Phase-Locked Loops & Their Application

OTHER IEEE PRESS BOOKS

Digital Signal Computers and Processors, *Edited by A. C. Salazar*
Systems Engineering: Methodology and Applications, *Edited by A. P. Sage*
Modern Crystal and Mechanical Filters, *Edited by D. F. Sheahan and R. A. Johnson*
Electrical Noise: Fundamentals and Sources, *Edited by M. S. Gupta*
Computer Methods in Image Analysis, *Edited by J. K. Aggarwal, R. O. Duda, and A. Rosenfeld*
Microprocessors: Fundamentals and Applications, *Edited by W. C. Lin*
Machine Recognition of Patterns, *Edited by A. K. Agrawala*
Turning Points in American Electrical History, *Edited by J. E. Brittain*
Charge-Coupled Devices: Technology and Applications, *Edited by R. Melen and D. Buss*
Spread Spectrum Techniques, *Edited by R. C. Dixon*
Electronic Switching: Central Office Systems of the World, *Edited by A. E. Joel, Jr.*
Electromagnetic Horn Antennas, *Edited by A. W. Love*
Waveform Quantization and Coding, *Edited by N. S. Jayant*
Communication Satellite Systems: An Overview of the Technology, *Edited by R. G. Gould and Y. F. Lum*
Literature Survey of Communication Satellite Systems and Technology, *Edited by J. H. W. Unger*
Solar Cells, *Edited by C. E. Backus*
Computer Networking, *Edited by R. P. Blanc and I. W. Cotton*
Communications Channels: Characterization and Behavior, *Edited by B. Goldberg*
Large-Scale Networks: Theory and Design, *Edited by F. T. Boesch*
Optical Fiber Technology, *Edited by D. Gloge*
Selected Papers in Digital Signal Processing, II, *Edited by the Digital Signal Processing Committee*
A Guide for Better Technical Presentations, *Edited by R. M. Woelfle*
Career Management: A Guide to Combating Obsolescence, *Edited by H. G. Kaufman*
Energy and Man: Technical and Social Aspects of Energy, *Edited by M. G. Morgan*
Magnetic Bubble Technology: Integrated-Circuit Magnetics for Digital Storage and Processing, *Edited by H. Chang*
Frequency Synthesis: Techniques and Applications, *Edited by J. Gorski-Popiel*
Literature in Digital Processing: Author and Permuted Title Index (Revised and Expanded Edition), *Edited by H. D. Helms, J. F. Kaiser, and L. R. Rabiner*
Data Communications via Fading Channels, *Edited by K. Brayer*
Nonlinear Networks: Theory and Analysis, *Edited by A. N. Willson, Jr.*
Computer Communications, *Edited by P. E. Green, Jr. and R. W. Lucky*
Stability of Large Electric Power Systems, *Edited by R. T. Byerly and E. W. Kimbark*
Automatic Test Equipment: Hardware, Software, and Management, *Edited by F. Liguori*
Key Papers in the Development of Coding Theory, *Edited by E. R. Berlekamp*
Technology and Social Institutions, *Edited by K. Chen*
Key Papers in the Development of Information Theory, *Edited by D. Slepian*
Computer-Aided Filter Design, *Edited by G. Szentirmai*
Laser Devices and Applications, *Edited by I. P. Kaminow and A. E. Siegman*
Integrated Optics, *Edited by D. Marcuse*
Laser Theory, *Edited by F. S. Barnes*
Digital Signal Processing, *Edited by L. R. Rabiner and C. M. Rader*
Minicomputers: Hardware, Software, and Applications, *Edited by J. D. Schoeffler and R. H. Temple*
Semiconductor Memories, *Edited by D. A. Hodges*
Power Semiconductor Applications, Volume II: Equipment and Systems, *Edited by J. D. Harnden, Jr. and F. B. Golden*
Power Semiconductor Applications, Volume I: General Considerations, *Edited by J. D. Harnden, Jr. and F. B. Golden*
A Practical Guide to Minicomputer Applications, *Edited by F. F. Coury*
Active Inductorless Filters, *Edited by S. K. Mitra*
Clearing the Air: The Impact of the Clean Air Act on Technology, *Edited by J. C. Redmond, J. C. Cook, and A. A. J. Hoffman*

Phase-Locked Loops & Their Application

Edited by

William C. Lindsey

Professor of Electrical Engineering
University of Southern California

Marvin K. Simon

Member of Technical Staff
Jet Propulsion Laboratory

A volume in the IEEE PRESS Selected Reprint Series,
prepared under the sponsorship of the
IEEE Communications Society.

The Institute of Electrical and Electronics Engineers, Inc. New York

PRINTED IN THE UNITED STATES OF AMERICA

IEEE International Standard Book Numbers: Clothbound: 0-87942-101-0
Paperbound: 0-87942-102-9

Library of Congress Catalog Card Number 77-073101

Sole Worldwide Distributor (Exclusive of the IEEE):

JOHN WILEY & SONS, INC.
605 Third Ave.
New York, NY 10016

Wiley Order Numbers: Clothbound: 0-471-04175-0
Paperbound: 0-471-04176-9

Contents

Introduction . . . 1

Part I: Basic Theory (Linear and Nonlinear) . . . 7

Piecewise Linear Analysis of Phase-Lock Loops, *C. R. Cahn* (*IRE Transactions on Space Electronics and Telemetry*, March 1962) . . . 8

Synchronous Communications, *J. P. Costas* (*Proceedings of the IRE*, December 1956) . . . 14

Design and Performance of Phase-Lock Circuits Capable of Near-Optimum Performance Over a Wide Range of Input Signal and Noise Levels, *R. Jaffe and E. Rechtin* (*IRE Transactions on Information Theory*, March 1955) . . . 20

Transient Analysis of Phase-Locked Tracking Systems in the Presence of Noise, *J. R. La Frieda and W. C. Lindsey* (*IEEE Transactions on Information Theory*, March 1973) . . . 31

Nonlinear Analysis of Generalized Tracking Systems, *W. C. Lindsey* (*Proceedings of the IEEE*, October 1969) . . . 42

Nonlinear Analysis of Correlative Tracking Systems Using Renewal Process Theory, *H. Meyr* (*IEEE Transactions on Communications*, February 1975) . . . 60

Functional Techniques for the Analysis of the Nonlinear Behavior of Phase-Locked Loops, *H. L. Van Trees* (*Proceedings of the IEEE*, August 1964) . . . 72

Phase-Locked Loop Dynamics in the Presence of Noise by Fokker–Planck Techniques, *A. J. Viterbi* (*Proceedings of the IEEE*, December 1963) . . . 90

A New Approach to the Linear Design and Analysis of Phase-Locked Loops, *C. S. Weaver* (*IRE Transactions on Space Electronics and Telemetry*, December 1959) . . . 107

Hangup in Phase-Lock Loops, *F. M. Gardner* (*IEEE Transactions on Communications*, October 1977) . . . 120

Part II: Acquisition . . . 125

The Influence of Time Delay on Second-Order Phase-Lock Loop Acquisition Range, *J. A. Develet, Jr.* (*Proceedings of the International Telemetering Conference*, September 1963) . . . 126

Phase-Lock Loop Frequency Acquisition Study, *J. P. Frazier and J. Page* (*IRE Transactions on Space Electronics and Telemetry*, September 1962) . . . 132

Phase-Locked Loop Pull-In Frequency, *L. J. Greenstein* (*IEEE Transactions on Communications*, August 1974) . . . 150

Acquisition Behavior of Generalized Tracking Systems in the Absence of Noise, *U. Mengali* (*IEEE Transactions on Communications*, July 1973) . . . 159

Part III: Threshold . . . 167

A Threshold Criterion for Phase-Lock Demodulation, *J. A. Develet, Jr.* (*Proceedings of the IEEE*, February 1963) . . . 168

Part IV: Stability . . . 177

The Effect of Time Delays on the Stability of Practical Phase Locked Receivers with Multiple Conversion by Utilizing Root Locus with Positive Feedback, *D. D. Carpenter* (*Conference Record, 1973 IEEE International Conference on Communications*, June 1973) . . . 178

Part V: Frequency Demodulation and Detection . . . 185

Demodulation of Wideband Frequency Modulation Utilizing Phase-Lock Technique, *R. C. Booton, Jr.* (*Proceedings of the 1962 National Telemetering Conference*, May 1962) . . . 186

Detection of Digital FSK and PSK Using a First-Order Phase-Locked Loop, *W. C. Lindsey and M. K. Simon* (*IEEE Transactions on Communications*, February 1977) . . . 210

Part VI: Tracking . . . 225

Delay-Lock Tracking of Binary Signals, *J. J. Spilker, Jr.* (*IEEE Transactions on Space Electronics and Telemetry*, March 1963) . . . 226

The Delay-Lock Discriminator—An Optimum Tracking Device, *J. J. Spilker, Jr. and D. T. Magill* (*Proceedings of the IRE*, September 1961) . . . 234

Part VII: Cycle Slipping and Loss of Lock . . . 249

Unlock Characteristics of the Optimum Type II Phase-Locked Loop, *R. W. Sanneman and J. R. Rowbotham* (*IEEE Transactions on Aerospace and Navigational Electronics*, March 1964) . . . 250

Simplified Formula for Mean Cycle-Slip Time of Phase-Locked Loops With Steady-State Phase Errors, *R. C. Tausworthe* (*IEEE Transactions on Communications*, June 1972) 260

Part VIII: Phase-Locked Oscillators 267

A Study of Locking Phenomena in Oscillators, *R. Adler* (*Proceedings of the IRE and Waves and Electrons*, June 1946) 268

Part IX: Operation and Performance in the Presence of Noise 275

Some Analytical and Experimental Phase-Locked Loop Results for Low Signal-to-Noise Ratios, *F. J. Charles and W. C. Lindsey* (*Proceedings of the IEEE*, September 1966) 276

The Response of a Phase-Locked Loop to a Sinusoid Plus Noise, *S. G. Margolis* (*IRE Transactions on Information Theory*, June 1957) 291

Part X: AGC, AFC, and APC Circuits and Systems 299

Theory of AFC Synchronization, *W. J. Gruen* (*Proceedings of the IRE*, August 1953) 300

The Lock-In Performance of an AFC Circuit, *G. W. Preston and J. C. Tellier* (*Proceedings of the IRE*, February 1953) 306

Automatic Phase Control: Theory and Design, *T. J. Rey* (*Proceedings of the IRE*, October 1960; Corrections in *Proceedings of the IRE*, March 1961) 309

The Response of an Automatic Phase Control System to FM Signals and Noise, *D. L. Schilling* (*Proceedings of the IEEE*, October 1963) 321

The Application of Linear Servo Theory to the Design of AGC Loops, *W. K. Victor and M. H. Brockman* (*Proceedings of the IRE*, February 1960) 332

Part XI: Digital Phase-Locked Loop 337

On Optimum Digital Phase-Locked Loops, *S. C. Gupta* (*IEEE Transactions on Communication Technology*, April 1968) 338

Performance of a First-Order Transition Sampling Digital Phase-Locked Loop Using Random-Walk Models, *J. K. Holmes* (*IEEE Transactions on Communications*, April 1972) 343

Part XII: Applications and Miscellaneous 357

Analysis of Synchronizing Systems for Dot-Interlaced Color Television, *T. S. George* (*Proceeedings of the IRE*, February 1951) 358

Application of the Phase-Locked Loop to Telemetry as a Discriminator or Tracking Filter, *C. E. Gilchriest* (*IRE Transactions on Telemetry and Remote Control*, June 1958) 366

Color-Carrier Reference Phase Synchronization Accuracy in NTSC Color Television, *D. Richman* (*Proceedings of the IRE*, January 1954) 382

Phase-Locked Loops with Signal Injection for Increased Pull-In Range and Reduced Output Phase Jitter, *P. K. Runge* (*IEEE Transactions on Communications*, June 1976) 410

On the Selection of Signals for Phase-Locked Loops, *J. J. Stiffler* (*IEEE Transactions on Communication Technology*, April 1968) 419

Author Index 425

Subject Index 427

Editors' Biographies 431

To

DOROTHY *ANITA*
JOHN *BRETTE*
JEFFREY

AND OUR PARENTS

Introduction

I. General

Perhaps more than in any other area, original sources relating to the development of phase-locked loop theory frequently tend to be neglected, misquoted, and sometimes are scarcely accessible to students and workers in the field. The primary reason for this is that the subject has been discussed in such a wide variety of contexts with far reaching applications that to uncover all of the published material one would have to search over an enormous number of engineering publications, journals, and books, many of which are written in foreign languages, spanning a time period of approximately 20 years.

In more leisured times, the best method of making original papers in a given field accessible was through voluminous "collected works" often published by learned societies in memory of their distinguished members and guests. While, today, such an approach is still within the realm of possibility, the Editors of this IEEE Press book feel that a collection of papers which form the nucleus of the subject would be much more valuable to those who wish to get an overview of and quickly come up to speed on the subject or who have contributed to the subject either by grappling with a loop in the laboratory or on paper.

Although, at first glance, the behavior of a loop employing the phase-lock principle might, because of its simple structure, appear easy to characterize, indeed this turns out not to be the case. In fact, its mysterious behavior is often so complicated to describe analytically that one turns to experimental observation, and even this at times is difficult, especially in the presence of noise. To this date, many fundamental unsolved problem areas still exist, particularly in the area of transient analysis and acquisition. In fact, a sound theory which describes steady-state behavior of higher order loops in noise only became available a few years back.

Phase-locked loops (PLL's) are used widely in modern communications, radar, telemetry, command, time and frequency control, ranging, and instrumentation systems. Because of the importance of the application of phase-locked loops, there has been a considerable amount of work done in this area. There are several books [1]-[7], each good in its own way, which look into the analysis, design, and application of the PLL in one way or the other. Some books [1], [2], [4], [5], [7] devote all or a majority of the material to this area.

II. Phase-Locked Loop Applications

In this section a number of applications which serve to motivate the reader as well as motivate the introduction of the PLL model itself are presented. In particular, problems pertaining to synchronization and carrier tracking, coherent demodulation of digital and analog signals, and frequency synthesis are included as examples. It is also worth noting here that integrated circuit technology has spawned a number of

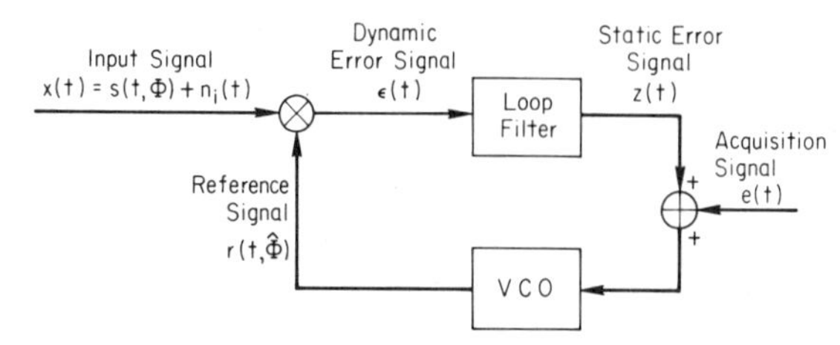

Fig. 1. Basic configuration of a phase-locked loop [5].

products that have and will continue to influence system design; however, few of these can rival the potential impact which monolithic circuits, including low-power medium-scale integration (MSI) and large-scale integration (LSI) digital logic elements, may have on the future uses of the PLL.

A. Carrier Tracking

In all applications pertaining to coherent telecommunications, it is necessary to reconstruct a carrier reference from a noise-corrupted version of the received signal. At least three methods are available: 1) tracking the instantaneous frequency and phase of the carrier component (if present) in the received signal spectrum 2) tracking instantaneous frequency and phase (of a phantom carrier) of a suppressed carrier signal, or 3) use of a hybrid of the first two methods.

Method 1) can be accomplished by designing a narrowband carrier tracking loop of the PLL or automatic phase control (APC) type. The basic configuration of such a loop is shown in Fig. 1. Here the input signal component $s(t, \Phi)$ and reference signal $r(t, \hat{\Phi})$ are characterized by $s(t, \Phi) = \sqrt{2}\, A(t) \sin \Phi(t)$ and $r(t, \hat{\Phi}) = \sqrt{2}\, K_1 \cos \hat{\Phi}(t)$, respectively. The additive function $n_i(t)$ represents background noise and $e(t)$ represents a signal acquisition voltage. In general, the *amplitude variations* $A(t)$ can be used to characterize any type of digital or analog amplitude modulation (AM), while the *phase function* $\Phi(t)$ can be used to characterize any type of digital or analog angle modulation, namely, frequency (FM) or phase (PM) modulation. The phase function $\hat{\Phi}(t)$ represents the loop estimate (or at least certain of its components) of $\Phi(t)$.

The PLL of Fig. 1 is essentially a closed-loop, electronic servomechanism in which the reference signal will acquire and track the signal component in the received signal $x(t) = s(t, \Phi) + n_i(t)$. Any phase error, say $\varphi(t) = \Phi(t) - \hat{\Phi}(t)$, between these two signals, which is within the bandwidth of the loop, is converted by the loop into a correction voltage that changes the instantaneous frequency and phase of the reference signal in order to make the loop track the input signal. In fact, in the absence of noise the product $s(t, \Phi) r(t, \hat{\Phi}) = AK_1 K_m \sin \varphi(t)$, appearing at the output of the multiplier with double frequency terms neglected, represents this voltage.

The implementation of this concept is quite simple (see Fig. 1). It can be accomplished with three basic parts: a *voltage-*

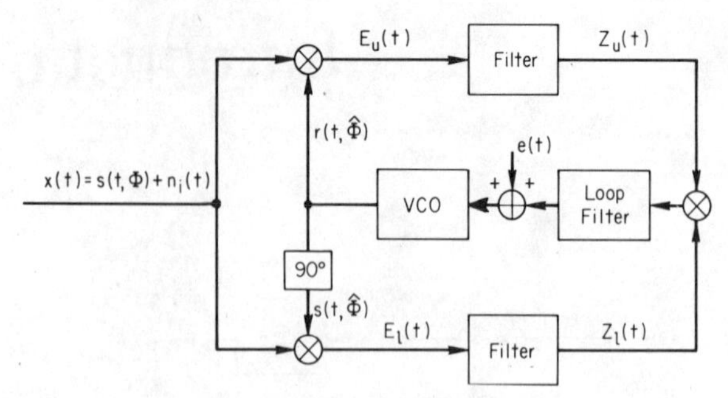

Fig. 2. The Costas loop for suppressed carrier tracking [5].

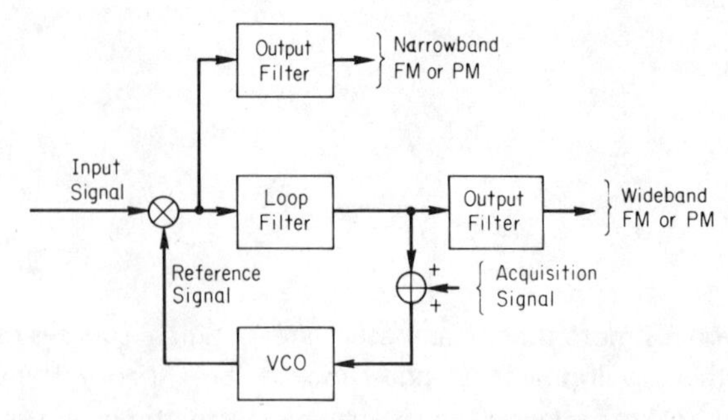

Fig. 3. Angle (digital or analog) demodulation by a PLL [5].

controlled oscillator (VCO), a *multiplier* or *phase detector*,[1] and a *loop filter*. The loop is so designed that when the average phase error is constant, the loop is *phase locked*. Should either of the signals change in phase, the loop filter output produces an average error voltage which is proportional in magnitude and direction to the original phase change. When applied to the VCO, this error voltage $z(t) + e(t)$ changes the frequency and phase in such a way as to retain phase lock with the input signal. In the simplest carrier tracking application, $A(t) = A$ and $\Phi(t) = \omega t + \theta(t)$ where ω is the *carrier radian frequency* and $\theta(t) = \Omega_0 t + \theta_0$ represents the *phase of the carrier to be tracked*. Furthermore, $\hat{\Phi}(t)$ is characterized by $\hat{\Phi}(t) = \omega_0 t + \hat{\theta}(t)$; here ω_0 is the *resonant* or *quiescent radian frequency* of the VCO and $\hat{\theta}(t)$ is the *loop estimate* of $\theta(t)$.

B. Suppressed Carrier Tracking

Many system applications require for one reason or another that the carrier component present in $s(t, \Phi)$ be completely suppressed at the transmitter. Even though the spectrum of the received signal $x(t)$ does not contain a carrier component, it is possible to reconstruct one. For example, suppressed carrier tracking can be accomplished by a variety of techniques, namely, the Costas loop, *N*th power loops, data-aided loops, decision-directed loops, delay-locked loops, hybrid loops, etc. In any case, the implemented loop is usually narrowband and employs the phase-lock principle.

A Costas loop, which provides for suppressed carrier tracking of the phase-shift keyed signal $s(t, \Phi) = \sqrt{2}\, A(t) \sin \Phi(t)$, with $A(t) = A$, $\Phi(t) = \omega_0 t + \theta$, $\theta = 0$ or π, during the time interval $(n-1)T \leqslant t \leqslant nT$, $n = 1, 2, \cdots$, is illustrated in Fig. 2.

C. Coherent Demodulation of Analog Signals

With appropriate choice of the loop circuit parameters, one can extract small angle modulation, phase modulation (PM), or frequency modulation (FM) of low index from the phase detector output of a narrowband PLL, or wideband FM or PM information from the loop filter output; see Fig. 3.

Amplitude modulation can be detected coherently by using a coherent amplitude detector driven from the voltage control oscillator (VCO) output as illustrated in Fig. 4. This circuit also provides amplitude detection for automatic gain control (AGC) or lock indication. Most PLL's incorporate phase

[1]The symbol ⊗ will be used to represent phase detectors which perform the basic operation of multiplication.

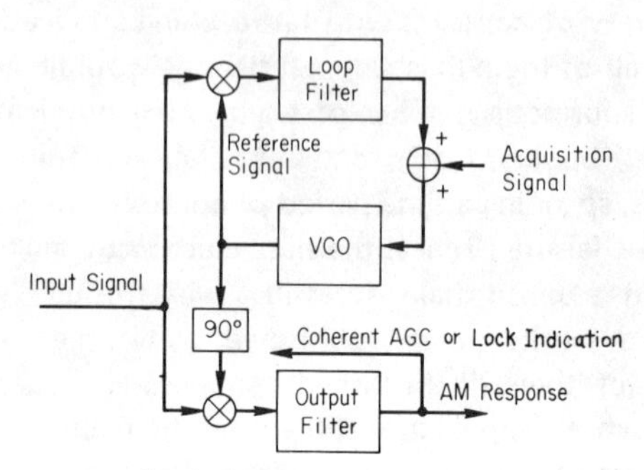

Fig. 4. Coherent amplitude (digital or analog) demodulation by a PLL [5].

detector circuits which are amplitude sensitive. Where close control of loop parameters is important, the signal level requires AGC processing. Examples of detection applications are frequency- and phase-shift keying, tracking, IF receivers with coherent AGC, and FM receivers.

D. Coherent Demodulation of Digital Signals

In the coherent demodulation of digital data, the signal $s(t, \Phi)$ frequently contains a sinusoidal component for carrier tracking purposes. When the digital modulation appears as a data-modulated subcarrier which has been phase modulated onto the transmitted carrier, then this modulation can be extracted at the phase detector output of Fig. 3. This output is applied to a subcarrier tracking loop followed by a digital data demodulator (output filter) for purposes of detecting the transmitted digital data stream. When the carrier is completely suppressed, for example, using phase-shift keyed modulation, then the noise-corrupted baseband modulation appears at the output of the lower phase detector (multiplier) of Fig. 2.

To coherently demodulate digital amplitude modulation which retains a carrier component in the transmitted signal, a 90° shift of the reference signal generated in Fig. 4 produces a coherent reference signal for coherent amplitude demodulation.

Since the reference signals in Figs. 3 and 4 are produced in the presence of noise, they are not perfect and give rise to the *noisy reference problem* in the theory of signal detection. Noisy reference signals produce deleterious effects on the data detection process when the PLL is operating synchronously and catastrophic effects when the synchronous reference signal fails.

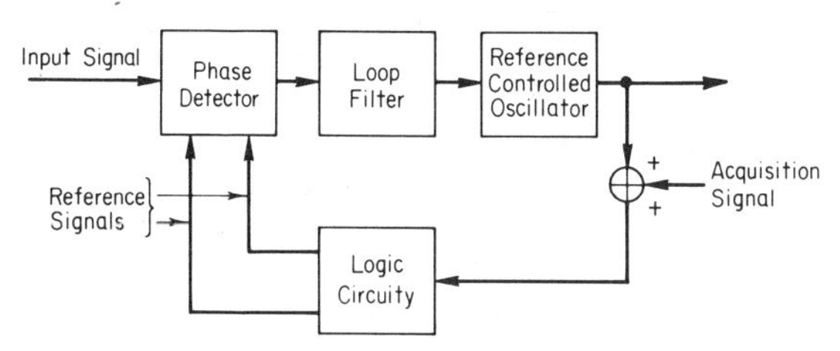

Fig. 5. Model of a class of symbol synchronization systems [5].

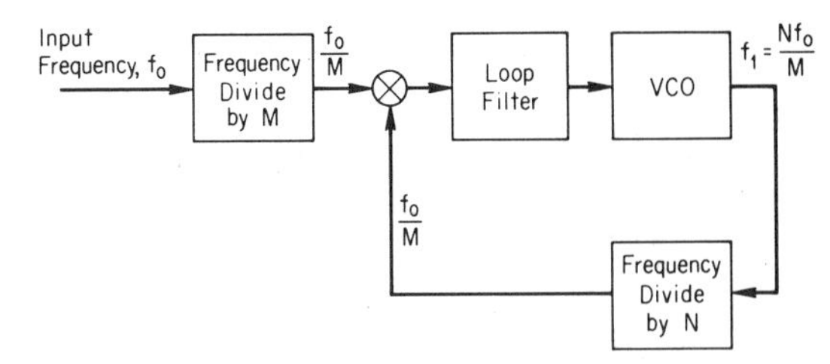

Fig. 6. PLL frequency multiplier/division (noise-free operation illustrated) [5].

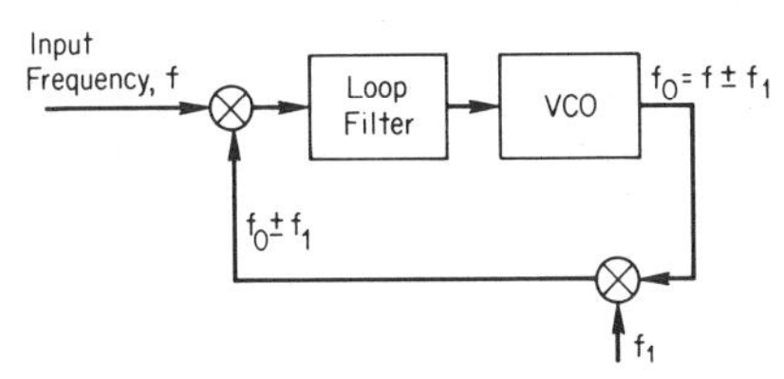

Fig. 7. PLL frequency translation (noise-free operation illustrated) [5].

E. Symbol (Bit) Synchronization Systems

In digital data transmission systems the usual first step in the demodulation/detection procedure is to establish synchronization (sync) between the transmitted and received carriers [6, ch. 9]. If subcarriers are used, the second step is to establish subcarrier sync. Of no less importance in the synchronization procedure is the establishment of *symbol (bit) sync*. Symbol synchronization has to do with determining the instants in time when the modulation may change states. One approach to providing this timing information is to utilize a separate communication channel solely for establishing symbol sync. More efficiently, however, are the techniques whereby this level of timing information can be obtained directly from the received data bearing signal. For this case a broad class of symbol synchronization systems, suggested by *maximum a posteriori* estimation techniques, can be represented by the diagram of Fig. 5. Specific phase detector characteristics included are the early–late gate type, those which incorporate an absolute value approach, a difference of squares approach, or a hybrid of these approaches, and the decision-directed type.

F. Frequency Synthesis

The PLL is an important building block in indirect frequency synthesis of spectrally pure signals. Frequency multiplication and/or division may be performed using a PLL in conjunction with divider elements as shown in Fig. 6. Frequency addition and subtraction may be effected by translation in Fig. 7. These operations may be combined in many ways to form programmable frequency synthesizers for spectrally pure signal generation and for special signal tracking and processing.

Phase-locked loops may someday replace present day envelope-type demodulators used in superheterodyne receivers for AM and FM reception by doing the job more efficiently and economically. PLL's may also replace banks of crystals in multichannel receivers and transmitters with a single crystal-controlled oscillator of high stability. They can provide precision control over motor speeds. In instrumentation they can be used in the implementation of variable time base electronic signal generators. In computers they can be used to synthesize and synchronize multiple clock frequencies from a single source. PLL's can be used as PM or FM modulators. They can be used to detect signal tones, demodulate angle modulation, measure Doppler and range, synchronize signals, track unstable signal sources, help provide automatic gain control (AGC), reconstruct signals, and generate or select precise signals for data transmission. Generally speaking, PLL's are used in the implementation of communication system modems, in telephone signaling and telemetry equipment, in tracking and navigation systems, in satellite and airborne systems in which Doppler effects must be avoided or measured, as well as in electronic test equipment; in fact, PLL's appear in most electronic applications that cover the frequency spectrum from very low to ultrahigh frequencies.

III. Model of a Phase-Locked Loop System

In this section we postulate the model of a phase-locked loop system, and as we can see, any of the aforementioned applications (among others) are represented as special cases of this model. A PLL system (Fig. 8) consists of five major subsystems: 1) the *phase detectors*, 2) the *loop filter*, 3) the *acquisition aid*, 4) the *waveform generator to be synchronized*, and 5) the *output filters*. For carrier tracking alone the output filters are not required. Frequently, the acquisition aid is not needed in the implementation of a PLL; however, when required, it assists in expediting the system to a rapid and satisfactory mode of operation. Once the signal is acquired, the aid usually is of no further assistance and is often disabled. For certain applications such as in phase-coherent communication receivers, a system is required which automatically adjusts the gain. To do this, an automatic gain control (AGC) system is employed which maintains a constant output signal level and provides linear operation throughout the system. Automatic gain control has many other uses; it keeps the output of automobile radio receivers at a pleasant listening level even when the received signal level changes drastically, as when driving among tall buildings. It also prevents sophisticated radar receivers from being saturated at high signal levels.

Choice of the phase detectors is a basic problem in signal design. Ultimately it is connected with the engineer's choice (if he has one) of designing the synchronizing (input) signal $s(t, \Phi)$ and the reference signal $r(t, \hat{\Phi})$ produced by the waveform generator to be synchronized. Here $\Phi(t)$ represents a time-dependent phase function of the input signal and $\hat{\Phi}(t)$ represents the system estimate of $\Phi(t)$ in the presence of the

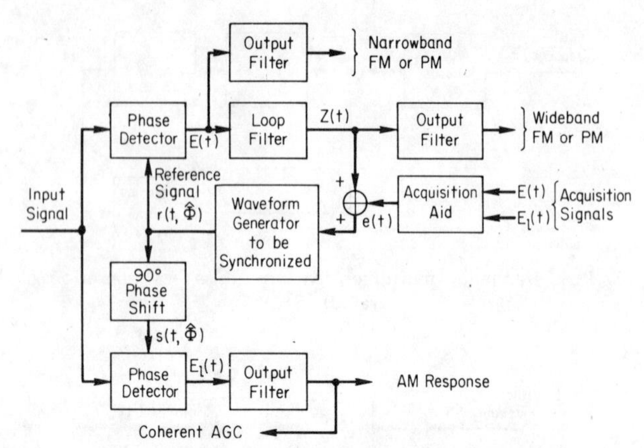

Fig. 8. Model of a class of phase-locked loops [5].

input noise $n_i(t)$. The phase function will be used to model the frequency and phase information to be extracted and/or tracked by the PLL. Any periodic signal (and some classes of nonperiodic signals) can carry this information.

IV. Operational Behavior and Performance Measures of PLL System

On the one hand, the operation of any PLL is conveniently discussed by considering its two modes of operation, namely, the *acquisition mode* (achieves the synchronous state) and the *synchronous or tracking mode* (retains the synchronous state). On the other hand, the performance of any PLL is characterized by certain performance measures which are statistical in nature when noise is present and deterministic when noise is absent. In order to introduce these performance measures, we shall discuss the operational behavior of a PLL from the viewpoint of rotating phasors. This point of view has the distinct advantage of being simple without losing generality in illustrating the operational behavior of a PLL for any application. There are considerable differences in the desirable parameter values of phase-locked loops for acquisition and tracking. For rapid acquisition the loop bandwidth needs to be as large as the channel noise will allow. On the other hand, during tracking, the bandwidth frequently needs to be as small as the oscillator instabilities and Doppler will permit. Frequently, the need for rapid sync acquisition and good tracking performance requires that two different loop bandwidths be implemented. If the loop bandwidth is reduced after acquisition by switching elements in the loop filter, the transients produced can bring the system out of lock. If the PLL is coupled with a coherent AGC, the latter will, depending upon the AGC time constant, keep the input amplitude high as long as the loop slips cycles and will reduce the gain after lock-up. Since the PLL bandwidth and damping are proportional to the input signal amplitude, the AGC can be used to control the PLL bandwidths and damping during acquisition as well as during tracking. The size of the attainable effect is limited by the dynamic range of the loop components.

Let the imaginary part of the phasor $\exp[i\Phi(t)]$ and the real part of the phasor $\exp[i\Phi(t)]$ be associated with the signals $s(t, \Phi)$ and $r(t, \hat{\Phi})$, respectively. These phasors, which rotate with instantaneous angular velocities $\omega(t) \triangleq \dot{\Phi}(t)$ rad/s and

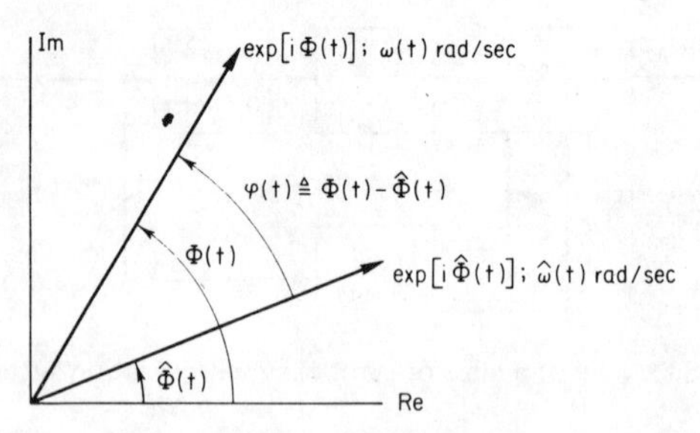

Fig. 9. Phasor diagram for describing operation of a PLL [5].

$\hat{\omega}(t) \triangleq \dot{\hat{\Phi}}(t)$ rad/s, respectively, are shown in Fig. 9. The *phase error* $\varphi(t) \triangleq \Phi(t) - \hat{\Phi}(t)$ has temporal characteristics which are strongly dependent on the application and mode of operation; however, the performance measures associated with the random process $\{\varphi(t)\}$ are essentially independent of the application. We note that when the phase function $\Phi(t)$ is characterized by a random process, the signal phasor $\exp[i\Phi(t)]$ and reference phasor $\exp[i\hat{\Phi}(t)]$ rotate randomly at rates proportional to $\dot{\Phi}(t)$ and $\dot{\hat{\Phi}}(t)$.

A. Acquisition Mode

Suppose the system enters the acquisition mode at $t = t_0$. Initially the phasors $\exp[i\Phi(t)]$ and $\exp[i\hat{\Phi}(t)]$ rotate at angular velocities $\omega \triangleq \omega(t_0)$ and $\omega_0 = \hat{\omega}(t_0)$ rad/s, respectively. Thus, initially $\exp[i\Phi(t)]$ rotates at rate $\Omega_0 \triangleq \dot{\varphi}(t_0) = \dot{\varphi}_0 = \omega - \omega_0$ rad/s relative to $\exp[i\hat{\Phi}(t)]$. The parameter $\Omega_0/2\pi$ is called the *initial frequency detuning* in the system.

Fig. 10 shows a typical behavior of $\varphi(t)$ and $\dot{\varphi}(t)$ for a second-order PLL operating in the acquisition mode without an acquisition aid in the absence of noise. From this figure it is clear that the behavior of $\varphi(t)$ and $\dot{\varphi}(t)$ during acquisition is highly dependent upon the initial state $y_0 \triangleq [\varphi(t_0), \dot{\varphi}(t_0)] = [\varphi_0, \dot{\varphi}_0]$. Let the synchronous mode be defined by the conditions $|\dot{\varphi}(t)| \leqslant \epsilon_\Omega$, $|\varphi(t) - 2n\pi| \leqslant \epsilon_\varphi$, n any integer, for $t = t_a$. Suppose these conditions are satisfied at $t_a = t_{acq} + t_0$ for the first time. The parameter t_{acq} is called the signal acquisition time (often called pull-in time or lock-in time) since it is the time required to reach the synchronous mode. When noise is present, the situation is much more complicated and we do not discuss all the details here; however, for this case t_{acq} becomes a random variable. Thus, an important performance measure for the acquisition mode is the set of statistical moments $T^n_{acq} = E(t^n_{acq})$; $E(\cdot)$ denotes expectation in the statistical sense. Obviously, T^n_{acq} depends upon the initial state of the system. When the initial state is random and characterized by a probability density function (pdf), then T^n_{acq} must be averaged over this density.

Shortly after acquisition, say $t > t_a$ seconds, the "steady-state" mean $\bar{\varphi}/2\pi = (\omega - \bar{\omega}_v)/2\pi$ represents the *mean residual frequency detuning*. Here $\bar{\omega}_v$ is the average radian frequency of the synchronized generator and $(\omega_0 - \bar{\omega}_v)/2\pi$ is the mean frequency shift of the waveform generator to be synchronized. When $\dot{\Phi}(t) = \omega$ is constant, the largest value of Ω_0, say $|\Omega_0|_m$, from which the system can reach the synchronous mode is called the *signal acquisition* (often called *pull-in* or *lock-in*) range, that is, $|\Omega_0|_m$ represents the maximum relative angular

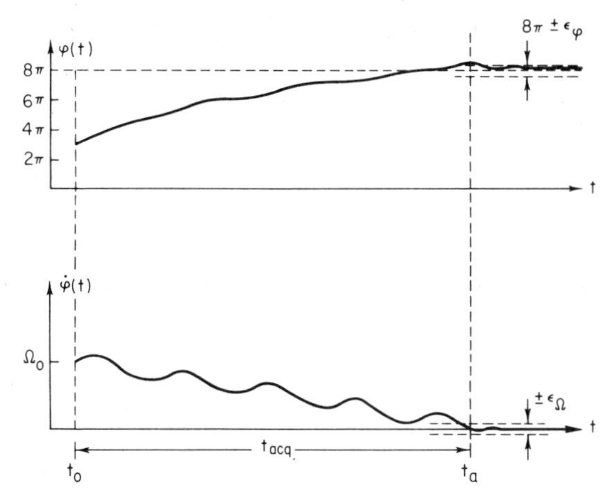

Fig. 10. Typical behavior of $\varphi(t)$ and $\dot{\varphi}(t)$ for a PLL in the acquisition mode (noise absent) [5].

velocity between phasors exp $[i\Phi(t)]$ and exp $[i\hat{\Phi}(t)]$ for which acquisition is possible. Since noise is present, characterization of the probability of acquisition in the interval $[t_0, t_a]$ is also of interest.

In the acquisition mode, a PLL is highly nonlinear. In fact, the so-called linear theory cannot be used to carry out a particular design nor can it be used to account for the system's operational behavior, that is, the acquisition problem is characteristically nonlinear.

B. Synchronous or Tracking Mode

Once the system is in the synchronous mode, a more or less steady-state condition exists and the system tends to automatically make adjustments so as to maintain $|\varphi(t) - 2n\pi| \leqslant \epsilon_\varphi$ and $|\dot{\varphi}(t)| \leqslant \epsilon_\Omega$ with high probability, that is, it tends to maintain the synchronous state. Due to the effects of noise $n_i(t)$ and instabilities in the waveform generator, random fluctuations in $\varphi(t)$ take place (see Fig. 11).

In the synchronous mode a number of performance measures are required to characterize and explain system behavior. The reason is that the performance measures that characterize this mode of operation are complicated to account for when both the system nonlinearity and noise are taken into consideration. If the system is linearized to simplify the problem, the performance measures which serve to characterize the fundamental behavior usually have no meaning, that is, operation in the synchronous mode is also characteristically nonlinear.

In this mode the time-dependent, conditional transition pdf $p(\varphi, t|y_0, t_0)$ of the absolute phase error $\varphi(t)$ is of interest. Furthermore, for some applications, for example, phase-coherent data transmission systems, the reduced phase error $\phi(t) \triangleq [\varphi(t) + \pi] \bmod 2\pi + (2n - 1)\pi$ (such as displayed by a phase meter) is of interest. Here n is considered to be any fixed integer. [See Fig. 12 for a typical sample function of $\varphi(t)$ and $\phi(t)$.] In loose terms, the phase error process $\{\phi(t)\}$ is called the *modulo-2π process*. Thus, the time-dependent, conditional transition pdf $p(\phi, t|y_0, t_0, n)$ of the phase error $\phi(t)$, which a phase meter would read, and steady-state conditional transition pdf $p(\phi|n)$ are of importance. In contrast, in two-way phase-coherent radar, sonar, and navigation systems,

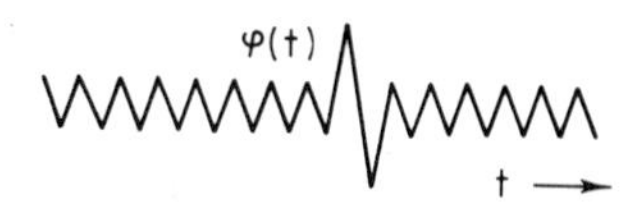

Fig. 11. Fluctuation of $\varphi(t)$ in the synchronous mode [5].

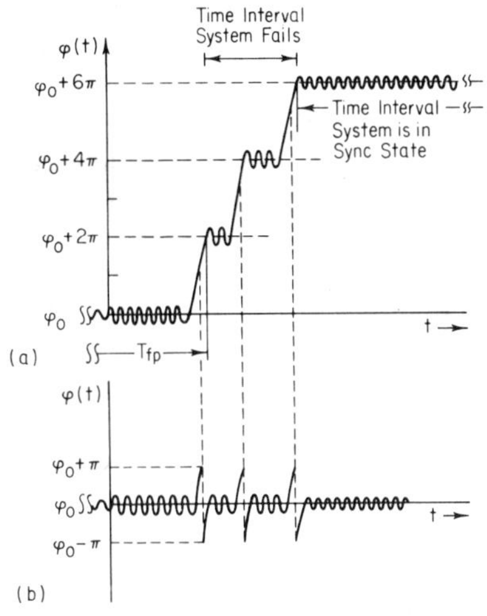

Fig. 12. Cycle slipping in PLL's. (a) Phase error $\varphi(t)$. (b) Reduced phase error $\phi(t)$ [5].

the phase of the reference waveform $r(t, \hat{\Phi})$ can help in providing for the measurement of range; therefore, a statistical characterization of the relative phase error $\varphi(t)$ is required in order to specify system performance. The reference $r(t, \hat{\Phi})$ is frequently used to measure Doppler shift. Where ϕ is the quantity of interest, some applications may require the mean and variance of ϕ to be less than a few degrees, while other applications allow the mean and standard deviation of ϕ to be as large as 30 or 45°.

With the passage of time, noise causes *synchronization failures* to occur. Here we define a synchronization (sync) failure as the change of φ from φ_0 to $\varphi_0 + 2\pi$ or $\varphi_0 - 2\pi$, i.e., $\Delta\varphi/2\pi = \pm 1$ cycle [see Fig. 12(a)]. This corresponds to the situation in Fig. 9 where, for any initial state $y_0 = [\varphi_0, \dot{\varphi}_0]$, the phasor exp $[i\Phi(t)]$ rotates $\pm 2\pi$ rad relative to the phasor exp $[i\hat{\Phi}(t)]$. In physical terms the generator to be synchronized drops or adds one cycle of oscillation in $r(t, \hat{\Phi})$ relative to $s(t, \hat{\Phi})$; or stated another way, the signal $s(t, \Phi)$ is shifted $\pm 2\pi/\omega$ seconds relative to the reference signal $r(t, \hat{\Phi})$. This phenomenon is called *cycle slipping*.

In the presence of noise and oscillator instabilities, cycle slipping leads to frequency and *phase diffusion* in the generator to be synchronized. Therefore, the average *cycle slipping rate* $\bar{S}$, i.e., the average number of cycles slipped per second, independent of the direction of rotation of exp $[i\Phi(t)]$ relative to exp $[i\hat{\Phi}(t)]$, is of interest. Since the time T_{fp} for a sync failure to occur for the first time is a random variable [see Fig. 12(a)], we are interested in its statistical description by means of its pdf or its moments, i.e., $E(T_{fp}^n)$. The first moment of this time is called the average time to *first loss of phase sync* or the *mean time to sync failure* and is denoted by $\tau(2\pi|\varphi_0) = E(T_{fp})$. Evaluation of this time is related to the so-called first-passage time problem studied in the theory of Markov processes.

The *probability of sync failure* $P(t)$ in the time interval $[t_0, t]$, as well as the *probability of slipping n cycles* $P(N = n)$ in the time interval $[t_0, t]$, are of interest. Since biases are frequently present in a PLL, the average number of clockwise (counterclockwise) rotations per second of the phasor exp $[i\Phi(t)]$ through 2π relative to the phasor exp $[i\hat{\Phi}(t)]$ and the respective probabilities are of interest. We shall refer to the dropping and adding of oscillations in $r(t, \hat{\Phi})$ relative to $s(t, \Phi)$ as cycle slips to the right and left, respectively. The average number of slips to the right (left) per second is characterized by $N_+ (N_-)$. Other parameters which require a statistical characterization by means of the nonlinear theory include the average time ΔT between cycle slipping events (sync failures) or the average time the system is in lock and the average time the system remains out of the synchronous state [see Fig. 12(a)]. Finally, a statistical description of the *frequency error* $\dot{\varphi}$ (together with $\bar{S} \triangleq N_+ + N_-$) serves to characterize the *frequency stability* of the synchronized generator.

V. Conclusions

This introduction was written based on the authors' experience [5] and [6] for the purpose of motivating the reader and providing an overview of a broad class of phase-locked loop systems. This approach avoids introducing any mathematical concepts prior to understanding their need; see Prelude to [5].

The development in the last few years of low-power medium-scale integration (MSI) and large-scale integration (LSI) digital elements have made digital mechanization of a PLL technically and economically feasible in many applications. Also, the availability of a low-cost PLL system in a monolithic circuit is emerging as a versatile building block, similar to the monolithic operational amplifier in the diversity of its applications. In the past, the primary use of the PLL has been in more sophisticated communication systems; however, with the rapid development of IC technology, PLL's are expected to be used widely in consumer electronics.

References

[1] F. M. Gardner, *Phaselock Techniques*. New York: Wiley, 1966.

[2] A. J. Viterbi, *Principles of Coherent Communication*. New York: McGraw-Hill, 1966.

[3] J. J. Stiffler, *Theory of Synchronous Communications*. Englewood Cliffs, NJ: Prentice-Hall, 1971.

[4] J. Klapper and J. T. Frankle, *Phase-Locked and Frequency Feedback Systems*. New York: Academic, 1972.

[5] W. C. Lindsey, *Synchronization Systems in Communication and Control*. Englewood Cliffs, NJ: Prentice-Hall, 1972.

[6] W. C. Lindsey and M. K. Simon, *Telecommunication Systems Engineering*. Englewood Cliffs, NJ: Prentice-Hall, 1973.

[7] A. Blanchard, *Phase-Locked Loops*. New York: Wiley, 1976.

[8] W. C. Lindsey and R. C. Tausworthe, "A bibliography of the theory and application of the phase-lock principle," Jet Propulsion Lab., Pasadena, CA, Tech. Rep. 32-1581, Apr. 1973.

Part I
Basic Theory
(Linear and Nonlinear)

Piecewise Linear Analysis of Phase-Lock Loops*

CHARLES R. CAHN†, MEMBER, IRE

Summary—The synchronizing performance of a phase-lock loop, in the absence of noise, is obtained by replacing the sinusoidal characteristic of the product demodulator, as a function of phase difference, by a triangular piecewise linear approximation. The analytical conditions for existence of a limit cycle (steady-state asynchronous mode) with a steady input frequency are derived. They may be solved numerically for any limit cycle, and a constrained minimization yields the minimum mistuning of the input carrier for which a limit cycle exists. Inside of this computed synchronization limit, an asynchronous mode does not exist.

The computation for a particular case (damping factor = 0.5) yields a synchronization limit very similar to previous analog computer results. It is found that for a relatively large ratio of noise bandwidth to hold-in range, the synchronization limit corresponds to a trajectory connecting adjacent unstable points in the phase plane. However, for a small ratio, the solution changes character, and the synchronization limit is found to become proportional to the square root of loop-noise bandwidth. This is in agreement with a conclusion reached by Gruen on the basis of experimental data and by Viterbi by another approximate method of analysis.

I. Introduction

A METHOD OF analyzing nonlinear systems is to replace the actual nonlinear characteristic by one constructed from straight-line segments. Then, within each segment, the system performance may be obtained by ordinary linear analysis methods, and a complete solution may be obtained by joining the various partial solutions. With this piecewise linear approach, some further understanding of the behavior of the original nonlinear system may be achieved than is possible by a completely linear approximation or by graphical or computational techniques. This approach may be applied to phase-lock loops to study their synchronizing performance in the absence of noise. In particular, the possible existence of limit cycles (periodic solutions) may be determined.

The block diagram of a phase-lock loop is shown in Fig. 1. In this figure ω denotes the free running frequency of the voltage-controlled oscillator, and λ_i and λ_0 are the input and output instantaneous phases, referred to the free-running frequency. The phase difference (or phase error) is $\lambda = \lambda_i - \lambda_0$. The loop filter, denoted by the transfer function $F(s)$, is used to extend the pull-in range of the loop beyond the closed-loop noise bandwidth.[1]

* Received by the PGSET, August 28, 1961. This paper was part of a study conducted for the Bendix Systems Division, Ann Arbor, Mich., and was accomplished under the over-all management and technical direction of the U. S. Army ADVENT Management Agency.

† The Bissett-Berman Corporation, 2941 Nebraska Ave., Santa Monica, Calif.

[1] W. J. Gruen "Theory of AFC synchronization," PROC. IRE, vol. 41, pp. 1043–1048; August, 1953.

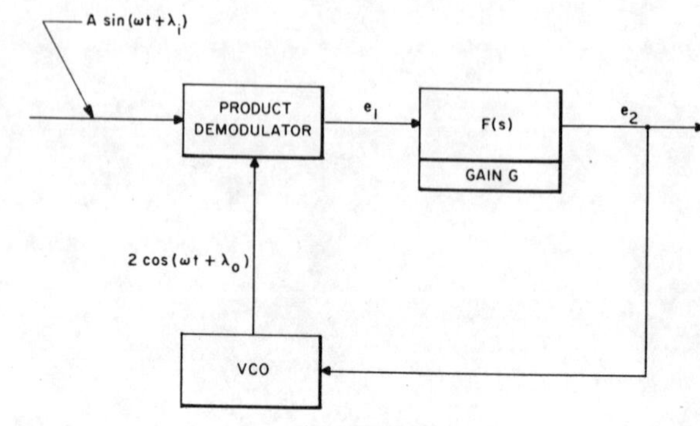

Fig. 1—Phase-lock loop.

A piecewise linear approximation to the characteristic of the product demodulator in the phase-lock loop is illustrated in Fig. 2, the sinusoidal variation with phase difference being approximated by a triangular variation. With the notation λ (mod 2π) denoting the phase difference expressed in the range $-\pi/2$ to $3\pi/2$ radians, the piecewise linear approximation is

$$\text{product demodulator output} = \lambda \ (\text{mod } 2\pi) \qquad -\pi/2 < \lambda(\text{mod } 2\pi) < \pi/2 \quad (\text{Domain } A)$$

$$= [\pi - \lambda(\text{mod } 2\pi)] \qquad \pi/2 < \lambda(\text{mod } 2\pi) < 3\pi/2. \quad (\text{Domain } B) \tag{1}$$

For convenience, the mod 2π notation is henceforth dropped and will be understood as implicit in all equations. The phase-lock loop equation then becomes

$$[s + KF(s)]\lambda = s\lambda_i \qquad -\pi/2 < \lambda < \pi/2 \tag{2}$$

and

$$[s - KF(s)]\lambda = s\lambda_i - KF(s)\pi \qquad \pi/2 < \lambda < 3\pi/2. \tag{3}$$

If $F(0) = 1$, the last term in (3) is simply $K\pi$.

The piecewise linear approximation presented above may, alternatively, be considered to represent exactly a phase-lock loop in which both inputs to the product demodulator are infinitely clipped. Previous work has shown that clipping of the signal input is desirable as an approximate method of maintaining optimum loop performance in the presence of noise when the signal or

Reprinted from *IRE Trans. Space Electron. Telemetry*, vol. SET-8, pp. 8–13, Mar. 1962

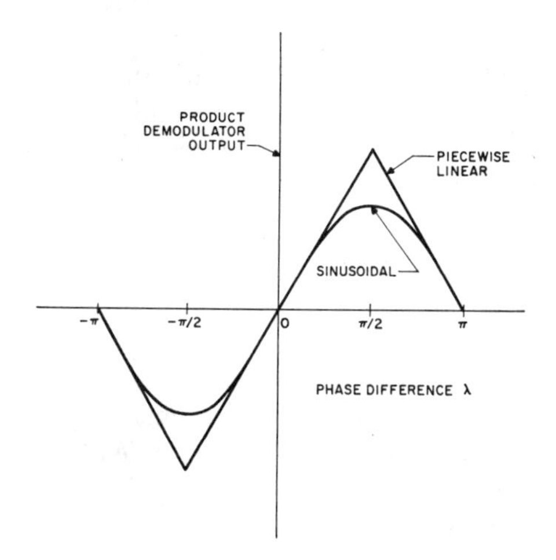

Fig. 2—Piecewise linear approximation to product demodulator characteristic.

noise level can vary over a wide range.[2] Also, it may be noted that the usual linear approximation for phase-locked loops is one portion of the piecewise linear approximation under discussion.

II. The Uncompensated Loop with a Constant Frequency Input

If the loop is uncompensated, $F(s) = 1$, and for a constant input frequency, (2) and (3) reduce to

$$(s + K)\lambda = \Delta\omega \qquad -\pi/2 < \lambda < \pi/2 \tag{4}$$

$$(s - K)\lambda = \Delta\omega \qquad \pi/2 < \lambda < 3\pi/2. \tag{5}$$

In either case the transient is exponential; however, (4) specifies a decreasing transient, and (5) an increasing transient. The solution of (4) is

$$\lambda(t) = \frac{\Delta\omega}{K} + \left[\lambda(0) - \frac{\Delta\omega}{K}\right]e^{-Kt} \qquad -\pi/2 < \lambda < \pi/2, \tag{6}$$

provided that $-\pi/2 < \lambda(0) < \pi/2$. Inspection of (6) shows that a stable point is reached at $\lambda = \Delta\omega/K$, unless $|\Delta\omega/K| > \pi/2$, in which case $\lambda(t)$ will ultimately exceed the range of validity for the equation.

If $\pi/2 < \lambda(0) < 3\pi/2$, the phase transient increases until λ enters the range of validity of (6). Then the stable point is reached, unless $|\Delta\omega/K| > \pi/2$, in which case a stable point does not exist. Thus if $|\Delta\omega/K| < \pi/2$, synchronization is reached within a single cycle, and if $|\Delta\omega/K| > \pi/2$, continuous cycle slipping occurs. With a sinusoidal product demodulator characteristic the corresponding boundary is $|\Delta\omega/K| = 1$; however, the behavior is otherwise very similar to the above.

[2] R. Jaffe and R. Rechtin, "Design and performance of phase-lock circuits capable of near-optimum performance over a wide range of input signal and noise levels," IRE Trans. on Information Theory, vol. IT-1, pp. 66–76; March, 1955.

III. The Second-Order Loop with a Constant Frequency Input

A. The Double-Time Constant Filter

If $F(s) = (1 + \tau_1 s)/(1 + \tau s)$, the second-order phase-lock loop is obtained. The piecewise linear approximation is now of real value, since the performance equation with a sinusoidal detector characteristic cannot be solved analytically. Again, a constant input frequency is considered, in which case the loop equation may be written, after algebraic manipulation, as

$$\left[\frac{\tau}{K}s^2 + \left(\tau_1\sqrt{\frac{K}{\tau}} + \frac{1}{\sqrt{K\tau}}\right)\sqrt{\frac{\tau}{K}}\,s + 1\right]\lambda = \frac{\Delta\omega}{K} \qquad -\pi/2 < \lambda < \pi/2 \tag{7}$$

and

$$\left[\frac{\tau}{K}s^2 - \left(\tau_1\sqrt{\frac{K}{\tau}} - \frac{1}{\sqrt{K\tau}}\right)\sqrt{\frac{\tau}{K}}\,s - 1\right]\lambda = \frac{\Delta\omega}{K} - \pi \qquad \pi/2 < \lambda < 3\pi/2. \tag{8}$$

These equations can, of course, be solved for the variation of the phase difference with time. However, for studying the synchronizing behavior, a phase-plane analysis is appropriate.

Making the substitutions[1]

$$\sqrt{K/\tau} = \omega_n \tag{9}$$

$$\tau_1\sqrt{K/\tau} + 1/\sqrt{K\tau} = \omega_n\tau_1 + \omega_n/K = 2\zeta \tag{10}$$

$$y = (s/\omega_n)\lambda, \tag{11}$$

the equations become

$$y\frac{dy}{d\lambda} + 2\zeta y + \lambda = \Delta\omega/K \qquad -\pi/2 < \lambda < \pi/2 \text{ (Domain } A), \tag{12}$$

$$y\frac{dy}{d\lambda} - \left(2\zeta - 2\frac{\omega_n}{K}\right)y - \lambda = \Delta\omega/K - \pi \qquad \pi/2 < \lambda < 3\pi/2 \text{ (Domain } B), \tag{13}$$

which may be solved analytically for the phase-plane trajectories. Note that the value of the phase difference λ is reduced modulo 2π to be in the indicated domains.

Eq. (12) and (13) may be solved after the substitution $y = vx$, where $x = \lambda - \Delta\omega/K$ in (12) and $x = \lambda + \Delta\omega/K - \pi$ in (13). For the moment, we shall indicate these solutions as

$$|x| = Af_A(v) \qquad -\pi/2 < \lambda < \pi/2 \quad \text{(Domain } A)$$
$$y = vx \qquad\qquad\qquad\qquad\qquad\qquad \tag{14}$$
$$x = \lambda - \Delta\omega/K$$

and

$$
\begin{aligned}
&| x | = Bf_B(v) \\
&\qquad \pi/2 < \lambda < 3\pi/2 \quad (\text{Domain } B) \\
&y = vx \\
&\qquad x = \lambda + \Delta\omega/K - \pi.
\end{aligned}
\tag{15}
$$

The coefficients A and B are the constants of integration and enter correctly, as may be seen from the fact that the differential equation is homogeneous when written in the independent variable x. That solutions passing by the singular point $x = 0$ of the equations are correctly joined will be seen later for a particular case and can be assured by properly selecting the form of the solutions.

The existence of a limit cycle (periodic solution for y as a function of λ) will now be presumed, and the conditions necessary for the limit cycle will be derived. Suppose that y remains positive. The solution is periodic if the value of y at $\lambda = 3\pi/2$ equals the value at $\lambda = -\pi/2$, and (14) and (15) may be pieced together on this hypothesis. Consider domain A first. At $\lambda = -\pi/2$, $x = -\pi/2 - \Delta\omega/K$, and the initial value of v is specified by the value of y at this point. Let this value be $v_0 < 0$. Then,

$$| -\pi/2 - \Delta\omega/K | = Af_A(v_0). \tag{16}$$

The trajectory is traced by letting $v \to -\infty$ and then return from $+\infty$ to a positive value v_1 at the point $\lambda = \pi/2$. Then,

$$\pi/2 - \Delta\omega/K = Af_A(v_1). \tag{17}$$

At this point the solution enters domain B. In crossing the boundary between the domains, the value of x changes from $\pi/2 - \Delta\omega/K$ to $-\pi/2 + \Delta\omega/K$, as may be seen from the definitions of x. Since y remains constant, it is seen that the value of v changes sign. Thus, at $\lambda = \pi/2$ in domain B, we have

$$\left| -\pi/2 + \frac{\Delta\omega}{K} \right| = Bf_B(-v_1). \tag{18}$$

The trajectory is traced again by letting $v \to -\infty$ and return to a positive value v_2 at $\lambda = 3\pi/2$, so that

$$\pi/2 + \Delta\omega/K = Bf_B(v_2). \tag{19}$$

As λ crosses into domain A, the value of x reverts to $-\pi/2 - \Delta\omega/K$, as in (16), hence v reverses signs again. Thus, a limit cycle is traced if $v_2 = -v_0$.

By simultaneous solution of (16) to (19), the constants A, B, and $\Delta\omega/K$ can be eliminated and the value of v_1 determined as a function of v_0. The result may be written in the form

$$\frac{f_A(v_0)}{f_B(-v_0)} = \frac{f_A(v_1)}{f_B(-v_1)}, \tag{20}$$

where $v_0 < 0$ and $v_1 > 0$. Eq. (20) is a transcendental equation, which can only be solved numerically for v_1 as a function of v_0. After any solution is obtained for v_1 and v_0, the corresponding value of $\Delta\omega/K$ is found by use of (16) and (17), as follows:

$$\frac{\pi/2 - \Delta\omega/K}{\pi/2 + \Delta\omega/K} = \frac{f_A(v_1)}{f_A(v_0)}. \tag{21}$$

By considering all solutions of (20), the minimum value of $| \Delta\omega/K |$ corresponding to a limit cycle can be determined. This minimum value will be a function of both ζ and ω_n/K.

The problem can be considered as a constrained maximization by which (21) is maximized subject to the constraint of (20). Taking the logarithms of the equations and differentiating partially with respect to v_1 and v_0, the equations

$$
\begin{aligned}
\frac{f_A'(v_1)}{f_A(v_1)} &= -\gamma \frac{f_B'(-v_1)}{f_B(-v_1)} \\
\frac{f_A'(v_0)}{f_A(v_0)} &= -\gamma \frac{f_B'(-v_0)}{f_B(-v_0)}
\end{aligned}
\tag{22}
$$

are obtained, where γ is a Lagrange multiplier, and the prime denotes differentiation with respect to the argument. It is seen that v_1 and v_0 are solutions of the same equation and are determined by γ. If a pair of values, v_0 and v_1, can be found satisfying (20), an extremal solution has been found.

B. Explicit Solution for $\zeta < 1$

To facilitate comparison with previous approximate and analog computer solutions, restriction to the case $\zeta < 1$ will be made. The explicit form of $f_A(v)$ is different for $\zeta = 1$ and $\zeta > 1$. Omitting the details of the integration, the functions $f_A(v)$ and $f_B(v)$ are

$$f_A(v) = \frac{1}{\sqrt{(v + \zeta)^2 + (1 - \zeta)^2}} \exp\left(-\frac{\zeta}{\sqrt{1 - \zeta^2}} \tan^{-1} \sqrt{1 - \zeta^2}/(v + \zeta)\right) \tag{23}$$

$$f_B(v) = \frac{1}{| v - \zeta_B - \sqrt{\zeta_B^2 + 1} |^{(1/2)(1 + \zeta_B/\sqrt{\zeta_B^2 + 1})} \, | v - \zeta_B + \sqrt{\zeta_B^2 + 1} |^{(1/2)(1 - \zeta_B/\sqrt{\zeta_B^2 + 1})}}, \tag{24}$$

where

$$\zeta_B = \zeta - \frac{\omega_n}{K}. \quad (25)$$

Performing the differentiations indicated in (22), we obtain after algebraic simplification

$$\frac{f_A'(v)}{f_A(v)} = -\frac{v}{(v+\zeta)^2 + (1-\zeta^2)}$$
$$\frac{f_B'(v)}{f_B(v)} = -\frac{v}{(v-\zeta_B)^2 - (\zeta_B^2+1)} \quad (26)$$

Substituting these simple expressions in (22), a quadratic in v is obtained. The two solutions of the quadratic are

$$v_0, \quad v_1 = -\zeta + \frac{\omega_n/K}{1-\gamma} \pm \sqrt{\left(\zeta - \frac{\omega_n/K}{1-\gamma}\right)^2 + \frac{2}{1-\gamma} - 1}. \quad (27)$$

In order that v_0 be negative and f_B be finite, the Lagrange multiplier must be in the range $0 < \gamma < 1$. It may be seen quite easily that $\gamma = 0$ yields v_0 and v_1 which satisfy (20), trivially, in the form $0 = 0$. In fact, this solution corresponds to the trajectory in the phase plane, connecting adjacent unstable (saddle) points of the loop. However, there may be another solution which yields a greater maximum and, therefore, a lower allowable mistuning.

Numerical calculation shows that for ω_n/K greater than about 0.25, the maximum actually does occur for $\gamma = 0$, and the corresponding synchronization limit curve is plotted in Fig. 3 for $\zeta = 0.5$.

For ω_n/K small, a simple expression for the maximum allowable mistuning can be obtained by a series expansion of the functions. The numerical solution shows that when ω_n/K is small, the synchronization limit occurs for $|v_0|$ and v_1 large. Hence, we can expand $f_A(v)/f_B(-v)$ in powers of $1/v$ and retain only the first few terms of the expansion. Actually, it is simpler to expand the logarithm of the ratio in a power series in $1/(v+\zeta)$. Terms no higher than the second power of $1/(v+\zeta)$ and first power of ω_n/K are retained. The result is

$$\log\left[\frac{f_A(v)}{f_B(-v)}\right] = -2\frac{\omega_n}{K}\frac{1}{v+\zeta} - \left(1 - 2\zeta\sqrt{1+\zeta^2}\frac{\omega_n}{K}\right)\frac{1}{(v+\zeta)^2}. \quad (28)$$

Using (20), the solution for v_1 in terms of v_0 is obtained as

$$\frac{1}{|v_0+\zeta|} - \frac{1}{v_1+\zeta} = \frac{2\omega_n/K}{1 - 2\zeta\sqrt{1+\zeta^2}\frac{\omega_n}{K}} \cong 2\omega_n/K, \quad (29)$$

again retaining only first order terms in ω_n/K.

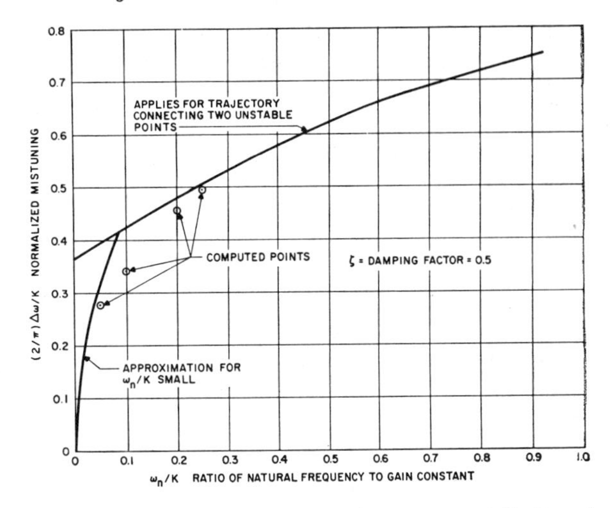

Fig. 3—Maximum allowable mistuning to prevent limit cycles.

The value of v_0 which maximizes (21) is desired. Expanding $f_A(v_1)/f_A(v_0)$ in a power series in $1/(v+\zeta)$ gives

$$\frac{f_A(v_1)}{f_A(v_0)} = \frac{|v_0+\zeta|}{v_1+\zeta}\left[1 - \zeta\left(\frac{1}{|v_0+\zeta|} + \frac{1}{v_1+\zeta}\right)\right]. \quad (30)$$

After substitution of (29) in (30), we obtain an expression only involving v_0, or

$$\frac{f_A(v_1)}{f_A(v_0)} = 1 + 6\zeta\frac{\omega_n}{K} - \frac{2\zeta}{|v_0+\zeta|} - 2\frac{\omega_n}{K}|v_0+\zeta|. \quad (31)$$

The maximum can be found by differentiating with respect to the variable $|v_0+\zeta|$, with the result

$$\left[\frac{1 - \frac{2}{\pi}\frac{\Delta\omega}{K}}{1 + \frac{2}{\pi}\frac{\Delta\omega}{K}}\right]_{\max} = 1 - 4\sqrt{\zeta\frac{\omega_n}{K}}, \quad (32)$$

correct to first order terms. Solving for $\Delta\omega/K$, gives

$$\left(\frac{2}{\pi}\frac{\Delta\omega}{K}\right)_{\min} = 2\sqrt{\zeta\frac{\omega_n}{K}}, \quad (33)$$

correct to first order terms. The approximate synchronization limit corresponding to (33) is plotted in Fig. 3. The points in the transition range between the two curves show that the series expansion yields a reasonable approximation to the true limit.

It will be noted that (33) displays the same functional dependence on ω_n and K discovered experimentally by Gruen[1] and presented as his 46. The analytical verification of his result is an interesting demonstration of the potential utility of the piecewise linear approximation employed in the derivation. A different approximate method of obtaining a relation similar to (33) is discussed by Viterbi.[3]

[3] A. J. Viterbi, "Acquisition and Tracking Behavior of Phase-Locked Loops," Jet Propulsion Lab., Pasadena, Calif., External Publication No. 673; 1959.

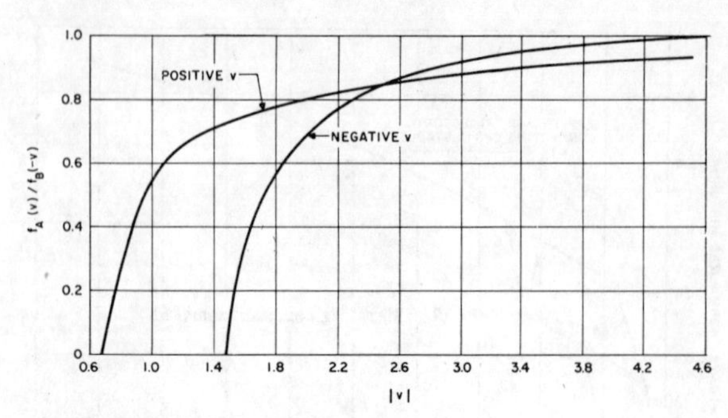

Fig. 4—Example of graph to discover all limit cycles ($\zeta = 0.5$ and $\omega_n/K = 0.1$).

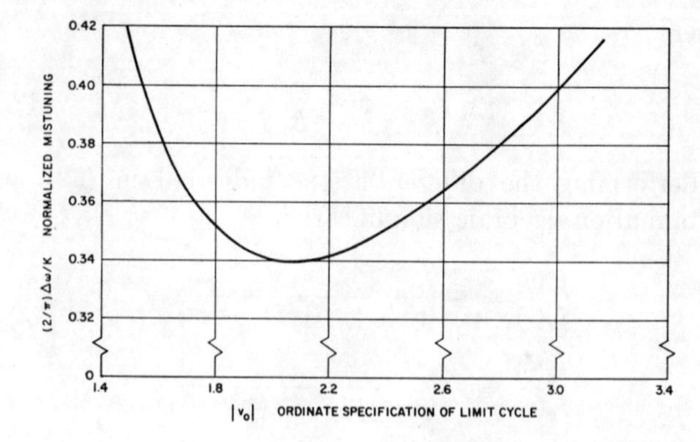

Fig. 5—Illustration of two limit cycle condition.

C. Discussion of Limit Cycles

When a single limit cycle exists for a specified value of mistuning in the second-order loop, the limit cycle is stable on both sides. That is, trajectories wind into the limit cycle both from above and from below. An example of a single stable limit cycle is seen in Fig. 16 of Viterbi's report.[3] In this figure, the mistuning is greater than the particular value which produces a trajectory starting from one unstable point and ending on the next.

The analysis in *A* and *B* of this section has shown that when ω_n/K is suitably small, a limit cycle can occur for a mistuning less than the particular value for which a trajectory connects two unstable points. However, the analysis was directed toward obtaining the *minimum* mistuning for which a limit cycle can exist, and, consequently, the fact that two limit cycles can exist for a given value of mistuning did not become apparent. When two limit cycles exist, it may be observed that if the upper limit cycle is stable, the lower one is unstable and merely represents the dividing line between those trajectories which wind toward the stable limit cycle and those which approach a stable point of the system.

To investigate this situation, a damping factor $\zeta = 0.5$ is again assumed. Furthermore, $\omega_n/K = 0.1$ is utilized, for which case Fig. 3 shows that the trajectory connecting two unstable points is not the limit cycle which occurs for the least mistuning. As discussed previously, the limit cycles for all values of mistuning are represented by the solutions to (20). These solutions may be obtained graphically by plotting the ratio $f_A(v)/f_B(-v)$ for both positive and negative values of v, as shown in Fig. 4. Any horizontal line which intersects both curves in this figure determines values of v_0 and v_1 satisfying (20), and these values can be substituted into (21) to obtain the corresponding mistuning. Fig. 5 shows the relation between allowable mistuning and v_0, which, it will be remembered, is a measure of the ordinate of the limit cycle at $\lambda = -\pi/2$. The curve in Fig. 5 displays a minimum, the value of which was previously plotted in Fig. 3.

In the range of frequencies between the minimum mistuning and that for which a trajectory connects two unstable points, two limit cycles are seen to exist. This range of mistuning is $0.34 < (2/\pi)(\Delta\omega/K) < 0.41$. Since $\omega_n/K = 0.1$, the range may also be specified as $3.4 < 2/\pi\ \Delta\omega/\omega_n < 4.1$. It is interesting to note that Figs. 15 and 16 in Viterbi's report[3] appear to be drawn for mistunings just outside the range where two limit cycles can occur. Fig. 15 is drawn for $\Delta\omega/\omega_n = 4.2$, and Fig. 16 is drawn for $\Delta\omega/\omega_n = 5.8$, which roughly agree with the above limits, considering that Viterbi's curves apply for a damping factor of 0.779. (The parameter 2ζ in Viterbi's report is defined as $\omega_n\tau_1$ rather than $\omega_n\tau_1 + \omega_n/K$, and his ζ is not the damping factor defined in (11).) It would be interesting to obtain phase-plane plots for an intermediate value of $\Delta\omega$, following Viterbi's analog computer method, to demonstrate the existence of two different limit cycles for the nonlinearized loop.

D. Time to Traverse Trajectories for Piecewise Linear System

As a further analytic note on the piecewise linear approximation, we note that the time to traverse a trajectory within a domain of linearity can be expressed by a relatively simple formula involving the parameter v presented previously. (A constant-frequency input signal is presumed.) The formula is obtained by starting from the integral expression

$$\omega_n(t_2 - t_1) = \int_{\lambda(t_1)}^{\lambda(t_2)} \frac{d\lambda}{y} \tag{34}$$

where $y = \lambda/\omega_n$, as before. Within a domain of linearity, λ and x, defined previously in (14) and (15), differ only by a constant, so that $d\lambda$ may be replaced by dx.

A given trajectory is specified in terms of the parameter v by (14) or (15). Omitting the subscript A or B, the integrand of (34) may be expressed as

$$\frac{d\lambda}{y} = \frac{dx}{vx} = \frac{1}{v}\frac{f'(v)}{f(v)}\,dv, \tag{35}$$

noting that the constant of integration and the polarity of x cancel out. Consequently, the time can be specified as an integral involving the function $f(v)$, which has an analytical representation. The functions for a constant applied frequency are given in (26). However, in using these functions, the singularity at $v = \infty$ must be duly considered.

In domain A, the elapsed time is

$$\omega_n(t_2 - t_1) = -\int_{v(t_1)}^{v(t_2)} \frac{dv}{(v + \zeta)^2 + (1 - \zeta^2)}$$

$$= \frac{1}{\sqrt{1 - \zeta^2}} \left[\tan^{-1} \frac{\sqrt{1 - \zeta^2}}{v(t_2) + \zeta} - \tan^{-1} \frac{\sqrt{1 - \zeta^2}}{v(t_1) + \zeta} \right] \quad (36)$$

which has been written in a form which is valid whether $v(t_1)$ and $v(t_2)$ have the same or opposite polarities. That is, when the trajectory passes through $x = 0$, v goes to $-\infty$ and returns from $+\infty$. In this case, the integral properly must be broken into two parts; however, the contributions at the infinite limits are zero as (36) is written, thereby avoiding any difficulty.

In domain B, the elapsed time is

$$\omega_n(t_2 - t_1) = -\int_{v(t_1)}^{v(t_2)} \frac{dv}{(v - \zeta_B)^2 - (\zeta_B^2 + 1)}$$

$$= \frac{1}{2\sqrt{\zeta_B^2 + 1}} \left[\log \left| \frac{\sqrt{\zeta_B^2 + 1} + v(t_2) - \zeta_B}{\sqrt{\zeta_B^2 + 1} - v(t_2) + \zeta_B} \right| - \log \left| \frac{\sqrt{\zeta_B^2 + 1} + v(t_1) - \zeta_B}{\sqrt{\zeta_B^2 + 1} - v(t_1) + \zeta_B} \right| \right]. \quad (37)$$

Because the logarithm approaches zero as $v \to \pm\infty$, (37) is also valid irrespective of the polarities of $v(t_1)$ and $v(t_2)$.

If a trajectory is computed analytically by joining analytical expressions at the breakpoints, a sequence of values of the parameter v will be a by-product. These values may be substituted in (36) and (37) to yield the time necessary to traverse each successive half cycle of the phase difference.

IV. Conclusions

Piecewise linearization of a phase-lock loop has proved effective for the analytical investigation of limit cycles in the second-order loop, in particular for computing the associated mistuning of the input carrier. Noting that a limit cycle represents an undesirable, asynchronous mode of operation of the phase-lock loop, a condition on the maximum allowable mistuning to insure no limit cycles may properly be interpreted as a synchronization limit, inside of which the phase-lock loop will pull in to a stable point irrespective of the initial point of the trajectory in the phase plane.

The computed synchronization limit is similar to the experimental curve obtained by Gruen for an actual phase-lock loop. However, an advantage of the piecewise linear analysis is the insight given into the role played by limit cycles in establishing the synchronization limit when the noise bandwidth is made much less than the hold-in range of the loop.

As noted in the discussion, the difference frequency associated with the limit cycle becomes large as the ratio of noise bandwidth to hold-in range approaches zero. In this case, if the initial frequency error is suitably bounded, the limit cycle will not be reached, and a greater mistuning is allowed than indicated by Fig. 3.

Synchronous Communications*

JOHN P. COSTAS†, ASSOCIATE MEMBER, IRE

Summary—It can be shown that present usage of amplitude modulation does not permit the inherent capabilities of the modulation process to be realized. In order to achieve the ultimate performance of which AM is capable synchronous or coherent detection techniques must be used at the receiver and carrier suppression must be employed at the transmitter.

When a performance comparison is made between a synchronous AM system and a single-sideband system it is shown that many of the advantages normally attributed to single sideband no longer exist. SSB has no power advantage over the synchronous AM (DSB) system and SSB is shown to be more susceptible to jamming. The performance of the two systems with regard to multipath or selective fading conditions is also discussed. The DSB system shows a decided advantage over SSB with regard to system complexity, especially at the transmitter. The bandwidth saving of SSB over DSB is considered and it is shown that factors other than signal bandwidth must be considered. The number of *usable* channels is not necessarily doubled by the use of SSB and in many practical situations *no increase* in the number of usable channels results from the use of SSB.

The transmitting and receiving equipment which has been developed under Air Force sponsorship is discussed. The receiving system design involves a local oscillator phase-control system which derives carrier phase information from the sidebands alone and does not require the use of a pilot carrier or synchronizing tone. The avoidance of superheterodyne techniques in this receiver is explained and the versatility of such a receiving system with regard to the reception of many different types of signals is pointed out.

System test results to date are presented and discussed.

Introduction

FOR A good many years, a very large percentage of all military and commercial communications systems have employed amplitude modulation for the transmission of information. In spite of certain well-known shortcomings of conventional AM, its use has been continued mainly due to the simplicity of this system as compared to other modulation methods which have been proposed. During the last few years, however, it has been felt by many responsible engineers that the increased demands being made on communications facilities could not be met by the use of conventional AM and that new modulation techniques would have to be employed in spite of the additional system complexity. Of these new techniques, single sideband has been singled out as the logical replacement for conventional AM and a great deal of publicity and financial support has been given SSB as a consequence.

Many technical reasons have been given to support the claim that SSB is better than AM and these points will be discussed in some detail later in this paper. In addition, many experiments have been performed which also indicate a superiority for SSB over AM. Some care must be taken, however, in drawing conclusions from the above statements. *We cannot conclude that SSB is superior to AM because we have no assurance whatever that conventional AM systems make efficient use of the modulation process employed.* In other words, AM as a modulation process may be capable of far better performance than that which is obtained in conventional AM systems. If an analysis is made of AM and SSB systems, it will be found that existing SSB systems are very nearly optimum with respect to the modulation process employed whereas conventional AM systems fall far short of realizing the full potential of the modulation process employed. In fact, it could honestly be said that we have been misusing rather than using AM in the past. Realization of the above situation raises some immediate questions: What are the equipment requirements of the optimum AM system? How does the performance of the optimum AM system compare with that of SSB? Which shows the greater promise of fulfilling future military and commercial communications requirements, optimum AM or SSB? The remainder of this paper will be devoted mainly to answering these questions.

* Original manuscript received by the IRE, August 31, 1956.
† General Electric Co., Syracuse, N. Y.

Synchronous Communications—The Optimum AM System

Receiver

Conventional AM systems fail to obtain the full benefits of the modulation process for two main reasons: Inefficient use of generated power at the transmitter and inefficient detection methods at the receiver. Starting with the receiver it can be shown that if maximum receiver performance is to be obtained the detection process must involve the use of a phase-locked oscillator and a synchronous or coherent detector. The basic synchronous receiver is shown in Fig. 1. The incoming signal is

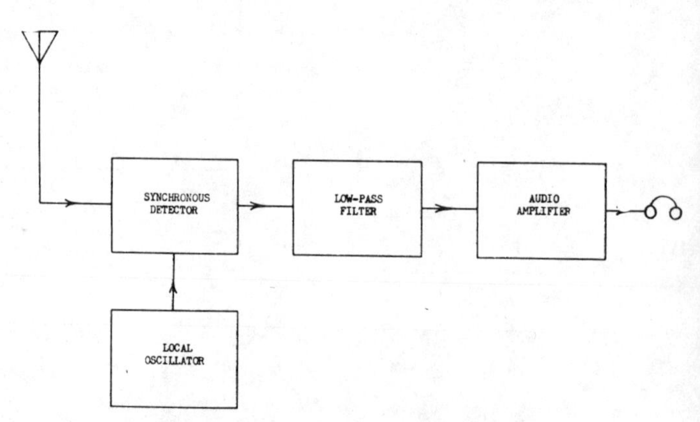

Fig. 1—Basic synchronous receiver.

mixed or multiplied with the coherent local oscillator signal in the detector and the demodulated audio output is thereby directly produced. The audio signal is then filtered and amplified. The local oscillator must be main-

Reprinted from *Proc. IRE*, vol. 44, pp. 1713–1718, Dec. 1956.

tained at proper phase so that the audio output contributions of the upper and lower sidebands reinforce one another. If the oscillator phase is 90° away from the optimum value a null in audio output will result which is typical of detectors of this type. The actual method of phase control will be explained shortly, but for the purpose of this discussion maintenance of correct oscillator phase shall be assumed.

In spite of the simplicity of this type of receiver, there are several important advantages worthy of note. To begin with, no IF system is employed which eliminates completely the problem of image responses. The opportunity to effectively use post-detector filtering allows extreme selectivity to be obtained without difficulty. The selectivity curve of such a receiver will be found to be the low-pass filter characteristic mirror-imaged about the operating frequency. Not only is a high order of selectivity obtained in this manner, but the selectivity of the receiver may be easily changed by low-pass filter switching. The carrier component of the AM signal is not in any way involved in the demodulation process and need not be transmitted when using such a receiver. Furthermore, detection may be accomplished at very low level and consequently the bulk of total receiver gain may be at audio frequencies. This permits an obvious application of transistors but more important it allows the selectivity determining low-pass filter to be inserted at a low-level point in the receiver which aids immeasurably in protecting against spurious responses from very strong undesired signals.

Phase Control: To obtain a practical synchronous receiving system some additions to the basic receiver of Fig. 1 are required. A more complete synchronous receiver is shown in Fig. 2. The first thing to be noted

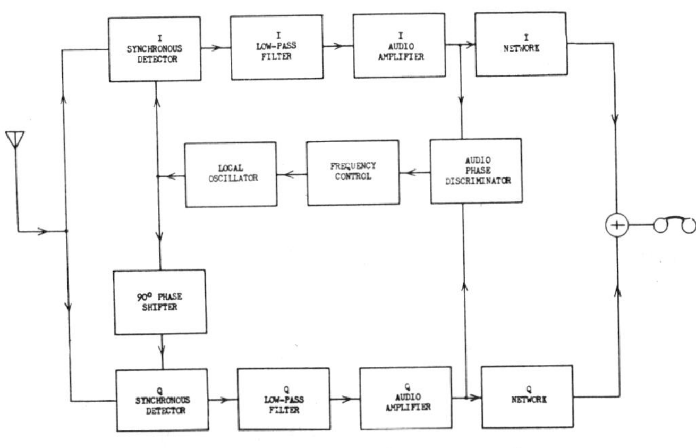

Fig. 2—Two-phase synchronous receiver.

about this diagram is that we have essentially two basic receivers with the same input signal but with local oscillator signals in phase quadrature to each other. To understand the operation of the phase-control circuit consider that the local oscillator signal is of the same phase as the carrier component of the incoming AM signal. Under these conditions, the in-phase or I audio amplifier output will contain the demodulated audio signal while the quadrature or Q audio amplifier will have no output due to the quadrature null effect of the Q synchronous detector. If now the local oscillator phase drifts from its proper value by a few degrees the I audio will remain essentially unaffected but there will now appear some audio output from the Q channel. This Q channel audio will have the same polarity as the I channel audio for one direction of local oscillator phase drift and opposite polarity for the opposite direction of local oscillator phase drift. The Q audio level is proportional to the magnitude of the local oscillator phase angle error for small errors. Thus by simply combining the I and Q audio signals in the audio phase discriminator a dc control signal is obtained which automatically corrects for local oscillator phase errors. It should be noted that phase control information is derived entirely from the sideband components of the AM signal and that the carrier if present is not used in any way. Thus since both synchronization and demodulation are accomplished in complete independence of carrier, suppressed-carrier transmissions may be employed.

It is unfortunate that many engineers tend to avoid phase-locked systems. It is true that a certain amount of stability is a prerequisite but it has been determined by experiment that for this application the stability requirements of single-sideband voice are more than adequate. Once a certain degree of stability is obtained, the step to phase lock is a simple one. It is interesting to note that this phase-control system can be modified quite readily to correct for large frequency errors when receiving AM due to Doppler shift in air-to-air or ground-to-air links.

It is apparent that phase control ceases with modulation and that phase lock will have to be reestablished with the reappearance of modulation. This has not proved to be a serious problem since lock-up normally occurs so rapidly that no perceptible distortion results when receiving voice transmission. It should be further noted that such a phase control system is inherently immune to carrier capture or jamming. In addition it has been found that due to the narrow noise bandwidth of the phase-control loop, synchronization is maintained at noise levels which render the channel useless for voice communications.

Interference Suppression: The post-detector filters provide the sharp selectivity which, of course, contributes significantly to interference suppression. However, these filters cannot protect against interfering signal components which fall within the pass band of the receiver. Such interference can be reduced and sometimes eliminated by proper combination of the I and Q channel audio signals. To understand this process consider that the receiver is properly locked to a desired AM signal and that an undesired signal appears, some of whose components fall within the receiver pass band. Under these conditions the I channel will contain the desired audio signal plus an undesired component due to the interference. The Q channel will contain only an interference component also arising from the presence of the interfering signal. In general the interference component in the I channel and the interference component in the Q channel are related to one another or they may be said to be correlated. Advantage may be taken of this correlation by treating the I and Q voltages with the I and Q networks and adding these network outputs. If properly done this process will reduce and sometimes eliminate the interfering signal from the receiver output as a result of destructive addition of the I and Q interference voltages.

The design of these networks is determined by the spectrum of the interfering signal and the details of network design may be found a report by the author.[1] Although such details cannot be given here it is interesting to consider one special interference case. If the interfering signal spectrum is confined entirely to one side of the desired signal carrier frequency the optimum I and Q networks become the familiar 90° phasing networks common in single-sideband work. Such operation does not however result in single-sideband reception of the desired signal since both desired signal sidebands contribute to receiver output at all times. This can be seen by noting that the Q channel contains no desired signal component so that network treatment and addition effects only the undesired audio signal components. The phasing networks are optimum only for the interference condition assumed above. If there is an overlap of the carrier frequency by the undesired signal spectrum the phasing networks are no longer optimum and a different network design is required for the greatest interference suppression.

This two-phase method of AM signal reception can aid materially in reducing interference. As a matter of fact it can be shown that the true anti-jam characteristics of AM cannot be realized unless a receiving system of the type discussed above is used. If we now compare the anti-jam characteristics of single sideband and suppressed-carrier AM properly received it will be found that intelligent jamming of each type of signal will result in a two-to-one power advantage for AM. The bandwidth reduction obtained with single sideband does not come without penalty. One of the penalties as we see here is that single sideband is more easily jammed than double sideband.

Transmitter

The synchronous receiver described above is capable of receiving suppressed-carrier AM transmissions. If a carrier is present as in standard AM this will cause no trouble but the receiver obviously makes no use whatever of the carrier component. The opportunity to employ carrier-suppressed AM transmissions can be used to good advantage in transmitter design. There are many ways in which to generate carrier-suppressed AM signals and one of the more successful methods is shown in Fig. 3. A pair of class-C beam power amplifiers are screen modulated by a push-pull audio signal and are driven in push-pull from an rf exciter. The screens are returned to ground or to some negative bias value by means of the driver transformer center tap. Thus in the absence of modulation no rf output results and during modulation the tubes conduct alternately with audio polarity change. The circuit is extremely simple and a given pair of tubes used in such a transmitter can easily match the average rf power output of the same pair of tubes used in SSB linear amplifier service. The circuit is self-neutralizing and the tune-up procedure is very much the same as in any other class-C rf power amplifier. The excitation requirements are modest and as an example the order of 8 w of audio are required to produce a sideband power output equivalent to a standard AM carrier output of 1 kw. Modulation linearity is good and the circuit is amenable to various feedback techniques for obtaining very low distortion which may be required for multiplex transmissions.

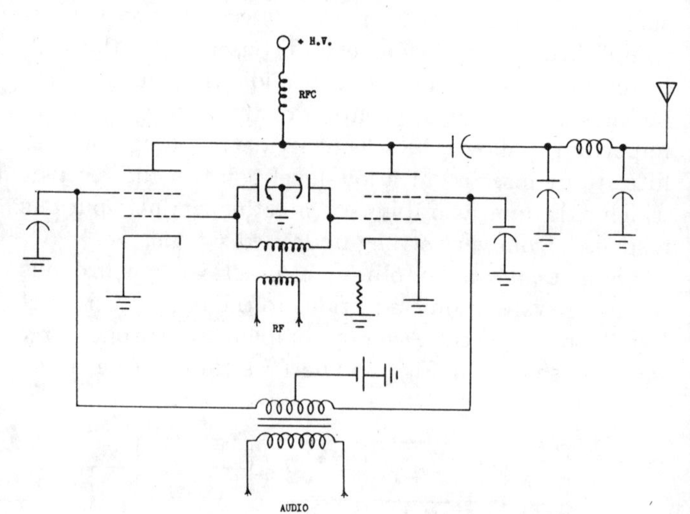

Fig. 3—Suppressed-carrier AM transmitter.

This transmitter circuit is by no means new. The information is presented here to indicate the equipment simplicity which can be realized by use of synchronous AM communications.

Prototype Equipment

A synchronous receiver covering the frequency range of 2–32 mc is shown in Fig. 4. The theory of operation of this receiver is essentially that of the two-phase synchronous receiver discussed earlier. This is a direct conversion receiver and the superheterodyne principle is not used. A rather unusual frequency synthesis system is employed to give high stability with very low spurious response. Only one crystal is used and this is a 100 kc oven-controlled unit.

This receiver will demodulate standard AM, suppressed-carrier AM, single sideband, narrow-band fm, phase modulation, and cw signals in an optimum man-

[1] J. P. Costas, "Interference Filtering," Mass. Inst. Tech. Res. Lab. of Elec., Tech. Rep. no. 185.

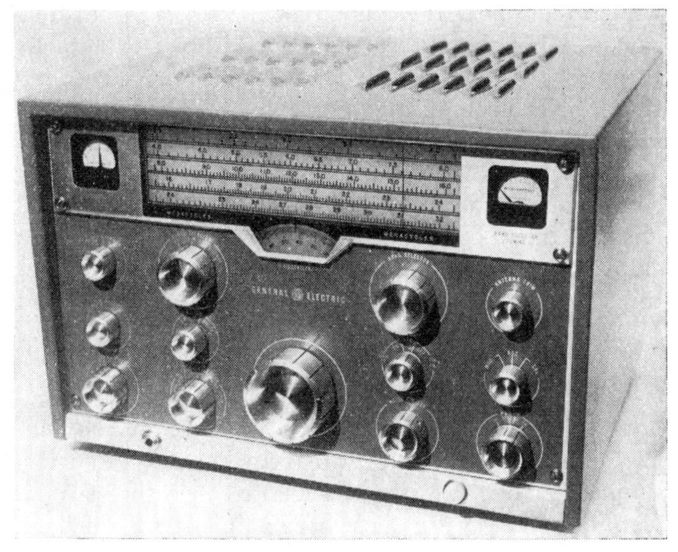

Fig. 4—The AN/FRR—48 (XW-1) synchronous receiver.

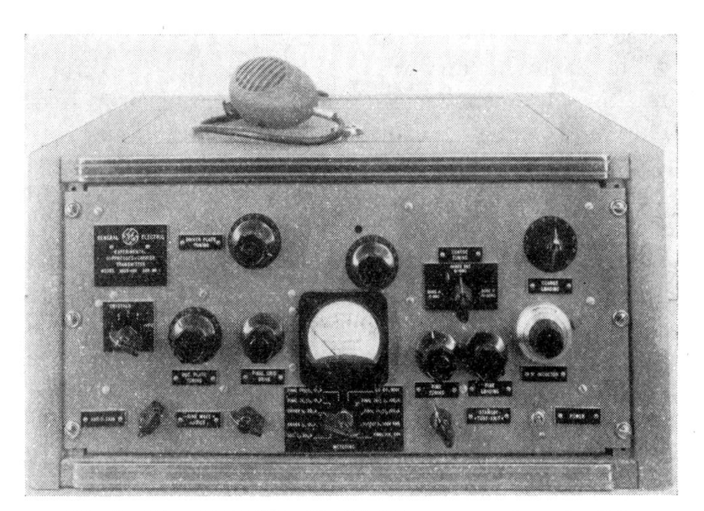

Fig. 5—The AN/FRT—29 (XW-1) suppressed-carrier AM transmitter.

ner. This versatility is a natural by-product of the synchronous detection system and no great effort is required to obtain this performance.

Fig. 5 shows a suppressed-carrier AM transmitter using a pair of 6146 tubes in the final. This unit is capable of 150-w peak sideband power output for continuous sine-wave modulation. The modulator is a single 12BH7 miniature double triode. Fig. 6 shows a transmitter capable of 1000-w peak sideband power output under continuous sine wave audio conditions. The final tubes are 4-250-A's and the modulator uses a pair of 6L6's. Both of these transmitters are continuously tunable over 2–30 mc.

A Comparison of Synchronous AM and Single Sideband

It is interesting at this point to compare the relative advantages and disadvantages of synchronous AM and single-sideband systems. Although single sideband has a clear advantage over conventional AM this picture is radically changed when synchronous AM is considered.

Fig. 6—The AN/FRT—30 (XW-1) suppressed-carrier AM transmitter.

Signal-to-Noise Ratio

If equal average powers are assumed for SSB and synchronous AM it can easily be shown that identical s/n ratios will result at the receiver. The additional noise involved from the reception of two sidebands is exactly compensated for by the coherent addition of these sidebands. The 9-db advantage often quoted for SSB is based on a full AM carrier and a peak power comparison. Since we have eliminated the carrier and since a given pair of tubes will give the same average power in suppressed-carrier AM or SSB service there is actually no advantage either way. If intelligent jamming rather than noise is considered there exists a clear advantage of two-to-one in average power in favor of synchronous AM.

System Complexity

Since the receiver described is also capable of SSB reception it would appear that synchronous AM and SSB systems involve roughly the same receiver complexity. This is not altogether true since much tighter design specifications must be imposed if high quality SSB reception is to be obtained. If AM reception only is considered these specifications may be relaxed considerably without materially affecting performance. The synchronous receiver described earlier may possess important advantages over conventional superheterodyne receivers but this point is not an issue here.

The suppressed-carrier AM transmitter is actually simpler than a conventional AM transmitter. It is of course far simpler than any SSB transmitter. There are no linear amplifiers, filters, phasing networks, or frequency translators involved. Personnel capable of operating or maintaining standard AM equipment will have no difficulty in adapting to suppressed-carrier AM. The military and commercial significance of this situation is rather obvious and further discussion of this point is not warranted.

Long-Range Communications

The selective fading and multipath conditions encountered in long-range circuits tend to vary the amplitude and phase of one sideband component relative to the other. This would perhaps tend to indicate an advantage for SSB but tests to date do not confirm this. Synchronous AM reception of standard AM signals over long paths has been consistently as good as SSB reception of the same signal. In some cases it was noted that the SSB receiver output contained a serious flutter which was only slightly discernible in the synchronous receiver output. Some attempt has been made to explain these results but as yet no complete explanation is available. One interesting fact about the synchronous receiver is that the local oscillator phase changes as the sidebands are modified by the medium since phase control is derived directly from the sidebands. In a study of special cases of signal distortion, it was found that the oscillator orients itself in phase in such a way as to attempt to compensate for the distortion caused by the medium. This may partially explain the good results which have been obtained. Perhaps another point of view would be that the synchronous receiver is taking advantage of the inherent diversity feature provided by the two AM sidebands.

Test results to date indicate that synchronous AM and single-sideband provide much the same performance for long-range communications. The AM system has been found on occasion to be better but since extensive tests have not been performed and a complete explanation of these results is not yet available it would be unfair to claim any advantage at this time for AM.

Spectrum Utilization

In theory, single-sideband transmissions require only half the bandwidth of equivalent AM transmissions and this fact has led to the popular belief that conversion to single sideband will result in an increase in usable channels by a factor of two. If a complete conversion to single sideband were made those who believe that twice the number of usable channels would be available might be in for a rather rude awakening. There are many factors which determine frequency allocation besides modulation bandwidth. Under many conditions it actually turns out that modulation bandwidth is not a consideration. This is a complicated problem and only a few of the more pertinent points can be discussed briefly here.

To begin with the elimination of one sideband is a complicated and delicate business. Any one of several misadjustments of the SSB transmitter will result in an empty sideband which is not actually empty. We are not thinking here of a telephone company point-to-point system staffed by career personnel, but rather we have in mind the majority of military and commercial field installations. This is in no way meant to be a criticism, but the technical personnel problem faced by the military especially in time of war is a serious one and this simple fact of life cannot be ignored in future system planning. Thus we must concede that single-sideband transmissions will in practice not always be confined to one sideband and that those who allocate frequencies must take this into consideration.

There may be those who would argue that SSB transmitting equipment can be designed for simple operation. This is probably true but in general operational simplicity can only be obtained at the expense of additional complexity in manufacture and maintenance. This of course trades one set of problems for another but if we assume ideal SSB transmission we are still faced with an even more serious allocation problem. We refer here to the problem of receiver nonlinearity which becomes a dominant factor when trying to receive a weak signal in the presence of one or more near-frequency strong signals. Under such conditions the single-signal selectivity curves often shown by manufacturers are next to meaningless. This strong undesired-weak desired signal situation often arises in practice especially in the military where close physical spacing of equipment is mandatory as in the case of ships or aircraft and where the signal environment changes due to the changing locations of these vehicles. Because of this situation allocations to some extent must be made practically independent of modulation bandwidth and the theoretical spectrum conservation of single sideband cannot always be advantageously used.

The problem of receiver nonlinearity is especially serious in multiple conversion superheterodyne receivers for obvious reasons. This was the dominant factor in choosing a direct conversion scheme in the synchronous receiver described earlier. Although this approach has given good results and continued refinement has indicated that significant advances over prior art can be obtained, it cannot be said however that the receiver problem is solved. This problem will probably remain a serious one until new materials and components are made available. This is a relatively slow process and it is not at all absurd to consider that by the time this problem is eliminated new modulation processes will have appeared which will eclipse both of those now being considered.

In short, the spectrum economies of SSB which exist in theory cannot always be realized in practice as there exist many important military and commercial communications situations in which no increase in usable channels will result from the adoption of single sideband.

Jamming

The reduction of transmission bandwidth afforded by single sideband must be paid for in one form or another. A system has yet to be proposed which offers nothing but advantages. One of the prices paid for this reduction in bandwidth is greater susceptibility to jamming as was previously mentioned. There is an understandable tendency at times to ignore jamming since the systems with which we are usually concerned provide us with ample worries without any outside aid. Jamming of course cannot be ignored and from a military point of view this raises a very serious question. If we concede for the moment that by proper frequency allocation single sideband offers a normal channel capacity advantage over AM, what will happen to this advantage when we have the greatest need for communications? It is almost a certainty that at the time of greatest need jamming will have to be reckoned with. Under these conditions any channel capacity advantage of SSB could easily vanish. A definite statement to this effect cannot be made of course without additional study but this is a factor well worth considering.

Conclusion

There is an undeniable need for improved communications and to date it appears that single sideband has been almost exclusively considered to supplant conventional AM. It has been the main purpose of this paper to point out that the improved performance needed can be obtained in another way. The synchronous AM system can compete more than favorably with single-sideband when all factors are taken into account.

Acknowledgment

Much of the work reported here was sponsored by the Rome Air Development Center of the Air Research and Development Command under Air Force contract AF 30(602) 584. The author wishes to acknowledge the support, cooperation, and encouragement which has been extended by the personnel of the Rome Air Development Center.

Design and Performance of Phase-Lock Circuits Capable of Near-Optimum Performance Over a Wide Range of Input Signal and Noise Levels*

R. JAFFE† AND E. RECHTIN†

Introduction

PHASE-LOCK LOOPS provide an efficient method for detection and tracking of narrow-band signals in the presence of wide-band noise. This paper explains how minimum-rms-error loops may be designed if the input-signal level, input-noise level, and a specification for transient performance are given. However, the system performance of such loops departs rapidly from the best obtainable performance if either the signal or the noise levels are different from the design levels, and if no compensating changes are made in the loop. A marked improvement results if the total input power is held constant, regardless of signal or noise levels. It will be demonstrated that a fixed-component loop preceded by a bandpass limiter yields near-optimum performance over a wide range of input signal and noise levels. The following topics will be discussed:

1. An outline of the theoretical design of minimum-rms-error, phase-lock loops when input-signal level, input-noise level, and a specification for transient error are given.

2. The effects of different input levels of signal and noise:
   a. On a system having a fixed-component loop that is optimum only for an original set of levels.
   b. On a system in which loop components maintain optimum performance when the new levels are given.

3. Characteristics of a bandpass limiter.

4. A comparison of the effect of different signal and noise levels:
   a. On a loop using a fixed filter preceded by an automatic-gain-control (AGC) system that holds the signal level constant.
   b. On a fixed-filter loop preceded by a bandpass limiter.
   c. On a variable-filter loop continually adjusted to be optimum.

5. Experimental verification of the fixed-component loop preceded by a bandpass limiter.

* Presented at the Western Electronics Show and Convention, Los Angeles, Calif., August 25–27, 1954. This paper presents the results of one phase of research carried out at the Jet Propulsion Laboratory, California Institute of Technology, under Contract No. DA-04-495-Ord 18, sponsored by the Department of the Army, Ordnance Corps.
† Jet Propulsion Laboratory, California Institute of Technology, Pasadena, Calif.

Description of the Loop

The elements of a typical phase-locked loop are shown in Fig. 1. Signal input to such a loop may be assumed to be $\sqrt{2}A \sin [\omega t + \theta_1(t)]$ where A is the rms signal amplitude, ω is the signal-center radian frequency, and $\theta_1(t)$ represents the information content of the signal. It is assumed that the noise input to the loop is narrowbanded about the signal-center frequency and has an essentially flat spectrum over the band; the noise input is represented by $N(t) \sin \omega t$. Output of the voltage-controlled oscillator (VCO) is assumed to have an rms amplitude of C, a center frequency identical to that of the signal, and a phase equal to $\theta_2(t)$.

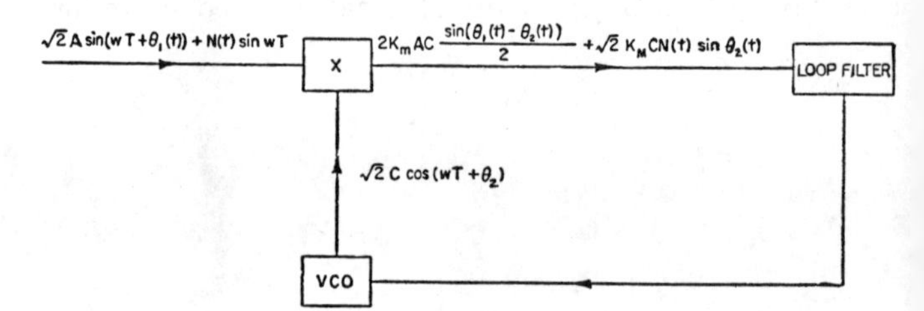

Fig. 1—Typical phase-lock loop.

Briefly, the operating principles of such a loop are as follows: The multiplier beats the signal input and the VCO output together, giving a low-frequency output proportional to the sine of $\theta_1 - \theta_2$. The loop filter accepts only this low-frequency term which is applied as a control voltage to the voltage-controlled oscillator, thereby forcing the VCO output phase θ_2 to be equal to the signal phase θ_1.

It is possible to make a phase-locked loop such that the loop need have a bandwidth only large enough to pass the *difference* between the signal frequency and the VCO estimate of the signal frequency. Since this difference frequency has considerably less variation than the actual signal frequency, the loop does not need nearly as large a bandwidth as would be needed if the loop were merely a tuned circuit placed between the system input and output, which would have to pass all frequencies over which the signal was expected to vary.

Since the bandwidth of the tracking loop is much smaller than that of a comparable nontracking filter, the amount of noise passed on to the output is proportionally smaller, and the loop accordingly picks up a greater resistance to interference.

Reprinted from *IRE Trans. Inform. Theory*, vol. IT-1, pp. 66–76, Mar. 1955.

Linearization of the Loop

To obtain a mathematical description of the loop suitable for subsequent analysis, it is desirable to approximate linearly the nonlinear operation of the actual loop.[1]

The low-frequency multiplier output is

$$2K_M AC \frac{\sin(\theta_1 - \theta_2)}{2} + N(t)\sqrt{2}\, C \sin \theta_2(t),$$

where K_M is the multiplier constant relating the multiplier voltage output both to the amplitude of its two inputs and to their phase difference in radians. Since the filter following the multiplier is low-pass, the high-frequency term in $2\,\omega t$ is disregarded. Under the assumption that the phase error $\theta_1 - \theta_2$ is small and that the input noise is uncorrelated with sine θ_2,

$$\sin(\theta_1 - \theta_2) \cong \theta_1 - \theta_2$$

and

$$N(t) \sin \theta_2 \cong \frac{N(t)}{\sqrt{2}}.$$

The multiplier output may therefore be represented approximately as

$$K_M C[A(\theta_1 - \theta_2) + N(t)].$$

The nonlinear operation of multiplication may now be linearized to subtraction, as shown in Fig. 2, together with associated changes of (1) making the phase input to the loop proportional to $\theta_1 + (N/A)$, (2) inserting an amplifier of gain A after the subtractor so that the voltage feeding the filter will be the same as in the nonlinearized configuration, and (3) representing the VCO as an integrator which relates its phase output to the integral of its input voltage by the VCO constant K_v. The phases around the loop and the transfer functions of the loop elements have been expressed in terms of their Laplace Transforms rather than as functions of time.

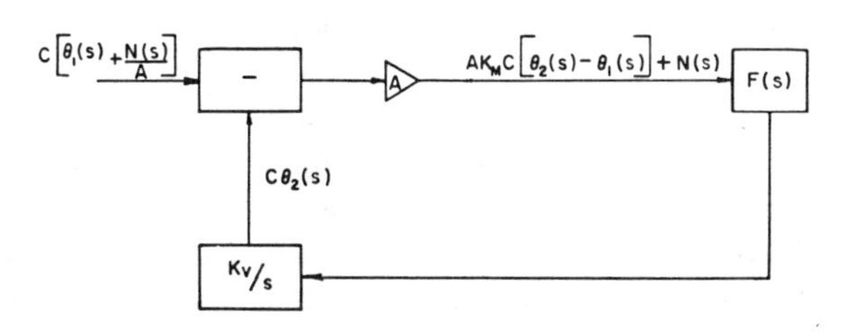

Fig. 2—Partial linearization of phase-lock loop.

Finally, the various gain coefficients may be combined into a single amplifier of gain K, as shown in Fig. 3, where K is evaluated in both radians and degrees. The product AK is defined to be the loop gain.

Criteria for Filter Design

The loop filter is important in insuring good loop operation and should be designed so that it performs two distinct functions:

1. It should minimize VCO phase-noise jitter due to noise interference.
2. It should maintain, at a specified amount, transient error in the VCO phase due to specified changes in signal phase.

Transient error is defined as the infinite-time integral of $(\theta_1 - \theta_2)^2$ and is caused solely by signal phase variations. Otherwise expressed,

$$E_T^2 = \int_0^\infty [\theta_1(t) - \theta_2(t)]^2\, dt.$$

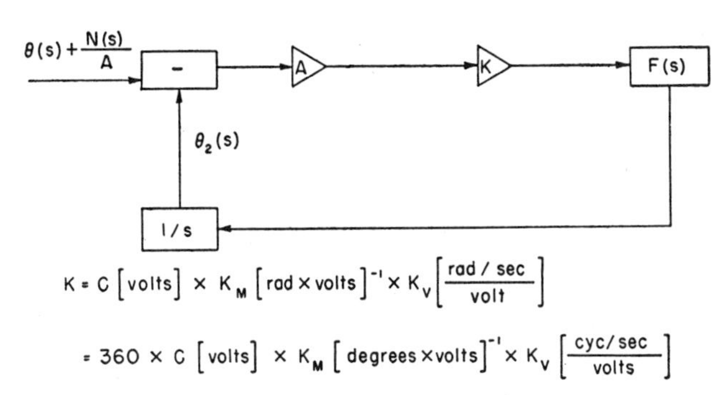

Fig. 3—Complete linearization of phase-lock loop.

Theory of Filter Design

If the rms phase jitter due to noise interference is denoted by σ_N, and if the transient error, as previously defined, is represented by E_T^2, the design criterion may be stated as follows:

$$\sigma_N^2 + \lambda^2 E_T^2 = \Sigma^2 = \text{minimum},$$

where λ is an undetermined constant, a Lagrangian multiplier used in the standard calculus-of-variations procedure. Theoretically, when the filter form has been computed (Appendix I), the value of λ may be determined from the requirement that the transient error to a specified signal must equal a specified value. Practically, λ is evaluated from the loop-bandwidth parameter B_0, which will be described in a subsequent paragraph.

Filters may be designed for many forms of signal phase inputs. It has been found that filters designed to have zero steady-state error to signal *frequency* changes (zero velocity-error loops) are relatively easy to mechanize and work well for a variety of inputs. This form of filter will be the only one discussed in the body of the paper, although the results of similar analyses concerning loops designed to follow phase steps and frequency ramps are discussed in Appendix II.

Derivation of the optimum filter is given in Appendix I. It is shown that such a filter for zero velocity-error loops has the transform

$$F(s) = \frac{B_0^2 + \sqrt{2}\, B_0 s}{A_0 K s},$$

where

$$B_0^2 \equiv \frac{\lambda(\Delta\omega)}{(N_0/A_0)} \sqrt{2\Delta f},$$

[1] The linearization appears quite legitimate, based on experimental evidence.

and where K represents the loop constant defined in Fig. 3, $\Delta\omega$ is the signal radian-frequency step, N_0 is the expected rms noise input of bandwidth Δf, and A_0 is the expected rms signal strength.

The output phase jitter and transient error of such an optimum loop, when the input noise and signal levels are those expected, are, respectively,

$$\sigma_N^2 = \frac{3}{2\sqrt{2}} B_0\left(\frac{N_0^2}{A_0^2 2\Delta f}\right) \text{rad}^2,$$

and

$$E_T^2 = \frac{(\Delta\omega)^2}{2\sqrt{2}\, B_0^3} \text{rad}^2 \text{ sec}.$$

It may be shown that B_0 is proportional to the expected loop bandwidth; B_0 is approximately equal to two times the loop noise bandwidth and three times the loop 3-db bandwidth. If the loop bandwidth is specified, B_0 may therefore be determined and λ may be evaluated from the definition of B_0.

It is important to note that the parameter B_0 varies with the expected input signal-to-noise ratio and that the value of the optimum filter varies both with B_0 and with the expected signal amplitude A_0.

This filter form is therefore optimum only for a particular input-signal level and a particular input-noise level. If the actual levels differ from their expected values, the filter is no longer optimum.

Performance of Loops Having Levels Other Than Design Levels

It may be shown, by integrating the actual loop transfer function, that the actual output phase jitter and transient error in a fixed-filter loop fed with signal and noise levels different from those for which it was designed are, respectively,

$$\sigma_N^2 = B_0\left(\frac{N^2}{A^2 2\Delta f}\right)\left[\frac{2(A/A_0) + 1}{2\sqrt{2}}\right] \text{rad}^2 \tag{1}$$

and

$$E_T^2 = \frac{(\Delta\omega)^2}{2\sqrt{2}\, B_0^3} \frac{1}{(A/A_0)^2} \text{rad}^2 \text{ sec}, \tag{2}$$

where A is actual signal level, N, actual noise level.

It also may be shown that an optimum variable filter with values continually adjusted with slowly changing signal and noise levels would have the transfer function

$$F(s) = \frac{B_0^2\left[\frac{(A/A_0)}{(N/N_0)}\right] + \sqrt{2}\, B_0\left[\frac{(A/A_0)}{(N/N_0)}\right]^{1/2} s}{KAs}; \tag{3}$$

and noise jitter and transient error, respectively, of

$$\sigma_N^2 = \frac{3B_0}{2\sqrt{2}} \frac{\sqrt{(N_0/A_0)}\,(N/A)^{3/2}}{2\Delta f} \text{rad}^2$$

and

$$E_T^2 = \frac{(\Delta\omega)^2}{2\sqrt{2}\, B_0^3} \frac{(N/A)^{3/2}}{(N_0/A_0)^{3/2}} \text{rad}^2.$$

Thus far, a filter has been derived that is optimum for fixed input levels, and its phase jitter and transient error have been obtained when the input levels matched and mismatched the design levels.

A variable filter has also been derived that was assumed to adjust itself to be optimum for different signal and noise inputs. Expressions for the phase jitter and transient error in this loop have been obtained.

The problem remaining is how to obtain optimum or near-optimum performance over a wide range of input signal and noise levels, yet achieve such performance using a filter than can easily be mechanized—instead of resorting to auxiliary servo loops that continually readjust the filter to keep it optimum.

One possible solution is to use an AGC based on the signal only. The AGC voltage may be obtained if the signal is multiplied by a 90-degree, phase-shifted version of the VCO output; the inputs to the multiplier will then be in phase, and the dc multiplier output will be proportional to the signal level and may be used for the AGC. The AGC method has several disadvantages: (1) It introduces additional components. (2) It introduces additional time constants which may cause two-loop oscillation. (3) It usually results in systems with less dynamic range.

Characteristics of a Bandpass Limiter

An alternate solution, which appeared promising, was to use a limiter preceding the loop. If the noise input to the system increased, the signal strength at the limiter output would decrease because of the limiter property of holding total output power constant. However, the noise bandwidth of the loop is directly dependent on the signal amplitude at the limiter output. Therefore, the increase in input noise, by forcing down the signal amplitude at the limiter output, would reduce the loop bandwidth and require that the phase jitter at the VCO output be a smaller percentage of the input noise. The system would therefore appear to be self-compensating.

The type of limiter to be considered performs perfect snap-action limiting, i.e., its output is assumed to be $+1$ for inputs greater than 0 and -1 for inputs less than 0; such a representation closely approximates an actual limiter fed with input signal and noise levels much greater than its limiting level. The limiter is followed by a filter which restricts the output to the zone centered about the input frequency and excludes the harmonics generated in the limiting action.

Davenport[2] proved that signal-to-noise ratio is essentially preserved in passing through a bandpass limiter. This fact has been experimentally verified at the Jet Propulsion Laboratory. Youla of this laboratory then proved that the total power output of a limiter in a given zone is constant. Mathematically stated:

$$N^2 + A^2 = L^2 = N_0^2 + A_0^2$$

and

$$\frac{N'}{A'} \cong \frac{N}{A},$$

[2] W. B. Davenport, "Signal-to-noise ratios in band-pass limiters," *Jour. of Appl. Phys.*, vol. 24, pp. 720–727; June, 1953.

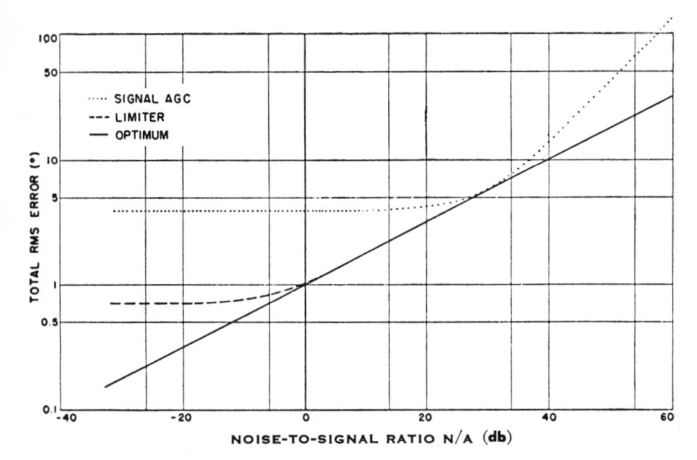

Fig. 4—Total phase error for various types of first-order loops.

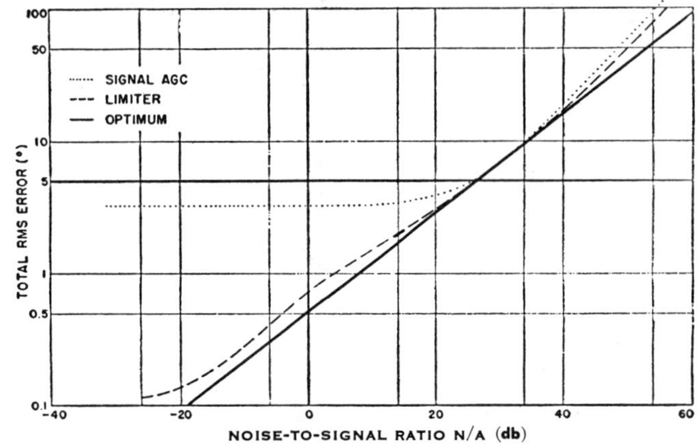

Fig. 5—Total phase error for various types of second-order loops.

where

N' and A' = limiter input noise and signal levels, respectively,
N and A = limiter output noise and signal levels, respectively,
L = total power output in a given zone (constant), and
N_0, A_0 = output levels of the limiter for which the loop was designed.

These results may be combined to give the ratio both of actual signal level to expected signal level and of actual noise level to expected noise level at the limiter output:

$$\left(\frac{A}{A_0}\right)^2 = \frac{1 + (N_0/A_0)^2}{1 + (N'/A')^2} \qquad \left(\frac{N}{N_0}\right)^2 = \frac{1 + (N_0/A_0)^2}{1 + (A'/N')^2}. \tag{4}$$

This relationship may in turn be applied to the expressions for the phase jitter and transient error [see (1) and (2)] in a fixed loop, giving the phase jitter and transient error in a fixed loop preceded by a limiter.

Comparison of AGC, Limiter, and Variable-Parameter Optimum Loops

It is now desirable to compare the behavior of the various loops discussed. The basis for comparison will be the relative phase jitter, transient error, and total error in the respective loops.

Total rms error is defined as

$$\Sigma = \sqrt{\sigma_N^2 + \lambda^2 E_T^2} \text{ rad},$$

and was the quantity minimized by the Wiener calculus-of-variations approach used in designing the filter.

Curves have been plotted for a loop using a fixed filter preceded by an AGC system that holds the signal level constant, for a fixed loop preceded by a limiter and for a variable-filter loop continually readjusted to be optimum. Expressions for loops designed to track input phase steps, frequency steps, and frequency ramps are derived in Appendix II and plotted in Figs. 4, 5, and 6.

Other theoretical curves for a zero velocity-error loop will now be discussed. For these plots the various parameters have been chosen to agree with the experimental

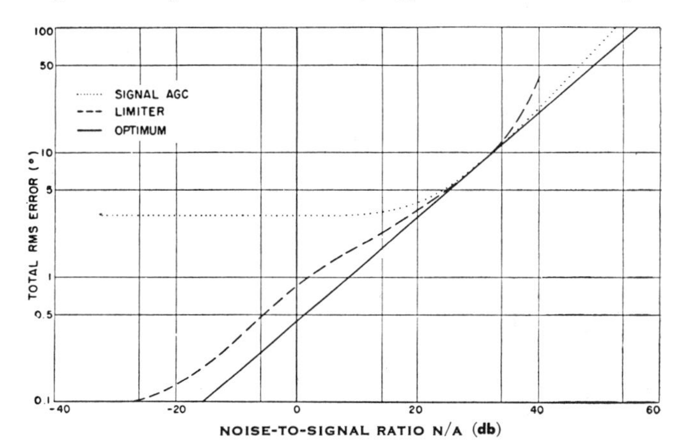

Fig. 6—Total phase error for various types of third-order loops.

and configuration to be described in a later section. Specifically, the input bandwidth of the noise was 900 cycles per second, and the loop noise bandwidth was $12\frac{1}{2}$ cycles per second (i.e., $B_0 = 25$); the loop was designed for an input noise-to-signal ratio of unity.

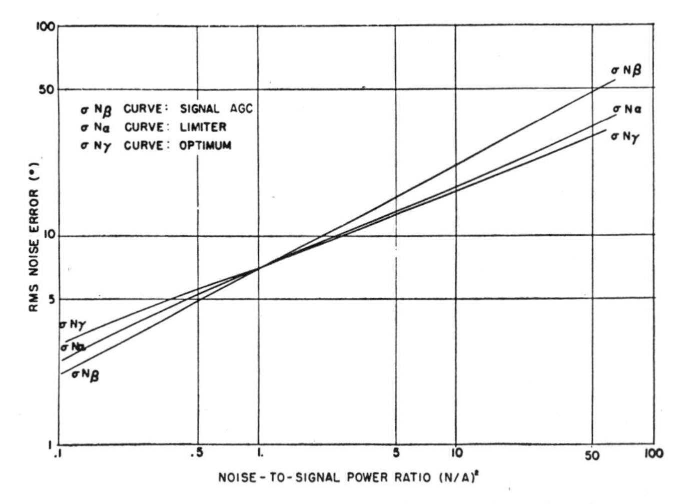

Fig. 7—Comparison of theoretical noise error in various loops.

Fig. 7 shows the rms phase-noise jitter to be expected in the different loops. All the curves coincide at the design point of unity input noise-to-signal ratio but diverge at other input ratios. The phase-noise jitter in a loop employing an optimum variable filter is not necessarily less

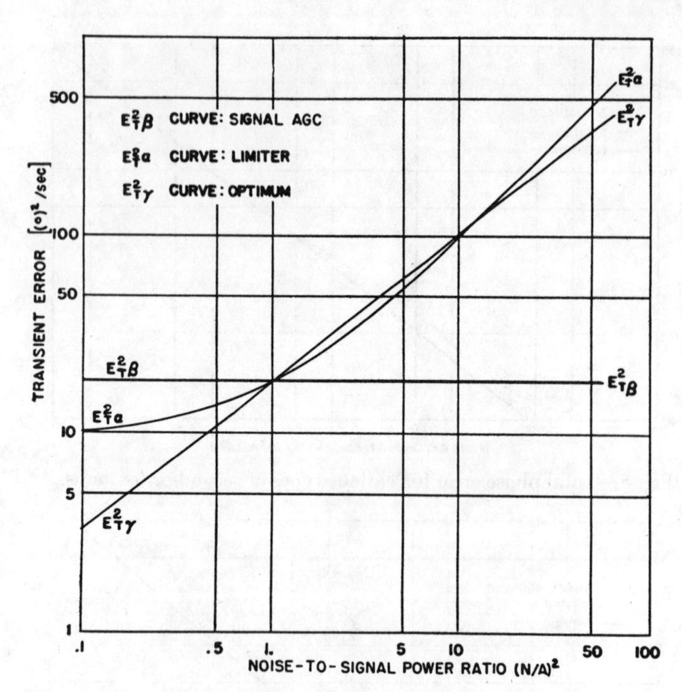

Fig. 8—Comparison of theoretical transient error in various loops.

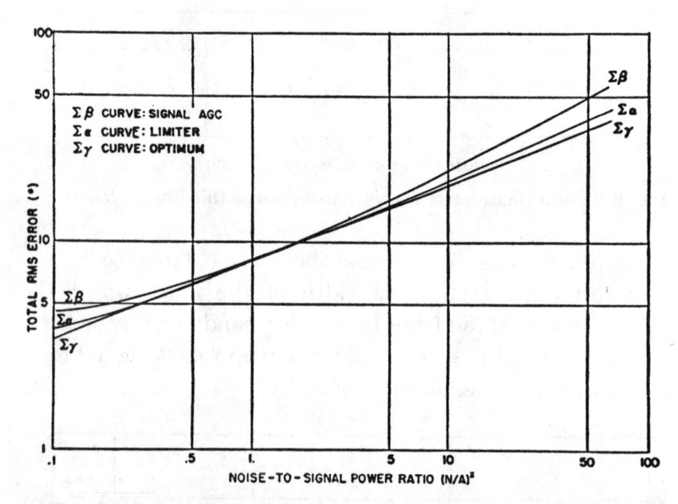

Fig. 9—Comparison of theoretical total error in various loops.

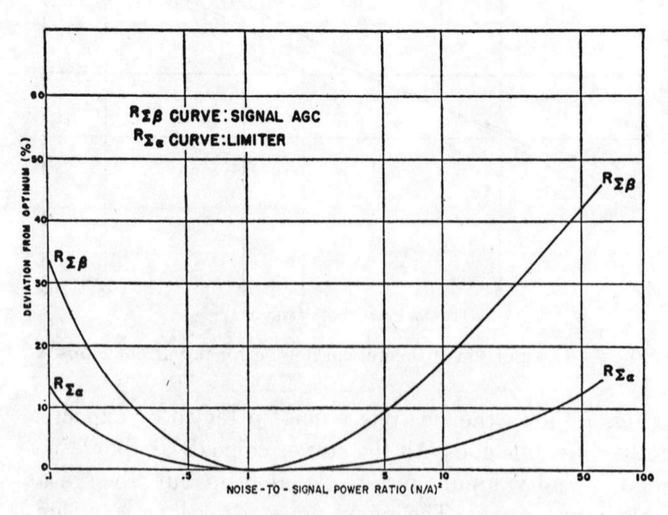

Fig. 10—Total error in signal-AGC and limiter-loop percentage deviation from optimum.

than the jitter in the other loops because the optimum loop is continually minimizing the total error rather than only the phase jitter or the transient error. Optimum performance may require that a slight increase in phase-noise jitter be allowed in order to obtain a considerable decrease in transient error.

Fig. 8 shows the transient error to be expected in the different loops. Since the transient error depends only on the loop gain in a fixed-filter loop tracking a given input frequency step, the AGC loop, which has constant input-signal amplitude, has a fixed loop gain and therefore a fixed amount of transient error.

Fig. 9 shows the total theoretically predicted rms error. However, it is easier to see the significant features of the loops if a plot is made showing the percentage by which the total error in the AGC and limiter loops deviates from the total error in the optimum loop (Fig. 10.) It is seen that the loop preceded by a limiter has a total error not more than 15 per cent greater than optimum, whereas the AGC loop has a total error exceeding that of an optimum loop by 45 per cent.

Experimental Configuration

It has been shown that a limiter loop is theoretically superior to an AGC loop and is nearly as good as a loop continually adjusted to be optimum by auxilliary servos. It remains to be shown that the theoretically derived results for such a limiter loop are experimentally verifiable.

Experimental results were obtained using equipment previously designed.[3] The experimental configuration is shown in Fig. 11. The noise was bandpassed to a 900-cycle

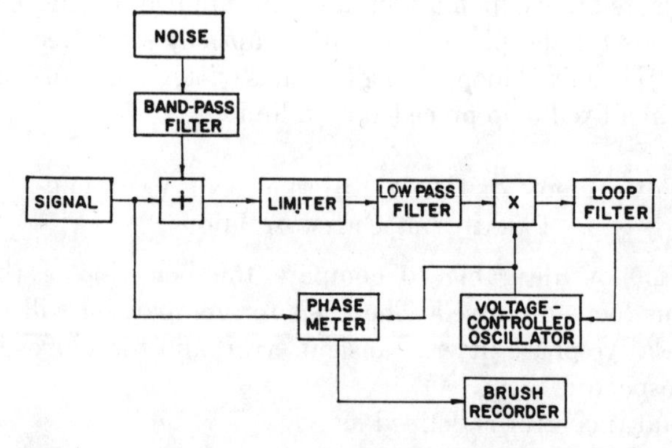

Fig. 11—Experimental configuration for limiter loop.

bandwidth centered at 5 kilocycles by a steep-sided filter and was then added to the 5-kilocycle signal. The sum was fed to a limiter consisting of a four-stage pentode amplifier having back-to-back silicon diodes across the input of each stage. The limiter output was passed through a steep-sided, 7-kilocycle, low-pass filter which removed all but the first zone of the limiter output. The output of this filter then served as the input to the phase-locked loop.

The loop itself consisted of (1) a multiplier using silicon diodes in a bridge circuit, (2) a multivibrator-type VCO

[3] Equipment used was designed by MacMillan of this laboratory.

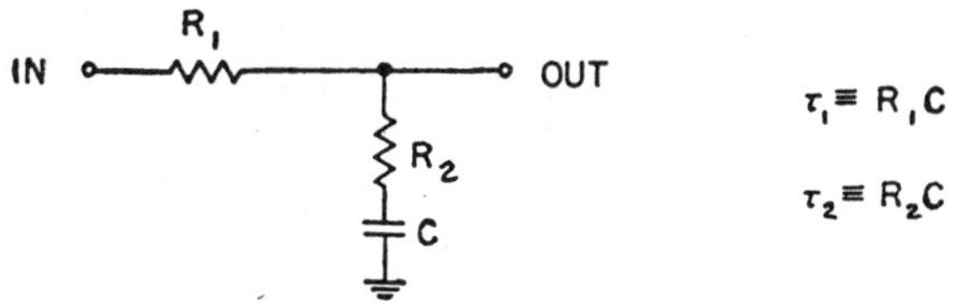

IN — R_1 — OUT; R_2, C to ground

$\tau_1 \equiv R_1 C$

$\tau_2 \equiv R_2 C$

$$\text{DESIRED } F(s) = \frac{B_0^2 + \sqrt{2}\, B_0 s}{K A_0} \qquad \text{ACTUAL } F(s) = \frac{\tau_2 s + 1}{(\tau_1 + \tau_2)s + 1}$$

$$\therefore \tau_1 = \frac{K A_0}{B_0^2} \qquad \cong \frac{\tau_2 s + 1}{\tau_1 s}$$

$$\tau_2 = \frac{\sqrt{2}}{B_0}$$

Fig. 12—Loop filter form.

having a low-pass filter on its output, which passed only the fundamental component of the multivibrator square-wave output, and (3) a loop filter as shown in Fig. 12.

The clean signal was compared with the VCO output in a phase meter, and the recorder output of the phase meter was connected to an oscillograph which was zeroed at the steady no-noise phase difference between the signal and the VCO. Any error in VCO tracking was then plotted directly on the oscillograph.

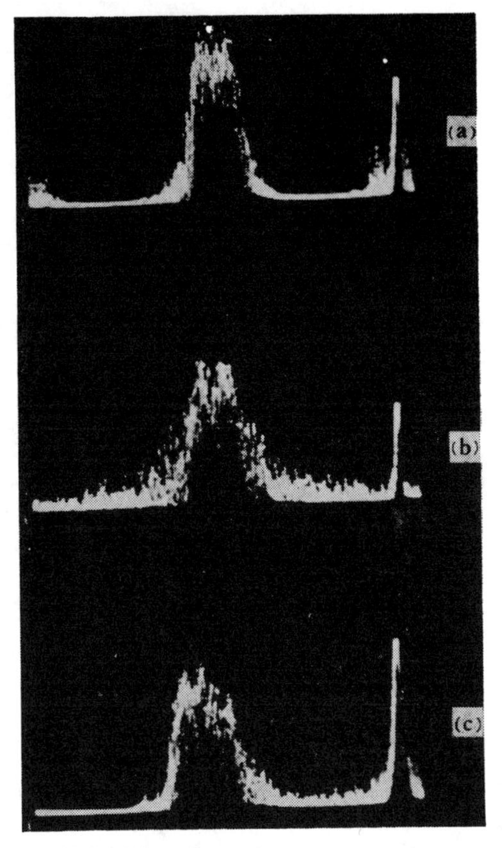

Fig. 13—Noise spectra at various points in limiter loop.

Experimental Results

Before discussing the data obtained from this experiment, it is interesting to look at the spectrum of the noise present in various parts of the loop. Figs. 13(a), 13(b), and 13(c), are, respectively, photographs of the noise spectrum at input to the limiter, at output of the limiter, and at output of the low-pass filter following the limiter.

The spectra are displayed on a logarithmic ordinate, with zero frequency at the right of the photograph. The input noise is centered at 5 kilocycles, with 900 cycles per second bandwidth. The first zone of the limiter output looks quite similar to the input noise except for the presence of some very low-level (considering the logarithmic ordinate) background noise generated in the limiter; also, the limiter input noise has been spread out in a manner similar to that predicted theoretically.[4] The last photograph shows the effect of the 7-kilocycle, low-pass filter in attenuating the limiter background noise past 7 kilocycles, implying that the higher-order noise zones generated in the limiter are cut off.

Experimental data on the noise error were obtained simply by adding a known amount of noise to the signal, and recording the VCO phase jitter on the recorder as shown in Figs. 14(a), 14(b), 14(c), for input noise-to-signal amplitude ratios of 1, 2, and 4, respectively. It is interesting to note how the frequency of the output noise jitter decreases as the input noise-to-signal amplitude ratio increases. This phenomenon implies that the loop bandwidth is decreasing with increasing input noise-to-signal ratio, a result theoretically predicted above.

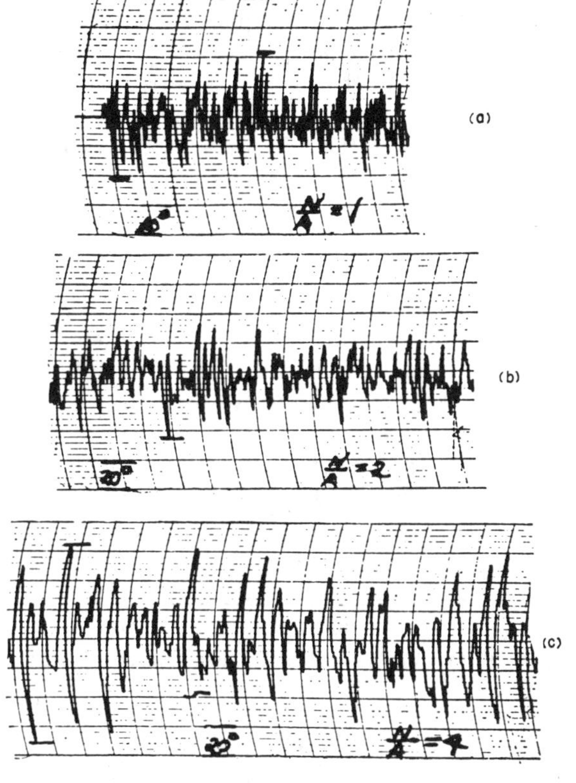

Fig. 14—Experimental phase jitter in limiter loop.

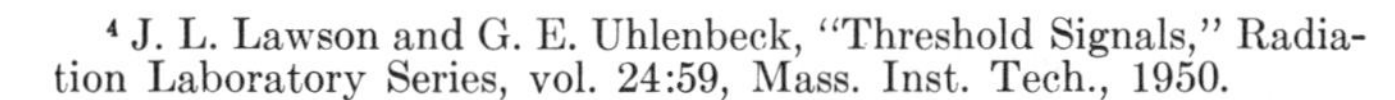

[4] J. L. Lawson and G. E. Uhlenbeck, "Threshold Signals," Radiation Laboratory Series, vol. 24:59, Mass. Inst. Tech., 1950.

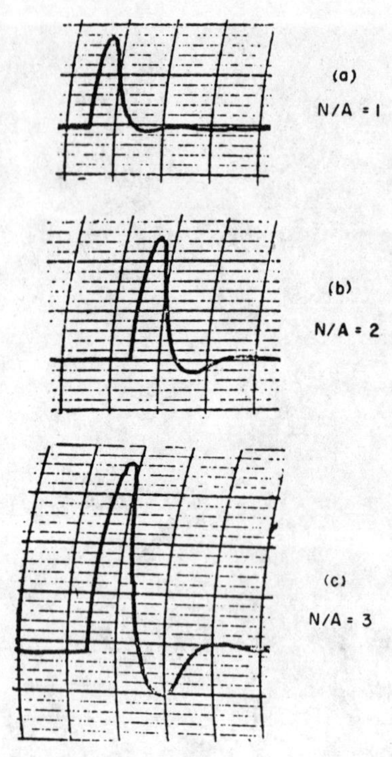

Fig. 15—Experimental transient error in limiter loop.

Experimental data on transient error were more difficult to obtain. In order to determine only the transient performance of the system, it was necessary to disconnect the noise source and reduce the signal output of the limiter in accordance with the limiter relationship already discussed (4). The transient error of the loop was then obtained by introducing a 2.5 cycle per second frequency step into the VCO and recording the loop phase error on the oscillograph. The step could have been introduced into either the signal source or the VCO; the VCO was chosen for convenience.

Figs. 15(a), 15(b), and 15(c) are recordings of the experimental transient error corresponding to input noise-to-signal ratios of 1, 2, and 3, respectively. The ordinate scale is 10 degrees for every heavy line. At the design level of unity signal-to-noise ratio, the Wiener theory requires about 5 per cent overshoot; as was expected, both the initial pip and the overshoot increase as the loop gain and bandwidth decrease.

Finally, the experimental results were combined to plot in Fig. 16, and in Figs. 17 and 18 (opposite) the experimental vs the theoretical values of noise error, transient error, and total error, respectively, in a phase-locked loop preceded by a limiter and employing a fixed filter that satisfied the Wiener optimum criterion for a design level of unity signal-to-noise ratio. The results are commensurate.

Conclusion

It has therefore been theoretically proved that a phase-lock loop preceded by a bandpass limiter approximates, over a wide range of input signal and noise levels, the optimum performance obtainable only with a variable filter that is continually readjusted to be optimum by an auxilliary servo system. It has been shown that, in addition to facility of mechanization, the limiter loop approximates optimum behavior considerably better than does an AGC-controlled loop in which the AGC is based upon the signal. Finally, the theoretical derivations leading to the expressions for error in the limiter loop have been experimentally verified and the effectiveness of a limiter phase-lock loop has been confirmed.

Appendix I

Derivation of Optimum Filter[5]

Let $Y(s)$ be the loop transfer function. By definition

$$\theta_2(s) = Y(s)\theta_1(s) \tag{5}$$

and

$$Y(s) = \frac{AKF(s)}{s + AKF(s)}. \tag{6}$$

The total noise power at the VCO output is

$$\sigma_N^2 = \frac{1}{2\pi j}\int_{-j\infty}^{+j\infty} |\, Y(s)\,|^2\, \Phi_N(s)\, ds,$$

where $\Phi_N(s)$ is the spectral density of the input phase noise. Assuming this spectral density to be flat, Figs. 2 and 3 show that

$$\Phi_N(s) = \Phi_N(0) = \frac{N^2}{A^2 2\Delta f}$$

where Δf is the input-noise bandwidth.

The transient error, previously defined as

$$E_T^2 = \int_0^\infty [\theta_2(t) - \theta_1(t)]^2\, dt$$

upon application of (5), may be written

$$E_T^2 = \frac{1}{2\pi j}\int_{-j\infty}^{+j\infty} |\, Y(s) - 1\,|^2\, |\,\theta_1(s)\,|^2\, ds.$$

Since the optimum filter is to be designed so that σ_N^2 is minimized under the constraint that E_T^2 have a specified value, it is desired that

$$\Sigma^2 = \sigma_N^2 + \lambda^2 E_T^2 = \text{minimum}. \tag{7}$$

The optimum filter is to be physically realizable, which for the purpose of this paper means that the loop transfer function has no poles in the right half of the s plane. Therefore, although the standard variational techniques are to be applied to minimize Σ^2, the integral expression for Σ^2 must be set up in such a manner that it yields a realizable filter.

Expansion of (7) gives

$$\Sigma^2 = \frac{1}{2\pi j}\int_{-j\infty}^{+j\infty} ds\; Y(s)Y(-s)\Phi_N(0) + \lambda^2[Y(s) - 1][Y(-s) - 1]\,|\,\theta_1(s)\,|^2. \tag{8}$$

[5] This derivation follows the approach to the Wiener filter theory given in unpublished notes by Rechtin of this laboratory.

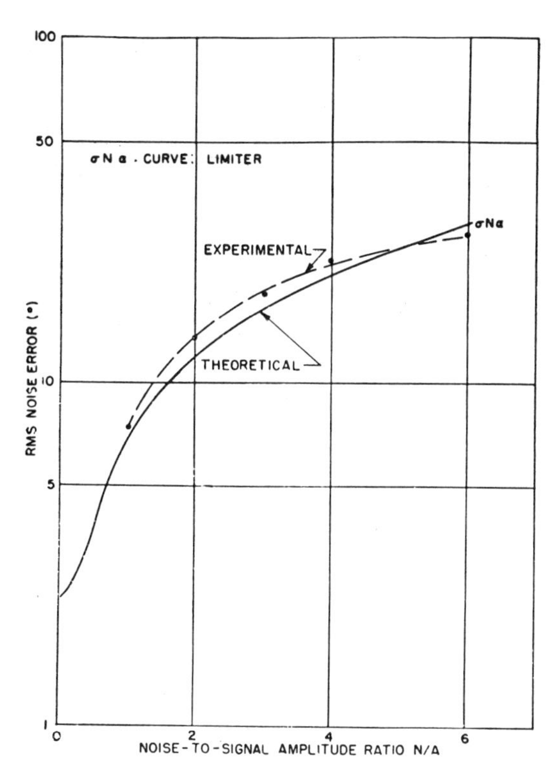

Fig. 16—Experimental vs theoretical noise error in limiter loop.

Applying the standard variational procedure to $Y(s)$, let

$$Y^*(s) = Y(s) + \epsilon\eta(s),$$

where $\epsilon\eta(s)$ is the variation to be minimized on the true optimum $Y(s)$, and $Y^*(s)$ is the estimate of $Y(s)$.

When $Y^*(s)$ is split into factors having positive and negative poles and substituted into (8),

$$\Sigma^2[Y(s) + \epsilon\eta(s)]$$
$$= \frac{1}{2\pi j}\int_{-j\infty}^{+j\infty} ds\,[Y(s) + \epsilon\eta(s)][Y(-s) + \epsilon\eta(-s)]\Phi_N(0)$$
$$+ \frac{1}{2\pi j}\int_{-j\infty}^{+j\infty} ds\,\lambda^2[1 - Y(s) - \epsilon\eta(s)]$$
$$\cdot[1 - Y(-s) - \epsilon\eta(-s)]\,|\,\theta_1(s)\,|^2.$$

Setting the variation of Σ^2 to zero at ϵ equals zero completes the standard variational procedure.

$$\frac{\partial\Sigma^2}{\partial\epsilon}[Y(s) + \epsilon\eta(s)]\bigg|_{\epsilon=0} = \frac{1}{2\pi j}\int_{-j\infty}^{+j\infty} ds\,\eta(s)$$
$$\cdot\{(Y(-s))(\lambda^2\,|\,\theta_1(s)\,|^2 + \Phi_N(0)) - \lambda^2\,|\,\theta_1(s)\,|^2\}$$
$$+ \frac{1}{2\pi j}\int_{-j\infty}^{+j\infty} ds\,\eta(-s)$$
$$\cdot\{(Y(s))(\lambda^2\,|\,\theta_1(s)^2\,| + \Phi_N(0)) - \lambda^2\,|\,\theta_1(s)\,|^2\} \quad (9)$$

In order to keep all terms in (9) split into factors having poles in either, but not both, the right or left half-plane, it is convenient to define

$$\lambda^2|\,\theta_1(s)\,|^2 + \Phi_N(0) \equiv |\,\psi(s)\,|^2 = \psi(s)\psi(-s). \quad (10)$$

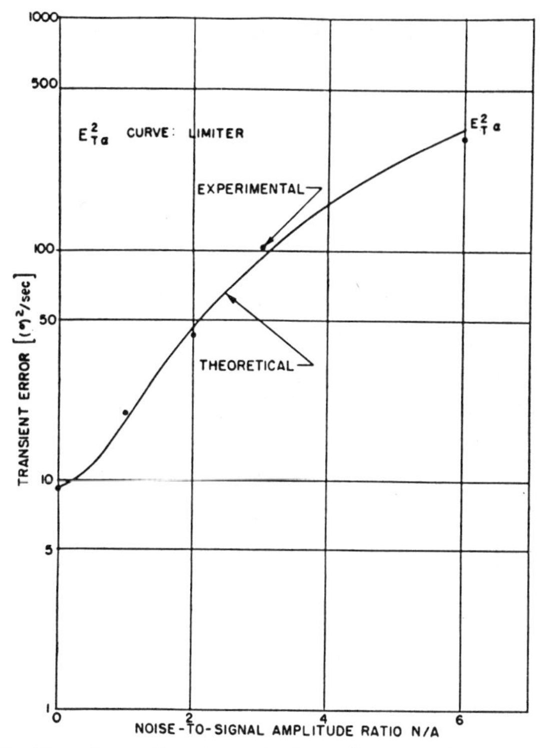

Fig. 17—Experimental vs theoretical transient error for limiter loop.

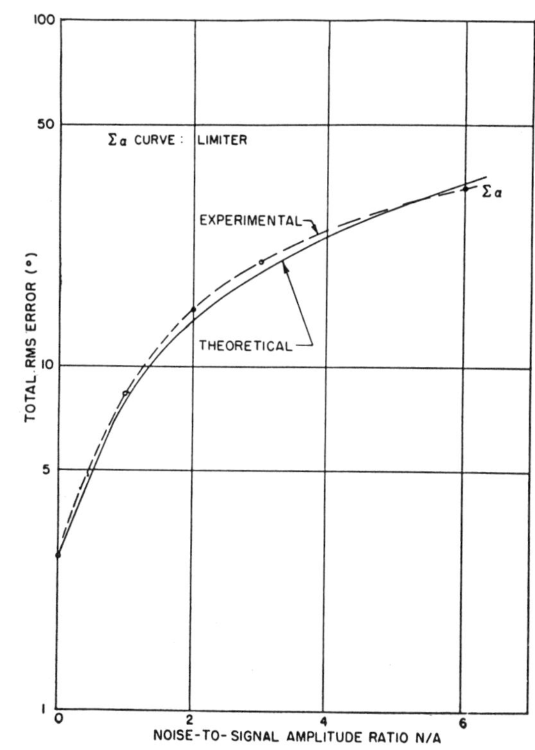

Fig. 18—Experimental vs theoretical total error in limiter loop.

It is now necessary to substitute (10) into (9) in such a way that the latter will be satisfied for realizable $Y(s)$, whereas the remainder of the integral will reduce to zero.

This can be effected by noting that

$$\int_{-j\infty}^{+j\infty} Z(s)W(s)\,ds = 0 \quad (11)$$

if $Z(s)$ and $W(s)$ are algebraic polynomials having poles only in the same half of the s plane.

Substitution of (10) into (9) and keeping together terms having similar poles

$$\int_{-j\infty}^{+j\infty} ds\ \eta(s)\psi(s)[Y(-s)\psi(-s) - \lambda^2 \mid \theta_1(s) \mid^2/\psi(s)]$$

$$+ \eta(-s)\psi(-s)[Y(s)\psi(s) - \lambda^2 \mid \theta_1(s) \mid^2/\psi(-s)] = 0. \quad (12)$$

It is now advisable to split

$$\frac{\mid \theta_1(s) \mid^2}{\psi(s)} = \left[\frac{\theta_1(s)^2}{\psi(s)}\right]_+ + \left[\frac{\theta_1(s)^2}{\psi(s)}\right]_-, \quad (13)$$

where the right-hand side of the equation is the partial fraction expansion of the left-hand side, having denominator terms in $+s$ and $-s$, i.e., having poles in the left and right halves of the s plane, respectively.

Substitution of (13) into (12) gives

$$\int_{-j\infty}^{-j\infty} ds\ \eta(s)\psi(s)\{Y(-s)\psi(-s) - \lambda^2[(\mid \theta_1(s) \mid^2/\psi(s))_+$$

$$+ (\mid \theta_1(s) \mid^2/\psi(s))_-]\}$$

$$+ \int_{-j\infty}^{+j\infty} ds\ \eta(-s)\psi(-s)\{Y(s)\psi(s) - \lambda^2[(\mid \theta_1(s) \mid^2/\psi(-s))_+$$

$$- (\mid \theta_1(s) \mid^2/(-s))_-]\} = 0.$$

Because of (11), the terms having similar poles drop out, leaving

$$\int_{-j\infty}^{+j\infty} ds\ \eta(s)\psi(s)\{Y(-s)\psi(-s) + \lambda^2(\mid \theta_1(s) \mid^2/\psi(s))_-\}$$

$$+ \int_{-j\infty}^{+j\infty} ds\ \eta(-s)\psi(-s)$$

$$\{Y(s)\psi(s) - \lambda^2(\mid \theta_1(s) \mid^2/\psi(s))_+\} = 0. \quad (14)$$

These integrals are identical except that one has an integrand in $+s$ and one, in $-s$. Therefore, the integrals are merely the negatives of each other, so that equating one of them to zero satisfies (14).

Setting the second integral to zero yields

$$Y(s) = \frac{\lambda^2}{\psi(s)}\left[\frac{\mid \theta_1(s) \mid^2}{\psi(s)}\right]_+, \quad (15)$$

which is the optimum loop transfer function composed of realizable terms. Note that all terms have poles in the left half-plane, thereby satisfying the requirement of physical realizability.

In this paper, (15) will be solved only for input frequency steps, although the results for other type inputs will be given.

Assume the input to be a frequency step

$$\theta_1(s) = \frac{\Delta\omega}{s^2}$$

where $\Delta\omega$ is the magnitude of the step in radians per second.

Therefore

$$\mid \theta_1(s) \mid^2 = \frac{(\Delta\omega)^2}{s^4}$$

and

$$\mid \psi(s) \mid^2 = \lambda^2 \mid \theta_1(s) \mid^2 + \Phi_N(0)$$

$$= \frac{\lambda^2(\Delta\omega)^2}{s^4} + \Phi_N(0) = \frac{B_0^4 + s^4}{s^4}\Phi_N(0),$$

where

$$B_0^4 \equiv \frac{\lambda^2\omega^2}{\Phi_N(0)}. \quad (16)$$

Factoring

$$\mid \psi(s) \mid^2 = \psi(s)\psi(-s)$$

$$= \Phi_N(0)\left[\frac{B_0^2 + \sqrt{2}\,B_0s + s^2}{s^2}\right]\left[\frac{B_0^2 - \sqrt{2}\,B_0s + s^2}{s^2}\right]$$

gives

$$\psi(s) = \sqrt{\Phi_N(0)}\left[\frac{B_0^2 + \sqrt{2}\,B_0s + s^2}{s^2}\right] \quad (17)$$

Substitution of (17) into the optimum-filter equation gives

$$Y(s) = \frac{\lambda^2s^2(\Delta\omega)^2}{\Phi_N(0)[B_0^2 + \sqrt{2}\,B_0s + s^2]}\left[\frac{1}{s^2[B_0^2 - 2B_0s - s^2]}\right]_+ \quad (18)$$

Separating the last term of (18) into partial fractions,

$$\frac{1}{s^2[B_0^2 + \sqrt{2}\,B_0s + s^2]} = \frac{1/B_0^2}{s^2} + \frac{\sqrt{2}/B_0^3}{s}$$

$$+ \frac{c}{s - B_0/\sqrt{2}\,(-1 + j)} + \frac{d}{s - B_0/\sqrt{2}\,(-1 - j)}.$$

The coefficients c and d are of no interest since the terms containing them have poles in the right half-plane and are not to be considered because of the definition of $[\ \ ]_+$ in (13).

The first two terms may be considered as having poles in the left half-plane, since they represent the limiting case of $1/(s - \epsilon)$ where ϵ is a small number representing the reciprocal of the dc gain. Since the pole approaches the origin from the left, in such a limiting process, the first two terms are included in the $[\ \ ]_+$ terms.

Therefore

$$Y(s) = \frac{\lambda^2(\Delta\omega)^2s^2}{\Phi_N(0)[B_0^2 + \sqrt{2}\,B_0s + s^2]}\left[\frac{1}{B_0^2s^2} + \frac{\sqrt{2}}{B_0^3s}\right],$$

which, with the aid of (16), reduces to

$$Y(s) = \frac{B_0^2 + \sqrt{2}\,B_0s}{B_0^2 + \sqrt{2}\,B_0s + s^2}.$$

As a check, it is interesting to note that $Y(s)$ is quasi-distortionless to a step-frequency change, i.e.,

$$\text{error}(s) \equiv \theta_1(s) - \theta_2(s) = \theta_1(s)[1 - Y(s)]$$

$$= \theta_1(s)\frac{s^2}{B_0^2 + \sqrt{2}\,B_0s + s^2},$$

so that, if $\theta_1(s) = \Delta\omega/(s^2)$, representing a step-frequency input, the infinite-time error is zero; i.e., the loop is quasi-distortionless to a step-frequency change:

$$\text{error}(s) = \frac{\Delta\omega}{s^2}\frac{s^2}{B_0^2 + \sqrt{2}\,B_0 s + s^2} = \frac{\Delta\omega}{B_0^2 + \sqrt{2}\,B_0 s + s^2}$$

$$\lim_{t\to\infty} \text{error}(t) = \lim_{s\to\infty} s \times \text{error}(s) = 0$$

Since the optimum loop transfer function is

$$Y(s) = \frac{B_0^2 + \sqrt{2}\,B_0 s}{B_0^2 + \sqrt{2}\,B_0 s + s^2}$$

and

$$Y(s) = \frac{AKF(s)}{s + AKF(s)}.$$

It follows that

$$F(s) = \frac{B_0^2 + \sqrt{2}\,B_0 s}{A_0 K s},$$

which was to be proved.

The noise error and transient error in such an optimum loop may be calculated as follows:

$$\sigma_N^2 = \frac{1}{2\pi j}\int_{-j\infty}^{+j\infty} |\,Y(s)\,|^2\,\Phi_N(0)\,ds$$

$$E_T^2 = \frac{1}{2\pi j}\int_{-j\infty}^{+j\infty} |\,\theta_1(s)\,|^2\,[1 - Y(s)]^2\,ds.$$

These integrals may be evaluated by involved contour integration and have been tabulated conveniently;[6] the evaluation of these integrals led to all explicit expressions for σ_N^2 and E_T^2 given in this paper.

Appendix II

Filters for Different Forms of Signal Inputs

Filter forms for fixed-component loops, fixed-component loops preceded by bandpass limiters, and variable-component loops will be given in this appendix, together with their associated transient responses and phase-noise jitters, for two other forms of signal inputs besides those discussed in the text. Derivation of these filters are similar to the derivation in Appendix I and are not repeated. Loops designed to track phase steps will be called first-order loops; loops designed to track frequency ramps will be called third-order loops. Second-order loops designed to track frequency steps are discussed in the text and will not be discussed in this Appendix.

(1) *First-order-loop*—If the signal input to the loop is assumed to be a phase step, i.e.,

$$\theta_1(t) = \Delta\theta \qquad \theta_1(s) = \frac{\Delta\theta}{s},$$

then the filter for the first-order, fixed-component loop has the transform

$$F(s) = \frac{B_1}{KA_0}$$

and has phase jitter and transient error, respectively, of

$$\sigma_N^2 = B_1\left(\frac{N^2}{2\Delta f A^2}\right)\left(\frac{A}{A_0}\right)$$

and (19)

$$E_T^2 = \frac{(\Delta\theta)^2}{2B_1(A/A_0)},$$

where B_1 is a quantity similar to B_0 in the previous discussion of the loop for tracking frequency steps, i.e.,

$$B_1 \equiv \frac{\lambda(\Delta\theta)}{N_0/A_0}\sqrt{2\Delta f}.$$

Substitution of (4) into (19) gives a total error in a first-order loop preceded by a bandpass limiter of

$$\Sigma^2 \equiv \sigma_N^2 + \lambda^2 E_T^2 = \left(\frac{B_1}{2}\right)\left(\frac{N^2}{A^2 2\Delta f}\right)\left(\frac{A}{A_0}\right) + \left(\frac{B_1}{2}\right)\left(\frac{N_0^2}{A_0^2 2\Delta f}\right)\left(\frac{A_0}{A}\right)$$

$$= \frac{B_1}{2}\left(\frac{N^2}{A^2 2\Delta f}\right)\left[\sqrt{\frac{1 + (N_0/A_0)^2}{1 + (N/A)^2}} + \frac{N_0^2}{A_0^2}\sqrt{\frac{1 + (N/A)^2}{1 + (N_0/A_0)^2}}\right].$$

It may be shown, using the same techniques employed in Appendix III for second-order loops, that the optimum variable-component filter for the first-order loop has the transform

$$F(s) = \frac{B_1}{KA_0}\frac{N_0}{N},$$

and that the total error in such a loop is

$$\Sigma^2 \equiv \sigma_N^2 + \lambda^2 E_T^2 = \left(\frac{B_1}{2}\right)\left(\frac{A/A_0}{N/N_0}\right)\left(\frac{N^2/A^2}{2\Delta f}\right) + \left[\frac{\lambda^2(\Delta\theta)^2}{2B_1}\right]\frac{N/N_0}{A/A_0}$$

$$= B_1\left(\frac{N_0}{A_0}\right)\left(\frac{N/A}{2\Delta f}\right).$$

Plots for the total error in first-order loops with (1) fixed components preceded by a signal-based AGC, (2) fixed components preceded by a bandpass limiter and (3) variable components are given in Fig. 4. Design parameters are

$$\frac{N_0}{A_0} = 1$$

$$B_0 = 10$$

$$\Delta f = 1 \text{ megacycle per second.}$$

The input-signal change was assumed to be a 1-radian phase step.

For comparison purposes, the total error in each of the three types of second-order loops is given in Fig. 5. The input-signal change was assumed to be a 6 radians per second (~ 1 cycle per second) frequency step.

(2) *Third-order-loop*—If the signal input to the loop is assumed to be a frequency ramp, ie.,

$$\theta_1(t) = \frac{(\Delta\alpha)t^2}{2} \qquad \theta_1(s) = \frac{\Delta\alpha}{s^3},$$

then the third-order, fixed-component loop filter has the transform

[6] "Servomechanisms," Radiation Laboratory Series, No. 25, pp. 369–370, Mass. Inst. Tech.; 1947.

$$F(s) = \frac{B_2^3 + 2B_2^2 s + 2B_2 s^2}{KA_0 s^2}$$

and has phase jitter and transient error, respectively, of

$$\sigma_N^2 = \left(\frac{N^2}{A^2}\right)\left(\frac{B_2}{2\Delta f}\right)\left[\frac{4(A/A_0) + 1}{4(A/A_0) - 1}\right]$$

and

$$E_T^2 = \frac{(\Delta\alpha)^2}{(A/A_0)[4(A/A_0) - 1]B_2^3}, \qquad (20)$$

where

$$B_2^3 \equiv \left[\frac{\lambda(\Delta\alpha)}{N_0/A_0}\right]\sqrt{2\Delta f}.$$

Substitution of (4) into (20) gives a total error in a third-order loop preceded by a band-pass limiter of

$$\Sigma^2 \equiv \sigma_N^2 + \lambda^2 E_T^2 = \left(\frac{B_2}{2}\right)\left[\frac{N^2}{A^2 2\Delta f}\right]\left[\frac{4(A/A_0) + 1}{4(A/A_0) - 1}\right] + \left(\frac{B_2}{2}\right)\left[\frac{N_0^2}{A_0^2 2\Delta f}\right]\frac{(A/A_0)}{[4(A/A_0) - 1]}$$

$$= B_2\left[\frac{N^2}{A^2 2\Delta f}\right]\left\{\left[\frac{4\sqrt{\frac{1 + (N_0/A_0)^2}{1 + (N/A)^2}} + 1}{4\sqrt{\frac{1 + (N_0/A_0)^2}{1 + (N/A)^2}} - 1}\right]\left(\frac{A}{A_0}\right)\right.$$

$$\left. + \frac{(A/A_0)^2}{(N/N_0)^2}\left[\frac{1}{\sqrt{\frac{1 + (N_0/A_0)^2}{1 + (N/A)^2}}\left[4\sqrt{\frac{1 + (N_0/A_0)^2}{1 + (N/A)^2}} - 1\right]}\right]\right\}.$$

It should be noted that the total error becomes infinite when the noise has reduced the signal level to one-fourth of the design level.

It may be shown that the optimum third-order, variable-component loop has the transform

$$F(s) = \frac{B_2^3\left[\frac{(A/A_0)}{(N/N_0)}\right] + 2B_2^2\left[\frac{(A/A_0)}{(N/N_0)}\right]^{2/3} s + 2B_2\left[\frac{(A/A_0)}{(N/N_0)}\right]^{1/3} s^2}{As^2},$$

and that the total error in such a loop is

$$\Sigma^2 = \sigma_N^2 + \lambda^2 E_T^2 = \left(\frac{5}{3}\right)\left(\frac{N}{A}\right)^{5/3}\left(\frac{N_0}{A_0}\right)^{1/3}\left(\frac{B_2}{2\Delta f}\right) + \frac{(\Delta\alpha)^2}{3B_2^5\left[\frac{A/A_0}{N/N_0}\right]^{5/3}}$$

$$= 2B_0\left(\frac{N^2}{A^2 2\Delta f}\right)\left[\frac{(N_0/N)}{(A_0/A)}\right]^{1/3}.$$

Plots for the total error in the three types of third-order loops are given in Fig. 6. The input-signal change was assumed to be a 1 radian per second² frequency ramp.

Appendix III

Optimum-Filter Design

The optimum filter is specified by

$$F(s) = \frac{B^2 + \sqrt{2}\, Bs}{KAs},$$

where

$$B^2 \equiv \frac{\lambda(\Delta\omega)}{(N/A)}\sqrt{2\Delta f}.$$

B varies in accordance with the signal- and noise-input levels and may be expressed in terms of B_0, the fixed-filter bandwidth parameter, as follows:

$$B^2 = B_0^2\left[\frac{(A/A_0)}{(N/N_0)}\right].$$

Therefore

$$F(s) = \frac{B_0^2\left[\frac{(A/A_0)}{(N/N_0)}\right] + \sqrt{2}\, B_0\left[\frac{(A/A_0)}{(N/N_0)}\right]s}{KAs}, \qquad (21)$$

giving transfer function for optimum variable filter (3).

All results described in this paper as pertaining to second-order, variable-filter loops were obtained by substituting (21) into the expression of $Y(s)$ in (6) and evaluating that which was required.

Transient Analysis of Phase-Locked Tracking Systems in the Presence of Noise

JAMES R. LA FRIEDA AND WILLIAM C. LINDSEY

Abstract—**This paper is concerned with the problem of obtaining time-dependent solutions to a class of Fokker–Planck equations that arise in the analysis and synthesis of a variety of first-order synchronization systems employing the phase-lock principle. These include the classical sinusoidal phase-locked loop, squaring and Costas loops, data-aided loops, hybrid loops, various symbol synchronizer mechanizations, and tunnel-diode oscillators.**

By analyzing the spectral properties of the associated time-dependent Fokker–Planck boundary value problem, eigenfunction expansions of the reduced modulo-2π phase-error-transition probability-density function are developed for a class of first-order synchronization systems. Whereas previous work treated the steady-state probability density function, this approach yields a complete statistical description of the phase-error process reduced modulo-2π.

I. Introduction

TWO of the most important practical and unsolved problems that remain in the analysis of phase-locked tracking loops in the presence of noise are the specifying of the system's signal acquisition and loss of lock characteristics. The reason why these problems remain largely unsolved has to do with two factors: 1) the dynamical equations that govern system response are characteristically nonlinear stochastic differential equations and 2) solving the signal acquisition and loss of lock problem is tantamount to obtaining and studying fundamental solutions to time-dependent Fokker–Planck (F–P) equations. Numerical solutions to time-dependent F–P equations can be obtained using the methods of finite differences and successive approximations; however, such approaches do not allow one to set forth a fundamental theory that can be used to give answers to other problems of interest. More frequently, these methods simply yield numerical solutions, and are subject to the formidable problems of discretizing equations, instability of solutions, and proving convergence of successive approximations.

This paper is primarily concerned with the problem of obtaining and studying analytical solutions to a class of time-dependent F–P equations that arise in the analysis and synthesis of a variety of first-order nonlinear tracking systems employing the phase-locked loop (PLL) principle. Such systems include the classical sinusoidal PLL [1], [2], squaring and Costas loops [3], data-aided loops [4], hybrid loops [5], various symbol synchronizers that are essentially digital PLL's [6], and tunnel-diode oscillators [7]. For this class of systems, the eigenfunction expansion method is used to obtain the principal time-dependent solution of the relevant F–P equation. The solution, which represents the transition probability-density function (pdf) of the phase-error process $\phi(t)$ reduced modulo-2π, completely characterizes the phase-error process. Such a transition pdf reveals, among other things, how the phase-error process undergoes diffusion and drift with the passage of time.

The problem of finding the transition pdf for a first-order phase-locked tracking loop is reduced to studying the spectral properties of an eigenvalue problem of Hill's equation [8], which is nonself-adjoint when loop detuning is present and self-adjoint when this detuning is absent. For the special case of a sinusoidal PLL, approximate analytic and numerical results have been obtained for the spectrum; and the transition pdf, the autocorrelation function, and the spectral density of the non-Gaussian phase-error process are illustrated for the nonlinear region of loop operation. Obtaining the transition pdf enables: 1) the time-dependent moments of the nonstationary non-Gaussian process $F[\phi(t)]$ to be evaluated (here $F[\cdot]$ represents any zero-memory time-invariant nonlinear transformation); 2) all steady-state joint pdf's of $\phi(t)$ and thus, the autocorrelation and spectral density of $F[\phi(t)]$ to be determined; 3) the transient behavior of loop response during signal acquisition to be evaluated; and 4) the validity and the limits of applicability of various quasi-linear loop theories [9]–[11] to be assessed.

II. Spectral Analysis of the Fokker–Planck Equation

To introduce the notation to be used in later analysis, we begin with the stochastic differential equation [1] for the phase-error process $\phi = \phi(t)$ of a first-order generalized tracking loop, i.e.,

$$\dot{\phi} = \Omega_0 - K[Ag(\phi) + n(t)]. \tag{1}$$

Here Ω_0 is the initial radian frequency detuning, A and K are constants depending upon the loop mechanization, $\{n(t)\}$ is a stationary white Gaussian noise process with zero mean and single-sided spectral density N_0 watts/hertz, and the function $g(\phi)$ is an odd symmetric periodic nonlinearity of class C^1. It follows from (1) that $\phi(t)$ is a temporally homogeneous continuous first-order Markov process, which is nonstationary for all $t < \infty$ and strictly stationary in the steady state. Such a process is completely

Manuscript received March 24, 1971; revised June 12, 1972. This work was supported in part by NASA Multidisciplinary Research Grant NGR-05-018-044 and NASA Contract NAS 7-100. This paper is based on a dissertation by J. R. La Frieda in fulfillment of the requirements for the Ph.D. degree, University of Southern California, Los Angeles, Calif.

J. R. La Frieda was with the Jet Propulsion Laboratory, California Institute of Technology, Pasadena, Calif. He is now with the Department of Electrical and Computing Engineering, University of Michigan, Ann Arbor, Mich. 48104.

W. C. Lindsey is with the Department of Electrical Fngineering, University of Southern California, Los Angeles, Calif. 90007.

Reprinted from *IEEE Trans. Inform. Theory*, vol. IT-19, pp. 155–165, Mar. 1973.

described by the transition probability density function $p \triangleq p(\phi,t \mid \phi_0,t_0)$, which satisfies the F–P equation

$$\frac{\partial}{\partial \phi}\left[\left\{[-\beta + \alpha g(\phi)] + \frac{\partial}{\partial \phi}\right\} p\right] = \frac{1}{D}\frac{\partial p}{\partial t} \quad (2)$$

with the sample space $\Omega = [-\pi,\pi]$. Here $\alpha = 4A/N_0K = A^2/N_0B_L$ is the signal-to-noise ratio in the loop bandwidth $B_L = AK/4$, $\beta = 4\Omega_0/N_0K^2 = \Omega_0\alpha/AK$, and $D = N_0K^2/4$ represents the diffusion coefficient. The transition pdf p is subject to the initial condition

$$\lim_{t \to t_0} p(\phi,t \mid \phi_0,t_0) = \delta(\phi - \phi_0) \quad (3)$$

and the linear homogeneous boundary conditions

$$B_1(p) = p(-\pi,t \mid \phi_0,t_0) - p(\pi,t \mid \phi_0,t_0) = 0$$

$$B_2(p) = \frac{\partial}{\partial \phi} p(\phi,t \mid \phi_0,t_0)\bigg|_{\pi} - \frac{\partial}{\partial \phi} p(\phi,t \mid \phi_0,t_0)\bigg|_{-\pi} = 0, \quad (4)$$

for all $t \in T \triangleq (t_0,t)$. The boundary conditions (4) follow directly from the periodicity of the extension of $p(\phi,t \mid \phi_0,t_0)$ over the entire ϕ axis, and from the fact that the derivative of a periodic function is also periodic.

A. Spectral Solutions to the Fokker–Planck Equation When $\Omega_0 \neq 0$.

To solve the two-point F–P boundary value problem (bvp) posed in (2)–(4), we apply the method of separation of variables, i.e., we look for a solution of the form

$$p(\phi,t \mid \phi_0,t_0) = s(t)u(\phi), \quad (5)$$

where $s(t)$ and $u(\phi)$ are functions dependent only on $t \in T$ and $\phi \in \Omega$, respectively. Letting $-\lambda$ be the separation constant and substituting (5) into (2) yields

$$\dot{s}(t) = -\lambda s(t) \quad (6)$$

and the regular, nonself-adjoint eingenvalue problem

$$Lu \triangleq u''(\phi) + [\alpha g(\phi) - \beta]u'(\phi) + \alpha g'(\phi)u(\phi) = \frac{-\lambda}{D} u(\phi)$$

$$B_1(u) = u(-\pi) - u(\pi) = 0$$

$$B_2(u) = u'(\pi) - u'(\pi) = 0. \quad (7)$$

Defining the even-symmetric periodic function

$$f(\phi) \triangleq \int^{\phi} g(x)\, dx, \quad (8)$$

and making the change of variables,

$$u(\phi) \triangleq y(\phi) \exp\left[-\frac{\alpha}{2}f(\phi) + \frac{\beta\phi}{2}\right], \quad (9)$$

(7) reduces to a general nonself-adjoint eigenvalue problem of the form of Hill's equation [8]

$$Ly \triangleq y''(\phi) - Q(\phi)y(\phi) = \frac{-\lambda}{D} y(\phi) \quad (10)$$

subject to the nonperiodic boundary conditions

$$B_1(y) = \exp[(-\beta\pi/2)]y(-\pi) - \exp[(\beta\pi/2)]y(\pi) = 0$$

$$B_2(y) = \exp[(-\beta\pi/2)]y'(-\pi) + \frac{\beta}{2}\exp[(-\beta\pi/2)]y(-\pi)$$

$$- \exp[(\beta\pi/2)]y'(\pi) - \frac{\beta}{2}\exp[(\beta\pi/2)]y(\pi) = 0, \quad (11)$$

where

$$Q(\phi) \triangleq -\frac{\alpha}{2}g'(\phi) + \frac{\alpha^2}{4}g^2(\phi) - \frac{\alpha\beta}{2}g(\phi) + \frac{\beta^2}{4} \quad (12)$$

is a periodic function that is "even symmetric" when $\beta = 0$.

Since the eigenvalue problem (7) and its formally self-adjoint form [(10) and (11)] are nonself-adjoint [12], the basic question that arises is whether (7) is well posed, i.e., whether a denumerable number of eigenvalues $\{\lambda_n, n = 0,1,2,\cdots,\}$ and corresponding eigenfunctions $\{u_n(\phi)\}$ exist and, in particular, whether the eigenfunctions form a complete system in the sense of mean-square convergence on the interval Ω. The answer to this question is affirmative in the self-adjoint case, i.e., when $\Omega_0 = 0$; however, for regular nonself-adjoint problems on a finite interval, it is the boundary conditions that determine if the eigenvalue problem is well posed and, hence, if a spectral representation for $p(\phi,t \mid \phi_0,t_0)$ is possible.

Because the boundary conditions (11) for the regular formally self-adjoint differential operator (10) can be shown [13] to be "regular" in the sense of Birkhoff [14] and "strongly regular" in the sense of Tamarkin [15], it follows that the nonself-adjoint problem (7) has a denumerable number of eigenvalues $\{\lambda_n, n = 0,1,2,\cdots,\}$ such that the corresponding eigenfunctions $\{u_n(\phi)\}$ form a complete biorthonormal set in $\mathscr{L}_2(\Omega)$. These eigenfunctions satisfy the biorthonormal condition

$$\langle v_n,u_m\rangle = \int_{-\pi}^{\pi} v_n(\phi)u_m(\phi)\, d\phi = \delta_{mn}$$

$$\triangleq \begin{cases} 1, & n = m \\ 0, & n \neq m \end{cases}, \quad (13)$$

where $\{v_n(\phi)\}$ is the set of eigenfunctions corresponding to the adjoint problem of (7), i.e.,

$$L^*v \triangleq v''(\phi) + [\beta - \alpha g(\phi)]v'(\phi) = \frac{-\lambda}{D} v(\phi)$$

$$B_1^*(v) = v(-\pi) - v(\pi) = 0$$

$$B_2^*(v) = v'(-\pi) - v'(\pi) = 0. \quad (14)$$

We remark that the eigenfunctions $\{u_n(\phi)\}$ not only form a basis for arbitrary functions in the domain of definition of the operator L, i.e.,

$$D(L) = \{u: B_i(u) = 0,\ i = 1,2\} \in \mathscr{L}_2(\Omega), \quad (15)$$

but they can be shown [13] to form a Riesz basis in the space $\mathscr{L}_2(\Omega)$ [16], i.e., a basis that is equivalent to a complete orthonormal basis in the sense that any function

$h \in \mathscr{L}_2(\Omega)$ has a unique expansion

$$h(\phi) = \sum_{n=0}^{\infty} c_n u_n(\phi), \qquad c_n \triangleq \langle h, v_n \rangle, \tag{16}$$

which converges to h in the norm of $\mathscr{L}_2(\Omega)$. Thus, because the $\{u_n(\phi)\}$ form a Riesz basis in $\mathscr{L}_2(\Omega)$, (which implies that the differential operator L of (7) is a spectral operator [17]) and because $p(\phi,t \mid \phi_0,t_0) \in \mathscr{L}_2(\Omega)$, for all fixed $t \in T$, except at $t = t_0$, $p(\phi,t \mid \phi_0,t_0)$ can be uniquely expanded in the generalized ($\mathscr{L}_2$ norm convergent) Fourier series

$$p(\phi,t \mid \phi_0,t_0) = \sum_{n=0}^{\infty} s_n(t) u_n(\phi), \tag{17}$$

where

$$s_n(t) \triangleq \int_{-\pi}^{\pi} p(\phi,t \mid \phi_0,t_0) v_n(\phi)\, d\phi. \tag{18}$$

In terms of the solution of (6), i.e.,

$$s_n(t) = s_n(t_0) \exp\left[-\lambda_n(t - t_0)\right] \tag{19}$$

it follows from (3) and (18) that the initial coefficients $s_n(t_0)$ are given by

$$s_n(t_0) = \int_{-\pi}^{\pi} \delta(\phi - \phi_0) v_n(\phi)\, d\phi = v_n(\phi_0), \tag{20}$$

so that (17) can be written as

$$p(\phi,t \mid \phi_0,t_0) = \sum_{n=0}^{\infty} v_n(\phi_0) u_n(\phi) \exp\left[-\lambda_n(t - t_0)\right]. \tag{21}$$

As it stands, (21) is the best approximation in the mean to $p(\phi,t \mid \phi_0,t_0)$; it is an exact solution only if the spectrum $\{\lambda_n, u_n(\phi)\}$, corresponding to a particular nonlinearity $g(\phi)$, is such that the series is uniformly convergent in the range $\phi \in \Omega$, $t > t_0$, and differentiable term-wise twice with respect to ϕ and once with respect to t. In the limit as $t \to t_0$, it follows from (3) and (21) that the Dirac delta function has the representation

$$\delta(\phi - \phi_0) = \sum_{n=0}^{\infty} v_n(\phi_0) u_n(\phi), \tag{22}$$

which can be shown [13] to converge weakly in the Schwartz sense to $\delta(\phi - \phi_0)$, by using the fact that the set $\{u_n(\phi)\}$ is complete in $\mathscr{L}_2(\Omega)$.

Integrating (7) over the sample space Ω gives

$$\left[u_n'(\phi) + (\alpha g(\phi) - \beta) u_n(\phi)\right] \Big|_{-\pi}^{\pi} = \frac{-\lambda_n}{D} \int_{-\pi}^{\pi} u_n(\phi)\, d\phi, \tag{23}$$

from which it follows that if $g(\phi)$ satisfies $g(\pi) = -g(-\pi) = 0$, then, for any n, either $\lambda_n = 0$ or

$$\int_{-\pi}^{\pi} u_n(\phi)\, d\phi = 0, \tag{24}$$

since the boundary conditions of (7) imply that (23) vanishes.

If we set $\lambda_0 = 0$, and solve the homogeneous problems

$$Lu_0 = 0 \qquad L^* v_0 = 0$$
$$B_i(u_0) = 0 \qquad B_i^*(v_0) = 0 \qquad i = 1,2, \tag{25}$$

then we find that $\lambda_0 = 0$ is an eigenvalue with corresponding bi-orthonormal eigenfunctions $v_0(\phi) = 1$ and

$$u_0(\phi) = C_1 \exp\left[\beta\phi - \alpha f(\phi)\right] \cdot \left[1 + C_2 \int_{-\pi}^{\phi} \exp\left[-\beta x + \alpha f(x)\right] dx\right],$$

where

$$C_1 = \left[\int_{-\pi}^{\pi} \exp\left[\beta\phi - \alpha f(\phi)\right] \cdot \left\{1 + C_2 \int_{-\pi}^{\pi} \exp\left[-\beta x + \alpha f(x)\right] dx\right\} d\phi\right]^{-1}$$

$$C_2 = \left[\exp(-2\beta\pi) - 1\right] \left\{\int_{-\pi}^{\pi} \exp\left[-\beta x + \alpha f(x)\right] dx\right\}^{-1}, \tag{26}$$

i.e., $\lambda_0 = 0$ yields the steady-state pdf $p(\phi) = u_0(\phi)$.

For $n > 0$, the spectrum cannot be found analytically; however, we show later (see Section IV-B) that the nonself-adjoint spectrum for $n > 0$ occurs in complex conjugate pairs, i.e., the complex eigenvalue λ_n, with positive real part, yields the periodic, complex eigenfunctions $u_n(\phi)$ and $\bar{v}_n(\phi)$, while its conjugate $\bar{\lambda}_n$ yields $\bar{u}_n(\phi)$ and $v_n(\phi)$. For example, considering the nonself-adjoint problems that result by setting $\alpha = 0$ in (7) and (14), we obtain the complete, biorthonormal spectrum

$$u_0(\phi) = (2\pi)^{-1} \qquad v_0(\phi) = 1 \qquad \lambda_0 = 0$$
$$u_n(\phi) = \bar{v}_n(\phi) = (2\pi)^{-\frac{1}{2}} e^{in\phi} \qquad \lambda_n = n^2 + i\beta n$$
$$\bar{u}_n(\phi) = v_n(\phi) = (2\pi)^{-\frac{1}{2}} e^{-in\phi} \qquad \bar{\lambda}_n = n^2 - i\beta n, \tag{27}$$

where

$$\langle u_n(\phi), v_m(\phi) \rangle = \langle \bar{u}_n(\phi), \bar{v}_m(\phi) \rangle = \delta_{nm}.$$

Expressing the spectrum of (7) and (14) in the real and imaginary parts

$$\lambda_n \triangleq \lambda_{nr} + i\lambda_{ni}$$
$$u_n(\phi) \triangleq u_{nr}(\phi) + iu_{ni}(\phi)$$
$$v_n(\phi) \triangleq v_{nr}(\phi) + iv_{ni}(\phi) \tag{28}$$

it follows from (21) that the transition pdf is of the form

$$p(\phi,t \mid \phi_0,t_0) = p(\phi) + \sum_{n=1}^{\infty} v_n(\phi_0) u_n(\phi) \exp\left[-\lambda_n(t - t_0)\right] + \sum_{n=1}^{\infty} \bar{v}_n(\phi_0) \bar{u}_n(\phi) \exp\left[-\bar{\lambda}_n(t - t_0)\right]$$

$$= p(\phi) + 2 \sum_{n=1}^{\infty} \{[v_{nr}(\phi_0) u_{nr}(\phi) - v_{ni}(\phi_0) u_{ni}(\phi)] \cos\left[\lambda_{ni}(t - t_0)\right] - [v_{nr}(\phi_0) u_{ni}(\phi) + v_{ni}(\phi_0) u_{nr}(\phi)] \cdot \sin\left[\lambda_{ni}(t - t_0)\right]\} \exp\left[-\lambda_{nr}(t - t_0)\right], \tag{29}$$

where the real functions $u_{nr},u_{ni},v_{nr},v_{ni} \in C^2(\Omega)$ satisfy the differential equations

$$u_{nr}'' + (\alpha g(\phi) - \beta)u_{nr}' + (\lambda_{nr} + \alpha g'(\phi))u_{nr} = \lambda_{ni}u_{ni},$$

$$u_{ni}'' + (\alpha g(\phi) - \beta)u_{ni}' + (\lambda_{nr} + \alpha g'(\phi))u_{ni} = -\lambda_{ni}u_{nr},$$

$$v_{nr}'' + (\beta - \alpha g(\phi))v_{nr}' + \lambda_{nr}v_{nr} = \lambda_{ni}v_{ni},$$

$$v_{ni}'' + (\beta - \alpha g(\phi))v_{ni}' + \lambda_{nr}v_{ni} = -\lambda_{ni}v_{nr}, \quad (30)$$

subject to periodic boundary conditions.

C. Spectral Solutions to the Fokker–Planck Equation when $\Omega_0 = 0$.

We now investigate the spectral properties of the F–P bvp when $\Omega_0 = \beta = 0$ in (2), (7), and (10), i.e., corresponding to the regular self-adjoint eigenvalue problem

$$Lu = u''(\phi) + \alpha g(\phi)u'(\phi) + \left[\frac{\lambda}{D} + \alpha g(\phi)\right] u(\phi) = 0$$

$$B_1(u) = u(-\pi) - u(\pi) = 0$$

$$B_2(u) = u'(-\pi) - u'(\pi) = 0 \quad (31)$$

and its formally self-adjoint form

$$Ly = y''(\phi) + \left[\frac{\lambda}{D} - \frac{\alpha}{2} g'(\phi) + \frac{\alpha^2}{4} g^2(\phi)\right] y(\phi) = 0$$

$$B_1(y) = y(-\pi) - y(\pi) = 0,$$

$$B_2(y) = y'(-\pi) - y'(\pi) = 0. \quad (32)$$

From the spectral theory of regular self-adjoint second-order differential operators, subject to periodic boundary conditions [12], it follows that the eigenvalues of (31) are real and form a sequence

$$0 = \lambda_0 < \lambda_1 < \lambda_2 < \cdots < \lambda_n < \cdots \quad (33)$$

such that the corresponding eigenfunctions $\{u_n(\phi), n \geq 0\} \in C^2(\Omega)$ are complete in $\mathscr{L}_2(\Omega)$ and satisfy the orthonormal condition

$$\langle u_n, u_m \rangle = \int_{-\pi}^{\pi} u_n(\phi)u_m(\phi) \exp [\alpha f(\phi)] \, d\phi = \delta_{nm}. \quad (34)$$

According to Floquet theory [12], Hill's equation will, in general, have only one periodic solution (and its constant multiples) of period π or 2π. If it should happen that two linearly independent (and therefore all) solutions of Hill's equation are of period π or 2π, then the eigenvalues $\{\lambda_n, n = 1,2,\cdots,\}$ are of multiplicity 2 and we call this an instance of *coexistence*, i.e., to each eigenvalue there coexists two periodic solutions of period π or 2π.

The case where $g(\phi) = \sin k\phi$, $k = 1,2$, is of special interest because of its practical significance and the fact that it interconnects with the coexistence theory of Ince's equation [8]. Namely, if we let $g(\phi) = \sin k\phi$, $k = 1,2$, in (31) and (32), and make the change of variables $\phi = 2x - \pi$, for $k = 1$, then we obtain:

$$u''(x) - (2\alpha \sin 2x)u'(x) + \left(\frac{4\lambda}{D} + 4\alpha \cos 2x\right) u(x) = 0, \quad k = 1 \quad (35)$$

$$y''(x) + \left(\frac{4\lambda}{D} - \frac{\alpha^2}{2} - 2\alpha \cos 2x + \frac{\alpha^2}{2} \cos 4x\right) y(x) = 0, \quad 0 \leq x \leq \pi \quad (36)$$

$$u''(\phi) + \alpha \sin 2\phi u'(\phi) + \left(\frac{\lambda}{D} + 2\alpha \cos 2\phi\right) u(\phi) = 0, \quad k = 2 \quad (37)$$

$$y''(\phi) + \left(\frac{\lambda}{D} - \frac{\alpha^2}{8} - \alpha \cos 2\phi + \frac{\alpha^2}{8} \cos 4\phi\right) y(\phi) = 0, \quad -\pi \leq \phi \leq \pi. \quad (38)$$

Equations (35) and (37) are recognized as special cases of Ince's three-parametric equation [18]

$$v''(z) + (\zeta \sin 2v)v'(z) + (\eta - q\zeta \cos 2v)v(z) = 0, \quad (39)$$

where ζ, η, and q are real constants. When Ince's equation is transformed into its formally self-adjoint form one obtains the Whittaker–Hill equation [18], i.e.,

$$\omega''(z) + \left(\eta - \frac{\zeta^2}{8} - (q+1)\zeta \cos 2z + \frac{\zeta^2}{8} \cos 4z\right) \cdot \omega(z) = 0, \quad (40)$$

for which (36) and (38) are special cases. Now the theory of Ince's equation and (24) enables us to prove the following conditions [13].

C − *1:* If $g(\phi) \in S \triangleq \{\sin k\phi, k = 1,2\}$, then each eigenvalue λ_n, $n > 0$, of (31) is of multiplicity 2, and for all values of α there coexists two linearly independent eigenfunctions $u_{ne}(\phi)$, $u_{no}(\phi)$, where $u_{ne}(\phi)$ and $u_{no}(\phi)$ are, respectively, for $k = 1,2$, even and odd periodic functions of period 2π and π, which satisfy the separated end-point boundary conditions

$$u_{ne}'(-\pi) = u_{ne}'(0) = u_{ne}'(\pi) = 0$$

$$u_{no}(-\pi) = u_{no}(0) = u_{no}(\pi) = 0. \quad (41)$$

C − *2:* If $g(\phi) \notin S$, then to each eigenvalue there corresponds a unique periodic eigenfunction.

Thus, following the development of (21) with $v_n(\phi)$ replaced by $u_n(\phi)$ and using the fact that a linear combination of $u_{ne}(\phi)$ and $u_{no}(\phi)$ is also an eigenfunction, the spectral solution of the F–P bvp corresponding to conditions $C - 1$ and $C - 2$ is, respectively,

$$p(\phi,t \mid \phi_0,t_0) = p(\phi) + \sum_{n=1}^{\infty} [u_{ne}(\phi_0)u_{ne}(\phi) + u_{no}(\phi_0)u_{no}(\phi)] \cdot \exp [\alpha f(\phi_0)] \exp [-\lambda_n(t - t_0)] \quad (42)$$

and

$$p(\phi,t \mid \phi_0,t_0) = p(\phi) + \sum_{n=1}^{\infty} u_n(\phi_0)u_n(\phi) \exp[\alpha f(\phi_0)] \cdot \exp[-\lambda_n(t - t_0)], \quad (43)$$

where $p(\phi) = u_0(\phi)$ is given by setting $\beta = 0$ in (26).

As a consequence of (33), it is easy to show, using the Weierstrass M test, that the infinite series in (42) and (43) converge uniformly in ϕ and t to $p(\phi,t \mid \phi_0,t_0)$, for all $\phi \in \Omega$ and $t > t_0$, and that these series remain uniformly convergent in the range stated when differentiated twice with respect to ϕ and once with respect to t. Thus we conclude that (42) and (43) are solutions of the F–P bvp, and that these solutions are continuous functions of ϕ and t, for all $\phi \in \Omega$ and $t > t_0$. We remark that the same conclusion holds for the infinite series (29) when $g(\phi) \in S$, since it can be verified that the real part λ_{nr} of the complex conjugate nonself-adjoint spectrum increases monotonically with respect to n (see Section IV-B).

As an example of the weak convergence of the spectral solutions to the Dirac delta function representation (22), note that if (27) is substituted into (29), then the transition pdf assumes the form

$$p(\phi,t \mid \phi_0,t_0) = \frac{1}{2\pi} + \frac{1}{\pi}\sum_{n=1}^{\infty} \cos n[\phi - \phi_0 - \Omega_0(t - t_0)] \cdot \exp[-Dn^2(t - t_0)], \quad (44)$$

and

$$\lim_{t\to t_0} p(\phi,t \mid \phi_0,t_0) = \frac{1}{2\pi} + \frac{1}{\pi}\sum_{n=1}^{\infty} \cos(n[\phi - \phi_0]) = \delta(\phi - \phi_0). \quad (45)$$

III. Application of the Theory to the Development of Loop Statistics

In this section the discussion shall be restricted to $g(\phi) \in S$, since this is the set of loop nonlinearities which has most practical significance.

A. Time-Variant Statistics

Using the spectral solution (42), all time-variant moments of the nonstationary non-Gaussian process $\psi(t) \triangleq F[\phi]$ may be evaluated, where $F[\cdot]$ is any memoryless time-invariant nonlinear transformation. For example, from (42) it follows that the time-variant mean of $\psi(t)$ may be written

$$E[\psi(t)] = \int_\Omega F(\phi)p(\phi,t \mid \phi_0,t_0)\, d\phi = \sigma_0 + \sum_{n=1}^{\infty} [u_{ne}(\phi_0)\sigma_{ne} + u_{no}(\phi_0)\sigma_{no}] \cdot \exp[\alpha f(\phi_0) - \lambda_n(t - t_0)], \quad (46)$$

where the correlation factors of $F[\phi]$ are

$$\sigma_0 = \int_\Omega F(\phi)p(\phi)\, d\phi \triangleq \langle F[\phi],p(\phi)\rangle$$

$$\sigma_{ne} \triangleq \langle F[\phi],u_{ne}(\phi)\rangle \qquad \sigma_{no} \triangleq \langle F[\phi],u_{no}(\phi)\rangle.$$

Letting $F[\phi] = 1$ and $\Omega = (-\phi_l,\phi_l)$ in (46), we obtain the probability that $|\phi| \le \phi_l$ at time $t - t_0$, i.e.,

$$\Pr[|\phi(t)| \le \phi_l; \phi_0] = \sigma_0 + \sum_{n=1}^{\infty} \sigma_{ne}u_{ne}(\phi_0) \exp[\alpha f(\phi_0) - \lambda_n(t - t_0)], \quad 0 \le \phi_l \le \phi_0 \le \pi, \quad (47)$$

which is of interest in computing the probability of signal acquisition at time $t - t_0$ given that $\phi = \phi_0$ at $t = t_0$.

B. Loop Statistics in the Steady State

Since $\phi(t)$ is stationary in the steady state, the transition pdf enables all nth-order steady-state joint pdf's to be evaluated. For example, the second joint pdf is the product of the first pdf at time $t_1, p(\phi_1)$, and the transition pdf at a later time $t_2, p(\phi_2,t_2 \mid \phi_1,t_1)$, so from (42) we have

$$p(\phi_2,t_2;\phi_1,t_1) = p(\phi_2)p(\phi_1) + h^{-1}(\alpha)\sum_{n=1}^{\infty} [u_{ne}(\phi_1)u_{ne}(\phi_2) + u_{no}(\phi_1)u_{no}(\phi_2)] \exp[-\lambda_n\tau], \quad (48)$$

where

$$h^{-1}(\alpha) \triangleq \int_\Omega \exp[-\alpha f(\phi)]\, d\phi, \qquad \tau \triangleq t_2 - t_1.$$

Using (48), the autocorrelation function and the spectral density of the stationary non-Gaussian process $\psi(t)$ are given, respectively, by

$$R_\psi(\tau) = \int_\Omega\int_\Omega F[\phi_1]F[\phi_2]p(\phi_2,\phi_1;\tau)\, d\phi_1\, d\phi_2 = \sigma_0^2 + h^{-1}(\alpha)\sum_{n=1}^{\infty} (\sigma_{ne}^2 + \sigma_{no}^2) \exp[-\lambda_n|\tau|] \quad (49)$$

and

$$S_\psi(\omega) = \int_{-\infty}^{\infty} R_\psi(\tau) \exp[-j\omega\tau]\, d\tau = \sum_{n=1}^{\infty} \frac{2\lambda_n(\sigma_{ne}^2 + \sigma_{no}^2)h^{-1}(\alpha)}{\omega^2 + \lambda_n^2}, \quad (50)$$

where $F[\phi_1] \triangleq F[\phi(t_1)]$ and $F[\phi_2] \triangleq F[\phi(t_2)]$.

C. Comparison With Statistical Linearization Methods

In the past, a great deal of effort has been expended on determining the behavior of the sinusoidal PLL in the nonlinear region of operation [9]–[11]. These analyses fail to take into account the nonlinear effect of the modulo-2π reduction of the phase-error process. Develet's [9] and Tausworthe's [11] analyses are based upon the assumption that

$$R_{\psi,\phi}(\tau) = C_\psi R_\phi(\tau)$$
$$R_\psi(\tau) = \tilde{C}_\psi R_\phi(\tau), \tag{51}$$

where C_ψ and $\tilde{C}_\psi$ are constants that depend only on the steady-state pdf $p(\phi)$ and the nonlinear transformation $\psi(t) = F[\phi(t)]$. A necessary and sufficient condition for (51) to hold is that the process $\phi(t)$ must belong to the class of separable random processes [19], i.e., the class for which the function

$$g(\phi_2,\tau) \triangleq \int_\Omega \phi_1 p(\phi_1,\phi_2;\tau)\, d\phi_1 \tag{52}$$

can be separated as

$$g(\phi_2,\tau) = g_1(\phi_2)g_2(\tau) \tag{53}$$

for all ϕ_2 and τ. Substituting (48) into (52) and arguing by contradiction, it is easy to show that

$$g(\phi_2,\tau) = h^{-1}(\alpha) \sum_{n=1}^{\infty} \langle \phi_1, u_{no}(\phi_1)\rangle u_{no}(\phi_2) \exp[-\lambda_n \tau] \tag{54}$$

is not separable; therefore, (51) does not hold in the nonlinear region of loop operation. From this contradiction it also follows that $p(\phi_1,\phi_2;\tau)$ does not posses a Barrett–Lampard expansion [20]. This is due to the fact that the class of processes that satisfy the Barrett–Lampard expansion is a subset of the class of separable processes [19].

IV. Analytic Approximations and Calculation of Eigenvalues

In this section we obtain analytic approximations and numerical values for the eigenvalues of the sinusoidal PLL. Analogous results can be produced for the squaring or Costas loop.

A. Analytic Approximations for $\beta = 0$

Substituting $g(\phi) = \sin\phi$ in (32) we obtain

$$y''(\phi) + \left(\lambda' - \frac{\alpha^2}{8} + \frac{\alpha}{2}\cos\phi + \frac{\alpha^2}{8}\cos 2\phi\right) y(\phi) = 0$$
$$y(-\pi) - y(\pi) = 0 \qquad y'(-\pi) - y'(\pi) = 0, \tag{55}$$

where $\lambda' \triangleq \lambda/D$. When $\alpha = 0$, (55) has periodic solutions $\cos n\phi$, $\sin n\phi$ with $\lambda' = n^2$, $n = 0,1,2,\cdots$. We now obtain perturbation solutions of (55), for the nonlinear region of operation $0 \le \alpha \le 6$, which are periodic and reduce to $\cos n\phi$ and $\sin n\phi$ as α approaches zero. Recalling the coexistence property of (31), we can write

$$\lambda_n'(\alpha) = n^2 + c_1\alpha + c_2\alpha^2 + c_3\alpha^3 + \cdots$$
$$y_{ne}(\phi) = \cos n\phi + \alpha f_1(\phi) + \alpha f_2(\phi) + \alpha f_3(\phi) + \cdots$$
$$y_{no}(\phi) = \sin n\phi + \alpha g_1(\phi) + \alpha^2 g_2(\phi) + \alpha^3 g_3(\phi) + \cdots, \qquad n > 0, \tag{56}$$

where the coefficients $c_1,c_2,\cdots$ and the functions $f_1(\phi)$, $f_2(\phi),\cdots,g_1(\phi),g_2(\phi),\cdots$ are to be determined from the conditions that the function $y_{ne}(\phi),y_{no}(\phi)$ are, respectively, even and odd, and normalized by making the coefficient of $\cos n\phi$ and $\sin n\phi$ be unity for all values of α. The advantage of this normalization procedure is that it yields perturbation solutions having a large radius of convergence.

To illustrate the method, we show the first few steps in the solution for y_{1o} and $\lambda_1'(\alpha)$. Letting $n = 1$ in (56) and substituting $y_{1o}(\phi),\lambda_1'(\alpha)$ in (55), we find that

$$-\sin\phi + \alpha g_1'' + \alpha^2 g_2'' + \cdots$$
$$+ [1 + c_1\alpha + c_2\alpha^2 + \cdots][\sin\phi + \alpha g_1 + \alpha^2 g_2 + \cdots]$$
$$+ \left[\frac{\alpha^2}{8}(\cos 2\phi - 1) + \frac{\alpha}{2}\cos\phi\right]$$
$$\cdot [\sin\phi + \alpha g_1 + \alpha^2 g_2 + \cdots] = 0. \tag{57}$$

The term independent of α in the preceding equation vanishes identically. Equating similar powers of α to zero yields a sequence of second-order linear differential equations with constant coefficients, for which we display the first three

$$Lg_1 = -c_1 \sin\phi - \tfrac{1}{2}\sin\phi\cos\phi$$
$$Lg_2 = -\left(c_1 + \frac{\cos\phi}{2}\right) g_1 - c_2\sin\phi - \tfrac{1}{8}(\cos 2\phi - 1)\sin\phi$$
$$Lg_3 = -[c_2 + \tfrac{1}{8}(\cos 2\phi - 1)]g_1 - (c_1 + \tfrac{1}{2})g_2 - c_3\sin\phi, \tag{58}$$

where in each the operator $L \triangleq (d^2/d\phi^2) + 1$. The general solution of the first equation is

$$g_1(\phi) = a_1\sin\phi + a_2\cos\phi + \frac{c_1}{2}\phi\cos\phi + \tfrac{1}{12}\sin 2\phi, \tag{59}$$

where a_1,a_2 are arbitrary constants. Since $g_1(\phi)$ must be odd, it follows that $a_2 = 0$. Since $g_1(\phi)$ is to be periodic, the term in $\phi\cos\phi$ must also vanish; hence $c_1 = 0$. Since $g_1(\phi)$ must contain no term in $\sin\phi$, $a_1 = 0$ also. Thus we have that $c_1 = 0$ and $g_1(\phi) = \sin 2\phi/12$. Substituting c_1 and $g_1(\phi)$ into the second equation and simplifying, we obtain

$$Lg_2 = (\tfrac{1}{6} - c_2)\sin\phi - \frac{\sin 3\phi}{12}. \tag{60}$$

Solving this equation for g_2 we find that

$$g_2(\phi) = a_3\sin\phi + a_4\cos\phi + \left(\frac{c^2}{2} - \frac{1}{12}\right)\phi\cos\phi + \frac{\sin 3\phi}{96}.$$

Applying the same restrictions on $g_2(\phi)$ as was done previously for $g_1(\phi)$, gives $a_3 = a_4 = 0$, $c_2 = \tfrac{1}{6}$, and

$g_2(\phi) = \sin 3\phi/96$. Consequently, to $0(\alpha^3)$ we have

$$\lambda_1'(\alpha) = 1 + \frac{\alpha^2}{6} + 0(\alpha^3),$$

$$y_{1o}(\phi) = \sin\phi + \frac{\alpha}{12}\sin 2\phi + \frac{\alpha^2}{96}\sin 3\phi + 0(\alpha^3). \quad (61)$$

Further terms are successively obtained by repeating the same process. The results obtained for the first three eigenvalues to $0(\alpha^5)$ are summarized in the following:

$$\lambda_1'(\alpha) = 1 + \frac{\alpha^2}{6} - \frac{\alpha^4}{2^5 \cdot 3^3} + 0(\alpha^5) \quad (62)$$

$$\lambda_2'(\alpha) = 4 + \frac{2\alpha^2}{15} + \frac{11\alpha^4}{8 \cdot 15^3} + 0(\alpha^5) \quad (63)$$

$$\lambda_3'(\alpha) = 9 + \frac{9\alpha^2}{70} + \frac{4689\alpha^4}{512 \cdot 35^3} + 0(\alpha^5). \quad (64)$$

Examining the first two terms of $\lambda_n'(\alpha)$, $n = 1,2,3$, we find that the eigenvalues of (55) and hence those of (31), with $g(\phi) = \sin\phi$, are approximated by

$$\lambda_n'(\alpha) \cong n^2 + \frac{n^2\alpha^2}{8n^2 - 2} + 0(\alpha^4). \quad (65)$$

The same result is also found to hold for $n > 3$. In the limit as $n \to \infty$, we note that (65) is consistent with the asymptotic behavior of the spectrum of (55) for large λ', viz.,

$$\lambda_n'(\alpha) = n^2 + \frac{\alpha^2}{8} \quad (66)$$

for which the corresponding asymptotic, normalized eigenfunctions of (31) are found to be

$$u_{ne}(\phi) = (\pi)^{-\frac{1}{2}} \cos n\phi \exp\left[\frac{\alpha}{2}\cos\phi\right]$$

$$u_{no}(\phi) = (\pi)^{-\frac{1}{2}} \sin n\phi \exp\left[\frac{\alpha}{2}\cos\phi\right]. \quad (67)$$

Because $\lambda_n'(\alpha)$ increases with respect to n^2 for fixed α, we note that the infinite expansion (42), and hence the expansions of the associated time-variant statistics that follow, can be adequately approximated by using a finite number of terms.

B. Perturbation Estimates for $\beta \neq 0$

We now obtain a first-order estimate for the eigenvalues of the nonself-adjoint problem (11) assuming $g(\phi) \in S$ by considering β as a perturbation on the self-adjoint spectrum $\{\lambda_n'(\alpha), u_0(\phi), u_{ne}(\phi), u_{no}(\phi)\}$ of (31). The nonself-adjoint or perturbed operator of (7) is rewritten as $L = L_0 + \beta(d/d\phi)$, where $L_0 = (d^2/d\phi^2) + (d/d\phi)[\alpha g(\phi)]$ is the unperturbed operator of (31). Since $\lambda_n'(\alpha)$, $n = 1,2,\cdots$, is twofold degenerate, for the zeroth-order perturbed eigenfunction we take the linear combination

$$u_{\beta n}(\phi) = a_n(\beta)u_{ne}(\phi) + b_n(\beta)u_{no}(\phi), \qquad n \geq 1,$$

and for the first-order approximation to the perturbed eigenvalue we let

$$\lambda_n'(\beta) = \lambda_n'(\alpha) + \beta\left.\left(\frac{\partial\lambda_n(\beta)}{\partial\beta}\right)\right|_{\beta=0}, \qquad n \geq 1, \quad (68)$$

so that as $\beta \to 0$, we have $a_n(\beta) \to a_n$, $b_n(\beta) \to b_n$, and hence $u_{\beta n}(\phi) \to u_n(\phi)$, $\lambda_n'(\beta) \to \lambda_n'(\alpha)$. Adding $\lambda_n'(\alpha)u_{\beta n}(\phi)$ to both sides of (7) and then differentiating with respect to β yields

$$\left.\left(\frac{\partial\lambda_n'(\beta)}{\partial\beta}\right)\right|_{\beta=0} = \pm\sqrt{\langle u_{no}', u_{ne}\rangle \cdot \langle u_{ne}', u_{no}\rangle}$$

so that (68) is

$$\lambda_n'(\beta) \cong n^2 + \frac{n^2\alpha^2}{8n^2 - 2} \pm \beta\sqrt{\langle u_{no}', u_{ne}\rangle\langle u_{ne}, u_{no}\rangle}. \quad (69)$$

As a check on (69), we note that (67) yields

$$\lambda_n'(\beta) = n^2 + \frac{n^2\alpha^2}{8n^2 - 2} \pm i\beta_n,$$

which is a result that is consistent with the exact values for the case $\alpha = 0$, $\beta \neq 0$, i.e., (23) and for the asymptotic case $\alpha \neq 0$, $\beta \neq 0$.

C. Calculation of Eigenvalues $\alpha \neq 0$, $\beta = 0$

We now present one iterative method for obtaining numerical values of the eigenvalues of (55). This method makes use of the fact that the Fourier coefficients of the even and odd eigenfunctions of (55) satisfy a five-term recurrence relation. Namely, if we let

$$y_e(\phi) = \sum_{n=0}^{\infty} A_n \cos n\phi \qquad y_o(\phi) = \sum_{n=0}^{\infty} B_n \sin n\phi \quad (70)$$

substitute these into (55), collect together like terms of $\cos n\phi$ and $\sin n\phi$, and equate each group to zero, then we obtain, for the even periodic eigenfunctions,

$$\left(\lambda' - \frac{\alpha^2}{8}\right)A_0 + \frac{\alpha}{4}A_1 + \frac{\alpha^2}{16}A_2 = 0, \qquad n = 0,$$

$$\frac{\alpha}{2}A_0 + \left(\lambda' - 1 - \frac{\alpha^2}{16}\right)A_1 + \frac{\alpha}{4}A_2 + \frac{\alpha^2}{16}A_3 = 0, \qquad n = 1,$$

$$\frac{\alpha^2}{8}A_0 + \frac{\alpha}{4}A_1 + \left(\lambda' - \frac{\alpha^2}{8} - 4\right)A_2 + \frac{\alpha}{4}A_3 + \frac{\alpha^2}{16}A_4 = 0, \qquad n = 2,$$

....

$$\left(\lambda' - \frac{\alpha^2}{8} - n^2\right)A_n + \frac{\alpha}{4}(A_{n-1} + A_{n+1}) + \frac{\alpha^2}{16}(A_{n-2} + A_{n+2}) = 0, \qquad n \geq 3. \quad (71)$$

While for the odd periodic eigenfunctions

$$(\lambda' - 1 - \tfrac{3}{16}\alpha^2)B_1 + \frac{\alpha}{4}B_2 + \frac{\alpha^2}{16}B_3 = 0, \qquad n = 1,$$

$$\frac{\alpha}{4}B_1 + \left(\lambda' - \frac{\alpha^2}{8} - 4\right)B_2 + \frac{\alpha}{4}B_3 + \frac{\alpha^2}{16}B_4 = 0, \qquad n = 2,$$

$$\frac{\alpha^2}{16}B_1 + \frac{\alpha}{4}B_2 + \left(\lambda' - \frac{\alpha^2}{8} - 9\right)B_3 + \frac{\alpha}{4}B_4 + \frac{\alpha^2}{16}B_5 = 0, \qquad n = 3,$$

$$\cdots\cdots\cdots$$

$$\left(\lambda' - \frac{\alpha^2}{8} - n^2\right)B_n + \frac{\alpha}{4}(B_{n-1} + B_{n+1}) + \frac{\alpha^2}{16}(B_{n-2} + B_{n+2}) = 0, \qquad n \geq 4. \quad (72)$$

Note that (71) and (72) constitute an infinite set of algebraic equations; hence, for an explicit computation of the Fourier coefficients A_n, B_n, and the eigenvalue λ', we actually have to solve an infinite set of equations. However, to circumvent this problem of dimensionality an approximation procedure was employed. Namely, the systems (71) and (72) were truncated at the fifth equation, thus yielding two systems of five equations in six unknowns, i.e., $A_0,\cdots,A_5$ and $B_1,\cdots,B_6$. The coefficients A_5,B_6 are set equal to zero and the coefficients A_4 and B_5 are set equal to unity, thus giving a nonhomogeneous system of five equations in four unknowns. One equation in each system of equations is therefore redundant, and it is used as an iteration error equation to find the eigenvalue and the approximate Fourier coefficients that characterize $y_e(\phi)$ and $y_o(\phi)$. For example, with $\alpha = 4$, $B_6 = 0$, and $B_5 = 1$, we obtain for the odd eigenfunction the nonhomogeneous system

$$\begin{aligned} (\lambda' - 4)B_1 + B_2 + B_3 &= 0, & n &= 1, \\ B_1 + (\lambda' - 6)B_2 + B_3 + B_4 &= 0, & n &= 2, \\ B_1 + B_2 + (\lambda' - 11)B_3 + B_4 &= -1, & n &= 3, \\ B_2 + B_3 + (\lambda' - 18)B_4 &= -1, & n &= 4, \\ B_3 + B_4 &= 27 - \lambda', & n &= 5. \end{aligned} \quad (73)$$

To find the eigenvalue λ', we use the first equation as our iteration control, i.e., we let

$$[\lambda'(\varepsilon) - 4]B_1 + B_2 + B_3 \triangleq E[\lambda'(\varepsilon)] \qquad (74)$$

be the iteration error, set $\lambda'(\varepsilon)$ equal to the approximate value $\lambda'(\alpha)$, solve the last four equations, compute the error $E[\lambda'(\varepsilon)]$, check its sign, and successively increase or decrease $\lambda'(\varepsilon)$ until the error is less than a fixed tolerance. In computing the eigenvalues λ_n' via (71) and (72), these equations were first truncated at $n = 5$ and then at $n = 6$. Increasing the dimensionality by one equation had the effect of changing only the fifth significant digit. Thus, for all practical purposes, five equations are sufficient in computing the eigenvalues. To obtain the eigenvalues λ_n', (71) and (72) were solved by means of the Gauss-elimination procedure, with an error tolerance of 10^{-5} to ensure reasonable accuracy. Computation time for obtaining 20 eigenvalues was less than 2 min on an SDS-930 computer. The relatively small amount of computation time required was due to the excellent initial estimate $\lambda_n'(\alpha)$. The corresponding normalized eigenfunctions $u_{ne}(\phi),u_{no}(\phi)$ were then obtained in increments of one degree in ϕ, by using a modified Runge–Kutta algorithm with an integration error upper bound of 10^{-6} to ensure reasonable accuracy.

V. Summary of Results $\alpha \neq 0$, $\beta = 0$

Considering the loop SNR region $0 \leq \alpha \leq 6$ where non-Gaussian statistics exist, the approximate analytic $\lambda_n'(\alpha)$ and the computed eigenvalues λ_n', $n = 1,\cdots,4$, for the sinusoidal PLL are tabulated in Table I. In this table, the values $\lambda_n'(\alpha)$, $n = 1,\cdots,4$, are given by (62) to (65), respectively. Comparing $\lambda_n'(\alpha)$ and λ_n', it is seen that the perturbation solutions $\lambda_n'(\alpha)$ yield excellent estimates of λ_n' not only for very small α, but throughout the nonlinear region of operation. In fact, the approximate eigenvalue with the least fidelity, i.e., $\lambda_1'(\alpha = 6)$, differs from its corresponding computed value with a maximum percentage error of 2 percent. The even and odd periodic eigenfunctions, u_{ne},u_{no}, $n = 1,\cdots,4$, of the sinusoidal PLL are illustrated in Figs. 1 and 2. It turns out that the asymptotic solutions, given by (66) and (67) yield excellent estimates of $\{\lambda_n',u_{ne},u_{no}\}$ not only for very large n but commencing at $n \geq 5$. This result is evident from the values given in Table II, where, for $n = 5$, $\alpha = 2,4$, it is seen that the computed eigenvalues differ from their corresponding asymptotic values by maximum percentage errors of 0.02 and 0.15 percent, respectively.

The transition pdf of the sinusoidal PLL is illustrated, at four different instants of time, in Figs. 3–5 for $\phi_0 = \pi/2$. In computing these functions, the first ten terms of (42) were used. From these figures it is clear that as α increases the transition time required for the steady-state pdf to be established decreases. The odd correlation factors σ_{no} for $F[\phi] = \phi$ and $F[\phi] = \sin\phi$ are tabulated in Table III. Recalling that the autocorrelation and spectral density functions of $F[\phi]$ depend on σ_n^2, from Table III it is evident that only the first few correlation factors are significant in evaluating these functions. The autocorrelation function

$$R_\phi(\tau) = \sum_{n=1}^{10} \frac{\sigma_{no}^2 \exp[-D\lambda_n'|\tau|]}{2\pi I_0(\alpha)} \qquad (75)$$

and the spectral density function

$$DS_\phi(f) = \frac{1}{\pi I_0(\alpha)} \sum_{n=1}^{10} \frac{\lambda_n'\sigma_{no}^2}{(\lambda_n')^2 + \left(\frac{2\pi f}{D}\right)^2} \qquad (76)$$

of the sinusoidal PLL are illustrated in Figs. 6 and 7, respectively. As a check on the accuracy of the eigenvalues given in Tables I and II and hence, on the corresponding eigenfunctions, the steady-state variance of ϕ, for $\alpha = 4$,

TABLE I
COMPUTED (λ_N') AND APPROXIMATE ($\lambda_N'(\alpha)$) EIGENVALUES OF PLL, ORDERS 1–4 FOR $\beta = 0$, $\alpha = 1 \cdots 6$

SNR α	N = 1 λ_N'	N = 1 $\lambda_N'(\alpha)$	N = 2 λ_N'	N = 2 $\lambda_N'(\alpha)$	N = 3 λ_N'	N = 3 $\lambda_N'(\alpha)$	N = 4 λ_N'	N = 4 $\lambda_N'(\alpha)$
1	1.16550	1.16551	4.13376	4.13374	9.12878	9.12878	16.12710	16.12698
2	1.64733	1.64816	4.54115	4.53985	9.51702	9.51770	16.50969	16.50794
3	2.39825	2.40610	5.24052	5.23300	10.17406	10.17444	17.15249	17.14286
4	3.33421	3.37060	6.27054	6.23763	11.11183	11.11182	18.0650	18.03174
5	4.35503	4.44557	7.66287	7.58796	12.35252	12.34779	19.25037	19.17460
6	5.39014	5.50000	9.39412	9.32800	13.92881	13.90504	20.73129	20.57143

TABLE II
PLL ASYMPTOTIC, PERTURBATION AND COMPUTED EIGENVALUES, ORDERS 5–10, $\beta = 0$, $\alpha = 2.4$

α = 2	ASYMPTOTIC VALUES	PERTURBATION VALUES	COMPUTED VALUES
N = 5	25.5	25.50504	25.50514
N = 6	36.5	36.50349	36.50358
N = 7	49.5	49.50256	49.50264
N = 8	64.5	64.50195	64.50202
N = 9	81.5	81.50154	81.50160
N = 10	100.5	100.50125	100.50130
α = 4	**ASYMPTOTIC VALUES**	**PERTURBATION VALUES**	**COMPUTED VALUES**
N = 5	27.0	27.02019	27.04007
N = 6	38.0	38.01398	38.02794
N = 7	51.0	51.01025	51.02051
N = 8	66.0	66.00784	66.01613
N = 9	83.0	83.00618	83.01268
N = 10	102.0	102.00500	102.01065

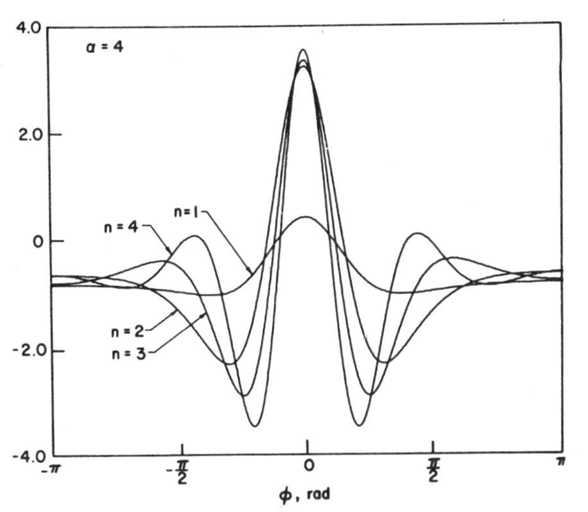

Fig. 1. Even periodic eigenfunctions of PLL, orders 1–4, $\alpha = 4$, $\beta = 0$.

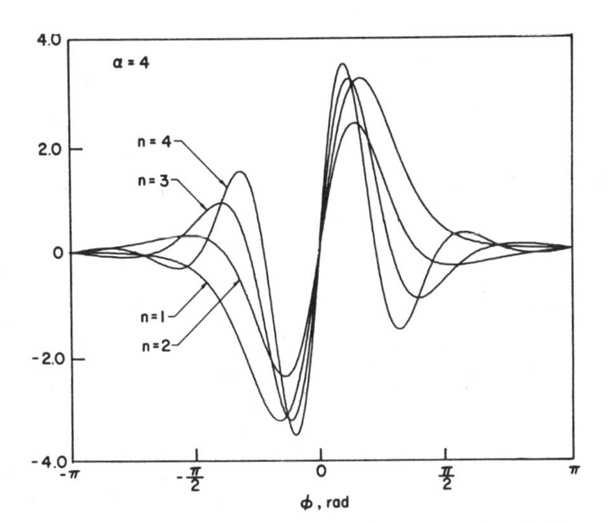

Fig. 2. Odd periodic eigenfunctions of PLL, orders 1–4, $\alpha = 4$, $\beta = 0$.

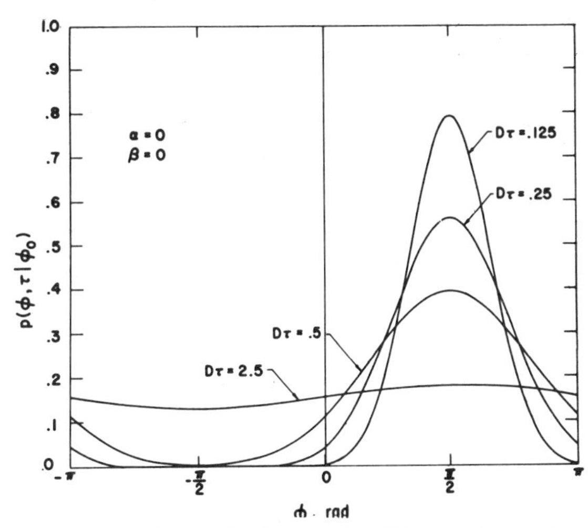

Fig. 3. Transition pdf of first-order PLL, $\alpha = 0$, $\beta = 0$.

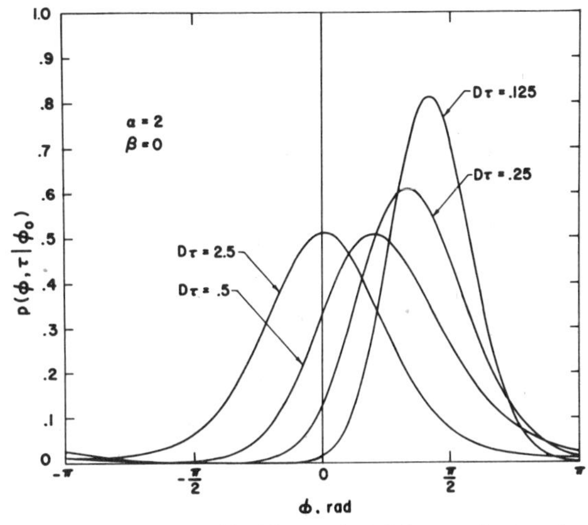

Fig. 4. Transition pdf of first-order PLL, $\alpha = 2$, $\beta = 0$.

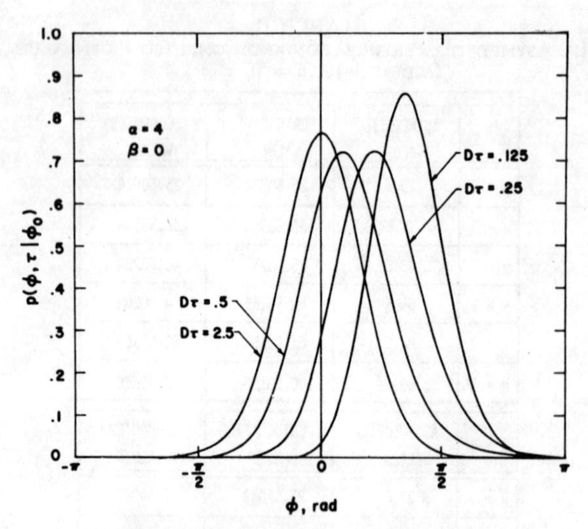

Fig. 5. Transition pdf of first-order PLL, $\alpha = 4$, $\beta = 0$.

TABLE III
PLL CORRELATION FACTORS FOR $F(\phi) = \phi$, $F(\phi) = \sin\phi$, ORDERS 1–10, $\alpha = 2.4$

	$\alpha = 2$		$\alpha = 4$	
N	$\sigma_{N,\phi}$	$\sigma_{N,\sin\phi}$	$\sigma_{N,\phi}$	$\sigma_{N,\sin\phi}$
1	3.19580	2.15247	4.55822	3.74383
2	-0.29830	0.59682	-0.52389	1.08307
3	0.42412	0.08942	0.26141	0.36753
4	-0.31248	0.00848	-0.09099	0.06908
5	0.25436	0.00054	0.09016	0.00954
6	-0.21361	0.00001	-0.07558	0.00105
7	0.18392	-	0.06583	0.00010
8	-0.16140	-	-0.05812	0.00002
9	0.14375	-	0.05201	-
10	-0.12953	-	-0.04701	-

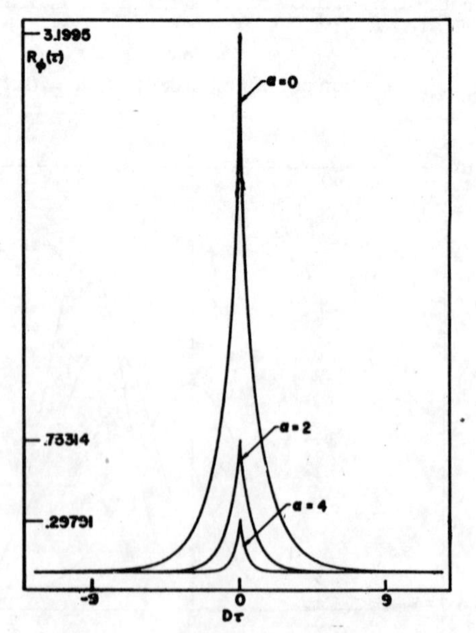

Fig. 6. Steady-state autocorrelation function of first-order PLL, $\beta = 0$, $\alpha = 0,2,4$.

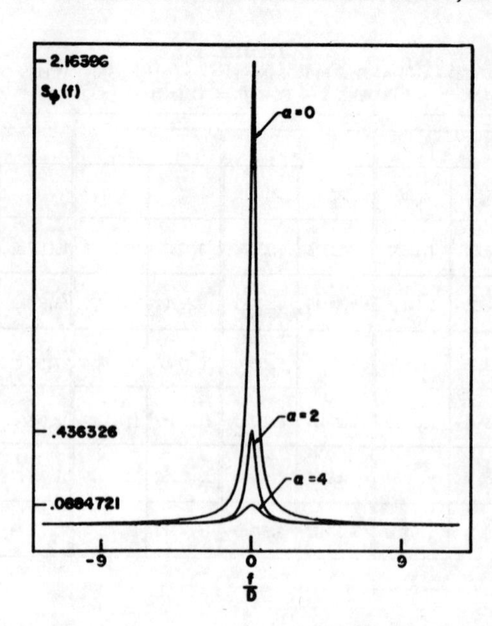

Fig. 7. Steady-state, normalized spectral density of first-order PLL, $\beta = 0$, $\alpha = 0,2,4$.

is found from (75) and Table III to be

$$\sigma_\phi^{\,2} = R_\phi(0) - R_\phi(\infty) = \sum_{n=1}^{10} \frac{\sigma_{no}^{\,2}}{2\pi I_0(\alpha)} = 0.29791$$

while from the known steady-state pdf $p(\phi)$, we have

$$\sigma_\phi^{\,2} = \int_{-\pi}^{\pi} \phi^2 p(\phi)\, d(\phi)$$

$$= \frac{\pi^2}{3} + 4 \sum_{n=1}^{\infty} \frac{(-1)^n I_n(\alpha)}{n^2 I_0(\alpha)} \cong 0.29822,$$

thus indicating excellent agreement with the nonlinear theory [1].

VI. CONCLUSIONS AND EXTENSIONS

The time-dependent F–P bvp corresponding to a general class of first-order phase-locked tracking systems has been analyzed and eigenfunction expansions have been developed for the reduced modulo-2π phase-error transition pdf. These expansions have been shown to be of practical value in obtaining either time-variant or steady-state statistics and in assessing the validity of statistical linearization methods.

We point out that the results presented here can be extended to find approximate solutions of the reduced modulo-2π transition pdf, $p(\phi,t \mid \phi_0,t_0)$, corresponding to higher order loops (see [13]). In addition, we note that the theory developed here has been extended to solving the first-passage time F–P bvp corresponding to a general class of first-order PLL's (see [21]). The results of this latter work yields a complete statistical description of the cycle-slipping behavior of first-order PLL's in the presence of noise.

ACKNOWLEDGMENT

The effective collaboration of Prof. L. R. Welch of the University of Southern California is greatly appreciated.

His helpful comments and various suggestions throughout the course of this work were invaluable.

References

[1] A. J. Viterbi, *Principles of Coherent Communications.* New York: McGraw-Hill, 1966, ch. 4.
[2] W. C. Lindsey, *Synchronization Systems in Communication and Control.* Englewood Cliffs, N.J.: Prentice-Hall, June 1972.
[3] J. P. Costas, "Synchronous communications," *Proc. IRE*, vol. 44, pp. 1713–1718, Dec. 1956.
[4] W. C. Lindsey and M. K. Simon, "Data-aided carrier tracking loops," *IEEE Trans. Commun. Technol.*, vol. COM-19, pp. 157–168, Apr. 1971.
[5] W. C. Lindsey, "Hybrid carrier and modulation tracking loops," in *1970 Proc. Int. Conf. Communications*, San Francisco, Calif., June 8–10, 1970; also see *IEEE Trans. Comm.* (Concise Paper), vol. COM-10, pp. 53–55, Feb. 1972.
[6] M. K. Simon, "Nonlinear analysis of an absolute value type of early-late gate bit synchronizer," *IEEE Trans. Comm. Technol.*, vol. COM-18, pp. 589–596, Oct. 1970.
[7] R. Adler, "A study of locking phenomena in oscillators," *Proc. IRE*, vol. 34, pp. 351–357, June 1946.
[8] W. Mangus and S. Winkler, *Hill's Equations.* New York: Wiley, 1966.
[9] J. A. Develet, Jr., "A threshold criterion for phase-lock demodulation," *Proc. IRE*, vol. 51, pp. 349–356, Feb. 1963.
[10] H. L. Van Trees, "Functional techniques for the analysis of the nonlinear behavior of phase-locked loops," *Proc. IEEE*, vol. 52, pp. 894–911, Aug. 1964.
[11] R. C. Tausworthe, "A method for calculating phase-locked loop performance near threshold," *IEEE Trans. Commun. Technol.*, vol. COM-15, pp. 502–506, Aug. 1967.
[12] E. A. Coddington and N. Levinson, *Theory of Ordinary Differential Equations.* New York: McGraw-Hill, 1955.
[13] J. R. La Frieda, "Transient analysis of nonlinear tracking systems," Ph.D. dissertation, Univ. Southern California, June 1970 (obtainable from Univ. Microfilms, Ann Arbor, Mich.); also [2, ch. 16].
[14] C. D. Birkhoff, "Boundary value and expansion problems of ordinary differential equations," *Trans. Amer. Math. Soc.*, vol. 9, pp. 373–395, 1908.
[15] J. D. Tamarkin, "Some points in the theory of differential equations," *Rend di Palermo*, vol. 34, pp. 354–3, 1912.
[16] V. P. Mikhailov, "On Riesz bases in $\mathscr{L}_2(0,1)$," *Dokl. Akad. Nauk SSSR*, vol. 144, pp. 981–984, 1962.
[17] N. Dunford, "A survey of the theory of spectral operators," *Bull. Amer. Math. Soc.*, vol. 64, pp. 214–217, 1958.
[18] F. M. Arscott, *Periodic Differential Equations.* New York: MacMillan, 1964.
[19] A. H. Nuttall, "Theory and application of the separable class of random processes," Massachusetts Inst. Technol., Cambridge, May 1958, ERL TR 343.
[20] J. F. Barrett and D. G. Lampard, "An expansion for some second-order probability distributions and its applications to noise problems," *IRE Trans. Inform. Theory*, vol. IT-1, pp. 10–15, Mar. 1955.
[21] J. R. La Frieda, "On the probability of cycle-slipping in first-order phase-locked loops," *Proc. Hawaii Int. Conf. System Sciences*, pp. 57–59, 1972.

Nonlinear Analysis of Generalized Tracking Systems

W. C. LINDSEY, MEMBER, IEEE

Abstract—This paper sets forth a rather general analysis pertaining to the performance and synthesis of generalized tracking systems. The analysis is based upon the theory of continuous Markov processes, in particular, the Fokker–Planck equation. We point out the interconnection between the theory of continuous Markov processes and Maxwell's wave equations by interpreting the charge density as a transition probability density function (pdf). These topics presently go under the name of probabilistic potential theory.

Although the theory is valid for $(N+1)$-order tracking systems with an arbitrary, memoryless, periodic nonlinearity, we study in detail the case of greatest practical interest, viz., a second-order tracking system with sinusoidal nonlinearity. In general we show that the transition pdf $p(y, t|y_0, t_0)$ is the solution to an $(N+1)$-dimensional Fokker–Planck equation. The vector $(y, t)=(\phi, y_1, \cdots, y_N, t)$ is Markov and ϕ represents the system phase error. According to the theory the transition pdf's $\{p(\phi, t|\phi_0, t_0), p(y_k, t_0|y_{k0}, t_0); k=1, \cdots, N\}$ of the state variables satisfy a set of second-order partial differential equations which represent equations of flow taking place in each direction of $(N+1)$-space. Each equation, and solution, is characterized by a potential function $U_k(y_k, t)$ which is related to the nonlinear restoring force $h_k(y_k, t)=-\nabla U_k(y_k, t)$; $k=0, 1, \cdots, N$. In turn the potential functions are completely determined by the set of conditional expectations $\{E(y_k, t|\phi), E(g(\phi), t|y); k=1, 2, \cdots, N\}$. It is conjectured that the potential functions represent the projections of the system Lyapunov function which characterizes system stability. This paper explores these relationships in detail.

I. INTRODUCTION

IN MODERN communication, radar tracking, missile guidance, and navigation systems, synchronization and tracking are generally accomplished by cross-correlating a locally generated reference signal with the received signal to produce a measurement of the error. In practice this reference signal is developed by means of a nonlinear device. A wide variety of nonlinear functions are available and the choice considerably influences system performance. For example, system performance could be mean-squared tracking error, moments of the mean time to first loss of synchronization or minimum acquisition time. The basic concept associated with a generalized tracking loop is best illustrated by considering first the familiar phase-locked loop (PLL) system of Fig. 1 and its well-known equivalent model [1] illustrated in Fig. 2. We assume that the phase detector is a multiplier and that the correlation function of the Gaussian additive noise process $\{n_i(t)\}$ is given by

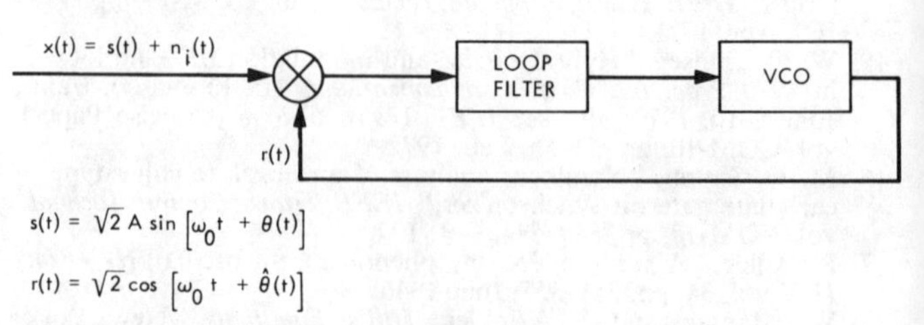

Fig. 1. Phase-locked loop mechanization.

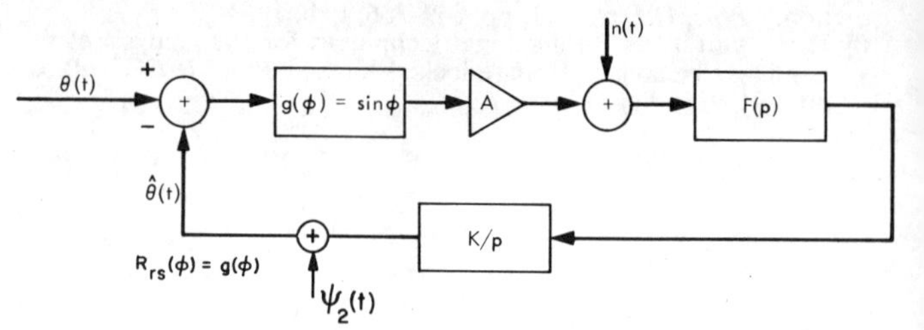

Fig. 2. Equivalent model for a PLL.

$$R_{n_i}(\tau) = \sigma_i^2 r(\tau) \cos \omega_0 \tau$$

where $\sigma_i^2 = N_0 W$ is the variance of $\{n_i(t)\}$, N_0 is the single-sided spectral density in watts/hertz, W is the bandwidth of the noise, and ω_0 is the center frequency of the narrow-band process $\{n_i(t)\}$. $r(\tau)=\text{sinc } \pi W\tau$ with sinc $x=\sin x/x$. In the limit as W approaches infinite, $R_{n_i}(\tau)$ approaches $N_0\delta(\tau)$ where $\delta(\tau)$ is the Dirac delta function. It has been shown [1]–[10] that we may describe system operation by the stochastic differential equation (see Fig. 1)

$$\dot{\phi}(t) = \dot{\theta}(t) - KF(p)[AR_{rs}(\phi) + n(t)] + \Delta\dot{\psi}(t); \quad p \triangleq \frac{d}{dt} \tag{1}$$

where $R_{rs}(\phi)=\sin \phi$. Here $\phi(t)=\theta(t)-\hat{\theta}(t)$ is the instantaneous phase error of the voltage control oscillator (VCO) with respect to the input signal $s(t)$, $F(p)$ is the transfer function of the loop filter in operator form where p is the Heaviside operator, $\Delta\dot{\psi}(t)=\dot{\psi}_1(t)-\dot{\psi}_2(t)$ represents the difference in the frequency instabilities of the transmitter oscillator $\{\dot{\psi}_1(t)\}$ and the VCO instabilities $\{\dot{\psi}_2(t)\}$, $P_c=A^2$ is the power in the input signal component, K is the open loop gain, and the process $\{n(t)\}$ has the correlation function

$$R_n(\tau) = \sigma^2 \text{ sinc } \pi W\tau$$

where $\sigma^2=N_0W/2$ so that for large W, $R_n(\tau)$ approaches $N_0\delta(\tau)/2$. In (1) $\theta(t)$ is the process to be tracked and $\hat{\theta}(t)$ is the tracker estimate of $\theta(t)$. In what follows we make the usual assumption that W is much greater than the bandwidth of the loop or, equivalently, that the correlation time

Manuscript received September 12, 1968; revised June 16, 1969. This paper presents the results of one phase of research carried out at the Jet Propulsion Laboratory, California Institute of Technology, Pasadena, Calif., under NASA Contract NAS 7-100. Part of the research was done under the Joint Services Electronics Program, Air Force Office of Scientific Research, at the University of Southern California, Los Angeles, Calif., under Grant AF AFOSR 69-1622.

The author is with the Department of Electrical Engineering, University of Southern California, Los Angeles, Calif. 90007.

Reprinted from *Proc. IEEE*, vol. 57, pp. 1705–1722, Oct. 1969.

$\tau_{\hat{\theta}}$ of the process $\{\hat{\theta}(t)\}$ is much greater than W^{-1} [6]. If this is the case then, for most practical purposes,

$$R_n(t_2 - t_1) \simeq \delta(t_2 - t_1) \int_{-\infty}^{\infty} R_n(\tau) d\tau = \frac{N_0}{2} \delta(\tau) = \frac{\sigma^2}{W} \delta(\tau).$$

In this sequel we also assume that the correlation time of the process $\{\Delta\psi(t)\}$ is a great deal smaller than the correlation time $\tau_{\hat{\theta}}$.

Frequently a PLL system must operate in conditions where external fluctuations due to additive noise are so intense that classical linear PLL theory does not adequately characterize loop performance nor explain loop behavior [1], [2]. The problem of investigating the effects of external noise on system accuracy has been carried through to completion for the linear model [1], [2]. As a consequence, certain approximate analyses have evolved for explaining and characterizing loop performance in the region of operation where direct linearization cannot be used, e.g., the quasi-linear analysis [3], the linear spectral method [2], [4], and the Volterra series method [5].

The statistical dynamics of a first-order PLL were given in [1], [7], and [9]. For a PLL system with integrating filter certain exact results have been reported [9]. An approximate solution for the joint probability density function of the phase error and phase-error rate was given in [10] for a system with a proportional-plus-integral control type loop filter. The mean time to first slip for zero detuning and initial conditions were evaluated by Viterbi [1], [8] for a first-order loop. Later, Tausworthe [11] derived a Fokker–Planck equation (for zero initial conditions) whose solution for the mean time to first slip was shown to agree with Viterbi's for the first-order loop. In addition, Tausworthe obtained an approximate solution for the mean time to first slip in a second-order PLL.

It can be shown that the stochastic differential equation of operation for the delay-locked loop (DLL) [12] with noise present, or the Nth-order tan-locked loop (TLL) when noise is absent [13], is identical with (1) except for the nonlinearity in the loop. In practice, a wide variety of nonlinearities can be synthesized from appropriate modifications of a DLL or a TLL, and hence, there formally exists the possibility of synthesizing the optimum nonlinearity having first determined the best choice of $F(p)$.

Since this is true, a generalized tracking loop like that of Fig. 3 is of interest. The equivalent loop model is shown in Fig. 4 while the stochastic differential equation of operation is given by (1) with $R_{rs}(\phi)$ replaced by $g(\phi)$. In what follows we shall assume (without loss in generality) that $g(\phi)$ is periodic. Physically speaking, $R_{rs}(\phi)=g(\phi)$ is the normalized cross-correlation function (double-frequency terms neglected) between the input signal component $s[t, \theta(t)]$ and the control signal $r(t)$. It will be of interest to do signal design on $s[t, \theta(t)]$ and $r(t)$, under appropriate constraints, and select that $F(p)$ which will optimize loop performance, i.e., we first seek the loop filter $F(p)$ which will optimize loop performance subject to some performance criterion

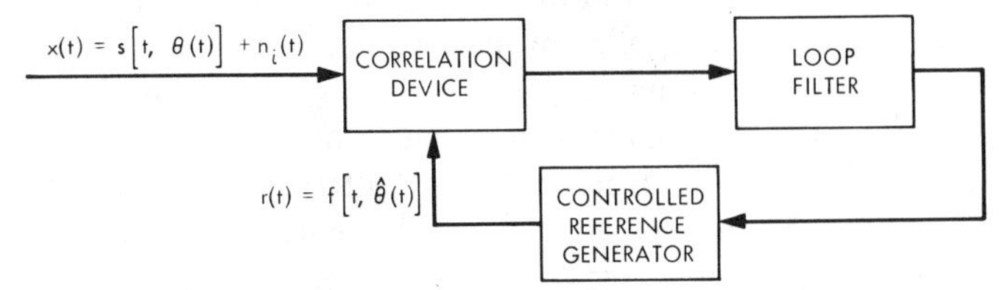

Fig. 3. Generalized tracking loop.

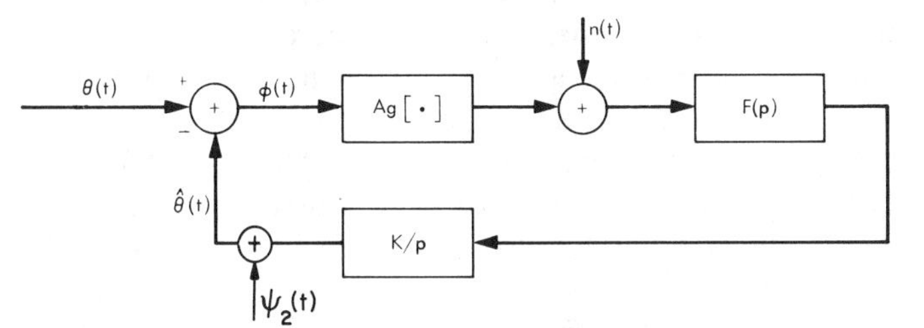

Fig. 4. Equivalent model for a generalized tracking loop.

and then optimize further by a choice of the nonlinearity.

An exact solution to the problem of analyzing and synthesizing an optimum tracker is formally possible on the basis of the theory of Markov processes [6]. The possibility of obtaining exact results enables 1) the nonlinear effects to be understood, 2) the limits of application of the various approximate methods and loop theories to be assessed, 3) one to perform stochastic optimization of the loop in the nonlinear region of operation, and 4) an optimum tracking theory to be developed. Thus a comparison of performance between the optimum tracker and any implemented suboptimum tracker, such as the PLL, can be made to see if the additional complexity required to mechanize the former is warranted. Also, it is just as satisfying to know what the optimum tracker is as it is to know what the channel capacity of a communication system is, even though each is unattainable in practice.

In summary, this paper presents rather general results for the transition probability density function (pdf) of the phase error occurring in the generalized tracking system of Fig. 3. In particular, a $(N+1)$-dimensional Fokker–Planck equation is derived whose solution is denoted by the multidimensional transition pdf $p(\mathbf{y}; t)=p(\mathbf{y}, t|\mathbf{y}_0, t_0)$ where $(\mathbf{y}, t)$ is a vector Markov process in the state variables $(\phi, y_1, \cdots, y_N)$. This Fokker–Planck equation is then reduced to a second-order partial differential equation (PDE) of flow whose solution is the transition pdf $p(\phi; t)$ of the phase-error process. The solution to this equation is shown to be completely characterized by the set of conditional expectations $\mathscr{S}_1=\{E(y_k, t|\phi), k=1, 2, \cdots, N\}$. In the steady state the PDE becomes an ordinary first-order differential equation for which the form of an exact solution is known. In addition, moments of the first passage time of the phase error are shown to be embedded in a knowledge of $\mathscr{S}_1$. Recursive formulas for these moments are derived and studied for the particular cases $N=0$ and $N=1$.

In addition, general formulas for determining the transition pdf's $p(y_k; t)$ of the state variables y_k, $k=1, 2, \cdots, N$

are given. It is shown that the transition pdf's $p(y_k;\, t)$, $k=1, 2, \cdots, N$ are solutions to second-order PDE's of flow. The transition pdf's are shown to be characterized by the set of conditional expectations $\mathcal{S}_2=\{E(g(\phi),\, t|y_k),\ k=1, 2, \cdots, N\}$. In the steady state these second-order PDE's become first-order ordinary differential equations for which the solutions are known. Recursive formulas for the moments of the first passage time of the projections y_k, $k=1, 2, \cdots, N$, are also derived.

The steady state expectations $\{E(y_k|\phi)\}$ are approximated by two methods. The first approximation is based upon the linear tracking theory while the second is based upon a modification and generalization of techniques due to Viterbi [1] and Holmes [14]. To check on the validity of these approximations $E(y_1|\phi)$ is measured via computer simulation methods for the case of greatest interest, i.e., $N=1$, $g(\phi)=\sin\phi$. The variance of the phase error is also computed for both approximations and the results are compared with measurements obtained earlier [10] via hardware simulation. It is further reasoned that a first-order tracker with nonlinearity $g(\phi)=\text{sgn}\,\phi$ minimizes the mean-square error over the class of linear filters and periodic nonlinearities when zero detuning exists within the loop.

For each projection of $\mathbf{y}$, we evaluate the steady state probability current and show that the probability field is irrotational when zero detuning exists in the loop. From these results we are then able to deduce the average number of phase jumps (cycles slipped) per unit of time in the steady state.

We close by concluding that specification of loop response, and hence, performance of a generalized tracking system in the nonlinear region of operation, is tantamount to possessing a knowledge of the set of expectations $\{E(y_k,\, t|\phi),\ E(g(\phi),\, t|y_k);\ k=1, 2, \cdots, N\}$. This knowledge may be obtained via computer simulation techniques [15], [16] or as is done here, one may approximate them; hence, stochastic optimization of the loop in the nonlinear region of operation is possible. The utility of the results becomes obvious when one considers the complexity of the equipment [10] and the amount of time required to measure only a few of the many statistical parameters at work in the loop, e.g., mean-square phase error or mean time to first loss of synchronization. On the other hand, the amount of time required to measure the transition pdf's and the moments thereof via computer simulation methods [15], [16] is also prohibitive. Thus, measurement or approximation of these conditional expectations when used in conjunction with the theory presented here is all that is required to analyze loop behavior and to carry out system synthesis in the nonlinear region of operation.

II. System Model

If we write the loop filter transfer function in the partial fraction expression

$$F(p) = F_0 + \sum_{k=1}^{N} \frac{1 - F_k}{1 + \tau_k p} \tag{2}$$

we have $F(p)=1$ if $F_0=1$ and $F_k=1$ for all $k=1, 2, \cdots, N$. For $N=1$, $F_0=F_1=\tau_2/\tau_1$, (2) reduces to the proportional-plus-integral control filter of considerable practical interest, i.e.,

$$F(p) = \frac{1 + \tau_2 p}{1 + \tau_1 p}. \tag{3}$$

If we substitute (2) into (1) and assume an input of the form $\theta(t) = \Omega_0 t + \theta$ where θ is a constant, we can write (1) as

$$\dot{\phi}(t) = \Omega_0 - F_0 K[Ag(\phi) + n(t)] - \sum_{k=1}^{N} \frac{1 - F_k}{1 + \tau_k p}[AKg(\phi) + Kn(t)] + \Delta\dot{\psi}(t) \tag{4}$$

where $\Omega_0 \triangleq \omega - \omega_0$ is the loop detuning. Introducing the state variable

$$y_k \triangleq -K\left[\frac{1 - F_k}{1 + \tau_k p}\right][Ag(\phi) + n(t)] \tag{5}$$

for $k=1, 2, \cdots, N$, we can replace (4) by the equivalent system of $(N+1)$ first-order stochastic differential equations,[1] i.e.,

$$\begin{aligned}
\dot{y}_0 &= \dot{\phi} = \Omega_0 - F_0 K[Ag(\phi) + n(t)] + \sum_{k=1}^{N} y_k + \Delta\dot{\psi}(t)\\
\dot{y}_1 &= -\frac{y_1}{\tau_1} - \frac{(1 - F_1)K[Ag(\phi) + n(t)]}{\tau_1}\\
&\ \ \vdots \qquad \vdots \qquad\qquad \vdots\\
\dot{y}_N &= -\frac{y_N}{\tau_N} - \frac{(1 - F_N)K[Ag(\phi) + n(t)]}{\tau_N}
\end{aligned} \tag{6}$$

where $y_0=\phi$. Written this way, it is clear from (6) that the coordinates $y_0, y_1, \cdots, y_N$ form components of a $(N+1)$-dimensional Markov vector $(\mathbf{y}, t)=(\phi, y_1, \cdots, y_N, t)$ since each component, $\dot{y}_k$, depends only upon the present values of $(\mathbf{y}, t)$ and a "white" Gaussian noise process. For convenience, we define the vector $\mathbf{y}_0' \triangleq (y_1, \cdots, y_N)$. Also in what follows we shall refer to a specific component of $\mathbf{y}$ as the projection of $\mathbf{y}$.

The most complete characterization of the state $\mathbf{y}$ of the tracking loop is its statistical description by means of the transition pdf, viz., $P(\mathbf{y},\, t|\mathbf{y}_0,\, t_0)$. In writing transition (conditional) pdf's, the conditioning variables will be written to the right of the vertical bar. The transition pdf $P(\mathbf{y};\, t) \triangleq P(\mathbf{y},\, t|\mathbf{y}_0,\, t_0)$, of course, can be formally determined by using the theory of Markov processes. In the next section we expose the procedure.

[1] By defining the Markov extension such that ϕ is one component of the vector process allows one to proceed with the analysis. Heretofore, the vector process has been defined such that ϕ is a weighted sum of the projections, thus leading to formidable mathematical difficulties if $N>1$. Also such an extension clearly depicts that the components of $\mathbf{y}_0'$ are Markov for a fixed ϕ. Further, if the τ_k's are much greater than one (narrowband loop) then it is clear that the y_k variables are slowly varying random processes.

III. The $(N+1)$-Dimensional Fokker–Planck Equation

Given the fact that the components of y form a vector Markov process, $P(y; t)$ satisfies the $(N+1)$-dimensional Fokker–Planck (F–P) equation [6], viz.,

$$\frac{\partial P(y;t)}{dt} + \sum_{k=0}^{N} \frac{\partial}{\partial y_k} \cdot \left\{ \left[K_k(y,t) - \frac{1}{2} \sum_{l=0}^{N} \frac{\partial}{\partial y_l} K_{lk}(y,t) \right] P(y;t) \right\} = 0 \quad (7)$$

where the intensity coefficients $K_k(y, t)$ and $K_{lk}(y, t)$ are defined by the formulas

$$\begin{aligned} K_k(y,t) &\triangleq \lim_{\Delta t \to 0} \frac{E[\Delta y_k | y]}{\Delta t} \\ K_{lk}(y,t) &\triangleq \lim_{\Delta t \to 0} \frac{E[\Delta y_l \Delta y_k | y]}{\Delta t} \end{aligned} \quad (8)$$

and the $E[\cdot|y]$ denotes mathematical expectation of the enclosed quantity given y. In the case of a stationary system, the coefficients $K_k(y, t)$ and $K_{lk}(y, t)$ do not depend explicitly on t.

In what follows the concept of probability current density

$$\mathscr{I}_k(y,t) \triangleq \left\{ \left[K_k(y,t) - \frac{1}{2} \sum_{l=0}^{N} \frac{\partial}{\partial y_l} K_{lk}(y,t) \right] P(y;t) \right\}; \quad k = 0, 1, \cdots, N \quad (9)$$

will be of use when evaluating the average number of phase jumps or cycles slipped per unit of time. Thus, we can write the F–P equation in the form of the equation of flow, i.e.,

$$\nabla \cdot \mathscr{I}(y,t) + \frac{\partial P(y;t)}{\partial t} = 0 \quad (10)$$

where the vector $\mathscr{I}(y, t) \triangleq [\mathscr{I}_0(y, t), \cdots, \mathscr{I}_N(y, t)]$ may be interpreted as a probability current density vector and ∇ is the differential operator for space, the del operator,

$$\nabla \triangleq \left[\frac{\partial}{\partial y_0}, \frac{\partial}{\partial y_1}, \cdots, \frac{\partial}{\partial y_N} \right].$$

We will also need to consider the probability current of the kth projection of $\mathscr{I}(y, t)$, i.e.,

$$\mathscr{I}_k(y_k, t) \triangleq \underbrace{\int \cdots \int}_{N\text{-fold}} \mathscr{I}_k(y,t) dy'_k \quad (11)$$

where $dy'_k \triangleq dy_0, dy_1, \cdots, dy_{k-1} dy_{k+1}, \cdots, dy_N$. The probability current of the kth projection describes the amount of probability crossing the hyperplane $y_k = y'_k$ in the positive direction per unit time. A geometric interpretation of (11) is possible. Define n_k to be a unit vector pointed along the positive direction of the y_k axis. Then $\mathscr{I}(y, t) \cdot n_k = \mathscr{I}_k(y, t)$ represents the probability current density flowing in the positive y_k direction, and $\mathscr{I}_k(y, t) dy'_k$ represents the amount of probability current flowing through the differential surface area dy'_k. Integrating over the surface gives the total probability current flowing through the hyperplane $y_k = y'_k$. Just as the equation of heat conduction involves a flow of heat, (7) involves a flow of probability.

Before we proceed we will give a graphic physical interpretation of the F–P equation which will prove useful and somewhat picturesque. For every sample function of a vector Markov process, the vector trajectory $y(t)$ can be thought of as the path of a point starting from $y(t_0)$ in an $(N+1)$-dimensional space $y = (\phi, y_1, \cdots, y_N)$. The position of this point at time t, i.e., $[y_0(t), \cdots, y_N(t)]$, may be envisioned as a Brownian particle undergoing diffusion in an $(N+1)$-space as a function of time. The set of sample functions of the process $y(t)$ is the ensemble of trajectories which move about in a random manner. The fraction of time that the particle spends in any region of the probability space R' is proportional to the total probability in that region.

The coefficients required for (7) can be straightforwardly evaluated by using (6) and (8). The differential equation whose solution describes the probability density $P(y; t)$ can then be written by mere substitution of these coefficients K_k and K_{lk} into (7). Such a computation yields

$$\begin{aligned} \frac{\partial P(y;t)}{\partial t} = &- \frac{\partial}{\partial \phi} \left\{ \left[\Omega_0 - F_0 A K g(\phi) + \sum_{k=1}^{N} y_k \right] P(y;t) \right\} \\ &+ \frac{F_0^2 N_0 K^2 \partial^2 P(y;t)}{4 \partial \phi^2} \\ &+ \sum_{k=1}^{N} \left(\frac{\partial}{\partial y_k} \left[\frac{y_k}{\tau_k} + \left(\frac{1 - F_k}{\tau_k} \right) A K g(\phi) \right] P(y;t) \right. \\ &\left. + \left\{ \frac{(1-F_k)^2 N_0 K^2}{4\tau_k^2} \frac{\partial^2}{\partial y_k^2} + \frac{F_0(1-F_k) N_0 K^2}{2\tau_k} \frac{\partial^2}{\partial \phi \partial y_k} \right\} P(y;t) \right) \\ &+ \sum_{k \neq l \neq 0}^{N} \sum^{N} \frac{(1-F_k)(1-F_l) K^2 N_0}{4 \tau_k \tau_l} \frac{\partial^2}{\partial y_k \partial y_l} P(y;t). \end{aligned} \quad (12)$$

In arriving at (12) we have assumed, without loss in generality, that $\Delta \psi(t) = 0$. Certain special cases of (12) are of great practical importance. For example, with $N = 1$ and $F_0 = F_1$, (12) becomes the F–P equation for the PLL system with the proportional-plus integral-control loop filter given in (3).

IV. Initial Conditions and Boundary Conditions

In order to obtain solutions to the F–P equation we have to supplement it with initial conditions and boundary conditions. For our purposes we specify the initial pdf $P(y; t)$ at time $t = t_0$ to be

$$\lim_{t \to t_0} P(y;t) = \prod_{k=0}^{N} \delta[y_k - y_k(t_0)] \quad (13)$$

and the subsequent evolution of this pdf is found from (12).

The boundary conditions themselves are determined by the physics of the problem. If $P[\phi, y'_0; t]$ is to be a pdf and since $K_k(y, t)$ and $K_{lk}(y, t)$ are periodic in ϕ, we have

$$\lim_{t \to \infty} P[\phi, y'_0; t] = \lim_{t \to \infty} P[\phi \pm 2n\pi, y'_0; t] = 0 \quad (14)$$

and in the steady state $P[\phi, y'_0; t]$ has an unbounded variance. This condition is directly traceable to the cycle-slipping (phase-jumps) phenomenon associated with generalized tracking systems. Thus to obtain a pdf with a finite variance in the steady state we require other considerations. Let

$$\tilde{p}(\phi,\ y'_0;\ t) \triangleq \sum_{n=-\infty}^{\infty} P[\phi + 2n\pi, y'_0; t]$$

for all ϕ. Note that $\tilde{p}$ is periodic in ϕ but, as such, it is not a probability density function since it is an infinite sum of density functions each with unit area. Thus, to obtain a solution which has the properties of a density function we define

$$p(\phi, y'_0; t) \triangleq \begin{cases} \tilde{p}(\phi, y'_0; t) \text{ in any } \phi \text{ interval } \phi\varepsilon[(2n-1)\pi, \\ \qquad (2n+1)\pi], n \text{ any fixed integer} \\ 0 \qquad \text{elsewhere} \end{cases} \tag{15}$$

where $p(y; t) \triangleq p(y, t|y_0, t_0)$.

To justify that $\tilde{p}(\phi, y'_0; t)$ is a solution, we note that $P(y; t)$ is a solution to (7) in the region R' for which $y_j = \pm\infty$, $j=0, \cdots, N$. The function $p(y; t)$ is defined in a region R which is the hyperslab formed by two hyperplanes located 2π radians apart. Hence, since each term in $\tilde{p}(\phi, y'_0; t)$ is a solution to (12) in R', the sum is also a solution in R. In cylindrical coordinates R may be regarded as a hypercylinder. We also assume that $p(y; t)$ is continuous, differentiable, and exists everywhere in R.

Since $p(y; t)$ is a transition pdf as defined in (15) the normalization condition, being the statement of conservation of probability,

$$\underbrace{\int \cdots \int}_{(N+1)\text{-fold}} p(y; t)y = 1 \tag{16}$$

holds for all t; consequently, $p(y; t)$ must approach zero faster than $y_k^{-(1+\varepsilon)}$, $\varepsilon > 0$, as y_k approaches infinity. Now (12) is a second-order partial differential equation of the parabolic type and its solution is determined by $2N+2$ independent boundary conditions. Noting these facts we have the following boundary conditions.

Along any edge of the surface Γ of the hyperslab R, i.e., the edges $y_k = \pm\infty$ for any and all $k=1, 2, \cdots, N$, we have the N boundary conditions

$$y_k p(\phi, y'_0; t)|_{y_k=\pm\infty} = 0; \qquad k = 1, 2, \cdots, N \tag{17}$$

since $p(y; t)$ approaches zero faster than $y_k^{-(1+\varepsilon)}$, $\varepsilon > 0$. As a consequence of (16) we also have that

$$p(\phi, y'_0; t)|_{y_k=\pm\infty} = 0$$

for all $k=1, 2, \cdots, N$. Since (16) holds for all t we have N other independent boundary conditions, viz.,

$$\frac{\partial}{\partial y_k} p(\phi, y'_0; t)|_{y_k=\pm\infty} = 0; \qquad k = 1, \cdots, N. \tag{18}$$

Now $\tilde{p}(\phi, y'_0; t)$ is periodic in ϕ because it is the sum of periodic functions; therefore

$$p(-\pi, y'_0; t) = p(\pi, y'_0, t). \tag{19}$$

It then follows from (19) that

$$\frac{\partial p(-\pi, y'_0; t)}{\partial y_k} = \frac{\partial p(\pi, y'_0; t)}{\partial y_k}; \qquad k = 1, 2, \cdots, N \tag{20}$$

which are not independent from the condition (19). Finally, if probability flow is to be conserved in all directions of the coordinate axes and since $\tilde{p}(\phi, y'_0; t)$ is periodic we have

$$\left.\frac{\partial p(\phi, y'_0; t)}{\partial \phi}\right|_{\phi=\pi} = \left.\frac{\partial p(\phi, y'_0; t)}{\partial \phi}\right|_{\phi=-\pi}. \tag{21}$$

Equations (17), (18), (19) and (21) define $2N+2$ independent boundary conditions. From (20) and (21) we can write in vector notation

$$\nabla p(y; t)|_{\phi=\pi} = \nabla p(y; t)|_{\phi=-\pi}.$$

In passing we point out the fact that if $\Omega_0 = 0$, the symmetry of (6) indicates[2] that $p(y; t) = p(-y, t)$.

It is interesting to note the merger of Maxwell's field equations with the theory of Markov processes, in particular, the Fokker–Planck equation. Using the Gauss theorem we note

$$\oint_R \nabla \cdot \mathscr{J} dR = \oint_\Gamma \boldsymbol{n} \cdot \mathscr{J} d\Gamma = -\frac{\partial}{\partial t} \oint_R p(y; t) dy = 0$$

where $\boldsymbol{n}$ is unit vector normal to the surface of Γ and directed positively outwards. From Maxwell's field equations we know that the divergence of the current density $\boldsymbol{J}$ is just the time rate of change of the charge density ρ. Also, the divergence of the flux density D is equal to ρ. Hence, if we interpret ρ as the transition pdf $\rho(y; t)$ and $\mathfrak{D}$ as a probability flux density then we may write $\nabla \cdot \mathfrak{D} = p(y; t)$, i.e., the net probability flux flowing out of a volume dR at time t is just equal to the probability of being in that volume at time t.

Interestingly enough if we integrate both sides of (12) with respect to y_j and $y_j(t_0)$ for all $j \neq k \neq 0$ and make use of the boundary conditions (17), (18), and (13) we arrive at[3]

$$\frac{\partial p}{\partial t} + \frac{\partial}{\partial \phi}[\mathscr{I}_0(\phi, y_k, t)] + \frac{\partial}{\partial y_k}[\mathscr{I}_k(\phi, y_k, t)] = 0 \tag{22}$$

where $p = p(\phi, y_k; t)$ and

$$K_0(\phi, y_k, t) = \Omega_0 + \sum_{j \neq k \neq 0}^{N} E(y_j, t|\phi, y_k) + y_k - AKF_0 g(\phi)$$

[2] The boundary condition given in (21) is required in finding the transition pdf's $p(y_k; t)$ but not for finding $p(\phi; t)$.

[3] Oscillator instabilities may be included here by replacing K_{00} by

$$K_{00} + \int_{-\infty}^{\infty} R(\tau) d\tau$$

where $R(\tau)$ is the correlation function of the random process $\Delta\psi(t) = \psi_1(t) - \psi_2(t)$; $\psi_1(t)$ represents the phase instabilities in $s[t, \theta(t)]$ and $\psi_2(t)$ represents the instabilities in the CRG output $r(t)$. To make this replacement valid and to remain within the framework of Markov process theory, the correlation time of $\Delta\psi(t)$ must be small when compared to the response time of the loop. Also, for brevity in notation, the initial values of the given variables in every conditional expectation have been suppressed.

$$K_k(\phi, y_k, t) = -\frac{1}{\tau_k}[y_k + (1 - F_k)AKg(\phi)]$$

$$K_{00} = \frac{N_0 F_0^2 K^2}{2}; \quad K_{0k} = K_{k0} = \frac{(1 - F_k)F_0 N_0 K^2}{2\tau_k};$$

$$K_{kk} = \frac{(1 - F_k)^2 N_0 K^2}{2\tau_k^2}; \qquad k \neq 0$$

with probability currents

$$\mathscr{I}_0(\phi, y_k, t) = \left\{\left[K_0(\phi, y_k, t) - \frac{K_{00}}{2}\frac{\partial}{\partial\phi} - \frac{K_{k0}}{4}\frac{\partial}{\partial y_k}\right]p\right\}$$

$$\mathscr{I}_k(\phi, y_k, t) = \left\{\left[K_k(\phi, y_k, t) - \frac{K_{kk}}{2}\frac{\partial}{\partial y_k} - \frac{K_{0k}}{4}\frac{\partial}{\partial\phi}\right]p\right\}.$$

V. Differential Equations for the Transition Probability Density Functions $p(y_k; t); k=0, 1, \cdots, N$

To find $p(y_k; t)$ we first need a differential equation whose solution is indeed $p(y_k; t)$. This is easily found by integrating both sides of (22) with respect to y_k and applying the appropriate boundary conditions. Without belaboring the details we obtain [17] the partial differential equation of flow in the kth direction

$$\nabla \cdot \mathscr{I}_k(y_k, t) + \frac{\partial p(y_k; t)}{\partial t} = 0 \tag{23}$$

with probability current

$$\mathscr{I}_k(y_k, t) = \left\{\left[K_k(y_k, t) - \frac{K_{kk}}{2}\frac{\partial}{\partial y_k}\right]p(y_k; t)\right\} \tag{24}$$

where, for $k = 0$,

$$K_0(\phi, t) = \Omega_0 - AKF_0 g(\phi) + \sum_{k=1}^{N} E(y_k, t|\phi) \tag{25a}$$

and for all $k \neq 0$,

$$K_k(y_k, t) = -\left[\frac{y_k + AK(1 - F_k)E(g(\phi), t|y_k)}{\tau_k}\right] \tag{25b}$$

where $E(y_k, t|\phi)$ is the conditional expectation of y_k given ϕ at time t and $E(g(\phi), t|y_k)$ is the conditional expectation of $g(\phi)$ given y_k at t.

It is convenient to introduce into (23) the nonlinear restoring force

$$h_k(y_k, t) \triangleq \frac{2K_k(y_k, t)}{K_{kk}} \tag{26}$$

and the potential function

$$U_k(y_k, t) = -\int^{y_k} h_k(x, t)dx \quad \text{or}$$
$$h_k(y_k, t) = -\nabla U_k(y_k, t) \tag{27}$$

so that (24) becomes

$$\mathscr{I}_k(y_k, t) = -\frac{K_{kk}}{2}\exp\left[-U_k(y_k, t)\right]\frac{\partial}{\partial y_k} \cdot \{p(y_k; t)\exp\left[U_k(y_k, t)\right]\}. \tag{28}$$

Assuming, in the limit as t approaches infinity that $p(y_k; t)$ approaches the steady state pdf $p(y_k)$, the stationary diffusion current is constant and obeys the law

$$\mathscr{I}_k = -\frac{K_{kk}}{2}\exp\left[-U_k(y_k)\right]\frac{\partial}{\partial y_k}\{p(y_k)\exp\left[U_k(y_k)\right]\}. \tag{29}$$

Solving (29) for $p(y_k)$ yields

$$p(y_k) = C_k \exp\left[-U_k(y_k)\right]\left\{1 + D_k\int_{l_k}^{y_k}\exp\left[U_k(x)\right]dx\right\} \tag{30}$$

where $D_k = -2\mathscr{I}_k/C_k K_{kk}$, and the lower limit $l_k = -\pi$ if $k=0$, and $l_k = -\infty$ if $k \neq 0$.

To evaluate the constants C_k and D_k for $k=0$ we make use of the boundary conditions. Since $p(\pi) = p(-\pi)$ from (19), we have from (30), with $y_0 = \phi$,

$$D_0 = \frac{\exp\left[-U_0(-\pi)\right] - \exp\left[-U_0(\pi)\right]}{\exp\left[-U_0(\pi)\right]\int_{-\pi}^{\pi}\exp\left[U_0(x)\right]dx}. \tag{31}$$

By means of the normalization condition the constant C_0 is easily determined. Furthermore, since $g(\phi)$ is periodic and since $g(\phi)$ is continuous we may also write [17]

$$p(\phi) = C_0' \exp\left[-U_0(\phi)\right] \cdot \int_{\phi}^{\phi + 2\pi}\exp\left[U_0(x)\right]dx \tag{32}$$

for any ϕ belonging to an interval of width 2π as defined in (15) and $U_0(\phi)$ is defined in (27) with $k=0$.

Equations (30) and (32) are remarkable in that they hold for all order loops and a broad class of nonlinearities. In fact, it is clear from (25) and (32) that the pdf of the phase error of an $(N+1)$-order loop is completely determined by the set of conditional expectations $\mathscr{S}_1 = \{E(y_k|\phi) \text{ for all } k\}$. Interestingly enough $E(y_k|\phi)$ is the minimum mean-square error estimate of y_k given ϕ.

For $k \neq 0$ the constant C_k is a normalization constant for $p(y_k)$ while the constant D_k is determined from the boundary condition $p(y_k) = 0$ at $y_k = \pm\infty$. Thus, from (30),

$$D_k = -\left[\int_{-\infty}^{\infty}\exp\left[U_k(x)\right]dx\right]^{-1} \tag{33}$$

When the D_k's are zero, i.e., $\mathscr{I}_k = 0$, the marginal pdf's reduce to

$$p(y_k) = C_k \exp\left[-U_k(y_k)\right] \tag{34}$$

for all $k=0, 1, \cdots, N$. When the conditions are such that (34) is true the kth projection is not allowed to penetrate the region of infinitely large positive or negative values, $k>0$.

In order to obtain explicit solutions for the $p(y_k)$ it appears that the conditional expectations must either be approximated or measured by the method of computer simulation. It is remarkable that the marginal pdf's $p(y_k)$ are determined by the set of conditional expectations $\{E(y_k|\phi), E(g(\phi)|y_k) \text{ for all } k\}$.

From (6) it is clear that in the steady state, the means of the coordinates y_k, $k \neq 0$, are

$$\bar{y}_k = -(1 - F_k)AK\overline{g(\phi)}; \qquad k = 1, 2, \cdots, N \tag{35}$$

where the overbar denotes statistical average. Further for small signal strengths, i.e., small AK, it is seen that the y_k's are zero mean Gaussian random variables having variance

$$\sigma_k^2 = (1 - F_k)^2 \frac{K^2 N_0}{4\tau_k}; \qquad k = 1, 2, \cdots, N. \tag{36}$$

For $k=0$, it is clear from (6) and (35) that the mean of the phase-error rate $\dot{\phi}$ is given by

$$\overline{\dot{\phi}} = \Omega_0 - AKF_0\overline{g(\phi)} + \sum_{k=1}^{N} \bar{y}_k = \Omega_0 - AKF(0)\overline{g(\phi)}. \tag{37}$$

To this end we have assumed that the transition pdf $p(\mathbf{y}; t)$ exists for all $t > t_0$. It is clear, however, that for certain loop designs the steady state pdf's may not exist since the loop may go unstable owing to a large loop gain or the form of $F(s)$ itself. It is the conjecture of this author that the set of potential functions $\{U_k(y_k, t), k=0, 1, \cdots, N\}$ form the basis from which one may construct a "Lyapunov function," say $V(\mathbf{y}, t)$, where $V(\mathbf{y}, t)$ represents the total potential at the point $\mathbf{y}(t)$. If for all $\mathbf{y} \in R$, $t \geq t_0$, this function behaves as required by Lyapunov's asymptotic stability theorem; that is, there exist three positive definite functions $W(\mathbf{y}, t)$, $W_1(\mathbf{y}, t)$, $W_2(\mathbf{y}, t)$ such that in R and for $t \geq t_0$ we can demonstrate that $W_1(\mathbf{y}, t) \geq V(\mathbf{y}, t) \geq W(\mathbf{y}, t)$; $\dot{V}(\mathbf{y}, t) \leq W_2(\mathbf{y}, t)$ and, in addition, that in R all the partial derivatives $\partial V/\partial y_k$ are bounded for $t \geq t_0$, i.e., $|\partial V/\partial y_k| \leq M$, $k=0, 1, \cdots, N$, then $p(\mathbf{y}; t)$ exists for all $t \geq t_0$. Obviously, the ability to demonstrate that $U_k(y_k, t)$ is bounded for all y_k and $t \geq t_0$ is directly related to the problem of showing that the conditional expectations $E[y_k, t|\phi]$ and $E[g(\phi), t|y_k]$ are bounded.

VI. Tracking Loops With $F_0=0$

Going back to (22), we see that when $F_0=0$, the reduced F–P equation degenerates, and the technique previously used fails. There is then no other alternative than to solve (12) with $F_0=0$ although it is possible to reduce it somewhat by integrating with respect to $y_1, \cdots, y_N$ and by using the boundary conditions (17) and (18).

This procedure yields the reduced F–P equation, viz.,

$$\frac{\partial}{\partial y_j}\left\{\left[\frac{y_j}{\tau_j} + \frac{AK(1-F_j)g(\phi)}{\tau_j}\right] p(\phi, y_j; t)\right\} + \frac{K_{jj}}{2}\frac{\partial}{\partial y_j} p(\phi, y_j; t)$$
$$- \frac{\partial}{\partial \phi}\left\{\left[\Omega_0 + y_j + \sum_{k \neq j \neq 1}^{N} E(y_k, t|\phi, y_j)\right] p(\phi, y_j; t)\right\}$$
$$= \frac{\partial p(\phi, y_j; t)}{\partial t}. \tag{38}$$

For the case $N=1$, $\Omega_0=0$, we obtain in the steady state,

$$p(\phi, \dot{\phi}) = C \exp\left[-\frac{\tau\rho}{2AK}\dot{\phi}^2 - \rho\int^{\phi} g(x)dx\right] \tag{39}$$

where $|\dot{\phi}| < \infty$, $\phi \in [(2n-1)\pi, (2n+1)\pi]$ for n any fixed integer, $\dot{\phi} = y_1$, and ρ is the signal-to-noise ratio existing in the loop bandwidth $W_L = 2b_L = AK/2$, i.e., $\rho = A^2/N_0 b_L = 2A^2/N_0 W_L$. The parameter C is a normalization constant.

Note that the density $p(\phi, \dot{\phi})$ can be written as $p(\phi)p(\dot{\phi})$; ϕ and $\dot{\phi}$ are therefore statistically independent random variables. Further, we note that for $\Omega_0=0$, $p(\phi)$ is identical to the expression for a first-order loop and that $\dot{\phi}$ is a Gaussian random variable with variance $N_0K^2/4\tau$.

VII. Evaluating the Expectations $E(y_k|\phi)$ and the Steady State Density $p(\phi)$ for an $(N+1)$-Order Tracker

The conditional expectation may be approximated with a great deal of accuracy by modifying and generalizing a method due to Viterbi [8] and Holmes [14]. We begin by multiplying both sides of (6) by $\exp(v/\tau_k)$ to obtain (see [17] for details)

$$\frac{d}{dv}[y_k(v)\exp(v/\tau_k)] = -\frac{(1-F_k)K}{\tau_k}$$
$$\cdot [Ag(\phi) + n(v)]\exp(v/\tau_k). \tag{40}$$

If we now take expectations of both sides conditioned upon $\phi(t)$ and interchange the order of expectation with differentiation we have

$$\frac{d}{dv}\{E[y_k(v)|\phi(t)]\exp(v/\tau_k)\} = -\frac{(1-F_k)K}{\tau_k}$$
$$\cdot\{AE[g\{\phi(v)\}|\phi(t)] + E[n(v)|\phi(t)]\}\exp(v/\tau_k). \tag{41}$$

Integrating both sides of (41) from t to infinity (the noise term drops out since it is independent of $\phi(t)$ in the future) and introducing the change of variables $v=(t+\tau)$ yields, in the steady state,

$$E(y_k|\phi) = \frac{AK(1-F_k)}{\tau_k}\int_0^{\infty} \exp(\tau/\tau_k)E[(g\{\phi(t+\tau)\}$$
$$- \overline{g\{\phi(t+\tau)\}})|\phi(t)]d\tau \tag{42}$$
$$- AK(1-F_k)\overline{g\{\phi(t+\tau)\}}$$

for all $k=1, 2, \cdots, N$. We note that $g[\phi(t+\tau)]$ is strictly stationary in the steady state since $\phi(t)$ reduced modulo 2π is strictly stationary in the steady state. The expectation under the integral sign of (42) may be estimated using the orthogonality principle (OP) to find the best $\rho_G(\tau)$ such that $E[\{g(\phi_2) - \overline{g(\phi_2)}\}|\phi_1]$ is estimated by $\rho_G(\tau)[g(\phi_1) - \overline{g(\phi_1)}]$ in the best linear mean-square sense. Here $\phi_2 = \phi(t+\tau)$, $\phi_1 = \phi(t)$, and $\rho_G(\tau)$ approaches zero as τ approaches minus infinity. First, we define the error ε,

$$\varepsilon \triangleq E_{\phi_1}[\{E_{\phi_2}(\{g(\phi_2) - \overline{g(\phi_2)}\}|\phi_1)$$
$$- \rho_G(\tau)[g(\phi_1) - \overline{g(\phi_1)}]\}^2]. \tag{43}$$

Let

$$y(\phi_1) \triangleq E_{\phi_2}[g(\phi_2) - \overline{g(\phi_2)}|\phi_1];$$
$$x(\phi_1) \triangleq [g(\phi_1) - \overline{g(\phi_1)}] \tag{44}$$

and write

$$\varepsilon = E_{\phi_1}[\{y(\phi_1) - \rho_G(\tau)x(\phi_1)\}^2]. \tag{45}$$

Thus the function $\rho_G(\tau)$ which produces the best linear mean-square estimate $E[g\{\phi(t+\tau)\} - \overline{g\{\phi(t+\tau)\}}|\phi_1]$ is easily shown, using the OP, to be given by[4]

$$\rho_G(\tau) = \frac{R_g(\tau) - (\bar{g})^2}{\sigma_G^2} = \frac{R_G(\tau)}{\sigma_G^2} \tag{46}$$

where for a stationary ϕ process, $\overline{g(\phi_1)} = \overline{g(\phi_2)} = \bar{g}$ and $G = g(\phi) - \bar{g}$. The minimum mean-square error $\varepsilon_m(\tau)$ is given by $\varepsilon_m^2(\tau) = \sigma_y^2 - \rho_G^2(\tau)\sigma_G^2$. Replacing the expectation $E[(g(\phi_2) - \bar{g})|\phi_1]$ by the best linear mean-square estimate $\rho_G(\tau)[g(\phi) - \bar{g}]$ in (42) we have

$$\hat{E}[y_k(t)|\phi(t)] = \frac{AK(1 - F_k)G(\phi)}{\tau_k} \int_0^\infty \rho_G(\tau) \exp(\tau/\tau_k) d\tau - AK(1 - F_k)\bar{g} \tag{47}$$

and the caret is used to denote the fact that we have used the linear estimate. If the loop is designed such that the correlation time of $\rho_G(\tau)$ is much less than τ_k then, to a good approximation,[5]

$$\hat{E}[y_k(t)|\phi(t)] \approx \frac{AK(1 - F_k)G(\phi)}{2\tau_k \sigma_G^2} S_G(0) - AK(1 - F_k)\bar{g} \tag{48}$$

for all $k = 1, 2, \cdots, N$. In (48), $S_G(0)$ is the spectral density of $G(\phi) = g(\phi) - \bar{g}$ at the origin.

At this point, in the development of a working theory, it appears that the goodness of the assumptions which lead one from (42) to (48) must be justified by direct measurement of $E(y_k|\phi)$. The measurement of $E(y_k|\phi)$ can readily be adapted for simulation on a digital computer. The fact that $E(y_1|\phi)$ is sinusoidal in the steady state for a sinusoidal PLL when $\Omega_0 = 0$, $N = 1$ has been verified by computer simulation techniques. Typical results from the simulation for a low signal-to-noise ratio case are shown in Fig. 5 along with a plot of $\hat{E}(y_1|\phi)$ to accentuate the agreement.

From (25) and (26) the steady state restoring force, for an $(N+1)$-order tracker, becomes

$$\hat{h}_0(\phi) = \beta(N) - \alpha(N)g(\phi) \tag{49}$$

where

$$\beta(N) = \frac{2}{K_{00}}\left[\Omega_0 - AK\bar{g} \sum_{k=1}^{N} (1 - F_k)\left(1 + \frac{S_G(0)}{2\tau_k \sigma_G^2}\right)\right]$$
$$\alpha(N) = \frac{2}{K_{00}}\left[AKF_0 - \frac{AKS_G(0)}{2\sigma_G^2} \sum_{k=1}^{N} \left(\frac{1 - F_k}{\tau_k}\right)\right]. \tag{50}$$

The use of (49) in (27) and (32) produces $p(\phi)$ for an (N+1)-order tracker. The factor $\beta(N) > 0$ is responsible for the asymmetry in $p(\phi)$; hence, $p(\phi)$ will be symmetric if the loop is designed such that $\beta(N) = 0$, i.e.,

$$\frac{\Omega_0}{AK} = \bar{g} \sum_{k=1}^{N} (1 - F_k)\left[1 + \frac{S_G(0)}{2\tau_k \sigma_G^2}\right]. \tag{51}$$

It is clear that for $\Omega_0 \neq 0$ this equation can never be satisfied by the design of a first-order system. The range of application of this theory must be determined by experimentation or methods of computer simulation.

[4] A random process $x(t)$ is said to be of the separable class if it is second-order stationary and satisfies $E[x(t+\tau)|x(t)] = \rho_x(\tau)x$ for all τ. The correlation function $\rho_x(\tau) = E[x(t+\tau)x(t)]/E(x^2)$.

[5] Throughout this paper the symbol $\approx$ is used to denote approximately equal to and the symbol $\cong$ implies asymptotically equal.

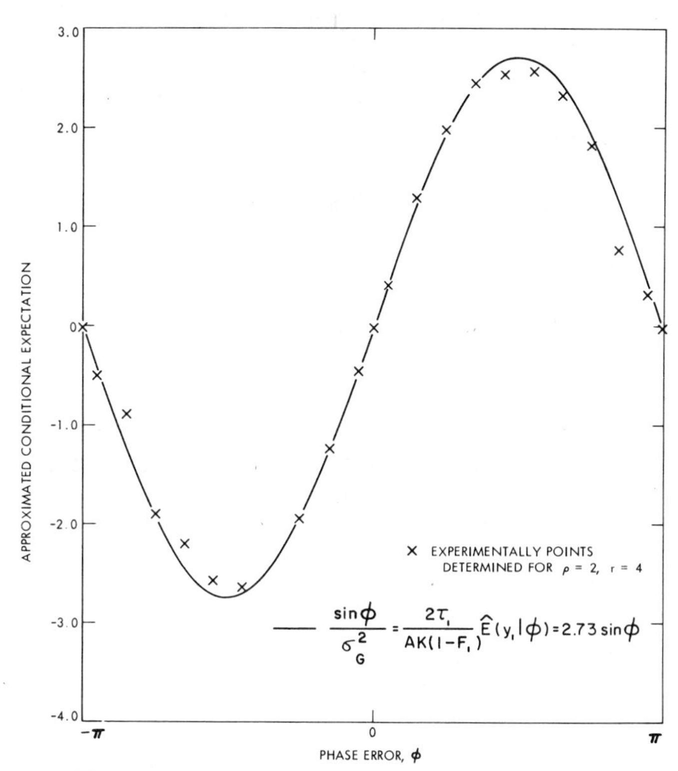

Fig. 5. Comparison of approximate conditional expectation with computer simulation results.

VIII. The Probability Density Function $p(\phi)$ for an $(N+1)$-Order, Sinusoidal PLL

Here $g(\phi) = \sin\phi$ and using (49) in (27) and (32) yields

$$p(\phi) = \frac{\exp[\beta\phi + \alpha\cos\phi]}{4\pi^2 \exp(-\pi\beta)|I_{j\beta}(\alpha)|^2} \cdot \int_\phi^{\phi+2\pi} \exp[-\beta x - \alpha\cos x] dx \tag{52}$$

where we have written β for $\beta(N)$ and α for $\alpha(N)$. The functions $I_\nu(x)$ are modified Bessel functions of imaginary order and of argument x. Equation (52) has been experimentally observed in the laboratory [10] for the case where $\Omega_0 = 0$.

The pdf $p(\phi)$ has the Fourier series representation

$$p(\phi) = \frac{1}{2\pi} \sum_{n=-\infty}^{\infty} C_n \exp(jn\phi); \qquad C_n = E[\exp(-jn\phi)] \tag{53}$$

where the circular-moments needed to define C_n are given by [17]

$$\overline{\cos n\phi} = \mathrm{Re}\left[\frac{I_{n-j\beta}(\alpha)}{I_{-j\beta}(\alpha)}\right]; \qquad \overline{\sin n\phi} = \mathrm{Im}\left[\frac{I_{n-j\beta}(\alpha)}{I_{-j\beta}(\alpha)}\right] \tag{54}$$

with Re[·] and Im[·] denoting, respectively, the real and the imaginary part of the bracketed quantities. Although we do not pursue the matter here we point out that the circular moments are related to the wedge functions $F_\nu(x)$ and $G_\nu(x)$ discussed in [18]. The wedge functions are defined by

$$F_\nu(x) \triangleq \frac{\pi}{2}\left[\frac{I_{j\nu}(x) + I_{-j\nu}(x)}{\sinh \pi\nu}\right] = \frac{\pi \mathrm{Re}\,[I_{j\nu}(x)]}{\sinh \pi\nu}$$

$$G_\nu(x) \triangleq \frac{j\pi}{2}\left[\frac{I_{j\nu}(x) - I_{-j\nu}(x)}{\sinh \pi\nu}\right] = \frac{-\pi\,\mathrm{Im}\,[I_{j\nu}(x)]}{\sinh \pi\nu} \tag{55}$$

and their name is derived from the fact that in potential theory they show a certain analogy to the solutions of Legendre's equation called "cone functions." We also write

$$I_{j\nu}(x) = \frac{\sinh \pi\nu}{\pi}[F_\nu(x) - jG_\nu(x)] \tag{56}$$

which relates the Bessel function to the wedge functions.

The mean value of the phase error may be found in terms of tabulated functions from (52) and the well-known Bessel function expansions for $\exp(\pm x \cos \phi)$. Without belaboring the details (see [17]) we have

$$\bar{\phi} = \int_{-\pi}^{\pi} \phi p(\phi) d\phi$$

$$\bar{\phi} = \frac{2 \sinh \pi\beta}{\pi |I_{j\beta}(\alpha)|^2} \sum_{m=1}^{\infty} \frac{m I_m(\alpha)}{m^2 + \beta^2} \cdot \left[\frac{I_0(\alpha)}{m} + \frac{I_m(\alpha)}{4m} + \sum_{\substack{k=1 \\ k \neq m}}^{\infty} \frac{2m(-1)^k I_k(\alpha)}{m^2 - k^2}\right]. \tag{57}$$

It is clear from (57) that with $\beta(N)=0$, $\bar{\phi}=0$. Furthermore, $\overline{\phi^2}$ is given by [17]

$$\overline{\phi^2} = \int_{-\pi}^{\pi} \phi^2 p(\phi) d\phi$$

$$\begin{aligned}\overline{\phi^2} = \frac{\sinh \pi\beta}{\pi |I_{j\beta}(\alpha)|^2} \Bigg\{ &\frac{I_0(\alpha)}{\beta}\left[\frac{\pi^2 I_0(\alpha)}{3} + 4 \sum_{k=1}^{\infty} \frac{(-1)^k I_k(\alpha)}{k^2}\right] \\ &+ 2\beta I_0(\alpha) \sum_{k=1}^{\infty} \frac{I_k(\alpha)}{k^2(\beta^2 + k^2)} \\ &+ 2\beta \sum_{k=1}^{\infty} \frac{(-1)^k I_k(\alpha)}{\beta^2 + k^2}\left[\left(\frac{\pi^2}{3} + \frac{1}{2k^2}\right) I_k(\alpha) \right. \\ &\left. + 4 \sum_{\substack{m=1 \\ m \neq k}}^{\infty} \frac{(-1)^m (k^2 + m^2) I_m(\alpha)}{(k^2 - m^2)^2}\right]\Bigg\}.\end{aligned} \tag{58}$$

The variance $\sigma_\phi^2 = \overline{\phi^2} - (\bar{\phi})^2$ is minimized when the loop is designed such that $\beta=0$ and α is maximized [17]. For $\beta=0$ we have, from (58), that

$$\sigma_\phi^2\Big|_{\min} = \frac{\pi^2}{3} + \frac{4}{I_0(\alpha)} \sum_{k=1}^{\infty} \frac{(-1)^k}{k^2} I_k(\alpha). \tag{59}$$

According to this theory, the moments of $\sin \phi$ and the phase-error rate $\dot{\phi}$ can be evaluated directly from (54) and the argument of stochastically equivalent processes [6] to yield [17]

$$\overline{\sin \phi} = \mathrm{Im}\left[\frac{I_{1-j\beta}(\alpha)}{I_{-j\beta}(\alpha)}\right]; \quad \overline{\sin^2 \phi} = \mathrm{Im}\left[\frac{I_{2-j\beta}(\alpha)}{I_{-j\beta}}\right]$$

$$\sigma_G^2 = \overline{\sin^2 \phi} - (\overline{\sin \phi})^2 \tag{60}$$

$$\sigma_{\dot{\phi}}^2 = \Omega^2 + \Delta^2 \sigma_G^2 + (F_1 K \sigma)^2$$

where

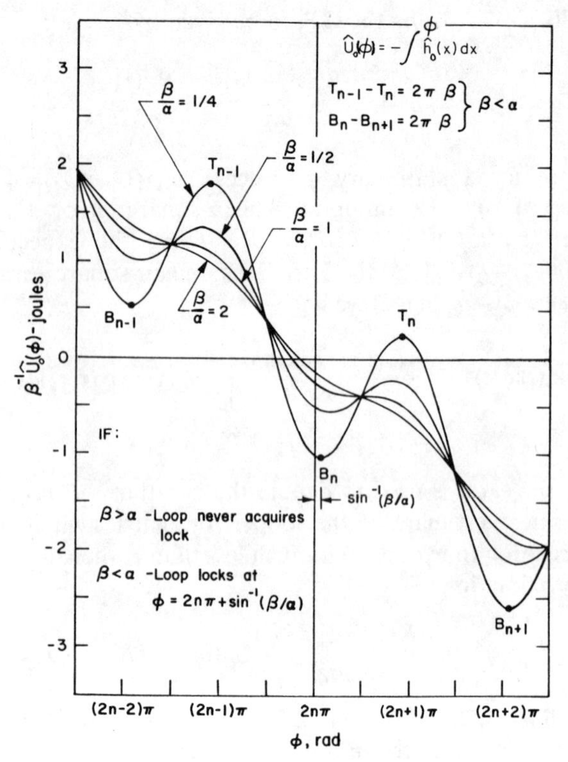

Fig. 6. Motion of the phase error from well to well as a function of the potential $\beta^{-1}\hat{U}_0(\phi)$.

$$\Omega = \Omega_0 - AK\bar{g} \sum_{k=1}^{N} (1 - F_k)\left[1 + \frac{S_G(0)}{2\tau_k \sigma_G^2}\right]$$

$$\Delta = AKF_0 - \frac{AKS_G(0)}{2\sigma_G^2}\left[\sum_{k=1}^{N}\left(\frac{1 - F_k}{\tau_k}\right)\right].$$

From (37) we see that, when $\beta=0$, the mean phase-error rate $\bar{\dot{\phi}} = \bar{\omega} - \omega_0 = \Omega_0$. Thus the mean frequency of the CRG $\bar{\omega}$ agrees with the frequency of $s[t, \theta(t)]$ on the average, i.e., $\bar{\omega} = \omega$.

It is convenient to give a physical interpretation of our solution in terms of the concept of "potential wells" containing a Brownian particle. As previously mentioned, the position of the trajectory $\mathbf{y}(t)$ at time t can be thought of as representative of the position of a particle undergoing Brownian motion. In fact, the motion of the ϕ projection along the ϕ axis can be interpreted as the motion of a particle in an external force field $\hat{h}_0(\phi)$ whose dependence on the position is nonlinear. The function $\hat{U}_0(\phi) = -\int^{\phi} \hat{h}_0(x)dx$ represents the potential at ϕ and $\{\hat{U}_0(\phi) - \hat{U}_0(\phi + 2\pi)\}$ represents the potential difference a distance of 2π radians apart.

Fig. 6 illustrates the normalized function $\beta^{-1}\hat{U}_0(\phi)$ versus ϕ for various values of β/α and for the case of a PLL. In fact all positions of possible phaselock are found when the restoring force $\hat{h}(\phi)$ is zero. According to this theory, if $\beta > \alpha$, phase lock is not possible since there exist no well bottoms B_n. The particle (phase error) will slip from well to well without coming to rest. If, however, $\beta = \alpha$, then the solutions to $\sin \phi = 1$ specify inflection points of $\hat{U}_0(\phi)$. If $\beta < \alpha$, well bottoms (B_n—stable points of *potential* minima) occur at $\phi = 2n\pi + \sin^{-1}(\beta/\alpha)$ and well tops (T_n—unstable points of potential maxima) occur at $\phi = (2n+1)\pi$

$+\sin^{-1}(\beta/\alpha)$ where n is any integer. According to this theory, the slipping occurs more rapidly the greater the β and the shallower the well depth. The depth of the wells is proportional to α. In other words, one wishes to design the loop such that for any external conditions, β is minimized and α is maximized, a fact which we have just observed analytically. The shape of the potential wells is obviously dependent upon the nonlinearity $g(\phi)$. Hence, we would intuitively expect that the "best" shape occurs when the well sides are straight, provided the loop is designed such that $\beta=0$. In fact, it is conjectured that an optimum signal acquisition device is one in which $g(\phi)$ is synthesized such that $\partial U_0(\phi, t)/\partial\phi$ is a symmetric square wave centered about zero for all t.

IX. Net Flow of Probability per Unit of Time Through the Hyperplane $y_k=y_k'$ and the Average Number of Cycles Slipped per Unit of Time for an $(N+1)$-Order Tracker

Due to the additive noise, discontinuities of oscillator (CGR) synchronization arises. The local oscillator may slip or gain a cycle of oscillations relative to the oscillations of the external signal $s[t, \theta(t)]$. In tracking applications the average number of cycles slipped per unit time is an important parameter as it is indicative of the error introduced into any Doppler measurement made to obtain velocity and changes in range.

To calculate the average number of cycles slipped per unit of time in the steady state, we make use of the concept of probability current (introduced earlier) in the ϕ direction. The average flow of probability through the hyperplane $\phi=\phi'$ in the positive ϕ direction per unit of time is easily found from (24), (30), and (52) to be given by

$$\mathscr{I}_0 = -\frac{K_{00}C_0D_0}{2}; \qquad \text{generalized tracker}$$
$$\mathscr{I}_0 = \frac{K_{00}\sinh\pi\beta}{4\pi^2|I_{j\beta}(\alpha)|^2}; \qquad (N+1)\text{-order PLL} \tag{61}$$

for all ϕ. This says that the flow of probability per unit time through the hyperplane $\phi=\phi'$ is constant.[6] If $\Omega_0=0$, then $D_0=0, \beta=0$. $\mathscr{I}_0(\phi)=0$ and there is no net flow of probability through the hyperplane $\phi=\phi'$.

In practice it is sometimes convenient to know the average number of cycles slipped per unit of time independent of direction. Denote the average number of cycles slipped to the right (positive ϕ direction) per unit of time by $\mathscr{I}_+$ and the average number of cycles slipped to the left (negative ϕ direction) per unit of time by $\mathscr{I}_-$. The parameter $\mathscr{I}_+$ represents the average number of trajectories of the ϕ projection traveling in the positive ϕ direction per unit time which pass through the hyperplane $\phi'=\phi$. Similarly $\mathscr{I}_-$ represents the average number of trajectories of the ϕ projection traveling in the negative ϕ direction per unit time which pass through the same hyperplane. Thus $\mathscr{I}_0=\mathscr{I}_+-\mathscr{I}_-$ represents the (net) average number of cycles slipped per unit time. The ratio of the currents can be obtained [17] using results [19] from the theory of statistical mechanics which specifies the rate of escape of Brownian particles over a potential barrier. For a cycle slip to occur it is necessary for the particle normally remaining in the plane of potential minimum at the position $\phi(t)$ to overcome the potential barrier represented by $\hat{U}_0(\phi)$. As a result of $\Omega_0\neq0$, the height of the barrier to the right is not equal to the height of the barrier to the left. If the height α is large, then Boltzmann's equation applies and

$$\frac{\mathscr{I}_+}{\mathscr{I}_-} = \exp[2\pi\beta] = \exp[\hat{U}_0(\phi) - \hat{U}_0(\phi+2\pi)] \tag{62}$$

for the $(N+1)$-order tracker. For $\beta=0$ we see from (62) that the to-the-left cycles slipped equal the to-the-right cycles slipped. Thus the total number of cycles slipped per unit of time independent of direction is $\bar{S}=\mathscr{I}_++\mathscr{I}_-$, viz.,

$$\bar{S} \approx \tfrac{1}{2}[C_0D_0K_0\coth\pi\beta]: \qquad \text{generalized tracker}$$
$$\approx \frac{K_{00}\cosh\pi\beta}{4\pi^2|I_{j\beta}(\alpha)|^2}; \qquad (N+1)\text{-order PLL.} \tag{63}$$

The quantity $D=(2\pi)^2\bar{S}$ is the diffusion coefficient representing the rate at which $\phi(t)$ is undergoing diffusion while Dt accounts for the fact that (14) is true. The expected value of the time intervals between cycle slipping events is given by $\Delta T\approx[\bar{S}]^{-1}$. Tikhonov [7], [9] seems to have been the first to apply this procedure to the analysis of a first-order PLL. According to Tikhonov, Stratonovich suggested this analogy [7].

The probability currents $\mathscr{I}_k$; $k=1, 2, \cdots, N$, may be easily computed from (29) and (30) in a manner similar to that described in this section. Thus the probability current flowing through the hyperplane $y_k=y_k'$ per unit of time in the positive direction of the y_k axis is $\mathscr{I}_k=-C_kD_kK_{kk}/2$ for all $k=1, 2, \cdots, N$. Again we find that the net flow of probability is constant and the field is irrotational when $D_k=0$. This says that there is no net flow of probability in the y_k direction.

X. The Probability Distribution $p(\phi)$ for the Second-Order Tracker

If we now consider the case where $N=1$, $F_0=F_1=\tau_2/\tau_1$ and define[7]

$$r \triangleq AKF_0\tau_2 \triangleq \frac{AK\tau_2^2}{\tau_1}; \qquad W_L \triangleq \frac{r+1}{2\tau_2} \text{ for } r\tau_1 \gg \tau_2 \tag{64}$$

[6] From Maxwell's field equations we know that the curl of the magnetic field intensity $\boldsymbol{H}$ is equal to the current density $\boldsymbol{J}$ in the steady state. Here the probability current $\mathscr{I}_0$ is analogous to $\boldsymbol{J}$ so that the curl of the ϕ projection of the electromagnetic field generated by electronic charges flowing in the PLL circuit is equal to $\mathscr{I}_0$. This curl is zero when the detuning is zero.

[7] The loop bandwidth W_L is defined as

$$W_L \triangleq 2b_L \triangleq \frac{1}{2\pi j}\int_{-j\infty}^{j\infty}|H(s)|^2 ds$$

where $H(s)$ is the closed transfer function when the loop is linearized [1], [2]. Using the linear theory one may compute the system damping behavior. For $r<4$ the system is underdamped, for $r=4$ the system is critically damped, and for $r>4$ the system is overdamped. Most tracking systems are designed such that $r=2$. This corresponds to $1/\sqrt{2}$ damping.

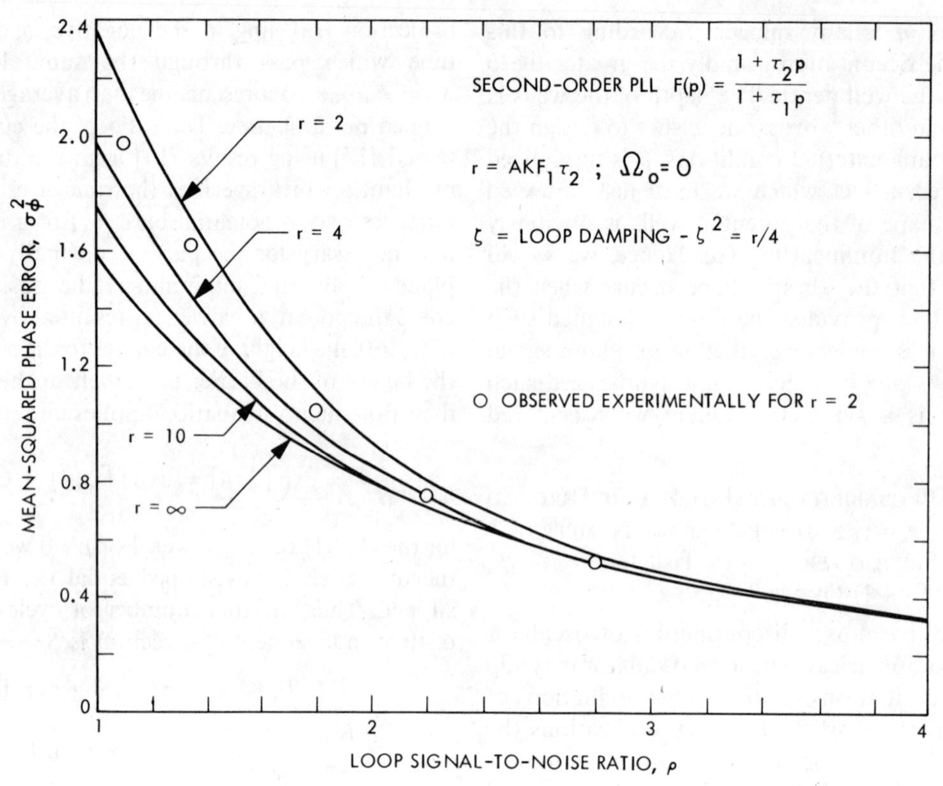

Fig. 7. Variance of the phase error versus loop signal-to-noise ratio ρ.

where r is related to the loop damping ($\zeta=\sqrt{r/4}$) and W_L is the loop bandwidth as defined from linear tracking theory. Thus $h_0(\phi)$ in (26) becomes

$$h_0(\phi) = -\left(\frac{r+1}{r}\right)\rho g(\phi) + \left(\frac{r+1}{r}\right)^2 \cdot \left[\frac{\rho\Omega_0}{2W_L} + \frac{\rho E(y_1|\phi)}{2W_L}\right] \tag{65}$$

and $\rho=2A^2/N_0W_L$, i.e., the signal-to-noise ratio in the loop bandwidth W_L. Using the linear least-squares estimate (48) to approximate $E(y_1|\phi)$ we have, from (49),

$$\hat{h}_0(\phi) = \beta - \alpha g(\phi)$$
$$\hat{U}_0(\phi) = -\beta\phi + \alpha\int^{\phi} g(x)dx \tag{66}$$

where

$$\beta = \beta(1) = \left(\frac{r+1}{r}\right)\frac{\rho}{F_1}\left[\frac{\Omega_0}{AK} - (1-F_1)\bar{g}\right]$$
$$\alpha = \alpha(1) = \left(\frac{r+1}{r}\right)\rho - \frac{1}{r\sigma_G^2} \tag{67}$$

and we have made use of the linear PLL theory to set $S_G(0)=N_0/2A^2=\sigma^2/2WA^2$. If the loop is designed such that $\beta\approx 0$, i.e., $\Omega_0\approx AK\bar{g}$ for the range of interest, then

$$p(\phi) \approx C_0' \exp\left[-\alpha(1)\int^{\phi} g(x)dx\right]. \tag{68}$$

The variance of ϕ for a sinusoidal PLL mechanization can be obtained by first solving for $\overline{\sin^2\phi}$ from

$$\overline{\sin^2\phi} = \int_{-\pi}^{\pi} \sin^2\phi\, p(\phi)d\phi = \frac{1}{\alpha}\left[\frac{I_1(\alpha)}{I_0(\alpha)}\right] \tag{69}$$

and the corresponding variance can be obtained from (68) and (69) by numerical integration on a digital computer. Fig. 7 illustrates the results for various values of r. The circles in Fig. 7 correspond to points obtained in the laboratory [10] by means of a hardware mechanization with $r=4$ and $g(\phi)=\sin\phi$. It is also instructive to approximate the conditional expectation using linear tracking theory. Thus [17],

$$\hat{E}(y_1|\phi) = (1-F_1)\left\{\frac{2rW_L}{(1+r)^2}\phi - \Omega_0\left(1+\frac{F_1}{1+r}\right)\right\}. \tag{70}$$

If $F_1 \ll 1$ we can write

$$\hat{h}_0(\phi) = -\left(\frac{r+1}{r}\right)\rho g(\phi) + \frac{\rho}{r}\phi + \frac{\Omega_0\rho}{AK} \tag{71}$$

so that

$$p(\phi) = C_0' \exp\left[-\left(\frac{r+1}{r}\right)\rho\int^{\phi} g(x)dx + \frac{\rho}{2r}\phi^2 + \frac{\Omega_0\rho}{AK}\phi\right] \tag{72}$$

which is independent Ω_0 for large AK. If we replace $g(x)$ by x then (72) is Gaussian and it is seen that the loop will always pull in lock so that $p(\phi)$ is symmetric about Ω_0/AK. For reasonable damping factors and Ω_0, a small F_0 produces very little mean-phase error. This, of course, is why second-order loops are used in practice. Contrast this with the solution, $N=0$, for a first-order loop with $\Omega_0\neq 0$ where it is noted that the ratio of (linear) mean-phase error for a second-order to loop to a first is $\bar{\phi}_2/\bar{\phi}_1 = F_0(r+1/r) \ll 1$ for equal loop bandwidths.

XI. The First Passage Time Model and Boundary Conditions on $P(\boldsymbol{y}; t)$

As discussed earlier, the variance of ϕ as determined from $P(\boldsymbol{y}; t)$ will be unbounded in the steady state. However, for $t<\infty$ the variance of ϕ as determined from (12) is bounded. In what follows, we work with $P(\boldsymbol{y}; t)$ when computing the moments of the first passage time of ϕ; however, when computing the moments of the projections of $\boldsymbol{y}_0' = (y_1, \cdots, y_N)$ one may work with the solution $p(\boldsymbol{y}; t)$. The domain R' spanned by the variables y_k is $R' = [|y_0| \le \infty, |y_1| \le \infty, \cdots, |y_N| \le \infty]$ and if $P(\boldsymbol{y}; t)$ is to be a probability density in R' we must have

$$\underbrace{\int \cdots \int}_{(N+1)\text{-fold}} P(\boldsymbol{y}; t) d\boldsymbol{y} = 1 \tag{73}$$

for all t. Further, the transition pdf of the kth projection of $\boldsymbol{y}$ is given by

$$P(\boldsymbol{y}_k; t) = \underbrace{\int \cdots \int}_{N\text{-fold}} P(\boldsymbol{y}; t) d\boldsymbol{y}_k'. \tag{74}$$

The flow of probability from the probability space R' through the surface Γ' surrounding R' into the infinite medium is characterized by the facts that along any edge of Γ'

$$y_j P(\boldsymbol{y}; t)|_{y_j = \pm\infty} = P(\boldsymbol{y}; t)|_{y_j = \pm\infty} = 0 \tag{75}$$

for all $j=0, 1, \cdots, N$. The fact that $y_j P(\boldsymbol{y}; t)$ in (75) approaches zero as y_j approaches infinity requires that $P(\boldsymbol{y}; t)$ approach zero faster than $[y_j]^{-(1+\varepsilon)}$, $\varepsilon > 0$. This is required in order for the law of conservation of probability to hold. Further, along any edge of Γ',

$$\frac{\partial}{\partial y_j} P(\boldsymbol{y}; t)|_{y_j = \pm\infty} = 0; \qquad j = 0, 1, \cdots, N. \tag{76}$$

In addition, when considering the first passage time of the kth projection y_k we must specify boundary conditions at the barriers $y_k = \pm y_{kl}$ for all $k=0, 1, \cdots, N$. By this we mean that we create a new process $q_k(t)$ such that, for all $k=0, 1, \cdots, N$,

$$q_k(t) = \begin{cases} y_k(t) & \text{if } y_k(\tau) < |y_{kl}| \text{ for every } \tau < t \\ \pm y_{kl} & \text{if } |y_k(\tau)| = y_{kl} \text{ for some } \tau \le t. \end{cases} \tag{77}$$

Such restrictions on the $\boldsymbol{y}(t)$ process place a restriction on the solution $P(\boldsymbol{y}; t)$ such that the normalization condition (73) no longer holds. (We discuss this point more fully in the next section.) We refer to the solution of (12) for the new process $\{q_k(t), k=0, \cdots, N\}$ as the restricted solution and denote it by $Q(\boldsymbol{y}, t)$. The boundary conditions (75) and (76) still hold if we replace $P(\boldsymbol{y}; t)$ by $Q(\boldsymbol{y}; t)$. In the next section we shall discuss further boundary conditions which must be imposed upon the solutions $Q(y_k, t)$ for all $k=0, 1, \cdots, N$.

The function $Q(y_k, t)$ is determined from (74) with $P(\boldsymbol{y}; t)$ replaced by $Q(\boldsymbol{y}, t)$. The reader unfamiliar with first passage time problems would do well in referring to [6] for an enlightening discussion.

XII. Moments of the First Passage Time of the kth Projection

In this section we shall be concerned with the problem of evaluating the moments of the random time $T(y_k)$ it takes for the kth projection of $\boldsymbol{y}(t)$ to exceed either of the barriers (limits) $y_k = \pm y_{kl}$ for the first time while the initial position at time $t = t_0$ is $y_k = y_{k0}$. Stated in another way, we wish to derive expressions for the moments of the expected time $E[T^n(y_k)]$ before one of the barriers is crossed for the first time.

We now show how to calculate the moments of the first-passage time $\tau^n(y_{kl}) = E[T^n(y_k)]$ by confining ourselves to the case where coefficients K_k and K_{lk} in the Fokker–Planck equation are time independent. Excluding from consideration any trajectory of the projection $y_k(t)$ as soon as it reaches the barrier values $-y_{kl}$ or y_{kl} for the first time we can describe the remaining trajectories by a probability density $Q(y_k, t)$ [6] such that $\Delta P = Q(y_k, t)\Delta y_k + 0[(\Delta y_k)^2]$. Here ΔP is the probability that the projection $y_k(t)$ assumes a value in the interval $[y_k, y_k + \Delta y_k]$ without ever having reached the barrier during the entire time interval (t_0, t). Then the integral

$$\psi_k(t) = \int_{-y_{kl}}^{y_{kl}} Q(y_k, t) dy_k; \qquad k = 0, 1, \cdots, N \tag{78}$$

gives the probability that $y_k(t)$ never reaches the barrier during the time interval (t_0, t). Initially, when no realization of $y_k(t)$ has yet managed to reach the barrier, the probability density is the same as the initial density (13) so that

$$\psi_k(t_0) = 1. \tag{79}$$

At subsequent times, the normalization condition

$$\int_{-\infty}^{\infty} Q(y_k, t) dy_k = 1 \tag{80}$$

is no longer valid, since with the passage of time more and more trajectories of the projection $y_k(t)$ are excluded from consideration as a result of having reached the barriers. Sooner or later all possible trajectories arrive at the barriers; hence,

$$\psi_k(\infty) = 0. \tag{81}$$

Inside the interval $[-y_{kl}, y_{kl}]$, the behavior of the probability density $Q(y_k, t)$ is described by the reduced Fokker–Planck equation since the kth projection of the trajectory $\boldsymbol{y}(t)$ cannot terminate inside $[-y_{kl}, y_{kl}]$. In fact, trajectories of the kth projection are excluded from consideration only when the barrier is reached. Furthermore, there is a nonzero probability current accumulating at the barrier which corresponds to those trajectories of the kth projection, which are absorbed. The value of this probability current at either barrier $y_k = y_{kl}$ or $y_k = -y_{kl}$ will, when used as the weight of a delta function, be that amount of probability required to normalize $Q(y_k, t)$. Therefore, the boundary conditions for the kth projection have to be altered in the following way. At time t slightly greater than t_0 there are practically no trajectories near the barriers, i.e., no trajectory has yet touched the barrier. But it is just these trajectories which are described by the probability density $Q(y_k, t)$; hence, we have the absorption conditions [6]

$$Q(\pm y_{kl}, t) = 0 \qquad \text{for all } t > t_0 = 0. \tag{82}$$

We define

$$Q(\pm y_{kl}) \triangleq \int_0^\infty Q(y_{kl}, t)dt = 0. \tag{83}$$

After calculating $Q(y_k, t)$ from (12), using (75) and (76) we can find the probability

$$\psi_k(t_0) - \psi_k(t) = 1 - \psi_k(t) \tag{84}$$

that the barrier is first reached during the time interval (t_0, t). Differentiating (84) with respect to t we obtain the probability density of the kth projection of the first passage time, viz.,

$$p_k(t) = -\frac{\partial \psi_k(t)}{\partial t}. \tag{85}$$

The nth moment of the first passage time of y_k is, therefore,

$$\tau^n(y_{kl}) = \int_0^\infty \int_{-y_{kl}}^{y_{kl}} -t^n \frac{\partial}{\partial t} Q(y_k, t)dy_k dt. \tag{86}$$

Integrating (86) by parts yields

$$\tau^n(y_{kl}) = -\int_{-y_{kl}}^{y_{kl}} \left[t^n \psi_k(t) \Big|_0^\infty - n \int_0^\infty t^{n-1} Q(y_k, t)dt \right] dy_k. \tag{87}$$

But from (79) we have that $\psi_k(t_0) = 1$ and from (81) we have $\psi_k(\infty) = 0$ so that (87) reduces to

$$\tau^n(y_{kl}) = n \int_{-y_{kl}}^{y_{kl}} \int_0^\infty t^{n-1} Q(y_k, t)dt dy_k \tag{88}$$

if we assume that $\psi_k(t)$ goes to zero faster than t^{-n} as t approaches infinity. Define

$$Q_{n-1}(y_k) \triangleq \int_0^\infty t^{n-1} Q(y_k, t)dt; \tag{89}$$

then the nth moment of the first passage time to y_k is given by

$$\tau^n(y_k) = n \int_{-y_{kl}}^{y_k} Q_{n-1}(z)dz \tag{90}$$

for all $k = 0, 1, \cdots, N$.

The functions $Q_{n-1}(y_k)$ in (89) may be determined from (23) and (24) by noting that each projection of y_k satisfies a partial differential equation of the flow from

$$\nabla \cdot \mathscr{I}_k(y_k, t) + \frac{\partial Q_k(y_k, t)}{\partial t} = 0. \tag{91}$$

If we substitute for $\mathscr{I}_k(y_k, t)$, from (24), into the above equation we easily write

$$\frac{K_{kk}}{2} \frac{\partial}{\partial y_k} \{ [h_k(y_k, t) - \frac{\partial}{\partial y_k} Q(y_k, t)] \} + \frac{\partial Q(y_k, t)}{\partial t} = 0 \tag{92}$$

where we have replaced $p(y_k; t)$ by the restricted form $Q(y_k, t)$. The conditional expectations which define $h_k(y_k, t)$ are now taken with respect to the restricted pdf $Q(y_k, t|\phi)$. Multiplying both sides of (92) by t^n and integrating between $t_0 = 0$ and infinity we obtain the differential equation

$$\begin{aligned} &\frac{d^2 Q_n(y_k)}{dy_k^2} - \frac{d}{dy_k}[h_k(y_k, \bar{t}) Q_n(y_k)] \\ &\qquad = \frac{2}{K_{kk}} \int_0^\infty t^n \frac{\partial Q(y_k, t)}{\partial t} dt \end{aligned} \tag{93}$$

where $Q_n(y_k)$ is defined in (89). In arriving at (93) we have made use of the mean value theorem [20] which says that[8]

$$\begin{aligned} E[y_k, \bar{t}|\phi] Q_n(\phi) &= \int_0^\infty t^n E[y_k, t|\phi] Q(\phi, t)dt \\ E[g(\phi), \bar{t}|y_k] Q_n(y_k) &= \int_0^\infty t^n E[g(\phi), t|y_k] Q(y_k, t)dt \end{aligned} \tag{94}$$

and $\bar{t}$ is a point such that $\bar{t} \varepsilon [0, \infty]$. Integrating by parts on the right-hand side of (93) yields

$$\frac{d^2 Q_n(y_k)}{dy_k^2} - \frac{d}{dy_k}[h_k(y_k, \bar{t}) Q_n(y_k)] = -\frac{2n Q_{n-1}(y_k)}{K_{kk}}. \tag{95}$$

This result may be integrated once so that

$$\begin{aligned} &\frac{dQ_n(y_k)}{dy_k} - [h_k(y_k, \bar{t}) Q_n(y_k)] \\ &\qquad = -\frac{2n}{K_{kk}} \int_{-y_{kl}}^{y_k} Q_{n-1}(x)dx + C_k(n) \end{aligned} \tag{96}$$

where $C_k(n)$ is a constant of integration to be determined. From (90) we write

$$\frac{dQ_n(y_k)}{dy_k} - h_k(y_k, \bar{t}) Q_n(y_k) = C_k(n) - \frac{2}{K_{kk}} \tau^n(y_k) \tag{97}$$

where $\tau^0(y_k) = u(y_k - y_{k0})$ is the unit step occurring at $y_k = y_{k0}$. Solving (97) for $Q_n(y_k)$ yields

$$\begin{aligned} Q_n(y_k) = \exp[-U_k(y_k, \bar{t})] \Big\{ E_k + \int_{-y_{kl}}^{y_k} \Big[C_k(n) - \frac{2}{K_{kk}} \tau^n(x) \Big] \\ \cdot \exp[U_k(x, \bar{t})] dx \Big\} \end{aligned} \tag{98}$$

$$\tau^0(x) = u(x - y_{k0})$$

[8] Let D be an open and connected domain and f and g be continuous and bounded on D with $g(p) \geq 0$ for all $p \varepsilon D$. Then, there is a point $\bar{p} \varepsilon D$ such that

$$\iint_D fg = f(\bar{p}) \iint_D g.$$

Recall that, for brevity in notation, the initial values of the given variables in every conditional expectation have been suppressed.

where E_k is a constant of integration. Since $Q(-y_{kl})=0$ from (83) we have that $E_k=0$ and since $Q(y_{kl})=0$ we find that

$$C_k(n) = \frac{2}{K_{kk}} C_k'(n) = \frac{2}{K_{kk}} \frac{\int_{-y_{kl}}^{y_{kl}} \tau^n(x) \exp\left[U_k(x, \bar{t})\right] dx}{\int_{-y_{kl}}^{y_{kl}} \exp\left[U_k(x, \bar{t})\right] dx}. \tag{99}$$

Using these values of $C_k'(n)$ and E_k in (98) and making use of (90) gives the recursive formula

$$\begin{aligned} \tau^n(y_{kl}) &= \frac{2}{K_{kk}} \int_{-y_{kl}}^{y_{kl}} \int_{-y_{kl}}^{y_k} \left[C_k'(n-1) - \tau^{n-1}(x)\right] \\ &\quad \cdot \exp\left[U_k(x, \bar{t}) - U_k(y_k, \bar{t})\right] dx dy_k \\ \tau^0(x) &= u(x - y_{k0}) \end{aligned} \tag{100}$$

for the nth moment of the first passage time for a generalized tracking loop; $k=0, 1, 2, \cdots, N$. The nth moment of the first slip time is obtained from (100) with $y_{0l}=\phi_l=2\pi$. Equation (100) generalizes earlier work due to Darling and Siegert [21] and Siegert [22].

XIII. Moments of the First Passage for a Sinusoidal Phase-Locked Loop; $g(\phi)=\sin\phi$

A. The First-Order Loop: $F_0=1$, $k=0$, $N=0$

In this case $g(\phi)=\sin\phi$. From (26) we have that

$$U_0(\phi) = -\frac{\rho\Omega_0}{AK}\phi - \rho\cos\phi \tag{101}$$

where $\rho=2A^2/N_0W_L$ and $W_L=AK/2$. Setting $\phi_l=2\pi$ in (100) we have

$$\begin{aligned} W_L\tau^n(2\pi) &= \frac{\rho}{2} \int_{-2\pi}^{2\pi} \int_{-2\pi}^{\phi} \left[C_0'(n-1) - \tau^{n-1}(x)\right] \\ &\quad \cdot \exp\left[\rho\left\{\cos\phi - \cos x) - \frac{\Omega_0}{AK}(\phi - x)\right\}\right] dx d\phi. \end{aligned} \tag{102}$$

This generalizes Viterbi's result [1]. For $n=1, \Omega_0=0, \phi_0=0$, $C_0'(0)=1/2$, and (102) reduces to Viterbi's result for the mean time to first slip.

B. The Second-Order Loop: $N=1$, $F_0=F_1=\tau_2/\tau_1$, $k=0$

From (25) and (27) the potential function becomes

$$\begin{aligned} U_0(x, \bar{t}) &= \left(\frac{r+1}{r}\right)\rho\cos\phi \\ &\quad + \left(\frac{r+1}{r}\right)^2 \frac{\rho}{2W_L}\left[\int^{\phi} E(y_1, \bar{t}|\phi) dy_1 + \Omega_0\phi\right] \end{aligned} \tag{103}$$

where r and ρ are defined in (64). For large signal-to-noise ratios the conditional expectation may be approximated using linear PLL theory. The fact that this may be done is due to the fact that for large signal-to-noise ratios ϕ rarely exceeds $\pm\pi$ and, when it does, the initial conditions do not effect this expectation. Thus, for $F_1 \ll 1$, we may use (70) in (103) to obtain

$$U_0(\phi, \bar{t}) \approx -\left(\frac{r+1}{r}\right)\rho\cos\phi - \frac{\rho}{2r}\phi^2 - \rho\frac{\Omega_0}{AK}\phi. \tag{104}$$

Consequently, the nth moment of the mean time to first slip is approximated by

$$\begin{aligned} W_L\tau^n(2\pi) &\approx \left(\frac{r+1}{r}\right)^2 \frac{\rho}{2} \int_{-2\pi}^{2\pi} \int_{-2\pi}^{\phi} \left[C_0'(n-1) - \tau^{n-1}(x)\right] \\ &\quad \cdot \exp\left[\rho\left(\frac{r+1}{r}\right)\{\cos\phi - \cos x\} \right. \\ &\quad \left. + \frac{\rho}{2r}(\phi^2 - x^2) + \frac{\rho\Omega_0}{AK}(\phi - x)\right] dx d\phi. \end{aligned} \tag{105}$$

This generalizes Tausworthe's result [11]. In fact for $n=1$, $\Omega_0=0$, $\phi_0=0$, $C_0'(0)=1/2$, (105) reduces to his result. As r approaches infinity (105) reduces to (102) as it should. For $n=0$, we have from (99)

$$C_0'(0) = \frac{\int_{\phi_0}^{2\pi} \exp(U_0(x, \bar{t})) dx}{\int_{-2\pi}^{2\pi} \exp(U_0(x, \bar{t})) dx}. \tag{106}$$

The product, $W_L\tau(2\pi)$, of (105) is plotted versus ρ in Figs. 8 and 9 for various values of r, with $\Omega_0/AK=0$ and $\phi_0=0$.

As an alternative the conditional expectations found using the orthogonality principle may be used to approximate $E(y_k, \bar{t}|\phi)$ with $\bar{t}=\infty$. For such a case we find from (48) and (100) that

$$\begin{aligned} W_L\tau^n(2\pi) &\approx \left(\frac{r+1}{r}\right)^2 \frac{\rho}{2} \int_{-2\pi}^{2\pi} \left[C_0'(n-1) - \tau^{n-1}(x)\right] \\ &\quad \cdot \exp\left[\alpha(\cos\phi - \cos x) - \beta(\phi - x)\right] dx d\phi \end{aligned} \tag{107}$$

where α and β are defined in (67). For $n=1$, $\phi_0=0$, $\beta=0$, (107) reduces to $\tau(2\pi)=1/\bar{S}$ where $\bar{S}$ is given in (63) with $\beta=0$. However, as one might suspect, the approximation is not extremely accurate owing to the fact that $E(y_k, \bar{t}|\phi)$ are not sinusoidal, for $t<\infty$.

Recall from (93) that the conditional expectations $E(y_k, \bar{t}|\phi)$ required here are to be taken with respect to the restricted probability density function $Q(\phi, y_k, t)$. It is, therefore, clear that the expectations needed for an accurate approximation of the expectations $E(y_k, \bar{t}|\phi)$ should differ considerably from those chosen on the basis of the orthogonality principle (OP) in the steady state. Indeed measurements discussed in [15] bear out this fact. Other results are given in [17] for the case where the orthogonality principle is used to produce estimates for the conditional expectations. There we produce estimates of the form $a_k\phi + b_k\phi^3$, and so forth.

XIV. Synthesis of Optimum Tracking Loops

A. Coherent Communications

The problem of choosing the "best" loop filter $F(p)$ as well as the "best" nonlinearity so as to provide for an optimum tracking loop depends upon what is meant by optimum. For example, during the signal acquisition mode

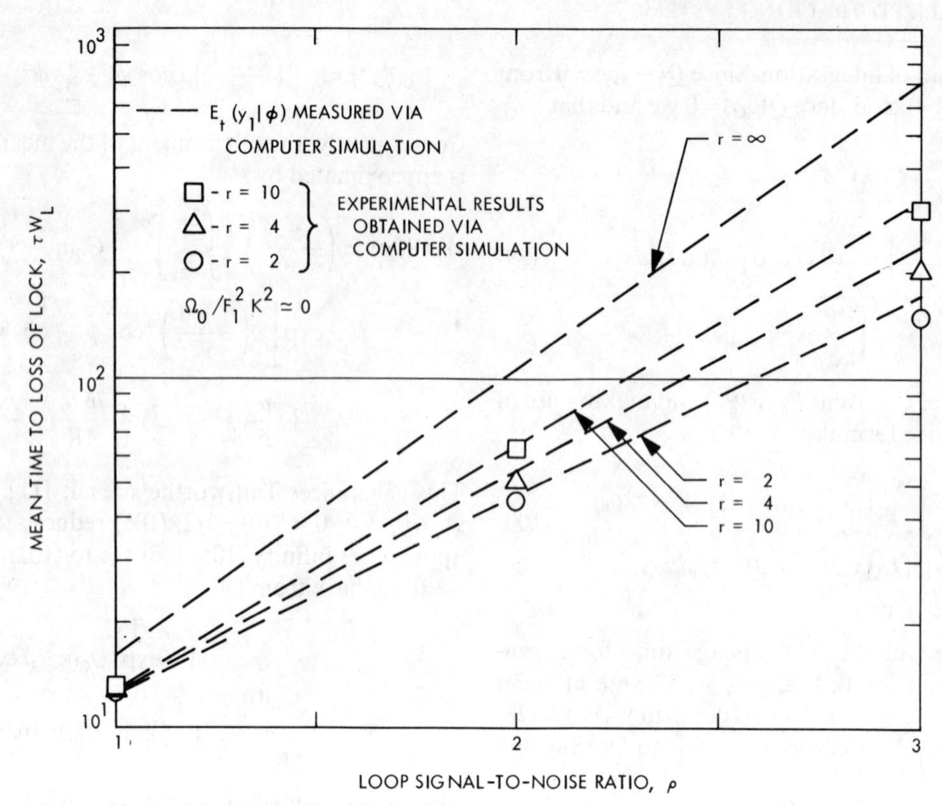

Fig. 8. Mean time to first slip bandwidth product versus ρ for various values of r.

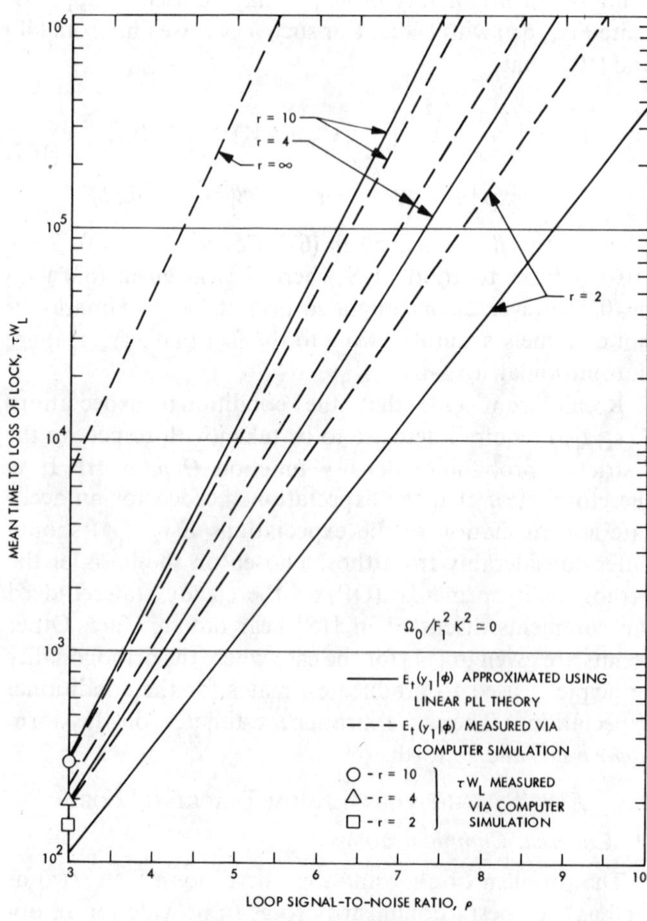

Fig. 9. Mean time to first slip bandwidth product versus ρ for various values of r.

the performance index is acquisition time. After the signal has been acquired, the problem becomes one of either tracking or data demodulation, or both, and the performance index changes. Hence, after acquisition a design based upon minimum acquisition time becomes suboptimum. For the case of phase-coherent communications one would want to minimize the mean-square phase error, i.e., minimize the functional

$$\min_{F(p),\, g(\phi)} G[F(p), g(\phi)] = \sigma_\phi^2 \tag{108}$$

subject to the linearity constraint on $F(p)$, and a gain constraint on the class of nonlinearities $\{g(\phi)\}$. In the case of tracking, one desires to maximize the expected time to loss of phase synchronization. In general, however, it is conceivable that such an optimization technique may be formidable or a solution to (108) may not even exist if $F(p)$ is restricted to being linear. Be that as it may, however, a few results are presently available for the case where one constrains the loop filter to be of the form $F(p)=1$ or $F(p)=1/(1+\tau_2 p)$ and then selects that nonlinearity $g(\phi)$ such that the mean-square error is minimum. This turns out to be equivalent [23], [24] to minimizing the area under the tail of the density $p(\phi)$. For these cases it can be shown [23], [24] that the optimum nonlinearity is

$$g(\phi) = \text{sgn}\,[\sin \phi] \tag{109}$$

when $F(p)=1$ and it is assumed that $\Omega_0=0$. The restoring force $h_0(\phi)$ becomes rectangular for this $g(\phi)$ since sgn $x=1$ for $x \geq 0$ and -1 for $x<0$.

In practice, it is desirable to design the loop such that

$p(\phi)$ is symmetric so that no bias is introduced in the phase measurement. From (52) it can be seen that any asymmetry in $p(\phi)$ is due to the $\beta(N)$. By proper design of the loop filter $\beta(N)$ can be made arbitrarily small for reasonable frequency offsets. This requires $N \geq 1$ so that

$$p(\phi) \approx C_0' \exp\left[-\alpha(N)\int^{\phi} g(\phi)d\phi\right] \tag{110}$$

if $\beta(N) \approx 0$. Paralleling the arguments due to Stiffler and Shaft [23], [24], we have, for the $(N+1)$-order tracker with $p(\phi)$ defined by (110), that the nonlinearity which minimizes the mean-squared phase error is also given by (109) for all N. Since $[\alpha(N)]^{-1}$ can be interpreted as a variance then σ_ϕ^2 is minimized over the choice of $F(p)$ when $\alpha(N)$ is maximized, i.e., the potential wells are deepest. This is accomplished with $N=1$, i.e., a second-order tracker. Hence we are lead to believe that, according to this theory, a second-order loop designed such that $\beta(1) \approx 0$ with $g(\phi)$ given by (109) is, for all practical purposes, that tracker which minimizes the variance of the phase error.

For the second-order tracker the minimum mean-square error obtainable is, from (109) and (110),

$$(\sigma_\phi^2)_{\min} = \frac{2[1 - \{1 + \pi\alpha + (\pi\alpha)^2/2\}\exp(-\pi\alpha)]}{\alpha^2[1 - \exp(-\pi\alpha)]} \tag{111}$$

where $\alpha = \alpha(1)$ is given is (67). For high signal-to-noise ratios, $\alpha = \rho$ and

$$(\sigma_\phi^2)_{\min} \cong \frac{2}{\rho^2}; \qquad \rho \gg 1. \tag{112}$$

At low values of ρ,

$$(\sigma_\phi^2)_{\min} \cong \frac{\alpha\pi^3}{3} \cdot \frac{1}{\exp(\pi\alpha) - 1}. \tag{113}$$

Thus, for large signal-to-noise ratios, the improvement in σ_ϕ^2 offered by the second-order tracker with optimum nonlinearity is $I = 10 \log_{10}(\rho/2)$ dB better than a second-order phase-locked loop with $g(\phi) = \sin\phi$. For equal phase variances and large signal-to-noise ratios, the second-order tracker with optimum nonlinearity requires that $\rho_{\text{PLL}} = \rho^2/2$ where ρ is the signal-to-noise ratio (see Fig. 10).

If for an $(N+1)$-order tracker, with $\Omega_0 = 0$, we desire to minimize the variance of ϕ then it is clear from (50) that the factor β is zero and $\alpha(N)$ is maximized when $N=0$. The minimum mean-square error is still given by (111) with $\alpha = \rho = 4A/N_0K$. Thus, we are lead to the conclusion that the optimum tracker is one for which the nonlinearity is given by (109) and $F(p)=1$ when $\Omega_0 = 0$. Hence a first-order system is optimum from the point of view of producing the minimum phase-error variance when $\Omega_0 = 0$. There are other approaches which may be used, aside from signal design, to vary $R_{rs}(\phi)$. We shall not discuss all of them in detail here; however, astute methods of instrumentation, as in Nth-order tan-lock [13] and delay locked loops [12] are possible.

B. Tracking Systems

In tracking applications, as opposed to phase-coherent communications, a more significant performance criterion

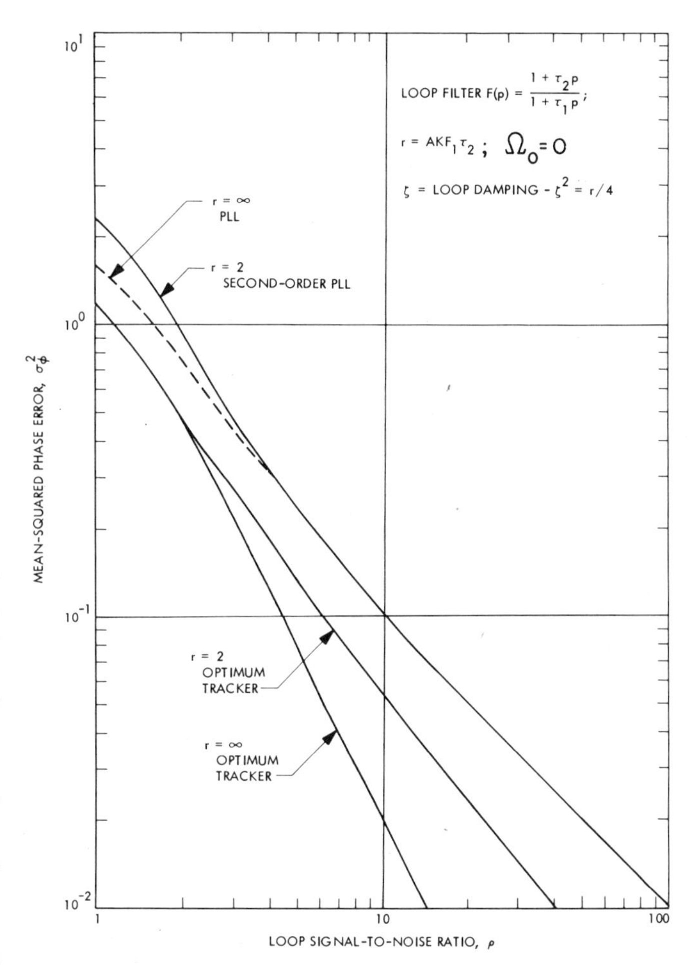

Fig. 10. Comparison of the variance of the phase error versus ρ for various r.

is the mean time to first loss of phase synchronization. For data demodulation with a tracker, mean-squared error assumes priority as a measure of performance. Furthermore, the criterion for loss of phase synchronization is arbitrary, i.e., the choice of ϕ_l.

Any attempt to perform stochastic optimization of the loop in the nonlinear region of operation for $N>0$ by selecting that $F(p)$ and $g(\phi)$ such that

$$\tau(y_k) = \max_{F(p),\, g(\phi)} E[T(y_k)] \tag{114}$$

is plagued by inexact knowledge of the expectations $E(g(\phi), \bar{t}|y_k)$ or $E(y_k, \bar{t}|\phi)$. For the case where $N=0$, $\Omega_0=0$, and $F_0=1$, it is possible to select that $g(\phi)$ for which $\tau(\phi_l)$ is a maximum. For this first-order tracker it is easy to show [24] that the nonlinearity which maximizes $\tau(\phi_l)$ is given by (109). The corresponding first passage time, as computed from (100) with $N=0$ and $g(\phi) = \text{sgn}[\sin\phi]$, is given by

$$\tau(\phi_l) = \frac{1}{2W_L}\left[\frac{1}{\rho}[\exp(\phi_l\rho) - 1] - \phi_l\right] \tag{115}$$

where $\rho = 4A/N_0K$. For small ρ, (115) approaches

$$\lim_{\rho \to 0} \tau(\phi_l) \cong \frac{\phi_l^2 \rho}{4W_L} \tag{116}$$

while for large ρ (115) simplifies to

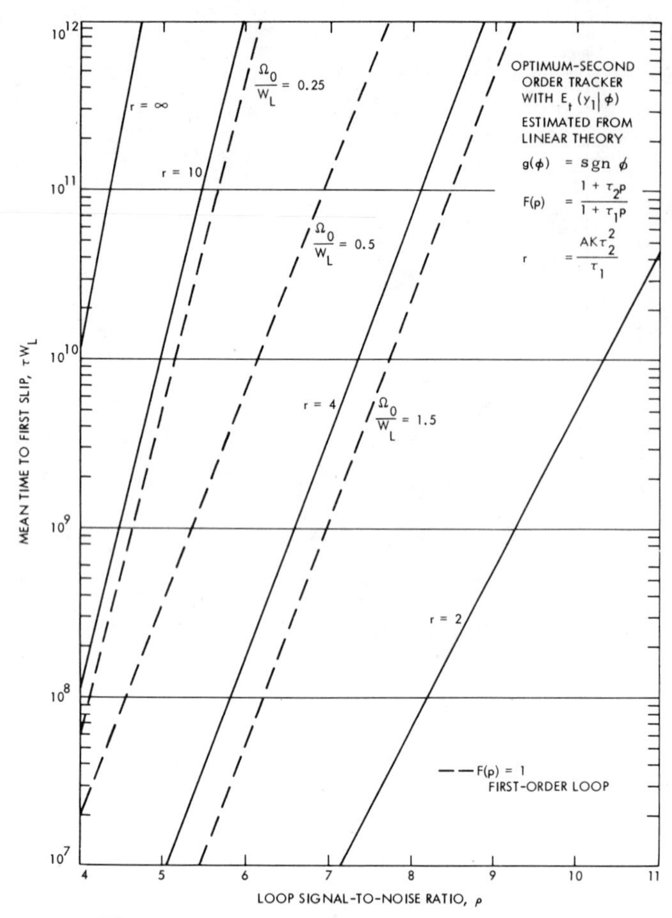

Fig. 11. Comparison of second-order tracker with a first-order tracker for $\Omega_0 \neq 0$.

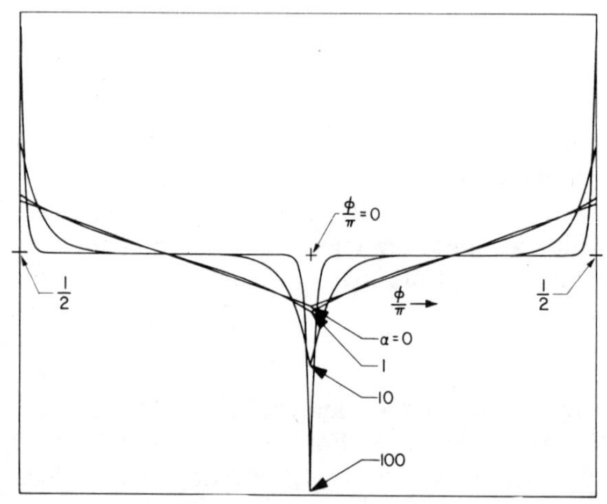

Fig. 12. Optimum reference waveform for various values of α.

$$\lim_{\rho\to\infty} \tau(\phi_l) \cong \frac{1}{2W_L}\left[\frac{1}{\rho}\exp(\rho\phi_l) - \phi_l\right]. \tag{117}$$

Using the linear tracking theory for an $(N+1)$-order loop it is easy to show that [11]

$$\sum_{k=1}^{N} E(y_k, \bar{t}|\phi) = AKF_0\left(1 - \frac{AKF_0}{2W_L}\right)\phi. \tag{118}$$

Employing a procedure similar to that used by Shaft [24] we have, within the approximation, that the optimum nonlinearity is again sgn $[\sin\phi]$ for $|\phi| \leq \phi_l$. The mean time to first loss of sync may be computed from (100). Omitting the details, we have, for $\Omega_0 = 0$, that

$$\tau(\phi_l) = \left(\frac{r+1}{r}\right)^2 \frac{\rho}{2W_L}\int_0^{\phi_l}\int_{\phi}^{\phi_l} \exp\left[\left(\frac{r+1}{r}\right)\rho(|\phi| - |x|) + \frac{\rho}{2r}(\phi^2 - x^2)\right] d\phi dx. \tag{119}$$

In the limit as r approaches infinity (119) reduces to (115) as it should. Numerical results comparing the performance of various trackers have been made in Fig. 11.

XV. The Optimum Reference Signal for a Square-Wave Input

Since a square-wave cross-correlation function is difficult to mechanize, there are various other approaches which can be used to vary $R_{rs}(\phi)$ which are interesting. For example, for a first-order loop [25], Layland constrained the signal $s[t, \theta(t)]$ to be a square wave (the case of digital communications) and maximized $p(0)$ with respect to $r(t)$ under a unit power constraint. If the same methodology is applied here for the optimum second-order tracker, then it is easy to show that [25] the optimum control signal is given by

$$\begin{aligned} r(\tau) = D'\Bigg\{&\int_{\tau}^{\pi/2} (\phi - \tau)w(\phi)d\phi + \left(\frac{\pi}{2} - \tau\right)w(\pi/2) \\ &\int_0^{\pi/2} \exp\left[-\frac{2\alpha}{\pi}\int_0^{\pi/2}(\pi/2 - \max\{\zeta,\eta\})r(\eta)\right]d\eta d\zeta \\ &+ w(\pi/2)\int_0^{\pi/2}(\pi/2 - \max\{\zeta,\eta\}) \\ &\cdot \exp\left[-\frac{2\alpha}{\pi}\int_0^{\pi/2}[\pi/2 - \max(\zeta,\eta)]r(\eta)d\eta\right]d\zeta\Bigg\} \end{aligned} \tag{120}$$

where D' is a normalizing constant for $\tau\varepsilon[0, \pi/2]$ and

$$w(\phi) = \exp\left[-\frac{2\alpha}{\pi}\int_0^{\phi}(\phi - \eta)r(\eta)d\eta\right]$$

which is independent of Ω_0 for the second-order tracker designed such that $\beta \approx 0$. (Note that $r(\tau)$ is implicitly a function of itself through $w(\phi)$.) Even though (120) is recursive and does not admit to a solution in closed form, it can be evaluated by employing an iterative numerical procedure on a digital computer. It is readily shown that the variance of the phase error becomes inversely proportional to $\sqrt{\alpha^{-3/2}}$ for large signal-to-noise ratios. Fig. 12 illustrates $r(\tau)$ for various values of α. In the limit as α approaches infinity, $r(\tau)$ becomes a delta function train and $R_{rs}(\phi)$ is given by (109).

XVI. Conclusions

The loop filter defined in (2) admits only those filters which have poles along the real axis in the left half plane provided the F_k's are less than one and the τ_k's are positive and real. If one allows for complex τ_k's and F_k's in (2) then a broader class of filter functions is admitted; however, the differential equations for the y_k's become complex and occur in conjugate pairs. On the basis of (3) the solution for $p(\phi)$ or $Q_n(\phi)$ is real since the sum of the conditional expectations $E(y_k|\phi)$ is real; however, the solutions for $p(y_k)$ or $Q_n(y_k)$ are no longer valid since they become complex and occur in conjugate pairs. It is possible to return to (6) and produce a new vector Markov process with real components by adding and subtracting the conjugate pairs of the differential equations for the y_k's. The solution to the revised F–P equation will then render solutions for the marginal densities which are real.

It is more interesting, although more difficult, to consider the general loop filter $F(s)=F_1+F'(s)$ where

$$F'(s) = \frac{N(s)}{D(s)} = \frac{\sum_{k=1}^{N} p_k s^{k-1}}{\sum_{k=1}^{N+1} q_k s^{k-1}} \tag{121}$$

where the q_n's and p_n's are real.

If one defines the vector Markov process

$$\begin{aligned} x_1 &= -\frac{K[Ag(\phi) + n(t)]}{D(s)} \\ x_2 &= \dot{x}_1 \\ &\vdots \\ x_N &= \dot{x}_{N-1} \end{aligned}$$

then from (1) we have

$$\begin{aligned} \dot{\phi}(t) &= \Omega_0 - KF_0[Ag(\phi) + n(t)] + \sum_{k=1}^{N} p_k x_k \\ \dot{x}_1 &= x_2 \\ \dot{x}_2 &= x_3 \\ &\vdots \\ \dot{x}_N &= -\frac{1}{q_{N+1}} \sum_{k=1}^{N} q_k x_k - \frac{K}{q_{N+1}}[Ag(\phi) + n(t)] \end{aligned} \tag{122}$$

where $\mathbf{x}=(\phi, x_1, \cdots, x_N)$ is a vector Markov process. The F–P equation for $Q(\mathbf{x}, t)$ or $p(\mathbf{x}; t)$ is easily set up and may be integrated, as before, over the projections of $\mathbf{x}_0'$ giving

$$\frac{\partial H}{\partial t} + \frac{K_{00}}{2} \frac{\partial}{\partial \phi}\left[h(\phi, t) - \frac{\partial}{\partial \phi}\right] H = 0 \tag{123}$$

when the boundary conditions are applied. Here $H=p(\phi; t)$ or $Q(\phi, t)$. For $N=1$, the above equation agrees with (23) and (92); however, for $N>1$ the expectations $E(x_k, t|\phi)$ are, in general, different from those defined in (25). The function $h_0(\phi, t)$ is given by

$$h(\phi, t) = \frac{2}{K_{00}}\left[\Omega_0 - KF_0 Ag(\phi) + \sum_{k=1}^{N} p_k E(x_k, t|\phi)\right]. \tag{124}$$

Acknowledgment

The author wishes to thank Drs. M. K. Simon and R. C. Tausworthe of the Jet Propulsion Laboratory and J. R. La Frieda of the University of Southern California for careful reading of the manuscript. Thanks are also due to Prof. C. L. Weber and L. R. Welch of the University of Southern California for offering various suggestions. Finally, the author thanks Dr. J. Layland of the Jet Propulsion Laboratory for his helpful comments and the use of Fig. 12.

References

[1] A. J. Viterbi, *Principles of Coherent Communication.* New York: McGraw-Hill, 1966, ch. 4.
[2] R. C. Tausworthe, "Theory and practical design of phase-locked receivers," Jet Propulsion Lab., Pasadena, Calif., TR 32-819, February 15, 1966.
[3] J. A. Develet, Jr., "A threshold criterion for phase-lock demodulation," *Proc. IEEE*, vol. 51, pp. 349–356, February 1963.
[4] R. C. Tausworthe, "A method for calculating phase-locked loop performance near threshold," *IEEE Trans. Communication Technology*, vol. COM-15, pp. 502–506, August 1967.
[5] H. L. Van Trees, "Functional techniques for the analysis of the nonlinear behavior of phase-locked loops," *Proc. IEEE*, vol. 52, pp. 894–911, August 1964.
[6] R. L. Stratonovich, *Topics in the Theory of Random Noise*, vol. 1. New York: Gordon and Breach, 1963.
[7] V. L. Tikhonov, "The effects of noise on phase-lock oscillation operation," *Automatika i Telemekhanika*, vol. 22, no. 9, 1959.
[8] A. J. Viterbi, "Phase-locked loop dynamics in the presence of noise by Fokker–Planck techniques," *Proc. IEEE*, vol. 51, pp. 1737–1753, December 1963.
[9] V. I. Tikhonov, "Phase-lock automatic frequency control application in the presence of noise," *Automatika i Telemekhanika*, vol. 23, 1960.
[10] F. J. Charles and W. C. Lindsey, "Some analytical and experimental phase-locked loop results for low signal-to-noise ratios," *Proc. IEEE*, vol. 54, pp. 1152–1166, September 1966.
[11] R. C. Tausworthe, "Cycle slipping in phase-locked loops," *IEEE Trans. Communication Technology*, vol. COM-15, pp. 417–421, June 1967.
[12] J. J. Spilker, "Delay-lock tracking of binary signals," *IEEE Trans. Space Electronics and Telemetry*, vol. SET-9, pp. 1–8, March 1963.
[13] J. C. Lindenlaub and J. J. Urhan, "Threshold study of phase-lock loop systems," Purdue University, Electronics Research Lab., Lafayette, Ind., Tech. Rept. TR-EE 66-19, December 1966.
[14] J. K. Holmes, "Stationary phase distribution for the second-order loop," Jet Propulsion Lab., Pasadena, Calif., SPS 37-49, vol. 3, 1967.
[15] R. C. Tausworthe and D. Sanger, "Experimental study of the first-slip statistics of the second-order phase-locked loop," Jet Propulsion Laboratory, Pasadena, Calif., SPS 37-43, vol. 3, pp. 76–80, 1967.
[16] R. W. Sanneman and J. R. Rowbotham, "Random characteristics of the type II phase-locked loop," *IEEE Trans. Aerospace and Electronic Systems*, vol. AES-3, pp. 604–612, July 1967.
[17] W. C. Lindsey, "Nonlinear analysis and synthesis of generalized tracking systems," pt. I and pt. II, University of Southern California, Dept. of Elec. Engrg., Los Angeles, Calif., Tech. Repts. 317 and 342, December 1968.
[18] S. P. Morgan, "Tables of Bessel functions of imaginary order and imaginary argument," California Institute of Technology, Pasadena, Calif., 1947.
[19] N. Wax, *Selected Papers on Noise and Stochastic Processes.* New York: Dover, 1954. pp. 65–70.
[20] R. C. Buck, *Advanced Calculus.* New York: McGraw-Hill, 1956, p. 58.
[21] D. A. Darling and A. J. F. Siegert, "The first passage time problem for a continuous Markov," *Am. Math. Stat.*, vol. 24, pp. 624–639, 1953.
[22] A. J. F. Siegert, "On the first passage time probability problem," *Phys. Rev.*, vol. 81, pp. 617–623, February 1951.
[23] J. J. Stiffler, "On the selection of signals for phase-locked loops," *IEEE Trans. Communication Technology*, vol. COM-16, pp. 239–244, April 1968.
[24] P. Shaft, "Optimum design of the nonlinearity in signal tracking loops," Stanford Research Institute, Menlo Park, Calif., May 1968.
[25] J. Layland, "On the optimum correlation function for a first-order tracking loop," Jet Propulsion Lab., Pasadena, Calif., SPS 37-50, vol. 4, pp. 284–286, April 1968.

Nonlinear Analysis of Correlative Tracking Systems Using Renewal Process Theory

HEINRICH MEYR

***Abstract*—A new method is presented which describes the behavior of an $(N + 1)$th-order tacking system in which the nonlinearity is either periodic [phase-locked loop (PLL) type] or a nonperiodic [delay-locked loop (DLL) type]. The cycle slipping of such systems is modeled by means of renewal Markov processes. A fundamental relation between the probability density function (pdf) of the single process and the renewal process is derived which holds in the transient as well as in the stationary state. Based on this relation it is shown that the stationary pdf, the mean time between two cycle slips, and the average number of cycles to the right (left) can be obtained by solving a single Fokker–Planck equation of the renewal process.**

The method is applied to the special case of a PLL and compared with the so-called periodic-extension (PE) approach. It is shown that the pdf obtained via the renewal-process approach can be reduced to agree with the PE solution for the first-order loop in the steady state only. The reasoning and its implications are discussed. In fact, it is shown that the approach based upon renewal-process theory yields more information about the system's behavior than does the PE solution.

INTRODUCTION

CORRELATIVE tracking systems, such as phase-locked loops (PLL's), Costas loops, and delay-locked loops (DLL's) have attracted much interest among researchers in the field of telecommunication and synchronization theory and are widely used in practice. A recently published bibliography [1] cites more than 800 papers and at present others are in the offing. Even though much has been published regarding their behavior and performance [2], several important problems have not been addressed nor even formulated. New applications arise, for example, in modern mass transportation systems which require precise velocity and distance measurement systems. One proposed solution to the problem [3]–[6] uses a DLL system to estimate the time difference between two versions of the same stochastic signal. It can be shown that this time difference is inversely proportional to the velocity. In the analysis of such a system [6], the intrinsic noise [7], [8] has been neglected for most communications applications; however, in mass transportation systems, it is of central importance and determines the limit of usefulness of such systems.

This paper consists of three sections. In Section I, background material and motivation for the problems to be treated are given. In Section II, a new method for the analysis of correlative tracking systems is presented. This method is based on the theory of renewal Markov processes and can be applied to systems with periodic nonlinearities as well as to systems with nonperiodic nonlinearities. It is shown that the stationary probability density function (pdf), the mean time between two cycles $E(\tau_L)$ of the phase process, and the average number of

Paper approved by the Associate Editor for Communication Theory of the IEEE Communications Society for publication after presentation at the Seventh Hawaii International Conference on System Science, 1974. Manuscript received June 16, 1974. This work was supported in part by the Office of Naval Research under Contract N-00014-67-A-0269-0022. The method described in this paper was originally developed for S-type loops in the author's Ph.D. dissertation, submitted to the Swiss Federal Institute of Technology, Zurich, Switzerland.

The author is with Hasler AG Bern, Bern, Switzerland.

Reprinted from *IEEE Trans. Commun.*, vol. COM-23, pp. 192–203, Feb. 1975.

cycles to the right (left) N_+,N_- for an $(N + 1)$th-order loop can be obtained by solving *one* Fokker–Planck (F–P) equation of a renewal process. This approach is contrasted with those employed earlier which did not use the renewal theory.

In Section III the theory developed in Section II is applied to PLL's. Up to now the stationary pdf of a PLL was obtained by the so-called periodic extension (PE) method, see for example [2], [9]. In order to obtain $E(\tau_L)$ the F–P equation with other boundary conditions had to be solved; the average number of cycles to the right (left) could be obtained exactly for a first-order system [2, ch.9]. We show here that $E(\tau_L),N_+,N_-$ are closely related to the probability currents first introduced by Lindsey [10] for a loop of arbitrary order. This paper closes with a comparison of the so-called PE solution and the "renewal process approach." It will be shown that the PE solution is identical to the renewal process solution only for a first-order loop. The reasoning and its implications are discussed in detail.

I. BACKGROUND AND PROBLEM CHARACTERIZATION

It is well known that the transition pdf of a correlative tracking system tends to zero in the steady state since the pdf has diffused over the entire interval $[-\infty,\infty]$. Thus, in the steady state the pdf possesses infinite variance. Since the steady-state behavior of the loop is of most practical interest, we must determine the transition pdf of an equivalent phase-error process that retains all the information of the statistical behavior yet yields a meaningful solution in the steady state. For a PLL, the infinite variance is due to the fact that the loop slips cycles. In the case of an S-type loop [defined as a tracking loop with nonperiodic nonlinearity [6] or with a tracking range much smaller than the period of the nonlinearity (DLL)], there always exists a probability that the loop falls out of lock, i.e., the error ϕ exceeds certain prescribed limits. In practice, therefore, an independent circuit (lock detector) supervises the loop and reinitializes acquisition as soon as the phase error exceeds a certain amount. Both phenomena, the cycle slipping of a PLL and the repeated acquisition of an S-type loop, can be conveniently modeled by regenerative (renewal) Markov processes.

Fig. 1 illustrates a typical trajectory of the phase-error process of a PLL, while Fig. 2 illustrates a plot of the trajectory reduced modulo 2π. The phase jumps for $\pm 2\pi$ whenever the trajectory hits one of the two barriers $\phi = (2n \pm 1)\pi$. Note that the number of phase jumps of the reduced modulo 2π process is *not* identical to the number of cycle slips the loop undergoes. Any crossing of the barriers at $(2n \pm 1)\pi$ causes a phase jump independent of whether or not the trajectory reaches either of the two barriers $\phi = (2n \pm 2)\pi$. (By definition, a loop has slipped one cycle when the phase error passes from ϕ_0 to $\phi_0 \pm 2\pi$.) A more appropriate model of the phase-error process for our purposes is the following.

At time $t = 0$ we start a random vector process

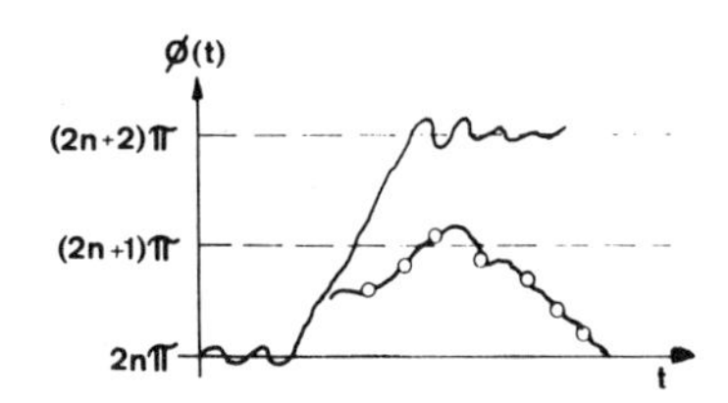

Fig. 1. Typical trajectories of the phase error.

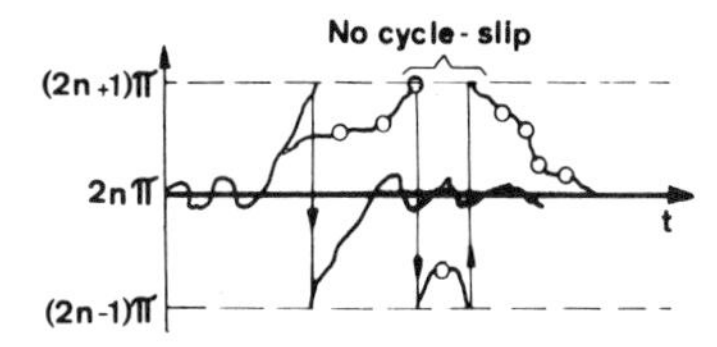

Fig. 2. Same trajectories reduced modulo-2π about the axes $(2n + 1)\pi$ and $(2n - 1)\pi$. Note the phase jumps of 2π.

$\boldsymbol{y} \triangleq [\phi,y_1,\cdots,y_n]$ in the hyperplane $\phi = \phi_0$ of the $(n + 1)$-dimensional space. The initial distribution of the random vector $\boldsymbol{y}_0 \triangleq [y_1,\cdots,y_n]$ is given by

$$q(\boldsymbol{y}_0\, t; = 0 \mid \phi = \phi_0; t = 0), \qquad -\pi \le \phi_0, \le \pi, \text{ arbitrary}.$$

As soon as the sample trajectory reaches one of the hyperplanes $\phi = \phi_0 \pm 2\pi$ for the first time it is absorbed. Immediately thereafter the trajectory jumps back to ϕ_0, thus making a phase jump of $\pm 2\pi$ and a new sample trajectory is started with the same initial distribution as at $t = 0$. Note that the interval is $[\phi_0 - 2\pi,\phi_0 + 2\pi]$ and therefore the number of phase jumps from the right (left) barrier to ϕ_0 equals exactly the number of cycles the loop slips to the right (left). See Fig. 3. Furthermore, because of the 2π periodicity of the PLL nonlinearity, the planes ϕ_0 and $\phi_0 \pm 2\pi$ are identical, provided the vector random process $\boldsymbol{y}$ is approximately Markovian and therefore the trajectory jumps for an amount of $\pm 2\pi$ in infinitely short time. This model exactly describes the behavior of a first-order PLL where $\boldsymbol{y}(t)$ is a scalar Markov process $\phi(t)$ with initial distribution

$$q(\phi; t = 0) = \delta(\phi - \phi_0).$$

For an $(N + 1)$th-order loop it has been noted [16], [17] that for low SNR the cycle slips occur in bursts and consequently the initial distribution is not the same for all single processes. However, for loop SNR's of practical interest, the mean time between two bursts is much longer than the mean burst length. Also, in [17] it is reported that for a Costas loop, multiple phase slips become unlikely for signal to noise that would be typically used in communications. Therefore the model introduced is well applicable.

The behavior of an S-type loop (DLL) is more complicated. The trajectory starts at $\phi = \phi_0$ at $t = 0$.[1] Due to the statistical nature of the supervising circuit (lock detector), which basically estimates the autocorrelation function, there exists for every ϕ a probability $(1 - S(\phi))$

[1] Assuming the system being in-lock, $\phi_{\min} \le \phi_0 \le \phi_{\max}$ where $\phi_{\min}$, $\phi_{\max}$ are two arbitrary limits defining the "in-lock" region.

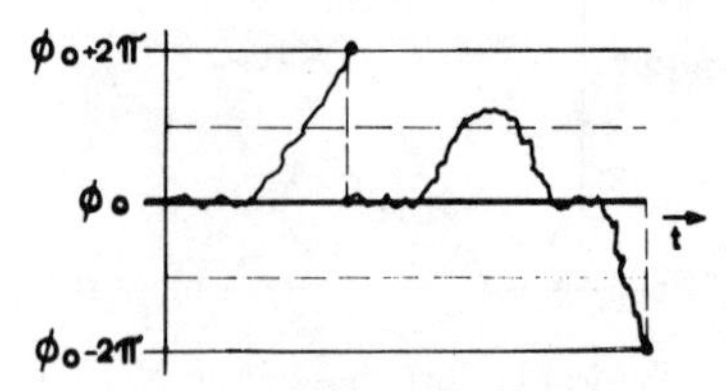

Fig. 3. Sample trajectory of the process defined in the 4π interval $[\phi_0 - 2\pi, \phi_0 + 2\pi]$.

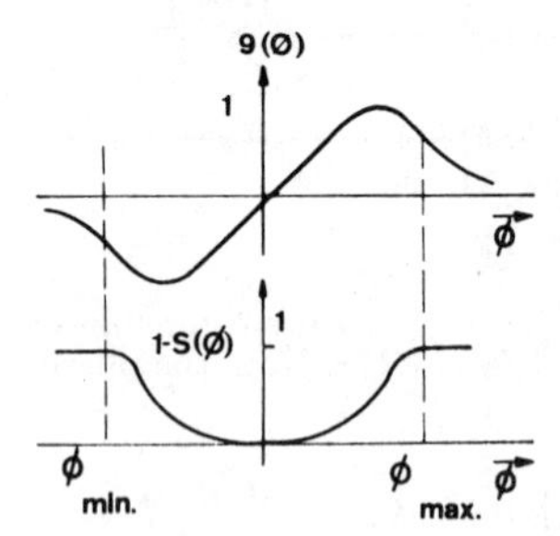

Fig. 4. Nonlinearity $g(\phi)$ and probability $(1 - S(\phi))$. $1 - S(\phi)$: probability that the lock detector gives a signal of being "out of lock" under the condition of ϕ being within range $[\phi,\phi + d\phi]$ at time t or equivalently it represents the probability that the sample trajectory is absorbed at time t in $[\phi,\phi + d\phi]$.

that the process is absorbed in the hyperslab $[\phi,\phi + d\phi]$. See Fig. 4. Since the mean time between two cycles for a reasonable loop is much greater than the reacquisition time, we assume that a new process is created immediately after the absorption of its predecessor at $\phi = \phi_0$, etc. (The extension to finite exponentially distributed acquisition times and an arbitrary initial distribution for ϕ_0 is straightforward.) The typical behavior of the phase error of the vector Markov process of an S-type loop is illustrated by Fig. 5. More precisely, we define the following for the kth process.

Initial distribution for the kth process:

$$\left.\begin{aligned} &q\Big[\phi; t = \sum_{i=0}^{k-1} \tau_{L,i}\Big] = \delta(\phi - \phi_0) \\ &q\Big[\mathbf{y}_0; t = \sum_{i=0}^{k-1} \tau_{L,i} \mid \phi = \phi_0; t = \sum_{i=0}^{k-1} \tau_{L,i}\Big] \end{aligned}\right\} \quad k \geq 1 \qquad (1)$$

where $\tau_{L,i}$ is the lifetime of the ith process ($\tau_{L,0} = 0$), and $\sum_{i=0}^{k-1} \tau_{L,i}$ is the starting point in time of the kth process.

II. STATIONARY PDF, MEAN TIME BETWEEN CYCLES, AND AVERAGE NUMBER OF CYCLES

A. Relation between the pdf's of the Single and Renewal Process

We assume the vector Markov processes defining the repeated process to be temporally homogeneous, i.e.,

$$q[\mathbf{y}; t - t_1 \mid \mathbf{y} = \mathbf{y}_1; t = t_1] = q[\mathbf{y}; t \mid \mathbf{y} = \mathbf{y}_1; t = 0] \qquad (2)$$

and to have the same initial distribution as in (1). At time t we define for the ith process the following events on a given sample space Ω:

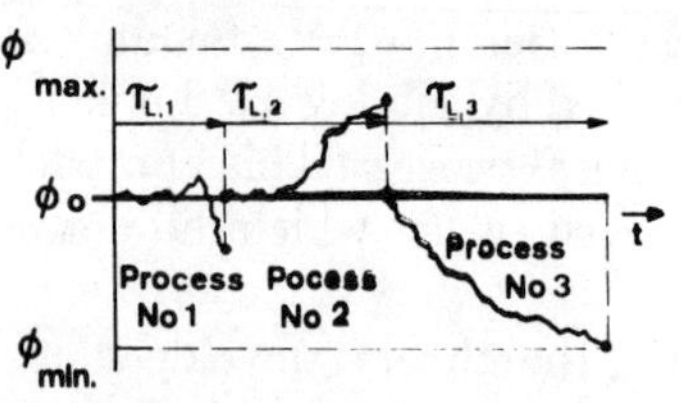

Fig. 5. Trajectory of the renewal process for an S-type loop.

α_i the sample trajectory of the ith process is within $[\phi,\phi + d\phi;\mathbf{y}_0,\mathbf{y}_0 + d\mathbf{y}_0]$ at time t

β_i the ith process is started in the time interval $[\tau,\tau + d\tau], \tau \epsilon [0,\infty]$

γ_i the sample trajectory of the ith process is absorbed at time t in the open interval $(\phi_{\min},\phi_{\max})$. (3)

These three events can be used to form both conditioned and (or) combined events:

$\alpha_i \cdot \beta_i$ the sample trajectory of the ith process is at time t in $[\phi,\phi + d\phi;\mathbf{y}_0,\mathbf{y}_0 + d\mathbf{y}_0]$ *and* was started in the time interval $[\tau, \tau + d\tau]$

$\alpha_i \cdot \bar{\gamma}_i$ the sample trajectory of the ith process is at time t in $[\phi,\phi + d\phi;\mathbf{y}_0,\mathbf{y}_0 + d\mathbf{y}_0]$ *and* is not absorbed in the interval $(\phi_{\min},\phi_{\max})$

$\alpha_i \mid \beta_i$ the sample trajectory of the ith process is at time t in $[\phi, \phi + d\phi;\mathbf{y}_0,\mathbf{y}_0 + d\mathbf{y}_0]$ under the *condition* that it was started in $[\tau, \tau + d\tau]$

$\bar{\gamma}_i \mid \alpha_i$ the sample trajectory of the ith process is not stopped at time t under the condition of being within $[\phi,\phi + d\phi;\mathbf{y}_0,\mathbf{y}_0 + d\mathbf{y}_0]$ where $\phi \epsilon (\phi_{\min}, \phi_{\max})$. (4)

Assigning probabilities to these events we arrive at[2,3]

$$P(\alpha_i \mid \beta_i) = \begin{cases} 0, & \text{for } t < \tau \\ q(\phi,\mathbf{y}_0;t - \tau) \cdot dV & \text{for } t \geq \tau \end{cases} \qquad dV = d\phi\, dy_1 \cdots dy_n$$

$$P(\bar{\gamma}_i \mid \alpha_i) = S(\phi)$$
$$P(\beta_i) = f_i(\tau) \cdot d\tau. \qquad (5)$$

For the definition of $P(\alpha_i \mid \beta_i)$ we have made use of the assumption that the process is temporally homogeneous. According to Bayes rule we can write

$$\begin{aligned} P(\alpha_i \cdot \beta_i) &= P(\alpha_i \mid \beta_i) \cdot P(\beta_i) \\ &= q(\phi,\mathbf{y}_0;t - \tau) \cdot f_i(\tau) \cdot d\tau \cdot dV. \end{aligned} \qquad (6)$$

By integrating over τ we arrive at

$$P(\alpha_i) = dV \cdot \int_0^t q(\phi,\mathbf{y}_0;t - \tau) \cdot f_i(\tau) \cdot d\tau. \qquad (7)$$

[2] These definitions hold for sufficiently small dV and $d\tau$. The treatment could be made more rigorous by using measure theory; however, none of these advanced tools will be needed.

[3] We assume that $S(\phi)$ vanishes outside an arbitrary, but finite, interval $(\phi_{\min}, \phi_{\max})$. With Pr 1 there exist no trajectories outside this interval.

Similar reasoning leads for $P(\alpha_i \cdot \bar{\gamma}_i)$ to

$$P(\alpha_i \cdot \bar{\gamma}_i) = P(\bar{\gamma}_i \mid \alpha_i) \cdot P(\alpha_i) = S(\phi) \cdot dV \int_0^t q(\phi, y_0; t - \tau) \cdot f_i(\tau)\, d\tau \tag{8}$$

and

$$P(\alpha_i \cdot \gamma_i) = [1 - S(\phi)] \cdot dV \int_0^t q(\phi, y_0; t - \tau) \cdot f_i(\tau)\, d\tau. \tag{9}$$

The sample space for the random variable γ_i contains only two points. Therefore

$$P(\alpha_i \cdot \bar{\gamma}_i) + P(\alpha_i \cdot \gamma_i) = P(\alpha_i). \tag{10}$$

The probability that at time t any trajectory is within $[\phi, \phi + d\phi; y_0; y_0 + dy_0]$ and is not absorbed is given by

$$P[\bigcup_{i=1}^{\infty} \alpha_i \cdot \bar{\gamma}_i] = S(\phi)\, dV \sum_{i=1}^{\infty} \int_0^t q(\phi, y_0; t - \tau) \cdot f_i(\tau)\, d\tau. \tag{11}$$

In arriving at (11) we have made use of the fact that the intersection of any two events $(\alpha_i \cdot \bar{\gamma}_i)$ and $(\alpha_m \cdot \bar{\gamma}_m)$, $i \neq m$, is empty. By defining

$$p(\phi, y_0; t)\, dV \triangleq P(\bigcup_{i=1}^{\infty} \alpha_i \bar{\gamma}_i), \tag{12}$$

the density function for the process with back jumps becomes

$$p(\phi, y_0; t) = S(\phi) \cdot \sum_{i=1}^{\infty} \int_0^t q(\phi, y_0; t - \tau) \cdot f_i(\tau)\, d\tau. \tag{13}$$

The starting point of the ith sample trajectory can be expressed by the sum of the lifetimes of the preceding $(i - 1)$ trajectories:

$$\tau = \tau_{L,1} + \cdots + \tau_{L,i-1} = \sum_{n=1}^{i-1} \tau_{L,n}; \qquad i \geq 2. \tag{14}$$

For any i, τ is the sum of $(i - 1)$ independent identically distributed random variables. (According to Feller a process which satisfies the property (14) is called a renewal process [11, p. 181].) Using a well known theorem from probability theory we get

$$f_i(\tau) = \underbrace{d_L(\tau) * d_L(\tau) * \cdots * d_L(\tau)}_{i - 1 \text{ fold convolution}}$$

$$f_1(\tau) = \delta(\tau) \tag{15}$$

where $d_L(\tau)$ is the density function of the life time $\tau_{L,i}$ of the individual processes defining the renewal process. Since all single processes are statistically equivalent we have dropped the index i in $d_L(\tau)$.

By Laplace transforming (13) we get

$$\int_0^{\infty} \exp(-st) \cdot p(\phi, y_0; t)\, dt = \int_0^{\infty} \exp(-st) \cdot \left[\sum_{i=1}^{\infty} S(\phi) \cdot \int_0^t q(\phi, y_0; t - \tau) \cdot f_i(\tau)\, d\tau\right] dt. \tag{16}$$

Assuming the validity of interchanging the order of summation and integration, the convolutional theorem of Laplace transform yields

$$p(\phi, y_0; s) = S(\phi) \cdot \sum_{i=1}^{\infty} q(\phi, y_0; s) \cdot f_i(s) \tag{17}$$

where $p(\phi, y_0; s)$ denotes the Laplace transform of $p(\phi, y_0; t)$. If we substitute in the integral

$$\int_0^{\infty} \exp(-st) f_i(t)\, dt \tag{18}$$

the density function $f_i(\tau)$ by the $(i - 1)$-fold convolution of (15), we find by repeated application of the Laplace convolution theorem

$$\int_0^{\infty} \exp(-st) \cdot f_i(t)\, dt = \left[\int_0^{\infty} \exp(-s\tau_L) \cdot d_L(\tau_L) \cdot d\tau_L\right]^{i-1} \tag{19[4]}$$

with

$$d_L(s) \triangleq \int_0^{\infty} \exp(-s\tau_L) \cdot d_L(\tau_L) \cdot d\tau_L$$

$$q(\phi, y_0; s) \triangleq \int_0^{\infty} \exp(-st) \cdot q(\phi, y_0; t)\, dt$$

$$p(\phi, y_0; s) \triangleq \int_0^{\infty} \exp(-st) \cdot p(\phi, y_0; t)\, dt. \tag{20}$$

The density functions of the single and renewal process are then related by

$$p(\phi, y_0; s) = S(\phi) \cdot q(\phi, y_0; s) \cdot [\sum_{i=2}^{\infty} d_L(s)^{i-1} + 1] = S(\phi) \cdot q(\phi, y_0; s) \cdot \frac{1}{1 - d_L(s)}. \tag{21}$$

Using the general definition of expectation

$$E[m(x)] \triangleq \int_{-\infty}^{\infty} m(x) \cdot p(x)\, dx$$

we can write

$$d_L(s) = \int_0^{\infty} \exp(-s\tau_L) \cdot d_L(\tau_L) \cdot s\tau_L = E[\exp(-s\tau_L)] \tag{22}$$

[4] For $i = 1$ we have $f_1(\tau) = \delta(\tau) \Rightarrow f_1(s) = 1$.

to arrive at our final result

$$p(\phi,y_0;s) = S(\phi)\cdot q(\phi,y_0;s)\cdot\left[\frac{1}{1-E[\exp(-s\tau_L)]}\right]. \quad (23)$$

Equation (23) represents the *fundamental relation* between the density functions of the single and repeated process in the transient as well as in the stationary state.

B. The F–P Equation of the Single Process

As long as the sample trajectory of the ith process has not been absorbed, its transition pdf obeys a F–P equation. Since all single processes, $i = 1,\cdots,\infty$, are assumed to be statistically identical, it is sufficient to study the one starting at $t = 0$ whose transition pdf obeys

$$\frac{\partial}{\partial t}q(\phi,y_0;t) + Lq(\phi,y_0;t) = 0$$

with

$$L \triangleq \sum_{k=0}^{N}\frac{\partial}{\partial y_k}K_k(y) - \tfrac{1}{2}\sum_{k=0}^{N}\sum_{j=0}^{N}\frac{\partial^2}{\partial y_k\partial y_j}K_{k,j}(y). \quad (24)$$

L is a spatial operator (forward diffusion operator) defined on some domain D with time-independent intensity coefficients in order to fulfill the temporal homogeneity of (1). The initial distribution is given by

$$q(\phi,y_0;t=0) = q(y_0;t=0\mid\phi;t=0)\cdot\delta(\phi-\phi_0). \quad (25)$$

Since the y_k's can take on all possible values from $-\infty$ to ∞, the boundary conditions for these random variables (rv's) take the form of conditions at $y_k = \pm\infty$. Due to the fact that

$$\int_D q(\phi,y_0;t)\,dV \le 1, \qquad t\in[0,\infty]$$

must hold, the function $q(\phi,y_0;t)$, viewed as a function of y_k with all other variables kept fixed, must vanish for $|y_k|\to\infty$. It will even be necessary that the stronger conditions

$$K_k(\phi,y_0)\cdot q(\phi,y_0;t)\,|_{y_k=\pm\infty} = 0, \qquad k = 1,\cdots,n \quad (26)$$

and

$$\frac{\partial}{\partial y_k}q(\phi,y_0;t)\,|_{y_k=\pm\infty} = 0 \quad (27)^5$$

be satisfied. According to the construction of the process and the definition of $S(\phi)$ we have the two additional boundary conditions for ϕ

$$q(\phi=\phi_{\min},y_0;t) = q(\phi=\phi_{\max},y_0;t) = 0. \quad (28)$$

The boundary conditions (28) are of the absorbing type. This means that the sample trajectories are stopped (absorbed) when they first reach one of the barriers located at $\phi = \phi_{\min},\phi_{\max}$ [11, ch. X.5], provided they arrive there and have not been absorbed inside the hyperslab $(\phi_{\min},\phi_{\max})$.

C. Differential Equation for the Stationary Distribution of the Renewal Process

We now develop the partial differential equation for the transition pdf $p(\phi,y_0;t)$ of the renewal process. Taking the Laplace transform of (24) we get

$$s\cdot q(\phi,y_0;s) - q(\phi,y_0;t=0) + Lq(\phi,y_0;s) = 0. \quad (29)$$

Replacing $q(\phi,y_0;s)$ by $p(\phi,y_0;s)$ using the result of (23) gives

$$L\frac{1}{S(\phi)}p(\phi,y_0;s) + s\frac{1}{S(\phi)}p(\phi,y_0;s) = \frac{q(\phi,y_0;t=0)}{1-E[\exp(-s\tau_L)]}. \quad (30)$$

To determine the stationary solution, assuming it exists, we make use of the final value theorem of Laplace transform which states

$$\lim_{t\to\infty}F(t) = \lim_{s\to0}(s\cdot F(s)),$$

therefore

$$\lim_{s\to0}\left\{L\frac{1}{S(\phi)}sp(\phi,y_0;s) + s^2(1/S(\phi))p(\phi,y_0;s)\right\} = \lim_{s\to0}\left[\frac{sq(\phi,y_0;t=0)}{1-E[\exp(-s\tau_L)]}\right]. \quad (31)$$

By expanding the term $\exp(-s\tau_L)$ in the right-hand side of (31) in a Taylor series

$$E[\exp(-s\tau_L)] = E\left[1 - s\tau_L + \frac{(s\tau_L)^2}{2!} + \cdots + \frac{(s\tau_L)^2}{n!} + \cdots\right], \quad (32)$$

the limit can easily be calculated

$$\lim_{s\to0}\left[\frac{s\cdot q(\phi,y_0;t=0)}{1-E[\exp(-s\tau_L)]}\right] = \lim_{s\to0}\left[\frac{q(\phi,y_0;t=0)}{1-1+(s/s)E(\tau_L)+\cdots+(s^k/s)[E(\tau_L{}^k)/k!]+\cdots}\right] = \frac{q(\phi,y_0;t=0)}{E(\tau_L)}. \quad (33)$$

By definition $E(\tau_L)$ is the *mean lifetime of a sample trajectory*. For $s\to0$ the left-hand side of (31) tends to

[5] The boundary condition (27) is sufficient for our purposes only if the intensity coefficients $K_{k,j}(\phi,y_0)$ remain bounded over the whole domain of definition of L.

$$\lim_{s\to 0}\left\{L\frac{1}{S(\phi)}(sp(\phi,y_0;s))+s^2\frac{1}{S(\phi)}p(\phi,y_0;s)\right\}$$
$$=L\frac{1}{S(\phi)}p(\phi,y_0;\,t=\infty) \quad (34)$$

and we arrive at our desired result

$$L\frac{1}{S(\phi)}p(\phi,y_0;\,t=\infty)=\frac{q(\phi,y_0;\,t=0)}{E(\tau_L)}. \quad (35)$$

It follows from (23) and (26)–(28) that the boundary conditions for (35) are identical to those of the single process.

It is interesting to note from (35) that $S(\phi)$ can always be absorbed in the spatial differential operator L

$$L\frac{1}{S(\phi)}\Rightarrow L^*.$$

This means that we can find for every renewal process a stochastically equivalent one with

$$S^*(\phi)=\begin{cases}1, & \phi\in(\phi_{\min},\phi_{\max})\\ 0, & \text{elsewhere.}\end{cases} \quad (36)$$

Note that $p(\phi,y_0;\,t=\infty)$ is found by solving a linear, inhomogeneous partial differential equation. Consequently there is no constant left to normalize $p(\phi,y_0;\,t=\infty)$ and we have to prove that $p(\phi,y_0,t=\infty)$ is indeed a pdf. Before we can do this we have to do some preparatory work.

D. The Mean Lifetime of a Sample Trajectory

In this section we wish to compute the mean lifetime of a trajectory. This problem is closely related, but not identical, to the well known "mean time to first passage" problem, see for example [2], [11]–[13].

The probability of finding the sample trajectory that was started at $t=0$ in the volume $[\phi,\phi+d\phi;y_0,y_0+dy_0]$ at time t without being absorbed is [see (8)]

$$P(\alpha_1\cdot\bar{\gamma}_1)=S(\phi)\cdot dV\cdot\int_0^t q(\phi,y_0;\,t-\tau)\cdot\delta(\tau)\,d\tau$$
$$=S(\phi)\cdot q(\phi,y_0;t). \quad (37)$$

The probability that the sample trajectory is anywhere within the domain D (and has not been absorbed) is found by integrating (37) over the whole domain D

$$W(t)=\int_D S(\phi)\cdot q(\phi,y_0;t)\,dV. \quad (38)$$

For $t=0$ we find

$$W(0)=S(\phi_0). \quad (39)$$

Note that $W(t)\le 1$ for $t\ge 0$ and therefore $q(\phi,y_0;t)$ is not a transition pdf because it does not have unit area [2], [9], [11]. The probability that the sample trajectory is stopped during the time interval $[0,t]$ is given by

$$1-W(t).$$

Differentiating with respect to t we obtain the pdf of the lifetime of the sample trajectory [see definition (15)]

$$d_L(t)\triangleq\frac{d}{dt}[1-W(t)]=-\frac{dW(t)}{dt}. \quad (40)$$

The mean lifetime $E(\tau_L)$ is then

$$E(\tau_L)=\int_0^\infty \tau\cdot d_L(\tau)\cdot d\tau=\int_0^\infty \tau\cdot\frac{-dW(\tau)}{d\tau}\cdot d\tau. \quad (41)$$

Integrating (41) by parts gives

$$E(\tau_L)=-W(\tau)\cdot\tau\,|_0^\infty+\int_0^\infty W(\tau)\,d\tau. \quad (42)$$

Under the assumption that $W(\tau)\to 0$, $\tau\to\infty$, we find by using (38)

$$E(\tau_L)=\int_0^\infty\int_D S(\phi)\cdot q(\phi,y_0;\tau)\,dV\,d\tau. \quad (43)$$

If we define [2], [9]

$$Q(\phi,y_0)=\int_0^\infty q(\phi,y_0;t)\,dt \quad (44)$$

and interchange the order of integration in (43) we can write $E(\tau_L)$ as

$$E(\tau_L)=\int_D S(\phi)\cdot Q(\phi,y_0)\,dV. \quad (45)$$

As long as the sample trajectory is not absorbed the behavior of $q(\phi,y_0;t)$ is described by the Fokker–Planck equation

$$\frac{\partial q(\phi,y_0;t)}{\partial t}+Lq(\phi,y_0;t)=0 \quad (46)$$

with the boundary conditions (26)–(28). By integrating (46) with respect to time we get

$$q(\phi,y_0;\,t=\infty)-q(\phi,y_0;\,t=0)+LQ(\phi,y_0)=0. \quad (47)$$

Since sooner or later all sample trajectories reach one of the two barriers located at $\phi=\phi_{\min},\phi_{\max}$ (or are absorbed inside the hyperslab $(\phi_{\min},\phi_{\max})$) the first term of (47) vanishes and there remains

$$LQ(\phi,y_0)=q(\phi,y_0;\,t=0). \quad (48)$$

Again the boundary conditions for (48) are the ones described in (26)–(28). Note the similarity of (48) and (35)!

E. Proof that $p(\phi,y_0;\,t=\infty)$ is Normalized to 1

We claim that the pdf of the renewal process is normalized to one, i.e.,

$$\int_D p(\phi,y_0;\,t=\infty)\,dV=1.$$

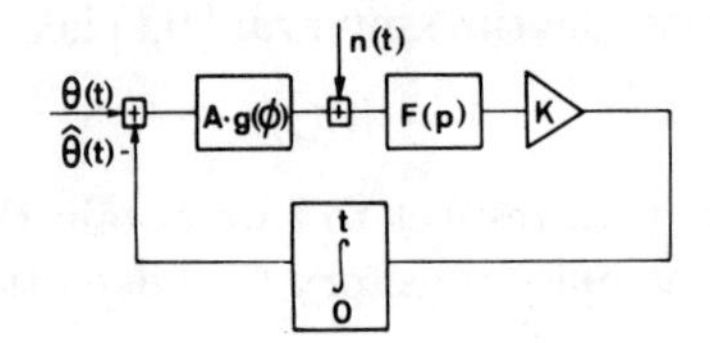

Fig. 6. Equivalent model for the PLL.

Proof: We define

$$Q^*(\phi,\boldsymbol{y}_0) = S(\phi)\cdot Q(\phi,\boldsymbol{y}_0). \tag{49}$$

Then instead of (45) and (48) we have

$$L\frac{1}{S(\phi)}Q^*(\phi,\boldsymbol{y}_0) = q(\phi,\boldsymbol{y}_0;\, t=0) \tag{50}$$

$$E(\tau_L) = \int_D Q^*(\phi,\boldsymbol{y}_0)\, dV. \tag{51}$$

The stationary pdf of the renewal process obeys the F–P equation (35). We compare (50) and (35). Both equations are identical, except for a constant factor $1/E(\tau_L)$ in the right-hand side and obey the same boundary conditions. But

$$\int_D Q^*(\phi,\boldsymbol{y}_0)\, dV = E(\tau_L). \tag{52}$$

Consequently

$$\int_D p(\phi,\boldsymbol{y}_0;t=\infty)\, dV = 1. \qquad \text{Q.E.D.}$$

III. APPLICATION OF THE THEORY TO TRACKING SYSTEMS WITH PERIODIC NONLINEARITY (PLL)

A. Differential Equation and Equivalent Model of the PLL

For the sake of the convenience and completeness, we repeat the well known differential equation of the loop [2, ch. 11]

$$\frac{d\phi}{dt} = -KF(p)[Ag(\phi)+n(t)] + \frac{d\theta}{dt}. \tag{53}$$

The mathematically equivalent model is shown in Fig. 6. Following Lindsey's approach [10] we write the loop filter transfer function in the partial fraction expansion

$$F(s) = F_0 + \sum_{k=1}^{N}\frac{1-F_k}{1-\tau_k s}. \tag{54}$$

For $N = 1$, $F_0 = F_1 = (\tau_2/\tau_1)$ $F(s)$ reduces to the proportional-plus-integral type filter used in second-order loops. If we assume for the moment

$$\theta(t) = \Omega_0 t + \theta_0 \tag{55}$$

we can replace (54) by the equivalent of $(N+1)$ first-order stochastic differential equations, that is

$$\frac{d\phi}{dt} = \Omega_0 - F_0K[Ag(\phi)+n] + \sum_{k=1}^{N} y_k$$

$$\frac{dy_1}{dt} = -\frac{y_1}{\tau_1} - \frac{(1-F_1)K[Ag(\phi)+n]}{\tau_1}$$

$$\vdots$$

$$\frac{dy_N}{dt} = -\frac{y_N}{\tau_N} - \frac{(1-F_N)K[Ag(\phi)+n]}{\tau_N}. \tag{56}$$

It is clear that the coordinates $\phi, y_1, y_2, \cdots, y_N$ form components of a $(N+1)$ vector Markov process if $n(t)$ can be approximated by a white Gaussian noise process [14, theorem 4.5]. The spatial operator L corresponding to (56) is easily found to be [2, p. 354]

$$L = \frac{\partial}{\partial\phi}\Big[\Omega_0 - F_0AKg(\phi) + \sum_{k=1}^{N} y_k\Big] - \frac{F_0^2N_0K^2}{4}\cdot\frac{\partial^2}{\partial\phi^2}$$
$$-\sum_{k=1}^{N}\Big\{\frac{\partial}{\partial y_k}\Big[\frac{y_k}{\tau_k} + \frac{1-F_k}{\tau_k}AKg(\phi)\Big]$$
$$+\frac{(1-F_k)^2N_0K^2}{4\tau_k^2}\cdot\frac{\partial^2}{\partial y_k^2} + \frac{F_0(1-F_k)N_0K^2}{2\tau_k}\frac{\partial^2}{\partial\phi\partial y_k}\Big\}$$
$$-\sum_{k\neq l\neq 0}^{N}\sum^{N}\frac{(1-F_k)(1-F_l)K^2N_0}{4\tau_k\tau_l}\frac{\partial^2}{\partial y_k\partial y_l}. \tag{57}$$

B. The F–P Equation of the Renewal Process

The F–P equation for the renewal process is a special case of the general result (35) with

$$S(\phi) = \begin{cases}1, & \phi \in (\phi_{\min},\phi_{\max})\\ 0, & \text{else}\end{cases} \tag{58}$$

$$\phi_{\min} = \phi_0 - 2\pi, \qquad \phi_{\max} = \phi_0 + 2\pi \tag{59}$$

and the boundary conditions (26)–(28). It is easy to show that L is periodic with a period of 2π and (35) can therefore be stated within any interval $[\phi_0 - 2\pi(1-k);\ \phi_0 + 2\pi(1+k)]$; $k = 0,\pm1,\cdots$.

In the stationary state the sample trajectory jumps back to a point $[\boldsymbol{y}_0;\boldsymbol{y}_0 + d\boldsymbol{y}_0]$ in the hyperplane $\phi = \phi_0$ with probability

$$q(\boldsymbol{y}_0;\, t=0 \mid \phi=\phi_0;\, t=0)\cdot dy_1\cdot dy_2\cdots dy_N. \tag{60}$$

Now if $p(\phi,\boldsymbol{y}_0;\, t=\infty)$ has to be the stationary pdf of an $(N+1)$th-order loop and if the initial distribution for all processes $q(\phi,\boldsymbol{y}_0;t - \sum_{n=0}^{i=1}\tau_{L,n})$, $i = 1,\cdots,\infty$, has to be the same, then the relation

$$q(\boldsymbol{y}_0;\, t=0 \mid \phi;\, t=0) = p(\boldsymbol{y}_0;\, t=\infty \mid \phi;\, t=\infty) \tag{61}$$

must hold and the *initial distribution is completely determined.*

If we integrate both sides of (35) with respect to y_j for all $j \neq k \neq 0$ and make use of the boundary conditions

(26), (27) we arrive at

$$\frac{\partial}{\partial\phi}J_0(\phi,y_k) + \frac{\partial}{\partial y_k}J_k(\phi,y_k) = \frac{p(y_k \mid \phi)\delta(\phi - \phi_0)}{E(\tau_L)}. \quad (62)^6$$

With the probability currents

$$J_0(\phi,y_k) = [K_0(\phi,y_k) - \tfrac{1}{2}K_{0,0}(\partial/\partial\phi) - \tfrac{1}{2}K_{0,k}(\partial/\partial y_k)]\cdot p(\phi,y_k)$$

$$J_k(\phi,y_k) = [K_k(\phi,y_k) - \tfrac{1}{2}K_{k,k}(\partial/\partial y_k) - \tfrac{1}{2}K_{k,0}(\partial/\partial\phi)]\cdot p(\phi,y_k) \quad (63)$$

and

$$K_0(\phi,y_k) = \Omega_0 - F_0AKg(\phi) + y_k + \sum_{l\neq k\neq 0}^{N} E[y_l \mid \phi,y_k]$$

$$K_k(\phi,y_k) = -\frac{1}{\tau_k}[y_k + (1 - F_k)AKg(\phi)]$$

$$K_{0,k} = K_{k,0} = \frac{(1 - F_k)F_0N_0K^2}{2\tau_k}$$

$$K_{k,k} = \frac{(1 - F_k)^2N_0K^2}{2\tau_k^2}$$

$$K_{0,0} = \frac{N_0F_0^2K^2}{2}. \quad (64)$$

To find the marginal pdf $p(y_k)$ or $p(\phi)$, we first need a differential equation whose solution is indeed $p(y_k)$ or $p(\phi)$. We first evaluate the differential equation for $p(\phi)$. This equation is easily found by integrating both sides of (62) over y_k and applying the appropriate boundary conditions. We obtain

$$\frac{d}{d\phi}[K_0(\phi)p(\phi)] - \tfrac{1}{2}K_{0,0}\frac{d^2}{d\phi^2}p(\phi) = \frac{\delta(\phi - \phi_0)}{E(\tau_L)} \quad (65)$$

with

$$K_0(\phi) = \Omega_0 - AKF_0g(\phi) + \sum_{k=1}^{N} E(y_k \mid \phi). \quad (65a)$$

The differential equation for $p(y_k)$ is then found to be

$$\frac{\partial}{\partial y_k}J_k(y_k) + [J_0(\phi_0 + 2\pi,y_k) - J_0(\phi_0 - 2\pi,y_k)] = \frac{p(y_k \mid \phi = \phi_0)}{E(\tau_L)} \quad (66)$$

with

$$J_k(y_k) = \left[K_k(y_k) - \frac{1}{2}\frac{\partial}{\partial y_k}K_{k,k}\right]p(y_k)$$

$$K_k(y_k) = -\frac{y_k + AK(1 - F_k)E[g(\phi) \mid y_k]}{\tau_k}.$$

Note the difference of the current $J_0(\phi,y_k)$ taken at $\phi = \phi_0 \pm 2\pi$ does not vanish, in general. This is in contrast to the PE solution.

C. Stationary Distribution Function $p(\phi)$

If we integrate both sides of (65) over ϕ we arrive at

$$K_0(\phi)p(\phi) - \tfrac{1}{2}K_{0,0}\frac{d}{d\phi}p(\phi) = \frac{1}{E(\tau_L)}[u(\phi - \phi_0) - D_0] \quad (67)$$

where $u(\phi - \phi_0)$ is the unit step function, D_0 is a constant, and

$$K_0(\phi) = \Omega_0 - AKF_0g(\phi) + \sum_{k=1}^{N} E(y_k \mid \phi).$$

The general solution of the linear inhomogeneous differential equation (67) is the sum of homogeneous and particular solution

$$p(\phi) = \frac{1}{K_{0,0}}\exp[-U_0(\phi)]\cdot\left\{D_1 + \frac{2}{E(\tau_L)} \cdot\int_{\phi_0-2\pi}^{\phi}[-u(\nu - \phi_0) + D_0]\exp[U_0(\nu)]\,d\nu\right\} \quad (68)$$

where $U_0(\phi)$ is the potential function [2, ch. 9] given by

$$U_0(\phi) = -\int_0^{\phi}\frac{2K_0(\nu)}{K_{0,0}}\,d\nu. \quad (69)$$

By making use of the boundary conditions

$$p(\phi = \phi_0 - 2\pi) = p(\phi = \phi_0 + 2\pi) = 0 \quad (70)$$

the constants D_1,D_0 are easily found to be

$$D_0 = \int_{\phi_0}^{\phi_0+2\pi}\exp[U_0(\phi)]\,d\phi \Big/ \int_{\phi_0-2\pi}^{\phi_0+2\pi}\exp[U_0(\phi)]\,d\phi,$$

$$D_1 = 0. \quad (71)$$

Equation (68) becomes then

$$p(\phi) = \frac{2}{E(\tau_L)K_{0,0}}\exp[-U_0(\phi)] \cdot\int_{\phi_0-2\pi}^{\phi}[D_0 - u(\nu - \phi_0)]\cdot\exp[U_0(\nu)]\,d\nu. \quad (72)$$

The result found can be given a graphic physical interpretation in terms of the probability current $J_0(\phi)$ which will prove useful. The right-hand side of (67) is by definition equal to the probability current $J_0(\phi)$,

$$J_0(\phi) = \frac{1}{E(\tau_L)}\cdot[u(\phi - \phi_0) - D_0]. \quad (73)$$

The probability current $J_0(\phi)$ is time independent (see

[6] In the following we omit for all functions in the stationary state the time variable t and agree to understand that $t = \infty$ whenever no explicit time reference is made.

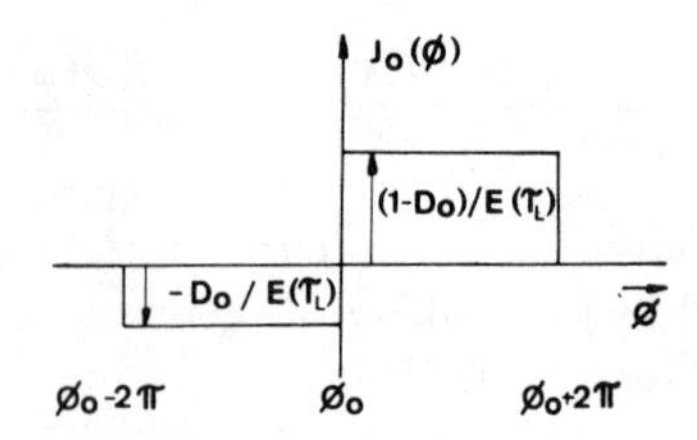

Fig. 7. Probability current $J_0(\phi)$ in the stationary state.

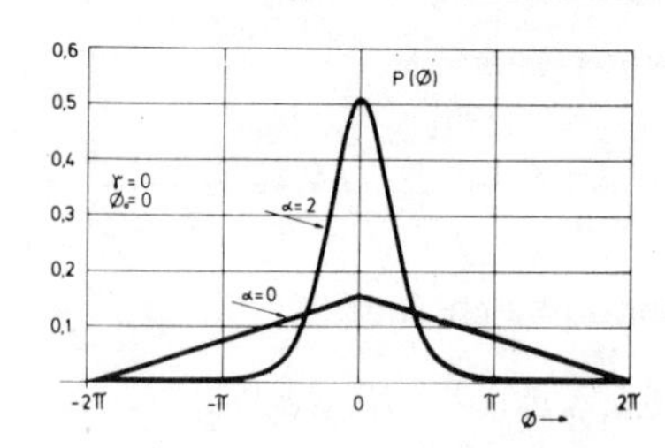

Fig. 8. PDF of the first-order sinusoidal PLL without detuning.

Fig. 7). The sample trajectories that are absorbed at the boundaries $\phi = \phi_0 \pm 2\pi$ are constantly replaced by the point source $\delta(\phi - \phi_0)/E(\tau_L)$ located at $\phi = \phi_0$ in a manner that on the average the same amount of trajectories ("particles") per unit of time are created and absorbed. The average number of sample trajectories or "particles" flowing to the right is given by $(1 - D_0)/E(\tau_L)$ whereas the average number of particles to the left is given by $D_0/E(\tau_L)$. Note that since the current $J_0(\phi)$ is a vectorial quantity the sign of the currents for positive and negative differences $(\phi - \phi_0)$ are different.

Example—First-order sinusoidal PLL: With $g(\phi) = \sin(\phi)$, $\phi_0 = 0$, and no loop detuning, the steady-state pdf is given by (Fig. 8)

$$p(\phi) = \frac{1}{2}\left(\frac{\alpha}{4B_L E(\tau_L)}\right)\cdot \exp(\alpha \cos \phi) \left[2\pi I_0(\alpha) - \int_0^{|\phi|} \exp(-\alpha \cos y)\, dy\right] \quad (74)$$

with

$$\alpha = \frac{A^2}{N_0 B_L}, \qquad B_L = \frac{AK}{4}, \qquad \gamma = \frac{\Omega_0}{4B_L} = 0 \quad (75)$$

where $I_0(\alpha)$ is the Bessel function. Note that in the limiting case $\alpha = 0$, the pdf is triangular and not uniform. This is due to the absorbing barriers and the 4π interval.

D. *Mean time Between Two Cycles $E(\tau_L)$ and Average Number of Cycles to the Right (Left) N_+, N_-*

In this section we study the interconnection of $E(\tau_L)$, $N_+(N_-)$ and the phase error rate $E(\dot\phi)$ and relate these quantities to the probability current. In tracking application the average number of cycles slipped per unit time $(N_+ + N_-)$ is an important parameter as it is indicative of the error introduced into any Doppler measurement made to obtain velocity and changes in range.

We obtain the average phase error rate in the stationary state by taking expected values on the first equation of (56)

$$E(\dot\phi) = \Omega_0 - F_0 KAE[g(\phi)] + \sum_{k=1}^{N} E(y_k). \quad (76)$$

$E(\dot\phi)$ can be considered as the average velocity of a sample trajectory in the stationary state.

If we integrate the reduced F–P equation (67) over the interval $[\phi_0 - 2\pi, \phi_0 + 2\pi]$ we get, using the boundary conditions (70),

$$\int_{\phi_0-2\pi}^{\phi_0+2\pi} K_0(\phi)p(\phi)\, d\phi = \frac{1}{E(\tau_L)}[-D_0 4\pi + 2\pi]. \quad (77)$$

After replacing $K_0(\phi)$ in (77) by its definition in (67) we get

$$\int_{\phi_0-2\pi}^{\phi_0+2\pi} K_0(\phi)p(\phi)\, d\phi = \Omega_0 - AKF_0 E[g(\phi)] + \sum_{k=1}^{N} E(y_k). \quad (78)$$

Since

$$E[g(\phi)] = \int_{\phi_0-2\pi}^{\phi_0+2\pi} g(\phi)p(\phi)\, d\phi,$$

$$E(y_k) = E_\phi[E(y_k \mid \phi)]. \quad (79)$$

Comparing (76), (77), and (78) we conclude

$$E(\dot\phi) = \frac{2\pi}{E(\tau_L)}(1 - 2D_0). \quad (80)$$

The velocity $E(\dot\phi)$ is a vectorial quantity. We split up

$$E(\dot\phi) = E(\dot\phi)_+ + E(\dot\phi)_- \quad (81)$$

and define

$$E(\dot\phi)_+ = \frac{2\pi(1 - D_0)}{E(\tau_L)} = J_0(\phi > \phi_0)\cdot 2\pi$$

$$E(\dot\phi)_- = -\frac{2\pi D_0}{E(\tau_L)} = J_0(\phi < \phi_0)\cdot 2\pi. \quad (82)$$

$E(\dot\phi)_+$ can be interpreted as the average velocity of all trajectories starting at ϕ_0 and ending at $\phi_0 + 2\pi$ while $E(\dot\phi)_-$ is the quantity corresponding to the trajectories arriving at $\phi_0 - 2\pi$. The mean time it takes a sample trajectory to get from ϕ_0 to $\phi_0 \pm 2\pi$ is given by the quotient of distance and velocity.

$$E(\tau_L)_+ = \frac{2\pi}{E(\dot\phi)_+} = \frac{E(\tau_L)}{1 - D_0}$$

$$E(\tau_L)_- = \left|\frac{2\pi}{E(\dot\phi)_-}\right| = \frac{E(\tau_L)}{D_0} \quad (83)$$

In practice one most often uses the reciprocal quantities $1/E(\tau_L)_\pm$. These quantities represent the average number of sample trajectories per unit of time arriving at the barrier $(\phi_0 + 2\pi),(\phi_0 - 2\pi)$ or equivalently the average number of cycles per unit of time the loop has been pushed to the right (left).

$$N_+ = \frac{1 - D_0}{E(\tau_L)}, \qquad N_- = \frac{D_0}{E(\tau_L)}. \tag{84}$$

Comparing (84) and (73), we can see that N_+ and N_- are identical to the absolute value of the probability current in the corresponding interval.

The mean time between two cycles $E(\tau_L)$ is independent of ϕ_0. This is intuitively clear and can be proven by simple reasoning. The mean time it takes a sample trajectory starting at $\phi = 0$ with a given distribution $(p(\boldsymbol{y}_0 \mid \phi = 0)$ to reach the hyperplane $\phi = 2\pi$ can be expressed as the expected value of the two random variables τ_1 and τ_2 (see Fig. 9).

$$E(\tau \mid 0)_+ = E(\tau_1)_+ + E(\tau_2)_+. \tag{85}$$

Similarly,

$$E(\tau \mid \phi_0)_+ = E(\tau_2)_+ + E(\tau_3)_+ \tag{86}$$

where $E(\tau_1)_+$ is the mean time a sample trajectory starting at $\phi = 0$ with $p(\boldsymbol{y}_0 \mid \phi = 0)$ needs to reach $\phi = \phi_0$, $E(\tau_2)_+$ is the mean time a sample trajectory starting at $\phi = \phi_0$ with $p(\boldsymbol{y}_0 \mid \phi = \phi_0)$ needs to reach $\phi = 2\pi$, and $E(\tau_3)_+$ is the mean time a sample trajectory starting at $\phi = 2\pi$ with $p(\boldsymbol{y}_0 \mid \phi = 2\pi)$ needs to reach $\phi = \phi_0 + 2\pi$.

The relation $E(\tau \mid \phi_0)_+ = E(\tau \mid 0)_+$ can hold if, and only if,

$$E(\tau_1)_+ = E(\tau_3)_+. \tag{87}$$

Using the stationarity of $p(\phi,\boldsymbol{y}_0)$ and the fact that the hyperplanes $\phi = 0$ and $\phi = 2\pi$ are "identical," i.e., $p(\boldsymbol{y}_0 \mid \phi = 0) = p(\boldsymbol{y}_0 \mid \phi = 2\pi)$, we can conclude that (87) is indeed true.

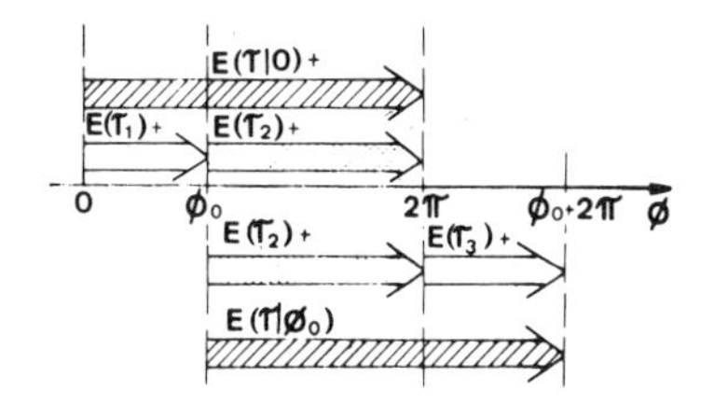

Fig. 9. Mean time between two cycles.

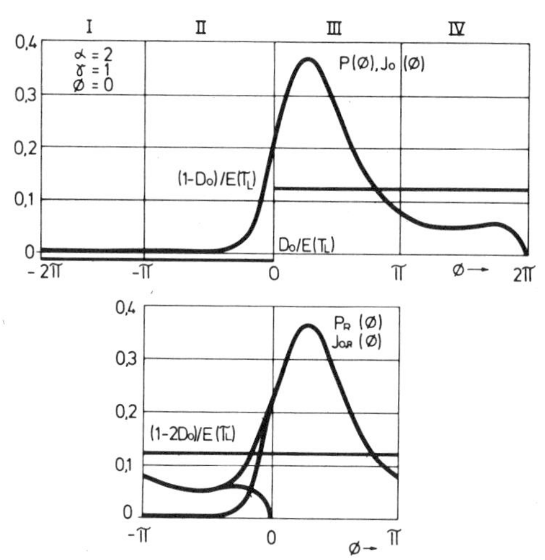

Fig. 10. PDF and probability current of the renewal process and corresponding functions for $[\phi_0 - \pi, \phi_0 + \pi]$.

E. Periodic Extension Versus Renewal Process Solution

We develop in this section the interconnection between the renewal process and the PE approach. For the moment we restrict our discussion to the first-order loop and give the general results at the end of this section.

The pdf $p(\phi)$ of the renewal process ranges over an interval of 4π. The pdf of the modulo 2π reduced phase error $\phi \in [\phi_0 - \pi,\phi_0 + \pi]$ (the range of a phase meter) can be easily obtained from $p(\phi)$. Since the events defining A_1,A_2,

A_1: The rv ϕ is within $[\phi,\phi + d\phi]$ or

$$[\phi - 2\pi,\phi - 2\pi + d\phi] \text{ for } \phi \in [\phi_0,\phi_0 + \pi]$$

A_2: The rv ϕ is within $[\phi,\phi + d\phi]$ or

$$[\phi + 2\pi,\phi + 2\pi + d\phi] \text{ for } \phi \in [\phi_0 - \pi,\phi_0],$$

are mutually exclusive we get the new pdf $p_R(\phi)$ by simply adding the probabilities of the respective intervals $d\phi$. This is equivalent to folding down the intervals in the range beyond $[\phi_0 - \pi,\phi_0 + \pi]$, as illustrated in Fig. 10.

By inspection of Fig. 10 we see without any calculation that the probability current $J_0(\phi)$ is now constant and equals the difference $(1 - 2D_0)/E(\tau_L)$. The values of $p_R(\phi)$ at $\phi = \phi_0 \pm \pi$ are by definition

$$\begin{aligned} p_R(\phi_0 - \pi) &= p(\phi_0 - \pi) + p(\phi_0 + \pi) \\ p_R(\phi_0 + \pi) &= p(\phi_0 + \pi) + p(\phi_0 - \pi). \end{aligned} \tag{88}$$

Therefore and due to the fact that $J_{0,R}$ is a constant we find

$$p_R(\phi_0 - \pi) = p_R(\phi_0 + \pi) \tag{89}$$

and

$$\frac{d}{d\phi} p_R(\phi_0 - \pi) = \frac{d}{d\phi} p_R(\phi_0 + \pi). \tag{90}$$

Interestingly enough these are the boundary conditions of the periodic extension solution. We show now that $p_R(\phi)$ and the PE solution $p_{\text{PE}}(\phi)$ are identical.

Proof: Let $p(\phi)$ be the unique solution of the differential equation (67) subject to the boundary conditions (70). By adding the differential equation for either $\phi + 2\pi$ or $\phi - 2\pi$,

$$K_0(\phi \pm 2\pi) \cdot p(\phi \pm 2\pi) - \tfrac{1}{2} K_{0,0} \frac{d}{d\phi} p(\phi \pm 2\pi) = \frac{u(\phi - \phi_0 \pm 2\pi) - D_0}{E(\tau_L)} \tag{91}$$

using 1) the linearity of (67), 2) the periodicity of the coefficients $K_0(\phi)$, $K_{0,0}$ we arrive at

$$K_0(\phi)[p(\phi) + p(\phi + 2\pi)] - \tfrac{1}{2}K_{0,0}\frac{d}{d\phi}[p(\phi) + p(\phi + 2\pi)] = \frac{1 - 2D_0}{E(\tau_L)}, \qquad \phi \in [\phi_0 - \pi, \phi_0]$$

$$K_0(\phi)[p(\phi) + p(\phi - 2\pi)] - \tfrac{1}{2}K_{0,0}\frac{d}{d\phi}[p(\phi) + p(\phi - 2\pi)] = \frac{1 - 2D_0}{E(\tau_L)}, \qquad \phi \in [\phi_0, \phi_0 + \pi]. \tag{92}$$

Therefore $p_R(\phi)$ is the normalized solution of the differential equation

$$K_0(\phi)p_R(\phi) - \tfrac{1}{2}K_{0,0}\frac{d}{d\phi}p_R(\phi) = \frac{1 - 2D_0}{E(\tau_L)}, \qquad \phi \in [\phi_0 - \pi; \phi_0 + \pi] \tag{93}$$

fulfilling the boundary conditions (89). In passing we note that (89), (90) guarantees $p_R(\phi)$ to be a periodic function in ϕ although $p(\phi)$ is not. The PE solution $p_{PE}(\phi)$ [2, ch. 9] is obtained by solving the differential equation

$$K_0(\phi)p_{PE}(\phi) - \tfrac{1}{2}K_{0,0}\frac{d}{d\phi}p_{PE}(\phi) = C_0 \tag{94}$$

where C_0 is a constant, subject to the boundary conditions

$$p_{PE}(-\pi) = p_{PE}(\pi)$$

and the normalization condition

$$\int_{-\pi}^{\pi} p_{PE}(\phi)\, d\phi = 1.$$

Consequently, since $p_R(\phi)$ and $p_{PE}(\phi)$ obey the same differential equation, the same periodic boundary conditions, and the same normalization condition, they must be identical. Q.E.D.

In the general case of an $(N + 1)$th order loop we have (48)

$$L_{\phi,\boldsymbol{y}_0}p(\phi,\boldsymbol{y}_0) = \frac{q(\phi,\boldsymbol{y}_0; t = 0)}{E(\tau_L)}. \tag{95}$$

After reducing the variables $y_1, \cdots, y_N$ and integrating once over ϕ we arrive at (67) where the intensity coefficients $K_0(\phi)$ and $K_{0,0}$ are given by (64) and (65a). In order to be able to proceed in the way just outlined for the first-order loop we must prove that the conditional expectations $E(y_k \mid \phi)$ are periodic.

Proof: Let $p(\phi,\boldsymbol{y}_0)$ be the unique solution of (95), subject to the boundary conditions (26)–(28). The operator $L_{\phi,\boldsymbol{y}_0}$ has the following properties:

A_1 time independent

A_2 linear

A_3 periodic in ϕ, i.e., $L_{\phi+2\pi k,\boldsymbol{y}_0} = L_{\phi,\boldsymbol{y}_0}$; $k = \pm 1 \cdots, \pm n$.

We add to (95) the solution for $\phi - 2\pi$, $\phi \in [\phi_0, \phi_0 + \pi]$

$$L_{\phi-2\pi,\boldsymbol{y}_0}p(\phi - 2\pi,\boldsymbol{y}_0) = \frac{q(\phi - 2\pi,\boldsymbol{y}_0; t = 0)}{E(\tau_L)} \tag{96}$$

which gives, using properties A_2 and A_3,

$$L_{\phi,\boldsymbol{y}_0}[p(\phi,\boldsymbol{y}_0) + p(\phi - 2\pi,\boldsymbol{y}_0)] = \frac{1}{E(\tau_L)}[q(\phi,\boldsymbol{y}_0; t = 0) + q(\phi - 2\pi,\boldsymbol{y}_0; t = 0)]. \tag{97}$$

By integrating out the state variables $y_1,y_2,\cdots,y_N$ using appropriate boundary conditions and integrating once over ϕ we arrive at

$$K_0(\phi)[p(\phi) + p(\phi - 2\pi)] - \tfrac{1}{2}K_{0,0}\frac{d}{d\phi}[p(\phi) + p(\phi - 2\pi)] = \frac{1 - 2D_0}{E(\tau_L)}. \tag{98}$$

On the other hand if we first integrate out the state variables $y_1,y_2,\cdots,y_N$ and then add the differential equations we get

$$K_0(\phi)p(\phi) + K_0(\phi - 2\pi)p(\phi - 2\pi) - \tfrac{1}{2}K_{0,0}\frac{d}{d\phi}[p(\phi) + p(\phi - 2\pi)] = \frac{1 - 2D_0}{E(\tau_L)}. \tag{99}$$

Now for every $\phi \in [\phi_0, \phi_0 + 2\pi]$ the equality

$$K_0(\phi)[p(\phi) + p(\phi - 2\pi)] = K_0(\phi)p(\phi) + K_0(\phi - 2\pi)p(\phi - 2\pi) \tag{100}$$

must hold. This is true if

$$K_0(\phi) = K_0(\phi - 2\pi). \tag{101}$$

If we add to (95) the solution for $\phi + 2\pi$; $\phi \in [\phi_0 - \pi; \phi_0]$ we find

$$K_0(\phi) = K_0(\phi + 2\pi). \tag{102}$$

Therefore, $K_0(\phi)$ is periodic and in view of (65a) also $E(y_k \mid \phi)$ is a periodic function. Q.E.D.

This contradicts some experimental results by Tausworthe [15] that indicated a linear term in $E(y_k \mid \phi)$ for a second-order loop. Other computer simulation results provided by Lindsey [10] did not show this term and gave a periodic $E(y_1 \mid \phi)$. However, it has to be kept in mind that the proof given is based on the validity of the model introduced in Section I. This model tacitly assumes the initial distribution $p(\boldsymbol{y}_0 \mid \phi = \phi_0)$ to be identical for all single processes, or equivalently, neglects the occurence of multiple cycle slips.

We also note that in the case of an $(N + 1)$th-order loop the PE equation is in disagreement with the equation obtained via renewal process approach. This can be seen if we compare the differential equations for the marginal distribution of y_k for both approaches

$$\underbrace{\frac{\partial}{\partial y_k} J_k(y_k) = 0,}_{\text{PE equation}}$$

$$\underbrace{\frac{\partial}{\partial y_k} J_k(y_k) + J_0(\phi_0 + 2\pi, y_k) - J_0(\phi_0 - 2\pi, y_k) = \frac{p(y_k \mid \phi_0)}{E(\tau_L)}}_{\text{Renewal process equation}}. \quad (103)$$

It is conjectured, however, that the total amount of probability current $J_0(\phi_0 + 2\pi, y_k) - J_0(\phi_0 - 2\pi, y_k)$ absorbed at the boundaries is compensated by the source $p(y_k \mid \phi_0)/E(\tau_L)$. Under this assumption the two equations (103) would be identical.

SUMMARY AND EXTENSION

A new method to describe the behavior of a correlative tracking system has been presented. It has been shown that the stationary pdf, the mean time between two cycles $E(\tau_L)$ and the average number of cycle slips $N_+(N_-)$ can be obtained by solving a single F–P equation for a renewal process. From a theoretical point of view it is very pleasing that correlative tracking systems with periodic or aperiodic nonlinearity can be treated in a unified mathematical framework.

A major feature of the method is that for the first time the actual behavior of the loop is mathematically modeled, thus giving new insight in the complicated structure of the underlying stochastic process. Several important results are derived. For instance, the average number of cycle slips to the right (left) $N_+(N_-)$, its relation with the mean time between two cycle slips $E(\tau_L)$, and the average phase error rate $E(\phi)$ can be obtained for an $(N + 1)$th-order loop with arbitrary nonlinearity. The mod 2π reduced pdf of a first-order loop has been shown to be identical with the one obtained by the so-called PE approach. This agreement, however, exists only in the stationary state while the transient solution of the two approaches are different. Since the renewal model describes the *actual behavior* of a first-order system we have to conclude that the PE solution does not adequately describe the transient behavior of such a system. In a forthcoming paper [18] more on this topic will be reported. In the case of an $(N + 1)$th-order system the renewal model introduced is valid only if multiple cycle slips can be neglected. More work has to be done in order to fully understand the complicated nature of the processes involved.

ACKNOWLEDGMENT

The author would like to thank Prof. W. C. Lindsey of the University of Southern California for his suggestion that the author's method for S-type loops be extended to loops with periodic nonlinearities, and for many interesting discussions and helpful comments. Many ideas expressed in this paper stem from these discussions. The author would also like to thank Dr. H. Mey, Head of the Research Division, Hasler Ltd., Berne, Switzerland, for his support and encouragement throughout this work, and Dr. E. Schultze, Hasler Research, for advice on the mathematical aspects of this paper.

REFERENCES

[1] W. C. Lindsey and R. C. Tausworthe, "A bibliography of the theory and application of the PLL," Jet Propulsion Laboratory, Pasadena, Calif., Tech. Rep. 32-1581, Apr. 1973.
[2] W. C. Lindsey, *Synchronization Systems in Communication and Control.* Englewood Cliffs, N. J.: Prentice-Hall, 1972.
[3] F. Mesch *et al.*, "Geschwindigkeitsmessung mit Korrelationsverfahren," Messtechnik, vol. 7, pp. 152–157, 1971; also vol. 8, pp. 163–168, 1971.
[4] R. Kuhne and H. Meyr, "Verfahren zur Messung Geschwindigkeit der Bewegung eines Korpers relativ zu einer Ungleichmassigkeiten aufweisenden Flache," Swiss Patent 6464-71, Apr. 5, 1971.
[5] H. Meyr, "Einrichtung zur Messung der Geschwindigkeit der Bewegung eines Korpers relativ zu einer Ungleichmassigkeiten aufweisenden Flache," Swiss Patent 6677-71, May 5, 1971.
[6] ——, "Untersuchung Korrelativer Tracking Systeme mit Hilfe der Fokker-Planck Methode," Ph.D. dissertation, Swiss Federal Institute of Technology, Zurich, Switzerland, 1973.
[7] J. J. Spilker, Jr., and D. T. Magill, "The delay-lock discriminator—An optimum tracking device," *Proc. IRE*, vol. 49, pp. 1403–1416, Sept. 1961.
[8] J. J. Spilker, Jr., "Delay-lock tracking of binary signals," *IEEE Trans. Space Electron. Telem.*, vol. JET-9, pp. 1–8, Mar. 1963.
[9] A. J. Viterbi, *Principles of Coherent Communications.* New York: McGraw-Hill, 1966.
[10] W. C. Lindsey, "Nonlinear analysis of generalized tracking systems," *Proc. IEEE*, vol. 57, pp. 1705–1722, Oct. 1969.
[11] W. Feller, *An Introduction to Probability Theory and Its Applications*, vol. II. New York: Wiley, 1966.
[12] R. L. Stratonovich, *Topics in the Theory of Random Noise.* New York: Gordon and Breach, 1963.
[13] E. B. Dynkin and A. A. Juschkewitsch, *Satze und Aufgaben uber Markoff'sche Prozesse.* Berlin, Germany: Springer-Verlag, 1969.
[14] A. H. Jazwinsky, *Stochastic Processes and Filtering Theory.* New York: Academic, 1970.
[15] R. C. Tausworthe, "Simplified formula for mean cycle-slip time of phase-locked loops with steady-state phase error," *IEEE Trans. Commun.*, vol. COM-20, pp. 331–337, June 1972.
[16] F. J. Charles and W. C. Lindsey, "Some analytical and experimental phase-locked loop results for low signal-to-noise ratios," *Proc. IEEE*, vol. 54, pp. 1152–1166, Sept. 1966.
[17] L. C. Palmer and S. A. Klein, "Phase slipping in phase-locked loop configurations that track biphase or quadriphase modulated carriers," *IEEE Trans. Commun.* (Concise Papers), vol. COM-20, pp. 984–991, Oct. 1972.
[18] W. C. Lindsey and H. Meyr, "Complete statistical description of the phase error generated by correlative tracking systems," to be published.

Functional Techniques for the Analysis of the Nonlinear Behavior of Phase-Locked Loops

HARRY L. VAN TREES, MEMBER, IEEE

Summary—In this paper we consider the analysis of a nonlinear feedback system. The purpose of the paper is twofold.

The first objective is to demonstrate the efficiency of the Volterra functional expansion technique as a method of analyzing nonlinear feedback systems. The techniques we demonstrate are valid for a large class of nonlinear systems. Several important advantages of the functional approach are as follows: 1) Random and deterministic inputs and disturbances are included. 2) All input-output relationships are explicit. One does not have to solve complicated differential equations. 3) Once one becomes facile with the properties of the expansion, the analysis of any particular nonlinear system is rapid and straightforward.

The second objective is to obtain some new and useful results for a device of practical importance. The particular nonlinear system that we will use as an example represents a phase-locked loop whose input signal is a phase-modulated sinewave which has been corrupted by additive noise. Two interesting cases of phase modulation are considered. In the first case the phase $\theta_1(t)$ is a deterministic function. In the second case the phase $\theta_1(t)$ is a sample function from a random process.

The results are presented as closed form analytic expressions. Several interesting cases are plotted as a function of the significant parameters.

I. Introduction

IN THIS PAPER we consider the analysis of the nonlinear system shown in Fig. 1. Our purpose is twofold.

1) To demonstrate the efficiency of nonlinear system analysis using Volterra functional expansion techniques, and
2) the system is an adequate model of a phase-locked loop. Consequently, the specific results will be of practical significance.

A. Volterra Functionals

The concept of functionals in the analysis of nonlinear systems is a generalization of the convolution integral used in linear system analysis. One desires to represent some arbitrary nonlinear system by a sequence of systems connected in parallel as shown in Fig. 2. Each one of the systems in the sequence has special properties which make it easy to analyze. The first system is a linear system. The output $y_1(t)$ is simply a convolution of the input $x(t)$ and the impulse response $h_1(t)$.

Manuscript received October 3, 1963; revised April 6, 1964. The work reported in this paper was supported by the United States Army, Navy and Air Force.

The author is with the Department of Electrical Engineering, Massachusetts Institute of Technology, Cambridge, Mass., and Lincoln Laboratory, Lexington, Mass.

Thus,

$$y_1(t) = \int_{-\infty}^{\infty} h_1(\tau)x(t-\tau)d\tau = \int_{-\infty}^{\infty} h_1(t-\tau)x(\tau)d\tau. \quad (1)$$

The second system is of a quadratic nature. The output $y_2(t)$ is a two-dimensional convolution of the input $x(t)$ and the impulse response $h_2(t_1, t_2)$. Thus,

$$y_2(t) = \int_{-\infty}^{\infty}\int_{-\infty}^{\infty} h_2(\tau_1, \tau_2)x(t-\tau_1)x(t-\tau_2)d\tau_1 d\tau_2$$

$$= \int_{-\infty}^{\infty}\int_{-\infty}^{\infty} h_2(t-\tau_1, t-\tau_2)x(\tau_1)x(\tau_2)d\tau_1 d\tau_2. \quad (2)$$

A simple example of a quadratic system is shown in Fig. 3. Since

$$Z(t) = \int_{-\infty}^{\infty} h(\tau)x(t-\tau)d\tau$$

and

$$y_2(t) = Z^2(t),$$

we can write

$$y_2(t) = \int_{-\infty}^{\infty} h(\tau_1)x(t-\tau_1)d\tau_1 \int_{-\infty}^{\infty} h(\tau_2)x(t-\tau_2)d\tau_2,$$

or

$$y_2(t) = \int_{-\infty}^{\infty}\int_{-\infty}^{\infty} h(\tau_1)h(\tau_2)x(t-\tau_1)x(t-\tau_2)d\tau_1 d\tau_2.$$

Here the two-dimensional kernel is simply

$$h_2(\tau_1, \tau_2) = h(\tau_1)h(\tau_2).$$

If $h(\tau)$ were a simple RC filter, the resulting two-dimensional kernel would be as shown in Fig. 4.

The third system is of a cubic nature. It is characterized by a three-dimensional kernel $h_3(t_1, t_2, t_3)$. The output $y_3(t)$ is

$$y_3(t) = \int_{-\infty}^{\infty}\int_{-\infty}^{\infty}\int_{-\infty}^{\infty} h_3(\tau_1, \tau_2, \tau_3)x(t-\tau_1)x(t-\tau_2) \cdot x(t-\tau_3)d\tau_1 d\tau_2 d\tau_3. \quad (3)$$

The total output $\hat{y}(t)$ is, in general, an infinite sum of $y_i(t)$.

$$\hat{y}(t) = \sum_{i=1}^{\infty} y_i(t). \quad (4)$$

Reprinted from *Proc. IEEE*, vol. 52, pp. 894–911, Aug. 1964.

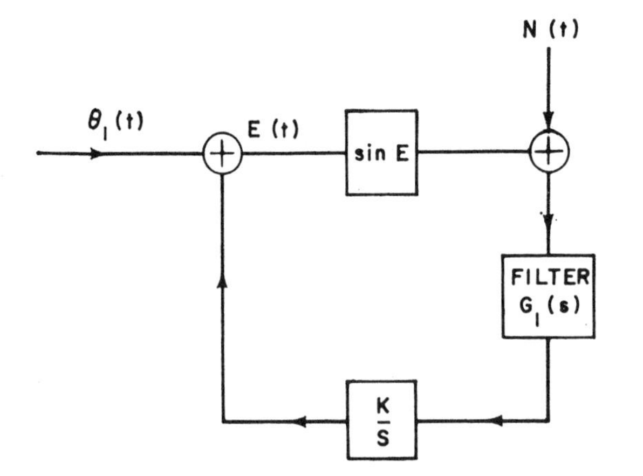

Fig. 1—A nonlinear feedback system.

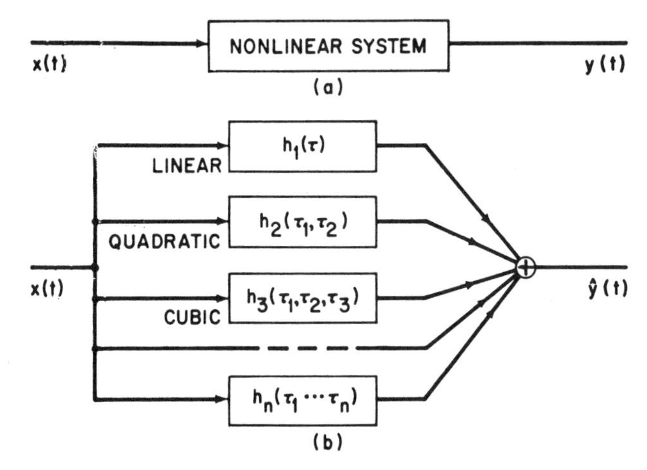

Fig. 2—Functional expansion of nonlinear system.

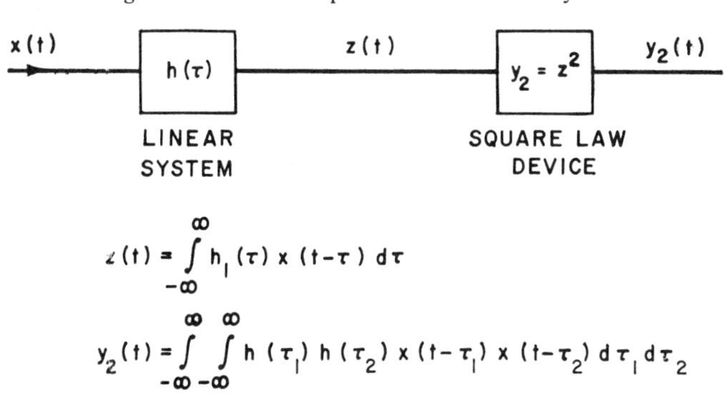

$$z(t) = \int_{-\infty}^{\infty} h_1(\tau)\, x(t-\tau)\, d\tau$$

$$y_2(t) = \int_{-\infty}^{\infty}\int_{-\infty}^{\infty} h(\tau_1)\, h(\tau_2)\, x(t-\tau_1)\, x(t-\tau_2)\, d\tau_1\, d\tau_2$$

Fig. 3—A simple quadratic system.

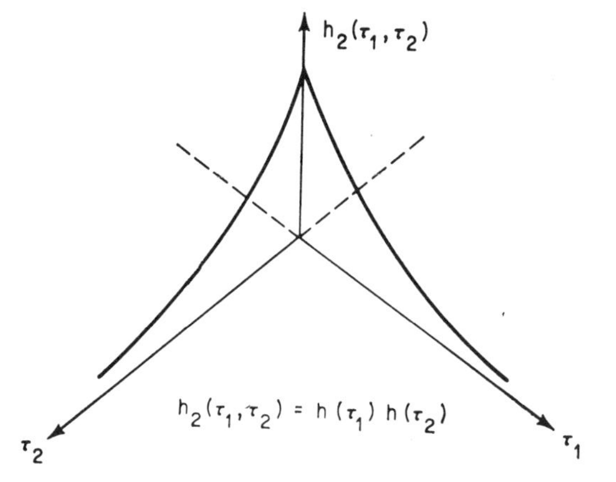

Fig. 4—A two-dimensional kernel.

The various impulse responses are called Volterra kernels after the mathematician who first studied this type of functional [1]. In the last several years considerable effort has been devoted to both the theoretical and practical problems of this type of representation. Discussion of primarily theoretical interest is contained in Wiener [2], Brilliant [3], and Chesler [4]. Research oriented in a more practical direction is contained in Barrett [5], George [6], and Van Trees [7]. We shall draw heavily on these references but will attempt to make our development self-contained.

A number of logical questions immediately arise.

1) What class of nonlinear systems may be represented in this manner?
2) How many terms in the infinite series must one use to obtain a good approximation to the actual system?
3) Given a nonlinear system, how does one find the appropriate kernels?
4) For what class of inputs can one find the properties of the output easily?

Rather than try to answer these questions in complete generality, we shall concern ourselves only with a specific example. For the general case one may consult the references. We will develop the construction of the kernels for the system in Fig. 1 and discuss the properties of the output for deterministic and random inputs.

B. Phase-Locked Loops

The idea of demodulation of angle-modulated signals using phase-locked loops or other feedback devices is well known.[1] Most early analyses were based on linear models. A derivation of a correct nonlinear model in the presence of noise is given in Van Trees [8]. The model, a nonlinear, randomly-time-varying system, appeared impossible to analyze exactly. Some results were obtained by approximating it by a linear, randomly-time-varying system. The simplified model in Fig. 1 was suggested by Develet [9]. This simplified model is well suited to the functional approach.

The functional approach may be viewed as a form of perturbation analysis. It was apparently first applied to nonlinear feedback systems by Wiener [10].[2]

Previous perturbation analyses are discussed in Margolis [13] and Schilling [14]. Our approach seems to be simpler than either of these, and we are able to obtain more extensive results. The use of perturbation techniques for solving nonlinear differential equations with deterministic inputs is well known (see, for example, Pipes [15]). Our approach, like other perturbation solutions, is a technique for obtaining an approximate solution.

[1] Since there are many articles motivating or describing the use of phase-locked loops, we will assume the reader is familiar with the general principles involved (see, for example, Jaffe and Rechtin [25], Gruen [26], or Van Trees [27]).

[2] This report is no longer readily available. Its essential content is available in Ikehara [11] and Deutsch [12].

An alternate method of attaining approximate solutions was developed by Booton [16]. Apparently a similar technique was developed independently at about the same time by Caughey [17]. This technique has been applied to the analysis of phase-locked loops in Develet [9].

Considerable effort has been expended in attempts to find exact solutions for nonlinear feedback systems excited by random imputs. For systems which are described by first-order differential equations whose forcing term is white Gaussian noise, the output is a simple Markov process. The behavior of the system can then be described by either the backward Kolmogorov equation or the Fokker-Planck equation.

Analysis of dynamic systems using these techniques has a long history. (See, for example, original papers by Uhlenbeck and Ornstein [18], Wang and Uhlenbeck [19], or Andronov, Pontryagin and Witt [20], and books by Bharucha-Reid [21] or Middleton [22].) The technique was first applied to the analysis of phase-locked loops by Tikhonov [23]. In this paper, he obtains an exact solution for the probability density for a loop without a filter.[3] The difficulty with this technique is that for loops with filters, one cannot solve the differential equation. In Section II, we will compare the result of the functional solution to the exact solution for the case of the first-order loop. We will use the model of Fig. 1 as our starting point.

II. Time Domain Analysis of First-Order System

The first case that we will consider is the system in Fig. 1 with the loop filter removed as shown in Fig. 5. This simple case serves two purposes. First, the technique is illustrated with a minimum of algebraic complexity. Second, for the special case of $\dot{\theta}_1(t)=\beta$, a constant, the answer may be compared to the solution in Tikhonov [23].

The differential equation describing the system of Fig. 3 is

$$\frac{dE(t)}{dt} + K \sin E(t) = \dot{\theta}_1(t) - KN(t) \equiv x(t) \qquad (5)$$

where we have defined $x(t)$ as an equivalent input. Observe that we are concerned with the behavior of the error signal. Now write

$$E(t) = E_1(t) + E_2(t) + \cdots = \sum_{i=1}^{\infty} E_i(t). \qquad (6)$$

Here $E_1(t)$ is the output of the linear system in Fig. 2. $E_2(t)$ is the output of the second-order system $h_2(\tau_1, \tau_2)$, and so forth.

We now want to find the analytic form of the various kernels. The easiest way to do this is to substitute (6) into (5) and sort the terms according to the order in which they involve $x(t)$. Expanding $\sin E(t)$ and performing the indicated substitution gives the following equation:

[3] A more widely available source of Tikhonov's result is contained in a recent article by Viterbi [24].

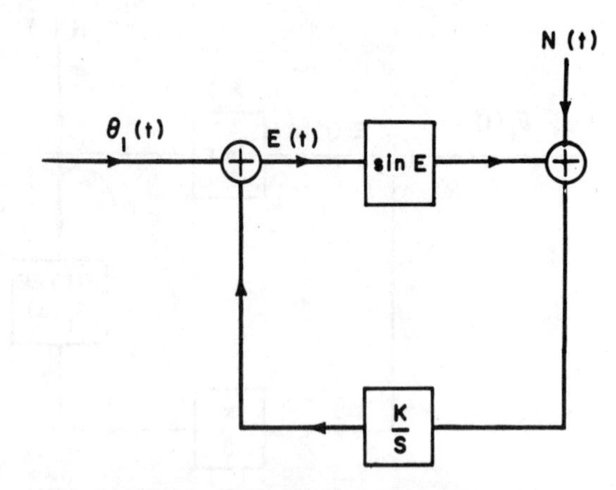

Fig. 5—Model of phase-locked loop without filter.

$$\begin{aligned}[\dot{E}_1(t) + \dot{E}_2(t) + \dot{E}_3(t) + \cdots] &+ K\Big[(E_1(t) + E_2(t) + E_3(t) + \cdots) \\ &- \frac{1}{3!}(E_1(t) + E_2(t) + E_3(t) + \cdots)^3 \\ &+ \frac{1}{5!}(E_1(t) + E_2(t) + \cdots)^5 + \cdots\Big] = x(t). \qquad (7)\end{aligned}$$

Equating terms of equal order in $x(t)$ gives the following set of equations:

$$\dot{E}_1(t) + KE_1(t) = x(t). \qquad (8)$$

All terms in (8) are first order in $x(t)$.

$$\dot{E}_2(t) + KE_2(t) = 0, \qquad (9)$$

$$\dot{E}_3(t) + KE_3(t) = \frac{K}{3!} E_1{}^3(t), \qquad (10)$$

$$\dot{E}_4(t) + KE_4(t) = \frac{K}{2} E_1{}^2(t)E_2(t), \qquad (11)$$

$$\dot{E}_5(t) + KE_5(t) = \frac{K}{2} E_1{}^2(t)E_3(t) - \frac{K}{5!} E_1{}^5(t). \qquad (12)$$

These equations may be solved sequentially.

$$\begin{aligned}E_1(t) &= \int_0^{\infty} e^{-K\tau}x(t-\tau)d\tau \\ &= \int_0^{\infty} h_1(\tau)x(t-\tau)d\tau \qquad (13)\end{aligned}$$

where we define

$$h_1(\tau) \equiv \begin{cases} e^{-K\tau} & \tau \geq 0 \\ 0, & \tau < 0. \end{cases} \qquad (14)$$

Clearly,

$$E_2(t) = 0 \qquad (15)$$

and, consequently, all higher-order even terms are zero.

$$E_3(t) = \int_0^\infty h_1(\tau) \cdot \frac{K}{3!} E_1^3(t-\tau) d\tau. \tag{16}$$

To write this as an explicit function of $x(t)$, we substitute (13) into (16). This gives,

$$E_3(t) = \int_0^\infty d\tau \int_0^\infty d\tau_1 \int_0^\infty d\tau_2 \int_0^\infty d\tau_3$$
$$\left[\frac{K}{3!} h_1(\tau)h_1(\tau_1)h_1(\tau_2)h(\tau_3)x(t-\tau-\tau_1) \cdot x(t-\tau-\tau_2)x(t-\tau-\tau_3)\right]. \tag{17}$$

In this form, it is easy to see that $E_3(t)$ depends in a third-order manner on $x(t)$. From (3), we know that we want a third-order relationship of the form

$$E_3(t) = \int_0^\infty d\tau_1 \int_0^\infty d\tau_2 \int_0^\infty d\tau_3 h_3(\tau_1, \tau_2, \tau_3) \cdot x(t-\tau_1)x(t-\tau_2)x(t-\tau_3). \tag{18}$$

In the time domain it is not necessary to find $h_3(\tau_1, \tau_2, \tau_3)$ explicitly. We shall see that in the transform domain the relationship between (17) and (18) follows easily. Similarly,

$$E_5(t) = \int_0^\infty h_1(\tau)\left[\frac{K}{2} E_1^2(t-\tau)E_3(t-\tau) - \frac{K}{5!} E_1^5(t-\tau)\right] d\tau. \tag{19}$$

Expressing in terms of $x(t)$ only, we have

$$E_5(t) = \frac{K^2}{12}\int_0^\infty d\tau_1 \int_0^\infty d\tau_2 \int_0^\infty \cdots \int_0^\infty d\tau_7$$
$$\cdot\left[h_1(\tau_1)h_1(\tau_2) \cdots h_1(\tau_7)x(t-\tau_1-\tau_2) \cdot x(t-\tau_1-\tau_3)x(t-\tau_1-\tau_4-\tau_5) \cdot x(t-\tau_1-\tau_4-\tau_6)x(t-\tau_1-\tau_4-\tau_7)\right.$$
$$+ \frac{K}{5!}\int_0^\infty d\tau_1 \int_0^\infty d\tau_2 \cdots \int_0^\infty d\tau_6 h_1(\tau_1) \cdots h_1(\tau_6) \cdot x(t-\tau_1-\tau_2)x(t-\tau_1-\tau_3)x(t-\tau_1-\tau_4)$$
$$\left.\cdot x(t-\tau_1-\tau_5)x(t-\tau_1-\tau_6)\right]. \tag{20}$$

Once again, the fifth-order relationship between $E_5(t)$ and $x(t)$ is clear. Eq. (20) is equivalent to the form

$$E_5(t) \int_0^\infty d\tau_1 \int_0^\infty d\tau_2 \cdots \int_0^\infty d\tau_5 h_5(\tau_1, \tau_2, \tau_3, \tau_4, \tau_5) \cdot x(t-\tau_1)x(t-\tau_2)x(t-\tau_3)x(t-\tau_4)x(t-\tau_5). \tag{21}$$

Higher-order kernels can be found in a similar manner. Observe that the kernels do *not* depend on the nature of the input. Thus once the functional relations are derived they are a property of the system. Now assume the series in (6) converges.[4] We have an explicit representation for $E(t)$.

Let us consider the approximate solution

$$E_{(5)}(t) \equiv E_1(t) + E_3(t) + E_5(t). \tag{22}$$

First, we will consider the simplest possible case. Let $\theta_1(t)=0$ (this corresponds to an incoming signal whose center frequency is the same as the VCO center frequency). Let $N(t)$ be a sample function from a *white, Gaussian* random process with a correlation function

$$R_N(\tau) = \frac{N_0}{2A^2} \delta(\tau). \tag{23}$$

The constant in the correlation function corresponds to a phase-locked loop whose received signal is a sinewave of rms value A corrupted by white, Gaussain noise with a double-sided spectral height $N_0/2$.

We want to evaluate the mean and variance of $E_{(5)}(t)$. By inspection we see that $\langle E_{(5)}(t)\rangle = 0$.

To evaluate $\langle E_{(5)}^2(t)\rangle$ we consider the double sum

$$\langle E_{(5)}^2(t)\rangle = \sum_{i=1,3,5} \sum_{j=1,3,5} < E_i(t)E_j(t) >. \tag{24}$$

In Appendix I, these terms are evaluated. One observes that a fundamental quantity in the solution is

$$\frac{KN_0}{4A^2} \equiv Z. \tag{25}$$

(Physically, this is the coherent noise-to-signal ratio.[5])

If we retain only terms in our answer which are third power or less in Z, we have

$$\langle E_{(5)}^2(t)\rangle \cong Z + \frac{1}{2} Z^2 + \frac{13}{24} Z^3. \tag{26}$$

The variance as a function of Z is plotted in Fig. 6. For comparison purposes the exact solution is shown, and the results predicted by a linear model are shown.

We now want to consider more interesting forms for $\theta_1(t)$. Before doing this, we want to demonstrate how the analysis proceeds in the transform domain.

[4] The convergence problem is certainly not trivial. However, most convergence proofs [2], [6] are involved and give conservative regions of convergence. Our primary concern is how well a small number of kernels approximate the actual system behavior. We will observe shortly that to evaluate the output of a kernel of higher order than five is too tedious to be of any practical value. It does not appear possible to obtain a closed-form solution in any very interesting cases. (In the case of a first-order loop without noise one can find the response to a frequency step, but this is not too interesting.) In general, the functional solution should be regarded as a method of obtaining approximate answers over a reasonable range of signal and noise levels.

[5] The noise bandwidth of the linearized loop is $K/2$. Thus, the noise power in this bandwidth is $KN_0/4$.

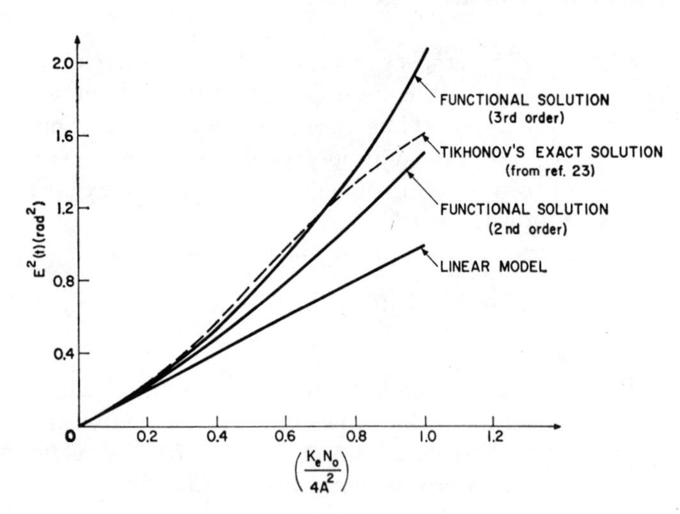

Fig. 6—Comparison of solutions.

III. Transform Domain Analysis of First-Order System

The advantages of transform domain analysis in the case of linear systems are well known.

For a function of one variable, the bilateral Laplace or exponential transform is

$$F(s) = \int_{-\infty}^{\infty} f(t)e^{-st}dt \tag{27}$$

$$f(t) = \frac{1}{2\pi j}\int_{-\infty}^{\infty} F(s)e^{st}ds. \tag{28}$$

In all cases of interest to us, the region of convergence includes the imaginary axis so that the Fourier transform also exists.

Similar relationships can be defined for functions of two variables

$$F(s_1, s_2) = \int_{-\infty}^{\infty}\int_{-\infty}^{\infty} f(t_1, t_2)e^{-s_1t_1-s_2t_2}dt_1dt_2, \tag{29}$$

$$f(t_1, t_2) = \left(\frac{1}{2\pi j}\right)^2 \int_{-\infty}^{\infty}\int_{-\infty}^{\infty} F(s_1, s_2)e^{+s_1t_1+s_2t_2}ds_1ds_2. \tag{30}$$

The relationship for n variables follows directly. Many of the familiar one-dimensional relationships in one dimension carry over to a n dimensions. We shall state several of the properties that we shall need in the sequel.

A. Review of Multidimensional Transform Properties

1) *Convolution:* In a linear system, convolution in the time domain corresponds to multiplication in the transform domain

$$\int_{-\infty}^{\infty} h_1(\tau)x(t-\tau)d\tau \leftrightarrow H_1(s)X(s). \tag{31}$$

The input-output relation of a second-order system is given by (2)

$$y_2(t) = \int_{-\infty}^{\infty}\int_{-\infty}^{\infty} h_2(\tau_1, \tau_2)x(t-\tau_1)x(t-\tau_2)d\tau_1d\tau_2.$$

This relationship is not quite a convolution. A two-dimensional convolution would be

$$\hat{y}_2(t_1, t_2) = \int_{-\infty}^{\infty}\int_{-\infty}^{\infty} h_2(\tau_1, \tau_2)x(t_1-\tau_1)x(t_2-\tau_2)d\tau_1d\tau_2. \tag{32}$$

Now, one can show [2], [6] that convolution corresponds to multiplication. Thus

$$\hat{Y}_2(s_1, s_2) = H_2(s_1, s_2)X(s_1)X(s_2). \tag{33}$$

To find the output of a quadratic system one could find the inverse transform of (33) and set $t_1 = t_2$.

Since

$$\hat{y}_2(t_1, t_1) = y_2(t_1). \tag{34}$$

An easy way is to perform the reduction in the transform domain. In other words, find $Y_2(s)$ from $\hat{Y}_2(s_1, s_2)$. A simple way to do this by inspection for rational functions is developed in George [6]. It is called association of variables. A table of associated transforms follows.[6]

$\hat{Y}_2(s_1, s_2)$	$Y_2(s_1)$
$\frac{A}{s_1+s_2+\alpha}$	$\frac{A}{s_1+\alpha}$
$\frac{B}{s_1+\alpha}\cdot\frac{C}{s_2+\beta}$	$\frac{BC}{s_1+\alpha+\beta}$
$F(s_1+s_2)$	$F(s_1)$
$\frac{A}{(s_1+\alpha)^n}\,\frac{B}{(s_2+\beta)^m}$	$AB\,\frac{(n+m-2)!}{(n-1)!(m-1)!}\,\frac{1}{(s+\alpha+\beta)^{n+m-1}}$

To reduce higher dimensional transforms, one applies the above relations successively.

2) *Cascades of Systems:*[7] Two types of cascades will be necessary.

a) A linear system $L_1(s)$ followed by an n-dimensional system with transform $H_n(s_1, s_2, s_3, \cdots, s_n)$. The transform of the cascade is

$$H_n(s_1, s_2, \cdots, s_n)L_1(s_1)L_1(s_2)L_1(s_3)\cdots L_1(s_n). \tag{35}$$

b) An n-dimensional system followed by a linear system $L_1(s)$. The transform of the cascade is

$$L_1(s_1+s_2+s_3+\cdots+s_n)H_n(s_1, s_2, s_3, \cdots, s_n). \tag{36}$$

[6] These relations are derived in George [6] on pp. 35–38. Since this reference is not widely available, a brief derivation is given in Appendix V.

[7] See George [6], p. 34, for further discussion.

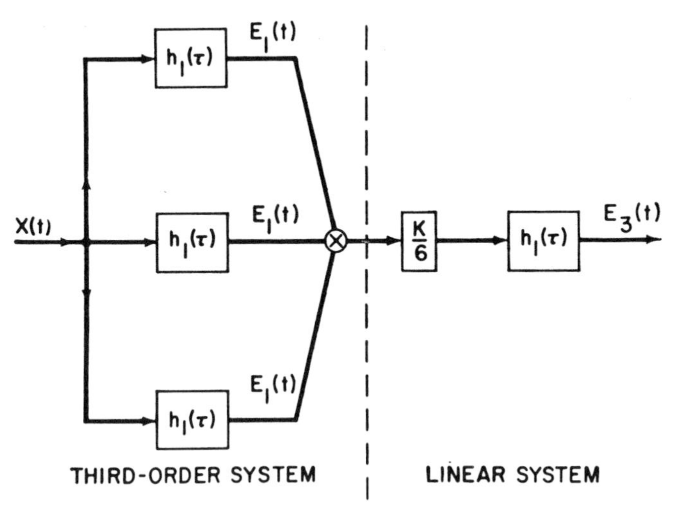

Fig. 7—Third-order kernel.

3) *Final Value Theorem:*[8]
If

$$h_2(t_1, t_2) \leftrightarrow H_2(s_1, s_2), \tag{37}$$

then

$$\int_{-\infty}^{\infty} h_2(t_1, t_2)dt_2 \leftrightarrow H_2(s_1, 0). \tag{38}$$

4) *Reversal:*
If

$$h_2(t_1, t_2) \leftrightarrow H_2(s_1, s_2), \tag{39}$$

then

$$h_2(-t_1, -t_2) \leftrightarrow H_2(-s_1, -s_2). \tag{40}$$

5) *Factorable Transforms:*
If

$$f(t_1, t_2) = f_1(t_1)g_1(t_2), \tag{41}$$

then

$$F_2(s_1, s_2) = F_1(s_1)G_1(s_2). \tag{42}$$

These properties are adequate for the problem of interest.

B. Kernels for Phase-Locked Loop

We may now express the kernels in the transform domain. Clearly, the linear kernel is

$$H_1(s) = \frac{1}{s + K} \cdot \tag{43}$$

To find the transform of (17), we observe that it could be viewed as the cascade shown in Fig. 7. The system to the left of the dotted line is a third-order system.

[8] See George [6] p. 41.

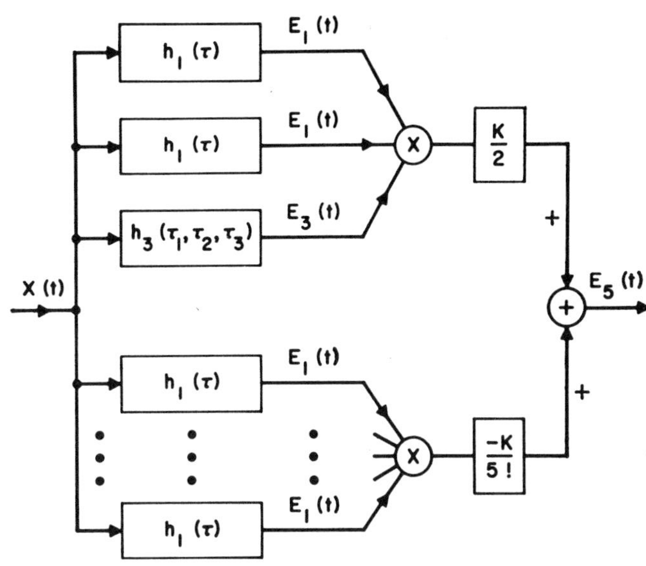

Fig. 8—Fifth-order kernel.

From the property in (41), its transform is

$$H_1(s_1)H_1(s_2)H_1(s_3) = \frac{1}{s_1 + K} \cdot \frac{1}{s_2 + K} \cdot \frac{1}{s_3 + K} \cdot \tag{44}$$

From the property in (36), the transform of the cascade is

$$\frac{K}{6} H_1(s_1 + s_2 + s_3)H_1(s_1)H_1(s_2)H_1(s_3)$$

$$= \frac{K}{6} \frac{1}{s_1 + s_2 + s_3 + K} \cdot \frac{1}{s_1 + K} \cdot \frac{1}{s_2 + K} \cdot \frac{1}{s_3 + K}$$

$$= H_3(s_1, s_2, s_3). \tag{45}$$

Similarly, the fifth-order kernel can be visualized as the system shown in Fig. 8.

The resulting transform is

$$H_5(s_1, s_2, s_3, s_4, s_5)$$

$$= H_1(s_1 + s_2 + s_3 + s_4 + s_5)$$

$$\cdot \left\{ \frac{K}{2} H_1(s_1)H_1(s_2)H_3(s_3, s_4, s_5) - \frac{K}{5!} \prod_{i=1}^{5} H_1(s_i) \right\}. \tag{46}$$

The first, third, and fifth kernels are now available in the transform domain. We will reconsider the problem of finding

$$\langle E_{(5)}{}^2(t) \rangle = \sum_{\tau=1,3,5} \sum_{j=1,3,5} \langle E_i(t)E_j(t) \rangle. \tag{47}$$

In the time domain analysis, the actual calculations were relegated to the Appendix since they were reasonably straightforward. In the transform domain, the techniques are less familiar so we will consider them in more detail.

C. Evaluation of Mean-Square Error

The first term of interest is $\langle E_1(t)E_3(t)\rangle$. From (101), we have

$$\langle E_1(t)E_3(t)\rangle = K^4\left(\frac{N_0}{2A^2}\right)^2 \int_0^\infty \int_0^\infty h_1(\tau)h_3(\tau, \tau_2, \tau_2)d\tau d\tau_2. \quad (48)$$

From the discussion in Section III-A, 1) we know that

$$h_3(\tau, \tau_2, \tau_2) \leftrightarrow H_3(s_1, s_2, s_3)\big|_{s_3 Ⓐ s_2}, \quad (49)$$

where s_3 Ⓐ s_2 means s_3 is associated with s_2. From Section III-A, 3) we know that

$$\int_0^\infty h_3(\tau, \tau_2, \tau_2)d\tau_2 \leftrightarrow H_3(s_1, s_2, s_3)\big|_{\substack{s_3 Ⓐ s_2 \\ s_2\to 0}}. \quad (50)$$

Finally, we observe that

$$\int_0^\infty h_3(\tau, \tau_2, \tau_2)d\tau_2 \equiv f(\tau) \quad (51)$$

is just a function of a single variable. From conventional transform theory

$$\int_0^\infty h_1(\tau)f(\tau)d\tau = \mathfrak{L}^{-1}[H_1(-s)F(s)]\big|_{t=0}. \quad (52)$$

Now let us perform the three indicated steps.

$$H_3(s_1, s_2, s_3) = \frac{K}{6}\,\frac{1}{s_1+s_2+s_3+K}\cdot\frac{1}{s_1+K}\cdot\frac{1}{s_2+K}\cdot\frac{1}{s_3+K}. \quad (53)$$

s_3 Ⓐ s_2 gives

$$H_3(s_1, s_2, s_3)\big|_{s_3 Ⓐ s_2} = \frac{K}{6}\,\frac{1}{s_1+s_2+K}\cdot\frac{1}{s_1+K}\cdot\frac{1}{s_2+2K}. \quad (54)$$

Letting $s_2 \to 0$ gives

$$H_3(s_1, s_2, s_3)\big|_{\substack{s_3 Ⓐ s_2 \\ s_2\to 0}} = \frac{1}{12}\,\frac{1}{(s_1+K)^2}. \quad (55)$$

Recalling that

$$H_1(s_1) = \frac{1}{s_1+K}$$

the third step is

$$\langle E_1(t)E_3(t)\rangle = \frac{K^4}{4}\left(\frac{N_0}{2A^2}\right)^2 \mathfrak{L}^{-1}\left[\frac{1}{-s+K}\cdot\left(\frac{1}{s+K}\right)^2\right]\Bigg|_{t=0} = \frac{K^2}{16}\left(\frac{N_0}{2A^2}\right)^2 = \frac{1}{4}Z^2 \quad (56)$$

which checks with (99). Now consider $\langle E_3^2(t)\rangle$. From (108) we have

$$\langle E_3^2(t)\rangle = K^3\left(\frac{N_0}{2A^2}\right)^3 \cdot\left\{9\int_0^\infty\int_0^\infty\int_0^\infty d\tau_1 d\tau_3 d\tau_5 h_3(\tau_1, \tau_1, \tau_3)h_3(\tau_3, \tau_5, \tau_5) + 6\int_0^\infty\int_0^\infty\int_0^\infty d\tau_1 d\tau_2 d\tau_3 h_3(\tau_1, \tau_2, \tau_3)h_3(\tau_1, \tau_2, \tau_3)\right\}. \quad (57)$$

Consider the first term. We observe that the integrals with respect to τ_1 and τ_5 are identical to (50). Therefore,

$$\langle E_3^2(t)\rangle_{(1)} = 9K^3\left(\frac{N_0}{2A^2}\right)^3 \mathfrak{L}^{-1}\left[\frac{1}{12}\,\frac{1}{(-s+K)^2}\cdot\frac{1}{12}\,\frac{1}{(s+K)^2}\right]\Bigg|_{t=0}. \quad (58)$$

For symmetrical arguments, it is perhaps easier to recall that

$$\mathfrak{L}^{-1}\left[\frac{1}{(-s+K)^2}\cdot\frac{1}{(s+K)^2}\right]_{t=0} = \frac{1}{2\pi j}\int_{-j\infty}^{+j\infty}\frac{1}{(-s+K)^2}\cdot\frac{1}{(s+K)^2}ds, \quad (59)$$

since the integral on the right is tabulated for rational functions up to tenth order.[9] Evaluating, we have

$$\langle E_3^2(t)\rangle_{(1)} = \frac{1}{64}K^3\left(\frac{N_0}{2A^2}\right)^3 = \frac{1}{8}Z^3. \quad (60)$$

The second term is

$$\langle E_3^2(t)\rangle_{(2)} = 6K^3\left(\frac{N_0}{2A^2}\right)^3 \cdot\int_0^\infty\int_0^\infty\int_0^\infty d\tau_1 d\tau_2 d\tau_3 h_3(\tau_1, \tau_2, \tau_3)h_3(\tau_1, \tau_2, \tau_3). \quad (61)$$

Now

$$\int_0^\infty\int_0^\infty\int_0^\infty d\tau_1 d\tau_2 d\tau_3 h_3(t_1+\tau_1, t_2+\tau_2, t_3+\tau_3)h_3(\tau_1, \tau_2, \tau_3) \leftrightarrow H_3(-s_1, -s_2, -s_3)H_3(s_1, s_2, s_3), \quad (62)$$

where we have used the convolution property.

The left side of (62) equals the desired expression when $t_1=t_2=t_3=0$. Thus, we associate s_3 with s_2, then associate s_2 with s_1. Next we take the inverse transform evaluated at $t=0$. [Equally well, one uses the tabulated integral of the form in (59)].

[9] This tabulation is available in James, Nichols, and Phillips [28].

$$H_3(-s_1,-s_2,-s_3)H_3(s_1,s_2,s_3)$$
$$=\frac{K^2}{36}\,\frac{1}{-s_1-s_2-s_3+K}\cdot\frac{1}{-s_1+K}\cdot\frac{1}{-s_2+K}$$
$$\cdot\frac{1}{-s_3+K}\cdot\frac{1}{s_1+s_2+s_3+K}\cdot\frac{1}{s_1+K}\cdot\frac{1}{s_2+K}\cdot\frac{1}{s_3+K}$$
$$=\frac{K^2}{36(2K)^3}\left\{\frac{1}{-s_1-s_2-s_3+K}\cdot\frac{1}{s_1+s_2+s_3+K}\right\}$$
$$\cdot\left\{\frac{1}{s_3+K}+\frac{1}{-s_3+K}\right\}\left\{\frac{1}{s_2+K}+\frac{1}{-s_2+K}\right\}$$
$$\cdot\left\{\frac{1}{s_1+K}+\frac{1}{-s_1+K}\right\}. \tag{63}$$

Now $s_3 Ⓐ s_2$ and then $s_2 Ⓐ S_1$ gives[10]

$$\frac{1}{288K}\left\{\frac{1}{-s+K}\cdot\frac{1}{s+K}\right\}\left\{\frac{6K}{(s+3K)(-s+3K)}\right\}. \tag{64}$$

Taking the inverse transform at $t=0$ gives

$$\langle E_3{}^2(t)\rangle_{(2)}=\frac{K^3}{8}\left(\frac{N_0}{2A^2}\right)^3$$
$$\mathcal{L}^{-1}\left[\frac{1}{-s+K}\cdot\frac{1}{s+K}\cdot\frac{1}{s+3K}\cdot\frac{1}{-s+3K}\right]_{t=0}$$
$$=\frac{1}{192}\left(\frac{KN_0}{2A^2}\right)^3=\frac{1}{24}Z^3. \tag{65}$$

Adding (60) and (65) gives

$$\langle E_3{}^2(t)\rangle=\frac{1}{6}Z^3, \tag{66}$$

which agrees with (106).

Using similar techniques, one can evaluate the higher-order terms in the transform domain.

Now consider the general case when $\dot\theta_1(t)$ is nonzero and/or there is a filter in the loop.

IV. Analysis of General Case

Now consider the general system shown in Fig. 1. A more convenient form for our purposes is shown in Fig. 9.

Here we have included the VCO gain in the loop filter by letting $G(s)=G_1(s)/K$.

First we want to find the kernels of the general system.

The differential equation is

$$\dot E(t)+G(p)\sin E(t)=x(t). \tag{67}$$

[10] One must keep the region of convergence in mind.

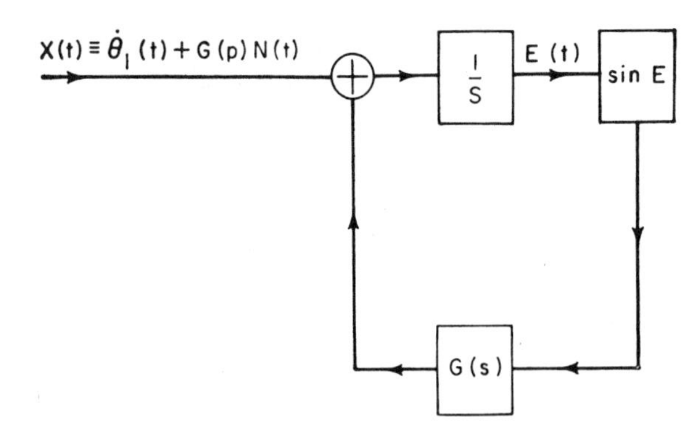

Fig. 9—Modified loop model.

One can show easily that the kernels are

$$H_1(s)=\frac{1}{s+G(s)} \tag{68}$$

$$H_3(s_1,s_2,s_3)=\frac{1}{3!}\,\frac{G(s_1+s_2+s_3)}{s_1+s_2+s_3+G(s_1+s_2+s_3)}\cdot\prod_{i=1}^{3}\frac{1}{s_i+G(s_i)} \tag{69}$$

$$H_5(s_1,s_2,s_3,s_4,s_5)=\frac{G(s_1+s_2+s_3+s_4+s_5)}{\sum_{i=1}^{5}s_i+G\left(\sum_{i=1}^{5}s_i\right)}\cdot\left\{\frac{1}{2}H_1(s_1)H_1(s_2)H_3(s_3,s_4,s_5)-\frac{1}{5!}\prod_{l=1}^{5}H_1(s_i)\right\}. \tag{70}$$

The higher-order kernels can be found similarly.

Now the simplicity of the analysis in the preceding sections rested on the fact that the input was a sample function from a *white*, Gaussian process. We would like to manipulate our system so that we are working with inputs of this type.

Let us assume that $\dot\theta_1(t)$ is a sample function from a *nondeterministic* [29] *nonwhite* Gaussian process. As before, $N(t)$ is a sample function of a *white* Gaussian process. In this case $x(t)$ could be formed as shown in Fig. 10.

Here $z_1(t)$ and $z_2(t)$ are sample functions from independent *white* Gaussian processes.

$$R_{z_1}(\tau)=R_{z_2}(\tau)=\delta(\tau). \tag{71}$$

The upper filter shapes the message. Since $\dot\theta_1(t)$ is nondeterministic, this is always possible. One chooses $L_{\dot\theta_1}(s)$ so that

$$a^2\,|\,L_{\dot\theta_1}(j\omega)\,|^2=S_{\theta 1}(j\omega). \tag{72}$$

In the lower loop the gain b is used to adjust the amplitude of the noise, and $G(s)$ provides the correct spectrum.

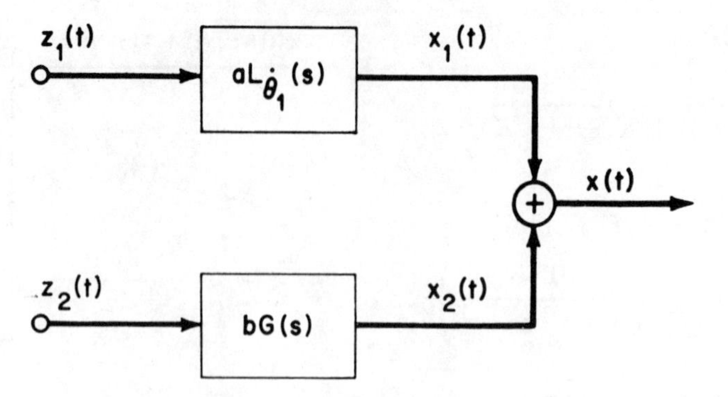

Fig. 10—Input forming filter.

We now consider the cascade of the shaping filter and the set of loop kernels and find the resulting kernels for the over-all system. Observe that we now have a two-input system.

The resulting linear term is[11]

$$E_1(s) = aL_{\theta_1}(s)H_1(s)Z_1(s) + bG_1(s)H_1(s)Z_2(s) \quad (73)$$

or

$$E_1(s) = K_{11}(s)Z_1(s) + K_{12}(s)Z_2(s), \quad (74)$$

where

$$K_{11}(s) \equiv aH_1(s)L_{\theta_1}(s) \quad (75)$$

and

$$K_{12}(s) \equiv bG_1(s)H_1(s). \quad (76)$$

For the third-order case the cascade kernel is obtained by a simple substitution

$$H_3(s_1, s_2, s_3)\left\{\prod_{i=1}^{3} X_1(s_i) + X_2(s_i)\right\}$$
$$= H_3(s_1, s_2, s_3)\big[a^2L_{\dot\theta_1}(s_1)L_{\dot\theta_1}(s_2)L_{\dot\theta_1}(s_3)Z_1(s_1)Z_1(s_2)Z_1(s_3)$$
$$+ 3a^2bL_{\dot\theta_1}(s_1)L_{\dot\theta_1}(s_2)G(s_3)Z_1(s_1)Z_1(s_2)Z_2(s_3)$$
$$+ 3ab^2L_{\dot\theta_1}(s_1)G(s_2)G(s_3)Z_1(s_1)Z_2(s_2)Z_2(s_3)$$
$$+ b^3G_1(s_1)G_1(s_2)G_1(s_3)Z_2(s_1)Z_2(s_2)Z_2(s_3)\big]. \quad (77)$$

Therefore, to specify the third-order cascade term, we need four kernels,

$$K_{3,30}(s_1, s_2, s_3) \equiv a^2L_{\dot\theta_1}(s_1)L_{\dot\theta_1}(s_2)L_{\dot\theta_1}(s_3)H_3(s_1, s_2, s_3) \quad (78)$$

$$K_{3,21}(s_1, s_2, s_3) \equiv 3a^2bL_{\dot\theta_1}(s_1)L_{\dot\theta_1}(s_2)G(s_3)H_3(s_1, s_2, s_3) \quad (79)$$

$$K_{3,12}(s_1, s_2, s_3) \equiv 3ab^2L_{\dot\theta_1}(s_1)G(s_2)G(s_3)H_3(s_1, s_2, s_3) \quad (80)$$

$$K_{3,03}(s_1, s_2, s_3) \equiv b^3G(s_1)G(s_2)G(s_3)H_3(s_1, s_2, s_3). \quad (81)$$

We now have the problem formulated in a manner that allows us to include a random-frequency modulation on the input carrier and an arbitrary filter in the loop. As one would suspect, the complexity of the calculation grows rapidly. For examples, we will consider the easiest, nontrivial cases.

[11] We take the liberty of using transform *notation*. The transform of a process does not exist in general.

A. Second-Order Loop—No Modulation

We consider the case

$$G(s) = 2d\left(1 + \frac{d}{s}\right) \quad (82)$$

and

$$\dot\theta_1(t) = 0. \quad (83)$$

All of the calculations are carried out in Appendix II. The following results are obtained:

Linear approximation

$$\langle E_1^2(t)\rangle = \frac{N_0}{2A^2}\cdot\frac{3d}{2} \equiv Z. \quad (84)$$

[This Z has the same physical significance as in (25). It is the coherent noise-to-signal ratio.]

Second-order approximation

$$\langle E_{(3)}^2(t)\rangle \cong Z + \frac{2}{3}Z^2. \quad (85)$$

The various approximations are plotted in Fig. 11.

B. First-Order Loop, Random Frequency Modulation

As a second example we consider the case of a first-order loop where the instantaneous frequency is a sample function from a Gaussian random process. In this case

$$G_1(s) = K. \quad (86)$$

The gain adjustment of the noise b^2 is

$$b^2 = \frac{N_0}{2A^2}\cdot \quad (87)$$

To characterize the modulation process, we need only to specify the correlation function.

$$R_{\dot\theta_1}(\tau) = a^2e^{-\alpha|\tau|}. \quad (88)$$

The calculations are carried out in Appendix III. The following results are obtained:

$$\langle E_1^2(t)\rangle = P\frac{1}{r(1+r)} + Qr \quad (89)$$

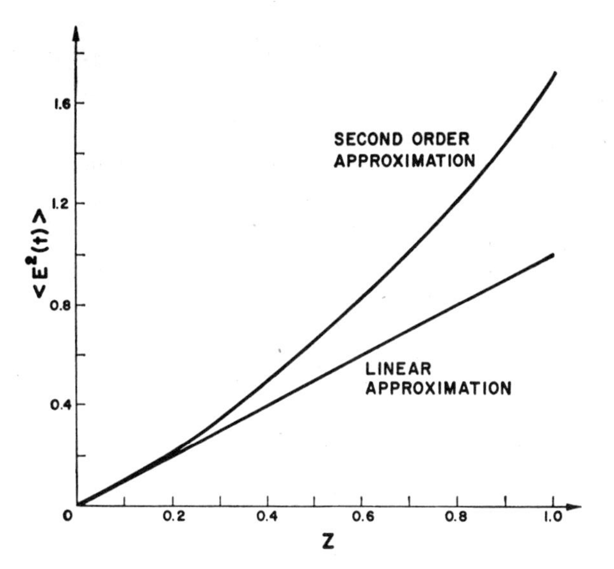

Fig. 11—Mean-square error in second-order loop.

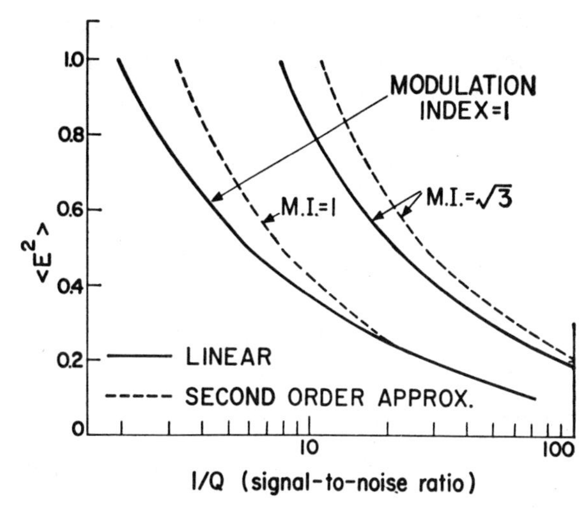

Fig. 12—Mean-square error as a function of modulation index and signal-to-noise ratio.

where we have defined

$$P = \frac{a^2}{\alpha^2}: \sim \frac{\text{mean-square value of instantaneous frequency}^{12}}{\text{square of effective bandwidth of process}} \tag{90}$$

$$Q = \frac{b^2\alpha}{2} = \frac{N_0}{2A^2}\cdot\frac{\alpha}{2}: \text{ coherent noise-to-signal ratio in modulation bandwidth} \tag{91}$$

$$r = \frac{K}{\alpha}: \text{ ratio of loop bandwidth to modulation-process bandwidth.} \tag{92}$$

Now one can choose r as a function of P and Q in order to minimize the linear approximation to the error. From (145) we observe that $r_{\min}$ is only a function of P/Q. In Fig. 12, we show the mean-square error as a function of Q. The parameter $P^{1/2}$ = modulation index determines which curve is appropriate.

The second-order approximation to the error is

$$\begin{aligned}\langle E^2(t)\rangle_{(2)} &= P\frac{1}{r(1+r)} + Qr \\ &+ 2\left[P^2\left[\frac{1+r-2r^2}{4r^2[1-r][1+r]^3}\right]\right. \\ &\left.+ PQ\left[\frac{3r^2+13r+6}{12[1+r]^3}\right] + Q^2\left[\frac{r^2}{4}\right]\right]\end{aligned} \tag{93}$$

or

$$\begin{aligned}\langle E^2(t)\rangle &= Q\left[\frac{P}{Q}\,\frac{1}{r(1+r)} + r\right] \\ &+ 2Q^2\left[\frac{P^2}{Q^2}f_1(r) + \frac{P}{Q}f_2(r) + \frac{r^2}{4}\right].\end{aligned} \tag{94}$$

[12] This corresponds to the square of the modulation index.

This can be interpreted most conveniently by assuming that we use $r = r_{\min}$ from Fig. 13 (page 908). The value of the mean-square error calculated using a second-order approximation is shown in Fig. 12.

V. Conclusions

The purpose of the paper was twofold. The first goal was to demonstate the efficiency of using functional expansion techniques in the analysis of nonlinear feedback systems. Several observations regarding the technique may be made.

One observes that the technique appears to be reasonably complex. However, if one looks at the calculations in detail, one sees that most of the operations involve the same type of operations. Therefore, after learning a few properties of multidimensional kernels and their transforms the analysis becomes straightforward. The properties derived are useful for a large class of nonlinear systems.

An obvious advantage is the ability to solve the problem for random inputs and noise. Aside from a few simple cases, no accurate answers were previously available for problems of this type[13] in the low signal-to-noise ratio region.

[13] The reviewer has suggested a comparison of the functional expansion technique with Booton's "equivalent gain" technique. The two techniques are compared in Appendix V for the case of a second-order loop with constant frequency input.

One should also observe that when one applies the Volterra functional expansion technique to a nonlinear system, we are essentially performing a perturbation analysis. The resulting solution will converge only for some region of the system inputs. Thus, in general, one can not predict instabilities using this technique.

The second goal was to obtain new results for an important nonlinear system. We considered a model of a phase-locked loop and solved three interesting cases.

1) First-order loop; constant frequency input.
2) Second-order loop; constant frequency input.
3) First-order loop; random frequency input.

Of these cases, only the first has been solved previously.

Finally one should observe that the paper was not intended to be encyclopedic. There are a number of interesting problems which can be considered using the techniques described.

1) Transient response to a step change in frequency or phase (see Barrett [5]).
2) Use of a nonlinear filter in the loop to optimize the system performance (see George [6]).
3) Adjustment of the parameters in a linear filter to minimize variance calculated using a second- or third-order approximation (a straightforward minimization).

We have attempted to include enough examples and detail so that the interested reader is equipped to solve new cases on his own.

Appendix I

We will evaluate the terms in (24). The terms of initial interest are: $\langle E_1^2(t)\rangle$, $\langle E_1(t)E_3(t)\rangle$, $\langle E_3^2(t)\rangle$, and $\langle E_1(t)E_5(t)\rangle$.

A. Evaluation of $\langle E_1^2(t)\rangle$

This is simply the linear approximation.

$$\langle E_1^2(t)\rangle = \int_0^\infty \int_0^\infty h_1(\tau_1)h_1(\tau_2)\langle N(t-\tau_1)N(t-\tau_2)\rangle d\tau_1 d\tau_2$$

$$= \int_0^\infty \int_0^\infty e^{-K\tau_1 - K\tau_2}\left(\frac{N_0}{2A^2}\right)\delta(\tau_1-\tau_2) = \frac{KN_0}{4A^2}. \quad (95)$$

The ratio $\frac{KN_0}{4A^2}$ will appear frequently. We define

$$\frac{KN_0}{4A^2} \equiv Z. \quad (96)$$

B. Evaluation of $\langle E_1(t)E_3(t)\rangle$

Using (13) and (17) we write

$$\langle E_1(t)E_3(t)\rangle = \frac{K}{3!}\int_0^\infty\int_0^\infty\int_0^\infty\int_0^\infty\int_0^\infty d\tau d\tau_1 d\tau_2 d\tau_3 d\tau_4 \cdot h_1(\tau)h_1(\tau_1)h_1(\tau_2)h_1(\tau_3)h_1(\tau_4) \cdot\langle x(t-\tau_4)x(t-\tau-\tau_1)x(t-\tau-\tau_2) \cdot x(t-\tau-\tau_3)\rangle. \quad (97)$$

Since the fourth moment of a Gaussian process factors, (97) becomes

$$\langle E_1(t)E_3(t)\rangle = \frac{K}{3!}\int_0^\infty\int_0^\infty\int_0^\infty\int_0^\infty\int_0^\infty d\tau d\tau_1 d\tau_2 d\tau_3 d\tau_4 \cdot h_1(\tau)h_1(\tau_1)h_1(\tau_2)h_1(\tau_3)h_1(\tau_4)\ 3K^4\left(\frac{N_0}{2A^2}\right)^2 \cdot\{u_0(\tau_4-\tau-\tau_1)u_0(\tau_2-\tau_3)\}(3), \quad (98)$$

where we have used the symmetry with respect to variables of interest. This reduces to

$$\langle E_1(t)E_3(t)\rangle = \frac{K^5}{2}\left(\frac{N_0}{2A^2}\right)^2\int_0^\infty h_1^2(\tau_2)d\tau_2\int_0^\infty h_1(\tau_4)d\tau_4 \cdot\int_0^\infty h_1(\tau_4-\tau)h_1(\tau)d\tau$$

$$= \frac{1}{4}\left(\frac{KN_0}{4A^2}\right)^2 = \frac{1}{4}Z^2. \quad (99)$$

We also observe that using (13) and (18) we could write

$$\langle E_1(t)E_3(t)\rangle = \int_0^\infty\int_0^\infty\int_0^\infty\int_0^\infty d\tau d\tau_1 d\tau_2 d\tau_3 h_1(\tau)h_3(\tau_1,\tau_2,\tau_3) \cdot\langle x(t-\tau)x(t-\tau_1)x(t-\tau_2)x(t-\tau_3)\rangle. \quad (100)$$

If $h_3(\tau_1,\tau_2,\tau_3)$ is expressed in a symmetrical form, this reduces to

$$\langle E_1(t)E_3(t)\rangle = K^4\left(\frac{N_0}{2A^2}\right)^2\int_0^\infty\int_0^\infty h_1(\tau)h_3(\tau,\tau_2,\tau_2)d\tau d\tau_2. \quad (101)$$

At the moment this expression is not useful since we have not solved for $h_3(\tau_1,\tau_2,\tau_3)$ explicitly. However, we will find it necessary for the transform-domain analysis.

C. Evaluation of $\langle E_3^2(t)\rangle$

From (17) we have

$$\langle E_3^2(t)\rangle = \frac{K^2}{36}\int_0^\infty\cdots\int_0^\infty d\tau_1 d\tau_2 d\tau_3 d\tau_4 d\tau_5 d\tau_6 d\tau_\alpha d\tau_\beta \cdot h_1(\tau_1)\cdots h_1(\tau_6)h_1(\tau_\alpha)h_1(\tau_\beta) \cdot\langle x(t-\tau_\alpha-\tau_1)x(t-\tau_\alpha-\tau_2)x(t-\tau_\alpha-\tau_3) \cdot x(t-\tau_\beta-\tau_4)x(t-\tau_\beta-\tau_5)x(t-\tau_\beta-\tau_6)\rangle. \quad (102)$$

The sixth-order moment factors, giving 15 terms. Recognizing symmetry, the expectation becomes

$$\langle x(t-\tau_\alpha-\tau_1)\cdots x(t-\tau_\beta-\tau_6)\rangle$$

$$= K^6\left(\frac{N_0}{2A^2}\right)^3$$

$$\cdot\{9u_0(\tau_1-\tau_2)u_0(\tau_\alpha+\tau_3-\tau_\beta-\tau_4)u_0(\tau_5-\tau_6)$$

$$+6u_0(\tau_\alpha+\tau_1-\tau_\beta-\tau_4)u_0(\tau_\alpha+\tau_2-\tau_\beta-\tau_5)$$

$$\cdot u_0(\tau_\alpha+\tau_3-\tau_\beta-\tau_6)\}. \tag{103}$$

$$\langle E_3^2(t)\rangle = \frac{K^8}{36}\left(\frac{N_0}{2A^2}\right)^3\left\{9\left[\int_0^\infty e^{-2K\tau_1}d\tau_1\right]^2\int_0^\infty e^{-K\tau_3}d\tau_3\right.$$

$$\cdot\int_0^\infty e^{-K\tau_\alpha}d\tau_\alpha\int_0^\infty e^{-K\tau_\beta}d\tau_\beta$$

$$\cdot e^{-K(\tau_\alpha+\tau_3-\tau_\beta)}u_{-1}(\tau_\alpha+\tau_3-\tau_\beta)$$

$$+6\int_0^\infty e^{-K\tau_\alpha}d\tau_\alpha\int_0^\infty e^{-K\tau_\beta}d\tau_\beta\int_0^\infty e^{-K\tau_1}d\tau_1\int_0^\infty$$

$$\cdot e^{-K\tau_2}d\tau_2\int_0^\infty e^{-K\tau_3}d\tau_3$$

$$\cdot e^{-K(\tau_\alpha-\tau_\beta+\tau_1)}u_{-1}(\tau_\alpha-\tau_\beta+\tau_1)$$

$$\cdot e^{-K(\tau_\alpha-\tau_\beta+\tau_2)}u_{-1}(\tau_\alpha-\tau_\beta+\tau_2)$$

$$\left.\cdot e^{-K(\tau_\alpha-\tau_\beta+\tau_3)}u_{-1}(\tau_\alpha-\tau_\beta+\tau_3)\right\} \tag{104}$$

$$\langle E_3^2(t)\rangle = \frac{K^8}{36}\left(\frac{N_0}{2A^2}\right)^3$$

$$\cdot\left\{\frac{9}{4K^2}\int_0^\infty e^{-2K\tau_3}d\tau_3\int_0^\infty e^{-2K\tau_\alpha}d\tau_\alpha(\tau_3+\tau_\alpha)\right.$$

$$+6\int_0^\infty e^{+2K\tau_\beta}d\tau_\beta\int_0^\infty e^{-4K\tau_\alpha}d\tau_\alpha$$

$$\cdot\left[\int_{\tau_\beta-\tau_\alpha}^\infty e^{-2K\tau_1}d\tau_1\right]^3$$

$$+6\int_0^\infty e^{-4K\tau_\alpha}d\tau_\alpha\int_0^{\tau_\alpha} e^{+2K\tau_\beta}d\tau_\beta$$

$$\left.\cdot\left[\int_0^\infty e^{-2K\tau_1}d\tau_1\right]^3\right\}. \tag{105}$$

Evaluating, we have

$$\langle E_3^2(t)\rangle = \frac{K^3}{48}\left(\frac{N_0}{2A^2}\right)^3 = \frac{1}{6}Z^3. \tag{106}$$

Once again, we observe that using (18) we could write

$$\langle E_3^2(t)\rangle = \int_0^\infty\cdots\int_0^\infty d\tau_1 d\tau_2 d\tau_3 d\tau_4 d\tau_5 d\tau_6 h_3(\tau_1,\tau_2,\tau_3)$$

$$\cdot h_3(\tau_4,\tau_5,\tau_6)\langle x(t-\tau_1)x(t-\tau_2)x(t-\tau_3)$$

$$\cdot x(t-\tau_4)x(t-\tau_5)x(t-\tau_6)\rangle. \tag{107}$$

Using symmetry, this becomes

$$\langle E_3^2(t)\rangle = K^3\left(\frac{N_0}{2A^2}\right)^3$$

$$\cdot\left\{9\int_0^\infty\int_0^\infty\int_0^\infty d\tau_1 d\tau_3 d\tau_5 h_3(\tau_1,\tau_1,\tau_3)\right.$$

$$\cdot h_3(\tau_3,\tau_5,\tau_5)+6\int_0^\infty\int_0^\infty\int_0^\infty d\tau_1 d\tau_2 d\tau_3$$

$$\left.\cdot h_3(\tau_1,\tau_2,\tau_3)h_3(\tau_1,\tau_2,\tau_3)\right\}. \tag{108}$$

We will use this form of the expectation in our transform domain analysis.

D. Evaluation of $\langle E_1(t)E_5(t)\rangle$

From (20) we have

$$\langle E_1(t)E_5(t)\rangle$$

$$=\frac{K^2}{12}\int_0^\infty\cdots\int_0^\infty d\tau\cdots d\tau_7 h_1(\tau)\cdots h(\tau_7)$$

$$\cdot\langle x(t-\tau)x(t-\tau_1-\tau_2)x(t-\tau_1-\tau_3)x(t-\tau_1-\tau_4-\tau_5)$$

$$\cdot x(t-\tau_1-\tau_4-\tau_6)x(t-\tau_1-\tau_4-\tau_7)\rangle$$

$$-\frac{K}{5!}\int_0^\infty\cdots\int_0^\infty d\tau d\tau_1\cdots d\tau_6 h_1(\tau)h_1(\tau_1)\cdots h_1(\tau_6)$$

$$\cdot\langle x(t-\tau)x(t-\tau_1-\tau_2)x(t-\tau_1-\tau_3)x(t-\tau_1-\tau_4)$$

$$\cdot x(t-\tau_1-\tau_5)x(t-\tau_1-\tau_6)\rangle. \tag{109}$$

Consider the first term in (109)

$$\langle E_1(t)E_5(t)\rangle_{(1)}$$

$$=\frac{K^2}{12}K^6\left(\frac{N_0}{2A^2}\right)^3$$

$$\cdot\int_0^\infty\cdots\int_0^\infty d\tau\cdots d\tau_7 h_1(\tau)\cdots h_1(\tau_7)$$

$$\cdot[6\delta(\tau-\tau_1-\tau_2)\delta(\tau_3-\tau_4-\tau_5)\delta(\tau_6-\tau_7)$$

$$+3\delta(\tau-\tau_1-\tau_4-\tau_5)\delta(\tau_2-\tau_3)\delta(\tau_6-\tau_7)$$

$$+6\delta(\tau-\tau_1-\tau_4-\tau_5)\delta(\tau_2-\tau_4-\tau_6)$$

$$\cdot\delta(\tau_3-\tau_4-\tau_7)]. \tag{110}$$

$$\langle E_1(t)E_5(t)\rangle_{(1)}$$

$$=\frac{K^8}{12}\left(\frac{N_0}{2A^2}\right)^3$$

$$\cdot\left\{6\int_0^\infty\cdots\int_0^\infty h_1(\tau_1+\tau_2)h_1(\tau_4+\tau_5)h_1(\tau_7)\right.$$

$$\cdot h_1(\tau_1)h_1(\tau_2)h_1(\tau_4)h_1(\tau_5)h_1(\tau_7)d\tau_1 d\tau_2 d\tau_4 d\tau_5 d\tau_7$$

$$+3\int_0^\infty\cdots\int_0^\infty h_1(\tau_1+\tau_4+\tau_5)h_1(\tau_3)h_1(\tau_7)$$

$$\cdot h_1(\tau_1)h_1(\tau_3)h_1(\tau_4)h_1(\tau_5)h_1(\tau_7)d\tau_1\cdots d\tau_7$$

$$+6\int_0^\infty\cdots\int_0^\infty d\tau_1\cdots d\tau_7 h_1(\tau_1)h_1(\tau_1+\tau_4+\tau_5)$$

$$\left.\cdot h_1(\tau_4+\tau_6)h_1(\tau_4+\tau_7)h_1(\tau_4)h_1(\tau_5)h_1(\tau_6)h_1(\tau_7)\right\}. \tag{111}$$

Evaluating we have

$$\langle E_1(t)E_5(t)\rangle_{(1)} = \frac{1}{4}Z^3. \tag{112}$$

Now consider the second term. All 15 terms are the same. Therefore,

$$\begin{aligned}\langle E_1(t)E_5(t)\rangle_{(2)} &= -\frac{K}{5!}(15)K^6\left(\frac{N_0}{2A^2}\right)^3 \\ &\cdot \int_0^\infty \cdots \int_0^\infty d\tau d\tau_1 \cdots d\tau_6 h_1(\tau)h_1(\tau_1)\cdots h_1(\tau_6) \\ &\cdot \{\delta(\tau-\tau_1-\tau_2)\delta(\tau_3-\tau_4)\delta(\tau_5-\tau_6)\}.\end{aligned} \tag{113}$$

This becomes

$$-\frac{K^7}{8}\left(\frac{N_0}{2A^2}\right)^3 \int_0^\infty \cdots \int_0^\infty h_1(\tau_1+\tau_2)h_1(\tau_4)h_1(\tau_6) \cdot h_1(\tau_1)h_1(\tau_2)h_1(\tau_4)h_1(\tau_6)d\tau_1 \cdots d\tau_6. \tag{114}$$

Evaluating, we have

$$\langle E_1(t)E_5(t)_{(2)} = -\frac{1}{16}Z^3. \tag{115}$$

One can show that the terms of $\langle E_3(t)E_5(t)\rangle$ are fourth order in Z. Therefore, an expression for the variance which is accurate to the third order in Z is

$$\begin{aligned}\langle E_{(5)}{}^2(t)\rangle &= Z + 2\left[\frac{1}{4}Z^2\right] + \frac{1}{6}Z^3 + 2\left[\frac{3}{16}Z^3\right] \\ &= z + \frac{1}{2}Z^2 + \frac{13}{24}Z^3.\end{aligned} \tag{116}$$

Appendix II

We consider the case

$$G(s) = 2d\left(1+\frac{d}{s}\right) \tag{117}$$

and

$$\dot{\theta}_1(t) = 0 \tag{118}$$

$$H_1(s) = \frac{1}{s+G(s)} = \frac{1}{s+2d\left(1+\frac{d}{s}\right)} = \frac{s}{s^2+2ds+2d^2}. \tag{119}$$

$$K_{12}(s) = bG_1(s)H_1(s) = b\left\{\frac{2ds+2d^2}{s^2+2ds+2d^2}\right\} \tag{120}$$

$$\langle E_1{}^2(t)\rangle = b^2\cdot\frac{1}{2\pi j}\int_{-j\infty}^{+j\infty} |K_{12}(s)|^2 ds = b^2\cdot\frac{3d}{2}\cdot \tag{121}$$

Setting $b^2 = N_0/2A^2$, we have the familiar linear result

$$\langle E_1{}^2(t)\rangle = \frac{N_0}{2A^2}\cdot\frac{3d}{2} \equiv Z. \tag{122}$$

This ratio will appear frequently so we will call it Z.[14] By analogy with (101) we have

$$\langle E_1(t)E_3(t)\rangle = \int_0^\infty\int_0^\infty 3k_{12}(\tau)k_{3,03}(\tau,\tau_1,\tau_1)d\tau d\tau_1 \tag{123}$$

where from (81)

$$K_{3,03}(s_1,s_2,s_3) = b^3G(s_1)G(s_2)G(s_3)H_3(s_1,s_2,s_3). \tag{124}$$

Following the approach of Section III-C we have

$$K_{3,03}(s_1,s_2,s_3) = b^3\frac{1}{3!}\frac{2d\left(\sum_{i=1}^3 s_i\right)+2d^2}{\left(\sum_{i=1}^3 s_i\right)^2+2d\left(\sum_{i=1}^3 s_i\right)+2d^2} \cdot \prod_{i=1}^3 \frac{2ds_i+2d^2}{s_i^2+2ds_i+2d^2} \tag{125}$$

or

$$K_{3,03}(s_1,s_2,s_3) = \frac{b^3}{6}\frac{2d\left(\sum_{i=1}^3 s_i\right)+2d^2}{(\sum s_i)^2+2d(\sum s_i)+2d^2} \cdot \prod_{i=1}^3\left\{\frac{d}{s_i+(d+jd)} + \frac{d}{s_i+(d-jd)}\right\}. \tag{126}$$

Associating s_3 with s_2, we have

$$\begin{aligned}K_{3,03}(s_1,s_2,s_3)\,|_{s_3 \text{Ⓐ} s_2} &= \frac{b^3}{6}\frac{2d(s_1+s_2)+2d^2}{(s_1+s_2)^2+2d(s_1+s_2)+2d^2} \\ &\cdot\left[\frac{2ds_1+2d^2}{s_1^2+2ds_1+2d^2}\right]\left[\frac{2d^2}{s_2+2d} + \frac{d^2}{s_2+2(d+jd)} + \frac{d^2}{s_2+2(d+jd)}\right].\end{aligned} \tag{127}$$

[14] It has the same physical significance as the Z defined in Appendix I. It is the coherent noise-to-signal ratio.

Now set $s_2=0$

$$K_{3,03}(s_1, s_2, s_3)\Big|_{s_2=0} = \frac{b^3}{6}\left(\frac{2ds_1 + 2d^2}{s_1^2 + 2ds + 2d^2}\right)^2 \cdot \frac{3}{2} d. \quad (128)$$

Therefore,

$$\langle E_1(t)E_3(t)\rangle = \frac{3b^4 d}{4} \mathcal{L}^{-1}\left\{\frac{-2ds + 2d^2}{(-s)^2 + 2d(-s) + 2d^2} \cdot \left[\frac{2ds + 2d^2}{s^2 + 2ds + 2d^2}\right]^2\right\}\Bigg|_{t=0}. \quad (129)$$

Using residues, we have

$$\langle E_1(t)E_3(+)\rangle = \frac{3b^4 d^2}{4} = 3\left(\frac{N_0}{2A^2}\right)^2 \frac{d^2}{4} = \left(\frac{N_0}{2A^2}\cdot\frac{3d}{2}\right)^2 \cdot \frac{1}{3} = \frac{1}{3} Z^2. \quad (130)$$

Thus the second-order approximation to $\langle E^2(t)\rangle$ is

$$\langle E^2(t)\rangle \cong Z + \frac{2}{3} Z^2. \quad (131)$$

Appendix III

First-Order Loop with Random Frequency Modulation

In this Appendix we evaluate the various terms for the conditions of Section IV-B.

From (86), (87), and (88) we have

$$G_1(s) = K \quad (132)$$

$$b^2 = N_0^* = \frac{N_0}{2A^2} \quad (133)$$

$$R_{\dot{\theta}_1}(\tau) = a^2 e^{-\alpha|\tau|}. \quad (134)$$

Now

$$S_{\dot{\theta}_1}(j\omega) = a^2 \frac{2\alpha}{\omega^2 + \alpha^2}. \quad (135)$$

From (72) we have

$$L_{\dot{\theta}_1}(j\omega) = \frac{\sqrt{2\alpha}}{j\omega + \alpha}. \quad (136)$$

From (73) the output of the linear term is

$$E_1(s) = aL_{\dot{\theta}_1}(s)H_1(s)Z_1(s) + bG_1(s)H_1(s)Z_2(s) = K_{11}(s)Z_1(s) + K_{12}(s)Z_2(s). \quad (137)$$

In our case

$$K_{11}(s) = \frac{a\sqrt{2\alpha}}{(s+\alpha)(s+K)} \quad (138)$$

and

$$K_{12}(s) = \frac{bK}{(s+K)}. \quad (139)$$

Since $Z_1(t)$ and $Z_2(t)$ are independent, we have

$$\langle E_1^2(t)\rangle = \frac{1}{2\pi j}\int_{-j\infty}^{+j\infty}\left\{\left|\frac{a\sqrt{2\alpha}}{(s+\alpha)(s+K)}\right|^2 + \left|\frac{bK}{(s+K)}\right|^2\right\} ds. \quad (140)$$

Evaluating, we have

$$\langle E_1^2(t)\rangle = \frac{a^2}{K(\alpha+K)} + \frac{b^2 K}{2}. \quad (141)$$

This can be rewritten as

$$\langle E_1^2(t)\rangle = \frac{a^2}{\alpha^2}\,\frac{1}{\frac{K}{\alpha}\left(1+\frac{K}{\alpha}\right)} + \frac{b^2\alpha}{2}\,\frac{K}{\alpha}. \quad (142)$$

We observe that $\alpha/2$ is the noise bandwidth of the phase process.

It is convenient to introduce the following parameters.

$$P = \frac{a^2}{\alpha^2}\left(\frac{\text{mean-square value of instantaneous frequency}}{\text{square of effective bandwidth of process}}\right)$$

$$Q = \frac{b^2\alpha}{2} \text{ (coherent noise-to-signal ratio in modulation bandwidth)}$$

$$r = \frac{K}{\alpha} \text{ (ratio of loop bandwidth to modulation-process bandwidth.)}$$

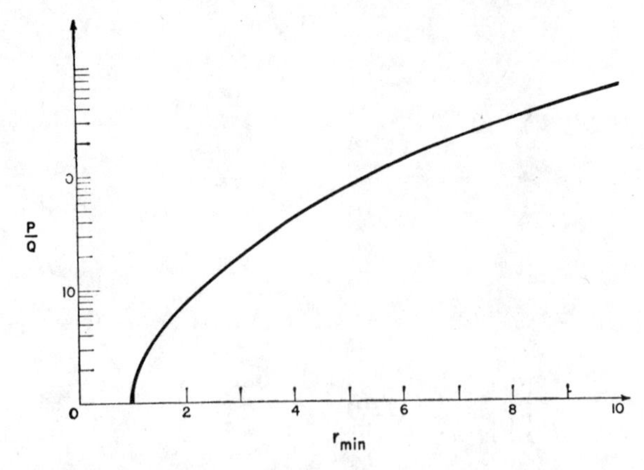

Fig. 13—Ratio of bandwidths for minimum error.

Then

$$\langle E_1{}^2(t)\rangle = P\frac{1}{r(1+r)} + Qr. \tag{143}$$

We can choose r to minimize $\langle E_1{}^2(t)\rangle$

$$\frac{\partial}{\partial r}\langle E_1{}^2(t)\rangle = P\frac{-(2r+1)}{(r^2+r)^2} + Q = 0, \tag{144}$$

or

$$r^4 + 2r^3 + r^2 - 2\frac{P}{Q}r - \frac{P}{Q} = 0 \tag{145}$$

The solution for minimum r is shown in Fig. 13. Now consider the first cross term

$$\begin{aligned}\langle E_1(t)E_3(t)\rangle = \langle\{&k_{11}(t)\otimes z_1(t) + k_{12}(t)\otimes z_2(t)\}\\ &\cdot\{k_{3,30}(t_1,t_2,t_3)\otimes z_1(t_1)z_1(t_2)z_1(t_3)\\ &+ k_{3,21}(t_1,t_2,t_3)\otimes z_1(t_1)z_1(t_2)z_2(t_3)\\ &+ k_{3,12}(t_1,t_2,t_3)\otimes z_1(t_1)z_2(t_2)z_2(t_3)\\ &+ k_{3,03}(t_1,t_2,t_3)\otimes z_2(t_1)z_2(t_2)z_2(t_3)\}\rangle. \end{aligned}\tag{146}$$

Since $z_1(t)$ and $z_2(t)$ are independent and the fourth-order moments factor, there are four nonzero terms.

$$\begin{aligned}\langle E_1(t)E_3(t)\rangle^{(1)} &\equiv \langle[k_{11}(t)\otimes z_1(t)][k_{3,30}(t_1,t_2,t_3)\\ &\qquad \otimes z_1(t_1)z_1(t_2)z_1(t_3)]\rangle\\ &= 3\int_0^\infty\int_0^\infty k_{11}(t_1)k_{3,30}(t_1,t_2,t_2)dt_1dt_2. \end{aligned}\tag{147}$$

This term is of the same form as (48). From (78), (136), and (69), we have

$$K_{3,30}(s_1,s_2,s_3) \equiv a^3\prod_{i=1}^{3}\frac{\sqrt{2\alpha}}{(s_i+\alpha)(s_i+K)}\,\frac{1}{3!}\,\frac{K}{\prod_{i=1}^{3}(s_i+K)}\cdot \tag{148}$$

Using the same approach as (48), we associate s_3 with s_2 and let $s_2=0$.

$$\begin{aligned}K_{3,30}(s_1,s_2,s_3) = \frac{a^3(2\alpha)^{3/2}K}{6(\alpha-K)^3}&\left\{\frac{1}{s_1+\alpha} - \frac{1}{s_1+K}\right\}\\ &\cdot\left\{\frac{1}{s_2+\alpha} - \frac{1}{s_2+K}\right\}\\ &\cdot\left\{\frac{1}{s_3+\alpha} - \frac{1}{s_3+K}\right\}\\ &\cdot\frac{1}{s_1+s_2+s_3+K}\cdot \end{aligned}\tag{149}$$

Then

$$K_{3,30}(s_1,s_2,s_3)\Big|_{s_3 \,\text{Ⓐ}\, s_2,\ s_2\to 0} = \frac{a^3(2\alpha)^{1/2}}{6(\alpha+K)}\,\frac{1}{(s_1+\alpha)(s_1+K)^2}, \tag{150}$$

$$\begin{aligned}\therefore \langle E_1(t)E_3(t)\rangle^{(1)} &= 3\mathfrak{L}^{-1}\left[\frac{a\sqrt{2}\,\alpha}{(-s+\alpha)(-s+K)}\cdot\frac{a^3(2\alpha)^{1/2}}{(s+\alpha)(s+K)^2 6(\alpha+K)}\right]\Bigg|_{t=0}\\ &= \frac{a^4[\alpha^2+\alpha K - 2K^2]}{4K^2[\alpha-K][\alpha+K]^3}\cdot \end{aligned}\tag{151}$$

$$\begin{aligned}\langle E_1(t)E_3(t)\rangle^{(2)} &\equiv \langle\{[k_{11}(t)\otimes z_1(t)][k_{3,12}(t_1,t_2,t_3)\\ &\qquad \otimes z_1(t)z_2(t_2)z_2(t_3)]\rangle\\ &= \int_0^\infty\int_0^\infty k_{11}(t_1)k_{3,12}(t_1,t_2,t_2)dt_1dt_2, \end{aligned}\tag{152}$$

which is exactly the same form as (148). From (80), (136), and (69) we have

$$\begin{aligned}K_{3,12}(s_1,s_2,s_3) &= ab^2(\sqrt{2\alpha}\,K^2)\left[\frac{1}{s_1+\alpha} + \frac{1}{s_2+\alpha} + \frac{1}{s_3+\alpha}\right\}\\ &\left\{\prod_{i=1}^{3}\frac{1}{(s_i+K)}\cdot\frac{K}{\left(\sum_{i=1}^{3}s_i\right)+K}\cdot\frac{1}{3!}\right\} \end{aligned}\tag{153}$$

$$K_{3,12}(s_1, s_2, s_3)\big|_{\substack{s_3 \text{Ⓐ} s_2 \\ s_2 \to 0}}$$

$$= \frac{ab^2\sqrt{2\alpha}\,K^2}{12(\alpha+K)} \frac{2s_1 + 3\alpha + K}{(s_1+\alpha)(s_1+K)^2}. \quad (154)$$

$$\langle E_1(t)E_3(t)\rangle^{(2)}$$

$$= \mathcal{L}^{-1}\left[\frac{a^2b^2(2\alpha)K^2}{12(\alpha+K)} \frac{(2s+3\alpha+K)}{(-s+\alpha)(-s+K)(s+\alpha)(s+K)^2}\right]$$

$$= +\frac{a^2b^2}{24} \frac{3\alpha^3 + 6\alpha^2K - 7\alpha K^2 - 2K^3}{(\alpha-K)(\alpha+K)^3}$$

$$= +\frac{PQ}{12} \frac{3 + 6r - 7r^2 - 2r^3}{[1-r][1+r]^3} \quad (155)$$

$$\langle E_1(t)E_3(t)\rangle^{(3)}$$

$$\equiv \langle [k_{12}(t) \otimes z_2(t)][k_{3,21}(t_1,t_2,t_3) \otimes z_1(t_1)z_1(t_2)z_2(t_3)]\rangle$$

$$= \int_0^\infty \int_0^\infty k_{12}(t_1)k_{3,21}(t_1, t_2, t_2)dt_1dt_2. \quad (156)$$

From (81), (136), and (69)

$$K_{3,21}(s_1,s_2,s_3) = \frac{a^2bK\cdot 2\alpha}{6}\left\{\left[\frac{1}{s_1+\alpha}\cdot\frac{1}{s_2+\alpha} + \frac{1}{s_1+\alpha}\cdot\frac{1}{s_3+\alpha} + \frac{1}{s_2+\alpha}\cdot\frac{1}{s_3+\alpha}\right]\left[\left(\prod_{i=1}^{3}\frac{1}{s_i+K}\right)\frac{K}{\left(\sum_{i=1}^{3}s_i\right)+K}\right]\right\}. \quad (157)$$

Then

$$K_{3,21}(s_1,s_2,s_3)\big|_{\substack{s_3 \text{Ⓐ} s_2 \\ s_2 = 0}}$$

$$= \frac{a^2bK^2\alpha}{3}\left[\frac{1}{2\alpha K(\alpha+K)}\left(\frac{s_1+3\alpha}{s_1+\alpha}\right)\frac{1}{(s_1+K)^2}\right]. \quad (158)$$

Then

$$= \mathcal{L}^{-1}\left[\frac{bK}{(-s+K)}\cdot\frac{a^2bK}{6(\alpha+K)}\frac{s+3\alpha}{(s+\alpha)(s+K)^2}\right]\Bigg|_{t=0}$$

$$= \frac{a^2b^2}{24}\frac{3\alpha+K}{(\alpha+K)^2} = \frac{a^2}{\alpha^2}\frac{b^2\alpha}{2}\frac{3+r}{(1+r)^2}$$

$$= \frac{1}{12}PQ\frac{(3+r)}{(1+r)^2}. \quad (159)$$

The last term is identical to (56)

$$\therefore \langle E_1(t)E_3(t)\rangle^{(4)} = \frac{K^2b^4}{16} = \frac{Q^2r^2}{4}. \quad (160)$$

The complete expression is

$$\langle E_1(t)E_3(t)\rangle = P^2\left\{\frac{1+r-2r^2}{4r^2[1-r][1+r]^3}\right\} + \frac{PQ}{12}\left\{\frac{3+6r-7r^2-2r^3}{(1-r)(1+r)^3}\right\} + \frac{PQ}{12}\frac{(3+r)}{(1+r)^2} + Q^2\frac{r^2}{4}. \quad (161)$$

$$\langle E_1(t)E_3(t)\rangle = P^2\left\{\frac{1+r-2r^2}{4r^2[1-r][1+r]^3}\right\} + PQ\left\{\frac{3r^2+13r+6}{12(1+r)^3}\right\} + Q^2\frac{r^2}{4}. \quad (162)$$

Appendix IV

Association of Transform Variables

In Section III-A, the following problem arose. The two-dimensional transform pair is

$$\hat{Y}_2(s_1,s_2) = \int_{-\infty}^{+\infty}\int_{-\infty}^{\infty}\hat{y}_2(t_1,t_2)e^{-s_1t_1-s_2t_2}dt_1dt_2 \quad (163)$$

and

$$\hat{y}_2(t_1,t_2) = \left(\frac{1}{2\pi j}\right)^2\int_{-\infty}^{+\infty}\int_{-\infty}^{+\infty}\hat{Y}_2(s_1,s_2)e^{+s_1t_1+s_2t_2}ds_1ds_2. \quad (164)$$

We would like to find

$$y_2(t_1) \equiv \hat{y}_2(t_1,t_1) \quad (165)$$

without actually finding the inverse transform indicated by (164). A procedure is derived in George.[15] We summarize it briefly. We will consider only the two-dimensional case since the extension is straightforward.

The transform of $y_2(t_1)$ is $Y_2(s_1)$,where

$$y_2(t_1) = \frac{1}{2\pi j}\int_{-\infty}^{\infty}Y_2(s)e^{+st_1}ds. \quad (166)$$

But, from (164) and (165), we have

$$y_2(t_1) = \hat{y}_2(t_1,t_1) = \frac{1}{(2\pi j)^2}\int_{-\infty}^{\infty}\int_{-\infty}^{\infty}ds_1ds_2\hat{Y}_2(s_1,s_2)e^{+(s_1+s_2)t_1}. \quad (167)$$

Now let $s_1+s_2=s$. Then,

$$y_2(t_1) = \frac{1}{2\pi j}\int_{-\infty}^{\infty}dse^{+st_1}\left\{\frac{1}{2\pi j}\int_{-\infty}^{\infty}\hat{Y}_2(s-s_2,s_2)ds_2\right\}. \quad (168)$$

Equating (166) and (168), we have

$$Y_2(s) = \frac{1}{2\pi j}\int_{-\infty}^{\infty}\hat{Y}_2(s-s_2,s_2)ds_2. \quad (169)$$

In general, this is still not too useful because it involves a convolution in the transform domain.

[15] See George [6], p. 34.

For systems with rational transforms, one can usually see the answer by inspection. Using an example similar to the one in George,[16] we have

$$Y_2(s_1, s_2) = \frac{A}{s_1 + s_2 + a} \cdot \frac{B}{s_1 + b} \cdot \frac{C}{s_2 + c}.$$

Then,

$$Y_2(s) = \frac{1}{2\pi j} \int_{-\infty}^{\infty} \frac{A}{s - s_2 + s_2 + a} \cdot \frac{B}{s - s_2 + b} \cdot \frac{C}{s_2 + c} ds_2$$

$$= \frac{A}{s + a} \cdot \frac{1}{2\pi j} \int_{-\infty}^{\infty} \frac{B}{s - s_2 + b} \cdot \frac{C}{s_2 + c} ds_2.$$

But the second integral is familiar

$$\frac{1}{2\pi j} \int_{-\infty}^{\infty} \frac{B}{-s_2 + (b + s)} \cdot \frac{C}{s_2 + c} ds_2 = \frac{BC}{s + b + c}.$$

Thus, we see that

$$\frac{A}{s_1 + s_2 + a} \text{ becomes } \frac{A}{s + a}$$

and

$$\frac{B}{s_1 + b} \cdot \frac{C}{s_2 + c} \text{ becomes } \frac{BC}{s + b + c}.$$

The other relations in Section III-A follow easily. For higher-order transforms, one simply associates variables successively.

Appendix V

Comparison with Equivalent Gain Model

Booton [16] suggests an approximate method of analyzing nonlinear systems. He assumed the input to the nonlinear no-memory element is Gaussian. He then replaces the actual nonlinearity by an equivalent gain.

For the nonlinear system of Fig. 1, the incremental gain for any value of $E(t)$ is $\cos E$. One takes the expected value of this incremental gain with the assumption E is a zero-mean Gaussian random variable with variance σ^2. (This is, of course, the mean-square error that we are looking for.) Denote the equivalent gain by k_e. In this case,

$$k_e = \int_{-\infty}^{\infty} \cos E \cdot \frac{1}{\sqrt{2\pi}\,\sigma} e^{-E^2/2\sigma^2} dE = e^{-\sigma^2/2}. \quad (170)$$

We apply this to the example of Section IV-A to obtain a comparison. Using straightforward techniques, the following equation is obtained.

$$Z = \frac{\frac{3}{2}\sigma^2}{e^{+\sigma^2/2} + \frac{1}{2} e^{+\sigma^2}} \quad (171)$$

[16] See George [6], p. 35.

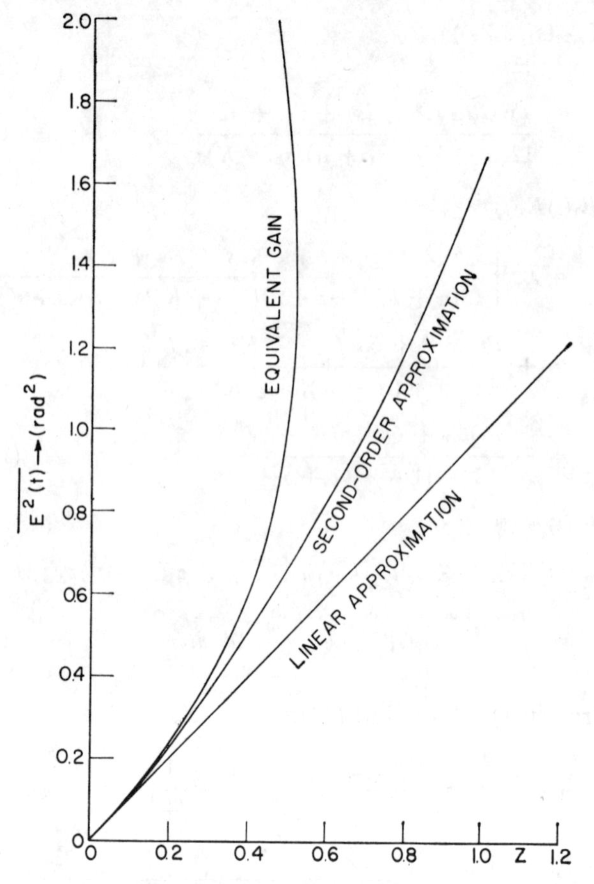

Fig. 14—Comparison of solutions.

where

$$Z = \frac{N_0}{2A^2} \cdot \frac{3d}{2} \quad (172)$$

is the signal-to-noise ratio in the process bandwidth. Eqs. (171) and (85) are plotted in Fig. 14.

For this particular system, the results agree within 25 per cent for $Z < 0.4 (S/N > 2.5)$. The maximum value of Z which satisfies (171) is approximately 0.52.

Acknowledgment

Most of the functional techniques used here were developed by various members of the Statistical Communications Theory Group at M.I.T., Cambridge, Mass., under the supervision of Y. W. Lee. Some of the specific ideas that were used (such as association of transform variables) are due originally to D. A. George. The actual model of the phase-locked loop which we used was brought to our attention by A. J. Viterbi of Jet Propulsion Laboratory, Pasadena, Calif. By making available his results prior to publication, he simplified our analysis.

References

[1] V. Volterra, "Theory of Functionals and of Integral and Integro-Differential Equations," Dover Publications, New York, N. Y.; 1959.

[2] N. Wiener, "Nonlinear Problems in Random Theory," M.I.T. Press, Cambridge, Mass.; 1959.

[3] M. B. Brilliant, "Theory of the Analysis of Nonlinear Systems," M.I.T. Research Laboratory of Electronics, Cambridge, Mass., Tech. Rept. No. 345; March 1958.

[4] D. A. Chesler, "Nonlinear Systems with Gaussian Inputs," M.I.T. Research Laboratory of Electronics, Cambridge, Mass., Tech. Rept. No. 366; February, 1960.
[5] J. F. Barrett, "The Use of Functionals in the Analysis of Nonlinear Physical Systems," Statistical Advisory Unit, Ministry of Supply, Great Britain, Rept. No. 1/57; 1957.
[6] D. A. George, "Continuous Nonlinear Systems," M.I.T. Research Laboratory of Electronics, Cambridge, Mass., Tech. Rept. No. 355; July, 1959.
[7] H. L. Van Trees, "Synthesis of Optimum Nonlinear Control Systems," M.I.T. Press, Cambridge, Mass.; 1962.
[8] ——, "A Threshold Theory for Phase-Locked Loops," M.I.T. Lincoln Laboratory, Lexington, Mass., Tech. Rept. No. 246; August, 1961.
[9] J. A. Develet, Jr., "A threshold criterion for phase-lock demodulation," Proc. IRE, vol. 51, pp. 349–356; February, 1963.
[10] N. Wiener, "Response of a Nonlinear Device to Noise," M.I.T. Radiation Laboratory, Cambridge, Mass., Rept. No. V-168; April, 1942.
[11] S. Ikehara, "A Method of Wiener in a Nonlinear Circuit," M.I.T. Research Laboratory of Electronics, Cambridge, Mass., Tech. Rept. No. 217; December 10, 1951.
[12] Ralph Deutsch, "Nonlinear Transformations of Random Processes," Prentice-Hall, Englewood Cliffs, N. J.; 1962.
[13] S. G. Margolis, "The response of a phase-locked loop to a sinusoid plus noise," IRE Trans. on Information Theory, vol. IT-3, pp. 136–142; June, 1957.
[14] D. L. Schilling, "The Response of an Automatic Phase Control System to FM Signals and Noise," Polytechnic Institute of Brooklyn, N. Y., Res. Rept. No. PIBMRI-1040-62; June, 1962.
[15] L. A. Pipes, "The reversion method for solving nonlinear differential equations," *J. Appl. Phys.*, vol. 23, pp. 202–207; February, 1952.
[16] R. C. Booton, Jr., "Nonlinear Control Systems with Statistical Inputs," M.I.T. Dynamic Analysis and Control Laboratory, Cambridge, Mass., Rept. No. 61; March, 1952.
[17] T. K. Caughey, "Response of Nonlinear Systems to Random Excitation," Lecture Notes, California Institute of Technology, Pasadena; 1953. (See "Equivalent linearization techniques," *J. Acoust. Soc. Am.*, vol. 35, pp. 1706–1712; November, 1963.)
[18] G. E. Uhlenbeck and L. S. Ornstein, "On the theory of Brownian motion," *Phys. Rev.*, vol. 36, pp. 823–841; Septmeber, 1930.
[19] M. C. Wang and G. E. Uhlenbeck, "On the theory of Brownian motion, II," *Rev. Mod. Phys.*, vol. 17, pp. 323–342; April–July, 1945.
[20] A. A. Andronov, L. S. Pontryagin, and A. A. Witt, "On the statistical investigation of a dynamical system," *J. Exp. and Theoretical Phys.* (*USSR*), vol. 3, p. 165; Febıuary, 1933.
[21] A. T. Bharucha-Reid, "Elements of the Theory of Markov Processes and Their Applications," McGraw-Hill Book Co., Inc., New York, N. Y.; 1960.
[22] D. Middleton, "An Introduction to Statistical Communicetion Theory," McGraw-Hill Book Co., Inc., New York, N. Y.; 1960.
[23] V. I. Tikhonov, "The effects of noise on phase-lock oscillation operation," *Automatika i Telemakhanika*, vol. 22; September, 1959.
[24] A. J. Viterbi, "Phase-locked loop dynamics in the presence of noise by Fokker-Planck techniques," Proc. IEEE, vol. 51, pp. 1737–1753; December, 1963.
[25] R. M. Jaffe and E. Rechtin, "Design and performance of the phase-lock circuits capable of near-optimum performance over a wide range of input signal and noise levels," IRE Trans. on Information Theory, vol. IT-1, pp. 66–76; March, 1955.
[26] W. J. Gruen, "Theory of AFC synchronization," Proc. IRE, vol. 41, pp. 1043–1048; August, 1953.
[27] H. L. Van Trees, "An Introduction to Feedback Demodulation," M.I.T. Lincoln Laboratory, Lexington, Mass., Group Rept. No. 64G-5; August 16, 1963.
[28] H. M. James, N. B. Nichols, and R. S. Phillips, "Theory of Servomechanisms," McGraw-Hill Book Co.Inc., New York, N.Y.; 1947.
[29] W. B. Davenport and W. Root, "Random Signals and Noise," McGraw-Hill Book Co., Inc., New York, N. Y.; 1958.

Phase-Locked Loop Dynamics in the Presence of Noise by Fokker-Planck Techniques*

ANDREW J. VITERBI†, SENIOR MEMBER, IEEE

Summary—**Statistical parameters of the phase-error behavior of a phase-locked loop tracking a constant frequency signal in the presence of additive, stationary, Gaussian noise are obtained by treating the problem as a continuous random walk with a sinusoidal restoring force. The Fokker-Planck or diffusion equation is obtained for a general loop and for the case of frequency-modulated received signals. An exact expression for the steady-state phase-error distribution is available only for the first-order loop, but approximate and asymptotic expressions are derived for the second-order loop. Results are obtained also for the expected time to loss of lock and for the frequency of skipping cycles. Some of the results are extended to tracking loops with nonsinusoidal error functions. Validity thresholds of widely accepted approximate models of the phase-locked loop are obtained by comparison with the exact results available for the first-order loop.**

Phase-Locked Loop Dynamics

The phase-locked loop is a communication receiver which operates as a coherent detector by continuously correcting its local oscillator frequency according to a measurement of the phase error. A block diagram of the device is shown in Fig. 1 with the pertinent input and output signals indicated. The output of the voltage-controlled oscillator (VCO) is a sinusoid whose frequency is controlled by the input voltage, $e(t)$; that is,

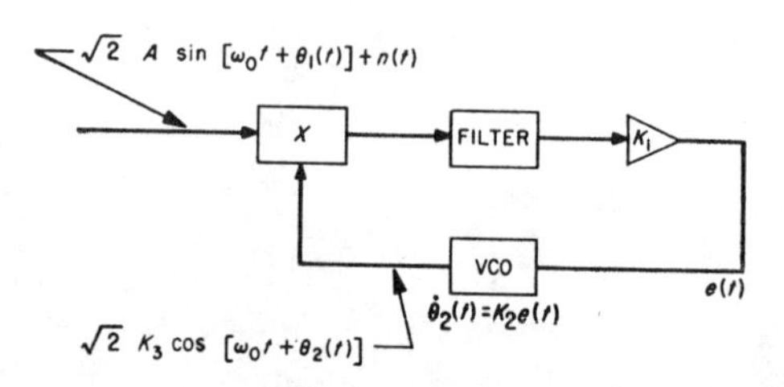

Fig. 1—Phase-locked loop.

$$\dot{\theta}_2(t) = \frac{d\theta_2(t)}{dt} = K_2 e(t), \tag{1}$$

so that when $e(t) = 0$, the oscillator frequency is ω_0. The received signal is a sinusoid of power A^2 watts and of arbitrary frequency and phase which may be either fixed or time-varying because of frequency or phase modulation at the transmitter. Thus, it may be represented by the expression

$$\sqrt{2}\,A \sin[\omega_0 t + \theta_1(t)]. \tag{2}$$

If the signal is a pure sinusoid with constant frequency ω and an initial phase θ, then

$$\theta_1(t) = (\omega - \omega_0)t + \theta. \tag{3}$$

Although we shall limit ourselves to this case in several of the following sections, we shall continue at present with the general formulation in which $\theta_1(t)$ is an arbitrary and possibly random process.

The received noise is assumed to be a stationary white Gaussian process of one-sided spectral density N_0 w/cps. We shall assume that the phase-locked loop is preceded by a band-pass filter with center frequency ω_0, a bandwidth β, which is very wide compared to the frequency region of interest, and a transfer function which is flat over this region. The only restriction on β is that it be less than or equal to $2\omega_0$. If we let $\beta = 2\omega_0$, the band-pass filter becomes a low-pass filter with this bandwidth. The noise process $n(t)$ over an arbitrary period of duration T can be expanded in a Fourier series whose coefficients are Gaussian variables which become independent in the limit as T approaches infinity [1]. By collecting the sine and cosine terms of the series, we can represent the noise process of infinite duration by the expression

$$n(t) = \sqrt{2}\,n_1(t) \sin \omega_0 t + \sqrt{2}\,n_2(t) \cos \omega_0 t, \tag{4}$$

where $n_1(t)$ and $n_2(t)$ are independent stationary wide-band Gaussian processes with flat spectra over the frequency range from zero to $\beta/2$. If we choose β to be $2\omega_0$ and restrict our interest to frequencies below ω_0 rad/sec, $n_1(t)$ and $n_2(t)$ may be regarded essentially as independent white Gaussian processes of one-sided spectral density N_0 w/cps.

Thus the product of input and reference signals is

$$\begin{aligned}
&2\{A \sin[\omega_0 t + \theta_1(t)] + n_1(t) \sin \omega_0 t + n_2(t) \cos \omega_0 t\} \\
&\quad \cdot \{K_3 \cos[\omega_0 t + \theta_2(t)]\} \\
&= K_3\{A \sin[\theta_1(t) - \theta_2(t)] - n_1(t) \sin \theta_2(t) + n_2(t) \cos \theta_2(t) \\
&\quad + A \sin[2\omega_0 t + \theta_1(t) + \theta_2(t)] + n_1(t) \sin[2\omega_0 t + \theta_2(t)] \\
&\quad + n_2(t) \cos[2\omega_0 t + \theta_2(t)]\}.
\end{aligned}$$

The double-frequency terms may be neglected since neither the filter nor the VCO will respond to these for

* Received April 17, 1963; revised manuscript received August 2, 1963. This paper presents the results of one phase of research carried out at the Jet Propulsion Laboratory, California Institute of Technology, under Contract No. NAS 7-100, sponsored by the National Aeronautics and Space Administration.

† University of California, Los Angeles, Calif., and consultant, Jet Propulsion Laboratory, Pasadena, Calif.

Reprinted from *Proc. IEEE*, vol. 51, pp. 1737–1753, Dec. 1963.

reasonable large ω_0. Then from Fig. 1 we see that

$$e(t) = K_1K_3F(s)\{A \sin [\theta_1(t) - \theta_2(t)] - n_1(t) \sin \theta_2(t) + n_2(t) \cos \theta_2(t)\} \quad (5)$$

where $F(s)$ is a rational function which represents in operational notation the effect of the linear filter in the loop. If we let

$$\phi(t) = \theta_1(t) - \theta_2(t) \quad (6)$$

and

$$K = K_1K_2K_3 \quad (7)$$

we obtain from (1) and (6)

$$\dot{\phi}(t) = \dot{\theta}_1(t) - K_2e(t),$$

and from (5) and (7) we have

$$\dot{\phi}(t) = \dot{\theta}_1(t) - KF(s)[A \sin \phi(t) - n_1(t) \sin \theta_2(t) + n_2(t) \cos \theta_2(t)]. \quad (8)$$

$\phi(t)$ defined by (6) is the instantaneous phase error or the phase difference between the received signal and the reference signal at the output of the phase-locked loop. If we let

$$n'(t) = - n_1(t) \sin \theta_2(t) + n_2(t) \cos \theta_2(t), \quad (9)$$

(8) reduces to

$$\dot{\phi}(t) = \dot{\theta}_1(t) - KF(s)[A \sin \phi(t) + n'(t)]. \quad (10)$$

This differential equation in operational form represents the dynamic operation of the phase-locked loop. It may also be written as the operational equation

$$\phi(t) = \theta_1(t) - \frac{KF(s)}{s} [A \sin \phi(t) + n'(t)], \quad (11)$$

which is represented by the block diagram or model of Fig. 2. It should be noted that (11) and the model of

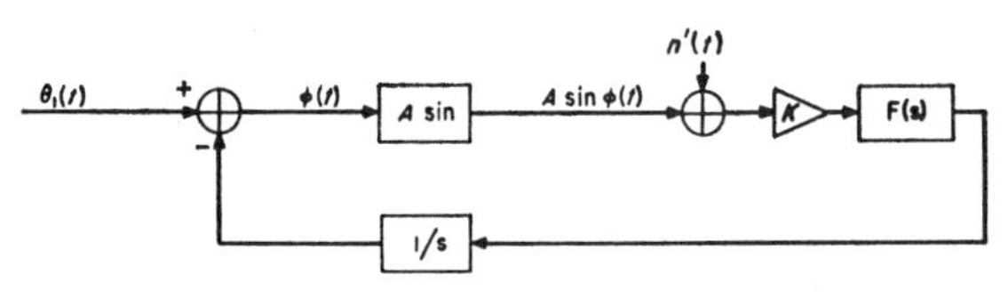

Fig. 2—Model of the phase-locked loop.

Fig. 2 are exact in all respects except for the fact that the terms centered about the double frequency, $2\omega_0$, have been assumed to be eliminated completely by the combination of the filter and the VCO.

Before we can proceed we must determine the statistics of the noise process $n'(t)$ defined by (9), which is the random driving function in the model. We can show that this is a white Gaussian process whenever the original noise process is Gaussian and white. We have taken $n(t)$ to be essentially white at least for frequencies less than twice the VCO quiescent frequency, ω_0, and have shown that consequently $n_1(t)$ and $n_2(t)$ are essentially white for frequencies up to ω_0. We have from above

$$n' = - n_1 \sin \theta_2 + n_2 \cos \theta_2, \quad (9)$$

and let us define similarly

$$n'' = n_1 \cos \theta_2 + n_2 \sin \theta_2. \quad (9a)$$

The process $\theta_2(t)$ depends on $\theta_1(t)$ and the noise, as is clear upon inspection of Fig. 2. Thus $\theta_2(t)$ may be nonstationary as in the case, for example, when $\theta_1(t)$ is given by (3). Although $\theta_2(t)$ is a function of the input noise, it can depend only on its past [*i.e.*, $n(t-\delta)$ where $\delta>0$], and since the noise is an essentially white process,[1] $n(t)$ is independent of $n(t-\delta)$. Therefore, $\theta_2=\theta_2(t)$ is independent of $n(t)$ and consequently also of $n_1(t)$ and $n_2(t)$, and we have as the joint probability density function of the three independent random variables: n_1, n_2, and θ_2,

$$p(n_1, n_2, \theta_2) = \frac{1}{2\pi\sigma^2} \exp(-n_1^2/2\sigma^2) \exp(-n_2^2/2\sigma^2)p(\theta_2),$$

where $\sigma^2 = N_0\omega_0/2\pi$ and $p(\theta_2)$ is an arbitrary nonstationary distribution. From this we can obtain the joint-probability density function of n', n'', and θ_2 by performing the linear transformation of (9) and (9a). The result is

$$p(n', n'', \theta_2) = p(n_1, n_2, \theta_2) \left| J\left(\frac{n_1, n_2, \theta_2}{n', n'', \theta_2}\right)\right|$$

$$= \frac{1}{2\pi\sigma^2} \exp(-n'^2/2\sigma^2) \exp(-n''^2/2\sigma^2)p(\theta_2),$$

since the absolute value of the Jacobian is unity. Hence we conclude that $n'(t)$ is a stationary process with exactly the same statistics as $n_1(t)$ and $n_2(t)$. It is Gaussian and essentially white at least over the frequency region up to ω_0 rad/sec, and its one-sided spectral density is N_0 w/cps.

The model of Fig. 2 first appeared without proof in a paper by Develet [2]. In the absence of noise, the model has been known for some time and analyzed by several authors beginning with Gruen [3]. Solutions of (11) in the absence of noise have been obtained for a number of filter-transfer functions and also for the case of received sinusoids with linearly time-varying fre-

[1] Actually, since $n(t)$ is "white" only for radian frequencies below $2w_0$, $n(t-\delta)$ is essentially independent of $n(t)$ only for $\delta>k\pi/w_0$, where k is a sufficiently large constant. On the other hand, since the combination of filter and VCO has a low-pass transfer function which is extremely narrow compared to $2w_0$, $\theta_2(t)$ is essentially independent of the input, $n(t)$, for $\delta<k\pi/w_0$, so that the net result is the same.

quency [4]. The general case in which additive noise is present has been treated by a variety of approximations. The first approach, by Jaffe and Rechtin [5], essentially replaced the sinusoidal nonlinearity of the model of Fig. 2 by a linear amplifier of gain A, which is the gain for arbitrarily small ϕ. Margolis [6] first analyzed the nonlinear operation in the presence of noise by perturbation methods obtaining a series solution of the differential equation of operation. By using the first few terms of the series he determined approximate moments of the phase error. Develet [2], who first proposed the operational model, applied Booton's quasi-linearization technique [7] to replace the sinusoidal nonlinearity by a linear amplifier whose gain is the expected gain of the device. Most recently, Van Trees [8] obtained a Volterra series representation of the closed-loop response by a perturbation method similar to the method employed by Margolis [6], but with the advantage of the simplified model he obtained more extensive results.

Unlike these analyses, Fokker-Planck or continuous random-walk techniques yield exact expressions for the statistics of the random process, $\phi(t)$. Unfortunately, expressions in closed form are available only for the first-order loop (*i.e.*, when the filter is omitted). For the general case a partial differential equation in ϕ and linear combinations of its time derivatives is derived, but a solution cannot be obtained in general. These techniques were first applied to this problem by Tikhonov [9], [10], who determined the steady-state probability distribution of ϕ for the first-order loop and an approximate expression for the distribution when the loop contains a one-stage RC filter.

All the analyses of this device have been concerned with the steady-state behavior. In this paper we shall obtain for the first-order loop not only its steady-state probability distribution and variance, but also the mean time to loss of lock, which is a transient phenomenon equivalent to a random-walk problem with absorbing boundaries. Also, we shall derive the Fokker-Planck equation for the general loop filter which produces zero mean error. We shall specialize this equation to the second-order loop and determine the form of the solution for the steady-state probability distribution of its phase error. In later sections we shall treat the effect of random modulation and tracking loops with other than sinusoidal error function. Finally we shall compare some of the results of the exact analysis with previous approximate results to determine the degree of validity of the approximate models.

First of all, in the next section a simple mechanical analog of the phase-locked loop is presented which provides a qualitative description of the operation of the device and an understanding of the nature of the statistical parameters required for its quantitative description.

The First-Order Loop and its Mechnical Analog

If the filter is omitted in Fig. 1, then $F(s)=1$ in (10). Furthermore, if we take the received signal to be a sinusoid of constant frequency and phase so that $\theta_1(t)$ is given by (3), we obtain the first-order differential equation

$$\dot{\phi}(t) = (\omega - \omega_0) - K[A \sin \phi(t) + n'(t)]. \quad (12)$$

Hence the term "first-order loop." Since $n'(t)$ is a white Gaussian process, the instantaneous change in ϕ represented by its derivative depends only on the present value of ϕ and the present value of the noise. Hence, $\phi(t)$ is a continuous Markov process, and we may use random-walk techniques to determine its probability distribution.

A mechanical analog is conducive to understanding the mechanism of this "random walk." Consider the pendulum of Fig. 3 which consists of a weightless ball

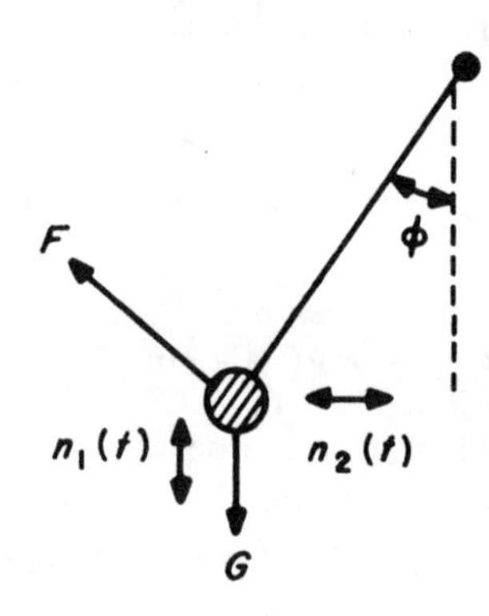

Fig. 3—Mechanical analog of the first-order loop.

attached by an infinitesimally thin weightless rod to a fixed point, and let the apparatus lie horizontally on a table top which is being randomly agitated. The pendulum is free to turn a full revolution about the point. Let the rod be initially at an angle ϕ with respect to the vertical axis and let an external force (such as a constant wind) be exerted on the ball in the vertical direction. Let the surface of the table be rough so that it produces a frictional force opposing motion of magnitude $f\dot{\phi}$. In addition, let the ball be equipped with an internal engine which exerts a constant force F along the axis of motion. The random agitation of the table produces a force on the ball which may be represented by two components which are stationary white Gaussian processes of zero means: $n_1(t)$ in the vertical direction, and $n_2(t)$ in the horizontal direction. Then by equating forces along the instantaneous axis of motion we obtain

$$f\dot{\phi} + G \sin \phi = F - n_1(t) \sin \phi - n_2(t) \cos \phi. \quad \textbf{(13)}$$

If we divide both sides of this equation by f and let

$$n_1 \sin \phi + n_2 \cos \phi = n',$$
$$F/f = \omega - \omega_0,$$
$$G/f = AK,$$
$$1/f = K,$$

we find that (13) is the same as (12). Also, the process $n'(t)$ defined here can be shown to be white and Gaussian by the same argument that was used in connection with the $n'(t)$ defined by (9) in the previous section. Thus the massless pendulum is the mechanical analog of a first-order loop with constant received frequency.

In the absence of the random forces it is clear that the pendulum approaches the equilibrium position

$$\phi_0 \sin^{-1} (F/G) = \sin^{-1}(\omega - \omega_0)/(AK), \qquad (14)$$

at which point the velocity is zero. Because this is a first-order system, there can be no overshoot. If $F>G$ or $(\omega-\omega_0)>AK$, there can be no equilibrium position, and the pendulum continues to revolve indefinitely, which corresponds to a loop that cannot achieve lock. When the random or noise forces are applied as well as the constant ones, the motion becomes a random walk, but when the noise variance is small, there is a strong tendency for the angle ϕ to approach and remain about this equilibrium position.

The complete statistical description of the random walk of the angle ϕ is given by its probability density as a function of time, $p(\phi, t)$. To understand qualitatively the behavior of this function, let us assume for the moment that the constant force $F=0$ and that initially (at $t=0$) the pendulum is at rest in the vertical position. Thus,[2] $p(\phi, 0)=\delta(\phi)$. With the passage of time, the effect of the random forces will be felt in the movement of the pendulum from the equilibrium position. The qualitative behavior of the probability density function is sketched in Fig. 4. Of course, the condition

$$\int_{-\infty}^{\infty} p(\phi, t) d\phi = 1$$

must always be met. After a sufficient amount of time, the random forces will push the pendulum around by more than half a revolution so that it will tend to return to the equilibrium position after a full cycle of rotation in either direction. This corresponds to the reference signal of the phase-locked loop advancing or retreating one cycle relative to the received signal. The average time for this occurrence depends on the signal-to-noise ratio. Thus after a sufficiently long period, the probability density will appear as a multimodal function, each mode being centered about equilibrium positions spaced 2π radians apart, the central mode being the largest with each successive maximum progressively smaller. After an even longer period equal to several times the average time between revolutions, the central mode of the probability density will have diminished, the modes to either side will have become almost as large, and more modes of significant magnitude will have appeared. The central mode will remain the largest since the pendulum may have revolved in either direction with equal probability. Finally, in the steady state an arbitrary number of revolutions will have occurred.

[2] In certain standard treatments of the continuous random walk problem, the probability density function is denoted $p(\phi, t|\phi_0, 0)$ which signifies that we are dealing with the function at time t given that $\phi=\phi_0$ at $t=0$. We avoid this cumbersome notation by specifying the initial condition on $p(\phi, t)$.

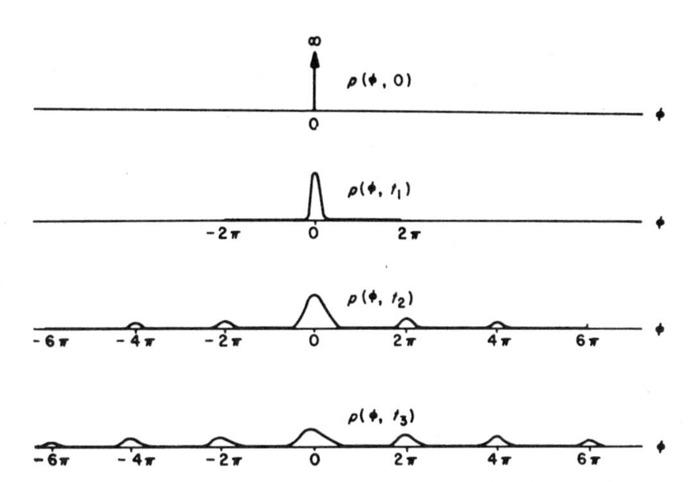

Fig. 4—Qualitative behavior of the probability density function for the first-order loop ($\omega=\omega_0$).

In the case for which F is not zero, or equivalently $\omega\neq\omega_0$, then clearly the pendulum will have a greater tendency to swing around in the sense corresponding to the direction of the force. Hence, the density function $p(\phi, t)$ will not be symmetrical. In either case, we are led to realize that the significant parameter, at least in the steady state, is the angle (or phase error) ϕ modulo 2π, since the number of revolutions of the pendulum which have occurred does not affect the present state of the system. In fact, although $p(\phi, t)$ yields a complete description of the statistical behavior, it would appear that a combination of the steady-state distribution of ϕ modulo 2π, and the frequency or average time between revolutions would yield a simpler and nearly complete representation. In the next two sections these parameters will be obtained quantitatively.

The Steady-State Phase-Error Probability Density for the First-Order Loop

A continuous random walk which is a Markov process is described by the statistical parameters of the incremental change of position as a function of the present position. Thus from (12) we note that in the infinitesimal increment of time Δt, the phase will change by an amount[3]

$$\Delta\phi = \int_t^{t+\Delta t} \dot{\phi}(t) dt = (\omega - \omega_0)\Delta t - (AK \sin \phi)\Delta t$$
$$-K \int_t^{t+\Delta t} n'(u) du.$$

[3] This assumes that $\phi(t)$ is a continuous process, which is justified by physical considerations.

Thus, since $n'(t)$ is a white Gaussian process with zero mean and

$$\overline{n'(u)n'(v)} = (N_0/2)\delta(u - v),$$

it follows that for a given position ϕ, $\Delta\phi$ is a Gaussian variable with mean

$$\overline{\Delta\phi} = [(\omega - \omega_0) - AK \sin \phi]\Delta t \tag{15}$$

and variance

$$\begin{aligned}\sigma^2_{\Delta\phi} &= \overline{(\Delta\phi)^2} - (\overline{\Delta\phi})^2 \\ &= K^2 \int_t^{t+\Delta t} \int_t^{t+\Delta t} \overline{n'(u)n'(v)}\, du dv \\ &= K^2(N_0/2)\Delta t. \end{aligned} \tag{16}$$

With the knowledge of the statistical parameters of the increment $\Delta\phi$, we may proceed to obtain $p(\phi, t)$. It was shown by Uhlenbeck and Ornstein [11] and Wang and Uhlenbeck [12] that for a continuous Markov process described by a first-order differential equation with a white Gaussian input, the instantaneous probability density $p(\phi, t)$ must satisfy the partial differential equation

$$\frac{\partial p(\phi, t)}{\partial t} = -\frac{\partial}{\partial \phi}[A(\phi)p(\phi, t)] + \frac{1}{2}\frac{\partial^2}{\partial \phi^2}[B(\phi)p(\phi, t)], \tag{17}$$

with the appropriate initial condition, where

$$A(\phi) = \lim_{\Delta t \to 0} \frac{1}{\Delta t} \overline{\Delta\phi}$$

$$B(\phi) = \lim_{\Delta t \to 0} \frac{1}{\Delta t} \overline{(\Delta\phi)^2},$$

provided

$$\lim_{\Delta t \to 0} \left(\frac{1}{\Delta t}\right) \overline{(\Delta\phi)^n} = 0 \qquad \text{for } n > 2.$$

Eq. (17) is known as the Fokker-Planck equation or the diffusion equation because it is a generalization of the equation for heat diffusion. From (15) and (16) we obtain for the first-order loop

$$A(\phi) = (\omega - \omega_0) - AK \sin \phi,$$

$$B(\phi) = K^2 N_0/2,$$

and it may be readily verified (using the property that the product of a set of Gaussian variables is the sum of products of pairs taken over all pairs of the variables) that

$$\lim_{\Delta t \to 0} \left(\frac{1}{\Delta t}\right) \overline{(\Delta\phi)^n} = 0 \qquad \text{for } n > 2.$$

Thus the Fokker-Planck equation holds for this case, and inserting the coefficients into (17) we obtain

$$\frac{\partial p}{\partial t} = \frac{\partial}{\partial \phi}[(AK \sin \phi + \omega_0 - \omega)p] + \frac{K^2 N_0}{4}\frac{\partial^2 p}{\partial \phi^2}. \tag{18}$$

If we take the initial value of ϕ to be ϕ_0, we have

$$p(\phi, 0) = \delta(\phi - \phi_0). \tag{19}$$

As was pointed out in the previous section, we are really interested in $p(\phi, t)$, where ϕ is taken modulo 2π. Therefore, we are tempted to solve (18) in the region—$\pi \leq \phi \leq \pi$ with appropriate boundary conditions. To do this properly we must first realize that since the coefficients of the Fokker-Planck equation in this case are periodic in ϕ, if $p(\phi, t)$ is a solution, then so is $p(\phi + 2\pi\, n, t)$ where n is any integer. Let us define the function

$$P(\phi, t) = \sum_{n=-\infty}^{\infty} p(\phi + 2\pi n, t).$$

Since each term of the series is a solution of (18), then $P(\phi, t)$ must also satisfy the Fokker-Planck equation

$$\frac{\partial P(\phi, t)}{\partial t} = \frac{\partial}{\partial \phi}[(AK \sin \phi + \omega_0 - \omega)P(\phi, t)] + \frac{K^2 N_0}{4}\frac{\partial^2 P(\phi, t)}{\partial \phi^2}, \tag{20}$$

with the initial condition

$$P(\phi, 0) = \sum_{n=-\infty}^{\infty} \delta(\phi - \phi_0 - 2\pi n).$$

Also, $P(\phi, t)$ must be periodic in ϕ since for any integer m

$$\begin{aligned} P(\phi + 2\pi m, t) &= \sum_{n=-\infty}^{\infty} p[\phi + 2(m + n), t] \\ &= \sum_{k=-\infty}^{\infty} p(\phi + 2k\pi, t) = P(\phi, t). \end{aligned}$$

Therefore, we may solve (20) over the interval of just one period $(-\pi \leq \phi \leq \pi)$ with the initial condition

$$P(\phi, 0) = \delta(\phi - \phi_0), \quad -\pi \leq \phi \leq \pi, \tag{21}$$

the boundary condition

$$P(\pi, t) = P(-\pi, t) \quad \text{for all } t, \tag{22}$$

and the normalizing condition

$$\int_{-\pi}^{\pi} P(\phi, t) = d\phi = 1 \quad \text{for all } t. \tag{23}$$

Although in principle the linear partial differential (20) with the conditions (21), (22), and (23) can be solved for $P(\phi, t)$, the procedure is somewhat complicated by the nonlinear behavior of the variable coefficients, and a closed-form solution cannot be obtained. On the other hand, the result of greatest interest is the steady-state distribution

$$P(\phi) = \lim_{t \to \infty} P(\phi, t). \tag{24}$$

By definition, the steady-state distribution is stationary.

Therefore,

$$\frac{\partial P(\phi)}{\partial t} = \lim_{t\to\infty} \frac{\partial P(\phi, t)}{\partial t} = 0. \tag{25}$$

Thus, in the steady state, the partial differential (20) reduces to an ordinary differential equation in $P(\phi)$. Letting

$$\alpha = (4A)/(KN_0) \tag{26}$$

and

$$\beta = [4(\omega - \omega_0)]/(K^2N_0), \tag{27}$$

we obtain

$$0 = \frac{d}{d\phi}\left[(\alpha \sin \phi - \beta)P(\phi) + \frac{dP(\phi)}{d\phi}\right]. \tag{28}$$

If we integrate once with respect to ϕ, we obtain a first-order linear differential equation which is readily solved as[4]

$$P(\phi) = C \exp (\alpha \cos \phi + \beta\phi) \left[1 + D\int_{-\pi}^{\phi} \exp - (\alpha \cos x + \beta x)dx\right] \quad -\pi \leq \phi \leq \pi. \tag{29}$$

To evaluate the constants, we must utilize the limiting form of the conditions (22) and (23); *i.e.*,

$$P(\pi) = P(-\pi) \tag{30}$$

and

$$\int_{-\pi}^{\pi} P(\phi)d\phi = 1. \tag{31}$$

Then using (30), we obtain

$$D = \frac{\exp(-2\beta\pi) - 1}{\int_{-\pi}^{\pi} \exp - (\alpha \cos x + \beta x)dx}, \tag{32}$$

and by means of (31), the constant C can be evaluated. In the special case $\beta = 0$ (which requires $\omega = \omega_0$; *i.e.*, when the frequency of the received signal is determined beforehand and the VCO quiescent frequency is tuned to this frequency so that the problem consists only of acquiring and tracking phase), we note from (32) and (27) that $D = 0$ so that

$$P(\phi) = \frac{\exp(\alpha \cos \phi)}{2\pi I_0(\alpha)} \quad -\pi \leq \phi \leq \pi, \tag{33}$$

[4] The results of (29), (32), and (33) were first obtained by V. I. Tikhonov [9]. Actually, these are a special case of an expression for a random-walk problem with arbitrary nonlinear restoring forces derived in A. A. Andronov, L. S. Pontryagin, and A. A. Witt [13].

since

$$C = \frac{1}{\int_{-\pi}^{\pi} \exp(\alpha \cos \phi)d\phi} = \frac{1}{2\pi I_0(\alpha)}.$$

The parameter α plays a very important role. From (26) we have

$$\alpha = \frac{(4A)}{(KN_0)} = \frac{(A^2)}{[N_0(AK/4)]}. \tag{34}$$

But A^2 is the received signal power, while $AK/4$ is an important parameter defined for the linearized model of the loop. If we replace the sinusoidal nonlinearity in the model of Fig. 2 by its gain A about $\phi = 0$, we obtain the linearized model. Then the variance of ϕ is obtained by using Parseval's theorem as:

$$\sigma_\phi^2 = \frac{1}{2\pi}\int_{-\infty}^{\infty} \frac{N_0}{2} \frac{K^2/\omega^2}{[1 + (A^2K^2/\omega^2)]} d\omega = \frac{N_0(AK/4)}{A^2} = \frac{1}{\alpha}. \tag{35}$$

The variance of ϕ is the same as the noise power at the output of an ideal low-pass filter of bandwidth $AK/4$ when the input is white noise of one-sided spectral density N_0. Hence, for the first-order filter, the loop bandwidth is defined as

$$B_L = AK/4, \tag{36}$$

so that (34) becomes

$$\alpha = (A^2)/(N_0B_L), \tag{37}$$

which is the SNR in the bandwidth of the loop.

Eq. (33) is plotted in Fig. 5 (p. 1744) for several values of α. It resembles a Gaussian distribution when the SNR, α, is large and becomes flat as α approaches zero. The asymptotic behavior of (33) for large α is of interest. Since for large α

$$I_0(\alpha) \sim (\exp \alpha)/(2\pi\alpha)^{1/2},$$

$$P(\phi) = \frac{[\exp(\alpha \cos \phi)]}{[2\pi I_0(\alpha)]} \sim \frac{\{\exp[\alpha(\cos \phi - 1)]\}}{(2\pi/\alpha)^{1/2}}.$$

Expanding $\cos \phi$ in a Taylor series, we obtain

$$P(\phi) \sim \frac{\exp\left[\frac{-\alpha\phi^2}{2}\left(1 - \frac{2\phi^4}{4!} + \frac{2\phi^6}{6!} \cdots\right)\right]}{(2\pi/\alpha)^{1/2}} \quad -\pi \leq \phi \leq \pi. \tag{38}$$

When α is large, $P(\phi)$ decays rapidly, so that the function is very small for all but very small values of ϕ. Thus the higher-order terms of the series representation of $\cos \phi$ have very little effect for moderate values of $P(\phi)$. Hence, the graph of $P(\phi)$ will appear to be nearly Gaussian for large α and, in this case, the results of the linear model are quite accurate.

The cumulative steady-state probability distribution

$$\text{Prob}\,(|\phi| < \phi_1) = \int_{-\phi_1}^{\phi_1} p(\phi)d\phi \qquad 0 < \phi_1 < \pi$$

is also of interest since it indicates the percentage of time during which the absolute value of the loop phase error ϕ is less than a given magnitude ϕ_1. This may be calculated when $\omega=\omega_0$ in the following manner. Expanding $P(\phi)$ of (33) in a Fourier series, we have

$$P(\phi) = \frac{\exp(\alpha \cos \phi)}{2\pi I_0(\alpha)} = \frac{1}{2\pi I_0(\alpha)}\left[I_0(\alpha) + 2\sum_{n=1}^{\infty} I_n(\alpha) \cos n\phi\right].$$

Then

$$\text{Prob}\,(|\phi| < \phi_1) = 2\int_0^{\phi_1} P(\phi)d(\phi) = \frac{\phi_1}{\pi} + \frac{2}{\pi}\sum_{n=1}^{\infty} \frac{I_n(\alpha)\sin n\phi_1}{nI_0(\alpha)}$$

for

$$0 < \phi_1 < \pi \quad \text{and} \quad \omega = \omega_0. \tag{39}$$

This series converges so rapidly that (39) could be calculated for several values of α without the use of a large-scale digital computer. The results are shown in Fig. 6. The variance of ϕ can be similarly obtained:

$$\sigma_\phi^2 = \int_{-\pi}^{\pi} \phi^2 \exp(\alpha\cos\phi)d\phi/2\pi I_0(\alpha) = \frac{1}{2\pi I_0(\alpha)}\int_{\pi}^{\pi}\phi^2\left[I_0(\alpha) + 2\sum_{n=1}^{\infty} I_n(\alpha)\cos n\phi\right]d\phi = \frac{\pi^2}{3} + 4\sum_{n=1}^{\infty}\frac{(-1)^n I_n(\alpha)}{n^2 I_0(\alpha)}. \tag{40}$$

This series converges even more rapidly than that of (39). It is plotted in Fig. 7 as a function of $1/\alpha$. Note that as the SNR, α, approaches zero, the variance approaches $\pi^2/3$, which is the variance of a random variable that is uniformly distributed from $-\pi$ to $+\pi$. Also shown in Fig. 7 is the variance of the phase error as determined from the linear model (35). It is evident that the linear model for the first-order loop with $\omega=\omega_0$ and no signal modulation produces results of reasonable accuracy (within 20 per cent) for $1/\alpha<1/4$ or when the loop SNR, $\alpha>4$.

For the general case ($\omega\neq\omega_0$), (29), (31), and (32) yield the entire distribution. However, analog or digital computation is required to evaluate the pertinent integrals. The case for which $(\beta/\alpha)=(\omega-\omega_0)/(AK)=\sin(\pi/4)$ is shown in Fig. 8. The constants as well as the distribution were obtained by means of the analog computer.

Mean Time to Loss of Lock and Frequency of Skipping Cycles

Since we have obtained only solutions for steady-state probabilities, a valuable statistic is the expected time required for the absolute value of the phase error to exceed some value ϕ_l when it is initially zero. When this occurs the loop will be said to have lost lock. Of particular interest is the case for which $\phi_l=2\pi$, which represents a loss or gain of a complete cycle, or for the mechanical analog, a complete revolution of the pendulum.

In the framework of the mechanical analog of the first-order loop, we may represent the out-of-lock boundaries by two knife edges at angles $+\phi_l$ and $-\phi_l$ relative to the vertical (Fig. 9). The pendulum is initially at an angle ϕ_0 relative to the vertical when the random external forces are applied. The first time that the angle reaches $\pm\phi_l$ the knife edges cut the rod and the process terminates.

This so-called first-passage time problem for Markov processes has been treated extensively in the literature [13]–[15]. It is possible to determine not only the first moment but all moments and even the distribution of the first-passage time for a Markov process described by a first-order differential equation with a white Gaussian driving function. However, computational difficulties render the form of the solution rather complex in all but the simplest cases. We shall employ a somewhat different method than previously used to obtain a simple expression for the expected time to the first occurrence of loss of lock. Closely related to the mean time to loss of lock is the frequency of skipping cycles. For the mechanical analog, this is the inverse of the expected time for the pendulum to swing a complete revolution in either direction. For the phase-locked loop, this represents the frequency of occurrence of the event that the loop VCO gains or drops a cycle relative to the received signal. In either case this corresponds to letting $\phi_l=2\pi$ in the determination of expected time. This is a very important parameter for tracking applications in which the phase-locked loop is used to measure the received Doppler frequency which is then integrated to determine relative range. An error of a full cycle will cause a significant error in the result.

We treat only the case of the first-order loop for which the VCO quiescent frequency is tuned to the received frequency ($\omega_0=\omega$) so that the equilibrium position is at $\phi=0$. This is also a good approximation to the steady-state behavior of the second-order loop with any value of $\omega-\omega_0$, but with very small integrator gain a, as will be discussed in a later section. For the first-order loop, when $\omega\neq\omega_0$, the same approach can be used, measuring phase error from the equilibrium position rather than from zero, but the results are in the form of integrals which require numerical calculation.

Let us assume that the loop is initially in lock so that

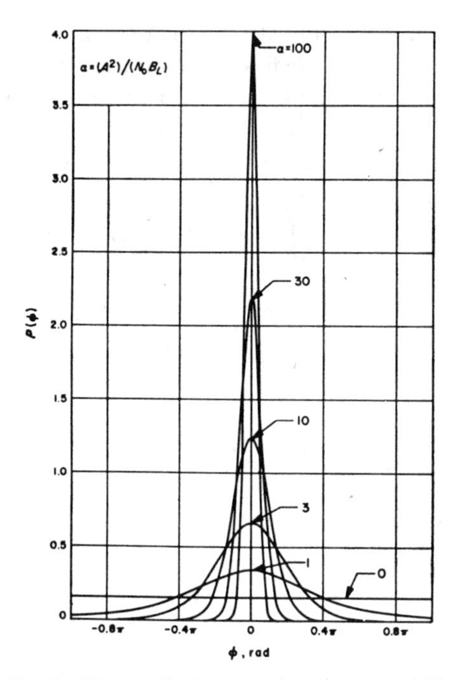

Fig. 5—First-order loop steady-state probability densities for $\omega = \omega_0$.

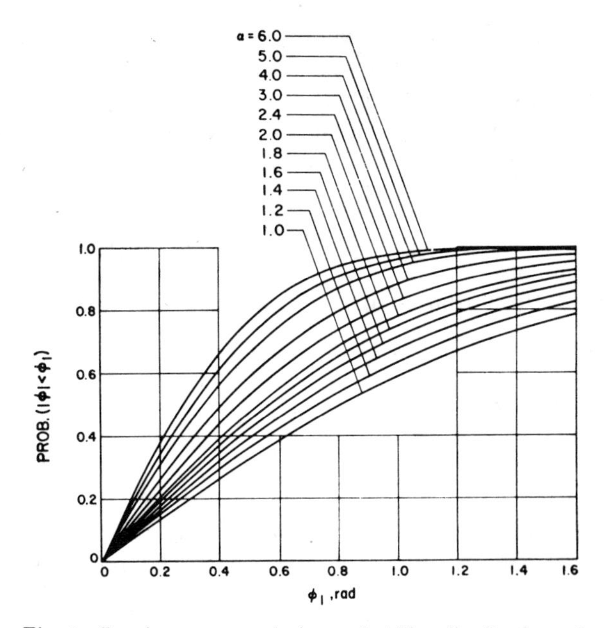

Fig. 6—Steady-state cumulative probability distributions of first-order loop for $\omega = \omega_0$.

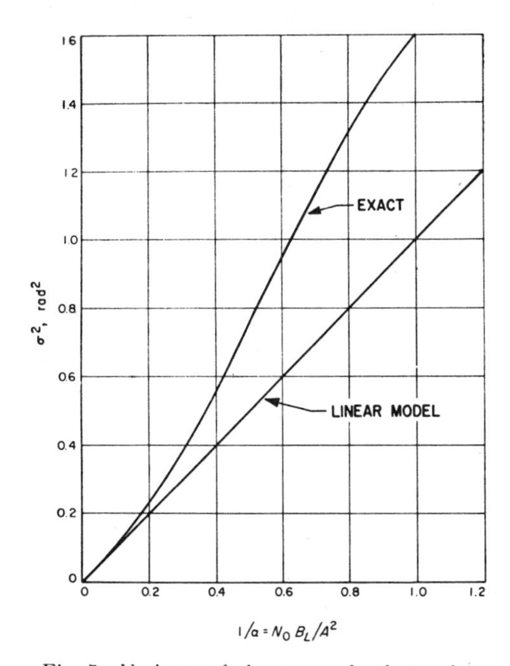

Fig. 7—Variance of phase-error for first-order loop where $\omega = \omega_0$.

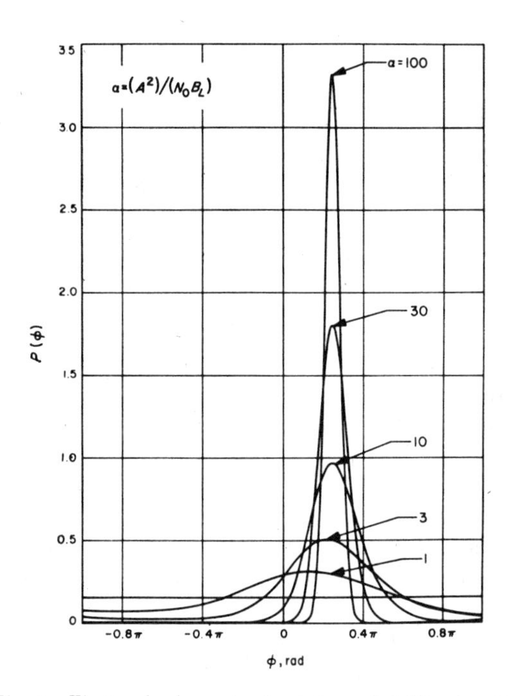

Fig. 8—First-order loop steady-state probability densities for $(\omega - \omega_0)/AK = \sin(\pi/4)$.

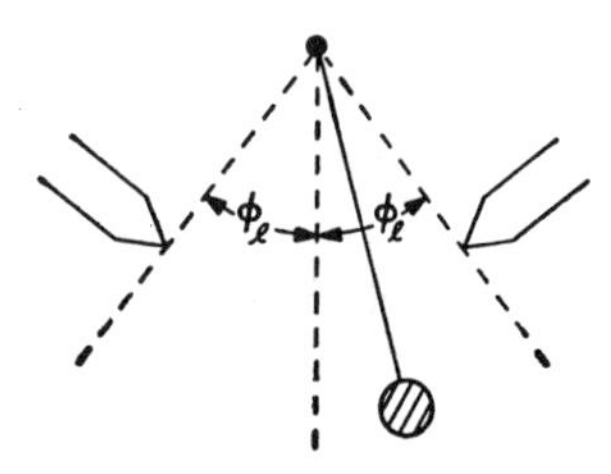

Fig. 9—Mechanical analog of first-passage time problem for first-order loop.

$\phi_0=0$. As long as the loop phase error (or the pendulum angle) remains within the limits $|\phi|<\phi_l$, the probability density of ϕ, which we denote here by $q(\phi, t)$, is described in the same manner as before by the Fokker-Planck equation

$$\frac{\partial q}{\partial t} = \frac{\partial}{\partial \phi}(AK \sin \phi q) + \frac{N_0 K^2}{4} \frac{\partial^2 q}{\partial \phi^2}, \tag{41}$$

with the initial condition

$$q(\phi, 0) = \delta(\phi) \qquad \text{for } |\phi| < \phi_l,$$

when

$$\omega = \omega_0.$$

We have denoted the probability density function of the phase error by $q(\phi, t)$ to distinguish it from the corresponding function $p(\phi, t)$ for the previous case without boundaries. As soon as $|\phi|$ reaches ϕ_l for the first time, the pendulum is removed from action so that

$$q(\phi, t) = 0 \qquad \text{for all } |\phi| \geq \phi_l. \tag{42}$$

Thus, in addition to the initial condition we have the boundary condition

$$q(\phi_l, t) = q(-\phi_l, t) = 0 \qquad \text{for all } t. \tag{43}$$

Solution of (41) over the interval $-\phi_l<\phi<\phi_l$ with the boundary conditions of (43) would yield the probability density $q(\phi, t)$. Its integral over the interval

$$\psi(t) = \int_{-\phi_l}^{\phi_l} q(\phi, t)d\phi \tag{44}$$

gives the probability that ϕ has not yet reached ϕ_l at time t. Note that as a consequence of (42) the limits on the integral of (44) could just as well be infinite. In fact,

$$\int_{-\infty}^{\infty} q(\phi, t)d\phi = \psi(t) \leq 1,$$

and this points out the fundamental difference between $q(\phi, t)$ and $p(\phi, t)$ of the previous section for which

$$\int_{-\infty}^{\infty} p(\phi, t)d\phi = 1 \qquad \text{for all } t.$$

In other words, $q(\phi, t)$ is not strictly a probability density function. In order to render it such we would have to normalize it by the probability that $|\phi|$ has never exceeded ϕ_l by the time t[*i.e.*, $\psi(t)$].

$\psi(t)$ must be a monotonically nonincreasing function of t, and from its definition it follows that the probability density function of the time required for ϕ to reach ϕ_l for the first time is $-[\partial\psi(t)/\partial t]$. Thus the expected time to reach the out-of-lock position ϕ_l is

$$T = \int_0^{\infty} -t\frac{\partial \psi(t)}{\partial t}dt = -[t\psi(t)]_0^{\infty} + \int_0^{\infty} \psi(t)dt. \tag{45}$$

If the nonincreasing function $\psi(t)$ approaches zero faster than $0(1/t)$, the first term on the right side of (45) is zero. This must be the case, for otherwise the integral of the second term would not exist. Then using (44) we obtain the expression for the expected time

$$T = \int_0^{\infty}\int_{-\phi_l}^{\phi_l} q(\phi, t)d\phi dt. \tag{46}$$

Now if we integrate both sides of (41) with respect to t over the infinite interval, we obtain

$$q(\phi, \infty) - q(\phi, 0) = \frac{\partial}{\partial \phi}[AK \sin \phi Q(\phi)] + \frac{N_0 K^2}{4} \frac{\partial^2 Q(\phi)}{\partial \phi^2}, \tag{47}$$

where

$$Q(\phi) = \int_0^{\infty} q(\phi, t)dt.$$

Clearly $q(\phi, \infty)=0$, and since ϕ is assumed initially at zero, $q(\phi, 0)=\delta(\phi)$. Therefore, we have

$$-\delta(\phi) = \frac{\partial}{\partial \phi}[AK \sin \phi Q(\phi)] + \frac{N_0 K^2}{4} \frac{\partial^2 Q(\phi)}{\partial \phi^2}, \tag{48}$$

with the boundary conditions

$$Q(\phi_l) = \int_0^{\infty} q(\phi_l, t)dt = 0$$

$$Q(-\phi_l) = \int_0^{\infty} q(-\phi_l, t)dt = 0, \tag{49}$$

which follow from (43). The solution of (48) may then be integrated with respect to ϕ over the interval $[-\phi_l, \phi_l]$ to obtain T, the expected time of (46). Taking the indefinite integral of both sides of (48), we obtain

$$C - u(\phi) = AK \sin \phi Q(\phi) + \frac{N_0 K^2}{4} \frac{\partial Q(\phi)}{\partial \phi}, \tag{50}$$

where $u(\phi)$ is the unit step function and C is a constant to be evaluated from the boundary conditions. The solution to the first-order differential equation is

$$Q(\phi) = D \exp(\alpha \cos \phi) + \exp(\alpha \cos \phi) \int_{-\phi_l}^{\phi} \frac{\exp(-\alpha \cos x)}{\gamma}[C - u(x)]dx, \tag{51}$$

where

$$\alpha = \frac{A^2}{N_0(AK/4)}$$

and

$$\gamma = \frac{N_0 K^2}{4} = \frac{AK}{\alpha} = \frac{4B_L}{\alpha}.$$

Application of the boundary conditions of (49) yields the values of the constants as $D=0$ and $C=1/2$. Thus,

$$Q(\phi) = \frac{\exp(\alpha\cos\phi)}{\gamma}\int_{-\phi_l}^{\phi}\exp(-\alpha\cos x)[\tfrac{1}{2}-u(x)]dx. \quad (52)$$

Integrating with respect to ϕ over the interval $[-\phi_l, \phi_l]$, we obtain the expression for the mean time to lose lock:

$$T = \int_{\phi_l}^{\phi_l} Q(\phi)d\phi = \frac{1}{\gamma}\int_{-\phi_l}^{\phi_l} d\phi \times \int_{-\phi_l}^{\phi}\exp\alpha(\cos\phi - \cos x)[\tfrac{1}{2} - u(x)]dx$$

$$= \frac{1}{\gamma}\int_0^{\phi_l}\int_{\phi}^{\phi_l}\exp\alpha(\cos\phi - \cos x)dxd\phi. \quad (53)$$

The domain of integration is the right isosceles triangle shown in Fig. 10. We can obtain a series representation of this double integral by expanding the integrands in Fourier series:

$$\exp(\alpha\cos\phi) = I_0(\alpha) + 2\sum_{m=1}^{\infty} I_m(\alpha)\cos m\phi$$

$$\exp(-\alpha\cos x) = I_0(\alpha) + 2\sum_{n=1}^{\infty}(-1)^n I_n(\alpha)\cos nx. \quad (54)$$

Then

$$T = \frac{1}{\gamma}\int_0^{\phi_l}\int_{\phi}^{\phi_l} \times \begin{bmatrix} I_0^2(\alpha) + 4I_0(\alpha)\sum_{n=2,4,6\cdots}(-1)^n I_n(\alpha)\cos n\phi \\ + 4\sum_{m=1}^{\infty}\sum_{n=1}^{\infty} -(1)^n I_m(\alpha)I_n(\alpha)\cos m\phi\cos nx \end{bmatrix} d\phi dx$$

$$= \frac{1}{\gamma}\left[\frac{I_0^2(\alpha)\phi_l^2}{2} + 4I_0(\alpha)\sum_{n=1}^{\infty}\frac{I_{2n}(\alpha)}{2n}\sin 2n\phi_l + 4\sum_{m=1}^{\infty}\sum_{n=1}^{\infty}(-1)^n I_m(\alpha)I_n(\alpha) \times \int_0^{\phi_l}\int_{\phi}^{\phi_l}\cos m\phi\cos nx dx d\phi\right], \quad (55)$$

where

$$\int_0^{\phi_l}\int_{\phi}^{\phi_l}\cos m\phi\cos nx dx d\phi = \begin{cases} \cos(n-m)\phi_l\left[\frac{1}{nm} + \frac{1}{n(n-m)}\right] - \frac{4\cos n\phi_l}{nm} - \frac{1}{(n-m)n} & \text{when } n = m \\ \left(\frac{1}{m^2} - \frac{\cos m\phi_l}{m^2}\right) & \text{when } n = m. \end{cases} \quad (56)$$

This expression may be computed without the aid of a large-scale digital computer because the sequence

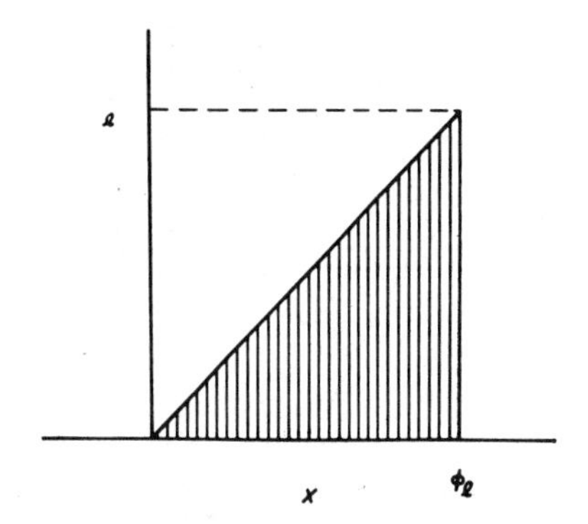

Fig. 10—Domain of integral T.

$I_n(\alpha)$, and consequently the above series, converges quite rapidly.

However, the most important result which we seek can be obtained in closed form. This is the frequency of skipping cycles, or, in other words, the inverse of the expected time between skipping cycles, which is $T(\phi_l=2\pi)$. It is clear from (55) and (56) that when $\phi_l=2\pi$,

$$T(2\pi) = \frac{2\pi^2}{\gamma}I_0^2(\alpha) = \frac{\pi^2\alpha I_0^2(\alpha)}{2B_L}, \quad (57)$$

where we have used

$$\gamma = \frac{N_0K^2}{4} = \frac{AK}{\alpha} = \frac{4B_L}{\alpha}$$

so that

$$\text{frequency of skipping cycles} = (2B_L)/\pi^2\alpha I_0^2(\alpha). \quad (58)$$

This parameter normalized by B_L is shown as a function of α in Fig. 11.

For large SNR, α,

$$I_0(\alpha) \sim (e^{\alpha})/(2\pi\alpha)^{1/2},$$

so that for large α,

$$\text{frequency of slipping cycles} \simeq [(4B_L)/\pi]e^{-2\alpha}. \quad (59)$$

Another parameter which is equally significant is the frequency of dropping or advancing half cycles ($\phi_l=\pi$). In the mechanical analog this corresponds to the pendulum arriving at the unstable equilibrium position and returning to the stable equilibrium position, either by the same route or by going around the full revolution. It is nearly intuitive that for a Markov process the frequency of this event is exactly double the frequency of skipping cycles. However, to show this rigorously, we note that the expected time for the pendulum to go from the equilibrium position $\phi=0$ to $\phi=\pi$ and to return is $T(\pi)+T'(\pi)$, where $T(\pi)$ is the expected time to go from 0 to $\pm\pi$, and $T'(\pi)$ is the expected time to go from π to either 0 or 2π. $T(\pi)$ is given by (55) with $\phi_l=\pi$, while we can show that

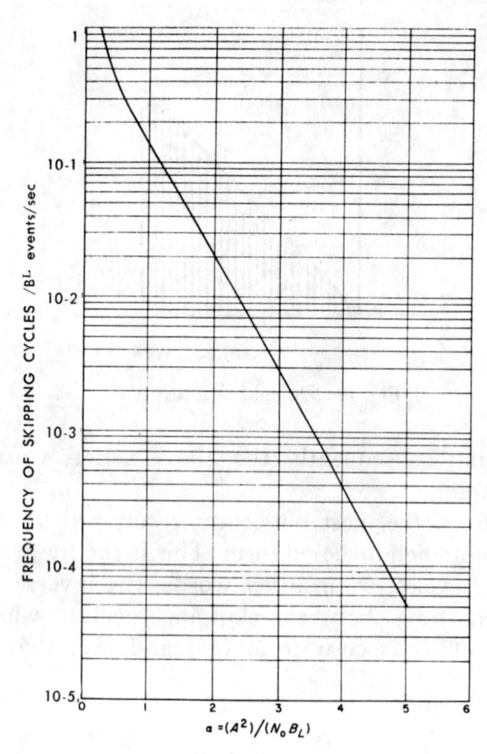

Fig. 11—Frequency of skipping cycles normalized by loop bandwidth for first-order loop where $\omega=\omega_0$.

$$T'(\pi) = \frac{1}{\gamma}\int_0^{\phi_l}\int_0^{\phi} \exp \alpha(\cos\phi - \cos x)dxd\phi.$$

The integrand is the same as that for $T(\pi)$, but the domain of integration is its complement with respect to the square of side π (Fig. 12). Therefore,

$$T(\pi) + T'(\pi) = \frac{1}{\gamma}\int_0^{\pi}\int_0^{\pi} \exp \alpha(\cos\phi - \cos x)dxd\phi$$

$$= (\pi^2/\gamma)I_0{}^2(\alpha) = [T(2\pi)/2], \tag{60}$$

and

$$\text{frequency of slipping half cycles} = (4B_L)/[\pi^2\alpha I_0{}^2(\alpha)]. \tag{61}$$

The Fokker-Planck Equation for Higher-Order Loops

Consider the phase-locked loop whose filter has the rational transfer function

$$F(s) = G(s)/H(s),$$

where $G(s)$ and $H(s)$ are polynomials such that $G(0)=1$, $H(0)=0$ and

$$\deg G(s) \le \deg H(s) = n - 1.$$

Then

$$G(s) = \sum_{k=0}^{n-1} a_k s^k; \qquad a_0 \not\equiv 0$$

$$H(s) = \sum_{k=1}^{n-1} b_k s^k; \qquad b_{n-1} \not\equiv 0. \tag{62}$$

This will be referred to as an nth-order loop. In this case, (11) which describes the operation of the loop becomes

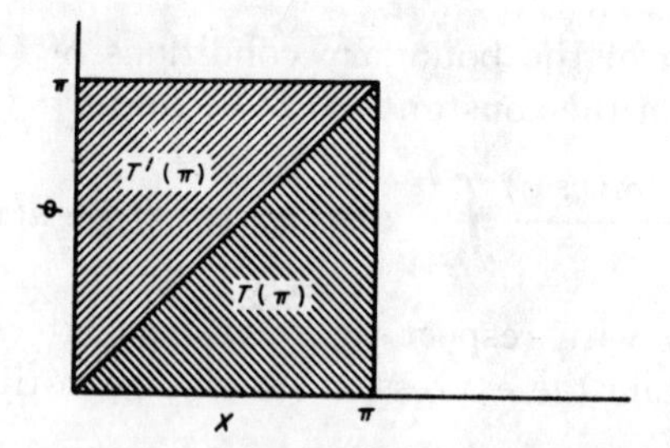

Fig. 12—Domains of integration for $T(\pi)$ and $T'(\pi)$.

$$sH(s)\phi = -KG(s)(A\sin\phi + n'), \tag{63}$$

since

$$s^k(\omega - \omega_0) = \frac{d^k}{dt}(\omega - \omega_0) = 0 \qquad \text{for } k \ge 1.$$

The reason for the pole at the origin of $F(s)$ is now clear. It eliminates the constant $(\omega-\omega_0)$ which causes the steady-state phase error in the first-order loop. Now let us define the random variable ϵ by the relation[5]

$$\phi = G(s)\epsilon. \tag{64}$$

Inserting this in (63), we obtain

$$sH(s)\epsilon = -K(A\sin\phi + n'), \tag{65}$$

which is an nth-order differential equation. Now let us define the n random variables $x_0, x_1, \cdots, x_{n-1}$ as

$$x_k = \frac{d^k\epsilon}{dt^k} \quad (k = 0, 1, \cdots, n-1.) \tag{66}$$

Inserting these for the derivatives of ϵ in (65) and using (62), we obtain

$$b_{n-1}\dot{x}_{n-1} + \sum_{k=1}^{n-2} b_k x_{k+1} = -K(A\sin\phi + n').$$

Also, we have

$$x_k = \frac{d}{dt}\frac{d^{k-1}\epsilon}{dt^{k-1}} = \dot{x}_{k-1},$$

so that we may express the derivatives $\dot{x}_k$ in terms of the variables x_k by the n differential equations

$$\begin{aligned}
\dot{x}_{n-1} &= -\sum_{k=1}^{n-2}\frac{b_k}{b_{n+1}}x_{k+1} - \frac{K(A\sin\phi + n')}{b_{n-1}} \\
\dot{x}_{n-2} &= x_{n-1} \\
&\vdots \\
\dot{x}_0 &= x_1
\end{aligned} \tag{67}$$

[5] This substitution which leads to the representation of ϕ as sum of the components of a Markov vector (67) was suggested by J. N. Franklin.

It follows also from (62), (64), and (66) that

$$\phi = \sum_{k=0}^{n-1} a_k x_k. \tag{68}$$

The random vector $(x_0, \cdots, x_{n-1})$ is a Markov vector since an incremental change depends only on the present state of the vector.

Wang and Uhlenbeck [12] have shown that for a vector Markov process, $x=(x_0, x_1, \cdots, x_{n-1})$, the Fokker-Planck equation is

$$\frac{\partial p(x)}{\partial t} = - \sum_{k=0}^{n-1} \frac{\partial}{\partial x_k} [A_k(x)p(x)] + \frac{1}{2} \sum_{k=0}^{n-1} \sum_{l=0}^{n-1} \frac{\partial^2}{\partial x_k \partial x_l} [B_{kl}(x)p(x)],$$

where

$$A_k(x) = \lim_{\Delta t \to 0} \frac{1}{\Delta t} \overline{(\Delta x_k)},$$

and

$$B_{kl}(x) = \lim_{\Delta t \to 0} \frac{1}{\Delta} \overline{(x_k)(x_l)} \tag{69}$$

with the initial condition

$$p(x, 0) = \prod_{k=0}^{n-1} \delta(x_k - x_{k,0}).$$

In our case,

$$A_k(x) = x_{k+1} \quad \text{for } k = 0, 1, \cdots, n-2$$

$$A_{n-1}(x) = - \sum_{k=1}^{n-2} \frac{b_k}{b_{n-1}} x_{k+1} - \frac{KA}{b_{n-1}} \sin \phi$$

$$B_{n-1,n-1}(x) = \lim_{\Delta t \to 0} \frac{1}{\Delta t} \frac{K^2}{b_{n-1}^2} \int_t^{t+\Delta t} \int_t^{t+\Delta t} \overline{n'(u)n'(v)} du dv = \frac{K^2 N_{0/2}}{b_{n-1}^2}$$

$$B_{k,l}(x) = 0 \quad \text{for all } k \neq n-1 \text{ and } l \neq n-1.$$

Thus, the Fokker-Planck equation for the nth-order loop is

$$\frac{\partial p(x, t)}{\partial t} = - \sum_{k=0}^{n-2} x_{k+1} \frac{\partial p(x, t)}{\partial x_k} + \frac{1}{b_{n-1}} \frac{\partial}{\partial x_{n-1}} \times \left[\left(\sum_{k=0}^{n-2} b_k x_{k+1} + KA \sin \phi \right) p(x, t) \right] + \frac{K^2 N_0}{4 b_{n-1}^2} \frac{\partial^2 p(x, t)}{\partial x_{n-1}^2}, \tag{70}$$

where

$$\phi = \sum_{k=0}^{n-1} a_k x_k.$$

Solution of this general case does not appear possible. However, in the next section some results are obtained for the second-order loop.

Steady-State Probability Distribution for the Second-Order Loop

The loop filter of greatest interest[6] is

$$F(s) = 1 + (a/s) = (s + a)/(s),$$

which requires a single integrator with gain a. In terms of the parameters of (62), $n=2$, $a_0=a$, $a_1=1$, $b_1=1$. Substituting these parameters in (67) and (70), we obtain the differential equations for the random variables

$$\dot{x}_1 = - K[(A \sin \phi + n'], \tag{71a}$$

$$\dot{x}_0 = x_1, \tag{71b}$$

and the Fokker-Planck equation

$$\frac{\partial p}{\partial t} = - x_1 \frac{\partial p}{\partial x_0} + \frac{\partial}{\partial x_1} [(AK \sin \phi) p] + \frac{K^2 N_0}{4} \frac{\partial^2 p}{\partial x_1^2}, \tag{72}$$

where

$$\phi = a x_0 + x_1. \tag{73}$$

If we restrict our attention to the steady-state probability distribution

$$p(x_0, x_1) = \lim_{t \to \infty} p(x_0, x_1, t).$$

Since

$$\lim_{t \to \infty} \frac{dp}{dt} (x_0, x_1, t) = 0,$$

we obtain

$$x_1 \frac{\partial p}{\partial x_0} = AK \frac{\partial}{\partial x_1} [(\sin \phi) p] + \frac{K^2 N_0}{4} \frac{\partial^2 p}{\partial x_1^2}.$$

With the substitution

$$z = a x_0, \tag{74}$$

we obtain an equation in $p(\phi, z)$ (note that the Jacobian of the transformation is a),

$$a(\phi - z) \left(\frac{\partial p}{\partial \phi} + \frac{\partial p}{\partial z} \right) = AK \frac{\partial}{\partial \phi} (\sin \phi p) + \frac{K^2 N_0}{4} \frac{\partial^2 p}{\partial \phi^2}. \tag{75}$$

Even this partial differential equation cannot be solved directly. However, since we are interested only in the density function of ϕ,

$$p(\phi) = \int_{-\infty}^{\infty} p(\phi, z) dz,$$

we may integrate both sides of (75) with respect to z

[6] Tikhonov [9], considered the RC low-pass filter whose transfer function is $1/(s+b)$. Its value is limited, however, since it does not reduce the mean phase-error to zero, as the perfect integrator does.

over the infinite line and obtain an ordinary differential equation in $p(\phi)$

$$a\left\{\frac{d(\phi p)}{d\phi} - \frac{d}{d\phi}\left[\int_{-\infty}^{\infty} zp(\phi, z)dz\right]\right\} = AK\frac{d}{d\phi}(\sin\phi p) + \frac{K^2N_0}{4}\frac{d^2p}{d\phi^2}. \quad (76)$$

But

$$\int_{-\infty}^{\infty} zp(\phi, z)dz = p(\phi)\int_{-\infty}^{\infty} zp(z\mid\phi)dz = p(\phi)E(z\mid\phi),$$

so that (76) becomes

$$0 = \frac{d}{d\phi}\left\{[AK\sin\phi - a\phi + aE(z\mid\phi)]p + \frac{K^2N_0}{4}\frac{dp}{d\phi}\right\}. \quad (77)$$

Unfortunately, it is not possible to determine exactly $E(z|\phi)$, which is a function of ϕ, without knowing $p(z, \phi)$, which would require solution of (75). However, its general form can be obtained as follows: from (73) and (74) we have $z=\phi-x_1$, so that

$$\begin{aligned} E[z(t)\mid\phi(t)] &= E[\phi(t) - x_1(t)\mid\phi(t)] \\ &= \phi(t) - E[x_1(t)\mid\phi(t)]. \end{aligned} \quad (78)$$

Integrating (71a) using (74), we have

$$x_1(\infty) - x_1(t) = -AK\int_t^{\infty}\sin\phi(\xi)d\xi - K\int_t^{\infty} n'(\xi)d\xi. \quad (79)$$

The expectation of the second term on the right side of (79) is zero, since $n'(t)=0$ for all t. Also,

$$E[x_1(\infty)\mid\phi(t)] = E[x_1(\infty)] = 0,$$

since it is clear that the mean of the process is zero. Therefore,

$$E[x_1(t)\mid\phi(t)] = AK\int_j^{\infty} E[\sin\phi(\xi)\mid\phi(t)]d\xi. \quad (80)$$

Combining (77), (78), and (80), and letting $\xi=t+\tau$, we obtain

$$0 = \frac{d}{d\phi}\left\{\frac{4A}{KN_0}\left(\sin\phi - a\int_0^{\infty} E[\sin\phi(t+\tau)\mid\phi(t)]d\tau\right)p(\phi) + \frac{dp(\phi)}{d\phi}\right\}. \quad (81)$$

At this point we are dealing with the random phase process ϕ. We may once again deal with ϕ modulo 2π, or equivalently define the function

$$P(\phi) = \sum_{n=-\infty}^{\infty} p(\phi + 2\pi n)$$

as we did for the first-order loop. Since the coefficients of (81) are periodic in ϕ, if $p(\phi)$ is a solution then so is $p(\phi+2\pi n)$ for all integers n, and hence so is $P(\phi)$. Thus we may replace $p(\phi)$ in (81) by $P(\phi)$. The magnitude of the expectation is always less than one, and becomes negligible for values of τ several times the inverse bandwidth of the spectrum of $\phi(t)$. This bandwidth is proportional to AK, as we found for the first-order loop. Therefore, the order of magnitude of the integral is inversely proportional to AK, and if $a\ll AK$, the second term in the coefficient of $P(\phi)$ is much smaller than the first. Neglecting this second term reduces (81) to the steady-state Fokker-Planck equation for the first-order loop (28) with $\omega=\omega_0$, whose solution is (33). Thus when the second integrator gain $a\ll AK$,

$$P(\phi) \simeq \frac{\exp(\alpha\cos\phi)}{2\pi I_0(\alpha)} \qquad -\pi\le\phi\le\pi. \quad (82)$$

On the other hand, for any value of a, when the SNR is large enough, $\phi(t)$ will be small for all time, so that $\sin\phi(t)\simeq\phi(t)$ and both $\phi(t)$ and $\sin\phi(t)$ will be nearly Gaussian processes. In this case, the expectation can be approximated by

$$\begin{aligned} \int_0^{\infty} E[\sin\phi(t+\tau)\mid\phi(t)]d\tau &\simeq \int_0^{\infty} E[\phi(t+\tau)\mid\phi(t)]d\tau \\ &= \left[\int_0^{\infty}\rho_\phi(\tau)d\tau\right]\sin\phi, \end{aligned} \quad (83)$$

where $\rho_\phi(\tau)$ is the normalized autocorrelation function of the stationary process $\phi(t)$. The integral can be obtained by using Parseval's theorem:

$$\int_0^{\infty}\rho_\phi(\tau)dt = \frac{1}{2\sigma^2}\int_{-\infty}^{\infty} R_\phi(\tau)d\tau = \frac{S_\phi(0)}{2\sigma^2},$$

where $R_\phi(\tau)$ is the unnormalized autocorrelation function, σ^2 the variance of ϕ, and $S_\phi(\omega)$ the spectral density. Since we have approximated $\sin\phi$ by ϕ, we may use the linearized version of Fig. 2 with the loop filter $F(s)=1+(a/s)$ inserted. Then

$$S_\phi(\omega) = \frac{N_0K^2}{2}\left|\frac{s+a}{s^2+AKs+aAK}\right|^2,$$

so that $S_\phi(0)=(N_0)/(2A^2)$.

$$\sigma^2 = \frac{1}{2\pi}\int_{-}^{} S_\phi(\omega)d\omega = \frac{N_0}{4A^2}(AK+a)$$

and

$$\int_0^{\infty}\rho_\phi(\tau)d\tau = 1/(AK+a).$$

Inserting this integral in (83) and substituting in (81) with $p(\phi)$ replaced by $P(\phi)$, we obtain

$$0 = \frac{d}{d\phi}\left\{\frac{4A}{KN_0}\left[\sin\phi\left(\frac{AK}{AK+a}\right)\right]P(\phi) + \frac{dP(\phi)}{d\phi}\right\},$$

whose solution with the boundary conditions of (30) and (31) is

$$P(\phi) \simeq \frac{\exp(\alpha' \cos \phi)}{2\pi I_0(\alpha')} \quad \text{for large } \alpha' \tag{84}$$

where the effective SNR, α', is given by

$$\alpha' = (A^2)/[N_0(AK + a)/(4)].$$

If we let $B_L = (AK+a)/4$, this is the same expression as that for the first-order loop with $\omega = \omega_0$. As would be expected, this expression for loop bandwidth for the second-order loop is that obtained from the linear model of the loop.

Random Modulation

Thus far we have considered only sinusoidal signals of constant frequency and phase. However, the exact method may be applied to signals with random frequency or phase modulation, provided the modulating process is stationary and Gaussian. We shall now derive the Fokker-Planck equation for a first-order loop and a specific random modulation, which will demonstrate the procedure in general.

The differential equation of operation for a first-order loop with arbitrary modulation is given by (10) as

$$\dot{\phi}(t) = \dot{\theta}_1(t) - K[A \sin \phi(t) + n'(t)]. \tag{10}$$

We shall assume that the VCO quiescent frequency is equal to the carrier frequency ($\omega = \omega_0$), and that the carrier signal is frequency modulated by the stationary Gaussian process $m(t)$ with a modulation index

$$K_F \frac{\text{rad/sec}}{\text{volt}} \cdot$$

Then

$$\dot{\theta}_1(t) = K_F m(t), \tag{85}$$

and (10) becomes

$$\dot{\phi}(t) = -AK \sin \phi(t) - Kn'(t) + K_F m(t). \tag{86}$$

Note that the modulation is an additional term in the driving function of the differential equation. Specifically we shall consider a Gaussian modulating signal whose spectrum[7] is

$$S_m(\omega) = \frac{M_0/2}{\omega^2 + \beta^2} \cdot \tag{87}$$

With this driving function $\phi(t)$ as given by (86) is no longer a Markov process, since the present value of $m(t)$ is correlated with the past and consequently $\dot{\phi}(t)$ is no longer independent of the past states of the system. However, we may proceed to express $m(t)$ in terms of an auxiliary white process, which will allow us to treat the problem in terms of a two-dimensional Markov process whose components are $\phi(t)$ and $m(t)$.

A stationary Gaussian process whose spectrum is given by (87) has the same statistics as the output of a first-order linear system (such as an RC low-pass filter) excited by white Gaussian noise. That is, we may represent $m(t)$ in terms of the white Gaussian process $\eta(t)$ of zero mean and of one-sided spectral density M_0, by the operational equation

$$m(t) = \frac{1}{s + \beta} \eta(t),$$

or equivalently by the differential equation

$$\dot{m}(t) = -\beta m(t) + \eta(t). \tag{88}$$

We may now treat the two first-order differential equations (86) and (88) as the defining equations for the two-dimensional Markov process $[\phi(t), m(t)]$ with white Gaussian driving functions $n'(t)$ and $\eta(t)$. Thus we may determine the Fokker-Planck equation for the two-dimensional probability density function in ϕ and m, $p(\phi, m, t)$, by evaluating the normalized moment coefficients of (69) and inserting these in the Fokker-Planck equation in the same way as was done for higher-order loops. The result is

$$\frac{\partial p}{\partial t} = \frac{\partial}{\partial \phi}\left[(AK \sin \phi - K_F m)p + \frac{K^2 N_0}{4} \frac{\partial p}{\partial \phi}\right] + \frac{\partial}{\partial m}\left[\beta m p + \frac{K^2 M_0}{4} \frac{\partial p}{\partial m}\right]. \tag{89}$$

The method used to derive (89) is easily generalized to the case of an nth-order loop with a received signal frequency modulated by a random Gaussian process with a rational spectrum, whose denominator polynomial is of degree $2k$, and whose numerator polynomial is of lower degree. The result in the general case is a Fokker-Planck equation for an $n+k$-dimension vector Markov process. The difficulty lies only in the solution of the partial differential equation. Even (89) which represents the simplest case ($n=k=1$) appears formidable. For the steady-state or stationary case in which the time derivative is set equal to zero, we have a problem of the same magnitude as for the second-order loop without modulation, for which we could obtain only approximate results.

Other Error Functions

While the majority of closed-loop tracking systems employ sinusoidal carriers and reference oscillators, occasionally for low frequencies and specific applications, square wave carriers are employed. In such cases the VCO is replaced by a multivibrator which, when the

[7] A Gaussian process with this spectrum is necessarily a Markov process which may be generated by driving a first-order linear system with white Gaussian noise; *cf.*, Wang and Uhlenbeck [12].

loop is locked, generates a square wave which is displaced by exactly a quarter period relative to the received signal. If the received signal power is A^2, the amplitude of the square wave must be A. Let the reference square wave be of amplitude K_3 and let all the other components and parameters of the loop be the same as for the sinusoidal case (Fig. 1). Then by taking the Fourier series expansion of the square waves and reproducing the analysis for sinusoidal loops, we find that the equation of operation becomes

$$\dot{\phi}(t) = \dot{\theta}_1(t) - KF(s)[Ah(\phi) + n'(t)], \qquad (90)$$

where $n'(t)$ has the same statistics as for sinusoidal loops and $h(\phi)$ is the triangular wave, one period of which is shown in Fig. 13. Thus the model of Fig. 2 describes this case when the sinusoidal nonlinearity is replaced by the function $h(\phi)$.

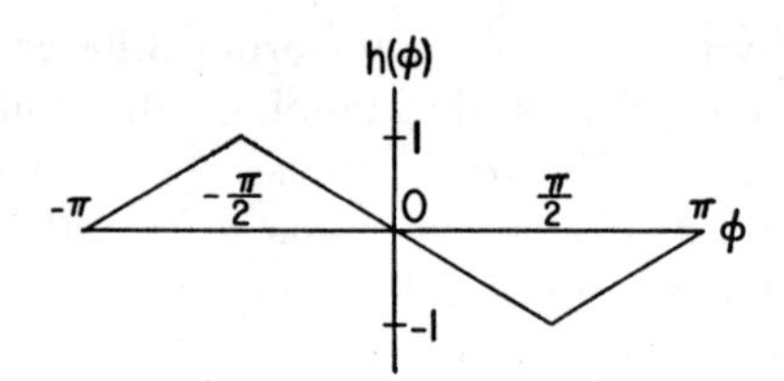

Fig. 13—Triangular error function.

Another system which has received much attention in a variety of applications including radar ranging is a tracking device often referred to as the delay-lock loop [16]–[18]. In one version of this device [18] the received signal alternates between $+A$ volts and $-A$ volts according to a binary code with a switching interval corresponding to the half period of the square wave in the above-mentioned case. The code is generated by a maximum-length linear shift register at the transmitter. The VCO in the receiver tracking loop consists of an identical shift register whose switching period is controlled by the loop filter output. The reference signal is derived by delaying the VCO-coder output by exactly two switching intervals and subtracting it from the undelayed coder output. The result of multiplying this by the received signal is an error function identical to Fig. 13 over the interval $-\pi<\phi<\pi$ and zero elsewhere [18]. However, the noise density is doubled by this procedure [so that the one-sided spectral density of $n'(t)$ is $2N_0$ in this case] and some self-noise is introduced into the system by the randomness of the code. However, the self-noise is negligible if the ratio N_0/A^2 is very large compared to the switching period [18]. We shall take this to be the case in the analysis which follows.

The various results obtained for sinusoidal loops can be generalized to arbitrary error functions $h(\phi)$ which are odd functions of ϕ, of which the cases just mentioned are particular examples. We shall consider only the results for the first-order loop. The steady-state probability density for a nonlinearity which is a periodic odd function of ϕ can be obtained by the same method as was used for the sinusoidal case [*cf.* (33)], and is

$$P(\phi) = C \exp[-\alpha g(\phi)] \qquad -\pi \le \phi \le \pi,$$

where

$$g(\phi) = \int^{\phi} h(x)dx, \qquad (91)$$

and

$$C = \frac{1}{\int_{-\pi}^{\pi} \exp[-\alpha g(\phi)]d\phi},$$

provided the timing periods of the received signal and the reference oscillator are initially synchronized.

The mean time to loss of lock has particular significance for the case of the coded signals of the delay-lock loop just discussed. For, assuming that the loop is initially in lock, if the phase error ever exceeds $\pm\pi$, there will be no deterministic restoring force tending to restore lock. We can show by an obvious generalization of the previous results that for an arbitrary odd error function (which need not be periodic), the mean time to loss of lock when the loop is initially in lock ($\phi_0=0$) is given by the expression [*cf.* (53)]

$$T(\phi) = \frac{1}{\gamma}\int_0^{\phi_l}\int_{\phi}^{\phi_l} \exp(-\alpha)[g(\phi) - g(x)]dxd\phi, \qquad (92)$$

where

$$\alpha = \frac{A^2}{N_0 B_L}, \qquad \gamma = \frac{4B_L}{\alpha}, \qquad B_L = AK/4,$$

and

$$g(\phi) = \int^{\phi} h(x)dx.$$

For the delay-lock tracking loop, we find by integrating the error function of Fig. 13 over a half period, that

$$-\alpha g(\phi) = \begin{cases} \dfrac{\alpha}{\pi}\left[\left(\dfrac{\pi}{2}\right)^2 - \phi^2\right] & 0 \le \phi \le \pi/2 \\ \dfrac{\alpha}{\pi}\left[-\left(\dfrac{\pi}{2}\right)^2 + (\phi-\pi)^2\right] & \pi/2 \le \phi \le \pi. \end{cases} \qquad (93)$$

We are most interested in the case $\phi_l=\pi$. However, the case $\phi_l=\pi/2$ is also worth consideration since it represents the mean time required for the loop to pass beyond the central linear region when initially $\phi=0$. The triangular domain of integration may be divided into three regions (Fig. 14) according to the regions of definition of $g(\phi)$ (93). Then referring to (93) and Fig. 14, we obtain from (92),

$$T(\pi) = A + B + C, \tag{94}$$

where

$$A = T(\pi/2) = \frac{1}{\gamma}\int_0^{\pi/2}\int_\phi^{\pi/2} \exp\left(\frac{\alpha}{\pi}\right)(-\phi^2 + x^2)dxd\phi$$

$$B = \frac{1}{\gamma}\int_{\pi/2}^{\pi}\int_\phi^{\pi} \exp(\alpha/\pi)[(\phi-\pi)^2 - (x-\pi)^2]dxd\phi$$

$$C = \frac{1}{\gamma}\int_0^{\pi/2}\int_{\pi/2}^{\pi} \exp\left(\frac{\alpha}{\pi}\right)\left[\frac{\pi^2}{2} - \phi^2 - (x-\pi)^2\right]dxd\phi.$$

By making the proper changes of variables we can show that

$$C = \frac{\alpha \exp\left(\frac{\alpha\pi}{2}\right)}{4B_L}\left[\int_0^{\pi/2} \exp -\left(\frac{\alpha\phi^2}{\pi}\right)d\phi\right]^2, \tag{95}$$

which is a tabulated integral, and

$$B = A = T(\pi/2)$$

$$= \frac{1}{\gamma}\int_0^{\pi/2}\int_0^{x} \exp\left(\frac{\alpha}{\pi}\right)(x^2 - \phi^2)d\phi dx$$

$$= \frac{\alpha}{4B_L}\left[\frac{\pi\alpha}{4} + \sum_{n=2}^{\infty}\frac{(\pi\alpha/4)^n}{n!}\frac{\prod_{i=1}^{n-1} 2i}{\prod_{i=1}^{n}(2i-1)}\right], \tag{96}$$

as can be shown by expanding the integrand in Taylor series. From (94), (95), and (96) we obtain the mean times $T(\pi)$ and $T(\pi/2)$ multiplied by B_L whose inverses are shown in Fig. 15. Similar results can be obtained for any odd error function by evaluating (92) for the proper $g(\phi)$.

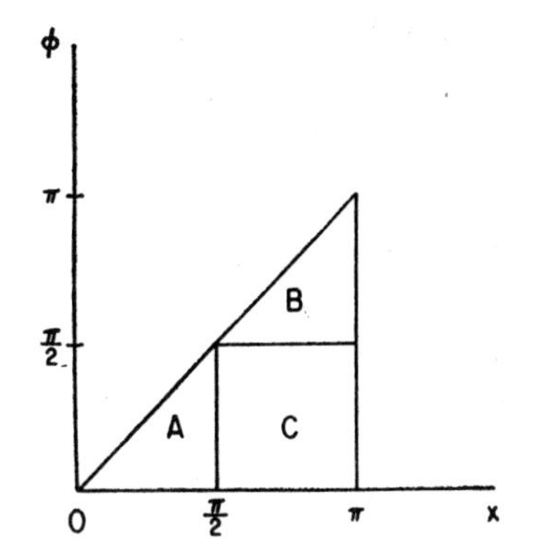

Fig. 14.—Domains of integrals of (94).

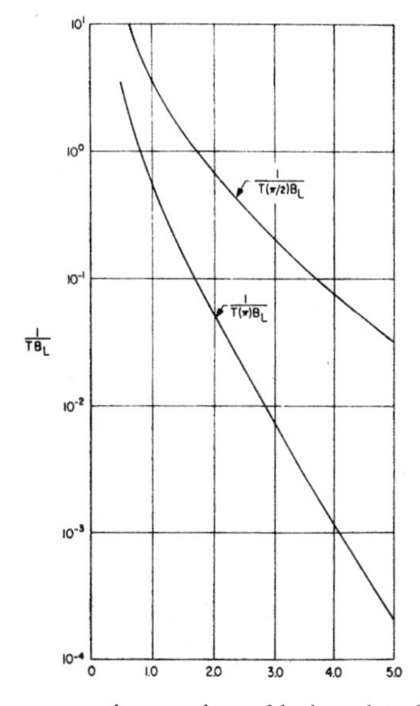

Fig. 15—Inverse-mean times to loss of lock and to first-passage from linear region (normalized by loop bandwidth).

Conclusions and Comparisons

This paper has dealt with the exact analysis of the nonlinear device described by the model of Fig. 2 or the nonlinear differential equation (10). The principal results, which can be generalized to any odd error function, have been:

1) The stationary (steady-state) probability density function, distribution, and variance for the first-order loop.
2) The expected time to loss of lock and frequency of skipping cycles for the first-order loop (this is particularly useful for constant-velocity Doppler tracking applications).
3) Approximate expressions for the stationary probability density of the second-order loop.
4) The partial differential (Fokker-Planck) equation for the probability density for higher-order loops, including the case in which the signal is frequency modulated by a stationary Gaussian process.

The limitations of the method become evident when we attempt to solve the partial differential equation for the higher-order cases. Although the equations are linear in the probability densities, since the coefficients are nonlinear functions of the dependent variables, no exact solution seems possible. Thus for the higher-order cases and particularly for modulated signals, we must rely on the approximate models mentioned in the first section of this paper which lend themselves to more direct methods. However, our exact results for the first-order loop are quite useful in determining the validity threshold of a particular model, which we shall define loosely as the value of SNR below which the model no longer yields useful results.

The most obvious parameter to consider for this comparison is the variance of the first-order loop. We have already seen (Fig. 7) that for the linear model (wherein $\sin\phi$ is replaced by ϕ), the variance as determined from the model underestimates the actual variance by more than 20 per cent when the SNR α is less than 4 (or 6 db), so that $\alpha = 4$ may be taken as the validity thresh-

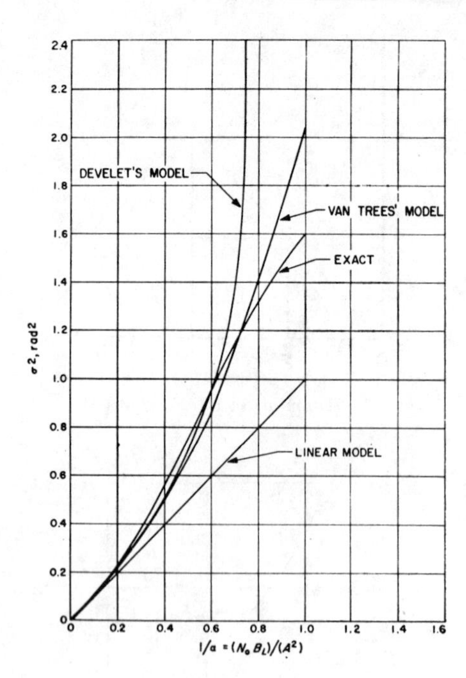

Fig. 16—Comparison of variance for first-order loop with results of approximate models.

old of the linear model. The variance for the exact model and the linear model are reproduced in Fig. 16 together with the variance obtained by using the models of Van Trees [8] and Develet [2]. Van Trees has shown that for the first-order loop the variance of the phase error can be written as a power series in $1/\alpha$. By calculating the first five terms of the Volterra functional expansion, Van Trees [8] found that the first three terms of the power series are

$$\sigma^2 = 1/\alpha + \frac{1}{2}(1/\alpha)^2 + \frac{13}{24}(1/\alpha)^3 \cdots \quad \text{(Van Trees)}.$$

This is shown in Fig. 16. Using the quasi-linearization technique, Develet [2] replaced the sinusoidal nonlinearity by its average gain under the assumption that the input distribution is approximately Gaussian. Since the gain of a sinusoidal nonlinearity for an input value x is $A \cos x$, the average gain when the input is Gaussian of zero mean and variance σ^2 is

$$\int_{-\infty}^{\infty} A \cos x \exp\left(\frac{-x^2}{2\sigma^2}\right) dx = A \exp\left(-\frac{\sigma^2}{2}\right).$$

Replacing the nonlinear element of Fig. 2 by this gain, we obtain by the usual linear analysis the variance of the phase error for the first-order loop:

$$\sigma^2 = (1/\alpha) \exp(\sigma^2/2) \qquad \text{(Develet)}.$$

The solution of this transcendental equation yields the value of the variance shown in Fig. 16. The maximum of $\sigma^2 \exp(-\sigma^2/2)$ is $2/e$ so that there can be no solution for $\alpha < e/2$, which means that the validity threshold of this model can be no lower than this value.

From Fig. 16 we note that the error in the Develet model is less than 10 per cent for $1/\alpha < 0.65$ or $\alpha > 1.54$, while the Van Trees approximation involving the first five Volterra kernels yields results of this accuracy for $1/\alpha < 0.80$ or $\alpha > 1.25$. Of course, with sufficient effort one can compute arbitrarily many terms of the Volterra series and consequently obtain arbitrarily many terms of the power series expansion of σ^2 thus extending the validity threshold of the model as far as may be desired. However, for higher-order loops, Van Trees' method becomes exceedingly complex and tedious, while Develet's method remains simple for all loop filters and even for modulated signals. In fact, using this method he has obtained fairly general results on the threshold of the phase-locked loop as a frequency modulation discriminator [2].

While we have certainly not exhausted the realm of application of closed-loop tracking devices, the results of this paper may serve as a guide for the analysis of a large class of such systems.

Acknowledgment

The author is indebted to Prof. J. N. Franklin and Dr. E. C. Posner for several valuable discussions during the preparation of this manuscript.

References

[1] W. B. Davenport, Jr., and W. L. Root, "Random Signals and Noise," McGraw-Hill Book Co., Inc., New York, N. Y.; 1958.
[2] J. A. Develet, Jr., "A threshold criterion for phase-lock demodulation," Proc. IRE, vol. 51, pp. 349–356; February, 1956.
[3] W. J. Gruen, "Theory of AFC synchronization," Proc. IRE, vol. 41, pp. 1043–1048; August, 1953.
[4] A. J. Viterbi, "Acquisition and Tracking Behavior of Phase-Locked Loops," *Proc. Symp. on Active Networks and Feedback Systems*, Polytechnic Inst. of Brooklyn, N. Y., vol. 10, pp. 583–619; April, 1960.
[5] R. M. Jaffe and E. Rechtin, "Design and performance of phase-lock circuits capable of near-optimum performance over a wide range of input signal and noise levels," IRE Trans. on Information Theory, vol. IT-1, pp. 66–76; March, 1955.
[6] S. G. Margolis, "The response of a phase-locked loop to a sinusoid plus noise," IRE Trans. on Information Theory, vol. IT-3, pp. 135–144; March, 1957.
[7] R. C. Booton, Jr., "The Analysis of Nonlinear Control Systems with Random Inputs," *Proc. Symp. on Nonlinear Circuit Analysis*, Polytechnic Inst. of Brooklyn, N. Y., pp. 369–391; April, 1953.
[8] H. L. Van Trees, "Functional Techniques for the Analysis of the Nonlinear Behavior of Phase-Locked Loops," presented at WESCON, San Francisco, Calif.; August 20–23, 1963.
[9] V. I. Tikhonov, "The effects of noise on phase-lock oscillation operation," *Automatika i Telemakhanika*, vol. 22, no. 9; 1959.
[10] ——, "Phase-lock automatic frequency control application in the presence of noise," *Automatika i Telemekhanika*, vol. 23, no. 3; 1960.
[11] G. E. Uhlenbeck and L. S. Ornstein, "On the theory of Brownian motion," *Phys. Rev.*, vol. 36, pp. 823–841; September, 1930.
[12] M. C. Wang and G. E. Uhlenbeck, "On the theory of Brownian motion II," *Rev. Mod. Phys.*, vol. 17, pp. 323–342; April–July, 1945.
[13] A. A. Andronov, L. S. Pontryagin, and A. A. Witt, "On the statistical investigation of a dynamical system," *J. Exp. Theoret. Phys. (USSR)*, vol. 3, p. 165; 1933.
[14] A. J. F. Siegert, "On the first passage time probability problem," *Phys. Rev.*, vol. 81, pp. 617–623; February, 1951.
[15] D. A. Darling and A. J. F. Siegert, "The first passage time problem for a continuous Markov process," *Ann. Math. Stat.*, vol. 24, pp. 624–639; 1953.
[16] J. J. Spilker, Jr., and D. T. Magill, "The delay-lock discriminator—an optimun tracking device," Proc. IRE, vol. 49, pp. 1403–1416; September, 1961.
[17] M. R. O'Sullivan, "Tracking system employing the delay-lock discriminator," IRE Trans. on Space Electronics and Telemetry, vol. SET-8, pp. 1–7; March, 1962.
[18] J. J. Spilker, Jr., "Delay-lock tracking of binary signals," IEEE Trans. on Space Electronics and Telemetry, vol. SET-9, pp. 1–8; March, 1963.

A New Approach to the Linear Design and Analysis of Phase-Locked Loops*

CHARLES SINCLAIR WEAVER†

Summary—Using the techniques and philosophy of control systems theory, the phase-locked loop is analyzed as a conventional feedback loop. The root-locus method yields graphs which specify how the transient response changes with signal strength. This method also reveals two thresholds which explain why a small change in signal strength or modulation may cause complete loss of detection. Charts show how the transients vary with various pole-zero patterns for both step and ramp inputs. The feedback equation shows why the phase-locked loop is an FM detector, and simplifies its design analysis to that of a simple audio network. The application of Wiener's criterion is simplified, and a new method of solution for the filter is presented which is applicable to almost any kind of signal. Because the phase-locked loop is nonlinear, there is no known solution for the filter except when the noise is white. The optimum transfer function may easily be reduced to the loop components. When used in an AM detector the phase-locked loop should be designed for minimum phase shift independent of the modulation.

Introduction

THE phase-locked loop provides a very attractive means of detecting a small signal in noise. This type of circuit is used increasingly by communications engineers in space-vehicle-to-earth data links, and wherever else the loss along the transmission path is very large or transmitter weight is at a premium. The phase-locked loop may even become the standard detector in the space-vehicle of the future.

Since the phase-locked loop can easily be linearized into a simple feedback control circuit, it seems obvious and natural that the precise analytical techniques of control systems engineering should be applied to this loop. With these techniques it is possible to calculate such a loop's transient response exactly, design for any given transient response, design loops for many kinds of signals and systems, predict the results of changes in signal strength and input noise, and derive Wiener filters for many kinds of signals. The Wiener criterion can be used only when the noise in the frequency range of interest is white. These techniques are mathematically simple, and yield easy-to-use design curves. Many such curves are included in this article.

Simple procedures are given for reducing the loop to its components, and for designing the loop as an FM detector. It is also shown that limiters do not always optimize the system for a wide range of signal-to-noise ratios.

I. Linearizing a Phase-Locked Loop

Before the root-locus and mean-squared-error analyses can be made, the phase-locked loop must be linearized. The linearization presented here differs from Jaffe and Rechtin's[1] because it is necessary to find the effect of the VCO phase upon the loop noise when designing for the Wiener criterion. A typical phase-locked loop (see Fig. 1) consists of a multiplier (phase discriminator), amplifier, and filter in the feedforward path, and a voltage-controlled oscillator (VCO) in the feedback path.

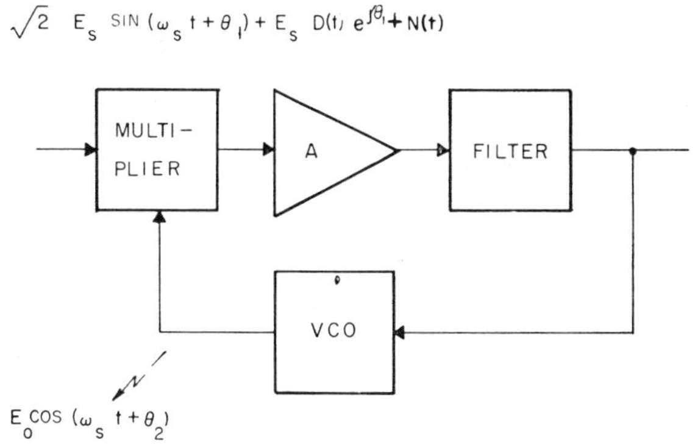

Fig. 1—A typical phase-locked loop.

The generalized incoming signal is $\sqrt{2}E_s \sin(\omega_s + \theta_1) + E_s D(t) e^{i\theta_1}$, where the first term represents the carrier and the second term the voltage in the modulation sidebands. The noise voltage is $N(t)$. The incoming signal, plus noise, is multiplied by the output of the VCO. Since the filter is low pass, we consider only the low-frequency components. The product component due to the carrier is

$$K_m E_0 E_s \sin(\theta_1 - \theta_2) \cong K_m E_0 E_s(\theta_1 - \theta_2) \qquad (1)$$

where K_m is the multiplier constant, and $(\theta_1 - \theta_2)$ is small. The component due to the noise may be found by expanding $N(t)$ in a complex fourier series:

$$N(t) = \sum_k C_k e^{i(\omega_k + \omega_s)t}.$$

* Manuscript received by the PGSET, July 1, 1959. The work reported in this paper was sponsored by the U. S. Air Force as part of Contract AF04(647)-97.

† Western Dev. Labs., Philco Corp., Palo Alto, Calif. Formerly with Government and Industrial Div., Philco Corp., Philadelphia, Pa.

[1] R. Jaffe and E. Rechtin, "Design and performance of phase-lock circuits capable of near-optimum performance over a wide range of input signal and noise levels," IRE Trans. on Information Theory, vol. IT-1, pp. 66–76; March, 1955.

Reprinted from *IRE Trans. Space Electron. Telemetry*, vol. SET-5, pp. 166–178, Dec. 1959.

Then

$$K_m\sqrt{2}N(t)E_0 \cos(\omega_s t + \theta_2)$$

$$= \sqrt{2}K_mE_0 \sum_k C_k e^{j(\omega_k+\omega_s)t}\left[\frac{e^{j(\omega_s t+\theta_2)} + e^{-j(\omega_s t+\theta_2)}}{2}\right]$$

$$= \frac{1}{\sqrt{2}} K_mE_0 \sum_k C_k e^{j\omega_k t}[e^{j(2\omega_s t+\theta_2)} + e^{-j\theta_2}]. \quad (2)$$

The low-frequency component will be

$$\frac{e^{-j\theta_2}}{\sqrt{2}} K_mE_0 \sum_k C_k e^{j\omega_k t} = K_mE_0 \frac{N(t)'}{\sqrt{2}}. \quad (3)$$

The component due to any AM side-bands may be similarly found by expanding in a complex Fourier series:

$$e^{j\theta_1}E_s D(t) = E_s \sum_i D_i e^{j(\omega_i+\omega_s)t}e^{j\theta_1}.$$

The low-frequency component will be

$$\frac{e^{j(\theta_1-\theta_2)}}{\sqrt{2}} K_mE_0E_s \sum_i D_i e^{j\omega_k t} = \frac{K_mE_sE_0 D(t)'}{\sqrt{2}}. \quad (4)$$

The sum of the low-frequency components is

$$K_mE_0\left[E_s(\theta_1 - \theta_2) + \frac{N(t)'}{\sqrt{2}} + \frac{E_s D(t)'}{\sqrt{2}}\right]. \quad (5)$$

The multiplier may now be converted to its linear equivalent, a summing junction, by setting the input equal to

$$\theta_1 + \frac{N(t)'}{E_s\sqrt{2}} + \frac{D(t)'}{\sqrt{2}}$$

(see Fig. 2).

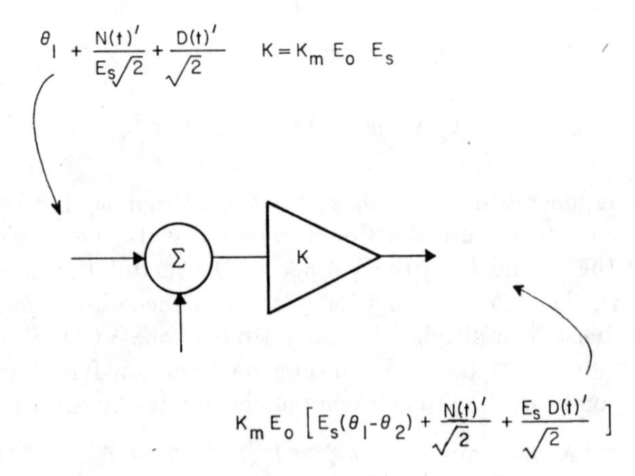

Fig. 2—Amplifier added after summing.

The VCO is assumed to change in frequency as the control voltage changes; the integral of the change in control voltage is proportional to the change in phase. The transfer function relating phase and voltage is

$$\frac{\theta_2}{E_c} = \frac{K_1}{s}. \quad (6)$$

The linearized loop is shown in Fig. 3. $K_2G(s)$ is the filter transfer function; and $G(s)$ is of the form

$$G(s) = \frac{\prod_i (s + a_i)}{\prod_k (s + b_k)}. \quad (7)$$

II. The Root-Locus Method and Feedback Equations

The linearized phase-locked loop may be considered as a conventional feedback loop, and analyzed by the root-locus method.[2-5] This analysis yields a simple graph showing how transient and frequency response change as parameters, such as signal strength and $G(s)$, are changed. The feedback equation relates the open- and closed-loop transfer functions, and is solved using a root-locus plot.

An "open loop" is one that has been broken open at "X," as in Fig. 4.

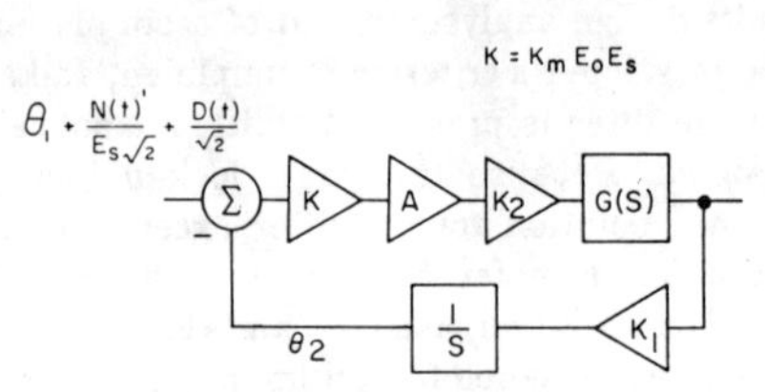

Fig. 3—The linearized phase-locked loop.

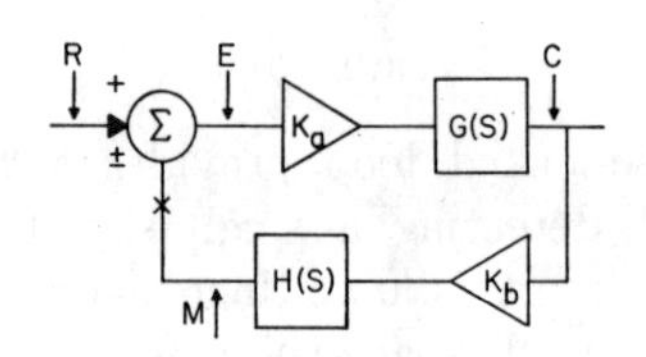

Fig. 4—The feedback loop broken open.

The open-loop transfer function is M/E. The closed-loop transfer function is C/R. The product of the gains of all the amplifiers in the loop of Fig. 3 is the open-loop gain K_{oL}. Design for transient response, and for minimum mean-squared-error, depends on K_{oL}. For the loop of Fig. 3,

$$K_{oL} = K_mE_0E_sAK_1K_2. \quad (8)$$

If the summer is negative, the feedback equation is

$$\frac{C(s)}{R(s)} = \frac{K_aG(s)}{1 + KG(s)H(s)}; \quad K = K_aK_b. \quad (9)$$

If the summer is positive, the sign in the denominator is reversed. Assume $KG(s)$ and $H(s)$ are polynomials of the form

[2] O. J. M. Smith, "Feedback Controls Systems," McGraw-Hill Book Co., Inc., New York, N. Y.; 1958.
[3] J. G. Truxal, "Automatic Feedback Control System Synthesis," McGraw-Hill Book Co., Inc., New York, N. Y.; 1955.
[4] "Reference Data for Radio Engineers," International Telephone and Telegraph Corp., New York, N. Y., 4th ed., pp. 354–358; 1956.
[5] J. A. Aseltine, "Transform Method in Linear System Analysis," McGraw-Hill Book Co. Inc., New York, N. Y.; p. 356; 1958.

$$K_aG(s) = K_a \frac{\prod_i (s + a_i)}{\prod_k (s + b_k)} = K_a \frac{Z_g}{P_g} \quad (10)$$

$$K_bH(s) = K_b \frac{\prod_i (s + \alpha_i)}{\prod_l (s + \beta_l)} = K_b \frac{Z_h}{P_h} \quad (11)$$

where Z denotes the product of the zeros, and P denotes the product of the poles. Then (9) is

$$G_{cL} = \frac{C}{R} = \frac{K_a \frac{Z_g}{P_g}}{1 + \frac{KZ_gZ_h}{P_gP_h}}$$

$$= K_a \frac{Z_gP_h}{P_gP_h + KZ_gZ_h}. \quad (12)$$

The closed-loop transfer function G_{cL} will have zeros in the same place as the zeros of $G(s)$ the feedforward transfer function. Also, the poles of $H(s)$ have become zeros in G_{cL}.

To find the poles in (12), set the denominator of (9) equal to zero. Then

$$KG(s)H(s) = -1 \quad (13)$$
$$= |1| \angle 180°$$

will give the location of all closed-loop poles, thus completely specifying the closed-loop transfer function. These poles will move on a locus as the gain is increased (K is K_{oL} in the phase-locked loop). To form the loci take all the points in the S plane where the angle of $G(s)$ $H(s)$ is 180°.

To find the S-plane phase angle at a particular point, add the angles from all zeros to the point, measuring counter-clockwise from a line parallel with the positive real axis. From this sum, subtract the total of the angles from all the poles to the point (see Fig. 5). The difference is the desired angle. The angles may be measured in several seconds by using a Spirule, a small plastic patented device manufactured by a company of the same name.

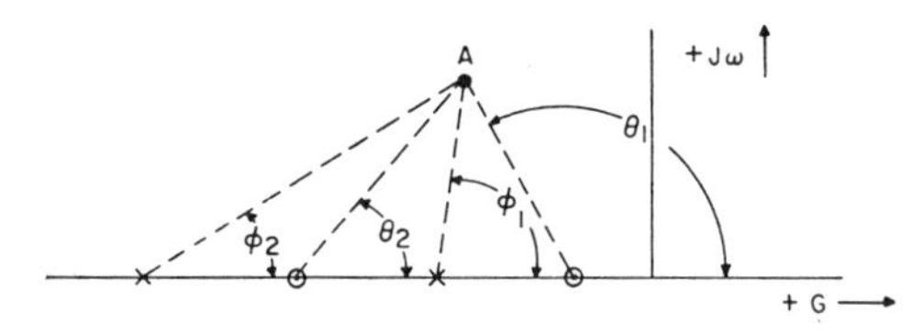

Fig. 5—Measuring the phase angle.

The root-locus can easily be plotted by applying the 6 following rules.

1) Loci start at open-loop poles at $K = 0$ and terminate at open-loop zeros (some at ∞) as $K \rightarrow \infty$. Thus, as gain increases, the closed-loop poles will move out along the loci away from the open-loop poles.

2) The loci, being complex conjugates, are symmetrical about the real axis.

3) All loci that go to infinity approach asymptotic lines whose angle with the real axis is

$$\frac{180° \pm 2n\pi}{\text{Number of Poles} - \text{Number of Zeros}}.$$

4) These asymptotes meet on the real axis at the "center of gravity" S.

$$S = \frac{\sum_i \alpha_i - \sum_j \alpha_j}{\text{No. Poles} - \text{No. Zeros}},$$

α_i is the distance of ith pole along the real axis, and α_j is the distance of the jth zero.

5) Any segment of the real axis that lies to the left of an odd number of real axis poles and zeros is a root-locus.

6) Any root-locus that leaves the real axis leaves it at right angles and near the center of a root-locus segment.

Once the root-locus has been plotted, the gain required to move a closed-loop pole to a specified point on the locus may be found by fulfilling the second condition of (13); namely, that

$$|H(s)G(S)| = 1/K. \quad (14)$$

The absolute magnitude of a complex number at a point is found by taking the ratio of the products of the distance from the zeros to the point, $\prod_j l_j$, to the product of the distances from the poles to the point, $\prod_k d_k$. Then

$$\left| H(s)G(s) \right| = \frac{\prod_j l_j}{\prod_k d_k}. \quad (15)$$

The root-locus plot indicates many of a phase-locked loop's transient characteristics. For example, if the loop's graph is like Fig. 6, the gain causing the loop to go unstable is the point where the poles cross the $j\omega$ axis. As the gain increases from zero, the damping in the loop will decrease, and as gain approaches the instability point, the system will tend to ring at every little disturbance (such as system noise). The noise bandwidth will decrease along with the damping. This demonstrates that response to noise is not a simple relationship (see Section V).

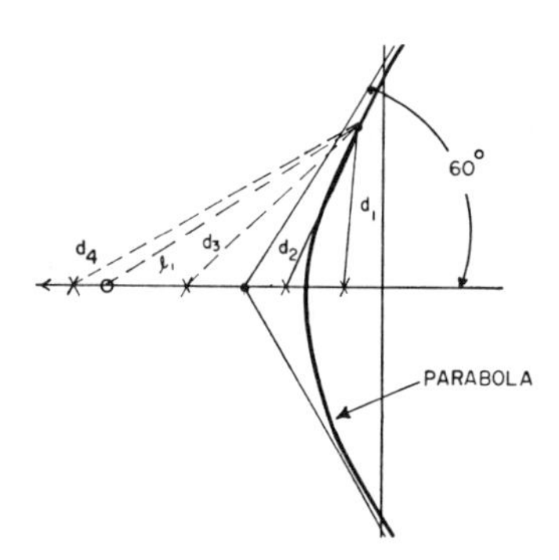

Fig. 6—Location of closed-loop poles. They will lie along the heavy curve.

III. Calculating Transient Response Due to Signal Input Only

The design example is for a system with step input. Design for step input permits exact specification of transient characteristics, and system requirements for step input are usually no more complex than for other inputs. Furthermore, it is often a fact (but not necessarily so) that a system having satisfactory transient response with step input will also function well with other inputs.

To design a phase-locked detector to respond to a frequency shift carrier or as a tracking filter, use the charts in Appendix I, which show the transient response of a linear system to various step and ramp inputs. Tables I and II give the formulas for the relations between step and ramp inputs and the pole and zero positions.

Suppose the transmission consists of one-millisecond carrier-shift pulses as shown in Fig. 7; f_1 will represent a 0 and f_2 will represent 1 in a binary code. This random code may be considered as a series of positive and negative step functions.

Eq. (12) shows that the poles in the feedback path appear as zeros in the closed-loop transfer function. According to Fig. 3, the only pole in the feedback path is $1/S$. Therefore, the closed-loop transfer function will automatically have a zero at the origin. We are neglecting $N(t)$ and $E_s\ D(t)$, so the only input will be $\theta_1(t)$. Since the signal is changing in steps of frequency, the phase change will be a ramp with a slope of $2\pi \times 6$ radians per millisecond (a step change from f_1 to f_2 is 6 cycles per millisecond). The transform of $\theta_1(t)$, is then

TABLE I

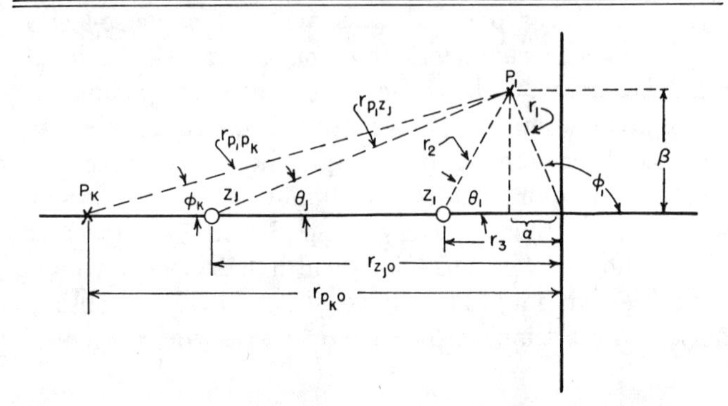

	Dominate Complex Pole-Pair Only	Dominate Complex Pole-Pair and a Zero	Contribution of Non-Dominate Poles and Zeros: Pole Correction Factor	Contribution of Non-Dominate Poles and Zeros: Zero Correction Factor
τ_r	$\frac{\phi_1}{\beta}$	$\frac{\phi - \theta_1}{\beta}$	Add $\frac{\phi_k}{\beta}$	Subtract $\frac{\theta_j}{\beta}$
τ_p	$\frac{\pi}{\beta}$	$\frac{\pi - \theta_1}{\beta}$		
τ_s	$\frac{1}{\alpha} \ln \frac{r_1}{\beta F}$	$\frac{1}{\alpha} \ln \frac{r_1 r_2}{\beta r_3 F}$	Add $\frac{1}{\alpha} \ln \left[\frac{r_{P_k O}}{r_{P_k P_1}}\right]$	Subtract $\frac{1}{\alpha} \ln \left[\frac{r_{Z_j O}}{r_{Z_j P_1}}\right]$
POR	$e^{-(\alpha/\beta)\pi}$	$\frac{r_2}{r_1} e^{-(\alpha/\beta)(\pi - \theta_1)}$	Multiply By $\frac{r_{P_k O}}{r_{P_k P_1}} e^{-(\alpha/\beta)\phi_k}$	Divide By $\frac{r_{Z_j O}}{r_{Z_j P_1}} e^{-(\alpha/\beta)\theta_j}$

$$\theta_1(s) = \frac{2\pi \times 6}{S^2}. \tag{16}$$

When $\theta_1(s)$ is multiplied by the closed-loop transfer function to obtain $C(s)$, the transfer function zero will cancel one of the poles of θ, (s). The transient response $C(s)$ would thus be *exactly* the same as the response of a system with identical poles and zeros minus the zero at the origin that was excited by a step input,

$$\theta(s) = \frac{12\pi}{S} = \frac{K_s}{S}. \tag{17}$$

This argument is also valid for the error voltage $E(s)$.

$$C(s) = K_a G(s) E(s) \tag{18}$$

$$E(s) = R(s) - K_b C(s) H(s), \tag{19}$$

then

$$\frac{E(s)}{R(s)} = \frac{1}{1 + KG(s)H(s)}$$

$$= \frac{P_g P_h}{P_g P_h + K Z_g Z_g} \tag{20}$$

where the notation is the same as in (12).

Two design requirements are that $C(s)$ have a settling time of less than a millisecond, and that $E(t)$ always be

TABLE II

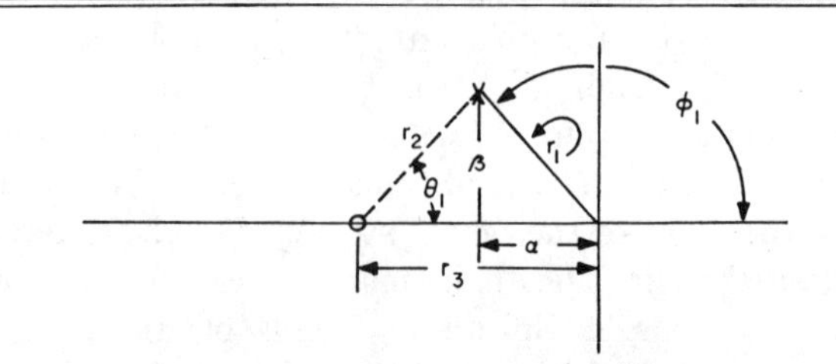

	Complex Pole-Pair	Complex Pole-Pair and Zero
τ_P	$\frac{\pi + \phi_1}{\beta}$	$\frac{\pi + \phi_1 - \theta_1}{\beta}$
τ_s	$\frac{1}{\alpha} \ln \frac{r_1}{\beta F} - \tau_P$	$\frac{1}{\alpha} \ln \frac{r_1}{\beta F} - \tau_p$
Peak Difference	$K_0 \frac{e^{-(\alpha/\beta)(\pi + \phi_1)}}{r_1}$	$K_0 \frac{r_2}{r_1 r_3} e^{-(\alpha/\beta)(\pi + \phi_1 - \theta_1)}$

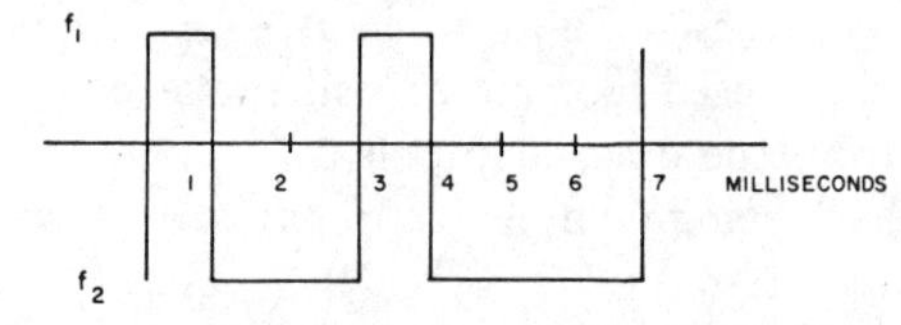

Fig. 7—Carrier-shift pulses; f_0 = carrier frequency, $f_1 - f_0 = f_0 - f_2 = 3$ kc.

less than 1.5 radians. The first ensures sufficient discrimination between pulses; and the second ensures that

$$E(t) = | \theta_1(t) - \theta_2(t) | \leqq \frac{\pi}{2}. \qquad (21)$$

If this difference is exceeded, the loop will fall out of synchronism, seriously distorting the detector's output. So, from (20)

$$E(s) = \frac{K_s}{S^2} \frac{P_g s}{P_g s + K_{oL} Z_g}. \qquad (22)$$

Using the final value theorem, the steady-state error-voltage is

$$E_{ss} = \lim_{S \to 0} SE(S) = \frac{K_s \prod_k b_k}{K_{oL} \prod_j a_j}. \qquad (23)$$

As shown in Appendix I, for a dominate pole-pair, and also for most nondominate configurations, the peak voltage is always less than twice the steady-state voltage. Then

$$E_p(t) \leqq 2Ess = \frac{2K_s \prod_k b_k}{K_{oL} \prod_j a_j} \leqq 1.5. \qquad (24)$$

The condition then is

$$\frac{K_s \prod_j b_k}{K_{oL} \prod_j a_j} \leqq 0.75. \qquad (25)$$

Although a much tighter upper bound could be put on $E_p(t)$, this one will usually ensure that the multiplier operates in the linear, low-distortion region. From (15)

$$\frac{K_s \prod_k \frac{b_k}{d_k}}{\prod_j \frac{a_j}{l_j} r_1} \leqq 0.75; \qquad k \neq 0, \qquad (26)$$

where r_1 is the distance from the complex pole to the origin. The importance of the inequality should be kept in mind when the Wiener criterion is applied (Section V).

As shown in Appendix I, if the closed-loop transfer function, $G_{cL}(s)$, has a zero at $S = -16 \times 10^3$ radians/msec, and a complex pole-pair at $S = [-4 \pm j7]\, 10^3$, $c(t)$ will be within four per cent of the steady-state value in one millisecond. Figs. 8 and 9 show how to adjust $G(s)$ to give the desired pole-zero pattern. Since a zero of $G(s)$ will be a zero of $G_{cL}(s)$, $G(S)$ must have a zero at -16×10^3. Next, measure the difference $\theta_1 - \phi_1$ with a Spirule; here it is $-90°$. Then place a pole on the negative real axis with an angle ϕ_2 equal to 90°. This will make the angle at the desired pole position be

$$\phi = \theta_1 - \phi_1 - \phi_2 = -180°.$$

In general, to make $\phi = -180°$, add a combination of poles and zeros to the loop.

Applying inequality (26) will give the required value of r_1,

$$12 \frac{\frac{r_5}{r_4}}{\frac{r_3}{r_2} \times 0.75} \leqq r_1. \qquad (27)$$

If each of the radii of Fig. 8 is multiplied by about three, the angular relationships will be preserved at the new pole position of Fig. 9. $G(S)$ now is

$$G(S) = \frac{(S + 48 \times 10^3)}{(S + 12 \times 10^3)}.$$

A simple lag network will realize this transfer function. From (10) and (9), the required gain is

$$K_{oL} = \frac{r_1 r_4}{r_2} = 12.1 \times 10^3. \qquad (28)$$

In this example, staying in the linear region required about three times as much bandwidth as that needed for adequate transient response. But reducing K_s to $4\pi \times 10^3$ would reduce the bandwidth to that needed for the

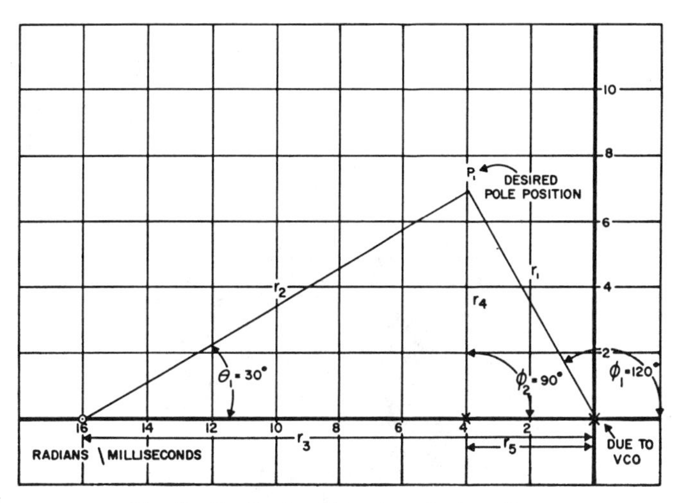

Fig. 8—Synthesis of loop transfer function.

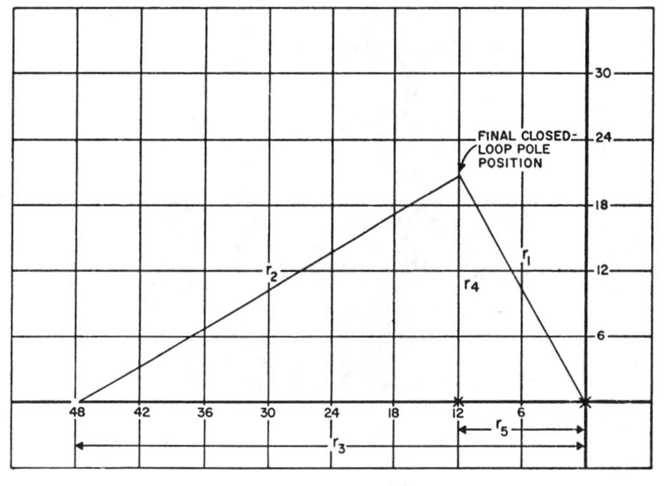

Fig. 9—Expanded bandwidth necessary for lock-in.

transient response. Reducing K_s lower would merely decrease the signal-to-noise ratio. The bandwidth must remain large to accommodate the signaling speed. There is minimum practical frequency shift for FSK detectors of this type.

As noted earlier, the change in K_{oL} is proportional to the change in signal voltage at the multiplier, and a change in signal will move the closed-loop poles along the root-locus. The root-locus plot in Fig. 10 shows the drastic changes in transient response as the signal strength varies. For pole-zero patterns with two poles between the origin and zero on the real axis, the root loci include a circle, centered on the zero, with a radius equal to the distance from the zero to a point on the real axis halfway between the two poles. When the condition of inequality (26) is not met, a lower gain may easily be found. Inequality (27) may be rewritten as

$$r_1 r_4 \geqq \frac{K_s r_5 r_2}{0.75 \times r_3}. \tag{29}$$

However,

$$r_1 r_4 = K_{oL} r_2;$$

then

$$K_{oL} \geqq \frac{K_s}{0.75} \times \frac{r_5}{r_3}, \tag{30}$$

$$E_s \geqq \frac{K_s r_5}{0.75 K_m E_0 A K_1 K_2 r_3}.$$

This is a general condition that holds for $G(S)$ of the above form (a form that appears frequently in the literature). Substitution of the numerical values of r_5 and r_3 in (30) shows that the phase-locked loop in the example is operating just in the region fulfilling the conditions of the inequality. A loss in signal strength will cause the loop to fall out of synchronism. This is an example of a lower threshold that has little to do with the signal-to-noise ratio (see Figs. 11 and 12).

When the gain is raised to 167×10^3, $G_{CL}(S)$ is no longer complex. The signal strength in this example need be increased by a factor of only fourteen to cover almost the complete range of transient response. This points out the necessity of close control over E_s. An excellent AGC

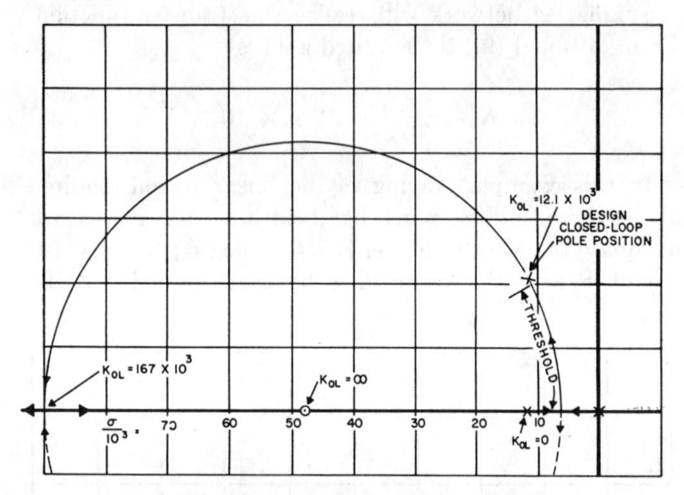

Fig. 10—Change of pole position with signal.

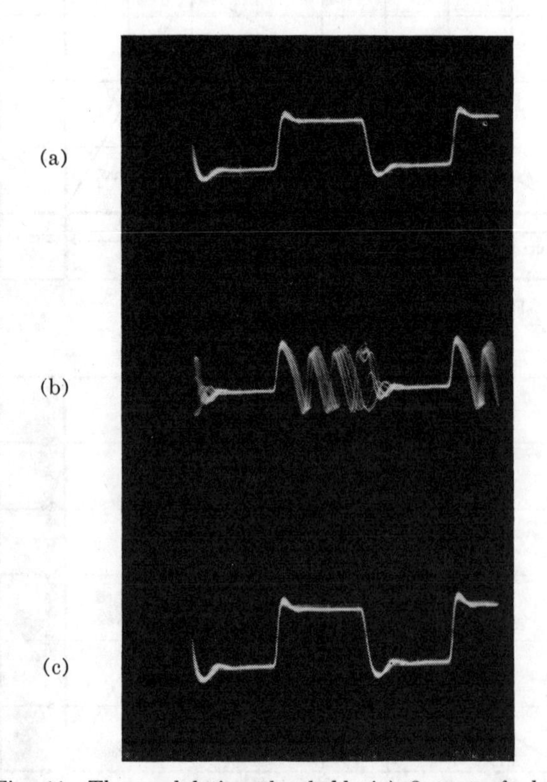

Fig. 11—The modulation threshold. (a) Output of phase-locked detector when the carrier is frequency modulated with a square wave. (b) Deviation increase 10 per cent. The loop falls out of lock. (c) Signal strength increased 10 per cent. The loop is again in lock.

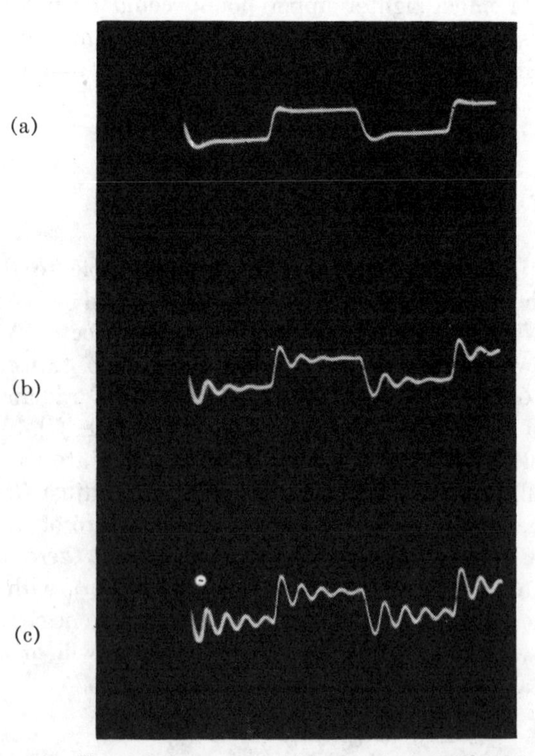

Fig. 12—Change of response with signal strength. (a) Response of the loop to a square wave. This loop is characterized by a complex pole-pair and a zero. (b) and (c) Response as the signal is varied over about 2:1.

or a good limiter—perhaps both—must be used. In Section V, it will be pointed out that the Wiener criteria may result in root loci that cross the imaginary axis, resulting in both an upper and lower threshold, and confining the gain to very narrow limits.

Conditions for Synchronism in General

When the input cannot exceed a specified maximum and the output has a steady-state value, a linear system will have the maximum difference between the input and the output (largest error voltage) if the input is a series of step functions. For nonperiodic inputs, the worst possible case would occur if the input changed its polarity from the maximum shift to one direction to the maximum shift in the other direction in such a manner that the two transients added. Then the peak error-voltage could be twice that of (24).

Applying inequalities (25), (26), and (30) will usually keep the noise-free signal in synchronism. Rewrite them as

$$\frac{D_m \prod_k b_k}{.37K_m E_0 A K_1 K_2 \prod_j a_j} \leq E_s, \tag{31}$$

$$\frac{D_m \prod_{k \neq 0} \frac{b_k}{d_k}}{\prod_j \frac{a_j}{l_j} \times r_1} \leq .37, \tag{32}$$

$$\frac{D_m r_5}{0.37K_m E_0 A K_1 K_2 r_3} \geq E_s, \tag{33}$$

where D_m is the maximum frequency deviation.

IV. Designing the Phase-Locked FM Detector

The root-locus method simplifies the design problem of the phase-locked FM detector to that of an exactly calculable audio network. The method also yields a simple explanation of phase-locked loop as an FM detector.

An FM carrier modulated by $f(t)$ will have the form

$$\sqrt{2}E_s \sin\left(\omega_{st} + D\int^t f(t)\,dt\right).$$

The input to the linearized phase-locked loop is then

$$\theta_1(t) = D\int^t F(t)\,dt$$

which transforms to

$$\theta_1(s) = \frac{DF(s)}{s}.$$

From (12)

$$C(S) = D\frac{F(S)}{S}K_a\frac{Z_g S}{P_g S + K_{oL}Z_g} \tag{34}$$

$$= K_a\frac{DF(S)Z_g}{P_g S + K_{oL}Z_g}.$$

The zero at the origin has acted as a differentiator, and the detected output is identical to that obtained from passing $f(t)$ through a filter with the transfer function,

$$\frac{DK_a Z_g}{P_g S + K_{oL}Z_g}.$$

The design of the detector (neglecting noise) reduces to picking an appropriate filter to pass $f(t)$, and applying the techniques discussed earlier to arrive at the desired configuration.

The noise power will be directly proportional to the bandwidth (if the relative shape of the bandpass remains the same), and thereby proportional to r_1. K_s and D_m are also proportional to r_1. However, the detected signal power is proportional to K_s^2 or D_m^2. The phase-locked loop thus improves signal-to-noise ratio much like a standard FM detector:

$$S/N \text{ improvement} = K\left(\frac{D_m}{f_m}\right)^2, \tag{35}$$

where f_m is the maximum modulation frequency. K is a function of the allowable distortion, and is normally taken as one.

As the bandwidth increases, the mean squared value of the noise, or the σ of the Gaussian distribution for the noise component (assuming white noise), increases proportionally. This means that the error voltage caused by noise has a higher probability of exceeding 90° as the bandwidth increases. So the phase-locked detector is like the standard FM detector in having an increase in the noise threshold as the deviation is increased.

For the usual case of a zero at the origin, a zero at $S = -r_3$ and two dominate complex poles, the distortion voltage may be found from Fig. 13. The ideal detector has the transfer function $r_3/r_1^2\, K_1 S$, where the notation is the same as in Fig. 8. The difference between this undistorted output and the output from the power block of Fig. 13 is the distortion voltage $q(t)$. It is equivalent in Figs. 13 and 14.

Then

$$\frac{Q(S)}{F(S)} = \frac{r_3}{r_1^2}K_1 - \frac{K_1(S + r_3)}{(S + \alpha + j\beta)(S + \alpha - j\beta)}$$

$$= \frac{r_3}{r_1^2}K_1\left[1 - \frac{\frac{S}{r_3} + 1}{\left(\frac{S}{\alpha + j\beta} + 1\right)\left(\frac{S}{\alpha - j\beta} + 1\right)}\right].$$

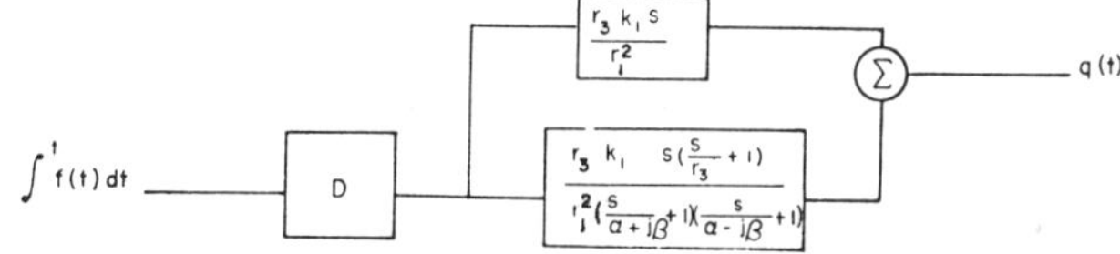

Fig. 13—Distortion voltage.

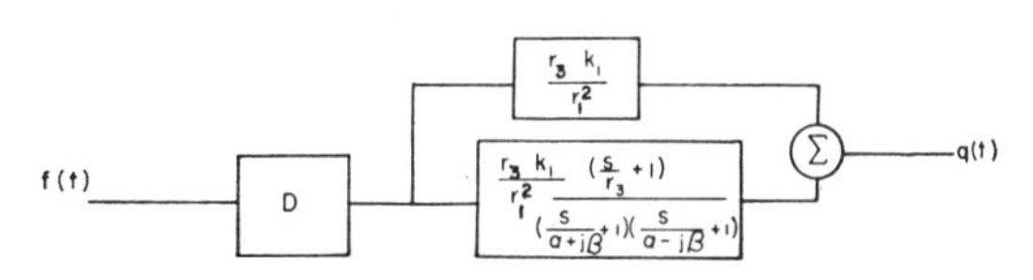

Fig. 14—Distortion voltage.

By Parcevell's equation, the percentage of distortion is

$$\frac{\int_0^\infty \left| 1 - \frac{\frac{S}{r_3} + 1}{\left(\frac{S}{\alpha + j\beta} + 1\right)\left(\frac{S}{\alpha - j\beta} + 1\right)} \right|^2 \Phi_f \, d\omega}{\int_0^\infty \Phi(\omega) \, d\omega} \times 100, \quad (36)$$

where $\Phi_f(\omega)$ is the power spectrum of $f(t)$. In theory, a phase-locked discriminator can be designed for any wide-band FM signal with any desired minimum of distortion. The noise spectrum will be discussed in Section V.

V. Design for Minimum Mean-Squared-Error

Since phase-locked loops and Wiener filters are often used where the signal-to-noise ratio is small, it seems useful to design phase-locked loops by using Wiener filter theory. The solution for G_0, the Wiener filter giving least mean-squared-error (best signal-to-noise ratio), has been derived by several authors.[6] It is

$$G_0(S) = \frac{1}{\Phi_{ii}^+} \mathcal{L}\mathfrak{F}^{-1}\left\{\frac{\Phi_{is}}{\Phi_{ii}^-} G_d(S)\right\}, \quad (37)$$

where $\mathcal{L}\mathfrak{F}^{-1}$ means the Laplace transform oft he inverse Fourier transform and $\Phi_{ii}(S)$ is the power spectrum of the signal plus noise. It may be represented by a collection of poles and zeros in the S plane that are symmetrical about both the real and imaginary axes. $\Phi_{ii}^+(S)$ is made up of all the poles and zeros of $\Phi_{ii}(S)$ in the left-half plane, and Φ_{ii}^- of corresponding poles and zeros in the right-half plane; Φ_{is} is the power spectrum of the signal Φ_{ss} plus the cross-power spectrum between the noise and the signal Φ_{ns}.

$$\Phi_{is} = \Phi_{ss} + \Phi_{ns}.$$

When there is no correlation between the signal and the noise, $\Phi_{ns} = 0$. $G_d(S)$ is the transfer function giving the desired output when the signal above is the input. For example, if the first derivative of the signal is needed, $G_d(S) = S$; if the signal is to be predicted, $G_d(S) = e^{st}$; and if the signal itself is to be the output, $G_d = 1$. When poles and zeros appear on the imaginary axis they will always appear in even numbers. Assign half to Φ_{ii}^+, and the remainder to Φ_{ii}^-. Eq. (37) may be applied by using S-plane diagrams. This technique may be illustrated by the example of differentiated white noise, and an uncorrelated signal in the form of a differentiated random ramp (a random step), with a power spectrum having two poles on the real axis (see Fig. 14).

The power spectrum for a series of random steps is[1]

$$\Phi_{ss}(\omega) = \frac{Kr}{T_s\left(\frac{1}{T_s^2} + \omega^2\right)}$$

$$= \frac{K_s}{(\omega^2 + \alpha^2)}; \quad \alpha^2 = \frac{1}{T_s^2}; \quad K_s = \frac{Kr}{T_s},$$

where T_s is the average time between discontinuities. This is a very commonly used design signal. (T_s and K_s usually are not known; and it is thought by the author that in most cases the receivers being designed merely have a narrow bandwidth and not Wiener optimum.)

$$\Phi_{nn} = A_0^2\omega^2 = -A_0^2S^2 \,|_{s=j\omega},$$

$$\Phi_{ii} = \Phi_{nn} + \Phi_{ss} = -\frac{K_s^2}{(S^2 - \alpha^2)}\bigg|_{s=j\omega} - A_0^2S^2\bigg|_{s=j\omega}$$

$$= -\frac{K_s^2 + A_0^2S^2(S^2 - \alpha^2)}{(S^2 - \alpha^2)}\bigg|_{s=j\omega} \quad (38)$$

So Φ_{ii} has poles identically equal to the poles of Φ_{ss}. The zeros are found by solving

$$-\frac{1}{S^2(S^2 - \alpha^2)} = \frac{A_0^2}{K_s^2}. \quad (39)$$

This brings to mind a root-locus. This locus of the zeros is shown in Fig. 15(b). The zero positions are found by letting

$$\left|\frac{1}{S^2(S^2 - \alpha^2)}\right| = \frac{A_0^2}{K_s^2}. \quad (40)$$

Inverting Φ_{ii}, as shown in Figure 15(c), reverses the poles and zeros; Φ_{ss}/Φ_{ii}^- is shown in (e). The constant multipliers K_s may now be neglected. They appear in

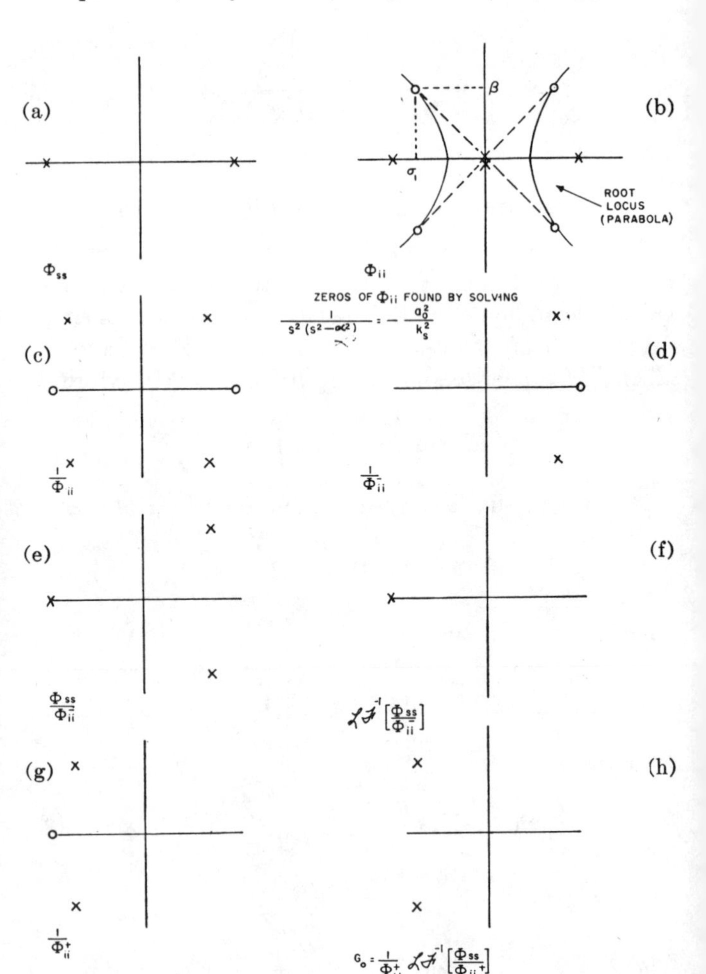

Fig. 15—S-plane solution for G_0.

[6] Truxal, *op. cit.*, p. 480.

the final solution as constants multiplying G_0, and do not change the signal-to-noise ratio.

In general $\mathcal{L}\mathfrak{F}^{-1}\,[\Phi_{ss}/\Phi_{i\bar{i}}]$ is the partial-fraction expansion of $\Phi_{ss}/\Phi_{i\bar{i}}$ about all poles in the right half plane. Here the only pole on the right is at $S = -\alpha$. Then

$$\mathcal{L}\mathfrak{F}^{-1}\left[\frac{\Phi_{ss}}{\Phi_{ii}^{-}}\right] = +\frac{K_1}{S+\alpha}. \tag{41}$$

Then, multiplying by $1/\Phi_{ii}^{+}$

$$G_0 = \frac{K_1}{(S+\sigma_1)^2 + \beta^2} \tag{42}$$

as shown in Figure 15(h).

As the signal-to-noise ratio increases, the poles of G_0 move out along the locus shown in Figure 15(b). Reference to a Bode plot shows that this is an increase in the bandwidth. This example used a simple signal spectrum; however, this technique also gives simple solutions for very complex spectra.

Eq. (5) shows the direct dependence of the equivalent noise upon the phase of the VCO. $N'(t)$ is the high-frequency envelope moved down to zero frequency with all Fourier components changed in phase by $\theta_2(t)$. Unfortunately, $\theta_2(t)$ is a function of $G(s)$, and $G(s)$ is not determined until G_0 is found. There is no solution for G_0 without Φ_{nn}. Only a noise spectrum that is white over the range of interest will give a solution. A white spectrum may be thought of as an infinitely large group of equal-amplitude sinusoids spaced an infinitesimal distance apart in frequency. If the phase or angular velocity is varied as a function of time, each sinusoid will be shifted in frequency an identical amount, so that the power distribution will remain the same. This will not be true if the original power distribution is not flat.

When the phase-locked loop is used as an FM detector, the transfer function may be thought of as a differentiator preceeding a network, $G_0(S)$. $G_0(S)$ is the filter that is to be designed for Wiener optimum (see Fig. 16).

$$SG_0(s) = G_{c1} = \frac{Z_g S}{P_g S + K_{oL} Z_g}. \tag{43}$$

$$G_0(S) = \frac{Z_g}{P_g S + K_{oL} Z_g}.$$

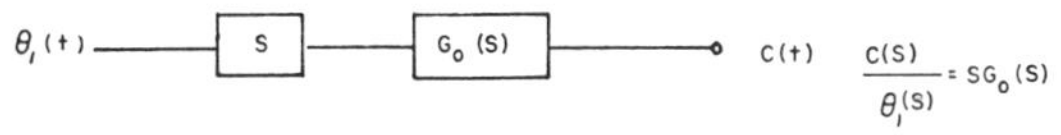

Fig. 16—Removing S-term from G_0.

The power spectrum of the phase input to the loop is

$$\Phi(\omega) = D^2 \left|\frac{1}{S}\right|^2_{s=j\omega} \Phi_f(\omega). \tag{44}$$

Then, if $\Phi(\omega)$ is passed through a differentiator, the power spectrum of the signal to $G_0(S)$ is

$$\begin{aligned}\Phi_{ss} &= |S^2|\,\Phi(\omega)\\ &= D^2\Phi_f(\omega),\end{aligned} \tag{45}$$

where Φ_{ss} is the power spectrum of $f(t)$ multiplied by D^2. The noise spectrum out of the differentiator is

$$\Phi_{nn} = -A_0^2 S^2 \,|_{s=j\omega}. \tag{46}$$

The design of the Wiener optimum phase-locked FM detector, as in Section IV, reduces to the synthesis of an audio frequency network with $f(t)$ as the signal, and a noise spectrum as given in (46).

If the loop is to be used to obtain minimum phase jitter, as in Fig. 17, S will not appear in the numerator of G_{cL}.

$$G_{oL} = \frac{K_{oL}\dfrac{G(s)}{S}}{L + K_{oL}\dfrac{G(s)}{S}} = \frac{K_{oL}Z_g}{P_g S + K_{oL}Z_g}. \tag{47}$$

In general, one loop cannot be designed as an optimum FM detector, and also to give minimum phase jitter. This will become obvious if $G(s)$ is derived separately for both cases.

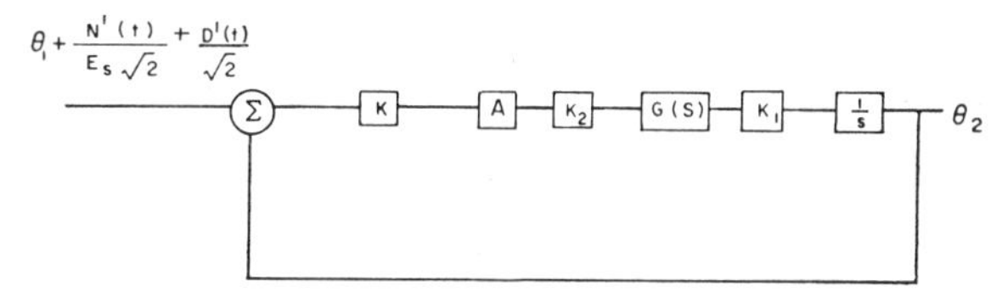

Fig. 17—The loop when phase is the output.

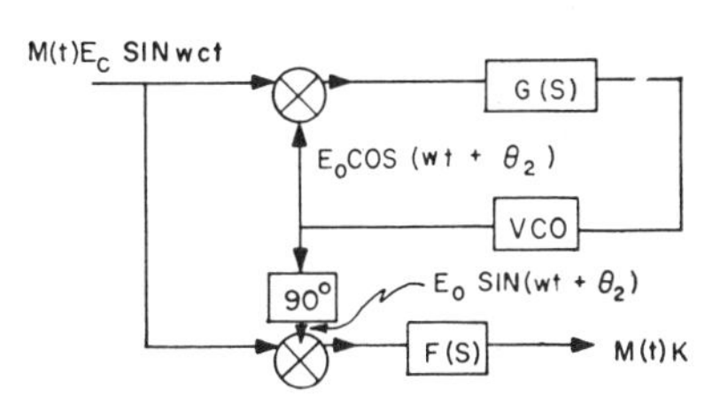

Fig. 18—AM detection.

The Phase-Locked AM Detector

If the output of the VCO is shifted 90° and multiplied with the signal (see Fig. 18), the demodulated amplitude modulation $M(t)$ will be obtained. Considering only the low-frequency component,

$$\begin{aligned}2K_m M(t)E_0 \sin \omega_c t E_0 \sin\,[\omega_c t + \theta_2(t)]\\ &= K_m E_c E_0 M(t) \cos\theta_2(t)\\ &= K_m E_c E_0 M(t)\left[1 - \frac{[\theta_2(t)]^2}{2}\right]\end{aligned} \tag{48}$$

when $\theta_2(t)$ is small. $\theta_2(t)$ is a function of $M(t)$, as can be seen from (5). The Wiener optimum for minimum *phase jitter* then, in general, cannot be found for AM. $\Phi_n(\omega)$ will be the power spectrum of the sideband voltage given in (2) and the spectrum of the voltage in (4). Both these spectra are functions of $\theta_2(t)$.

One possible design procedure could be as follows. If the carrier frequency drifts, design $G(s)$ to make the closed-loop pass band as narrow as the required transient response will allow—the loop must track efficiently. It is hoped that the side bands and most of the noise are removed. Some noise will appear as phase jitter.

However, if the band pass of $F(s)$ must be large to pass the demodulated signal compared to the phase-locked loop pass band, then most of the noise arriving at $F(s)$ will have traveled via the signal channel.

The noise voltage $N(t)'$ given in (2) is a "stationary random process," when $\theta_2(t)$ is constant. This means that

$$\overline{N(t)^2} = \overline{N(t \pm \tau)}$$

or, $N(t)'$ has the same power spectrum when $N(t)$ is multiplied by either a sine or a cosine. If the spectrum of $N(t)$ is

$$\Phi(\omega) = A_0^2$$

then the spectrum of the noise out of the multiplier is

$$\Phi_{nn} = K_m^2 E_0^2 2 A_0^2. \tag{49}$$

The signal spectrum will be

$$\Phi_{ss} = K_m^2 E_0^2 \Phi_{mm}(\omega), \tag{50}$$

where $\Phi_{mm}(\omega)$ is the spectrum of the modulating voltage,

$$F(s) = G_0,$$

and of course G_0 is found from (37).

Noise Bandwidth and the Limiter

A limiter ahead of the loop not only maintains a fixed transient response as the signal strength varies, but in certain special cases automatically reduces the bandwidth if the signal decreases. A band-pass limiter tends to have the same signal-to-noise ratio at its output as it does at its input. Further, it tends to keep the total power output constant. Jaffe and Rechtin have shown that the following relationships hold:

$$\frac{E_{si}}{E_{s0}} = \sqrt{\frac{1 + (N_0 \backslash E_{s0})^2}{1 + (N_i \backslash E_{si})^2}} \tag{51}$$

$$= \text{Ratio of Change of } K_{0L},$$

$$\left(\frac{N_i}{N_0}\right)^2 = \frac{1 + (N_0 \backslash E_{s0})^2}{1 + (E_{si} \backslash N_i)^2}; \tag{52}$$

where N_0^2 = design level noise, N_i = new noise power, E_s = design signal level, and E_{si} = new signal level.

Under certain conditions the bandwidth will decrease as the signal strength decreases. The Bode plot and root locus make this apparent. A good example is Fig. 10. However, if $G(s)$ had two poles somewhere near the origin, rule 4) of the root-locus method shows that two poles will cross the $j\omega$ axis as K_{0L} is increased. As these poles approach the axis, the noise bandwidth is decreased. In this region, a decrease signal strength would result in a greater noise bandwidth, not less. The root-locus should be consulted to determine the precise effect of the limiter and the *AGC*.

When the number of poles of $G(s)$ exceeds the number of zeros by two, the root-locus eventually crosses into the right-half plane. This also may occur if $G(s)$ has only two poles near the origin and the zero sufficiently far to the left. When E_s is high enough to cause the poles to move across the axis, the loop goes unstable and stops detecting. This is an example of an upper threshold that has little to do with noise.

Appendix I
Response of Step and Ramp Input

Tables I and II summarize the transient response $C(t)$ to step and ramp inputs $R(t)$ as a function of pole-zero patterns. First we must derive these responses (see Fig. 19).

Rise Time: τ_r, the time for $C(t)$ to first reach the value that will eventually be steady-state, C_{ss};

Peaking Time: τ_p, the time for $C(t)$ to reach the first peak;

Settling Time: τ_s, the time for the envelope of the response to reach and remain less than a specified fraction F of the final steady-state value;

Peak-Overshoot Ratio: POR, the ratio of the maximum value of $C(t)$ to C_{ss}.

Exact relations between these definitions and the pole-zero locations may be derived (see Fig. 20). Let the transfer

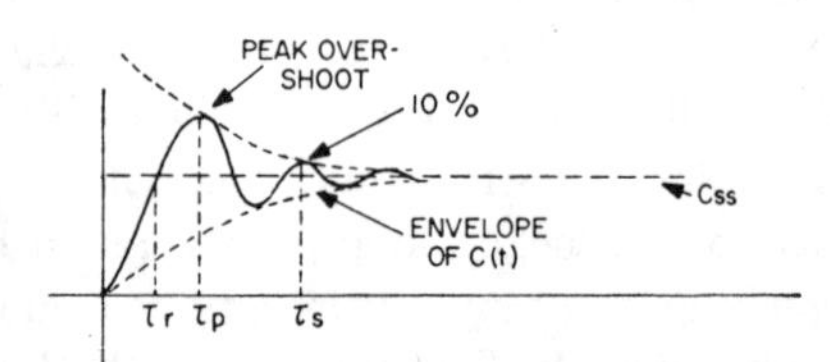

Fig. 19.

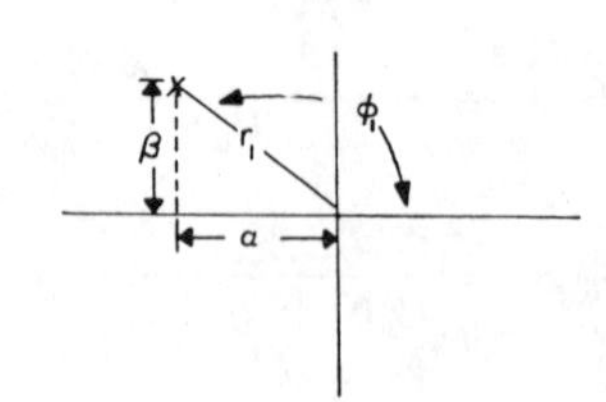

Fig. 20.

function be

$$G_{cL}(S) = \frac{K}{(S + \alpha)^2 + \beta^2}$$

from a table of transforms

$$C(t) = K_0 + \frac{K_0 r_1}{\beta} e^{-\alpha t} \sin \beta t - \phi$$

$$K_0 = \frac{K}{r_1^2} = C_{ss};$$

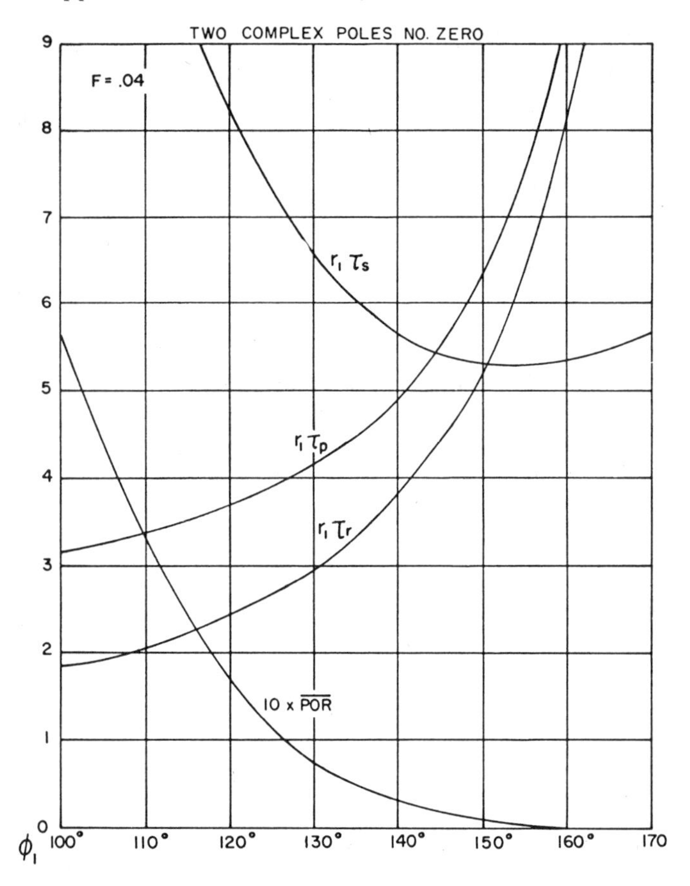

Fig. 21.

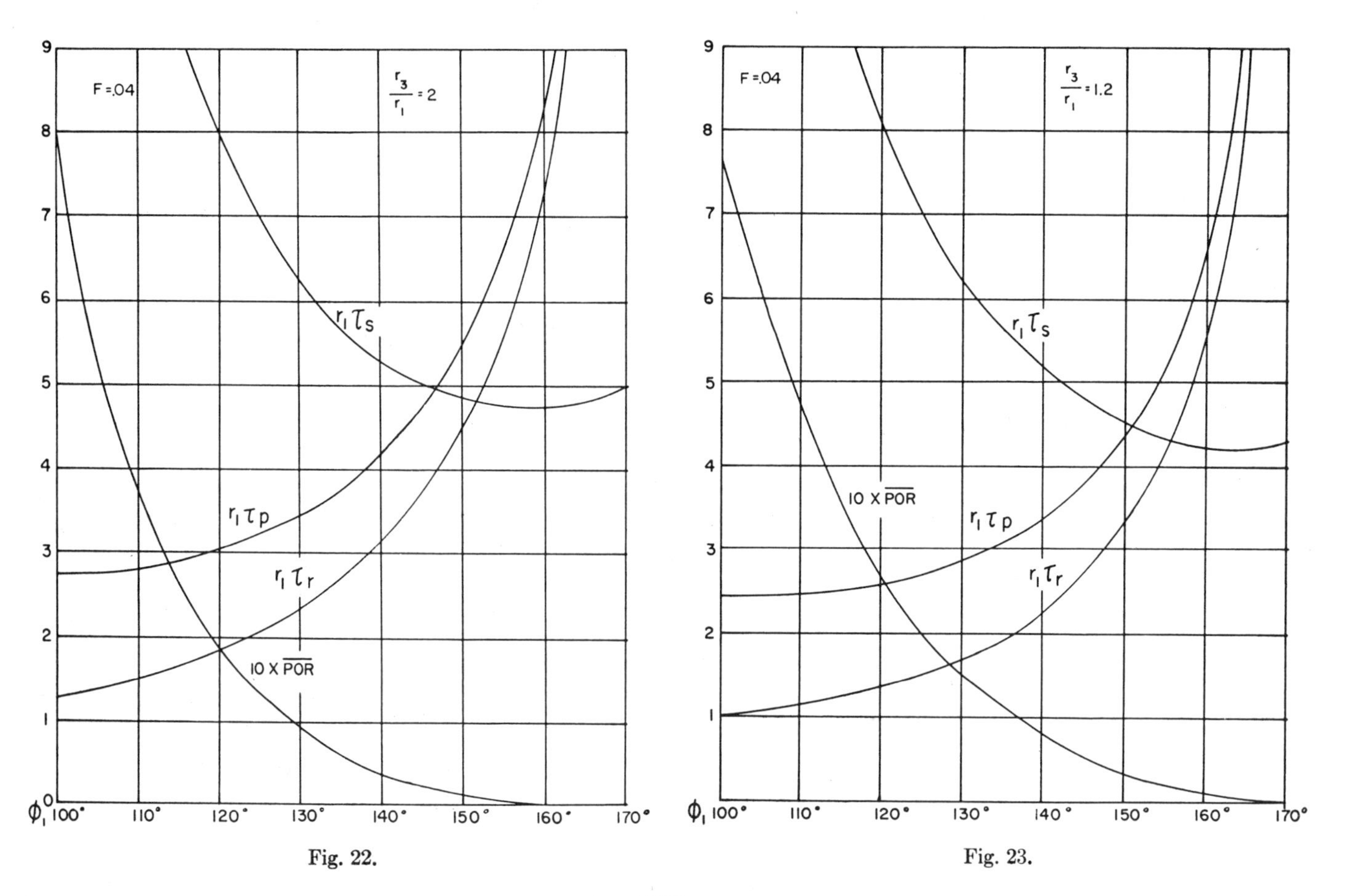

Fig. 22.

Fig. 23.

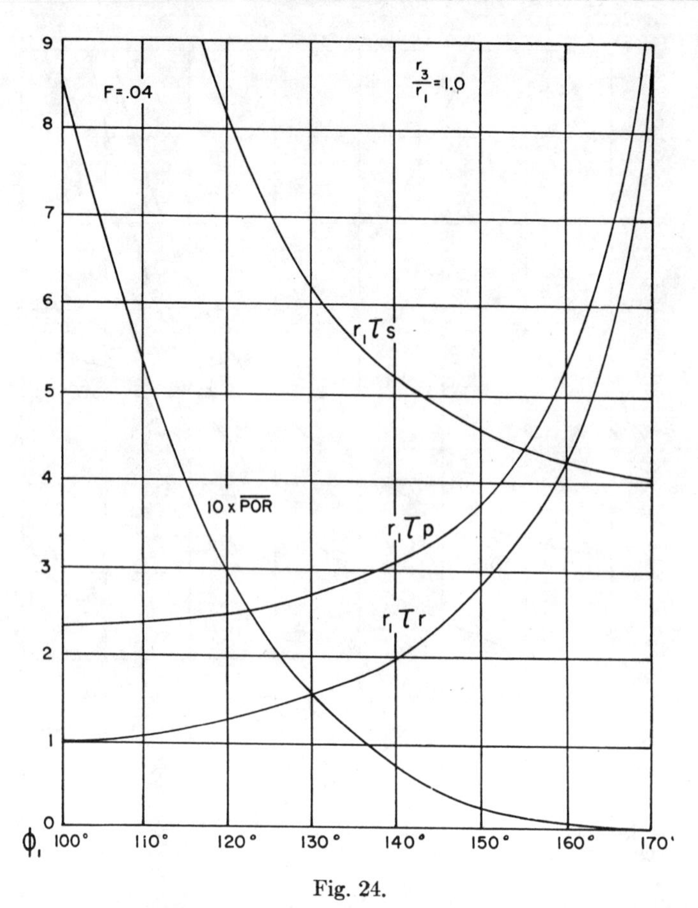

Fig. 24.

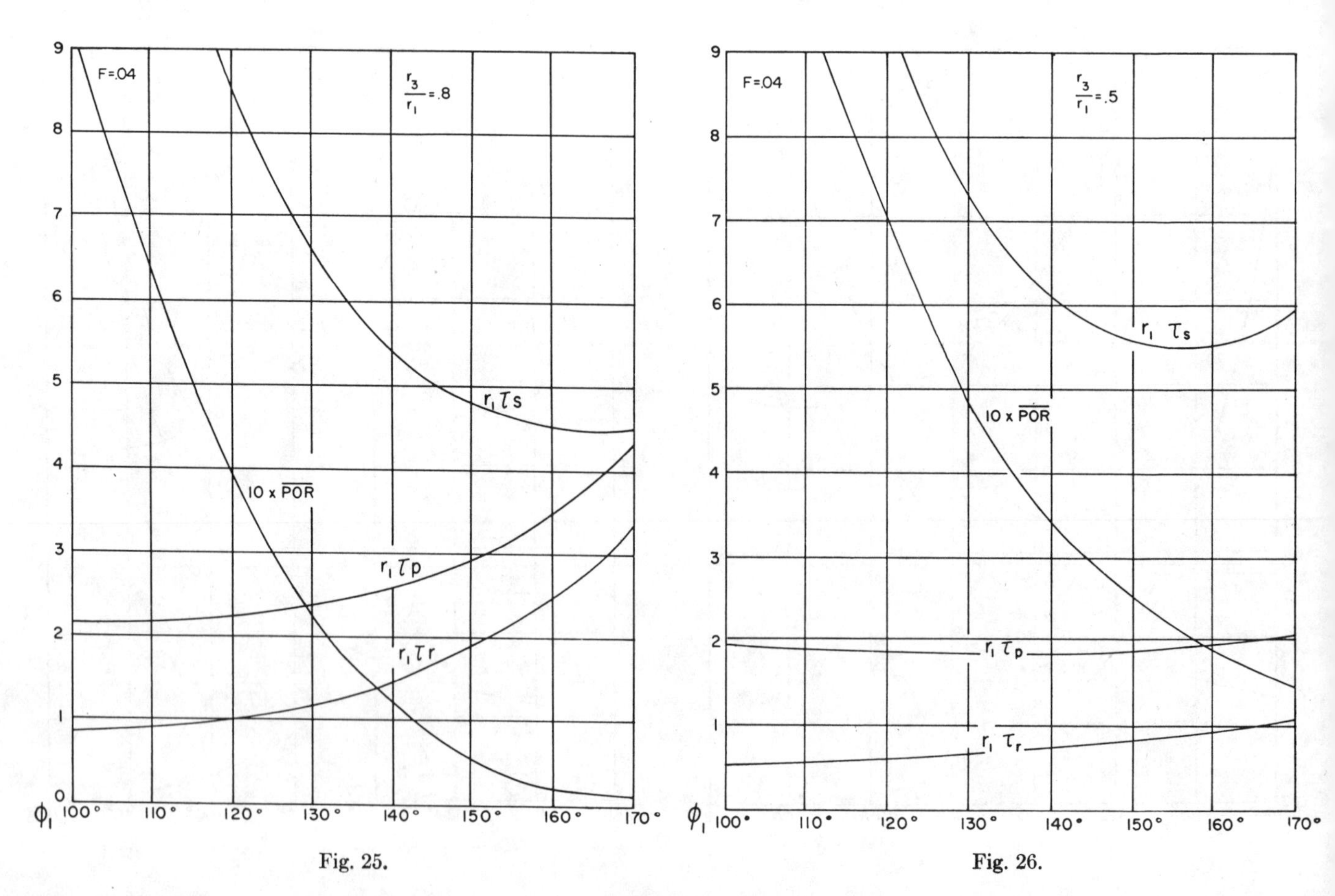

Fig. 25.

Fig. 26.

here C_{ss} is the final value of $C(t)$

$$C'(t) = \frac{K_0 r_1}{\beta} e^{-\alpha t} \sin \beta t$$

hen

$$\tau_r = \frac{\phi_1}{\beta}.$$

τ_p occurs when $C'(t)$ first equals zero or

$$\tau_p = \frac{\pi}{\beta}.$$

Setting the envelope of $C(t)$ equal to FK_0 and solving for time gives

$$\tau_s = \frac{1}{\alpha} \ln \frac{r_1}{\beta F}.$$

Finally,

$$\text{POR} = \frac{C(\tau_p) \quad C_{ss}}{C_{ss}} = \frac{\alpha}{\beta}\pi.$$

If the loop is represented by a complex pole-pair and a zero, similar results may be obtained in the same manner. If there are additional poles on the real axis for enough to the left of the origin (large α's) to cause only a small effect on the transient response, and additional zeros, $C(t)$ may be found graphically[7] and treated similarly.

For a ramp input (as when tracking doppler shifts):

Peaking time: τ_p, time for the difference between $C(t)$ and the steady-state ramp component of $C(t)$, to reach maximum.

Peak difference: PD, difference between $C(\tau_p)$ and the ramp.

Setting time: τ_s, time for the envelope to fall to a fraction F of the peak difference.

A more complete derivation of these tables will appear in a forthcoming Philco WDL Report.

Appendix II
Block Diagram Reduction

Block Diagram Reduction

Complex phase-locked loops may be reduced to single loops if the pass band of the phase-locked loop is small compared to the pass bands of any preceding amplifier.

The multiple loop of Fig. 27 is a good example of a more complicated configuration. K_n is the mixer constant, and G is the IF amplifier gain. The right-hand loop can be linearized, and since the IF amplifier just changes the amplitude of the input, this gain, along with K_n, E_m, E_s, and E_0 may be moved inside the loop. Fig. 27 is then changed to Fig. 28 which may be redrawn as in Fig. 29.

Consider two cascaded summing points as in Fig. 30. If $b = kc$, (a) and (b) are equivalent. Then Fig. 28 becomes as shown in Fig. 31.

This analysis may be carried on for any additional number of loops. The only effect of the frequency multi-

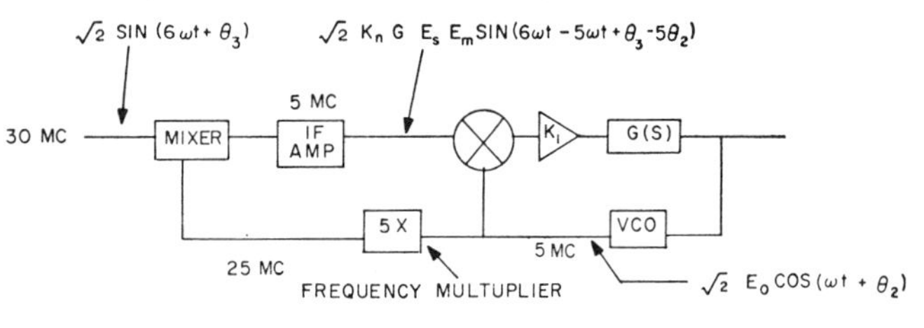

Fig. 27—Multiple loop.

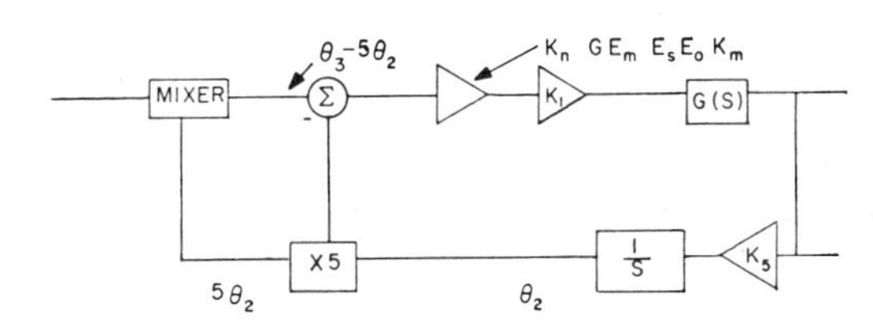

Fig. 28.

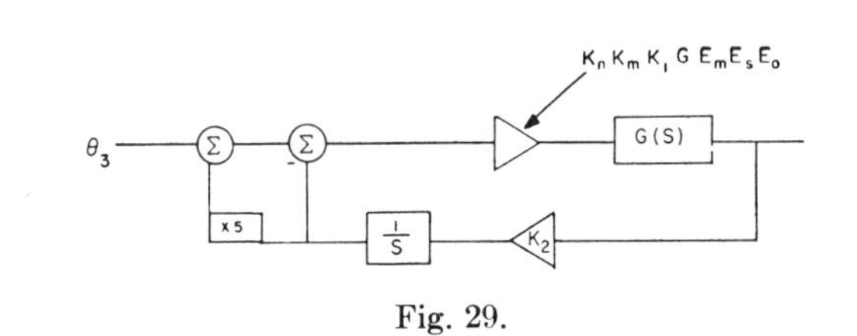

Fig. 29.

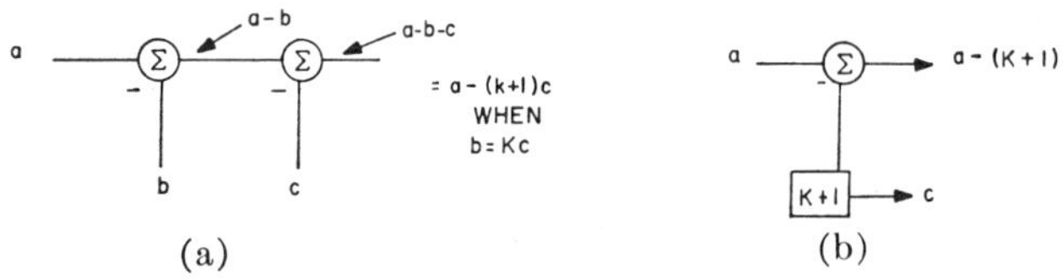

(a) (b)

Fig. 30.

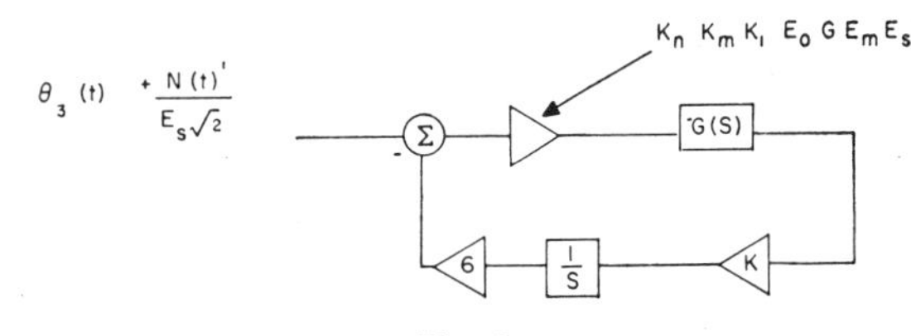

Fig. 31.

plier is to increase the loop gain by six. K_1 and K_2 are design parameters so that the additional loop provides no particular theoretical advantage and, in practice, probably means a considerable increase in complexity. Notice that G appears inside the single loop with or without the frequency multiplier in feedback. Savant[8] summarizes several additional block-diagram substitutions theorems that may be of value.

There is no known satisfactory way of reducing an IF amplifier with a transfer function other than gain. Attempts at this reduction result in highly complex and nonlinear relationships. This is the reason that the amplifiers preceding the loop must be flat over the range of operation of the phase-locked detector, if the response is to be easily calculable.

[7] Aseltine, *op. cit.*, pp. 101–105.

[8] C. J. Savant, "Basic Feedback Control System Design," McGraw-Hill Book Company, Inc., New York, N. Y., p. 19; 1958.

Hangup in Phase-Lock Loops

FLOYD M. GARDNER, SENIOR MEMBER, IEEE

***Abstract*—A phase-lock loop occasionally will take a long time to settle to equilibrium. Phase dwells at a large error for a prolonged interval. This phenomenon has been dubbed "hangup."**

The periodic nature of phase detectors is responsible for hangup, which occurs near the reverse-slope, unstable null. Restoring force is small in the vicinity of the reverse null, and noise causes the loop to equivocate about the null.

Hangup is very troublesome when fast acquistion is needed with high reliability. One example is synchronization of digital communications.

Hangups can be avoided if a large restoring force is applied for large phase errors and if equivocation is prevented.

An implementation of an antihangup circuit is proposed.

INTRODUCTION

Phase locked loops (PLL's) occasionally exhibit unduly prolonged phase transients during initial acquisition of lock. The loop appears to stick, temporarily, at a large value of phase error before settling to its normal tracking condition of small error. This phenomenon has been dubbed the "hangup" effect.

Hangup is associated with the reverse-slope null of the phase detector (PD). The problem is caused by the small output voltage that occurs in the vicinity of the reverse null and by equivocation back and forth across the null.

Dwell time at the hangup point is finite (the loop cannot remain indefinitely at the unstable null), but even a short dwell time may be excessive for some applications. An example arises in burst-mode digital communications where rapid acquistion is demanded for efficient usage of channel time. Failure of a PLL to settle within an allotted synchronization preamble prevents correct detection of an identification preamble and causes loss of the entire burst.

Hangup has been obscure because most PLL users do not require fast phase acquisition (frequency acquisition is typically very much slower), because the loop always settles eventually, and because hangup is an event of low probability. It simply has not been noticed by most users. Its existence came as an unwelcome surprise to builders of TDMA (Time Division Multiple Access) modems.

Recent practice [1-6] in TDMA synchronizers has been to substitute narrow-band, tuned filters for a PLL in order to avoid the hangup effect. The resulting equipment tends to be overly complex and suffers from other unwanted characteristics. A hangup-free, phase-locked synchronizer would be preferable, if one could be devised.

In this concise paper we examine the causes of hangup and then formulate a pair of conditions that are necessary for prevention of hangup. It is thought that the two conditions together are sufficient, but experimental confirmation remains to be obtained.

Methods are given for implementing the conditions and a particular implementation is proposed.

ATTRIBUTES OF HANGUP

Almost all phase detectors have a periodic characteristic of output voltage vs. phase error; examples are shown in Fig. 1. Because of the periodicity, the characteristic must have two nulls per cycle. One null has a slope that provides negative feedback for the loop and is the stable null for equilibrium tracking. The other null has a reverse slope that provides unstable, positive feedback so stable tracking about this null is not possible.

Experiments have revealed that if the initial phase error in a PLL is very close to the reverse null, then the loop can dwell in the vicinity of the null for a prolonged time; this is the hangup effect. It is exemplified by the curves of Fig. 2 which show typical phase transients in a first-order PLL with sinusoidal phase detector. The phase trajectories originating near 180° remain in that vicinity for a long time before decaying towards equilibrium at 0°. (The first-order loop is not unique; it was chosen for an example only because its phase transient is easily obtained by integrating the nonlinear differential equation of the loop [7].)

The loop cannot dwell at the reverse null forever; it must eventually move away and converge towards the normal equilibrium null (assuming that no conditions exist that prevent eventual locking). But, if fast phase acquisition is needed, the hangup interval can be excessively long and can severely degrade the performance of fast TDMA systems.

Paper approved by the Editor for Data Communication Systems of the IEEE Communications Society for publication after presentation at the URSI/USNC Conference, Stanford, CA, June 24, 1977. Manuscript received April 26, 1977. This work was performed for the European Space Agency, Noordwijk, The Netherlands.

The author is with the Gardner Research Company, Palo Alto, CA 94301.

Reprinted from *IEEE Trans. Commun.*, vol. COM-25, pp. 1210–1214, Oct. 1977.

Hangup is a natural feature of phase acquisition of a PLL. It has been obscured in the past only because most PLL users are not concerned with extremely fast phase acquisition. It has no relation to frequency pull-in, which is known to be a slow process.

On the contrary, all loops considered here will be assumed to encounter negligible initial frequency error so that there is no frequency acquisition to be achieved. The PLL is only required to rephase its VCO, rapidly, at the start of signal reception.

It has been suggested that an initial frequency offset should prevent hangup because the phase would not be able to dwell at the reverse null. This is incorrect. In a first-order loop, a frequency offset merely shifts the location of the nulls, but both are still present. For example, see [8, p. 42]. Hangup will still occur at the shifted reverse null.

In a second-order loop, hangup will occur if the state trajectory of the loop (phase and phase rate) comes near a saddle point in the phase plane. (See [9, chapter 3] for explanation of phase-plane analysis of a PLL). Hangup will occur if the initial state is close enough to a separatrix; it does not depend upon the less likely condition of initial state being close to the saddle point.

Others have suggested that a "kickoff" pulse be introduced into the loop simultaneously with the introduction of the signal in order to push the loop out of hangup. Undoubtedly that would work if the loop were indeed in hangup, but might push the loop into hangup if its initial state were not in the hangup region. If the kickoff pulse was applied blindly, it could not be expected to provide any net benefit. If a kickoff pulse is applied only if the loop is in hangup, then it might very well be helpful. The anti-hangup methods discussed later could be considered to be a variation on selectively applied kickoff schemes.

It has also been suggested that noise should tend to drive the loop out of its hangup region. In fact, it has been found [4, 10] that noise worsens hangup; an explanation will be deferred to the next section.

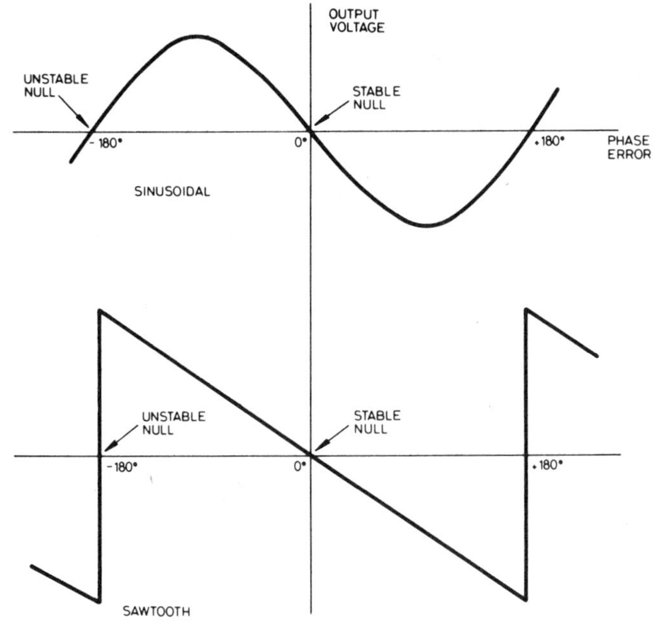

Fig. 1. Phase-Detector Characteristics.

CAUSES OF HANGUP

The reverse null of the phase detector is clearly implicated as a source of hangup. However, what is the exact physical mechanism by which hangup occurs? Two causes have been identified: weak restoring force and directional equivocation.

The rate of phase correction in a PLL is proportional to the output voltage from the phase detector; that error voltage can be visualized as a restoring force. If the error voltage is very small (as it is near the reverse null of a sinusoidal phase detector), then the restoring force is also small and the loop converges very slowly towards equilibrium. The curves of Fig. 2 illustrate the point amply.

Recognition of this effect leads to the following.

Anti-Hangup Rule #1: Provide a large restoring force, particularly in the vicinity of the reverse null.

A sinusoidal phase detector has a small restoring force around the reverse null, whereas a sawtooth PD has a large output voltage around its reverse null (which ideally becomes a discontinuity). One might expect that the sawtooth characteristic would eliminate hangup.

To the intense disgust of those who followed this expectation, it was found that hangup still occurs with any practical

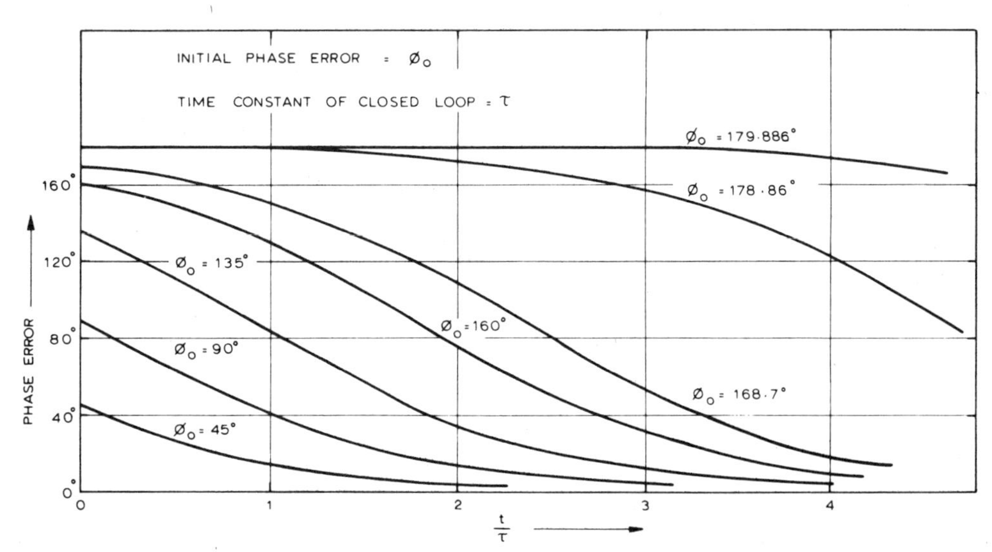

Fig. 2. Transient Phase Error (In first-order loop with sinusoidal phase detector).

sawtooth phase detector. The probability of hangup may be reduced, but not to zero.

Noise and other disturbances, such as modulation of the signal, cause perturbations that degrade the ideal performance of the sawtooth PD, or of any PD with a discontinuous characteristic. One explanation [11, 12] is that the disturbances smooth out the discontinuity, converting it to an unstable null with finite reverse slope. The average restoring force in the vicinity of the discontinuity is small because of the noise, despite the fact that the noiseless characteristic appears to promise a large restoring force.

For an alternative viewpoint, imagine a PLL with sawtooth PD and an initial phase error very close to the discontinuity. The PD developes an output voltage that instructs the loop to move, say, clockwise and the loop begins to obey, at high speed. But, before much phase correction can accumulate, a noise event occurs that throws the instantaneous phase error to the other side of the discontinuity, whereupon the PD instructs the loop to move counterclockwise. This motion is maintained until another noise event pushes the apparent phase error back across the discontinuity and the correction direction is reversed once again.

This equivocation is possible only in the near vicinity of the reverse null; noise events large enough to reverse the direction are very rare if the phase error is sufficiently removed from the reversal point. Nonetheless, the loop sometimes does fall into an initial phase close to the reverse null and it can equivocate if there is enough noise or other disturbance. The effect clearly worsens with increasing noise.

Recognition of such behavior leads to the following.

Anti-Hangup Rule #2: Avoid equivocation, expecially in the vicinity of an unstable null.

Each rule, by itself, is necessary but not sufficient to avoid hangup. It is tempting to think that the two rules, taken together, are sufficient, but that conclusion has not been demonstrated, either by analysis or by experiment. The author has been unable to uncover any other causes of hangup and he advances the conjecture that the two rules, together, are sufficient. The remainder of this concise paper is devoted to implementations of the rules.

PREVENTION OF HANGUP

The following procedure, if implemented, obeys the antihangup rules and is expected to prevent hangup.

1. Establish a direction of correction by an initial measurement.

2. Apply a restoring force (large, unless the loop is already close to equilibrium) in the chosen direction.

3. Maintain the chosen direction without reversal until equilibrium phase has been reached. It is far better to move steadily in the "wrong" direction than to be unable to decide which way to turn and not move at all.

This last point may be amplified by recognizing that the phase error of the PLL lies on a circle with equilibrium at 0° and hangup at 180° (for a simple phase detector). If the initial phase falls near 180°, it makes very little difference which way the correction is made; the settling time provided by a large restoring force is nearly the same for either direction around the circle. The important consideration is to avoid indecision at 180°. As with many situations, the actual decision made is immaterial; the crucial action is to decide quickly and then carry through.

Several antihangup implementations have been identified and others will be found as the subject is studied further. If the loop is time-discrete, if it performs iterative corrections, and if it remembers its previous estimate of phase error while computing its next correction [13], then equivocation will be avoided if the system rejects any "correction" that carries the estimated phase error over a hangup boundary. It would appear that only fairly complex digital systems can take advantage of this method.

Another approach [14] is to watch for a zero-crossing of the signal and then force-reset the phase of the VCO into coincidence with the crossing. The resetting is readily accomplished in low-frequency systems that employ a string of digital counters; resetting of counters is a simple matter. Noise must be small in order to assure reliable recognition of a zero crossing.

Suppose that phase error could be measured unambiguously and accurately. It would then be possible to apply an impulse to the VCO control line: an impulse with an area that exactly cancelled the measured phase error. A control voltage applied to the VCO alters the frequency. The integral of the frequency change (the area of the impulse) is the accumulated phase change caused by the control. If such a technique could be implemented, it would provide rapid, hangup-free acquisition of phase.

There are numerous obstacles to the implementation of this technique, some practical and some inherent. Difficulties include the following.

1) Ordinary phase detectors are ambiguous; any particular output voltage can correspond to either of two phases.

2) The ambiguity can be resolved by employing two phase detectors in quadrature, thereby establishing the correct quadrant. Since the PD is nonlinear, some linearizing scheme must be employed.

3) Phase detector output is proportional to signal amplitude. Some means would have to be devised to normalize the measurement in order to remove the effect of variable amplitude.

4) Noise is omnipresent and causes error in the phase measurement. Filtering of the PD output is required so as to reduce the effects of noise, but some noise-caused error is inescapable.

Filtering requires that a nonzero time interval be devoted to the measurement. If the phase error is time varying, then the filtering will cause a dynamic error in the measurement.

With persistence, additional problems can be found, but these should suffice for now. The above-listed difficulties are a serious barrier, partly because we are asking for accurate measurement from inaccurate instruments and partly because we are attempting a single-step, open-loop correction. If we could devise a scheme that worked with inaccurate measurements and if the corrections were closed-loop or iterative, then relatively simple equipment and successful operation could be expected. The penalty would be a longer acquisition time than promised, but not delivered, by the single-step method. A closed-loop implementation is described in the next section.

A SIMPLE ANTIHANGUP METHOD

The block diagram of a straightforward antihangup scheme is shown in Fig. 3. It operates by applying a large restoring force to the loop whenever magnitude of phase error exceeds 90° and shutting off the force whenever the error magnitude is less than 90°. Sequential logic prevents reversal of the direction while the force is turned on.

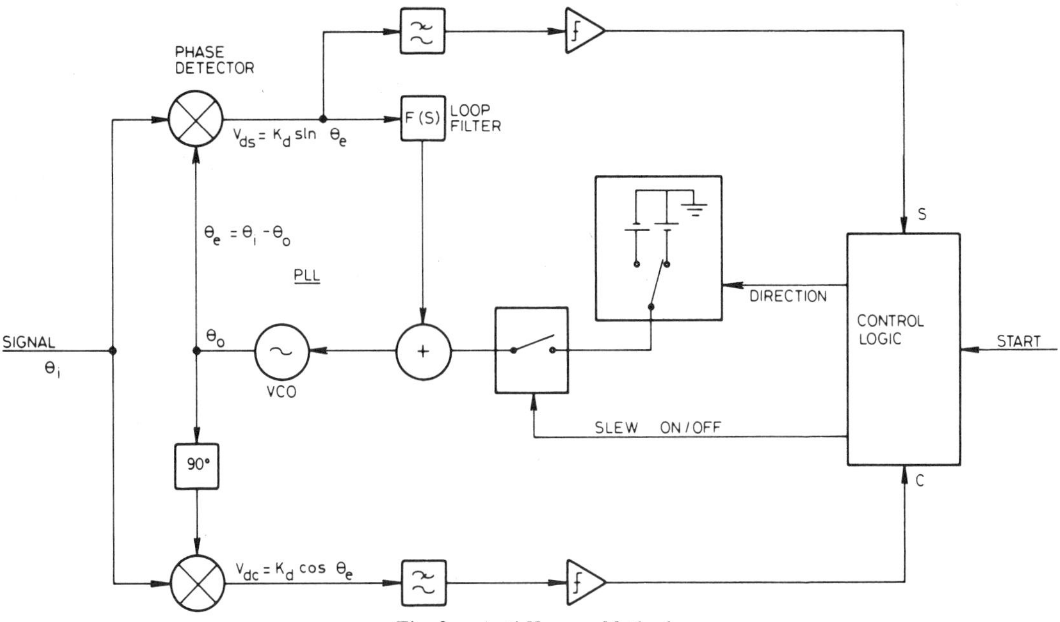

Fig. 3. Anti-Hangup Method.

The restoring force is applied as a large voltage of the correct polarity to the control line of the VCO. This voltage acts to slew the phase (that is, offset the frequency) of the VCO. Amplitude of the slew voltage must be large enough to overcome any possible output from the phase detector. If the voltage were too small, there would be some phase at which the PD output cancelled the slew voltage and the loop could hang up at that point.

The loop phase detector provides an output proportional to the sine of the phase detector while the quadrature-driven PD generates an output proportional to the cosine. Outputs of the PD's are filtered to reduce noise and then clipped to produce bilevel signals [15] (denoted S and C). The quadrant of the phase error can be determined by simple logic operations on S and C. Only quadrant information is needed; there is no attempt at accurate measurement of the phase error itself.

Bilevel output S is used to set the direction of phase slew. The direction is free to change as long as the slew is shut off but is held constant while the slew is turned on. A memory device (a simple latch) is needed to hold the direction.

The cosine is positive in the right half circle and negative in the left; since $C = \text{sign} (\cos \theta_e)$, it is used to turn the slew on and off.

Because of the low-pass filters, S and C zero crossings will be delayed from the quadrant-boundary crossings of the phase error. That means that the slew will stay on for some time after the phase error has crossed into the right half circle; it stays on until the low-pass filter has discharged sufficiently to allow C to change sign.

That lag is beneficial, if not prolonged unduly. The slew provides a faster convergence towards equilibrium than does the simple PLL by itself. The delay would only be harmful if the loop were carried beyond equilibrium before the slew shut off. Of course, the delay would be disastrous if the phase error overshot by 90° or more; the slew would never shut off under those conditions.

When the signal appears initially in the left half circle, we want to be able to select a slew direction before applying the slew. To accomplish that, it is necessary to recognize that a new signal has appeared and that slew must be inhibited while S is determined. That is to say, some kind of startup procedure is needed. It is symbolized by the arrow labeled START in Fig. 3, and is implemented in a manner determined by the specific application. A TDMA system will usually provide Start of Burst pulses that would serve the startup purpose very well.

We noted earlier that the penalty for an incorrect direction of slew was small. If a starting scheme proved to be overly complex or difficult, a random direction could be selected without seriously increasing the acquisition time.

COMMENTS

The purpose of this concise paper has been to bring the hangup phemomenon to light and to propose methods for its avoidance. It is thought that the principles and methods are sound, but experimental confirmation has not been attempted yet.

Controlled slewing as a means of hangup prevention has been worked out for a first-order PLL or, what amounts to the same thing, a second-order loop with very large damping. Behavior of a lightly damped second-order loop is not clear. Behavior during frequency pull-in is also obscure. In the application that motivated this work, the initial frequency error was negligible so these questions did not arise. Note that frequency acquisition tends to be a slow process so that one might not be overly concerned with phase hangup if a frequency error must be corrected first.

The primary application of antihangup methods is likely to be in carrier and clock synchronizers for digital communications where rapid startup is demanded. The regenerators and phase detectors for synchronizer application are significantly more complex than the simple sinusoidal phase detectors shown here. Work is under way to adapt the antihangup principles to synchronizer circuits.

REFERENCES

[1] M. Asahara, N. Toyonaga, S. Sasaki, and T. Miyo, "A 4-Phase PSK Modem for an SCPC Satellite Communication System", *Conference Record, ICC'75,* pp. 12-28 to 12-32, San Francisco: June 1975.

[2] M. Asahara, H. Nakamura, and T. Sugiura, "8-Phase and 16-Phase High Speed PSK Modems for PCM-TDMA Satellite Communication", *Fujitsu Scientific and Technical Journal,* pp. 59-89, Dec. 1972.

[3] *Communications Presented at the International Colloquim on Digital Communications by Satellite;* Paris, 1972, Editions Chiron. C. Denance, "Study of a PSK Modem for a TDMA System", pp. 247-255, (in French), M. Ogawa and M. Ohkawa, "A New 8-Phase Modem System for TDMA", pp. 256-265. T. Sekizawa, M. Asahara, H. Nakamura and T. Sugiura, "8-Phase and 16-Phase High Speed PSK Modems for Satellite Communication", pp. 266-276. Y. Matsuo, S. Sugimoto, and S. Yokoyama, "A Study on Carrier and Bit-Timing Recovery for Ultra-high-Speed PSK-TDMA Systems", pp. 277-286.

[4] F. M. Gardner, *Carrier and Clock Synchronization for TDMA Digital Communications,* European Space Agency: Noordwijk, Netherlands, Report 744976, June 1974. Republished as ESA TM-169, Dec. 1976.

[5] —, "Comparison of QPSK Carrier Regenerator Circuits for TDMA Application", *Conference Record, ICC'74,* paper 43B, Minneapolis, June 1974.

[6] —, "Clock Recovery for QPSK-TDMA Receivers", *Conference Record, ICC'75,* pp. 28-11 to 28-15, San Francisco, June 1975.

[7] D. Richman, "Color-Carrier Reference Phase Synchronization Accuracy in NTSC Color Television", *Proc. IRE, 43,* pp. 106-133, Jan. 1954.

[8] F. M. Gardner, *Phaselock Techniques,* Wiley: New York, 1966.

[9] A. J. Viterbi, *Principles of Coherent Communication,* McGraw-Hill: New York, 1966.

[10] S. L. Goldman, "Second-Order Phase-Lock Loop Acquisition in the Presence of Narrow-Band Gaussian Noise", *IEEE Trans., COM-21,* pp. 297-300, April 1973, Correction: *COM-21,* p. 1434, Dec. 1973.

[11] K. Nozaka, T. Muratani, M. Ogi, and T. Shoji, "Carrier Synchronization Techniques of PSK-Modem for TDMA Systems", *Intelsat/IEE Conf. on Digital Satellite Communication,* pp. 145-165, London, Nov. 1969.

[12] A. H. Pouzet, "Characteristics of Phase Detectors in Presence of Noise", *ITC Proceedings,* pp. 818-826, Los Angeles: 1972.

[13] R. D. Gitlin and J. Salz, "Timing Recovery in PAM Systems", *BSTJ, 50,* pp. 1645-1670, May-June 1971.

[14] J. A. C. Bingham, "Simultaneous and Rapid Regeneration of Carrier and Data Timing From Polyphase Signals", *Conf. Rec. ICC'76,* pp. 44-23 to 44-26, Philadelphia, June 1976.

[15] K. Hiroshige, "A Simple Technique for Improving the Pull-in Capability of Phase-Lock Loops", *IEEE Trans., SET-11,* pp. 40-46, March 1965.

Part II
Acquisition

THE INFLUENCE OF TIME DELAY ON SECOND-ORDER PHASE-LOCK LOOP ACQUISITION RANGE*

By JEAN A. DEVELET, Jr.

Summary.—The influence of time delay on second-order phase-lock receiver acquisition range is discussed. The time delay for the most part is considered lumped in the predetection i.f. amplifier and/or a radio frequency interference (r.f.i.) rejection filter at the VCO output. From the aforementioned specific ways in which time delay may be introduced, a general receiver model is extracted. A simple analysis of this model is presented which gives the pull-in range and false-lock points for the special case of high receiver gain.

Introduction.—In certain applications of phase-lock loops to radio reception, the introduction of time delay within the feedback path is unavoidable. Two ways in which time delay may be introduced are illustrated in Fig. 1.

For extremely narrow-band highly sensitive receivers (≈ -170 dBm or less), an i.f. amplifier of many stages is unavoidable. In addition, if the signal/noise ratio at point X (Fig. 1) is to be kept reasonable (e.g. > -20 dB), a narrow-band crystal filter may be required somewhere in the i.f. chain†. These factors can contribute to non-negligible time delay in the loop if the VCO connection is made to the first mixer. This type connection is highly desirable in narrow-band systems to reduce i.f. bandwidth requirements.

If severe r.f.i. requirements are imposed, one may need an extremely selective filter at the VCO output to prevent spurious signals generated in multiplier chains from radiating to other equipments by the conductive path shown in Fig. 1. This selectivity can only be obtained by many sections of a filter network which in turn introduces unwanted time delay.

Whether the aforementioned sources of time delay are significant in a particular application or some other source is of primary concern, one may construct a simple model of a phase-lock loop incorporating time delay for purposes of estimating the effects on pull-in range.

It is realized, of course, that excessive time delay can cause loop instability for a given set of parameters; however, since it will be shown that pull-in range is substantially independent of loop bandwidth, it will be assumed that closed loop bandwidths are sufficiently narrow with respect to the time delay present that a stable situation prevails.

Analysis.—Consider the simplified model (Fig. 2) for the second-order receiver with time delay. The various quantities which appear in Fig. 2 are defined as follows:

- E = received signal amplitude, v
- ω_s = received signal radian frequency
- K_a = loop filter d.c. gain
- α, β = loop filter time constants, sec
- K_v = VCO constant, rad/v/sec
- ω_0 = VCO center or rest frequency, rad/sec
- ω_1 = VCO instantaneous frequency, rad/sec
- ϕ = modulation index imposed on VCO due to $\omega_s \neq \omega_1$, rad
- τ = composite loop time delay, sec

The analysis may now proceed utilizing a technique previously utilized by Gruen[1]. Gruen's method is essentially a final value analysis which seeks possible stable points of equilibrium ignoring the dynamics of how the system proceeds to these points.

Normally, without time delay, there are, at the most, two stable states: one at the signal frequency if mistuning is sufficiently small, and the other in the vicinity of the VCO rest frequency if the mistuning is very large. The introduction of time delay yields the possibility of many more equilibrium frequencies.

The cause of these equilibrium frequencies can be seen by consideration of Fig. 2. If the VCO is mistuned, the predominant output of the product demodulator is a sinusoid which, after passing through the loop filter, frequency modulates the VCO. This frequency modulation causes sidebands. If the modulating sinusoid is of such a frequency that one of these sidebands coincides with the signal frequency, a d.c. voltage appears at the output of the product demodulator. This d.c. voltage may be of sufficient magnitude to maintain the VCO offset and thus create a stable point of equilibrium.

The author is with Aerospace Corporation, El Segundo, California.

*This research was performed while the author was affiliated with the Space Technology Laboratories, Inc.

†Phase detector balance becomes very critical for small predetection signal-to-noise ratios.

Reprinted with permission from *Proc. Int. Telemetering Conf.*, vol. 1, Sept. 23–27, 1963, pp. 432–437, Institution of Electrical Engineers, London, England.

Let us now seek the possible equilibrium frequencies. Consider the VCO at rest at a stable frequency, ω_1. The sinusoidal portion of the product demodulator output is given by

$$V_\varepsilon = E \sin (\omega_s - \omega_1)\, t \quad . \; . \; . \; . \; . \; . \; . \; . \; . \; . \; . \; (1)$$

Assuming $(\omega_s - \omega_0) >> 1/\beta$ or $1/\alpha$, the loop filter output resulting from the voltage of eqn. (1) causes frequency modulation of the VCO given by

$$\omega(t) = \frac{K_a K_v E \alpha}{\beta} \sin (\omega_s - \omega_1)\, t \quad . \; . \; . \; . \; . \; . \; . \; . \; . \; . \; (2)$$

where $\omega(t)$ is the frequency deviation of the VCO due to the input signal, rad/sec.

The phase modulation corresponding to $\omega(t)$ is

$$\phi(t) = \phi \cos (\omega_s - \omega_1)t = \frac{-K_a K_v E \alpha}{\beta(\omega_s - \omega_1)} \cos (\omega_s - \omega_1)\, t \quad . \; . \; . \; . \; . \; . \; . \; (3)$$

From eqn. (3), it can be seen that

$$\phi = \frac{-K_a K_v E \alpha}{\beta(\omega_s - \omega_1)}, \text{ rad} \quad . \; . \; . \; . \; . \; . \; . \; . \; . \; . \; . \; (4)$$

Eqn. (3) is an important result. It indicates that the VCO has sidebands, one of which is identical in frequency with the incoming signal. Introducing the time delay, τ, we obtain as the reference input to the product demodulator

$$v_r(t) = 2 \cos \left[\omega_1 (t - \tau) - \frac{K_a K_v E \alpha}{\beta(\omega_s - \omega_1)} \cos (\omega_s - \omega_1)(t - \tau) \right] \quad . \; . \; . \; . \; . \; (5)$$

Finally, a more exact expression for the output voltage, V_ε, than given by eqn. (1) is

$$V_\varepsilon = E \sin \left[(\omega_s - \omega_1)t + \omega_1 \tau + \frac{K_a K_v E \alpha}{\beta(\omega_s - \omega_1)} \cos (\omega_s - \omega_1)(t - \tau) \right] \quad . \; . \; . \; . \; (6)$$

For small values of the index, ϕ, eqn. (6) may be simplified. This simplification may be made in the practical situation of high loop gain. Thus

$$V_\varepsilon \cong E \sin (\omega_s - \omega_1)t + \frac{K_a K_v E^2 \alpha}{2\beta\, (\omega_s - \omega_1)} \cos (\omega_s - \omega_1)\, \tau \quad . \; . \; . \; . \; . \; . \; . \; (7)$$

In eqn. (7), the fixed phase shift, $\omega_1 \tau$, was neglected, since the VCO will introduce an equal and opposite shift cancelling $\omega_1 \tau$ giving consistent relationships forced by the closed loop.

It is noted from eqn. (7), that eqn. (1) is essentially correct except for the small d.c. perturbation which is the driving force to pull the loop to a stable point of equilibrium.

Imagine that the system is now at rest at the frequency ω_1, i.e., the VCO offset from the rest frequency ω_0 is maintained by the d.c. voltage in eqn. (7) and, vice versa, the d.c. voltage in eqn. (7) is maintained by the VCO offset. The equation of equilibrium becomes

$$\omega_1 - \omega_0 = \frac{K^2_v K^2_a E^2 \alpha}{2\beta\, (\omega_s - \omega_1)} \cos (\omega_s - \omega_1)\, \tau \quad . \; . \; . \; . \; . \; . \; . \; . \; . \; (8)$$

Eqn. (8) is the principal result of this paper. It determines the possible stable equilibrium frequencies $\omega_1 = \omega_k$ in which the final rest condition of the receiver may be found.

Eqn. (8) is identical to Gruen's result[2] with the exception that the $\cos (\omega_s - \omega_1)\tau$ term appears due to time delay. The effect of this term is interesting and will be the subject of subsequent discussion.

Let us now simplify the notation of eqn. (8) and bring in conventional definitions for the loop parameters[1]:

$\Delta\omega_0 = \omega_s - \omega_0$, the initial signal offset from VCO rest frequency, rad/sec
$\Delta\omega = \omega_s - \omega_1$, VCO offset from signal frequency, rad/sec
$\omega^2_n = EK_aK_v/\beta$, loop natural frequency (rad/sec)^2
$K = EK_aK_v$, overall d.c. loop gain, rad/sec
$\xi =$ loop damping ratio
$\alpha = (2\xi/\omega_n)-(1/K)$, natural result of above definitions

Eqn. (8) may now be written

$$\tau\Delta\omega_0 - \tau\Delta\omega = \frac{(2\xi\omega_n K - \omega^2_n)\tau^2}{2\Delta\omega\tau}\cos\Delta\omega\tau \qquad (9)$$

Introducing the normalizing variables

$\phi_0 = \tau\Delta\omega_0$, rad
$\theta = \tau\Delta\omega$, rad
$A = (2\xi\omega_n K - \omega^2_n)\tau^2/2$, rad

Eqn. (9) becomes

$$Y = \frac{\theta_0 - \theta}{A} = \frac{\cos\theta}{\theta} \qquad (10)$$

The condition on the validity of eqn. (10) is, as previously mentioned, that ϕ given by eqn. (4) be much less than unity. In terms of the newly defined parameters, this condition will be met by most high gain loops ($\omega_n/K \to 0$) and for values of $\theta >> 2\xi\omega_n\tau$. It would be wise in actual calculation, using the results of this simplified analysis, to check that this condition does exist.

Eqn. (10) is a transcendental equation with multiple solutions, θ_k, which in general cannot be attained explicitly. In Fig. 3, Y versus θ is plotted, illustrating conditions which may prevail for various θ_0 and A.

The intersection of the curves of the left- and right-hand portions of eqn. (10) gives the desired θ_k. Not all θ_k are stable equalibrium points. Physical reasoning based on assuming a small perturbation away from an intersection indicates the only stable solutions are those in which the ordinate of the straight line becomes greater than the ordinate of the $\cos\theta/\theta$ function for values of θ slightly less than θ_k. Since stable and unstable points alternate, it is clear that the system will eventually drive to a point of stable equilibrium.

It is interesting and significant that the first solution (intersection), if it exists, is always an unstable point of equilibrium. The loop will either be driven until $\theta_k = \theta_1$, a stable point, or until $\theta = 0$, which corresponds to acquisition of the desired signal. Strictly speaking, the foregoing analysis does not hold to $\theta = 0$ for finite time delay τ, but only down to those values of θ for which the relation $\theta >> 2\xi\omega_n\tau$ is satisfied.

Another interesting point exhibited by Fig. 3 is that the VCO may actually be driven away from the signal to reach equilibrium. This phenomenon is not predicted by previous analyses[1, 3] which do not take time delay into account.

Having provided a general result, eqn. (10), from which all stable solutions may be calculated, graphically or by iteration, let us examine certain limiting cases which yield more explicit results.

Case I: Large A, Arbitrary θ_0.—For this special case, it is clear from Fig. 3 that the stable points of possible equilibrium are

$$\theta_k = 0$$

or

$$\theta_k = \frac{(4k-1)\pi}{2} \quad (k=1, 2, 3, \ldots) \qquad (11)$$

Eqn. (11) represents all possible stable points, and the unperturbed equilibrium point will be the closest point to θ_0.

Translating eqn. (11) into the possible stable frequency states, we have

$$\Delta f_k = 0$$

or

$$\Delta f_k = \frac{4k-1}{4\tau}, \text{c/s} \quad (k=1, 2, 3, \ldots) \quad \ldots\ldots\ldots\ldots \quad (12)$$

where Δf_k = stable-lock frequency-offset from the signal frequency, c/s
τ = loop time delay, sec

Note, for large A, the system will not pull in unless $\Delta_{60} < 1/4\tau$ c/s

Case II: Small A, Arbitrary θ_0.—For this special case, Fig. 3 indicates that only two stable points exist: one is at $\theta = 0$ (signal lock) and the other is slightly higher or lower than θ_0 depending on whether cos θ_0 is negative or positive, respectively. Thus, for small A, it is concluded that the unperturbed equilibrium point is approximately $\Delta\omega_0$ itself, and pull-in will not occur.

Case III: Critical A, $\theta_0 < \pi/2$.—Probably the most significant situation is one in which $\theta_0 < \pi/2$ and a relation is sought between A and θ_0 in order to predict pull-in range. If one assumes θ_0 is small (small time delay for a fixed $\Delta\omega_0$), then explicit analytical results may be obtained without recourse to exact solution of eqn. (10). These results represent a perturbation of those obtained in Refs. 1 and 3. The perturbation gives the effects of small time delay on pull-in range.

For small θ_0, eqn. (10) may be approximated as

$$\frac{\theta_0 - \theta}{A} \cong \frac{1-(\theta^2/2)}{\theta} \quad \ldots\ldots\ldots\ldots\ldots \quad (13)$$

Eqn. (13) has two solutions. The interesting case which defines pull-in range is when the straight line is just tangent to the cos θ/θ curve in Fig. 3. This point may be found by setting the discriminant in the quadratic given by eqn. (13) equal to zero. One thereby obtains

$$\theta_0 \approx \left[4A\left(1 - \frac{A}{2}\right)\right]^{1/2} \quad \ldots\ldots\ldots\ldots \quad (14)$$

Substitution of the various loop parameters in eqn. (14) gives the usual formulation for maximum pull-in range in a second-order receiver including a time delay correction.

$$\Delta\omega_0 \approx \left\{2\left(2\xi\omega_n K - \omega^2{}_n\right)\left[1-\left(2\xi\omega_n K - \omega^2{}_n\right)\frac{\tau^2}{4}\right]\right\}^{1/2}, \text{rad/sec} \quad \ldots\ldots \quad (15)$$

Note, eqn. (15) reduces to the familiar[1]

$$\Delta\omega_0 = \sqrt{2(2\xi\omega_n K - \omega^2{}_n)}, \text{rad/sec} \quad \ldots\ldots\ldots\ldots \quad (16)$$

for $\tau = 0$

Conclusions.—Previous results [1, 3] on pull-in range for second-order phase-lock receivers have been modified to include receivers with delay for the special but important case of $\omega_n/K \to 0$ and $\theta >> 2\xi\omega_n\tau$.

A general transcendental equation, eqn. (10), has been derived which can be used to calculate exact equilibrium frequencies given the necessary parameter values.

Certain limiting cases were then investigated yielding the following interesting results.

First, high-gain receivers with delay yield a multiplicity of stable or false-lock points, specifically every $(4k-1)/4\tau$ c/s away from the signal frequency. Upon locking to an offset frequency, the loop will remain stable and never lock to the signal if all parameters remain constant. In order to lock to the signal, parameters must be disturbed or the initial offset must be less than $1/4\tau$ c/s.

Second, if one uses the technique of very slowly changing the VCO center frequency ω_0, to approach and finally lock to the signal, one finds that as the loop delay becomes larger, false-lock forces overshoot of the signal to finally achieve acquisition. In fact, it may be impossible with a specific circuit design to tune the VCO high (low) enough to acquire. This situation may best be overcome by turning off the signal until the VCO is tuned to within $1/4\tau$ c/s of the signal frequency and then reinserting the signal. If this is not possible, one might try reducing the loop bandwidth, ω_n, during the acquisition mode.

Third, small time delays yield little difference over effects already noted in previous analyses[1, 3]. The loop will stay either very close to the initial offset from the signal or at a critical offset will progress toward the signal to eventually achieve lock. This critical offset is given by eqn. (15).

It must be remembered that the final value analysis utilized here assumes that loop time delay was not large enough to cause dynamic instability. That is, it was assumed a final value eventually is reached.

Finally, it is noted that arbitrary phase shift versus frequency characteristics around the closed loop may be analyzed in a similar fashion. This more general problem was not undertaken in the interest of simplicity and physical interpretation.

Acknowledgment.—The author wishes to acknowledge the fact that Sidney Bergman of Space Technology Laboratories, Inc., Redondo Beach, California, first indicated that delay effects had been observed in the laboratory and postulated the mechanism herein analyzed. In addition, the novel technique of loop analysis put forth by Gruen was an invaluable aid in obtaining physical insight as well as quantitative results. The comments of Dr. R. C. Booton, Jr., of Space Technology Laboratories, Inc., aided significantly in clarifying the presentation of this material.

REFERENCES

(1) GRUEN, W. J.: 'A Simple Derivation of Pull-In Range and Pull-In Time for Second-Order Frequency and Phase Control Loops', Thompson Ramo Wooldridge, Inc., Canoga Park, California, March 7th, 1962, p. 1.

(2) *Ibid.*, p. 6.

(3) VITERBI, A. J.: 'Acquisition and Tracking Behavior of Phase-Locked Loops', Jet Propulsion Laboratory, Pasadena, California, External Publication 673, July 14th, 1959, p. 1.

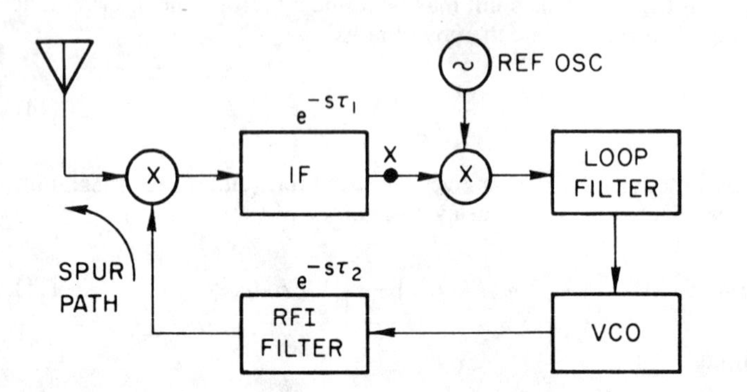

Fig. 1.—Introduction of time delay in phase-lock reception.

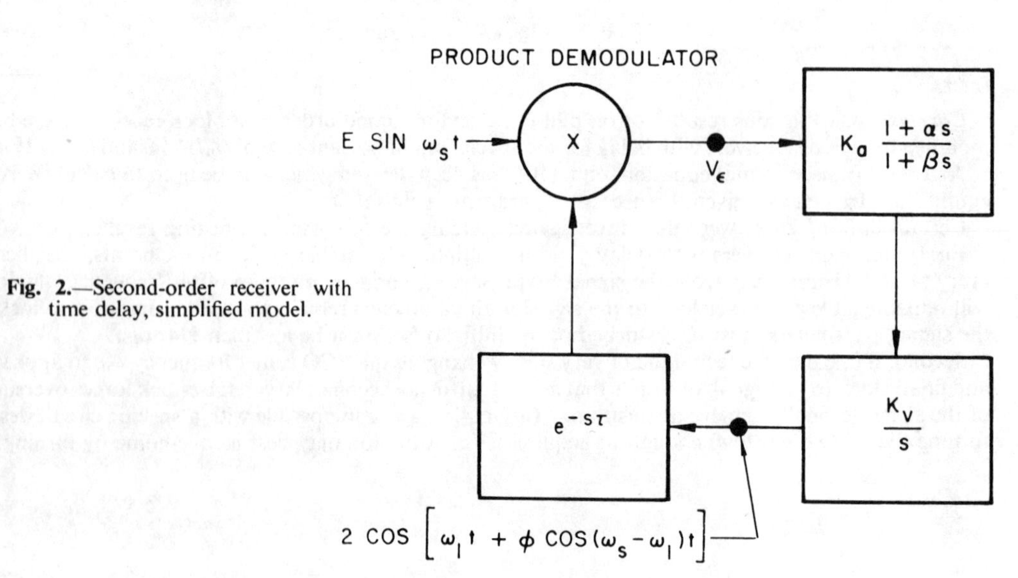

Fig. 2.—Second-order receiver with time delay, simplified model.

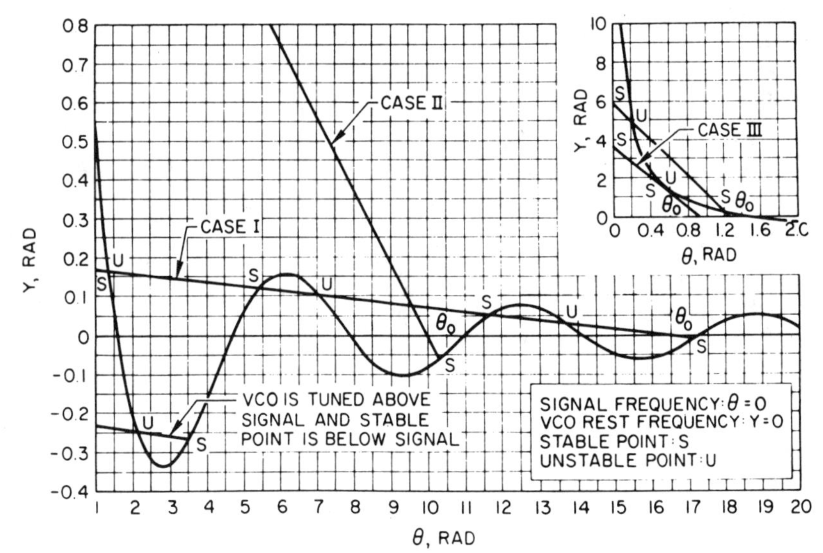

Fig. 3.—Y versus θ for various values of A.

Phase-Lock Loop Frequency Acquisition Study*

J. P. FRAZIER† MEMBER, IRE, AND J. PAGE†, MEMBER, IRE

Summary—The ability of a phase-lock loop using a proportional plus integral control filter to acquire a noisy signal when the local oscillator is being swept was determined empirically by means of a low-frequency, GEESE[1] model of such a system.

The effects of the damping factor and natural frequency on the frequency acquisition properties of linear loops (as distinguished from a loop in which the IF signal is limited) are considered in this study. In addition, consideration is given to a loop in which the IF signal is "hard" limited and the loop designed to maximize the sweep rate under the constraint that the probability of acquisition is equal to or greater than 90 per cent for a given SNR.

The rms phase jitter in the output signal was measured as a method of verifying the standard analytic approach to predicting phase jitter.

The results of the study are as follows:

1) The range of damping factors from 0.5 to 0.85 yields near optimum acquisition performance.
2) Although a drop in loop gain produces lower phase jitter for a given $(S/N)_{IF}$, it degrades the over-all ability of the loop to acquire and track a signal.
3) A "hard" limiter in the IF can be effectively used as a gain control to enhance loop performance.
4) Using an empirical formula derived from experimental results, the VCO sweep rate for 90 per cent probability of acquisition can be predicted accurately, given the $(SNR)_{IF}$ and loop parameters.

I. Introduction

PHASE-LOCKED oscillators are discussed extensively in the literature; [1]–[6] however, the nonlinear nature of phase-locked oscillators in an unlocked condition, *i.e.*, before and during frequency acquisition (lock-on), precludes a comprehensive analysis of their acquisition capability. The ability of a phase-lock loop to acquire a noisy signal when the local oscillator is being swept has been determined empirically using a low-frequency GEESE[1] model of such a system.

This report considers the effects of the natural frequency ω_n and the damping factor ζ on the frequency acquisition properties of a linear loop (as distinguished from a loop in which the IF signal is limited). In addition, consideration is given to a loop in which the IF signal is "hard" limited (infinite clipping) and the loop designed to maximize the sweep rate under the constraint that the probability of acquisition is equal to, or greater than, 90 per cent of a given signal-to-noise (SNR).

The rms phase jitter in the output signal also was measured as a method of verifying the standard analytical approach to predicting phase jitter.

The results are presented in graphical form, and from these results in empirical formula is derived which predicts the sweep rate for 90 per cent probability of acquisition (P_a) as a function of the SNR and loop parameters.

* Received December 15, 1961.
† Defense Systems Department, GE Company, Syracuse, N. Y.
[1] General Electric Electronic System Evaluator.

II. General Loop Description

The block diagram of the particular phase-lock loop considered is shown in Fig. 1. The incoming signal is mixed with the local oscillator signal and filtered in the IF amplifier. In the locked condition, the IF signal is of exactly the same frequency as the reference oscillator, and only a phase difference exists between them. The phase detector senses any phase error, and its output is a voltage proportional to the sine of the phase difference. This error voltage is filtered and applied to the VCO to control the VCO frequency. The dynamic characteristics of the loop are determined primarily by the characteristics of the control filter $H(s)$.

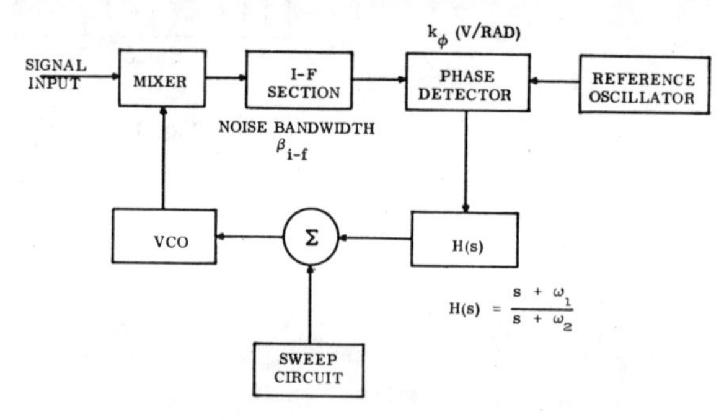

Fig. 1—System block diagram.

The input and reference signals are CW signals from commercial oscillators, the mixer and the phase detector are regular analog multipliers, and the IF section is made up of two simulated, synchronously-tuned stages with a single-sided noise bandwidth β_{IF}. The VCO is an audio oscillator which was designed especially for this type of work and which has a linear output frequency vs input voltage characteristic. The sweep circuit was nothing more than an analog integrator. The control filter, $H(s)$, is a proportional plus integral filter for which $\omega_1/\omega_2 \approx 10^3$.

During the tests, the input signal was fixed in frequency and amplitude, and the noise level was increased. The VCO then was swept at various rates, and the probability of acquisition thus was the number of times that the loop locked in one hundred attempts. The sweep was not disabled as soon as acquisition occurred.

The experimental results include the effects on the ability of this system to acquire and track a noisy signal for various damping factors and bandwidth ratios, and for limiting the IF signals.

III. Loop Performance

In the particular loop under consideration, for a given equivalent double-sided bandwidth β_n, [see (19)] the natural frequency ω_n is a maximum for $\zeta = \frac{1}{2}$. Conversely,

Reprinted from *IRE Trans. Space Electron. Telemetry*, vol. SET-8, pp. 210–227, Sept. 1962.

for a given natural frequency, the noise bandwidth is minimum when $\zeta = \frac{1}{2}$. Most of the performance data in this report was taken on systems with a damping factor ζ equal to 0.5. The comparative data on systems with damping factors indicate that for $\zeta > \frac{1}{2}$ there is no marked change in acquisition characteristics; but for $\zeta = \frac{1}{3}$, the acquisition performance is significantly poorer than for $\zeta \geq \frac{1}{2}$.

When the IF signal is limited, the loop can be designed so that, at some low SNR, a high probability of acquisition is obtainable for a given sweep rate. The limiter performs a gain-control function which provides good dynamic performance. In addition, when using the limiter and "matching"[2] the gain at some minimum expected SNR, the loop operates as well at the match point as would a linear loop. (Any increase in SNR will improve acquisition performance over that obtainable with a linear loop.)

The general results are presented in graphical form in Figs. 2–6 (pp. 211–212). Fig. 2 illustrates the degradation in system performance as a function of SNR for a linear system with $\zeta = \frac{1}{2}$. In Fig. 3 the performance of 4 systems with a fixed noise bandwidth but different damping factors is compared. The over-all system performance is generally better for $\zeta = \frac{1}{2}$, inasmuch as the decrease in P_a is not as rapid with increasing sweep rate and noise does not degrade the performance as much. The apparent improvement in maximum sweep rate for higher damping factors, and likewise the decrease for lower damping factors, is tied to the relative stability of the loops.

The results of tests on a linear system and on two systems with limiting (one matched with no noise and the other matched for $(S/N)_{IF} = -2$ db) are shown in Fig. 4. This figure illustrates the severe degradation of system performance for an uncompensated system, as a function of the SNR when limiting occurs. This poor performance is due primarily to the reduction in the natural frequency ω_n and damping factor ζ of the system as the signal is attenuated in the limiter. The "matched" system at the match point, $(S/N)_{IF} = -2$ db, shows performance comparable to a linear loop and an improvement in performance for the higher SNR's. This is to be expected when it is realized that increasing the SNR in the IF causes an increase in loop gain. The total increase in loop gain is fixed when the match point is specified.

Using certain assumptions, the rms phase jitter can be calculated for any system of these types when the loop noise bandwidth is properly accounted for [see Fig. 10 and (21), (25), (28)]. Fig. 5 is a plot of the measured phase jitter, in which these results are compared to the calculated values. The differences between calculated and measured values are within 20 per cent.

[2] "Matching" is the procedure whereby the loop gain is set to give a minimum noise bandwidth $\zeta_0 = 1/2$ at one SNR in the IF, $(S/N)_{IF}$. Fig. 10 shows the factor α by which the signal voltage out of a limiter varies as a function of the SNR into the limiter, $(S/N)_{IF}$. The point at which $\zeta_0 = 1/2$ is set will be designated α_0.

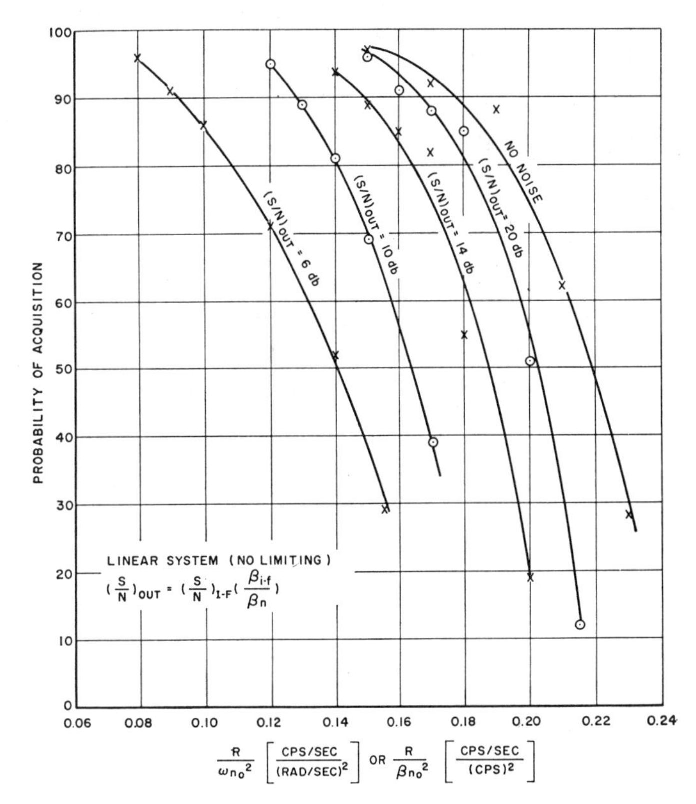

Fig. 2—Probability of acquisition vs normalized sweep rate ($\zeta = \frac{1}{2}$).

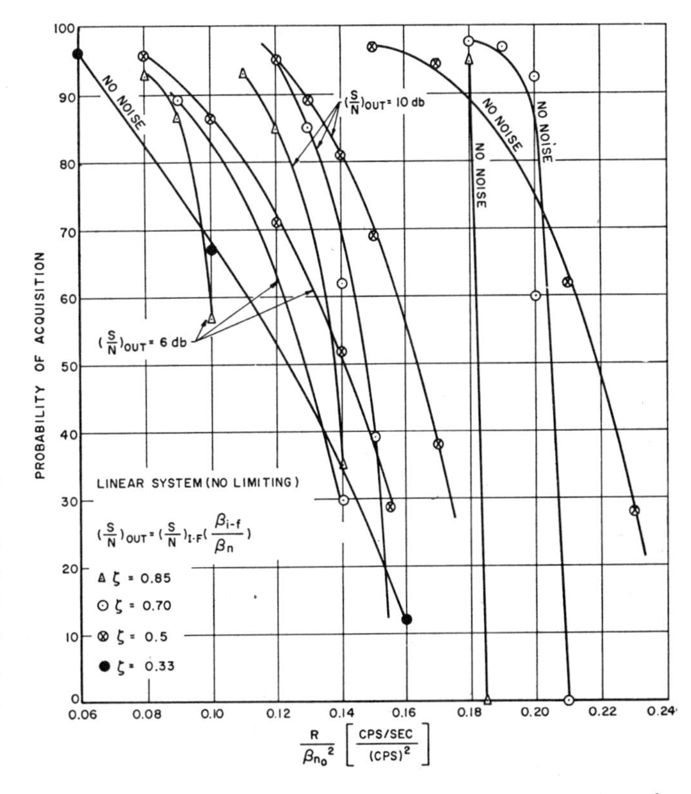

Fig. 3—Probability of acquisition vs normalized sweep rate for various damping factors.

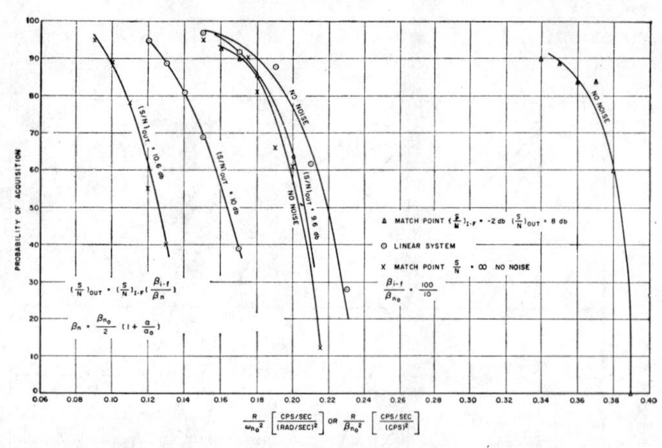

Fig. 4—Probability of acquisition vs normalized sweep rate, $R/\beta_{n_o}^2$, a comparison of a linear system with two systems having limiting in the IF.

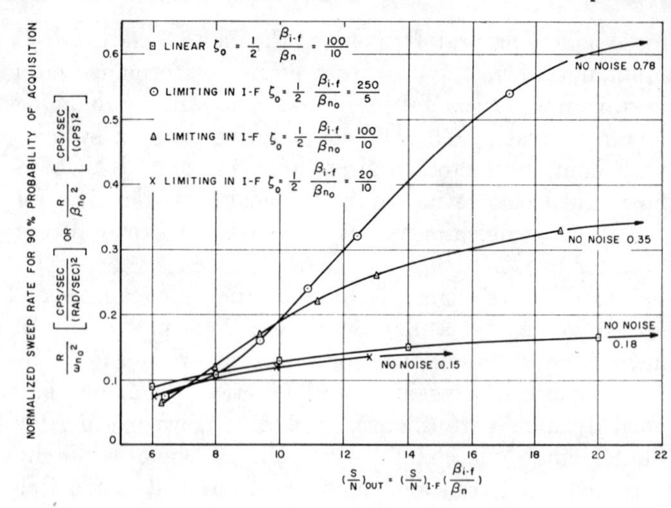

Fig. 6—Normalized sweep rate for P_a = 90 per cent vs $(S/N)_{out}$ for various systems.

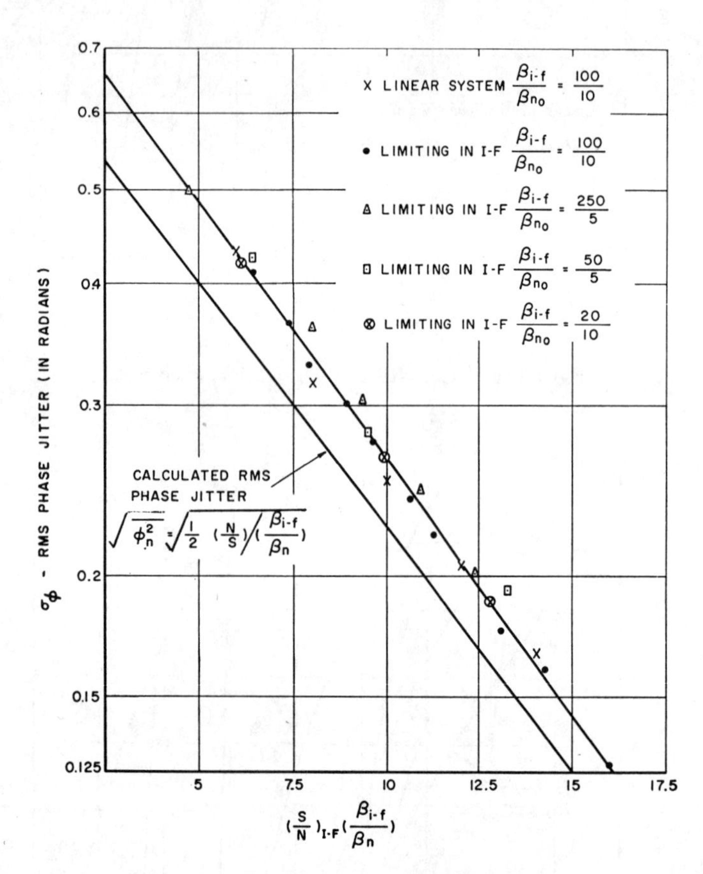

Fig. 5—RMS phase jitter vs $(S/N)_{out}$, db, for the various systems considered.

The results of these tests can best be summarized as shown in Fig. 6, which is a plot of the normalized maximum sweep rate for 90 per cent P_a vs $(S/N)_{out}$. It is worth noting that the curves converge near $(S/N)_{out}$ = 8 db, the point at which the loops with limiting IF's were matched to perform like the linear loop.

It is also interesting to note that the system with a bandwidth ratio β_{IF}/β_n = 2 never performs quite as well as the linear loop; this performance (like that for $\zeta = \frac{1}{3}$) is because the nonlinear loop has a relatively poorer stability characteristic (the IF break points affect the loop response).

From these results it was possible to develop a formula to predict the sweep rate for P_a = 90 per cent. The accuracy of this formula in predicting the 90 per cent point is within 10 per cent of the measured values, and holds for damping factors equal to, or greater than, 0.5 ($\zeta \geq \frac{1}{2}$).

$$R_{90}(\text{cps/sec}) = \frac{(\pi/2 - 2.2\sigma_0)\left(0.9\dfrac{\alpha}{\alpha_0}\right)\omega_{n_o}^2}{2\pi(1+\delta)}, \tag{1}$$

where

σ_0 = calculated rms output phase jitter,
δ = overshoot (see Appendix III and Fig. 24),
α = signal suppression factor in limiter,
α_0 = suppression factor for the $(S/N)_{IF}$ for which loop is "matched,"
ω_{n_o} = natural frequency of loop at match point.

The general form of this equation was first suggested by Dr. L. J. Neelands.[3]

IV. Loop Analysis

Phase Jitter

The loop under consideration is shown in Fig. 7. The input signal is a sine wave having an rms voltage A and angular frequency, ω_s. In addition, a noise signal having Gaussian statistical properties is present. The noise spectrum is band-limited but flat over a frequency range which is much broader than the IF noise bandwidth. The VCO signal was a constant rms voltage of B, a frequency of $(\omega_s - \omega_i)$, and a phase of ϕ_0.

The mixer and phase detector in the block diagram model are perfect multipliers. They therefore possess the characteristic that, if one of the signals is clean and the other consists of a sine wave plus noise, the input SNR is preserved in the output and the noise spectrum retains its shape, although shifted in frequency. The output of

[3] Consultant, GE Company, Instrumentation and Guidance Product Section, Advance Systems Engineering, Syracuse, N. Y.

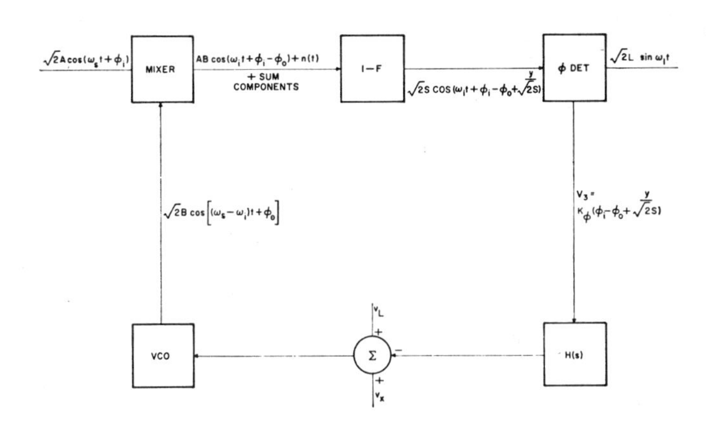

Fig. 7—Phase-locked loop block diagram.

the mixer is therefore a sine wave at the intermediate frequency, with a noise spectrum symmetrical about it, plus a sum-frequency signal with an associated noise spectrum.

The higher frequency components are removed in the IF section, which has a noise bandwidth, β_{IF}. The bandwidth of the IF must be much greater than the loop bandwidth or the IF filter time constants will effect the loop response and many of the assumptions to be made in the analysis will no longer be valid. A large ratio of bandwidths is generally the case in an actual system. The output of the IF now consists of a signal having a power S and band-limited noise of bandwidth β_{IF} centered about the intermediate frequency. The total signal can be written as the sum of signal plus noise.

$$V_{IF} = \sqrt{2S} \cos (\omega_i t + \phi_i - \phi_0) + n(t).$$

The noise can be written in terms of its in-phase and quadrature components:

$$V_{IF} = \sqrt{2S} \cos (\omega_i t + \phi_i - \phi_0) + x \cos (\omega_i t + \phi_i - \phi_0) - y \sin (\omega_i t + \phi_i - \phi_0), \quad (2)$$

where x and y are independent, Gaussian, random variables having zero means. Since x and y are independent, the average noise power, or mean-squared value, reduces to

$$N = \overline{x^2} = \overline{y^2}. \quad (3)$$

The SNR in the IF now can be written as

$$\frac{S}{N} = \frac{S}{\overline{x^2}} = \frac{S}{\overline{y^2}}. \quad (4)$$

The effect of input noise or, correspondingly, IF noise, on the phase of the VCO signal can be investigated by first expressing the effect of the noise on the IF signal as a random phase modulation of a relatively clean IF signal. The response of the loop to phase variations of the input signal then can be used to find the VCO phase jitter due to the input noise. Eq. (2) can be written

$$V_{IF} = (\sqrt{2S} + x) \cos (\omega_i t + \phi_i - \phi_0) - y \sin (\omega_i t + \phi_i - \phi_0). \quad (5)$$

The quadrature part of the noise produces the phase variation. A vector addition of (5) results in the desired phase-modulated signal. For large SNR's,[4] the signal can be written

$$V_{IF} \approx \sqrt{2S} \cos (\omega_i t + \phi_i - \phi_0 + y/\sqrt{2S}). \quad (6)$$

The mean-squared phase jitter σ^2 is now found:

$$\sigma^2 \approx \frac{\overline{y^2}}{2S} = \left(\frac{1}{2}\right)\left(\frac{N}{S}\right). \quad (7)$$

Although the assumption of a high SNR in the above derivation appears too restrictive for general application, the experimental results indicate that (7) can be used to predict output mean-squared phase jitter to within 3 db with IF SNR's of less than -10 db.

The phase detector compares the phase of the IF signal with the phase of a reference oscillator and provides the error-sensing capabilities of the loop. The operation of this particular loop differs from the loop shown in Fig. 8 only to the extent that a heterodyning stage is introduced. Since the heterodyning action of Fig. 7 preserves the phase relationship between the input signal and VCO signal, the basic operation of the two loops is the same, and the results of this study can be applied to the loop shown in Fig. 8.

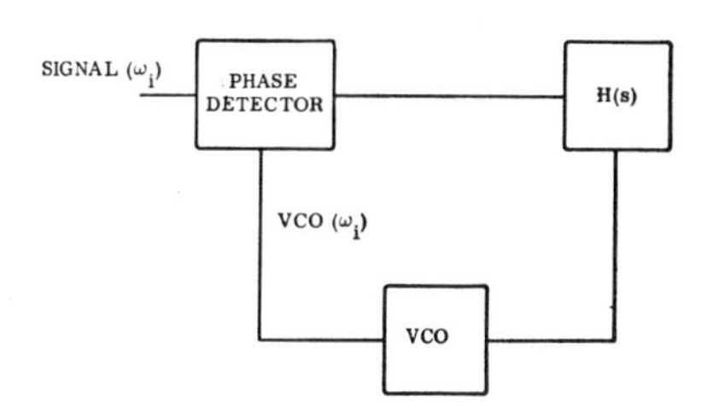

Fig. 8—Alternate form of phase-locked loop.

The loop filter $H(s)$ is a proportional plus integral filter which shapes the closed-loop characteristics. It provides the means for obtaining the optimum compromise between phase error due to noise and phase error due to phase variations of the input signal or the VCO signal other than those due to noise. These latter errors hereafter will be called transient errors. The summing amplifier is used to set the VCO center frequency and to supply the VCO sweep voltage. The frequency sweep must be employed whenever it is desired to lock onto a signal whose frequency cannot be predicted within the loop pull-in range.

Definition of Phase Errors

The loop phase error is defined as the difference in phase between the VCO signal and the incoming signal.

[4] Another approach to the analysis would be to describe the spectrum of the noise out of the phase detector, independent of the signal phase. This spectrum could be treated analytically as a disturbing function entering the loop after the ϕ detector and the ϕ jitter then calculated due to this disturbance. The same expressions for phase jitter will be obtained using this approach as will be derived in (13) and (18). However, this latter analysis will not be dependent upon a high SNR if the IF as is the analysis in the text.

The loop phase errors can be of the two types mentioned previously—namely, transient phase errors, and phase jitter due to noise. To investigate the magnitude of these errors, an equivalent loop transfer function must be derived relating the VCO output phase to the phase of the incoming signal. This is done in Appendix III, where a loop transfer function is derived for the case where the loop is in the locked condition.

The loop transfer functions can be summarized as follows:

$$\frac{\phi_{VCO}(s)}{\phi_{SIG}(s)} = \frac{K(\omega_2/\omega_1)(s+\omega_1)}{s^2 + [\omega_2 + K(\omega_2/\omega_1)]s + K\omega_2}$$

$$= \frac{K(\omega_2/\omega_1)(s+\omega_1)}{s^2 + 2\zeta\omega_n s + \omega_n^2}, \tag{8}$$

where

ω_2 and ω_1 are the critical frequencies of the loop filter $H(s) = (s+\omega_1)/(s+\omega_2)$,

K = loop velocity gain constant,

$\omega_n^2 = K\omega_2$,

$\zeta \approx \frac{1}{2}\sqrt{\frac{K\omega_2}{\omega_1^2}}$.

The phase error response is

$$\frac{\phi_{error}(s)}{\phi_{sig}(s)} = \frac{s(s+\omega_2)}{s^2 + 2\zeta\omega_n s + \omega_n^2}. \tag{9}$$

A given constant Doppler offset of $\Delta\omega$ radians per second produces a phase ramp having the transform $\Delta\omega/s^2$. Using the Laplace transform Final Value Theorem, the steady-state phase error in the loop can be found from (9) to be $\Delta\omega/K$. This steady-state error will be present after the loop pulls in and locks in frequency. In other words, the loop tracks a constant-frequency signal with zero frequency error and with a phase error equal to $2\pi\ \Delta f/K$ radians. Appendix III shows that the loop tracks a frequency ramp of R radians/sec^2 for a short period of time with an error R/ω_n^2. This error increases at a rate of ω_2 radians/sec [see (45)] until the total loop error exceeds $\pi/2$ radians. The loop then will lose lock. Since the ratio ω_n/ω_2 is normally in the order of 1000, the $\omega_2 t$ term does not appreciably affect the acquisition capabilities of the loop.

The phase jitter in the loop can be determined from Fig. 7. The output of the phase detector is

$$K_\phi[\phi_i - \phi_0 + (y/\sqrt{2S})],$$

where K_ϕ is the phase detector gain and the remaining terms are as defined previously. The output phase is related to the phase detector output as follows:

$$\phi_0(s) = K_\phi[\phi_i - \phi_0 + (y/\sqrt{2S})]H(s)\frac{K_{VCO}}{s}, \tag{10}$$

where K_{VCO} is the VCO gain in radians/sec/volt. In Appendix III, the open loop transfer function is defined as

$$G(s) = K_\phi \frac{K_{VCO}}{s} H(s). \tag{11}$$

Therefore

$$\phi_0(s) = G(s)[\phi_i - \phi_0 + (y/\sqrt{2S})]. \tag{12}$$

A block diagram (Fig. 9) derived from (12) illustrates the relationships between the phase error, the signal phase, and the phase jitter in the IF. This block diagram can be solved for $\phi_E(s)$ by deriving a set of simultaneous equations.

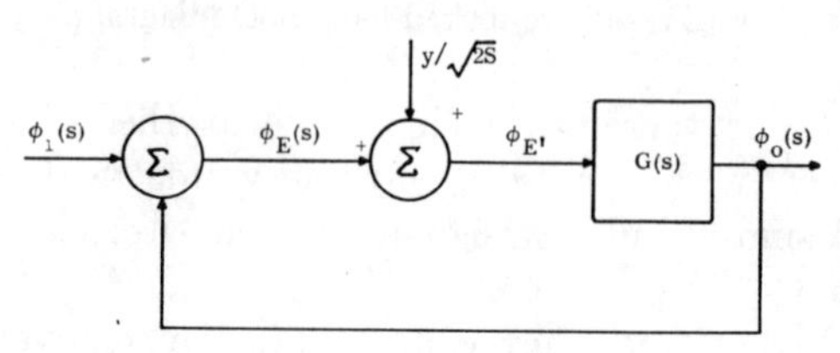

Fig. 9—Simplified block diagram showing important phase relationships.

The final expression for loop phase error is found from Fig. 9 to be

$$\phi_E(s) = \phi_1(s)\left[\frac{1}{1+G(s)}\right] - \frac{y}{\sqrt{2S}}\left[\frac{G(s)}{1+G(s)}\right]$$

$$= \phi_1(s)\left[\frac{1}{1+G(s)}\right] - \frac{y}{\sqrt{2S}}F(s). \tag{13}$$

The first part of (13) is the transient error term derived in Appendix III. The second term defines the loop phase jitter.

It will be recalled that $y/\sqrt{2S}$ is a random variable, having zero mean, which defines the IF phase variation due to noise. The mean-squared value of these variations was found to be related to the SNR by

$$\sigma^2 = \frac{\overline{y^2}}{2S} = \frac{1}{2}\left(\frac{N}{S}\right). \tag{14}$$

The IF equivalent noise bandwidth β_{IF} is related to the power density spectrum by

$$\beta_{IF} = \frac{N}{\Phi_N(\omega_i)} \quad \text{or} \quad \Phi(\omega_i) = N/\beta_{IF} \tag{15}$$

where $\Phi_N(\omega_i)$ is the maximum density located at the IF center frequency. An equivalent phase-jitter power density is

$$\Phi_{IF}(\omega) = \frac{\Phi_N(\omega_i)}{2S}, \tag{16}$$

since the noise power and phase jitter are related by (14). Eq. (13) defines the relationship between the IF phase jitter and the loop phase jitter. The output phase-jitter power density is therefore related to the input spectrum by

$$\Phi_0(\omega) = |F(j\omega)|^2\ \Phi_{IF}(\omega) = \frac{|F(j\omega)|^2\ \Phi_N(\omega)}{2S}.$$

The mean-squared phase jitter can be found by evaluating the integral

$$\sigma_0^2 = \int_{-\infty}^{\infty} \Phi_0(\omega)\, d\omega = \int_{-\infty}^{\infty} \frac{\Phi_N(\omega)}{2S} |F(j\omega)|^2\, d\omega. \qquad (17)$$

If the IF noise power is distributed uniformly over a bandwidth appreciably wider than the loop bandwidth, (15) can be substituted into (17), and the constants can be brought outside the integral. Therefore,

$$\sigma_0^2 = \left(\frac{1}{2}\right)\left(\frac{N}{S}\right)\left(\frac{1}{\beta_{IF}}\right) \int_{-\infty}^{\infty} |F(j\omega)|^2\, d\omega$$
$$= \left(\frac{1}{2}\right)\left(\frac{N}{S}\right)\left(\frac{\beta_n}{\beta_{IF}}\right), \qquad (18)$$

where β_n is the equivalent noise bandwidth of the loop. The integral was evaluated for this transfer function, and β_n was found to be [3]

$$\beta_n = \frac{\pi\omega_n}{2\zeta} [1 + (2\zeta - \omega_n/K)^2] \approx \frac{\pi\omega_n}{2\zeta} (1 + 4\zeta^2). \qquad (19)$$

The damping factor ζ for minimum noise bandwidth can be found by differentiating (19) and setting the results equal to zero. The resulting damping factor is 0.5. In other words, for a loop having a given undamped natural frequency, the loop has a minimum amount of phase noise if the damping factor is 0.5. Conversely, it can be shown that, for a given β_n, the undamped natural frequency can be maximized if ζ is 0.5.

In summary (13) states that the loop phase error arises from two sources: 1) noise, and 2) Doppler error and its derivatives. From Appendix III it was found that the phase errors caused by Doppler shifts and Doppler rate were inversely proportional to the loop gain K and ω_n^2, respectively. However, since $\omega_n^2 = K\omega_2$, both types of phase error are reduced in direct proportion to increases in loop gain. The second type of error, that due to noise, was found to be a random variable having zero mean and a mean-squared value given by (18). Combining (23) and (25),

$$\beta_n \approx \pi\left(\omega_1 + \frac{\omega_n^2}{\omega_1}\right) = \pi\left(\omega_1 + K\frac{\omega_2}{\omega_1}\right)$$

shows that the mean-squared value of phase jitter increases linearly with loop gain.

Gain Control

In Appendix III it is shown that the loop gain is dependent on the signal voltage. In the preceding section it was pointed out that the ability of the loop to track a signal having uncertain and varying Doppler shifts is directly dependent on loop bandwidth and, therefore, on signal voltage. To operate the loop at its optimum at all times, it would be necessary to design the loop as an adaptive servo. This would entail the identification of signal dynamics and SNR and then the minimization (13) by a gain adjustment. The complexity of the adaptive loop would depend on the order of the Doppler characteristics which are of importance and also on how much of the signal dynamics can be assumed to be known. In general, this approach would lead to more problems in system complexity and instability than it would solve in phase-error minimization.

A coherent AGC could be implemented, using the output voltage of a phase detector which senses the phase difference between the IF and a 90° phase-shifted version of the reference. The average value of the phase-detector output will be directly proportional to the signal strength and should, therefore, result in a system which keeps the signal power constant. Although it is relatively unimportant, it should be pointed out that this AGC would be sensitive to the signal Doppler characteristics. Doppler errors result in loop phase errors which will also appear in the AGC phase detector. For any loop phase error, the AGC voltage would drop, indicating an apparent signal-level decrease which actually does not occur. The main drawback of the coherent AGC, other than the addition of an extra feedback loop, is that it is insensitive to SNR. Theoretically, the transient error [the first term in (13)] will remain constant regardless of the input SNR. However, the phase jitter represented by the second term will increase rapidly with decreasing SNR. The result is a phase-locked loop which can operate over a limited range of SNR's. Jaffe and Rechtin [1] point out that the dynamic range of the loop can be improved over the AGC system and made to approach the ideal adaptive system if a band-pass limiter is inserted in the IF.

Band-Pass Limiter

The limiter considered is an ideal limiter, the output voltage of which can be only plus or minus the limit level. Davenport [2] has analyzed the SNR characteristics of this type of limiter when it is followed by a band-pass filter that removes the harmonics generated by the limiting action. The SNR is essentially preserved in the band-pass limiter; the maximum degradation is -1 db for low SNR's, and the maximum improvement is 3 db for high SNR's.

The total power output of a limiter in a given frequency zone has been shown to be constant [1] and equal to the signal power with no noise.

$$P_0 = (8/\pi^2)E, \text{ if } \pm E \text{ are the limit levels.}$$

Any noise added to the signal will therefore result in an output signal suppression. The magnitude of this suppression can be found as follows:

$$P_0 = S_{out} + N_{out},$$

where P_0 is the constant power output. The ratio of output signal power, S_{out}, to signal power with no noise is therefore

$$\frac{S_{out}}{P_0} = \frac{S_{out}}{S_{out} + N_{out}} = \frac{1}{1 + \dfrac{N_{out}}{S_{out}}}.$$

Thus, for an S_{out}/N_{out} of zero db, the signal voltage will be down 3 db from its value with no noise. This result, when combined with Davenport's results, allows the signal voltage out of the band-pass limiter to be found as a function of the input SNR and the limit level. The result is shown in Fig. 10.

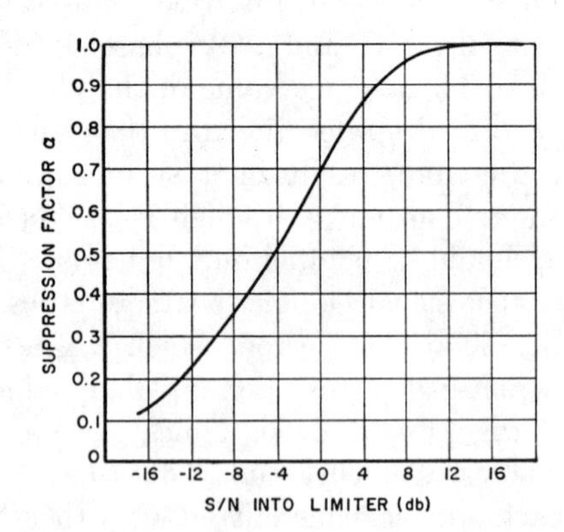

Fig. 10—Output signal voltage suppression factor in an ideal band-pass limiter as a function of input signal-to-noise ratio.

The action of the limiter as the SNR drops can now be related to a decrease in signal voltage, the resultant drop in loop gain and a lowering of the loop noise bandwidth. The mean-squared phase jitter is therefore reduced as shown in (18).

For SNR's greater than the design ratio, the limiter output signal power is increased, thus raising the loop gain. Although the noise bandwidth is increased, the improvement in loop dynamics is such that a more optimum minimization of (13) is obtained than is provided by a loop whose gain is controlled by a coherent AGC.

V. Analog Computer Simulation

General Description

The GEESE analog computer model which was used to obtain the experimental results is an actual phase-locked loop. As such, all statements made, and equations derived, in the Section IV apply directly to the computer model. Fig. 1 therefore serves as a block diagram of the computer model. The only change in notation required is that the noise was introduced into the model IF rather than before the mixer (see Fig. 32). This was done in order to prevent computer overloads and also to aid in controlling the noise spectral shape. The operation of the loop is unchanged if a perfect multiplier is assumed for a balanced mixer. The loop under study, and its computer model, are the same in the following respects:

1) The signal and oscillator voltages are sine waves. The gain of the VCO's and phase detectors are expressed in radians per second per volt and in volts per radian, respectively.
2) The computer phase detector and balanced mixer are multipliers. This type of device is usually assumed in an analysis of the loop, although in practice a balanced mixer can be made to only approach this ideal.
3) The noise characteristics and analyses of both loops are identical.

The two loops differ as follows:

1) The IF filter and $H(j\omega)$ time constants are obtained in the model with high-gain, operational amplifiers and their associated input and feedback networks. A computer filter will have the same transfer function as the actual filter of which it is a model. However, the flexibility of having easily and accurately controlled time constants is incorporated in the model.
2) A time/frequency scaling is used in deriving the model. For example, if a time scale factor of 10^3 is used, the following relationships would exist between the actual loop and its model.

Factor	Actual	Model
ω_1	10×10^3 radians/sec	10 radians/sec
ω_2	10 radians/sec	0.01 radian/sec
K	10^7 sec^{-1}	10^4 sec^{-1}
β_n	10×10^3 cps	10 cycles/sec

Note that the open-loop gain must be scaled in order to maintain the same relative stability in the model as that in the original loop. If the above constants are substituted into (23) and (24), it will be seen that both loops have a damping factor of 0.5 and that the undamped natural frequency of the model is reduced by the scale factor 10^{-3}.

The analog computer simulation was programmed by straightforward application of the equations given in Appendix II. The signal and VCO frequencies were chosen to be 2750 and 1250 cps, respectively, for an intermediate frequency of 1500 cps. The IF filter was composed of two identical stages of simulated single-tuned filters. The Q of the filters was adjusted to give the desired noise bandwidth, β_{IF}, obtained from the following expression [3]:

$$\beta_{IF} = 1.22\beta_{3\,db}.$$

$\beta_{3\,db}$ is the 3-db bandwidth of the combined two-stage filter, and has the same units as β_{IF}. When limiting was introduced into the loop, it was added after the IF amplifiers and before the phase detector. The limit level was set to limit 20 db into the signal, *i.e.*, the IF sine waves were clipped at 1/10 of the peak value. The limiter output was not filtered before application to the phase detector. This is permissable, since the loop itself filters out the high frequencies generated in the limiting process.

β_{IF} for the loop with no limiting was $2\pi(100)$ radians/sec, and the loop bandwidth, β_n, was $2\pi(10)$ radians/sec. ω_n and ω_1 are thus uniquely determined by (24) and (25) for any specified damping factor. In the investigation of the loop with different damping factors, the loop gain K

TABLE I
TABLE OF CALCULATED VALUES

Parameter	$(S/N)_{IF}$	α	α/α_0	ω_n^2	ζ	β_n	$(S/N)_{out}$	σ_c	$R/\omega_{n_0}^2$	$R/\omega_{n_0}^2$	σ_m
In Terms of	db	—	—	$(rad/sec)^2$	—	rad/sec	db	rad	$\frac{cps/sec}{(rad/sec)^2}$	$\frac{cps/sec}{(rad/sec)^2}$	rad
From		Fig. 10	—	Equations (23), (25), (26)	Equations (24), (25), (27)	Equations (25), (28)	Equation (20)	Equation (22)	Calculated Equation (1)	Measured Fig. 21	Measured Fig. 5
	∞	1	3.33	83.3	0.912	21.6π	∞	0	0.748	0.781	0
	+3	0.84	2.80	70.0	0.84	19π	17.2	0.098	0.541		
	−3	0.55	1.83	45.8	0.675	14.2π	12.4	0.168	0.299	0.322	0.201
	−5	0.46	1.53	38.3	0.62	12.7π	10.9	0.206	0.226	0.237	0.245
	−7	0.38	1.27	31.7	0.565	11.3π	9.4	0.237	0.169	0.163	0.306
	−9	0.30	1	25	0.50	10π	8.0	0.283	0.116		0.365
	−11	0.24	0.8	20	0.447	9π	6.4	0.344	0.077	0.075	0.420

Table illustrating the method for predicting maximum sweep rate for SNR'S other than the design point SNR. Loop designed to acquire with a $(S/N)_{IF} = -9$ db and a $\beta_{IF}/\beta_n = 250/5$.

was kept constant at 10^4 in determining ω_2 by (23). An example is as follows (for a complete listing see Appendix III):

$\zeta = 1/2$	$\zeta = 1/\sqrt{2}$
$\omega_n = \omega_1 = 10$ radians/sec	$\omega_n = 9.43$ radians/sec
$K = 10^4 \ \sec^{-1}$	$\omega_1 = 6.67$ radians/sec
$\omega_2 = 10^{-2}$ radians/sec	$K = 10^4 \ \sec^{-1}$
	$\omega_2 = 8.9 \times 10^{-3}$ radians

While investigating the effects of limiters in the IF, it was necessary to cover a large range of S/N_{IF} and β_{IF}/β_n ratios. In so doing, both the IF and loop bandwidths in the model were changed. The β_{IF}/β_n ratios used were 100/10, 50/5, 250/5, 20/10, where the bandwidth units are in cps. These numbers are designated on the data as a reference for comparison. It is the ratio, not the absolute value of bandwidth, that is important. This is true, because a change in β_n is merely a change in scale factor, provided all loop parameters are scaled accordingly. This scale factor is removed in the normalized curves, as can be seen by comparing Figs. 19 and 20.

Matching Procedure

When calculating the required loop gain for a loop with IF limiting, the limiter signal suppression must be taken into account. This effect is described in Section IV, and α, the ratio of limiter gain to limiter gain with no noise, is plotted in Fig. 10. In order to investigate the loop operation for SNR's both above and below the design SNR, the model loop was always designed for minimum $\beta_n(\zeta = \frac{1}{2})$ at a S/N_{out} of 8 db. For an example of the matching procedure, refer to Table I. The problem was to have the loop provide a 17-db improvement in S/N for $\beta_{IF} = 250$ cps and a $S/N_{IF} = -9$ db. The loop was therefore designed for $\beta_n = 5$ cps and $\zeta_0 = 0.5$, where the subscript "0" denotes a design parameter at the design SNR. The loop time constants then were calculated, as was done for the nonlimiting loop. The gain of the loop dc amplifier was increased by the reciprocal of the limiter suppression factor for $(S/N)_{IF} = -9$ db, to provide the desired $K_0 = 10^4$ at $(S/N)_{IF} = -9$ db. In this case, $\alpha_0 = 0.3$ (see Fig. 10), so that the gain of the loop with no noise would be equal to 3.33 K_0. Once this part of the design was completed, (26)–(28) and Fig. 10 could be used to predict the changes in loop parameters for other $(S/N)_{IF}$ ratios. In addition, using (20) and (21) and Fig. 24, the loop phase jitter and sweep rate for 90 per cent acquisition also could be predicted. The results of these predictions also are shown in Table I.

VI. RESULTS

General

In the experimental evaluation of the phase-lock loop, the frequency of the local oscillator was swept through the pull-in range one hundred times and the number of times that the loop locked up then became the probability of acquisition. The sweep was not discontinued once acquisition occurred, and when the loop maintained lock for a specified amount of time, this was counted as acquisition. (A few strip-chart records of acquisition records are shown in Figs. 11–16.)

The phase-noise measurements are described in Appendix IV. The σ_0^2 measurements for each SNR were made five separate times and then averaged. The duration of each measurement was equal to five time constants

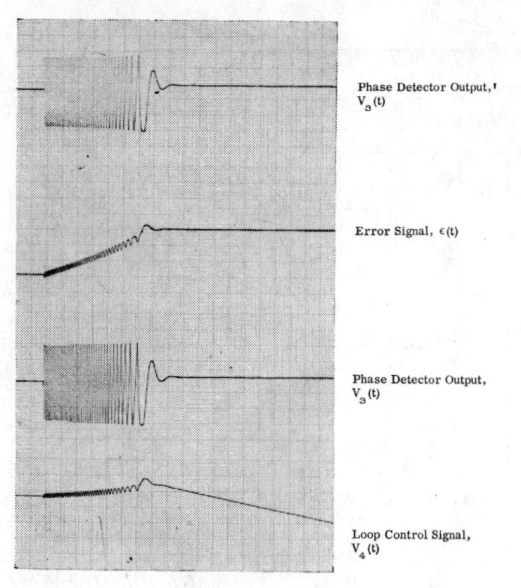

Fig. 11—Acquisition record-loop acquires, no noise, $R/\beta_{n_0}{}^2 = 0.0375$ cps/sec/(cps)^2.

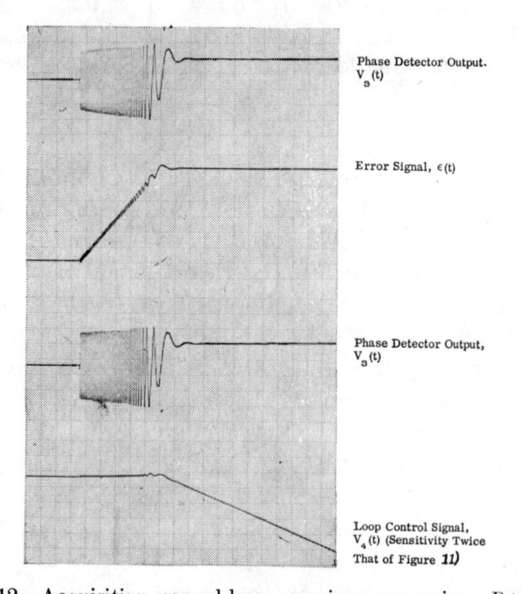

Fig. 12—Acquisition record-loop acquires, no noise, $R/\beta_{n_0}{}^2 = 0.15$ cps/sec/(cps)^2.

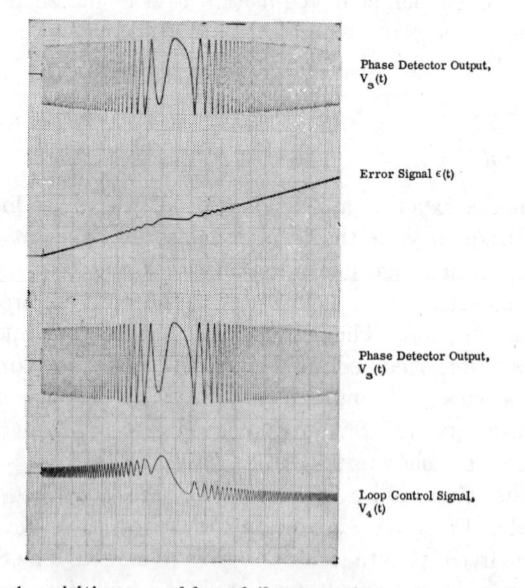

Fig. 13—Acquisition record-loop fails to acquire, no noise, $R/\beta_{n_0}{}^2 = 0.195$ cps/sec/(cps)^2.

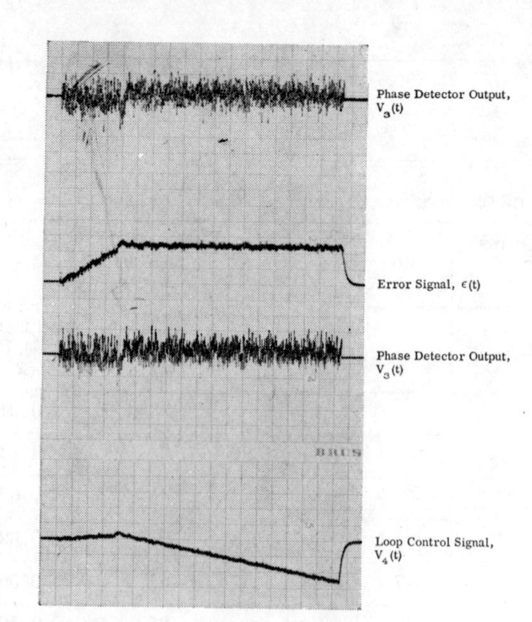

Fig. 14—Acquisition record-loop acquires, $(S/N)_{out} = 4$ db $R/\beta_{n_0}{}^2 = 0.075$ cps/sec/(cps)^2.

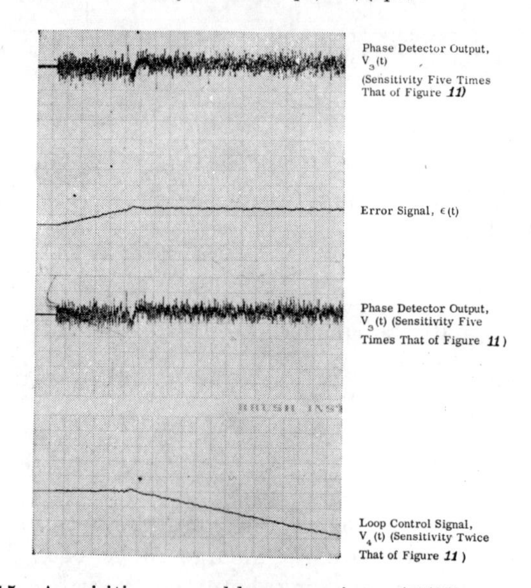

Fig. 15—Acquisition record-loop acquires, $(S/N)_{out} = 7$ db, $R/\beta_{n_0}{}^2 = 0.075$ cps/sec/(cps)^2.

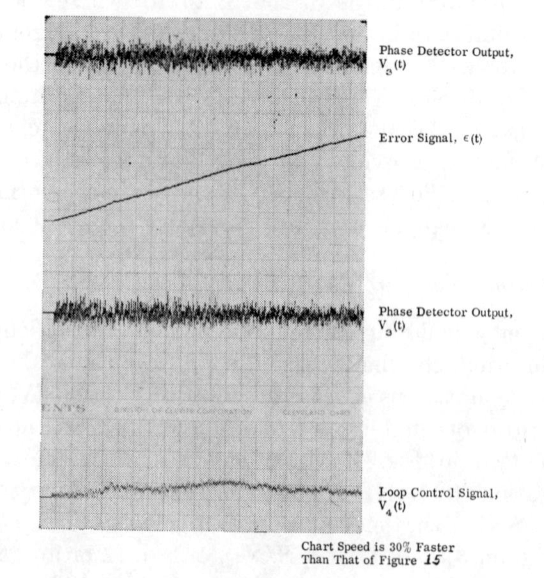

Fig. 16—Acquisition record-loop fails to acquire, $(S/N)_{out} = 7$ db, $R/\beta_{n_0}{}^2 = 0.0975$ cps/sec/(cps)^2.

of the integrating filter, and it was verified that longer measurements would not change the time average significantly.

The experimental results are plotted in the curves showing the probability of acquisition vs normalized sweep rate for a number of different types of loops, with the SNR as a parameter. The noise was introduced into the IF, and the SNR at the intermediate frequency was measured at the output of the IF amplifier section. The types of loops considered were a linear loop with various damping factors, and loops with limiting in the IF, some of which were "matched" to improve acquisition performance.

The experimental results were repeatable with good accuracy. Many of the early tests were repeated as a means of establishing a high confidence level in the validity of the tests. Any tests, the results of which departed radically from those expected (as, for example, $\zeta = \frac{1}{3}$ and the radical drop in P_a for the no-noise case in Fig. 15), were repeated to verify the results.

Measurement of the IF Signal-to-Noise Ratio, $(S/N)_{IF}$

The measurements of the SNR at the output of the IF were made as follows:

1) The rms noise level at the IF output was measured when the loop was unlocked, to give $\sqrt{N}$.
2) The rms signal level was measured when the loop was locked, to give $\sqrt{S}$.
3) The rms value of signal-plus-noise was measured when the loop was locked, to verify the relationship $\sqrt{N+S} = \sqrt{P_T}$. Throughout each test, the signal-plus-noise voltage out of the IF was monitored continuously.
4) The signal-to-noise ratio is then S/N.

All $(S/N)_{IF}$ measurements were made with a Ballantine Model 320 true rms meter.

Linear System (No Limiting, $\zeta_0 = \frac{1}{2}$), Fig. 2

Fig. 2 presents the data taken on a linear loop with minimum noise bandwidth ($\zeta_0 = \frac{1}{2}$). This is the basic system under consideration, and the SNR at the output ranged down to 6 db, the minimum considered. For the linear case the noise bandwidth, the damping factor, and the loop bandwidth were constant throughout the tests.

Limiting in the IF (Match Point, No Noise), Fig. 17

In this particular loop, the IF signal was limited with the limit level set 20 db into the signal. The loop was designed to have a damping of $\frac{1}{2}$ ($\zeta_0 = \frac{1}{2}$) with no noise in the IF, *i.e.*, it was "matched" at the no-noise point. As Fig. 17 shows, the loop gain is degraded, and the acquisition properties of this loop are considerably poorer than those of the linear loop as the SNR in the IF (or at the output, for that matter) decreases.

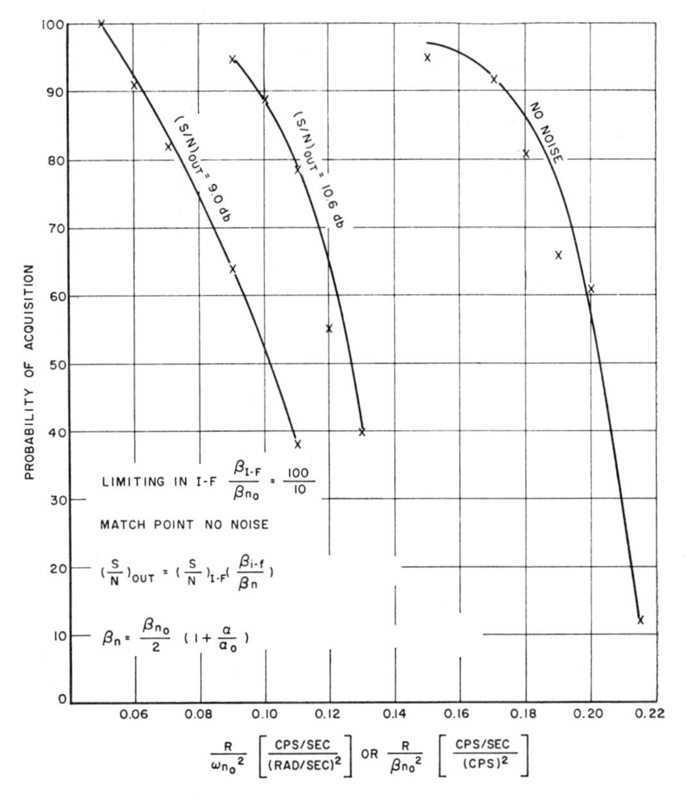

Fig. 17—Probability or acquisition vs normalized sweep rate ($\zeta = \frac{1}{2}$), limiting in the IF.

Linear System (Various Damping Factors), Figs. 3 and 18

In taking the data to compare loop performance for various damping factors, it was decided to keep the noise bandwidth of the loop constant, while changing the loop damping. This required changing ω_n, since (25) shows that ω_n is determined by the noise bandwidth and damping factor. Figs. 3 and 18 are taken from the same data and differ only in the manner in which the abscissa variable is normalized. The sweep rate has been normalized with respect to β_{n0}^2 in Fig. 3 and ω_{n0}^2 in Fig. 18. Figure 3 gives a more meaningful presentation of the results since, if S/N_{IF}, β_{IF}, and S/N_{out} are specified, then β_n is fixed. When it is recognized that ω_{n0} is smaller for damping factors other than $\frac{1}{2}$, it can be seen that, while the two curves appear to be different, they actually present the same data. This question of normalization does not arise in the other graphs, since $\zeta_0 = \frac{1}{2}$ in all cases. Under these conditions, (31) shows that β_n expressed in cycles per second has the same numerical values as ω_n expressed in radians per second.

One conclusion that can be drawn from either figure, is that a damping factor of $\frac{1}{3}$ provides very poor acquisition performance and that this degradation is a function of the relative stability of the loop.

A rough glance at Fig. 3 indicates that with no noise in the system, the higher the damping factor, the higher the allowable sweep rate for 90 per cent acquisition. However, for the higher sweep rates, the acquisition performance degrades more rapidly for the higher damping factors. It should be noted that, even with no noise, the

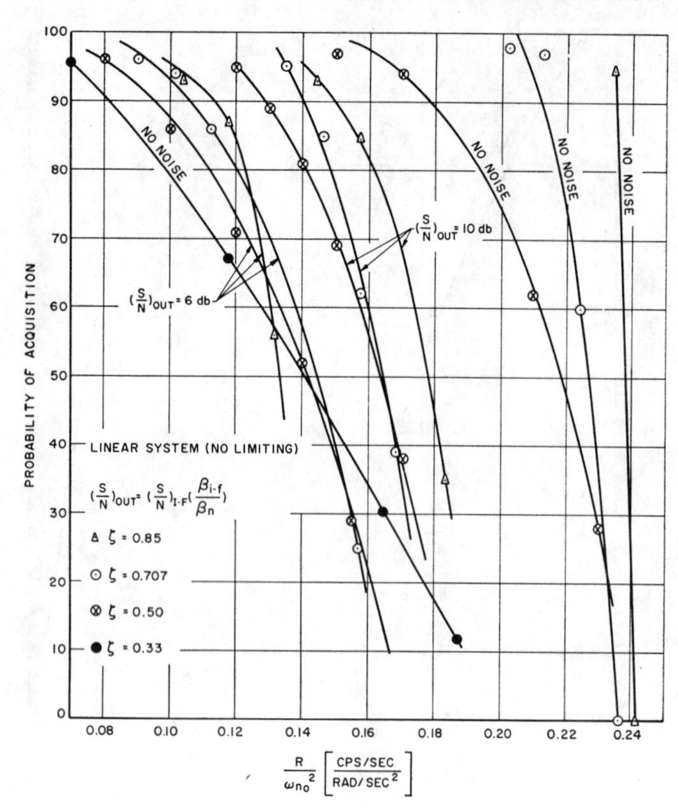

Fig. 18—Probability of acquisition vs normalized sweep rate, $R/\omega_{n_0}^2$, for various damping factors.

damping factor of $\frac{1}{2}$ has a higher sweep rate for a probability of 70 per cent. Although the higher damping factors give a somewhat better performance with no noise, the loop performance degrades more rapidly with noise.

It would seem that if the bandwidth requirements are specified, a minimum noise bandwidth ($\zeta = \frac{1}{2}$) gives better performance.

System Comparison, Fig. 4

This graph compares the performance of three different systems. The first is the linear system with a minimum noise bandwidth ($\zeta_0 = \frac{1}{2}$). The acquisition performance is plotted for this loop for the no-noise case and for a $(S/N)_{out}$ of 10 db. The second is a system with a limiting IF which has the proper gain setting, with no noise in the system ($\zeta_0 = \frac{1}{2}$ with no noise in the system). This loop has considerably poorer characteristics in the presence of noise, due to the signal suppression in the limiter. The third system here is a system with a limiting IF which is "matched" to have a $\zeta_0 = \frac{1}{2}$ at $(S/N)_{out} =$ 8 db. This system not only performs better for no noise but shows a marked improvement for noise in the system. This illustrates rather markedly the advantage of limiting and "matching" to enhance over-all system performance.

Limiting in the IF ($\beta_{IF}/\beta_{n_0} = 100/10$). Fig. 19

These graphs are for a system with a bandwidth ratio of 10, which is matched at $(S/N)_{out} = 8$ db. This particular loop performs about the same as a linear system for a $(S/N)_{out} = 8$ db, and as the SNR increases, the acquisition performance improves.

This is illustrative of what limiting can do to enhance over-all system performance. The loop is designed to "match" at some low SNR, and for any higher SNR the acquisition capability increases.

Limiting in the IF ($\beta_{IF}/\beta_{n_0} = 50/5$), Fig. 20

This graph shows the loop performance for a bandwidth ratio of 10 (as in Fig. 19) but with a loop equivalent noise bandwidth of 5 cps, as compared to 10 cps in the previous figure. For all practical purposes, the performance is the same.

Limiting in the IF ($\beta_{IF}/\beta_{n_0} = 250/5$) Fig. 21

In this case the bandwidth ratio was increased so that the input to the limiter had a much lower SNR. This allowed a very low match point ($\alpha_0 = 0.3$) so that, as the S/N increased, the maximum sweep rate with no noise increased tremendously. This indicates that when a limiting loop is "matched," or designed to give a specified performance at some low SNR, its acquisition performance can only get better for higher SNR's.

Limiting in the IF ($\beta_{IF}/\beta_{n_0} = 20/10$) Fig. 22

It should be noted here that the basic system with no noise has poorer acquisition properties than the linear system (Fig. 2). This particular configuration required a "match" at $(S/N)_{IF} = 5$ db ($\alpha_0 = 0.9$), and there was no appreciable enhancement of system performance at higher SNR. The addition of the time constants contributed by the narrow IF bandwidths causes the loop to be basically more unstable, and, consequently, this affected the over-all acquisition performance.

Phase-Jitter Measurements, Fig. 5

The measured phase jitter for the various loops tested are plotted here for comparison with the calculated values of phase jitter [see (18)]. The measured jitter was within 20 per cent of the calculated value for all cases, and for the higher SNR's the measurements were within 15 per cent. As seen from the average line through the measured points, the errors between measured and calculated values begin to diverge for the lower SNR's.

Normalized Maximum Sweep Rate vs $(S/N)_{out}$, a System Comparison, Fig. 6

The data presented here is a summary of the normalized sweep rates for 90 per cent acquisition, as a function of $(S/N)_{out}$. A graphical comparison is made of four different loops. The linear loop with constant signal strength, δ, ω_n, and β_n is compared with three loops using a limiter as a gain control; β_{IF}/β_{n_0} in these three loops is 50, 10, and 2. The loops were "matched" for a $(S/N)_{out} =$ 8 db. The match points were therefore $(S/N)_{IF} = -9$, -2, and 5 db, respectively. Changes in $(S/N)_{IF}$ were made in 2-db increments, and the change in β_n with α was taken into account when calculating $(S/N)_{out}$.

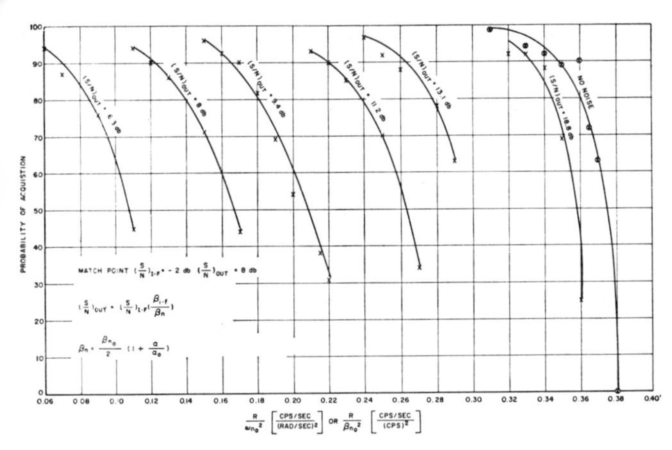

Fig. 19—Probability of acquisition vs normalized sweep rate, ($\zeta_0 = \frac{1}{2}$), limiting in the IF, $\beta_{IF}/\beta_{n_0} = 100/10$.

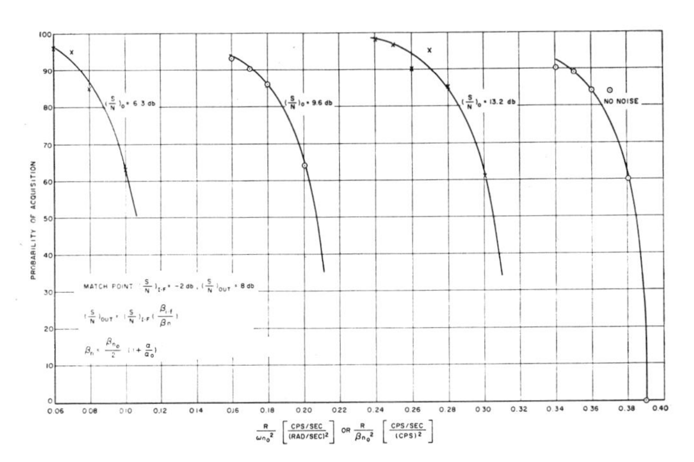

Fig. 20—Probability of acquisition vs normalized sweep rate ($\zeta_0 = \frac{1}{2}$), limiting in the IF, $\beta_{IF}/\beta_{n_0} = 50/5$.

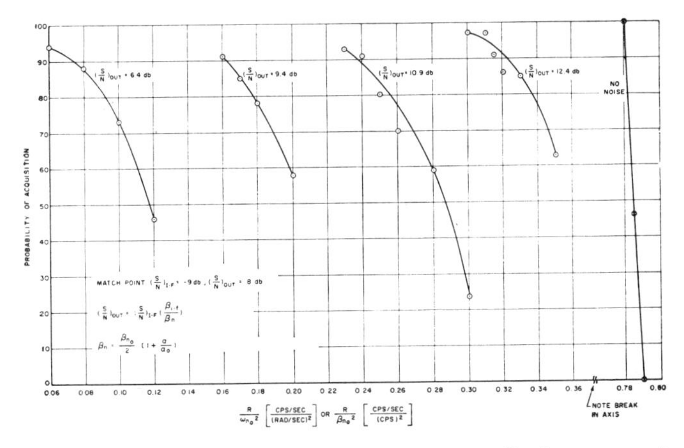

Fig. 21—Probability of acquisition vs normalized sweep rate ($\zeta_0 = \frac{1}{2}$), limiting in the IF, $\beta_{IF}/\beta_{n_0} = 250/5$.

Starting at $(S/N)_{out} = 8$ db, where all loops are equivalent when normalized with respect to ω_{n_0}, and comparing the curves as the $(S/N)_{IF}$ is varied in 2-db increments, the following can be seen:

1) As $(S/N)_{IF}$ increases there is less improvement in $(S/N)_{out}$ for the limited loop as compared to the linear loop; but as $(S/N)_{IF}$ decreases, the $(S/N)_{out}$ for the limited loop is higher than that for the linear loop.

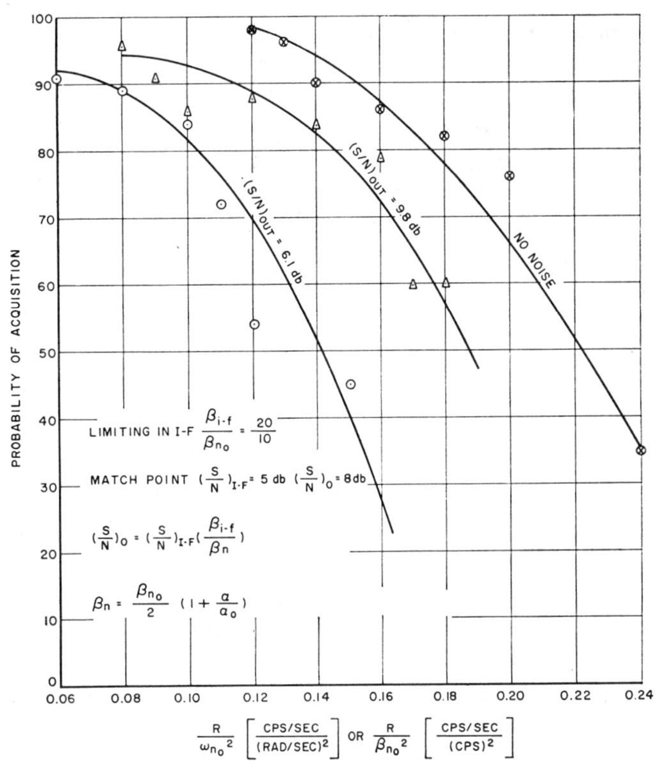

Fig. 22—Probability of acquisition vs normalized sweep rate ($\zeta_0 = \frac{1}{2}$), limiting in the IF, $\beta_{IF}/\beta_{n_0} = 20/10$.

2) As $(S/N)_{IF}$ increases, the acquisition performance of the limited loop is better than that of the linear loop; but as $(S/N)_{IF}$ decreases, the linear loop is superior to the limited loop in acquisition performance.

The effect of less improvement in $(S/N)_{out}$ for the limiting loop, with $(S/N)_{IF}$ higher than the match point, is more than offset by improved loop dynamics. The over-all result is that, with respect to acquisition capability, the limiter loops out-perform the linear loop.

VII. Conclusions

It has been shown that for a specified noise bandwidth, a phase-locked loop having a damping factor of $\frac{1}{2}$ yields optimum acquisition performance. Acquisition is defined here as locking the phase of a frequency-swept VCO onto the phase of an incoming signal that is perturbed by noise. The range of damping factors from 0.5 to 0.85 yields near optimum performance, while decreasing the damping factor below 0.5 causes rapidly deteriorating performance.

The dependence of over-all loop performance on loop gain and, therefore, on signal strength has been of prime interest. The effects of a drop in loop gain are significant in two respects. First, a lower gain results in decreased noise bandwidth. Secondly, the loop damping factor and natural frequency are adversely affected by a low loop gain. The over-all effect is a somewhat lower phase jitter for a given $(S/N)_{IF}$ but a decreased ability of the loop to acquire and track a signal having changing Doppler characteristics. This latter effect is much the more important in a loop of this configuration when the VCO is being swept for the purpose of phase lock.

The indicated need for an effective gain control led to the analytical investigation and computer implementation of a loop having a band-pass limiter. The results of the study show that limiting, in itself, does not degrade the loop acquisition or steady-state performance. On the contrary, it can serve as an effective gain control which allows the system to be designed for optimum operation at the minimum expected IF SNR, and yet does not seriously degrade loop performance for higher SNR's. Other advantages of the limiter as a gain control are as follows:

1) No significant time constant is involved.
2) An additional feedback loop is not introduced.

Eq. (1), derived from the computer results, has been shown to yield an estimate, accurate to within 10 per cent, of the sweep rate for 90 per cent probability of acquisition. For loop damping factors less than 0.5, the accuracy of prediction decreases until, for a $\zeta = \frac{1}{3}$, loop instability results in a degradation of performance much more severe than the simple equation would predict.

Appendix I

List of Symbols

ω_s	Signal angular frequency
ω_i	Reference oscillator angular frequency
ϕ_i	Signal phase angle
ϕ_0	VCO phase angle
ϕ_E	Total loop phase error
S	Signal power in the IF
N	Noise power in the IF
x	In-phase component of IF noise
y	Quadrature component of IF noise
σ	RMS phase jitter in the IF
σ_0	RMS phase jitter in the output signal
K_ϕ	Phase detector gain
K_{VCO}	VCO gain
K	Velocity gain constant
ω_n	Loop undamped natural frequency
ζ	Loop damping factor
δ	Per cent overshoot in phase error for a VCO frequency ramp
α	Signal suppression factor in limiter
$\Delta\omega$	Doppler error in radians per second
R	VCO linear sweep rate in cycles per second per second
β_{IF}	Equivalent IF single-sided noise bandwidth
β_n	Equivalent loop double-sided noise bandwidth
$\Phi_n(\omega)$	Noise power density spectrum in IF
$\Phi_{IF}(\omega)$	Phase jitter power density spectrum in the IF
$\Phi_0(\omega)$	Output phase jitter power density spectrum
$H(s)$	Loop filter transfer function
$G(s)$	Open loop transfer function
$F(s)$	Closed loop transfer function

The subscript "0" on loop parameters ($\omega_{n_0}, \beta_{n_0}, \zeta_0, K_0, \alpha_0$) denotes the specific value of that parameter for which the loop was designed to operate at one particular input SNR.

Appendix II

List of Useful Equations

Introduction

The following equations are useful in the analysis of phase-locked loops having a proportional plus integral network as a loop filter. Refer to Appendix I for nomenclature.

Power Relationships

By definition:

$$S/N_{out} = S/N_{IF}(\beta_{IF}/\beta_n). \tag{20}$$

$$\sigma_{out}^2 = \left(\frac{1}{2}\right)\left(\frac{N}{S_{IF}}\right)\left(\frac{\beta_n}{\beta_{IF}}\right). \tag{21}$$

$$\sigma_{out}^2 = \left(\frac{1}{2}\right)\left(\frac{N}{S_{out}}\right). \tag{22}$$

General Loop Parameters

$$\omega_n^2 = K\omega_2. \tag{23}$$

$$\zeta = \frac{1}{2}\sqrt{\frac{K\omega_2}{\omega_1^2}} = \left(\frac{1}{2}\right)\left(\frac{\omega_n}{\omega_1}\right). \tag{24}$$

$$\beta_n = \frac{\pi\omega_n}{2\zeta}(1 + 4\zeta^2) = \pi\left(\omega_1 + \frac{\omega_n^2}{\omega_1}\right). \tag{25}$$

Variation of Loop Parameters with Limiter Suppression Factor

$$\left(\frac{\omega_n}{\omega_{n_0}}\right)^2 = \frac{\alpha}{\alpha_0}. \tag{26}$$

$$\left(\frac{\delta}{\delta_0}\right)^2 = \frac{\alpha}{\alpha_0}. \tag{27}$$

$$\frac{\beta_n}{\beta_{n_0}} = \frac{1 + \frac{\alpha}{\alpha_0}}{2} \quad \text{for} \quad \zeta_0 = 0.5$$
$$= \frac{1 + 2\frac{\alpha}{\alpha_0}}{3} \quad \text{for} \quad \zeta_0 = 0.707. \tag{28}$$

For $\zeta = 0.5$

$$\omega_n = \omega_1. \tag{29}$$

$$\frac{K}{\omega_1} = \frac{\omega_1}{\omega_2}. \tag{30}$$

$$\beta_n = 2\pi\omega_n = 2\pi\omega_1. \tag{31}$$

Appendix III

Closed-Loop Responses in a Phase-Locked Loop

Derivation of Closed-Loop Transfer Function

In studying the acquisition capabilities of a phase-locked loop, one of the most important factors is the phase error in the phase detector as a function of a driving voltage applied to the VCO. The reason is that the loop will lose lock if the phase error exceeds $\pm\pi/2$ radians. For any given steady-state signal, the loop has only one stable operating point every 2π radians in the phase detector. The quiescent phase error can lie anywhere in the range of $\pm\pi/2$ radians. The tendency of the loop to lose lock under transient or noise conditions is directly dependent on the steady-state phase error in the phase detector. The nearer the error is to $\pm\pi/2$ radians, the easier it is to cause an unlock. After the transient dies out, it is possible for the loop to pull back into lock, provided the new steady-state conditions allow a phase error of less than $\pm\pi/2$ radians. Otherwise the phase detector output will continue to cycle and no stable operating point can be obtained.

The basic loop to be considered in this analysis is shown in Fig. 7. The mixer and phase detector both are product devices which produce only the sum and difference frequencies when sine waves are applied to them. $H(s)$ is the loop filter and, in our case, is a proportional plus integral network.

In order to linearize the loop equations, the following assumptions are made:

1) The phase of the reference oscillator signal is arbitrarily chosen as zero, except for the $\omega_i t$ term. This is permissible for only one of the signals.
2) The lean frequency of the VCO is adjusted by V_L so that its frequency is $\omega_s - \omega_i$.
3) Since the mixer does not affect the loop phase relationships, both it and the IF can be eliminated, and any gain associated with them is reflected in the magnitude $\sqrt{2S}$.
4) The following assumption can lead to a smal error in the phase-detector gain for large phase errors, but it must be made if linear equations are to be obtained. The assumption is that the phase-detector gain K_ϕ is proportional to $(\phi_i - \phi_0)$ rather than $\sin(\phi_i - \phi_0)$. The maximum error introduced by the assumption is a factor of $\pi/2$ for phase errors of $\pm\pi/2$ radians. In the experimental test, the steady-state phase error was kept at zero so that the phase detector would be operating in its most linear region. This also allowed maximum excursions of phase error before the loop becomes unlocked.
5) The amplitude of the sine waves was held constant so that the loop would be time invariant.

The loop now can be redrawn as shown in Fig. 23, and the loop transfer functions can be derived using linear differential equations having constant coefficients.

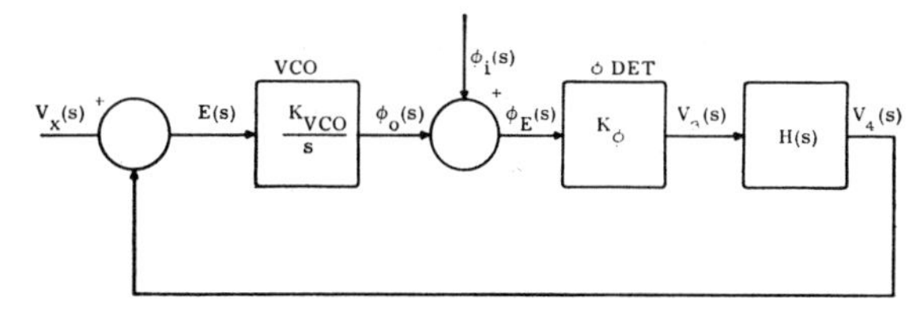

Fig. 23—Linearized block diagram, phase-locked loop.

A one-volt change in $E(t)$ produces a change in VCO frequency of K_{VCO} radians per second. Since we are interested in the phase change, the transfer function of the VCO is written K_{VCO}/s, indicating the integration process relating phase to frequency. A one-radian phase error, ϕ_E, produces an output of K_ϕ volts in the phase detector. The gain of the phase detector is therefore K_ϕ volts/radian. $H(s)$ is a proportional plus integral filter having the transfer function, $(1 + \tau_1 s)/(1 + \tau_2 s)$. A gain, $K_3 = \omega_1/\omega_2$, is added so that $H(s) = (s + \omega_1)/(s + \omega_2)$. The above relationships are summarized as follows:

$$K_1 = K_{VCO}K_\phi, \qquad K_3 = \omega_1/\omega_2, \qquad \phi_E = \phi_i - \phi_0.$$

Since the phase error is not zero for a loop frequency error, steady-state error signals can be present. These can be eliminated at the beginning of a test by nulling V_4 with V_L. Thus, the phase errors caused by the application of a test signal, can be determined by finding the response of ϕ_0 to that particular signal.

The open-loop transfer function now can be written

$$G(s) = \frac{K_1 H(s)}{s} = \frac{K_{VCO}K_\phi(s + \omega_1)}{s(s + \omega_2)} = \frac{K_{VCO}K_\phi K_3(1 + s\tau_1)}{s(1 + s\tau_2)}. \quad (32)$$

The velocity gain constant is $K = \lim_{s\to 0} sG(s)$. Therefore,

$$K = K_{VCO}K_\phi K_3 = K_1 K_3 = K_1(\omega_1/\omega_2).$$

The closed-loop response, $F(s) = V_4/V_x = \phi_0/\phi_i$, can be expressed as

$$F(s) = \frac{G(s)}{1 + G(s)} = \frac{K_1(s + \omega_1)}{s(s + \omega_2) + K_1(s + \omega_1)}$$

$$F(s) = \frac{K_1(s + \omega_1)}{s^2 + (\omega_2 + K_1)s + K_1\omega_1},$$

or using $K = K_1(\omega_1/\omega_2)$,

$$F(s) = \frac{K(\omega_2/\omega_1)(s + \omega_1)}{s^2 + [\omega_2 + K(\omega_2/\omega_1)]s + K\omega_2}. \quad (33)$$

In terms of ζ and ω_n,

$$F(s) = \frac{K(\omega_2/\omega_1)(s + \omega_1)}{s^2 + 2\zeta\omega_n s + \omega_n^2}, \quad (34)$$

where

$$\omega_n^2 = K\omega_2 = K_1\omega_1 \tag{35}$$

$$\zeta = \frac{\omega_2 + K_1}{2\omega_n} = \frac{\omega_2 + K_1}{2\sqrt{K_1\omega_1}}$$

$$\doteq \frac{1}{2}\sqrt{\frac{K_1}{\omega_1}} = \frac{1}{2}\sqrt{\frac{K\omega_2}{\omega_1^2}} = \frac{1}{2}\left(\frac{\omega_n}{\omega_1}\right). \tag{36}$$

Phase-Error Response to a Ramp in Frequency

Of interest here is the phase-error response of ϕ_0 to V_x, since the phase error is limited to a maximum of $\pi/2$ radians if lock is to be maintained.

If a ramp in frequency in the VCO is desired, $V_x(t)$ will be a ramp. Let the desired ramp in frequency be R' rad/sec/sec. This requires a $V_x(t) = (R'/K_{VCO})$ volts/second.
Therefore,

$$V_x(s) = (R'/K_{VCO})(1/s^2). \tag{37}$$

The error response can be written

$$\frac{E(s)}{V_x(s)} = \frac{1}{1 + G(s)}; \tag{38}$$

also,

$$\frac{\phi_0(s)}{E(s)} = \frac{K_{VCO}}{s}. \tag{39}$$

Combining (38) and (39),

$$\frac{\phi_0(s)}{V_x(s)} = \frac{K_{VCO}}{s(1 + G(s))}. \tag{40}$$

Combining (40) and (37) to find the phase error,

$$\phi(s) = \frac{R'}{s^3[1 + G(s)]} = \frac{R'}{s^3}\left[\frac{1}{1 + \frac{K_1(s + \omega_1)}{s(s + \omega_2)}}\right]$$

$$= \frac{R'}{s^3}\left[\frac{s(s + \omega_2)}{s^2 + (\omega_2 + K_1)s + K_1\omega_1}\right]$$

$$\phi_0(s) = \frac{R'(s + \omega_2)}{s^2(s^2 + 2\zeta\omega_n s + \omega_n^2)}, \tag{41}$$

where ω_n and ζ are the same as in (35) and (36).
Finding the inverse transform of (41) and using the fact that $\omega_2 \ll \zeta\omega_n$,

$$\frac{\phi_0(t)}{R'} = \frac{1}{\omega_n^2}(\omega_2 t + 1)$$

$$+ \frac{1}{\omega_n^2\sqrt{1 - \zeta^2}}\epsilon^{-\zeta\omega_n t}\sin(\omega_n\sqrt{1 - \zeta^2}\,t + \Psi) \tag{42}$$

where

$$\Psi = \tan^{-1}\frac{\sqrt{1 - \zeta^2}}{\zeta} \quad \text{and} \quad \pi < \Psi < 3\pi/2. \tag{43}$$

Since the tangent function repeats every π radians, maximum and minimum phase errors will be obtained when $\omega_n\sqrt{1 - \zeta^2}\,t$ increases by π radians. Therefore,

$$t = \frac{n\pi}{\omega_n\sqrt{1 - \zeta^2}} \quad (n = 0, 2, 4, \cdots)$$

are times for minimums, and

$$t = \frac{n\pi}{\omega_n\sqrt{1 - \zeta^2}} \quad (n = 1, 3, 5, \cdots) \tag{44}$$

are times for maximums.

The maximum phase error in the phase detector can be found by substituting (44) into (42), since $\phi_E(t) = -\phi_0(t)$ under the initial conditions assumed in the loop.

$$\phi_E \max = -\frac{R'}{\omega_n^2}\left[\frac{\omega_2\pi}{\omega_n\sqrt{1 - \zeta^2}} + 1 + \epsilon^{-\zeta\pi/\sqrt{1-\zeta^2}}\right]. \tag{45}$$

If the term envolving ω_2 is neglected (it is less than 1 per cent), the overshoot is equal to

$$\epsilon^{-\zeta\pi/\sqrt{1-\zeta^2}}.$$

A graph of this function appears in Fig. 24.

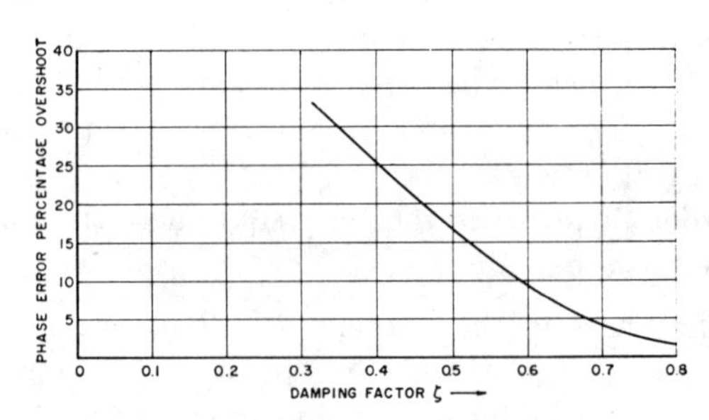

Fig. 24—Phase error percentage overshoot as a function of damping factor for a linearly changing input frequency.

The loop responses to a frequency ramp were taken on the computer and are shown in Figs. 25–28 for different damping factors.

Two other important factors affecting the ability of the loop to lock when the VCO is being swept can be seen from (42).

1) The maximum sweep rate R is directly proportional to ω_n^2.
2) The ability of the loop to acquire and track a sweep is further limited if noise is present in the system. This is true because any phase jitter due to noise will be added to the steady-state phase error due to the ramp, making it possible for a noise spike to cause an unlock.

Loop Response to a Voltage Step in the VCO

The most useful check on the operation of the computer model was to apply a voltage step to the VCO at V_x and monitor the loop response V_4. Eq. (34) was derived

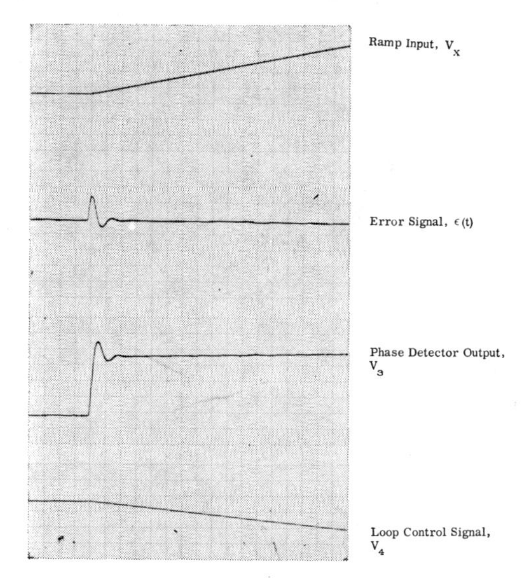

Fig. 25—Ramp response, $\zeta = \frac{1}{3}$.

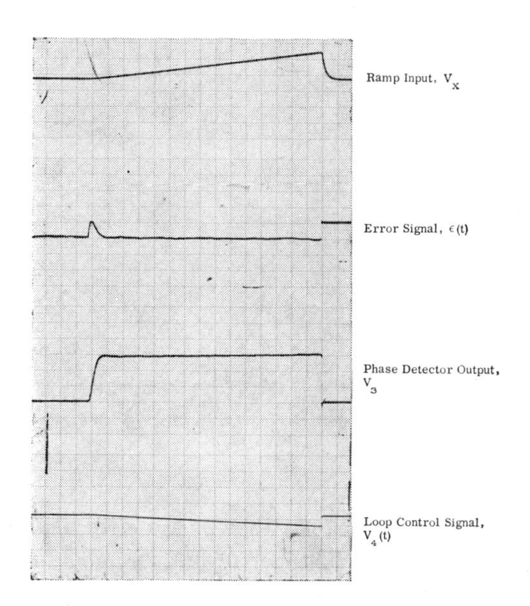

Fig. 26—Ramp response, $\zeta = 1/\sqrt{2}$.

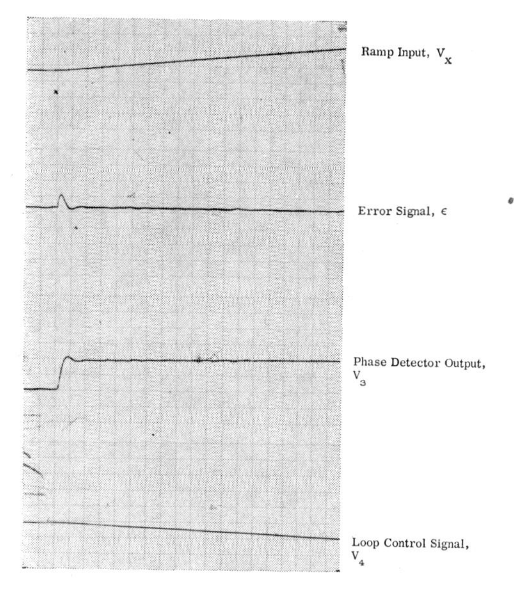

Fig. 27—Ramp response, $\zeta = \frac{1}{2}$.

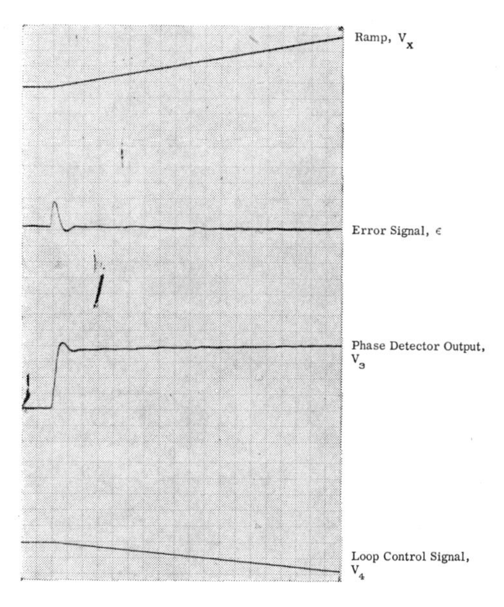

Fig. 28—Ramp response, $\zeta = \frac{1}{2}$, ramp rate = twice that of Fig. 27.

as the closed-loop transfer function relating the output and signal phases. However, an inspection of Fig. 23 will show that the transfer function $F(s)$ is also equal to $V_4(s)/V_x(s)$. If V_x is a voltage step of amplitude A, $V_4(t)$ can be found by taking the inverse transform of

$$V_4(s) = \left(\frac{A}{S}\right)\left(\frac{\omega_2}{\omega_1}\right)\frac{K(s+\omega_1)}{s^2 + 2\zeta\omega_n s + \omega_n^2}. \tag{46}$$

The resulting time function is

$$V_4(t) = A\left[1 - \frac{\epsilon^{-\zeta\omega_n t}}{\sqrt{1-\zeta^2}}\sin\left(\omega_n\sqrt{1-\zeta^2}\,t + \Psi\right)\right], \tag{47}$$

where

$$\Psi = -\tan^{-1}\frac{\sqrt{1-\zeta^2}}{\zeta} \quad \text{and} \quad \frac{\pi}{2} < \Psi < \pi.$$

The speed of response and the overshoot are therefore related to loop gain through (35) and (36).

Eq. (47) was solved for several damping factors, and the results are tabulated as follows:

Parameter	Damping Factor, ζ			
	1/2	$1/\sqrt{2}$	1/3	0.805
ω_n (radians)	10	9.43	9.25	8.75
ω_1 (radians)	10	6.67	13.9	5.15
ω_2 (radians)	10^{-2}	8.9×10^{-3}	8.55×10^{-3}	7.65×10^{-3}
K sec^{-1}	10^4	10^4	10^4	10^4
Step Overshoot (per cent)	30	21	42	16.6
Time to First Minimum	0.6 sec	0.706	0.643	—

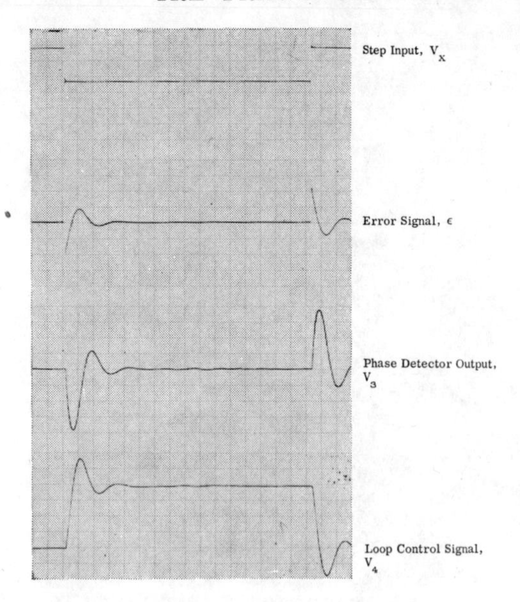

Fig. 29—Step response, $\zeta = \frac{1}{3}$.

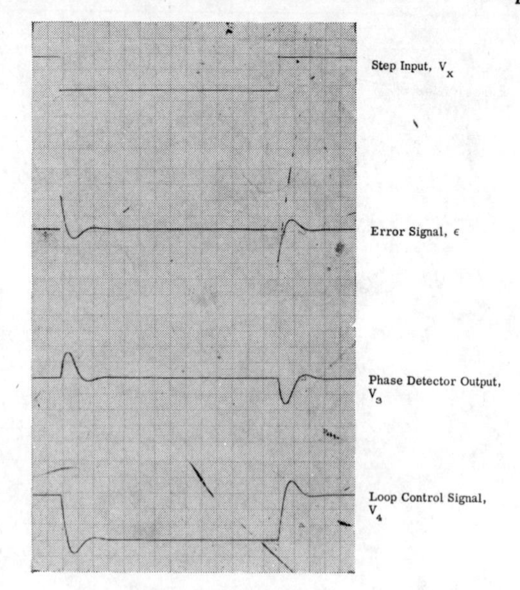

Fig. 31—Step response, $\zeta = \frac{1}{2}$.

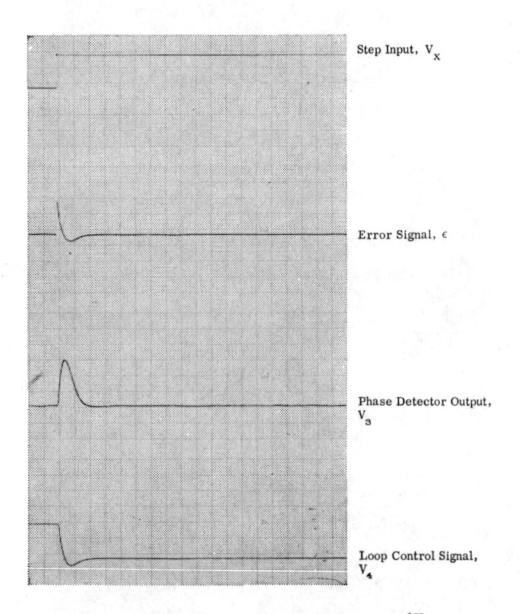

Fig. 30—Step response, $\zeta = 1/\sqrt{2}$.

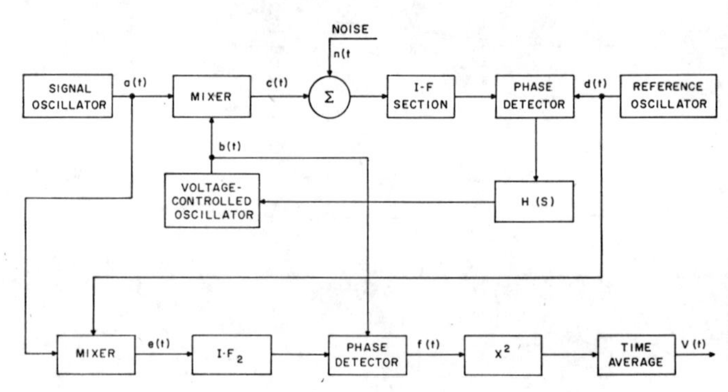

Fig. 32—Basic phase-lock loop with phase-jitter measurement network.

Since the per cent of overshoot is dependent on ζ only, an error in loop gain is reflected through the damping factor as a change in overshoot. The loop step response for various damping factors is shown in Figs. 29–31. These figures are recordings taken from the computer model, and they include signals monitored at other points in the loop.

Appendix IV

Phase-Jitter Measurements

In the evaluation of the phase-lock loop, it proved helpful to actually measure the rms phase jitter of the locked oscillator signal. This phase jitter represents errors, and it is generally the limiting factor in loop performance. Under certain conditions, this phase error can be calculated so that these phase-error measurements can serve as a check on system performance. In the tests on the loop with limiting in the IF, these measurements contributed to a greater understanding of loop performance.

The basic loop and the phase-measuring circuit are shown in Fig. 32. The lower portion of the diagram indicates the circuitry used in measuring the phase errors.

The mixer in the lower loop derives the difference frequency component from the input and reference signals. In the locked condition, this difference is exactly the locked-oscillator frequency. IF_2 selects this difference frequency, which is equal to the locked oscillator frequency, and feeds it to the auxiliary phase detector. The output of this phase detector is a voltage linearly proportional to the phase difference between the two inputs. This voltage is then squared and averaged. The output of the averaging device is proportional to the mean-square value of the phase variations between the input and the output signal. Initial phase difference can be made equal to zero with no noise in the system.

The loop voltages are as follows:

$$a(t) = A \cos(\omega_s t + \phi_i) \tag{48}$$

$$b(t) = B \cos[(\omega_s - \omega_i)t + \phi_0] \tag{49}$$

$$c(t) = \frac{AB}{2} \cos(\omega_i t + \phi_i - \phi_0) - \frac{AB}{2} \cos[(2\omega_s - \omega_i)t + \phi_i + \phi_0] \tag{50}$$

$$d(t) = L \sin \omega_i t \tag{51}$$

$$e(t) = \frac{AL}{2} \sin[(\omega_s - \omega_i)t + \phi_i] + \frac{AL}{2} \sin[(\omega_s + \omega_i)t + \phi_i]. \tag{52}$$

Since IF_2 removes the high-frequency component in $e(t)$, the output of the phase detector is

$$f(t) = k_\phi \, \Delta\phi, \tag{53}$$

where

$$\Delta\phi = \phi_0 - \phi_i.$$

The voltage, $f(t)$ then is linearly proportional to the phase difference between the input signals and the locked oscillator signal. For the phase detector used in this application, the output had a slope that was linear with phase and equal to $200/\pi$ volts per radian. (See Fig. 33)

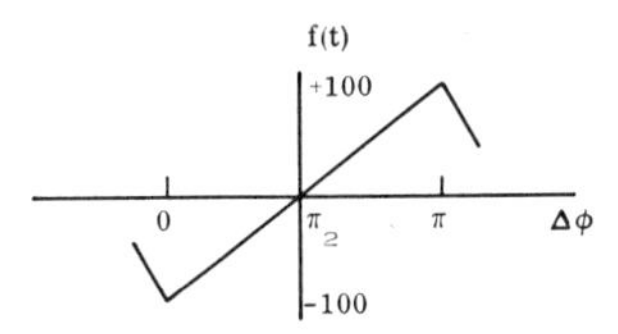

Fig. 33—Characteristic of the phase detector in the phase-jitter measuring network.

In this particular case, the phase variations are generally a function of the SNR at the output of the IF amplifier. Regardless of the type of cause of the phase variations, voltage $f(t)$ is proportional to the phase variations.

The squaring and time-averaging circuit performs the squaring, integrating, and time-averaging parts of the rms definition.

$$V(t) = \lim_{T\to\infty} \frac{1}{T} \int_0^T f^2(t)\, dt. \tag{54}$$

The output of the time-averaging filter is the mean-squared value of a voltage proportional to phase difference. The output of the phase detector can be nulled so that the steady-state $\phi_0 - \phi_i$ is very nearly zero. Under this condition, the voltage variations out of the phase detector are caused by phase jitter due to noise. Therefore,

$$f(t) = k_\phi \phi_0, \tag{55}$$

$$f(t)^2 = k_\phi^2 \phi_0^2, \tag{56}$$

and out of the time-averaging circuit,

$$\overline{f(t)^2} = \overline{k_\phi^2 \phi_0^2} = k_\phi^2 \overline{\phi_0^2} \tag{57}$$

Then for any one measurement from the time-averaging circuit, the mean-squared phase variation is

$$\overline{\phi_0^2} = \overline{f^2(t)}/k_\phi^2, \tag{58}$$

and the rms phase jitter is

$$\sigma_0 = \sqrt{\overline{\phi_0^2}} = \sqrt{\overline{f(t)^2}/k_\phi^2}. \tag{59}$$

In this particular application, the integrating time constant in the averaging filter was 100 seconds, and the noise bandwidth of the system was from 0 to 10 cps. The measurements of $\sqrt{\overline{\phi_0^2}}$ were within 15 per cent of the calculated value for high SNR's. See (18).

References

[1] R. Jaffe and R. Richtin, "Design and performance of phase-lock circuits capable of near-optimum performance over a wide range of input signal and noise levels," IRE Trans. on Information Theory, vol. IT-1, pp. 66–76; March, 1955.

[2] W. B. Davenport, Jr., "Signal-to-noise ratios in band-pass limiters," *J. Appl. Phys.*, vol. 24, pp. 720–727; June, 1953.

[3] E. M. Robinson, "Acquisition Capabilities of Phase-Locked Oscillators in the Presence of Noise," GE Co., Syracuse, N. Y., TIS No. R60DSD11; September, 1960.

[4] R. Leck, "Phase-lock AFC loop," *Electronic and Radio Engr.*; April and May, 1957.

[5] A. J. Viterbi, "Acquisition and Tracking Behavior of Phase-Locked Loops," Jet Propulsion Lab., Pasadena, Calif., External Publication No. 673; July 14, 1959.

[6] H. T. McAleer, "A new look at the phase-locked oscillators," Proc. IRE; June, 1959.

Phase-Locked Loop Pull-In Frequency

LARRY J. GREENSTEIN, MEMBER, IEEE

Abstract—A computerized procedure for obtaining the pull-in frequency of a phase-locked loop is described. The procedure, which consists of solving for the limit cycle of the out-of-lock loop and finding the frequency offset below which no solution exists, is quite general with respect to phase detector function and loop filter. It is applied in this study to the case of a second-order loop with a sinusoidal phase detector.

The results of the study give normalized pull-in frequency X as a function of the damping factor ζ and normalized natural frequency β of the linearized closed loop. A review of several previously published investigations of this case is given, and the limitations identified point the way to the more exact solution described here. In addition, an experimental study is described which produces excellent agreement with the new results, and a curve-fitting procedure is outlined which leads to a highly accurate functional description of X in terms of β and ζ.

I. INTRODUCTION

A STUDY of pull-in frequency in second-order phase-locked loops (PLL's) with sinusoidal phase detectors is described. This subject has already been treated theoretically by various authors [1]–[8], with varying results. The approach used here is based on a computerized solution of the periodic frequency error when the loop is out of lock. The pull-in frequency (ω_p) is solved for as the frequency offset $\Delta\omega$ below which no periodic solution exists.[1]

The computer solutions are made efficient by the use of fast Fourier transform (FFT) algorithms and can be obtained with arbitrarily high accuracy. What is more, the method itself is completely general and can be used for any phase detector function (e.g., sawtooth, tanlock, rectangular) and loop filter transfer function. This is in contrast to previously reported solutions which are restricted in terms of either filter type, phase detector function, or both.

Despite the generality of the method of solution, the focus of this paper is on second-order loops (i.e., those containing lag/lead or lag-only loop filters) with sinusoidal phase detector functions. Fig. 1(a) shows a block diagram of such a PLL, and Fig. 1(b) shows the equivalent linearized version for the case where $\phi(t) \ll 1$. The effective transfer function between $\theta_i(t)$ and $\theta_0(t)$ in the latter case is

Paper approved by the Associate Editor for Communication Theory of the IEEE Communications Society for publication without oral presentation. Manuscript received December 10, 1973; revised February 11, 1973.

The author is with Bell Laboratories, Holmdel, N. J. 07733.

[1] The offset $\Delta\omega$ is the excess of the input frequency above the voltage-controlled oscillator (VCO) rest frequency and, because of symmetry, we assume $\Delta\omega > 0$ throughout with no loss in generality.

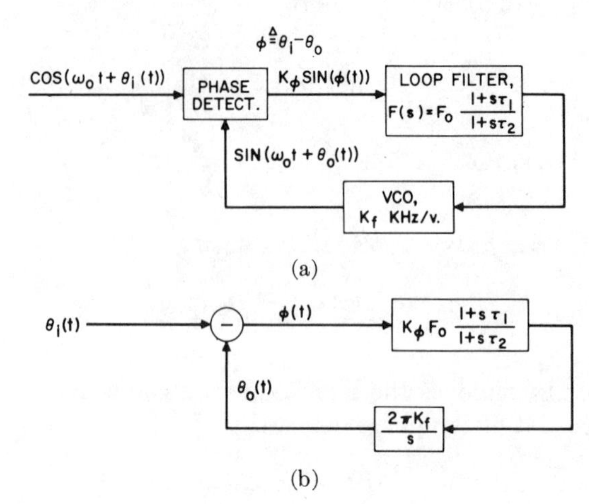

Fig. 1. Second-order PLL. (a) Block diagram. (b) Equivalent linearized version.

$$\frac{\Theta_0(s)}{\Theta_i(s)} \equiv H(s) = \frac{K(1+s\tau_1)}{s^2\tau_2 + (1+K\tau_1)s + K} \tag{1}$$

where

$$K \triangleq 2\pi K_f K_\phi F_0 \tag{2}$$

F_0 being the dc gain of the loop filter. Following the convention of many authors, we define the loop damping factor ζ, natural frequency ω_n, and normalized natural frequency β as follows:

$$\zeta \triangleq \frac{1+K\tau_1}{2(K\tau_2)^{1/2}} \tag{3}$$

$$\omega_n \triangleq (K/\tau_2)^{1/2} \tag{4}$$

$$\beta \triangleq \omega_n/K = 1/(K\tau_2)^{1/2}. \tag{5}$$

In terms of the preceding quantities, τ_1 and τ_2 can be written as

$$\tau_1 = \frac{1}{K}\left(\frac{2\zeta}{\beta} - 1\right) \tag{6}$$

and

$$\tau_2 = \frac{1}{K\beta^2} \tag{7}$$

and the linear transfer function defined by (1) can be written as

$$H(s) = \frac{\beta^2 + (2\zeta\beta - \beta^2)(s/K)}{(s/K)^2 + 2\zeta\beta(s/K) + \beta^2}. \tag{8}$$

In the absence of noises (input and VCO) and nonlineari-

Reprinted from *IEEE Trans. Commun.*, vol. COM-22, pp. 1005–1013, Aug. 1974.

ties (amplifier and VCO control circuit), the PLL locking frequency (that $\Delta\omega$ below which the locked loop remains locked) is K r/s. In the absence of the preceding perturbations and extraneous circuit time constants, the PLL pull-in frequency (that $\Delta\omega$ below which pull-in always occurs) is a function of just K, τ_1, and τ_2, expressible in the form

$$\omega_p = K \text{ fcn } (\beta,\zeta). \tag{9}$$

Defining the normalized pull-in frequency to be

$$X \triangleq \omega_p/K = \text{fcn } (\beta,\zeta) \tag{10}$$

our objective reduces to 1) finding numerical solutions for X in terms of β and ζ, and 2) finding a functional description that fits these solutions accurately for all β and ζ of interest. The values of ζ of interest lie in the range $0 < \zeta \leq 1.0$, higher values corresponding to an overdamped loop; the values of β of interest lie in the range $0 < \beta \leq 2\zeta$, higher values corresponding to negative values of the lead time constant (6).

Section II presents some previously published theoretical results for $X(\beta,\zeta)$ and attempts to explain the differences among them. It is found that the solutions that are most accurate for $\beta \ll 2\zeta$ (corresponding to so-called high-gain loops) are the least useful in the vicinity of $\beta = 2\zeta$, which represents the special case of a lag-only loop filter ($\tau_1 = 0$).

Section III describes a computerized procedure that can be used with any loop filter and phase detector function to find pull-in frequency. The application of this procedure to the particular case at hand is discussed in some detail, and a family of curves (Fig. 10) is used to summarize the results.

Section IV describes a study in which the new theoretical results are compared with laboratory measurements of pull-in frequency. The agreement (Fig. 12) is found to be excellent.

Section V discusses a search for a two-variable function $\tilde{X}(\beta,\zeta)$ to describe the computed theoretical results. The function obtained, (47), is found to be accurate over all ($\beta \leq 2\zeta$, $\zeta \leq 1.0$) to within ± 2 percent.

II. REVIEW

Fig. 2 shows a generalized phase-locked loop with a fixed frequency offset $\Delta\omega$. Under the conditions stipulated for $g(\phi)$ and $F(s)$, K is the locking frequency. In the out-of-lock steady state ($\Delta\omega > \omega_p$), the variations $a(t)$ and $b(t)$ are periodic, while $c(t)$ is a combination of a periodic variation and (if $b(t)$ has a dc component) a ramp function.

For the particular case where $g(\phi) = \sin\phi$ and $F(s) = (1 + s\tau_1)/(1 + s\tau_2)$, ω_p has been derived by several authors. The collective work of Richman [2], Rey [3], Moschytz [4], Viterbi [5], Lindsey [6], and Mengali [7] provides a representative sampling, and their methods and results (expressed in the notation of Section I) are summarized here. Fig. 3 shows X versus β for each of these authors for $\zeta = 0.1$; Figs. 4 and 5 provide the same comparisons for $\zeta = 0.5$ and 0.9, respectively; and Fig. 6

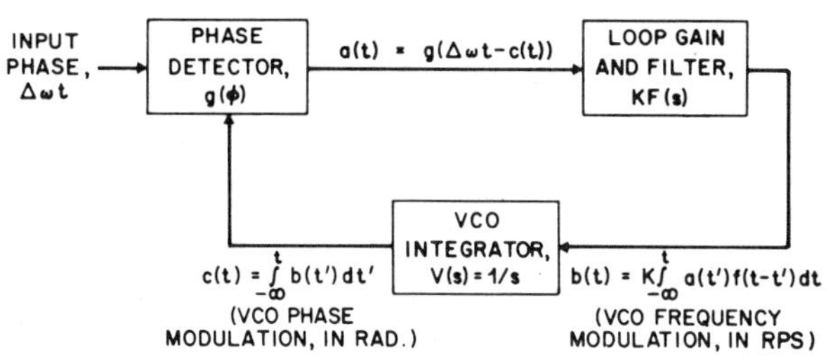

$g(\phi)$ has period 2π, odd symmetry, and peak value unity.
$F(0) = 1$.
$f(t) = \mathcal{L}^{-1}\{F(s)\}$.
K = locking frequency in r/s.

Fig. 2. Generalized PLL.

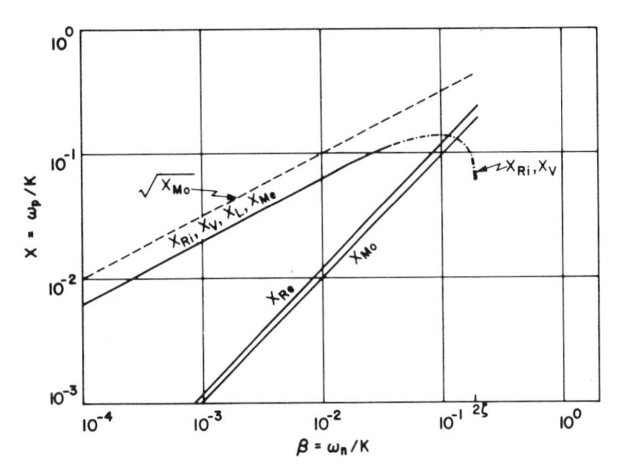

Fig. 3. Published theoretical results for $\zeta = 0.1$.

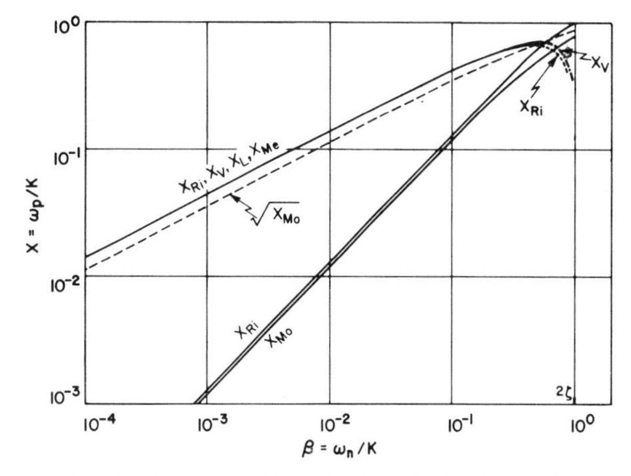

Fig. 4. Published theoretical results for $\zeta = 0.5$.

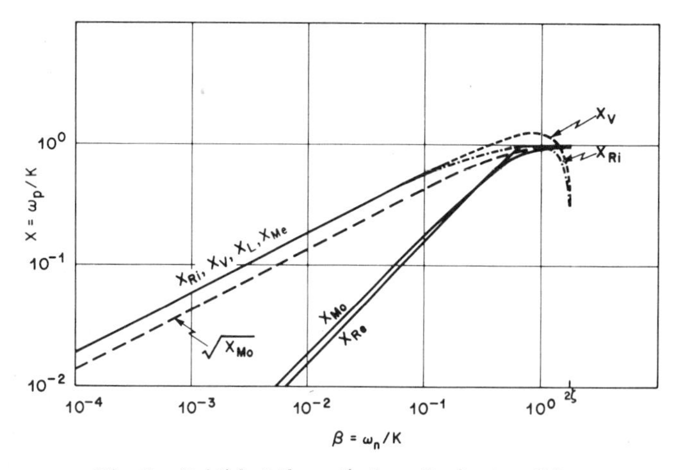

Fig. 5. Published theoretical results for $\zeta = 0.9$.

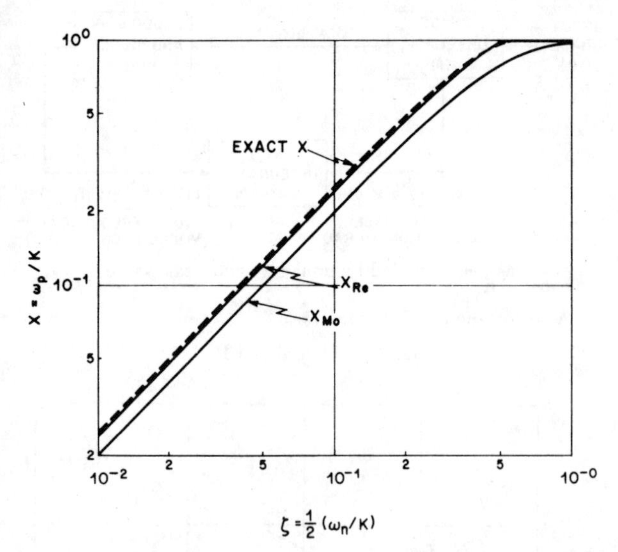

Fig. 6. Theoretical results for lag-only loop filter ($\beta = 2\zeta$).

compares results from Rey, Moschytz, and an exact solution derived here for the lag-only filter case, $\beta = 2\zeta$.

The method of Rey invokes the periodicity of $b(t)$ in the out-of-lock loop, and assumes that it can be approximated by just the dc term plus first harmonic. A further assumption is that pull-in occurs when the peak value of $b(t)$ just equals the frequency offset $\Delta\omega$. Given the first assumption, dc and first harmonic components of $b(t)$ can be found which are in balance around the loop, i.e., where the sine of the resulting phase difference between input and VCO, when passed through the loop filter, yields the same dc and first harmonic components that $b(t)$ had to start with. Given the second assumption, ω_p can be found as that value of $\Delta\omega$ for which $\max_t \{b(t)\} = \Delta\omega$. Rey's resulting solution can be expressed as follows:

$$X_{\mathrm{Re}} = \min\left\{1,\; 0.096\left[\frac{1+[(1+yD^2)(1+y(D-E)^2)/(1+yE)^2]^{1/2}}{[1.202+2.202y(D-E)^2]^{1/2}}\right\}\right. \qquad (11)$$

where

$$D = (2\zeta - \beta)/\beta \qquad (12)$$

$$E = 1/\beta^2 \qquad (13)$$

and y is solution of the equation

$$y = 0.9312\left(\frac{1+yD^2}{1+yE^2}\right)\frac{1+y(D-E)^2}{1.202+2.202y(D-E)^2}. \qquad (14)$$

The min { , } operation in (11) takes cognizance of the fact that pull-in frequency cannot exceed locking frequency, i.e., $X \leq 1$. The preceding result can be easily evaluated on a digital computer, and this was done to obtain the curves labeled X_{Re} in Figs. 3, 4, and 5.

The method of Moschytz is essentially the same as that of Rey. The primary difference is that Moschytz does not include the effect of VCO phase modulation ($c(t)$ in Fig. 2) on the phase detector output. Thus, $a(t)$ is assumed to be just sin $\Delta\omega t$, and ω_p is solved for as that $\Delta\omega$ for which the resulting $b(t)$ has a peak value equal to $\Delta\omega$. The resulting solution is then

$$X_{\mathrm{Mo}} = \beta[(1+\{2\zeta(\zeta-\beta)\}^2)^{1/2} + 2\zeta(\zeta-\beta)]^{1/2}. \qquad (15)$$

Moschytz observed experimentally that, under some conditions at least, X is more accurately approximated by the square root of X_{Mo}. This approximation is represented by the dotted curves of Figs. 3 through 5.

The theoretical solutions of Rey and Moschytz are found by the present study to be fairly accurate for $\beta \approx 2\zeta$, but highly inaccurate for $\beta \ll 2\zeta$. The major problem appears to be the assumption that pull-in occurs when $\max_t \{b(t)\} = \Delta\omega$. (The derivation of Moschytz has the additional weakness of ignoring the dc component of $b(t)$.) The exact solutions for $b(t)$ obtained in the course of this study are found to have peak values near $\Delta\omega$ only when $\beta \approx 2\zeta$, but not when $\beta \ll 2\zeta$. This finding (which can also be obtained using phase plane plots) explains the declining accuracy of (11) and (15) as β decreases from the vicinity of 2ζ.

By contrast, virtually exact results for X in the region $\beta \ll 2\zeta$ are produced by the methods of Richman, Viterbi, Lindsey, and Mengali, but their results are not valid for $\beta \approx 2\zeta$. Richman's result is a by-product of his derivation of frequency pull-in time T_F, wherein ω_p is obtained as the value of $\Delta\omega$ beyond which $T_F = \infty$. The derivation of T_F rests on some approximations that assume a relatively long lead time constant, i.e., $K\tau_1 \gg 1$, which can be shown from (6) to correspond to $\beta \ll 2\zeta$. Richman's solution is

$$X_{\mathrm{Ri}} = [2(2\zeta\beta - \beta^2) - (2\zeta\beta - \beta^2)^2]^{1/2}. \qquad (16)$$

The curves of Figs. 3 through 5 for this solution demonstrate the breakdown in validity as β approaches 2ζ (dotted portions).

Viterbi, Lindsey, and Mengali, in methods not dissimilar from that of Richman, use phase plane descriptions to derive pull-in frequency. Viterbi derives the equation relating the instantaneous phase and frequency differences between the input and VCO signals, and estimates the change in frequency difference per cycle change in phase. He then finds ω_p as the value of $\Delta\omega$ below which that change is always negative, regardless of initial conditions. An implicit assumption, however, is that the frequency difference between the input and VCO signals changes relatively little in a cycle.[2] Since this is not the case for $\beta \approx 2\zeta$, Viterbi's result is accurate only for $\beta \ll 2\zeta$. The solution

$$X_V = [2(2\zeta\beta - \beta^2)]^{1/2} \qquad (17)$$

can be seen ((16) and Figs. 3 through 5) to be very similar to that of Richman.

[2] This is akin to assuming that, when $\Delta\omega = \omega_p^+$, the periodic variation of $b(t)$ does not come close to $\Delta\omega$. Because Rey's solution is based on what amounts to a contrary assumption, it is no surprise that his results are most accurate precisely where Viterbi's are not, and vice versa.

Using a similar approach, Lindsey derives a formulation for the case of arbitrary periodic $g(\phi)$. He simplifies the differential equation relating the frequency and phase differences by postulating (like Richman) a condition equivalent to $K\tau_1 \gg 1$. The resulting solution is therefore applicable only for $\beta \ll 2\zeta$ and, furthermore, necessitates machine computations in general. However, under a further stipulation corresponding to $2\beta\zeta \ll 1$, Lindsey's solution reduces to a simple form which, for sinusoidal $g(\phi)$, is identical to Viterbi's, i.e.,

$$X_L \approx X_V = [2(2\zeta\beta - \beta^2)]^{1/2}, \qquad \beta \ll 2\zeta \ll \frac{1}{\beta}. \quad (18)$$

Mengali also treats the case of arbitrary periodic $g(\phi)$. His approach is to expand the expression for the instantaneous frequency difference into a power series with respect to $(1/\bar{\nu})$, where $\bar{\nu}$ is the frequency difference at the start of the phase cycle. His approximation consists of discarding third-order terms and higher, which makes the resulting solution decreasingly valid as β approaches ζ, especially as ζ gets large. In the region of greatest validity, Mengali's solution for sinusoidal $g(\phi)$ is the same as Lindsey's,[3] i.e.,

$$X_{\mathrm{Me}} \doteq X_L \approx X_V = [2(2\zeta\beta - \beta^2)]^{1/2}, \qquad \beta \ll 2\zeta. \quad (19)$$

The results of Lindsey and Mengali differ only in the region where β approaches $\zeta - (\zeta^2 - 1)^{1/2}$, with $\zeta > 1$, in which case Lindsey's solution appears to be more valid.

The special case $\beta = 2\zeta$ deserves a separate evaluation of the published results. Rey and Moschytz give the solutions for this case explicitly, and they are

$$X_{\mathrm{Re}}(\beta = 2\zeta) = \min\{1, (3\zeta[(0.423 + 0.12\zeta^4)^{1/2} - 0.992\zeta^2]^{1/2})\} \quad (20)$$

and

$$X_{\mathrm{Mo}}(\beta = 2\zeta) = 2\zeta[(1 + 4\zeta^4)^{1/2} - 2\zeta^2]^{1/2}. \quad (21)$$

The various approximations of Richman, Viterbi, Lindsey, and Mengali are all inappropriate in this circumstance, and so their solutions do not apply. The preceding two results are plotted in Fig. 6, along with a dotted curve representing exact results from the present study.

To summarize, the solutions of Richman, Viterbi, Lindsey, and Mengali are asymptotically valid in the region $\beta \ll 2\zeta$, corresponding to the widely used high-gain PLL, while the results of Rey and Moschytz are approximately correct for the special case $\beta = 2\zeta$, corresponding to the important case of lag-only loop filtering. The general approach reported here yields exact results for all (β, ζ) and thus permits the validity regions of previously published solutions to be determined.

[3] The extrapolations of Lindsey's and Mengali's solutions into the region $\beta \rightarrow 2\zeta$ are omitted from Figs. 3 through 5 because they are invalid and differ little from those for Richman and Viterbi.

III. THEORETICAL STUDY

The method of solution to be described is similar to that of Woodbury [8], who derives a periodic solution for $a(t)$, Fig. 2, when the loop is out of lock, and then solves for ω_p as the $\Delta\omega$ below which the periodic frequency is not real. To obtain a solution analytically, however, Woodbury approximates $a(t)$ by just the dc term plus the first two harmonics. Moreover, the resulting pull-in frequency still requires a computer solution. By accepting the necessity of a computer solution to begin with, the method described here obtains a more exact determination of $a(t)$ and correspondingly exact solutions for ω_p.

A. General Analysis

Fig. 2 depicts the pertinent relationships for a generalized PLL. With the loop out of lock ($\Delta\omega > \omega_p$), $a(t)$ and $b(t)$ are periodic functions of time, and so is $c(t)$ except for a ramp function, as we shall see.

To begin the analysis, we write $a(t)$ as the complex Fourier series

$$a(t) = \sum_{k=-\infty}^{\infty} A_k \exp(jk\omega_0 t) \quad (22)$$

where ω_0 is the fundamental frequency, and

$$A_k \triangleq \frac{\omega_0}{2\pi} \int_0^{2\pi/\omega_0} a(t) \exp(-jk\omega_0 t)\, dt. \quad (23)$$

The filtered signal $b(t)$ is then

$$b(t) = K \sum_{k=-\infty}^{\infty} \underbrace{A_k F(s = jk\omega_0)}_{B_k} \exp(jk\omega_0 t) \quad (24)$$

and the integrated signal $c(t)$ is[4]

$$c(t) = KA_0 t + \underbrace{\sum_{k=-\infty;\,k\neq 0}^{\infty} \underbrace{KA_k[F(jk\omega_0)/jk\omega_0]}_{\Theta_k} \exp(jk\omega_0 t)}_{\theta(t)}. \quad (25)$$

Finally, the phase detector output is

$$a(t) = g\left((\Delta\omega - KA_0)t - K \sum_{k=-\infty;\,k\neq 0}^{\infty} A_k \frac{F(jk\omega_0)}{jk\omega_0} \cdot \exp(jk\omega_0 t)\right). \quad (26)$$

From physical considerations, the loop can be said to oscillate at the difference between the input frequency and the average VCO frequency. This leads to the relationship

$$\Delta\omega - KA_0 = \omega_0. \quad (27)$$

Combining (27) with (26), equating (26) with (22), and using the normalizations

[4] As defined here, $\theta(t)$ is the periodic phase modulation of the VCO.

$$W_0 = \omega_0/K \tag{28a}$$

$$\Delta W = \Delta\omega/K \tag{28b}$$

and

$$T = Kt \tag{28c}$$

we obtain the nonlinear equation describing the out-of-lock loop:

$$\sum_{k=-\infty}^{\infty} A_k \exp(jkW_0T) = g\left(W_0T - \sum_{k=-\infty;k\neq0}^{\infty} A_k \cdot \frac{F(s/K = jkW_0)}{jkW_0} \exp(jkW_0T)\right). \tag{29}$$

Given the phase detector and filter functions $g(\cdot)$ and $F(\cdot)$, the A_k that solve (29) are functions solely of W_0. If these A_k exist and can be found, then the frequency offset ΔW that produces the periodic frequency W_0 can be obtained from (27), i.e.,

$$\Delta W = W_0 + A_0(W_0). \tag{30}$$

In general, the variation of ΔW with W_0 will have one of two forms. These are depicted in Fig. 7 for the case of a second-order loop with sinusoidal phase detection. In Fig. 7(a), where $\zeta = 0.1$ and $\beta = 0.05$, ΔW exhibits a minimum with respect to W_0, and there is also a W_0 below which no solution to (29) (i.e., no limit cycle) exists. The minimum ΔW represents the highest offset frequency for which the PLL can sustain a limit cycle; below that frequency, the loop converges to a locked condition. Therefore, ω_p/K is easily identified in Fig. 7(a).

In Fig. 7(b), where $\zeta = 0.1$ and $\beta = 0.2$, the minimum of ΔW occurs at the same W_0 below which no solution to (29) exists. Here again, the minimum ΔW clearly corresponds to ω_p/K, but its precise location on the W_0 axis is computationally harder to identify. The reason is that the practical determination of whether or not a limit cycle exists is somewhat fuzzy, as later discussion shows. Fortunately, the minimum in cases such as Fig. 7(b) is always sufficiently broad that ω_p/K can be pinpointed with high accuracy.

B. Computational Approach

Computerized methods are necessary both to solve (29) for the A_k and to find the minimum of ΔW with respect to W_0, as given by (30). In solving (29), each of the periodic variations $a(T)$ and $\theta(T)$[5] is represented in the computer by an array of N time samples that accurately describes its variation over a cycle. Thus, for example, $a(T)$ is represented by an array

$$\{a_n\} \equiv (a_0, a_1, \cdots a_n, \cdots a_{N-1}) \tag{31}$$

where

[5] The functions a and θ are given by (22) and (25), except that now we are invoking the normalizations $T = Kt$ and $W_0 = \omega_0/K$.

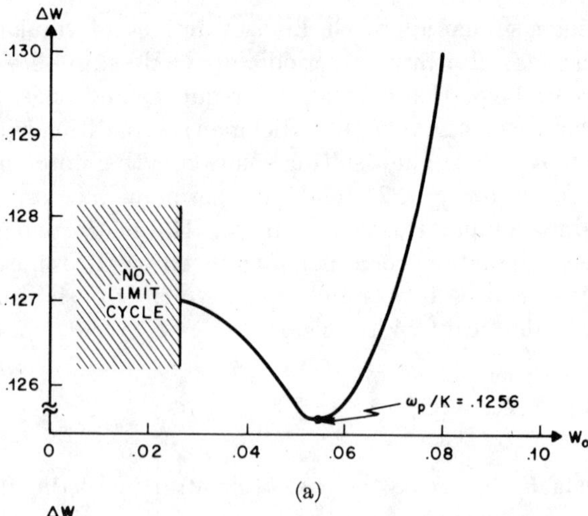

(a)

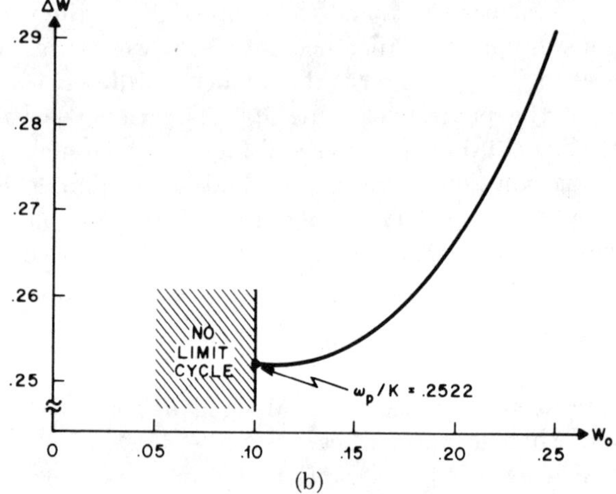

(b)

Fig. 7. Variations of ΔW with W_0. (a) $\zeta = 0.1$; $\beta = 0.05$. (b) $\zeta = 0.1$; $\beta = 0.2$.

$$a_n = a\left(T = \frac{2\pi n}{NW_0}\right), \qquad n = 0, N-1. \tag{32}$$

The value of N must be sufficiently large that all harmonics of $a(T)$ beyond the first $(N/2 - 1)$ are negligible. In that case, the Fourier series coefficients of $a(t)$ can be accurately computed using the discrete Fourier transform (DFT) [9, sec. 6.2] of $\{a_n\}$, i.e.,[6]

$$\{A_k\} \doteq \frac{1}{N}[\text{DFT}\{a_n\}] \triangleq \{(1/N) \sum_{n=0}^{N-1} a_n \exp(-j2\pi kn/N); \quad k = 0, N-1\} \tag{33}$$

If, in addition, N is an integral power of 2, the DFT and its inverse [9, sec. 6.2] can be very efficiently implemented on a computer using FFT algorithms [9, sec. 6.4–6.6].

The solution of (29) for given W_0 (and the resulting solution of ΔW in (30)) uses the iterative procedure described in Fig. 8. The periodic component of the VCO integrator output is characterized by the Fourier coefficient array $\{C_k\}$, while the Fourier coefficient array $\{\Theta_k\}$ repre-

[6] The elements of this array from $k = N - 1$ to $k = (N/2) + 1$ represent the negative-frequency components of the spectrum and are the complex conjugates of the elements from $k = 1$ to $k = (N/2) - 1$, respectively.

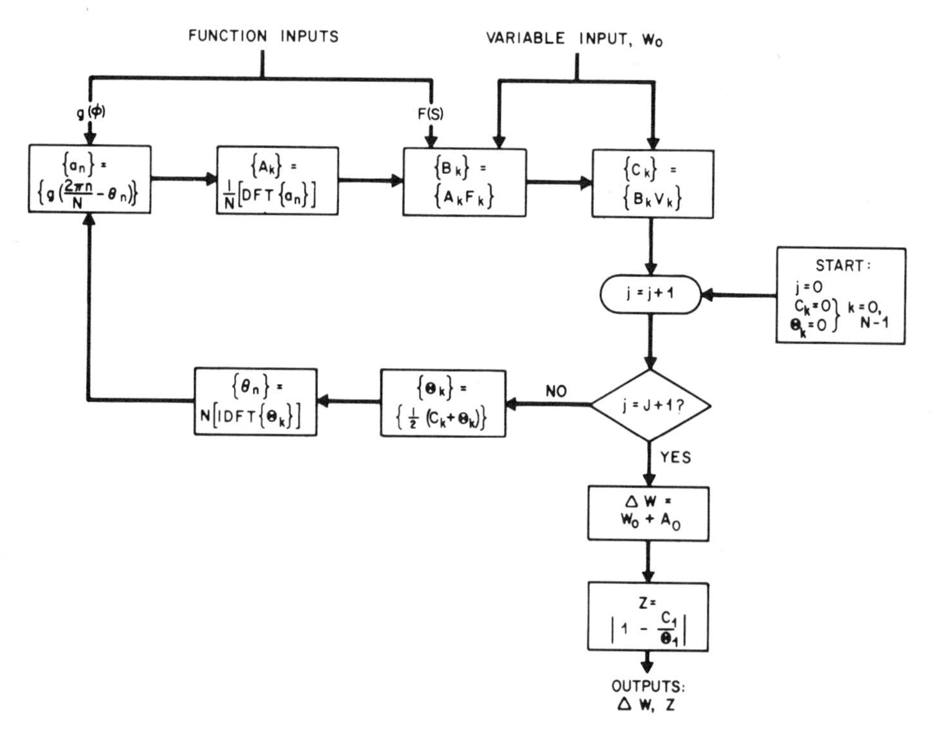

Fig. 8. Procedure for finding ΔW.

sents the periodic phase modulation entering the phase detector. In a stable limit cycle, of course, these two arrays are one and the same. At the start, all C_k and Θ_k are taken to be zero. The two arrays are averaged to produce a new $\{\Theta_k\}$, and the inverse DFT operation produces an estimate of the periodic phase modulation, represented by the array $\{\theta_n\}$. The phase detector output is then computed as $\{a_n\} = \{\sin((2\pi n/N) - \theta_n)\}$, and a DFT operation produces the Fourier coefficient array $\{A_k\}$. This array is processed by complex multiplications representing the loop filter and VCO integrator responses ($F(S)$ and $V(S) = 1/S$, where $S \equiv s/K$) at the harmonic frequencies. Thus,

$$B_k = A_k F_k, \qquad k = 0, N-1 \tag{34}$$

and[7]

$$C_k = B_k V_k, \qquad k = 1, N-1 \tag{35}$$

where, because of the way the DFT is defined, the proper relationships for the arrays F_k and V_k are

$$F_k = \begin{cases} F(S = jkW_0), & 0 \le k < N/2 \\ 0, & k = N/2 \\ F(S = -j(N-k)W_0), & N/2 < k \le N-1 \end{cases} \tag{36}$$

and

$$V_k = \begin{cases} V(S = jkW_0) = \dfrac{1}{jkW_0}, & 1 \le k < N/2 \\ 0, & k = N/2 \\ V(S = -j(N-k)W_0) = -\dfrac{1}{j(N-k)W_0}, & N/2 < k \le N-1. \end{cases} \tag{37}$$

Having obtained a new estimate of $\{C_k\}$, the preceding process is repeated until $j = J + 1$. As j gets large, the relative changes in the complex arrays $\{A_k\}$, $\{B_k\}$, $\{C_k\}$, and $\{\Theta_k\}$ will diminish with each iteration, provided a solution to (29) (i.e., a PLL limit cycle) exists. An equivalent result is that $\{C_k\}$ and $\{\Theta_k\}$, which should be identical, will converge. If this convergence does not occur as j increases, then there is no limit cycle for the given periodic frequency W_0.

In implementing the procedure of Fig. 8, it is necessary to choose N (the number of time samples per cycle and twice the number of periodic signal harmonics) and J (the number of iterations used to solve for the limit cycle). For the second-order PLL with sinusoidal phase detection, it has been found that $N = 32$ and $J = 40$ give an agreeable balance between high accuracy and program efficiency. In general, there is no adequate substitute for trying different N (and J) and observing when the arrays are sufficiently large (and convergent) to give good results.

In addition to choosing N and J, a test is needed to determine, after J iterations, whether or not a limit cycle exists. In cases like the one depicted in Fig. 7(b), such a

[7] Note that $\theta(t)$, (25), has only harmonic components. Since the C_k are computed as part of the solution for the θ_k, the component for $k = 0$ has no meaning and is not computed.

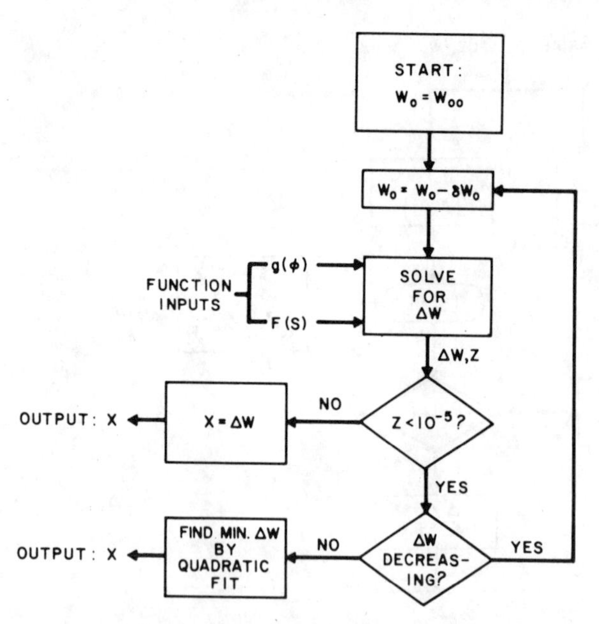

Fig. 9. Procedure for finding X.

determination is critical in deciding where to stop the search over W_0 and accept the nearest valid solution of ΔW as ω_p/K. For the second-order PLL with sinusoidal phase detection, a good measure of whether a limit cycle exists is whether the final quantity

$$Z \triangleq \left| 1 - \frac{C_1}{\Theta_1} \right| \tag{38}$$

is less than some low number, say, 10^{-5}. For $J = 40$, Z is typically 10^{-7} or less near a distinct minimum like the one in Fig. 7(a). Whereas this test has been studied for just one class of PLL's, it should be suitable for any PLL in which the first harmonic is the dominant component of $\theta(t)$ near pull-in.

Deciding what threshold to compare Z against is a bit problematical, since it may be hard to say in a given instance whether no limit cycle exists or convergence is just slow. Fortunately, in cases such as Fig. 7(b), the value of ΔW near the minimum changes imperceptibly as the computed Z increases from 10^{-7} to 10^{-3}. Thus, while the distinction between a valid limit cycle and no limit cycle may be blurred using this method, there is broad leeway insofar as accurately identifying ω_p/K is concerned.

Fig. 9 shows the computerized procedure in which the iterative solution of Fig. 8 is embedded. The periodic frequency W_0 is decreased in suitably small increments (δW_0) from some value (W_{00}) above the range of interest. For each W_0, ΔW and Z are determined and, if Z exceeds 10^{-5}, X ($= \omega_p/K$) is taken to be the most recently computed ΔW. If Z is less than 10^{-5}, and ΔW starts to increase, a quadratic fit to the three most recent values of ΔW is used to locate the minimum, which is taken to be ω_p/K.

C. Results

Fig. 10 shows some results for X versus β with ζ as a parameter. These curves are all obtained from the com-

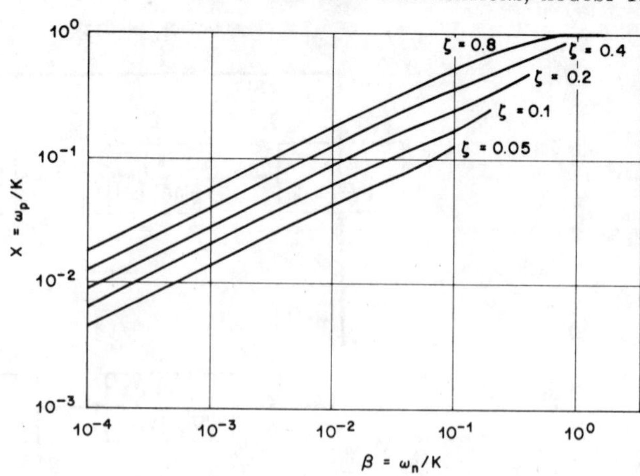

Fig. 10. New theoretical results for $X(\beta, \zeta)$.

puterized method of solution just described, as is the dotted curve in Fig. 6 for the case $\beta = 2\zeta$. The agreements with the results of Richman, Viterbi, Lindsey, and Mengali [i.e., $X \approx (4\zeta\beta)^{1/2}$] for $\beta \ll 2\zeta$, and with those of Moschytz and Rey for $\beta \approx 2\zeta$, are apparent.

IV. EXPERIMENTAL STUDY

A. Configuration

The variation of X with β and ζ was measured in the laboratory using the arrangement shown in Fig. 11. The VCO used was a Damon 6836 crystal-controlled device having a rest frequency of 140.0 MHz, an FM deviation sensitivity K_f of 30.4 KHz/V and a linear deviation range of ± 140 KHz. For frequency modulation rates up to 100 KHz, the frequency deviation response of this VCO is constant to within ± 1 dB.

The operational amplifier was used to provide the dc filter gain F_0 (Fig. 1). The measured values of gain, input resistance, and 3-dB frequency were 1018, 130 KΩ, and 5 MHz, respectively.

The phase detector used was a Vari-L (DBM 100E) double balanced mixer. The low-frequency output of this device is $K_\phi \sin \phi$, where K_ϕ is a function of the two input powers. For this experiment, the input signal level was adjusted to make $K_\phi = 0.07$ V.

The variable elements in the experiment were 1) the input signal attenuation, to control K_ϕ; 2) the loop attenuation (R_a and R_b), to control the dc loop gain and filter source impedance; and 3) the lag/lead filter (R_c, R_d, and C), to control the values of τ_1 and τ_2. In principle, the experiment could be performed with fixed settings of the loop attenuation and filter capacitance, with R_c and R_d alone being varied to cover the desired region of (τ_1, τ_2) (or (β, ζ), using (3) and (5)). The enlarged array of experimental variables was necessitated by several practical considerations, e.g., upper limits on ω_p dictated by the VCO deviation range, lower limits on ω_p dictated by the achievable measurement resolution, upper limits on R_c dictated by amplifier drift problems, etc. In all,

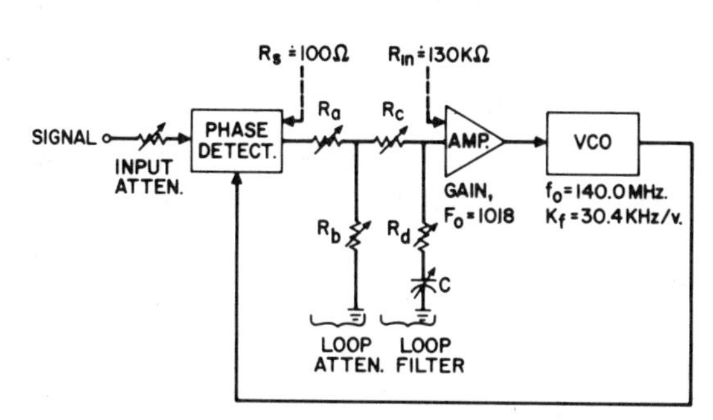

Fig. 11. Experimental configuration.

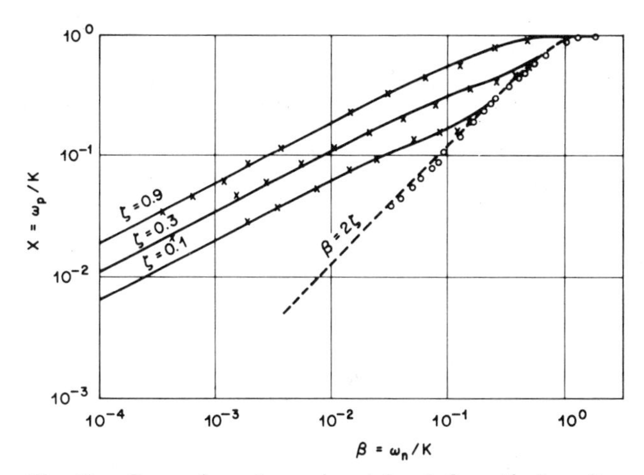

Fig. 12. Comparison of experimental and theoretical results.

five combinations of loop attenuation and filter capacitance were used to cover the (β,ζ) range of interest.

B. Procedure

The equations used in this study were (3), (5), (10),

$$K = 2\pi K_f K_\phi F_0 \frac{R_b}{R_a + R_b} \left\{ \frac{R_{in}}{R_{in} + R_c + R_a R_b/(R_a + R_b)} \right\} \tag{39}$$

and

$$\tau_2 = C \left\{ R_d + \frac{R_{in}[R_c + R_b(R_a + R_s)/(R_a + R_b + R_s)]}{R_{in} + R_c + R_b(R_a + R_s)/(R_a + R_b + R_s)} \right\}. \tag{40}$$

For each combination of loop attenuation and filter capacitance, the procedure used was the following: For each of several values of R_d, R_c was adjusted (using (3), (39), and (40)) to achieve a specified value of ζ. For that combination of R_c and R_d, ω_p was determined (where measurable) by reducing $|\Delta\omega|$ towards zero on each side of the VCO rest frequency and averaging the two values for which pull-in occurred. The quantities β and X were then computed using (5) and (10).

For the special case $\beta = 2\zeta$ ($R_d = 0$), the experiment is simpler because X is a function of just one parameter. The procedure used in this case was to measure ω_p for a representative range of R_c values and compute K, τ_2, ζ and X using the equations just mentioned.

C. Results

The solid curves of X versus β in Fig. 12 are the theoretical results for $\zeta = 0.1$, 0.3, and 0.9. The experimental data for these cases are given by the crosses, and the agreement between theory and measurement is seen to be generally quite good. Comparable agreement was also obtained for $\zeta = 0.5$ and 0.7. The dotted curve of Fig. 12 gives the theoretical variation for the case $\beta = 2\zeta$, and the corresponding experimental data are given by the circles; in this case also, agreement is quite good. Among other factors, these favorable results can be attributed to special care in performing the measurements and careful experimental designs to minimize the effects of noises, nonlinearities, and extraneous time constants.

V. MATHEMATICAL DESCRIPTION

To maximize the utility of the new results for $X(\beta,\zeta)$, an attempt was made to find a functional description $\tilde{X}(\beta,\zeta)$ that is accurate for all $(\beta \leq 2\zeta, \zeta \leq 1.0)$. The first step in the search was the dual observation that 1) $X \approx (4\zeta\beta)^{1/2}$ for $\beta \ll 2\zeta$; and 2) as β approaches 2ζ, the curves of X versus β have the appearance of asymptotic exponentials. These observations suggest a representation for $X(\beta,\zeta)$ of the form

$$X(\beta,\zeta) = 1 - \exp\{-(4\zeta\beta)^{1/2} Y(\beta,\zeta)\} \tag{41}$$

where $Y \doteq 1$ when $\beta \ll 2\zeta$.

The second step in the search was to compute and plot the function $Y(\beta,\zeta)$ using the computed values of $X(\beta,\zeta)$ and the formula (derived from (41)),

$$Y(\beta,\zeta) = \frac{1}{(4\zeta\beta)^{1/2}} \ln \frac{1}{1 - X(\beta,\zeta)}. \tag{42}$$

The third step was the observation that Y versus β, for $\beta \leq 2\zeta$ and $\zeta \leq 1.0$, appears to be well fitted by a function of the form

$$Y(\beta,\zeta) = \exp\{u_1(\zeta)\beta^{1/2} + u_2(\zeta)\beta + u_3(\zeta)\beta^2\}. \tag{43}$$

The final step was to compute u_1, u_2, and u_3 for a representative set of ζ values and to find suitable curve fits for these variations with ζ. The results are

$$u_1(\zeta) \doteq 0.5(1 + \zeta)(1 - e^{-6\zeta}) \tag{44}$$

$$u_2(\zeta) \doteq -0.275/\zeta \tag{45}$$

and

$$u_3\,\zeta(\doteq (0.115 - 0.07\zeta^2 + 1.5\zeta^4)/\zeta^2. \tag{46}$$

Combining these results with (41) and (43) yields

$$X(\beta,\zeta) \doteq 1 - \exp\{-(4\zeta\beta)^{1/2} \exp[0.5(1 + \zeta)(1 - e^{-6\zeta}) \cdot \beta^{1/2} - 0.275(\beta/\zeta) + (0.115 - 0.07\zeta^2 + 1.5\zeta^4)(\beta/\zeta)^2]\} \equiv \tilde{X}(\beta,\zeta). \tag{47}$$

This functional approximation has been compared with exact values computed for 175 densely packed points in the region $(\beta \leq 2\zeta,\ \zeta \leq 1.0)$. The results show that $\tilde{X}(\beta,\zeta)$ is accurate to within ± 2 percent for all (β,ζ) of interest.

For the special case $\beta = 2\zeta$, (47) reduces to

$$\tilde{X}(\beta = 2\zeta) = 1 - \exp\{-2.58\zeta \exp[(1+\zeta)(1-e^{-6\zeta}) \cdot (\zeta/2)^{1/2} - 0.28\zeta^2 + 6\zeta^4]\}. \quad (48)$$

This approximation has an accuracy of 1.5 percent for all ζ, while the functionally different approximation of Rey in (20) has an accuracy of 6 percent.

VI. CONCLUSION

The computerized method of solution for PLL pull-in frequency presented here is the most general and exact known to the author. The efficiency of the procedure depends primarily on the phase detector function $g(\phi)$. If $g(\phi)$ has sharp discontinuities (such as the sawtooth and rectangular functions), accurate limit cycle solutions may require representation by more than the 16 harmonics used here for sinusoidal detectors. The resulting increase in N will, of course, increase the cost of the computer runs.

It should also be mentioned that, in contrast to the methods of Richman, Viterbi, Lindsey, Mengali, and others, this method does not lend itself to the determination of frequency pull-in time, which is an important parameter in many cases. At the same time, it provides a simple and accurate way to study the behavior of PLL's in the out-of-lock condition, over and above merely determining pull-in frequency.

ACKNOWLEDGMENT

The author is deeply indebted to S. Michael for his collaboration in the experimental study, and to W. Mammel for invaluable help in executing the computer programs.

REFERENCES

[1] W. J. Gruen, "Theory of AFC synchronization," *Proc. IRE*, vol. 41, pp. 1043–1048, Aug. 1953.
[2] D. Richman, "Color-carrier reference phase synchronization accuracy in NTSC color television," *Proc. IRE*, vol. 43, pp. 106–133, Jan. 1954.
[3] T. J. Rey, "Automatic phase control: Theory and design," *Proc. IRE*, vol. 48, pp. 1760–1771, Oct. 1960.
[4] G. S. Moschytz, "Miniaturized RC filters using phase-locked loop," *Bell Syst. Tech. J.*, vol. 44, pp. 823–870, May–June 1965.
[5] A. J. Viterbi, *Principles of Coherent Communication.* New York: McGraw-Hill, 1966, ch. 3.
[6] W. C. Lindsey, *Synchronization Systems in Communication and Control.* Englewood Cliffs, N. J.: Prentice-Hall, 1972, sec. 10-2.3, 10-2.4.
[7] U. Mengali, "Acquisition behavior of generalized tracking systems in the absence of noise," *IEEE Trans. Commun.*, vol. COM-21, pp. 820–826, July 1973.
[8] J. R. Woodbury, "Phase-locked loop pull-in range," *IEEE Trans. Commun. Technol.*, vol. COM-16, pp. 184–186, Feb. 1968.
[9] B. Gold and C. M. Rader, *Digital Processing of Signals.* New York: McGraw-Hill, sec. 6.2, and 6.4–6.6.

Acquisition Behavior of Generalized Tracking Systems in the Absence of Noise

UMBERTO MENGALI

Abstract—**This paper describes a new method for analyzing the acquisition behavior of a second-order generalized tracking system in the absence of noise. In particular it allows closed-form expressions of the pull-in range and of the frequency acquisition time for any type of loop nonlinearity. The results of the present analysis are compared with those already known in the literature and with others obtained via the numerical solution of the system equation.**

I. Introduction

As IS KNOWN [1], a broad class of synchronization and tracking systems used in various fields of telecommunication engineering are described by the same basic model, mathematically identical to that of the conventional phase-locked loop except for the phase-detector characteristic that can be other than sinusoidal. For example, the delay-lock loop [2] has a triangular characteristic, and sawtooth phase comparators are used to provide synchronization in digital communication networks [3] and [4]. Other loop nonlinearities are those of the tanlock [5] and of the modified tanlock receiver [6]. The list could continue at some length (see [1, p. 116]) but even these few examples are enough to give substance to the category of the so-called generalized tracking systems [7], that is, of the tracking loops with an arbitrary periodic phase-detector characteristic $g(\varphi)$.

An important problem in the study of these systems is their description during the frequency acquisition operation. As expected, a considerable number of papers has been published on this subject. To indicate some of the major efforts in this field, and limiting ourselves to second-order loops, let us begin with a theory suggested by Richman [8].

Essentially, Richman's method is based on the fact that, during the acquisition, a tracking system can be treated as a first-order loop with a slowly varying voltage added to the output of the loop filter. Richman obtained an equation for this voltage that accounts for the loop acquisition mechanism. Integrating this equation yields the frequency acquisition time T_f, whereas the pull-in range is obtained as that value of the loop detuning for which T_f approaches infinity. This method, originally limited to a sinusoidal $g(\varphi)$, has been extended later to other nonlinearities. In particular, Byrne [9] studied the case of a sawtooth function and Meer [10] considered the sawtooth and the triangular characteristics. Recently Shakhtarin [11] explored the case of the tanlock-type function.

Further extentions are difficult, however, because of two basic obstacles. The first is that the differential equation for the acquisition voltage does not appear integrable in closed form except for the case of a sinusoidal nonlinearity. Second, for an arbitrary $g(\varphi)$, even obtaining the differential equation is a formidable task.

Other interesting approaches have been suggested [12]-[14] for the determination of T_f, but they are limited to specific nonlinearities. In [1] and [15] overbounds are given to T_f for any $g(\varphi)$. In closing, we will cite Rey [16] and Woodbury [17], who studied a new method for computing the pull-in range in the sinusoidal case, Lindsey [1], and Yevtyanov and Snedkova [18], who extended it to an arbitrary characteristic.

This paper presents a general theory for the computation of the frequency acquisition time, and of the pull-in range of second-order tracking loops with arbitrary periodic characteristics. The significant feature of this theory is that both the acquisition time and the pull-in range can be expressed in closed form as functions of the relevant system parameters.

Paper approved by the Communication Theory Committee of the IEEE Communications Society for publication without oral presentation. Manuscript received January 31, 1973. This work was supported in part by the Centro di Studi per i Metodi e i Dispositivi di Radiotrasmissione of the C. N. R.

The author is with the Department of Electrical and Electronics Engineering, University of Pisa, Pisa, Italy.

Reprinted from *IEEE Trans. Commun.*, vol. COM-21, pp. 820-826, July 1973.

Our exposition will proceed as follows. In Section II an approximate solution is found of the system equation during a 2π increase of the loop phase error. This solution is then used for the determination of the pull-in range (Section III) and of the frequency acquisition time (Section IV). Finally, in Section V, the conclusions are drawn.

II. Description of a Cycle Skip

In the absence of noise the describing equation of the loop is [7]

$$p\varphi = -AK\, F(p) g(\varphi) + \Omega, \tag{1}$$

where $p \triangleq d/dt$ is the "Heaviside" operator, φ the phase error, A^2 the power of the synchronizing signal, K the open loop gain, $F(p)$ the loop filter transfer function, $g(\varphi)$ the loop nonlinearity, and Ω the radian frequency difference between the input signal and the natural frequency of the voltage-controlled oscillator. Here $g(\varphi)$ is assumed as an arbitrary periodic odd function of φ:

$$g(\varphi) = g(\varphi + 2\pi)$$
$$g(\varphi) = -g(-\varphi) \tag{2}$$

and $F(p)$ corresponds to a second-order imperfect integrating filter

$$F(p) = \frac{1 + \tau_2 p}{1 + \tau_1 p} = F_0 \frac{p + 1/\tau_2}{p + F_0/\tau_2}, \tag{3}$$

where $F_0 \triangleq \tau_2/\tau_1$ is the ratio between the ac gain and the dc gain of the filter.

Substituting (3) into (1) and normalizing the time by setting $\tau \triangleq AKF_0 t$ yields

$$\frac{d^2\varphi}{d\tau^2} + \left[\frac{F_0}{r} + g'(\varphi)\right] \frac{d\varphi}{d\tau} + \frac{g(\varphi)}{r} = \frac{\gamma F_0}{r} \tag{4}$$

in which $g'(\varphi) \triangleq dg/d\varphi$, $r \triangleq AKF_0\tau_2$, and $\gamma \triangleq \Omega/AKF_0$ is the normalized detuning.

Equation (4) is the differential equation to be studied. Before proceeding, however, we point out that, in the limit, for $F_0 \to 0$ and AKF_0 approaching some nonzero constant, it becomes

$$\frac{d^2\varphi}{d\tau^2} + g'(\varphi) \frac{d\varphi}{d\tau} + \frac{g(\varphi)}{r} = 0, \tag{5}$$

which is the differential equation of a loop with a perfect integrating filter.

Next, let us recall briefly the phase-plane technique often used to describe the acquisition mode of operation of the loop (in particular, see [12]). To this end, setting $v \triangleq d\varphi/d\tau$ in (4) yields

$$v\frac{dv}{d\varphi} + v\left[\frac{F_0}{r} + g'(\varphi)\right] + \frac{g(\varphi)}{r} = \frac{\gamma F_0}{r}. \tag{6}$$

In Fig. 1 a typical phase-plane plot of (6) is drawn with φ reduced modulo 2π. As we can see, starting from A, the point

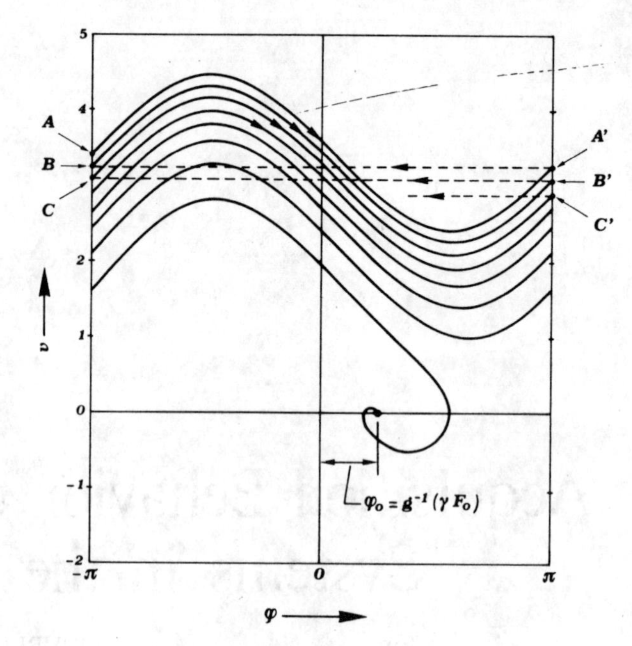

Fig. 1. Typical phase-plane plot of (6).

(φ, v) follows the trajectory AA' until A'. Then, instantaneously, it skips back to B, which has the same ordinate as A' and begins to run along $BB' \cdots$, etc. The oscillatory motion from left to right, to left, to right $\cdots$ etc., indicates that the phase error is cycling through multiples of 2π. In the case of Fig. 1, little by little the instantaneous frequency error v decreases (the trajectories get lower and lower) until the last cycle in the pull-in process is reached. At this point the frequency lock is achieved and (φ, v) continues toward the equilibrium point $\varphi_0 = g^{-1}(\gamma F_0)$, where $g^{-1}(\cdot)$ is the inverse function of $g(\cdot)$.

Let us now seek the equation of an arbitrary trajectory in the phase plane. To this end let us denote with $\bar{v}$ the value of v at the beginning of this trajectory. The ordinate of any other point will be a function both of φ and of $\bar{v}$, say $v = v(\varphi, \bar{v})$. Obviously, it must be

$$v(-\pi, \bar{v}) = \bar{v}, \qquad \text{for any } \bar{v}.$$

Alternatively, we can set

$$v = v(\varphi, \bar{v}) = \bar{v} + w(\varphi, \bar{v}) \tag{7}$$

with $w(\varphi, \bar{v})$ a new function of φ and $\bar{v}$ such that

$$w(-\pi, \bar{v}) = 0, \qquad \text{for any } \bar{v}. \tag{8}$$

Under certain conditions $w(\varphi, \bar{v})$ can be expressed in the form of an asymptotic expansion [19]

$$w(\varphi, \bar{v}) = w_0(\varphi) + \frac{w_1(\varphi)}{\bar{v}} + \frac{w_2(\varphi)}{\bar{v}^{-2}} + \cdots, \tag{9}$$

where, because of (8), the functions $w_n(\varphi)$ satisfy

$$w_n(-\pi) = 0, \qquad n = 0, 1, 2, \cdots. \tag{10}$$

In this way our problem is reduced to finding the expressions of $w_n(\varphi), n = 0, 1, 2, \cdots$. In general, for a given φ, when n increases, the functions $w_n(\varphi)$ decrease at first, then increase again making the above series divergent. Nevertheless, (9) can be used for calculation because the error incurred by ending the series at any term is of the order of the next term omitted (a characteristic property of asymptotic series). Thus if $\bar{v}$ is sufficiently large, few terms on the right-hand side of (9) are expected to give a result with a small error, and the larger the $\bar{v}$, the smaller the error.

The functions $w_n(\varphi)$ can be determined as follows. Inserting (9) into (7) and the latter into (6) produces

$$f_1(\varphi)\bar{v} + f_0(\varphi) + \frac{f_{-1}(\varphi)}{\bar{v}} + \frac{f_{-2}(\varphi)}{\bar{v}^2} + \cdots = 0, \tag{11}$$

where

$$\begin{aligned} f_1(\varphi) &= w_0'(\varphi) + \frac{F_0}{r} + g'(\varphi) \\ f_0(\varphi) &= w_1'(\varphi) + w_0(\varphi)w_0'(\varphi) + w_0(\varphi)\left[\frac{F_0}{r} + g'(\varphi)\right] \\ &\quad + \frac{g(\varphi)}{r} - \frac{\gamma F_0}{r} \\ f_{-1}(\varphi) &= w_2'(\varphi) + w_0(\varphi)w_1'(\varphi) + w_1(\varphi)w_0'(\varphi) + w_1(\varphi) \\ &\quad \cdot \left[\frac{F_0}{r} + g'(\varphi)\right] \\ f_{-2}(\varphi) &= \cdots \\ &\vdots \end{aligned} \tag{12}$$

We ask if it is possible to satisfy (11) by setting

$$f_n(\varphi) = 0, \qquad n = 1, 0, -1, -2, \cdots$$

This entails equating the right-hand sides of (12) to zero. In doing this, one obtains

$$\begin{aligned} w_0'(\varphi) &= -g'(\varphi) - \frac{F_0}{r} \\ w_1'(\varphi) &= \frac{-g(\varphi)}{r} + \frac{\gamma F_0}{r} \\ w_2'(\varphi) &= -w_0(\varphi)\left[-\frac{g(\varphi)}{r} + \frac{\gamma F_0}{r}\right] \\ w_3'(\varphi) &= \cdots \\ &\vdots \end{aligned} \tag{13}$$

that is, a system of simultaneous linear differential equations of the first order that can be solved in succession in a straightforward way. Assuming $g(-\pi) = 0$ (a condition always satisfied in cases of practical interest) and using the initial conditions (10), one gets

$$\begin{aligned} w_0(\varphi) &= -g(\varphi) - \frac{F_0}{r}(\varphi + \pi) \\ w_1(\varphi) &= -\frac{1}{r}\int_{-\pi}^{\varphi} g(x)dx + \frac{\gamma F_0}{r}(\varphi + \pi) \\ w_2(\varphi) &= \frac{\gamma F_0}{r}\int_{-\pi}^{\varphi} g(x)dx - \frac{1}{r}\int_{-\pi}^{\varphi} g^2(x)dx \\ &\quad + \frac{\gamma F_0^2}{r^2}\int_{-\pi}^{\varphi}(x + \pi)dx - \frac{F_0}{r^2}\int_{-\pi}^{\varphi}(x + \pi)g(x)dx \\ w_3(\varphi) &= \cdots \\ &\vdots \end{aligned} \tag{14}$$

which, along with (9) and (7), provides the equation of the trajectories in Fig. 1.

Admittedly, our result pertains only to those lying in the upper half-plane. However a one-to-one correspondence exists between these and those of the lower half-plane. In fact, it is a simple matter to verify that if a function $v_u(\varphi)$ satisfies (6) with $\gamma = \bar{\gamma}$ and $v_u(-\pi) = \bar{v}$, then $v_1(\varphi) \triangleq -v_u(-\varphi)$ satisfies (6) with $\gamma = -\bar{\gamma}$ and $v_1(\pi) = -\bar{v}$ (see Fig. 2). For this reason, in the sequel we will limit our analysis to the case $\bar{v} > 0$.

III. Pull-In Range

In this section we will determine an approximate formula for the frequency acquisition range of a generalized tracking loop. Our procedure is based on the fact that the pull-in range represents the range of values of the normalized detuning γ for which periodic solutions of (6) are impossible. On the other hand, if a periodic solution exists, it must be $v(\pi, \bar{v}) = \bar{v}$ for some value of $\bar{v}$ and vice versa.

Hence, the pull-in range can be determined by finding the values of γ such that the frequency error decay $\Delta v \triangleq v(\pi, \bar{v}) - \bar{v}$ cannot be zero, whatever $\bar{v}$ is. This argument leads us to compute Δv. This can be done by substituting (14) into (9) and the latter into (7). The result is

$$\Delta v = -\frac{2\pi F_0}{r}\left[1 - \frac{\gamma}{\bar{v}} + \frac{1}{\bar{v}^2}Q(\gamma)\right] + 0\left(\frac{1}{\bar{v}^3}\right) \tag{15}$$

where

$$Q(\gamma) \triangleq \frac{\langle g^2 \rangle}{F_0} + \frac{\langle \varphi g \rangle}{r} - \frac{\pi\gamma F_0}{r}, \tag{16}$$

where $\langle\cdot\rangle$ denotes the average of the enclosed quantity on the interval $-\pi \leqslant \varphi \leqslant \pi$ and $0(1/\bar{v}^3)$ goes to zero as $1/\bar{v}^3$ or faster when $\bar{v}$ becomes large.

As a preliminary to the consideration of (15) observe that in all cases of practical interest, $\langle \varphi g \rangle$ is positive (as a rule, for $-\pi \leqslant \varphi \leqslant \pi$ it is $g(\varphi) \gtrless 0$ according to whether $\varphi \gtrless 0$).

It follows that for $\gamma < 0$ it is $Q(\gamma) > 0$ and, therefore, neglecting $0(1/\bar{v}^3)$ henceforth, the right-hand side of (15) is always negative. Instead, if γ is positive and so large that

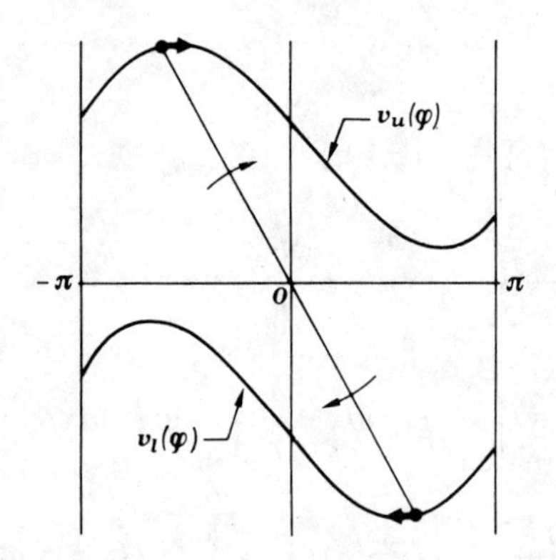

Fig. 2. One-to-one correspondence between the trajectories of the two half-planes.

$Q(\gamma) < 0$, certainly Δv will be zero for some $\bar{v}$. Finally, if we suppose $\gamma > 0$ but $Q(\gamma) > 0$, and set $\Delta v = 0$ in (15), we obtain

$$\bar{v}^2 - \gamma \bar{v} + Q(\gamma) = 0. \tag{17}$$

Straightforward algebra indicates that the real solutions of (17) are positive. Hence, if we want $\Delta v \neq 0$ for $\bar{v} > 0$, we must require that (17) has complex solutions, i.e.,

$$\gamma^2 - 4Q(\gamma) < 0$$

or, using (16),

$$\gamma < 2\left[\frac{\langle g^2 \rangle}{F_0} + \frac{\langle \varphi g \rangle}{r} + \left(\frac{\pi F_0}{r}\right)^2\right]^{1/2} - \frac{2\pi F_0}{r}. \tag{18}$$

We can summarize the previous discussion by saying that $\Delta v = 0$ is impossible for $\bar{v} > 0$ if (18) holds true and, at the same time, $Q(\gamma) > 0$, i.e.,

$$\gamma < \frac{r}{\pi F_0}\left(\frac{\langle g^2 \rangle}{F_0} + \frac{\langle \varphi g \rangle}{r}\right). \tag{19}$$

On the other hand, condition (19) turns out[1] to be always less restrictive than (18). Therefore (take into account that we limited our attention to the trajectories with $\bar{v} > 0$), the normalized pull-in range is simply

$$|\gamma| \leqslant \gamma_m \triangleq 2\left[\frac{\langle g^2 \rangle}{F_0} + \frac{\langle \varphi g \rangle}{r} + \left(\frac{\pi F_0}{r}\right)^2\right]^{1/2} - \frac{2\pi F_0}{r}. \tag{20}$$

It is worthwhile comparing this result with those obtained by other authors. In Fig. 3 the continuous line gives $\Omega_m/AK \triangleq F_0\gamma_m$ versus F_0 as obtained by Richman [8] for the sinusoidal phase-detector characteristic whereas the dashed lines represent (20) for some values of r. An analogous comparison is shown in Fig. 4 for the case of the sawtooth function

$$\text{saw}(\varphi) \triangleq \varphi, \qquad |\varphi| < \pi.$$

Here the continuous line gives the result of Yevtyanov and Snedkova [18]. Both figures indicate fair agreements in the range of greater practical interest, say $F_0 \leqslant 0.2$. Finally, let us suppose $F_0 << 1$. In this case (20) yields

[1]To show this, compare the right-hand sides of (18) and (19) setting $a \triangleq (\langle g^2 \rangle/F_0) + (\langle \varphi g \rangle/r)$ and $b \triangleq r/\pi F_0$.

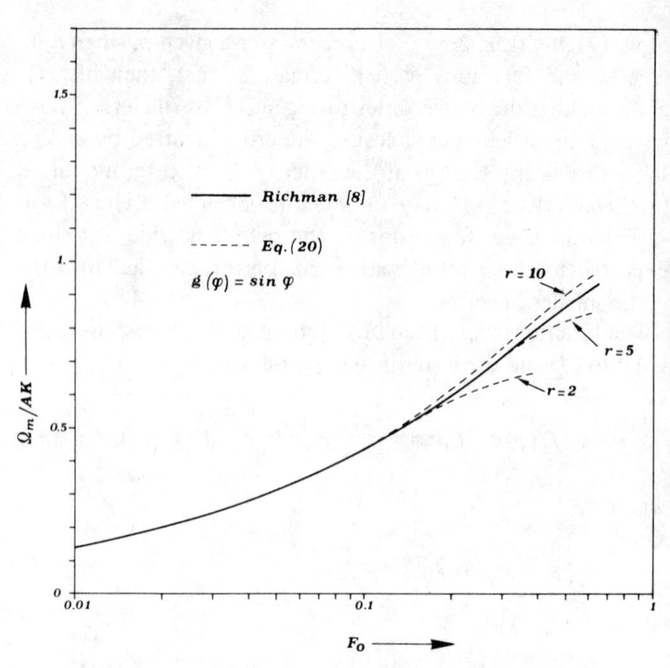

Fig. 3. Pull-in range versus F_0 for a sinusoidal nonlinearity.

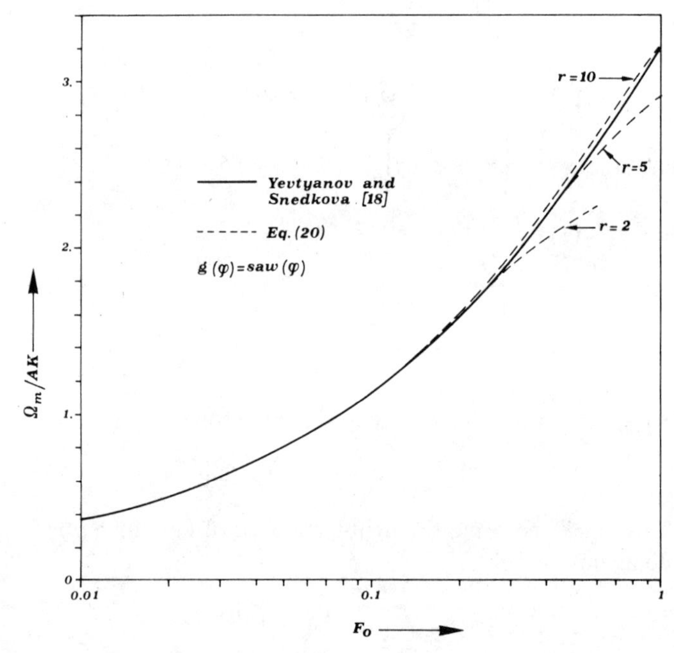

Fig. 4. Pull-in range versus F_0 for a sawtooth nonlinearity.

$$\frac{\Omega m}{AK} \simeq 2\sqrt{F_0 \langle g^2 \rangle}$$

that is, the same result Lindsey obtained under the same hypotheses [1, p. 463].

IV. Frequency Acquisition Time

In this section we will compute the frequency acquisition time. To this purpose, we first determine the time $\Delta\tau$ that point (φ, v) takes to traverse from $-\pi$ to $+\pi$ a trajectory in Fig. 1. Observing that, by definition, $d\tau = d\varphi/v$, we can write

$$\Delta\tau = \int_{-\pi}^{\pi} \frac{d\varphi}{v}$$

or, using (7) and (9),

$$\Delta\tau = \frac{1}{\bar{v}} \int_{-\pi}^{\pi} \left[1 + \frac{w_0(\varphi)}{\bar{v}} + \frac{w_1(\varphi)}{\bar{v}^2} + \cdots\right]^{-1} d\varphi. \quad (21)$$

In order to obtain tractable formulas it is expedient to represent the integrand in (21) as a series expansion in descending powers of $\bar{v}$. In doing this one gets

$$\left[1 + \frac{w_0(\varphi)}{\bar{v}} + \frac{w_1(\varphi)}{\bar{v}^2} + \cdots\right]^{-1} = 1 - \frac{w_0(\varphi)}{\bar{v}} + \frac{w_0^2(\varphi) - w_1(\varphi)}{\bar{v}^2} + \cdots \quad (22)$$

and (22) collecting with (1), (14), and (21) yields

$$\Delta\tau = \frac{2\pi}{\bar{v}}\left[1 + \frac{\pi F_0}{\bar{v} r} + \frac{P}{\bar{v}^2}\right] + 0\left(\frac{1}{\bar{v}^4}\right), \quad (23)$$

where

$$P \triangleq \langle g^2 \rangle + \frac{1}{r}(2F_0 - 1)\langle \varphi g \rangle - \frac{\pi\gamma F_0}{r} + \frac{1}{3}\left(\frac{2\pi F_0}{r}\right)^2. \quad (24)$$

Next, we turn our attention to the dependence of v on the time. In Fig. 5 a typical graph of v versus τ, say $v = v(\tau)$, is drawn. Here the points $A, B, C, \cdots$, etc., correspond to those with the same label in Fig. 1. Call $\bar{v} = \bar{v}(\tau)$ the piecewise linear path joining these points. If $\bar{v}$ is large enough, the decay Δv in (15) and the time $\Delta\tau$ in (23) are so small that $\bar{v} = \bar{v}(\tau)$ can be confused with the solution of the equation

$$\frac{d}{d\tau}\bar{v}(\tau) = \frac{\Delta v}{\Delta\tau} \quad (25)$$

with initial condition $\bar{v}(0) = v(0)$ (see Fig. 5). Therefore, if we define τ_f, the normalized frequency acquisition time as the time $v(\tau)$ takes to go to zero, Fig. 5 shows that τ_f can be computed as the instant in which $\bar{v}(\tau)$ crosses some positive level λ. Experimental evidence indicates approximately $\lambda = 1.3\, g_m$, where g_m denotes the maximum of $g(\varphi)$.

Thus τ_f can be obtained as follows. Assuming that for $\tau = 0$ the voltage-controlled oscillator oscillates at its natural frequency, we can write

$$v(0) = \left.\frac{d\varphi}{d\tau}\right|_{\tau=0} = \frac{1}{AKF_0}\left.\frac{d\varphi}{dt}\right|_{t=0} = \frac{\Omega}{AKF_0} = \gamma.$$

Then, inserting (15) and (23) into (25) [neglect $0(1/\bar{v}^3)$ and $0(1/\bar{v}^4)$] and integrating between the limits $\bar{v} = \gamma$ at $\tau = 0$ and $\bar{v} = \lambda$ at $\tau = \tau_f$, without belaboring the details one obtains

$$\tau_f = \frac{r}{F_0}\left\{\ln\left(\frac{\gamma}{\lambda}\right) + \left[\frac{2\pi F_0}{r} + \gamma\left(1 + \frac{P}{Q}\right)\right] \cdot \left[\tan^{-1}\left(\frac{\gamma}{R}\right) - \tan^{-1}\left(\frac{2\lambda - \gamma}{R}\right)\right]\Big/ R + \frac{1}{2}\left(1 - \frac{P}{Q}\right)\ln\left[\frac{\lambda^2 Q}{\gamma^2(\lambda^2 - \gamma\lambda + Q)}\right]\right\} \quad (26)$$

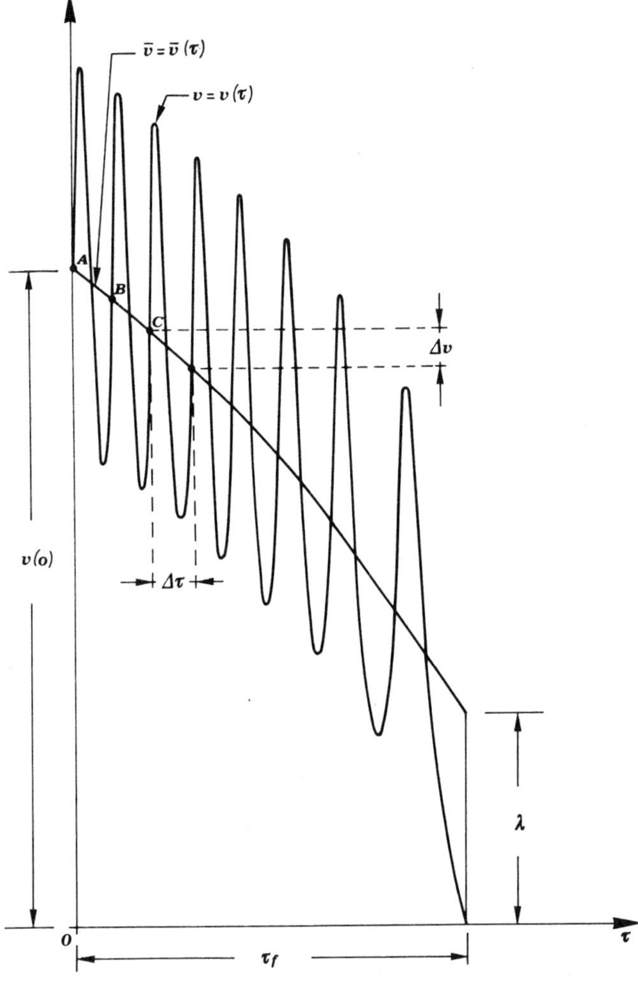

Fig. 5. Typical graph of v versus τ.

where $Q \triangleq Q(\gamma)$ as defined in (16) and

$$R \triangleq (4Q - \gamma^2)^{1/2}$$

Equation (26) pertains to the case of an imperfect integrating loop filter. As noted in Section II, however, we can extend this result to a loop with an ideal integrating filter letting $F_0 \to 0$ while F_0AK approaches some nonzero constant. Carrying out these operations in (26), yields

$$\tau_f = \frac{r}{\langle g^2 \rangle}\left[\frac{\gamma^2 - \lambda^2}{2} + \left(\langle g^2 \rangle - \frac{\langle \varphi g \rangle}{r}\right)\ln\left(\frac{\gamma}{\lambda}\right)\right]. \quad (27)$$

Let us compare these results with those of previous authors. In Fig. 6 (26) is plotted as a continuous line for various loop parameters and $\lambda = 1.3\, g_m$. Here the abscissa is $\Delta f / B_L \triangleq \Omega / 2\pi B_L$, with B_L the one-sided equivalent noise bandwidth of the loop

$$B_L \triangleq \frac{r + 1}{4\tau_2(1 + F_0/r)}$$

and $T_f B_L$ is the ordinate, with $T_f \triangleq \tau_f / F_0AK$ the denormalized acquisition time. The dashed lines represent Richman's formula. As we can see, the two theories are in satisfactory agreement. The same conclusions hold true for Fig. 7 in which (26)

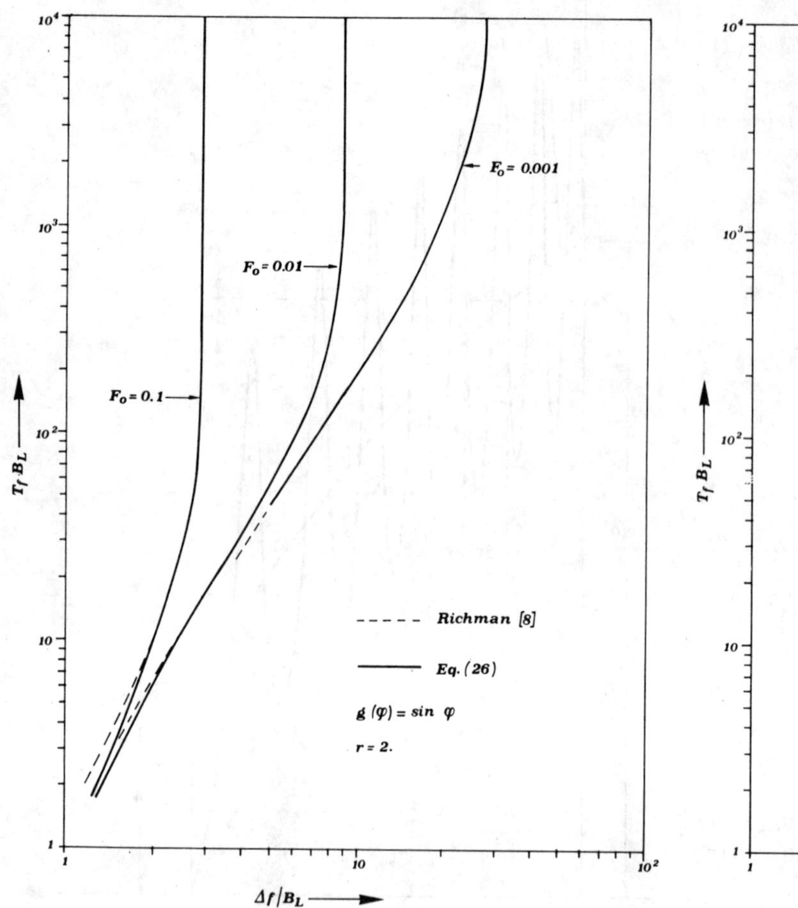

Fig. 6. Acquisition time versus $\Delta f/B_L$ for a sinusoidal nonlinearity.

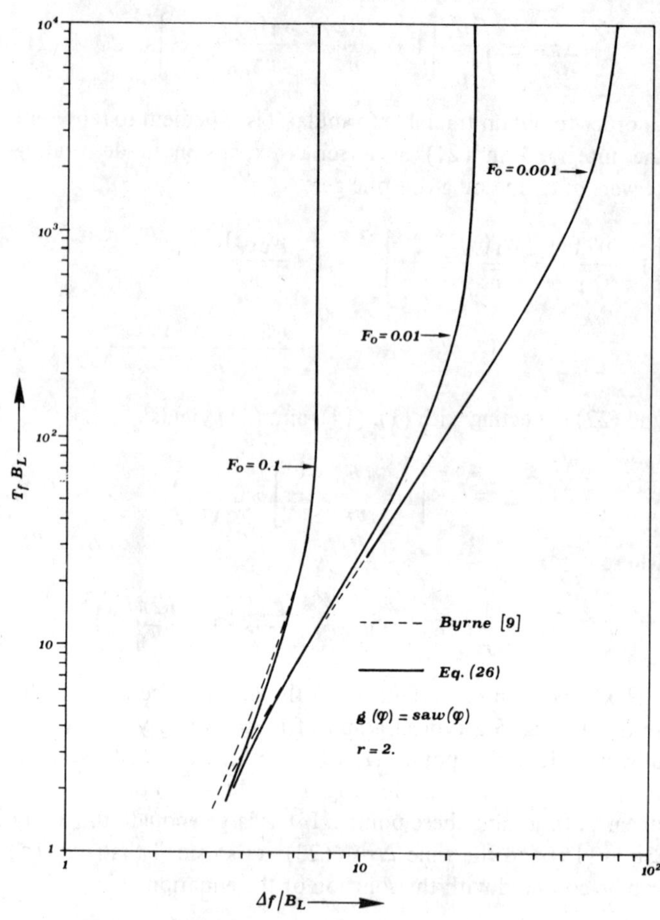

Fig. 7. Acquisition time versus $\Delta f/B_L$ for a sawtooth nonlinearity.

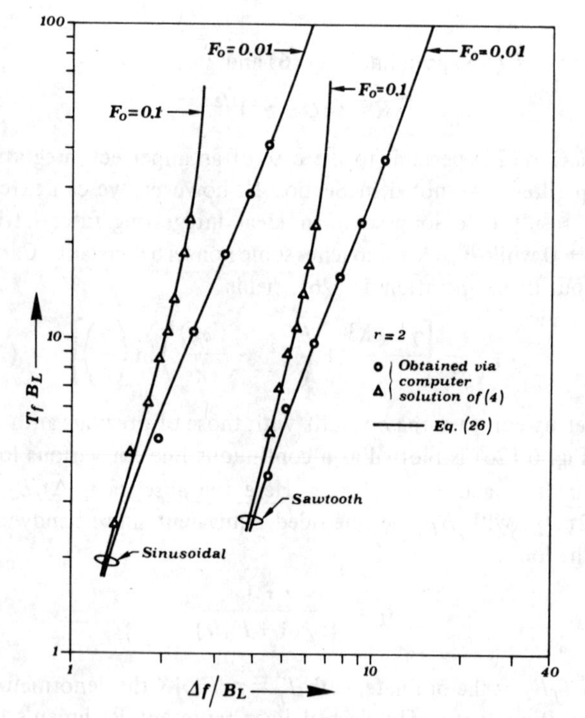

Fig. 8. Comparison with exact values.

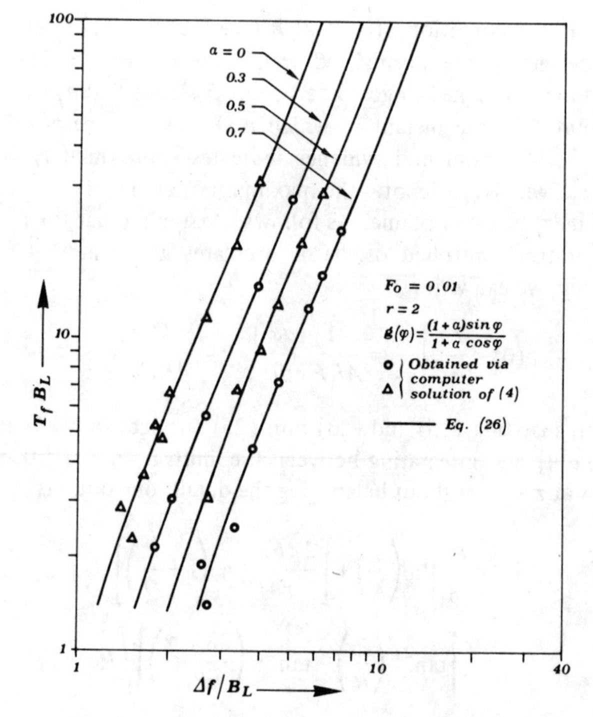

Fig. 9. Acquisition time for a Tanlock nonlinearity.

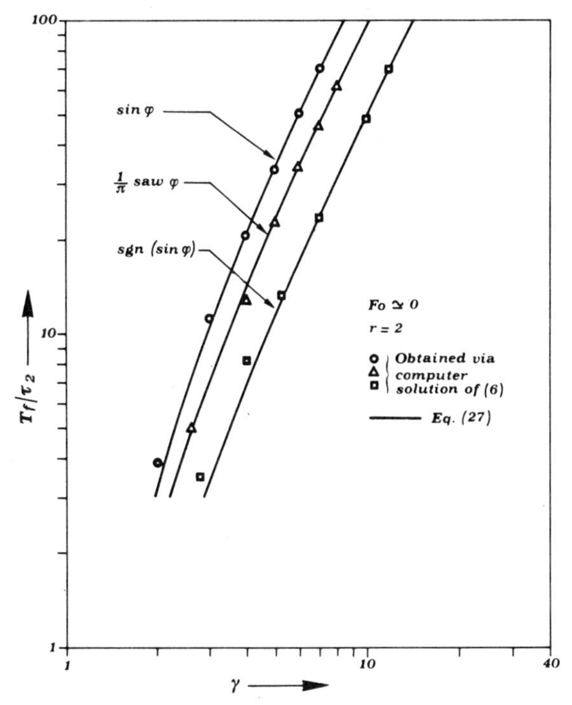

Fig. 10. Acquisition time in the case of a perfect integrating filter.

is compared with the numerical results obtained by Byrne for the sawtooth function.

Fig. 8 illustrates plots of (26) for the sinusoidal and the sawtooth characteristics. Superimposed are the exact values obtained by numerical solution of (4). Analogous results are given in Fig. 9 in the case of a tanlock-type nonlinearity. Finally let us consider the case of a perfect integrating loop filter. For $\gamma >> 1$ Viterbi [12] has found $\tau_f = r\gamma^2$ when $g(\varphi) = \sin\varphi$ and Meer [10] $\tau_f = 3/2\pi^2\, r\gamma^2$ and $\tau_f = 6/\pi^2\, r\gamma^2$ for the sawtooth and triangular functions, respectively. It is easy to verify that all these results satisfy the formula

$$\tau_f = \frac{1}{2\langle g^2 \rangle} r\gamma^2$$

to which (27) is reduced when γ is large.

In Fig. 10 (27) is plotted for some shapes of $g(\gamma)$. Superimposed are the exact values obtained via the numerical solution of (6).

V. Conclusions

A method for estimating the pull-in range and the frequency acquisition time of a second-order loop with any type of nonlinearity has been presented.

A merit of this method is that, for the first time, a closed-form expression has been obtained of the acquisition time of a generalized tracking system. The results of the present analysis have been found in fair agreement with those previously known for some types of nonlinearities and with the exact results obtained via numerical solution of the system equation.

References

[1] W. C. Lindsey, *Synchronization Systems in Communication and Control.* Englewood Cliffs, N. J.: Prentice-Hall, 1972.
[2] J. J. Spliker, "Delay-lock tracking of binary signals," *IEEE Trans. Space Electron. Telem.*, vol. SET-9, pp. 1–8, Mar. 1963.
[3] J. S. Mayo, "Experimental 224 Mb/s PCM terminals," *Bell Syst. Tech. J.*, pp. 1813–1841, Nov. 1965.
[4] F. J. Witt, "An experimental multiplexer-demultiplexer using pulse stuffing synchronization," *Bell Syst. Tech. J.*, pp. 1843–1885, Nov. 1965.
[5] L. M. Robinson, "Tanlock: A phase lock loop of extended tracking capability," in *Proc. Nat. Winter Conf. Military Electronics*, Feb. 1962, Los Angeles, Calif., pp. 396–421.
[6] J. J. Uhran, Jr., and J. C. Lindenlaub, "Experimental results for phase-lock loop systems having a modified *n*th-order tanlock phase detector," *IEEE Trans. Commun. Technol.*, vol. COM-16, pp. 787–795, Dec. 1968.
[7] W. C. Lindsey, "Nonlinear analysis of generalized tracking systems," *Proc. IEEE*, vol. 57, pp. 1705–1722, Oct. 1969.
[8] D. Richman, "Color-carrier reference phase synchronization accuracy in NTSC color television," *Proc. IRE*, vol. 42, pp. 106–133, Jan. 1954.
[9] C. J. Byrne, "Properties and design on the phase controlled oscillator with a sawtooth comparator," *Bell Syst. Tech. J.*, pp. 559–603, Mar. 1962.
[10] S. A. Meer, "Analysis of phase-locked loop acquisition: A quasi-stationary approach," in *1966 IEEE Int. Conv. Rec.*, vol. 14, pt. 7, pp. 85–106.
[11] B. I. Shakhtarin, "Certain characteristics of nonlinear phase synchronization systems," *Telecommun. Radio Eng.*, pp. 126–129, Apr. 1971.
[12] A. J. Viterbi, "Acquisition and tracking behavior of phase-locked loops," *Proceedings of the Symposium on Active Network and Feedback Systems.* New York: Polytechnic Press, 1960.
[13] F. G. Splitt, "Design and analysis of linear phase-locked loop of wide dynamic range," *IEEE Trans. Commun. Technol.*, vol. COM-14, pp. 432–440, Aug. 1970.
[14] E. N. Protonotarios, "Pull-in time in second-order phase-locked loop with a sawtooth comparator," *IEEE Trans. Circuit Theory*, vol. CT-17, pp. 372–378, Aug. 1970.
[15] P. D. Shaft and R. C. Dorf, "Minimization of communication-signal acquisition time in tracking loops," *IEEE Trans. Commun. Technol.*, vol. COM-16, pp. 495–499, June 1968.
[16] T. J. Rey, "Automatic phase-control: Theory and design," *Proc. IRE*, vol. 48, pp. 1760–1771, Oct. 1960.
[17] J. R. Woodbury, "Phase locked loop pull-in range," *IEEE Trans. Commun. Technol.*, vol. COM-16, pp. 184–186, Feb. 1968.
[18] S. I. Yevtyanov and V. K. Snedkova, "Pull-in range in phase-lock automatic frequency control and phase-detector characteristic," *Telecommun. Radio Eng.*, pp. 113–116, May, 1970.
[19] E. T. Whittaker and G. N. Watson, *Modern Analysis.* New York: Cambridge Univ. Press, 1940, ch. 8.

Part III
Threshold

A Threshold Criterion for Phase-Lock Demodulation*

JEAN A. DEVELET, JR.†, MEMBER, IRE

***Summary*—An analytical threshold criterion has been developed for the general phase-lock receiver utilizing Booton's quasi-linearization technique. This criterion is established for arbitrary information and noise spectral densities. The information is assumed phase- or frequency-encoded on the received signal. Explicit results are centered around the case of additive white Gaussian noise.**

The principal nonlinearity is assumed to be the phase detector which is represented as a product device.

Threshold curves are derived for three types of signals:

1) **Bandlimited phase-encoded white Gaussian signals, optimal receiver;**
2) **Bandlimited phase-encoded white Gaussian signals, second-order receiver;**
3) **Frequency-encoded white signals, optimal receiver.**

The phase-encoded white Gaussian signal threshold is then compared with Shannon's results. It was found that the optimal receiver threshold occurs 10 $\log_{10}(e)$ or 4.34 db above Shannon's limit.

The second-order receiver was found to threshold 2 to 3 db above the optimal receiver in the region of normally encountered output signal-to-noise power ratios.

Frequency-encoded white signals represent the character of residual noise in a communication link oscillator system. Residual frequency noise is induced by the ever present thermal noise in oscillator circuits. This particular thermal-induced noise cannot be removed entirely. For this fundamental noise process maximum receiver sensitivities are derived.

An interesting result of quasi-linearization is that, for the signals considered, previous solutions of the Wiener-Hopf equation may be applied with only slight modifications.

I. Introduction

PREVIOUS ANALYSES of phase-lock receiver performance have been based on linearized models of the actual transfer function.[1–4] Since the phe-

* Received June 25, 1962; revised manuscript received, September 26, 1962.

† Aerospace Corporation, El Segundo, Calif. Formerly with Space Technology Laboratories, Inc., a Subsidiary of Thompson Ramo Wooldridge, Inc., Redondo Beach, Calif.

[1] W. J. Gruen, "Theory of AFC synchronization," PROC. IRE, vol. 53, pp. 1043–1048; August, 1953.

[2] B. D. Martin, "A Coherent Minimum-Power Lunar Probe Telemetry System," Jet Propulsion Lab., California Inst. Tech., Pasadena, Calif., External Publication No. 610, pp. 1–72; August, 1959.

[3] C. E. Gilchriest, "Application of the phase-locked loop to telemetry as a discriminator or tracking filter," IRE TRANS. ON TELEMETRY AND REMOTE CONTROL, vol. TRC-4, pp. 20–35; June, 1958.

[4] R. Jaffe and E. Rechtin, "Design and performance of phase-lock circuits capable of near-optimum performance over a wide range of input signal and noise levels," IRE TRANS. ON INFORMATION THEORY, vol. IT-1, pp. 66–76; March, 1955.

Reprinted from *Proc. IEEE*, vol. 51, pp. 349–356, Feb. 1963; and from *Proc. IEEE*, vol. 51, p. 580, Apr. 1963.

nomenon of loop threshold is caused by the effects of nonlinearities, it is understandable that the linear models referenced cannot yield a threshold criterion.

This paper describes the application of the *quasi-linearization* technique of Booton[5] to the determination of an analytic threshold criterion. The analysis will be restricted to the case of Gaussian signals corrupted by additive white Gaussian noise. Extension to other situations is possible but not treated in this paper.

II. Analysis

Consider the phase-lock receiver of Fig. 1.

The low-frequency output of the product demodulator can be shown to be

$$v_e(t) = -E\sin\epsilon(t) - X(t)\sin[m(t)+\epsilon(t)] + Y(t)\cos[m(t)+\epsilon(t)] \quad (1)$$

(Signal)

where

$E=\sqrt{2S_{if}}$, received signal amplitude, volts;
$\epsilon(t)$ = instantaneous loop error, radians;
$m(t)$ = instantaneous signal modulation, a Gaussian random variable, radians;
$X(t)$, $Y(t)$ = uncorrelated white Gaussian noise variables of average power $\Phi_{if}BW_{if}$, watts.

For the normal phase-lock receiver situation the predetection bandwidth, BW_{if}, may be assumed much larger than the phase-lock loop bandwidth. In this situation $Y(t)$ and $X(t)$ have a correlation time[6] much shorter than $m(t)$ or $\epsilon(t)$. Under these conditions, it can be shown that the noise portion of (1) can be represented by a Gaussian variable $N(t)$ which has the same correlation function as $Y(t)$ or $X(t)$. Thus (1) becomes

$$v_e(t) = -E\sin\epsilon(t) + N(t). \quad (2)$$

Eq. (2) may now be utilized to obtain an analytical representation of the general receiver of Fig. 1. Fig. 2 portrays this representation.

The representation of Fig. 2 can be seen to be a simple servomechanism with the exception of the nonlinear element E sin [ ]. Booton[5] has provided an approximation technique which can be used to replace the nonlinear element by an equivalent gain, K_A. This technique essentially determines the *average* gain of the nonlinear device under the expected operating conditions. K_A may be found utilizing an averaging procedure which is a slight variation of Booton's equation (58).[7] This variation was obtained by an integration by parts and is in a

[5] R. C. Booton, Jr., "Nonlinear Control Systems with Statistical Inputs," Mass. Inst. Tech., Cambridge, Mass., Rept. No. 61, pp. 1-35; March 1, 1952.

[6] The time in seconds it takes the correlation function to drop to a small value compared to the value at zero seconds.

[7] *Ibid.*, p. 21.

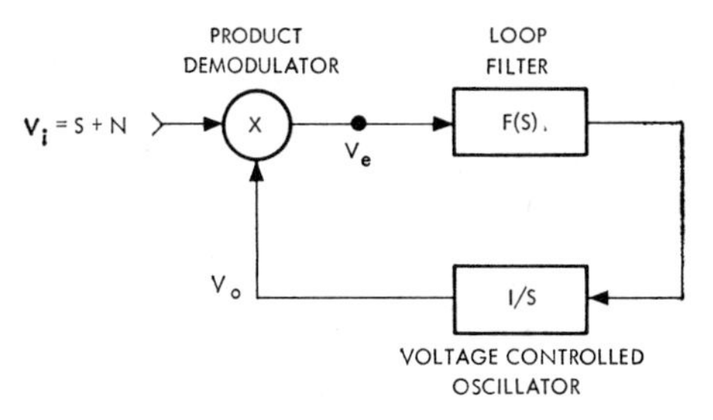

$$V_i = E\sin[wt+m(t)] + x(t)\sin wt + y(t)\cos wt$$
$$V_0 = 2\cos[w(t)+\epsilon(t)]$$

Fig. 1—General phase-lock receiver.

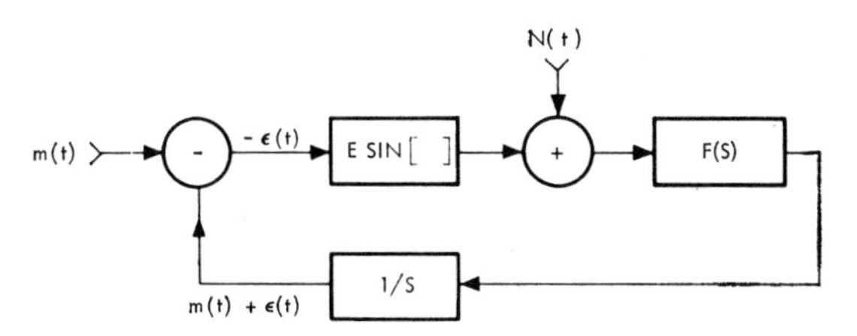

Fig. 2—Analytical receiver representation.

more convenient form:

$$K_A = \int_{-\infty}^{\infty} g'(x)p_1(x)dx \quad (3)$$

where

K_A = equivalent element gain,
$g'(x) = E\cos x$,
$p_1(x)$ = probability density of $\epsilon(t)$ which must be Gaussian to conform to Booton's criteria.

Letting $\overline{\epsilon(t)^2}=\sigma^2$, substituting into (3), and integrating yields

$$K_A = E\exp(-\sigma^2/2). \quad (4)$$

The *quasi-linear* receiver representation obtained by linearizing the E sin [ ] transfer function is now shown in Fig. 3.

Denoting the signal and noise *one-sided* power-spectral densities of $m(s)$ and $N(s)$ as $\Phi_m(\omega)$ rad²/cps and $\Phi_n(\omega)$ watts/cps respectively, it is a simple matter to show

$$\sigma^2 = \underbrace{\int_0^\infty \Phi_m(\omega)\left|1-\frac{\phi_0}{\phi_i}(\omega)\right|^2 df}_{\text{Modulation Error}} + \underbrace{\int_0^\infty \frac{\exp(\sigma^2)\Phi_n(\omega)}{2S_{if}}\left|\frac{\phi_0}{\phi_i}(\omega)\right|^2 df}_{\text{Noise Error}} \quad (5)$$

where

$$\frac{\phi_0}{\phi_i}(\omega) = \frac{E\exp\left(-\frac{\sigma^2}{2}\right)F(s)/s}{1+E\exp(-\sigma^2/2)F(s)/s}.$$

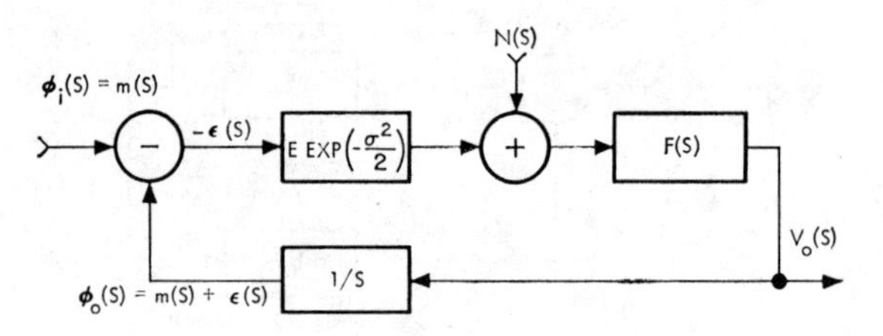

Fig. 3—Quasi-linear phase-lock receiver.

Eq. (5) is a key relationship. As a consequence of the exp (σ^2), it implicitly contains a threshold criterion for the receiver model of Fig. 3. It will be shown more explicitly in what follows how this criterion is obtained.

Eq. (5) can be solved for the received signal power S_{if} as follows:

$$S_{if} = \frac{\dfrac{\exp \sigma^2}{2} \displaystyle\int_0^\infty \Phi_n(\omega) \left| \frac{\phi_0}{\phi_i}(\omega) \right|^2 df}{\sigma^2 - \displaystyle\int_0^\infty \Phi_m(\omega) \left| 1 - \frac{\phi_0}{\phi_i}(\omega) \right|^2 df}. \tag{6}$$

It is desired now to choose a function ϕ_0/ϕ_i such that for a specified σ, $\Phi_m(\omega)$, and $\Phi_n(\omega)$, S_{if} is minimized. This function will yield maximum receiver sensitivity. In order to find this optimum function let

$$\frac{\phi_0}{\phi_i}(\omega) = A(\omega) e^{i\theta(\omega)}. \tag{7}$$

Substitution of (7) in (6) yields

$$S_{if} = \frac{\dfrac{\exp(\sigma^2)}{2} \displaystyle\int_0^\infty \Phi_n(\omega) A^2(\omega) df}{\sigma^2 - \displaystyle\int_0^\infty \Phi_m(\omega)[1 - 2A(\omega)\cos\theta(\omega) + A^2(\omega)] df} \cdot \tag{8}$$

Clearly for minimum S_{if}, $\theta(\omega)$ should be chosen equal to zero. Thus,

$$S_{if} = \frac{\dfrac{\exp(\sigma^2)}{2} \displaystyle\int_0^\infty \Phi_n(\omega) A^2(\omega) df}{\sigma^2 - \displaystyle\int_0^\infty \Phi_m(\omega)[1 - A(\omega)]^2 df}. \tag{9}$$

$A(\omega)$ may be found by standard variational methods. That is, add to $A(\omega)|_{\text{opt}}$ the function $\epsilon\eta(\omega)$. Eq. (9) is then differentiated with respect to ϵ and the derivative and ϵ set equal to zero. Since $\eta(\omega)$ is arbitrary, an equation for $A(\omega)|_{\text{opt}}$ for all ω is obtained. The result of this mathematical manipulation is

$$A(\omega)|_{\text{opt}} = \frac{\Phi_m(\omega)}{\Phi_m(\omega) + \dfrac{\Phi_n(\omega)\exp(\sigma^2)}{2S_{if_{\min}}}}. \tag{10}$$

Eq. (10) is an interesting result. It is the identical result one obtains if the minimization of σ^2 in (5) is carried out. Since minimization of S_{if} and σ^2 is achieved by the same transfer function, the general results of Wiener are applicable with the exception of the fact that the noise power-spectral density is replaced by

$$\frac{\Phi_n(\omega)\exp(\sigma^2)}{2S_{if_{\min}}}.$$

Further consideration of Wiener-Hopf solutions will use results of previous workers in the field. In particular only the case of white Gaussian noise will be treated. That is $\Phi_n(\omega) = 2\Phi_{if}$ where Φ_{if} is the *one-sided* predetection power-spectral density of the receiver.

General Solution for White Gaussian Noise

Yovits and Jackson in their important paper 1955[8] found a particularly useful form of the optimum *realizable* transfer function for the situation of white noise. Their results for *linear* filters are repeated below utilizing the terminology previously developed,[9] as follows:

$$\left| 1 - \frac{\phi_0}{\phi_i}(\omega) \right|^2_{\text{opt}} = \frac{\Phi_n}{\Phi_n + \Phi_m(\omega)}. \tag{11}$$

The corresponding minimum following error is

$$\sigma^2 = \Phi_n \int_0^\infty \log_\epsilon \left(1 + \frac{\Phi_m(\omega)}{\Phi_n} \right) df. \tag{12}$$

In order to treat the performance of the quasi-linear receiver model one need just replace Φ_n in (11) and (12) by Φ_{if} exp $(\sigma^2)/S_{if}$. The following fundamental relations are thereby obtained:

$$\left| 1 - \frac{\phi_0}{\phi_i}(\omega) \right|^2_{\text{opt}} = \frac{\Phi_{if}\exp(\sigma^2)}{\Phi_{if}\exp(\sigma^2) + S_{if}\Phi_m(\omega)}, \tag{13}$$

$$\sigma^2 = \frac{\Phi_{if}\exp(\sigma^2)}{S_{if}} \int_0^\infty \log_\epsilon \left(1 + \frac{S_{if}\Phi_m(\omega)}{\Phi_{if}\exp(\sigma^2)} \right) df. \tag{14}$$

Eq. (14) is one of the principal results of this paper. Given a receiver output quality constraint and a receiver noise density Φ_{if}, (14) will not yield a bounded solution for σ^2 if the received signal power S_{if} is too small. The value of S_{if} below which the solution for σ^2 ceases to exist is the threshold for phase-lock demodulation obtained from the quasi-linear model.

III. Applications

The previous results will now be applied to three situations which are important in communication engineering. In all cases the background noise will be assumed white and Gaussian. The signal modulations and transfer functions to be considered are the following:

[8] M. C. Yovits and J. L. Jackson, "Linear filter optimization with game theory considerations," 1955 IRE National Convention Record, pt. 4, pp. 193–199.

[9] *Ibid.*, pp. 195–196.

1) Reception of a bandlimited phase-encoded white Gaussian signal spectrum with the optimal transfer function,
2) Reception of a bandlimited phase-encoded white Gaussian signal spectrum with a second-order transfer function,
3) Reception of a frequency-encoded white signal spectrum with the optimal transfer function.[10]

The first case is important to illustrate optimal communication of information by use of a phase-lock receiver.

The second situation is of practical interest since a second-order loop is easily realized and is amenable to measurements verifying the theory which has been developed.

The last situation is important in that oscillators in a communication link are not perfectly stable and must be "tracked" in phase in order to continue receiving the signal. The results here point to fundamental sensitivity limitations of phase-lock receivers given the oscillator system is perturbed by a random walk phase process.

Bandlimited Phase-Encoded White Gaussian Signal, Optimal Receiver

Let the signal power-spectral density in (14) be given by

$$\begin{aligned} \Phi_m(\omega) &= \Phi_m \qquad 0 \le f \le f_m, \\ \Phi_m(\omega) &= 0 \qquad f_m < f. \end{aligned} \tag{15}$$

Integration of (14) yields

$$\sigma^2 = \frac{\Phi_{if} \exp(\sigma^2)}{S_{if}} f_m \log_\epsilon \left[1 + \frac{\Phi_m S_{if}}{\Phi_{if} \exp(\sigma^2)} \right]. \tag{16}$$

Since $\Phi_{if} \exp(\sigma^2)/S_{if}$ is the *one-sided* phase noise power-spectral density in the receiver output,[11] and $\Phi_m f_m$ is the mean-square signal power in the receiver output, one may rewrite (16) in a more meaningful form exhibiting signal-to-noise power ratios input and output. Thus,

$$\left(\frac{S}{N}\right)_i = \frac{\exp(\sigma^2)}{2\sigma^2} \log_\epsilon \left[1 + \left(\frac{S}{N}\right)_0 \right] \tag{17}$$

where

$$\left(\frac{S}{N}\right)_i = \frac{S_{if}}{2\Phi_{if} f_m};$$

the input signal-to-noise power ratio referred to twice the information bandwidth.

$$\left(\frac{S}{N}\right)_0 = \frac{\Phi_m f_m}{\dfrac{\Phi_{if}}{S_{if}} \exp(\sigma^2) f_m} = 2\left(\frac{S}{N}\right)_i \frac{{\sigma_m}^2}{\exp(\sigma^2)};$$

the output signal-to-noise power ratio referred to the information bandwidth. σ_m = modulation index, radians.

Eq. (17) has a minimum value for $(S/N)_i$ at $\sigma = 1.0$ for a fixed system output quality $(S/N)_0$. Substitution of $\sigma = 1.0$ radian in (17) yields the threshold result depicted in Fig. 4.

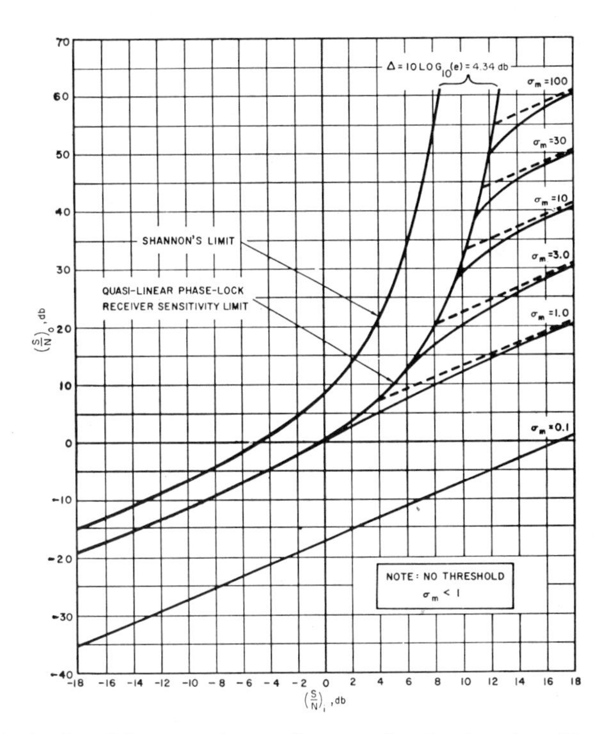

Fig. 4—Quasi-linear receiver performance for the situation of bandlimited white Gaussian phase-encoded signals with the optimal transfer function.

The curves above threshold represent system output quality vs signal-to-noise power ratio input asymptotically converging on the conventional high $(S/N)_i$ relation. These are constructed with the modulation index $\sigma_m = \sqrt{\Phi_m f_m}$ as a parameter. Since

$$\left(\frac{S}{N}\right)_0 = \frac{2{\sigma_m}^2 \left(\dfrac{S}{N}\right)_i}{\exp(\sigma^2)},$$

and σ is a function of $(S/N)_i$ governed by (17), a simultaneous solution of this relation and (17) yields σ and hence $(S/N)_0$ vs $(S/N)_i$. It is interesting to note the curvature as one approaches threshold. Note also for $\sigma_m < 1.0$ the quasi-linear model yields no threshold.

The threshold criterion depicted by Fig. 4 is significantly below a standard FM or PM discriminator as would be expected since more optimal demodulation is employed. It is, however, higher than that predicted by use of Shannon's results. This is as it should be, for above threshold information may be conveyed by the phase-lock communication system and the rate of communication should be bounded by a communication theoretical result such as Shannon's. More discussion of this appears in Section IV.

[10] This situation corresponds to a random walk in signal phase.
[11] Inspection of the transfer function $V_0(s)/N(s)$ in Fig. 3 reveals the fact $\exp(\sigma^2)$ is a factor in the output noise also.

Bandlimited Phase-Encoded White Gaussian Signal, Second-Order Receiver

The previous section treated the case of the optimal receiver which requires an infinite number of elements to realize. The second-order receiver though not optimum is easy to realize in hardware.

As with the optimal receiver, let the signal power-spectral density be given by (15), as follows

$$\Phi_m(\omega) = \Phi_m \qquad 0 \le f \le f_m,$$
$$\Phi_m(\omega) = 0 \qquad f_m < f.$$

Return to the fundamental equation (5) is necessary since we are not dealing with an optimal receiver for the signal given by (15).

The transfer function $\phi_0/\phi_i(\omega)$ has been derived by Gruen for the second-order receiver.[1] Gruen's results are restated below for convenience.

$$\frac{\phi_0}{\phi_i}(s) = \frac{\omega_n^2\left[1 + \left(\frac{2\zeta}{\omega_n} - \frac{1}{K}\right)s\right]}{\omega_n^2 + 2\zeta\omega_n s + s^2} \tag{18}$$

where

ω_n = loop natural frequency, rad/sec,
ζ = loop damping ratio,
K = loop gain, sec.$^{-1}$.

Eq. (18) may be obtained from (5) by substituting for $F(s)$ the transfer function of a lag-lead filter.

$$F(s) = K_0 \frac{1 + \tau_1 s}{1 + \tau_2 s} \tag{19}$$

where

K_0 = a dc gain,
τ_1; τ_2 = time constants of the lag-lead filter, seconds.

Various methods of realizing $F(s)$ have been discussed thoroughly by Martin.[2] In addition to the above substitution for $F(s)$, the following parameters are defined after Gruen:[12]

$$\omega_n^2 = \frac{K}{\tau_2},$$
$$\tau_1 = \frac{2\zeta}{\omega_n} - \frac{1}{K},$$
$$K = E \exp\left(-\frac{\sigma^2}{2}\right) K_0. \tag{20}$$

where

ω_n = loop natural frequency, rad/sec,
ζ = loop damping factor.

Since (20) shows that the loop natural frequency, and damping vary with σ the loop error, all analysis to follow will be carried out in terms of the zero σ (high $(S/N)_i$) values of these quantities, ω_{n0} and ζ_0 respectively. The receiver design will center around optimization of ω_{n0} to achieve maximum sensitivity.

[12] Gruen, *op. cit.*, p. 1045.

Eqs. (15) and (18) may now be substituted in (5) and the two integrals evaluated. The following simplifying assumptions will be made to yield meaningful results without resort to computer integration of the modulation error in (5):

$$\frac{2\zeta}{\omega_n} \gg \frac{1}{K},$$
$$\frac{2\pi f_m}{\omega_n} \ll 1.$$

The first assumption is usually the case in practical receivers. The second assumption restricts the validity of our calculations to the more interesting region of high signal-to-noise ratios in the receiver output. With the above substitutions and assumptions, it can be shown that (5) becomes, for the second-order receiver

$$\sigma^2 = \frac{(2\pi)^4 \Phi_m f_m^5 \exp(\sigma^2)}{5\omega_{n0}^4} + \frac{\Phi_{if}\exp(\sigma^2)\left[1 + 4\zeta_0^2\exp\left(-\frac{\sigma^2}{2}\right)\right]}{8S_{if}\zeta_0}\omega_{n0}. \tag{21}$$

Assuming a fixed damping ζ_0, Φ_{if}, Φ_m and f_m, (21) may be solved for S_{if} and minimized with respect to ω_{n0}. Thus, maximum receiver sensitivity is achieved. Performing this manipulation one obtains as the minimum receiver input signal to noise-power ratio defined as in the optimal receiver,

$$\left(\frac{S}{N}\right)_i = \frac{S_{if}}{2\Phi_{if}f_m} = \frac{\exp\left(\frac{6}{5}\sigma^2\right)}{2\sigma^2} \cdot \left\{\frac{5\pi\left[1 + 4\zeta_0^2\exp\left(-\frac{\sigma^2}{2}\right)\right]}{16\zeta_0}\right\}^{4/5}\left(\frac{S}{N}\right)_0^{1/5}, \tag{22}$$

where the relation for signal-to-noise output power ratio is given by the same relation as in the optimal receiver, *i.e.*,

$$\left(\frac{S}{N}\right)_0 = \frac{2\sigma_m^2\left(\frac{S}{N}\right)_i}{\exp(\sigma^2)}. \tag{23}$$

As in (17), (22) has a minimum value at a particular value of σ which depends on system output quality $(S/N)_0$ and receiver damping ratio ζ_0. Only one particular value of $\zeta_0 = 1/\sqrt{2}$ will be chosen in what follows.[13]

[13] Since the previous section has shown the second-order receiver is not optimal for the type of signal treated here, the author does not feel justified in using other than a commonly encountered value for ζ_0.

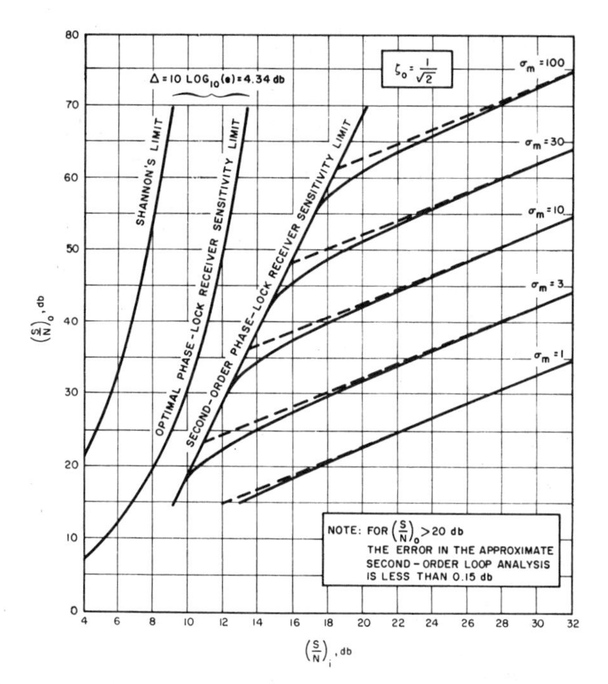

Fig. 5—Quasi-linear receiver performance for the situation of band-limited white Gaussian phase-encoded signals with a second-order transfer function.

For $\zeta_0 = 1/\sqrt{2}$ the value of σ which yields minimum $(S/N)_i$ and hence minimum receiver threshold is 1.01 radians. Substitution of $\sigma = 1.01$ radians into (22) yields the following threshold relation:

$$\left(\frac{S}{N}\right)_i = 4.08\left(\frac{S}{N}\right)_0^{1/5}. \tag{24}$$

Fig. 5 graphs (24). Note that only values for $(S/N)_0 > 20$ db are considered. This is consistent with the approximate evaluation of the modulation error portion of (5).

As with the optimal receiver, the curves above threshold represent system output quality vs signal-to-noise power ratio input asymptotically converging on the conventional high $(S/N)_i$ relation. These are constructed with the modulation index σ_m as a parameter. Loop error, σ, may be eliminated for purposes of the graphs in Fig. 5 by simultaneous solution of (22) and (23).

Shannon's lower limit and the optimal phase-lock receiver sensitivity are shown for comparison in Fig. 5. It is interesting to note that for output signal-to-noise ratios of practical interest (≈ 20 to 40 db) that the second-order receiver is only 2 to 3 db poorer than the optimal receiver or 6 to 7 db poorer than Shannon's limit.

Frequency-Encoded White Signals, Optimal Receiver

In a coherent communication system, oscillator (clock) stability is of great importance especially when accurate Doppler measurements or low information rates are to be conveyed through the link.

In general it is not easy to describe the effect of oscillator noise on coherent systems because the shape of the oscillator system phase power-spectral density is not known. In one specific instance, however, the effect is amenable to calculation, that is, when the resulting random process imposed on the received signal phase is caused by thermal noise. In this special instance Edson[14] has shown that the *frequency* modulation $[m \cdot (t)$ Fig. 2$]$ has a white power-spectral density. Develet[15] then obtained by simple integration and Fourier transformation the *one-sided* phase power-spectral density, Φ_m, given by

$$\Phi_m = \frac{2}{\tau_c \omega^2}, \qquad \frac{\text{rad}^2}{\text{cps}} \tag{25}$$

where

τ_c = coherence time of the oscillator system; the time in seconds it takes the phase drift to build up to one (1) radian rms,

ω = radian frequency, rad/sec.

This power-spectral density can be identified with $m(t)$ a random walk process.

One seeks now the fundamental receiver sensitivity in the presence of additive white Gaussian noise given that the received signal has a phase power-spectral density governed by (25). Substitution of (25) in (14) yields

$$\sigma^2 = \frac{\Phi_{if} \exp(\sigma^2)}{S_{if}} \int_0^\infty \log_\epsilon \left[1 + \frac{2S_{if}}{\omega^2 \tau_c \Phi_{if} \exp(\sigma^2)}\right] df. \tag{26}$$

Integration gives the simple result

$$S_{if} = \frac{\Phi_{if}}{2\tau_c}\left[\frac{\exp(\sigma^2)}{\sigma^4}\right]. \tag{27}$$

Eq. (27) may be differentiated with respect to σ to find the minimum value of S_{if}. Thus,

$$S_{if}\big|_{\min} = \frac{\Phi_{if}}{\tau_c}\left(\frac{e^2}{8}\right) \tag{28}$$

where

$$\sigma_{\min} = \sqrt{2} \text{ radian.}$$

Considering that the receiver noise density is given by Boltzmann's constant times the equivalent receiver temperature, (28) may be restated as

$$S_{if}\big|_{\min} = \frac{KT_{\text{eq}}}{\tau_c}\left(\frac{e^2}{8}\right) \tag{29}$$

[14] W. A. Edson, "Noise in oscillators," *Proc. IRE*, vol. 48, pp. 1454–1466; August, 1960.

[15] J. A. Develet, Jr., "Fundamental Sensitivity Limitations for Second Order Phase-Lock Receivers," presented at URSI Spring Meeting, Washington, D.C., May 4, 1961, STL Tech. Note 8616-0002-NU-000; June 1, 1961.

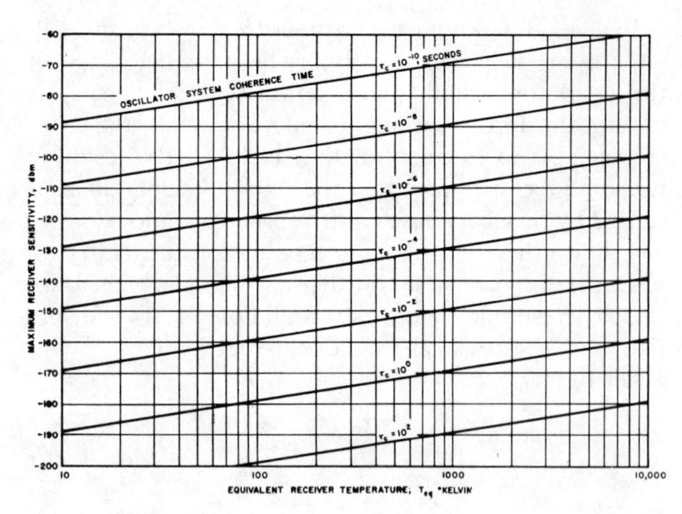

Fig. 6—Receiver sensitivity vs equivalent receiver temperature.

where

$K = 1.38 \times 10^{-23}$, Joules/°K,

T_{eq} = equivalent receiver temperature, °K.

Fig. 6 plots phase-lock receiver sensitivity as a function of T_{eq} with τ_c as a parameter. It represents a fundamental sensitivity limitation for phase-lock reception given by the *quasi-linear* receiver model, for reception of signals disturbed by a random walk phase process.

IV. Comparison with Shannon's Results

Let us now derive Shannon's limit as depicted in Fig. 4. Consider the communication channel of Fig. 7. It is clear that the error-free information rate in bits/sec of the video portion of the link cannot be greater than the maximum error-free information rate which can be passed through the RF channel. This fact allows one to use Shannon's theorem to bound the output signal to noise-power ratio $(S/N)_0$ for a given input signal to noise-power ratio $(S/N)_i$. The phase-lock demodulator cannot enhance the information transfer process.

The RF signal modulation considered in the derivation of Fig. 4 was bandlimited, phase-encoded, white, and Gaussian. Since it is difficult to specify RF bandwidth occupancy for such a phase modulation at intermediate modulation indices ($\sigma_m \approx 1.0$), only an extreme case will be considered which yields an absolute lower bound on RF signal-to-noise power ratio for a given output signal-to-noise power ratio. It will be assumed that the RF channel has infinite bandwidth. Shannon has shown that this condition corresponds to the case of maximum information flow per watt of RF channel signal power in the presence of white noise.

For this case, the use of Shannon's result[16] gives the upper bound on information flow in the RF channel as

$$C_{\max} = \frac{S_{if}}{\Phi_{if} \log_e 2}, \quad \text{bits/sec.} \tag{30}$$

[16] C. E. Shannon, "Communication in the presence of noise," Proc. IRE, vol. 37, pp. 10–21; January, 1949.

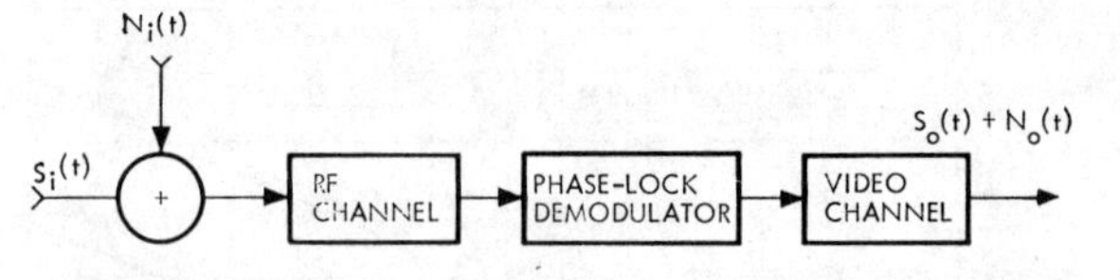

Fig. 7—Communication channel.

Since the signals are bandlimited and the noise is white and additive in the quasi-linear receiver model output, the information rate passing out of the demodulator cannot exceed

$$C_0 = f_m \frac{\log_e \{1 + (S/N)_0\}}{\log_e 2}, \quad \text{bits/sec.} \tag{31}$$

In addition, $C_{\max} \geq C_0$ because the phase-lock demodulator cannot enhance the RF channel's ability to transfer information, but can only equal or degrade this ability. Therefore

$$\left(\frac{S}{N}\right)_i = \frac{S_{if}}{2\Phi_{if} f_m} \geq \frac{1}{2} \log_e \left\{1 + \left(\frac{S}{N}\right)_0\right\}. \tag{32}$$

Eq. (32) represents a lower bound on required signal-to-noise power input and is graphed in Fig. 4. A significant point is the consistency of Booton's quasi-linear approximate model and Shannon's result. Note the constant difference factor, $10 \log_{10}(e)$, for all values of $(S/N)_0$. A difference factor is to be expected, for in real-time phase-lock demodulation no *a priori* knowledge of the signals to be transmitted is available as in Shannon demodulation.

V. Conclusions

For the first time a simple approximate theory for phase-lock threshold has been developed.

The quasi-linearization technique of Booton was the principal analytic tool in developing this threshold criterion.

It was shown that all previous Wiener-Hopf solutions in the linear filter domain apply in this situation with one minor adjustment. The noise power-spectral density is replaced by $\Phi_{if} \exp(\sigma^2)/S_i$ wherever it appears in previous results. It must be kept in mind that only one particular servomechanism (the phase-lock loop) and nonlinearity (the product demodulator) was treated in this analysis.

Certain interesting examples were shown to illustrate the usefulness of the procedure. In particular for the special case of communication with white Gaussian phase-encoded signals in the presence of additive white Gaussian noise the optimal phase-lock receiver threshold performance was found to be 4.34 db poorer than Shannon's limit. This degradation in performance is explained by the fact that no *a priori* knowledge of the signals to be transmitted is available in real-time phase-lock demodulation. This contrasts to the Shannon demodulator in which knowledge of the set of transmitted signals is available to the receiver. Apparently *a priori* knowledge is worth 4.34 db in signal power.

Limited measurements which have been conducted on a wideband second-order loop at Space Technology Laboratories, Inc., Redondo Beach, California, indicate very close agreement with predictions based on the quasi-linear theory for loop rms phase error, σ, out to the predicted threshold.[17]

It is important to note that the threshold analysis in this paper was based on a mean-square signal, noise, and loop error criterion. Threshold is defined as the input signal-to-noise power ratio at which, for a given quality constraint, the loop error becomes unbounded. If, however, short-term statistics of the receiver output are important to the observer, an entirely new criterion may require development. In any event, it is considered this paper presents the *lower bound* on sensitivity, because with any new criterion a bounded mean-square loop error will certainly be a prerequisite.

[17] R. C. Booton, Jr., "Demodulation of wideband frequency modulation utilizing phase-lock techniques," *Proc. Nat'l Telemetering Conf.*, Washington, D. C., May 23–25, 1962, vol. 2, Fig. 11, p. 14.

ACKNOWLEDGMENT

The author wishes to acknowledge the stimulating discussions with Dr. R. C. Booton, Jr., of Space Technology Laboratories, Inc., Redondo Beach, Calif., and Dr. C. R. Cahn of the Bissett-Berman Corp., Santa Monica, Calif., which aided in my solution of this problem.

CORRECTION

Jean A. Develet, Jr., author of "A Threshold Criterion for Phase-Lock Demodulation," which appeared on pages 349–356 of the February, 1963, issue of PROCEEDINGS, has called the following to the attention of the *Editor*.

On page 350 [p. 169 in this reprint volume] in (1) the word "Signal" should be deleted.

Also on page 350 [p. 169], in Fig. 1, the second line of the equation should read $V_0 = 2 \cos [w(t) + m(t) + \epsilon(t)$ instead of $V_0 = 2 \cos [w(t) + \epsilon(t)]$.

On page 354 [p.173 in this volume], column 2 line 8 $m^{1\prime}(t)$ should read $\dot{m}(t)$.

Part IV
Stability

THE EFFECT OF TIME DELAYS ON THE STABILITY OF PRACTICAL PHASE LOCKED RECEIVERS WITH MULTIPLE CONVERSION BY UTILIZING ROOT LOCUS WITH POSITIVE FEEDBACK

Daniel D. Carpenter
TRW Systems Group
Redondo Beach, Calif.

ABSTRACT

A practical root locus technique is given for determining the stability of phase-locked receivers utilizing multiple frequency conversions. An approximate phase transfer function is derived for the predetection crystal filter in the loop and for each IF amplifier stage that follows a frequency conversion. The transfer function of the IF amplifiers have the form $H(s) = -(ds-1)$. The effect of $H(s)$ in the last IF stage is to produce positive feedback for all nonzero delays, d. However, the loop is not necessarily unstable due to the positive feedback. The particular case of the Pioneer spacecraft receiver with triple conversion is used to demonstrate the method with root locus sketches.

INTRODUCTION

The basic stability problem for a practical phase-locked receiver with multiple frequency conversion results from the time delays in the loop and the predetection filter. However, the effects of IF amplifiers stages and the predetection crystal filter are usually ignored in obtaining the closed loop linear transfer function. But in order to develop a suitable practical model for stability analysis, the interaction of these effects must be considered.

Lindsey[1] discusses several interesting practical cases in which delays enter into phase-lock loop problems. For the general effect of a transportation delay, d, Lindsey modifies the open loop transfer function by exp(-ds) and proceeds to derive the conditions for the critical delay to prevent oscillation. Shpolyanskiy[2], et al, use this transfer function to represent the delay of an IF amplifier in a phase-locked loop. They also give a condition for the critical delay and indicate that it is the result of applying the Mikhailov stability criterion. However, for determining the transient response and degree of stability, this transfer function complicates root locus and other types of stability analysis. The problem becomes more difficult when an attempt is made to incorporate more than one IF amplifier response for the cases with multiple conversion in the phase locked loop.

In order to obtain a useful stability analysis technique for phase-locked receiver with multiple conversions, a phase transfer function is derived for the IF amplifier stages and the crystal filter that allows the direct application of root locus transient stability analysis.[3] The phase transfer function for an IF stage is given by the form

$$H(s) = 1-ds$$

and is based on a deriation using an instantaneous frequency argument. The delay d is equivalent to the slope of the phase transfer function for the IF stage. Also derived is the phase transfer function of a "single pole" crystal filter given by

$$H_c(s) = \frac{1}{1+T_c s}$$

By writing the phase transfer function of an IF stage in the form

$$H(s) = -(ds-1)$$

the direct application of root locus techniques with positive feedback can be made. The effect is to produce cases with positive feedback with a right half plane zero at 1/d. It is demonstrated that this does not necessarily imply instability.

Based on these results the case of the Pioneer spacecraft receiver, with triple conversion, is used to demonstrate the method by means of root locus diagrams.[3] First, the effects of the crystal filter and coupling networks are considered. These effects give rise to negative feedback, but instability can result. Next, the additional effects of time delay in the first and third IF amplifiers are considered. It is shown that the first IF delay is relatively unimportant when compared with the third IF delay. In fact, for any non-zero delay d_3 for the third IF positive feedback always results with the loop still being very stable for reasonably small d_3. The interpretations are given by an interesting set of positive feedback root locus sketches.

STABILITY ANALYSIS

Figure 1 gives a block diagram of the Pioneer spacecraft phase locked receiver to be used for the analysis. This particular receiver utilizes three frequency conversions and the analytical model developed is general enough to demonstrate how the stability analysis can be applied to other cases. The input phase to the loop is $\Theta_i(t)$ and the multiplied version of the VCO phase for comparison is $\Theta_o(t)$. The various voltages at points in the loop are indicated with the phase functions being given in operator form. In general linearizing the loop equation by assuming $\sin \Phi_3(t) \cong \Phi_3(t)$ will yield the loop transfer function. This can also be accomplished by inspection from the linearized block diagram of Figure 2. The individual phase transfer functions, derived in the appendices, are

$$H_1(s) = 1-d_1 s, \quad d_1 = \text{time delay of } 1^{st} \text{ IF stage}$$

$$H_2(s) = \frac{1}{1+T_c s}, \quad \frac{1}{T_c} = 1{,}000\ \pi \text{ rad/sec}$$

$$H_3(s) = 1-d_3 s, \quad d_3 = \text{time delay of } 3^{rd} \text{ IF stage}$$

The general transfer function of the loop is

$$\frac{\hat{\Theta}_o(s)}{\hat{\Theta}_i(s)} = \frac{MK_\ell\, H_1(s)H_c(s)H_3(s)\, F(s)/s}{1\ K_\ell H_c(s)H_3(s)\ [MH_1(s)+Q/H_c(s)+N]\ \frac{F(s)}{s}}$$

where M = 108, N = 2, Q = 1/2. Substituting for the individual phase transfer functions yields

$$\frac{\hat{\Theta}_o(s)}{\hat{\Theta}_i(s)} = \frac{108\ K_\ell \frac{(1-d_3 s)}{(1+T_c s)} \left[s\left(\frac{T_c}{2} - 108\ d_1\right) + 110.5\right] \frac{F(s)}{s}}{1+K_\ell \frac{(1-d_3 s)}{(1+T_c s)} \left[s\left(\frac{T_c}{2} - 108\ d_1\right) + 110.5\right] \frac{F(s)}{s}}$$

Reprinted from *Conf. Rec., 1973 IEEE Int. Conf. on Commun.*, vol. II, June 11-13, 1973, pp. 33-14-33-19.

For stability we are concerned with the denominator

$$D(s) \equiv 1 + KH(s)$$

$$\equiv 1 + K_\ell \frac{(1-d_3 s)}{(1+T_c s)} \left[s\left(\frac{T_c}{2} - 108\, d_1\right) + 110.5\right] \frac{F(s)}{s}$$

In order to interpret positive and negative feedback by root locus methods we use the following rule. Change all negative coefficients of s, in the single pole and zero form $\pm s \pm a$, to positive values such that the resulting sign of KH(s) determines the overall sense of feedback. Thus 1+KH(s) represents negative feedback and 1-KH(s) represents positive feedback. This allows the usual root locus angular conventions to be applied directly to the positive feedback case.

This procedure yields the form

$$D(s) = 1 - K_\ell \frac{(d_3 s-1)}{(1+T_c s)} \left[s\left(\frac{T_c}{2} - 108\, d_1\right) + 110.5\right] \frac{F(s)}{s}$$

indicating the case of positive feedback provided $T_c > 216\, d_1$. However, this is easily established since

$$d_1 < \frac{T_c}{216} = \frac{1}{216{,}000\,\pi} = 1.47\ \mu s$$

where d_1 will be at least two orders of magnitude less than 1μ sec. Thus we have positive feedback with a right half plane zero at $s = 1/d_3$. It is noticed that for this case we have positive feedback for all non zero values of delay d_3. It is obvious that for d_1 and T_c set to zero and d_3 very small the form of the loop approaches that of the conventional case and can be made stable. Thus, although we have positive feedback there is no reason to suppose that we necessarily have instability. Before we discuss the interesting root locus sketches with delay it is instructive to consider the cases without delay first. These cases demonstrate the additional effects of the crystal filter and the coupling networks on the conventional transfer function. Parameter values that are used in the analysis are defined as follows:

Natural frequency at threshold	ω_n = 800 rad/sec
Damping ratio at threshold	ζ = .707
Limiter suppression factor	K_α = 1 (no noise)
VCO slope gain	K_{vco} = 2π x350 rad/sec/v
Phase detector gain	K_ϕ = 7.5 volts/rad
Loop gain at threshold	$K_\ell = K_\alpha K_\phi K_{vco}$ = 16,450
Loop filter, phase lag network	$F(s) = \frac{1+T_2 s}{1+T_1 s}$, T_1 = 2.82 sec, T_2 = 1.8 msec
Crystal filter -3 dB r.f. bandwidth	B_{-3} = 1 KHz

Negative Feedback Cases. The transfer function of the loop without the time delays d_1 and d_3 reduces to

$$\frac{\hat{\theta}_o(s)}{\hat{\theta}_i(s)} = \frac{\frac{108\, K_\ell F(s)}{s(1+T_c s)} \left[\frac{T_c}{2} s + 110.5\right]}{1 + \frac{K_\ell F(s)}{s\,(1+T_c s)} \left[\frac{T_c}{2} s + 110.5\right]}$$

which represents negative feedback for all values of the parameters. Without the crystal filter we have the basic loop transfer function

$$\frac{\theta_o(s)}{\theta_i(s)} = \frac{108 \times 110.5\, K_\ell \frac{F(s)}{s}}{1 + 110.5\, K_\ell \frac{F(s)}{s}}$$

Using the parameters listed in Figure 2 the denominator for this transfer function becomes

$$D_1(s) = 1 + \frac{K(s+566)}{s(s+.352)} \doteq 1 + \frac{K(s+566)}{s^2}$$

where K is the gain for root locus. The root locus for this case is sketched in Figure 3. For the value of loop gain K_ℓ = 16,450 the closed loop poles are indicated at ω_n = 800 rad/sec , ζ = .707, corresponding to the threshold values.

Adding the effect of the crystal filter we obtain from above the expression for the denominator

$$D_2(s) = 1 + \frac{K(s+566)\ (s+695{,}000)}{s^2(s+3142)}$$

The root locus for this case is sketched roughly in Figure 4. The effect of the crystal filter is to produce a pole directly -3142 and an unimportant zero at -695,000. The result of the crystal filter is to reduce the previous value of ζ to .65 and increase ω_n to 1075. In addition a closed loop pole is produced at -1,750 which is essentially unimportant. Thus placing the pole of the crystal filter at only 3 times the open loop zero at -566 does not appreciably change the stability. Usual rules of thumb would tend to place this pole somewhat farther out.

Figure 2 indicates the transfer function of the loop filter together with the coupling networks from the phase detector and to the VCO. The transfer function F(s) is modified to give

$$F_m(s) = \frac{(1+s/566)}{(1+s/333{,}300)(1+s/.352)(1+s/59{,}000)}$$

Incorporating this change in the previous case yields the root locus sketch in Figure 5. The effect of the two additional poles at -59,000 and -333,300 change the plot only slightly for values of gain that are of interest, i.e. near the origin. However, it is seen that for very large gains the system can become unstable with closed loop poles in the right half plane.

The above cases indicate the general effects of additional filtering in the loop for the negative feedback cases. Since the values of the parameters used were based on later design values the stability indicated in the root locus sketches did not reflect adverse conditions that could result.

Positive Feedback Case. It was established for time delays in the first and third IF stages that the stability of the loop was governed by the positive feedback characteristic equation given by

$$D_3(s) = 1 - K_\ell \frac{(d_3 s-1)}{(1+T_c s)} \left[s\left(\frac{T_c}{2} - 108\, d_1\right) + 110.5\right] \frac{F(s)}{s}$$

It was established that for the example used in the paper, that the coefficient of s in the bracketed term was positive, thus giving rise to the positive feedback case. Since, the delay d for the first IF stages in most other applications be very small, the delay d_3 in the third IF will produce the positive feedback case. Thus for a well designed loop the effect of d_1 will be very small. It is noted that, independent of the value of d_3, the effect of d_1 is to move the crystal filter induced zero at $s = -221/T_c$ to the position $s = -221/T_c - 216\,d_1$, where $d_1 << T_c/216 = 1.47$ μsec with the change being very small. Since, for the present case, this zero is relatively unimportant to the effect of d_1 is can be neglected. Thus, for any non zero delay d_3 we have the case of positive feedback governed by $d_3(s)$. A set of positive feedback root locus sketches is given in Figures 6a, 6b, and 6c for large and small delays. For a positive feedback root locus diagram we note the two basic rule changes from those for negative feedback. These two new rules are:

1. A locus exists on the real axis if the number of real poles and zeros to the right is even.

2. The angle of the symptotes is $n\pi/(p-z)$, $n = 0,2,4,\ldots$, which changes the determination of the angle of departure.

Figure 6a indicates a rough root locus sketch with a small delay d_3. For reasonable value of gain the locus near the origin is similar to the previous cases with negative feedback. However, for large values of gain the locus moves along the negative real axis as in the negative feedback towards the closed loop pole at $s = -\infty$. But due to the position feedback it suddenly appears to snap back to the right half plane and approachs the open loop zero at $s = 1/d_3$ for $K \to \infty$. Thus for small values of d_3 and moderate gain the positive feedback case holds for $d_3 > o$ and is very stable, with closed loop ω_n and ζ similar to the cases for negative feedback.

Increasing the value of the delay d_3 moves the right half plane zero at $1/d_3$ closer towards the origin until a point is reached where it has more attraction than the remote zero at -695,000. The effect is shown in the rough sketch of Figure 6b. Here the system becomes unstable for much smaller values of gain. It is interesting to note that as the gain increases the locus moves out on the positive real axes until a value of gain is reached such that it suddenly reappears on the left hand side real axis and approaches the zero at $s = -695{,}000$ such that apparently the system now becomes stable as $K \to \infty$. Figure 6c indicates the slight change due to the coupling networks. The effect is small as it should be for a proper design.

An accurate root locus diagram has been plotted near the origin for gain values of interest. For reasonably small values of delays, d_3, the changes in ζ and ω_n were very slight. In fact the departure of the loci near the origin were such that the loci for $d_3 = .2$ msec and .13 msec still produced damping ratios of $\zeta \cong .5$ and .4 for moderate gain values.

CONCLUSIONS

A practical root locus technique has been developed to determine the stability of phase-locked loops containing sources of time delays such as IF amplifier stages that follow frequency conversion. An approximate phase transfer function is derived for the predetection crystal filter that is usually in the loop. Also an approximate phase transfer function was derived for each IF amplifier stage in the form $H(s) = -(ds-1)$. For an IF amplifier stage preceeding the loop phase detector with delay d, the resulting loop equation gives rise to positive feedback. However, the loop is usually stable for reasonable values of delay d. In order to demonstrate the method the Pioneer spacecraft receiver with triple conversion was used, root locus diagrams were sketched for the negative feedback cases to show the general effect of a single pole crystal filter and the loop filter coupling networks. The positive feedback cases resulted from delays in the last IF stage. Root locus diagrams were sketched as a function of d_3 to indicate the changes in transient responses that are possible.

The method developed could also be applied, for example, to AFC and frequency synthesizer loops. Additional work that could be done is to determine how well the transfer function $H(s) = -(ds-1)$ approximates IF amplifier stages with a large number of poles and zeros for small delay d.

APPENDICES

I. Phase Transfer Function for IF Amplifier Stages.

Near band-center the voltage transfer function is assumed to be constant in gain and linear in phase. Therefore, we have over a bandwidth B

$$H(i\omega) = |H(i\omega)| = e^{-id(\omega-\omega_o)} \quad \text{for } |f-f_o| \leq B/2$$

where "d" is the time delay of the filter and can be taken as the slope of the actual phase transfer function $H(i\omega)$ at the center frequency f_o. The phase transfer function can now be obtained by using an "instantaneous frequency" argument as follows: Consider a phase modulated input voltage to $H(i\omega)$ in the form

$$v_i(t) = A \sin [\omega_o t + \theta(t)]$$

If $\theta(t)$ is slowly varying with respect to the response of $H(i\omega)$ then we can use the instantaneous phase derivation $\omega_o + \theta'(t)$ as the "frequency" and apply the definition of a transfer function to obtain the output voltage as

$$v_o(t) = A\, M[\omega_o t+\theta'(t)] \sin [\omega_o t+\theta(t)+\psi(\omega_o+\theta'(t))]$$

where in general

$$H(i\omega) = M(\omega)\, e^{i\psi(\omega)}$$

For the above case we have, setting $|H(i\omega_o)| = 1$,

$$v_o(t) = A \sin [\omega_o t+\theta(t) - d\theta'(t)]$$

or in operator notation, $p \equiv d/dt(\cdot)$,

$$v_o(t) = A \sin [\omega_o t+(1-dp)\theta(t)]$$

Thus, the output phase $\theta_o(t)$ is given by

$$\theta_o(t) = (1-dp)\theta(t)$$

so that the phase transfer function has the form

$$\frac{\hat{\theta}_o(s)}{\hat{\theta}_i(s)} = 1-ds$$

II. Phase Transfer Function of a Single Pole Crystal Filter.

A crystal filter is usually used in a phase locked loop with frequency conversion as the bandwidth restricting element in the loop. The crystal filter helps reduce the dynamic range of noise at the input to the loop phase detector. A "single pole" crystal is often used and can be modeled as the voltage transfer function

$$H(s) = \frac{\hat{v}_o(s)}{\hat{v}_i(s)} = \frac{2\,\zeta\omega_o s}{s^2+2\zeta\omega_o s+\omega_o^2}\ ,\ Q \doteq \frac{1}{2\zeta} = \frac{f_o}{B_{-3}}$$

Where B_{-3} is the -3 dB bandwidth of H(s) in Hz. "Single pole" is used to refer to the baseband properties. Although there are several ways to derive a phase transfer function, we use a direct physically motivated method. The method is to first apply a sinewave at $t = 0$ to H(s). After the turn-on transients die out we let the phase of this sinewave experience a step of "a" radians at $t = T$. Thus, by initializing the output time to $t = T$ we have upon suitably interpreting the output, the phase step response of H(s) from which the phase transfer function is obtained.

The input voltage is defined as

$$v_i(t) = \sin \omega_o t \quad , \quad t \leq t < T$$
$$= \sin (\omega_o t + a) \quad t \geq T$$

By using Laplace transforms, the output voltage $v_o(t)$ is obtained. By dropping the transient terms due to turn-on and using the fact that $\zeta \doteq 0.5 \times 10^{-4}$ we have the output voltage at the time of the phase step in the form

$$v_o(t) = \sqrt{1+a(1-e^{-\zeta\omega_o t})^2}\ \sin\left[\omega_o t + \tan^{-1} a(1-e^{-\zeta\omega_o t})\right]$$

For "a" small we have the output phase due to a step

$$\mathrm{Tan}^{-1}\, a(1-e^{-\zeta\omega_o t}) \doteq a(1-e^{-\zeta\omega_o t})$$

Therefore, the phase transfer function can be written

$$H_c(s) = \frac{\hat{\theta}_o(s)}{\hat{\theta}_i(s)} = \frac{1}{1+T_c s} = \frac{1}{1+s/\zeta\omega_o} = \frac{1}{1+s/\pi B_{-3}}$$

REFERENCES

1. Lindsey, W. C., Synchronization Systems In Communication and Control, Prentice-Hall, Englewood Cliffs, New Jersey, 1972.

2. Shpolyansky, V. A., Tyufyakin, L.S., Kovsakov, P.P, "Automatic Phase Control With Delay", Automation and Remote Control, 22, No. 3 (1961) pp. 1303-1316.

3. Carpenter, D. D., "Pioneer Receiver Stability Analysis", Space Technology Labs., IOC 9332.3-139, November 8, 1963.

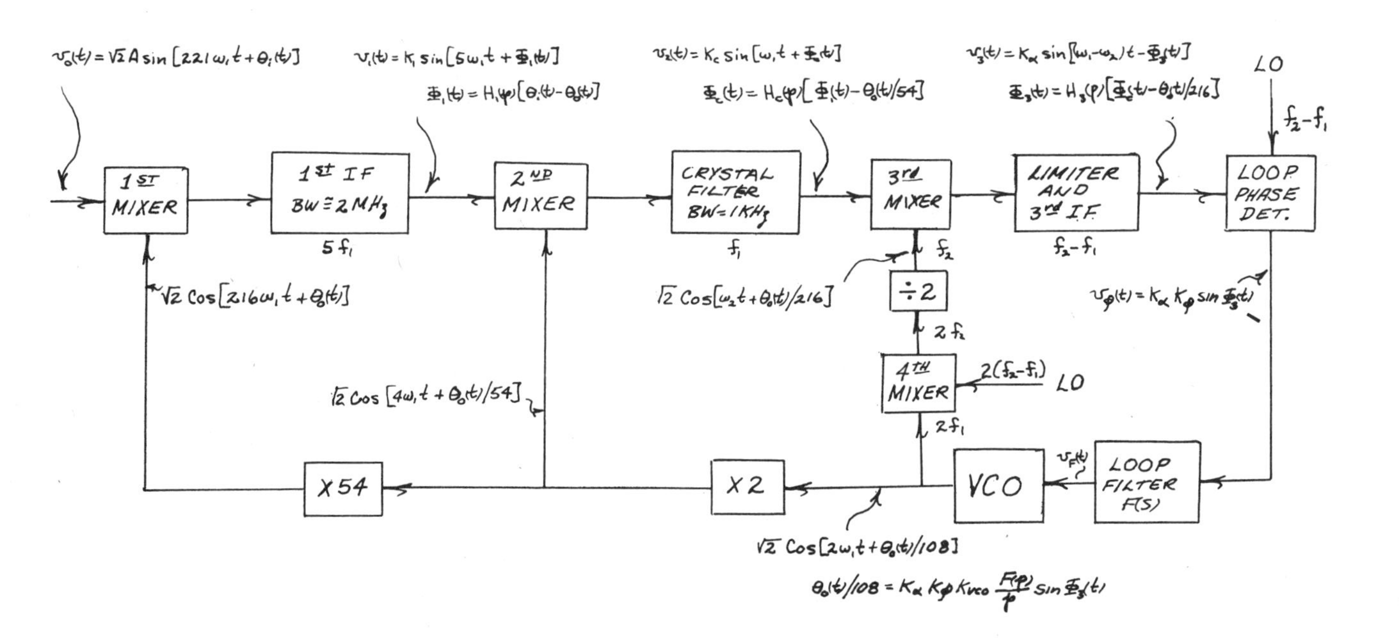

ANALYTICAL MODEL FOR PIONEER RECEIVER
FIGURE 1

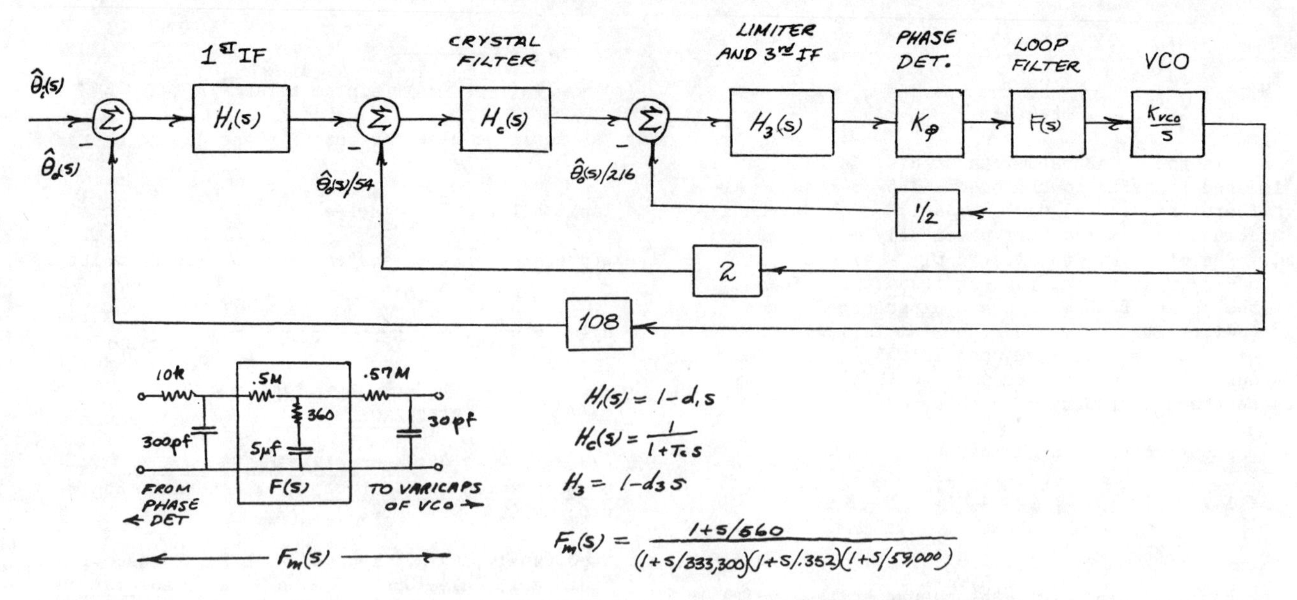

LINEARIZED PHASE TRANSFER FUNCTION
FIGURE 2

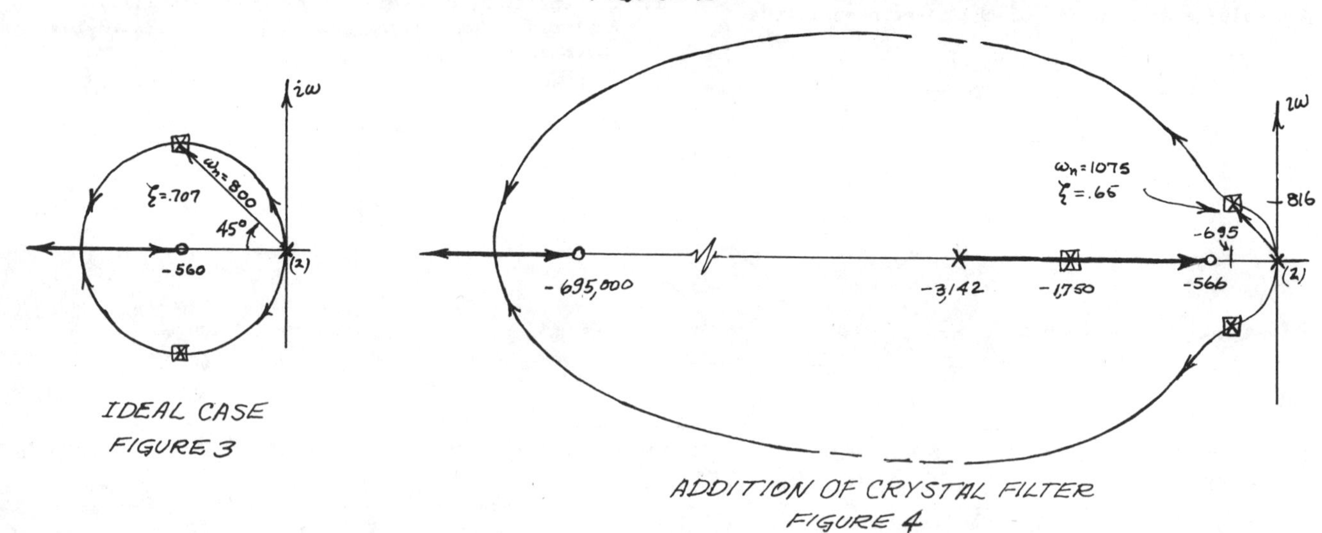

IDEAL CASE
FIGURE 3

ADDITION OF CRYSTAL FILTER
FIGURE 4

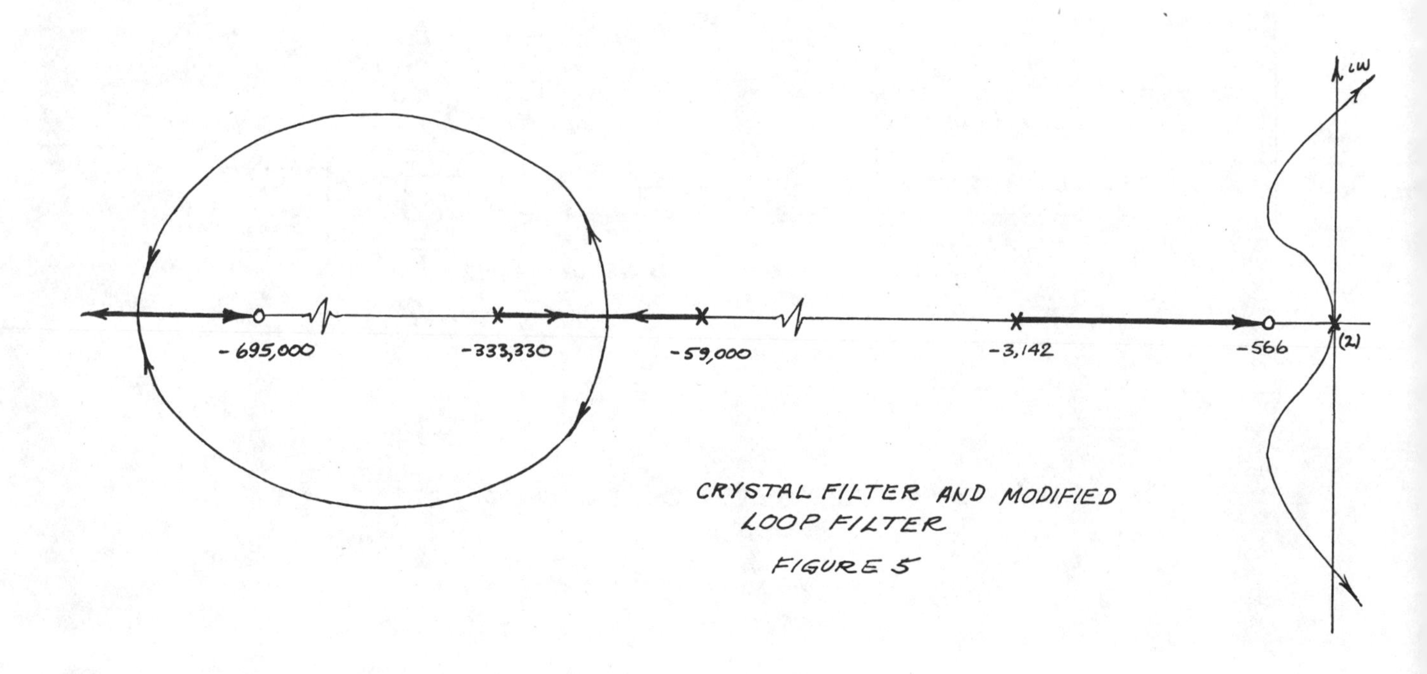

CRYSTAL FILTER AND MODIFIED LOOP FILTER
FIGURE 5

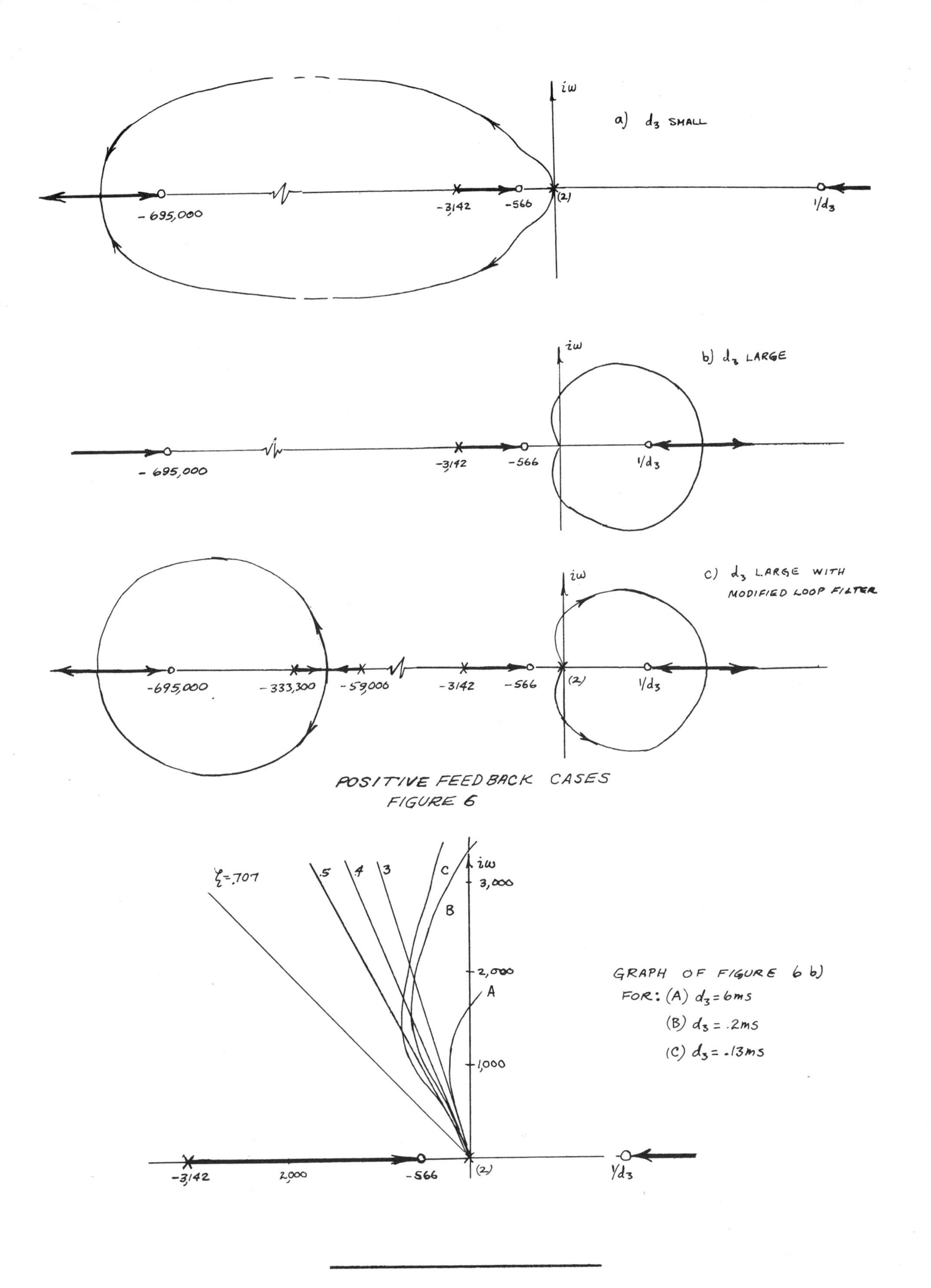

POSITIVE FEEDBACK CASES
FIGURE 6

Part V
Frequency Demodulation and Detection

DEMODULATION OF WIDEBAND FREQUENCY MODULATION UTILIZING PHASE-LOCK TECHNIQUE

Richard C. Booton, Jr.
Space Technology Laboratories, Inc.
Redondo Beach, California

Introduction

Soon will take place the first experimental demonstrations of telephone and television communication by means of active satellite repeaters. The first spacecraft for this purpose will of necessity be limited in transmitter power capability. Now and for some time the communication system design must center around achieving maximum information flow per watt of satellite radiated power, even at the expense of another precious quantity, bandwidth.

The modulation technique chosen for the early satellite relays has been wideband analog frequency (or phase) modulation. This encoding has the capability of yielding high output signal-to-noise ratio with small spacecraft transmitted power. However, full advantage cannot be taken of the power reduction capability of this modulation form with a standard-discriminator demodulator. Accordingly, considerable effort has been devoted to the development of more efficient demodulation schemes. Two examples are the BTL FM-feedback demodulator and the phase-lock demodulator to be described in the present paper.

The demodulator described here was developed by STL for NASA/GSFC under RELAY Contract NAS5-1302. The design objective was a demodulator capable of handling frequency-modulated television signals. The baseband signals were to be standard 525 line video, with a bandwidth of approximately 3 Mc, plus an aural subcarrier at 4.5 Mc. A demodulator with the capability of handling such a TV signal can handle between 120 and 240 FDM telephone channels.

Evaluation of several methods that reduce the required signal, including the FM-feedback discriminator, led to a decision to utilize the phase-lock technique. Extensive experience has been gained with phase-lock circuits although in different applications. Much of the STL experience was obtained in connection with the design of receivers for satellite and deep space applications. Such applications usually require extremely sensitive receivers with very narrow bandwidths. Although wideband demodulation was a different application, the phase-lock technique was sufficiently promising that it was selected for the development of a wideband demodulator. The prime advantage of the phase-lock technique is the extreme simplicity of the circuits which makes possible low loop time delays and, accordingly, high loop bandwidths.

Mathematical Description

Figure 1 shows an abbreviated block diagram of the wideband phase-lock demodulator developed to handle TV signals. The phase-lock loop itself is shown in the center of the figure and consists of three units: a voltage-controlled oscillator that generates a replica of the signal input, a phase detector to measure the phase relation between the signal input and the VCO output, and a third unit for amplification and filtering to control the loop characteristics. Because of the feedback action the phase of the VCO output follows the phase of the signal input, at least within the dynamic capability of the loop.

The dynamic characteristics of the loop were selected to be second order, although higher order loops have been investigated. A mathematical description of a linearized second-order loop is shown in Figure 2. The phase detector determines the difference between the input phase and the VCO output phase. The action of the phase detector is indicated as a simple subtraction of phases because the nonlinear effects of the actual phase detector have been neglected. The loop filter is shown as a linear block that furnishes gain and a simple compensation to insure satisfactory closed loop characteristic. The principal effect of the VCO is indicated as an integration from the voltage input to the phase of the VCO output. In addition to the usual second order parameters, pure time delays are shown both in the forward path and in the feedback path. Although these delays are shown as pure time delays, they might in fact be a convenient approximation to higher order effects whose characteristic frequencies are so high that the dominant effect over the loop bandwidth is simply a phase shift linear with frequency. The two equations in Figure 2 show the mathematical form for the transfer function from the phase input to the VCO output and the transfer function from the phase input to the phase error.

If it were possible to eliminate the time delays, achievement of any value of loop bandwidth would be a simple matter of selecting appropriate values for the constant K, α and β. The presence of the time delays $\tau 1$ and $\tau 2$, however, restricts the bandwidths that can be satisfactorily achieved. Figure 3 indicates a Nyquist diagram for a circuit such as that shown in Figure 2 and indicates the tendency of the time delays to cause system instability. For a fixed total time delay, attempting to increase the gain K beyond a certain value results in an unstable system. Analysis indicated that in order to

Reprinted from *Proc. 1962 National Telemetering Conf.*, vol. II, May 23–25, 1962, pp. 1–24.

achieve the desired loop bandwidth of approximately 7.5 Mc the total time delay had to be reduced to approximately 30 nanoseconds.

Physical Description

The physical description of the demodulator is given next. First, one should point out that an appreciable amount of conventional receiving equipment is required in addition to the demodulator. Figure 4 shows a complete receiving system. The output of the antenna-mounted preamplifier, mixer, and IF amplifier is the input to the rack-mounted portion of the receiver. The rack-mounted equipment includes the demodulator, amplifiers, filters, and equalizers. A standard FM discriminator is provided also. A decision was made that all of this equipment could be mounted on one chassis and Figure 5 shows an artist's conception of how the receiver would appear physically. Note that the phase demodulator itself consists of only two modules. Figure 6 shows a photograph of the actual receiver as completed. Figures 7 and 8, respectively, show the physical appearance of the phase detector-filter and the VCO. As an illustration of the other modules, Figure 9 is a photograph of an IF amplifier.

Experimental Measurements with Sinusoidal Baseband Signal

A series of measurements have been performed to determine the extent to which the performance of the phase-lock demodulator could be predicted by theoretical means. This comparison should indicate the accuracy with which the loop parameters are known. A simple situation was chosen to simplify the analysis. The experimental setup is shown in Figure 10. A sinusoidal baseband signal of 50 kc was used to frequency modulate a 120 Mc IF signal to a peak-to-peak deviation of 20 Mc. Noise was added to this FM signal and the sum was amplified and applied as the input to the phase-lock demodulator. The noise in a 1 Mc band was measured using a low pass filter and a true rms meter. The signal component was measured using a harmonic analyzer with a bandwidth of 6 cps. The experimental results are shown by the circles in Figure 11. A quasilinear analysis[1,2] has been applied for an approximate evaluation of the expected signal-to-noise ratio in the output of the phase-lock demodulator. This analysis can be applied down to an input signal-to-noise ratio of approximately 0 db. The quasilinear analysis indicates that this value of input should correspond to a threshold in some sense. The analysis is not applicable below this point but one would expect the output signal-to-noise ratio to decrease rapidly as it does experimentally. For comparison, Figure 11 also shows the expected signal-to-noise ratio output from a standard discriminator with an input identical to that of the phase-lock demodulator.

Experimental TV Data

The principal use for which the present demodulator was designed is the demodulation of TV signals. Extensive measurements have been performed to determine the IF signal-to-noise ratio required to give satisfactory picture quality. As a reference point, the output of the phase-locked demodulator is compared with the output of a conventional FM discriminator. Figure 12 shows the experimental setup for such a comparison. A closed circuit camera together with the associated television circuitry derives a video signal which is used to frequency modulate a carrier at the normal IF frequency of 120 Mc. Standard program material transmitted by a Los Angeles TV station has also been used. Carefully shaped and measured wideband noise is added to the frequency modulated signal and the sum is amplified and limited at the IF frequency. The noise bandwidth is the minimum bandwidth capable of satisfactorily transmitting the signal. The output of the IF amplifier-limiter is an accurate simulation of the signal plus noise that would appear at the input to the demodulator in a communication satellite experiment. This simulated IF signal is used to drive in parallel a phase-locked demodulator and a conventional discriminator each of which is used to control a monitor. The quantity of direct interest is the IF signal-to-noise ratio required to give satisfactory performance. Figure 13 compares the output of two monitors when no noise is added to the signal. The following figure, Figure 14, compares the two monitor pictures with a 12 db IF signal-to-noise ratio and illustrates that at high signal-to-noise ratios the performance of the phase-locked demodulator and the standard discriminator are approximately the same. The following figures, Figures 15 through 21, compare the two monitor pictures with signal-to-noise ratios of 8 db, 4 db, 2 db, 0 db, -1 db, -2 db, and -3 db. As the signal-to-noise ratio is decreased both monitors show steadily worsening picture quality with the important difference that the standard discriminator output degrades more rapidly than the phase-lock demodulator output. Figure 15, for example, shows the situation for a signal-to-noise ratio of 8 db which is just below the threshold of 10-12 db commonly assumed for a standard discriminator. The discriminator output is beginning to show the effects of the noise whereas the phase-lock picture is essentially unaffected by the noise. Comparison of the monitor pictures for the two demodulators involves a subjective comparison, because of the somewhat different nature of the noise on the picture, as described below. Most observers, however, seem to agree that threshold conditions for the two demodulators occur with the standard discriminator requiring approximately 5 db more signal-to-noise ratio; e.g., the phase-lock monitor output for the -1 db signal-to-noise ratio has approximately the same picture quality as the standard discriminator output with a signal-to-noise ratio of +4 db.

The noise output of the standard discriminator appears as a fairly uniform graininess in the picture. The principal effect of noise on the phase-lock demodulator, however, is strikingly different. At random intervals the phase-lock demodulator will lose lock. After loss of lock the baseband output tends to go to full black or full white for a very short interval. After this short interval the demodulator locks up again. The losses in lock thus appear in the monitor picture as short black and white horizontal lines.

Conclusions

The design effort discussed in this paper was directed toward the development of a discriminator that could be used with a high deviation frequency modulated signal with a baseband signal consisting of a 525 line video plus a 4.5 Mc aural subcarrier, and reduce the threshold by approximately 6 db below the threshold of a standard discriminator. The principal design problem overcome during the development was the decrease of the loop time delay sufficiently that a loop bandwidth of approximately 7.5 Mc could be achieved without stability problems. This problem was solved satisfactorily by reduction of the unit time delays to the point where the total time delay is less than 30 nanoseconds. Although extensive measurements have been made to verify the dynamic properties of the loop, perhaps the best illustration that the development was satisfactory lies in the quality of the monitor pictures as illustrated in this paper.

Acknowledgment

The support and encouragement of NASA/GSFC, specifically D. G. Mazur and R. J. Mackey, is gratefully acknowledged. Special appreciation is due the members of the STL Communication Laboratory, in particular J. A. Develet, Jr., C. W. Stephens, K. H. Hurlbut, and W. A. Garber, for the development described in this paper.

References

1. R. C. Booton, Jr., "The Analysis of Nonlinear Control Systems with Random Inputs," Proceedings of the Symposium on Nonlinear Circuit Analysis, Polytechnic Institute of Brooklyn, 23-24 April 1953.

2. Application of the quasilinear analysis to the phase-lock loop was suggested and performed by J. A. Develet, Jr.

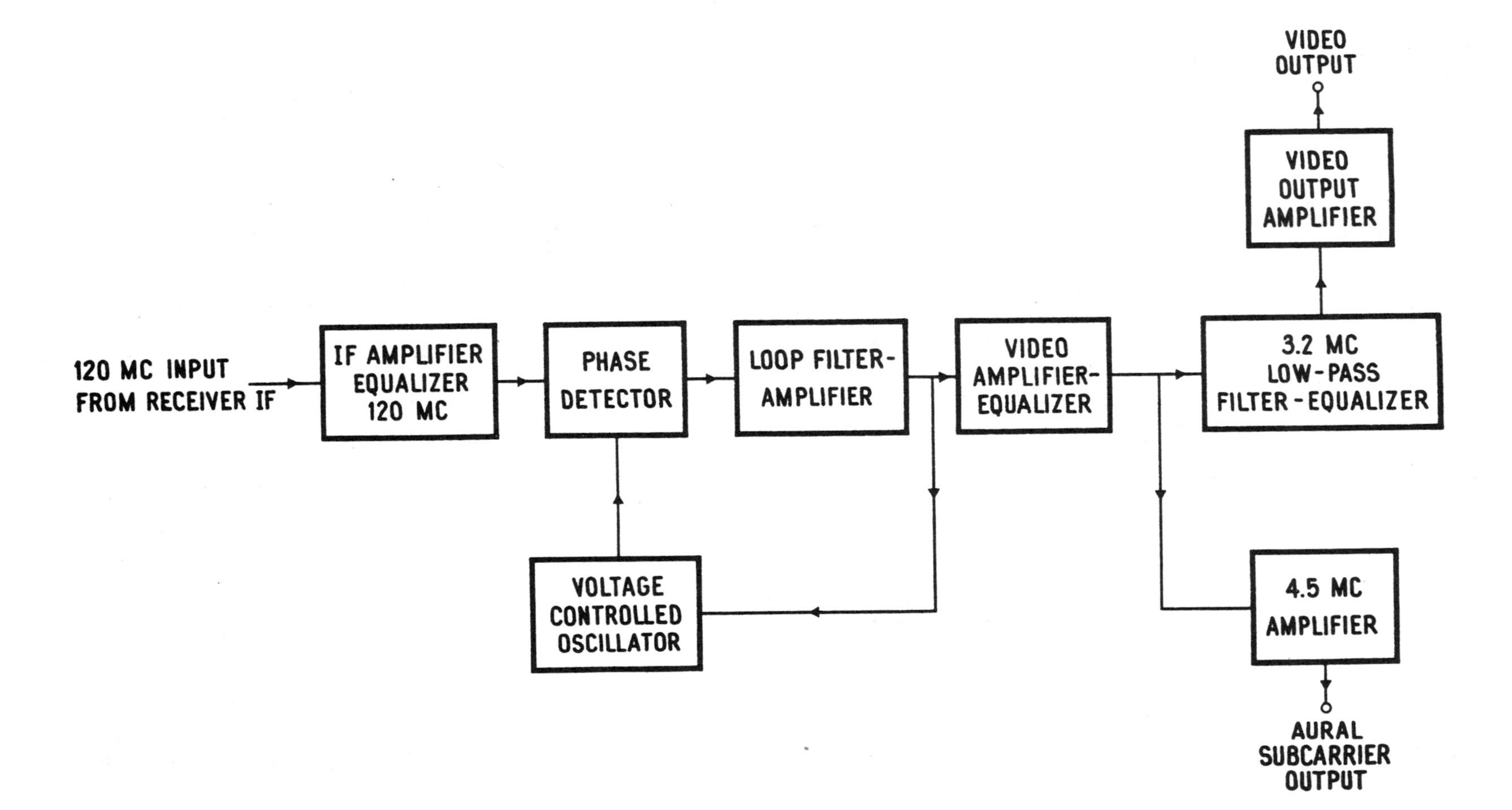

Figure 1. Block Diagram of Demodulator for TV

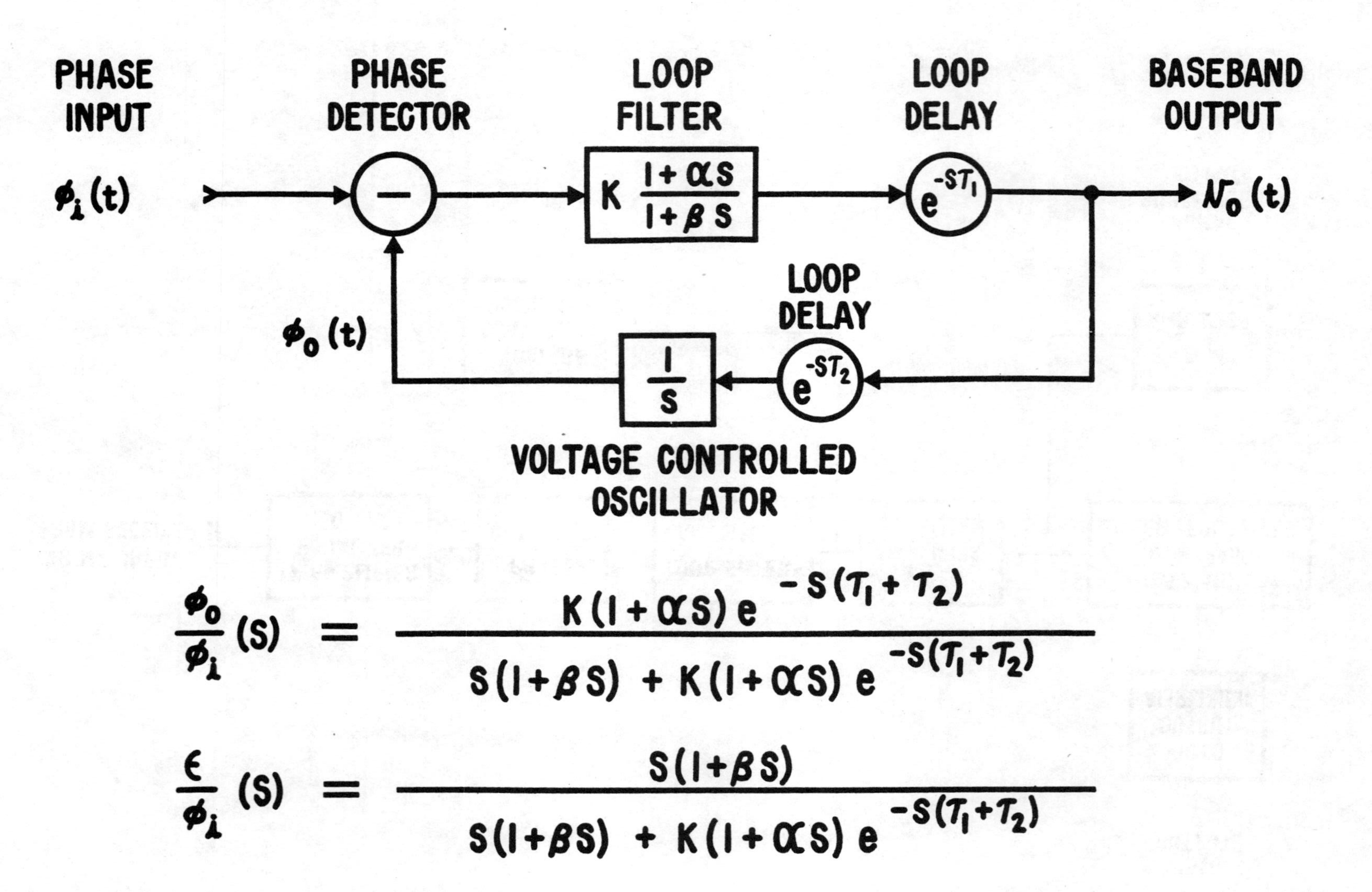

Figure 2. Linearized Second-Order Phase-Lock Loop

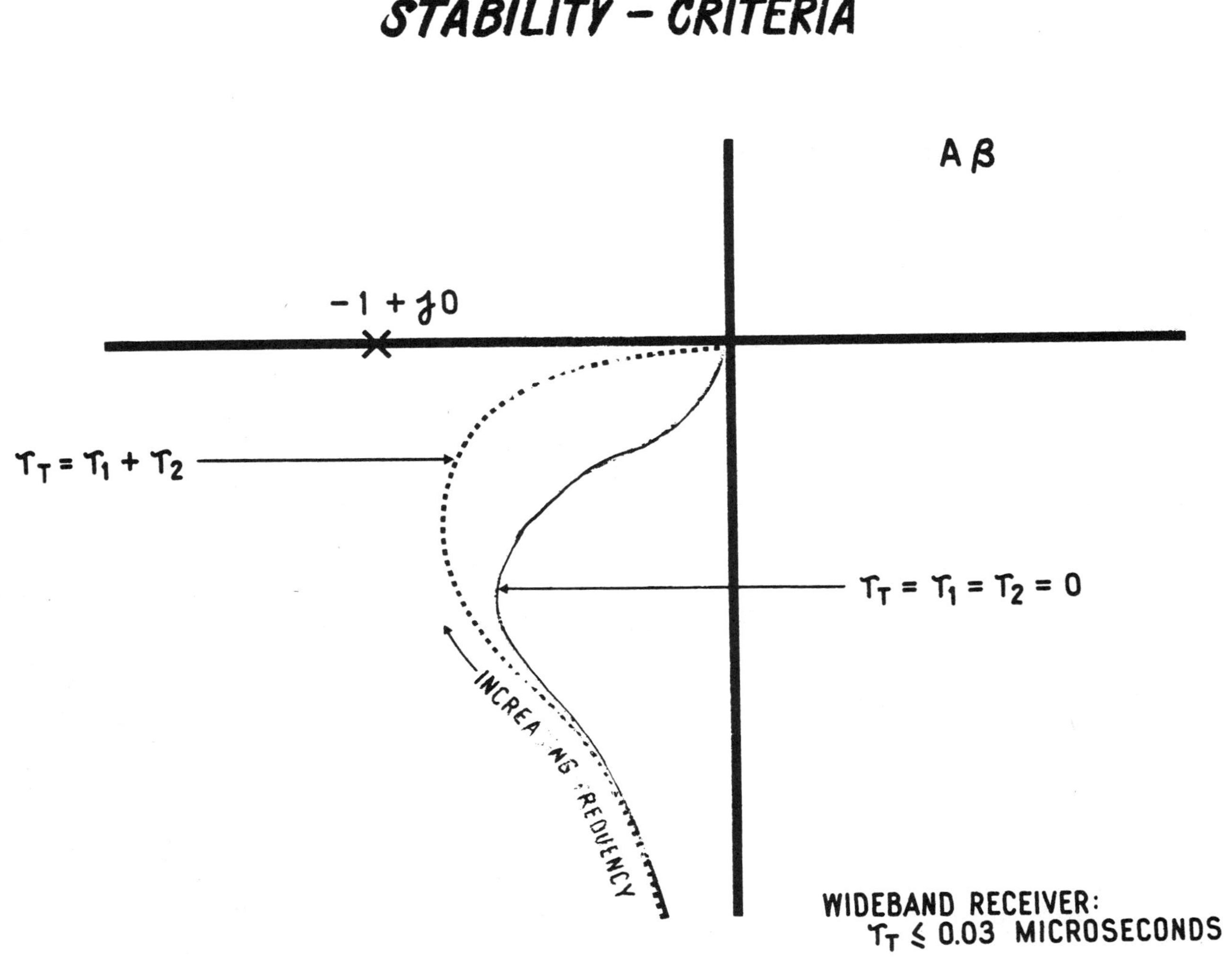

Figure 3. Nyquist Diagram

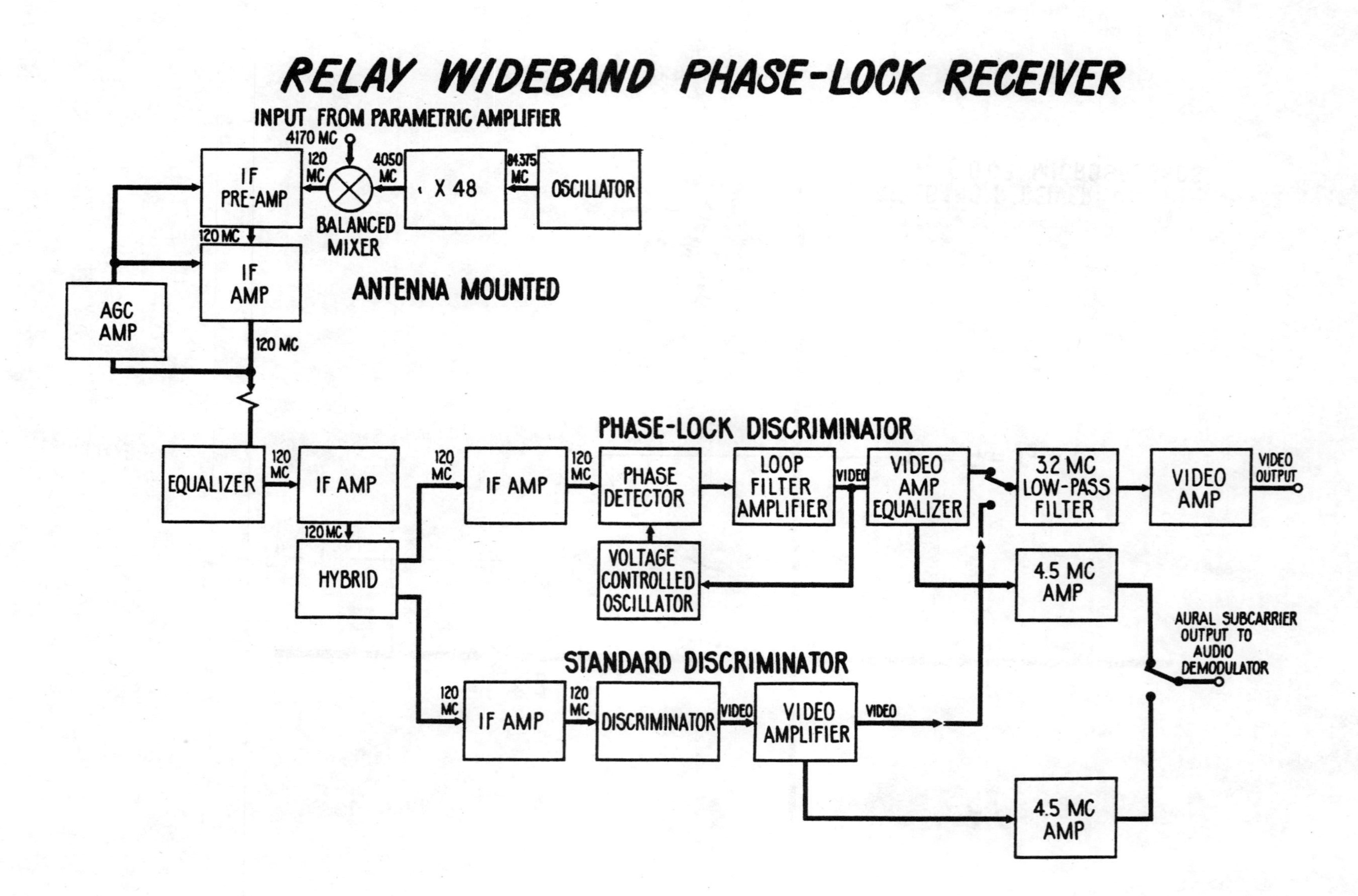

Figure 4. Complete Receiver Block Diagram

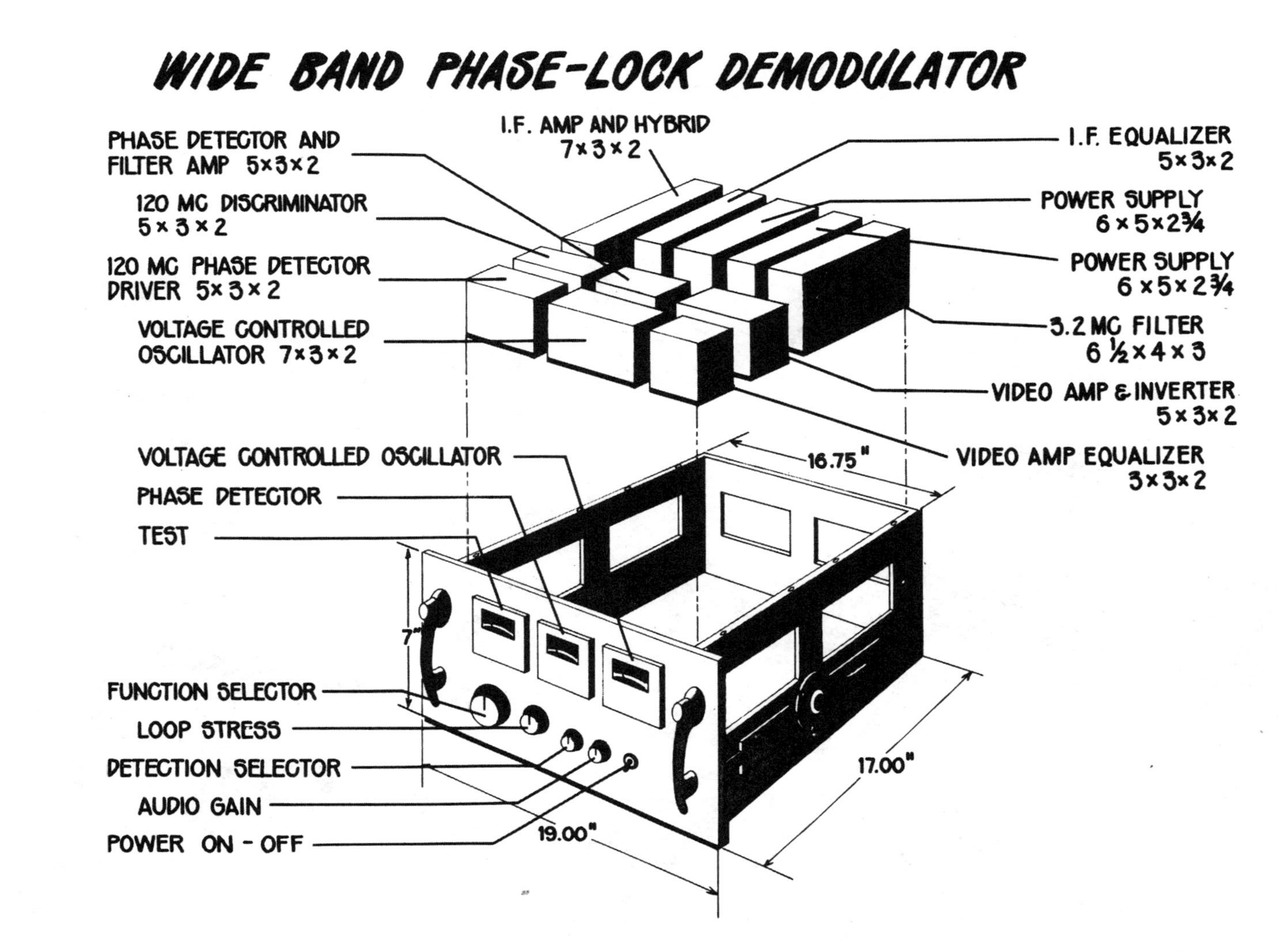

Figure 5. Artist's Conception of Complete Receiver

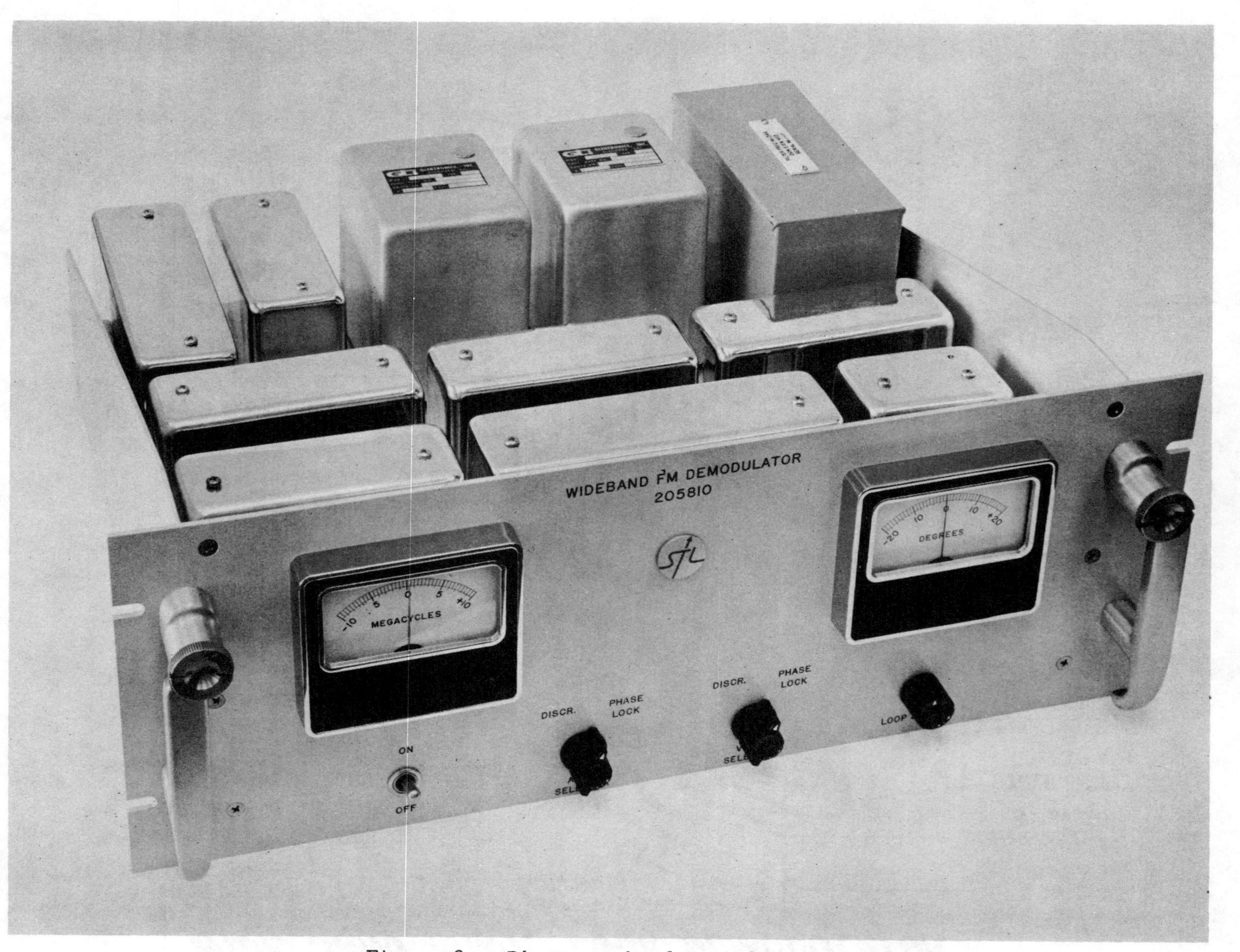

Figure 6. Photograph of Actual Receiver

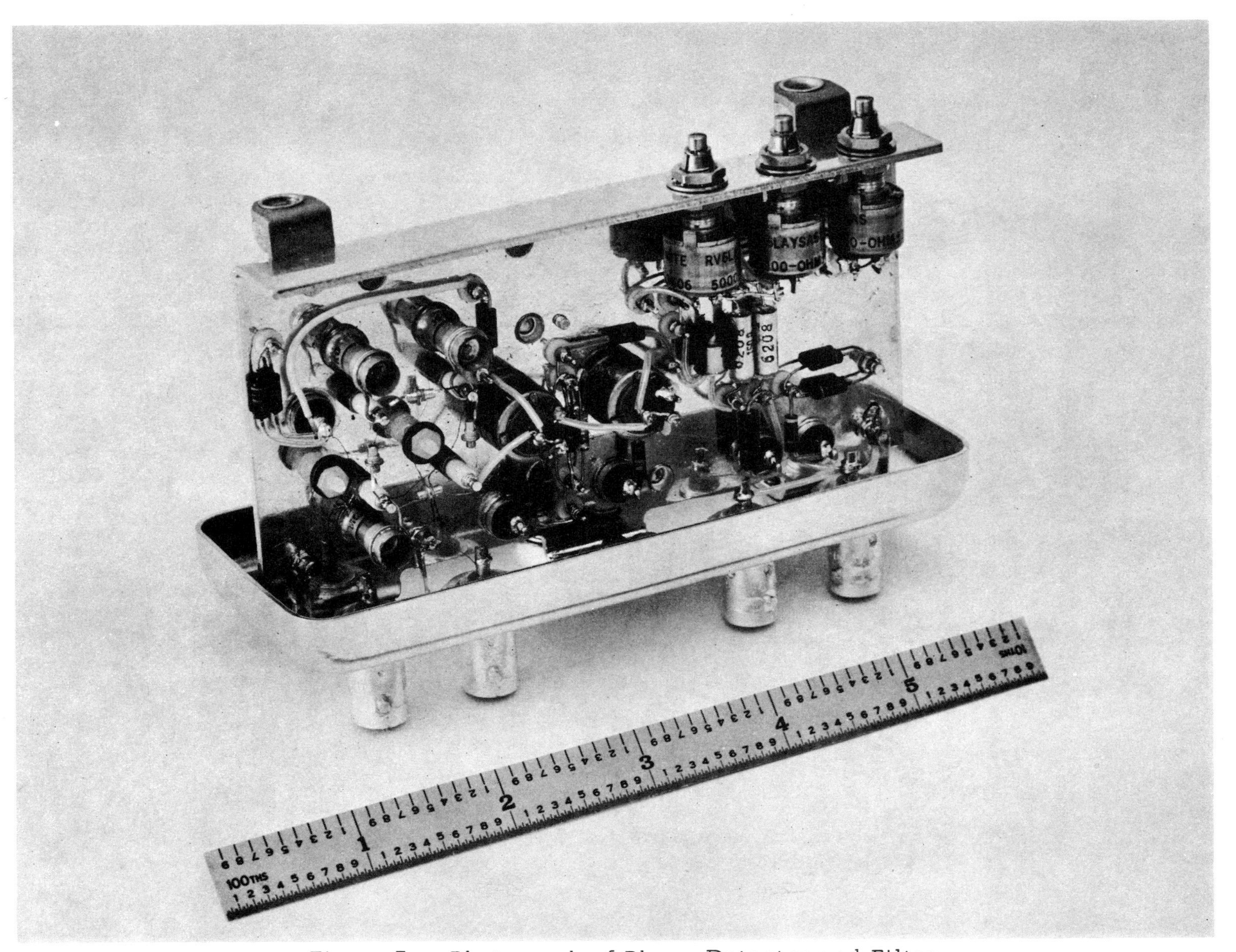

Figure 7. Photograph of Phase Detector and Filter

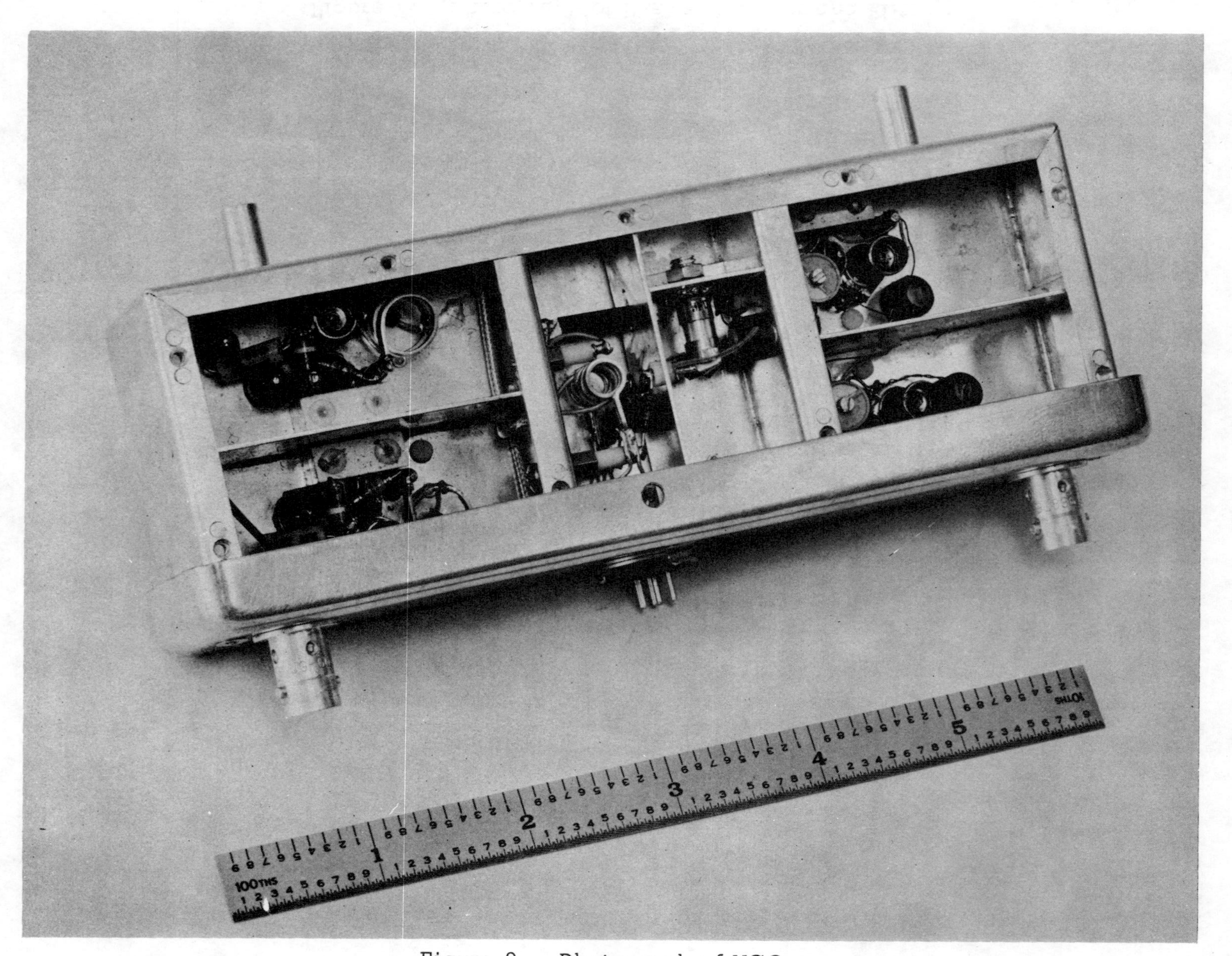

Figure 8. Photograph of VCO

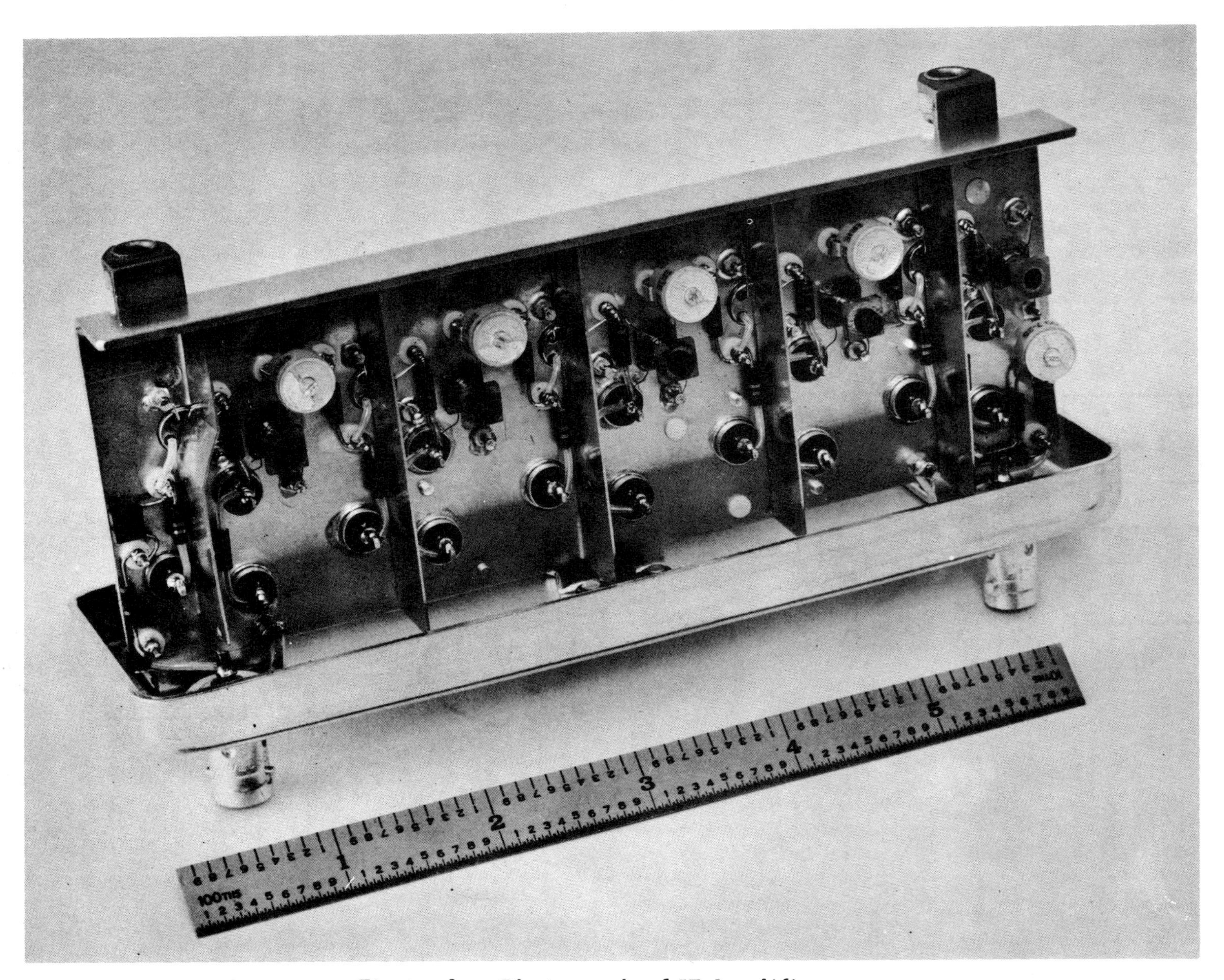

Figure 9. Photograph of IF Amplifier

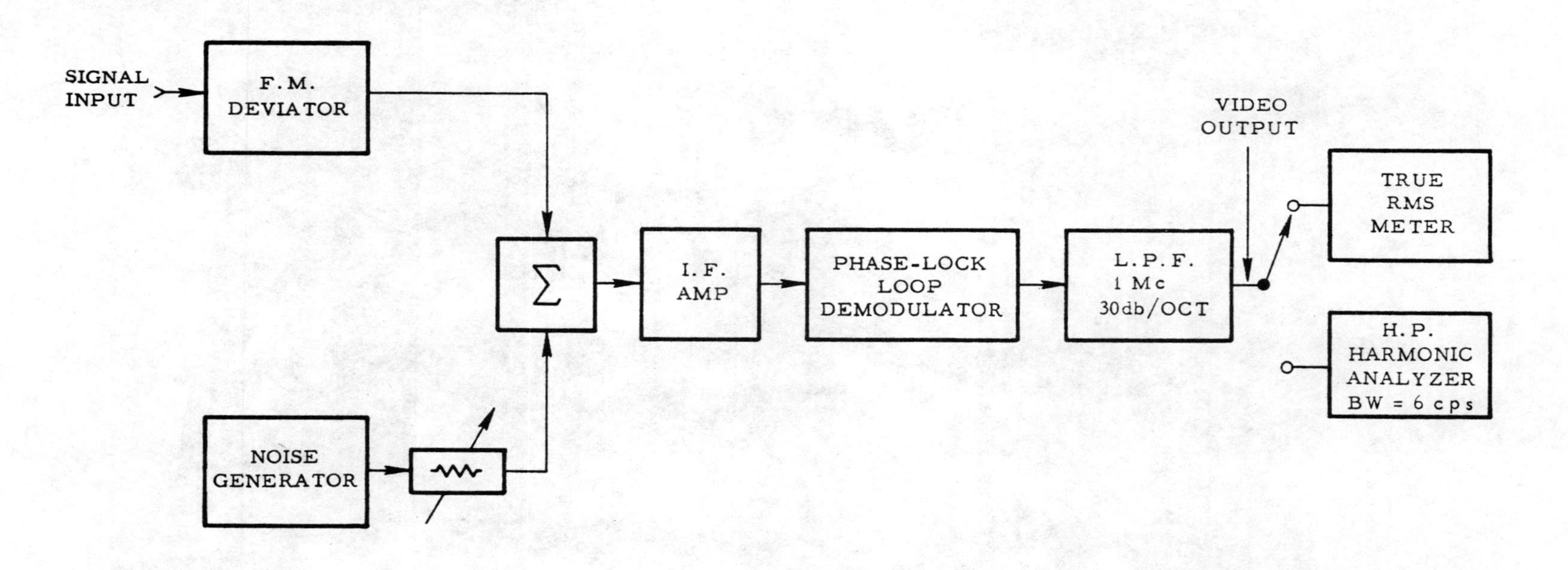

Figure 10. Experimental Set-up for Sinusoidal Signal Measurement

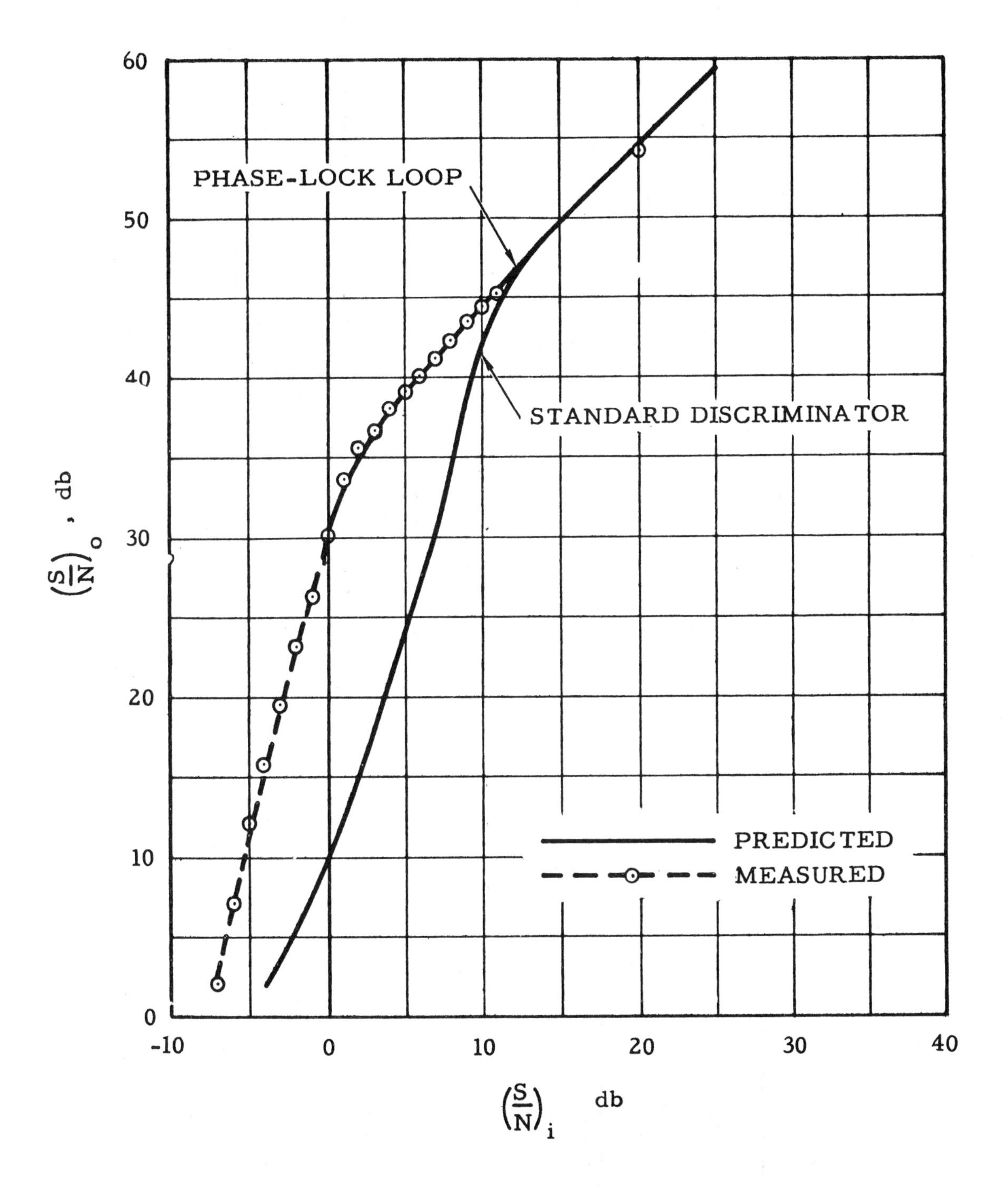

Figure 11. Experimental Measurements of Signal to Noise Ratio

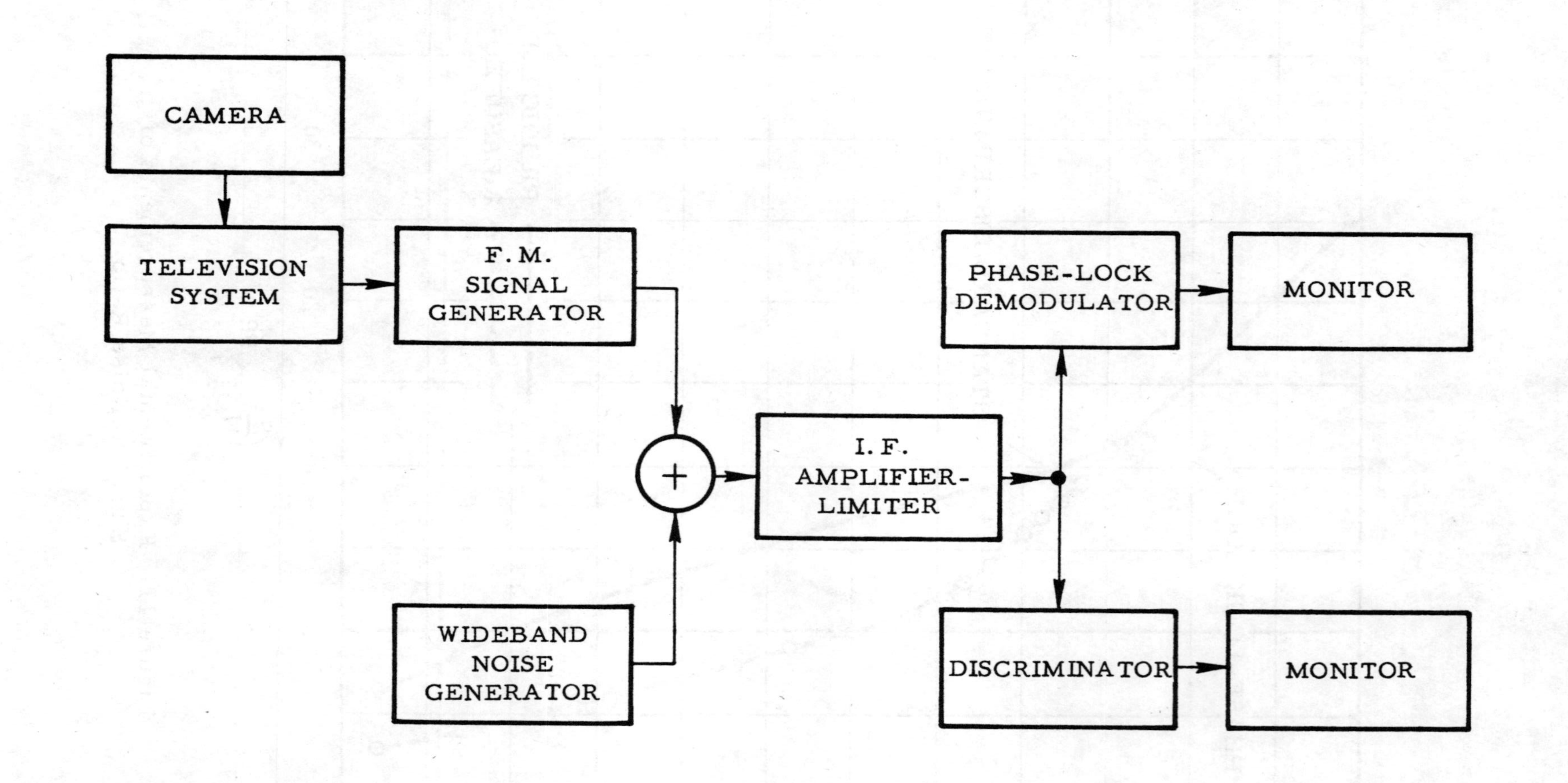

Figure 12. Experimental Set-up for TV Measurements

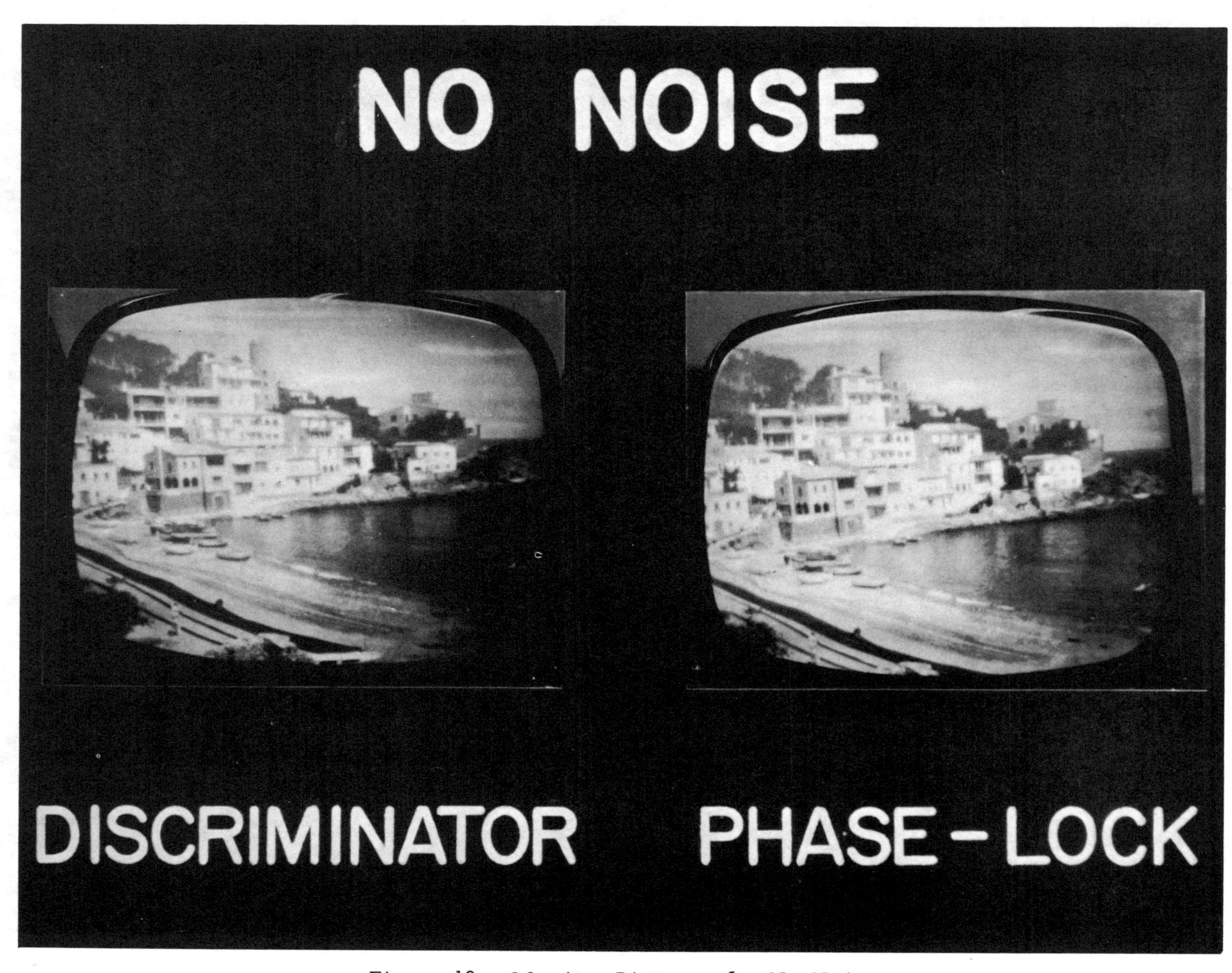

Figure 13. Monitor Pictures for No Noise

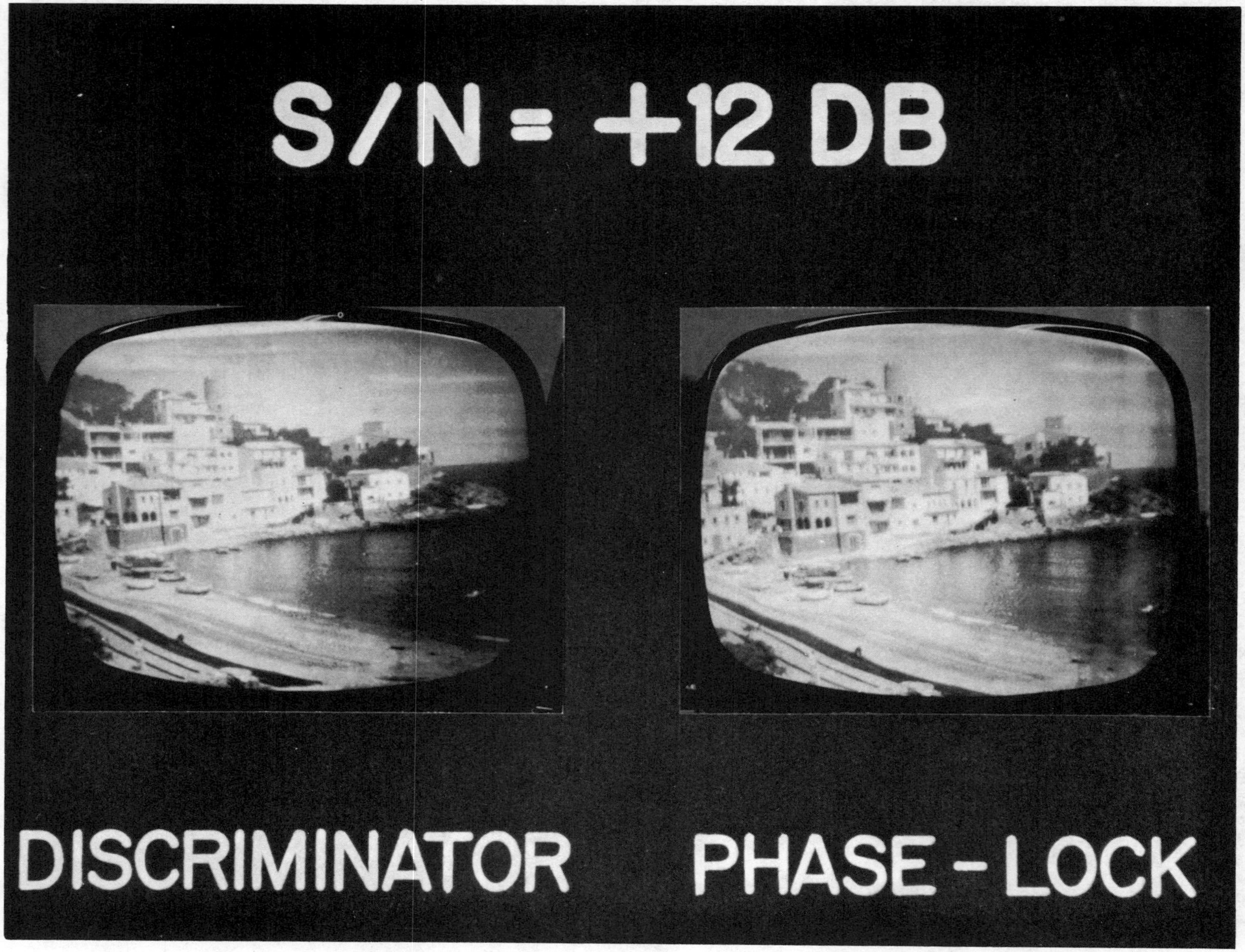

S/N = +12 DB

DISCRIMINATOR

PHASE - LOCK

Figure 14. Monitor Pictures for S/N = + 12 DB

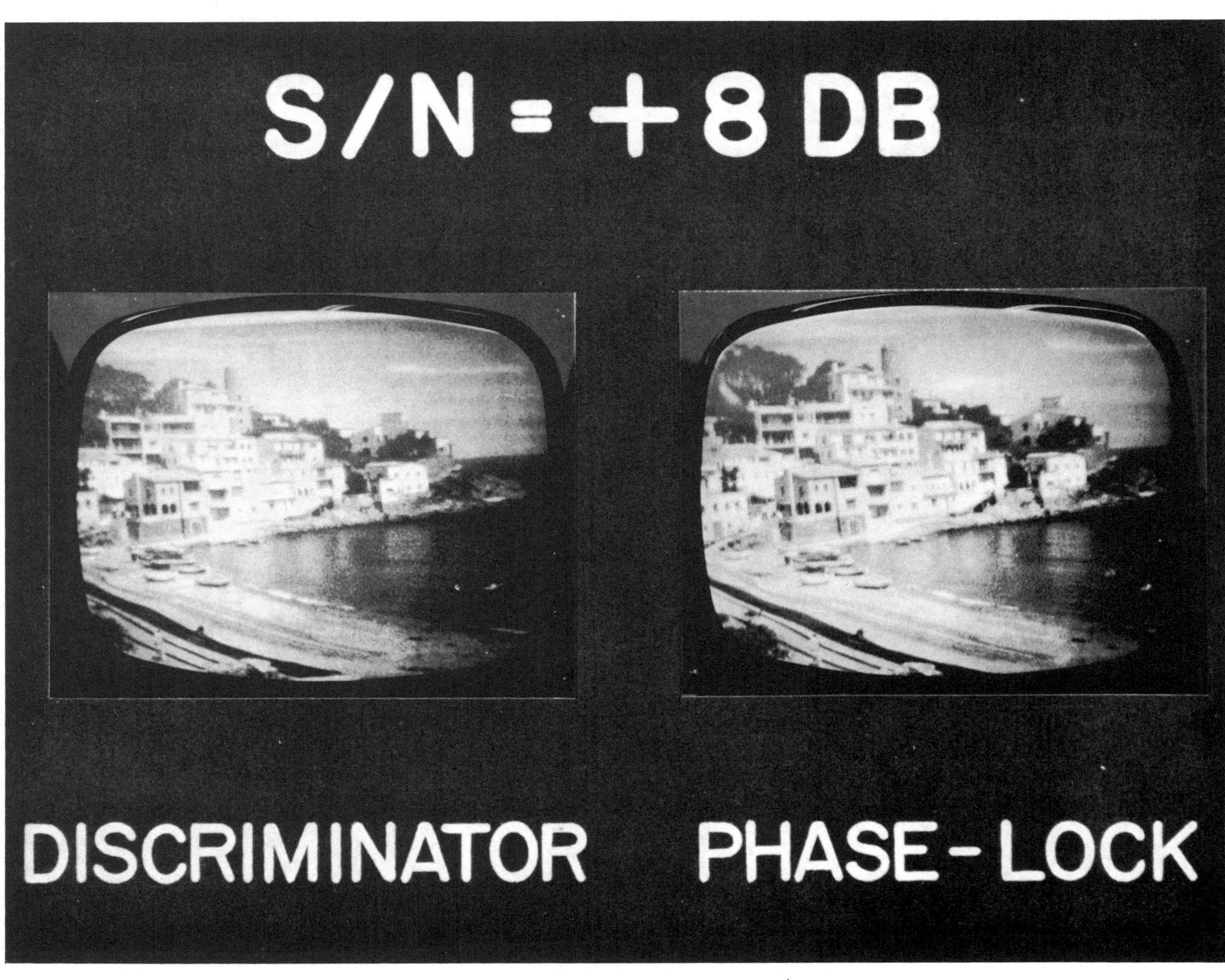

Figure 15. Monitor Pictures for S/N = + 8 DB

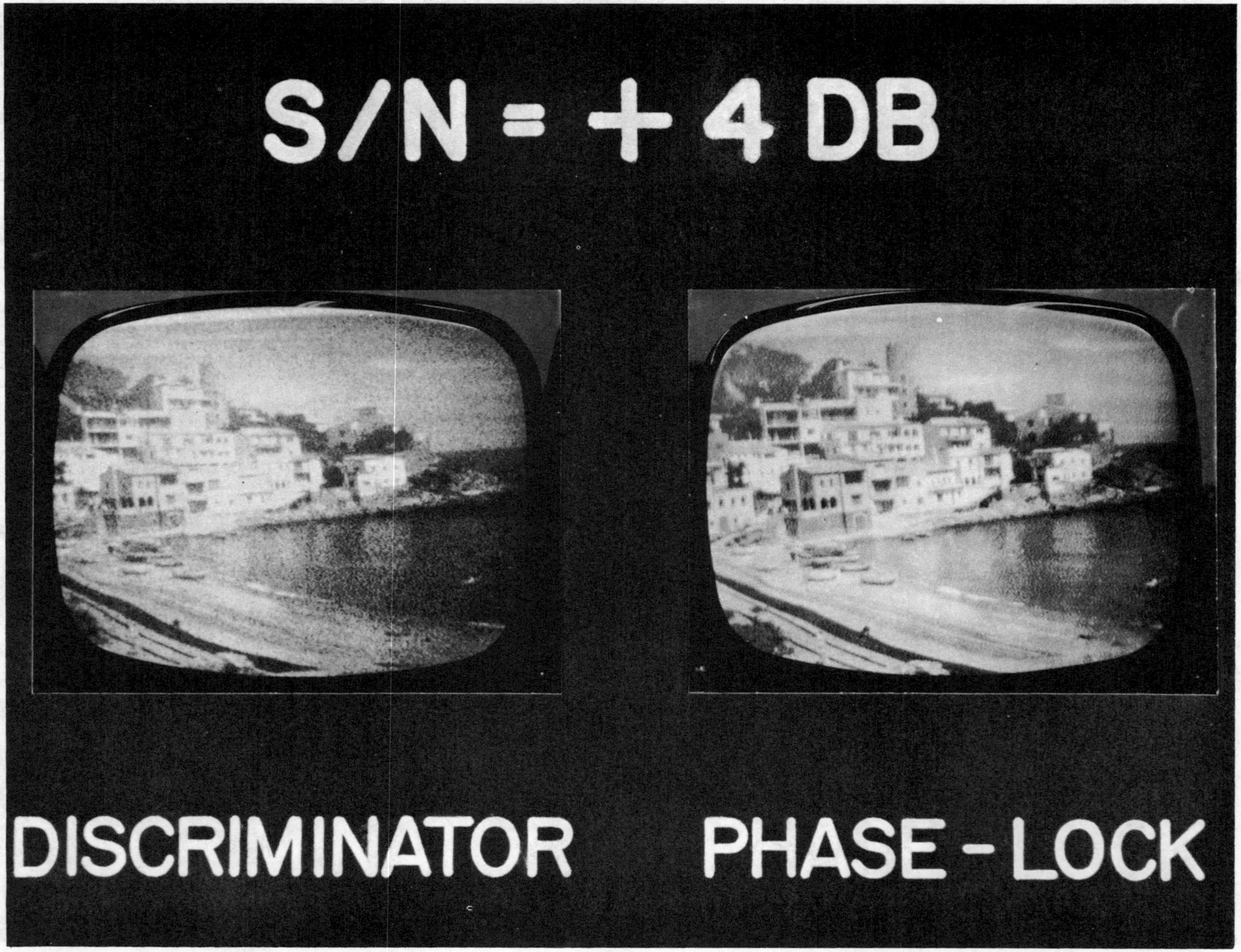

Figure 16. Monitor Pictures for S/N = + 4 DB

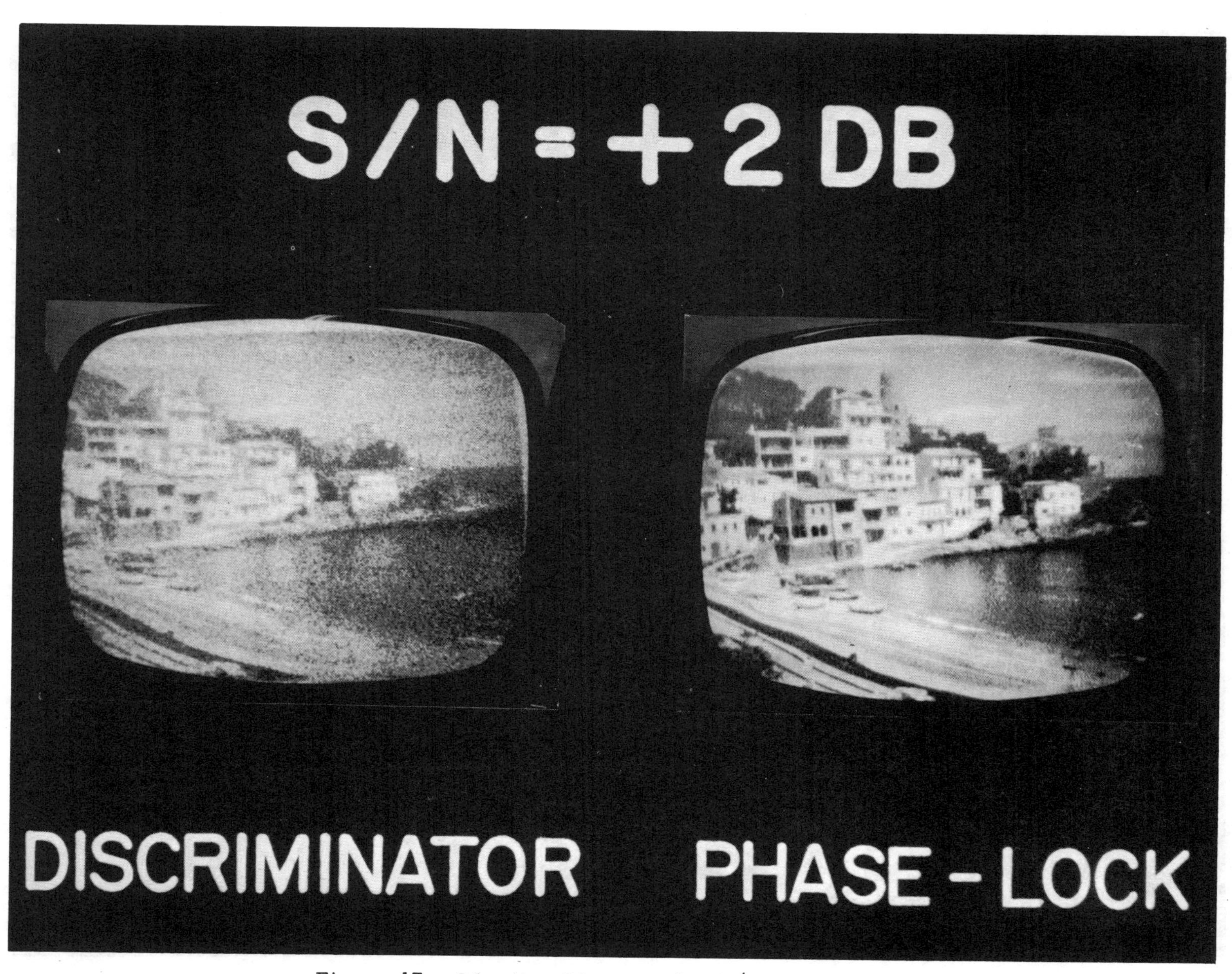

Figure 17. Monitor Pictures for S/N = + 2 DB

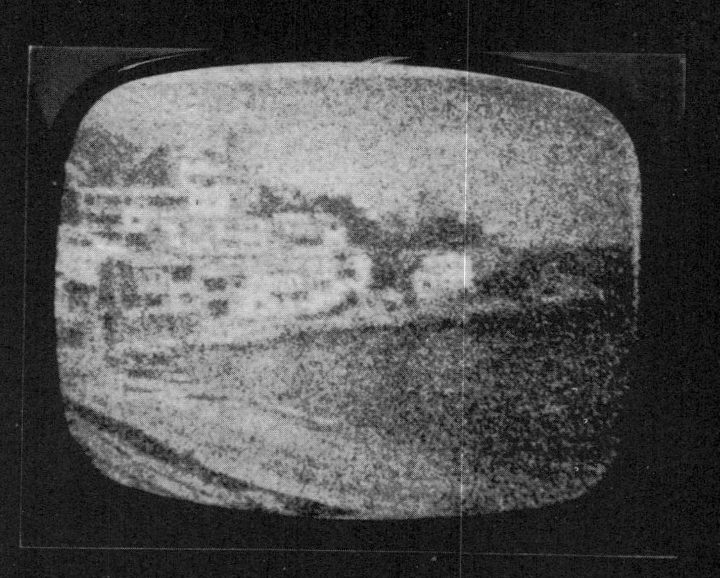

Figure 18. Monitor Pictures for S/N = +0 DB

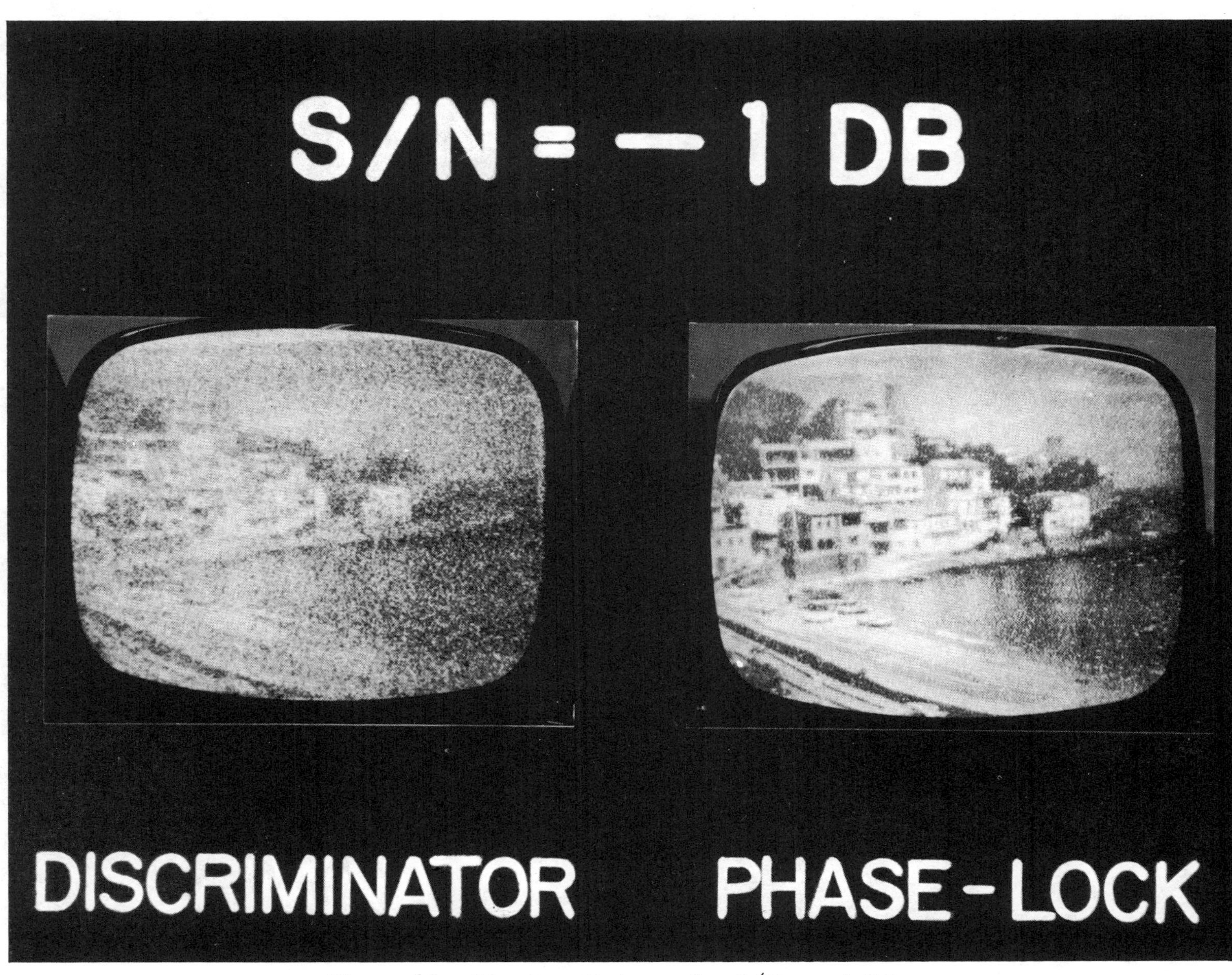

Figure 19. Monitor Pictures for S/N = - 1 DB

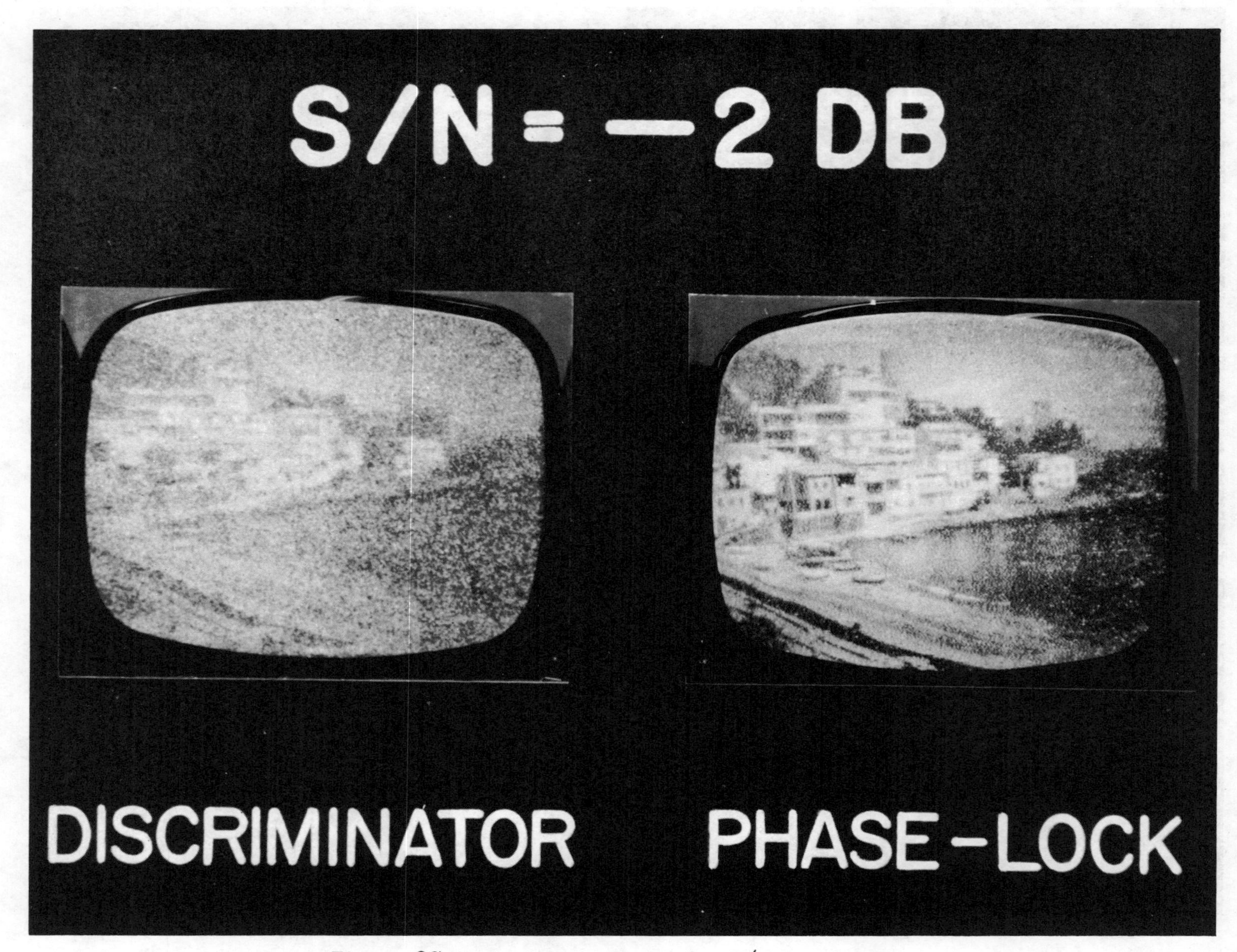

Figure 20. Monitor Pictures for S/N = - 2 DB

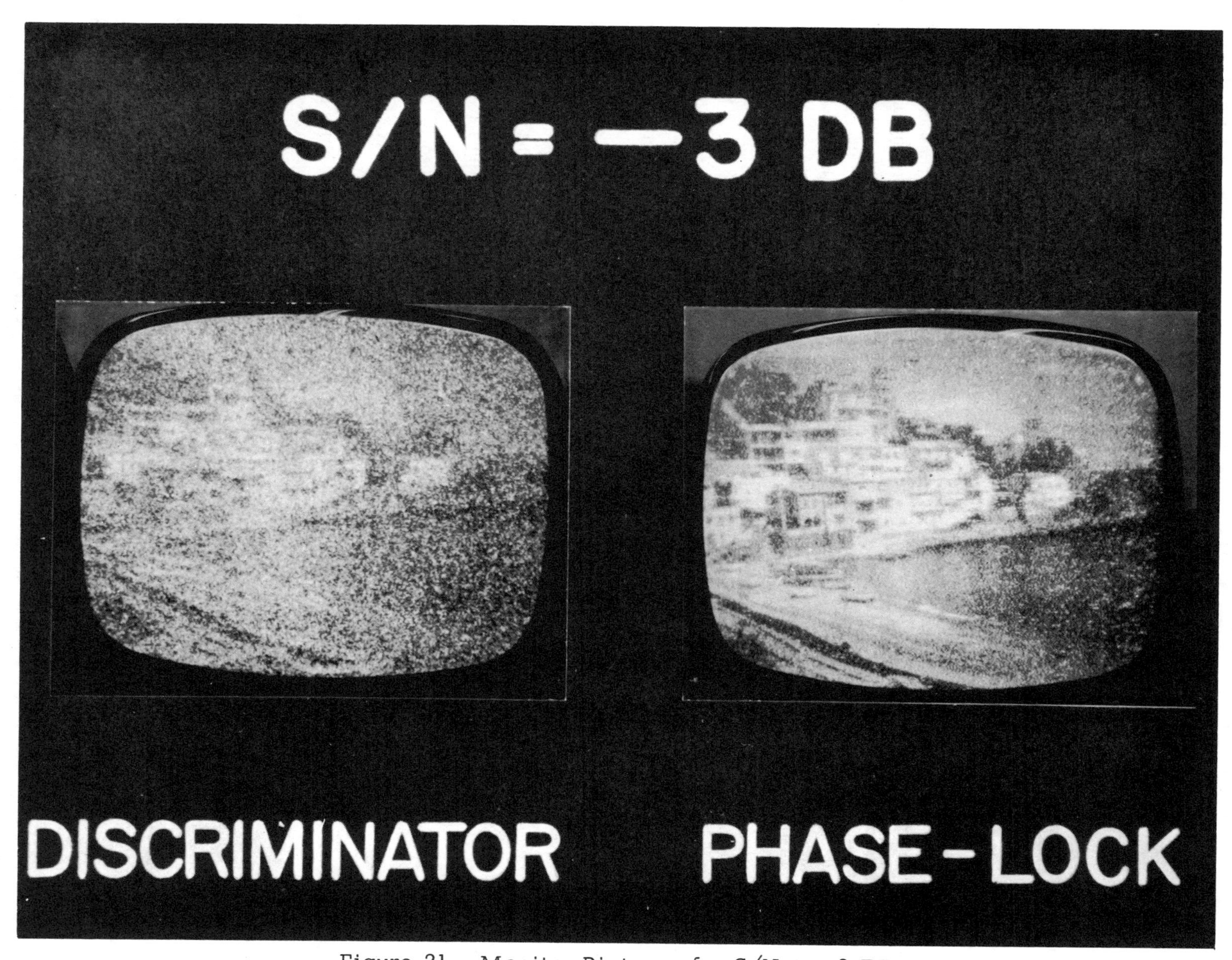

Figure 21. Monitor Pictures for S/N + - 3 DB

Detection of Digital FSK and PSK Using a First-Order Phase-Locked Loop

WILLIAM C. LINDSEY, FELLOW, IEEE, AND MARVIN K. SIMON, SENIOR MEMBER, IEEE

Abstract—A classical problem in digital frequency-shifted keyed (FSK) demodulation is the evaluation of the bit error probability performance when an estimator-correlator that incorporates a phase-locked loop (PLL) is employed. Although some attention has been devoted to this problem in the past, an accurate account of the mechanism which produces decision errors has not yet been advanced. This paper examines a special case, viz., a first-order PLL preceded by a wideband IF filter, of the above problem using a new approach which is based upon the renewal Markov process theory and the Meyr distribution. In particular, the ad hoc approach of invoking the Gaussian assumption on the decision variable and patching it with a correction term based on Rice's click theory is not used. Rather, the effective noise is properly characterized by unfolding the renewal Markov process associated with the loop phase error. As a slight extension of the results, the performance of the above PLL detector operating on low data rate PSK is given and demonstrated to be approximately 3 dB superior to that of FSK reception.

The theory and analysis presented herein apply to the special case where the bandwidth of the IF filter preceding the first-order PLL is required to be several times the data rate because of frequency uncertainties due to channel Doppler and oscillator instabilities, but the frequency deviation to data rate ratio may be chosen small (if desired) to optimize system error probability performance. In addition to presenting results for the case where the oscillator instabilities are assumed absent and channel Doppler is prefectly tuned out at the receiver oscillator, the effects of small residual Doppler on bit error probability performance is considered. In all cases tested, excellent agreement was obtained between theory and computer simulation results.

I. INTRODUCTION

DETECTION of digital FSK is an old and interesting topic from which various analyses as well as hardware developments have proceeded. These developments can be bisected into two categories commonly called *wideband* FSK and *narrowband* FSK. Wideband FSK is the term used to imply the fact that the receiver IF filter bandwidth is much greater than the data rate. Narrowband FSK implies that the IF bandwidth is on the order of the data rate. In addition, two receiver mechanizations have been considered for these two categories: one employs discriminator detection while the other uses the phase-locked loop (PLL) concept. Both detection methods are valid; however, in this paper we restrict our investigation to evaluating the detection capabilities of a first-order PLL operating in the face of a wideband IF filter, say, greater than four times the data rate. The case of basic interest that involves a narrowband IF filter followed by a second-order loop appears to remain a formidable problem.

Paper approved by the Editor for Communication Theory of the IEEE Communications Society for publication after presentation at the 1975 National Telecommunications Conference. Manuscript received July 14, 1975; revised June 10, 1976. This paper presents the results of one phase of research carried out at the Jet Propulsion Laboratory, California Institute of Technology, Pasadena, CA. This work was supported by the National Aeronautics and Space Administration under Contract NAS 7-100 and the Office of Naval Research under Contract N-00014-67-A-0269-0022.

W. C. Lindsey is with the with the Department of Electrical Engineering, University of Southern California, Los Angeles, CA 90007.

M. K. Simon is with the Jet Propulsion Laboratory, Pasadena, CA 91103.

There are several reasons, theoretical as well as practical, why the following development considers the performance of wideband IF, PLL detection of digital FSK. First of all there are situations arising in practice which would require a wideband IF section. These include the cases where the additive channel noise is not a problem or where the channel frequency uncertainty (due to Doppler and oscillator instabilities) is large (as might be the case in low rate command systems or space relay links) and a cheap and simple receiver is the added constraint [1]. Moreover, in low data rate links (say, a few bits per second), frequency discriminators suffer from the problem of performance degradations due to frequency drifts caused by aging, temperature, etc. Second, to the authors' knowledge, no previous papers have attacked this problem from a strict theoretical point of view although a few experimental results have been reported. The reason for this stems from the fact that the advance in the theory of PLL, which serves as the key to explaining the noise mechanism causing bit errors, has only recently been made. Third, by introducing only a few valid and astute assumptions, one can arrive at a theory which is useful for design purposes; however, as far as the authors can tell, a theoretical treatment of the FSK detection problem associated with a narrowband IF filter is formidable when a PLL is used. In this paper an approximate mathematical account is given for the noise mechanism which creates bit errors in the detection of digital FSK or PSK using a first-order loop preceeded by a wideband IF filter.

Considerable theoretical and experimental work has been accomplished for the case of detecting FSK using a narrowband IF filter followed by a frequency discriminator or a discriminator and a low-pass filter. In the next section, these and other contributions are placed into proper perspective. When this is done, one can see that only one paper strictly addresses the problem of wideband FSK detection and this is from the experimental viewpoint.

II. REVIEW OF THE LITERATURE

In the past few years, interest in the problem of demodulation of FSK type signals, e.g., (PAM/FM, MFSK, CPFSK, and MSK) has increased owing to the fact that a number of

Reprinted from *IEEE Trans. Commun.*, vol. COM-25, pp. 200–214, Feb. 1977.

applications which require simplicity in hardware have arisen as well as decoding algorithms [2]-[4] which allow exploitation of the phase continuity associated with the transmitted oscillations. In many of the applications, overall uncertainty of the received carrier frequency is small whereas in other applications this uncertainty (due to channel Doppler and transmitter and receiver frequency instabilities) cannot be neglected. In either case, the communication system design engineer has the option of employing coherent or noncoherent detection of the received information. Depending upon the application and system cost constraints, tradeoffs exist between systems which operate fully coherent (e.g., PSK) versus those which operate noncoherently (FSK).

It is well known that noncoherent detection of FSK and CPFSK signals requires the implementation of matched filters followed by envelope detectors [4], [5]. In order to realize the matched filter operation in hardware some form of frequency estimation and control is required in the receiver. The purpose of this paper is to present a theory for use in the design of a digital FSK receiver which employs a PLL to set up the desired matched filter as the arriving signal frequency switches. The principle of operation is described as follows: A PLL is used to rapidly estimate the signal frequency and phase. The loop filter output is processed via the usual correlation method, viz., an integrate and dump circuit is used. Thus, the receiver structure functions as an "estimator-correlator" in that the loop quickly estimates the carrier phase and frequency of the arriving tone and correlates the loop filter output voltage in an integrate and dump circuit. In applications where the initial frequency uncertainty due to channel Doppler and oscillator instability is of concern, some form of frequency estimator (AFC) can be used in the receiver to combat the offending loop stress. This can be accomplished by averaging the loop filter output and using this to reduce the stress thereby improving system performance. This method is particularly useful when the data stream transition density is sufficient to give a good frequency estimate. Of course, Manchester coded FSK [6]-[8] can always be employed to overcome the data transition density problem at the expense of lowering the system noise immunity and increasing system bandwidth.

Over the past decade a number of authors have been concerned with the theoretical performance of digital FSK when demodulated by a frequency discriminator. In the early 1960's, Smith [9] pointed out that the performance of optimum detection of PCM/FM as determined by Kotel'nikov [10] could be achieved (within 1 dB) by an ordinary FM discriminator when the bandwidth of the IF filter preceding the discriminator is made equal to the data rate. Further analytical estimates of this fact were made by Shaft [11]. His approach followed that of Meyerhoff and Mazer [12] who in turn based their technique on the classical work of Rice [13]. Data detection was based upon threshold comparison of a single sample of the discriminator output taken at the end of each data pulse. Experimental verification (with reasonably close agreement) of the above analyses is given in the reported measurements made by Radiation, Inc. [14]. Aeronutronics, Inc. [15], and EMR, Inc. [16]. The impulse-like properties of FM noise were first described quantitatively by Cohn [17] in his doctoral dissertation and later in a published conference paper [18]. Following a similar but somewhat different approach, Rice [19] presented a theory which characterized the statistics of the impulsive behavior (so-called "clicks" or "pops") of the noise at the output of an ideal limiter-discriminator combination. These contributions now allowed one to include the effects of post-detection filtering which had heretofore been ignored. Based then upon these works, Mazo and Salz [20] presented an approximate theory to describe the theoretical performance of digital FSK when demodulated by a frequency discriminator followed by a low-pass filter. Shortly after, Schilling, Hoffman, and Nelson [21] considered the same problem and gave further experimental and analytical support to the fact that in certain regions FM discriminator low-pass filter detection can be made to closely approach that of matched filter detection. With regard to the discriminator detection of wide-band PCM/FM, wherein the IF bandwidth must be made many times as wide as the PCM data rate to accommodate carrier frequency uncertainties due to Doppler or instabilities, the following contributions are noted. McRae [22] determined analytical expressions for error rates for the standard limiter-discriminator followed by a low-pass filter. Chen [23] first extended the works by Meyerhoff and Mazer [12] and Shaft [11] to the wideband case and later included the effect of low-pass filtering at the output [24]. Finally, Tjhung and Wittke [25] reported analytical and experimental results for the case where the FM signal is distorted by a restricted predetection filter bandwidth. Although Bennett and Salz [26] had taken this effect into account in their analysis, their receiver model did not include a low-pass filter following the discriminator.

To the authors' knowledge very little information concerning the demodulation of digital FSK by a PLL has been given in the literature. Whereas Schilling, Billig, and Kermisch [27] have addressed the problem and given preliminary experimental results, the noise mechanism which causes decision error is not adequately modeled.

The present paper differs from this earlier work [27] in that the Gaussian assumption regarding the decision variable, or one of its "components," is not invoked, a first-order PLL is assumed and the IF filter is wideband. The probability density function (pdf) of the decision variable proceeds on the basis of the *Meyr distribution* [28] which accounts for the direction in which the probability mass associated with the phase error is wrapped around the phase cylinder (probability space). This distribution is used to generate the pdf of the phase error increment which disturbs the correlator output. The cycle slipping probabilities derived in [29] as well as their direction are then used to unfold the probability mass which has been wrapped around the phase cylinder due to loop slipping during the bit interval. Using this approach, the Gaussian assumption, which is never valid in the signal-to-noise ratio region of practical interest, is not invoked, although the Poisson assumption is made regarding the cycle-slipping distribution. In addition, frequency uncertainties due to resid-

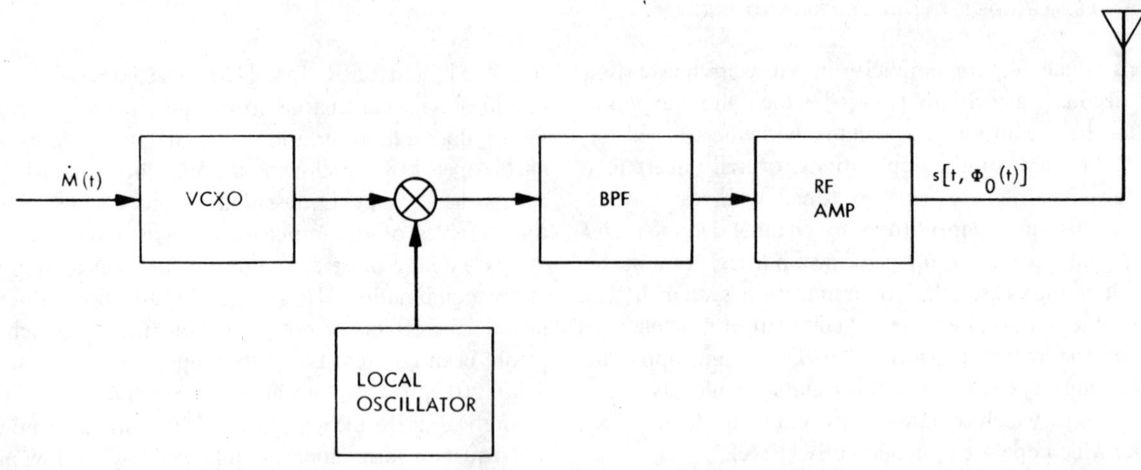

Fig. 1. Simplified transmitter mechanization.

ual Doppler effects, etc., are included as well as a determination of demodulator design philosophy as functions of system data rate, deviation ratio, and signal-to-noise level. Computer simulations are given to verify the region of validity of our Poisson assumption.

In the next section, a mathematical model of the system under consideration is described; this is followed by a presentation of the chief theoretical results which are then evaluated and presented for use in predicting system design, performance.

III. SYSTEM MODEL

The transmitted oscillation is characterized by

$$s[t,\Phi_0(t)] = \sqrt{2}A \sin \Phi_0(t) \tag{1}$$

where $\Phi_0(t)$ represents the total instantaneous phase and A^2 the average transmitted power. A simplified mechanization of the transmitter is illustrated in Fig. 1. The instantaneous frequency $\dot{\Phi}_0(t)$ of $s[t,\Phi_0(t)]$ is characterized by

$$\dot{\Phi}_0(t) = 2\pi[f_0 + \Delta f \dot{M}(t)], \qquad \dot{M}(t) = \sum_{k=-\infty}^{\infty} d_k p(t-kT) \tag{2}$$

where f_0 represents the nominal carrier frequency, Δf represents the one-sided frequency deviation, d_k is, in general, a random variable and $p(t - kT)$ characterizes the kth transmitted pulse of duration T s. Although the case emphasized herein is that in which d_i is a discrete binary random variable, i.e., PCM/FM, the theory which follows can be applied to modulation schemes of the PAM/FM type. In the latter case, mean-squared error is the performance measure as opposed to bit error probability.

The received signal

$$x(t) = s[t,\Phi(t)] + n_i(t), \qquad \Phi(t) = \Phi_0(t) + \theta(t) \tag{3}$$

is assumed to be a noise-corrupted, Doppler shifted, phase-shifted version of the transmitted signal in which $n_i(t)$ is additive white Gaussian noise of single-sided spectral density N_0 W/Hz, and $\theta(t) \triangleq \theta_0 + d(t)$ with $d(t)$ the channel Doppler and θ_0 an arbitrary phase angle. For this received signal model, it is well known [4, ch. 9] that the optimum receiver for noncoherent detection consists of matched filters followed by envelope detectors, samplers, and a threshold comparison circuit. Although the configuration mentioned is theoretically optimum in the sense of processing the received data bit-by-bit (assuming perfect synchronization) its implementation into practice requires frequency estimation circuits due to the presence of oscillator instabilities and channel Doppler. A popular mechanization of a receiver, which has the combined capability of estimating the frequency needed for setting up the matched filter and detection of the data, is to employ a PLL followed by an integrate and dump circuit. Fig. 2 illustrates a simplified realization of the circuit.[1] The PLL is designed such that early in the bit interval, it locks to the frequency and phase of the arriving noise contaminated oscillation, $x(t)$, thus providing a reference estimate $r(t,\hat{\Phi})$ of the arriving signal component $s(t,\Phi)$. This reference estimate is then used to produce the information-bearing signal $z(t)$, into the correlator (Fig. 2), the sign of whose output provides a decision regarding the data bit. In what follows, we shall assume that $p(t)$ is a rectangular pulse, i.e., $p(t) = 1, 0 \leqslant t \leqslant T$; $p(t) = 0$, otherwise. In cases where this assumption cannot be invoked, a multiplier driven by $p(t)$ and $z(t)$ must precede the integrated and dump circuit.

IV. MATHEMATICAL MODEL OF THE RECEIVING SYSTEM

The equation of operation of the estimator part of the receiver is well-known having been derived in many papers and textbooks on the subject of PLL (e.g., [29, ch. 3]). The form of this equation which is most useful to the problem at hand, however, may not be immediately obvious. Thus, we shall briefly trace through the derivation keeping the details to a minimum. We begin by defining the total phase error $\varphi(t)$ by

[1] We point out that the classical noncoherent FSK detector which assumes *a priori* knowledge of the transmitted frequencies does not apply here. The optimum receiver for the case of unknown frequencies can be implemented, to a first approximation, as in Fig. 2.

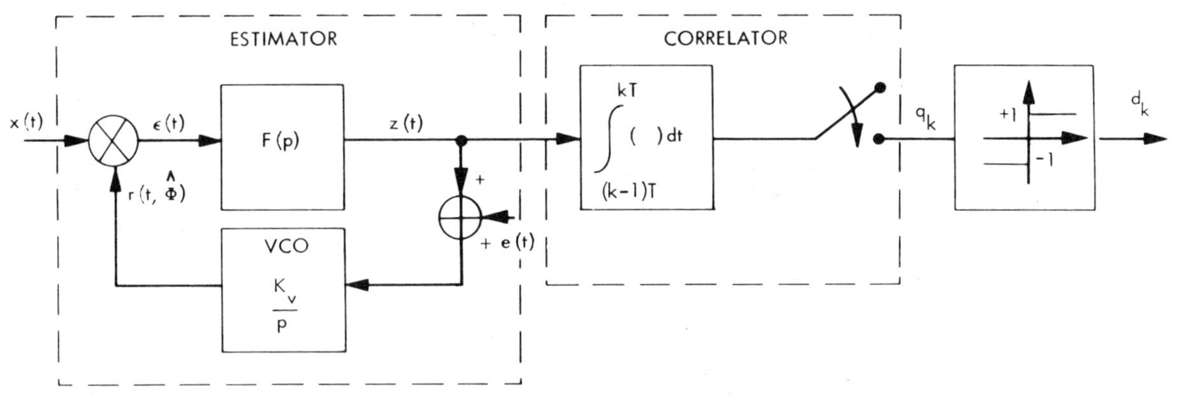

Fig. 2. Estimator/correlator employing a phase-locked loop.

$$\varphi(t) \triangleq \Phi(t) - \hat{\Phi}(t) \tag{4}$$

where the instantaneous phase $\hat{\Phi}(t)$ of the receiver VCO output $r(t)$ is

$$\hat{\Phi}(t) = \omega_0 t + \hat{\theta}(t), \qquad \omega_0 \triangleq 2\pi f_0. \tag{5}$$

From Fig. 2 we observe that

$$\hat{\theta}(t) = \frac{K_v z(t)}{p} + \frac{K_v e(t)}{p} \tag{6}$$

where $p = d/dt$ is the Heaviside operator and K_v denotes the receiver VCO gain. The voltage $e(t)$ represents a center frequency tuning voltage whose purpose is to eliminate loop stress due to channel Doppler and transmitter instabilities. Since combining (2) and (3) gives

$$\Phi(t) = \omega_0 t + \theta_0 + d(t) + 2\pi\Delta f M(t) \tag{7}$$

then

$$\begin{aligned} \dot{\varphi}(t) &= \dot{\Phi}(t) - \dot{\hat{\Phi}}(t) \\ &= \dot{d}(t) + 2\pi\Delta f \dot{M}(t) - K_v z(t) - K_v e(t) \end{aligned} \tag{8}$$

so that the input $z(t)$ to the correlator of Fig. 2 is

$$z(t) = \frac{1}{K_v}\,[\dot{d}(t) + 2\pi\Delta f \dot{M}(t) - \dot{\varphi}(t) - K_v e(t)]. \tag{9}$$

Assuming, without loss in generality, that $K_v = 1$, then the decision voltage q_k generated during the kth bit interval $(k-1)T \leqslant t \leqslant kT$ is given by

$$\begin{aligned} q_k &= \int_{(k-1)T}^{kT} z(t)\,dt \\ &= d(kT) - d[(k-1)T] + 2\pi\Delta f\{M(kT) - M[(k-1)T]\} \\ &\quad + \varphi[(k-1)T] - \varphi(kT) + \int_{(k-1)T}^{kT} e(t)\,dt \end{aligned} \tag{10}$$

where we have assumed perfect bit synchronization. For the present let us assume that the Doppler residual

$$\Delta d = d(kT) - d[(k-1)T] \tag{11}$$

can be neglected and that $e(t) = 0$. Thus, the decision variable q_k reduces to

$$q_k = 2\pi\Delta f T d_k - \underbrace{\{\varphi(kT) - \varphi[(k-1)T]\}}_{\text{noise increment}} \tag{12}$$

where we have also used the fact that

$$d_k T = \int_{(k-1)T}^{kT} \dot{M}(t)\,dt = M(kT) - M[(k-1)T]. \tag{13}$$

Since d_k is a discrete binary random variable taking on values plus one or minus one, that at $t = kT$ the receiver of Fig. 2 announces

$$\hat{d}_k = \begin{cases} 1, & \text{if } q_k \geqslant 0 \\ -1, & \text{if } q_k < 0. \end{cases} \tag{14}$$

As is usually the case in statistical detection theory, the decision variable q_k is composed of two components; one having to do with the signal (in this case, $2\pi\Delta f T d_k$) and the other having to do with the additive noise. However, we note that the noise increment in (12), which perturbs the signal component, is non-Gaussian so that an accurate assessment of system performance requires a statistical characterization of the random variable $\varphi(kT) - \varphi[(k-1)T]$. This is the subject of the next section. Before going on, however, to this all important part of the paper we conclude this section with a few brief remarks and illustrations which will hopefully provide insight into what follows.

In an effort to explain the loop's ability to track a frequency which switches randomly at the data rate $R = 1/T$, the differential equation of loop operation in the absence of noise has been solved by means of a digital computer. Fig. 3 illustrates the loop phase error $\varphi(t)$ as a function of time for various values of $\delta \triangleq 1/B_L T$ with the normalized frequency deviation $\Delta f T$ held constant. Here B_L represents the single-

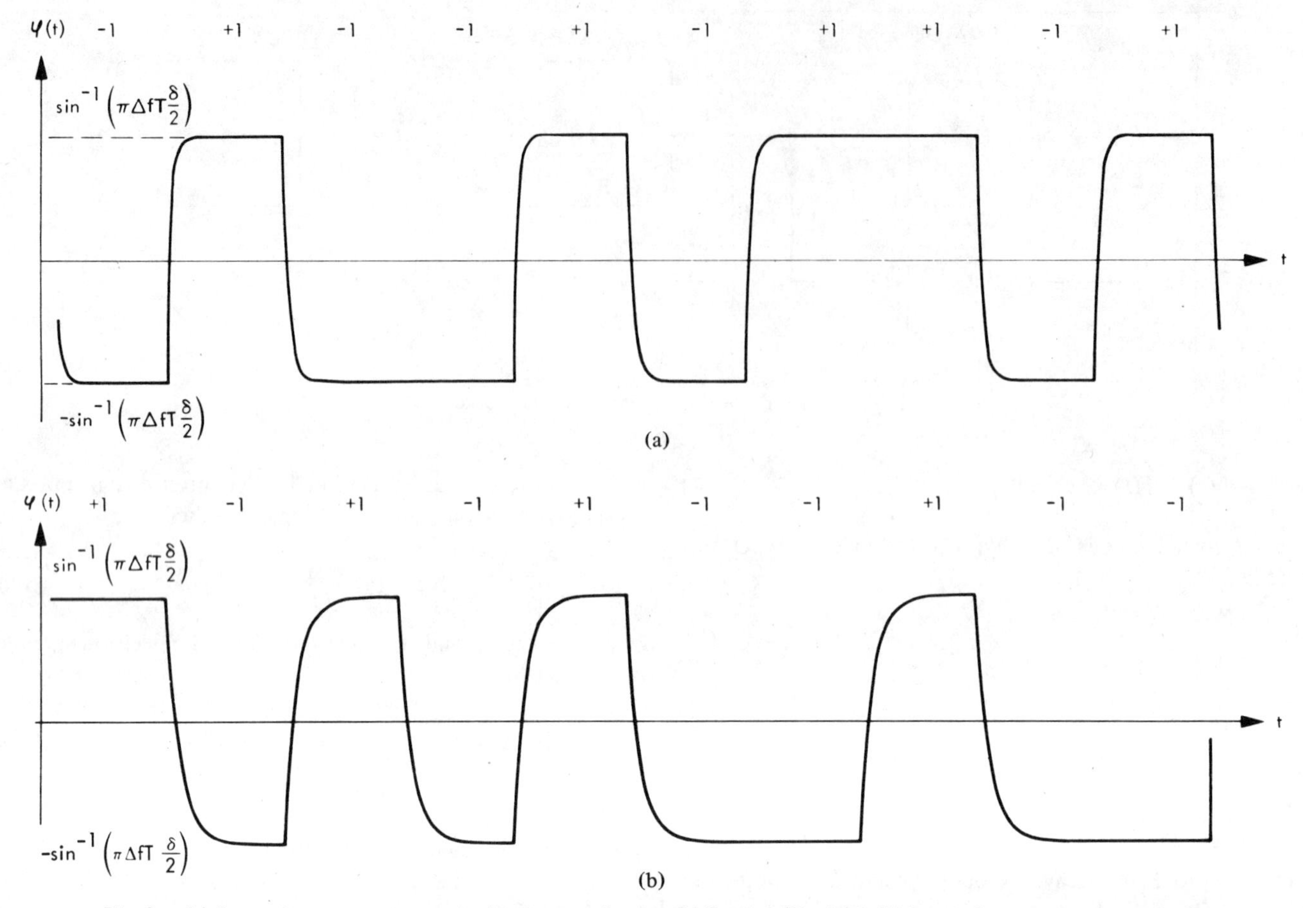

Fig. 3. (a) Loop phase error response in the absence of noise; $B_LT = 5, \Delta fT = 0.25$. (b) Loop phase error response in the absence of noise; $B_LT = 2, \Delta fT = 0.25$.

sided loop noise bandwidth, and a first-order loop, i.e., $F(p) = K_p$, is assumed. We observe that during each bit interval the loop locks to the steady-state phase error ϕ_{ss} which for a first-order loop with radian frequency detuning Ω_0 is given by $\phi_{ss} = \sin^{-1}(\Omega_0/AK)$ with K denoting the total product of the loop component gains. Recalling that for a first-order loop $B_L = AK/4$ and noting that $\Omega_0 = 2\pi\Delta f d_k$, we get

$$\phi_{ss} = \sin^{-1}\left(\frac{2\pi\Delta f d_k}{4B_L}\right) = d_k \sin^{-1}\left(\frac{\pi\Delta fT\delta}{2}\right) \tag{15}$$

so that the sign of ϕ_{ss} is directly dependent on the data sequence $\{d_k\}$. The fact that, for all practical purposes, the loop reaches a steady-state condition prior to the end of the bit interval is important in what follows, viz., we shall use this fact to justify the assumption that $\varphi(kT)$, $\varphi[(k-1)T]$, and hence the noise increment $\varphi(kT) - \varphi[(k-1)T]$ can be characterized by their steady-state pdf's. Furthermore, for $B_LT > 1$, the random variables $\varphi(kT)$ and $\varphi[(k-1)T]$ are approximately independent; thus pdf of the increment $\varphi_d = \varphi(kT) - \varphi[(k-1)T]$ is found from the convolution of the pdf's of $\varphi(kT)$ and $\varphi[(k-1)T]$, respectively. Finally, we note that although the above arguments have been advanced for the noise-free case, one would anticipate no significant degradation due to the transient behavior during acquisition in noise since, in practical situations, this transient interval represents a small fraction of the bit interval and furthermore $\Delta f \ll f_0$.

Although the theory given here assumes a first-order loop, we hasten to point out that the statistics of the phase error are not drastically altered when an integrating type of loop filter is used provided its time constant is of the order of the bit duration T. However, the cycle-slipping properties are considerably altered thus presenting analytical difficulties (see [29, ch. 10]) when attempting to treat this case.

V. STATISTICS OF THE ADDITIVE NOISE INCREMENT

This section discusses and develops results which lead to a statistical characterization of the additive noise increment given in (12). We begin by discussing the statistics of the phase error renewal process. Meyr [28] has found the solution to the Fokker–Planck equation [herein called the renewal process solution and denoted by $\phi_r(t)$] which accounts for the direction in which the probability mass associated with the phase error trajectories has been deposited on the phase cylinder (probability space). If ϕ_0 denotes the initial value of phase error at some arbitrary starting time instant t_0, then using the Meyr distribution $p_r(\phi_r)$ [28, eq. 72], defined over the 4π interval $(-2\pi + \phi_0, 2\pi + \phi_0)$, it is easy to show, after some manipulation, that

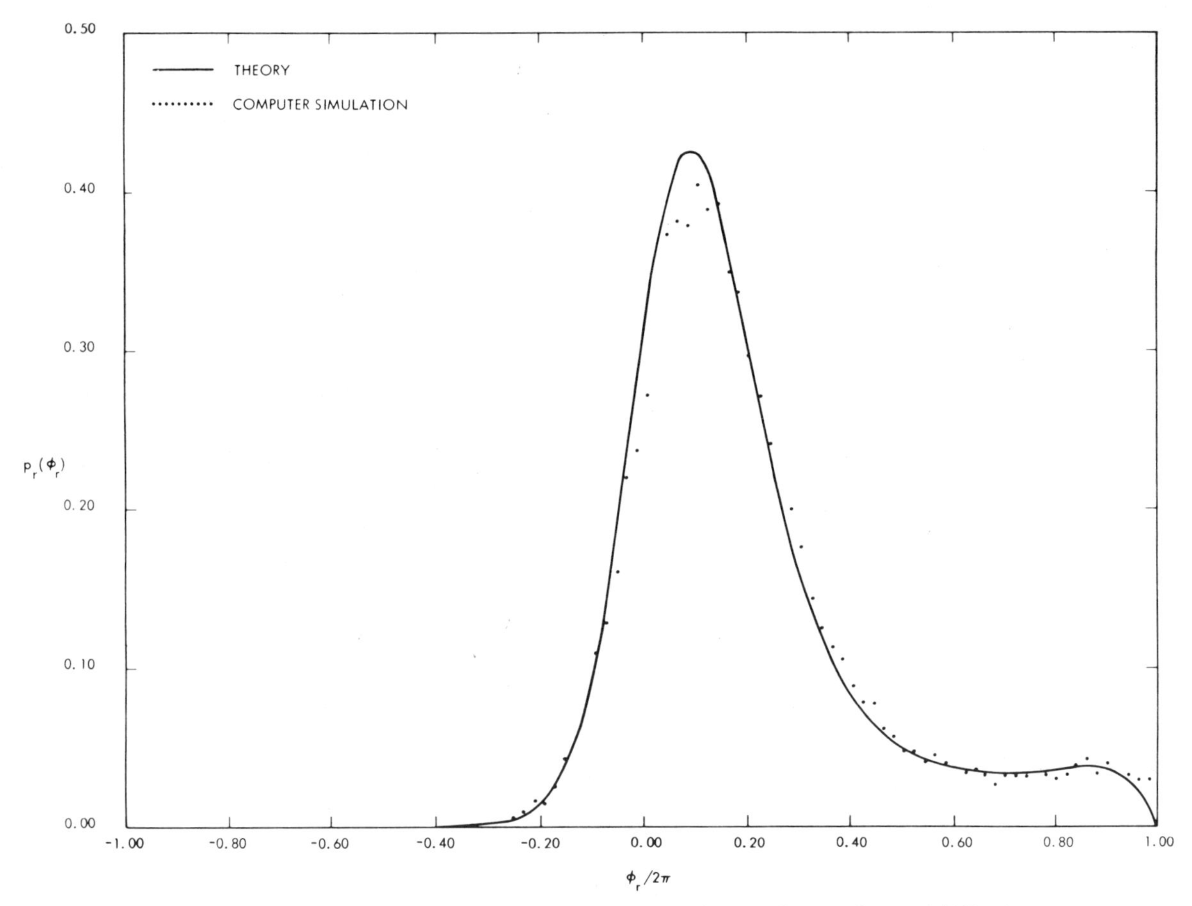

Fig. 4. A plot of the pdf of the renewal process $\phi_r(t)$ at time $t = kT$; $\rho = 2$ dB, $\Omega_0/AK = 0.7$.

$$p_r(\phi_r) = \begin{cases} \dfrac{\exp(\alpha \cos \phi_r + \beta \phi_r)}{4\pi^2 \,|\, I_{j\beta}(\alpha) \,|^2} \left\{ e^{-\beta\pi} \displaystyle\int_{-2\pi+\phi_r}^{\phi_0} \exp(-\alpha \cos y - \beta y)\, dy \right\}, \\ \qquad \phi_0 \leqslant \phi_r \leqslant 2\pi + \phi_0 \\ \dfrac{\exp(\alpha \cos \phi_r + \beta \phi_r)}{4\pi^2 \,|\, I_{j\beta}(\alpha) \,|^2} \left\{ e^{-\beta\pi} \displaystyle\int_{-2\pi+\phi_0}^{\phi_r} \exp(-\alpha \cos y - \beta y)\, dy \right\}, \\ \qquad -2\pi + \phi_0 \leqslant \phi_r \leqslant \phi_0 \end{cases} \tag{16}$$

where $I_{j\beta}(\alpha)$ is the modified Bessel function of imaginary order $j\beta$ and real argument α, and for a first-order loop,

$$\alpha = \rho \triangleq \frac{A^2}{N_0 B_L} = R_d \delta, \qquad R_d = \frac{A^2 T}{N_0}$$
$$\beta \triangleq \rho \frac{\Omega_0}{AK} = d_k \frac{\rho\pi\Delta f T\delta}{2}. \tag{17}$$

The parameter ρ is the well-known loop signal-to-noise ratio, and β/ρ is the normalized loop detuning. For purposes of numerical evaluation, the function $|\, I_{j\beta}(\alpha)^2 \,|$ can be expressed in integral form [29, ch. 9], viz.,

$$|\, I_{j\beta}(\alpha) \,|^2 = \frac{e^{-\beta\pi}}{\pi} \int_0^{\pi} \exp(-2\beta z) I_0(2\alpha \sin z)\, dz. \tag{18}$$

A typical plot of $p_r(\phi_r)$ versus ϕ_r is illustrated in Fig. 4 for the special case of $\phi_0 = 0$, $\rho = 2$, and $\Omega_0/AK = 0.7$. For $\beta = 0$ and $\phi_0 = 0$, (16) simplifies to

$$p_r(\phi_r) = \frac{\exp(\alpha \cos \phi_r)}{2\pi I_0(\alpha)} \left\{ \frac{\displaystyle\int_{|\phi_r|}^{2\pi} \exp(-\alpha \cos y)\, dy}{\displaystyle\int_0^{2\pi} \exp(-\alpha \cos y)\, dy} \right\} \tag{19}$$

or in series form

$$p_r(\phi_r) = \frac{\exp(\alpha \cos \phi_r)}{4\pi^2 I_0^2(\alpha)} \left[(2\pi - |\, \phi_r \,|) I_0(\alpha) - 2 \sum_{k=1}^{\infty} (-1)^k \frac{I_k(\alpha)}{k} \sin k \,|\, \phi_r \,| \right], \quad |\, \phi_r \,| \leqslant 2\pi \tag{20}$$

The pdf of the renewal process solution, $\phi_r(t)$, is related to that of the periodic extension solution $\phi_p(t)$, which is defined on the 2π interval $(-\pi + \phi_0, \pi + \phi_0)$, by [29]

$$p_p(\phi_p) = \begin{cases} p_r(\phi_p) + p_r(\phi_p + 2\pi), & -\pi + \phi_0 \leqslant \phi_p \leqslant \phi_0 \\ p_r(\phi_p) + p_r(\phi_p - 2\pi), & \phi_0 \leqslant \phi_p \leqslant \phi + \phi_0 . \end{cases} \tag{21}$$

Furthermore, when $\phi_0 = 0$, the pdf of the periodic extension solution reduces to the well-known Tikhonov distribution [31] for the so-called "modulo-2π reduced" phase error ϕ, viz.,

$$p(\phi) = \frac{\exp(\alpha \cos \phi + \beta\phi)}{4\pi^2 \exp(-\pi\beta) \mid I_{j\beta}(\alpha) \mid^2} \cdot \int_{\phi}^{\phi+2\pi} \exp(-\alpha \cos y - \beta y)\, dy, \qquad |\phi| \leqslant \pi \tag{22}$$

The relation between the renewal process solution and the actual loop phase error $\varphi(t)$ is crucial to the characterization of the noise increment in (12). In particular, for small δ (where the assumption that $\varphi(kT)$ and $\varphi[(k-1)T]$ have reached their steady-state values is valid), we have

$$\varphi_d \triangleq \varphi(kT) - \varphi[(k-1)T] = \underbrace{\phi_r(kT) - \phi_r[(k-1)T]}_{\phi_\Delta} + 2\pi N \tag{23}$$

where

$$N = n_+ - n_- \tag{24}$$

is a random variable with n_+ and n_- denoting the number of times the sample trajectories of the renewal process arrive at the barriers $2\pi + \phi_0$ and $-2\pi + \phi_0$, respectively, in the time interval $(k-1)T \leqslant t \leqslant kT$. The quantities n_+ and n_- are also synonomous with the number of cycles the loop has been pushed to the right and to the left, respectively, in the above T-second interval. To determine then the pdf of φ_d, one first computes the convolution of $p_r\{\phi_r[(k-1)T]\}$ and $p_r[\phi_r(kT)]$ and *unfolds* this pdf by the set of discrete cycle-slipping probabilities characterizing N.

To characterize the cycle-slipping probabilities, we shall assume that the cycle slips which tend to increase and decrease φ by 2π radians form independent Poisson processes with rates of occurrence N_+ and N_-, respectively [29, ch. 9]. The region of validity of this assumption is verified later by computer simulation. Then the probability that a net number, say $N = n$, of cycle slips occur in a given T-second interval is given by

$$P_n \triangleq \Pr\{N = n\} = \left(\frac{N_+}{N_-}\right)^{n/2} \cdot \exp(-\bar{S}T) I_n [2T\sqrt{N_+ N_-}], \qquad n = 0, \pm 1, \pm 2, \cdots \tag{25}$$

where

$$\frac{N_+}{N_-} = \exp(2\pi\beta)$$

$$N_\pm = \frac{B_L \exp(\pm\pi\beta)}{\pi^2 \alpha \mid I_{j\beta}(\alpha) \mid^2}$$

$$\bar{S} \triangleq N_+ + N_- = \frac{2B_L \cosh \pi\beta}{\pi^2 \alpha \mid I_{j\beta}(\alpha) \mid^2} . \tag{26}$$

At this point, it is worthwhile to remind the reader that in computing the set of cycle-slipping probabilities which characterize $\varphi(t)$ in the interval $(k-1)T \leqslant t \leqslant kT$, the data bit d_k which was transmitted in this interval determines the sign of β [see (17)] to be used in (26). For example, if $d_k = +1$, then from (17) β is positive, and from (26) $N_+/N_- > 1$ i.e., the loop tends to slip more cycles to the right than to the left. The same fact holds true when computing the pdf's of $\phi_r[(k-1)T]$ and $\phi_r(kT)$, that is, the sign of β to be used in (16) is determined by the data digits d_{k-1} and d_k, respectively. Thus, when $d_k = d_{k-1}$, the pdf of $\phi_r(kT) - \phi_r[(k-1)T]$ is the convolution of (16) with itself, while for $d_k = -d_{k-1}$, the pdf is obtained by the convolution of (16) and its mirror image. Letting $\phi_\Delta \triangleq \phi_r(kT) - \phi_r[(k-1)T]$, then the pdf $p_\Delta(\phi_\Delta)$ of ϕ_Δ is illustrated in Fig. 5 for the two cases $d_k = d_{k-1}$ and $d_k = -d_{k-1}$ and the same loop parameters as in Fig. 4.

Using the set of discrete probabilities given in (25), the pdf of N can be expressed as a weighted sum of delta functions. Convolving this pdf with $P_\Delta(\phi_\Delta)$ gives the pdf $p_d(\varphi_d)$ of the noise increment φ_d, viz.,

$$p_d(\varphi_d) = \sum_{n=-\infty}^{\infty} P_n p_\Delta(\varphi_d + 2\pi n), \qquad -\infty \leqslant \varphi_d \leqslant \infty \tag{27}$$

where without any loss in generality, we have assumed $\phi_0 = 0$. We note that since the pdf $p_\Delta(\phi_\Delta)$ is distributed on a 4π interval, then the unfolding procedure characterized by (27) causes an overlap of probability masses [see Fig. 6(a)]. Fig. 6(b) is an attempt of explaining this phenomenon by showing two different sample trajectories of the actual phase error process $\varphi(t)$ which have different statistics but which both contribute to the pdf $P_\Delta(\phi_\Delta)$ at the same value of ϕ_Δ. In part (a) of the figure, the loop undergoes a single cycle slip to the right, whereas in part (b) of this figure, the loop does not slip a cycle within the same T-second interval.

Fig. 7 typically illustrates the diffusion of the noise increment φ_d along the real line as ρ is varied at fixed Ω_0/AK for the case $d_k = d_{k-1} = 1$. An important point to note here is that the pdf $P_d(\varphi d)$ as computed from (27) is a *continuous* function of φ_d on the entire real line and thus represents a proper unwrapping of the probability mass associated with the phase error. The reason behind this stems from the fact that $p_r(\phi_r)$ as computed from (16) is zero at its endpoints $\phi_r = \pm 2\pi$ and thus the convolution pdf $p_\Delta(\phi_\Delta)$ is correspondingly zero at its endpoints $\phi_\Delta = \pm 4\pi$ (see Figs. 4 and 5). This approach of unfolding the convolved renewal process pdf is in direct contrast to characterizing ϕ_Δ as a truncated Gaussian random variable [27] or the convolution of the Tikhonov pdf [30] with itself either of which when unfolded results in a piecewise con-

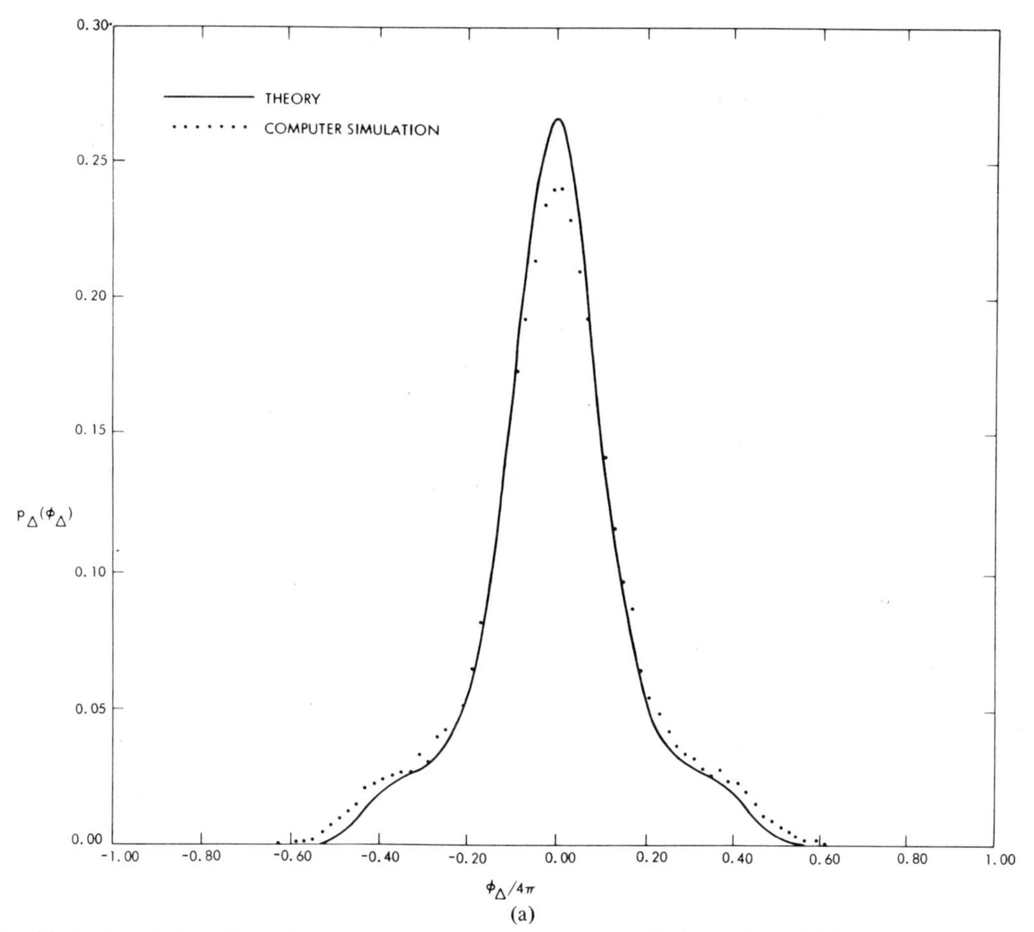

(a)

Fig. 5. (a) A plot of the pdf of the renewal process increment $\phi_\Delta = \phi_r(kT) - \phi_r[(k-1)T]$; $d_k = d_{k-1}$, $\rho = 2$ dB, $\Omega_0/AK = 0.7$. (b) A plot of the pdf of the renewal process increment $\phi_\Delta = \phi_r(kT) - \phi_r[(k-1)T]$; $d_k = -d_{k-1}$, $\rho = 2$ dB, $\Omega_0/AK = 0.7$.

tinuous pdf for φ_d implying discontinuous trajectories for the $\varphi(t)$ process.

VI. ERROR PROBABILITY PERFORMANCE WHEN DOPPLER IS ABSENT

From (12) and (14) and the definition of the noise increment φ_d given in (23), we can immediately write down an expression for the system error probability, viz.,[2]

$$\begin{aligned} P_E &= p_{+1} \Pr\{\hat{d}_k = -1 \mid d_k = +1\} \\ &\quad + p_{-1} \Pr\{\hat{d}_k = +1 \mid d_k = -1\} \\ &= p_{+1} \Pr\{\varphi_d > 2\pi\Delta fT\} + p_{-1} \Pr\{\varphi_d < -2\pi\Delta fT\} \\ &= p_{+1} \int_{2\pi\Delta fT}^{\infty} p_d(\varphi_d)\, d\varphi_d + p_{-1} \int_{-\infty}^{-2\pi\Delta fT} p_d(\varphi_d)\, d\varphi_d \end{aligned} \tag{28}$$

[2] When d_k is a random variable of the continuous type (the case of PAM/FM), system performance is accounted for by evaluation of the mean-squared error

$$\sigma^2 = E\{E[(q_k - 2\pi\Delta fTd_k)^2 \mid d_k]\}$$

from which the signal-to-noise ratio SNR $\triangleq \sigma_{dk}^2/\sigma^2$ serves as the system performance measure.

where p_{+1} and p_{-1} are, respectively, the *a priori* probabilities of $d_k = +1$ and $d_k = -1$. Since we have previously observed that the convolved pdf $p_\Delta(\phi_\Delta)$ can take on one of two forms (see Figs. 4 and 5) depending upon whether $d_{k-1} = d_k$ or $d_{k-1} = -d_k$, then from (27) the same holds true for $p_d(\varphi_d)$. Thus, each of the integrals in (28) must be broken up into two parts with weighting coefficients dependent on the data transition density $p_t \triangleq \Pr\{d_k \neq d_{k-1}\}$, viz.,

$$\begin{aligned} \int_{2\pi\Delta fT}^{\infty} p_d(\varphi_d)\, d\varphi_d &= p_t \int_{2\pi\Delta fT}^{\infty} p_d^{-+}(\varphi_d)\, d\varphi_d + (1-p_t) \int_{2\pi\Delta fT}^{\infty} p_d^{++}(\varphi_d)\, d\varphi_d \\ \int_{-\infty}^{-2\pi\Delta fT} p_d(\varphi_d)\, d\varphi_d &= p_t \int_{-\infty}^{-2\pi\Delta fT} p_d^{+-}(\varphi_d)\, d\varphi_d \\ &\quad + (1-p_t) \int_{-\infty}^{-2\pi\Delta fT} p_d^{--}(\varphi_d)\, d\varphi_d \end{aligned} \tag{29}$$

where the superscript notation on $p_d(\varphi_c)$ denotes the successive polarities of d_{k-1} and d_k. Clearly, dependent upon the given data source, the *a priori* probabilities p_{+1}, p_{-1} and the transition probability p_t may or may not be related. For example, for independent input data bits,

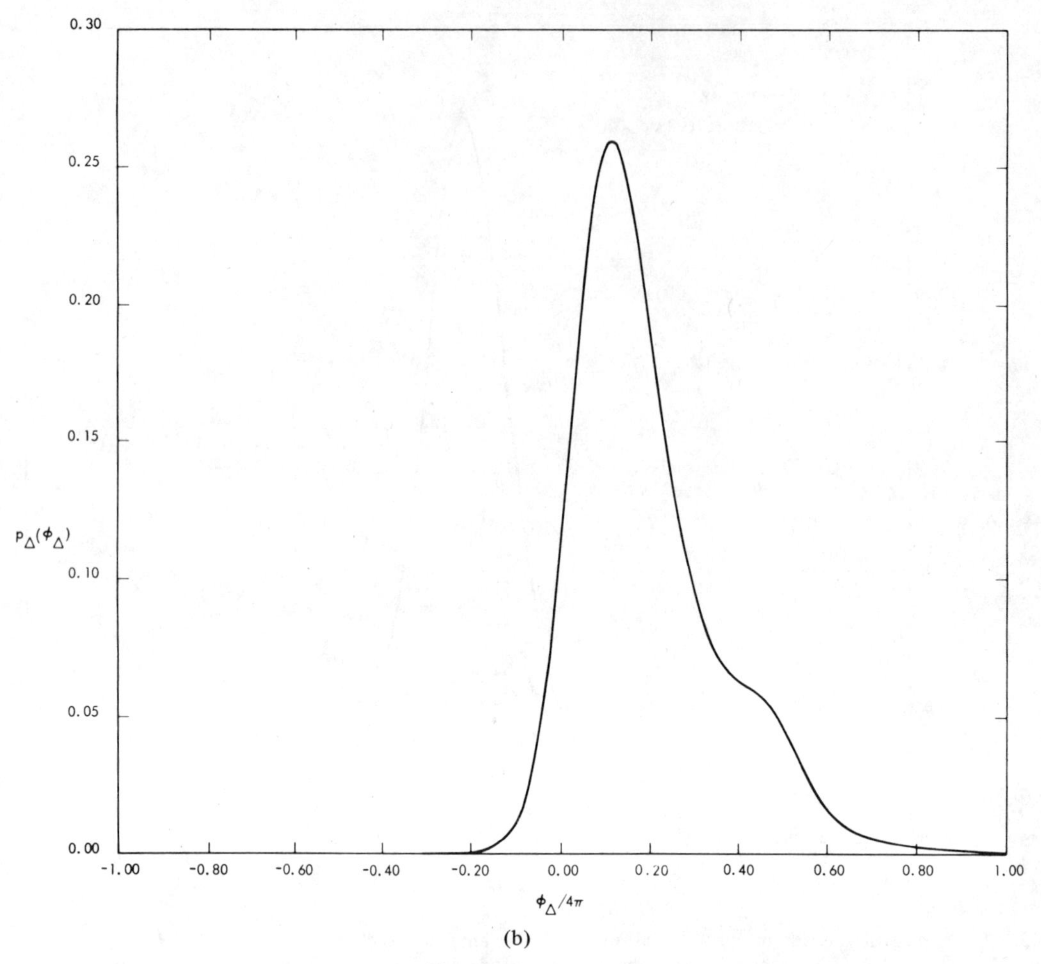

(b)

Fig. 5. Continued.

$$p_t = 2p_{+1}p_{-1} \tag{30}$$

whereas for a first-order Markov source,

$$p_{+1} = p_{-1} = \frac{1}{2} \tag{31}$$

for any nonzero value of p_t.

The numerical results presented in this paper will be for the case of random data with equiprobable bits in which case (30) applies with $p_{+1} = p_{-1} = p_t = 1/2$. Furthermore, the following symmetry properties hold:

$$p_d^{-+}(\varphi_d) = p_d^{+-}(-\varphi_d)$$
$$p_d^{++}(\varphi_d) = p_d^{++}(-\varphi_d) = p_d^{--}(\varphi_d) = p_d^{--}(-\varphi_d). \tag{32}$$

Thus, the error probability of (28) simplifies, for this special case, to

$$P_E = \frac{1}{2}\int_{2\pi\Delta fT}^{\infty} p_d^{-+}(\varphi_d)\, d\varphi_d + \frac{1}{2}\int_{2\pi\Delta fT}^{\infty} p_d^{++}(\varphi_d)\, d\varphi_d. \tag{33}$$

Fig. 8 is a plot of the error probability P_E as computed (by numerical integration) from (33) vs signal-to-noise ratio $R_d \triangleq A^2T/N_0$ for fixed values of δ and $\Delta fT = 0.25$, i.e., the deviation ratio corresponding to orthogonal FSK. Notice that when $\delta = 1$, the ability of the loop to rapidly set up the frequency and phase estimates of the data pulse is breaking down. In addition, the independence assumption concerning the random variables $\varphi[(k-1)T]$ and $\varphi(kT)$ used in finding the pdf of the noise increment is also becoming suspect. Thus, we represent the theoretical curve corresponding to this case by dashed lines. Also illustrated in this figure is the error probability performance of coherent FSK (detected by a matched filter receiver) given by

$$P_E = \frac{1}{2}\,\text{erfc}\left(\frac{\sqrt{R_d}}{2}\right) \tag{34}$$

where

$$\text{erfc}\, \chi \triangleq \frac{2}{\sqrt{\pi}}\int_{\chi}^{\infty} \exp(-y^2)\, dy. \tag{35}$$

One observes from this figure the large penalty in signal-to-noise ratio that must be paid for the inability of the receiver to have exact knowledge of the transmitter oscillator frequencies or the perturbations of them caused by the channel. This lack of knowledge is manifested in the necessity of having to design a large IF bandwidth to data rate ratio, whereupon the usual

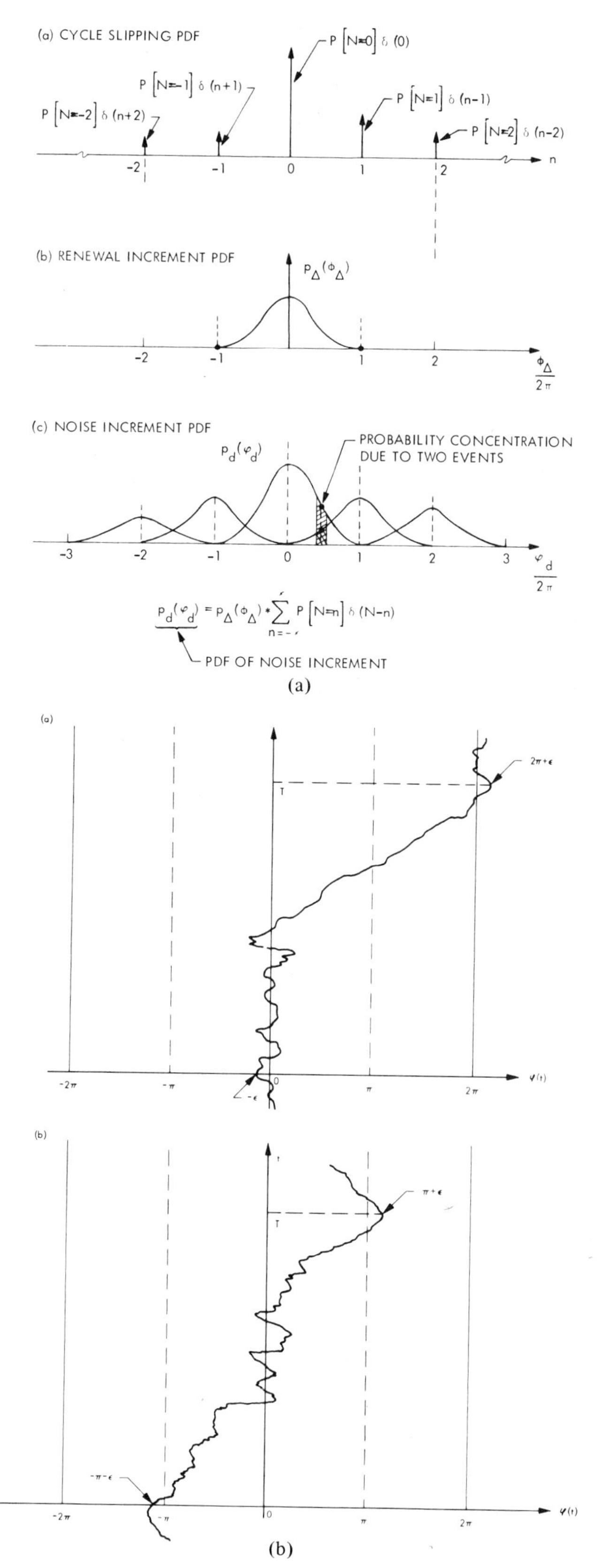

Fig. 6. Generation of the probability density function of the noise increment process. Two trajectories having different statistical properties which both contribute to $p_d(\varphi_d)$ at $\varphi_d = 2\pi + 2\epsilon$.

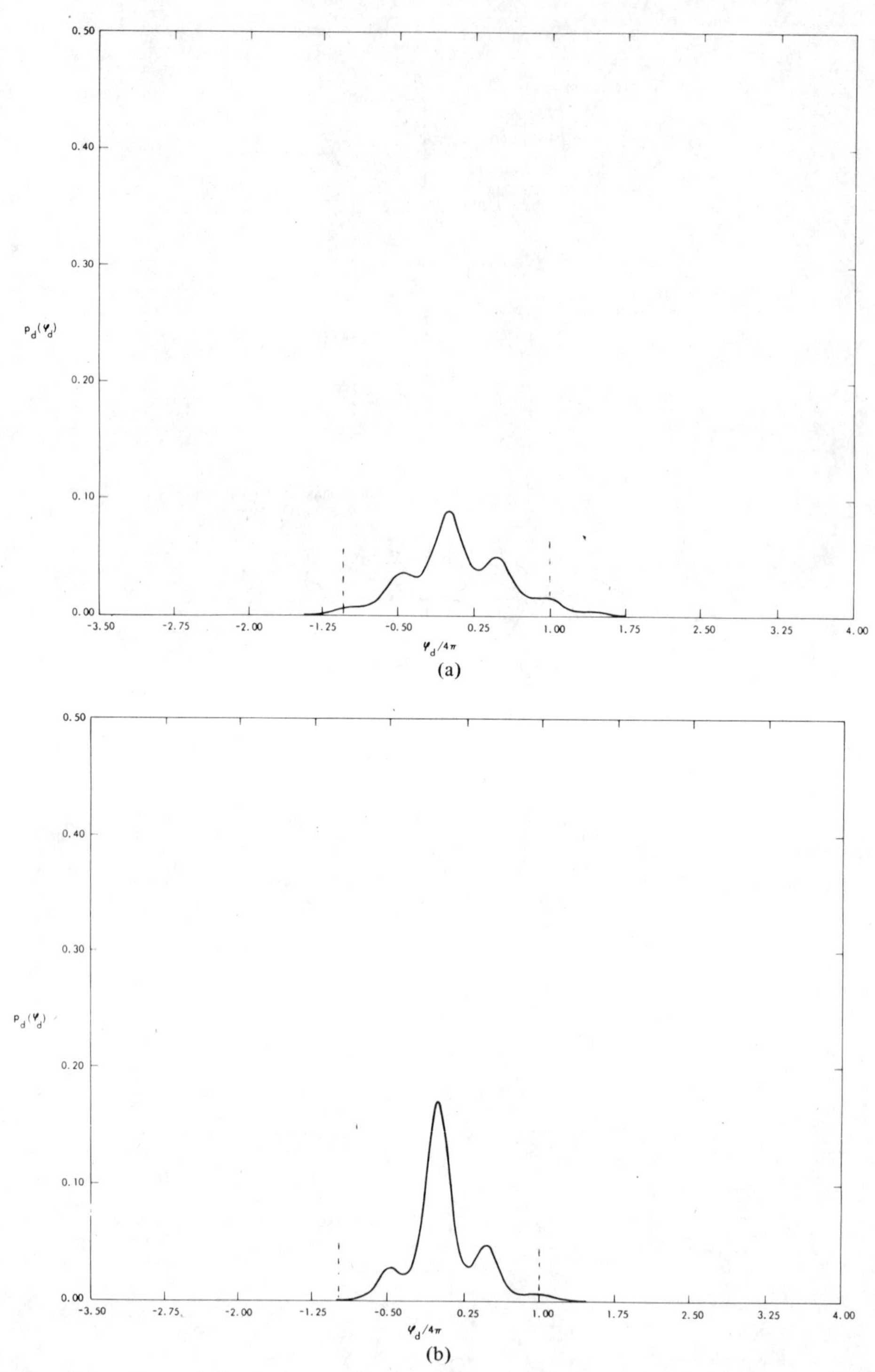

Fig. 7. A plot of the pdf $p_d(\varphi_d)$ of the noise increment; $\delta = 0.2$, $\Delta fT = 0.25$, $R_d = 6$ dB. (b) A plot of the pdf $p_d(\varphi_d)$ of the noise increment; $\delta = 0.2$, $\Delta fT = 0.25$, $R_d = 8$ dB. (c) A plot of the pdf $p_d(\varphi_d)$ of the noise increment; $\delta = 0.2$, $\Delta fT = 0.25$, $R_d = 10$ dB. (d) A plot of the pdf $p_d(\varphi_d)$ of the noise increment; $\delta = 0.2$, $\Delta fT = 0.25$, $R_d = 12$ dB.

bandlimited "white" noise assumption made at the PLL input remains valid.

One measure of the value of any theory is the degree of accuracy with which it agrees with simulation results. The equation of operation of a first-order PLL was simulated on an XDS 930 computer. The actual loop phase error was computed in the presence of noise, as a function of time (in discrete steps Δt) and by proper setting of the boundaries at $\pm 2\pi$, the renewal process was generated. The value of the renewal process phase error was recorded at discrete T-second intervals ($T/\Delta t$ was kept fixed at typically a value of 100) and a histogram of the relative frequency of occurrence of 100 values[3] of phase error over the interval $(-2\pi, 2\pi)$ was taken. This histogram then corresponds to a simulated pdf of $\phi_r(t)$ and is super-

[3] Actually, these values are "bins" of width $4\pi/100$ and the relative frequency of occurrence for each bin is determined by the number of times the phase error (at the end of each T-second interval) falls in that bin.

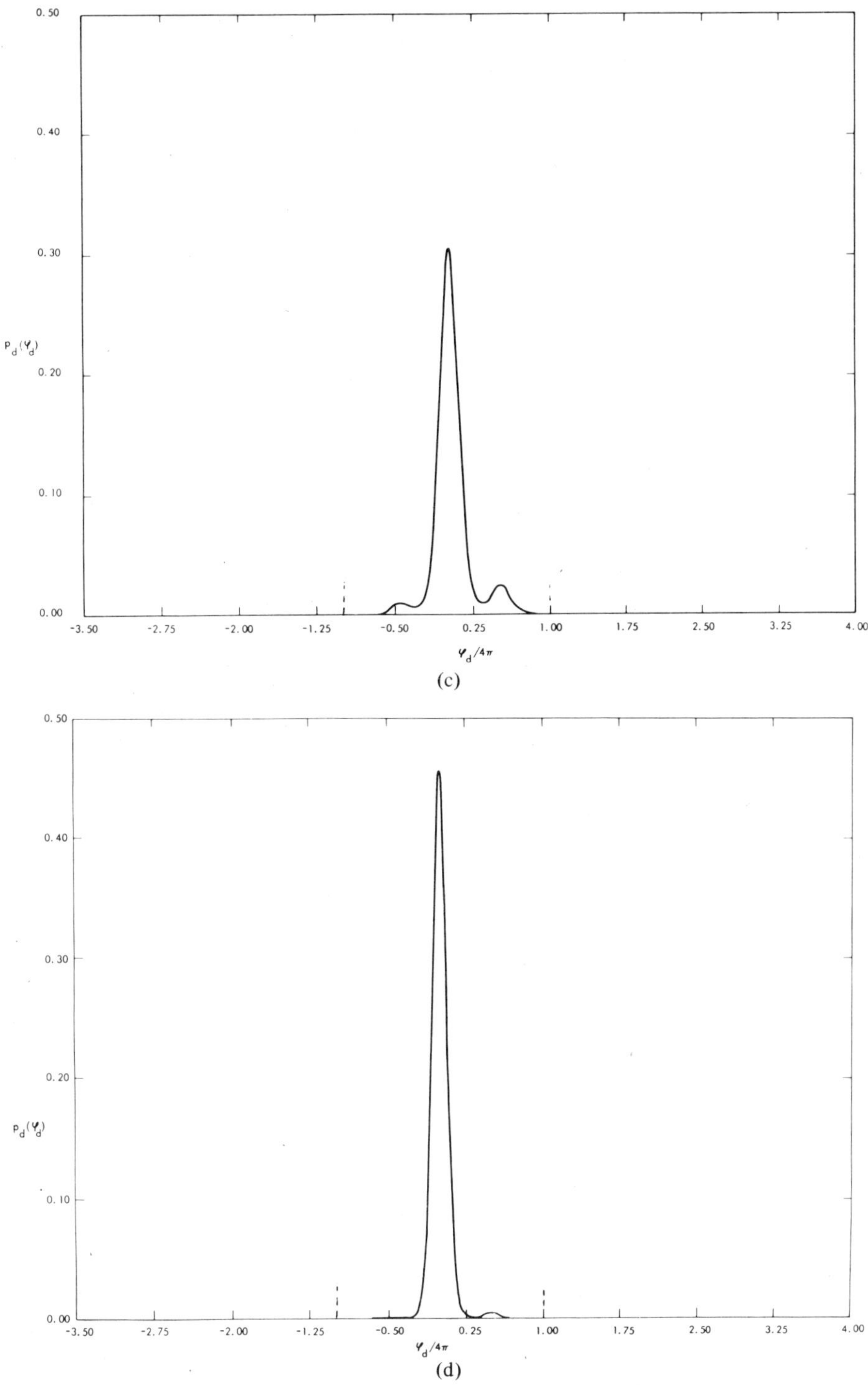

(c)

(d)

Fig. 7. Continued.

imposed on the theoretical result in Fig. 4. The closeness of fit is readily observed. Furthermore, the difference of adjacent T-second samples of the simulated renewal process phase error was recorded along with a count of the actual net number of cycle slips which occurred in each T-second interval. A histogram of the relative frequency of occurrence of 100 values of this difference phase error over the interval $(-4\pi, 4\pi)$ was recorded and is superimposed on the corresponding theoretical result in Fig.5(a).[4]

[4] Thus far, the input signal for these simulation results was assumed to correspond to all one's data, i.e., a CW signal.

At this point, the PLL simulation was augmented by the inclusion of a random data source. Using the actual cycle slip count made in each T-second interval to characterize the random variable N of (23) along with the recorded values of $\phi_r(kT)$ and $\phi_r[(k-1)T]$, the decision variable q_k of (12) was computed in each interval, and its sign recorded as the corresponding data estimate. Thus, in this manner, simulation results for average error probability performance were obtained. These simulation results are superimposed on the theoretically computed curves of Fig. 8. The agreement, one can observe, is excellent.

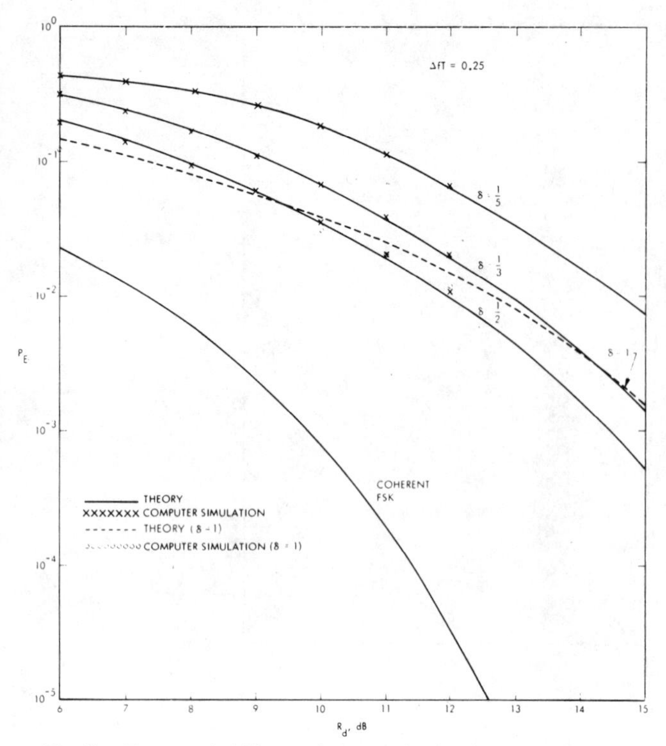

Fig. 8. Error probability performance in the absence of Doppler.

Finally, the error probability performance as a function of normalized frequency deviation ΔfT is illustrated in Fig. 9 for $\delta = 1/3$ and $R_d = 4, 6, 10$. One observes from these results that an optimum ΔfT (in the sense of minimizing P_E) exists whose value is in the neighborhood of that corresponding to optimum detection of PCM/FM [11], i.e., $\Delta fT = 0.358$.

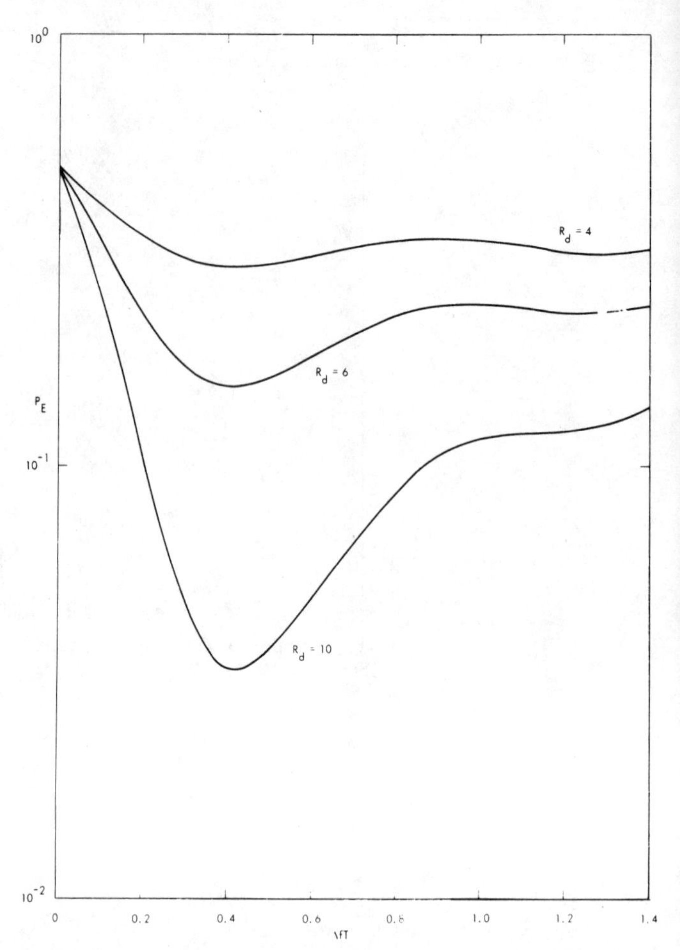

Fig. 9. A plot of error probability performance as a function of normalized frequency deviation; $\delta = 1/3$.

VII. ERROR PROBABILITY PERFORMANCE IN THE PRESENCE OF DOPPLER RESIDUALS

We recall that in arriving at the expression for the decision variable q_k given in (12), we assumed that the Doppler residual Δd of (11) could be neglected. If we now wish to take its effect into account, then from (10) we obtain the modified result

$$q_k = \Delta d + 2\pi\Delta fTd_k - \varphi_d \tag{36}$$

where we have also made use of the definition of the noise increment φ_d given in (23). Since the data estimate $\hat{d}_k$ is still determined by the decision criterion of (14), then using the definition of error probability given in (28) we get

$$\begin{aligned} P_E &= p_{+1} \Pr\{\varphi_d > \Delta d + 2\pi\Delta fT\} \\ &\quad + p_{-1} \Pr\{\varphi_d < \Delta d - 2\pi\Delta fT\} \\ &= p_{+1} \int_{\Delta d + 2\pi\Delta fT}^{\infty} p_d(\varphi_d)\, d\varphi_d \\ &\quad + p_{-1} \int_{-\infty}^{\Delta d - 2\pi\Delta fT} p_d(\varphi_d)\, d\varphi_d. \end{aligned} \tag{37}$$

Also, as in (29), we must split up each integral of (37) into two parts. Again for the special case of random equiprobable bits, (37) simplifies to

$$\begin{aligned} P_E &= \frac{1}{4} \int_{-\Delta d + 2\pi\Delta fT}^{\infty} p_d{}^{-+}(\varphi_d)\, d\varphi_d \\ &\quad + \frac{1}{4} \int_{\Delta d + 2\pi\Delta fT}^{\infty} p_d{}^{-+}(\varphi_d)\, d\varphi_d \\ &\quad + \frac{1}{4} \int_{-\Delta d + 2\pi\Delta fT}^{\infty} p_d{}^{++}(\varphi_d)\, d\varphi_d \\ &\quad + \frac{1}{4} \int_{\Delta d + 2\pi\Delta fT}^{\infty} p_d{}^{++}(\varphi_d) d\varphi_d. \end{aligned} \tag{38}$$

Fig. 10 plots P_E as computed from (27), (32), and (38) versus R_d for $\Delta fT = 0.25$, $\delta = 1/3$, and several values (including zero) of normalized Doppler residual $\Delta d/2\pi\Delta fT \triangleq f_r/\Delta f$. Note that $f_r \triangleq \Delta d/2\pi T$ represents the equivalent Doppler residual frequency which is not tuned out by $e(t)$ of Fig. 2.

VIII. BIT ERROR RATE PERFORMANCE FOR PSK MODULATION

We now turn our attention to the case of PSK modulation. As is well known, detection of suppressed carrier PSK can be

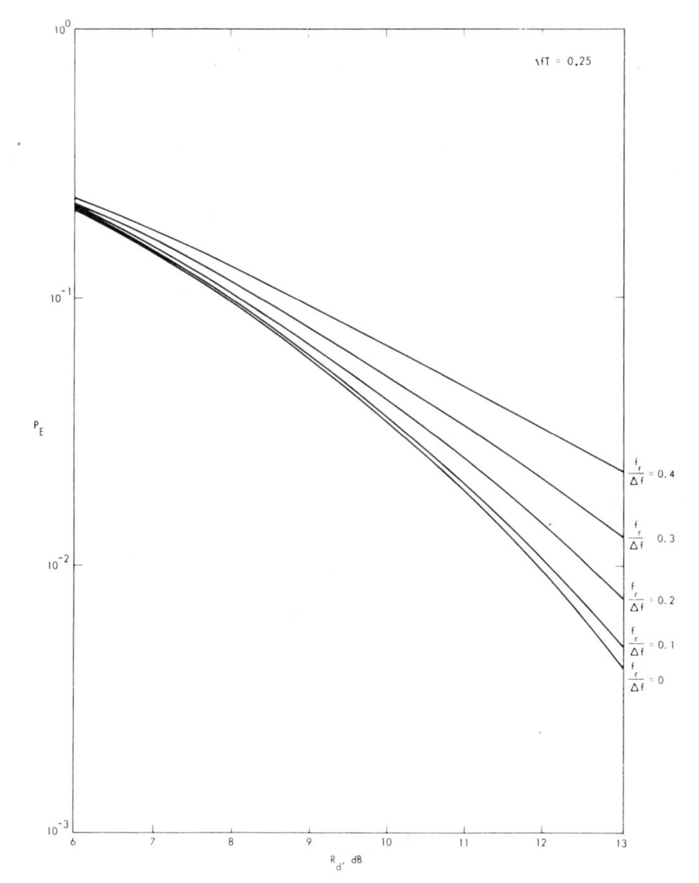

Fig. 10. Error probability performance in the presence of Doppler residuals; $\delta = 1/2$.

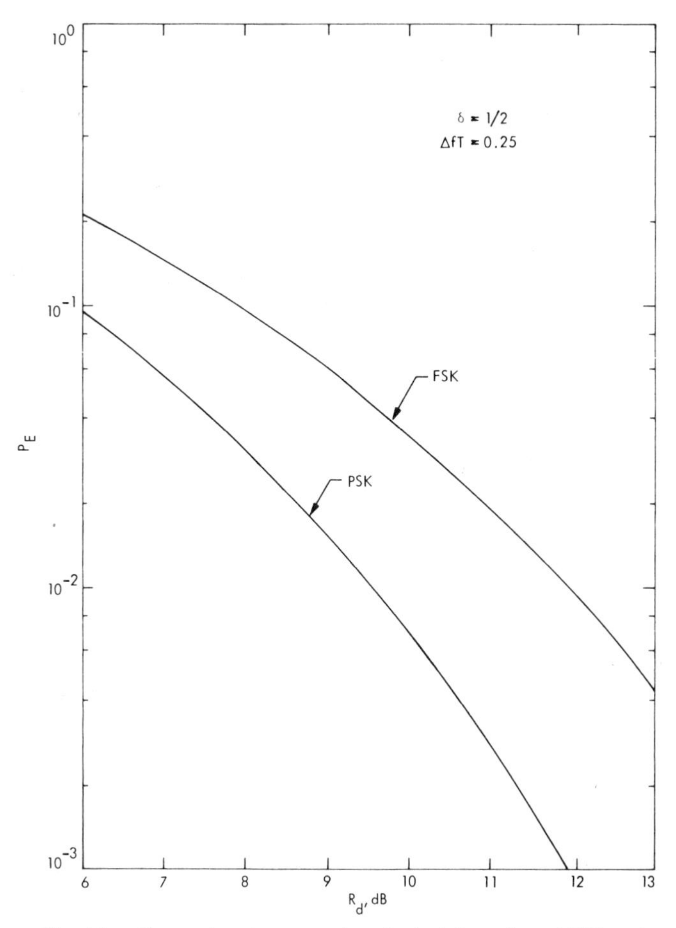

Fig. 11. Comparison between phase-locked detection of PSK and FSK.

accomplished using a Costas or squaring loop [4] when $B_L T \ll 1$. However, there are many low data rate applications for which $B_L T$ is on the order of unity. One might think that for a given data rate $B_L T$ can always be made much less than unity by narrowing the loop bandwidth: but, this bandwidth can only be reduced to a value which is ultimately determined by the channel Doppler or oscillator instabilities. Therefore, for the case where $B_L T \cong 1$, an alternate approach to PSK detection is of interest.

Again, the PLL configuration of Fig. 1 can be used as a demodulator and the resulting performance found using the same approach discussed earlier for FSK. For the purpose of being concise, we avoid presenting the mathematical details and merely give the graphical results illustrated in Fig. 11. As one might anticipate, PSK, when detected in the above manner, offers approximately a 3 dB advantage over FSK.

IX. CONCLUSIONS

This paper has presented a mathematical model for use in explaining the noise mechanism which creates bit errors in a demodulator/data detector configured as a performance of systems which employ a class of digital FM first-order PLL preceeded by a wideband IF filter. Unlike earlier investigations, the assumption of a Gaussian decision variable patched with a correction term based on Rice's click theory is not used; instead, the noise mechanism which accounts for decision errors is modelled on the basis of the Meyr distribution and renewal Markov process theory. Particular emphasis has been devoted to characterizing the performance of binary FSK systems designed to accommodate large frequency uncertainties caused by channel Doppler. It would be interesting to consider the case where the channel Doppler is small and investigate performance when a narrow IF filter precedes the estimator-correlator. Unfortunately, the theory as presented in this paper does not apply to this case, and thus, further investigation of this problem is required. However, extension of the results presented herein to the case of MFSK as well as PAM/FM is rather straightforward.

ACKNOWLEDGMENT

The authors are indebted to Mrs. M. Easterling for her computer programing assistance which was all important in arriving at the numerical results presented herein. The authors also wish to thank M. G. Pelchat for his comments on this paper.

REFERENCES

[1] C. A. Hinrichs, "Digital simulation of a communication link for pioneer Saturn Uranus atmospheric entry probe," McDonnell Douglas Astronautics Co. East, Tech. Rep. NASA CR-137600, NASA Contract NAS2-7935, Feb. 1975.

[2] M. G. Pelchat, R. C. Davis, and M. B. Luntz, "Coherent demodulation of continuous phase binary FSK signals," in *Proc. 1971 Int. Telemetry Conf.*

[3] R. DeBuda, "Coherent demodulation of frequency-shift keying

with low deviation ratio," *IEEE Trans. Commun.*, vol. COM-20, pp. 429–435, June 1972.

[4] W. C. Lindsey and M. K. Simon, *Telecommunication Systems Engineering.* Englewood Cliffs, NJ: Prentice-Hall, 1973.

[5] W. P. Osborne and M. B. Luntz, "Coherent and noncoherent detection of CPFSK," *IEEE Trans. Commun.*, vol. COM-22, pp. 1023–1036, Aug. 1974.

[6] N. M. Shehadeh and R. F. Chiu, "Transmission characteristics of split-phase PCM codes," NASA Final Rep., pt. 1, Contract NAS 9-9270, Mar. 1970.

[7] H. P. Hartmann, "Spectrum of Manchester coded FSK," *IEEE Trans. Commun.*, vol. COM-20, pp. 1001–1004, Oct. 1972.

[8] C. Chen and H. P. Hartmann, "Comments on 'spectrum of Manchester coded FSK,'" *IEEE Trans. Commun.*, vol. COM-22, pp. 270–271, Feb. 1974.

[9] E. F. Smith, "Attainable error probabilities in demodulation of random PCM-FM waveforms," *IRE Trans. Space Electron. Telemetry*, vol. SET-8, pp. 290–297, Dec. 1962.

[10] V. A. Kotel'nikov, *Theory of Optimum Noise Immunity* (transl. R. A. Silverman). New York: McGraw-Hill, 1959.

[11] P. D. Shaft, "Error rate of PCM-FM using discriminator detection," *IEEE Trans. Space Electron. Telemetry*, vol. SET-9, pp. 131–137, Dec. 1963.

[12] A. A. Meyeroff and W. M. Mazer, "Optimum binary FM reception using discriminator detection and IF shaping," *RCA Rev.*, vol. 22, pp. 698–728, Dec. 1961.

[13] S. O. Rice, "Properties of a sine wave plus random noise," *Bell Syst. Tech. J.*, vol. 27, pp. 109–157, Jan. 1948.

[14] D. D. McRae and F. A. Perkins, "PCM synchronization and multiplexing study," Radiation Inc., Melbourne, FL, Final Rep., June 14, 1963.

[15] *Telemetry System Study*, vol. 2, *Experimental Evaluation Program*, Aeronutronic Systems, Inc., Newport Beach, CA, U.S. Army Signal Res. Develop. Labs., Final Rep. Contract DA-36-039 SC-73182, Dec. 18, 1959.

[16] L. R. Brown, "Experimental determination of signal-to-noise relationships in PCM/FM and PCM/PM transmission," Electromechanical Research, Inc., Sarasota, FL, Oct. 20, 1961.

[17] J. Cohn, "A new analysis of FM threshold reception," Ph.D. dissertation, Northwestern Univ., Evanston, IL, 1957.

[18] —, "A new approach to the analysis of FM threshold reception," *Proc. Nat. Electron. Conf.*, 1958, pp. 221–236.

[19] S. O. Rice, "Noise in FM receivers," in *Time Series Analysis*, M. Rosenblatt, Ed. New York: Wiley, 1963, ch. 25.

[20] J. E. Mazo and J. Salz, "Theory of error rates for digital FM," *Bell Syst. Tech. J.*, vol. XLV, pp. 1511–1535, Nov. 1966.

[21] D. L. Schilling, E. Hoffman, and E. A. Nelson, "Error rates for digital signals demodulated by an FM discriminator," *IEEE Trans. Commun. Technol.*, vol. COM-15, pp. 507–517, Aug. 1967.

[22] D. McRae, "Error rates in wide band FSK with discriminator demodulation," *Proc. Int. Telemetering Conf.*, pp. 48–77, 1967.

[23] C. Chen, "Note on discriminator detection of PCM/FM," *IEEE Trans. Aerospace Electron. Syst.*, pp. 478–479, May 1968.

[24] —, "Discriminator detection of wide-band PCM/FM," *IEEE Trans. Aerospace Electron. Syst.*, pp. 126–127, Jan. 1969.

[25] T. T. Tjhung and P. H. Wittke, "Carrier transmission of binary data in a restricted band," *IEEE Trans. Commun. Technol.*, vol. COM-18, pp. 295–304, Aug. 1970.

[26] W. R. Bennett and J. Salz, "Binary data transmission by FM over a real channel," *Bell Syst. Techn. J.*, vol. XLII, pp. 2387–2426, Sept. 1963.

[27] D. L. Schilling, J. Billig, and D. Kermisch, "Error rates in FSK using the phase-locked loop demodulator," *Conf. Rec. 1st IEEE Ann. Commun. Conv.*, June, 1965, pp. 75–81.

[28] H. Meyr, "Nonlinear analysis of correlative tracking systems using renewal process theory," *IEEE Trans. Commun.*, vol. COM-23, pp. 192–203, Feb. 1975.

[29] W. C. Lindsey, *Synchronization Systems in Communication and Control.* Englewood Cliffs, NJ: Prentice-Hall, 1972.

[30] V. I. Tikhonov, "Phase-lock automatic frequency control operation in the presence of noise," *Automation and Remote Control*, vol. 21, pp. 209–214, 1960. (Transl. from *Automatika i Telemekhaniki*, vol. 21, Mar. 1960.)

Part VI
Tracking

Delay-Lock Tracking of Binary Signals*

J. J. SPILKER, JR.†, SENIOR MEMBER, IRE

Summary—**This paper presents the theory of operation and an evaluation of performance of a delay-lock tracking system for binary signals. The delay-lock discriminator is a nonlinear feedback system which employs a form of cross correlation in the feedback loop and continuously estimates the relative delay between a reference signal and a delayed version of that signal which is perturbed with additive noise. Binary maximal-length, shift-register sequences are used as the signal because they can easily be regenerated with any desired delay and they possess desirable autocorrelation functions.**

Problems of target search and acquisition are studied. The system performance in the presence of additive Gaussian noise is discussed. Computations are made of the effect of amplitude-limiting the received data on the system noise performance.

Introduction

The DELAY-LOCK discriminator has been described previously[1,2] as a device for tracking the delay difference between two correlated waveforms. The discriminator is a nonlinear feedback system which employs a form of cross correlation in the feedback loop. This device tracks the delay of a broad-band signal much in the same manner that a phase-lock discriminator tracks the phase of a sinusoidal signal. In this paper a modified version of the delay-lock discriminator is described which is particularly designed to track binary signals generated from feedback shift-registers. One of the main reasons for the interest in this type of signal is that it is easy to regenerate the signal in the discriminator for use in the cross-correlation operations with any desired amount of delay. Furthermore, certain classes of these sequences possess desirable autocorrelation properties; *e.g.*, the maximal-length linear shift-register sequences[3,4] have "two-level" autocorrelation functions and can be designed to have periods which are long enough so that ambiguities are of no concern.

In an actual tracking operation the binary sequence is applied to an RF carrier and transmitted to the target, where it is returned to the tracking equipment either as a radar reflection or via a transponder. The RF signal is then demodulated, and the demodulated binary sequence plus additive noise is fed into the delay-lock discriminator in order to obtain the estimate of the delay between the original and received sequences. This paper is concerned only with the delay estimation operation after demodulation of the RF carrier.

The analysis of this feedback system is presented in four main parts:

1) Derivation of the system equations.
2) Evaluation of noise effects, caused by receiver noise and system self-noise.
3) Determination of transient performance—acquisition and pull-out effects.
4) Computation of noise effects with amplitude-limited inputs.

Description of the Discriminator

A block diagram of the modified delay-lock discriminator is shown in Fig. 1. The received signal originated from an n-stage maximal-length linear feedback shift-register FSR, and has a period $M\Delta = (2^n - 1)\Delta$, where Δ is the digit width. This binary signal, plus additive Gaussian white noise, is fed into the cross-correlation network, where a comparison is made with time-displaced versions of the same pseudorandom binary signal as used in the transmitter. As is shown in a later section, the implementation of this cross correlator can be substantially simplified if the received data is first converted to binary form by means of an amplitude-limiter. Then the multiplication of ± 1 binary digits can be replaced by a modulo 2 adder.

The output of the cross-correlator network in Fig. 1 is,[5]

$$k\,\delta s[t + \hat{T}(t)]\{\sqrt{P_s}\,s[t + T(t)] + n(t)\}, \tag{1}$$

where $\delta s(t) \triangleq s(t + \Delta) - s(t - \Delta)$, and P_s is the received

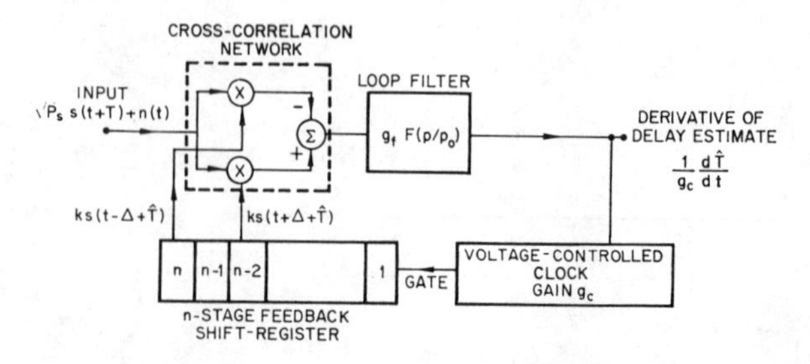

Fig. 1—Delay-lock discriminator for binary shift-register sequences.

* Received October 16, 1962. This work was supported by the Lockheed Independent Research Program..

† Communication Sciences Dept., Philco Western Development Laboratories, Palo Alto, Calif. Formerly with Lockheed Missiles and Space Company.

[1] J. J. Spilker, Jr. and D. T. Magill, "The delay-lock discriminator—an optimum tracking device," Proc. IRE, vol. 49, pp. 1403–1416; September, 1961.

[2] M. R. O'Sullivan, "Tracking systems employing the delay-lock discriminator," IRE Trans. on Space Electronics and Telemetry, vol. SET-8, pp. 1–7, March, 1962.

[3] B. Elspas, "The theory of autonomous linear sequential switching networks," IRE Trans. on Circuit Theory, vol. CT-6, pp. 45–60; March, 1959.

[4] S. W. Golomb and L. R. Welch, "Nonlinear Shift-Register Sequences," Jet Propulsion Lab., Calif. Inst. of Tech., Pasadena, Calif., Memorandum No. 20-149; 1957.

[5] Notice that $\delta s/2\Delta$ is reminiscent of the expression for the differentiated signal which is used in the cross-correlation operation in footnote reference 1:

$$\frac{ds}{dt} = \lim_{\Delta \to 0} \frac{s(t + \Delta) - s(t - \Delta)}{2\Delta}.$$

Reprinted from *IEEE Trans. Space Electron. Telemetry*, vol. SET-9, pp. 1–8, Mar. 1963.

signal power. As will be shown, this product contains a low-pass spectral component which serves to keep the discriminator accurately tracking the target delay once the system has been "locked on." The low-pass filter which follows the cross-correlation network, is designed on the basis of the expected dynamics of target motion, and its purpose is to remove as much noise and other interference as possible. The output of the loop filter is an estimate of the time derivative of delay (proportional to the target's radial velocity) and is used to control the clock rate of the FSR.

The delay estimate $\hat{T}$ can be easily obtained by sensing the time instants when the transmitter and receiver FSRs go through a particular state, *e.g.*, the "all ones" state, and determining the time difference between these time instants.

System Equations

In order to analyze the response of the system, the signal cross-correlation term in (1) is expressed as

$$k\,\delta s(t+\hat{T})s(t+T)\sqrt{P_s} \triangleq k\sqrt{P_s}[D_\Delta(\epsilon)+n_s(t,\epsilon)], \qquad (2)$$

where $\epsilon \triangleq T - \hat{T}$ is the delay error. The term which is not explicitly dependent on time, namely, $D_\Delta(\epsilon)$, the discriminator characteristic, has been separated from the remainder $n_s(t, \epsilon)$, the self-noise term. The discriminator characteristic can easily be shown to be[6]

$$\begin{aligned} D_\Delta(\epsilon) &= E[\delta s(t+\hat{T})s(t+T)] \\ &= R_s(\epsilon-\Delta) - R_s(\epsilon+\Delta), \end{aligned} \qquad (3)$$

where $R_s(\ )$ is the signal autocorrelation function.

In this paper the binary signal has amplitudes $s = \pm 1$. The autocorrelation function[7,8] and the discriminator characteristic for this maximal-length sequence are plotted in Fig. 2. As can be seen, $D_\Delta(\epsilon)$ varies linearly with ϵ for $|\epsilon| < \Delta$ and is zero for $2\Delta \leq |\epsilon| \leq (M-2)\,\Delta$. Since $s(t)$ is periodic with period $M\Delta$, $D_\Delta(\epsilon)$ has the same periodicity.

The self-noise term in the output of the multiplier network can be expressed as

$$\begin{aligned} n_s(t,\epsilon) &= s(t+\Delta+\hat{T})s(t+T) \\ &\quad - s(t-\Delta+\hat{T})s(t+T) - D_\Delta(\epsilon) \\ &= s(t+\Delta+\hat{T}) \oplus s(t+T) \\ &\quad - s(t-\Delta+\hat{T}) \oplus s(t+T) - D_\Delta(\epsilon), \end{aligned} \qquad (4)$$

where the symbol $\oplus$ represents modulo 2 addition. Multiplication of ± 1 terms is equivalent to modulo 2 addition and complementation of 0, 1 digits.

[6] The symbol $E(\ )$ represents the ensemble average operator. The sequence $s(t)$ is assumed to have an equidistributed random time origin.

[7] S. W. Golomb, "Sequences with Randomness Properties" Glenn Martin Co., Baltimore, Md., Final Rept., Contract No. SC-54-36611; 1955.

[8] B. Elspas, "A Radar System Based on Statistical Estimation and Resolution Considerations," Stanford Electronics Lab., Stanford Univ., Calif., Rept. No. 361-1; 1955. See especially Appendix D.

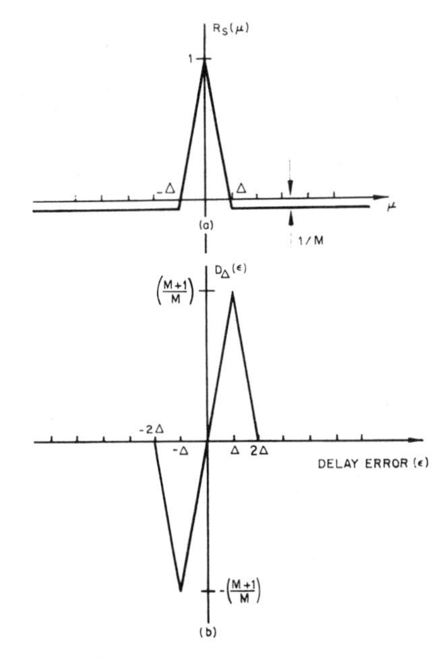

Fig. 2—Autocorrelation function $R(\mu)$, and discriminator characteristic $D_\Delta(\epsilon)$, for maximal-length binary shift-register sequences.

In order to determine the effect of the self-noise on the system performance, it is necessary to determine the power spectrum of $n_s(t, \epsilon)$. This computation is performed for values of delay error $\epsilon = 0, \pm m\Delta$. It has been known for some time that maximal-length linear binary sequences possess the "cycle and add" property,[9,10] *i.e.*,[11] $s(t+i\Delta) \oplus s(t+j\Delta) = s(t+r\Delta)$. This relationship can be used to simplify the form of (4):

$$\begin{aligned} n_s(t, \epsilon = m\Delta) &= s(t+T+j\Delta) - s(t+T+(j+1)\Delta), \\ &\qquad \text{for } m = 0, M, \text{ etc.} \\ &= \pm\,[s(t+T+n\Delta) - 1/M], \\ &\qquad \text{for } m = \pm 1, M \pm 1, \text{ etc.} \\ &= s(t+T+r\Delta) - s(t+T+q\Delta), \\ &\qquad \text{otherwise,} \end{aligned} \qquad (5)$$

where j, n, r, q are integers which are dependent on the delay error $\epsilon = m\Delta$; for no value of m is $r = q$. Hence, it is seen that for $\epsilon = \pm\Delta$, the self-noise power spectrum is the same as the signal power spectrum $G_s(f)$. The signal power spectral density has components at frequencies which are integer multiples of $1/M\Delta$, *i.e.*,

[9] S. W. Golomb, "Sequences with the Cycle and Add Property," Jet Propulsion Lab., Calif. Inst. of Tech., Pasadena, Calif., Section Rept. No. 8-573; 1957.

[10] R. C. Titsworth and L. R. Welch, "Modulation by Random and Pseudo-Random Sequences," Jet Propulsion Lab., Calif. Inst. of Tech., Pasadena, Calif., Progress Rept. No. 20-387; June, 1959; See especially sec. 3.

[11] In other words, if S is the state vector of the FSR at time $t + i\Delta$, and T is the transformation matrix of the FSR, then $(I + T^{(j-i)})S = T^{(r-i)}S$ where I is the identity matrix. The relationship between i, j, r can be found using the characteristic polynomial of T, *i.e.*, $\phi(\lambda) = |T - \lambda I|$. In particular, one can use the Caley-Hamilton theorem, which shows that $\phi(T) = 0$, and the periodicity relation, $T^M = I$, to obtain the relationship between these integers.

$$G_s(f) = \frac{1}{M\Delta} G(f) \sum_{\nu=-\infty}^{\infty} \delta(f - \nu/M\Delta),$$

where $\delta(f - f_0)$ is the Dirac delta function defined by the operator equation, $\int H(f)\delta(f - f_0)\, df = H(f_0)$. The quantity $G(f)$, which can be considered as the "envelope" of the signal spectrum, is plotted in Fig. 3(a).

$$G(f) = \Delta\left(\frac{\sin \pi\Delta f}{\pi\Delta f}\right)^2.$$

For other values of delay error the self-noise is expressed as the difference between two versions of the same sequence having different time origins, and the self-noise spectrum is

$$\begin{aligned} G_{n_s}(f, \epsilon = m\Delta) &= 2G_s(f)[1 - \cos 2\pi f\Delta], \\ &\quad \text{if} \quad m = 0, M, \quad \text{etc.} \\ &= 2G_s(f)[1 - \cos 2\pi(r - q)f\Delta], \\ &\quad \text{if} \quad 1 < |\, m \,| < M - 1, \quad \text{etc.} \end{aligned} \tag{6}$$

The self-noise spectrum for $\epsilon = \Delta$ is depicted in Fig. 3(b). Notice that it is a general characteristic of the spectra in (6), that they have a null at the origin, and as a result the self-noise is more easily removed by the loop filter.

The noise input to the discriminator is assumed to be white Gaussian noise. Because of this spectral density, it produces a noise component in the output of the cross-correlation network which is also white,

$$k\,\delta s(t + \hat{T})n(t) \triangleq kn_n(t). \tag{7}$$

The spectral density of $n_n(t)$ is $G_{n_n}(f) = P_d N_0$, where P_d is the average power of $\delta s(t + \hat{T})$ and N_0 is the input noise spectral density. For values of $\hat{T}$ which are fixed or vary little in a time interval Δ, the average power in $\delta s(t + \hat{T}) \triangleq s(t + \hat{T} + \Delta) - s(t + \hat{T} - \Delta)$ is $P_d = 2$. Hence, $G_{n_n}(f) = 2N_0$ wsec.

Thus, the output of the cross-correlator network can be written using (1), (2) and (7) as

$$k\sqrt{P_s}[D_\Delta(\epsilon) + n_s(t, \epsilon) + n_n(t)/\sqrt{P_s}]. \tag{8}$$

By referring to Figs. 1 and (8), it can be seen that the system equation (in operator notation) is

$$\begin{aligned} p\hat{T} &= kg_c g_f \Delta\sqrt{P_s}F(p/p_0) \\ &\quad \cdot [D_\Delta(\epsilon) + n_s(t, \epsilon) + n_n(t)/\sqrt{P_s}], \end{aligned} \tag{9}$$

where g_f is the loop-filter gain-constant, p_0 is the loop-filter frequency-constant, and g_c is the gain of the voltage-controlled clock. Define g_0 as the dc loop gain, *i.e.*, a steady delay error of ϵ sec, $|\, \epsilon \,| < \Delta$ produces a $g_0\epsilon/\Delta$ cps change in clock frequency or a $g_0\epsilon$ sec delay change/sec. The dc loop gain is given by $g_0 = kg_f g_c \sqrt{P_s}\,(M + 1)/M$. The system equation (9) can then be written as

$$\begin{aligned} p\hat{T} &= g_0\Delta F(p/p_0)\left(\frac{M}{M + 1}\right) \\ &\quad \cdot [D_\Delta(\epsilon) + n_s(t, \epsilon) + n_n(t)/\sqrt{P_s}]. \end{aligned} \tag{10}$$

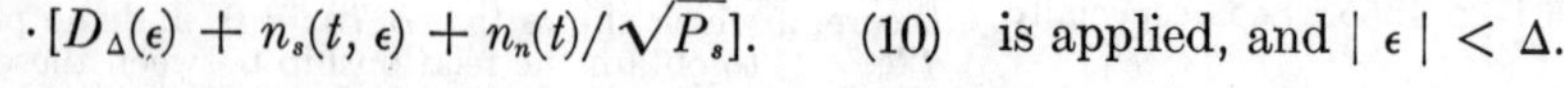

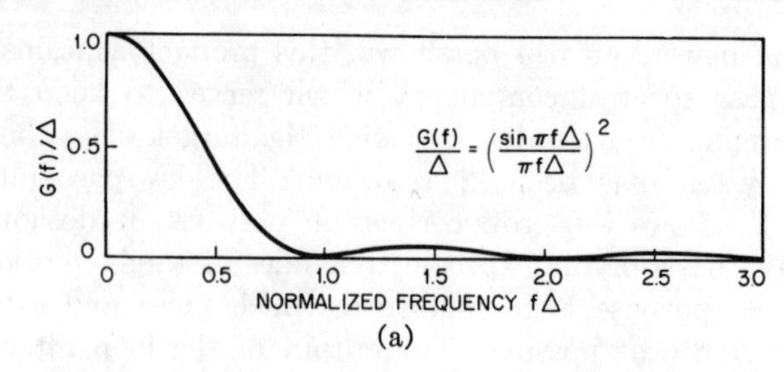

(a)

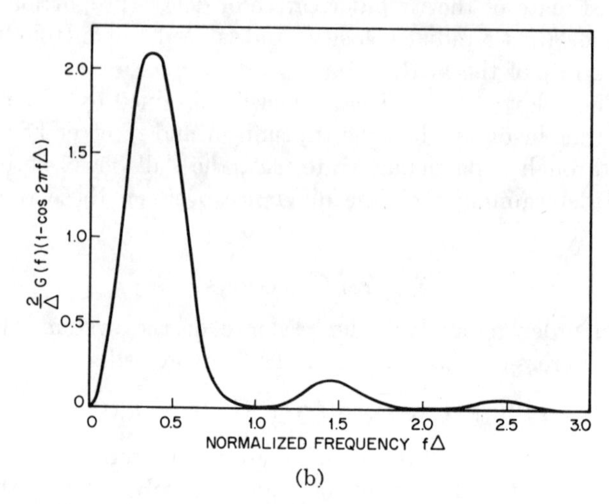

(b)

Fig. 3—"Envelopes" of power spectral densities. (a) Binary sequence spectral density. (b) Self-noise spectral density for $\epsilon = \Delta$.

Noise Performance

In this section the effects of receiver noise and self-noise on system accuracy are determined. The discriminator is assumed to be in the locked-on state, *i.e.*, $|\, \epsilon \,| < \Delta$, so that the system is operating in the region where $D_\Delta(\epsilon) = [(M + 1)/M](\epsilon/\Delta)$.

Define the normalized loop gain $g = g_0/p_0$. The linearized system equation can then be obtained from (10) as

$$\hat{T}/\Delta = g\,\frac{F(p/p_0)}{(p/p_0)}\left[\frac{\epsilon}{\Delta} + \frac{n_s(t, \epsilon) + n_n(t)/\sqrt{P_s}}{(M + 1)/M}\right].$$

Since $\epsilon = T - \hat{T}$, (10) can be written as

$$\hat{T}/\Delta = H(p/p_0)\left[T/\Delta + \frac{n_s(t, \epsilon) + n_n(t)/\sqrt{P_s}}{(M + 1)/M}\right], \tag{11}$$

where $H(p/p_0)$ is the linearized closed-loop transfer function

$$H(p/p_0) \triangleq \frac{gF(p/p_0)}{p/p_0 + gF(p/p_0)}.$$

The linear feedback network depicted in Fig. 4(a) can be shown to have the same system equation as (11), and therefore it provides performance equivalent to that of Fig. 1 when the equivalent input

$$\frac{T(t)}{\Delta} + \frac{n_s(t, \epsilon) + n_n(t)/\sqrt{P_s}}{(M + 1)/M}$$

is applied, and $|\, \epsilon \,| < \Delta$.

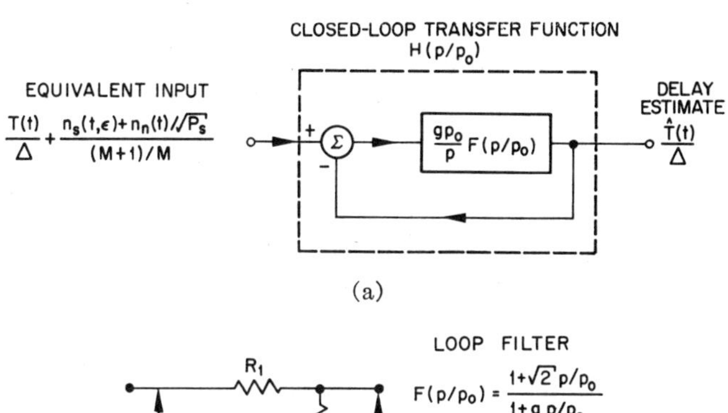

(a)

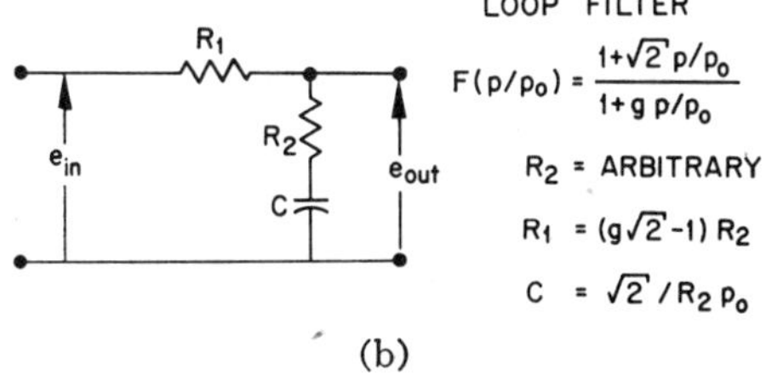

(b)

Fig. 4—Linearized equivalent circuit for the delay-lock discriminator, valid for $|\epsilon| < \Delta$. (a) Block diagram. (b) Loop-filter referred to in the analysis.

Consider that the closed-loop transfer function $H(p/p_0)$ has the form

$$H(p/p_0) = \frac{1 + \sqrt{2}p/p_0}{1 + \sqrt{2}p/p_0 + (p/p_0)^2}.$$

This transfer function has been shown optimum for ramp inputs of delay in the presence of white noise in that it minimizes the total squared transient error plus the mean-squared error caused by interfering noise.[12] The relationship between p_0 and the transient response is discussed further in the next section. The loop-filter shown in Fig. 4(b),

$$F(p/p_0) = \frac{1 + \sqrt{2}p/p_0}{1 + gp/p_0}, \tag{12}$$

can be used to approximate this closed-loop response as closely as desired by making the normalized gain constant g large.

The mean-square delay error caused by the receiver noise can be evaluated from (7) and (11) to be

$$\sigma_{n_n}^2 = \frac{\Delta^2}{P_s}\left(\frac{M}{M+1}\right)^2 \int_{-\infty}^{\infty} G_{n_n}(f)\,|H(j\omega/p_0)|^2\,df$$

$$= 2.12\,\Delta^2\left(\frac{M}{M+1}\right)^2 \frac{N_0 p_0}{P_s}. \tag{13}$$

For example, if the receiver noise temperature is 900°K, $p_0 = 1$ radian/sec, $M \gg 1$, $\Delta = 10^{-6}$ sec and $P_s = 10^{-15}$ watt, then $N_0 = 6.2 \times 10^{-21}$ wsec, and the rms delay error caused by noise is $\sigma_{n_n} = 2.5 \times 10^{-9}$ sec.

The self-noise spectrum has been shown to depend upon the delay error ϵ. Hence, the mean-square value of delay error caused by the self-noise also depends on ϵ. An upper bound on the mean-square delay error can be found by assuming the most pessimistic spectral density function, *i.e.*, the self-noise spectrum obtained for $\epsilon = \pm\Delta$. The mean-square self-noise delay error is then

$$\sigma_{n_s}^2 = \Delta^2 \int_{-\infty}^{\infty} G_{n_s}(f)\,|H(j\omega/p_0)|^2\,df$$

$$= \frac{\Delta^2}{M}\sum_{\nu=-\infty}^{\infty} \frac{1 + 2(2\pi\nu/M\Delta p_0)^2}{1 + (2\pi\nu/M\Delta p_0)^4}\left(\frac{\sin \nu\pi/M}{\nu\pi/M}\right).$$

Under conditions where the number of digits in a period M is large and the loop filter constant p_0 is small compared to the signal bandwidth ($\approx\frac{1}{2}\Delta$), this expression simplifies to

$$\sigma_{n_s}^2 \cong \Delta^2\left(\frac{p_0\Delta}{2}\right). \tag{14}$$

For the parameters just considered, $p_0 = 1$ radian/sec, $\Delta = 10^{-6}$ sec, (14) gives an rms self-noise delay error of $\sigma_{n_s} = 7.07 \times 10^{-10}$ sec.

When the delay error fluctuations become too large, the discriminator encounters a threshold effect. This effect is caused by the fact that large delay errors force the discriminator to operate, at least temporarily, in a region of negative loop gain, since $D(\epsilon)$ has a negative slope for $\Delta < |\epsilon| < 2\Delta$.

Experimental measurements made thus far indicate that the system threshold occurs when the total rms delay error is about $\sigma_n = 0.30\Delta$. For delay errors having $\sigma_n \leq 0.30\Delta$, experiments showed a negligible probability of losing the locked-on state in the absence of transient errors. For delay error fluctuations which have a Gaussian amplitude statistic and $\sigma_n = 0.30\Delta$, the probability that $|\epsilon| > \Delta$ in the linearized model is only 0.00087. This statement assumes that transient errors caused by fluctuations in T are small compared to Δ. Assuming that the self-noise error is negligible, the system threshold can be computed from (13) and is given approximately by

$$\frac{P_s}{p_0 N_0} = 23.5\left(\frac{M}{M+1}\right)^2. \tag{15}$$

Transient Performance

In this section two problems concerning the transient performance of the discriminator in the absence of noise are investigated:

1) How rapidly can a given region be searched and the target acquired for a given closed-loop noise bandwidth, $B_n = 1.06\,p_0$ cps?
2) Once the target has been acquired, what is the maximum change in velocity that can be tolerated without losing the locked-on condition?

In order to carry out these objectives most readily, we rewrite (10) neglecting the noise effects and making use of the following time normalizations $x \triangleq \epsilon/\Delta$, $y \triangleq T/\Delta$, $y - x = \hat{T}/\Delta$, $\tau \triangleq p_0 t$, $s \triangleq p/p_0 = d/d\tau$, $D(x) \triangleq D_\Delta(\epsilon)$

$$s(y - x) = gF(s)D(x). \tag{16}$$

[12] R. Jaffe and E. Rechtin, "Design and performance of phase-locked circuits capable of near-optimum performance over a wide range of input signal levels," IRE Trans. on Information Theory, vol. IT-1, pp. 66–72; March, 1955.

The integral plus proportional control loop-filter of (12) now becomes

$$gF(s) = \frac{1 + \sqrt{2}s}{1/g + s}.$$

Thus we obtain the operator equation

$$(1/g + s)s(y - x) = (1 + \sqrt{2}s)D(x).$$

This system equation can be rewritten in time derivative notation

$$\dot{y}/g + \ddot{y} = \dot{x}/g + \ddot{x} + D(x) + \sqrt{2}D'(x)\dot{x}, \qquad (17)$$

where $\dot{x} \triangleq dx/d\tau$, $D'(x) \triangleq dD/dx$, etc.

The phase-plane method of solving this second-order differential equation is to compute the trajectories in x, $\dot{x}$ space which describe the solution of this equation for the desired sets of initial conditions. Define a new variable $\gamma \triangleq \ddot{x}/\dot{x} = d\dot{x}/dx$, which from (17) is given by

$$\frac{d\dot{x}}{dx} = \gamma[x, \dot{x}, \dot{y}, \ddot{y}]$$

$$= -\frac{D(x) + [\sqrt{2}D'(x) + 1/g]\dot{x} - \dot{y}/g - \ddot{y}}{\dot{x}}. \qquad (18)$$

Computer solutions to these trajectories can be obtained by approximating the differential equation (18) with the difference equation

$$\dot{x}_{n+1} - \dot{x}_n \triangleq \dot{x}\left(x_0 + \sum_{j=0}^{n} \delta_j\right) - \dot{x}\left(x_0 + \sum_{j=0}^{n-1} \delta_j\right)$$

$$\cong \gamma(\dot{x}_n, x_n)\delta_n, \qquad (19)$$

where x_0, $\dot{x}_0$ are the initial values of x, $\dot{x}$, and δ_j is the jth increment in x. Satisfactory solutions can be obtained by letting the computer function as an adaptive device, so that the size of the interval is made variable. In this particular sequential computation[13] the interval size is taken as[14]

$$|\delta_j| = \frac{\delta}{1 + |\gamma(\dot{x}_j, x_j)|}.$$

The value δ to be used is 0.02.

Consider the search and acquisition problem where the normalized loop gain $g = \infty$. Assume that the target search velocity is a constant, $\dot{y}(t) = \dot{y}$, $\ddot{y} = 0$. The variable γ now becomes

$$\gamma(x, \dot{x}) = -\frac{D(x) + \sqrt{2}D'(x)\dot{x}}{\dot{x}}.$$

[13] R. Bellman, "Adaptive Control Processes: A Guided Tour," Princeton University Press, Princeton, N. J., ch. 4; 1961.

[14] Notice that the distance moved in the x, $\dot{x}$ plane in one increment is

$$\Delta r_n = \sqrt{(x_{n+1} - x_n)^2 + (\dot{x}_{n+1} - \dot{x}_n)^2}$$

$$= \delta\left(\frac{1 + \gamma_n^2}{1 + 2|\gamma_n| + \gamma_n^2}\right)^{1/2}.$$

Hence, the distance moved is bounded by $\delta/\sqrt{2} \leq r_n \leq \delta$. Thus, the maximum distance moved is δ, yet needlessly small increments are not used for small or moderate $|\gamma|$.

The acquisition trajectories for these conditions are plotted[15] in Fig. 5. If the delay error is decreasing from the left, the system does not respond until $x = -2$. As can be seen, if the search velocity $|\dot{y}| \leq 2.2$ the system locks on, and the state variable converges to the origin. As an example, consider $\Delta = 10^{-6}$ sec, $p_0 = 10$ rps. Then we have $\dot{y} = (dT/dt)/p_0\Delta = 10^5\, dT/dt$. For electromagnetic propagation $c = 3 \times 10^5$ km/sec and the maximum search velocity (two-way propagation time) permitted is $v = (\frac{1}{2})\, 2.2cp_0\Delta = 3.3$ km/sec. From (15) it can be seen that the threshold value of signal power is $P_s = 21.2N_0v/c\Delta = 235.3N_0$ watt in this example ($M \gg 1$).

Notice, however, that even if the system fails to lock-on in this particular interval of x, the velocity error is less at the end of the transient than at the beginning. Since $D(x)$ is periodic every M, the system always locks on eventually, because with $g = \infty$ there is no decay in x outside of the intervals $|x| < 2$. It may, however, pass through stable regions of x many times before locking on. In practice this behavior might rely on unrealistic storage times in the filter as is implicit in the assumption that $g = \infty$. Furthermore, the time required for lock-on might be intolerable for search velocities $|\dot{y}| > 2.2$.

The effect of target velocity transients can also be ascertained from Fig. 5. If the system is initially locked on, $x = 0$, $\dot{x} = 0$ and the target suddenly changes velocity to $\dot{y}$, the system response is described by that portion of the trajectories which begins at $x = 0$, $\dot{x} = \dot{y}$. As can be seen, normalized velocity transients of $\dot{y} = 3.38$ can be tolerated without losing the locked-on state.

Acquisition trajectories have also been obtained for finite loop gain systems with $g = 10$. The equation for γ in this example is

$$\gamma(x, \dot{x}, \dot{y}) = -\frac{D(x) + (\sqrt{2}D'(x) + 0.1)\dot{x} - 0.1\dot{y}}{\dot{x}}.$$

Because of the finite loop gain, the maximum steady-state clock frequency change is $g = 10p_0$. Thus, for this reason alone, the system would never lock on for $|\dot{y}| > 10$.

The trajectories for these conditions are shown in Fig. 6. As can be seen, the fact that the loop gain is reduced from ∞ to 10 has very little effect on the maximum tolerable search velocity; $|\dot{y}| \leq 2.2$ is still permitted. Notice however, that the curves which lock on converge to $x = 0.1\dot{y}$ or $\epsilon = x\Delta = 0.1\Delta\dot{y}$ rather than to $x = 0$. For $\Delta = 10^{-6}$ sec and $\dot{y} = 1$, the steady-state lock-on point is $\epsilon = 0.1$ μsec. In practice, by measuring $\dot{y}$ one can correct for this steady-state bias error.

Although it is not shown on the trajectory curves, the value of x, for the curves which do not lock on, decays with a time constant $g/p_0 = 10/p_0$ sec. Thus, if the period of the sequence is large so that $M \gg 1$, the value of $\dot{x}$ will have decayed almost to $\dot{y}$ when $x = M - 2$.

[15] The trajectory computations were performed by Lt. C. S. Mulloy on a CDC-1604 computer and are presented as part of his report entitled "Digital Analysis of the Delay-Lock Discriminator," U. S. Naval Postgraduate School, Monterey, Calif., 1962.

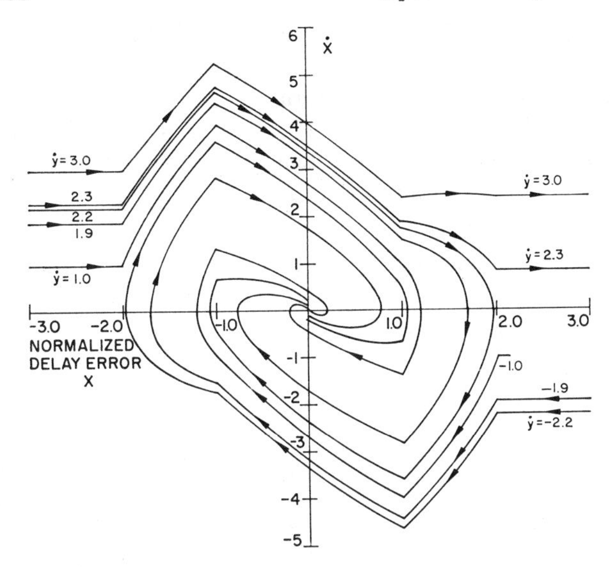

Fig. 5—Acquisition trajectories for loop gain $g = \infty$ plotted for various values of $\dot{y}$, the normalized search velocity.

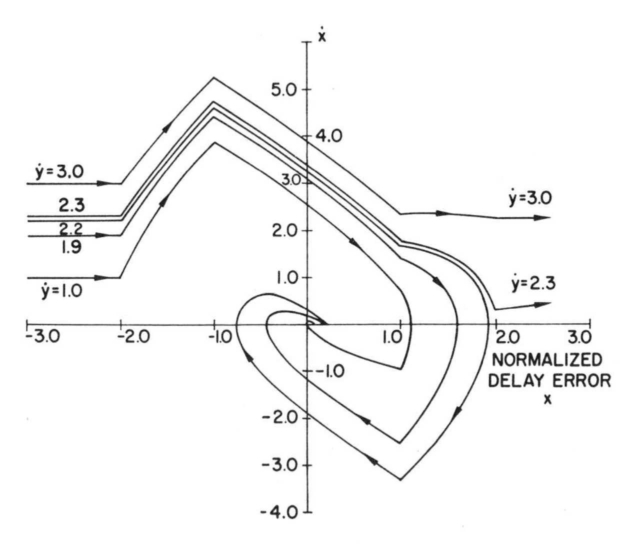

Fig. 6—Acquisition trajectories for loop gain $g = 10$ plotted for various values of $\dot{y}$, the normalized search velocity.

As well as knowing whether or not the system will lock on for a given search velocity, it is also important to know how long the transient lasts. The time response of the system can be obtained from computer solutions to the difference equation

$$t_{n+1} - t_n \triangleq t\left(x_0 + \sum_{i=0}^{n} \delta_i\right) - t\left(x_0 + \sum_{i=0}^{n-1} \delta_i\right)$$

$$\cong \frac{\delta}{(1 + |\gamma_n|)\dot{x}(x_n)}.$$

The results of this computation are shown in Fig. 7 as plots of x and $\dot{x}$, for $g = \infty$ and the maximum tolerable search velocity $\dot{y} = 2.2$. The time required for the transient to subside within $|x| < 0.1$ is $\tau = 5.6$. For $p_0 = 10$ radian/sec this lock-on time is about 0.56 sec.

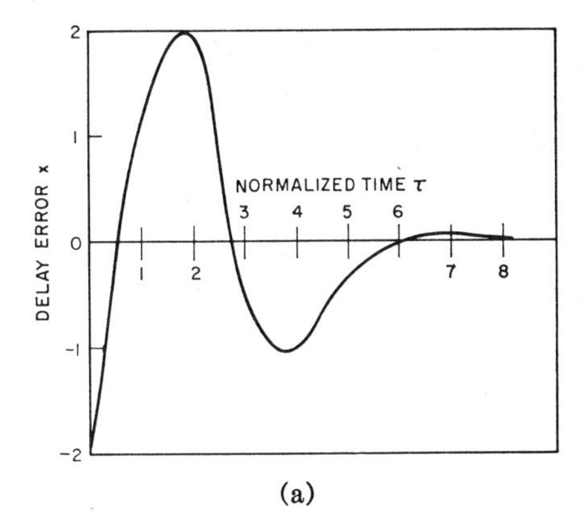

(a)

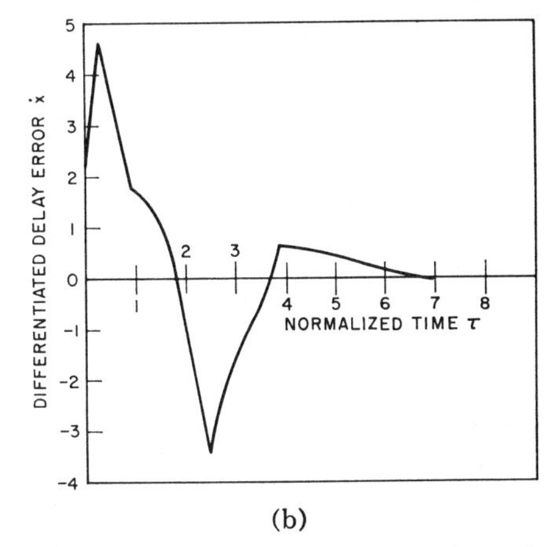

(b)

Fig. 7—Acquisition transient for $g = \infty$ and maximum search velocity, $y = 2.2$. (a) Normalized delay error response. (b) Differentiated delay error response.

Effect of Amplitude-Limiting the Received Data

Considerable simplification in the circuitry is possible if the received data are amplitude-limited before entering the cross-correlation network. The input to the cross-correlation network then has the binary form $u(t) = A \text{ sgn } [\sqrt{P_s}\, s(t) + n(t)]$, where A is the limiter output amplitude and sgn x is the sign function

$$\text{sgn } x = \begin{cases} 1, & \text{if } x \geq 0 \\ -1, & x < 0. \end{cases}$$

Thus, the waveforms fed into the discriminator have $\pm A$ amplitude and, as a result, the multiplication circuits of Fig. 1 can be replaced by modulo 2 adders as shown in Fig. 8. It is apparent that this change is inconsequential for a noise-free input. The purpose of this section is to determine how much this limiting action influences the noise performance when the input signal-to-noise ratio SNR is small.

The first step in the solution of this problem is to compute the cross correlation between the output of the limiter $u(t + \mu)$ and the signal component $s(t)$. It is this cross correlation which determines the useful signal component in the correlator network output through its relationship to the discriminator characteristic. Consider that band-limited white Gaussian noise enters the limiter.

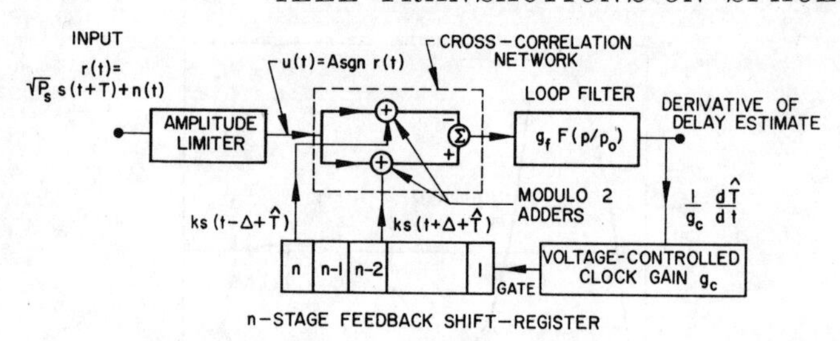

Fig. 8—Block diagram of the delay-lock discriminator operating on amplitude-limited inputs.

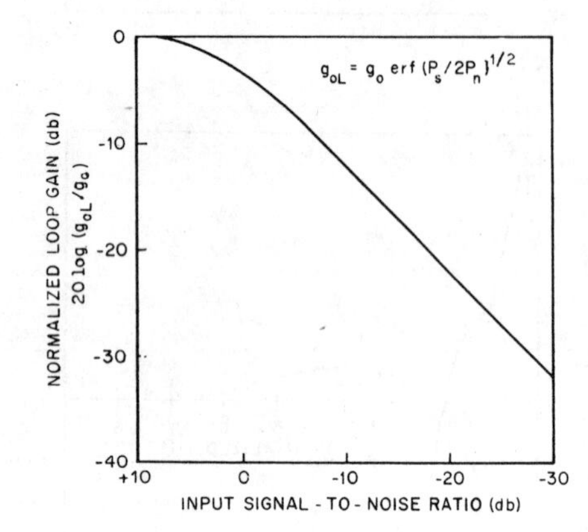

Fig. 9—Plot of normalized loop gain with limiting vs input SNR.

The noise is low pass having a maximum frequency B and average power P_n. It is shown in the Appendix that the cross correlation is related to the signal autocorrelation function by the equation,

$$E[u(t+\mu)s(t)] \triangleq AR_{us}(\mu) = AR_s(\mu) \operatorname{erf} \sqrt{P_s/2P_n}, \tag{20}$$

where erf x is the error function. For small input SNR, (20) becomes

$$AR_{us}(\mu) = \sqrt{\frac{2}{\pi}\frac{P_s}{P_n}}\, AR_s(\mu).$$

It is interesting to compare this cross-correlation function with that which is obtained without the limiter when the average total input power is held constant at the same level A^2. This type of operation is obtained if the discriminator is preceded by an "ideal" AGC amplifier which has constant average output power A^2. The cross correlation for this situation is

$$AR_s(\mu)\sqrt{\frac{P_s}{P_s+P_n}}.$$

By comparing this expression with (20), it is apparent that at small values of input SNR, the cross correlation with amplitude-limited inputs is smaller by a factor of $\sqrt{2/\pi}$. Since both expressions are proportional to the square root of the input SNR for small SNR, the loop gain of the discriminator g_0 decreases in the same proportion. More generally, the loop gain with limiting, g_{0L}, varies with SNR as

$$g_{0L} = g_0 \operatorname{erf} \sqrt{P_s/2P_n},$$

where g_0 is the loop gain for noise-free inputs. This expression is plotted in Fig. 9.

It is next necessary to determine the spectral density of the noise output of the cross-correlator network. It is known that at low SNR the autocorrelation function of the noise in the limiter output is[16]

$$R_{n0}(\mu) = \frac{2}{\pi} A^2 \sin^{-1} \frac{R_n(\mu)}{P_n}.$$

Hence, the noise spectrum in the limiter output is[17]

$$G_{n0}(f) = \frac{2}{\pi}\frac{A^2}{P_n}$$

$$[G_n(f) + 0.167\,{}_3G_n(f)/P_n^2 + 0.075\,{}_5G_n(f)/P_n^4 + \cdots].$$

Assume that the noise bandwidth is large compared to the effective correlator-filter bandwidth. Then the noise effects in the cross-correlator output are dependent mainly upon $G_{n0}(0)$, and this spectral density has the value

$$\begin{aligned} G_{n0}(0) &= 2\int_0^\infty \frac{2}{\pi} A^2 \sin^{-1}\,[R_n(\mu)/P_n]\, d\mu \\ &= \frac{4}{\pi} A^2 \int_0^\infty \sin^{-1}\left[\frac{\sin 2\pi B\mu}{2\pi B\mu}\right] d\mu \\ &= \frac{2}{\pi}\frac{A^2}{\pi B}\int_0^\infty \left[\frac{\sin x}{x} + \frac{1}{6}\left(\frac{\sin x}{x}\right)^3 \right. \\ &\quad \left. + \frac{3}{40}\left(\frac{\sin x}{x}\right)^5 + \cdots\right] dx = \frac{A^2}{2B}\left(\frac{2.2}{\pi}\right). \end{aligned} \tag{21}$$

The degradation in system performance at low SNR caused by the limiter action can now be determined by taking the ratio of the square of the signal component amplitude at the cross-correlator output to the noise spectral density $G_{n0}(0)$. This ratio is then compared with the corresponding ratio obtained with an ideal AGC amplifier having average output power A^2. For operation with the limiter, this ratio can be obtained from (20), (21) as

$$\frac{R_{su}^2(\mu)/R_s^2(\mu)}{G_{n0}(0)} = \frac{P_s}{P_n/2B}\left(\frac{1}{1.1}\right). \tag{22}$$

Operation with the ideal AGC yields a ratio which is larger by only a factor of 1.1 or 0.4 db. Hence, the use of an amplitude limiter is attractive since it allows the use of binary logic for the cross-correlation circuitry at only a small expense in the theoretical noise performance.

Discussion

Tracking systems operating on signals generated from maximal-length shift-register codes have received attention for some time, particularly with respect to radio

[16] R. Price, "A useful theorem for nonlinear devices having Gaussian inputs," IRE Trans. on Information Theory, vol. IT-4, pp. 69–72; June, 1958.

[17] The notation ${}_3G_n(f)$ represents the convolution in the frequency domain $G_n(f)*G_n(f)*G_n(f)$.

astronomy[18,19] and satellite tracking.[20] The purpose of this paper has been to compute and analyze the performance of a delay-lock discriminator which has been designed to operate on this type of binary signal. Solutions to the phase-plane trajectories have been obtained for the search and acquisition transients. The effects of additive noise and system self-noise on the measurement accuracy have been ascertained.

In addition, it has been shown that the process of amplitude-limiting the received data produces very little degradation in the system noise performance while permitting substantial simplification in the circuitry, *i.e.*, binary logic can be used in place of analog multipliers and dc amplifiers. In practice, deficiencies in performance of the analog circuitry would more than make up for this theoretical difference.

With either binary or analog inputs to the discriminator, if the average input power is fixed, the loop gain decreases in proportion to the square root of the SNR. Hence the loop gain should be made large enough to satisfy the search velocity requirements at the lowest SNR likely to be encountered.

It should be pointed out that certain improvements in noise performance can result if the loop filter is made adaptive. One can sense whether or not the system is locked-on by means of an external correlator. When the loop is unlocked and the discriminator is in its search mode, the loop filter can be designed to permit target acquisition with the desired search velocity. Once the target has been acquired, the closed-loop bandwidth can be decreased, and the noise performance can thereby be improved.

As has already been mentioned, it is possible to generate sequences having arbitrarily long periods by using enough shift-register stages. For extremely long period sequences, however, a straightforward search procedure may lead to an intolerably long acquisition time. In this situation it is better to use a composite sequence made up of several sequences each of relatively prime periods, and to track each sequence with a separate delay-lock discriminator. In this way the period of the composite sequence is the product of the periods, whereas the acquisition times are simply related to the periods of the individual sequences. If n sequences are properly combined, the power available for the acquisition of each sequence is $1/n$th of the total power. Easterling[20] has described one interesting means for combining an odd number of binary sequences; the sequences are simply combined in a majority logic.

Appendix

The purpose of this Appendix is to compute the cross correlation between the limiter output $u(t+\mu)$ and the reference signal component $s(t)$. This cross correlation can be written

$$E\left[z(t) \triangleq \frac{u(t+\mu)s(t)}{A}\right] = R_{us}(\mu)$$

$$= Pr[z(t) = 1] - Pr[z(t) = -1],$$

where $u(t) = A\,\text{sgn}\,[\sqrt{P_s}\,s(t) + n(t)]$. In order to simplify the notation, define $s(t) = s_0$, $s(t+\mu) = s_\mu$, $n(t+\mu) = n_\mu$. Define the signal probabilities $Pr\ (s=1) = p = (M+1)/2M$, and $Pr\ (s = -1) = q = 1 - p$. The probability that $z = 1$ can then be written

$$\begin{aligned} P_r(z = 1) &= pPr(n_\mu > -\sqrt{P_s}s_\mu \mid s_0 = 1) \\ &\quad + qPr(n_\mu < -\sqrt{P_s}s_\mu \mid s_0 = -1) \\ &= Pr(n_\mu > -\sqrt{P_s})[pPr(s_\mu = 1 \mid s_0 = 1) \\ &\quad + qPr(s_\mu = -1 \mid s_0 = -1)] \\ &\quad + Pr(n_\mu > \sqrt{P_s})[pPr(s_\mu = -1 \mid s_0 = 1) \\ &\quad + qPr(s_\mu = 1 \mid s_0 = -1)], \end{aligned} \tag{23}$$

where use has been made of the fact that the probability density of n is symmetric about the origin. The probability that $z = -1$ can be obtained in a similar manner

$$\begin{aligned} &Pr(z = -1) \\ &= Pr(n_\mu < -\sqrt{P_s})[pPr(s_\mu = 1 \mid s_0 = 1) \\ &\quad + qPr(s_\mu = -1 \mid s_0 = -1)] \\ &\quad + Pr(n_\mu < \sqrt{P_s})[pPr(s_\mu = -1 \mid s_0 = 1) \\ &\quad + qPr(s_\mu = 1 \mid s_0 = -1)]. \end{aligned} \tag{24}$$

The autocorrelation function of the signal is expressed by

$$\begin{aligned} R_s(\mu) &= E[s(t)s(t+\mu)] \\ &= p[Pr(s_\mu = 1 \mid s_0 = 1) - Pr(s_\mu = -1 \mid s_0 = 1)] \\ &\quad + q[Pr(s_\mu = -1 \mid s_0 = -1) - Pr(s_\mu = 1 \mid s_0 = -1)]. \end{aligned}$$

Hence, by using (23), (24), the cross-correlation function $R_{us}(\mu)$ can be written

$$R_{us}(\mu) = R_s(\mu)Pr(|\,n_\mu\,| < \sqrt{P_s}).$$

Under the assumption that the noise has stationary Gaussian amplitude statistics and has a mean-square value P_n, we obtain

$$R_{us}(\mu) = R_s(\mu)\ \text{erf}\ \sqrt{P_s/2P_n}.$$

Acknowledgment

The author would like to acknowledge the contributions of Lt. C. S. Mulloy, U.S.N., of the U.S. Naval Postgraduate School for programming the solutions to acquisition trajectories and for valuable comments made during the course of this study. Special appreciation also is expressed to R. A. Dye and D. T. Magill of Lockheed Missiles & Space Company, who contributed many useful suggestions and have provided experimental verification of much of this theory.

[18] R. Price and P. E. Green, Jr., "Signal Processing in Radar Astronomy-Communication via Multipath Media," M.I.T. Lincoln Lab., Lexington, Mass., Rept. No. 234; October, 1960.

[19] W. K. Victor, R. Stevens, and S. W. Golomb, "Radar Exploration of Venus," Goldstone Observatory Report for March–May, 1961, Jet Propulsion Lab., Calif. Inst. Tech., Pasadena, Rept. No. 32-132; August, 1961.

[20] M. F. Easterling, "A skin-tracking radar experiment involving the COURIER satellite," IRE Trans. on Space Electronics and Telemetry, vol. SET-8, pp. 76–84; June, 1962.

The Delay-Lock Discriminator—An Optimum Tracking Device*

J. J. SPILKER, JR.†, MEMBER, IRE, AND D. T. MAGILL†, MEMBER, IRE

Summary—**The delay-lock discriminator described in this paper is a statistically optimum device for the measurement of the delay between two correlated waveforms. This new device seems to have important potential in tracking targets and measuring distance, depth, or altitude. It operates by comparing the transmitted and reflected versions of a wide-bandwidth, random signal. The discriminator is superior to FM radars in that it can operate at lower power levels; it avoids the so-called "fixed error," and it is free of much of the ambiguity inherent in such periodically modulated systems. It can also operate as a tracking interferometer.**

The discriminator is a nonlinear feedback system and can be thought of as employing a form of cross-correlation along with feedback. The basic theory of operation is presented, and a comparison is made with the phase-lock FM discriminator. Variations of performance with respect to signal spectrum choice, target velocity, and signal and interference power levels are discussed quantitatively. The nonlinear, "lock-on" transient and the threshold behavior of the discriminator are described. Performance relations are given for tracking both passive and actively transmitting targets. Results of some experimental measurements made on a laboratory version of the discriminator are presented.

Introduction

In many problems of position measurement, interferometry, and tracking, it is necessary to measure the delay difference between two versions of the same signal, *e.g.*, the transmitted signal and the returned signal reflected from a target. In the domain of pulse radar, emphasis in recent years has been placed on the improvement of positioning accuracy in the presence of noise, and this effort has led to the development of advanced, matched-filter and pulse-compression techniques.[1,2]

The purpose of this paper is to present an improved delay estimation technique which operates on wide-bandwidth, continuous signals in the presence of interfering noise. The delay-lock discriminator, which is described herein, provides an optimum, continuous measurement of delay by operating on a wide-bandwidth, random, continuous signal. Throughout most of this paper, the signal is considered to be either filtered Gaussian random noise, or a sine wave randomly modulated in frequency. The signals are usually nonperiodic. Operation with pulsed signals is also possible, although this possibility is not treated specifically.

The delay-lock discriminator is shown as it might be used in tracking Fig. 1. This tracking problem differs from the conventional pulse radar problem in that only a single target is to be tracked by each discriminator. (There may, however, be several discriminators.) The target is tracked continuously as a function of time rather than at periodic intervals. (Dispersive effects in the target return are to be neglected in this discussion.)

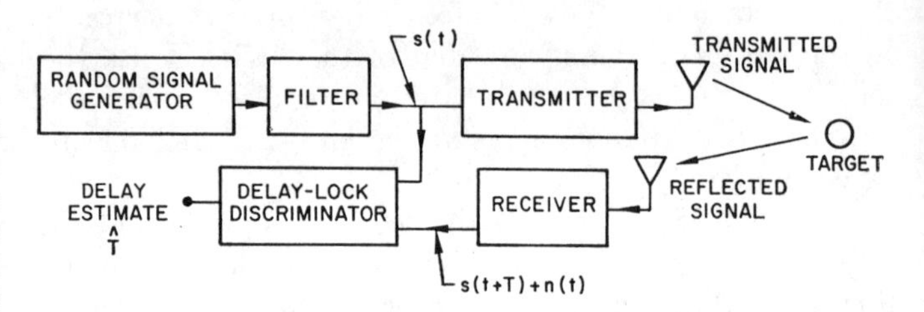

Fig. 1—Use of the delay-lock discriminator in tracking.

Although this radar uses a continuous signal, it differs from ordinary FM radars[3,4] in that, first, it avoids the so called "fixed error"; secondly, it is free of much of the ambiguity inherent in such periodically modulated systems; and finally, it can operate with multiple targets present at the same bearing from the antenna, thus providing range discrimination. The discriminator can be used in a second type of application: it can operate on a signal transmitted from the target which arrives via two separate receiving antennas. The delay difference in the two received signals is then measured by a method similar to that of a tracking interferometer.

Random-signal distance-measuring systems in themselves are not new. It is well known that when cross-correlations are made between the transmitted and received waveforms, the time difference can be accurately ascertained, if the received signal is sufficiently free of interference. These techniques have limitations, however, in that if the target is moving rapidly, the cross-correlation operation has limited useful integration time.

A somewhat different form of distance measuring technique employing random signals has been described by B. M. Horton[5] and was proposed for use as an altim-

* Received by the IRE, March 23, 1961; revised manuscript received, June 23, 1961.

† Commun. and Controls Res., Lockheed Missiles and Space Co., Palo Alto, Calif.

[1] C. E. Cook, "Pulse compression—key to more efficient radar transmission," Proc. IRE, vol. 18, pp. 310–316; March, 1960.

[2] Matched Filter Issue, IRE Trans. on Information Theory, vol. IT-6, pp. 310–413; June, 1960.

[3] D. G. C. Luck, "Frequency Modulated Radar," McGraw-Hill Book Co., Inc., New York, N. Y.; 1949.

[4] M. A. Ismail, "A precise new system of FM radar," Proc. IRE, vol. 44, pp. 1140–1145; September, 1956.

[5] B. M. Horton, "Noise-modulated distance measuring system," Proc. IRE, vol. 47, pp. 821–828; May, 1959.

Reprinted from *Proc. IRE*, vol. 49, pp. 1403–1416, Sept. 1961.

eter. This technique simply involves the direct multiplication of the transmitted and received signals, followed by a frequency-discrimination operation. Basically, this is a special type of correlation technique which is capable of operating over a relatively small range of delay. However, this system has a limitation on the dynamic range of delay which for many purposes would be overly restrictive.

Correlation techniques can be extended to cope better with time varying delays as shown in Fig. 2. A single element in the simple cross-correlation process is shown in Fig. 2(a). The fixed delay T_m is one of a large set of delays to be tested for maximum cross-correlation. (Notice that the time shift T is negative for a real delay.) The delay which produces the largest cross-correlation voltage V_m is considered the best estimate over a given interval of time. However, the integration time τ is limited to relatively short periods of time over which the delay $T(t)$ does not fluctuate enough to change the cross-correlation significantly. This restriction on integration time can, in some situations, cause severe limitations on the accuracy of the delay estimate in the presence of interference.

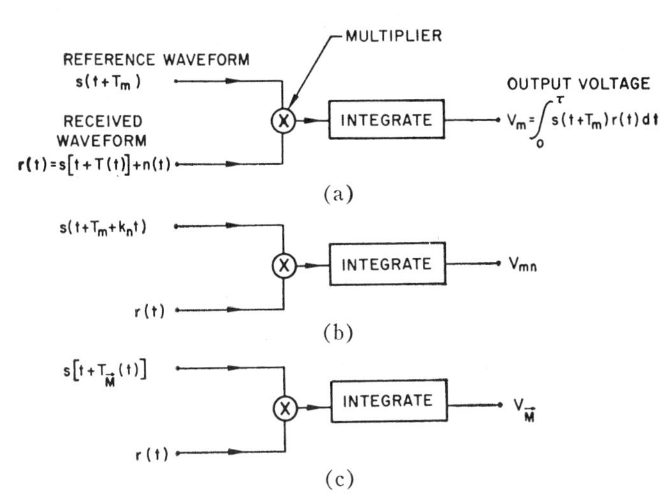

Fig. 2—Use of cross-correlation in delay estimation. Reference waveforms have: (a) Fixed delays T. (b) Fixed, plus linearly varying delays T_m+k_nt. (c) Time varying delay, $T_{\vec{M}}(t)$ chosen from a set of time functions τ sec long with bandwidth W.

A modified cross-correlation process is shown in Fig. 2(b). Here comparisons are made between the received signal and a two-dimensional set of fixed, plus linearly varying delays, and the integration time can be increased to time intervals over which the time delay is well approximated by a member of this set. Matched filter analogs to this technique have been discussed in the literature.[6]

The final stage of accuracy that can be achieved is shown in Fig. 2(c), where a multidimensional set of delay functions $T_{\vec{M}}(t)$ of length τ sec and bandwidth W are used as comparison delay functions. It is clear, however, that to use large dimensions for $\vec{M}$ would be unfeasible in most practical problems.

The delay-lock discriminator provides an approximation to this last technique by generating its own comparison delay function $T_{\vec{M}}(t)$ through the use of cross-correlation and error feedback.

Delay-Lock Discriminator

A block diagram of the delay-lock discriminator is shown in Fig. 3. As the figure shows, the discriminator is basically a nonlinear feedback system employing a multiplier, linear filter, and a controllable delay line.[7] The controllable delay line can have a number of implementations, *e.g.*, a ferrite-core delay line with magnetically controlled permeability and delay, or a servo-controlled electric or ultrasonic delay line. The ultrasonic lines are preferable for delays in the millisecond range or greater. In practice, it is often desirable to have an automatic gain control or limiter to maintain constant power at the discriminator input.

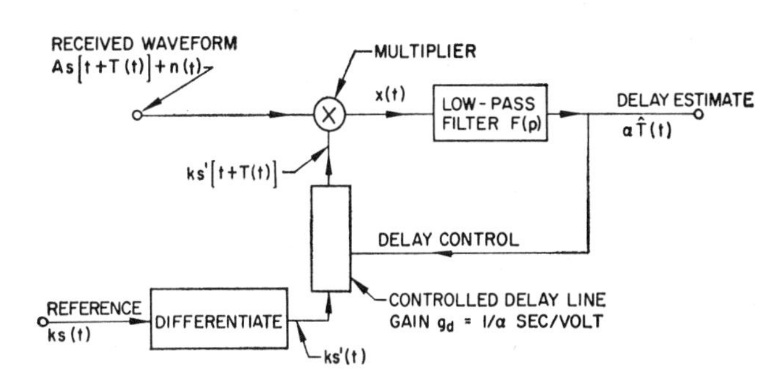

Fig. 3—Block diagram of the delay-lock discriminator. The symbols k, α, are constants.

Through an analysis similar to that used by Lehan and Parks,[8] this discriminator, or a slightly modified version of it, can be shown to be optimum in that it provides the maximum likelihood (*a posteriori*, most probable) estimate of the delay. This derivation has been made by Spilker in unpublished work under the assumption of Gaussian random delay and interfering noise. In general, the truly optimum discriminator contains a second feedback loop which serves to reduce the effects of intrinsic or self-noise described in this section. It should be pointed out, however, that the maximum likelihood discriminator taking the form shown in Fig. 3 requires a nonrealizable loop filter $F(p)$. In this paper the configuration of elements of Fig. 3 is retained, but the loop filter is constrained to be realizable and is optimized for an important class of target delay functions;

[6] R. M. Lerner, "A matched filter detection system for complicated Doppler shifted signals," IRE Trans. on Information Theory, vol. IT-6, pp. 373–385; June, 1960.

[7] The output of a passive, lossless, delay line with an input $f(t)$ is actually $\sqrt{1+dT(t)/dt}\,f[t+T(t)]$ rather than just $f[t+T(t)]$ as shown in Fig. 3. The square root term is present to keep the output energy equal to the input energy. However, in most practical problems we have the relationship $dT(t)/dt \ll 1$, and this effect can be ignored.

[8] F. W. Lehan and R. J. Parks, "Optimum demodulation," 1953 IRE National Convention Record, pt. 8, pp. 101–103.

delay functions which can be approximated by a series of ramps fall into this class.

The operation of the discriminator can be analyzed by examining the multiplier output $x(t)$. The delay error may be defined as $\epsilon(t) = T(t) - \hat{T}(t)$. We can then write the Taylor series for the delayed signal

$$s(t+T) = s(t+\hat{T}) + \epsilon s'(t+\hat{T}) + \frac{\epsilon^2}{2} s''(t+\hat{T}) + \cdots$$

where the primes refer to differentiation with respect to the argument, and all derivatives of $s(t)$ are assumed to exist.[9] Initially, the delay error $\epsilon(t)$ is assumed to be small so that the Taylor series expansion of $s(t+T)$ about $s(t+\hat{T})$ converges rapidly. The multiplier output then has the series expansion

$$\frac{x(t)}{k} = A\Big[s(t+\hat{T})s'(t+\hat{T}) + \epsilon(t)[s'(t+\hat{T})]^2 + \frac{\epsilon^2(t)}{2!} s''(t+\hat{T})s'(t+\hat{T}) + \cdots\Big] + n(t)s'(t+\hat{T}). \quad (1)$$

For convenience, $s(t)$ is normalized to have unity power, and thus the received signal power is $P_s = A^2$. The term $(s')^2$ has a nonzero average value which will be defined as P_d, the power in the differentiated signal, and is dependent only upon the shape of the signal spectrum. We can then write $[s'(t)]^2 \triangleq P_d + s_2(t)$ where $s_2(t)$ has a zero mean. By making use of this last definition, we can rewrite (1) as

$$\frac{x(t)}{k} = AP_d\epsilon(t) + n_e(t), \quad (2)$$

where the first term is the desired error correcting term, and the second term $n_e(t)$ is an equivalent noise term caused by the interfering noise $n(t)$ and the remainder of the infinite series (distortion and intrinsic noise effects). If ϵ is small, $n_e(t)$ has little dependence upon $\epsilon(t)$.

The delay tracking behavior is evident from (2). Suppose that the input delay $T(t)$ is suddenly increased by a small amount. The error $\epsilon(t)$, assumed initially small, will also suddenly increase; the multiplier output will increase, and therefore the delay estimate $\hat{T}(t)$ will increase and tend to track the input delay. The discriminator output is indeed an estimate of the delay.

The representation of the multiplier output given in (2) permits the use of the partially linearized equivalent network shown in Fig. 4. The closed-loop transfer function $H(p)$ is

$$H(p) = \frac{F(p)}{1 + kAP_dF(p)/\alpha} \quad (3)$$

[9] RC filtered white noise for example is nondifferentiable. It seems, however, that for most physical systems parasitic effects cause the signal functions to be differentiable. See S. O. Rice, "Mathematical Analysis of Random Noise," in "Noise and Stochastic Process," edited by N. Wax, Dover Publications, New York, N. Y., pp. 193–195; 1954.

where p is the complex frequency variable. This representation is equivalent to that shown in Fig. 3 because the input to the loop filter $F(p)$ is the same in both instances. The delay estimates thus obtained are identical.

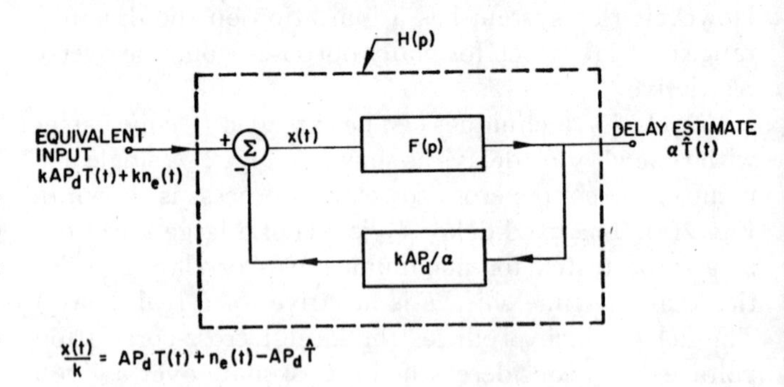

Fig. 4—Partially-linearized equivalent circuit for the delay-lock discriminator.

Notice that the equivalent transfer function $H(p)$ is still nonlinear because it is dependent upon the input signal amplitude A. In the initial part of this discussion, A is assumed constant, and $H(p)$ is assumed linear. In a later paragraph, the effect of AGC or limiting the input signal on the loop transfer function is discussed.

The equivalent input noise $n_e(t)$ is dependent upon $\hat{T}(t)$. However, it can be seen that under conditions of small delay error, this effect can be neglected. This linearized equivalent circuit, then, has its greatest use under conditions of small delay error, *i.e.*, "locked-on" operation. Notice that if $n(t)$ is "white," the interfering noise component of $n_e(t)$ is also white.

To provide a relatively simple yet useful and rather general analysis of the discriminator operation, the signal $s(t)$ will be assumed to have the form of a random frequency, modulated sine wave

$$s(t) = \sqrt{2}\sin[\omega_0 t + \phi(t)] = \sqrt{2}\sin\left[\omega_0 t + \int_0^t \omega_i(t')dt'\right]. \quad (4)$$

The spectrum of this signal can have a wide range of shapes[10] depending upon the statistics of $\omega_i(t)$, but for convenience in calculation, the spectrum of $s(t)$ will be taken to be rectangular with bandwidth B_s and center frequency f_0 as shown in Fig. 5. (It is assumed that $B_s < 2f_0$ and that $\omega_i(t)$ has a zero average value.) Then we can write the expressions:

$$s'(t) = \sqrt{2}\,\omega_s(t)\cos[\omega_0 t + \phi(t)]$$

$$P_d = (2\pi)^2\left[f_0^2 + \frac{1}{3}\left(\frac{B_s}{2}\right)^2\right], \quad (5)$$

[10] D. Middleton, "An Introduction to Statistical Communication Theory," McGraw-Hill Book Co., Inc., New York, N. Y., pp. 604–625; 1960.

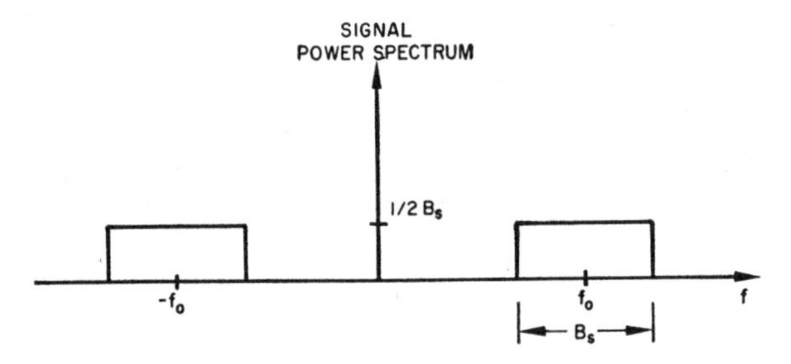

Fig. 5—Power spectral density of $s(t)$.

where we have defined $\omega_s(t)=\omega_0+\omega_i(t)$.

Define the quantities $a_n=E[s'(t)s^{(n)}(t)]$. Note that $a_n=0$ if n is even.

In general, the input to the linearized equivalent circuit can be written as the sum of the equivalent inputs to the discriminator, signal and three types of interference noise terms,

$$\text{Signal term} = kAP_{d\epsilon}(t)$$

$$\text{Noise term} = kn_e(t) = k[n_d(t) + n_i(t) + n_n(t)] \quad (6)$$

where $n_d(t)$ represents a nonlinear distortion term (it is small for small ϵ); $n_i(t)$ is an intrinsic or self-noise term, which is dependent upon the carrier characteristics, and $n_n(t)$ is an external interference term, which is dependent upon external noise at the discriminator input. By making use of (4) and (5), these noise terms can be evaluated as

$$n_d(t) = A\left[a_3\frac{\epsilon^3(t)}{3!} + a_5\frac{\epsilon^5(t)}{5!} + \cdots\right]$$

$$\begin{aligned} n_i(t) &= A\Big\{\epsilon(t)[(s'(t+\hat{T}))^2 - a_1] \\ &\quad + \frac{\epsilon^2(t)}{2!}s'(t+\hat{T})s''(t+\hat{T}) \\ &\quad + \frac{\epsilon^3(t)}{3!}[s'(t+\hat{T})s'''(t+\hat{T}) - a_3] + \cdots\Big\} \\ n_n(t) &= n(t)s'(t+\hat{T}) \\ &= \sqrt{2}\, n(t)\omega_s(t+\hat{T})\cos[\omega_0 t + \phi(t+\hat{T})]. \end{aligned} \quad (7)$$

The distortion terms are taken as those terms of the form ϵ^n for $n\neq 1$.

The terms in the multiplier output with spectra centered about $\omega=2\omega_0$ have been neglected because they will be assumed to be above the passband of the loop filter. This is not possible for low-pass spectra, of course.

The importance of the intrinsic noise term in determining the performance of the discriminator is dependent upon how much of its spectrum passes through the low-pass loop filter. Notice that the intrinsic noise terms are present even if the interference $n(t)$ is absent. It can be seen from (6) and (7) that the intrinsic noise effect is relatively small for this type of signal if

$$|\omega_i(t)|/\omega_0 \le B_s/2\omega_0 < 1,$$

and the bandwidth of the instantaneous frequency $\omega_i(t)$ is large compared to the closed-loop bandwidth. It should be pointed out that the operation of the discriminator is not restricted to the use of fixed envelope signals. However, the intrinsic noise contributions will generally increase if envelope fluctuations of the signal are allowed.

Comparison with the Phase-Lock FM Discriminator

The operation of the delay-lock discriminator is analogous, in several respects, to the operation of the phase-lock FM discriminator (see Fig. 6 for a diagram of the phase-lock loop). It is desirable to investigate the differences and similarities of these two devices.

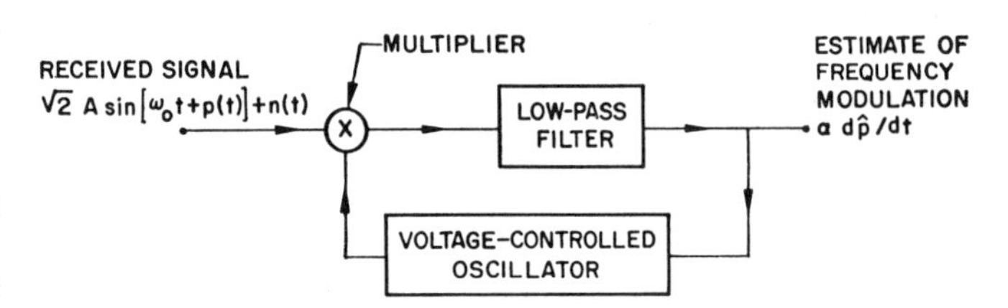

Fig. 6—Block diagram of the phase-lock discriminator.

For pure sine wave carriers (unmodulated carrier bandwidth of zero), delay modulation has a corresponding modulation in phase, *i.e.*,

$$\sin[\omega_0 t + \phi(t)] = \sin\omega_0[t + T(t)] \text{ if } \phi(t) = \omega_0 T(t).$$

Thus, if pure sine wave carriers are used, the delay line and its reference carrier input can be replaced by a differentiator and voltage-controlled oscillator. The differentiator can be lumped into the loop filter of the phase-lock loop. Theoretically, therefore, for the special case of a pure sine wave carrier, the delay-lock discriminator functions exactly as a phase-lock loop.[11]

The delay-lock discriminator normally operates with a wide bandwidth signal when used as a tracking device, and with this type of signal there is no longer a direct correspondence with the phase-lock discriminator operation. As might be expected, however, there are analogous features in both discriminators. For example, the delay-lock discriminator has a threshold error and lock-on performance which are analogous to those in the phase-lock loop.

Discriminator Operating Curve

Thus far, it has been indicated that the discriminator will tend to track the delay variations of an incoming signal provided that the delay error magnitude, $|\epsilon|=|T-\hat{T}|$, is small. In this section we seek to determine how small this error must be and what occurs as the error becomes larger.

[11] In practice, the delay-lock discriminator uses a delay line with restricted dynamic range of delay. Thus, it can be operated only with the sine wave carriers having a limited peak phase deviation.

Assume that $s(t)$ is a stationary (wide sense), ergodic, random variable with zero mean, and that the delays $T(t)$ and $\hat{T}(t)$ are constant or slowly varying with time. Under these conditions, the loop filter when properly optimized forms the average of the multiplier output to obtain:

$$E[x(t)] = E\{[As(t+T) + n(t)]ks'(t+\hat{T})\}$$
$$= -kAR_s'(T - \hat{T})$$

where $n(t)$ and $s(t)$ are assumed independent, and $R_s'(\tau) = d/d\tau[R_s(\tau)]$, the derivative of the autocorrelation function of $s(t)$. The important component in the multiplier output is not always linearly dependent upon the delay error, but, more generally, is functionally dependent upon the error through the differentiated autocorrelation function, and thereby causes changes in the effective loop gain.

The multiplier output can be written using (6) and (7) as

$$\frac{x(t)}{k} = -AR_s'[\epsilon(t)] + n_i(t) + n_n(t) \tag{8}$$

where we have used the relationship

$$R_s'(\epsilon) = \sum_{n=1}^{\infty} a_n \epsilon^n / n!.$$

A further general statement can be made with respect to the effective loop gain for small $|\epsilon|$. The correction component of the multiplier output for small $|\epsilon|$ is $kA\epsilon(t)a_1$ where a_1, in general, is given by

$$a_1 = \int_{-\infty}^{\infty} \omega^2 G_s(f) df$$

and depends only on the shape of the signal spectrum.

Threshold Error

To illustrate the nonlinear behavior of the discriminator, some exemplary signal spectra are shown in Fig. 7 along with their corresponding discriminator characteristics. If $s(t)$ is taken to have a rectangular band-pass spectrum as shown in Fig. 7(a), then, in the region $|\epsilon| < \frac{1}{4}f_0$, the discriminator curve is approximately linear and has a positive slope. However, if the error exceeds the threshold error[12] ϵ_T, the point at which the slope of the discriminator curve first becomes zero, the slope becomes negative, and further small incremental increases in ϵ in this region produce decreases in $\hat{T}$. Thus, the discriminator is unlocked and temporarily unstable with respect to small noise perturbations. Notice that in Fig. 7(a) there are several possible positive-slope lock-on regions, a characteristic of band-pass signal spectra. The effective loop gain, however, dependent upon the magnitude of the slope in these regions, decreases considerably as the delay error moves several inverse bandwidths away from the origin.

[12] In general, the threshold error ϵ_T in the fundamental lock-on region about $\epsilon = 0$ is given by the smallest value of ϵ which can satisfy the equation

$$\int \omega^2 G_s(\omega) \cos \omega\epsilon d\omega = 0.$$

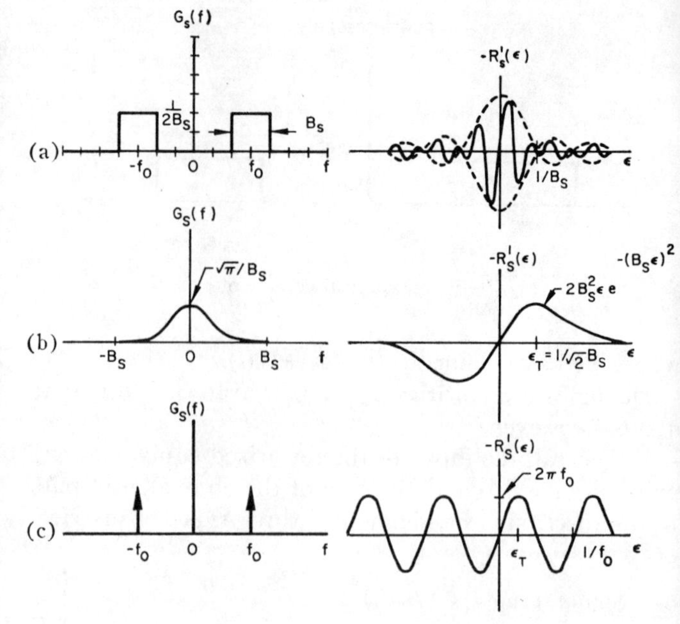

Fig. 7—Signal-power spectral density (unity-signal power) and the corresponding discriminator characteristics. (a) Rectangular band-pass spectrum $\epsilon_T = 1/4f_0$. (b) Gaussian low-pass spectrum $G(f) = (\sqrt{\pi}/B_s) \exp -(\pi f/B_s)^2$, $\epsilon_T = 1\sqrt{2}B_s$. (c) Pure sine-wave signal.

A Gaussian shape of low-pass signal spectrum is shown in Fig. 7(b). The discriminator curve for this signal spectrum has only one lock-on region, a characteristic which is also obtained using white noise passed through low-pass filters with poles on the negative real axis in the p plane. The threshold error for this Gaussian spectrum is $\epsilon_T = 1/\sqrt{2}B_s$. These signals with low-pass spectra are, of course, assumed to be detected (AM, FM, or PM) versions of the actual transmitted RF waveform.

The last spectrum, Fig. 7(c), corresponds to a pure sine wave carrier and phase-lock loop type of operation. Obviously, there are an unlimited number of indistinguishable lock-on regions here, and each has the same loop gain and threshold error. The use of this type of carrier in a tracking problem has limitations unless the delay variations are restricted to values less than $1/f_0$.

Dynamic Range

The dynamic range of the delay-lock discriminator is defined to be the maximum delay excursion of the controlled delay line, and is determined by the largest delay line control input voltage and the delay line gain. The maximum control input voltage in turn is determined by the signal amplitude, the loop filter dc gain, and the peak value of the discriminator curve. Thus, the dynamic range[13] ΔT is

$$\Delta T = kg_d AF(0)R'_{s\,\text{peak}} = gR'_{s\,\text{peak}}/P_d \tag{9}$$

where $g \triangleq kg_d P_d AF(0)$ is the dc loop gain. This value of ΔT relates to the fundamental lock-on region. The values for other regions, if they exist, will be correspondingly less.

Delay Ambiguities and Interference from Other Targets

If a signal having a band-pass spectrum is used, there will exist ambiguities, in many situations, as to which lock-on zone the discriminator is using. An exception to this statement occurs if the ambiguity can be resolved by other means (such as knowledge of the exact target position at a certain instant of time, as might be the situation in tracking a rocket from its firing position). A means for resolving this ambiguity could be to control externally the bias on the controlled delay line and to observe some characteristic of the discriminator curve, *e.g.*, its slope or peak amplitude in a given lock-on region. The problem, then, is analogous to the resolution problem of radar.

Woodward[14] has defined a measure of time ambiguity for radar signals called the time resolution constant T_c. This constant is a measure of the width of the envelope of the discriminator characteristic; and for a rectangular signal spectrum, this time ambiguity has the value $T_c = 1/B_s$. It is difficult to determine the correct lock-on region from others that are separated in delay from it by less than $1/B_s$.

Of course, if a properly chosen low-pass signal spectrum is used, multiple lock-on regions will not exist, and hence ambiguities of this sort do not occur.

Considerations of a similar nature arise when one attempts to compute the interference caused by the presence of multiple targets. Suppose that the discriminator is locked on to a target with delay T, and an interfering target comes into view with delay T_i and returned signal amplitude A_i. Then the multiplier output in the discriminator is $-[AR'_s(T-\hat{T})+A_iR_s(T_i-\hat{T})]$, and the discriminator will operate so as to minimize the sum of these two terms rather than the desired term $-AR'_s(T-\hat{T})$ alone. If the relative effect of the interfering target is to be small, then it is necessary to have the ratio $|A_iR'_s(T_i-\hat{T})/AR_s(T-\hat{T})|$ small for the desired accuracy maximum error $|T-\hat{T}|$. Thus, if a small effect only is to be caused by the second target, it must be separated from the desired target by a delay $|T_i-T| \gg 1/B_s$. It is also desirable that $R_s(\tau)$ decrease rapidly with increasing τ to make up for the differences in path attenuations from the target returns caused by a relatively close undesired target. Spectra with gradual cutoffs are therefore desirable because of the rapid falloffs of $R_s(\tau)$ for large τ, *e.g.*, if $G_s(\omega) \sim \exp -(\omega/2B_s)^2$, then $R_s(\tau) \sim \exp -(B_s\tau)^2$.

Lock-on Performance

Before a target can be tracked, the discriminator must lock on to the target delay so that the discriminator is operating in its linear region. This operation can be performed in practice by manually or automatically sweeping the bias on the delay line control throughout the expected range of the target delay. An alternative approach is to set the delay to correspond to the perimeter of some circular region surrounding the radar. Then targets will be tracked as they enter this region. In this subsection a short analysis is made of the nonlinear lock-on transient when the signal is first applied to the discriminator.

Two discriminator curves are shown in Fig. 8, one for a band-pass spectrum, the other for a low-pass spectrum. Both spectra have Gaussian shapes. If we assume that the received signal has a fixed delay T, and that the quiescent discriminator delay is zero, the steady-state conditions of the discriminator must then satisfy the equation

$$-gR'_s(T-\hat{T})/P_d = \hat{T} = (T-\epsilon). \tag{10}$$

It can be seen that solutions to this equation are given by the intersections of $-R'_s(\epsilon)$ and the straight line in Fig. 8. Recall that only the positive slope regions are stable zones with respect to noise perturbations.

A typical lock-on transient for the signal with a low-pass spectrum is as follows: when the input signal is first applied to the discriminator at $t=0$, the error $\epsilon(t)$ has its initial value $\epsilon(0+) = T$. As a result of this error, the loop filter input takes on a positive value, and $\hat{T}$ will begin to increase from zero and rise towards T. To describe the exact behavior of the loop, the loop filter must be specified.

If a simple low-pass RC filter is used as the loop filter, the lock-on transient is described by a first-order nonlinear differential equation. Referring to Fig. 3 and (8), and neglecting noise effects, one readily finds the differential equation to be

$$\left[\frac{1}{\omega_f}\frac{d\hat{T}}{dt} + \hat{T}\right] = \frac{F(0)}{\alpha}x(t) = -gR'_s(T-\hat{T})/P_d \tag{11}$$

where $\omega_f \triangleq 1/RC$. If $T(t) = T$ is a constant, and $\hat{T}(0) = 0$, then the transient response can be obtained by integrating

$$dt = d\hat{T}/\omega_f[-gR'_s(T-\hat{T})/P_d - \hat{T}(t)]. \tag{12}$$

[13] Notice that this dynamic range restriction is different from that encountered with phase-lock discriminators. Here it is the maximum delay excursion which is limited, whereas, with the phase-lock loop, the maximum frequency excursion is the quantity limited. The reason for this difference is that in the phase-lock loop, the multiplier output controls the *frequency* of the VCO.

[14] P. M. Woodward, "Probability and Information Theory with Applications to Radar," McGraw-Hill Book Co., Inc., New York, N. Y., pp. 115–118; 1953.

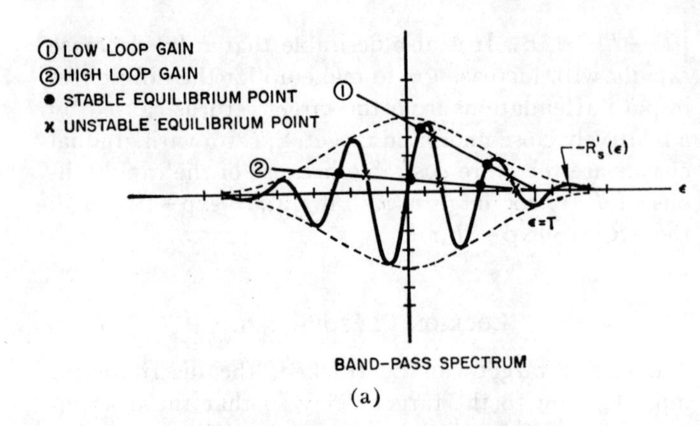

(a)

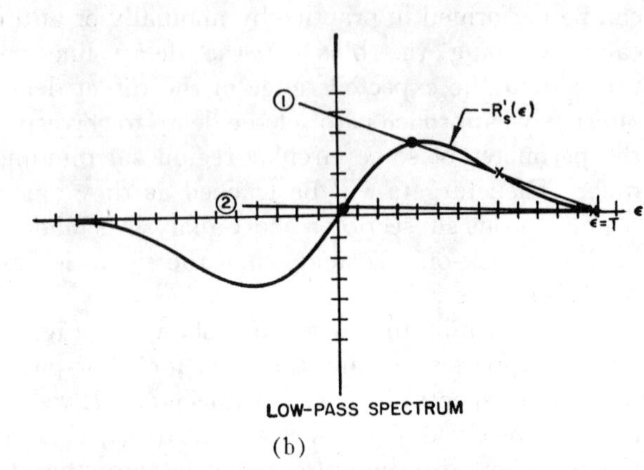

(b)

Fig. 8—Possible steady-state conditions. (a) Band-pass signal spectrum. (b) Low-pass signal spectrum.

Notice that the slope $d\hat{T}/dt$ is proportional to the difference between $-R_s'(T-\hat{T})$ and the straight line as shown in Fig. 8. Thus the slope becomes zero whenever the two curves cross. Of course if there are any zero slope points in negative discriminator slope regions, they are still unstable because of noise considerations neglected in (12).

If the Gaussian low-pass spectrum of Fig. 7(b) is assumed for the signal, and the loop gain is sufficient so that only one zero slope point exists, then the time τ required for the error to change from $\epsilon(0)=T$ to $\epsilon(\tau)=\epsilon_T$, the threshold condition, is given by

$$\tau = \int_0^\tau dt = \int_0^{T-\epsilon_T} \frac{d\hat{T}}{\omega_f[-gR_s'(T-\hat{T})/P_d - \hat{T}]}$$

$$\approx \int_0^{T-\epsilon_T} \frac{P_d\, d\hat{T}}{-\omega_f g R_s'(T-\hat{T})} \tag{13}$$

where the last expression assumed $-gR_s'(T-\hat{T}) \gg P_d\hat{T}$ in the region of interest. For the Gaussian spectrum (13) becomes

$$\tau = \hat{T}\int_{\epsilon_T} \frac{P_d e^{(B_s\epsilon)^2} d\epsilon}{\omega_f g 2B_s^2\epsilon} = \frac{P_d}{2B_s^2\omega_f g}\left[\ln y + \sum_{n=1}^{\infty} \frac{y^{2n}}{2n(n!)}\right|_{\epsilon_T B_s}^{TB_s}$$

where $y \triangleq B_s\epsilon$.

Now, by making use of the series representation

$$\frac{1}{2y^2}(e^{y^2}-1) = \frac{1}{2}\sum_{n=0}^{\infty}\frac{y^{2n}}{(n+1)!}$$

and the approximation $n(n!) \approx (n+1)!$ for $n \gg 1$, then for $y>1$ we have

$$\tau \approx \frac{P_d}{4B_s^2\omega_f g}\left[\frac{1}{y^2}(e^{y^2}-1-y^2)\right|_{\epsilon_T B_s}^{TB_s} \tag{14}$$

Thus, with sufficiently large loop gain and the absence of interfering noise, the discriminator will eventually lock on even if the initial delay error is large. However, if $\epsilon(0)=T\gg 1/B_s$, the lock-on time will become extremely large and interfering noise effects will become of dominant importance.

Referring to (13) one sees that low-pass signals with autocorrelation functions which decrease rapidly with delay for large delays (desirable because of the effects of multiple targets) have lock-on times which increase extremely rapidly with initial delay error for large initial errors, *e.g.*, from (14),

$$\tau \sim e^{(TB_s)^2}/(TB_s)^2$$

for large TB_s with the Gaussian signal spectrum.

Accuracy of the Discriminator

In this section we return to the investigation of linear discriminator operation and the linearized equivalent representation shown in Fig. 4. The objective of this section is to determine the accuracy of the discriminator and the threshold value of input SNR. Intrinsic noise effects are assumed negligible compared to those caused by other error terms. Both band-pass and low-pass signal spectra are considered. The input signal amplitude is assumed fixed.

The target delay to be used is a ramp of delay beginning at $t=0$ and corresponds to a sudden change in velocity, *i.e.*,

$$T(t) = 0 \qquad \text{for } t<0$$
$$= \frac{2v}{c}t \qquad t \geq 0,$$

where v is the target velocity, and c the velocity of light. The Laplace transform of the delay is $T(p)=2v/cp^2$. Although real targets, of course, cannot change velocity instantaneously in this manner, they can approximate this ramp well enough to make the results of this analysis useful. This sudden ramp of delay is also important in studying the discriminator response when the return from a constant velocity target is suddenly applied to the input. Furthermore, the general behavior of the transient errors and the steady-state errors with velocity inputs are of interest by themselves. The linearized analysis used here applies only if the delay error at the beginning of the transient $\epsilon(0)$ is much less than the threshold error.

Two loop filters are shown in Fig. 9. The first of these, a simple integrator, produces a closed-loop transfer function [obtained from (3)] which is given by

$$H(p) = \frac{\alpha}{kAP_d}\left(\frac{1}{1 + p/p_0}\right). \tag{15}$$

This filter has zero steady-state error to step inputs of delay, but a finite nonzero steady-state error to ramp inputs. The second loop filter, shown in Fig. 9(b), is composed of an integrator and an RC filter. The closed-loop transfer function for this filter is

$$H(p) = \frac{\alpha}{kAP_d} \frac{1 + \sqrt{2}\, p/p_0}{1 + \sqrt{2}\, p/p_0 + (p/p_0)^2}. \tag{16}$$

This loop filter has been shown optimum for ramp inputs in the presence of white noise, in that it minimizes the total squared transient error plus the mean square error caused by interfering noise.[15] The frequency p_0 would then be chosen by relative weighting of the two types of errors. The frequency here will be chosen from other considerations, namely, to keep the peak transient error below a set value. This filter produces zero steady-state error in response to a ramp input.

The transient error is defined as the delay error $T(t) - \hat{T}(t)$ for a given delay function $T(t)$ in the absence of discriminator interference $n_e(t)$. The transient error for the simple integrator type of loop filter [Fig. 9(a)] with a ramp of delay as the input is shown in Fig. 10(a). The corresponding closed-loop frequency response is shown in Fig. 10(b). Notice that the error rises to a final steady-state value $\epsilon_t(\infty) = 2v/cp_0$ for a target radial velocity v, and a corresponding steady-state target position error $2v/p_0$. It is obviously desirable to have $\epsilon_t(\infty) < \epsilon_T$ and the position error small enough to obtain the required position accuracy. Suppose, then, that we choose p_0 to obtain the desired small steady-state transient error $\epsilon_t(\infty)$, *i.e.*, $p_0 = 2v/c\epsilon_t(\infty)$.

For this value of p_0, what is the lowest input SNR for which we can keep the delay errors below the threshold value ϵ_T most of the time? If the equivalent input noise, $n_e(t)$, is assumed Gaussian, and produces an rms error σ_{ϵ_n} in the delay estimate, then a reasonable condition for the discriminator to be said to operate above threshold is that $\sigma_{\epsilon_n} \leq \epsilon_T/3$. For delay errors which are approximately Gaussian (transient errors are assumed much less than ϵ_T), this condition corresponds to a probability of $|\epsilon| \geq \epsilon_T$ of less than or equal to 0.27 per cent at any instant of time.

The rms value of noise error for white noise inputs can be found from the expression

$$\sigma_{\epsilon_n}^2 = \int_{-\infty}^{\infty} k^2 G_{n_n}(f)\, |H(j\omega)/\alpha|^2 df = G_{n_n}(0) p_0 / 2(AP_d)^2$$

[15] R. Jaffe and E. Rechtin, "Design and performance of phase-locked circuits capable of near optimum performance over a wide range of input signal levels," IRE TRANS. ON INFORMATION THEORY, vol. IT-1, pp. 66–72; March, 1955.

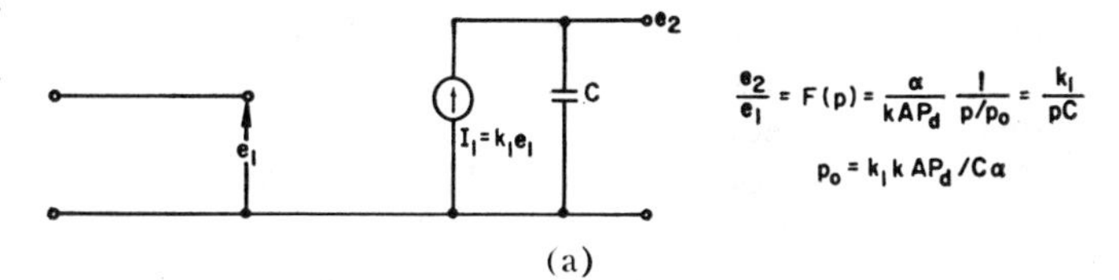

(a)

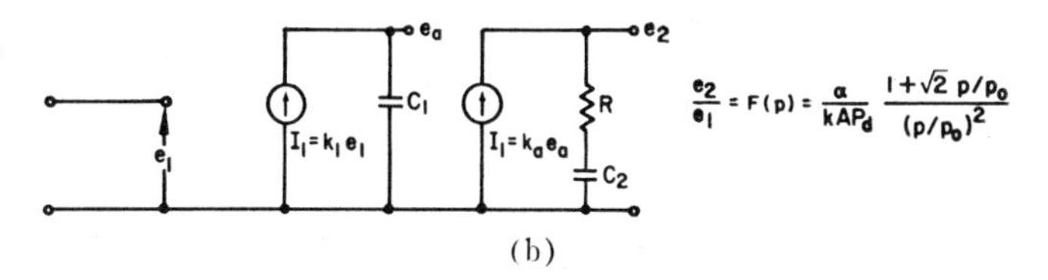

(b)

Fig. 9—Two loop filters and their transfer functions.

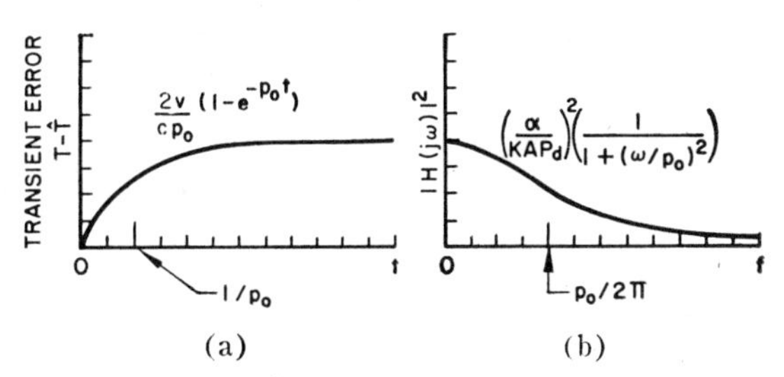

(a) (b)

Fig. 10—Discriminator performance with the loop filter of Fig. 9(a). (a) Transient error in response to a ramp of delay. (b) Closed-loop frequency response.

where $G_{n_n}(f)$ is the power spectral density of the noise term $n_n(t)$. For the value of p_0 chosen, we have

$$\sigma_{\epsilon_n}^2 = \frac{G_{n_n}(0)v}{c\epsilon_t(\infty)(AP_d)^2}. \tag{17}$$

The threshold occurs, then, when $G_{n_n}(0)$ has the value $c\epsilon_T^2\epsilon_t(\infty)(AP_d)^2/9v$. The power spectral density $G_{n_n}(0)$ is in turn related to the input noise spectral density. For white input noise $n(t)$ and a signal spectrum which is rectangular or Gaussian in shape, the spectrum of $G_{n_n}(f)$ is also white.

For white interfering noise $n(t)$ with power P_n in a bandwidth $2B_s$ (both positive and negative frequency regions are used throughout this paper), the amplitude of this power spectral density is[16]

$$G_{n_n}(f) = P_d G_s(f) * G_n(f)$$

and

$$G_{n_n}(0) = P_d P_n / 2B_s. \tag{18}$$

This last relation is valid regardless of the shape of the spectrum of $s(t)$ and has assumed that the spectrum of $s'(t+\hat{T})$ is the same as that of $s'(t)$.

Now by combining (17) and (18), the threshold input SNR can be found

$$(\mathrm{SNR})_{\text{threshold}} = \left[\frac{A^2}{P_n}\right]_{\text{threshold}} = \frac{4.5(v/c)}{B_s P_d \epsilon_T^2 \epsilon_t(\infty)}. \tag{19}$$

[16] The use of the asterisk indicates convolution in the frequency domain.

It is seen that, in general, the threshold SNR increases as the transient error ϵ_t is made smaller for fixed velocity v, just as expected.

To evaluate this expression, the power spectrum of $s(t)$ must be specified so that ϵ_T and P_d can be determined. If $s(t)$ has a Gaussian low-pass spectrum,

$$G_s(f) = \frac{\sqrt{\pi}}{B_s} \exp - (\pi f/B_s)^2, \qquad \sigma \text{ is } \frac{B_s}{\sqrt{2\pi}}$$

for this spectrum, then

$$\epsilon_T = 1/\sqrt{2}B_s, \qquad P_d = (2\pi\sigma)^2 = 2B_s^2.$$

Thus the threshold SNR is

$$(\text{SNR})_{\text{threshold}} = \frac{4.5(v/c)}{B_s\epsilon_t(\infty)} \tag{20}$$

As an example, suppose $B_s = 1$ Mc, which makes $\epsilon_T = 0.707$ μsec, $\epsilon_t(\infty) = 0.1$ μsec (98.4 ft. transient error), $v = 2000$ mph, and $(v/c = 2.99 \times 10^{-6})$, then the threshold SNR is 1.36×10^{-4} or -38.6 db.

The transient error for the loop filter depicted in Fig. 9(b) in response to the same ramp input of delay $T(t) = 2(v/c)t$ is shown in Fig. 11(a). The closed-loop frequency response is shown in Fig. 11(b). The peak transient error for this filter is $\epsilon_t(t_{\text{peak}}) = 0.91\ (v/cp_0)$ and occurs at time $t_{\text{peak}} = 1.11/p_0$. Because the transient error is significant over a limited time interval only (about $2t_{\text{peak}}$) and has a limited rise time, it can be seen that in order to have the peak transient error from a real target be well approximated by that given in Fig. 11(a), the actual change in target velocity must occur over a time interval less than t_{peak}. In other words, the maximum target velocity transient considered here is the velocity change that can occur in a period of time t_{peak}.

The peak transient error will be set at $\epsilon_T/3$, *i.e.*, $p_0 = 2.72v/c\epsilon_T$. Threshold will be said to occur when the delay error caused by noise ϵ_n has an rms value $\sigma_{\epsilon_n} = \epsilon_T/3$. For Gaussian ϵ_n, this condition corresponds to a probability of $|\epsilon| \geq \epsilon_T$ of 0.27 per cent when there is no transient error. The probability of $|\epsilon| \geq \epsilon_T$ at peak transient error is 2.3 per cent.

The mean square delay error caused by a white interfering noise input can be found using (16) as[17]

$$\sigma_{\epsilon_n}{}^2 = \int_{-\infty}^{\infty} k^2 G_{n_n}(f) \,|\, H(j\omega)/\alpha \,|^2 df$$

$$= 1.06 G_{n_n}(0) p_0/(AP_d)^2 = (\epsilon_T/3)^2. \tag{21}$$

Now by using (18), (21) and the relation for p_0, the threshold input SNR can be found as

$$(\text{SNR})_{\text{threshold}} = \left(\frac{A^2}{P_n}\right)_{\text{threshold}} = \frac{13.0(v/c)}{B_s P_d \epsilon_T{}^3}. \tag{22}$$

[17] This integral has been evaluated using D. Bierens de Haan, "Nouvelles tables d'intégrales définés," Hafner Publishing Co., New York, N. Y., p. 47; 1957.

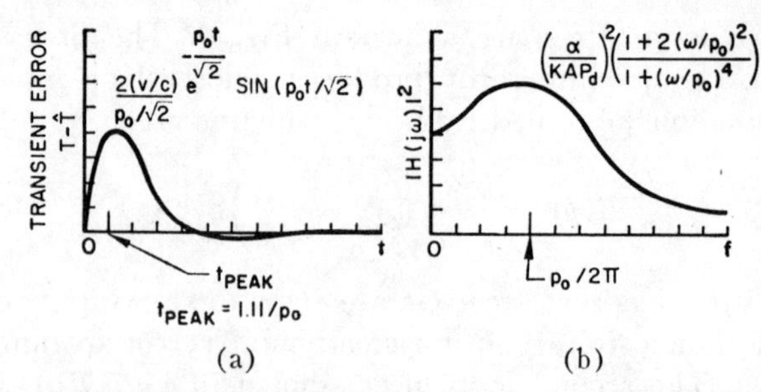

Fig. 11—Discriminator performance with the loop filter of Fig. 9(b). (a) Transient error in response to a ramp of delay. (b) Closed-loop frequency response.

If the signal spectrum has a Gaussian shape as before $\sigma = B_s/\sqrt{2\pi}$, then (22) becomes

$$(\text{SNR})_{\text{threshold}} = 18.4(v/c). \tag{23}$$

As a second example, suppose $v = 2000$ mph $(v/c = 2.99 \times 10^{-6})$. Then the threshold (SNR) is 5.5×10^{-5} or -43 db, an improvement of more than 4 db over that provided by the first filter.

Tracking an Actively Transmitting Target

One of the more important applications of the delay-lock discriminator is to track a target which is itself transmitting a wide-bandwidth, random signal. Information on the target position can be obtained by estimating the delay difference $T(t)$ between the signals as they arrive at the two antennas as shown in Fig. 12. The signal received from one antenna is fed into the discriminator as the reference, and the other received signal is fed into the input. By comparing the delay differences for three such pairs of antennas, the target position (including range) can be determined as the intersection point of three hyperboloids.[18] Two pairs of antennas are sufficient to provide angular information.

As it concerns the operation of the delay-lock discriminator, this problem differs from the one just discussed only in that the noise-perturbed signal received in one antenna is used as the reference. As a result, a corresponding degradation in accuracy at low input SNR is to be expected. By referring to Fig. 13 and (2), one can write the low-frequency terms of the multiplier output as

$$x(t) = A_1 A_2 P_d \epsilon(t) + n_e(t) \tag{24}$$

where $n_e(t)$ is the equivalent linearized interference and has the representation

$$n_e(t) = A_1 A_2 [n_d(t) + n_i(t)] + A_1 s(t + T) n_2'(t + \hat{T})$$
$$+ A_2 s'(t + \hat{T}) n_1(t) + n_1(t) n_2'((t + \hat{T}), \tag{25}$$

[18] Actually, there are two intersections of the three hyperboloids, one on each side of the plane of the antennas. However, if the antennas are on the ground, it is usually easy to decide which point is correct.

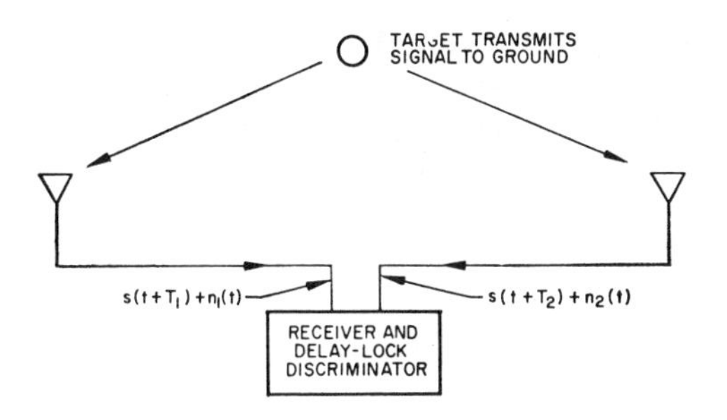

Fig. 12—Tracking a target which is transmitting a wide bandwidth signal.

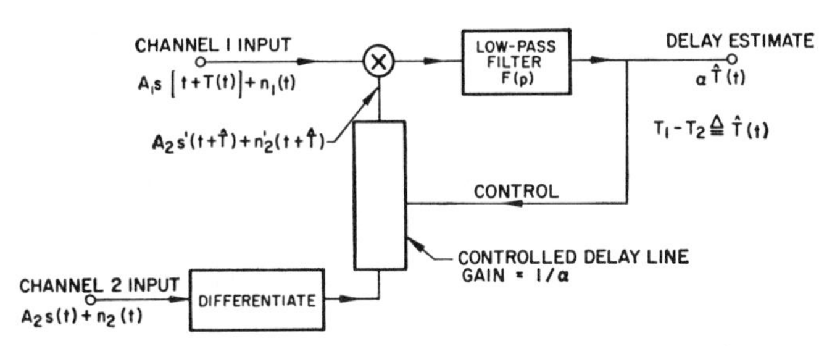

Fig. 13—Operation of the delay-lock discriminator with a noise perturbed reference.

where $n_d(t)$, $n_i(t)$ are the distortion and intrinsic noise components, respectively. Notice that in addition to the noise terms of (6) there are now two additional noise terms, another $S \times N$ term (signal times noise term) and a $N \times N$ term.

In many situations the dominant noise terms are generated in the two receiver amplifiers, and the noise terms $n_1(t)$ and $n_2(t)$ are independent of one another. If the spectra of the signal and these noise components have bandwidths much greater than the closed loop bandwidth of the discriminator, then $n_e(t)$ can be considered to have an approximately white spectrum, and the results of the previous section can be used to obtain the performance of this discriminator. Of course, if $n_1(t)$ and $n_2(t)$ contain components which are not independent, any dc and low-frequency noise components which might then exist must be taken into account.

If, for example, we take the signal and independent noise components to have rectangular spectra with bandwidths B_s, then the equivalent noise spectrum in the low-frequency region is

$$G_{n_n}(0) = \frac{1}{2B_s}(A_1^2 P_{dn_2} + A_2^2 P_d P_{n_1} + P_{dn2} P_{n_1})$$
$$= \frac{A_1^2 A_2^2 P_d}{2B_s}(r_1 + r_2 + r_1 r_2) \qquad (26)$$

where P_{n1}, P_{dn2} are the average powers in $n_1(t)$, $n_2'(t)$, and r_1, r_2 are the noise-to-signal power ratios on channels 1 and 2, respectively.

If intrinsic noise effects are negligible, then threshold input SNR can be obtained by combining (26) with either (17) or (21), depending on which loop filter is used. Notice that constant k in (15) and (16) now becomes A_2. Consider that the loop filter of Fig. 9(b) is used. Then by using (21), (26) and assuming equal SNR on both channels, the threshold SNR can be evaluated as

$$(\text{SNR})_{\text{threshold}} = h + \sqrt{h + h^2} \qquad (27)$$

where

$h \triangleq 13.0(v/c)/B_s P_d \epsilon_T^3$, and v is the radial velocity difference to the antennas. For $h \ll 1$, this relation becomes $(\text{SNR})_{\text{threshold}} = \sqrt{h}$. As an example, suppose that $v = 2000$ mph. If the signal spectrum is low-pass and rectangular, then $\epsilon_T \cong \frac{1}{2} B_s$, $P_d = (2\pi B_s)^{2/3}$, and the threshold SNR is 4.85×10^{-3} or -23 db.

Effect of AGC or Limiting on the Discriminator Performance

As pointed out earlier, to control the discriminator loop gain it is desirable to feed the received data through a limiter or an amplifier with strong AGC. The use of an ideal AGC serves to maintain a constant average input power for the discriminator and has a relatively simple effect. Thus, the ideal AGC acts as a variable attenuator which produces no distortion of the input, but attenuates the amplitude of the signal component in its output. Its only effect is to vary the discriminator loop gain as a function of the input SNR. If the total input power to the discriminator is A_t^2 and only one channel has noise added and AGC control, then the effective loop gain is

$$\text{loop gain} = g = \frac{k A_t P_d g_d F(0)}{\sqrt{1 + P_n/P_s}}. \qquad (28)$$

The change in the loop gain, however, is important because it affects both the dynamic range of the discriminator and the closed-loop bandwidth. Here we encounter the problem: if a time invariant loop filter is to be used, what value of the loop gain should be assumed in designing the loop filter?

Once the discriminator has locked onto the signal, the most critical phenomenon is the occurrence of the threshold or loss of lock condition. While the discriminator is operating well above threshold at a fixed SNR, there is a linear relationship—the closed-loop transfer function $H(j\omega)$—between the true delay and the delay estimate. Thus we can follow the discriminator with a linear filter with a transfer function $H_2(j\omega)$, and the product $H_T(j\omega) = H(j\omega)H_2(j\omega)$ can be chosen so that it is optimum in some sense, *e.g.*, $H_T(j\omega)$ can be chosen as the realizable Wiener filter which allows some particular value of delay (negative, if a predictor is desired) in forming the estimate of the target delay. Consequently, one reasonable approach to this is to optimize the loop filter using the threshold value of loop gain, and then to choose a time invariant filter $H_2(j\omega)$, so that $H_T(j\omega)$ is optimum in the Wiener sense at large input SNR.

Now, if we consider that the single noisy channel has a limiter preceding the discriminator, it can be shown that the dominant effect is the change in the loop gain. However, the exact dependence of loop gain on input SNR is not always exactly the same as with perfect

AGC, because the limiter input and output SNR are not always equal. For example, Davenport[19] has shown that the signal power output of an ideal band-pass limiter for sine wave plus Gaussian noise inputs is related to the total limiter output power by the expression

$$P_{s\,\mathrm{out}} = \frac{P_T}{1 + b(P_n/P_s)}$$

where P_T is the band-pass limiter output power, $\pi/4 \leq b \leq 2$, and b depends on the input SNR P_s/P_n. The loop gain varies with $P_{s\,\mathrm{out}}$ roughly in the same manner as with AGC.

If the input to an ideal limiter is a Gaussian signal plus independent Gaussian noise, then the discriminator operating curve can be obtained using the results of Bussgang[20] which show that the cross-correlation between the input and output of the limiter is proportional to the autocorrelation function of the input. The discriminator operating curve can thus be shown to be

$$-R'(\tau) = \sqrt{P_s P_T}\,\frac{-R_s'(\tau)}{\sqrt{1+r}}$$

where P_T is the limiter output power, P_s is the signal power, and $R_s(\tau)$ is the autocorrelation function of the signal which has been normalized to unity power. In this way the loop gain is changed exactly as it was with the AGC. The loop gain change may not be the only effect, because in passing through the limiter the noise statistics change and harmonics of the signal are generated. In practice, however, the limiting operation can usually be done at an IF or RF frequency (before detection if $s(t)$ is to be low-pass) so that harmonic content is removed by band-pass filters and is of little concern.

It is sometimes convenient to limit both received data channels at the multiplier inputs; the reference channel is differentiated and delayed before amplitude limiting. With this type of operation the multiplier inputs are both binary random variables, and the multiplier circuit can be implemented by using an AND circuit. If both inputs to the receiver have stationary Gaussian statistics and the noise-to-signal power ratios on the two channels are r_1 and r_2, then the discriminator characteristic can be shown to be[21]

$$\frac{2}{\pi} P_T \sin^{-1}\left[R_s'(\tau)/\sqrt{(1+r_1)(1+r_2)R_s(0)R_s''(0)}\right]$$

where P_T is the output power of each of the limiters, and $R_s(\tau)$ is the autocorrelation function of the signal. The ratio $R_s'(0)/\sqrt{R_s(0)R_s''(0)}$ is less than unity as can be shown using the Schwarz inequality. This discriminator characteristic is of roughly the same shape as obtained without limiting since $\sin^{-1}x \approx x$ for $|x| < 1$. The peak value of the loop gain varies in proportion to $[(1+r_1)(1+r_2)]^{-1/2}$.

If a band-pass limiter is used on both received input channels, then using Price's[22] (6) we can show that the cross-correlation is given by

$$R_{12}(\tau) = \frac{\left(\frac{\pi}{8}\right) P_T R_s(\tau)}{\sqrt{(1+r_1)(1+r_2)}} \left\{ 1 + \sum_{m=1}^{\infty} \frac{\left[\left(\frac{1}{2}\right)\left(\frac{3}{2}\right)\cdots\left(\frac{2m-1}{2}\right)\right]^2}{m!(m-1)!} \cdot \frac{\rho_s^{2m}(\tau)}{\sqrt{(1+r_1)(1+r_2)}} \right\}$$

where $R_s(\tau) \triangleq \rho_s(\tau)\cos[\omega_0\tau + \lambda(\tau)]$, *i.e.*, $\rho_s(\tau)$ is the envelope of $R_s(\tau)$. The discriminator characteristic, $-R_{12}'(\tau)$, is not greatly different in shape from $-R_s'(\tau)$. The loop gain is again attenuated by both noise-to-signal ratios r_1 and r_2.

Experimental Version of the Discriminator

A laboratory model of the delay-lock discriminator has been constructed and tested. The objective of the experimental work was to demonstrate the basic principles of operation and to provide experimental verification of some of the theory of linear operation. Ferrite-core delay lines with magnetically controlled permeability[23] were used in these particular experiments as the variable delay elements.

A block diagram of the experimental delay-lock discriminator is shown in Fig. 14. In this experimental equipment, the reflection and transmission from the target were simulated by another delay line similar to the one used in the discriminator. An AGC amplifier was provided in the signal input channel to maintain constant input power to the discriminator. The loop filter $F(p)$ consisted of a RC low-pass filter with a time constant of 8.8 msec.

The phase delay vs control current characteristic of the delay lines used in the experimental discriminator is shown in Fig. 15. Note the nonlinearity and hysteresis effect present even for relatively small delay variations. Additional measurements have shown the slope of the curve (*i.e.*, the delay line gain) to vary as a function of carrier frequency approximately ±10 per cent of the value shown. The group delay (slope of the phase shift vs frequency curve) is expected to vary by about this same amount from the values shown.

The amplitude spectrum of the carrier measured at the output of the AGC amplifier is shown in Fig. 16(a).

[19] W. B. Davenport, Jr., "Signal-to-noise ratios in bandpass limiters," *J. Appl. Phys.*, vol. 24, pp. 720–727; June, 1953.

[20] J. J. Bussgang, "Cross-correlation functions of amplitude-distorted Gaussian signals," Mass. Inst. Tech., Cambridge, Mass., RLE TR No. 216, pp. 4–13; March, 1952.

[21] If x_1 and x_2 are limited forms of $y_1 \triangleq s + n_1$ and $y_2 \triangleq s' + n_2$, respectively, then it can be shown that $R_{x_1x_2}(\tau) = P_T\{4\Pr[y_1(t)>0, y_2(t+\tau)>0] - 1\}$, where $\Pr(y>0)$ is the probability that $y>0$. This probability can be evaluated by integrating the bivariate normal distribution to obtain the above result.

[22] R. Price, "A note on the envelope and phase-modulated components of narrow-band Gaussian noise," IRE Trans. on Information Theory, vol. IT-1, pp. 9–12; September, 1955.

[23] H. W. Katz and R. E. Schultz, "Miniaturized ferrite delay lines," 1955 Natl. IRE Convention Record, pt. 2, pp. 78–86.

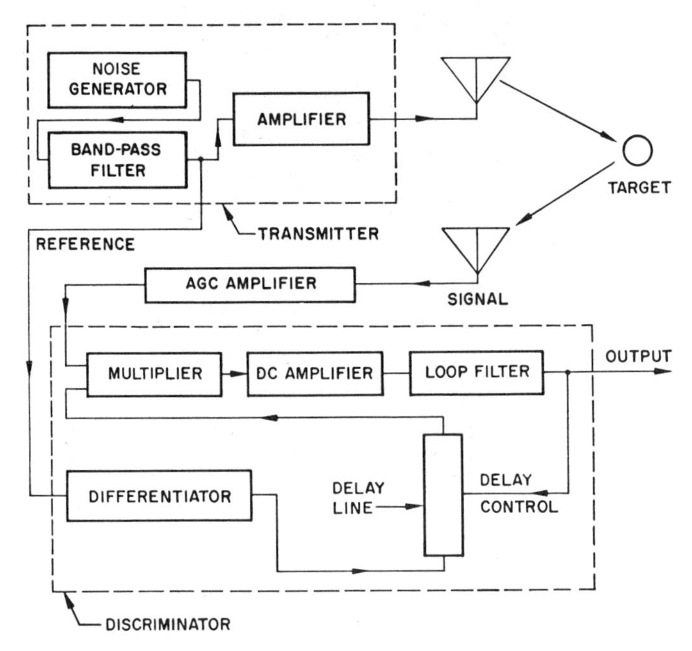

Fig. 14—Experimental system-block diagram.

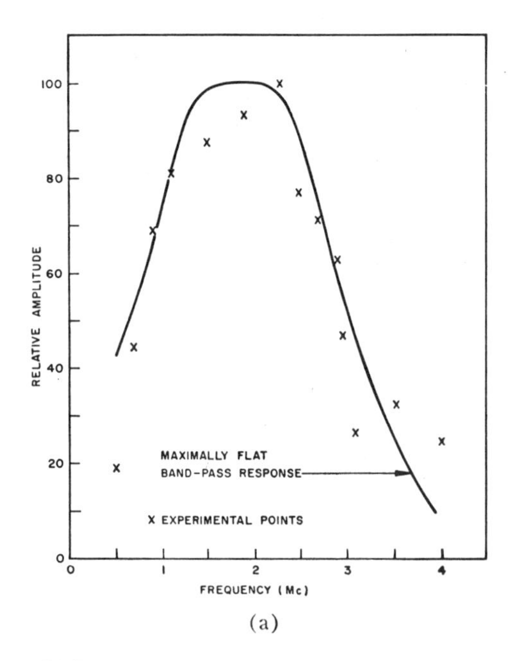

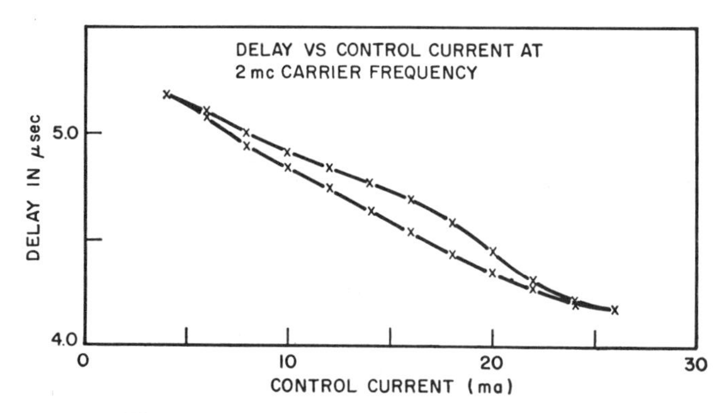

Fig. 15—Discriminator delay-line characteristic.

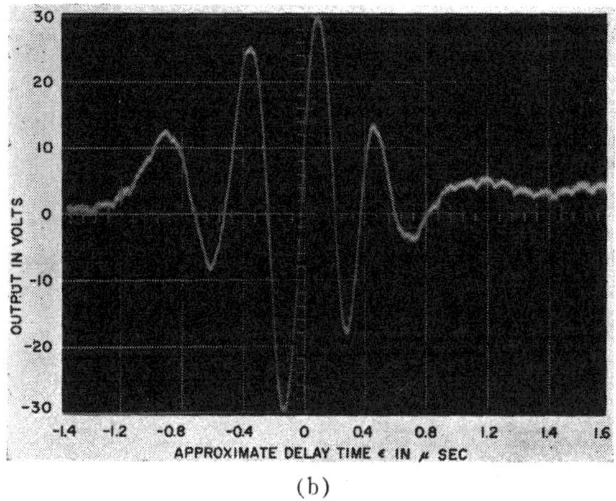

Fig. 16—(a) Amplitude spectrum of carrier. (b) Oscillogram of discriminator characteristic.

A maximally flat band-pass spectrum with the same 3 db points as the experimental data is plotted for comparison. This spectrum has a center frequency of 1.85 Mc.

The discriminator characteristic obtained from the experimental system is shown in Fig. 16(b). The horizontal axis is labelled "approximate delay" since much nonlinear distortion was produced by the delay line. Note that the delay variation presented is about $2\frac{1}{2}$ times that shown in Fig. 15. It should be pointed out, however, that the operating range for the measurements presented is the main lock-on region at the center of the oscillogram. This portion of the discriminator characteristic, which is quite linear, extends ± 0.12 μsec about the $\epsilon=0$ position. This wave corresponds to a center frequency of approximately 2.1 Mc for a symmetrical band-pass spectrum. Thus, the major linear region of the characteristic extends over a range of delay error ϵ that agrees to within 13 per cent of the value predicted by the maximally flat band-pass approximation to the experimental spectrum.

Measured open-loop and closed-loop amplitude responses are plotted in Figs. 17(a) and 17(b), respectively. The measured open-loop response coincides well with the theoretical response of an RC low-pass loop filter with a cutoff frequency of 18 cps. Using a linearized equivalent circuit similar to Fig. 4, we see that for such a simple filter the only effect of the feedback will be to multiply the cutoff frequency by a factor of $1+g$. The measured loop gain was $g=11$. The theoretical closed loop response plotted in Fig. 17(b) is that of a single real-axis pole with a cutoff frequency of 216 cps. The measured closed-loop response, plotted also in Fig. 17(b), again matches the theoretical curve closely.

Another check on the theory of linear operation of the delay-lock discriminator can be made by measuring the transient error for a triangular wave input. Figs. 18(a) and 18(b) show the discriminator responses for 13 ma, peak-to-peak inputs of frequencies 24 cps and 240 cps, respectively. Since a triangular wave is a summation of an infinite number of ramps, the transient error to a triangular wave can be found from the error to a ramp.

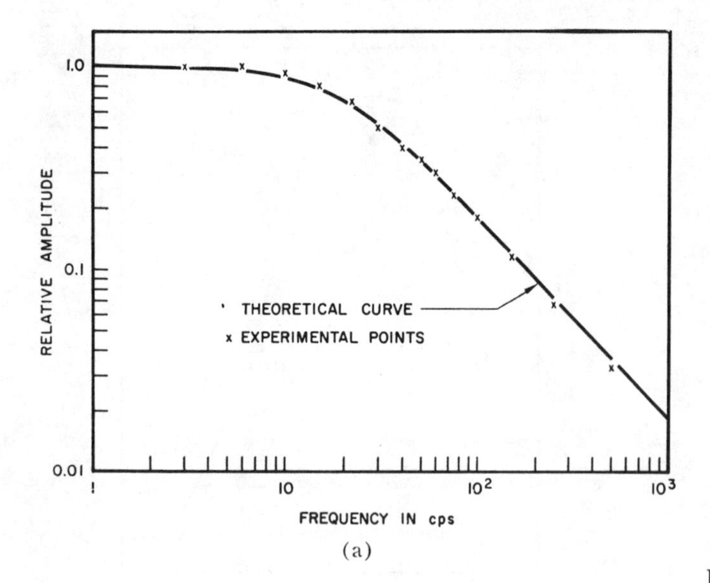

(a)

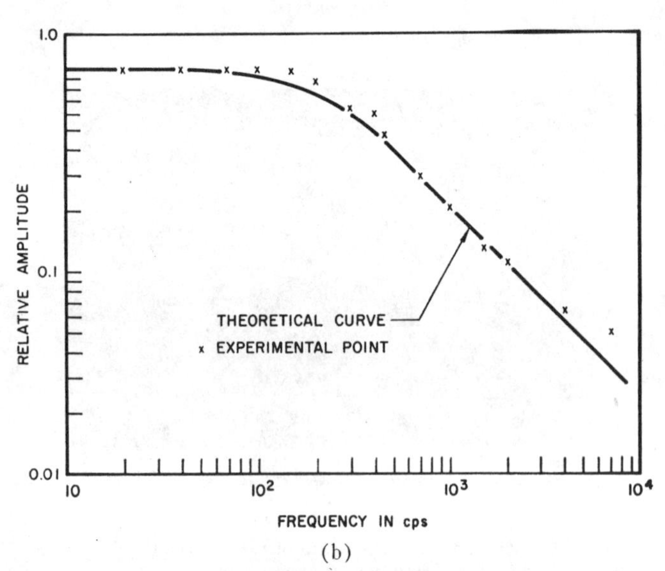

(b)

Fig. 17—(a) Open-loop amplitude response. (b) Closed-loop amplitude response.

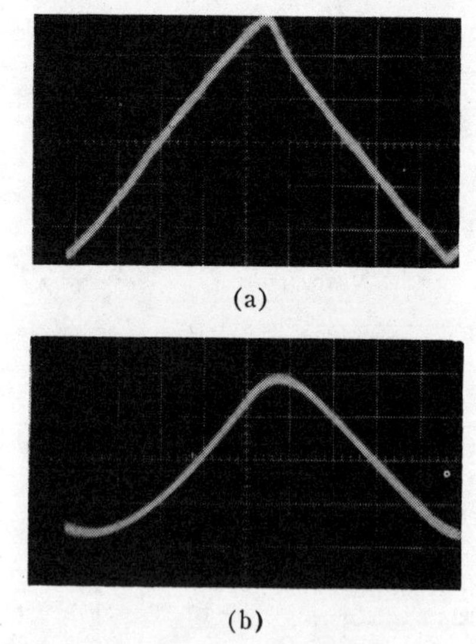
(a)

(b)

Fig. 18—Oscillograms of discriminator delay-line, control current for triangular wave control current in modulating delay line. (a) 24 cps input. Vertical scale is 2 ma per division, while the horizontal scale is 5 msec per division. (b) 240 cps input. Vertical scale is 2 ma per division, while the horizontal scale is 500 μsec per div.

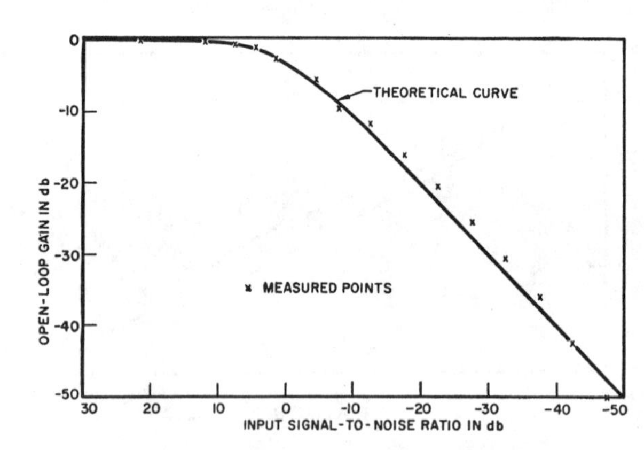

Fig. 19—Loop gain vs input signal-to-noise power ratio.

From elementary control theory, using a linearized equivalent circuit similar to Fig. 4, it is possible to find a simple expression for the delay error $\epsilon(t)$ to a ramp input of delay.

$$\epsilon(t) = \frac{at}{(1+g)} + \frac{agRC}{(1+g)^2}(1 - e^{-(1+g)t/RC}) \quad (29)$$

where the input is

$$T(t) = at, \quad t \geq 0$$
$$= 0, \quad t < 0.$$

If we define ϵ' as the peak-to-peak output error in response to a triangular wave input, then it can be shown that[24]

$$\epsilon' = 2\epsilon(t = T_0/2) \quad (30)$$

[24] Choose the time origin so that the input function is an even function of time.

where T_0 is the period of the triangular wave input. Further calculations using (29) and (30) predict a peak-to-peak amplitude of 11.6 ma for the 24 cps input, and a peak-to-peak amplitude of 8.0 ma for the 240 cps input. These predictions are in good agreement with the oscillograms of Figs. 18(a) and 18 (b).

The measured loop gain as a function of input, SNR power ratio, is plotted in Fig. 19. A theoretical curve based on (28) is plotted in the same figure and corresponds closely with the experimental points. This curve of loop gain is an indirect indication of closed-loop discriminator dynamic range and closed-loop bandwidth. A version of this discriminator, with a higher dc loop gain than that described here, has operated at input SNR ratios as low as -40 db.

Discussion

On the basis of these results, the delay-lock discriminator appears to have good potential in tracking rapidly moving targets while using very low, received, SNR ratios. It is especially suited to tracking problems where the initial target position is known or where tracking is to begin only when the target enters a fixed perimeter. However, by the use of search techniques, targets of unknown initial position can be tracked.

By properly choosing the signal spectrum shape and bandwidth, good performance can be obtained with respect both to reducing the ambiguity in target position and discriminating against undesired targets.

In practice, where tracking is required over moderately long distances, the use of servocontrolled ultrasonic delay lines seems attractive. Delays in the millisecond range are attainable using such lines, and the linearity of delay vs control voltage can be made quite good. However, the response of the servosystem has to be taken into account in computing the closed-loop response. The presence of this servomotor within the loop may require some modification of the loop filter depending on the speed of response desired. Other delay techniques using such devices as magnetic recorders or shift registers might also be useful where long delays are desired.

The delay line also restricts the signal frequency spectrum that can be used because of its delay-bandwidth product limitations. At present, quartz delay lines can function at frequencies up to 100 Mc. However, the state of the art prevents direct discriminator operation at frequencies much above this with delays in the millisecond region. If transmission frequencies above this are to be used (a likely requirement) and delays are large, the transmitted signal can be formed by amplitude or frequency-modulating an RF sine wave with low-pass random energy. The low-pass random waveform can then be synchronously (or nonsynchronously) detected at the receiver, and the detected signal fed into the delay-lock discriminator. If synchronous detection is to be used, it should be noted that the phase of the local oscillator used for detection must follow the phase modulation of the carrier caused by the reflection from the moving target.

It is also possible to devise modified versions of the delay-lock discriminator which can operate directly on FM deviated signals and use video delay lines. Nonsynchronous forms of the discriminator can provide delay estimates which are free of the possible ambiguities caused by the fine structure of the signal autocorrelation function. In essence, this type of operation is made possible by ignoring the fine structure and working only with the envelope of the autocorrelation function. If the delay-lock discriminator is to be used in an interferometer, the delay variations generally are in the microsecond region or less, and the frequency limitations of the delay lines become greatly relaxed.

Further work is being carried out on the problems of locking-on and unwanted target discrimination. For example, reflections from undesired targets can be discriminated against in both range and velocity by making the closed-loop bandwidth relatively small. Then, if the undesired target passes rapidly enough through the range of the target to which the discriminator is locked, the interfering transients which result occur too rapidly to affect materially the discriminator output. Adaptive filtering techniques seem to be appropriate here; one loop filter can be employed during the lock-on transient, and another can be used after lock-on is established.

List of Symbols

A = signal amplitude
B_s = signal bandwidth (cps)
c = velocity of light (or of sound if sonic propagation is considered)
$e = 2.718$
E = expected value of a random variable
f = frequency (cps)
f_0 = center frequency of the signal
$F(p)$ = loop filter transfer function
g = loop gain
$G_s(f)$, $G_n(f)$ = signal, noise, power, spectral densities
h = a constant
$H(p)$ = linearized equivalent transfer function
k = reference signal amplitude
$n(t)$ = input noise waveform
$n_e(t)$ = equivalent noise
p = complex frequency
p_0 = filter cutoff frequency in rad/sec
P_s, P_n = signal, noise-average power
r = input noise-to-signal power ratio
$R_s(\tau)$, $R_n(\tau)$ = signal, noise autocorrelation functions
$s(t)$ = signal waveform (unity power)
t = the variable time
$T(t)$ = delay
$\hat{T}(t)$ = estimate of delay
v = velocity of the target
$x(t)$ = multiplier output
y = a variable
α = relative amplitude of the delay estimate
$\delta(f)$ = Dirac delta function
ΔT = dynamic range of the discriminator
$\epsilon(t)$ = delay error $T(t) - \hat{T}(t)$
ϵ_T = threshold delay error
$\rho(\tau)$ = envelope of the normalized autocorrelation function
σ = standard deviation of a random variable
τ = a variable representing time
$\phi(t)$ = phase function
ω = angular frequency
$\omega_i(t)$ = instantaneous angular frequency

Acknowledgment

The authors would like to acknowledge the valuable comments and suggestions of their associates in Communications Research at Lockheed Missiles and Space Company. Special thanks are expressed to M. R. O'Sullivan for his interesting and rewarding comments.

Part VII
Cycle Slipping and Loss of Lock

Unlock Characteristics of the Optimum Type II Phase-Locked Loop

R. W. SANNEMAN, MEMBER, IEEE, AND J. R. ROWBOTHAM, MEMBER, IEEE

***Summary*—This paper describes a study of the statistical unlock characteristics of the commonly used optimum type II phase-locked loop taking into account the nonlinearity of the phase detector. The unlock behavior is investigated by simulating the loop with random noise inputs on a digital computer. A criterion of loop unlock is evolved using the phase portrait of the loop and consists of crossing of separatrices by loop error and a measure of loop error rate. Approximately 100 statistically independent trials were made for each of a number of combinations of initial phase error, initial frequency error, and input noise amplitude. From these computer runs, cumulative frequency distributions of unlock time and curves of mean time to unlock were obtained and are presented. The phase portrait of the loop containing time information also is included in the paper, along with a plot of capture characteristics for large initial frequency offsets.**

INTRODUCTION

PHASE-LOCKED loops are employed in certain low-frequency radio navigation systems, such as MUTNS[1] and in coherent communication receivers to filter a received signal and in some instances to determine the phase of a received signal with respect to a reference. The presence of noise in the received signal produces noise in the loop output which, if sufficiently large, can cause the loop to unlock because of the nonlinearity of the phase detector. In certain applications, loop unlock produces a serious if not catastrophic degradation in system performance. The deterministic acquisition and tracking behavior of the phase-locked loop, taking into account the nonlinear characteristics of the phase detector, were investigated by Viterbi[2] while the statistical aspects of frequency acquisition, taking into account the nonlinearity, were investigated by Frazier and Page.[3] In this paper, the results of an investigation of the statistics associated with unlock of a phase-locked loop are described. The statistical information was obtained from a simulation of the phase-locked loop on a digital computer.

The type II phase-locked loop derived by Jaffe and Rechtin,[4] which is optimum in the minimum mean-squared error sense to an input consisting of a step in frequency plus white Gaussian noise, is most commonly employed in receivers and is the type considered in the investigation. This loop will be referred to as the opti-

Manuscript received May 13, 1963. This work was sponsored by Motorola, Inc., Scottsdale, Ariz.

The authors are with Motorola Inc., Western Military Electronics Center, Scottsdale, Ariz.

[1] Multiple User Tactical Navigation System, a classified precise navigation system configured for the tactical military environment (developed by Motorola, Inc.).

[2] A. J. Viterbi, "Acquisition and tracking behavior of phase-locked loops," *Proc. Symp. on Networks and Feedback Systems*, Polytechnic Institute of Brooklyn, N. Y., pp. 583–619; April, 1960.

[3] J. P. Frazier and J. Page, "Phase-lock loop frequency acquisition study," IRE TRANS. ON SPACE ELECTRONICS AND TELEMETRY, vol. SET-8, pp. 210–227; September, 1962.

[4] R. Jaffe and E. Rechtin, "Design and performance of phase-lock circuits capable of near-optimum performance over a wide range of input signal and noise levels," IRE TRANS. ON INFORMATION THEORY, vol. IT-1, pp. 66–76; March, 1955.

Reprinted from *IEEE Trans. Aerosp. Navig. Electron.*, vol. ANE-11, pp. 15–24, Mar. 1964.

mum type II loop. Before the statistics of loop unlock could be determined, it was necessary to establish a criterion of loop unlock. This criterion was established by means of a phase portrait of the loop, obtained by digital simulation. Random noise inputs to the loop for determining unlock statistics were produced in the simulation by generating a pseudo-random binary sequence in the digital computer.

MODEL FOR SIMULATION

The optimum type II phase-locked loop derived by Jaffe and Rechtin[3] has an integration and lead in the compensation network resulting in a damping ratio of 0.707. A block diagram of the loop is shown in Fig. 1. The symbols used in the block diagram and the derivation below are defined as follows:

$x_i(t)$ = input signal plus noise
$x_o(t)$ = output signal plus noise
$c(t)$ = phase detector output
$\phi_i(t)$ = phase of input signal
$\phi_o(t)$ = phase of output signal
$\phi_i(o)$ = initial phase of input signal
$\phi_o(o)$ = initial phase of output signal
$n_i(t)$ = input noise
$n_s'(t)$ = first component of Rice equivalent representation of noise
$n_c'(t)$ = second component of Rice equivalent representation of noise
$n_s(t) = (K_pS_o/2)n_s'(t)$
$n_c(t) = (K_pS_o/2)n_c'(t)$
$\epsilon(t) = \phi_i(t) - \phi_o(t)$ = loop error
$\dot{\epsilon}(t)$ = time rate of change of $\epsilon(t)$
$a(t)$ = phase detector nonlinearity output
$v_2(t)$ = output of integration branch of compensation network
$v_1(t)$ = output of gain branch of compensation network
$v(t)$ = output of compensation network
$v_2(o)$ = initial output of integration branch of compensation network
ω_n = undamped resonant frequency of loop
K_p = phase detector gain
ω = carrier frequency of input and output signals
S_i = rms amplitude of input signal
S_o = rms amplitude of output signal.

The input signal plus noise can be expressed as

$$x_i(t) = \sqrt{2}\, S_i \sin(\omega t + \phi_i) + n_i(t). \tag{1}$$

The output signal can be expressed as

$$x_o(t) = \sqrt{2}\, S_o \cos(\omega t + \phi_o). \tag{2}$$

The phase detector is assumed to be an ideal multiplier with gain K_p. Its output is

$$c(t) = K_p x_i(t) x_o(t) = 2K_pS_oS_i \sin(\omega t + \phi_i)\cos(\omega t + \phi_o) + \sqrt{2}\, K_pS_on_i(t)\cos(\omega t + \phi_o). \tag{3}$$

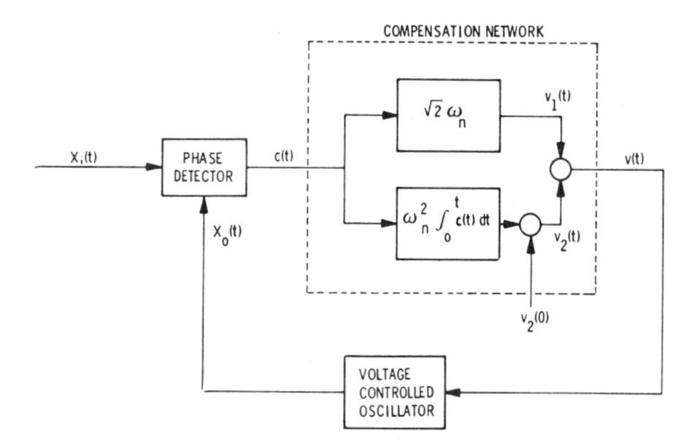

Fig. 1—Optimum type II phase-locked loop.

The noise $n_i(t)$ is assumed to be band-passed white noise and can be represented by the following Rice equivalent:[5]

$$n_i(t) = n_s'(t)\sin\omega t + n_c'(t)\cos\omega t. \tag{4}$$

$n_s'(t)$ and $n_c'(t)$ are independent noises having low-pass spectra and mean-squared values equal to the mean-squared values of $n_i(t)$. Substituting this expression for $n_i(t)$ into the expression for $c(t)$,

$$\begin{aligned} c(t) = {} & 2K_pS_oS_i\sin(\omega t + \phi_i)\cos(\omega t + \phi_o) \\ & + \sqrt{2}\, K_pS_on_s'(t)\sin\omega t\cos(\omega t + \phi_o) \\ & + \sqrt{2}\, K_pS_on_c'(t)\cos\omega t\cos(\omega t + \phi_o). \end{aligned} \tag{5}$$

Making use of the identities for the product of sines and cosines,

$$\begin{aligned} c(t) = {} & K_pS_oS_i\sin(\phi_i - \phi_o) + K_pS_oS_i\sin(2\omega t + \phi_i + \phi_o) \\ & - \frac{K_pS_o}{\sqrt{2}} n_s'(t)\sin\phi_o + \frac{K_pS_o}{\sqrt{2}} n_s'(t)\sin(2\omega t + \phi_o) \\ & + \frac{K_pS_o}{\sqrt{2}} n_c'(t)\cos\phi_o + \frac{K_pS_o}{\sqrt{2}} n_c'(t)\cos(2\omega t + \phi_o). \end{aligned} \tag{6}$$

The phase-locked loop is a low-pass filter with respect to phase and is assumed to have a bandwidth that is much narrower than the carrier frequency, ω. For this reason, the noise components containing the frequency 2ω will not pass the loop and may be deleted from the input without affecting the output. The phase detector output then becomes

$$c(t) = K_pS_oS_i\sin(\phi_i - \phi_o) - \frac{K_pS_o}{\sqrt{2}} n_s'(t)\sin\phi_o + \frac{K_pS_0}{\sqrt{2}} n_c'(t)\cos\phi_o. \tag{7}$$

The factor K_p converts the dimension of power into radians of phase. By letting

$$K_pS_oS_i = 1, \tag{8}$$

[5] S. A. Rice, "Mathematical analysis of random noise," *Bell Sys. Tech. J.*, vol. 54, No. 7.

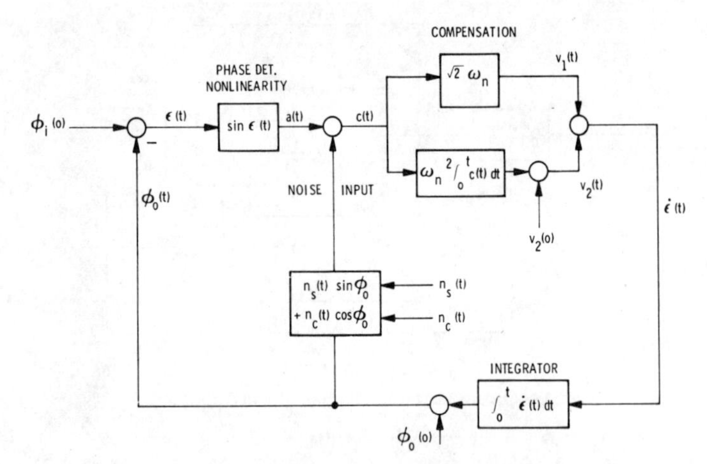

Fig. 2—Optimum type II phase-locked loop in terms of phase.

$$\frac{K_pS_o}{\sqrt{2}}\, n_s'(t) = -\, n_s(t), \tag{9}$$

and

$$\frac{K_pS_o}{\sqrt{2}}\, n_c'(t) = n_c(t). \tag{10}$$

The dimension of $c(t)$ in (7) is radians of phase, where $n_s(t)$ and $n_c(t)$ have the dimension of radians of phase. Eq. (7) becomes

$$c(t) = \sin(\phi_i - \phi_o) + n_s(t)\sin\phi_o + n_c(t)\cos\phi_o. \tag{11}$$

This equation expresses the output of the phase detector in terms of the phase of the loop input signal, the phase of the voltage controlled oscillator (VCO) output, and the random noise input. Noting this expression, the form of the compensation in Fig. 1, and the fact that the VCO is an integrator of phase, a block diagram of the optimum type II phase-locked loop can be constructed and is depicted in Fig. 2. In the simulation, the phase of the input signal, $\phi_i(t)$, is assumed to be a constant, $\phi_i(o)$. The compensation filter output therefore equals the time rate of change of the error, $\epsilon(t)$.

Initial conditions $\phi_o(0)$ and $v_2(0)$ are inserted at the output of each integrator.

Digital Simulation

The optimum type II phase-locked loop was simulated on an LGP-30 digital computer by performing digitally the computations shown in Fig. 2. On a given iteration n, each computation shown is performed starting with the error $\epsilon[(n-1)P]$ from the previous iteration and ending with the error $\epsilon[nP]$, where P is the time for one iteration or the sample period. The integrations shown are performed by adding to a stored sum, the product of the integrand on a given iteration and the sample period, P. The simulation becomes more accurate as the sample period P diminishes but the computation time per unit of real time increases making it necessary to set the sample period, P, to effect a compromise between simulation accuracy and consumed computer time. One check of the accuracy of the digital simulation for a given sample period is to compare the frequency response of the actual loop with that of the digitally simulated loop. In order to make this comparison, a linearized model of the loop is assumed, where the phase detector characteristic is represented as a gain of unity. This model is accurate for small loop error where the error is approximately equal to the sine of the error.

The transfer function from either phase or noise input to the integrator output of the linear model of the actual loop is

$$Y(s) = \frac{\dfrac{\sqrt{2}}{\omega_n}s + 1}{\dfrac{s^2}{\omega_n{}^2} + \dfrac{\sqrt{2}}{\omega_n}s + 1}. \tag{12}$$

The frequency response is

$$Y(\omega) = \frac{1 + j\sqrt{2}\dfrac{\omega}{\omega_n}}{1 - \left(\dfrac{\omega}{\omega_n}\right)^2 + j\sqrt{2}\dfrac{\omega}{\omega_n}}. \tag{13}$$

The magnitude and phase of the frequency response are shown in Fig. 3 as the curves labeled "actual."

A corresponding frequency response can be derived for the digitally simulated loop. The integrations indicated are performed by rectangular integration according to the rule,

$$y_o[nP] = Py_i[nP] + y_o[(n-1)P] \tag{14}$$

where y_i is the integrand, y_o is the integral, n is the iteration number and P is the sampling period. For P sufficiently small, (14) provides a close approximation of the integration

$$y_0 = \int_o^{nP} y_i(t)dt.$$

The Z transform of (14) is

$$Y_o(z) = PY_i(z) = z^{-1}Y_o(z) \tag{15}$$

and the pulse transfer function of the integration is

$$\frac{Y_o(z)}{Y_i(z)} = P\frac{z}{z-1} \tag{16}$$

where $Y_o(z)$ is the Z transform of $y_o(t)$ and $Y_i(z)$ is the Z transform of $y_i(t)$. The pulse transfer block diagram of the linearized digitally simulated loop is shown in Fig. 4. It should be noted that the phase detector characteristic is assumed to be of unity gain only for the purpose of comparison of the frequency response of the actual and simulated loops. The phase detector nonlinearity, as shown in Fig. 2, was included in the actual digital simulation.

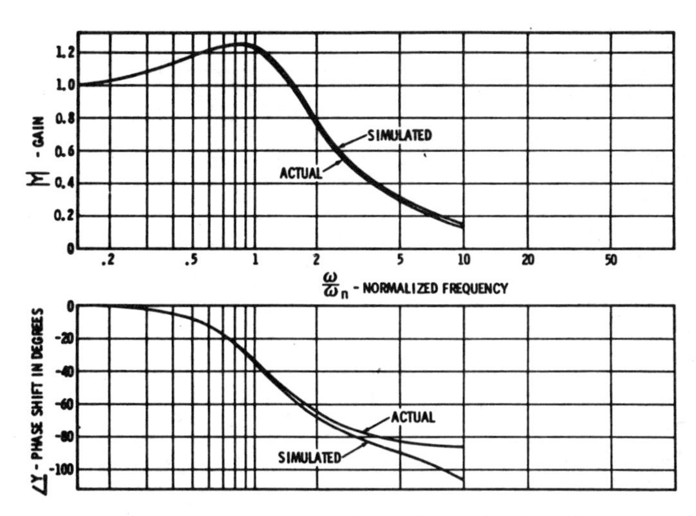

Fig. 3—Frequency response of actual and simulated loops.

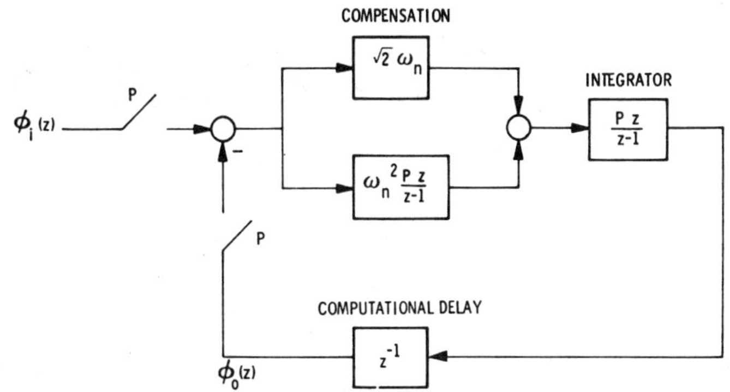

Fig. 4—Pulse transfer function block diagram of the linearized simulated loop.

The z^{-1} term in the feedback path represents a one iteration delay in the computations required in the compensation and integration. The pulse transfer function of the simulated loop is

$$Y(z) = \frac{k(k + \sqrt{2})z - \sqrt{2}\,k}{z^2 + (k^2 + \sqrt{2}\,k - 2)z - \sqrt{2}\,k + 1} \tag{17}$$

where $k = \omega_n P$.

The frequency response is obtained by making the substitution, $z = e^{jP\omega}$, in (17) and using series approximations for the exponential terms. The frequency response of the digitally simulated loop is

$$Y(\omega) + \frac{N_r + jN_i}{D_r + jD_i}, \tag{18}$$

where

$$N_r = 1 - \frac{k + \sqrt{2}}{2} ka^2 + \frac{k + \sqrt{2}}{24} k^3 a^4 \tag{19}$$

$$N_i = (k + \sqrt{2})a - \frac{k + \sqrt{2}}{6} k^2 a^3 \tag{20}$$

$$D_r = 1 - \left(1 + \frac{\sqrt{2}}{2} k + \frac{k^2}{2}\right) a^2 + \left(\frac{7}{12} + \frac{\sqrt{2}}{24} k + \frac{k^2}{24}\right) k^2 a^4 \tag{21}$$

$$D_i = (\sqrt{2} + k)a - \left(1 + \frac{\sqrt{2}}{6} k + \frac{k^2}{6}\right) ka^3 \tag{22}$$

and

$$a = \frac{\omega}{\omega_n}.$$

A sample period P of $1/16\omega_n$ was selected for the simulation on the basis of the response given by (18)–(22). For this value of P corresponding to a value of k of 1/16, the amplitude and phase frequency responses of the digitally simulated loop are plotted in Fig. 3 as the curves labeled "simulated." It is seen that the gain of the simulation closely approximates that of the actual loop, while the phase shift differs by no more than 20 degrees up to ω/ω_n equal to 10. As a check, sinusoids of various frequencies were fed into the simulated loop having a sampling period P of $(1/16)(1/\omega_n)$. The responses of the loop at these frequencies checked those plotted in Fig. 4.

Phase Plane Diagram and Loop Capture Characteristics

In determining a suitable criterion of loop unlock, the phase portrait of the optimum type II phase-locked loop is utilized. Viterbi[1] has included in his work such a diagram which was not available when this study was begun. For this reason and because it was felt that a phase portrait with time information included would be of use in phase-locked loop analysis, data for a phase portrait of the loop was obtained by means of the loop digital simulation discussed in the previous paragraph. The phase portrait is plotted in Fig. 5 and consists of a family of trajectories of error $\epsilon(t)$, and normalized error rate $\dot{\epsilon}(t)/\omega_n$, describing the response of the loop to any set of initial values of $\epsilon(t)$ and $\dot{\epsilon}(t)/\omega_n$ along the trajectories. $\epsilon(t)$ is plotted along the abscissa and $\dot{\epsilon}(t)/\omega_n$ along the ordinate. The responses were obtained by starting the simulation described in the previous paragraphs at various pairs of initial conditions and allowing the simulation to continue until $\epsilon(t)$ and $\dot{\epsilon}(t)/\omega_n$ came to rest at zero. Values of $\epsilon(t)$ and $\dot{\epsilon}(t)/\omega_n$ were recorded at given times.

The phase plane diagram has two families of curves, phase trajectories and the isocrones. The isocrones, shown as dashed lines in Fig. 5, are time reference curves such that the distance between isochrones on any phase trajectory represents a specified increment of time. This time increment is equal to $1/4\omega_n$ seconds in the phase plane diagram of Fig. 5.

There are two families of isochrones continuous within regions of π radians along the abscissa that are repeated symmetrically throughout the diagram. One

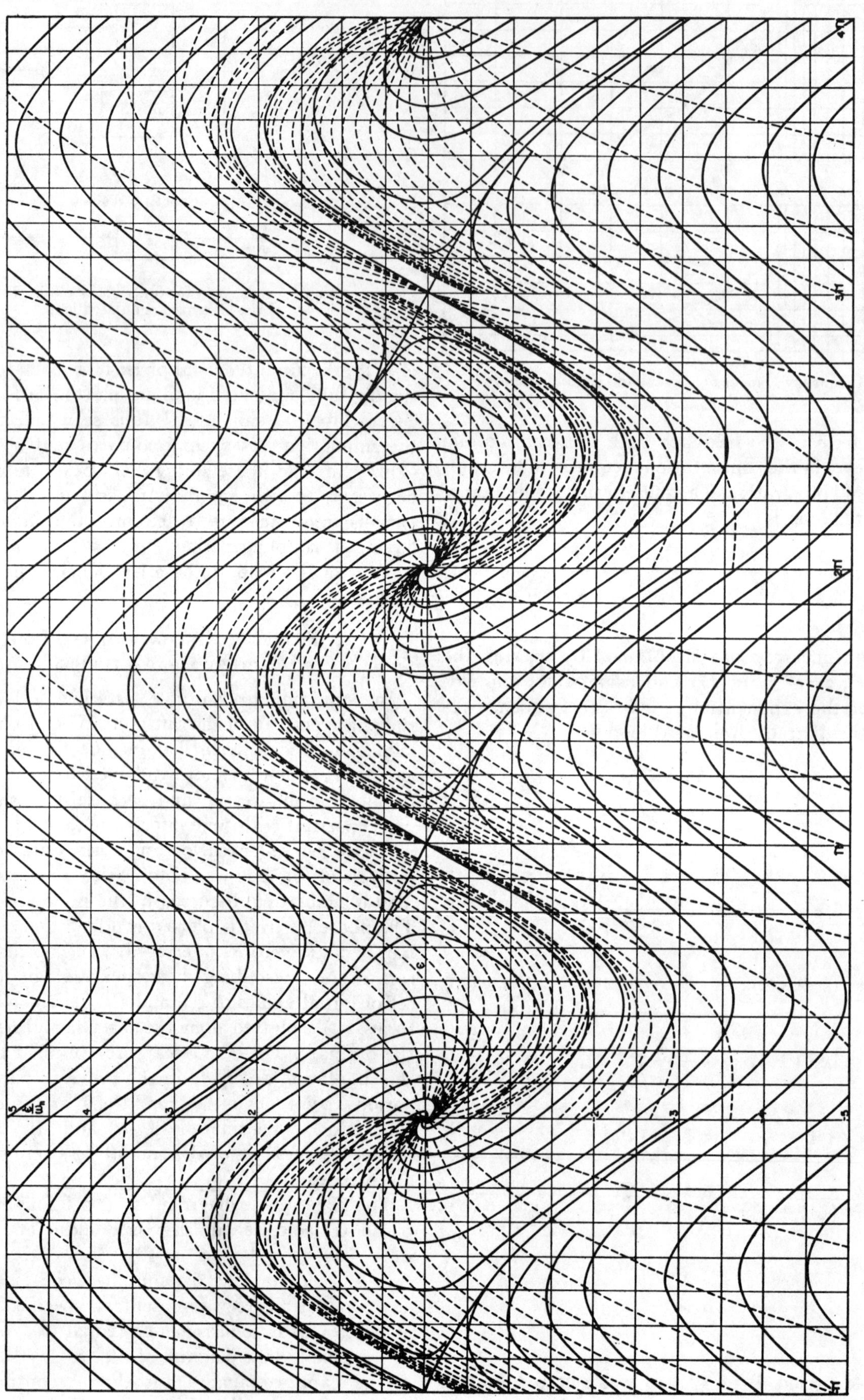

Fig. 5—Phase portrait of loop.

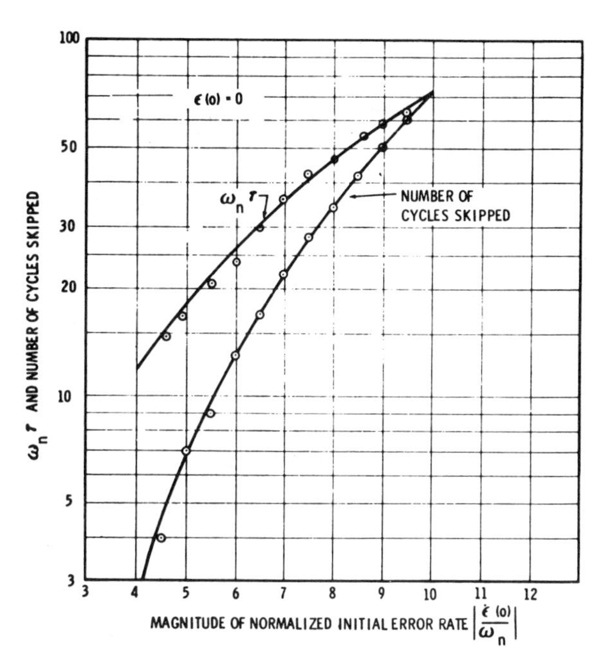

Fig. 6—Capture characteristics of loop.

family, referenced to the ordinate axis at $\epsilon=0$, is continuous in the region $0\leq\epsilon\leq\pi$ for values of $\dot{\epsilon}/\omega_n$ above the separatrix running from the singularity at $\epsilon=\pi$ radians back towards the equilibrium point at $\epsilon=0$. The other family of isochrones in this region is referenced to the negative ordinate axis at $\epsilon=\pi$ radians. Inspection of the phase plane diagram shows the symmetry of these families of isochrones as they are drawn in adjacent regions.

Location of the initial conditions of error and error rate defines the starting point of a phase trajectory. A trajectory not starting on one of the plotted trajectories can be readily sketched by interpolation between adjacent trajectories on the diagram. The trajectory direction is always to the right for error rates that are positive. Complete transient response of the loop to an initial error and error rate can be determined from the trajectory, where time is determined by counting the number of isochrones intersected by the trajectory. In cases where the trajectory extends beyond the region where one family of isochrones is continuous, some interpolation of the time increment at the boundary is usually necessary. As an example, for the initial conditions of $\dot{\epsilon}(o)/\omega_n=3.6$ and $\epsilon(o)=-\pi/4$ radians, the phase error will settle to within 5 degrees of the equilibrium state in approximately $33/8\omega_n$ seconds. The maximum phase error during the transient is 90 degrees.

Fig. 6, showing the capture characteristics of the optimum phase-locked loop, can be used in those cases where the initial normalized error rate is beyond the range of the phase plane diagram. The capture characteristics shown assume that there is no initial phase error. The settling time, τ, shown in Fig. 6 is defined as the time in which the phase error settles to within 0.2 radian of its equilibrium state.

Generation of Random Inputs

The noise inputs to the simulated loop shown in Fig. 2, $n_s(t)$ and $n_c(t)$, are to be independent Gaussian noises having the spectral characteristics of the circuits preceding the phase-locked loop. One such noise can be obtained by generating a pseudorandom binary sequence[5] in the digital computer by means of a shift register arrangement and digitally filtering the sequence. The digital filter output has the spectral characteristics of the filter and is nearly Gaussian provided that the filter bandwidth is considerably less than the width of the spectrum of the sequence. In addition, the repetition length of the sequence must be longer than measurement times in the simulation if the noise is to appear random. A second noise, which is independent of the first over times of interest, can be generated by half adding certain bits of the shift register to obtain another pseudorandom binary sequence. This second sequence also is passed through a digital filter to obtain the desired spectral characteristics.

Two pseudorandom binary sequences were generated in the digital computer by the shift register arrangement shown in Fig. 7. A 30-bit shift register was programmed since the digital computer word length was 30 bits plus a sign and spacer bit. A maximal length sequence was generated by half adding the 23rd and 30th bit, feeding it back to the input of the word, and shifting all bits of the word one bit to the right. The first binary sequence was taken from the half adder output. The second was obtained by half adding the 17th and 23rd bits. The sequences generated have a length of $2^{30}-1$ or approximately 10^9 bits and closely approximate random binary sequences having a mean of 1/2 and independence between bits. The generated sequences contain only ones and zeros which are converted into sequences having a zero mean and an rms value of A by converting one to A and zero to $-A$. The time between bits is taken as Q in the simulation and each bit effectively is held for this period to form a pseudorandom square wave sequence of period Q and amplitude A or $-A$. The power spectrum of this sequence has a form $(\sin L\omega/L\omega)^2$ where L is a constant and is relatively flat near zero frequency.

Referring to Fig. 2, the noises on $\epsilon(t)$ and $v_2(t)$ determine the unlock behavior of the loop and it is only necessary that these noises be accurately simulated regardless of how $n_s(t)$ and $n_c(t)$ are generated. The noise transfer characteristics of the loop from $n_s(t)$ and $n_c(t)$ to both $v_2(t)$ and $\epsilon(t)$ have cutoff frequencies approximating the loop bandwidth. The noises on $\epsilon(t)$ and $v_2(t)$ have the correct spectral characteristics if $n_s(t)$ and $n_c(t)$ are pseudorandom pulse sequences with no digital filtering but with a bandwidth wide enough so that the spectrum of the pulse sequence is relatively flat over the band-

[5] T. G. Birdsall and M. P. Ristenbatt, "Introduction to Linear Shift-Register Generated Sequences," University of Michigan Research Institute, Ann Arbor, Tech. Rept. No. 90; October, 1958.

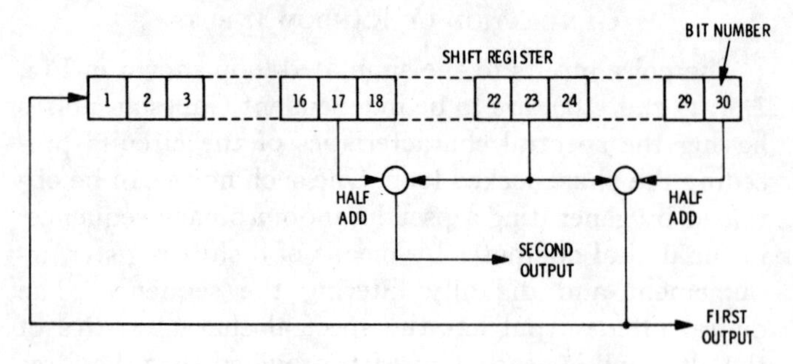

Fig. 7—Pseudorandom sequence generator.

width of the phase-locked loop. The probability distributions of the noises on $\epsilon(t)$ and $v_2(t)$ are approximately Gaussian. It was found experimentally that a bandwidth ratio of 10 to 1 between the spectrum of the pulse sequence and the bandwidth of the phase-locked loop is sufficient to yield nearly Gaussian distributions on these noises.

The autocorrelation function of the pulse sequences representing $n_s(t)$ and $n_c(t)$ is

$$\begin{aligned} R(\tau) &= \lim_{T\to\infty} \frac{1}{2T}\int_{-T}^{T} n_s(t)n_s(t+\tau)dt \\ &= A^2\left(1-\frac{|\tau|}{Q}\right) \qquad \text{for } |\tau| \le Q \\ &= \text{zero, elsewhere.} \end{aligned} \tag{23}$$

The spectral density of the pulse sequence is

$$\begin{aligned} G(\omega) &= \frac{2}{\pi}\int_0^\infty R(\tau)\cos\omega t d\tau \\ &= \frac{QA^2}{\pi}\left[\frac{\sin\frac{Q\omega}{2}}{\frac{Q\omega}{2}}\right]^2 . \end{aligned} \tag{24}$$

The normalized spectral density is

$$\frac{G(\omega)}{G(o)} = \left[\frac{\sin\frac{Q\omega}{2}}{\frac{Q\omega}{2}}\right]^2 = \left[\frac{\sin m\frac{\omega}{\omega_n}}{m\frac{\omega}{\omega_n}}\right]^2 \tag{25}$$

where

$$G(o) = \frac{QA^2}{\pi} = \text{spectral density at zero frequency} \tag{26}$$

and

$$m = \frac{Q\omega_n}{2} . \tag{27}$$

In order that the bandwidth of the noise spectrum be 10× the bandwidth of the phase-locked loop, m was chosen equal to 1/16 so that Q equals $1/8\omega_m$. The spectral density of the noise for m equal to 1/16 is plotted in Fig. 8. The noise bandwidth of the optimum type II phase-locked loop is[3] 1.89 ω_n and it is seen from Fig. 8 that the spectrum of the input noise is relatively flat over this bandwidth. It should be noted also that the period of the pulse sequence Q is twice the loop sample period $1/16\omega_m$, and one noise sample is held for two loop sample periods.

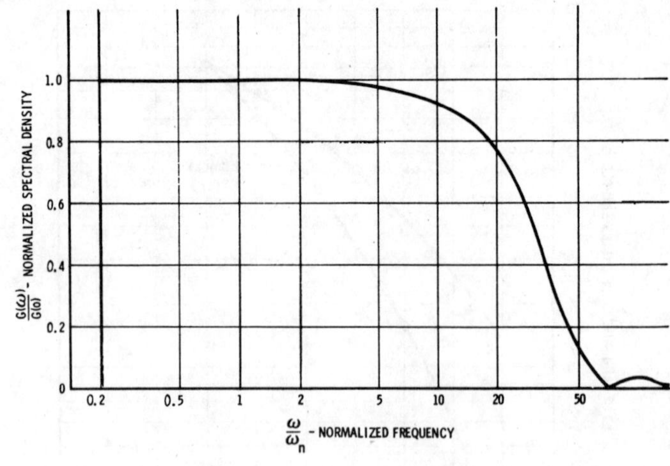

Fig. 8—Normalized spectral density of pseudorandom sequence.

Criterion of Loop Unlock

Before the statistics of time to unlock can be investigated, a criterion of loop unlock must be established. This is essential to any study of loop unlock behavior. It is logical to conjecture that loop unlock is dependent upon a combination of error and error rate and therefore that the phase plane diagram can be used in defining unlock. A problem arises because the phase plane diagram defines the response of the system to initial conditions and does not take into account random or nonrandom driving functions of any kind. Noise represents a random driving function.

The phase plane diagram of the loop is plotted in Fig. 5 and the region around one of the equilibrium states with separatrices is sketched in Fig. 9. A logical choice of an unlock criterion would be when a trajectory defined by some measures of error and error rate crosses the separatrices for a given time. Referring to Fig. 2, consider the error $\epsilon(t)$ and normalized error rate $\dot{\epsilon}(t)/\omega_n$ as the measures of error and error rate to be used. The error $\epsilon(t)$ is filtered by the loop and contains only low-frequency components but the error rate $\dot{\epsilon}(t)$ contains the high-frequency components of the input noise passing through the branch of the compensation having gain $\sqrt{2}\omega_n$ and is of considerable amplitude. A trajectory of $\epsilon(t)$ and $\dot{\epsilon}(t)/\omega_n$ on the phase plane would jitter back and forth across the separatrices at a relatively high rate and any single crossing of the separatrices does not indicate that $\epsilon(t)$ and $\dot{\epsilon}(t)/\omega_n$ tend toward another equilibrium state in the long term. Other choices for the error and rate measures are the error $\epsilon(t)$ and the error rate $v_2(t)/\omega_n$. The error rate $v_2(t)$ is filtered to a narrow

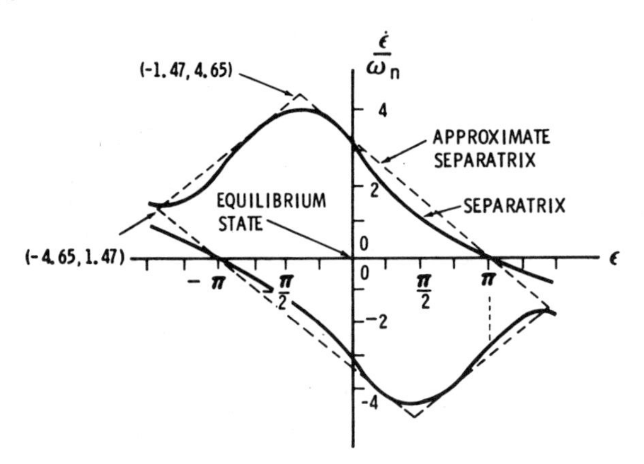

Fig. 9—Separatrices of phase portrait.

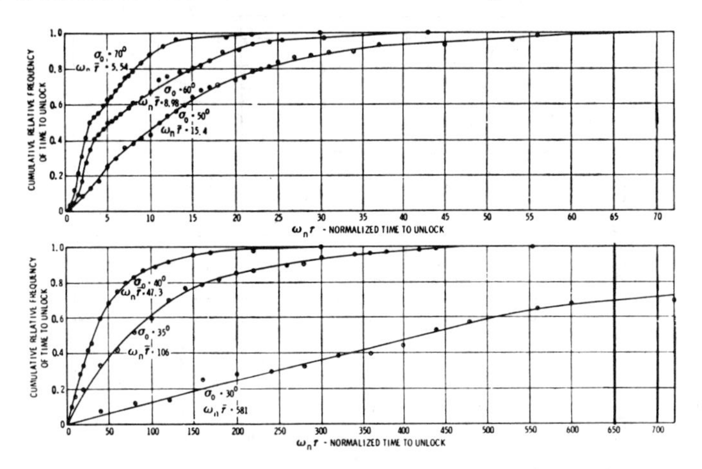

Fig. 10—Cumulative frequency distributions of normalized time to unlock for $\epsilon(0)=0$ and $\dot{\epsilon}_0(0)/\omega_n=0$.

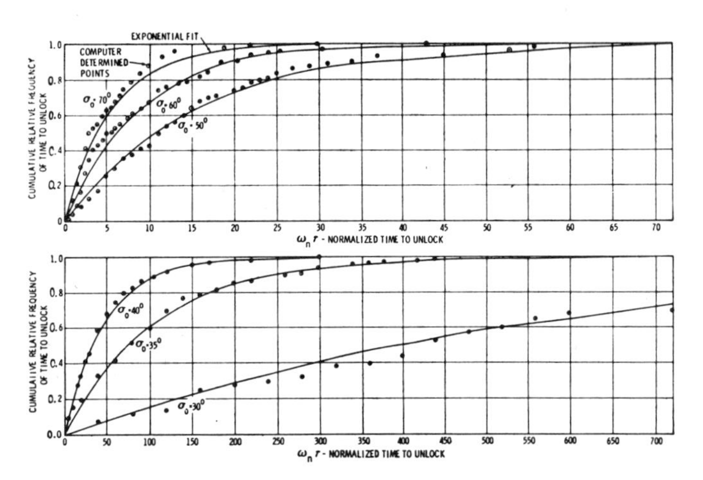

Fig. 11—Exponential fits of cumulative frequency distributions for $\epsilon(0)=0$ and $\dot{\epsilon}_0(0)/\omega_n=0$.

bandwidth but does not contain the narrow-band noise component that comes through the upper branch of the compensation from $a(t)$. This component of noise affects the long term behavior of the loop and hence unlock. Measures of error and error rate which eliminate all of the above objections are the error $\epsilon(t)$ and the normalized error rate

$$\frac{v_2(t)+\sqrt{2}\,\omega_n a(t)}{\omega_n}=\frac{v_2(t)}{\omega_n}+\sqrt{2}\,a(t).$$

The rate $v_2(t)+\sqrt{2}\omega_n a(t)$ contains the desired low-frequency components of noise. In addition, it becomes equal to $\dot{\epsilon}(t)$ when the noise inputs $n_s(t)$ and $n_c(t)$ are removed and it does not change instantaneously when the noise inputs are removed instantaneously. These three factors make this rate the logical choice for use in defining unlock on the phase portrait. A criterion of loop unlock was conjectured as follows. Unlock occurs when the error $\epsilon(t)$ and the error rate $\dot{\epsilon}_o(t)/\omega_n$ defined by $\dot{\epsilon}_o(t)/\omega_n=v_2(t)/\omega_n+\sqrt{2}a(t)$ cross a separatrix of the phase plane diagram for longer than a time $3/\omega_n$. The time factor is introduced into the criterion to prevent small, high-frequency components on either $\epsilon(t)$ or $\dot{\epsilon}_o(t)$ from indicating loop unlock. To facilitate computer mechanization, the separatrices were approximated by the straight lines as shown in Fig. 9. The coordinates of the end points are shown in the figure.

Thirty trials were made using the simulated loop shown in Fig. 2 with random inputs of sufficient magnitude to cause crossing of the separatrices by $\epsilon(t)$ and $\dot{\epsilon}_o(t)/\omega_n$. In each case it was found that when a separatrix was crossed for more than a time $3/\omega_n$, the trajectory described by $\epsilon(t)$ and $\dot{\epsilon}_o(t)/\omega_n$ remained outside the separatrices for a period much longer than $3/\omega_n$, hovering around another equilibrium point, verifying that the criterion for unlock is valid.

Determination of Unlock Statistics

Computer simulations of the operation of the optimum type II phase-locked loop starting with a given set of initial conditions and given levels of noise input were made using the methods of digital simulation and noise generation given in the previous paragraphs. In the simulation, the loop sample interval, P, was set equal to $1/16\omega_n$ and the pulse interval of the pseudo-random pulse sequence Q was set equal to $1/8\omega_n$. Approximately 100 trials were made for each of a variety of combinations of initial conditions and input noise levels. The initial conditions are the initial values of $\epsilon(t)$ and $\dot{\epsilon}_o(t)/\omega_n$, each of which was varied between ± 2 radians. The levels of the noise inputs were taken as those that would give rms noise outputs of an optimum type II phase-locked loop having a linear rather than sinusoidal phase detector characteristic. This rms noise output is

$$\sigma_0=\sqrt{G(o)\int_0^\infty |\,Y(\omega)\,|^2 d\omega}. \tag{28}$$

Substituting (26) and (27) into (28) and performing the indicated operations,

$$\sigma_o = 1{:}03A\sqrt{Q\omega_n}. \tag{29}$$

The quantity $Q\omega_n$ was set equal to 1/8 so that the pulse amplitude $A=2.79\sigma_o$.

For each trial, the simulation was allowed to continue until the unlock criterion was satisfied—that is, when a

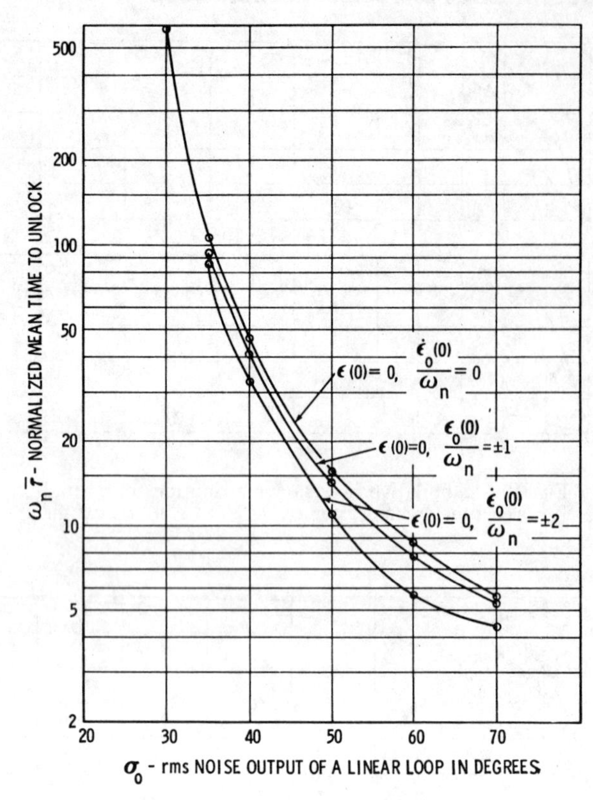

Fig. 12—Normalized mean time to unlock vs rms noise output.

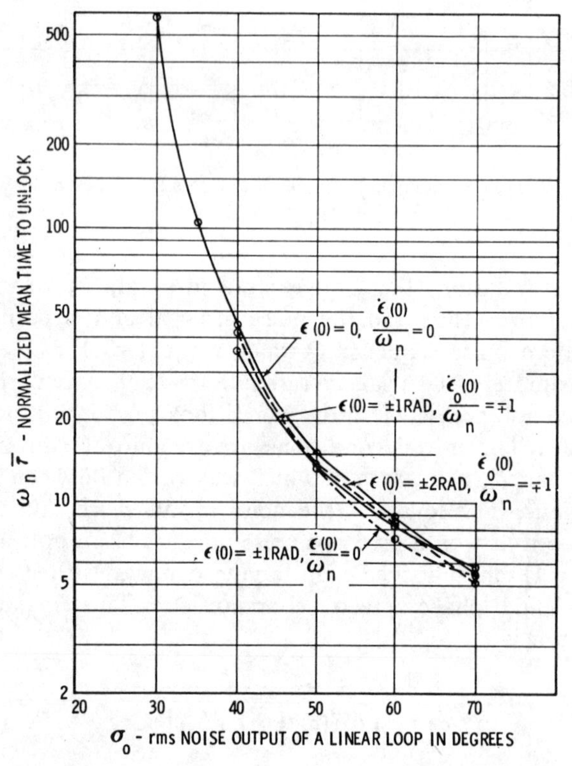

Fig. 13—Normalized mean time to unlock vs rms noise output.

trajectory defined by $\epsilon(t)$ and $\dot{\epsilon}_o(t)/\omega_n$ crossed a separatrix in the phase plane. The time from the start of a trial until unlock was recorded and a new trial then was begun. The random noise sequence never repeated during the entire study, so that each trial was statistically independent of all other trials. The times to unlock for each

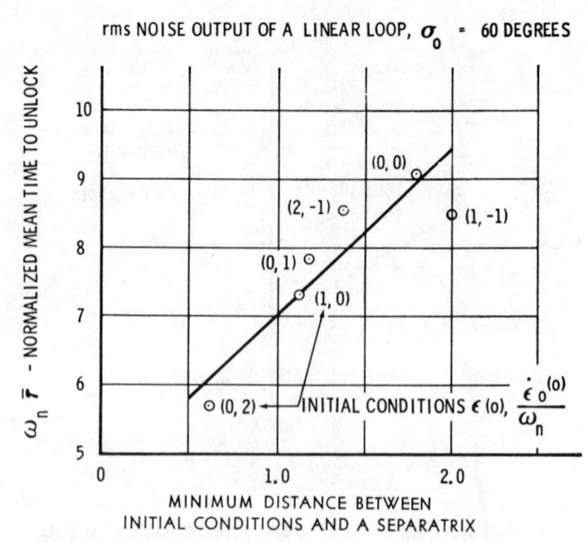

Fig. 14—Normalized mean time to unlock vs minimum distance between initial conditions and a separatrix.

set of approximately 100 trials for given initial conditions and input noise amplitude were plotted into cumulative relative frequency distributions of normalized time to unlock, $\omega_n\tau$. The distributions for zero initial conditions and various input noise amplitudes are plotted in Fig. 10. An ordinate of the distribution is obtained by taking the total number of unlocks occurring in the time corresponding to the abscissa and dividing by the total number of trials. The resulting frequency distributions approximate the cumulative probability distribution of time to unlock. The mean normalized time to unlock was calculated for each set of trials by summing the unlock times and dividing by the number of trials. For each distribution, it was found that a function of the form

$$P(\omega_n\tau) = 1 - e^{-(\omega_n\tau)/\omega_n\bar{\tau}} \tag{30}$$

closely fits the distribution, where

$P(\omega_n\tau)$ = cumulative relative frequency of time to unlock, $\omega_n\tau$
$\doteq$ cumulative probability of time to unlock
$\bar{\tau}$ = mean time to unlock.

In Fig. 11, this function is shown plotted along with the data obtained from the computer simulation for zero initial conditions.

Curves of the normalized mean time to unlock vs the rms noise output of a linear loop having the same input noise as the simulated loop are plotted in Figs. 12 and 13. It would be expected that the mean time to unlock would decrease as the minimum distance between the initial conditions and a separatrix in the phase plane decreases. Indeed, this is the case and is demonstrated by the plot in Fig. 14 for each set of initial conditions used in the study. The ordinate of a circled point is the normalized mean time to unlock for that point while the abscissa is the minimum distance between a separatrix and the initial conditions for the point. Although correlation exists between the mean time to unlock and the minimum distance between the initial conditions, the spread of mean time to unlock is not very great for the range of initial conditions tried.

The following two examples are given to illustrate the use of the data presented.

Example I

An optimum type II phase-locked loop having a noise bandwidth of 0.5 cps is locked onto an input signal so that no phase or frequency error exists in the absence of noise. The S/N ratio at the input of the loop is such that the rms noise at the loop output calculated for a linear loop is 50 degrees. It is desired to find the mean time to unlock and the probability of unlock in 10 seconds.

The undamped resonant frequency of the optimum type II phase-locked loop is related to the noise bandwidth by $\omega_n = 1.89B_L$ where

ω_n = undamped resonant frequency in radians/second
B_L = loop noise bandwidth in cps.

Thus $\omega_n = (1.89)\ (0.5) = 0.945$ radians/second. From Fig. 12, $\omega_n\bar{\tau} = 15.5$ for $\epsilon(o) = o$, $\dot{\epsilon}_o(o)/\omega_n = o$ and $\sigma_o = 50°$ and the mean time to unlock is $\bar{\tau} = 15.5/\omega_n = 16.4$ seconds.

For a time to unlock τ of 10 seconds, $\omega_n\tau = (0.945)(10) = 9.45$. From Fig. 10 for $\sigma_o = 50°$ and $\omega_n\tau = 9.45$, the cumulative relative frequency of time to unlock in 10 seconds, which is equivalent to the probability of unlock in 10 seconds, is 0.44.

Example II

The S/N ratio at the input of an optimum type II phase-locked loop having a noise bandwidth of 0.002 cps is to be limited to a value which will make the probability of unlock less than 0.01 in 600 seconds. It is desired to find the rms noise on the output of the loop, calculated on a linear basis, necessary to meet these conditions assuming zero initial phase and frequency error with no noise input.

The probability of 0.01 is too low to be used in Fig. 11. Instead, (30) is utilized.

$$P(\omega_n\tau) = 1 - e^{-(\omega_n\tau)/\omega_n\bar{\tau}}.$$

From (30), $\tau/\bar{\tau} = 0.01$ for $P(\omega_n\tau) = 0.01$ and $\bar{\tau} = \tau/0.01 = 60{,}000$ seconds. Also, $\omega_n = 1.89B_L = (1.89)(0.002) = 0.00378$ radians/second and $\omega_n\bar{\tau} = (0.00378)(60{,}000) = 227$. From Fig. 12 for $\epsilon(o) = o$, $\dot{\epsilon}_o(o)/\omega_n = o$ and $\omega_n\bar{\tau} = 227$, the maximum output noise, σ_o, to meet the above conditions is 32 degrees.

Conclusions

A study of the statistics associated with unlock of an optimum type II phase-locked loop with a signal plus random noise input is described in this paper. The loop was simulated on a digital computer with the random input noise simulated by generating pseudo-random binary sequences in a shift register programmed into the computer. A criterion of loop unlock was evolved with the aid of a phase portrait of the loop which was obtained by means of the digital simulation. Simulations of loop operation with a signal plus random input noise were made and approximately 100 statistically independent trials were run for each set of a number of values of initial phase error, initial frequency error, and input noise amplitude. Cumulative frequency distributions of time to unlock and curves of mean time to unlock were obtained for the various input conditions. The cumulative frequency distributions could be approximated closely by the function $1 - e^{-a}$ where a is the ratio of time to unlock to mean time to unlock. Correlation was found between the mean time to unlock and the minimum distance in the phase plane diagram between the initial conditions and a separatrix.

Acknowledgment

The authors wish to express their appreciation to E. J. Groth, Chief Engineer, Motorola, Inc., and other members of the Preliminary Design and Analysis Group for their many helpful suggestions.

Simplified Formula for Mean Cycle-Slip Time of Phase-Locked Loops With Steady-State Phase Error

ROBERT C. TAUSWORTHE, MEMBER, IEEE

Abstract—Previous work has shown that the mean time from lock to a slipped cycle of a phase-locked loop is given by a certain double integral. Accurate numerical evaluation of this formula for the second-order loop has proved extremely vexing because the difference between exponentially large quantities is involved. This article simplifies the general formula to avert this problem, provides a useful approximation to a needed conditional expectation, and produces an asymptotic formula for the mean slip time that is moderately accurate even at low loop SNR (less than 7 dB) and small steady-state phase errors (less than 0.3 rad). The approximations extend to higher order loops, as well.

Paper approved by the Communication Theory Committee of the IEEE Communications Society for publication without oral presentation. It presents the results of one phase of research carried out at the Jet Propulsion Laboratory, California Institute of Technology under Contract NAS 7-100, sponsored by the National Aeronautics and Space Administration. Manuscript received December 4, 1970; revised November 15, 1971.

The author is with the Jet Propulsion Laboratory, California Institute of Technology, Pasadena, Calif.

I. Introduction

A NUMBER of authors (e.g., [1], [2]) have derived a formula for the mean time to a first cycle slip of a phase-locked loop by considering an associated two-dimensional differential equation

Reprinted from *IEEE Trans. Commun.*, vol. COM-20, pp. 331–337, June 1972.

describing the time-dependent loop error statistics. Lindsey [2] gives a recursive formula for higher moments of this slip time, extended to loops of a generalized type and arbitrary order. Lindsey's formula, however, contains in it an unknown function, a time-dependent conditional expectation that, for loops of degree greater than or equal to two, has never been evaluated. The author [3] was able to approximate this function in a way that gave extremely good agreement with experimental data, so that, formally at least, the cycle-slip statistics were considered solved.

Evaluation of the formula for the mean first-slip time (a double integral) to give actual numerical answers has, until now, presented quite a problem. Theoretical manipulations, such as series expansions or an asymptotic formula, have not been forthcoming. Direct evaluation by numerical integration involved differences of exponentially large quantities and require very careful programming in multiprecision arithmetic with very small error tolerances to obtain reasonable answers at even moderate loop signal-to-noise ratios (SNR) [4]. Consequently, such evaluations were very slow and costly to obtain, and of use only at lower loop SNR.

This article presents a method in which a much-reduced precision program can be used to obtain the mean first-cycle slip time for a loop of arbitrary degree tracking at a specified SNR and steady-state phase error. It also presents a simple approximate formula that is asymptotically tight at higher loop SNR. As a limiting case, the asymptotic formula agrees, for a first-order loop at zero offset, with the known formula given in Viterbi [1].

II. The Mean Cycle-Slip Time Formula

To introduce the notation to be used, we start with the usual representation [1] of the phase error process $\phi = \phi(t)$,

$$\dot{\phi} = \dot{\theta} - AKF(p)\left[\sin\phi + \frac{n(t)}{A}\right], \tag{1}$$

where $\theta = \theta(t)$ is the instantaneous phase offset between the incoming sinusoid and the VCO at rest, A is the rms signal amplitude, $n(t)$ is a baseband white Gaussian noise with one-sided spectral density N_0, K is the loop gain, $p = d/dt$ is the Heaviside operator, and $F(s)$ is the loop filter transfer function.

To illustrate the method, we shall work out the second-order loop result when $\dot{\theta} = \Omega_0$. The loop filter is assumed to take the form

$$F(s) = \frac{1 + \tau_2 s}{1 + \tau_1 s} = F + \frac{(1 - F)}{1 + \tau_1 s}. \tag{2}$$

The coefficient F is defined in terms of the two time constants, $F = F(\infty) = \tau_2/\tau_1$. When $F = 1$, the loop degenerates to one of the first order.

A state-vector representation $\boldsymbol{\phi} = (\phi_1, \phi_2)$ of the process of particular use in this case is

$$\phi_1 = \phi$$

$$\phi_2 = -\frac{AK(1 - F)}{1 + \tau_1 p}\left\{\sin\phi + \frac{n(t)}{A}\right\}. \tag{3}$$

The time derivative $\dot{\boldsymbol{\phi}}$ of $\boldsymbol{\phi}$ is described then by

$$\dot{\phi}_1 = \Omega_0 - AKF\left[\sin\phi + \frac{n(t)}{A}\right] + \phi_2$$

$$\dot{\phi}_2 = -\frac{AK(1 - F)}{\tau_1}\left[\sin\phi + \frac{n(t)}{A}\right] - \frac{1}{\tau_1}\phi_2. \tag{4}$$

The last equation shows quite clearly that the representation of the process by $\boldsymbol{\phi}$ is a vector Markov process, because incremental changes in $\boldsymbol{\phi}$ are dependent only upon present values of the system state and a white Gaussian disturbing function. Extension to high-order loops alters (3) and (4) only by the inclusion of extra state variables ϕ_i, one for each additional pole in $F(s)$.

Based on this model, use of Lindsey's [2] general formula for the mean time to a first cycle slip yields the following expression to be studied:

$$T = \frac{4}{N_0(KF)^2}\int_{-2\pi}^{2\pi}\int_{x}^{2\pi}[u(y - \phi_0) - D]e(x, y)\,dy\,dx. \tag{5}$$

The function $u(\cdot)$ is the ordinary unit-step, and $\phi_0 = \phi(0)$. The constant D is given by

$$D = \int_{\phi_0}^{2\pi} e(0, y)\,dy \Big/ \int_{-2\pi}^{2\pi} e(0, y)\,dy \tag{6}$$

and the function $e(x, y)$ is

$$e(x, y) = \exp\left[\frac{4}{N_0(KF)^2}\int_{y}^{x} E(\dot{\phi} \mid \phi)\,d\phi\right]. \tag{7}$$

Finally, the conditional expectation required in (7) is

$$E(\dot{\phi} \mid \phi) = \int_{0}^{\infty}\int_{-\infty}^{+\infty} \dot{\phi}\, p(\phi_2, t \mid \phi)\,d\phi_2\,dt. \tag{8}$$

A simple interpretation of $E(\dot{\phi} \mid \phi)$ is possible: it is the average velocity at a given value of ϕ, averaged over the entire slip. There is no direct dependence on the t parameter.

An expression may also be written for higher order moments of the first-slip time [2, Sec. XII]; however, it is necessary first to evaluate a function $E_{\phi_2}(\dot{\phi}, \bar{t} \mid \phi)$, where the expectation is over ϕ_2 only and $\bar{t}$ results from an application of the mean-value theorem to satisfy

$$E(\dot{\phi}t^{m-1} \mid \phi) = E_{\phi_2}(\dot{\phi}, \bar{t} \mid \phi)E(t^{m-1} \mid \phi) \tag{9}$$

when averaged over the slip. The value m is the slip-time moment index. For all but the first-order moment, there is a $\bar{t}$ dependence to be accounted for.

In retaining the generality of his result, Lindsey [2] neglected to point out the intrinsic simplicity of (8), which readily lends itself toward measurement and accurate estimation of T.

As the reader may appreciate, the formal evaluation of $E(\dot{\phi} \mid \phi)$ is not possible without knowing the probability density of the error process; however, approximation is possible. Lindsey [5], for example, has tried the orthogonality principle, but the resulting values for T do not agree well with observed data. (The orthogonality principle does, however, produce a useful accurate estimate of the steady-state phase error statistics.)

III. Approximation of $E(\dot{\phi} \mid \phi)$

Another method [3] approximates the conditional expectation $E(\dot{\phi} \mid \phi)$, and therefore T, to a high experimental accuracy. The form of $\dot{\phi}$ given in (4) suggests

$$E(\dot{\phi} \mid \phi) = -AKF \sin \phi + B(\phi), \tag{10}$$

where $B(\phi)$ is a function to be determined, say as a series, $B(\phi) = b_0 + b_1\phi + \cdots$. Experimental evidence [6], [7] indicates that the first two terms of $B(\phi)$ are in strong dominance.

The two coefficients b_0 and b_1 may therefore be solved for by observing the behavior of ϕ during times when ϕ is yet small enough to apply linear theory and have stationary statistics. According to linear theory, once the initial transients have died (i.e., the loop has reached steady state),

$$\phi = \frac{\Omega_0}{AK} - \frac{1}{A \cos \phi_{ss}} \int_0^\infty l(\tau)n(t-\tau)\, d\tau$$

$$L(s) = \int_0^\infty l(\tau)e^{-st}\, d\tau. \tag{11}$$

Here $L(s)$ represents the well-known loop transfer function linearized about the steady-state phase error point ϕ_{ss}. As we have agreed to start observing the slip beginning at steady-state lock, we take $\phi_0 = \phi_{ss}$. We shall find it convenient in what follows to define

$$\hat{\rho} = \frac{4A}{N_0KF}$$

$$r = AKF\tau_2$$

$$h(x) = \frac{1}{AKF}\int^x E(\dot{\phi} \mid \phi)\, d\phi \tag{12}$$

it being understood that the indefinite integral is to be evaluated only at its upper limit. The parameter r is very closely related to the second-order loop damping factor, as $\zeta = r^{1/2}(1 + F/r)/2$.

To obtain two equations from which b_0 and b_1 may be extracted, values for $E(\dot{\phi})$ and $E(\phi\dot{\phi})$ can be computed in two ways: 1) directly from (10), 2) by using the linear loop theory valid at low angular disturbances,

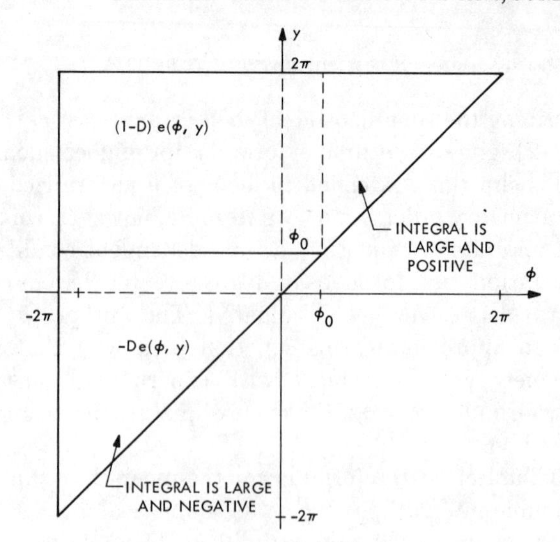

Fig. 1. The region of integration for T, showing subregions with positive and negative integrand.

$$E(\dot{\phi}) = E_\phi[E(\dot{\phi} \mid \phi)] \approx -AKFE(\sin \phi) + b_0 + b_1\phi_0$$

$$\approx -\frac{1}{A \cos \phi_0} E\left[\int_0^\infty l'(\tau)n(t-\tau)\, d\tau\right] = 0$$

$$E(\phi\dot{\phi}) = E_\phi[E(\dot{\phi}\phi \mid \phi)]$$

$$\approx -AKFE(\phi \sin \phi) + b_0\phi_0 + b_1E(\phi^2)$$

$$\approx E\left\{\frac{1}{A^2 \cos^2 \phi_0} \int_0^\infty \int_0^\infty l'(\tau_1)l'(\tau_2)\right.$$

$$\left.\cdot n(t-\tau_1)n(t-\tau_2)\, d\tau_1\, d\tau_2\right\}$$

$$= \frac{N_0 l^2(0)}{4A^2 \cos^2 \phi_0} = \frac{N_0}{4A^2 \cos^2 \phi_0}[\lim_{s\to\infty} sL(s)]^2. \tag{13}$$

Solution for the unknowns yields the following approximate form for $E(\dot{\phi} \mid \phi)$:

$$E(\dot{\phi} \mid \phi) = AKF\left[\sin \phi_0 - \sin \phi + \frac{(1-F)\cos \phi_0}{1 + r \cos \phi_0}(\phi - \phi_0)\right]. \tag{14}$$

Defining the coefficients

$$u \triangleq \sin \phi_0 - v\phi_0$$

$$v \triangleq \frac{(1-F)\cos \phi_0}{1 + r\cos \phi_0} \tag{15}$$

puts $h(x)$ and the function $e(x, y)$ into the form

$$h(x) = ux + \cos x + \tfrac{1}{2}vx^2$$

$$e(\phi, y) = \exp[\hat{\rho}h(\phi) - \hat{\rho}h(y)]. \tag{16}$$

At this point, the values of T are formally attainable. However, because of the exponential character of (16), the integration required to evaluate T must be sufficiently precise not to allow significant error when its positive and negative parts are combined.

IV. Integration of T

As can be seen in (5), the integration for T has positive and negative components that are exponentially related to $\hat{\rho}$. Attempts at evaluating the formula directly at large $\hat{\rho}$ show that the difference between these positive and negative parts is generally much less than the accuracy of the numerical methods conveniently available. Hence, precise evaluations of T have been extremely difficult and costly to obtain.

We now develop a formula that excludes integration over negative integrands, and thus retains the accuracy of the numerical integration method.

Fig. 1 shows the region of integration for T, as given in (5). The formula for T can alternatively be written as

$$T = \frac{\hat{\rho}}{AKF}\left\{-D \int_{-2\pi}^{\phi_0} \int_{-2\pi}^{y} e(\phi, y)\, d\phi\, dy + (1 - D) \int_{\phi_0}^{2\pi} \int_{-2\pi}^{y} e(\phi, y)\, d\phi\, dy\right\}. \tag{17}$$

The latter integral over $(-2\pi, y)$ can be broken into two parts $(-2\pi, \phi_0)$ and (ϕ_0, y) as indicated by the dashed line in Fig. 1. When the integrals (6) for D and $1 - D$ are inserted, (17) can be manipulated to give

$$T = \frac{\hat{\rho}}{AKF}\left\{D \int_{-2\pi}^{\phi_0} \int_{y}^{\phi_0} e(\phi, y)\, d\phi\, dy + (1 - D) \int_{\phi_0}^{2\pi} \int_{\phi_0}^{y} e(\phi, y)\, d\phi\, dy\right\}. \tag{18}$$

This rearrangement of the integration has dismissed the negative portion of T altogether, so that (18) can be evaluated straightforwardly by numerical integration routines having only moderate precision. The new region of integration is shown in Fig. 2.

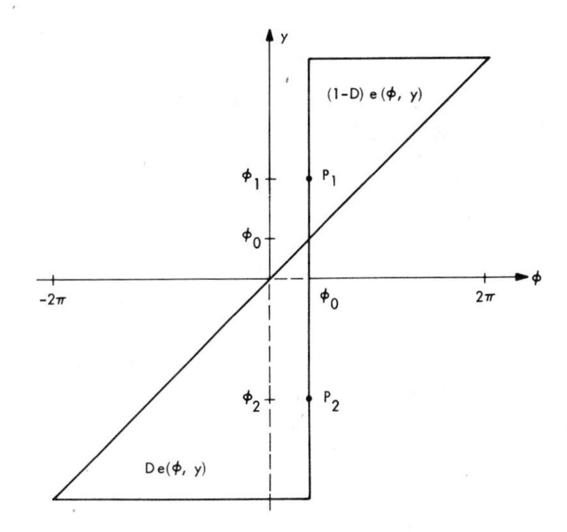

Fig. 2. Transformed integration formula for T having positive integrand only.

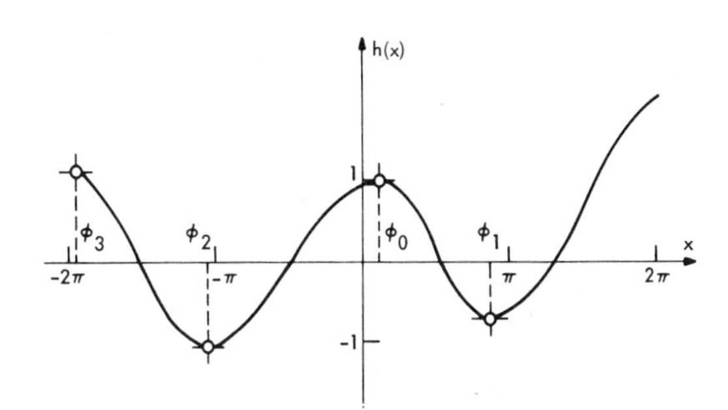

Fig. 3. The exponent function $h(x)$.

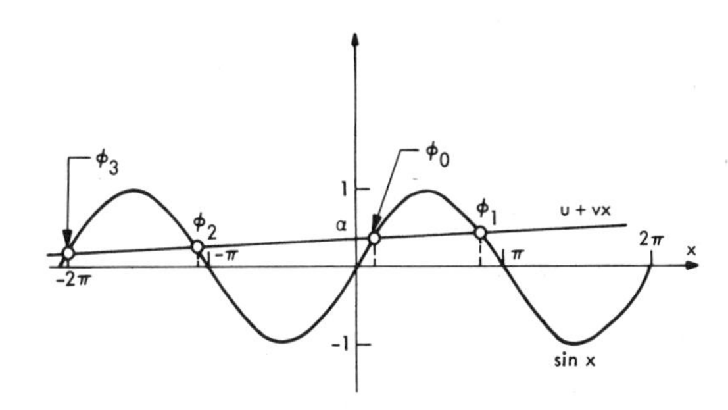

Fig. 4. Solution method for maxima and minima of $h(x)$.

V. Asymptotic Evaluation of T

Due to the exponential character of $e(\phi, y)$ as a function of $\hat{\rho}$, the dominant contributions to T occur in neighborhoods of the relative maxima of $e(\phi, y)$, and thereby, at the relative maxima of the function $h(\phi) - h(y)$. At such a point, call it $P^* = (\phi^*, y^*)$, the coordinates ϕ^* and y^* are maxima and minima of $h(x)$, respectively, on the region $-2\pi \leq \phi \leq y \leq 2\pi$. These are illustrated in Fig. 3, denoted ϕ_0, ϕ_1, ϕ_2, and ϕ_3; the maximum nearest the origin and both minima are solutions to $h'(\phi_i) = 0$, viz.,

$$u + v\phi_i = \sin \phi_i \tag{19}$$

shown in Fig. 4. It is of interest to point out at such points that $E(\dot{\phi} \mid \phi) = 0$; hence, these are precisely the lock point ϕ_0 and the first-slip saddle points ϕ_1 and ϕ_2.

The remaining maximum of interest ϕ_3 may either be a solution to $h'(\phi) = 0$, if one exists (when $u \geq 2\pi v$), or $\phi_3 = -2\pi$ otherwise.

There are only two relative maximum points within the new bounds of integration, labeled $P_1 = (\phi_0, \phi_1)$ and $P_2 = (\phi_0, \phi_2)$ in Fig. 2. The Taylor series of $h(x)$ about the maximum points P_1 and P_2 allow the integrand, say in the neighborhood of P_1, to be written as

$$e(\phi, y) = e(\phi_0, \phi_1) \exp\left\{\tfrac{1}{2}\hat{\rho}h''(\phi_0)(\phi - \phi_0)^2 + \cdots\right\} \cdot \exp\left\{-\tfrac{1}{2}\hat{\rho}h''(\phi_1)(y - \phi_1)^2 + \cdots\right\}. \tag{20}$$

A similar expression arises at P_2. Since $h''(x) = (v - \cos x)$ is negative at ϕ_0 and positive at ϕ_1, integration of $e(\phi, y)$ about P_1 is very nearly related to the normal probability integral. When $\hat{\rho}$ is even moderately large, the quadratic nature of (20) extends far enough that the y-integral limits in (18) can be replaced by $(-\infty, +\infty)$ with only a small error. The limits on ϕ can similarly be replaced by $(-\infty, 0)$ and $(0, \infty)$ in the two regions. The result is an asymptotic formula for T, which we denote as $\hat{T}$;

$$\hat{T} = \frac{\pi}{AKF(\cos\phi_0 - v)^{1/2}} \cdot \left[\frac{De(\phi_0, \phi_2)}{(v - \cos\phi_2)^{1/2}} + \frac{(1 - D)e(\phi_0, \phi_1)}{(v - \cos\phi_1)^{1/2}}\right]. \quad (21)$$

The validity of this formula depends only on the one condition that $\hat{\rho}$ is large enough that any deviations from the quadratic character postulated in (20) occur when $e(\phi, y) \ll e(\phi_0, \phi_{1,2})$ and thus cause negligible errors.

The same technique can be applied to the computation of D. We find, for example,

$$\int_{\phi_0}^{2\pi} e(0, y)\, dy \approx e(0, \phi_1) \exp\left[\frac{\hat{\rho}}{2} h''(\phi_0)\phi_0^2\right] \cdot \int_{-\infty}^{+\infty} \exp\left[-\frac{\hat{\rho}}{2} h''(\phi_1)(y - \phi_1)^2\right] dy$$

$$= e(0, \phi_1) \exp\left[\frac{\hat{\rho}}{2} h''(\phi_0)\phi_0^2\right]\left[\frac{2\pi}{\hat{\rho} h''(\phi_1)}\right]^{1/2} \quad (22)$$

so that the asymptotic value of D is

$$\hat{D} = \frac{e(0, \phi_1)[v - \cos\phi_2]^{1/2}}{e(0, \phi_1)[v - \cos\phi_2]^{1/2} + e(0, \phi_2)[v - \cos\phi_1]^{1/2}}. \quad (23)$$

Insertion of this $\hat{D}$ into (21) gives the final asymptotic formula for $\hat{T}$. It is convenient to introduce at this point the parameters

$$\rho = \frac{A^2}{N_0 b_L} = \hat{\rho}\left(\frac{r + F}{r + 1}\right)$$

$$b_L = \frac{r(r + 1)}{4\tau_2(r + F)}. \quad (24)$$

The parameter b_L is the loop baseband single-sided noise bandwidth, as referred to $\phi_0 = 0$, and ρ represents the baseband SNR inside this bandwidth. The actual passband around the carrier is twice this wide because of the zero-beat heterodyning effect of the loop. Hence, we may take

$$w_L = 2b_L \quad (25)$$

as the loop carrier-bandwidth. The expressions for $\hat{T}$ are

$$\hat{T}w_L = \left(\frac{r + 1}{r + F}\right)\frac{\pi(1 - \hat{D})e(\phi_0, \phi_1)}{[(\cos\phi_0 - v)(v - \cos\phi_1)]^{1/2}}$$

$$= \left(\frac{r + 1}{r + F}\right)\frac{\pi\hat{D}e(\phi_0, \phi_2)}{[(\cos\phi_0 - v)(v - \cos\phi_2)]^{1/2}}. \quad (26)$$

The angles ϕ_0, ϕ_1, and ϕ_2 are all solutions to $h'(\phi) = u + v\phi - \sin\phi = 0$ and are independent of $\hat{\rho}$.

VI. Evaluation for the First-Order Loop With Frequency Offset

The first-order loop results in our theory by setting $F = 1$. Then, we have $v = 0$, $u = \sin\phi_0$, and $\hat{\rho} = \rho$. Further, the solutions to $h'(\phi) = 0$ are

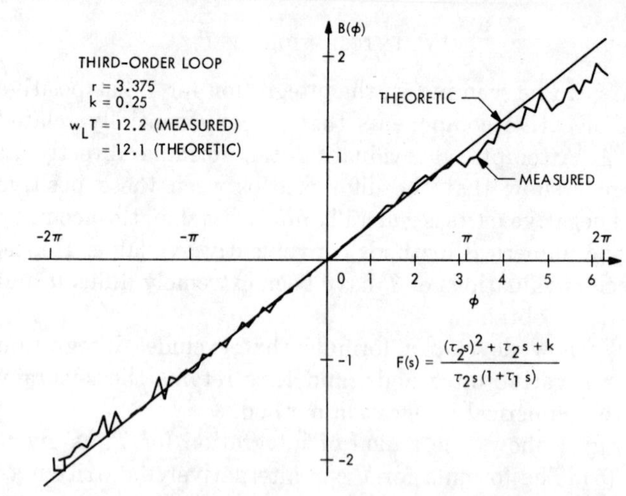

Fig. 5. The function $B(\phi)$ for a third-order loop at unity SNR.

$$\phi_0 = \arcsin(u)$$

$$\phi_1 = \pi - \phi_0$$

$$\phi_2 = -\pi - \phi_0. \quad (27)$$

These values inserted into (23) and (26) yield

$$\hat{T}w_L = \frac{\pi \exp[2\rho(\phi_0 \sin\phi_0 + \cos\phi_0)]}{2\cos\phi_0 \cosh(\pi\rho\sin\phi_0)}$$

$$= \frac{\pi}{2}e^{2\rho}, \qquad \phi_0 = 0. \quad (28)$$

The limiting case $\phi_0 = 0$, is the correct asymptotic result contained in Viterbi [1].

VII. Extension to Higher Order Loops

To extend the formulas for T and $\hat{T}$ to loops of higher order it is necessary merely to recompute $E(\dot{\phi} \mid \phi)$ and to verify that any assumptions are experimentally justified. By reviewing the steps, it is clear that $E(\dot{\phi} \mid \phi)$ can be written in the same form as (10), perhaps with more terms in $B(\phi)$. But again, experimental evidence, such as that shown in Fig. 5 for the third-order loop, shows that it is sufficient only to keep b_0 and b_1 in significant consideration. Moreover, the two equations given in (13) are general, applying to all loops in which higher order b_i may be neglected.

As a conclusion, since (14) represents $E(\dot{\phi} \mid \phi)$ for the third-order loop—and probably those of higher order, as well, although this has not been verified—then the formulas for T and $\hat{T}$ derived for the second-order loop also hold for the third-order loop (and probably the others) as well. The only reinterpretation necessary is that ϕ_0 results from accelerations as well as frequency offset terms of the input wave.

VIII. Evaluation and Comparison of Results

To calibrate the effectiveness of the asymptotic formula and the accuracy of the improved integral formula, test cases were run on a digital computer. The results were also compared with previous experimental data [6]–[10] for further verification that the approximation $E(\dot{\phi} \mid \phi)$

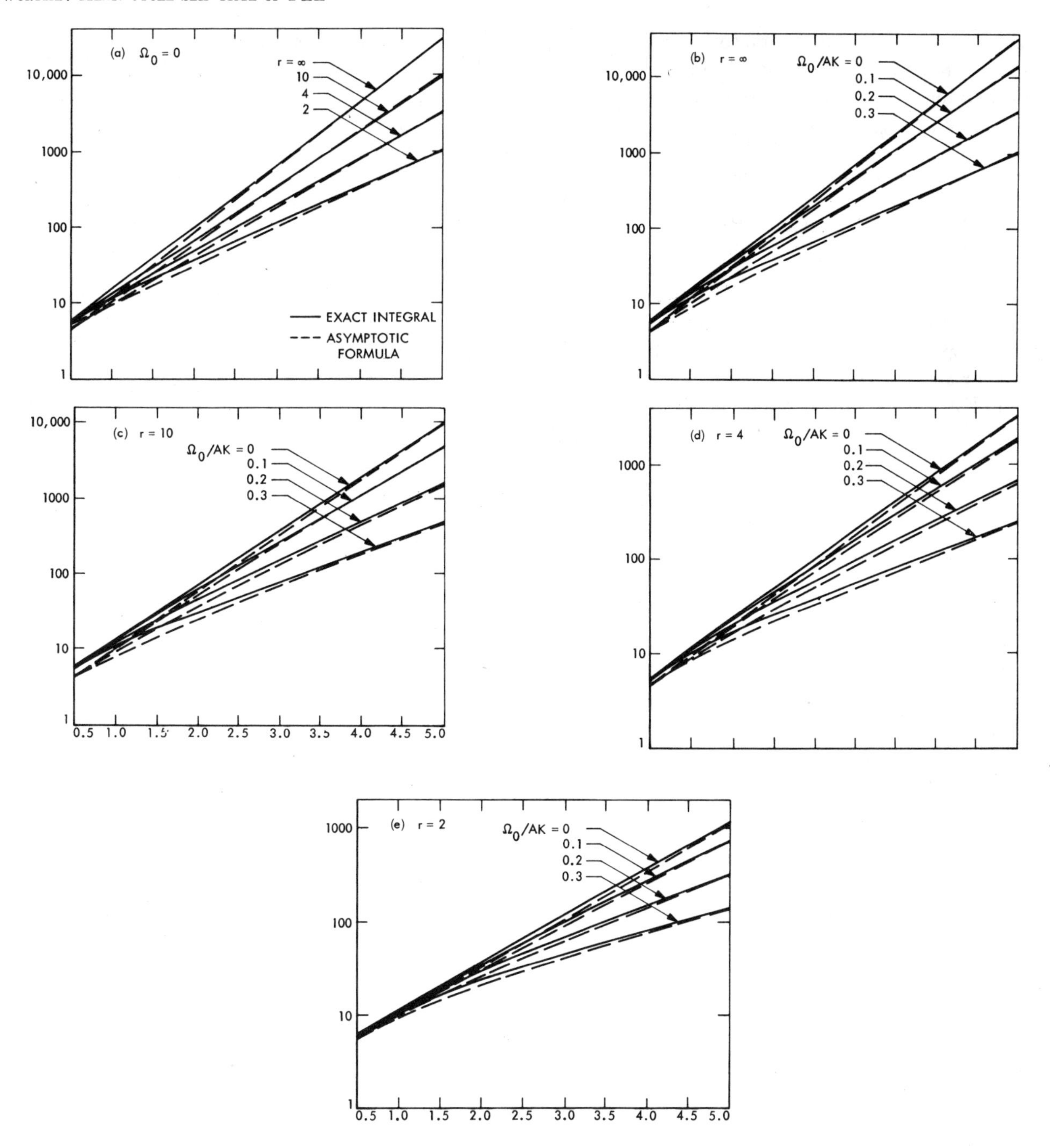

Fig. 6. Comparison of the integral and asymptotic formulas for mean slip time T as a function of loop SNR ρ for various loop damping and detuning parameters r, Ω_0/AK. Vertical scale: Tw_L; horizontal scale: loop SNR, ρ.

was accurate. Computations show, as illustrated in Fig. 6(a)–(e), that the asymptotic formula is accurate for the range of parameters considered within 35 percent, for ρ as low as $\rho = 0.5$, and increases in accuracy to about 10 percent at $\rho = 5$.

The curves in Fig. 6(a)–(e) clearly illustrate the asymptotic exponential character of both T and $\hat{T}$ as a function of ρ. The exponential coefficient ξ can be found from the asymptotic behavior of $\hat{T}$ and fit to a convenient approximating form, such as

$$\xi = \left(\frac{r+1}{r+F}\right)[h(\phi_0) - h(\phi_1)]$$

$$\phi_0 = \arcsin(u)$$

$$\phi_1 = \pi - 2.6414/r + 1.8318/r^2 - u(1.0156 - 1.1295/r + 1.0/r^2)$$

$$\phi_2 = -\pi + 2.6414/r + 1.8318/r^2 - u(1.0156 - 1.5266/r + 1.6825/r^2). \tag{29}$$

A somewhat more accurate asymptotic form for $\hat{T}$ than (26) is suggested by extrapolating the known exact slip-time formula for the first-order loop with $\phi_0 = 0$, viz.,

$$Tw_L = \pi^2 \rho I_0^2(\rho) \tag{30}$$

to include the effects of ϕ_0 and r. Such a form is

$$\hat{T}w_L = \left\{\pi^2\left(\frac{r+1}{r+F}\right)\frac{\xi(1-\hat{D})\cdot}{[(\cos\phi_0 - v)(v - \cos\phi_1)]^{1/2}}\right\}\rho I_0^2(\tfrac{1}{2}\xi\rho). \tag{31}$$

Evaluations of this last formula show errors no larger than about 15 percent for $r \geq 2$, $\phi_0 \leq 0.3$, and $\rho \geq 1$. For the first-order loop, the result is exact at $\phi_0 = 0$, and within 9 percent for $\rho \geq 1$ and $\phi_0 \leq 0.3$; for $\rho \geq 5$, it lies within 4 percent. For the second-order loop, $r \geq 2$ and $\phi_0 = 0$, the results are within 11 percent at $\rho = 1$, and taper to 3 percent at $\rho = 5$. For all $\phi_0 \leq 0.3$, $r \geq 2$, the formula is correct to 10 percent for $\rho \geq 5$.

REFERENCES

[1] A. J. Viterbi, *Principles of Coherent Communications*. New York: McGraw-Hill, 1966, pp. 96–103.

[2] W. C. Lindsey, "Nonlinear analysis of generalized tracking systems," *Proc. IEEE*, vol. 57, pp. 1705–1722, Oct. 1969.

[3] R. C. Tausworthe, "Cycle slipping in phase-locked loops," *IEEE Trans. Commun. Technol.*, vol. COM-15, pp. 417–421, June 1967.

[4] W. C. Lindsey, private communication.

[5] ——, "Nonlinear analysis and synthesis of generalized tracking systems, part II," Electron. Sci. Lab., Univ. Southern California, Rep. USCEE-342, Apr. 1969.

[6] D. Sanger and R. C. Tausworthe, "An experimental study of the first-slip statistic of the second-order phase-locked loop," Jet Propul. Lab., Space Programs Sum. 37–43, vol. 4, Feb. 1967.

[7] J. Holmes, "A simulation of first-slip times versus the static phase offsets for first- and second-order phase locked loop," Jet Propul. Lab., Space Programs Sum. 37–58, vol. 2, pp. 29–32, July 1969.

[8] R. W. Sanneman and J. R. Rowbotham, "Unlock characteristics of the optimum type II phase-locked loop," *IEEE Trans. Aerosp. Navig. Electron.*, vol. ANE-11, pp. 15–24, Mar. 1964.

[9] B. M. Smith, "The phase-locked loop with filter: Frequency of slipping cycles," *Proc. IEEE* (Lett.), vol. 54, pp. 296–297, Feb. 1966.

[10] F. J. Charles and W. C. Lindsey, "Some analytical and experimental phase-locked loop results for low signal-to-noise ratios," *Proc. IEEE*, vol. 54, pp. 1152–1166, Sept. 1966.

Part VIII
Phase-Locked Oscillators

A Study of Locking Phenomena in Oscillators*

ROBERT ADLER†, ASSOCIATE, I.R.E.

Summary—**Impression of an external signal upon an oscillator of similar fundamental frequency affects both the instantaneous amplitude and instantaneous frequency. Using the assumption that time constants in the oscillator circuit are small compared to the length of one beat cycle, a differential equation is derived which gives the oscillator phase as a function of time. With the aid of this equation, the transient process of "pull-in" as well as the production of a distorted beat note are described in detail.**

It is shown that the same equation serves to describe the motion of a pendulum suspended in a viscous fluid inside a rotating container. The whole range of locking phenomena is illustrated with the aid of this simple mechanical model.

I. Introduction

THE BEHAVIOR of a regenerative oscillator under the influence of an external signal has been treated by a number of authors. The case of synchronization by the external signal is of great practical interest; it has been applied to frequency-modulation receivers[1,2] and carrier-communication systems,[3] and formulas, as well as experimental data, have been given for the conditions required for synchronization.[4–8] The other case, arising when the external signal is not strong enough to effect synchronization, is of practical importance in beat-frequency oscillators. Here the tendency toward synchronization lowers the beat frequency and produces strong harmonic distortion of the beat note.[4–8]

It is the purpose of this paper to derive the rate of phase rotation of the oscillator voltage at a given instant from the phase and amplitude relations between the oscillator voltage and the external signal at that

* Decimal classification: R133×R355.91. Original manuscript received by the Institute, October 2, 1945; revised manuscript received January 16, 1946.

† Zenith Radio Corporation, Chicago, Illinois.

1 C. W. Carnahan and H. P. Kalmus, "Synchronized oscillators as frequency-modulation receiver limiters," *Electronics*, vol. 17, pp. 108–112; August, 1944.

2 G. L. Beers, "A frequency-dividing locked-in oscillator frequency-modulation receiver," Proc. I.R.E., vol. 32, pp. 730–738; *Elec. Eng.* December, 1944.

3 D. G. Tucker, "Carrier frequency synchronization," *Post Office Elec. Eng.*, vol. 33, pp. 75–81; July, 1940.

4 E. V. Appleton, "The automatic synchronization of triode oscillators," *Proc. Camb. Soc.*, vol. 21, pp. 231–248; 1922–1923.

5 D. G. Tucker, "The synchronization of oscillators," *Elec. Eng.*, vol. 15, pp. 412–418, March, 1943; pp. 457–461, April, 1943; vol. 16, pp. 26–30; June, 1943.

6 D. G. Tucker, "Forced oscillations in oscillator circuits," *Jour. I.E.E.* (London), vol. 92, pp. 226–234; September, 1945.

7 S. Byard and W. H. Eccles, "The locked-in oscillator," *Wireless Eng.*, vol. 18, pp. 2–6; January, 1941.

8 H. G. Möller, "Über Störungsfreien Gleichstromempfang mit den Schwingaudion," *Jahr. für Draht. Teleg.*, vol. 17, pp. 256–287; April, 1921.

Reprinted from *Proc. IRE and Waves and Electrons*, vol. 34, pp. 351-357, June 1946.

instant; in other words, to find a differential equation for the oscillator phase as a function of time. This equation must be expected to describe the case of synchronization where any transient disturbance vanishes in time, giving way to a steady state in which phase difference between oscillator and external signal is constant. It must also give frequency and wave form of the beat note, in case no synchronization occurs. To cover both cases, it must contain a parameter which decides whether or not the transient term will vanish in time, thus producing an equivalent to the criteria for synchronization derived by other methods. Finally, the equation must suggest a mechanical analogy simple enough to give a clear picture of what actually happens in an oscillator when an external signal is impressed upon it.

In the following analysis, it is assumed that the impressed signal and the free oscillation are of similar frequency. Locking effects at submultiple frequencies are analogous in many respects, but the analysis does not apply directly.

II. Conditions for Bandwidth and Time Constants

In attempting to derive the rate of phase rotation at a given instant from no other data but phase and amplitude relations at that same instant, we assume implicitly that there are no aftereffects from different conditions which may have existed in the past. The value of such an assumption lies in the fairly simple analysis which it permits. But our experience with practical oscillators warns us that it may not always be justified. In this section we will study the requirements which an oscillator must meet so that our analysis may be applicable.

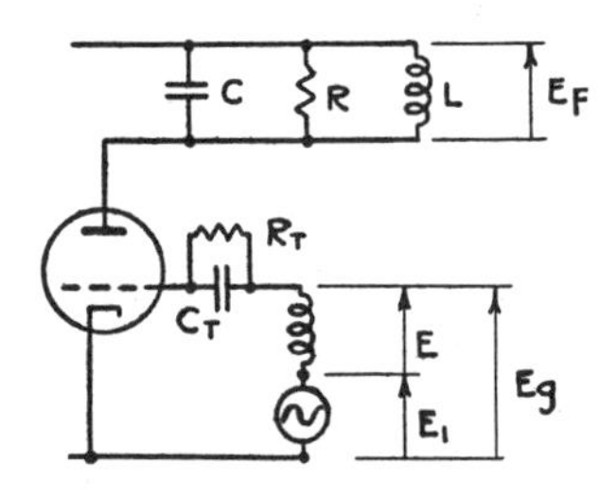

Fig. 1—Oscillator circuit.

If an oscillator is disturbed but not locked by an external signal, we observe a beat note—periodic variations of frequency and amplitude. If these variations are rapid, a sharply tuned circuit in the oscillator may not be able to respond instantaneously, or a capacitor may delay the automatic readjustment of a bias voltage. In either case, our assumption would be invalid. To validate it, we shall have to specify a minimum bandwidth for the tuned circuit and a maximum time constant for the biasing system. To establish these limits, let us study the circuit shown in Fig. 1, with the understanding that the impressed signal is not strong enough to cause locking. We will use the following symbols:

Angular frequencies:

ω_0 = free-running frequency

ω_1 = frequency of impressed signal

$\Delta\omega_0 = \omega_0 - \omega_1$ = "undisturbed" beat frequency

ω = instantaneous frequency of oscillation

$\Delta\omega = \omega - \omega_1$ = instantaneous beat frequency.

Voltages:

E_p = voltage across plate load

E = voltage induced in grid coil

E_1 = voltage of impressed signal

E_g = resultant grid voltage

Q = figure of merit of plate load L, C, R.

If the oscillator were undisturbed, the only frequencies present[9] would be ω_0 and ω_1, producing a beat frequency $\Delta\omega_0$. Actually, a lower beat frequency is observed, so that the value of ω averaged over one complete beat cycle is shifted toward ω_1. We cannot yet predict, however, how large the excursions of the momentary value of ω might be. We may think of ω as of a signal which is frequency modulated with the beat note $\Delta\omega$; this beat note is known to contain strong harmonics if the oscillator is almost locked, so that ω can be represented by a wide spectrum of frequencies extending to both sides of its average value.

If the plate circuit is to reproduce variations of ω without noticeable delay, each half of the pass band must be wide compared to the "undisturbed" beat frequency. For a single tuned circuit we can write

$$\frac{\omega_0}{2Q} \gg \Delta\omega_0. \tag{1}$$

Without reference to any specific type of circuit, we can say that the frequency of the external signal should be near the center of the pass band.

Up to this point, we have assumed that the circuit of Fig. 1 operates as a linear amplifier. But it is well known[10] that some nonlinear element must be present to stabilize the amplitude of any self-excited oscillator. Curved tube characteristics may produce a nonlinear relation between grid voltage and plate current, distorting every individual cycle of oscillation ("instantaneous" nonlinearity); plate-current saturation is an example for this case. On the other hand, a nonlinear element may control the transconductance as the amplitude varies, thus acting like an automatic volume control; the relation between grid voltage and plate current may then remain linear over a period of many cycles. Oscillators stabilized by an inverse-feedback circuit containing an incandescent lamp provide perhaps the best example for this type. The combination of C_T and R_T in the circuit of Fig. 1 functions also as a controlling element of the automatic-volume-control type; at the same time, some nonlinearity of the "instantaneous" type will generally be present in this circuit.

We want the instantaneous amplitudes of the plate

[9] "Frequency" always means the angular frequency.

[10] B. van der Pol, "The nonlinear theory of electric oscillations," Proc. I.R.E., vol. 22, pp. 1051–1086; September, 1934.

current and of the voltage E fed back to the grid to be the same as if the total grid voltage E_r at that instant had been stationary for some time; earlier amplitudes should have no noticeable aftereffects. How fast the amplitudes vary depends on the beat frequency. The amplitude control mechanism should, therefore, have a time constant which is short compared to one beat cycle.[11] (For the circuit of Fig. 1 this time constant would be of the order $T = C_T R_T$.) Since the shortest possible beat cycle corresponds to the "undisturbed" beat frequency $\Delta\omega_0$, we can write

$$T \ll \frac{1}{\Delta\omega_0}. \tag{2}$$

If the oscillator contains only amplitude limiting of the "instantaneous" type, this condition is inherently satisfied. An oscillator of the pure automatic-volume-control type will show the same locking and synchronizing effects as long as it fulfills[12] condition (2). But when the amplitude control mechanism acts too slow to accommodate the beat frequency, phenomena of an entirely different character appear. Such an oscillator would fall outside the scope of the mathematical analysis presented in the following, but its special characteristics merit brief discussion.

In an oscillator of the pure automatic-volume-control type, let us represent all elements outside the tuned circuit L, C, R by a negative admittance connected in parallel with L, C, R. The numerical value of this negative admittance is proportional to the gain in the oscillator tube. Over a long period of time, the automatic-volume-control mechanism will so adjust the gain that the negative admittance becomes numerically equal to the positive loss admittance of L, C, R. At this point the net loss vanishes and the prevailing amplitude is maintained indefinitely, as if the tuned circuit had infinite Q.

Now, let an external signal of slightly different frequency be superimposed upon this oscillation, so that the resulting amplitude varies periodically. Then if the automatic-volume-control mechanism acts so slowly that no substantial gain adjustments can be made within one beat cycle, that value of negative admittance which resulted in zero net loss will be retained. In other words, the system acts as if the Q of the plate circuit were still infinitely large. An external signal E_1 with a frequency very close to ω_0 will then produce a large near-resonant amplitude, increasing further the closer ω_1 approaches ω_0. This magnified signal of frequency ω_1, superimposed on the original signal of frequency ω_0, which is still maintained, produces amplitude modulation of a percentage much greater than would correspond to the ratio E_1/E.

Evidently, similar effects could be observed if the tuned circuit had of itself a Q high enough to violate condition (1). This suggests an alternative way of stating that condition. The tuned circuit will "memorize" phase and amplitude for a period of the order T', its "decay time." This period must be short compared to a beat cycle[11]

$$T' \ll \frac{1}{\Delta\omega_0}. \tag{1a}$$

For a simple tuned circuit, $T' = (2Q/\omega_0)$ hence $(\omega_0/2Q) \gg \Delta\omega_0$ which is the same as (1).

If an oscillator fulfills both conditions (1) and (2), the amplitude modulation arising from a given signal E_1 is solely determined by the ratio E_1/E and by the shape of the amplitude-limiting or automatic-volume-control characteristic. Most oscillators operate in a fairly flat part of this characteristic, so that the amplitude actually varies less than in proportion to E_1/E. Keeping this in mind, we further assume a weak external signal

$$E_1 \ll E \tag{3}$$

so that the amplitude variations of E will also be small compared to E itself.

A surprisingly large number of practical cases meet all three conditions.

III. Derivation of the Phase as a Function of Time

Let Fig. 2 be a vector representation of the voltages in the grid circuit as they are found at a given instant.

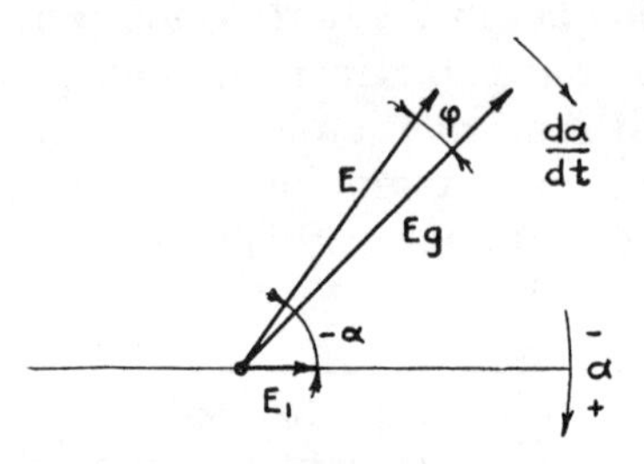

Fig. 2—Vector diagram of instantaneous voltages.

Furthermore, let E_1 be at rest with respect to our eyes; any vector at rest will therefore symbolize an angular frequency ω_1, that of the external signal, and a vector rotating clockwise with an angular velocity $(d\alpha/dt)$ shall represent an angular frequency $\omega_1 + (d\alpha/dt)$, or angular beat frequency of

$$\Delta\omega = \frac{d\alpha}{dt} \tag{4}$$

relative to the external signal.

It is important to keep in mind that this vector diagram shows beat frequency and phase. Many high-frequency oscillations may occur during a small shift of the vectors. We call $(d\alpha/dt)$ the instantaneous angular beat frequency; we would count $(1/2\pi)(d\alpha/dt)$ beats per second if this speed of rotation were maintained. Actually, $(d\alpha/dt)$ may vary and a complete beat cycle may never be accomplished.

With no external signal impressed, E_g and E must coincide: the voltage E returned through the feedback

[11] $1/\Delta w_0$, or the time required for one radian of a beat cycle, is used in the following.

[12] For synchronization on a subharmonic of the impressed signal, nonlinearity of the "instantaneous" type is necessary.

circuit must have the same amplitude and phase as the voltage E_g applied to the grid. Those nonlinear elements which limit the oscillator amplitude will adjust the gain so that $|E| = |E_g|$; but the phase can only coincide at one frequency, the free-running frequency ω_0. At any other frequency the plate load would introduce phase shift between E_g and E. Fig. 3 shows a typical curve of phase shift versus frequency for a single tuned circuit as assumed in Fig. 1. The amount of lead or lag of the voltage drop across such a circuit with respect to the current flowing through it is plotted. For our oscillator

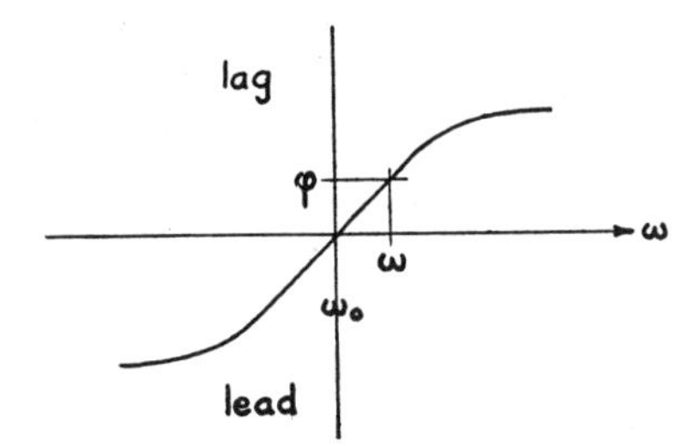

Fig. 3—Phase versus frequency for a simple tuned circuit.

circuit, we may take the curve to represent the lead or lag of E with respect to E_g as a function of frequency.

Let now an external voltage E_1 be introduced, and let Fig. 2 represent the voltage vectors at a given instant during the beat cycle. Evidently, the voltage E returned through the feedback circuit is now no longer in phase with the grid voltage E_g; the diagram shows E lagging behind E_g by a phase angle ϕ.

No such lag could be produced if the oscillator were still operating at its free frequency ω_0. We conclude that the frequency at this instant exceeds ω_0 by an amount which will produce a lag equal to ϕ in the plate circuit.

With $E_1 \ll E$ according to our third condition, inspection of Fig. 2 yields

$$\phi = \frac{E_1 \sin(-\alpha)}{E} = -\frac{E_1}{E}\sin\alpha. \tag{5}$$

The instantaneous frequency ω follows from Fig. 3. But our first condition implies that the pass band of the plate circuit is so wide that all frequencies are near its center. So we are using only a small central part of the ϕ versus ω curve which approaches a straight line with the slope

$$A = \frac{d\phi}{d\omega}\cdot \tag{6}$$

Then, if ω_0 is the free frequency, the phase angle for another frequency ω close to it will be

$$\phi = A(\omega - \omega_0). \tag{7}$$

The instantaneous beat frequency $\Delta\omega$ is the difference between ω and the impressed frequency ω_1. Setting again $\Delta\omega_0 = \omega_0 - \omega_1$, we have

$$\phi = A(\omega - \omega_0) = A[(\omega - \omega_1) - (\omega_0 - \omega_1)] = A[\Delta\omega - \Delta\omega_0]. \tag{8}$$

Now, substituting (5) on the left and (4) on the right, we find

$$-\frac{E_1}{E}\sin\alpha = A\left[\frac{d\alpha}{dt} - \Delta\omega_0\right] \tag{9a}$$

and substituting

$$B = \frac{E_1}{E}\cdot\frac{1}{A}$$

we obtain

$$\frac{d\alpha}{dt} = -B\sin\alpha + \Delta\omega_0. \tag{9b}$$

Adding the impressed frequency ω_1 on both sides, we may also write

$$\omega = -B\sin\alpha + \omega_0. \tag{9c}$$

This means physically that the instantaneous frequency is shifted from the free-running frequency by an amount proportional to the sine of the phase angle existing at that instant between the oscillator and the impressed signal. The shift is also proportional to the impressed signal E_1, but inversely proportional to the oscillator grid amplitude E and to the phase versus frequency slope A of the tuned system employed.

For a single tuned circuit, textbooks give

$$\tan\phi = 2Q\frac{\omega - \omega_0}{\omega_0} \tag{10}$$

and for small angles we can write

$$\phi = 2Q\frac{\omega - \omega_0}{\omega_0}\cdot \tag{10a}$$

Hence, substituting into (6)

$$A = \frac{2Q}{\omega_0} \tag{10b}$$

and

$$B = \frac{E_1}{E}\,\frac{\omega_0}{2Q}\cdot \tag{10c}$$

Equation (9b) reads, therefore, for a single tuned circuit,

$$\frac{d\alpha}{dt} = -\frac{E_1}{E}\,\frac{\omega_0}{2Q}\sin\alpha + \Delta\omega_0. \tag{11}$$

The possibility of a steady state is immediately apparent; $(d\alpha/dt)$ must then be zero, so that in the steady state

$$0 = -\frac{E_1}{E}\,\frac{\omega_0}{2Q}\sin\alpha + \Delta\omega_0 \tag{12a}$$

or

$$\sin\alpha = 2Q\frac{E}{E_1}\cdot\frac{\Delta\omega_0}{\omega_0}\cdot \tag{12b}$$

This gives the stationary phase angle between oscillator and impressed signal. Since $\sin\alpha$ can only assume values between +1 and −1, no steady state is possible if the right side of (12b) is outside this range. This gives the condition for synchronization

$$\left| 2Q \frac{E}{E_1} \cdot \frac{\Delta\omega_0}{\omega_0} \right| < 1 \tag{13a}$$

or

$$\frac{E_1}{E} > 2Q \left| \frac{\Delta\omega_0}{\omega_0} \right| . \tag{13b}$$

Because of its practical importance for receiver applications, another form of this condition shall be considered. E is the voltage which the oscillator (Fig. 1) produces across its grid coil; but if a locked oscillator is used to replace an amplifier, the voltage E_p across the plate circuit is the one that matters, since (E_p/E_1) represents the total gain. Now the tuned circuit is equivalent to a plate load $R_p = Q\sqrt{(L/C)}$, so that for a given transconductance g_m

$$E_p = E \cdot g_m \cdot Q \sqrt{\frac{L}{C}} \cdot$$

Combining this with (13b), we obtain

$$\frac{E_p}{E_1} < \left| \frac{\omega_0}{2\Delta\omega_0} \right| \cdot g_m \sqrt{\frac{L}{C}} \cdot \tag{13c}$$

It is interesting to note that Q, the only circuit constant entering into (13b) where the grid voltage E is of interest, cancels out in (13c) where the plate voltage E_p is determined.

For an oscillator which contains a plate load other than a simple tuned circuit, the condition for synchronization may be written

$$\frac{E_1}{E} > | A\Delta\omega_0 | \tag{13d}$$

whereby $A = (d\phi/d\omega)$ for the particular type of plate load.

IV. Approximation for the Pull-In Process

Turning now to the transient solution of the differential equation (9b), we examine first the case $\Delta\omega_0 = 0$. This means that the free-running frequency equals that of the impressed signal and that locking will eventually occur for any combination of voltages and circuit constants as evidenced by all forms of (13).

The equation

$$\frac{d\alpha}{dt} = -B \sin \alpha \tag{14a}$$

shows what happens when the external signal E_1 is suddenly switched on with an initial lag α_1 behind the free-running oscillator. Equation (14a) is quite similar to

$$\frac{d\alpha}{dt} = -B\alpha \tag{14b}$$

and actually goes over into this form when α is small. Equation (14b) has the familiar solution

$$\alpha = \alpha_1 \epsilon^{-Bt} \tag{14c}$$

and this means physically that the oscillator phase "sinks" toward that of the impressed signal, first approximately, and later accurately as a capacitor discharges into a resistor. The speed of this process, according to (10c) which defines B, is proportional to the ratio of impressed voltage to oscillator voltage and to the bandwidth of the tuned circuit.

If the free-running frequency is not equal to that of the impressed signal, but close enough to permit locking for a given combination of constants according to (13), the manner in which the steady state is reached must still resemble a capacitor discharge. It is particularly worth noting that the final value α_∞ is always approached from one side in an aperiodic fashion. The accurate solution for this case will be given later.

V. Phenomena Outside the Locking Range

To obtain a general solution giving α as a funcion of time, it is necessary to integrate (9b). We first substitute

$$K = \frac{\Delta\omega_0}{B} \tag{15a}$$

which means for a single tuned circuit

$$K = 2Q \frac{E}{E_1} \frac{\Delta\omega_0}{\omega_0} \cdot \tag{15b}$$

By comparing with (13a) and (13d), we find that the condition for synchronization can now be written

$$| K | < 1. \tag{15c}$$

Substituting into (9b) we obtain

$$\frac{d\alpha}{dt} = -B(\sin \alpha - K). \tag{16}$$

Integration gives

$$\tan \frac{\alpha}{2} = \frac{1}{K} + \frac{\sqrt{K^2 - 1}}{K} \tan \frac{B(t - t_0)}{2} \sqrt{K^2 - 1} \tag{17a}$$

or

$$\alpha = 2 \tan^{-1} \left[\frac{1}{K} + \frac{\sqrt{K^2 - 1}}{K} \tan \frac{B(t - t_0)}{2} \sqrt{K^2 - 1} \right] \tag{17b}$$

wherein t_0 is an integration constant.

Let us now assume that the condition for synchronization is not fulfilled, so that $|K| > 1$. This makes $\sqrt{K^2-1}$ real. With continually increasing t, the term $[B(t-t_0)/2]\sqrt{K^2-1}$ will pass through $\pi/2$, $3\pi/2$, etc., and the tangent on the right side of (17a) will become $+\infty$, $-\infty$, etc. in succession; at these instants $\alpha/2$ must also be $\pi/2$, $3\pi/2$, etc., although it will assume values different from $[B(t-t_0)/2]\sqrt{K^2-1}$ during the intervals.

So, while $[B(t-t_0)/2]\sqrt{K^2-1}$ increases uniformly with time, $\alpha/2$ will grow at a periodically varying rate; but the total length of a period must be the same for both. The average angular beat frequency—the actual number of beats in 2π seconds—is therefore

$$\overline{\Delta\omega} = B\sqrt{K^2 - 1} \tag{18a}$$

or, substituting from (15a),

$$\overline{\Delta\omega} = \Delta\omega_0 \frac{\sqrt{K^2 - 1}}{K}. \tag{18b}$$

$\Delta\omega_0$ is that beat frequency which would appear if the oscillator maintained its free frequency; $\sqrt{K^2-1}/K$ approaches unity for large values of K, far from the point where locking occurs; but it drops toward zero when this point $(K=1)$ is approached.

Fig. 4 shows a plot of the average beat frequency $\overline{\Delta\omega}$ versus the undisturbed beat frequency $\Delta\omega_0$ as computed from (18b).

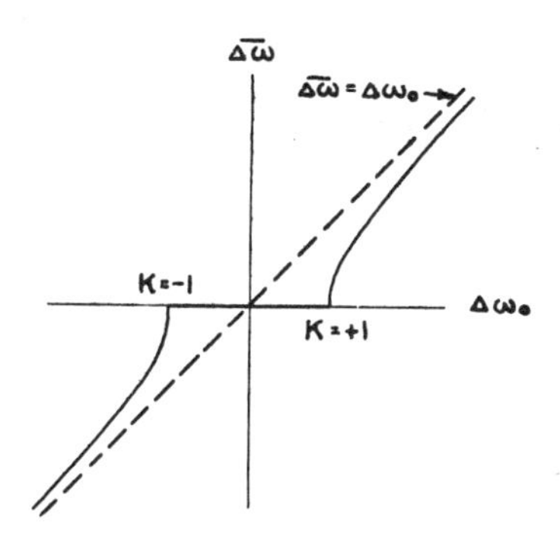

Fig. 4—Reduction of beat frequency due to locking.

In the intervals between the arguments $\pi/2$, $3\pi/2$, etc., the two angles in (17a) cannot be the same because of the factor $\sqrt{K^2-1}/K$ with which one tangent is multiplied, and the addition of $1/K$. For large values of K, $1/K$ vanishes and the factor approaches unity, so that the rate of increase of $\alpha/2$ with time will vary by a smaller percentage as the beat frequency increases; but (16) shows that $d\alpha/dt$ must still vary between $B(K-1)$ and $B(K+1)$. Now, $BK=\Delta\omega_0$, according to (15), and B represents the highest difference $\Delta\omega_{max}$ for which locking can occur ($K=1$ for $B=\Delta\omega_0$). So the instantaneous beat frequency $\Delta\omega$ will vary periodically between $\Delta\omega_0-\Delta\omega_{max}$ and $\Delta\omega_0+\Delta\omega_{max}$ as long as $\Delta\omega_0$ exceeds $\Delta\omega_{max}$.

$\Delta\omega_{max}$ itself is determined by (13). It is

$$\Delta\omega_{max} = \frac{\omega_0}{2Q}\frac{E_1}{E} \tag{19a}$$

or

$$\Delta\omega_{max} = \frac{\omega_0}{2}\frac{E_1}{E_p} g_m \sqrt{\frac{L}{C}} \tag{19b}$$

for a single-tuned circuit, and

$$\Delta\omega_{max} = \frac{1}{A}\cdot\frac{E_1}{E} \tag{19c}$$

for any type of plate load for which $A = d\phi/d\omega$.

If K is only slightly above unity, the factor $\sqrt{K^2-1}/K$ falls far below unity, and the phase angle between E_1 and E increases at an extremely nonuniform rate. Inspection of the vector diagram in Fig. 2 gives the resultant grid voltage $E_g=E-E_1 \cos\alpha$. To illustrate the wave form of the resultant beat note

Fig. 5—Wave form of beat note for cos $\alpha(t)$.

the function cos $\alpha(t)$ is plotted in Fig. 5. Operation very close to locking is assumed. Other wave forms are possible in beat-frequency oscillators where the beat note is produced in a separate detector; a constant phase shift may then be added to α on the way to the detector. Fig. 6 shows an example with a phase shift of $\pi/2$: the

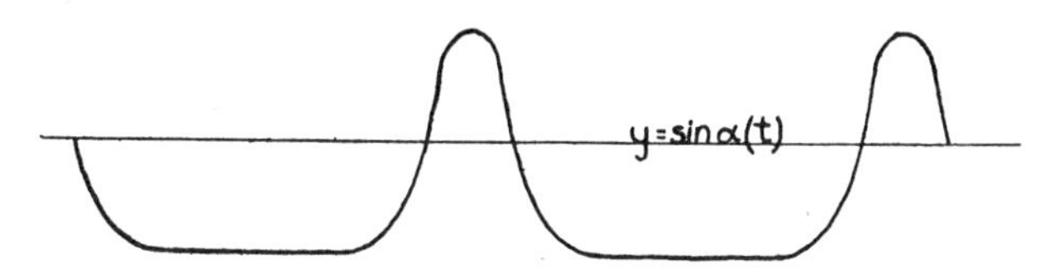

Fig. 6—Wave form of beat note for cos $\left[\alpha(t)+\frac{\pi}{2}\right]$.

function plotted is cos $[\alpha(t)+\pi/2]$ which equals $-\sin \alpha(t)$.

VI. Accurate Analysis of the Pull-In Process

To make the discussion of (17a) complete, we may finally apply it to the case of an oscillator pulling into the locked condition, $|K|<1$. The term $\sqrt{K^2-1}$ then becomes $j\cdot\sqrt{1-K^2}$. By use of the relation $\tanh x = -j \tan jx$, equation (17a) is transformed[13] into

$$\tan\frac{\alpha}{2} = \frac{1}{K} - \frac{\sqrt{1-K^2}}{K}\tanh\frac{B(t-t_0)}{2}\sqrt{1-K^2}. \tag{20a}$$

The integration constant t_0 permits one to fit the equation to the initial phase difference α_1, which exists when the external signal is switched on.

As t increases, the functions tanh and coth go assymptotically toward unity. The steady state must therefore be given by

$$\tan\frac{\alpha}{2(\infty)} = \frac{1-\sqrt{1-K^2}}{K}. \tag{20b}$$

Using (16) we identify K with $\sin\alpha_\infty$ for the steady state. Hence $\sqrt{1-K^2}=\cos\alpha_\infty$ and (20b) becomes $(1-\cos\alpha_\infty)/\sin\alpha_\infty$, which is indeed equal to $\tan(\alpha_\infty/2)$ by a trigonometrical identity.

VII. A Mechanical Model

In conclusion, let us construct a mechanical model to

[13] Equation (20a) holds for $\sin\alpha_1>k$. Otherwise, substitute coth for tanh.

illustrate the processes which we have derived. To provide a full analogy, the model must follow the same differential equation (9b)

$$\frac{d\alpha}{dt} = -B \sin \alpha + \Delta\omega_0.$$

Let us forget $\Delta\omega_0$ for the moment. A pendulum in a viscous fluid would follow the remaining equation if α is taken to mean the angle between the pendulum and a vertical line. If we assume the viscosity of the fluid to be so great that we need not consider the inertia of the pendulum, the angular speed of the pendulum $d\alpha/dt$ is proportional to the force which causes it to move. We may shape the pendulum so that one unit of force will produce one unit of speed. Now, if B is the weight of the pendulum, the force acting to return it to its rest position will indeed be $-B \sin \alpha$.

To include the term $\Delta\omega_0$, we must add a constant force. We may also bring $\Delta\omega_0$ over to the left side of the equation; since $d\alpha/dt$ stands for angular speed, $-\Delta\omega_0$ on the left would mean a constant backward rotation of the pendulum with respect to the liquid. Constant forward rotation of the liquid with respect to the pendulum would produce the same force, and we choose this interpretation for our model shown in Fig. 7.

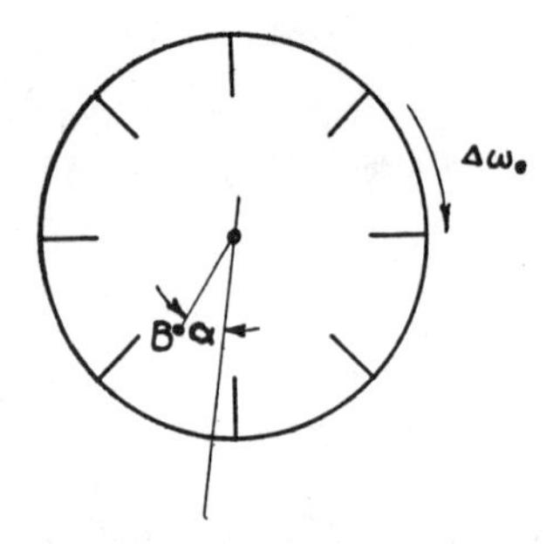

Fig. 7—Mechanical model: pendulum in a rotating container filled with viscous liquid.

The viscous liquid is enclosed in a drum rotating with an angular speed $\Delta\omega_0$. Again we assume that the viscosity of the liquid is so great that it will follow the rotation of the drum completely. Let us also assume that the rotation of the liquid is not noticeably affected by inserting the pendulum.

Remembering now that the vertical direction represents the phase of the impressed signal, while the position of the pendulum indicates the relative phase of the oscillator grid voltage, we can go through the whole range of phenomena by rotating the drum with various speeds, corresponding to the undisturbed beat frequencies $\Delta\omega_0$.

At low drum speed, the pendulum will come to rest at a definite angle α_∞ which will increase as the drum speed rises. If disturbed, the pendulum will "sink" back; it will never go past the rest position since inertia effects are absent.

If we lift the pendulum clockwise to any point below $\alpha_1 = \pi - \alpha_\infty$, it will come back counterclockwise; but if we lift it past this limit, or over to the right, it will return clockwise. This is the reason why there are two different transient solutions for (20a).

At a certain critical drum speed $\Delta\omega_{max} = B$ the pendulum will stand horizontal; if the drum is further accelerated, it will "unlock" and begin to go around, moving fast on the right but very slow on the left and completing a much smaller number of revolutions than the liquid.

But as we increase the speed further, the fast whirling fluid takes the pendulum along, irrespective of the weight. The motion appears much more uniform, and the speed of the pendulum becomes nearly equal to that of the drum: the average beat frequency $\overline{\Delta\omega}$ is approaching the undisturbed value $\Delta\omega_0$.

Acknowledgment

C. W. Carnahan and H. P. Kalmus, in the course of their work on locked oscillators for frequency-modulation receivers,[1] assembled a great deal of information regarding the behavior of such oscillators inside and outside the locking range. To study these phenomena further, they built a 1000-cycle oscillator which permitted direct observation of phase and amplitude variations on the oscilloscope. They investigated the influence of time constants and, among other effects mentioned in this paper, observed the large amplitude modulation which occurs when the time constant of the grid bias is large (case of "infinite Q" noted in Section II). Discussion of these experiments laid the groundwork for the analysis presented here, and the author gratefully acknowledges this important contribution.

Part IX
Operation and Performance in the Presence of Noise

Some Analytical and Experimental Phase-Locked Loop Results for Low Signal-to-Noise Ratios

F. J. CHARLES, MEMBER, IEEE, AND W. C. LINDSEY, MEMBER, IEEE

***Abstract*—This paper is concerned with the nonlinear behavior of the second-order phase-locked loop (PLL) in the presence of noise. The loop filter is of the proportional-plus-integral control type. This filter corresponds to the one generally employed for carrier tracking purposes in the implementation of phase-coherent communication systems.**

The paper is composed essentially of two parts: the first part presents analytical results which pertain to the probability distribution of the phase-error. Since these analytical results are approximations, valid only for certain regions of signal-to-noise ratio, they are complemented by experimental results obtained from simulation of the PLL system in the laboratory. The experimental techniques used to measure the statistical properties of the loop behavior and the corresponding results comprise the second part of the paper.

Approximate analytical expressions for the distribution of the system phase-error are first obtained by using the Fokker-Planck apparatus and, secondly, by assuming that the PLL behaves as a very narrow band-pass filter. The range of signal-to-noise ratios for which these approximations are valid is obtained by graphically comparing the analytical expressions to experimentally derived phase-error distributions. In addition, measurements relative to the variance of the phase-error are compared to those predicted by the linear PLL theory and the variance as computed from the approximate solutions.

Finally, experimental results relative to the probability distribution of the time intervals between cycle-slipping events are given for signal-to-noise ratios in a range where the linear PLL theory does not apply. In particular, the maximum length of time the loop may be expected to remain in-lock is illustrated graphically as a function of signal-to-noise ratio in the loop bandwidth.

I. INTRODUCTION

SINCE an exact analysis of the practical second-order phase-locked loop in the presence of noise remains, at this time, a mathematically intractable problem, the design of phase-coherent communication systems must, for the most part, be predicated on the basis of a linear or quasi-linear PLL theory. However, in space communications, where a significant monetary value is associated with any particular mission, there exists an incentive to develop an exact nonlinear theory of phase-locked loops. Such a theory could be employed to predict accurately, prior to launch, the performance of the communcation system towards the end of mission lifetime.

Over the past few years several techniques for analyzing the behavior of phase-locked loops in the presence of noise have evolved. Rather than expounding on the history and development of PLL theory in this paper, we refer the interested reader to material contained in [1]. Suffice it to say, however, that most of the theory which has been developed to date is based upon linearization techniques, series representations, or the so-called spectral approximation introduced by Tausworthe [2]. Even though these methods give valid results and design trends which may be used in practice, emphasis has been placed on developing a theory which specifies the variance of the phase-error as a function of the signal-to-noise ratio existing in the bandwidth of the loop. Such a theory is valid insofar as it goes; however, there are situations which arise in practice where the variance of the phase-error is not sufficient in predicting the performance of a communication system. For example, the probability distribution of the phase-error is needed when attempting to characterize the performance of communication systems employing noisy phase references [3], [4] or when one attempts to allocate, in some optimal manner, the total transmitter power between a carrier and one or more subcarriers [5]. Another situation which arises in practice occurs when one attempts to measure the received Doppler frequency. Here, knowledge of the cycle-slipping phenomenon of the loop is important because the Doppler measurement is certainly in error if the loop has slipped cycles during the measurement. In this particular case, the variance of the phase-error contributes nothing in specifying the cycle-slipping phenomenon.

The most promising analytical approach used to develop an exact nonlinear theory of phase-locked loops is based on the Fokker-Planck method [1], [6], and [7]. This particular method leads to an expression for the probability density of the nonstationary phase-error which, in the steady-state, gives an unbounded variance. Of course, this behavior of the variance is a result of the cycle-slipping phenomenon associated with phase-locked loops. However, Viterbi [1] and Tikhonov [6], [7] were succesful in applying this method to the analysis of the first-order loop by recognizing that the probability distribution of the phase-error reduced modulo 2π is stationary and possesses a bounded variance.

We note that the modulo 2π representation of the phase-error is equivalent to ignoring the fact that the loop actually slips cycles. Consequently, another probability distribution, viz., the probability distribution of time intervals between cycle slipping events, is needed. These two distributions, viz., the probability distribution of the time intervals between cycle-slipping events and the joint probability dis-

Manuscript received January 26, 1966; revised May 10, 1966; and June 30, 1966. This paper presents the results of one phase of research carried out at the Jet Propulsion Laboratory, California Institute of Technology, under Contract NAS 7-100, sponsored by the National Aeronautics and Space Administration.

The authors are with the Jet Propulsion Laboratory, California Institute of Technology, Pasadena, Calif.

Reprinted from *Proc. IEEE*, vol. 54, pp. 1152–1166, Sept. 1966.

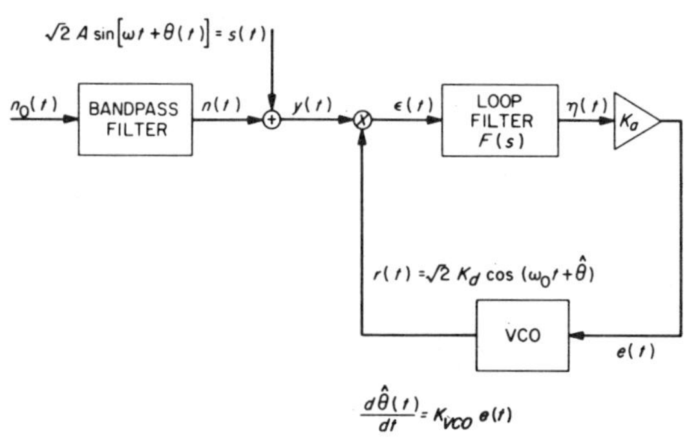

Fig. 1. Phase-locked system.

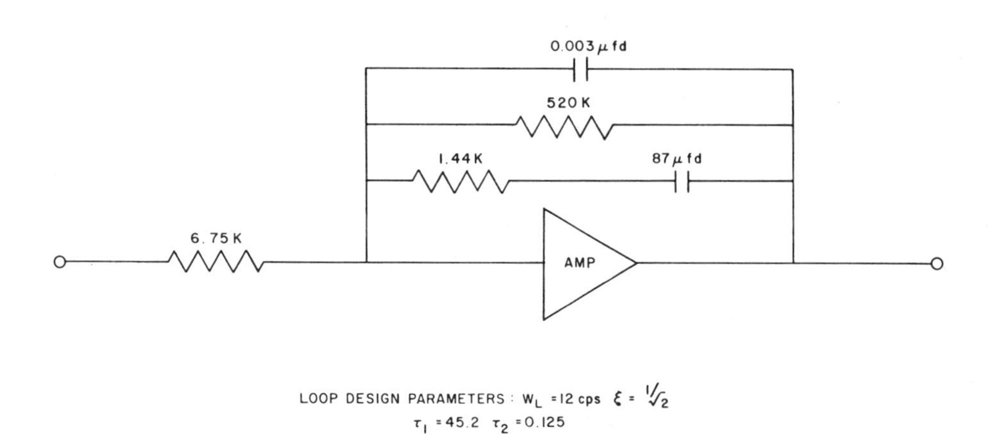

Fig. 2. Loop filter mechanization.

tribution of the phase-error and phase-error rate, are sufficient in characterizing the *fundamental* behavior of the PLL.

As a rule, in the treatment of nonlinear devices in the presence of noise, approximate solutions must be accepted because of the mathematical difficulties involved and, quite frequently, these theoretical simplifications are adaptable to practical results. For cases in which the approximate solutions might be suspect, their particular value to the design engineer depends upon a comparison with experimental results. Thus, the experimental results may be used to complement the analytical approximations and vice versa. In this paper, we present experimental and approximate analytical results which pertain to the probability distribution of the phase-error reduced modulo 2π, the second moment of this distribution and experimental results relative to the probability distribution of the time intervals between cycle-slipping events. These distributions (and their properties) can be used to predict more accurately the performance of phase-coherent communication systems and, in addition, may prove to be valuable in postulating useful loop models which account for the cycle-slipping phenomenon.

II. The Equation of Operation

Figure 1 illustrates the basic PLL system under consideration. The input signal $s(t)$ to the PLL is assumed to be a sinusoid with average power $P=A^2$ watts, a frequency of ω rad/s and a phase of $\theta(t)=\theta$ radians. The perturbing noise $n(t)$ is obtained by passing a stationary white Gaussian noise process through the band-pass filter. The random process $n_0(t)$ is assumed to posses a double-sided spectral density of N_0 watts/(c/s). In practice, the band-pass filter corresponds to the IF filter which always precedes the PLL and it is assumed to have a bandwidth W which is large compared to the bandwidth of the loop.

We are concerned here with the behavior of the PLL system of Fig. 1 when the loop filter is of the proportional-plus-integral control type, viz.,

$$F(p)=\frac{1+\tau_2 p}{1+\tau_1 p}. \tag{1}$$

In (1), p denotes the Heaviside operator. The particular mechanization of the loop filter under study is illustrated in Fig. 2. With this particular loop filter it can be easily shown [1] that the equation of operation is given by

$$\ddot{\phi}(t)+[a+b\cos\phi(t)]\dot{\phi}(t)+c\sin\phi(t)=\Omega_0/\tau_1+N(t) \tag{2}$$

where

$$a=\frac{1}{\tau_1},\qquad b=\frac{AK\tau_2}{\tau_1},\qquad c=\frac{AK}{\tau_1}$$

and

$$N(t)=-\frac{K}{\tau_1}\cdot\left[1+\tau_2\frac{d}{dt}\right]\cdot n'(t). \tag{3}$$

In (2) we note that $\Omega_0 = \omega - \omega_0$ is the initial detuning and in (3) the process $n'(t)$ corresponds to the low-pass version of $n(t)$.

It is common practice to simplify (2) by assuming $|\phi| < 1$ radian and to use, for design purposes, the linearized equation of operation obtained by substituting the approximations $\cos\phi \approx 1$ and $\sin\phi \approx \phi$. Using this approach it is convenient to define a number of parameters which relate the PLL to the more familiar filter, viz., the natural undamped frequency ω_n, the relative damping ratio ξ, and the equivalent noise bandwidth W_L [two-sided, see (22)]. In terms of the loop parameters it can be readily shown that

$$\omega_n = \sqrt{c} = \sqrt{\frac{AK}{\tau_1}}$$
$$\xi = \frac{a+b}{2\sqrt{c}} \approx \frac{1}{2}\tau_2\sqrt{\frac{AK}{\tau_1}}$$
$$W_L = \frac{c+(a+b)^2}{2(a+b)} \approx \frac{1}{2}\left(\frac{AK}{\tau_1}\tau_2 + \frac{1}{\tau_2}\right). \quad (4)$$

The approximations in (4) are valid if the inequality $a \ll b$ is satisfied. In most practical designs the loop is underdamped so that the loop filter zero $1/\tau_2$ is located at or near the natural frequency ω_n. Consequently, the above inequality reduces to $1/\tau_1 \ll AK$ which is not difficult to satisfy in practice. The parameter values associated with the experimental loop considered in this paper are $\omega_n = 11.3$ rad/s, $\xi = 1/\sqrt{2}$, $W_L = 12$ c/s.

III. The Two-Dimensional Fokker-Planck Equation

The analytical simplification introduced by linearizing the differential equation of operation can only be justified for situations involving high signal-to-noise ratios. For the case of low signal-to-noise ratios a more realistic approach to the analysis of the PLL is to use the Fokker-Planck method. This approach makes it necessary to characterize the properties of the process $N(t)$, which we recall is defined by (3), in order to compute the coefficients of the Fokker-Planck equation. In particular, the autocorrelation function of $N(t)$ is required to determine the coefficient defined by [8, p. 453]

$$D = \lim_{\Delta t \to 0} \frac{1}{\Delta t} \int_{t_1}^{t_1+\Delta t} \int_{t_2}^{t_2+\Delta t} \overline{N(t_1)N(t_2)}\, dt_1 dt_2. \quad (5)$$

Writing the correlation function $R_N(t_2 - t_1)$ as the Fourier transform of the spectrum of $N(t)$ we obtain

$$R_N(\tau) = \frac{N_0}{2\pi}\left(\frac{K}{\tau_1}\right)^2 \int_{-\pi W}^{\pi W} [1 + (\omega\tau_2)^2] \cos\omega\tau d\omega, \quad \tau = t_2 - t_1 \quad (6)$$

where $N_0/2\pi$ is the spectral density of the process $n'(t)$ and is constant over the arbitrarily wide region $|\omega| \le 2\pi(W/2)$. Under these conditions and recalling that

$$\lim_{W\to\infty} \int_{-\pi W}^{\pi W} \cos\omega\tau d\omega = 2\pi\delta(\tau), \quad \lim_{W\to\infty} \frac{\sin \pi W\tau}{\tau} = \pi\delta(\tau) \quad (7)$$

we can approximate $R_N(\tau)$ by

$$R_N(\tau) = N_0\left(\frac{K}{\tau_1}\right)^2 \left\{ [1 + (\pi W\tau_2)^2]\delta(\tau) + 2W\left(\frac{\tau_2}{\tau}\right)^2 \left[\cos \pi W\tau - \frac{\sin \pi W\tau}{\pi W\tau}\right]\right\}. \quad (8)$$

Since the last term in (8) contributes only a small amount to the computation of D in (5), we ignore this term; the approximate form for the correlation function of $N(t)$ is given by

$$R_N(t_2 - t_1) \approx N_0\left(\frac{K}{\tau_1}\right)^2 [1 + (\pi W\tau_2)^2]\delta(t_2 - t_1). \quad (9)$$

Using this expression to compute the coefficient D in (5) we obtain

$$D = N_0\left(\frac{K}{\tau_1}\right)^2 [1 + (\pi W\tau_2)^2]. \quad (10)$$

The remaining coefficients can be readily computed and the resulting two-dimensional Fokker-Planck equation (in the steady-state) takes the form given by [8]

$$-\frac{\partial}{\partial\phi}[\dot\phi p(\phi, \dot\phi)] + \frac{\partial}{\partial\dot\phi}[a\dot\phi + b\dot\phi g_\phi(\phi) + cg(\phi)]p(\phi, \dot\phi) + \frac{1}{2}D\frac{\partial^2}{\partial\dot\phi^2}[p(\phi, \dot\phi)] = 0 \quad (11)$$

where $g(\phi)$ is an arbitrary nonlinearity which possesses a continuous first derivative. In our case [see (2)], $g(\phi) = \sin\phi$ and $g_\phi(\phi) = \partial/\partial\phi g(\phi)$. We have also assumed for convenience zero initial detuning, i.e., $\Omega_0 = 0$.

IV. Approximate Solutions to the Fokker-Planck Equation

In the preceding discussion, we obtained the Fokker-Planck equation whose steady-state solution is the joint probability distribution $p(\phi, \dot\phi)$. In this section we present approximate solutions to the nonlinear partial differential equation (11) and give evidence which attests to the fact that our approximation gives logically noncontradictory results for several important cases.

We assume a steady-state solution of the form

$$f(\phi, \dot\phi) = K' \exp\left[m\dot\phi^2 + n\int \dot\phi g_\phi(\phi) d\dot\phi + q\int g(\phi)d\phi + k\int g(\phi)g_\phi(\phi)d\phi\right]. \quad (12)$$

This $f(\phi, \dot\phi)$ is then substituted into the partial differential equation (11) and if the values of m, n, q, and k are selected such that

$$q = -\frac{ac}{D}, \quad k = -\frac{bc}{D}$$
$$m = -\frac{a}{2D}; \quad n = -\frac{b}{D} \quad (13)$$

we find that all terms in the result drop out except a residual term

$$R(\phi, \dot{\phi}) = \frac{b}{D} \dot{\phi}^2 g_{\phi\phi}(\phi)$$
$$= \frac{P\tau_2}{N_0 G} \cdot \frac{\tau_1}{1 + (\pi W \tau_2)^2} \cdot \dot{\phi}^2 g_{\phi\phi}(\phi) \quad (14)$$

where $G = AK$ is the equivalent loop gain, and $P = A^2$ is the observed signal power. This residual term $R(\phi, \dot{\phi})$ is zero if $b = 0$, or, in terms of the loop filter time constants, if $\tau_2 = 0$ or $\tau_1 = \tau_2 = 0$. For the typical situation occurring in practice we note that

$$|R(\phi, \dot{\phi})| \ll \left(\frac{P}{2N_0 W}\right) \dot{\phi}^2 \quad (15)$$

where $(P/2N_0 W)$ is the signal-to-noise ratio in the IF bandwidth. Intuitively, we expect that the residual term approaches zero for low values of signal-to-noise ratio even though the rate of change of ϕ becomes large. This, of course, is the region where the linear PLL theory does not hold.

In terms of the loop parameters, we have

$$p(\phi, \dot{\phi}) = K' \exp\left[-\dot{\phi}^2 \left(\frac{\beta_1 + \beta_2 \cos\phi}{2}\right) + \alpha_1 \cos\phi + \alpha_2 \cos 2\phi\right] \quad (16)$$

where

$$\alpha_1 = \frac{2\sqrt{P}}{N_0 K} \cdot \frac{1}{1 + (\pi W \tau_2)^2}; \quad \alpha_2 = \frac{P}{2N_0} \cdot \frac{\tau_2}{1 + (\pi W \tau_2)^2}$$
$$\beta_1 = \frac{2\tau_1}{N_0 K^2} \cdot \frac{1}{1 + (\pi W \tau_2)^2}; \quad \beta_2 = \frac{2\sqrt{P}}{N_0 K} \cdot \frac{\tau_1 \tau_2}{1 + (\pi W \tau_2)^2} \quad (17)$$

and K' is the normalizing constant. The occurrence of the "two-ϕ" term is due to the nonlinear damping term in (2).

Our solution is, therefore, exact for several interesting cases.

Case I: If we let $\tau_2 = 0$, this corresponds to the RC circuit loop filter $F(p) = (1 + \tau_1 p)^{-1}$. For this case the exact solution is given by

$$p(\phi, \dot{\phi}) = \frac{\exp\left[-\frac{\beta_1}{2}\dot{\phi}^2 + \alpha_1 \cos\phi\right]}{2\pi\sqrt{(2\pi\beta_1)^{-1}}\, I_0(\alpha_1)}; \quad \begin{matrix} |\phi| \le \pi \\ |\dot{\phi}| \le \infty \end{matrix} \quad (18)$$

where $I_0(x)$ is an imaginary Bessel function of order zero and argument x. Note that this distribution of the phase-error $p(\phi)$ is independent of τ_1.

Case II: If we let $\tau_1 = \tau_2 = 0$ we obtain the loop filter $F(p) = 1$ and we find that

$$p(\phi) = \frac{\exp(\alpha_1 \cos\phi)}{2\pi I_0(\alpha_1)}; \quad |\phi| \le \pi. \quad (19)$$

This corresponds to the result derived by Viterbi [1] and Tikhonov [6]. Note that this distribution is identical to the distribution of the phase-error given in (18).

Case III: We linearize $g(\phi)$, i.e., $g(\phi) = \phi$; and the distribution of the phase-error becomes

$$p(\phi) = \frac{\exp(-\phi^2/2\alpha^{-1})}{\sqrt{2\pi\alpha^{-1}}} \quad (20)$$

which is valid for high signal-to-noise ratios. The parameter α of this distribution plays a very important role when the linear PLL theory is involved. In fact, it is easy to show, using Parseval's theorem and the linearized model [1] of the loop, that the phase variance is given by

$$\sigma_\phi^2 = \frac{N_0}{A^2} \frac{1}{2\pi j} \int_{-j\infty}^{j\infty} |H(s)|^2 ds = \frac{N_0 W_L}{A^2} = \frac{1}{\alpha} \quad (21)$$

where $H(s)$ is the closed-loop transfer function and

$$W_L = \frac{1}{2\pi j} \int_{-j\infty}^{j\infty} |H(s)|^2 ds \quad (22)$$

is the equivalent noise bandwidth of the loop.

In order to observe the behavior of the approximate solution in the region of small signal-to-noise ratios (i.e., $R(\phi, \dot{\phi}) \approx 0$) we now write (16) in terms of the loop parameters,

$$p(\phi, \dot{\phi}) = K' \exp\left[-\frac{2\tau_1}{N_0 K^2}\dot{\phi}^2 - \frac{2\sqrt{P}\tau_1\tau_2}{N_0' K}\dot{\phi}^2 \cos\phi\right] \cdot \exp\left[+\frac{2\sqrt{P}}{N_0' K}\cos\phi + \frac{P\tau_2}{2N_0'}\cos 2\phi\right] \quad (23)$$

where

$$N_0' = N_0[1 + (\pi W \tau_2)^2] \approx N_0 (\pi W \tau_2)^2.$$

We note that for weak signal conditions, i.e., small $\sqrt{P}$, the phase-error ϕ becomes uniformly distributed while its derivative $\dot{\phi}$ becomes Gaussian. This is a rather striking result when compared to the large signal-to-noise ratio situation where, we recall from the linear PLL theory that both ϕ and $\dot{\phi}$ are Gaussian variables. Finally, we note from (23) that, for weak signal conditions, the variance of $\dot{\phi}$ is $\sigma_{\dot{\phi}}^2 = (N_0' K^2)/4\tau_1$. Further, we see from (23) that there is some indication that $p(\phi, \dot{\phi})$ possesses a multimodal structure for low signal-to-noise ratios.

Before abandoning the Fokker-Planck method we will examine the stochastic differential equation of operation for further simplification. Rewriting (2) in terms of the loop parameters, we have

$$\frac{\tau_1}{G\tau_2}\ddot{\phi} + \left(\frac{1}{G\tau_2} + \cos\phi\right)\dot{\phi} + \frac{1}{\tau_2}\sin\phi(t) = \frac{\tau_1}{G\tau_2}N(t). \quad (24)$$

If we further assume that $\ddot{\phi}$ is small in comparison to $\dot{\phi}$ (this is equivalent to assuming that $P[|\phi| > \pi/2] \approx 0$) and that $(1/G\tau_2) < 1$, we have the stochastic fluctuation equation

$$(\cos\phi)\dot{\phi} + \frac{1}{\tau_2}\sin\phi = \frac{\tau_1}{G\tau_2}N(t) \quad (25)$$

whose Fokker-Planck solution is given by

$$p(\phi) = K' \cos\phi \exp\left[-\frac{\alpha}{2}\sin^2\phi\right]; \quad |\phi| \le \frac{\pi}{2}. \quad (26)$$

As we shall see, the approximation obtained in the next section contains this term. We will further observe that, for $\alpha > 6$ dB, this distribution [i.e., (26)] is a good approximation to the distribution of the phase-error which has been obtained in the laboratory.

Thus, at this point, we have two approximations, viz., (23) and (26), which are predicated on the Fokker-Planck theory. As we shall see, when the experimental results are discussed, the approximation given in (23) is valid for $\alpha < 0$ dB, i.e., in this region $R(\phi, \dot{\phi})$ is small, while the approximation given in (26) is valid for $\alpha > 6$ dB. Further, we shall see that for $\alpha > 9$ dB (20) adequately characterizes the distribution $p(\phi)$.

V. The PLL as a Narrow Band-Pass Filter

The purpose of this section is to present an analytical expression for the phase-error distribution which is valid (in the engineering sense) for signal-to-noise ratios $\alpha > 0$ dB. The method of approximation involves the concept that the PLL is a very narrow band-pass filter whose output, i.e., the VCO output, is a cosine wave. The phase modulation on the VCO output is a *narrow-band process* and, in this case, is the system phase-error since the input signal is unmodulated. By a *narrow-band process*, we mean a stationary random process whose realizations are close to being sinusoidal oscillations of some fixed frequency ω_0 for time intervals equal to a large number of periods $2\pi/\omega_0$; clearly, such a process has mean value zero.

With this concept in mind we assume that the phase-locked loop may be replaced by a rectangular band-pass filter with center frequency ω_0 and bandwidth W_L cycles. Thus, the output of the filter (assuming the input is an unmodulated sine wave of power P watts plus white noise) will be a sinusoid of frequency ω_0 plus a random disturbance which possesses mean squared value $2W_L N_0 = \sigma^2$. The distribution of the phase of this process is easily determined by the method of characteristic functions and is given by [8]

$$p(\phi) = \frac{1}{2\sqrt{\pi}}\left[\frac{\exp\left(-\frac{\alpha}{2}\right)}{\sqrt{\pi}} + \sqrt{\frac{\alpha}{2\pi}}\cos\phi\exp\left(-\frac{\alpha}{2}\sin^2\phi\right) \cdot \left\{1 + \exp\left(-\frac{\alpha}{2}\right)\Phi\left(\sqrt{\frac{\alpha}{2}}\cos\phi\right)\right\}\right]; \quad \phi \le \pi. \quad (27)$$

For small α this becomes uniformly distributed which is in agreement with the Fokker-Planck theory (23) while for larger α the distribution becomes

$$p(\phi) = \sqrt{\frac{\alpha}{2\pi}}\cos\phi\exp\left(-\frac{\alpha}{2}\sin^2\phi\right); \quad |\phi| \le \frac{\pi}{2} \quad (28)$$

which agrees with the Fokker-Planck solution given by (26). Of course, for large values of α, the distribution becomes Gaussian with zero mean and variance α^{-1}

VI. Mechanical Analog of the Second-Order Phase-Locked Loop

The form of the second-order differential equation (2) suggests a mechanical analog which may be used to illustrate the general behavior of the second-order PLL. Consider, for example, the pendulum system shown in Fig. 3. The pendulum consists of a bob of mass M attached by an infinitesimally thin, weightless rod of length L to a shaft which is free to rotate with the same angular velocity as the bob. We assume that the shaft turns a fan in a manner which does not require energy from the pendulum system.

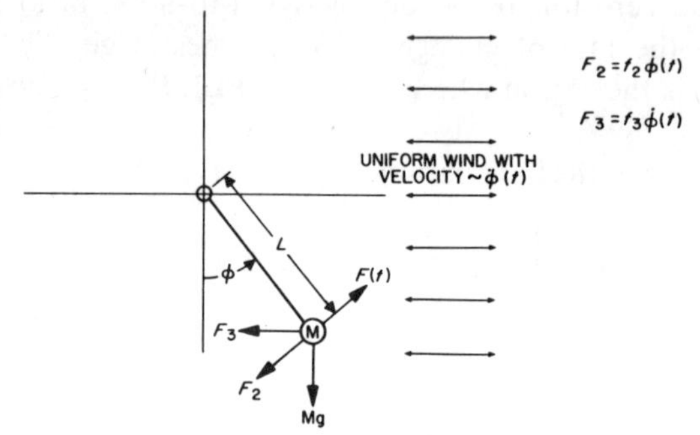

Fig. 3. Mechanical analog of PLL.

The fan is situated so that the wind it generates blows uniformly across the plane of the pendulum as indicated in Fig. 3. We assume that the wind blows from right to left when the pendulum revolves in the counterclockwise direction and blows in the opposite direction when the motion of the pendulum is reversed. Since the velocity of the wind is proportional to the angular velocity of the pendulum, an external force, denoted by $F_3 = f_3\dot{\phi}(t)$, is applied to the bob. We further assume that a motor inside the bob causes a random tangential force $F(t)$ to be exerted on the bob as shown in Fig. 3. In addition, suppose that the pendulum system is parallel to a fixed surface with which the bob is always in contact. Then there is exerted on the bob a frictional force $F_2 = f_2\phi(t)$. Finally, a vertical gravitational force Mg acts on the bob.

The resultant force F_r, producing the torque or moment of force with respect to the center of rotation and causing the angular motion of the bob, is proportional to the angular acceleration, i.e., $F_r = ML\dot{\phi}(t)$. The expression for the resultant force F_r can be readily derived by referring to Fig. 3 and the equation of motion for the pendulum becomes

$$\ddot{\phi}(t) + [a + b\cos\phi(t)]\dot{\phi}(t) + c\sin\phi(t) = D_1(t) \quad (29)$$

where

$$a = \frac{f_2}{ML}; \quad b = \frac{f_3}{ML}; \quad c = g/L; \quad D_1(t) = \frac{F(t)}{ML}.$$

A comparison of (3) and (29) shows the following correspondence between the parameters of the phase-locked loop and the pendulum system; viz.,

$$a = \frac{1}{\tau_1} = \frac{f_2}{ML}; \quad b = \frac{G\tau_2}{\tau_1} = \frac{f_3}{ML}; \quad c = \frac{G}{\tau_1} = \frac{g}{L}$$

$$D_1(t) = \frac{\Omega_0}{\tau_1} + N(t). \tag{30}$$

In order to relate the behavior of the pendulum to the operation of the phase-locked loop we note that the angular displacement of the bob is analogous to the loop phase-error. It is convenient, therefore, to represent the angular position of the bob by a point on the sinusoidal trajectory shown in Fig. 4 and obtained by considering the horizontal projection of the swinging pendulum. Consequently, the instantaneous phase-error in the loop is described graphically by the random motion of a point on the sinusoidal trajectory. If the random force $F(t)$ acting on the pendulum has a nonzero mean $\overline{F(t)}$, corresponding to an initial frequency error in the loop, the angular positions given by $\sin^{-1}[\overline{F(t)}/Mg]$ (or equivalently in terms of the loop parameters $\sin^{-1}[\Omega_0/G]$) represent equilibrium points.

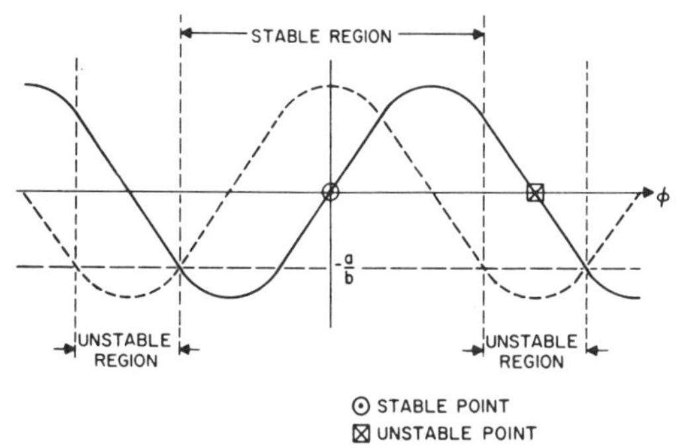

Fig. 4. Phase-error trajectory.

It can be shown [9] that the equilibrium points ϕ_s which satisfy the inequality $\cos\phi_s > -a/b = -1/G\tau_2$ are stable equilibrium points whereas the values ϕ_s' for which $\cos\phi_s' \leq -a/b$ are conditionally stable points. It is now possible to partition the sinusoidal trajectory into stable and unstable regions (see Fig. 4). If the phase-error is such that the point on the trajectory lies within a stable region and the noise is removed from the system the point will return, under the influence of the internal system forces, to the equilibrium position within the stable region. In contrast, if the point enters an unstable region its velocity will increase and it will be forced to travel through the unstable region towards the next stable region. However, because of the overshoot in the system and the presence of noise, the point may travel through several cycles in succession (corresponding to rotation of the pendulum) before settling down once again in a stable region. This phenomenon is commonly referred to as *cycle-slipping* and is an indication that the loop is "out of lock."

From a study of this pendulum model a number of important and *fundamental* conclusions can be drawn relative to the behavior of the PLL. We note that the occurrence of unstable regions along the trajectory is a result of the tangential component of the wind force F_3 changing the direction in which it acts on the bob whenever the phase-error exceeds $\pm\pi/2$ radians. It is evident that decreasing the ratio $a/b = f_2/f_3 = 1/G\tau_2$ by increasing the proportionality constant f_3 of the wind force has the effect of reducing the random variations of the phase-error provided $|\phi| < \cos^{-1}(-a/b)$ radians. However, this can be done only at the expense of decreasing the width of the stable regions which results in more frequent occurrences of the cycle-slipping events. We also note that, since the time required for the pendulum to move through a fixed angular displacement is proportional to $\sqrt{L}$, increasing the pendulum's length will tend to reduce $\dot{\phi}(t)$. Consequently, the length of the time intervals between the occurrences of cycle-slipping events will increase as the pendulum is made longer.

Thus, from these observations, we conclude that in order to minimize the random variations of the phase-error and the tendency of the loop to slip cycles, it is necessary (for this particular type of loop) to minimize the ratios $1/G\tau_2$ and $g/L = G/\tau_1$. For a given product $G = AK$ this amounts to maximizing the loop filter parameters τ_1 and τ_2 within the constraints of the particular situation. These observations are in agreement with those observed from the Fokker-Planck solutions.

VII. Experimental Techniques to Measure the Statistical Parameters of Phase-Locked Loops Operating in Noise

Most of the experimental work previously performed on phase-locked loops, with a few exceptions [10]–[13], has been concerned with the response of the loop to deterministic signals. The reason for this is, in part, due to the difficulty and complexity associated with the measurement of the statistical parameters of the loop operating in noise. In this section we will limit the discussion to a description of the experimental techniques used and in the following sections present results obtained employing these techniques.

A. Phase-Error Measurements

The essentials of the instrumentation technique used to measure the probability distribution of the phase-error can be briefly outlined by first noting that, for the case of an unmodulated signal in additive noise and zero initial frequency offset, the phase error $\phi(t)$ is simply the negative of the phase-jitter (or noise) $\hat{\theta}(t)$ on the VCO output signal. Consequently, in order to measure the phase-error distribution it was necessary to demodulate coherently the noisy VCO reference signal to obtain a voltage proportional to the instantaneous phase-jitter $\hat{\theta}(t)$ modulo 2π. The demodulator shown in Fig. 5 consisted of a low-pass flip-flop phase detector whose transfer characteristic was linear over the range $|\phi| \leq \pi$ radians. A special purpose amplitude distribution analyzer (ADA) was then used to obtain 800 uniformly spaced samples of the amplitude distribution (corresponding to increments of $\pi/400$ radians in ϕ) of the phase detector output. This amplitude distribution can be readily converted to the phase-error distribution modulo 2π by simply changing the abscissa scale units from volts to

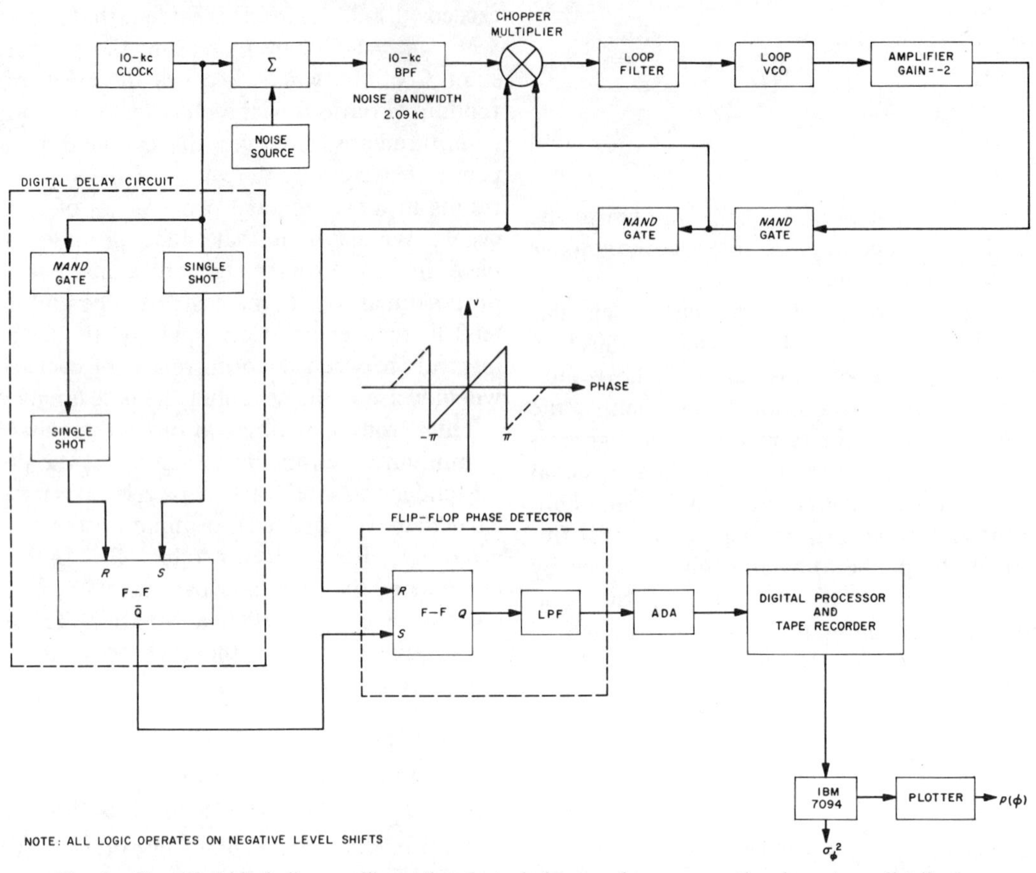

Fig. 5. Simplified block diagram illustrating the technique used to measure the phase-error distribution.

radians in accordance with the linear transfer characteristic of the phase detector. The samples of the distribution, in binary coded decimal form were then processed and stored on magnetic tape in a format acceptable to an IBM 7094 computer. The output of the computer consisted of a point plot of the phase-error distribution $p(\phi)$ and a plot of the corresponding cumulative distribution $P[\phi \leqq \phi_1]$. In addition a print-out of the variance σ_ϕ^2 was obtained.

B. Cycle-Slipping Measurements

A simplified block diagram illustrating the technique used to register the occurrence of a cycle-slipping event is shown in Fig. 6. The purpose of this instrumentation was to generate a sequence of narrow pulses as a function of time where each pulse indicates that the phase-error $\phi(t)$ has changed by 2π radians. Such a time sequence can be recorded on magnetic tape and processed by a digital computer to obtain the statistics of the cycle-slipping phenomenon. Since the technique used to generate the sequence of narrow pulses may not be immediately evident, we will now describe it in some detail.

Let us consider first that the phase-locked loop is operating normally in the absence of noise. The digital delay circuit is adjusted so that the set and reset inputs to Flip-Flop 1 are exactly in phase. We note that when this condition is satisfied, the set and reset inputs to Flip-Flop 5 will be exactly out of phase so that it will behave as a phase detector with a low-pass characteristic as shown in Fig. 5. When noise is added to the system, the outputs of Flip-Flop 1 can be used as a measure of the phase-error ϕ. For instance, when the phase-error is positive the pulse width of the Q output is proportional to the magnitude of ϕ modulo 2π. In a similar manner the $\bar{Q}$ output may be used when the phase-error is negative.

It is convenient to restrict our attention initially to the case where ϕ is positive and consider the sequence of events which occur as the magnitude of ϕ increases. We observe that the differentiator circuit following the low-pass phase detector provides a pulse whenever $|\phi| = n\pi$, $n = 1, 2, \cdots$. The polarity of this pulse can be used to indicate whether ϕ is increasing in the positive direction (negative pulse) or in the negative direction (positive pulse). For the case under consideration, the negative pulse from the differentiator triggers a single-shot which in turn sets Flip-Flop 2. As the phase-error approaches $+2\pi$ radians, a narrow pulse obtained by delaying the trailing edge of the set input to Flip-Flop 1 by 2π radians enables the NAND gate No. 1 and a single cycle-slipping event is registered. Of course, it is necessary to route this event pulse back to Flip-Flop 2 and reset it immediately so that NAND gate 1 is disabled and consequently prevents any further event indications until the phase error has again changed by 2π radians.

In a similar manner, the $\bar{Q}$ output of Flip-Flop 1 and Q output of Flip-Flop 3 are in a set position when the phase error exceeds $-\pi$ radians. Since the remaining input to NAND gate 2 is a narrow pulse, obtained by delaying the

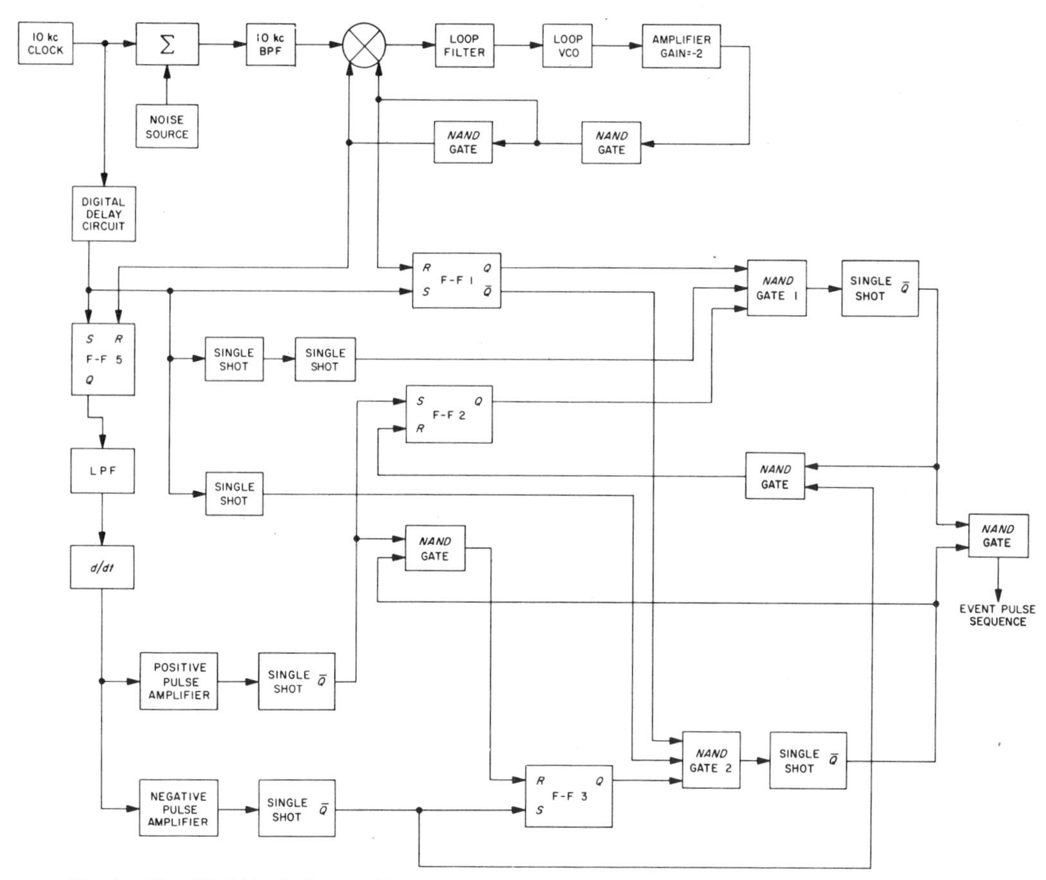

Fig. 6. Simplified block diagram illustrating the technique used to generate the event pulse sequence.

trailing edge of the set input to Flip-Flop 1 by a small increment, a pulse is generated as ϕ approaches -2π radians. The desired sequence is finally obtained by combining the outputs of NAND gates 1 and 2. Although it is impossible with this particular implementation to indicate uniquely when ϕ has changed by exactly 2π radians, it is possible using standard digital logic to obtain an indication when ϕ has changed by 99.3 percent of 2π radians. This is considered more than sufficient to indicate the occurrence of a cycle-slipping event.

Since the quantities which determine the statistics of cycle-slipping are represented by the random time intervals separating the leading edges of adjacent event pulses, it is evident that the recording of the pulse sequence on magnetic tape must necessarily include a time base or reference. The simplest way to achieve this is to synchronize the event pulses to some known clock which we will refer to as bit sync. A block diagram of the event synchronizer is shown in Fig. 7. The first event pulse sets Flip-Flop 1 and conditions the set level (SL) of Flip-Flop 2 so that it will set on the next negative level shift from the input single-shot. The NAND gate 1 will set Flip-Flop 3 on the trailing edge of the single-shot $\bar{Q}$ output starting a pulse (i.e., the new event pulse) which is terminated by the next leading edge of the single-shot $\bar{Q}$ output. A logic level diagram is included in Fig. 7 to illustrate these operations.

We note that although timing "noise" is introduced in the process of synchronizing the event pulse sequence, it can be neglected provided the clock frequency (which in our experiment was chosen as 2.0 kc/s) is much larger than the highest rate of cycle slipping. In the simplest terms, the synchronized event pulse sequence can now be recorded on magnetic tape as a sequence of binary "ones" and "zeros" by observing the synchronized sequence at bit sync time and writing a "1" if an event pulse is present and a "0" if no pulse is present. Since each binary digit on the tape corresponds to a time interval equal to the clock period Δt seconds, the time between cycle-slipping events can be determined by counting the number of consecutive "zeros" between adjacent "ones" recorded on the tape, adding one and multiplying by Δt. A machine language computer program was written to process the tape record and a FORTRAN subroutine was used to plot the distributions of the time intervals between cycle-slipping events.

VIII. Experimental Data for the Second-Order Phase-Locked Loop

The experimental phase-locked loop system was shown in block diagram form in Fig. 1. The input signal was a 10 kc/s unmodulated sinusoid carefully tuned to match the frequency of the VCO. The input band-pass filter, used to bandlimit the noise, had a measured one-sided noise bandwidth of 2.09 kc/s. The loop was designed to give a linearized loop noise bandwidth of $W_L = 12$ c/s. Additional loop parameters were shown in Fig. 2. For convenience, the signal level was maintained constant through the experi-

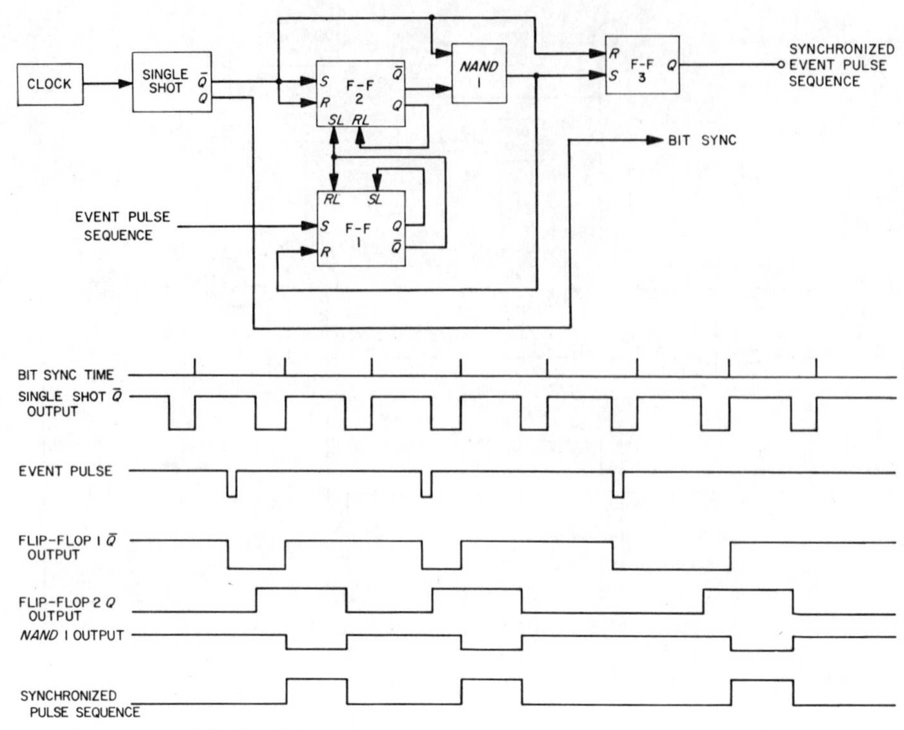

Fig. 7. Event pulse synchronizer and logic level diagram.

ment and the signal-to-noise ratio was varied by changing the noise power level.

A. The Phase-Error Distributions

The procedure used to measure the probability distribution of the phase-error consisted of setting a predetermined signal-to-noise ratio at the input to the loop and then performing the phase-error measurement as described in Section VII. It is convenient to specify the conditions under which the data was taken in terms of α, i.e., the signal-to-noise ratio in the noise bandwidth W_L of the loop. In particular, we consider three analytical models and determine the region of signal-to-noise ratio over which each model adequately describes the phase-error distribution relative to that measured in the laboratory. The three models to be considered are:

1) bandpass filter loop model (BPF); (27),
2) first-order or RC circuit loop model; (18) and (19),
3) linear model (LM); (20).

Illustrated in Fig. 8(a), for various values of the parameter α, is the experimental phase-error distribution (i.e., the point plot) with the distribution, (27), superimposed. We note that the BPF model 1) appears to be a good approximation to the distribution of the phase-error for values of $\alpha > 1.5$ dB. The ability with which model 2) characterizes the phase-error distribution when we consider $\alpha = \alpha_1$ is illustrated in Fig. 8(b). It appears that this model gives the best approximation to the distribution of the phase-error for the range of signal-to-noise ratios considered, viz., $\alpha > 0$ dB. Finally, Fig. 8(c) illustrates the manner in which the linear PLL model approximates the distribution of the phase-error. We observe that the linear model 3) appears to be a valid representation of the phase-error distribution for values of $\alpha > 9$ dB.

The cumulative distributions of the measured phase-error are shown in Fig. 9. Notice from this figure that $P[|\phi| > (\pi/2)]$ is essentially zero if $\alpha > 6$ dB. This observation may be used to justify the claim that the Fokker-Planck approximation given by (26) is valid for signal-to-noise ratios such that $\alpha > 6$ dB. Finally, it is clear from Fig. 9 that for $\alpha < 0$ dB the phase-error distribution tends to become uniformly distributed. This is in direct agreement with the Fokker-Planck solution, (23), and the BPF solution, (27).

In Fig. 10 the variance of each of the distributions obtained from the analytical models 1)–3) is compared graphically to the variance of the measured distribution. It is evident that a lower bound on the variance of the phase-error is given by the variance as determined from the linear PLL theory. However, the linear theory produces adequate results provided the signal-to-noise ratio $\alpha > 9$ dB. We also observe that the most accurate estimate of the variance of the phase error is given by the first-order or RC circuit loop model 2) for $\alpha > 0$ dB. Finally, the expression for the variance of the phase-error, as determined from the BPF model, produces reasonably adequate results for $0 \leq \alpha \leq 6$ dB and for $\alpha > 6$ dB is quite an accurate representation of the variance.

B. The Cycle-Slipping Distributions

The cycle-slipping phenomenon is best discussed by considering two regions of loop operation, viz., the region for which $\alpha > 6$ dB and the region for which $0 < \alpha < 6$ dB. For the particular loop design used to obtain the experimental data, the parameters were chosen such that $a/b \approx 0$ (i.e., $AK >> 1/\tau_2$). Consequently, the stable region indicated in

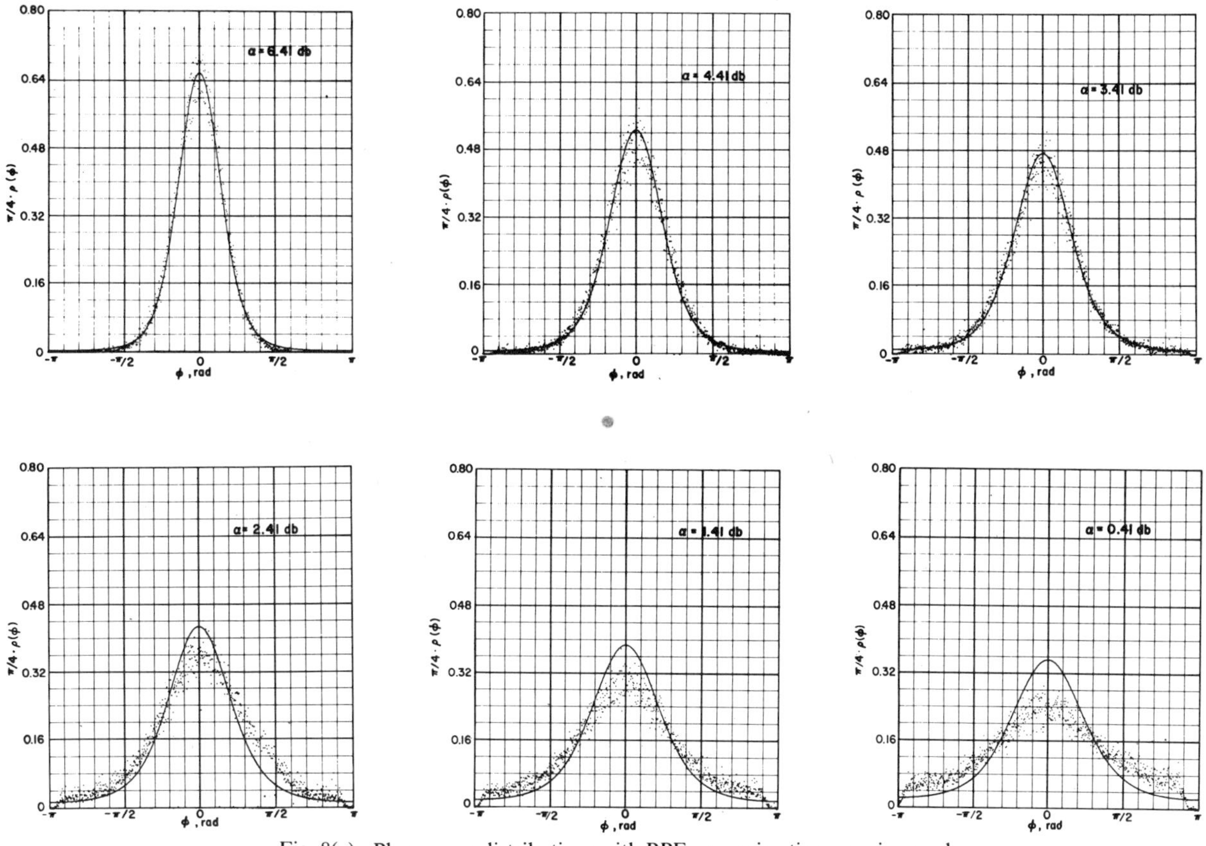

Fig. 8(a). Phase-error distributions with BPF approximation superimposed.

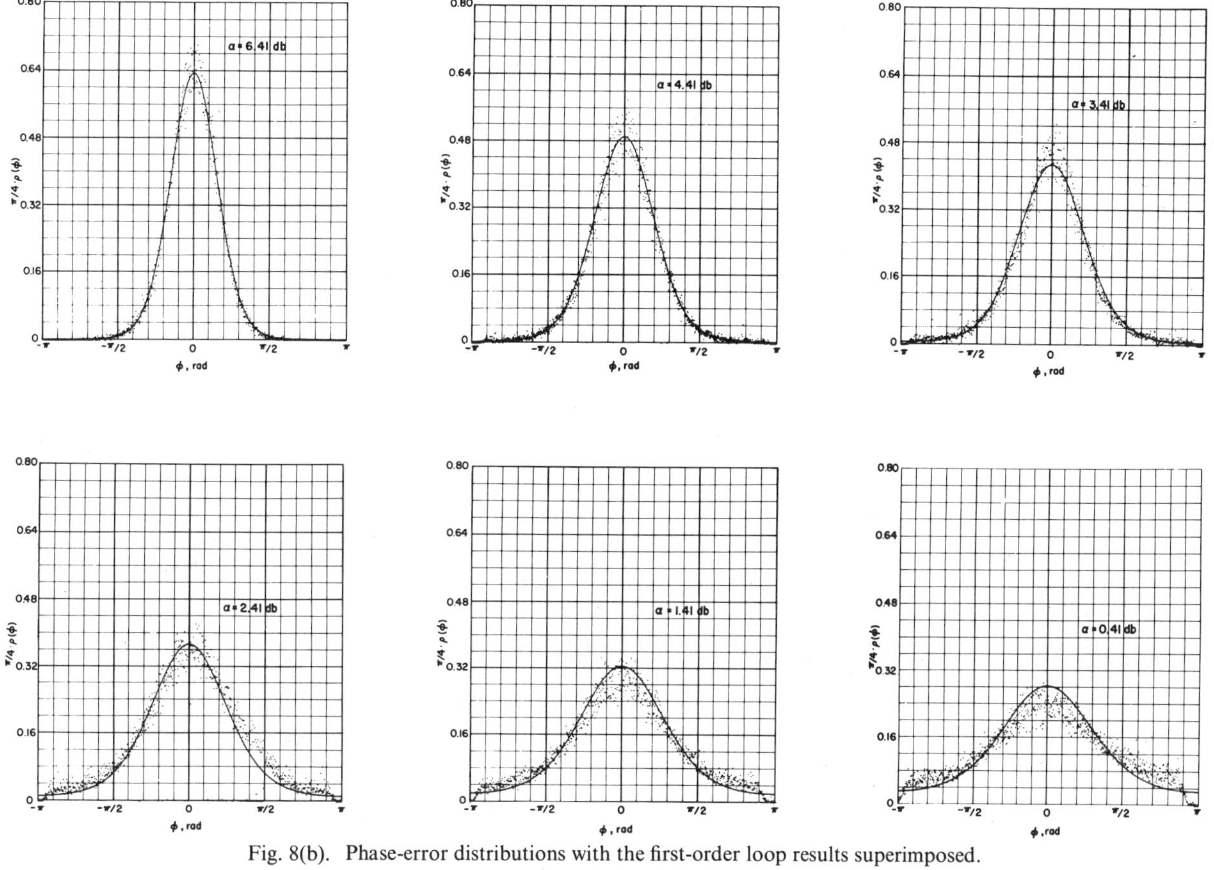

Fig. 8(b). Phase-error distributions with the first-order loop results superimposed.

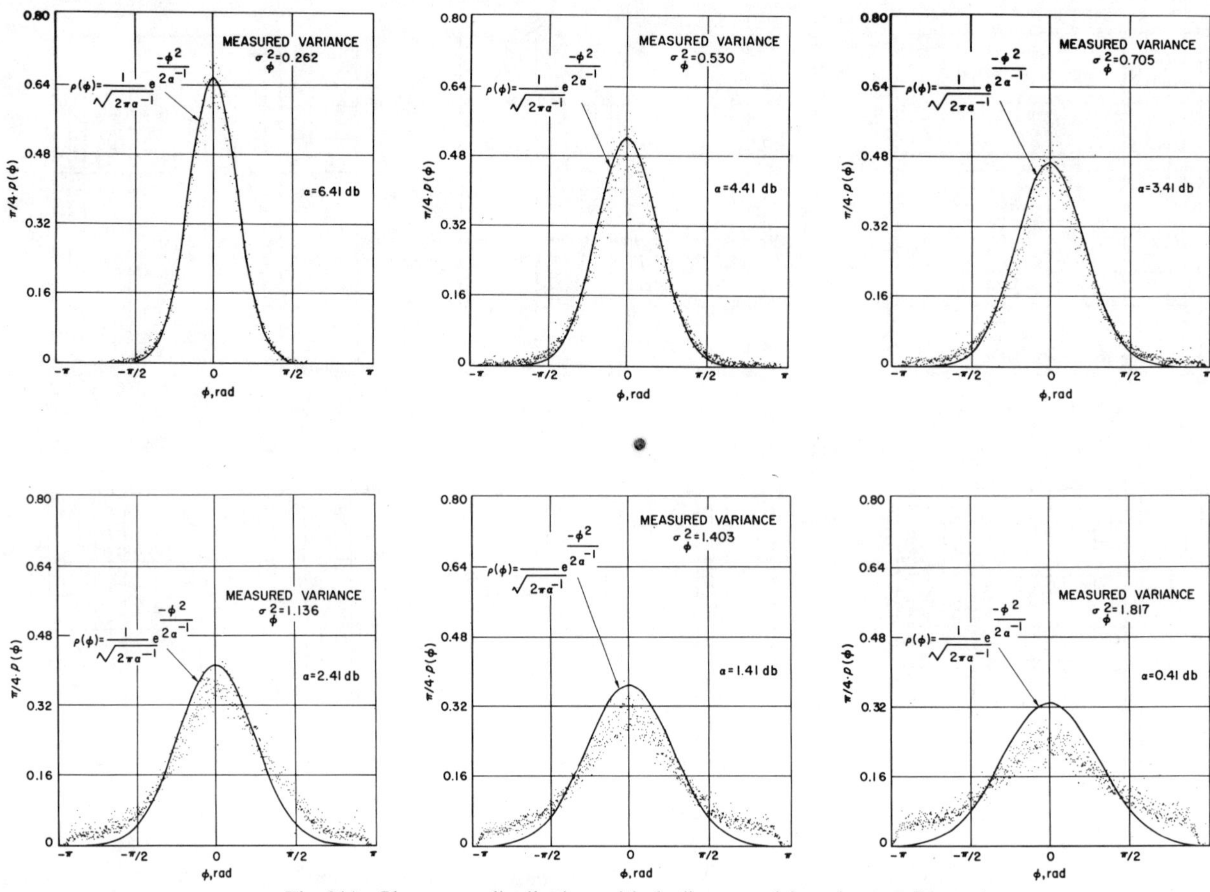

Fig. 8(c). Phase-error distributions with the linear model results superimposed.

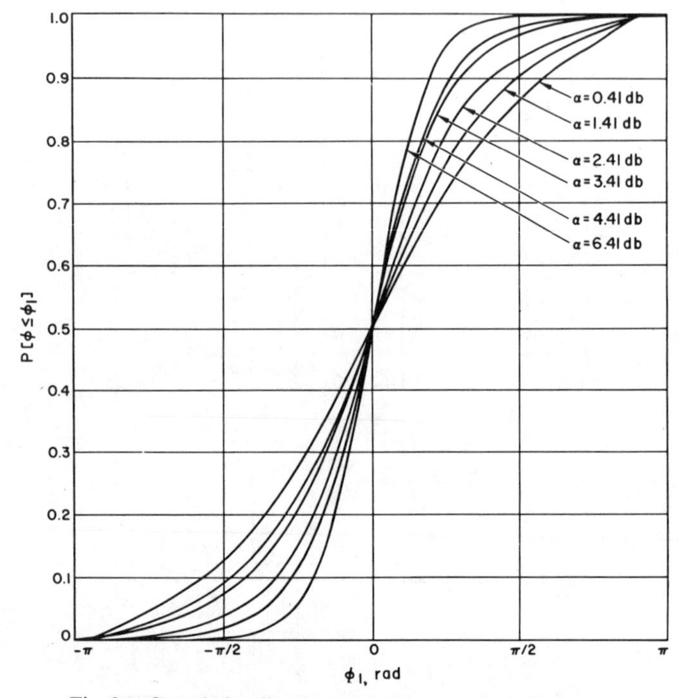

Fig. 9. Cumulative distributions of the measured phase-error.

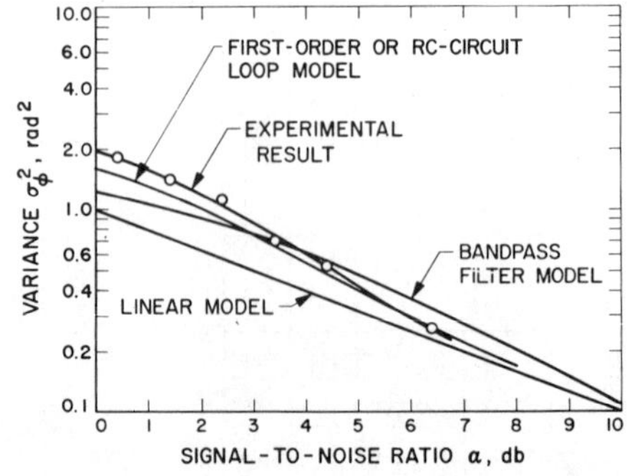

Fig. 10. Experimental and analytical results relative to the variance of the phase-error.

Fig. 4 is essentially bounded by $\pm\pi/2$. With this in mind we observe from the cumulative distribution plots of the phase-error shown in Fig. 9 that, for $\alpha > 6$ dB, the probability of $|\phi|$ being greater than $\pi/2$ is extremely small. This implies that, for such a range of α, the loop tends to operate for a relatively long period of time without slipping cycles. The truth of this statement is well supported indeed by empirical observations in the laboratory.

Let us now consider the event pulse sequence representative of the cycle-slipping phenomenon. We recall that in this sequence each pulse marks the time of occurrence of a cycle-slipping event. As we have indicated previously, once the loop slips a cycle it will, due to the inertia of the system, tend to slip several cycles in succession. Consequently, for $\alpha > 6$ dB we would expect a typical event pulse sequence to consist of widely separated pulse bursts containing relatively closely spaced event pulses. The lengths of the pulse bursts correspond to the time intervals the loop is "out-of-lock" while the lengths separating the bursts correspond to the time intervals the loop remains "in lock."

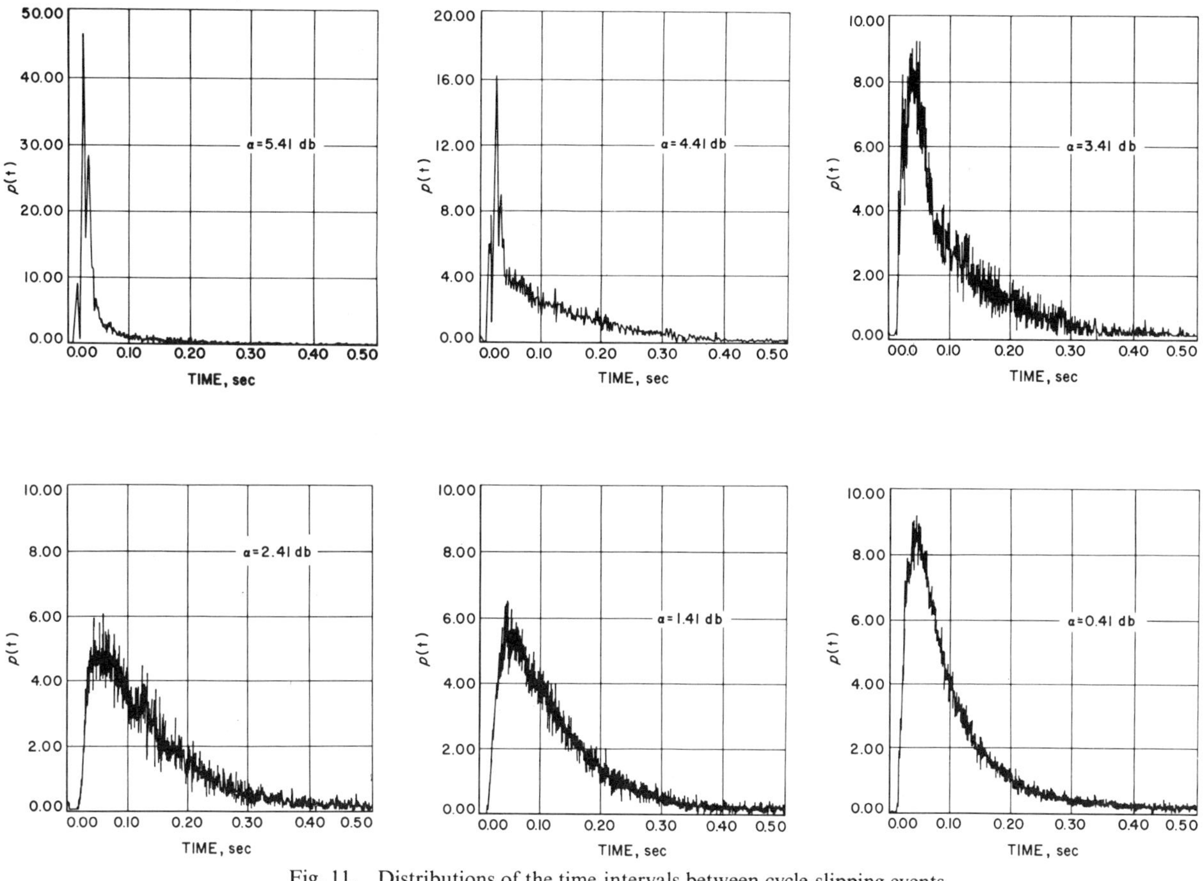

Fig. 11. Distributions of the time intervals between cycle-slipping events.

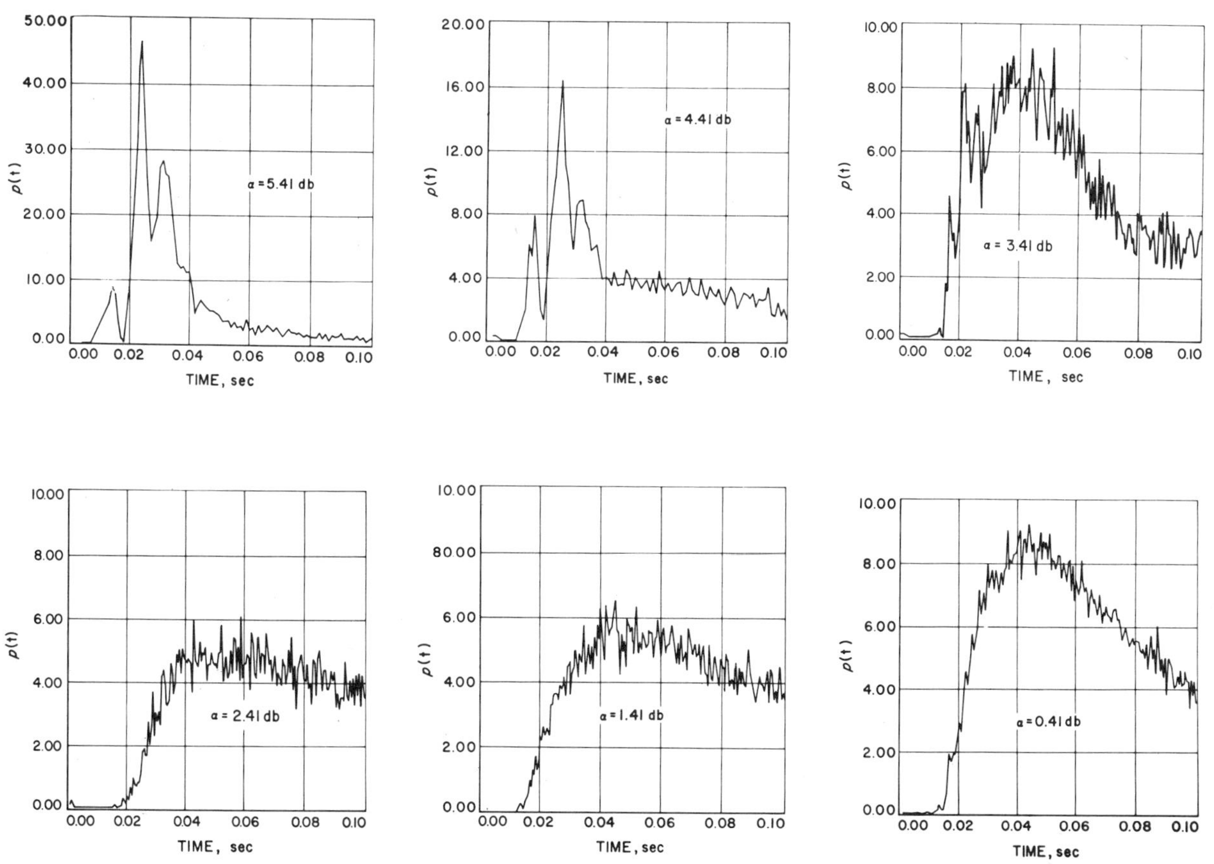

Fig. 12. Distributions of the time intervals between cycle-slipping events (expanded time scale).

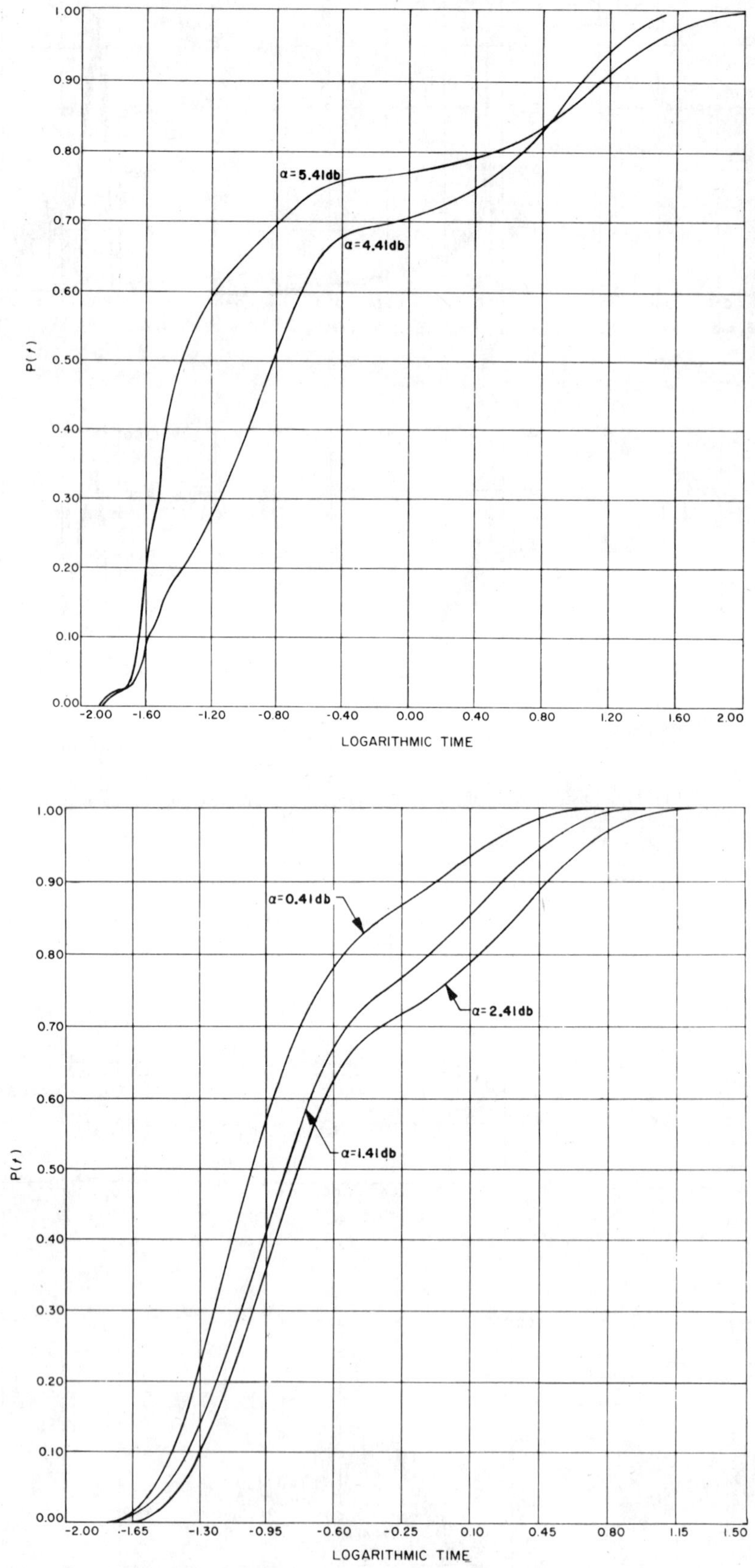

Fig. 13. Cumulative distributions of the time intervals between cycle-slipping events.

The experimental results which pertain to the cycle-slipping phenomenon are shown in Figs. 11 and 12. Shown in Fig. 11 for various values of α are the probability distributions $p(t)$ of the time interval t seconds between cycle-slipping events. These same distributions are shown in Fig. 12 on an expanded horizontal scale illustrating the fine grain structure of the distributions. Figure 13 shows the cumulative probability $P(t \leq t_1)$ for the values of α considered.

Consider first the distribution of the time intervals shown in Fig. 11 for $\alpha = 5.41$ dB. This particular distribution we associate with the time intervals separating the pulses contained within the bursts; it does not provide any essential information concerning the separation of pulse bursts or the length of time the loop remains in lock. This fact can be verified in at least two ways. First, we recognize from the cumulative distributions of the phase-error that at this particular signal-to-noise ratio the frequency with which the loop slips cycles is still reasonably small and, consequently, the loop remains in lock for relatively long periods of time. In addition, we observe from the cumulative distribution shown in Fig. 13 that for $\alpha = 5.41$ dB, the time intervals lying in the range 0.40 to 1.0 second occur with essentially zero probability. However, the probability that time intervals less than 0.40 second occur is approximately 0.77, whereas the probability that time intervals greater than 1.0 second occur is approximately 0.23. Consequently, we can conclude that for the particular loop design under consideration there exists two disjoint sets of time intervals for $\alpha > 5.5$ dB, viz., one set containing short time intervals related to the separation of pulses within the burst and a second set containing longer time intervals related to the length of time the loop stays in lock. The difficulty involved in distinguishing between these two sets of time intervals for lower values of α is clearly evident from an examination of the remaining cumulative distributions shown in Fig. 13. In fact, for $\alpha = 0.41$ dB the burst nature of the event pulse sequence has totally disappeared.

From a practical point of view it would be desirable to consider only the time intervals separating the pulse bursts and, as a result, obtain the statistics of the time intervals during which the loop remains in lock. Although this information is contained in the cumulative distributions, it is impossible to extract it for $\alpha < 5.5$ dB. However, with a minor modification of the instrumentation technique the distribution of the time intervals during which the loop remains "in-lock" can be measured. This modification simply amounts to starting the experiment under a known set of conditions and then automatically stopping the experiment with the first event pulse, recording the time it takes to occur, and then repeating the procedure a sufficient number of times.[1] With this distribution available, the corresponding cumulative distribution in Fig. 13 can be modified to obtain the cumulative distribution of the time intervals separating pulses within the bursts. Although this later distribution is of no real importance for design purposes, it does provide information regarding the fundamental behavior of the loop. The complexity of the "out-of-lock" behavior of the loop is immediately apparent from the trimodal structure of the distributions in Fig. 12 for $\alpha = 5.41$ dB and $\alpha = 4.41$ dB. It is conjectured that the position of the peaks in these distributions are directly related to the natural frequency of the loop, however, the exact relationship has not yet been determined.

[1] This experiment is currently being carried out to obtain the distribution of the first-passage times. From this distribution the mean time to loss of lock may be evaluated. This mean time corresponds to the expected time to loss of lock which Viterbi obtained for the first-order loop using the Fokker-Planck equation with an absorbing boundary.

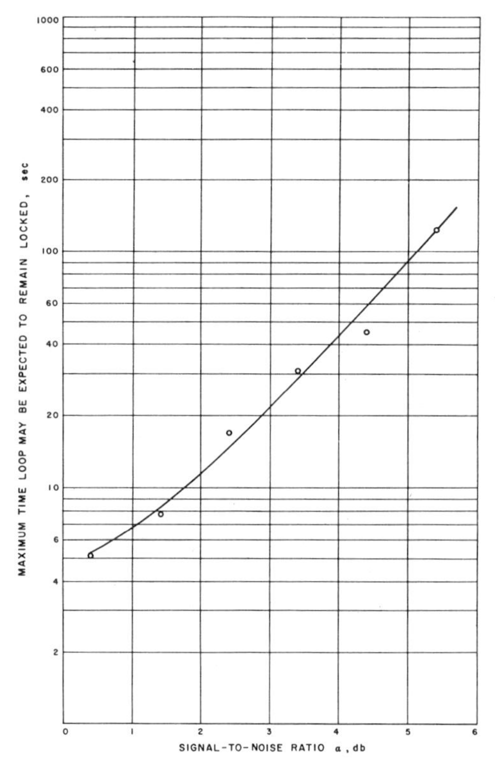

Fig. 14. The maximum time the loop may be expected to remain in lock vs. signal-to-noise ratio.

Although the cumulative distributions in Fig. 13 are at present more of academic than practical interest, they do provide a result useful in practical applications; viz., the maximum length of time the loop (with the particular set of parameters indicated in Fig. 2) may be expected to stay "in-lock." An estimate of these time intervals was obtained from the cumulative distributions in Fig. 13 by noting the value of the abscissa for which the cumulative distribution becomes approximately unity. These time values are shown in Fig. 14 as a function of α and may be considered as representing an upper bound on the mean time to loss of lock for the second-order PLL.

IX. Conclusions

This paper has been concerned with the *fundamental* behavior of the second-order phase-locked loop of greatest practical interest. This behavior is characterized by the probability distribution of the phase-error reduced modulo 2π and the probability distribution of the time intervals between cycle-slipping events.

In particular, it has been demonstrated through the use of laboratory measurements and approximate analytical solutions that the distribution of the phase-error is adequately characterized, in a signal-to-noise ratio region where the loop is usually expected to operate as a carrier tracking filter, by

$$p(\phi) = \frac{1}{2\sqrt{\pi}}\left[\frac{\exp(-\alpha/2)}{\sqrt{\pi}} + \sqrt{\frac{\alpha}{2\pi}}\cos\phi\exp\left(-\frac{\alpha}{2}\sin^2\phi\right)\left\{1 + \exp(-\alpha/2)\Phi\sqrt{\frac{\alpha}{2}}\cos\phi\right)\right\}\right]; \quad \begin{matrix}|\phi| \le \pi \\ \alpha \ge 1.5 \text{ dB.}\end{matrix} \tag{31}$$

This distribution has been obtained using the band-pass filter model of the loop. For large α, say $\alpha > 6$ dB, this distribution becomes

$$p(\phi) = \sqrt{\frac{\alpha}{2\pi}}\cos\phi\exp\left(-\frac{\alpha}{2}\sin^2\phi\right); \quad \begin{matrix}|\phi| < \frac{\pi}{2} \\ \alpha > 6 \text{ dB.}\end{matrix} \tag{32}$$

For $\alpha > 6$ dB the Fokker-Planck solution to the second-order stochastic differential equation of the loop is approximated by

$$p(\phi) = K\cos\phi\exp\left(-\frac{\alpha}{2}\sin^2\phi\right); \quad \begin{matrix}\alpha > 6 \text{ dB} \\ |\phi| \le \pi/2\end{matrix} \tag{33}$$

which is in direct agreement with the model obtained using the band-pass filter approximation (31) and the results obtained in the laboratory.

Finally, the distribution of the phase-error is approximated more simply by

$$p(\phi) = \frac{\exp(\alpha\cos\phi)}{2\pi I_0(\alpha)}; \quad \begin{matrix}|\phi| \le \pi \\ \alpha > 0 \text{ dB}\end{matrix} \tag{34}$$

which is the solution to the Fokker-Planck equation when the loop filter possesses unit gain at all frequencies or when the loop filter is an RC circuit [see (18)]. For $\alpha > 9$ dB, the distribution of the phase-error is essentially Gaussian with variance α^{-1}. The usefulness of these results becomes evident when one attempts to study the performance of coherent communication systems which employ noisy reference signals. A second area of application of the results arises in optimum power allocation analysis.

The cycle-slipping behavior of the loop is the most difficult to model analytically. In fact, for design purposes, it would appear that the distribution of the time-intervals between cycle-slipping events is of no great practical importance; however, in terms of understanding the *fundamental behavior* of the second-order loop this distribution is of utmost importance. The so-called "first-passage time," i.e., mean time to loss of lock, is of greater interest in practical designs and can be measured using a slight modification to the instrumentation techniques employed in observing the distribution of time intervals between cycle-slipping events. One observation, relative to the cycle-slipping behavior of the loop, pertains to the nature with which the loop actually slips cycles. It appears, from the empirical observations, that there is a sharp dichotomy where the loop begins to slip cycles frequently and where the loop slips cycles infrequently. This occurs in a region where 5 dB $< \alpha <$ 6 dB. In essence, this means that the error introduced in a Doppler measurement is largely due to thermal noise or oscillator instabilities if $\alpha > 6$ dB.

References

[1] A. J. Viterbi, "Phase-locked loop dynamics in the presence of noise by Fokker-Planck techniques," *Proc. IEEE*, vol. 51, pp. 1737–1753, December 1963.

[2] R. C. Tausworthe, "Theory and practical design of phase-locked receivers," Jet Propulsion Labs., Pasadena, Calif., JPL Tech. Rept. 32-819, vol. I, February 1966.

[3] W. C. Linsdey, "The detection of PSK signals with a noisy phase reference," *1965 Proc. Nat'l Telemetering Conf.*, pp. 50–53.

[4] ——, "Optimal design of one-way and two-way coherent communication links," to be published in IEEE Trans. on Communication Technology.

[5] J. Stiffler, "Bit and subcarrier synchronization in a binary PSK communication system," presented at the 1964 Nat'l Telemetering Conf., Los Angeles, Calif.

[6] V. I. Tikhonov, "Influence of noise on phase-locked oscillator operation," *Automatika i Telemekhanika*, vol. 20, September 1959.

[7] ——, "Phase-lock automatic frequency control operation in the presence of noise," *Automatika i Telemekhanika*, vol. 21, p. 301, March 1960.

[8] D. Middleton, *Introduction to Statistical Communication Theory*. New York: McGraw-Hill, 1960.

[9] F. J. Charles, "A second-order phase-locked loop study," M.S. thesis, Syracuse University, Syracuse, N. Y., November 1965.

[10] J. P. Frazier and J. Page, "Phase-lock loop frequency acquisition study," *IRE Trans. on Space Electronics and Telemetry*, vol. SET-8, pp. 210–227, September 1962.

[11] R. W. Sanneman and J. R. Rowbotham, "Unlock characteristics of the optimum type two phase-locked loop," *IEEE Trans. on Aerospace and Navigational Electronics*, vol. ANE-11, pp. 15–24, March 1964.

[12] D. L. Schilling, "The response of an automatic phase control system to FM signals and noise," *Proc. IEEE*, vol. 51, pp. 1306–1316, October 1963.

[13] D. L. Schilling and J. Billig, "On the threshold extension capability of the PLL and FDMFB," *Proc. IEEE* (*Correspondence*), vol. 52, pp. 621–622, May 1964.

The Response of a Phase-Locked Loop to a Sinusoid Plus Noise*

STEPHEN G. MARGOLIS†

Summary—**The phase-locked loop is a practical device for separating a sinusoidal signal from additive noise. In this device the incoming signal-plus-noise is multiplied by a noise-free sinusoid generated by a voltage-controlled oscillator (vco). The filtered product is used to lock the phase of the vco output to that of the incoming signal, thus producing a relatively clean version of the incoming signal in which the noise manifests itself as a small phase modulation. Analysis of this noise-produced phase modulation is complicated by the presence of the multiplier at the input to the loop. This paper presents a perturbation method which reduces this inherently nonlinear servo analysis problem to the analysis of a series of linear systems, the first of which is related to the linear model used by previous authors. The perturbation technique permits the phase modulation resulting from an arbitrary noise input to be computed to any desired accuracy. This analysis is particularly useful in predicting loop performance when it is used as a narrow-band receiver in a phase-comparison angle-measuring system.**

List of Symbols

THE FOLLOWING is a list of the symbols used in this paper:

A = peak amplitude of sinusoidal input, volts.

A_1 = peak amplitude of sinusoidal input to reference channel, volts.

$A_m(\theta_s)$ = peak amplitude of sinusoidal input to measurement channel, volts.

C = capacitance used in compensating network, farads.

$f(t)$ = random time-function, dimensionless.

$g(t)$ = random time function defined in (30c) dimensionless.

$H(S)$ = transfer function of phase-locked loop, relating small transient changes in input phase angle to the resulting changes in output phase angle.

$h(\tau)$ = inverse L-transform of $H(S)$.

$J = \frac{2N}{A}$ = noise-to-signal ratio, dimensionless.

$K = \frac{GA}{2} \times \frac{(1 \text{ radian})}{\text{volt-second}}$ = loop gain constant, radians/second.

N = noise amplitude factor, volts.

N_1 = noise amplitude factor for reference channel, volts.

$N_m(\theta_N)$ = noise amplitude factor for measurement channel, volts.

R_1, R_2 = resistances used in compensating network, ohms.

S = complex frequency, radians/second.

t = time, seconds.

T = averaging interval, seconds.

$T_1 = (R_1 + R_2)C$ = larger time-constant of compensating network seconds.

$T_2 = R_2C$ = smaller time-constant of conpensating network, seconds.

$v_0(t)$ = output of voltage-controlled oscillator, volts.

$v_1(t)$ = input to phase-locked loop, volts.

$v_2(t)$ = output of multiplier, volts.

$v_4(t)$ = input to control terminal of voltage-controlled oscillator, volts.

$v_5(t)$ = output of angle-measuring system, volts.

$v_m(t)$ = input to measurement channel from antenna, volts.

$v_r(t)$ = input to measurement channel from reference channel, volts.

W = noise spectral density, $\frac{(\text{volts})^2\text{-second}}{\text{radian}}$.

$\Delta\omega$ = half-bandwidth of narrow-band noise, radians/second.

$\Delta\omega' = \frac{\Delta\omega}{K}$ = normalized half-bandwidth of narrow-band noise, dimensionless.

θ_N = angle between reference line and line joining noise-source location and antenna location, radians.

θ_s = angle between reference line and line joining signal-source location and antenna location, radians.

$\lambda(t) = \lambda_{vco} - \lambda_i$ = noise-produced phase error, radians.

$\lambda_1(t), \lambda_2(t) \cdots \lambda_n(t)$ = components of $\lambda(t)$, radians.

$\lambda_i(t)$ = phase angle of input sinusoid, radians.

$\Lambda_i(s)$ = L-transform of $\lambda_i(t)$.

$\lambda_{vco}(t)$ = phase angle of voltage-controlled oscillator output, radians.

$\Lambda_{vco}(s)$ = L-transform of $\lambda_{vco}(t)$.

σ_1 = rms noise level at input to reference channel, volts.

σ_f = rms value of $f(t)$, dimensionless.

σ_r = rms noise level at input to measurement channel, volts.

τ = dummy variable used in convolution integral, seconds.

$\omega_0 = \omega_c - \omega_1$ = difference between center-frequency of noise band and angular frequency of input sinusoid, radians/second.

$\omega_0' = \frac{\omega_0}{K}$ = normalized difference frequency, dimensionless.

ω_1 = angular frequency of input sinusoid; center-frequency of voltage controlled oscillator, radians/second.

ω_c = center-frequency of narrow noise band, radians/second.

ω_N = closed-loop bandwidth of compensated phase-locked loop, radians/second.

* Manuscript received by the PGIT, November 23, 1956. This paper was presented at IRE-WESCON, Los Angeles, Calif.; August 21-24, 1956. Results are presented of one phase of research carried out at Jet Propulsion Lab., Calif. Inst. Tech., under contract no. DA-04-495-Ord 18, sponsored by the Dept. of the Army, Ordnance Corps.

† Westinghouse Electric Corp., Pittsburgh, Pa. Formerly with Jet Propulsion Lab., Calif. Inst. Tech., Pasadena, Calif.

Reprinted from *IRE Trans. Inform. Theory*, vol. IT-3, pp. 136-142, June 1957.

Introduction

The basic function of a phase-locked loop is to lock the phase of the output of a voltage-controlled oscillator (vco)[1] to the phase of an incoming sinusoidal signal. A properly designed loop is capable of performing this function even in the presence of noise power which greatly exceeds the signal power.[2] The noise manifests itself as a random phase modulation of the vco output. This random phase modulation is usually treated by deriving an equivalent transfer function for the loop, by which transient changes in input phase can be related to the resulting changes in output phase; the incoming signal-plus-noise is usually approximated by a signal phase-modulated by noise, and the output phase modulation is then found by passing the input phase modulation through the equivalent transfer function. This paper takes a somewhat more direct approach to the problem, and thus avoids the necessity of converting a signal-plus-noise to a signal phase-modulated by noise; consequently avoiding defining equivalent phase angle when the noise power greatly exceeds the signal power. The results are easily interpreted in terms of the noise-free behavior of the loop, which will be reviewed first.

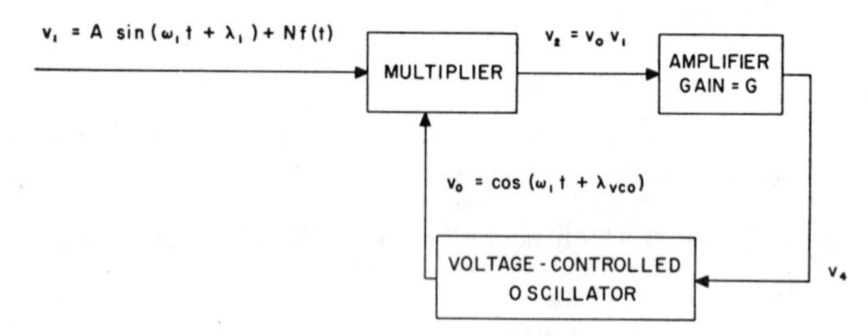

Fig. 1—Basic structure of a phase-locked loop.

Fig. 1 shows the basic structure of a phase-locked loop with a noisy input. Assuming for the moment that $N = 0$ (*i.e.*, the input signal is free from noise) and neglecting terms with frequency $2\omega_1$, the output of the amplifier is

$$v_4 = \frac{GA}{2} \sin (\lambda_i - \lambda_{vco}) \doteq \frac{GA}{2} (\lambda_i - \lambda_{vco}) \quad \text{for} \quad (\lambda_i - \lambda_{vco}) \ll \frac{\pi}{2}. \tag{1}$$

It is assumed that the sensitivity of the vco is 1 radian per second per volt,[3] so that

$$\dot{\lambda}_{vco} = v_4 \times \frac{1 \text{ radian}}{\text{volt-second}}. \tag{2}$$

Combining (1) and (2), the equation relating the vco phase to the input phase is[4]

$$\dot{\lambda}_{vco} + K\lambda_{vco} = K\lambda_i \tag{3}$$

where

$$K = \tfrac{1}{2}GA \times \frac{1 \text{ radian}}{\text{volt-second}}.$$

The steady-state solution is $\lambda_{vco} = \lambda_i$. The response of this simple loop to small changes in the input phase λ_i can be found by L-transforming (3). The result is

$$H(S) = \frac{\Lambda_{vco}(S)}{\Lambda_i(S)} = \frac{K}{S + K} \tag{4}$$

which defines the equivalent transfer function for the loop. Here Λ_{vco} and Λ_i are the transforms of λ_{vco} and λ_i, respectively.

In the simple loop the constant K determines both the bandwidth of the transfer function and the range of frequencies over which the loop will lock (pull-in range).[5] These parameters may be controlled separately by including an RC compensating network, connected between

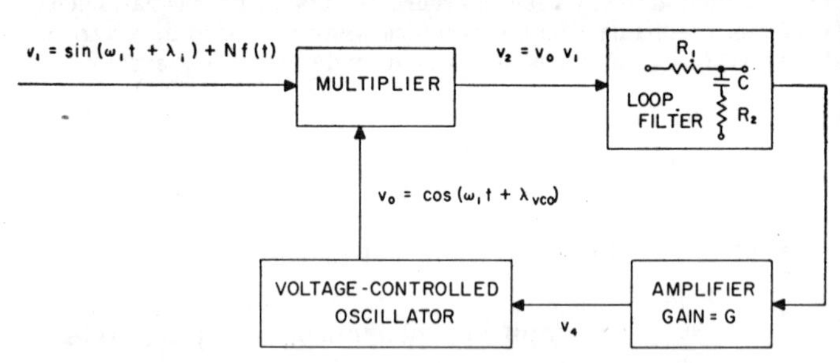

Fig. 2—Phase-locked loop with RC compensation.

the multiplier output and the vco input, as shown in Fig. 2. For the compensated loop.[5,6]

$$H(S) = \frac{\Lambda_{vco}(S)}{\lambda_i(S)} = \frac{\dfrac{KT_2}{T_1} S + \dfrac{K}{T_1}}{S^2 + \left(\dfrac{1}{T_1} + \dfrac{KT_2}{T_1}\right)S + \dfrac{K}{T_1}} \tag{5}$$

where

$$T_1 = (R_1 + R_2)C$$

$$T_2 = R_2C$$

and

$$K = \tfrac{1}{2}GA \times \frac{1 \text{ radian}}{\text{volt-second}},$$

as before. Here again, the steady-state solution is $\lambda_{vco} = \lambda_i$ and the pull-in range is $\pm K$ radians per second, but the bandwidth of the transfer function ω_N is equal to $\sqrt{K/T_1}$. For properly damped response to changes in λ_i it is usual to choose

$$\frac{KT_2}{T_1} \doteq 1.5\omega_N$$

[1] In this paper, the term voltage-controlled oscillator refers to an oscillator in which the deviation of the output *frequency* from its nominal valve is proportional to a control voltage.

[2] R. Jaffee and E. Rechtin, "Design and performance of phase-locked circuits capable of near optimum performance over a wide range of input signal and noise levels," IRE Trans., Vol. IT-1, p. 66; March, 1955.

[3] The actual numerical value of the vco sensitivity can be absorbed into the gain G.

[4] By definition, K has the units radians/second.

[5] W. J. Gruen, "Theory of afc synchronization," Proc. IRE, vol. 41, pp. 1043-1048; September, 1953. See pp. 1045-1046.

[6] P. F. Ordung, J. E. Gibson, and B. J. Shinn, "Closed Loop Automatic Phase Control," presented at AIEE Summer and Pacific General Meeting, Los Angeles, Calif.; June 21-25, 1954.

and to make

$$\frac{KT_2}{T_1} \gg \frac{1}{T_1}.$$

The Analysis of Noise in Simple Loops

Referring again to Fig. 1, the input to the simple loop is

$$v_1 = A \sin(\omega_1 t + \lambda_i) + Nf(t). \tag{6}$$

For convenience in the subsequent analysis, the noise has been written as the product of a constant N and a time-varying part $f(t)$. The vco output is

$$v_0 = \cos(\omega_1 t + \lambda_{vco}). \tag{7}$$

The multiplier forms the product $v_0 v_1$, and this is

$$v_2 = \frac{A}{2}\left[\frac{2N}{A} f(t) \cos(\omega_1 t + \lambda_{vco}) + \sin(\lambda_i - \lambda_{vco})\right]. \tag{8}$$

Here again, periodic terms with frequency $2\omega_1$ have been ignored. In practice, the vco does not respond to these terms when they appear at its input. The amplifier output is

$$v_4 = Gv_2 \tag{9}$$

and because v_4 controls the frequency of the voltage-controlled oscillator,

$$\dot{\lambda}_{vco} = v_4 \times \left(\frac{1}{\text{volt-second}}\right). \tag{10}$$

Eqs. (8)-(10) combine to give

$$\dot{\lambda} + K \sin \lambda = KJf(t) \cos(\omega_1 t + \lambda_i + \lambda) \tag{11}$$

where $K = \frac{1}{2}GA$, $J = 2N/A$, and λ has been written for $\lambda_{vco} - \lambda_i$. Eq. (11) describes the behavior of the simple loop without any approximations other than those implicit in the use of an ideal multiplier to represent a physical component.

In cases of practical interest the phase modulation λ is small—certainly less than $\pm 90°$. In addition, if the statistical properties of the time function $f(t)$ are fixed, the amplitude of λ must increase if N, which may be regarded as the amplitude of the noise, is increased. Thus if λ is written as a power series[7]

$$\lambda(t) = \lambda_0(t) + J\lambda_1(t) + J^2\lambda_2(t) + J^3\lambda_3(t) + \cdots \tag{12}$$

the solution of (11) can be reduced to solving for the coefficients $\lambda_0(t)$, $\lambda_1(t)$, $\lambda_2(t)$, $\lambda_3(t)$, $\cdots$ each of which is a random time function. This is easily done by substituting the series (12) into (11) and replacing $\sin \lambda$ by $(\lambda - \lambda^3/6 + \cdots)$ and $\cos \lambda$ by $(1 - \lambda^2/2 + \cdots)$ whenever they appear and then equating the coefficients of equal powers of J. The result is the series of linear equations,

$$\lambda_0 = 0 \tag{13}$$

$$\dot{\lambda}_1 + K\lambda_1 = Kf(t) \cos(\omega_1 t + \lambda_i) \tag{13a}$$

$$\dot{\lambda}_2 + K\lambda_2 = K[-\lambda_1 f(t) \sin(\omega_1 t + \lambda_i)] \tag{13b}$$

$$\dot{\lambda}_3 + K\lambda_3 = K\left[\frac{\lambda_1^3}{6} - \frac{\lambda_1^2}{2} f(t) \cos(\omega_1 t + \lambda_i) - \lambda_2 f(t) \sin(\omega_1 t + \lambda_i)\right] \tag{13c}$$

$$\dot{\lambda}_n + K\lambda_n = K[\textit{a function of } \lambda_{n-1}, \lambda_{n-2}, \cdots, \lambda_1].$$

These equations have a simple interpretation. Eq. (13a) together with the series (12) implies that a first approximation to the effect of noise on the loop can be found by analyzing the system shown schematically in Fig. 3(a). The noise $f(t)$ is multiplied by $\cos(\omega_1 t + \lambda_i)$ and the resulting low-frequency noise after passing through a low-pass filter with transfer function $K/(S + K)$ and an attenuator with a transmission equal to J gives the phase modulation $\lambda(t)$.

Fig. 3(b) is a schematic representation which includes the effect of the first three terms in the series (12). An even more detailed physical model can be constructed by including four terms in the series (12) as shown in Fig. 3(c). Despite the seeming complexity of these models, the computation is straightforward in that there is no feedback; all signals flow from left to right. In addition, each equation contains only one unknown.

The physical model shown in Fig. 3(b) is useful in finding

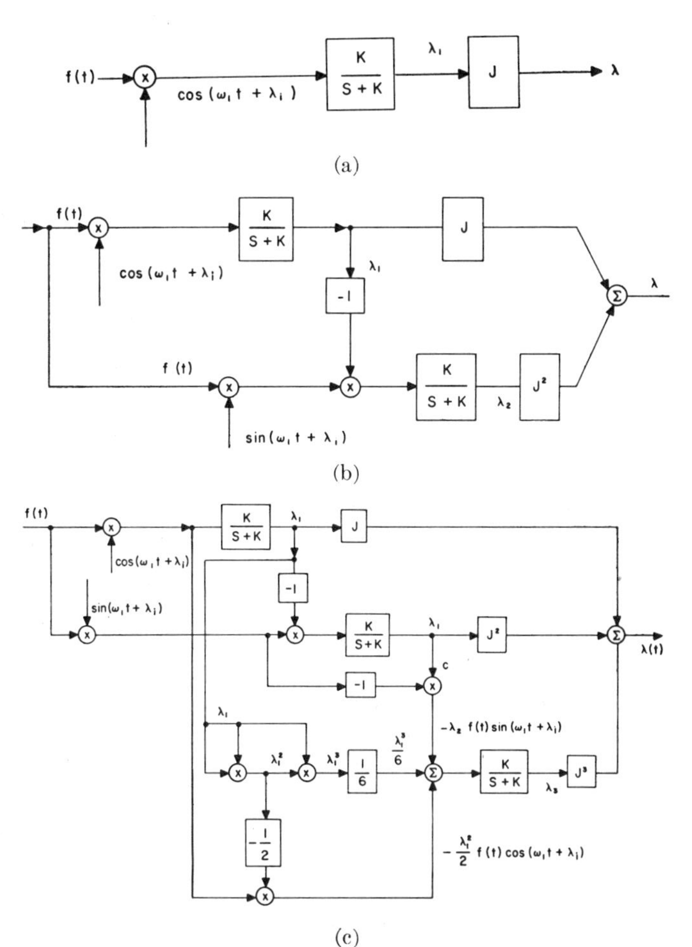

Fig. 3—(a) First approximation to effects of noise on a simple loop. (b) Model for second approximation to effects of noise on a simple loop. (c) Model for third approximation to effects of noise on a simple loop.

[7] The use of this series expansion was suggested by L. R. Welch of the Jet Propulsion Lab., Pasadena, Calif.

any steady-state phase error which may be produced when the noise $Nf(t)$ is multiplied by the phase-modulated vco output. Any correlation between these waveforms would be expected to produce a dc component in the output of the multiplier which would eventually manifest itself as a steady-state phase offset. The physical model makes it clear that λ_1 contains no steady component, but that λ_2 may contribute a steady-state phase error owing to the multiplication of the time function $f(t) \sin(\omega_1 t + \lambda_i)$ by λ_1, which is just a filtered version of $f(t) \cos(\omega_1 t + \lambda_i)$.

Since the impulse response of a system with transfer function $K/(S + K)$ is $K\epsilon^{-K\tau}$ the time function λ_1 is the convolution of $K\epsilon^{-K\tau}$ and $f(t) \cos(\omega_1 t + \lambda_i)$, *i.e.*,

$$\lambda_1 = \int_0^\infty K\epsilon^{-K\tau} f(t - \tau) \cos [\omega_1(t - \tau) + \lambda_i] \, d\tau \tag{14}$$

and the average value of λ_2 is just the average value of $-\lambda_1 f(t) \sin(\omega_1 t + \lambda_i)$ or

$$\lambda_{2\text{avg}} = -\lim_{T\to\infty} \frac{1}{2T} \int_{-T}^{T} \left\{ \int_0^\infty K\epsilon^{-K\tau} f(t - \tau) \cdot \cos [\omega_1(t - \tau) + \lambda_i] \, d\tau \right\} f(t) \sin(\omega_1 t + \lambda_i) \, dt. \tag{15}$$

Since $f(t)$ is assumed to be a random time function with no periodic components, (15) reduces to

$$\lambda_{2\text{avg}} = -\frac{K}{2} \int_0^\infty \epsilon^{-K\tau} \phi_{ff}(\tau) \sin \omega_1 \tau \, d\tau \tag{16}$$

where ϕ_{ff} has its usual definition,

$$\phi_{ff} = \lim_{T\to\infty} \frac{1}{2T} \int_{-T}^{T} f(t) f(t - \tau) \, d\tau. \tag{17}$$

For the specific case in which $f(t)$ is white noise which has been passed through a filter with an ideal rectangular band-pass characteristic, as shown in Fig. 4,

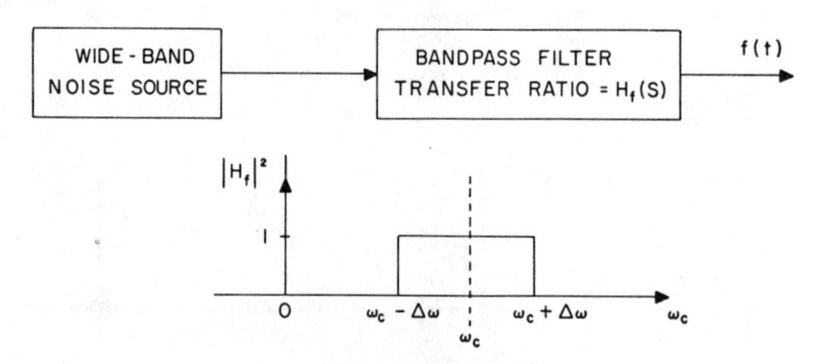

Fig. 4—Generation of $f(t)$.

$$\phi_{ff} = 4W \frac{\sin \Delta\omega\tau}{\tau} \cos \omega_c \tau \tag{18}$$

$$= \sigma_f^2 \frac{\sin \Delta\omega\tau}{\Delta\omega\tau} \cos \omega_c \tau \tag{18a}$$

where W is the noise power per unit bandwidth. For this type of noise,

$$\lambda_{2\text{avg}} = -\frac{K\sigma_f^2}{2\Delta\omega} \int_0^\infty \frac{\epsilon^{-K\tau}}{\tau} \sin \Delta\omega\tau \cos \omega_c \tau \sin \omega_1 \tau \, d\tau. \tag{19}$$

The definite integral (19) can be found in integral tables;[8] its value is

$$\lambda_{2\text{avg}} = -\frac{K\sigma_f^2}{16\Delta\omega} \ln \frac{K^2 + (\omega_1 + \omega_c + \Delta\omega)^2}{K^2 + (\omega_1 + \omega_c - \Delta\omega)^2} - \frac{K\sigma_f^2}{16\Delta\omega} \ln \frac{K^2 + (\omega_1 - \omega_c + \Delta\omega)^2}{K^2 + (\omega_1 - \omega_c - \Delta\omega)^2}. \tag{20}$$

If the noise is confined to a narrow band, $\omega_1 + \omega_c \gg \Delta\omega$. In this case, the first term in (20) becomes small compared to the second term; evaluation of the second term is simplified by the introduction of the normalized variables

$$\omega_0' = \frac{\omega_c - \omega_1}{K}$$

and

$$\Delta\omega' = \frac{\Delta\omega}{K}.$$

For narrow-band noise

$$\lambda_{2\text{avg}} \cong \frac{\sigma_f^2}{16\Delta\omega'} \ln \frac{1 + (\omega_0' + \Delta\omega')^2}{1 + (\omega_0' - \Delta\omega')^2} \tag{21}$$

which is an odd function of ω_0'. The resulting phase error is

$$J^2\lambda_{2\text{avg}} \cong \frac{1}{4} \frac{N^2\sigma_f^2}{A^2\Delta\omega'} \ln \frac{1 + (\omega_0' + \Delta\omega')^2}{1 + (\omega_0' - \Delta\omega')^2}. \tag{22}$$

The ratio $N^2\sigma_f^2/A^2$ is [from (6) and (18)] just the square of the ratio of rms noise to peak signal, measured at the input to the loop. The phase bias is thus a function of: $N^2\sigma_f^2/A^2$, the ratio of mean-square noise to the square of signal amplitude; $\Delta\omega'$, the ratio of noise bandwidth (see Fig. 4) to loop bandwidth K; and ω_0', the ratio of ω_0 to loop bandwidth. Here ω_0 is the difference between the signal frequency ω_1 and the noise center frequency, ω_c. Fig. 5 is a plot of $J^2\lambda_2$ vs ω_0' for $N^2\sigma_f^2/A^2 = 1$ and $\Delta\omega' = 1$.

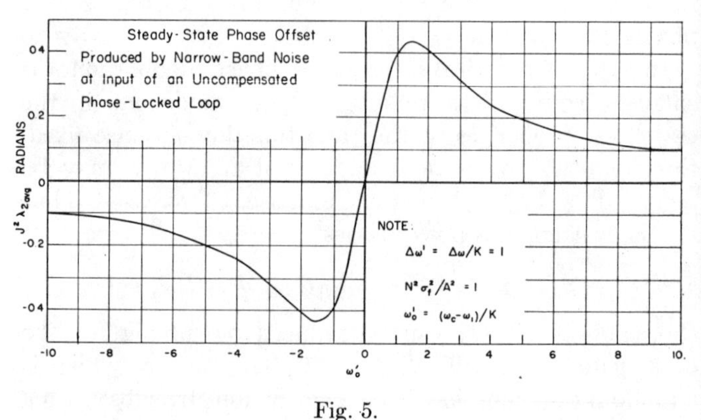

Fig. 5.

The phase error vanishes when $\omega_0' = 0$; *i.e.*, when the center-frequency of the noise band coincides with ω_1.

Noise in Loops with RC Compensation

The analysis of practical phase-lock circuits, which include an RC compensation network as shown in Fig. 2,

[8] W. Grobner and N. Hofreiter, "Integraltafel; Zweiter Teil, Bestimmte Integrale," Vienna, Springer-Verlag; 1950.

follows the same general plan as that used to analyze the uncompensated loop. As before, ignoring components in v_2 with frequency $2\omega_1$

$$v_1 = A \sin(\omega_1 t + \lambda_i) + N f(t) \tag{23}$$

$$v_0 = \cos(\omega_1 t + \lambda_{vco}) \tag{24}$$

$$v_2 \cong \frac{A}{2}\left[\frac{2N}{A} f(t) \cos(\omega_1 t + \lambda_{vco}) + \sin(\lambda_i - \lambda_{vco})\right] \tag{25}$$

and

$$\dot{\lambda} = v_4 \times \left(\frac{1}{\text{volt-second}}\right). \tag{26}$$

Because of the inclusion of the RC network, however,

$$v_4 + T_1 \dot{v}_4 = G(v_2 + T_2 \dot{v}_2) \tag{27}$$

$$\text{where } T_1 = (R_1 + R_2)\,C \text{ and } T_2 = R_2 C. \tag{28}$$

Eq. (27) becomes

$$\ddot{\lambda} + \frac{1}{T_1}\dot{\lambda} = \frac{K}{T_1}\Big\{J f(t) \cos(\omega_1 t + \lambda_i + \lambda) - \sin\lambda + T_2 \frac{d}{dt}\left[J f(t) \cos(\omega_1 t + \lambda_i + \lambda) - \sin\lambda\right]\Big\}. \tag{29}$$

Substitution of the power series (12) into (29) and collection of the coefficients of equal powers of J lead to a series of equations which resembles the series (13):

$$\lambda_0 = 0 \tag{30}$$

$$\ddot{\lambda}_1 + \left(\frac{1}{T_1} + \frac{KT_2}{T_1}\right)\dot{\lambda}_1 + \frac{K}{T_1}\lambda_1 = \frac{K}{T_1} f(t) \cos(\omega_1 t + \lambda_i) + \frac{KT_2}{T_1}\frac{d}{dt}\left[f(t) \cos(\omega_1 t + \lambda_i)\right] \tag{30a}$$

$$\ddot{\lambda}_2 + \left(\frac{1}{T_1} + \frac{KT_2}{T_1}\right)\dot{\lambda}_2 + \frac{K}{T_1}\lambda_2 = -\frac{K}{T_1}\lambda_1 f(t) \sin(\omega_1 t + \lambda_i) - \frac{KT_2}{T_1}\frac{d}{dt}\left[\lambda_1 f(t) \sin(\omega_1 t + \lambda_i)\right] \tag{30b}$$

$$\ddot{\lambda}_3 + \left(\frac{1}{T_1} + \frac{KT_2}{T_1}\right)\dot{\lambda}_3 + \frac{K}{T_1}\lambda_3 = \frac{K}{T_1} g(t) + \frac{KT_2}{T_1}\frac{d}{dt} g(t)$$

where

$$g(t) = \left[\frac{\lambda_1^3}{6} - \frac{\lambda_1^2}{2} f(t) \cos(\omega_1 t + \lambda_i) - \lambda_2 f(t) \sin(\omega_1 t + \lambda_i)\right] \tag{30c}$$

These equations permit the effect of noise to be interpreted in terms of a series of block diagrams. The interpretation of (30a) is shown in Fig. 6(a). A first approximation to the phase-modulation is found by multiplying the noise $f(t)$ by the unmodulated vco output $\cos(\omega_1 t + \lambda_i)$ and passing the resulting waveform through a filter with transfer function

$$H(S) = \frac{\dfrac{KT_2}{T_1} S + \dfrac{K}{T_1}}{S^2 + \dfrac{1}{T_1} + \dfrac{KT_2}{T_1} S + \dfrac{K}{T_1}}$$

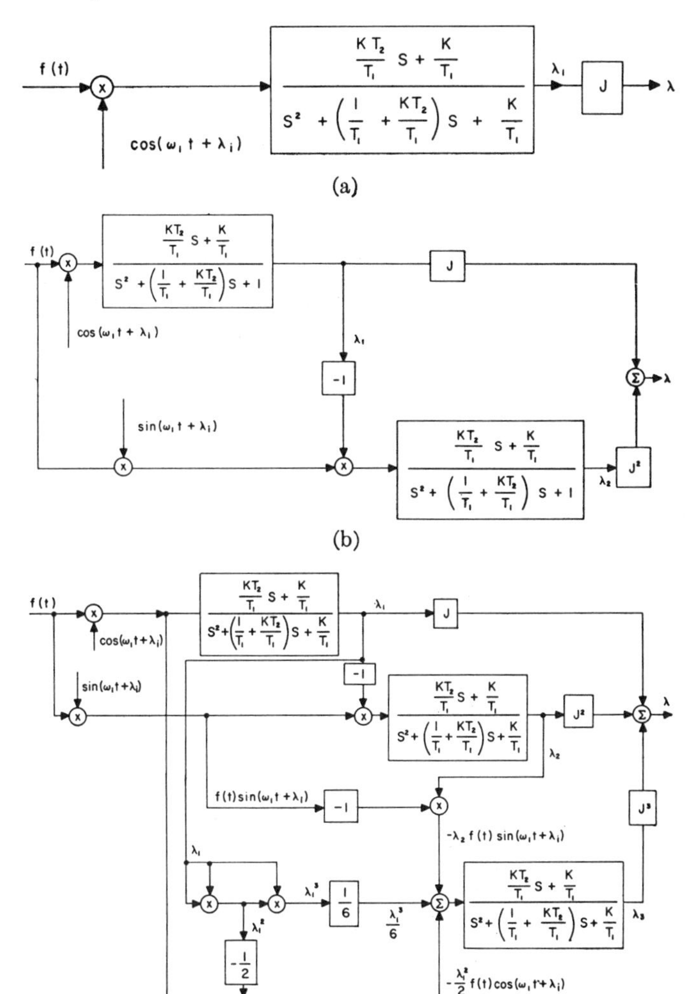

Fig. 6—(a) Model for first approximation to effects of noise on a compensated loop. (b) Model for second approximation to effects of noise on a compensated loop. (c) Model for third approximation to effects of noise on compensated loop.

and an attenuator with transmission J. More detailed models based on (30b) and (30c) are shown in Figs. 6(b) and 6(c). As before, all signals flow from left to right; there is no feedback.

The Use of a Phase-Locked Loop in the Reference Channel of an Angle-Measuring System

Fig. 7 is a simplified block diagram of a simultaneous-lobing angle-measuring system used to measure the azimuth angle of a remote signal source. Synchronous detection is used to discriminate against noise which may be produced by a second remote source. The relatively noise-free reference signal required by the synchronous detector is provided by a phase-locked loop.

The reference channel is assumed to be fed by a non-directional antenna. On the other hand, the measurement channel is fed by an antenna designed to produce an output A_m proportional to the azimuth angle θ_s, with a phase reversal when the azimuth passes through zero, as shown in Fig. 8. The input from the antenna to the reference channel is

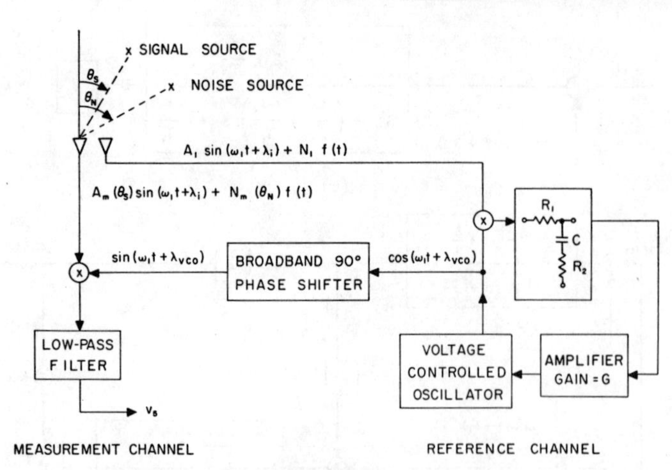

Fig. 7—Phase-locked loop used to provide reference channel for angle-measuring system.

$$v_1 = A_1 \sin(\omega_1 t + \lambda_i) + N_1 f(t) \tag{31}$$

and the antenna and reference inputs to the measurement channel are

$$v_m = A_m(\theta_s) \sin(\omega_1 t + \lambda_i) + N_m(\theta_N) f(t) \tag{32}$$

$$v_r = \sin(\omega_1 t + \lambda_{\text{vco}}). \tag{33}$$

In the absence of noise, ($N_1 = N_m = 0$) the steady-state output of the vco will be $\cos(\omega_1 t + \lambda_i)$. The reference input to the angle-measuring multiplier will then be $\sin(\omega_1 t + \lambda_i)$, making the dc component of the multiplier

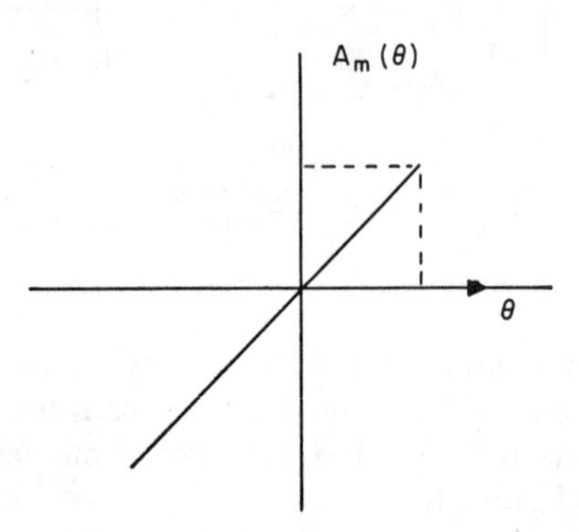

Fig. 8—Typical form for $A_m(\theta)$.

output just $A_m/2$. The factor A_m has the form given in Fig. 8, so that the output of the multiplier measures the angle of arrival of the signal $\sin(\omega_1 t + \lambda_i)$ provided there is no noise. Noise at the input to the reference channel produces a phase-jitter of the vco output. If this phase jitter results in a component in the vco output which is correlated with the noise in the measurement channel, v_5 will contain a spurious dc component owing to the noise. The results of the previous sections will be used to compute this spurious component.

To find the output of the angle-measuring multiplier when noise is present at the inputs of both channels, the product $v_r v_m$ is formed and is expanded in a power series in J. In this application both J and λ are defined for the reference channel by $J = 2N_1/A_1$ and $\lambda = \lambda_{\text{vco}} - \lambda_i$. When the coefficients of J are collected in the resulting expression for the output of the angle-measuring multiplier, the result is

$$v_5 = \frac{A_m(\theta)}{2} + N_m f(t) \sin(\omega_1 t + \lambda_i) + JN_m \lambda_1 f(t) \cos(\omega_1 t + \lambda_i). \tag{34}$$

The multiplier is assumed to contain a filter which rejects periodic components at the frequency $2\omega_1$. The first term in (34) represents the desired output. The second term has no dc component and can be made as small as desired by low-pass filtering of v_5. The term $JN_m\lambda_1 f(t) \cos(\omega_1 t + \lambda_i)$ contributes a dc error; the dc arises from the multiplication of λ_1, a filtered version of $f(t) \cos(\omega_1 t + \lambda_i)$, by $f(t) \cos(\omega_1 t + \lambda_i)$ itself. Thus the steady-state error in v_5 is

$$v_{5\text{avg}} = \lim_{T\to\infty} \frac{1}{2T} \int_{-T}^{T} \left\{ JN_m \int_0^\infty h(\tau) f(t-\tau) \cdot \cos[\omega_1(t-\tau) + \lambda_i]\, d\tau \right\} f(t) \cos(\omega_1 t + \lambda_i)\, dt \tag{35}$$

and if $f(t)$ contains no periodic components,

$$v_{5\text{avg}} = \frac{JN_m}{2} \int_0^\infty \phi_{ff}(\tau) h(\tau) \cos \omega_1 \tau \, d\tau \tag{36}$$

where $h(\tau)$ is the inverse transform of $H(S)$ and ϕ_{ff} is the autocorrelation of $f(t)$. For noise with an autocorrelation function

$$\phi_{ff}(\tau) = \sigma_f^2 \frac{\sin \Delta\omega\tau}{\Delta\omega\tau} \cos \omega_1 \tau \tag{37}$$

(produced by passing white noise through an ideal band-pass filter of bandwidth $2\Delta\omega$ centered at ω_1)

$$v_{5\text{avg}} = \frac{JN_m}{4} \sigma_f^2 \int_0^\infty \frac{\sin \Delta\omega\tau}{\Delta\omega\tau} (1 + \cos 2\omega_1\tau) h(\tau)\, d\tau. \tag{38}$$

If $H(S)$ has the form given by (5) with parameters chosen for good response to transient charges in input phase, *i.e.*,

$$H(S) = \frac{\sqrt{2}\omega_N S + \omega_N^2}{S^2 + \sqrt{2}\omega_N S + \omega_N^2} \tag{39}$$

the value of $h(\tau)$ is

$$h(\tau) = \sqrt{2}\omega_N \epsilon^{-(\sqrt{2}/2)\omega_N\tau} \cos \frac{\sqrt{2}}{2} \omega_N \tau \tag{40}$$

using this specific form for $h(\tau)$ in (36) gives

$$v_{5\text{avg}} = \frac{JN_m\sigma_f^2}{4\Delta\omega} \sqrt{2}\omega_N \cdot \int_0^\infty \frac{\sin \Delta\omega\tau}{\tau} \epsilon^{-(\sqrt{2}/2)\omega_N\tau} \cos \frac{\sqrt{2}}{2} \omega_N \tau \, d\tau + \frac{JN_m\sigma_f^2}{4\Delta\omega} \sqrt{2}\omega_N \cdot \int_0^\infty \frac{\sin \Delta\omega\tau}{\tau} \epsilon^{-(\sqrt{2}/2)\omega_N\tau} \cos \frac{\sqrt{2}}{2} \omega_N \tau \cos 2\omega_1\tau \, d\tau. \tag{41}$$

In the practical case where the noise bandwidth $\Delta\omega$ and the loop bandwidth ω_N are small compared to the carrier frequency ω_1, the second term may be neglected in comparison to the first term. Using the known value of the first term gives[9]

$$v_{5\text{avg}} = \frac{\sqrt{2}}{2}\frac{\sigma_1}{A_1}\sigma_m\frac{\omega_N}{\Delta\omega}\left[\tfrac{1}{2}\arctan\frac{\sqrt{2}\dfrac{\Delta\omega}{\omega_N}}{1-\left(\dfrac{\Delta\omega}{\omega_N}\right)^2}\right] \quad (42)$$

when

$$\frac{\Delta\omega}{\omega_N} \leq 1$$

$$v_{5\text{avg}} = \frac{\sqrt{2}}{2}\frac{\sigma_1}{A_1}\sigma_m\frac{\omega_N}{\Delta\omega}\left[\frac{\pi}{2}+\tfrac{1}{2}\arctan\frac{\sqrt{2}\dfrac{\Delta\omega}{\omega_N}}{1-\left(\dfrac{\Delta\omega}{\omega_N}\right)^2}\right] \quad (42a)$$

when

$$\frac{\Delta\omega}{\omega_N} \geq 1.$$

For $\Delta\omega \ll \omega_N$ (noise bandwidth much less than loop bandwidth)

$$v_{5\text{avg}} \doteq \frac{1}{2}\frac{\sigma_1}{A_1}\sigma_m \quad (43)$$

and for $\Delta\omega \gg \omega_N$ (noise bandwidth much greater than loop bandwidth)

$$v_{5\text{avg}} \doteq \frac{\sqrt{2}}{2}\frac{\pi}{2}\frac{\sigma_1}{A_1}\sigma_m\frac{\omega_N}{\Delta\omega} \quad (44)$$

where $\sigma_1 = N_1\sigma_f$ and $\sigma_m = N_m\sigma_f$ are the rms noises in reference and measuring channels, respectively. The error in the measurement of the null point in the antenna pattern produced by the presence of correlated noise at the inputs of both the reference and measuring channels is thus a function of: σ_1/A_1, the ratio of rms noise to peak signal in the reference channel; σ_m, the rms noise in the measurement channel; and $\omega_N/\Delta\omega$, the ratio of loop bandwidth to noise bandwidth. A plot of $v_{2\text{avg}}$ vs $\Delta\omega/\omega_N$ for $(\sigma_1/A_1)\sigma_m = 1$ is given in Fig. 9. The voltage $v_{5\text{avg}}$ is converted to a phase error by dividing it by the calibration constant of the measurement channel, K_A volts-per-angular mil.

[9] The principal branch of the arctan function, running from arctan $-\infty = -\pi/2$ through arctan $0 = 0$ to arctan $+\infty = \pi/2$ should be used in evaluating (42) and (42a).

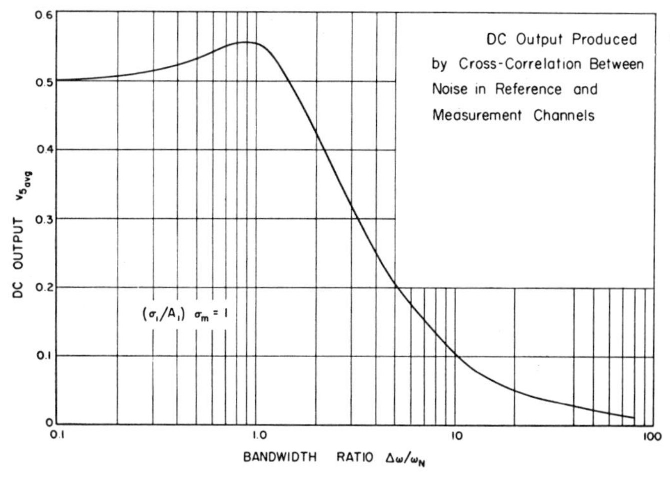

Fig. 9.

Conclusion

A method has been derived for determining the phase modulation produced in the output of a phase-locked loop by noise at its input. The method has been shown to be useful in predicting the steady-state phase offset produced by cross-correlation effects within a single loop and in predicting the errors which may occur when the loop is used to provide a reference channel for an angle-measuring system. The analysis presented in this paper permits the phase modulation resulting from input noise to be computed to any desired accuracy (at the cost of increasing analytic difficulty); a physical interpretation has been included as a guide to choice of a model sufficiently complex to explain the phenomenon sought, yet sufficiently simple to permit a solution to be found.

Part X
AGC, AFC, and APC Circuits and Systems

Theory of AFC Synchronization*

WOLF J. GRUEN†, MEMBER, IRE

Summary—**The general solution for the important design parameters of an automatic frequency and phase-control system is presented. These parameters include the transient response, frequency response and noise bandwidth of the system, as well as the hold-in range and pull-in range of synchronization.**

I. Introduction

AUTOMATIC FREQUENCY and phase-control systems have been used for a number of years for the horizontal-sweep synchronization in television receivers, and more recently have found application for the synchronization of the color subcarrier in the proposed NTSC color-television system. A block diagram of a general AFC system is shown in Fig. 1.

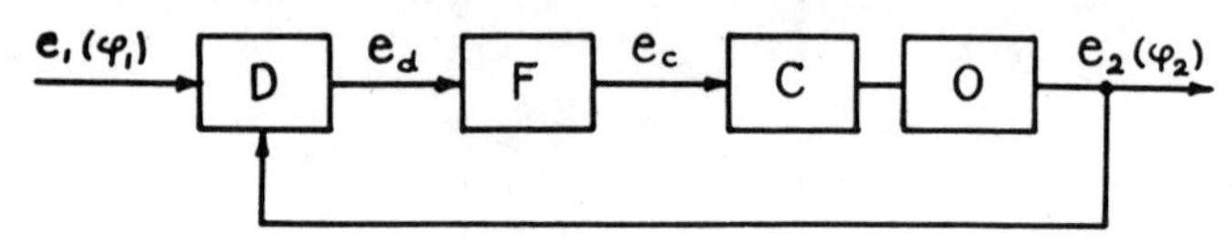

Fig. 1—Block diagram of A.F.C. loop.

The phase of the transmitted synchronizing signal e_1 is compared to the phase of a local oscillator signal e_2 in a phase discriminator D. The resulting discriminator output voltage is proportional to the phase difference of the two signals, and is fed through a control network F to a frequency-control stage C. This stage controls the frequency and phase of a local oscillator O in accordance with the synchronizing information, thereby keeping the two signals in perfect synchronism. Although in practice the transmitted reference signal is often pulsed and the oscillator comparison voltage non-sinusoidal, the analysis is carried out for sinusoidal signal voltages. The theory, however, can be extended for a particular problem by writing the applied voltages in terms of a Fourier series instead of the simple sine function. An AFC system is essentially a servomechanism, and the notation that will be used is the one followed by many workers in this field. An attempt will be made to present the response characteristics in dimensionless form in order to obtain a universal plot of the response curves.

II. Derivation of the Basic Equation

If it is assumed that the discriminator is a balanced phase detector composed of peak-detecting diodes, the discriminator-output voltage can be derived from the vector diagram in Fig. 2. For sinusoidal variation with time, the synchronizing signal e_1 and the reference signal e_2 can be written

$$e_1 = E_1 \cos \phi_1 \tag{1}$$

and

$$e_2 = E_2 \sin \phi_2. \tag{2}$$

ϕ_1 and ϕ_2 are functions of time and, for reasons of simplicity in the later development, it is arbitrarily assumed that ϕ_1 and ϕ_2 are in quadrature when the system is perfectly synchronized, that is when $\phi_1 = \phi_2$.

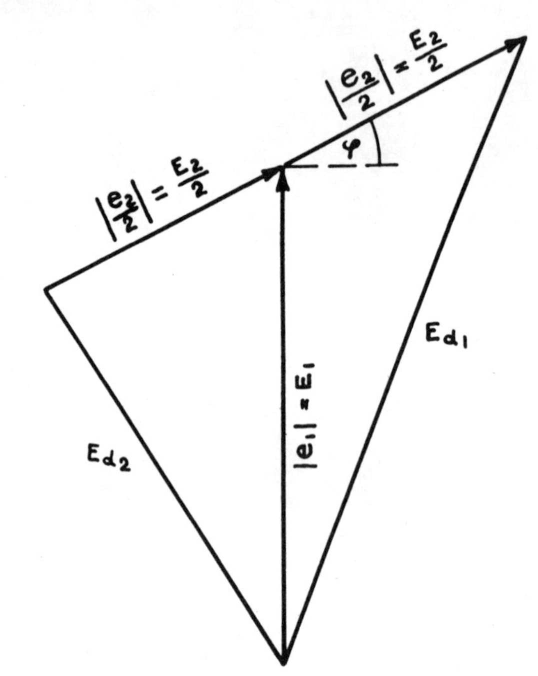

Fig. 2—Discriminator vector diagram.

While one of the discriminator diodes is fed with the sum of e_1 and $e_2/2$, the other is fed with the difference of these two vectors as shown in Fig. 2. The resulting rectified voltages E_{d1} and E_{d2} can be established by simple trigonometric relations. Defining a difference phase

$$\phi \equiv \phi_1 - \phi_2, \tag{3}$$

one obtains

$$E_{d1}^2 = E_1^2 + \frac{E_2^2}{4} + E_1 E_2 \sin \phi \tag{4}$$

and

$$E_{d2}^2 = E_1^2 + \frac{E_2^2}{4} - E_1 E_2 \sin \phi. \tag{5}$$

The discriminator output voltage e_d is equal to the dif-

* Decimal classification: R583.5. Original manuscript received by the Institute, August 21, 1952; revised manuscript received February 25, 1953.

† General Electric Co., Syracuse, N. Y.

Reprinted from *Proc. IRE*, vol. 41, pp. 1043–1048, Aug. 1953.

ference of the two rectified voltages, so that

$$e_d = E_{d1} - E_{d2} = \frac{2E_1E_2}{E_{d1} + E_{d2}} \sin \phi. \tag{6}$$

If the amplitude E_1 of the synchronizing signal is larger than the amplitude E_2 of the reference signal, one obtains

$$E_{d1} + E_{d2} \cong 2E_1. \tag{7}$$

The discriminator output voltage then becomes

$$e_d = E_2 \sin \phi \tag{8}$$

and is independent of the amplitude E_1 of the synchronizing signal. As ϕ_1 and ϕ_2 are time-varying parameters, it should be kept in mind that the discriminator time constant ought to be shorter than the reciprocal of the highest difference frequency $d\phi/dt$, which is of importance for the operation of the system.

Denoting the transfer function of the control network F as $F(p)$, the oscillator control voltage becomes

$$e_c = F(p)E_2 \sin \phi. \tag{9}$$

Assuming furthermore that the oscillator has a linear-control characteristic of a slope S, and that the free-running oscillator frequency is ω_0, the actual oscillator frequency in operational notation becomes

$$p\phi_2 = \omega_0 + Se_c. \tag{10}$$

Substituting (3) and (9) into (10) then gives

$$p\phi + SE_2F(p) \sin \phi = p\phi_1 - \omega_0. \tag{11}$$

The product SE_2 repeats itself throughout this paper and shall be defined as the gain constant

$$K \equiv SE_2. \tag{12}$$

K represents the maximum frequency shift at the output of the system per radian phase shift at the input. It has the dimension of radians/second.

Equation (11) can be simplified further by measuring the phase angles in a coordinate system which moves at the free-running speed ω_0 of the local oscillator. One obtains

$$\boxed{p\phi + F(p)K \sin \phi = p\phi_1}. \tag{13}$$

This equation represents the general differential equation of the AFC feedback loop. $p\phi$ is the instantaneous-difference frequency between the synchronizing signal and the controlled-oscillator signal and $p\phi_1$ is the instantaneous-difference frequency between the synchronizing signal and the free-running oscillator signal.

Equation (13) shows that all AFC systems with identical gain constants K and unity d.c. gain through the control network have the same steady-state solution, provided that the difference frequency $p\phi_1$ is constant. If this difference frequency is defined as

$$\Delta\omega \equiv p\phi_1 = \omega_1 - \omega_0, \tag{14}$$

the steady-state solution is

$$\sin \phi = \frac{\Delta\omega}{K} \tag{15}$$

This means the system has a steady-state phase error which is proportional to the initial detuning $\Delta\omega$ and inversely proportional to the gain constant K. Since the maximum value of sin ϕ in (15) is ± 1, the system will hold synchronism over a frequency range

$$|\Delta\omega_{\text{Hold-in}}| \leqq K. \tag{16}$$

Equations (15) and (16) thus define the static performance limit of the system.

III. Linear Analysis

An AFC system, once it is synchronized, behaves like a low-pass filter. To study its performance it is permissible, for practical signal-to-noise ratios, to substitute the angle for the sine function in (13). Then, with the definition of (3), one obtains

$$p\phi_2 + KF(p)\phi_2 = KF(p)\phi_1. \tag{17}$$

This equation relates the output phase ϕ_2 of the synchronized system to the input phase ϕ_1. It permits an evaluation of the behavior of the system to small disturbances of the input phase, if the transfer function $F(p)$ of the control network is specified.

a. $F(p) = 1$

This is the simplest possible AFC system, and represents a direct connection between the discriminator output and the oscillator control stage. Equation (17) then becomes

$$p\phi_2 + K\phi_2 = K\phi_1. \tag{18}$$

If the initial detuning is zero, the transient response of the system to a sudden step of input phase $|\phi_1|$ is

$$\frac{\phi_2}{|\phi_1|}(t) = 1 - e^{-Kt}. \tag{19}$$

Likewise, the frequency response of the system to a sine-wave modulation of the input phase is

$$\frac{\phi_2}{\phi_1}(j\omega) = \frac{1}{1 + j\dfrac{\omega}{K}}. \tag{20}$$

The simple AFC system thus behaves like an RC-filter and has a cut-off frequency of

$$\omega_c = K \text{ [radians/sec]}. \tag{21}$$

George[1] has shown that the m.s. phase error of the system under the influence of random interference is proportional to the noise bandwidth, which is defined as

[1] T. S. George, "Synchronizing systems for dot interlaced color TV," Proc. I.R.E.; February, 1951.

$$B = \int_{-\infty}^{+\infty} \left| \frac{\phi_2}{\phi_1}(j\omega) \right|^2 d\omega. \tag{22}$$

The integration has to be carried out from $-\infty$ to $+\infty$ since the noise components on both sides of the carrier are demodulated. Inserting (20) into (22) then yields

$$B = \pi K \text{ [radians/sec]}. \tag{23}$$

It was shown in (15) that for small steady-state phase errors due to average frequency drift, the gain constant K has to be made as large as possible, while now for good noise immunity, i.e., narrow bandwidth, the gain constant has to be made as small as possible. A proper compromise of gain then must be found to insure adequate performance of the system for all requirements. This difficulty, however, can be overcome by the use of a more elaborate control network.

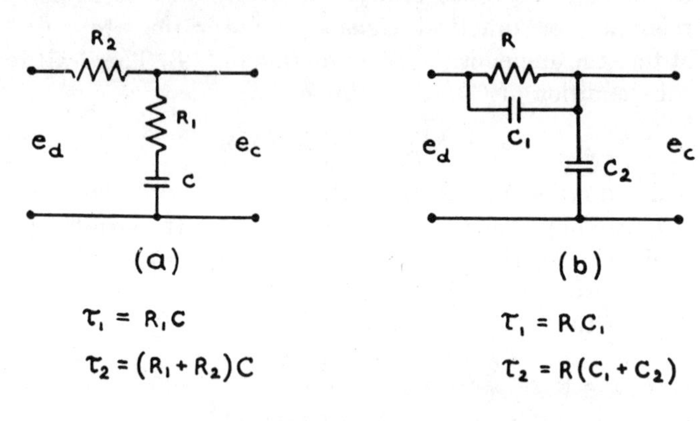

Fig. 3—Proportional plus integral control networks.

b. $F(p) = \dfrac{1 + \tau_1 p}{1 + \tau_2 p}$

Networks of this type are called proportional-plus integral-control networks[2] and typical network configurations are shown in Fig. 3. Inserting the above transfer function into (17) yields

$$p^2\phi_2 + \left(\frac{1}{\tau_2} + K\frac{\tau_1}{\tau_2}\right) p\phi_2 + \frac{K}{\tau_2}\phi_2 = K\frac{\tau_1}{\tau_2}p\phi_1 + \frac{K}{\tau_2}\phi_1. \tag{24}$$

ϕ_1 and ϕ_2 are again relative phase angles, measured in a coordinate system which moves at the free-running speed of the local oscillator. To integrate (24), it is convenient to introduce the following parameters

$$\omega_n^2 \equiv \frac{K}{\tau_2} \tag{25}$$

and

$$2\zeta\omega_n \equiv \frac{1}{\tau_2} + K\frac{\tau_1}{\tau_2}. \tag{26}$$

ω_n is the resonance frequency of the system in the absence of any damping, and ζ is the ratio of actual-to-critical damping. In terms of the new parameters the time constants of the control network are

$$\tau_1 = \frac{2\zeta}{\omega_n} - \frac{1}{K} \tag{27}$$

and

$$\tau_2 = \frac{K}{\omega_n^2}. \tag{28}$$

With these definitions (24) becomes

$$p^2\phi_2 + 2\zeta\omega_n p\phi_2 + \omega_n^2\phi_2 = \left(2\zeta\omega_n - \frac{\omega_n^2}{K}\right)p\phi_1 + \omega_n^2\phi_1. \tag{29}$$

The transient response of the system to a sudden step of input phase $|\phi_1|$ is found by integration of (29) and the initial condition for the oscillator frequency is obtained from (10). The transient response then is

$$\frac{\phi_2}{|\phi_1|}(t) = 1 - e^{-\zeta\omega_n t}\left[\cos\sqrt{1-\zeta^2}\,\omega_n t - \frac{\zeta - \dfrac{\omega_n}{K}}{\sqrt{1-\zeta^2}}\sin\sqrt{1-\zeta^2}\,\omega_n t\right]. \tag{30}$$

For $\zeta<1$ the system is underdamped (oscillatory), for $\zeta=1$ critically damped and for $\zeta>1$ overdamped (non-oscillatory). In order to avoid sluggishness of the system, a rule of thumb may be followed making. $4<\zeta<1$[2]. The transient response of (30) can be plotted in dimensionless form if certain specifications are made for the ratio ω_n/K. As the time constant τ_1 of the control network must be positive or can at most be equal to zero, the maximum value for ω_n/K is found from (27), yielding

$$\left.\frac{\omega_n}{K}\right|_{\max} = 2\zeta. \tag{31}$$

In this case the control network is reduced to a single time constant network ($\tau_1=0$). On the other hand, if for a fixed value of ω_n the gain of the system is increased towards infinity, the minimum value for ω_n/K becomes

$$\left.\frac{\omega_n}{K}\right|_{\min} = 0. \tag{32}$$

Fig. 4 shows the transient response of the system for these two limits and for a damping ratio of $\zeta=0.5$.

[2] G. S. Brown and D. P. Campbell, "Principles of Servomechanisms," John Wiley & Sons Publishing Co., New York, N. Y.; 1948.

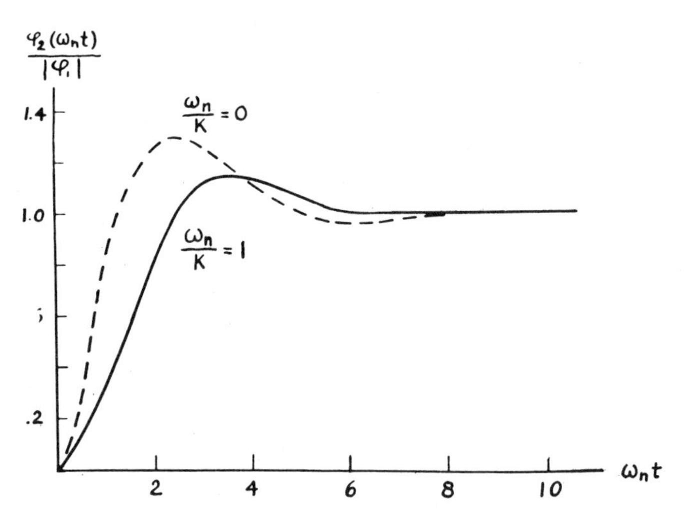

Fig. 4—Transient response for $\zeta=0.5$.

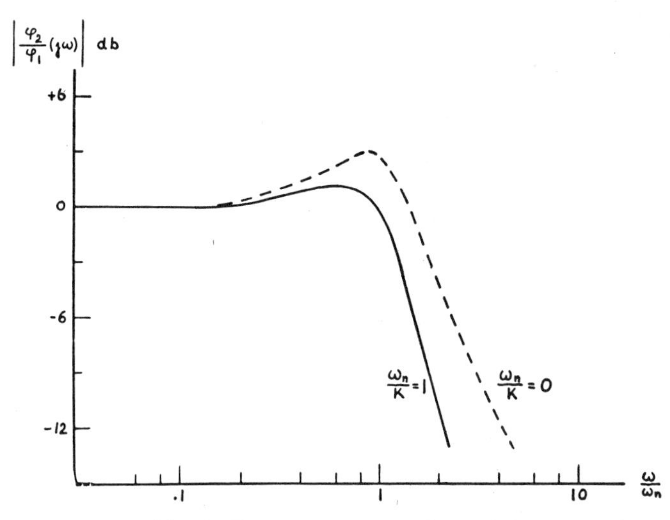

Fig. 5—Frequency response for $\zeta=0.5$.

The frequency response of the system is readily found from (24) and one obtains

$$\frac{\phi_2}{\phi_1}(j\omega) = \frac{1 + j2\zeta\frac{\omega}{\omega_n}\left(1 - \frac{\omega_n}{2\zeta K}\right)}{1 + j2\zeta\frac{\omega}{\omega_n} - \left(\frac{\omega}{\omega_n}\right)^2}. \qquad (33)$$

Its magnitude is plotted in Fig. 5 for the two limit values of ω_n/K and for a damping ratio $\zeta=0.5$. The curves show that the cut-off frequency of the system, for $\zeta=0.5$, is approximately

$$\omega_c \cong \omega_n \text{ [radians/sec.]}. \qquad (34)$$

If ϕ_1 and ϕ_2 in (33) are assumed to be the input and output voltage of a four-terminal low-pass filter, the frequency response leads to the equivalent circuit of Fig. 6.

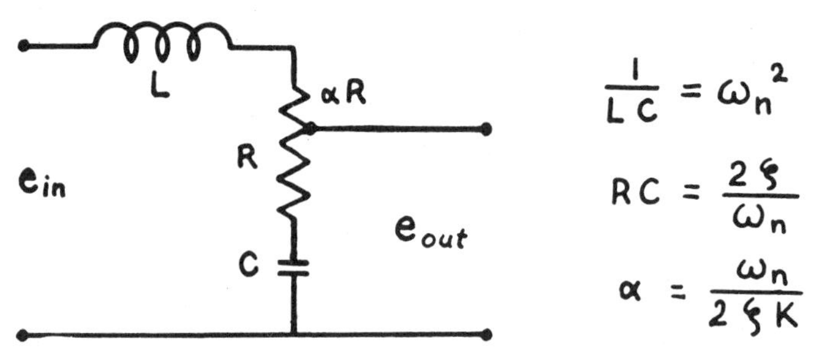

Fig. 6—Equivalent low-pass filter.

The noise bandwidth of the system is established by inserting (33) into (22) and one obtains

$$B=\omega_n\int_{-\infty}^{+\infty}\frac{1+4\zeta^2\left(\frac{\omega}{\omega_n}\right)^2\left[1-\frac{\omega_n}{2\zeta K}\right]^2}{1-(2-4\zeta^2)\left(\frac{\omega}{\omega_n}\right)^2-\left(\frac{\omega}{\omega_n}\right)^4}\,d\left(\frac{\omega}{\omega_n}\right). \qquad (35)$$

The integration, which can be carried out by partial fractions with the help of tables, yields

$$B = \frac{4\zeta^2 - 4\zeta\frac{\omega_n}{K} + \left(\frac{\omega_n}{K}\right)^2 + 1}{2\zeta}\pi\omega_n. \qquad (36)$$

For small values of ω_n/K, it is readily established that this expression has a minimum when $\zeta=0.5$. Hence, the noise bandwidths for the limit values of ω_n/K and $\zeta=0.5$ become

$$B\Big|_{(\omega_n/K)\to 1} = \pi\omega_n = \pi K \text{ [radians/sec]} \qquad (37)$$

and

$$B\Big|_{(\omega_n/K)\to 0} = 2\pi\omega_n \text{ [radians/sec]}. \qquad (38)$$

The above derivations, as well as the response curves of Figs. 4 and 5, show that the bandwidth and the gain constant of the system can be adjusted independently if a double time-constant control network is employed.

c. *Example*

The theory is best illustrated by means of an example. Suppose an AFC system is to be designed, having a steady state phase error of not more than 3° and a noise bandwidth of 1,000 cps. The local oscillator drift shall be assumed with 1,500 cps.

The required gain constant is obtained from (15), yielding

$$K = \frac{\Delta\omega}{\sin\phi} = \frac{2\pi\cdot 1{,}500}{\sin 3^\circ} = 180{,}000 \text{ radians/sec.}$$

Since K is large in comparison to the required bandwidth, the resonance frequency of the system is established from (38).

$$\omega_n = \frac{B}{2\pi} = \frac{2\pi\cdot 1{,}000}{2\pi} = 1{,}000 \text{ radians/sec.}$$

The two time constants of the control network, assuming a damping ratio of 0.5, are determined from (27) and (28) respectively

$$\tau_1 = \frac{2\zeta}{\omega_n} - \frac{1}{K} = \frac{1}{1{,}000} - \frac{1}{180{,}000} \cong 10^{-3} \text{ sec},$$

and

$$\tau_2 = \frac{K}{\omega_n^2} = \frac{180{,}000}{1{,}000^2} = 0.18 \text{ sec.}$$

These values K, τ_1, and τ_2 completely define the AFC system. A proper choice of gain distribution and control-network impedance still has to be made to fit a particular design. For example, if the peak amplitude of the sinusoidal oscillator reference voltage is $E_2=6$ volts, the sensitivity of the oscillator control stage must be $S=30{,}000$ radians/sec/volt to provide the necessary gain constant of 180,000 radians/sec. Furthermore, if the capacitor C for the control network of Fig. 3(a) is assumed to be 0.22 uf, the resistors R_1 and R_2 become 4.7 kΩ and 820 kΩ respectively, to yield the desired time constants.

IV. Non-Linear Analysis

While it was permissible to assume small phase angles for the study of the synchronized system, thereby linearizing the differential (13), this simplification cannot be made for the evaluation of the pull-in performance of the system. The pull-in range of synchronization is defined as the range of difference frequencies, $p\phi_1$, between the input signal and the free-running oscillator signal, over which the system can reach synchronism. Since the difference phase ϕ can vary over many radians during pull-in, it is necessary to integrate the nonlinear equation to establish the limit of synchronization.

Assuming that the frequency of the input signal is constant as defined by (14), (13) can be written

$$p\phi + F(p)K \sin \phi = \Delta\omega. \tag{39}$$

Mathematically then, the pull-in range of synchronization is the maximum value of $\Delta\omega$ for which, irrespective of the initial condition of the system, the phase difference ϕ reaches a steady state value. To solve (39), the transfer function of the control network again must be defined.

a. $F(p) = 1$

The pull-in performance for this case has been treated in detail by Labin.[3] With $F(p)=1$ (39) can be integrated by separation of the variables and it is readily found that the system synchronizes for all values of $|\Delta\omega| < K$. The condition for pull-in then is

$$|\Delta\omega|_{\text{Pull-in}} < K. \tag{40}$$

Large pull-in range and narrow-noise bandwidth thus are incompatible requirements for this system.

b. $F(p) = \dfrac{1+\tau_1 p}{1+\tau_2 p}$

Inserting this transfer function into (39) and carrying out the differentiation yields

$$\frac{d^2\phi}{dt^2} + \left[\frac{1}{\tau_2} + K\frac{\tau_1}{\tau_2}\cos\phi\right]\frac{d\phi}{dt} + \frac{K}{\tau_2}\sin\phi = \frac{\Delta\omega}{\tau_2}. \tag{41}$$

This equation can be simplified by inserting the coefficients defined in (25) and (26), and by dividing the resulting equation by ω_n^2. This leads to the dimensionless equation.

$$\frac{d^2\phi}{\omega_n^2 dt^2} + \left[\frac{\omega_n}{K} + \left(2\zeta - \frac{\omega_n}{K}\right)\cos\phi\right]\frac{d\phi}{\omega_n dt} + \sin\phi = \frac{\Delta\omega}{K}. \tag{42}$$

A further simplification is possible by defining a dimensionless difference frequency

$$y \equiv \frac{d\phi}{\omega_n dt} \tag{43}$$

and one obtains a first order differential equation from which the dimensionless time $\omega_n t$ has been eliminated. It follows

$$\frac{dy}{d\phi} = \frac{\frac{\Delta\omega}{K} - \sin\phi}{y} - \frac{\omega_n}{K} - \left(2\zeta - \frac{\omega_n}{K}\right)\cos\phi. \tag{44}$$

There is presently no analytical method available to solve this equation. However, the equation completely defines the slope of the solution curve $y(\phi)$ at all points of a $\phi-y$ plane, except for the points of stable and unstable equilibrium, $y=0$; $\Delta\omega/K=\sin\phi$. The limit of synchronization can thus be found graphically by starting the system with an infinitesimal velocity Δy at a point of unstable equilibrium, $y=0$; $\phi=\pi-\sin^{-1}\Delta\omega/K$, and finding the value of $\Delta\omega/K$ for which the solution curve just reaches the next point of unstable equilibrium located at $y=0$; $\phi=3\pi-\sin^{-1}\Delta\omega/K$. The method is discussed by Stoker[4] and has been used by Tellier and Preston[5] to find the pull-in range for a single time constant AFC system.

To establish the limit curve of synchronization for given values of ζ and ω_n/K, a number of solution curves have to be plotted with $\Delta\omega/K$ as parameter. The limit of pull-in range in terms of $\Delta\omega/K$ then can be interpolated to any desired degree of accuracy. The result, obtained in this manner, is shown in the dimensionless graph of Fig. 7, where $\Delta\omega/K$ is plotted as a function of ω_n/K for a damping ratio $\zeta=0.5$. Since this curve represents the stability limit of synchronization for the system, the time required to reach synchronism is infinite when starting from any point on the limit curve. The same applies to any point on the $\Delta\omega/K$-axis, with exception of the point $\Delta\omega/K=0$, since this axis describes a system having either infinite gain or zero bandwidth, and neither case has any real practical significance. The practical pull-in range of synchronization, therefore, lies inside the solid boundary. The individual points

[3] Edouard Labin, "Theorie de la synchronization par controle de phase," *Philips Res. Rep.*, (in French); August, 1941.

[4] J. J. Stoker, "Non-linear vibrations," *Interscience;* New York, 1950.

[5] G. W. Preston and J. C. Tellier, "The Lock-in Performance of an A.F.C. Circuit," Proc. I.R.E.; February, 1953.

entered in Fig. 7 represent the measured pull-in curve of a particular system for which the damping ratio was maintained at $\zeta = 0.5$. For small values of ω_n/K this

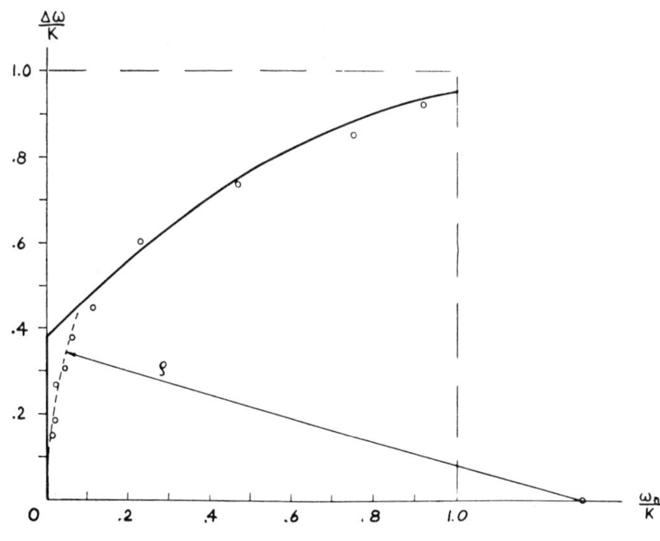

Fig. 7—Pull-in range of synchronization for $\zeta = 0.5$.

pull-in curve can be approximated by its circle of curvature which, as indicated by the dotted line, is tangent to the $\Delta\omega/K$-axis and whose center lies on the ω_n/K-axis. The pull-in range thus can be expressed analytically by the equation of the circle of curvature. If its radius is denoted by ζ, the circle is given by

$$\left(\frac{\omega_n}{K} - \zeta\right)^2 + \left(\frac{\Delta\omega}{K}\right)^2 = \zeta^2. \tag{45}$$

Hence, for $(\omega_n/K) \to 0$, the pull-in range of synchronization is approximately

$$\left| \Delta\omega_{\text{Pull-in}} \right|_{(\omega_n/K)\to 0} < \sqrt{2\zeta\omega_n K - \omega_n^2} \cong \sqrt{2\zeta\omega_n K}. \tag{46}$$

ζ can be interpreted as a constant of proportionality which depends on the particular design of the system, and which increases as the system gets closer to the theoretical limit of synchronization.

Equation (46) shows that the pull-in range for small values of ω_n/K is proportional to the square root of the product of the cut-off frequency ω_n and the gain constant K. Since the bandwidth of a double time constant AFC system can be adjusted independently of the gain constant, the pull-in range of such a system can exceed the noise bandwidth by any desired amount.

V. Conclusions

The performance of an AFC system can be described by three parameters. These are the gain constant K, the damping ratio ζ and the resonance or cut-off frequency ω_n. These parameters are specified by the requirements of a particular application and define the over-all design of the system. It has been shown that among the systems with zero, single and double time constant control networks, only the latter fulfills the requirement for achieving good noise immunity, small steady-state phase error and large pull-in range.

The Lock-In Performance of an AFC Circuit*

G. W. PRESTON†, ASSOCIATE, IRE AND J. C. TELLIER†, SENIOR MEMBER, IRE

Summary—**The lock-in condition of the phase-detector reactance tube-controlled oscillator afc system has been investigated. For the case of RC coupling between phase detector and the reactance tube, an explicit relation is obtained for the lock-in condition which involves the filter time constant, the initial frequency error, the phase detector constant and the reactance tube controlled oscillator sensitivity constant.**

IN CERTAIN compatible color-television systems, hue and saturation are transmitted as phase and amplitude modulation of a sine wave, the color "subcarrier." In the particular system for which specifications have been established by the National Television System Committee, the frequency of the subcarrier is approximately 3.89 mc. In order to recover the phase modulation at the receiver without ambiguity, a phase reference must be available. For this purpose, a color "reference" signal is transmitted as a burst of color carrier, just after the horizontal synchronizing pulse. The common afc system shown in block diagram form in Fig. 1 is often employed at the receiver to extract a relatively noise-free reproduction of the burst from the transmitted signal.

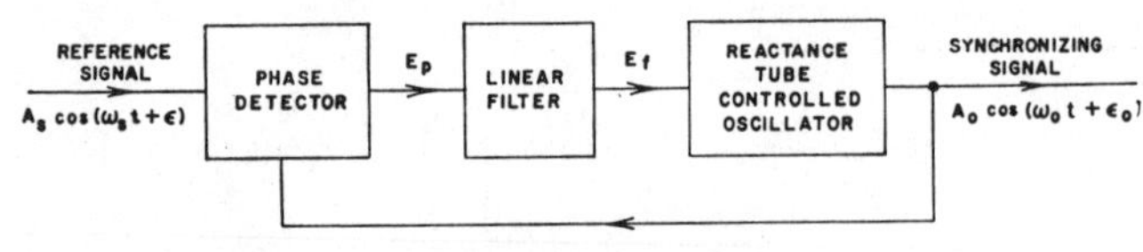

Fig. 1—Block diagram.

For the purposes of the present discussion, this burst is assumed to have been separated by gating from the synchronizing pulses, and filtered somewhat to reduce noise but without producing any significant amount of phase shift. It is permissible, moreover, to regard the synchronizing signal as being continuous rather than intermittent since the band pass of the circuit, as a whole, will always be narrow compared to the line-repetition frequency, which is 15,750 cps.

The phase-detector reactance, tube-controlled oscillator afc system functions in the following manner. The phase detector has a voltage output E_p, which is proportional to the cosine of the difference $\phi_s-\phi_0$ in the phases of the reference signal and the controlled oscillator and to twice the signal voltage amplitude, $E_p=2A_sK_1 \cos(\phi_s-\phi_0)$. This voltage is applied (disregarding for the moment the intervening linear filter) to the grid of the reactance-tube controlled oscillator. The oscillator in turn assumes a frequency which differs froms its normal value Ω (obtained when the reactance tube control voltage is zero) by an amount which is essentially proportional to the control voltage;

$$\omega_0 = \Omega + K_2E_p = \Omega + 2A_1K_1K_2 \cos(\phi_s - \phi_0).$$

If the reference signal frequency is $\omega_a+\Omega$, and if synchronization is assumed, it is seen that in the equilibrium condition there must exist a static phase difference

$$\left(\phi_s - \phi_0 - \frac{\pi}{2}\right)_{\text{static}} = \sin^{-1}\left(\frac{\omega_a}{2A_sK_1K_2}\right).$$

The noise performance of this circuit has been discussed by George[1] although his analysis involves a linearizing assumption which is valid in the steady state for IF signal-to-noise ratios of practical interest. George shows that the mean-square phase fluctuation, due to noise, of the oscillator depends upon the transient response of the coupling network and that it is also proportional to $A_sK_1K_2$. The static phase shift, on the

* Decimal classification: R361.215. Original manuscript received by the Institute, November 16, 1951; revised manuscript received August 27, 1952.

† Philco Corp., Philadelphia 34, Pa.

[1] T. S. George, "Analysis of synchronizing systems for dot-interlaced color television," PROC. I.R.E., vol. 39, pp. 124–131; February, 1951.

Reprinted from *Proc. IRE*, vol. 41, pp. 249-251, Feb. 1953.

other hand, varies nearly inversely as $A_sK_1K_2$. It is desirable to minimize the static phase error by maximizing K_1K_2 provided the noise performance requirements can be satisfied by making the band pass of the coupling network sufficiently narrow. However, as the band pass is narrowed, a point is reached at which the oscillator may fail to synchronize following an initial frequency difference. The purpose of the present investigation was to determine constraints upon the filter characteristic within which synchronization will occur. An abrupt frequency difference between the reference voltage and the oscillator occurs in practice when switching from one station to another.

With respect to lock-in, this circuit will, in general, exhibit a peculiarity common to many types of nonlinear oscillatory systems; *it has more than one steady-state condition*. Specifically, following a step change in the reference frequency, the controlled oscillator may become synchronized to the new frequency, or it may reach the state of oscillation (with periodic perturbations) at its normal frequency Ω. Which of these final conditions is assumed depends upon the initial phase and frequency errors. If, however, suitable constraints are observed in the design of the linear filter, only the former steady-state condition will be possible, regardless of initial conditions.

By numerical and graphical methods, this constraint for a simple RC coupling network, has been found to be

$$\frac{1}{RC} \geqq 0.64 \frac{\omega_a^2}{K_1K_2} \tag{1}$$

approximately, provided $\omega_a \leqq 0.5K_1K_2$.

The methods used in obtaining this condition will be discussed after a mathematical statement of the problem has been made. Unfortunately, qualitative considerations of nonlinear systems of this type contribute little to a clear understanding of their behavior.

The conventional discriminator-type balanced phase detector has a voltage output which is proportional to the difference between the absolute values of the sum and difference of the complex reference and oscillator voltages.

Thus, if the reference voltage is $A_s \exp\{j\int^t \omega_s dt\}$ and the oscillator voltage is $A_0 \exp\{j\int^t \omega_0 dt\}$, the output of the balanced phase detector is (assuming $(A_0/A_s)^2$ is small compared to unity)

$$E_p(t) = 2A_sK_1 \cos\left[\int^t (\omega_0 - \omega_s)dt\right] \tag{2}$$

approximately, in which K_1 is a constant characteristic of the phase detector.

In the particular case of low-pass RC coupling, the control voltage E_f is given by

$$\frac{dE_f}{dt} + \alpha E_f = \alpha E_p, \tag{3}$$

in which $1/\alpha = RC$.

Letting $\phi \equiv \phi_0 - \phi_s = \int^t(\omega_0 - \omega_s)dt$, then

$$\frac{d\phi}{dt} = \omega_0 - \omega_s \quad \text{and} \quad \frac{d^2\phi}{dt} = \frac{d\omega_0}{dt} = K_2\frac{dE_f}{dt}, \tag{4}$$

where $d\omega_s/dt$ has been neglected since the synchronizing burst is crystal controlled at the transmitter and will drift very slowly.

It follows from (2), (3), and (4) that

$$\frac{d^2\phi}{dt^2} + \alpha\frac{d\phi}{dt} - 2\alpha A_sK_1K_2 \cos\phi = -\alpha\omega_a.^2 \tag{5}$$

The equation is highly nonlinear owing to the presence of cos ϕ, and it is clear that no methods are presently available for obtaining a solution.

The appearance of this equation can be improved somewhat by making the substitutions,

$$\lambda = \sqrt{\frac{\alpha}{2A_sK_1K_2}}, \quad \gamma = \frac{\omega_b}{2A_sK_1K_2} \quad \text{and}$$

$$\tau = \sqrt{2\alpha A_sK_1K_2}\, t, \tag{6}$$

which result in

$$\frac{d^2\phi}{d\tau^2} + \lambda\frac{d\phi}{d\tau} - \cos\phi = -\gamma. \tag{7}$$

This equation may be reduced to a first-order one with the elimination of the independent variable τ by the substitutions $p = d\phi/d\tau$ and hence $pdp/d\phi = d^2\phi/d\tau^2$; this gives

$$p\frac{dp}{d\phi} + \lambda p - \cos\phi = -\gamma. \tag{8}$$

The disappearance of the variable $\tau = \sqrt{2\alpha A_sK_1K_2}t$ is no disadvantage since, in practice, the time required for look-in to occur is not significant. At lock-in, $p = 0$ and $\phi = \arccos\gamma$. Within the range of its validity, (1) gives approximately the necessary constraints upon λ and γ so that the above values are boundary conditions on *all* solutions of (8).

The behavior exhibited by this type of nonlinear oscillatory system is perhaps somewhat easier to visualize in terms of an analogue, the ideal damped pendulum acted upon by a constant torque and by the force of gravity. Then λ is analogous to the angular displacement of the bob measured from the downward vertical and p is analogous to the angular momentum.

With no torque, the pendulum is in stable equilibrium at $\phi = 0$, $p = 0$, and is in unstable equilibrium at $\phi = \pi/2$, $p = 0$. The effect of a torque γ is a shift of the stable and unstable equilibrium points by the angle arcsin γ; the stable equilibrium point is shifted in the direction of the applied torque, the unstable equilibrium point in the opposite direction.

[2] This equation is discussed at some length by J. J. Stoker, "Nonlinear Vibrations," Interscience, New York, N. Y., pp. 70–80; 1950, in connection with the pull-out torque of synchronous motors.

When the torque is increased to unity, the points of stable and unstable equilibrium coalesce at $\phi=-\pi/2$.

The applied torque supplies energy to the pendulum uniformly for each degree of rotation, whereas the energy dissipated by damping per degree of rotation is proportional to the angular velocity. The gravitational field is conservative, of course, and there is an oscillatory flow of energy from kinetic to potential. Clearly, at very high angular velocities, the dissipated energy will exceed the energy supplied by the torque and the pendulum will slow down. However, at low enough velocities, the energy gained from the applied torque per degree of rotation must exceed the dissipated energy, and the condition of rest is not stable unless the maximum gravitational torque ($\phi=\pm\pi/4$) exceeds the applied torque.

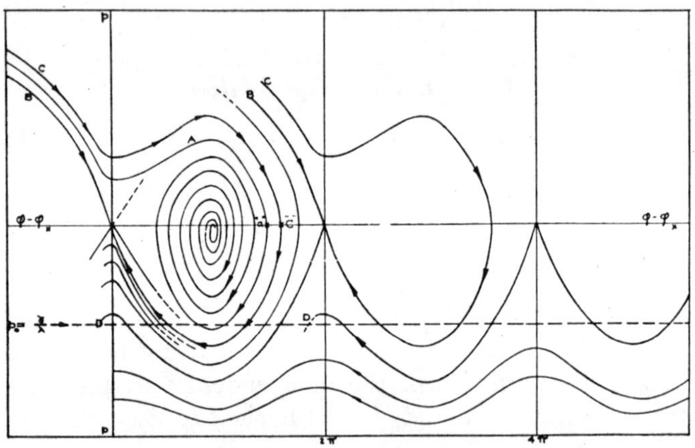

Fig. 2—Representative solution curves for λ less than λ_c—free-hand sketch.

At any instant, the phase error ϕ and the frequency error, which is proportional to p, define a point in the p, ϕ plane; and since p and ϕ are implicit functions of time, this representative point will move in time and generate a curve $p(\phi)$ which is a solution of (8) with the given initial condition. Moreover (8), when solved for $dp/d\phi$, defines uniquely the slope of the solution curve which passes through any given point, except at those points $p=0$, $\phi=\arccos(-\gamma)$, usually referred to as "saddle points." As a consequence, no two solution curves can intersect anywhere on the p, ϕ plane except at the saddle points.

Let us suppose that the initial angular velocity is large and opposite in sense to the applied torque. Then p will decrease toward zero, while ϕ steadily increases until the p axis is reached. If the initial conditions have been chosen so that the pendulum comes to rest, it will stop short of a "saddle point" and fall back, stopping and reversing its motion again at some point short of its previous turning point, and so on, oscillating about the intermediate stable equilibrium point with steadily decreasing amplitude until its energy is fully dissipated by damping; or the pendulum may stop and reverse its direction only once and eventually reach terminal rotation speed under the applied torque.

There will be one solution curve B (Fig. 2), which terminates at an unstable equilibrium point from the left, and one C from the right. Finally, the curve D represents the solution curve when the pendulum is given an infinitesimal angular impulse while in unstable equilibrium.

Since the solution curves cannot intersect, the curves A, B, C, and D divide the p, ϕ plane into two sets of regions, those to the left and above the lines C, but to the right of the B lines (the stable region), and those to the right of the C lines, but to the left of the B and D lines (the unstable region). All solution curves in the stable region spiral into a stable equilibrium point, while all solution curves in the unstable region eventually oscillate about the axis $p=-\gamma/\lambda$.

Thus, given values of γ and λ, depending upon the initial conditions of motion and displacement, the pendulum will eventually either damp out or will approach continuous rotation.

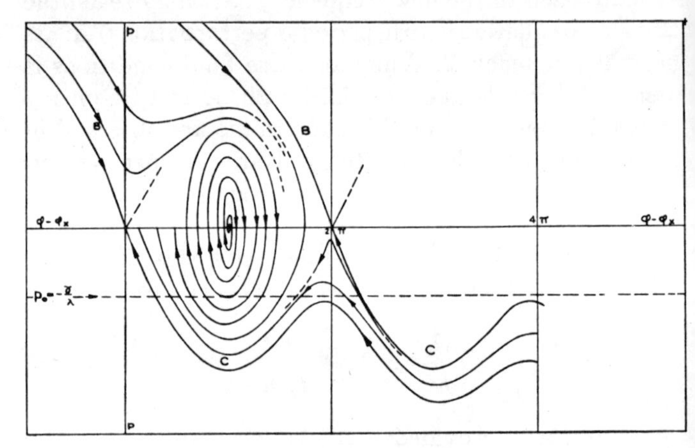

Fig. 3—Representative solution curves for λ slightly greater than λ_c—free-hand sketch.

However, for a given value of γ, as λ is increased, the point c at which the curve C crosses the p axis, moves to the right, until for a certain value λ, the point c reaches the next point of unstable equilibrium. We call this value of λ, λ_c the "second critical damping factor."[3] For λ greater than λ_c, the curve C no longer crosses p axis, and the unstable regions have completely disappeared (Fig. 3).

The values of the "second critical damping factor" for several values of γ were determined by cut and try methods. The boundary solution curves were traced by differential analysis to determine whether or not they cross the p axis.

For γ less than 0.5, the relationship is quite linear and given very nearly by $\lambda_c=0.8\gamma$; however, this line cannot be extrapolated with any degree of confidence.

Inserting the original parameters, we get for the lock-in condition (1).

Acknowledgment

The authors gladly acknowledge the assistance of Mrs. Florence Katz, who performed the computations and prepared the manuscript.

[3] *Ibid.*, referring to investigations by Lyon and Edgerton in which "critical damping factor" is distinct from this one.

Automatic Phase Control: Theory and Design*

T. J. REY†, SENIOR MEMBER, IRE

Summary—**Automatic Phase Control (APC) systems with sinusoidal reference are analyzed, and design criteria are presented. The limit of the pull-in range is derived; the result is believed to be novel, and is found to be in agreement with experiment.**

Glossary

a = lead parameter of RC filter (Fig. 6)
B = noise bandwidth (17)
G = forward transfer function of filterless APC system at lock (6)
J_0 = Bessel function of order 0
$\pm K$ = limits of the control range (frequencies are in radians per second)
$N(\omega) = N_\omega \exp(-i\psi_\omega)$ = filter transfer function ($N(0)=1$)
t = real time
v = control voltage
$\hat{V}$ = amplitude of the first-order beat
Y = closed-loop transfer function of locked APC system
W = open-loop transfer function of locked APC system
μ = modulation index
ν = frequency of FM in locked system
τ = time constant of RC filter (Fig. 6)
φ = a small change in ϕ
ϕ = oscillation phase relative to reference (1)
ω = frequency, error frequency in closed loop
ω_s = reference frequency
Ω = tuning error in open loop
$\hat{\Omega}$ = maximum pull-in frequency

I. Introduction

A. General

AUTOMATIC Phase Control (APC) serves to synchronize an oscillator with a sinusoidal reference signal of low power. It differs from AFC systems wherein a frequency discriminator supplies the control signal so that a frequency error remains. APC has developed from a method of motor tuning in which the oscillation and the reference are combined to generate a field that rotates at the error frequency.[1-3]

The electronic system contains a phase detector (*e.g.*, a balanced mixer), a frequency modulator (*e.g.*, a reactance tube) and a low-pass filter between these parts (Fig. 1). The filtered beat between the oscillation and the reference signal controls the modulator, and tends to compensate the open-loop error frequency Ω. The desired control signal is dc proportional to $\sin \phi_\infty = \Omega/K$; the limit of the control range is $\pm K$, where

* Received by the IRE, November 27, 1959; revised manuscript received April 15, 1960. Presented in summary at the NEREM Meeting in Boston, Mass., November 19, 1958. The work reported in this paper was performed by Lincoln Laboratory, a center for research operated by Massachusetts Institute of Technology, with the joint support of the U. S. Army, Navy, and Air Force.
† M.I.T. Lincoln Lab., Lexington, Mass.

[1] D. G. Tucker, "Carrier frequency synchronization," *P.O. Elec. Engrs. J.*, vol. 33, p. 75–81; July, 1940.
D. G. Tucker, "The synchronization of oscillators," *Electronic Engrg.*, vol. 16, pp. 26–30; June, 1943.
[2] W. J. Gruen, "Theory of AFC synchronization," PROC. IRE, vol. 41, pp. 1043–1048; August, 1953.
[3] E. Labin: "Théorie de la synchronization par contrôle de phase," Philips Res. Repts., vol. 4, pp. 291–315; August, 1949.

Reprinted from *Proc. IRE*, vol. 48, pp. 1760–1771, Oct. 1960; and from *Proc. IRE*, vol. 49, p. 590, Mar. 1961.

K = maximum beat voltage $\hat{V} \times$ modulator sensitivity,
$\hat{V}$ = smaller of the inputs to the detector $\times$ conversion gain $\times$ dc gain between detector and modulator.

The angle ϕ_∞ accounts for the static phase difference between the reference and the locked oscillation (Fig. 2). When $\Omega > K$, the system is in its asynchronous state and Ω must be reduced to the value $\hat{\Omega}$ for lock to occur. When the "pull-in ratio" $(\hat{\Omega}/K) < 1$, a hysteresis effect (sometimes called "pulling effect") exists since, if Ω drifts slowly, the state does not change when Ω enters the interval between $\hat{\Omega}$ and K from either side (Fig. 3).

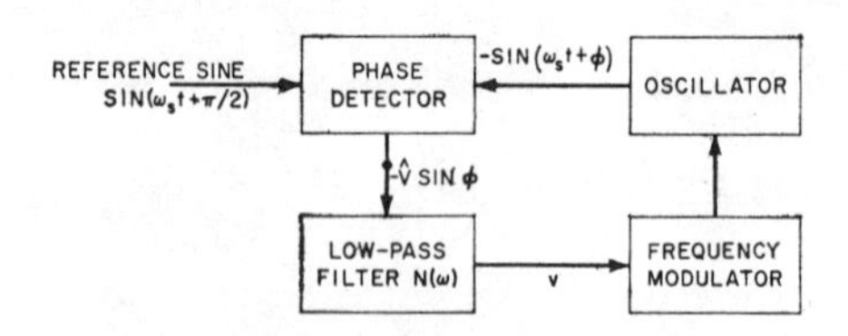

Fig. 1—Block diagram of APC system.

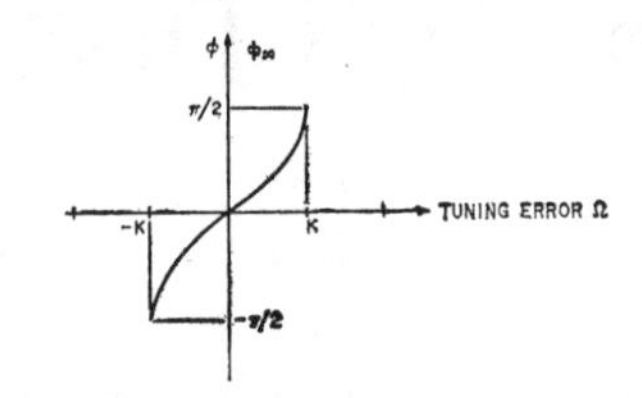

Fig. 2—Phase vs tuning error characteristic.

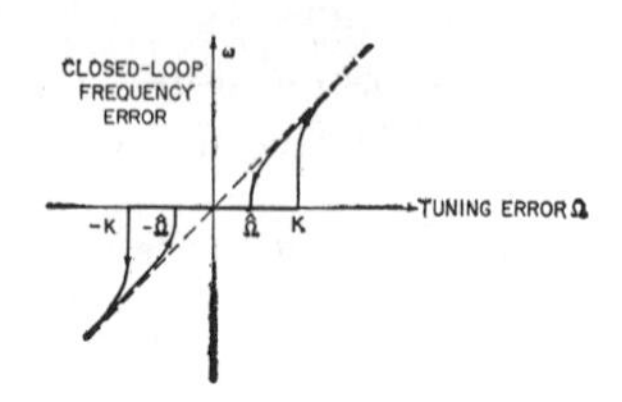

Fig. 3—Hysteresis effect.

The objects of synchronization are varied and sometimes conflicting, *e.g.*,

1) Power in excess of the reference
2) Spectral purity in excess of the reference (frequency synthesis)
3) Phase close to the reference (TV, Radio interferometry)
4) Tracking of frequency changes (*e.g.*, satellites)[4,5]
5) Spectral purification of the oscillation (microwaves)[6–8]

[4] R. Leek, "Phase-lock AFC loop, tracking signals of changing frequency," *Electronic and Radio Engr.*, vol. 34, pp. 141–146, 177–183; April/May, 1957.
See also R. Jaffe and E. Rechtin, Jet Propulsion Lab., California Institute of Technology, Progress Rept. No. 20–243; December, 1954.
[5] L. V. Berkner, "Annals of the International Geophysical Year," Pergamon Press, New York, N. Y., vol. 6, pp. 410–416; 1958.
[6] M. Peter and M. W. Strandberg, "Phase stabilization of microwave oscillators," Proc. IRE, vol. 43, pp. 869–873; July, 1955.
[7] G. Winkler, "Progress in Phase-Lock Techniques," *Proc. of the 11th Annual Symp. on Frequency Control*, U. S. Signal Corps, Fort Monmouth, N. J., pp. 335–336; May, 1957.
[8] I. L. Bershtein and V. L. Sibiriakov, "Phase stabilization of microwave generators," *Radioteckh. i Elektron.*, vol. 2, pp. 944–945; July, 1957.

Further possibilities include:

6) A source of FM centered on the reference frequency (Section III G)
7) A phase or frequency discriminator (Sections III, D and IV, B).

The object of this paper is the preparation of a design basis in terms of system parameters.

Section II deals with a filterless system to exemplify such basic subjects as phase pull-in, asynchronous degeneration, and the effects of small perturbations at lock.

Section III establishes the equivalent circuit that allows for a filter in the control system. Design equations (19) and (22) follow by optimization with regard to detector noise. Different criteria hold when phase perturbations are considered (Section III, D and E). Section III concludes with the pull-out problem and with self-maintaining FM in the locked state.

Section IV hinges on the demonstration of the asynchronous limit cycle. Its degenerative properties allow $(\hat{\Omega}/K)$ to be determined (Section IV-C), and hence, to be compared with experimental data (Section IV-D). Agreement is satisfactory, as is the result that pull-in range $\hat{\Omega}$ and noise bandwidth B are roughly equal in a system optimized with regard to B.

Section V reviews some complications.

B. *Analytical*

Let

$\sin(\omega_s t + \pi/2)$ = the time dependence of the reference signal,
$-\sin(\omega_s t + \phi)$ = the time dependence of the oscillation.

Then their first-order beat has the form

$$V = -\hat{V}\sin\phi. \tag{1}$$

The modulator control voltage v is the output of the filter N fed with V, and determines the instantaneous angular frequency

$$\dot{\phi} = \Omega + Kv/\hat{V}, \tag{2}$$

where $\dot{\phi} = d\phi/dt$, Ω = open-loop tuning error, and the modulation sensititivy $K/\hat{V}$ is regarded as a constant.

Since exact analysis is precluded by the nonlinearity of (1), previous theory of the asynchronous state has been confined mainly to numerical or graphical analysis with regard to the phase plane. The present treatment is based on three observations:

1) The closed system tends toward a steady state or limit cycle wherein the control voltage approaches a harmonic series.

2) The dc and the fundamental component of the control voltage suffice for an approximate analysis of two important cases:

a) the synchronous state,

b) the steady asynchronous state, including that which corresponds to a tuning error at the pull-in limit ($\Omega=\hat{\Omega}$).

3) Operational methods can be applied readily to case a), but not to case b).

II. The Filterless Case

The analysis is simplified if the filter passes the beat of (1) without modification while rejecting all other components in the detector output; then $v=V$ and

$$\dot{\phi} = \Omega - K \sin \phi, \tag{3a}$$

$$= K(\sin \phi_\infty - \sin \phi), \tag{3b}$$

where

$$\phi_\infty = \arc\sin \Omega/K,$$

$$|\phi_\infty| < \pi/2 \quad \text{and} \quad \phi \text{ is real if } |\Omega| < K. \tag{4}$$

The circle diagram of Fig. 4 illustrates the differential equation, and suggests a driven pendulum subjected to a driving torque (Ω), to gravity ($K \sin \phi$) and to viscous damping ($\dot{\phi}$); the inertia is negligible in the absence of a filter [but see (23a)].

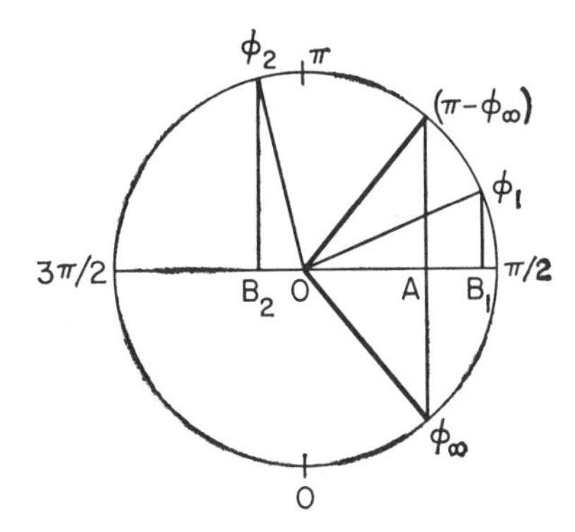

Fig. 4—Phase regions.

When $|\Omega| < K$, the "velocity" $\dot{\phi}$ is proportional to $(0A - 0B)$, where A and B are the projections on the diameter through $(\pi/2)$ of ϕ_∞ and ϕ, respectively. Thus,

1) $\dot{\phi}$ is negative when $\phi(=\phi_1)$ is on the short arc between ϕ_∞ and $(\pi-\phi_\infty)$;

2) $\dot{\phi}$ is positive when $\phi(=\phi_2)$ is on the long arc between ϕ_∞ and $(\pi-\phi_\infty)$.

Since $\dot{\phi}=0$ only when $\phi=\phi_\infty$ or $(\pi-\phi_\infty)$, equilibrium is stable at ϕ_∞ but unstable at $(\pi-\phi_\infty)$. The phase cannot pass through the equilibrium values, and must reach ϕ_∞ with a total phase change $<2(\pi-\phi_\infty)$ from anywhere; this process of initial stabilization is called "phase pull-in." The exact integral of (3b) is considered in the Appendix, but can be approximated by linearization; thus

$$\sin \phi \sim \phi \quad \text{for} \quad -\pi/2 < \phi < \pi/2,$$

then

$$(\phi - \phi_\infty) \sim \phi_0 \exp(-KT),$$

$$\sin \phi \sim (\pi - \phi) \quad \text{for} \quad \pi/2 < \phi < 3\pi/2,$$

then

$$(\phi - \phi_\infty) \sim \phi_0 \exp KT. \tag{5}$$

The decay or growth is roughly exponential with time constant $1/K$. However, $\dot{\phi}$ is not discontinuous at $\pm\pi/2$, although the approximation suggests it.

The case $\Omega > K$ merits attention. The system is then asynchronous, of course, but the average frequency error is less than Ω. Eq. (3a) in the form

$$\dot{\phi} = K\left(\frac{\Omega}{K} - \sin \phi\right)$$

shows that

$$0 < \dot{\phi} < \Omega \quad \text{where} \quad \phi = \pi/2,$$

$$\Omega < \dot{\phi} \quad \text{where} \; \phi = -\pi/2.$$

Hence, ϕ dwells longer on the arc from 0 to π than on the supplementary arc, $\sin \phi$ has a positive mean, and the control signal $-\hat{V} \sin \phi$ has a dc component of the correct polarity for reducing the tuning error. The asynchronous state can be said to have degenerative properties.

It will now be shown that the locked system is described by a forward transfer function

$$G = K/i\nu \tag{6}$$

and a feedback function

$$\cos \phi_\infty = \sqrt{1 - \Omega^2/K^2} \tag{7}$$

with reference to the perturbation φ of the controlled phase that results from a small voltage disturbance

$$E = A\hat{V} \exp i\nu t, \qquad |A| \ll 1.$$

The proof is as follows. When the loop is opened between detector and modulator, and E together with the lock-maintaining bias $-\hat{V} \sin \phi_\infty$ is applied to the latter, the oscillation phase is perturbed by

$$\varphi_0 = \frac{K}{i\nu} \frac{E}{\hat{V}}.$$

The perfect integration is due to FM, and (6) follows on taking $V\varphi_0$ as the output. However, the detector output is

$$-\hat{V} \sin(\phi_\infty + \varphi_0) \sim -\hat{V} \sin \phi_\infty - \hat{V}\varphi_0 \cos \phi_\infty$$

$$\text{if } |\varphi_0| \ll 1, \tag{8}$$

and reveals the feedback function of (7) that depends on tuning error but not on the perturbing frequency.

A sudden perturbation φ_0 of the oscillation phase is easily shown to decay as

$$\varphi = \varphi_0 \exp(-Kt \cos \phi_\infty). \quad (9)$$

Locked system properties are considered further in the next section.

III. The Synchronous State

A. The Stable State

The low-pass filter will be described by the transfer function

$$N(\omega) = N_\omega \exp[-i\psi_\omega], \; N_\omega \text{ and } \psi_\omega \text{ real}, \; N(0) = 1. \quad (10)$$

In the case of the locked loop, allowance for the filter is made by modifying the forward function to NG; the over-all transfer function $NG \cos \phi_\infty$ must satisfy Nyquist's criterion for stability. (Evidently NG and $\cos \phi_\infty$ play the parts of the "μ" and "β" of feedback theory.) The stably locked system is a form of regulator; with regard to small perturbations, it is represented by an equivalent circuit comprising linear transfer functions and a differencing element (Fig. 5). The signal E injected into the filter perturbs the oscillation phase by the amount

$$\varphi = YE/\hat{V},$$

where Y is the transfer function

$$Y = NG/(1 + NG \cos \phi_\infty). \quad (11)$$

In many instances, the stability condition is related to the frequency ν at which the filter has a 90° lag through

$$\frac{K}{\nu} N_\nu \cos \phi_\infty < 1 \quad \text{where} \quad \psi_\nu = \frac{\pi}{2} \cdot \quad (12)$$

The static phase error ϕ_∞ of (4) is not affected by N if the system is stable.

B. RC Filter with Compensation

A particularly suitable filter has the configuration of Fig. 6 and the transfer function

$$N(\omega) = \frac{1 + i\omega a\tau}{1 + i\omega\tau}, \qquad 0 \leq a \leq 1; \quad (13)$$

the limit $a = 0$ corresponds to a simple RC filter, and the limit $a = 1$ corresponds to no filter. By substituting in (11),

$$Y = K \frac{1 + i\nu a\tau}{K \cos \phi_\infty + i\nu(1 + aK\tau \cos \phi_\infty) - \nu^2\tau}, \quad (14a)$$

or

$$Y = \frac{1 + i\nu(2\zeta K/\omega_n)}{1 + i\nu(2\zeta/\omega_n) - (\nu/\omega_n)^2} \sec \phi_\infty, \quad (14b)$$

where

$$\omega_n^2 = \frac{K}{\tau} \cos \phi_\infty, \quad 2\zeta = (K\tau \cos \phi_\infty)^{-1/2} + a(K\tau \cos \phi_\infty)^{1/2},$$

$$k = \frac{aK\tau \cos \phi_\infty}{1 + aK\tau \cos \phi_\infty}, \qquad 2\zeta\omega_n = \frac{1}{\tau} + aK \cos \phi_\infty, \quad (15a)$$

or conversely,

$$K \cos \phi_\infty = \frac{\omega_n}{2\zeta(1 - k)}, \qquad \frac{1}{\tau} = 2\zeta\omega_n(1 - k),$$

$$a = 4\zeta^2 k(1 - k). \quad (15b)$$

Now, a stable linear feedback system can be represented by a passive circuit; corresponding to the locked system with the compensated RC filter is the LRC circuit of Fig. 7. It has resonant frequency ω_n, and damping ratio ζ; the output terminals 1, 3 are across C_{eq} and kR_{eq} in series, and deliver the output $\varphi\hat{V} \cos \phi_\infty$ rather than $\varphi\hat{V}$

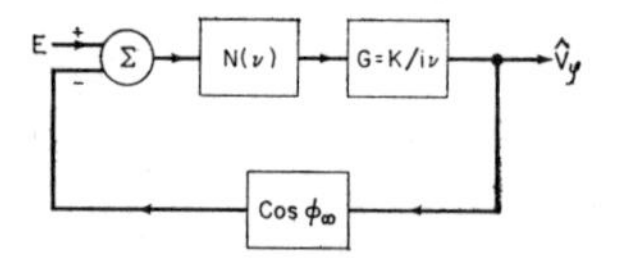

Fig. 5—Locked system equivalent circuit.

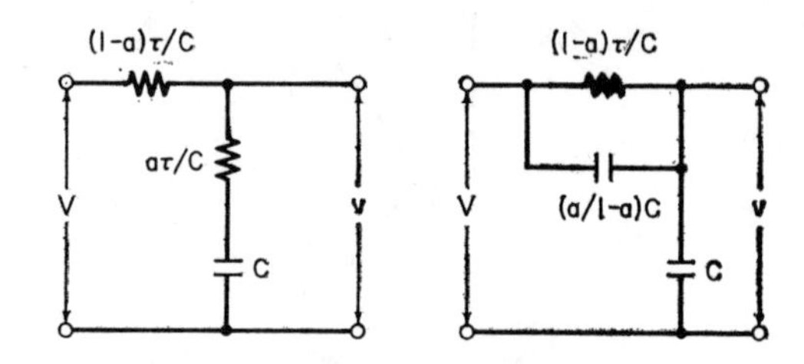

Fig. 6—Filters with transfer function $(1+i\omega a\tau)/(1+i\omega\tau)$.

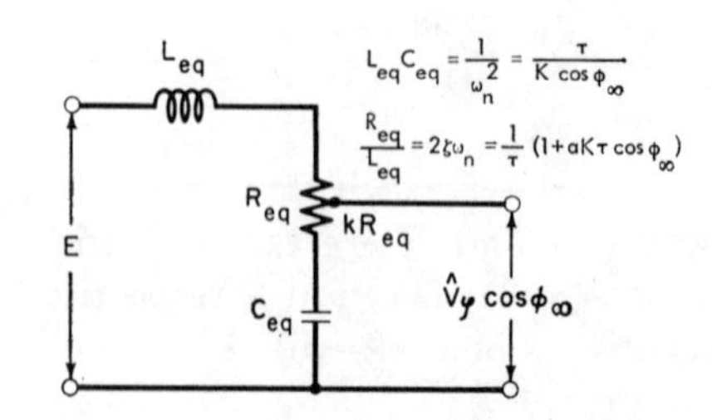

Fig. 7—Equivalent passive circuit with filter of Fig. 6.

If E is a small step at time $t = 0$, the subsequent phase perturbation follows from (15b) as

$$\varphi = \frac{E}{\hat{V}} \sec \phi_\infty [1 - g(t)],$$

$$g(t) = (\exp - \zeta\omega_n t) \left[\cos \sqrt{1 - \zeta^2}\omega_n t - \frac{(2k - 1)\zeta}{\sqrt{1 - \zeta^2}} \sin \sqrt{1 - \zeta^2}\omega_n t \right]. \quad (16)$$

The cases $K\tau = 1.6$ and 32, $\Omega = 0$, and a chosen according to (21a), are illustrated for both the sinusoidal and step responses by Fig. 8.

A certain design criterion will now be discussed.

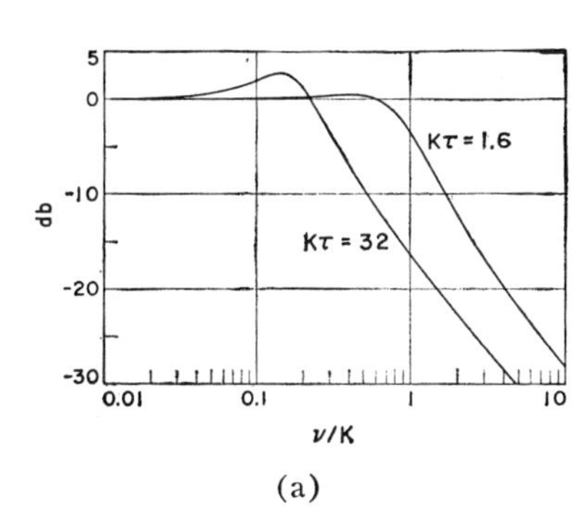

(a)

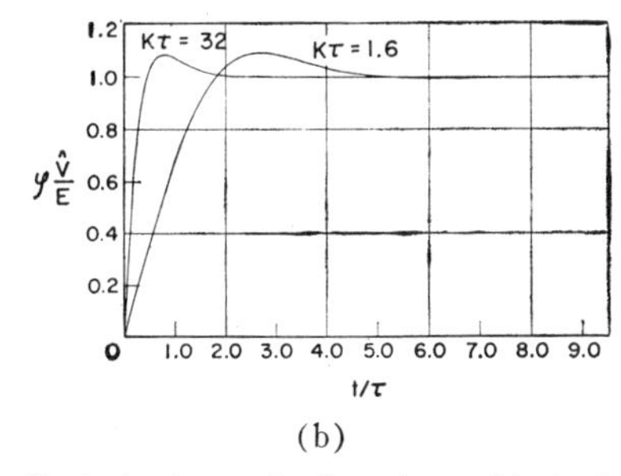

(b)

Fig. 8—Optimized transfer functions of locked system; (a) sinusoidal response and (b) step response.

C. The Noise Bandwidth

Let white noise of mean square (MS) value $\overline{\delta e^2} = \overline{A^2}\hat{V}^2 \delta\nu$ per unit bandwidth be injected into the detector output. The resultant phase perturbation is random, with MS value

$$\overline{\varphi^2} = \overline{A^2}B, \tag{17a}$$

where (the lower limit of integration being $-\infty$ for noise at the detector input)

$$B = \int_0^\infty |Y|^2 d\nu. \tag{17b}$$

If $\overline{\varphi^2}$ is small, clipping caused by nonlinearity is unimportant; (17a) then shows the relative contribution of injected noise to the oscillation spectrum, and can be expressed in the form

$$\text{in-band spurious contribution} = 10 \log_{10} \overline{A^2}B \text{ db}. \tag{17c}$$

The "integrated noise bandwidth" B is an index of noise performance; the spectral distribution of φ is $|Y|$, of course, and the 3-db bandwidth is found from (14a), *e.g.*, as

$$K \cos \phi_\infty \quad \text{for} \quad a = 1,$$

and as

$$\omega_n\{1 - 2\zeta^2 + [1 + (1 - 2\zeta^2)^2]^{1/2}\}^{1/2} \quad \text{for} \quad a = 0.$$

The value of B for a system without filter or with a simple RC filter is, with (14) for $a = 0$ or 1 and on integrating (17b) by partial fractions or in the complex plane,

$$B = \pi K / 2\sqrt{1 - \Omega^2/K^2}. \tag{18a}$$

At the edge of the control range $\Omega = K$, the feedback vanishes, B is infinite, and operation is unstable because of noise; B has its least value $\pi K/2$ when $\Omega = 0$. Given $\Omega = \Omega_0$, the corresponding B is minimized on taking

$$K = \sqrt{2}\Omega_0;$$

then

$$B = \pi\Omega_0 (= \pi K/\sqrt{2}). \tag{18b}$$

The time constant τ affects B only indirectly, by precluding pull-in if Ω exceeds $\hat{\Omega}$ (Fig. 3). A system with simple RC filter is then optimized with regard to detector and predetector noise by

$$K = \sqrt{2}\Omega_0, \qquad K\tau = 2.6, \tag{19}$$

on using (35) and $\Omega_0 = \hat{\Omega}$. As Ω drifts towards K, B is doubled when $\Omega \sim 1.33\Omega_0$; for greater drifts, it is necessary to increase K and $K\tau$ beyond the above values.

In the case of a system with compensated RC filter, (18a) is replaced by

$$B = \left(\frac{1 + a^2 K\tau \cos \phi_\infty}{1 + aK\tau \cos \phi_\infty}\right) \frac{\pi K}{2} \sec \phi_\infty \tag{20a}$$

$$= \left(\frac{1}{2\zeta} + 2k\zeta\right) \frac{\pi\omega_n}{2} \sec^2 \phi_\infty. \tag{20b}$$

It is readily shown[2] that B has a broad minimum if

$$\zeta = \tfrac{1}{2}\sqrt{1 + (\omega_n / K \cos \phi_\infty)^2}$$

so that

$$B = \pi\omega_n \quad \text{if} \quad \zeta = 1/2 \quad \text{and} \quad \omega_n \ll K.$$

The last condition entails a low pull-in ratio (Section IV, C), ensures a small static phase error in the absence of drift and corresponds to the lead parameter

$$a = (\sqrt{K\tau} - 1)/K\tau.$$

The increase of B with the tuning error is evident from (20a). However, B is minimized by the lead parameter a if

$$(1/a) = 1 + \sqrt{1 + K\tau \cos \phi_\infty}, \tag{21a}$$

so that

$$B_{\min} = (\pi/\tau)(\sqrt{1 + K\tau \cos \phi_\infty} - 1) \sec^2 \phi_\infty. \tag{21b}$$

The lowest curve of Fig. 9 represents (21a) if the abscissae are read as $K\tau \cos \phi_\infty$. Eq. (21b) is also shown there, normalized with regard to $a=0$ on dividing by $(\pi K/2 \cos \phi_\infty)$; the middle curve is valid for $\phi_\infty=0$. The slight improvement for $K\tau \sim 1$ becomes more marked as $K\tau$ increases; it is somewhat less at tuning errors for which B is not optimized, as indicated by the upper curve that has been computed from (20a) and (21a) on taking $\Omega=\sqrt{3}K/2$ for B but $\Omega=0$ for a.

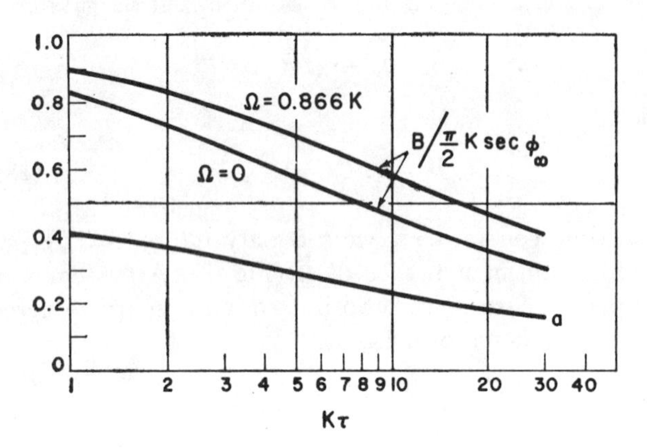

Fig. 9—Normalized noise bandwidth and a for $a=(1+\sqrt{1+K\tau})^{-1}$.

Given $\Omega=\Omega_0$, the system is optimized further by differentiating (21a) with respect to K, and substituting

$$K = \sqrt{2}\Omega_0\sqrt{1 + \epsilon}.$$

The resulting equation

$$\epsilon = (\Omega_0\tau/4)(1 - \epsilon)^2\sqrt{1 + 2\epsilon};$$

is solved approximately by

$$3\epsilon = \left[\sqrt{(1 + 4/\Omega_0\tau)^2 + 6} - (1 + 4/\Omega_0\tau)\right];$$

clearly,

$$\sqrt{2} \leq (K/\Omega_0) < 1.8,$$

the upper limit being approached when $\Omega\tau_0 \gg 4$.

System optimization is thus complicated by the effect of the lead parameter a on both B and $\hat{\Omega}/K$. Two limiting cases are of interest, $K\tau \gg 1$ [see the paragraph following (20b) above], and $K\tau < 10$, say. In the latter case, ϕ_∞ can be appreciable so that a may be chosen for $\cos \phi_\infty = 1/\sqrt{2}$ and K from (19). Also, in view of (35), nearly optimal relations are

$$K = \sqrt{2}\Omega_0, \qquad K\tau = 6.7, \qquad a = 0.39. \tag{22}$$

The remarks following (19) still apply with regard to drift.

D. Perturbations of the Reference Phase

A small signal E injected into the filter is equivalent to the perturbation ϵ of the reference phase $(\omega_s t + \pi/2)$, if

$$E = \epsilon \hat{V} \cos \phi.$$

In the closed loop, the detector output is then

$$E/(1 + NG \cos \phi_\infty),$$

and the oscillation phase perturbation is

$$\varphi = \epsilon \cdot Y \cos \phi_\infty,$$

where Y is defined by (11); the perturbation vanishes where $\Omega=K$, *i.e.*, where B is infinite.

The effects of changes in the reference frequency are of interest; it will be shown that the controlled frequency settles at the correct value if the reference frequency suffers a small impulse or step; however, the response to the ramp $\dot{\omega}_s$ tends to the error limit $\dot{\omega}_s/K \cos \phi_\infty$.

A small step Ω_s at time $t=0$ of the reference frequency ω_s is identical with the ramp function of phase

$$\epsilon = \Omega_s t, \qquad t \geq 0.$$

The controlled frequency is then subjected to an evanescent error that follows from (16) as

$$-\ddot{\varphi} + \Omega_s = \Omega_s g(t);$$

the change φ of the controlled phase has the steady-state component $(\Omega_s/K \cos \phi_\infty)$, as can be shown by integration or by static considerations alone.

If the reference frequency changes at the constant rate $\dot{\omega}_s(=d\omega/dt)$, then

$$\epsilon = \dot{\omega}_s t^2/2, \qquad t \geq 0,$$

so that

$$\dot{\varphi} - \dot{\omega}_s = -\dot{\omega}_s g(t),$$

i.e., the rate of following the frequency tends towards its correct value. However, steady error components are given by

$$(-\varphi + \dot{\omega}_s t) \rightarrow \dot{\omega}_s/K \cos \phi_\infty$$

for the frequency, and by

$$\left(-\varphi + \frac{\dot{\omega}_s t^2}{2}\right) \rightarrow \frac{\dot{\omega}_s t}{K \cos \phi_\infty} - \dot{\omega}_s \frac{\zeta^2 - 2k}{\omega_n^2}$$

for the phase. The situation is complicated if φ or noise is substantial.[4]

E. Inherent Perturbations

Small phase perturbations arising within the oscillator (Section V, B) are modified by the factor $(1+NG \cos \phi_\infty)^{-1}$.

In the case of the compensated RC filter, reductions occur at perturbation frequencies below

$$\omega_c = \left\{K \cos \phi_\infty/\tau\left[2(1 - a) - a^2 K\tau \cos \phi_\infty\right]\right\}^{1/2}$$

if ω_c is real, or at all frequencies if

$$(2/a) < 1 + \sqrt{1 + 2K\tau \cos \phi_\infty}.$$

When a is given by (21a), the cutoff frequency equals the resonance frequency ω_n of (15a). The closed-loop perturbation φ is related to the open-loop perturbation $\varphi_{o.l.}$ by the equivalent circuit of Fig. 7 if $\hat{V}\varphi_{o.l.}$ is applied between terminals 1, 2 and $\hat{V}\varphi$ is measured across terminals 1 and 3.

The integrated effect of the loop depends upon the nature of the uncontrolled oscillation spectrum. For random phase perturbations confined to the band (ω_a, ω_b) with spectral power distribution ω^{-2}, the uncontrolled MS deviation of phase $\overline{\varphi_{o.l.}^2}$ and frequency $\overline{\omega^2}$ are related through

$$\overline{\varphi_{o.l.}^2} = \overline{\omega^2}/\omega_a\omega_b.$$

The normalized MS deviation in the closed loop is then given by

$$\frac{\overline{\varphi^2}}{\overline{\varphi_{o.l.}^2}} = \frac{\omega_a\omega_b}{\omega_b - \omega_a}\int_a^b \frac{(1+\omega^2\tau^2)d\omega}{[(K\cos\phi_\infty - \omega^2\tau)^2 + \omega^2(1+aK\tau\cos\phi_\infty)^2]}$$

where $\omega_a < \omega_b$. The deviation is least for no filter ($a=1$). and $(K/\tau) > \omega^2$ is desirable.

F. Pull-Out

The perturbations considered so far were assumed to be too small to destroy the locked state. An example of another type is the following:

The error Ω_0 is corrected by $\phi_0 = \text{arc sin } \Omega_0/K$ when the additional error Ω_1 is applied suddenly (*e.g.*, owing to a step voltage $\hat{V}\Omega_1/K$ at the frequency modulator); what is the greatest value $\hat{\Omega}_1$ that will not throw the system out of synchronism? The condition $|\Omega| \leq K$ (where $\Omega = \Omega_0 + \hat{\Omega}_1$) is implied, of course.

On allowing for the compensated RC filter that has the differential equation

$$v + \tau\dot{v} = V + a\tau V,$$

and combining with (1), (2) and (4), the system equation is

$$\ddot{\phi} + \dot{\phi}\left(\frac{1}{\tau} + aK\cos\phi\right) + \frac{K}{\tau}\sin\phi = \frac{\Omega}{\tau}. \quad (23a)$$

In the filterless case ($a=1$), Section II shows that the new equilibrium

$$\phi_\infty = \sin\frac{(\Omega_0 + \Omega_1)}{K}$$

is reached without overshoot, and the static condition $|\Omega| = K$ answers the problem.

This is not so if $a=0$. The problem is then identical with that of the pull-out torque for synchronous motors; (23a) is the torque equation if $\ddot{\phi}$ is due to inertia, $(\dot{\phi}/\tau)$ arises from the slip between the cage and the rotating field, $(K/\tau)\sin\phi$ is due to the angle between the fields of rotor and armature, and (Ω/τ) represents the torque load.

An apparent approach is to linearize as in Section II, D; (23a) then becomes

$$\ddot{\phi} + 2\omega_n\dot{\phi} + \omega_n^2\phi = \omega_n^2\tan\phi_\infty \quad (23b)$$

which can be solved for ϕ in terms of the initial values $\phi = \phi_0$, $\dot{\phi}_0 = \Omega_1$. Since ϕ_∞ is a point of stable equilibrium, it is obvious that if $|\phi - \phi_\infty|$ is never appreciable, the system will settle at ϕ_∞. However, the linearized equation does not provide the solution $\hat{\Omega}_1$.

The two equilibria ϕ_∞ and $(\pi - \phi_\infty)$ were found in Section II; $(\pi - \phi_\infty)$ is unstable also in the presence of a filter, although either point can then be traversed, in general. If the perturbations are such that $(\pi - \phi_\infty)$ is not traversed, there can be no pullout; the critical condition follows as

$$\ddot{\phi} = \dot{\phi} = 0 \quad \text{where} \quad \phi = \pi - \phi_\infty. \quad (24)$$

Since

$$\ddot{\phi} = \frac{d\dot{\phi}}{dt} = \dot{\phi}\frac{d\dot{\phi}}{d\phi},$$

(23a) can be rewritten as a relation between the variables ϕ and $\dot{\phi}$. This leads to the "phase plane" concept of Nonlinear Mechanics, *i.e.*, the graph of $\dot{\phi}$ vs ϕ. Methods of constructing and interpreting such trajectories are well known.[9-11] Clearly, when $\dot{\phi} > 0$, the representative point $(\dot{\phi}, \phi)$ is in the upper half plane and moves to the right as time increases; conversely for $\dot{\phi} < 0$. The pull-out problem is represented by Fig. 10(a);[9] the critical trajectory passes through ϕ_0 with infinite slope and cuts the abscissa again at $(\pi - \phi_\infty)$. Since equilibrium is unstable there, $\dot{\phi}$ increases again and the system pulls out into the asynchronous limit cycle for $(\Omega_0 + \Omega_1)$ (see Section IV, B); alternatively, $\dot{\phi}$ reverses sign and the system settles at the stable equilibrium point ϕ_∞.

The case $a=0$ has been treated by numerical approximation[11] and, more extensively, by integraph[12] solution of (23a). The result depends upon both Ω_0 and the damping term $(K\tau)^{-1/2}$. The lowest value

$$\frac{\Omega_0 + \Omega_1}{K} \sim 0.72$$

was obtained for $\Omega_0 = 0$, $1/K\tau = 0$; the static limit unity is obtained with sufficient damping $(1/\sqrt{K\tau} > 0.11)$.

[9] Z. Jelonek, O. Celinski and R. Syski, "Pulling effect in synchronized systems," *Proc. IEE*, vol. 101, pt. 4, pp. 108–117; November, 1953.

[10] Z. Jelonek and C. I. Cowan, "Synchronized systems with time delay in the loop," *Proc. IEE*, Monograph No. 229R, March, 1957, or vol. 104C, pp. 388–397; September, 1957.

[11] J. J. Stoker, "Nonlinear Vibrations," Interscience Publishers, Inc., New York, N. Y., pp. 70–98; 1950.

[12] W. V. Lyon and H. E. Edgerton, "Transient torque-angle characteristics," *Trans. AIEE*, vol. 49, pp. 686–699; April, 1930.

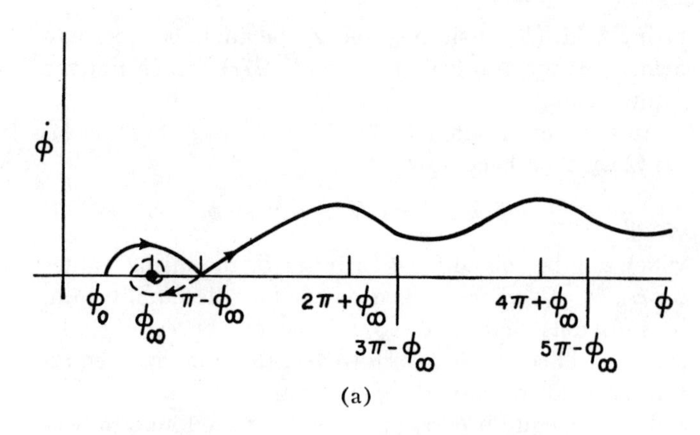

(a)

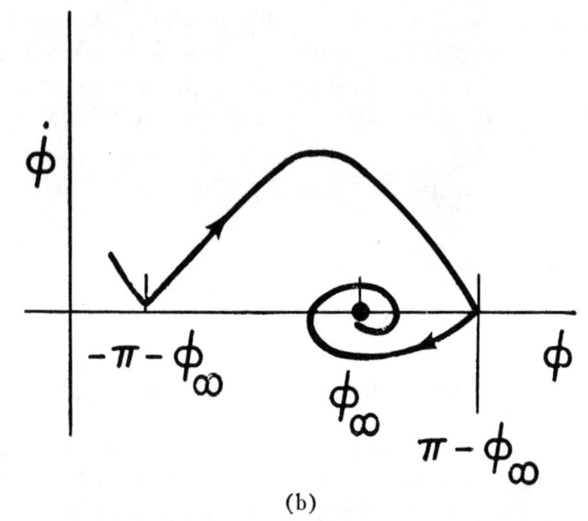

(b)

Fig. 10—(a) Pull-out trajectory. (b) Pull-in trajectory.

As $a>0$ increases the damping term, the situation for $0<a<1$ is intermediate between that for $a=0$ and $a=1$.

G. Instability

The existence of the locked state is compatible with a violation of the stability condition (12). Self-maintained oscillations then occur in the loop and modulate the phase. The oscillations are limited by nonlinearity and are almost sinusoidal when their amplitude is small. A suitable form of the phase is

$$\phi = \phi_\infty + \mu \sin \theta, \qquad \dot{\theta} = \nu, \tag{25}$$

where the oscillation frequency ν and the deviation index μ will be determined with the aid of (1) and (2), and the Fourier-Bessel series

$$\exp (i\mu \sin \theta) = J_0(\mu) + 2iJ_1(\mu) \sin \theta + 2J_2(\mu) \cos 2\theta + \text{terms in } 3\theta \text{ etc.};$$

the control voltage V follows as the output of the filter N fed with $-\hat{V} \sin \phi$, as

$$\frac{-v}{\hat{V}} = a_0 + b_1 \sin (\theta - \psi_\nu) + a_2 \cos (2\theta - \psi_{2\nu}) + \cdots$$

where

$$a_0 = J_0(\mu) \sin \phi_\infty, \qquad b_1 = 2J_1(\mu) \cos \phi_\infty N_\nu,$$
$$a_2 = 2J_2(\mu) \sin \phi_\infty N_{2\nu}, \cdots. \tag{26}$$

Combining with (2) but ignoring terms of frequency 2ν etc., then

$$\nu\mu \cos \theta = \Omega - KJ_0(\mu) \sin \phi_\infty - KN_\nu 2 \cos \phi_\infty J_1(\mu) \sin (\theta - \psi_\nu).$$

Assuming ν and μ to be constant, this equation is satisfied separately by the dc and by the coefficients of sin θ and cos θ if

$$\sin \phi_\infty = \frac{\Omega}{KJ_0(\mu)} \quad \cos \psi_\nu = 0,$$

$$\mu = -\frac{K}{\nu} N_\nu \cos \phi_\infty 2J_1(\mu). \tag{27}$$

Thus, if the criterion (12) is violated, synchronism is subject to periodic FM of frequency ν and deviation index μ such that

$$\frac{\mu^2}{8} \sim 1 - \frac{\nu}{N_\nu K \cos \phi_\infty} \quad \text{if} \quad \mu^2 \ll 1. \tag{28}$$

The gravest of the suppressed terms in (26) has frequency 2ν and relative amplitude

$$\frac{a_2}{b_1} = \frac{J_1(\mu)}{J_2(\mu)} \tan \phi_\infty \frac{N_{2\nu}}{N_\nu}.$$

Now $N_{2\nu} < N_\nu$ and $(J_1/J_2) \sim (\mu/4)$ if μ is small; then if $|\phi_\infty| < \pi/4$, the ignored term is slight and can be allowed for, *e.g.*, by a variable part of μ of frequency 3ν. Such higher approximations will not be considered further.

IV. Nonsynchronous States

A. The Hysteresis Effect

Hysteresis occurs if the boundary between the locked and the unlocked state depends upon the approach (Fig. 3). In the locked state, Ω can drift slowly through the interval $\pm K$. However, it is well known that in the asynchronous state, $|\Omega|$ can fall well below K before lock occurs. The effect can be estimated as follows: When the loop is opened at the frequency modulator, the filter output is AC of frequency Ω and peak amplitude $\hat{V} N_\omega$ as distinct from the direct $p-d$ $(\Omega/K)\hat{V}$ at lock. Hence, the equality

$$\hat{\Omega} = KN_{\hat{\Omega}}$$

defines a first approximation to the locking ratio $\hat{\Omega}/K$; for a simple RC filter, this is equivalent to

$$\frac{\hat{\Omega}}{K}\sqrt{K\tau} \sim \left[\left(1 + \frac{1}{4K^2\tau^2}\right)^{1/2} - \frac{1}{2K\tau}\right].$$

The better approximations derived in Sec. IV-C yield somewhat higher values by allowing for asynchronous degeneration in the closed loop.

B. Steady Asynchronism

In the asynchronous state, the phase ϕ differs from the open-loop value Ωt because of modulation by the control voltage. An expression to be considered is

$$\phi = \theta + \gamma + \mu \sin \theta, \qquad \dot{\theta} = \omega \tag{29}$$

where γ is a phase angle and μ is a deviation index.

A steady asynchronous state is described by ω = constant, while γ and μ are constant or harmonic with period $2\pi/\omega$ (or a multiple thereof). Eq. (29) then includes a ϕ component that grows linearly with time, but assigns purely periodic forms to $\sin \phi$ and the derivatives $\dot{\phi} - \omega, \ddot{\phi}$ etc. Since these, and not ϕ, enter into the system equation [*e.g.*, (23)], it follows that (29) is in the nature of a particular solution. On using it, instead of (25) for a harmonic balance as in Section III-G:

$$\begin{aligned} \omega &= \Omega + KJ_1(\mu) \sin \gamma \\ \mu\omega \sin \psi_\omega &= + KN_\omega[J_0(\mu) - J_2(\mu)] \cos \gamma \\ \mu\omega \cos \psi_\omega &= - KN_\omega[J_0(\mu) + J_2(\mu)] \sin \gamma. \end{aligned} \tag{30}$$

For a very small deviation index, these relations simplify to

$$\omega \sim \frac{\Omega}{2}\left(1 \underset{(-)}{+} \sqrt{1 - \frac{2K^2}{\Omega^2}(N \cos \psi)_\omega}\right),$$

$$|\mu\omega| \sim KN_\omega, \qquad \gamma \sim \psi_\omega \underset{(+)}{-} \frac{\pi}{2} \cdot \tag{31}$$

The upper signs are chosen for continuity with the properties of the filterless case (Section II); thus, when $\Omega > K$, then $\omega > 0$, $\mu > 0$ and ϕ is a minimum when

$$\theta = \pi, \qquad \phi = \pi + \gamma \sim \frac{\pi}{2} + \psi_\omega;$$

when $\Omega < -K$, then $\omega < 0$, $\mu < 0$ and $|\phi|$ is a minimum when

$$\theta = 0, \qquad \phi = \gamma \sim -\frac{\pi}{2} + \psi_\omega. \tag{31a}$$

The control voltage has the mean value $[(\omega - \Omega)/K]\hat{V}$ here, compared with $(-\Omega/K)\hat{V}$ at lock; the relation is reminiscent of a frequency discriminator.

C. Pull-In

The pull-in problem can be stated as follows: Given the asynchronous state for $\Omega > K$, Ω is reduced very slowly; at what value $\Omega = \hat{\Omega}$ does synchronism occur?

The answer depends on the nature of the system. For the filterless case, the identity $\hat{\Omega} = K$ was found (Section II). A more general result is suggested by the condition that the control voltage should just reach the required value $-\hat{V}\hat{\Omega}/K$. This is consistent with the determination of the critical limit cycle, that is, the unstable asynchronous state that is reached as μ grows very slowly.

In view of (29), the stationary condition

$$\dot{\phi} = \ddot{\phi} = 0$$

is attained when

$$\mu = 1, \qquad \theta = \pi; \tag{32}$$

implied stationary conditions are

$$\dot{\mu} = \dot{\gamma} = \dot{\omega} = 0. \tag{33}$$

When the system equation is of the second order, the stationary conditions imply the phase[13] $(\pi - \phi_\infty)$ and an asynchronous state that is critical since $(\pi - \phi_\infty)$ is a point of unstable equilibrium. The plot [Fig. 10(b)] of $\dot{\phi}$ vs ϕ touches the abscissae in cusps at $[(2n+1)\pi - \phi_\infty]$; $\dot{\phi}$ fluctuates between 0 and $2\omega_t$, where ω_t denotes the value of ω for $m = 1$.

Eq. (24) was introduced by the pull-out problem (Section III, F) where it represents a transient condition that terminates in either ϕ_∞ or in the stable asynchronism described by (29) for $|\mu| < 1$.

In the context of the slow approach to $|\mu| = 1$ from stable asynchronous states with $|\mu| < 1$, (24) defines a critical limit cycle that is nearly attained over many complete cycles of ϕ until a slight perturbation launches the phase transient that terminates in ϕ_∞ [Fig. 10(b)].

The pull-out frequency is never less than the pull-in frequency, in accordance with the practice of running synchronous motors up to speed on no-load.

On solving (30) for $\mu = 1$ (and writing $N_{\omega_t} = N_t$, etc.),

$$\frac{\omega_t}{KN_t} = \frac{0.965}{\sqrt{2.202 - \cos^2 \psi_t}} \tag{34a}$$

and

$$\frac{\hat{\Omega}}{\omega_t} = 1 + \frac{\cos \psi_t}{2N_t}, \tag{34b}$$

providing that $|\Omega/K| < 1$.

The first equation allows ω_t to be computed, given the value of K and the function N. Its nature usually requires a numerical method of solution; the pull-in ratio $\hat{\Omega}/K$ then follows from (34b).

This has been computed for the system with the RC filter of Fig. 6 for $(1/a) = 1 + \sqrt{1 + K\tau}$, and plotted in Fig. 11(a) against the independent variable $K\tau$. Hysteresis is seen to set in when $K\tau \sim 1.65$ and corresponds to $\phi_K = 0.46$; the pull-in ratio falls as $K\tau$ increases further, as was to be expected. The corresponding ratio[14] $B_{\min}/\hat{\Omega}$ has also been entered in Fig. 11(a), and appears to be fairly constant. The ratio $K\sqrt{2(N \cos \psi)_{\hat{\Omega}}}/\hat{\Omega}$ [see (31)] is shown as well, and lies between the limits

[13] Actually, (31a) only indicates that the phase is in the quadrant of $(\pi - \phi_\infty)$ when $\dot{\phi}$ is a minimum.

[14] This $B_{\min}$ has been both optimized and computed for $\Omega = 0$.

1.05 and 1.19. The ratio $\hat{\Omega}/K$ is shown in Fig. 11(b) for $(1/a)=1+\sqrt{1+2^{-1/2}K\tau}$.

Noise depresses the ratio near unity since lock is unstable where $\Omega=K$, and raises it somewhat where damping is small ($K\tau \gg 1$) and random perturbations of the asynchronous state modify the unperturbed values of $\dot{\phi}$ and $\ddot{\phi}$.

The ratio $\hat{\Omega}/K$ can be increased artificially by increasing the in-phase component of the filter transfer function transiently, *e.g.*, by short circuiting the top arm of the filter of Fig. 6, by shunting this arm with two diodes connected back to back, or by rocking the tuning capacitor; an equivalent of the latter expedient is the injection of low-frequency ac into the modulator, the resulting deviation being reduced drastically when the system locks. However, these expedients are questionable if they tend to make $\Omega=K$ or to introduce spurious signals.

For the system with a simple RC filter, (34) reduces to the explicit form

$$(\omega_t\tau)^2 = \sqrt{0.423(K\tau)^2 + 0.075} - 0.273, \quad \hat{\Omega} = 1.5\omega_t, \quad (35)$$

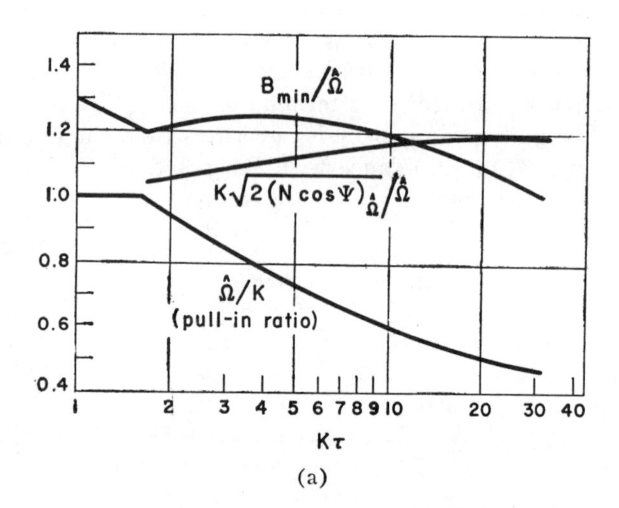

(a)

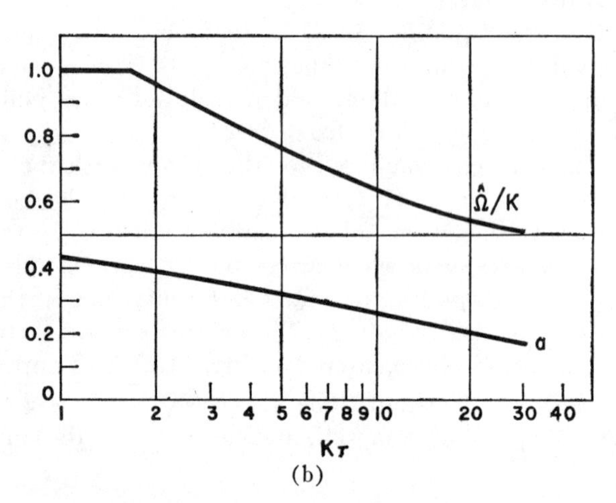

(b)

Fig. 11—(a) Pull-in ratio, approximate criterion and $B/\hat{\Omega}$ for $a=(1+\sqrt{1+K\tau})^{-1}$. (b) Pull-in ratio and a for

$$a = (1 + \sqrt{1 + 2^{-1/2}K\tau})^{-1}.$$

or

$$\frac{\hat{\Omega}}{K}\sqrt{K\tau} \sim 1.21\sqrt{1 - \frac{0.446}{K\tau}} \quad \text{if} \quad K\tau \gg 1.$$

D. Experimental Verification

Fig. 12 presents a plot of (35) and the results[9] of experiment and of graphical analysis.

Hysteresis sets in where $K\tau \sim 0.95$, $\psi_{\hat{\Omega}}=0.77$. The quantity

$$\frac{K}{\hat{\Omega}}\sqrt{2(N\cos\psi)_{\hat{\Omega}}} = \frac{K}{\hat{\Omega}}\sqrt{\frac{2}{1+\hat{\Omega}^2\tau^2}}$$

is also plotted, and lies between 1.02 and 0.97. Agreement between theory and experiment is good; the discrepancies at great values of $K\tau$ are attributed to experimental error by the experimenters.[9]

For the system with the filter of Fig. 6 in which $\zeta=0.5$, (34) has been computed for independent variable ω_n/K, using (15). This is plotted in Fig. 13, together with the results[2] of experiment and of graphical analysis. Quantity $K\sqrt{2(N\cos\psi)_{\hat{\Omega}}}/\hat{\Omega}$ is also plotted, and lies between 0.75 and 1.19. It should be noted that, since $\zeta=0.5$, the abscissa $\omega_n/K=1$ implies $a=0$, $K\tau=1$; then $\hat{\Omega}/K \sim 0.96$ agrees with this point of Fig. 12, as it should.

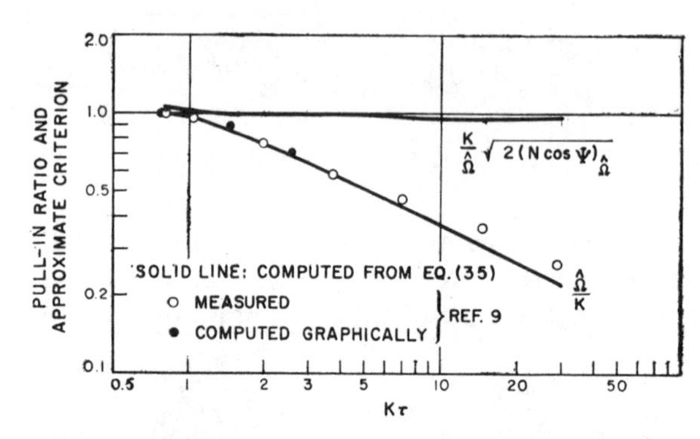

Fig. 12—Pull-in ratio of system with simple RC filter.

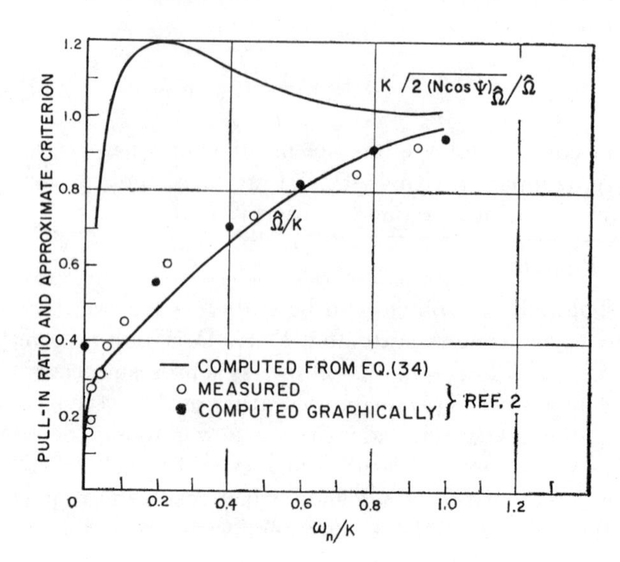

Fig. 13—Pull-in ratio for $\zeta=0.5$.

An approximation to the pull-in ratio is suggested by the vanishing of the surd in (31):

$$\frac{K}{\hat{\Omega}}\sqrt{2(N\cos\psi)_{\hat{\Omega}}} = 1. \tag{36}$$

Agreement with (34) is excellent for the system with a simple RC filter. However, the computations for other systems (Figs. 11 and 13) show that (36) provides only a rough estimate there.

Experience with systems containing a simple RC filter and a delay line, respectively, has led to the conjecture[10] that hysteresis sets in where $\psi_K \sim \pi/4$, but this conflicts with the value 0.46 in the case illustrated by Fig. 11(a). Moreover, the present theory must be modified when N is a delay line, perhaps to allow for the term of frequency 2ω in the $\sin\phi$ series. The results of applying (34) to this case differ significantly from the experimental data (Fig. 14).

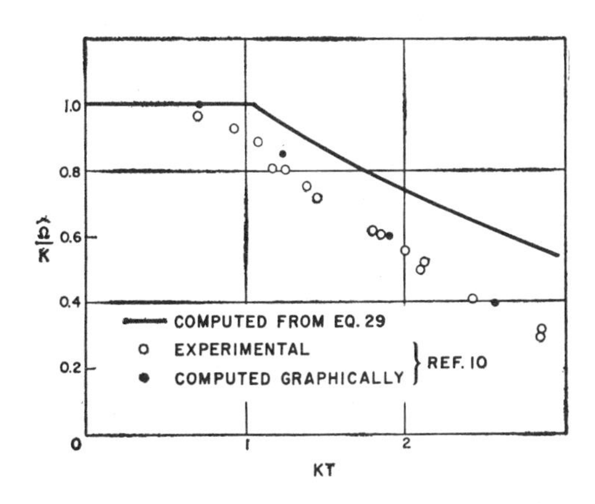

Fig. 14—Pull-in ratio for system with delay T in the loop.

Finally, it should be noted that APC circuits will synchronize at tuning errors Ω such that

$$\hat{\Omega} < |\Omega| < K$$

when the loop is first closed, if the initial conditions happen to be more favorable than those considered in Section III, C.

V. Imperfections

A. The Detector

The specification of a physical APC system in terms of K and N is usually incomplete, though effects that are represented by linear transfer functions can be absorbed in N; *e.g.*, if there is a time constant τ_1 in addition to the filter function of (13), the locked system remains stable if

$$\frac{K\tau_1}{1+\tau_1/\tau} < 1 + aK\tau. \tag{37}$$

However, the representation of the detector as a source of the beat signal is often complicated by the fact that dc is generated when the frequency of the weaker signal is equal to that of the stronger *or* to one of its harmonics. The detector then acts as a sampler, and a plurality of synchronous states is inherent in the system. Stability is limited by an inequality of type

$$K < \omega_s/\pi, \tag{38}$$

where ω_s now denotes the frequency of the stronger signal. However, if both Ω and K are small compared with ω_s, the sampling aspect is significant only with regard to the noise bandwidth B in relation to the pre-detector bandwidth B_p, let us say. The predetector noise spectrum is folded into the low-frequency band by beating with the harmonics of the strong signal, and magnifies the contribution of B_p to the B as defined in Section III, C by a factor of order

$$B_p \bigg/ \frac{\omega_s}{2} \tag{39}$$

if this is large.

The subject of APC with pulse reference will be considered at another time.

B. Inherent Noise

Spurious modulation also occurs at power-line frequency and its harmonics, *e.g.*, by leakage to the diode cathodes where these are ac heated.

It was shown in Section III, E that inherent phase noise is reduced only to a limited extent by APC; this can convert incoherent amplitude noise into phase noise. Inherent oscillator noise depends partly on the response to injected ac; the relation between this and the output of an oscillator with APC is interesting, and will be considered in a further paper. However, synchronization by APC and by injection are very different processes. The advantage of APC is that the oscillator and the control circuit may be designed quite independently; the quality factor Q of the elements that determine the oscillation frequency when the APC loop is open does not affect the ease of automatic tuning.

Appendix—Solutions of (3)

Approximate solutions of the synchronization equation in the filterless case have been presented above [(5) and (9)]. Periodic solutions are best approximated by (29). Since $N(\omega) \equiv 1$, (30) simplifies to

$$\gamma = -\frac{\pi}{2}, \quad \frac{\omega}{K} = \frac{J_0(\mu)+J_2(\mu)}{\mu}, \quad \frac{\Omega}{K} = \frac{\omega}{K} + J_1(\mu). \tag{40}$$

The example $\mu = 0.8$, *i.e.*, $\Omega/K = 1.521$, has been computed and is presented in Fig. 15(a), in normalized form. The ordinates $-\sin\phi$ represent the instantaneous beat frequency or control signal outside synchronism.

The exact solution of (3) is obtained by separating the variables and integrating with the aid of the successive substitutions

$$x = \tan\frac{\phi}{2}, \qquad \zeta^2 = \left(x - \frac{K}{\Omega}\right)^2 \bigg/ \left(1 - \frac{K^2}{\Omega^2}\right),$$

or by using tables (*e.g.*, Dwight, 436.00). The solutions have the indefinite forms

$$\frac{\sqrt{K^2 - \Omega^2}}{2}\, t = \frac{1}{2} \ln \left(\frac{\sqrt{K^2 - \Omega^2} - K + \Omega \tan \phi/2}{\sqrt{K^2 - \Omega^2} + K - \Omega \tan \phi/2}\right) \quad \text{for } \Omega < K; \quad (41)$$

or

$$\frac{\sqrt{\Omega^2 - K^2}}{2}\, t = \arctan \left(\frac{\Omega \tan (\phi/2) - K}{\sqrt{\Omega^2 - K^2}}\right) \quad \text{for } K < \Omega. \quad (42)$$

The last equation is illustrated by Fig. 15(b) for $\Omega/K = 1.521$. Both the approximate and the exact solutions demonstrate asynchronous degeneration, the beat periods being 1.33 $2\pi/\Omega$ and 1.31 $2\pi/\Omega$, respectively.

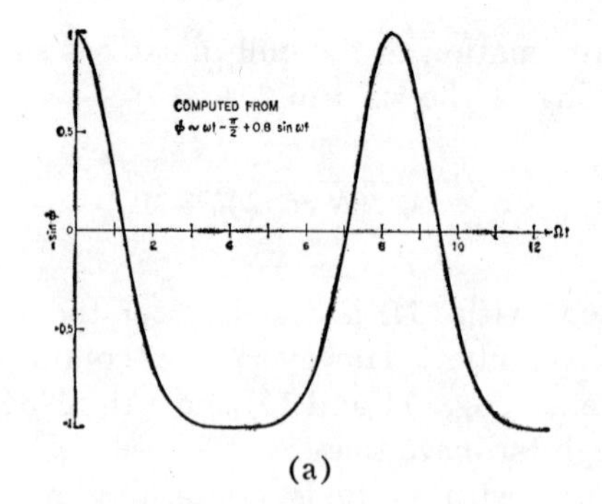

(a)

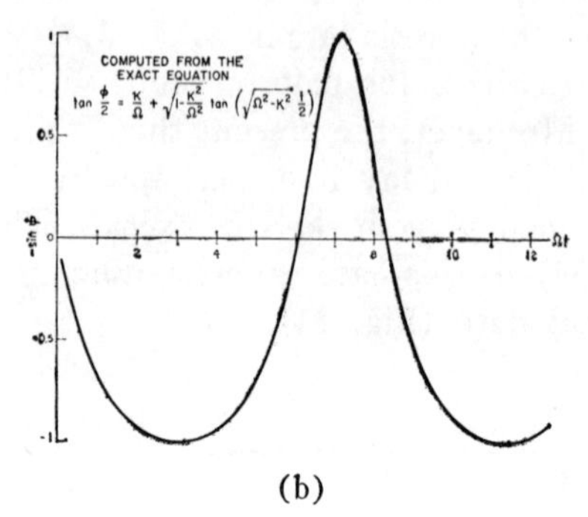

(b)

Fig. 15—Asynchronous control $p-d$ (normalized); (a) approximate, (b) exact.

CORRECTION

T. J. Rey, author of "Automatic Phase Control: Theory and Design," which appeared on pages 1760–1771 of the October, 1960 issue of PROCEEDINGS, has called the following to the attention of the *Editor*.

[*Note:* The page numbers given below are pages in this reprint volume.]

Page	*Column*	*Line*	*For*	*Read*
311	1	14 from top	ϕ	ϕ_∞
311	1	6 from bottom	$<2(\pi-\phi_\infty)$	$<(\pi \pm 2\phi_\infty)$
313	1	7 from top	unit bandwidth	bandwidth $\delta\nu$
314	2	18 from top	$-\ddot{\varphi}+\Omega_s$	$-\dot{\varphi}+\Omega_s$
314	2	26 from top	$\dot{\varphi}-\dot{\omega}_s$	$\ddot{\varphi}-\dot{\omega}_s$
314	2	11 from bottom	φ	$\epsilon-\varphi$
315	1	6 from top	1 and 3	2 and 3
315	1	13 from bottom	$a_T V$	$a_T \dot{V}$
315	2	6 from top	$2\omega_n\dot{\phi}$	$2\zeta\omega_n\dot{\phi}$
315	1	7 from bottom	sin	$\sin^{-1}$
316	2	10 from top	$\cos \psi_\nu$	, $\cos \psi_\nu$
316	2	last	[ ]	$[\ ]^{1/2}$
317	2	5 from bottom	ϕ_K	ψ_K
319	1	Fig. 14	Eq. 29	Eq. 34

The Response of an Automatic Phase Control System to FM Signals and Noise*

DONALD L. SCHILLING†, MEMBER, IEEE

***Summary*—An Automatic Phase Control (APC) System is analyzed in order to determine its response to frequency modulated signals, and narrow-band Gaussian noise. Emphasis is placed on the system's response to signals having "ramp" type FM.**

The response of the APC System to an FM signal is obtained using a perturbation and a piecewise linear technique. The response of the system to noise is obtained using an iteration technique. The complete response to an FM signal and noise is then discussed using an iteration technique and extending the results previously obtained.

INTRODUCTION

THE APC SYSTEM is the basic element in many communication and radar systems. The resulting second-order nonlinear differential equation also represents the motion of a pendulum when acted on by an applied force and a nonlinear friction force. The device can be used to synchronize two oscillators, to track an accelerating satellite or other target, or as an FM receiver.

In contradistinction to the analyses performed by previous investigators, the assumptions of linearizing the system, and of assuming a signal-to-noise (S/N) ratio greater than unity, will not be made.[1]

THEORETICAL RESULTS

The APC System analyzed in this paper is shown in Fig. 1(a). The signal voltage,

$$e_c(t) = S \sin \phi_1(t)$$
$$= S \sin\left(\omega_1 t + \alpha \int_0^t e_m(\lambda)d\lambda\right), \quad (1a)$$

and the output voltage of the voltage controlled oscillator (VCO)

$$e_v(t) = \cos \phi_2(t)$$
$$= \cos\left(\omega_2 t + \int_0^t G_2 e_0(\lambda)d\lambda\right) \quad (1b)$$

are multiplied in the phase detector. The resulting difference frequency signal,

* Received January 29, 1963; revised manuscript received July 9, 1963. This paper represents a partial summary of the author's Ph.D. dissertation, "The Response of an APC System to FM Signals and Noise," Polytechnic Institute of Brooklyn, N. Y. The work reported here was partially supported at the Polytechnic Institute of Brooklyn under Contract No. AF-18(600)-1505 and Grant No. AFOSR-62-295.

† Department of Electrical Engineering, Polytechnic Institute of Brooklyn, Brooklyn, N. Y.

[1] E. Kreindler, "The Theory of Phase Synchronization of Oscillators with Application to the DOPLOC Tracking Filter," Electronics Res. Labs., School of Engrg, Columbia University, New York N. Y., Tech. Rept. T-1/157; August, 1959.

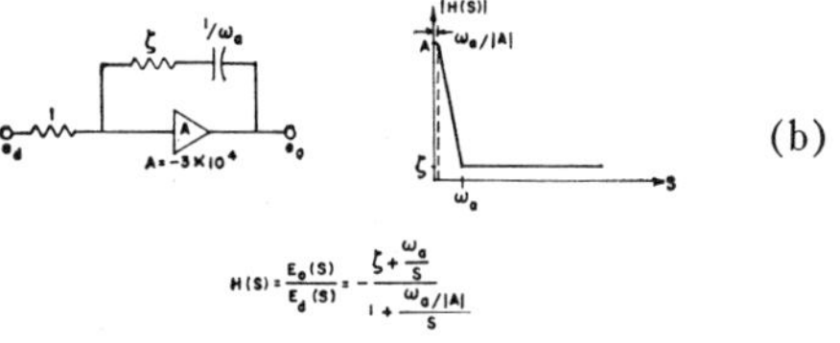

Fig. 1—An automatic phase control system. (a) An APC System. (b) An active phase lag filter.

$$e_d(t) = SG_1 \sin(\phi_2(t) - \phi_1(t))$$
$$= SG_1 \sin\left(\Omega t + \int_0^t (G_2 e_0(\lambda) - \alpha e_m(\lambda))d\lambda\right) \quad (2a)$$

represents the input to the filter. In this equation

$$\Omega = \omega_2 - \omega_1$$

G_1 represents the gain of the phase detector, and
G_2 represents the sensitivity of the voltage controlled oscillator (rad/sec/volt).

The voltage output of the filter

$$e_0(t) = \int_0^t h(t-\lambda)e_d(\lambda)d\lambda \quad (2b)$$

corrects the frequency of the VCO. Thus the frequency of the VCO attempts to follow the instantaneous frequency of the input signal.

The Response of an APC System to an FM Signal

The type of filter used greatly influences the performance of an APC System. It limits the frequency range of operation of the system and restricts the type of input signals with which it can synchronize.

To enable the system to synchronize to a signal whose frequency varies linearly with time, when the signal is embedded in noise, an active phase lag filter, as shown in Fig. 1(b), is employed. Using this filter in the system shown in Fig. 1(a) the second-order nonlinear differential equation describing the performance of the system becomes

$$\frac{d^2\phi}{dt^2} + \omega_n\zeta \cos\phi \frac{d\phi}{dt} + \omega_n\omega_a \sin\phi = -\alpha\frac{de_m(t)}{dt}$$
$$(t \ll A/\omega_a) \quad (3)$$

Reprinted from *Proc. IEEE*, vol. 51, pp. 1306–1315, Oct. 1963.

where $\phi = \phi_2 - \phi_1$, $\omega_n = G_1 G_2 S$ rad/sec, and $\alpha e_m(t)$ represents the frequency modulation of the input signal. It should be noted that the loop gain (or sensitivity) of the system ω_n can be considered equal to unity, and its actual magnitude considered as part of the magnitude of the parameters of the phase lag filter.

Eq. (3) can be rewritten in normalized form,

$$\frac{d^2\phi}{d\tau^2} + \epsilon \cos\phi \frac{d\phi}{d\tau} + \sin\phi = -a\frac{de_m(\tau)}{a\tau} \tag{4}$$

where

$$\tau = \sqrt{\omega_a \omega_n}\, t$$

$$\epsilon = \sqrt{\omega_n/\omega_a}\, \zeta$$

and

$$a = \alpha/\omega_n \omega_a.$$

If $\epsilon = a = 0$, this equation becomes the well-known pendulum equation. If a is not zero, the equation represents the motion of a pendulum with an applied force. The term, $\epsilon \cos\phi d\phi/d\tau$ represents a nonlinear friction force, which damps the motion of the pendulum as long as $|\phi| < \pi/2$. When $|\phi| > \pi/2$, the coefficient of friction $\epsilon \cos\phi$ becomes negative. If one assumes that the phase error ϕ is always much less than 1 radian, (4) can be linearized and a linear analysis performed.

It is of interest to approximately determine the time required for the system to complete a cycle of its motion, given a set of initial conditions. The time τ can be written as

$$\tau_{bc} = \int_b^c \frac{d\phi}{\phi'} \tag{5}$$

where

$$\phi' = \frac{d\phi}{d\tau}.$$

As previously noted, if $a = \epsilon = 0$, (4) reduces to the pendulum equation. A first integral can be obtained, and ϕ' found. If when $\phi(0) = 0$, $\phi'(0) = \Omega_n$, ϕ' becomes

$$\phi' = \sqrt{\Omega_n^2 - 4\sin^2\frac{\phi}{2}} \tag{6}$$

and

$$\tau_{ab} = \int_a^b \frac{d\left(\frac{\phi}{2}\right)}{\sqrt{\left(\frac{\Omega_n}{2}\right)^2 - \sin^2\frac{\phi}{2}}}. \tag{7a}$$

The time τ that it takes a pendulum to move $\frac{1}{4}$ of a cycle is[2]

$$\tau = K(k^2) \tag{7b}$$

where $K(k^2)$ is the complete elliptic integral of the first kind, and when $\Omega_n < 2$,

$$k^2 = \left(\frac{\Omega_n}{2}\right)^2 = \sin^2\left(\frac{\phi_{\max}}{2}\right). \tag{7c}$$

When $\Omega_n > 2$, the motion of the pendulum is no longer bounded. The pendulum now continually rotates in one direction. If some damping were present, the path of the pendulum would be perturbed, and the pendulum would rotate until, when $\phi = 2n\pi$, the velocity of the pendulum, Ω_n, became less than 2.

For example, if $0 \le |\Omega_n| \le 1.8$, $0 \le \phi_{\max} \le 128°$, and the time T (in seconds) required by the pendulum to complete a full period is bounded by,[2]

$$2\pi \le (\omega_a \omega_n)^{1/2} T \le 9. \tag{7d}$$

The concept of perturbing the motion of the pendulum by a friction force, and an applied force will be discussed in detail in the following section.

A Perturbation Solution:[3] The equation describing the APC System can be rewritten in the form

$$\phi'' + \sin\phi = -\epsilon\cos\phi\phi' - ae_m'. \tag{8}$$

If ϵ and a are each less than unity, a solution using the perturbation technique can be found. It has been shown, using a phase plane analysis[4] that a should be less than one-half for locking to occur. No such fundamental restriction on ϵ exists. Although choosing ϵ less than unity restricts the range of usefulness of the solution, the solution obtained using the perturbation technique is valid even when the phase error $\phi(\tau)$ and the frequency error $\phi'(\tau)$ are large. This should be contrasted to the results that can be obtained using a linear analysis, where although ϵ can take on any value, the solution is restricted to small values of $\phi(\tau)$ and small values of $\phi'(\tau)$.

A solution to (8), can be written in the form

$$\phi(\tau) = \Phi_0(\tau) + \epsilon\Phi_1(\tau) + a\Phi_2(\tau) + 0(\epsilon^2, a^2, \epsilon a). \tag{9}$$

Substituting (9) into (8), and neglecting second-order terms, one obtains

$$\Phi_0'' + \sin\Phi_0 = 0, \tag{10a}$$

$$\Phi_1' + \frac{\sin\Phi_0}{\Phi_0'}\Phi_1 = -\frac{1}{\Phi_0'}\int_0^\tau \Phi_0'^2 \cos\Phi_0 d\tau \tag{10b}$$

and

$$\Phi_2' + \frac{\sin\Phi_0}{\Phi_0'}\Phi_2 = -\frac{1}{\Phi_0'}\int_0^\tau e_m'(\tau)\Phi_0'(\tau)d\tau. \tag{10c}$$

[2] L. M. Milne-Thomson, "Jacobian Elliptic Function Tables," Dover Publications, Inc., New York, N. Y.; 1950.

[3] J. J. Stoker, "Nonlinear Vibrations," Interscience Publishers, Inc., New York, N. Y.; 1950.

[4] A. J. Viterbi, "Acquisition and Tracking Behavior of Phase-Locked Loops," Jet Propulsion Lab., California Institute of Technology, Pasadena, External Publication No. 673; July, 1959.

Eq. (10a) is the equation of the pendulum. Eqs. (10b), (10c), and the differential equations resulting from higher-order terms, are all first-order linear differential equations, with time-varying coefficients. The solution to this type of equation is well-known.

These equations can be solved in terms of elliptic functions and integrals.[5] The solution for the case of a frequency ramp modulated input signal is shown in Fig. 2. In this figure the parameters chosen were

$$a = 0.157, \qquad \epsilon = 0.25,$$

and

$$\left.\frac{d\phi}{d\tau}\right|_{\tau=0} = \Omega_n = 1.6, \qquad \text{when } \phi(0) = 0.$$

An interesting characteristic of the perturbation solution, as seen in the figure, is that not only does the period decrease with time τ but the negative half-cycle takes longer to complete than does the preceding positive half-cycle. The reason for this is that during the positive half-cycle $\phi_{\max} < \pi/2$. The coefficient of friction $\epsilon \cos \phi$, although small, is still positive. During the negative half-cycle, due to the offset caused by the frequency ramp, the friction term has less effect on the result. In this example $|\phi_{\max}| > \pi/2$, and the friction term is actually negative over part of the negative half-cycle. Thus it takes more time to complete this half-cycle.

The perturbation solution approaches the solution obtained using the simple linear analysis as the phase error decreases. Although the perturbation technique yields more accurate results than the linear analysis, it is more difficult to use. A simple technique, which takes into account the occurrence of "negative damping," has also been investigated. This is a piecewise-linear analysis, and is here described.

A Piecewise-Linear Analysis: A piecewise-linear solution was obtained using Fig. 3 as a model. Using this solution the effects of various types of frequency modulation were investigated. In addition, the maximum initial frequency error, to achieve locking, was determined. This result is most important when $a \neq 0$. When $a = 0$ (*i.e.*, the frequency of the input signal and the initial frequency of the VCO differ by Ω), the APC System will always eventually lock (synchronize) to the input signal.[4]

Using (4) and Fig. 3, the piecewise-linear equations become

$$\frac{d^2\phi}{d\nu^2} + \epsilon_1 \frac{d\phi}{d\nu} + \phi = -a_1 \frac{de_m}{d\nu}, \qquad -\frac{\pi}{2} \leq \phi \leq \frac{\pi}{2} \qquad \text{(11a)}$$

[5] D. L. Schilling, "The Response of an Automatic Phase Control System to Frequency Modulated Signals and Noise," Ph.D. dissertation, Polytechnic Institute of Brooklyn, N. Y., Research Rept. No. PIBMRI 1040-62; June, 1962.

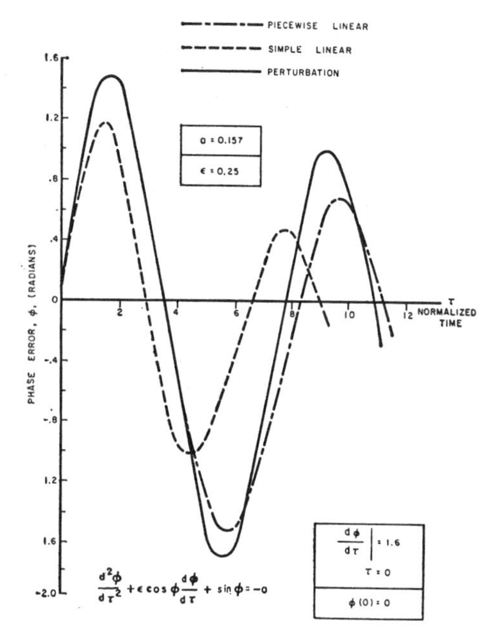

Fig. 2—A comparison of the approximate solutions used to obtain the response of the APC System to a frequency ramp modulated signal.

and

$$\frac{d^2\phi}{d\nu^2} - \epsilon_1 \frac{d\phi}{d\nu} - \phi = \pi - a_1 \frac{de_m}{d\nu}, \qquad -\frac{3\pi}{2} \leq \phi \leq -\frac{\pi}{2} \qquad \text{(11b)}$$

where

$$\nu = \sqrt{\frac{2}{\pi}}\,\tau = \sqrt{\frac{2}{\pi}\,\omega_a\omega_n}\,t$$

$$\epsilon_1 = \sqrt{\frac{2}{\pi}}\,\epsilon = \sqrt{\frac{2}{\pi}\left(\frac{\omega_n}{\omega_a}\right)}\,\zeta$$

and

$$a_1 = \sqrt{\frac{\pi}{2}}\,a = \sqrt{\frac{\pi}{2}}\,\frac{\alpha}{\omega_a\omega_n}.$$

Using the case of ramp modulation, a comparison between the results obtained using the perturbation, piecewise linear, and simple linear techniques was made. These results are shown in Fig. 2. In this example the parameters chosen were

$$a = 0.157, \qquad \epsilon = 0.25$$

and

$$\left.\frac{d\phi}{d\tau}\right|_{\tau=0} = \Omega_n = 1.6, \qquad \text{when } \phi(0) = 0.$$

The simple linear solution damps out most quickly. The frequency of the damped oscillations obtained using the piecewise linear analysis is smaller than that obtained using the simple linear analysis. It is clear from the

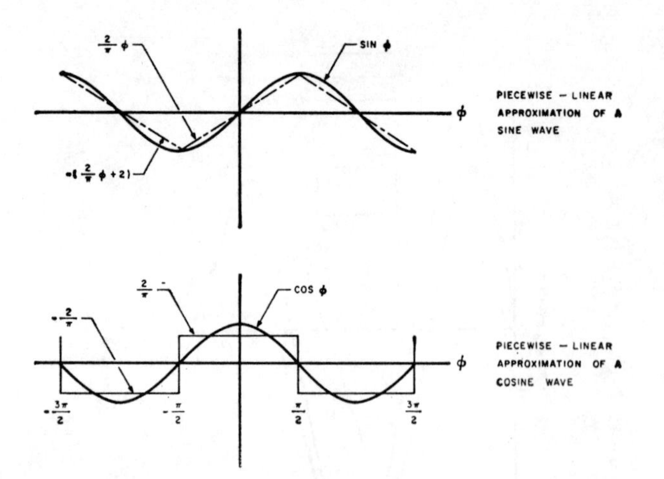

Fig. 3—A piecewise linear approximation.

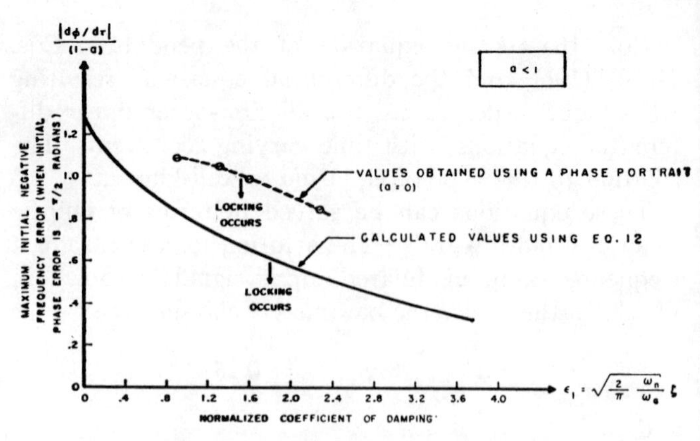

Fig. 4—The necessary and sufficient conditions for the APC System to lock to a frequency ramp modulated signal.

figure that the piecewise linear analysis approximates the solution obtained using the perturbation technique more closely than does the simple linear solution.

Using (11a) and (11b) the set of necessary and sufficient conditions for the APC System to synchronize (lock) to a frequency ramp modulated signal can be shown to be

$$a < 1$$

and

$$\left|\frac{d\phi}{d\tau}\right| < \begin{cases} \sqrt{\dfrac{\pi}{2}}\left(1 - \dfrac{\epsilon_1}{2}\right)(1-a), & \epsilon_1 < 2 \text{ (underdamped)} \\ \sqrt{\dfrac{\pi}{2}}(\sqrt{2}-1)(1-a), & \epsilon_1 = 2 \text{ (critically damped)} \\ \sqrt{\dfrac{\pi}{2}}\,\dfrac{(1-a)}{\epsilon_1}, & \epsilon_1 > 2 \text{ (overdamped)} \end{cases}$$

when

$$\phi(0) = -\frac{\pi}{2} \text{ radians.} \tag{12}$$

These results are plotted in Fig. 4. It is seen from this figure that when the initial phase error is $-\pi/2$, the maximum frequency error, for locking to occur, decreases as the normalized damping coefficient ϵ_1 increases. This is due to the fact that when the phase error ϕ becomes more negative than $-\pi/2$ radians, the damping force becomes negative. However, locking is still possible when the phase error exceeds $-\pi/2$ radians.

As shown in Fig. 4, similar results can be obtained using a phase plane analysis. The approximation of a constant coefficient of damping, in the piecewise linear analysis, results in the differences between the two curves. These results should be contrasted to the results obtained using a simple linear analysis. Using the simple linear analysis it can be shown that locking will always occur; a result which is obviously incorrect when $a \neq 0$.

The Response of an APC System to Colored Gaussion Noise

Since the APC System is a nonlinear closed loop device, its response to noise cannot simply be determined by using ordinary linear feedback theory, and a Spectral Density analysis. To determine the statistics of the noise at e_d and at e_0, an iteration technique was utilized.

The Iteration Technique: The iteration technique employed in this analysis can be explained using Fig. 5. This figure depicts the phase detector as a multiplier followed by an RC filter. The bandwidth ω_0 of this filter is much larger than the frequency error $\dot{\phi}$ obtained in the signal analysis. Thus, in the previous section we neglected the effect of this filter except to eliminate second-harmonic distortion. In this section we consider the system's response to noise. The noise is shaped by the RC filter since the noise has a relatively wide bandwidth (compared to $\dot{\phi}$).

In order to obtain a first approximation of the solution, the phase jitter present in the output of the VCO is neglected. Then,

$$e_{v_0}(t) = \cos \omega_2 t.$$

The first approximation of the statistics of the difference frequency voltage $e_{d_1}(t)$ can be determined once the statistics of the noise are specified. Having determined $e_{d_1}(t)$, the statistics of $e_{0_1}(t)$ can be calculated, since $e_{0_1}(t)$ and $e_{d_1}(t)$ are related by the transfer function of the active phase lag filter. Since the phase lag filter involves an integration, the statistics of $e_0(t)$ will be nonstationary. The first approximation of the output phase jitter $\theta_{2_1}(t)$ is then

$$\theta_{2_1}(t) = G_2 \int_0^t e_{0_1}(\lambda)d\lambda.$$

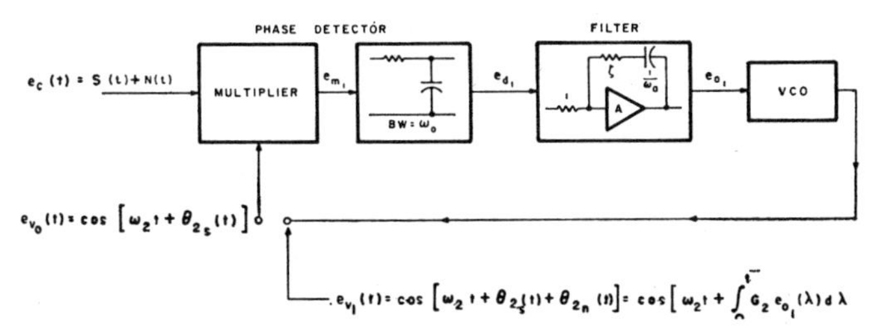

Fig. 5—The $(i+i)$ iteration

Using successive iterations, as illustrated in Fig. 5 the statistics of $e_{d_i}(t)$, $e_{0_i}(t)$ and $\theta_{2_i}(t)$ can be determined.

Characterization of the Input Noise: The results obtained below refer to an input consisting of narrow-band Gaussian noise. The noise is derived by passing white Gaussian noise through a single tuned IF filter, having a center frequency ω_2 and a bandwidth ψ. The noise emanating from this filter has an autocorrelation function given by

$$R_n(\tau) = \sigma_n{}^2 e^{-\psi|\tau|} \cos \omega_2 \tau$$

and a mean value of zero.

$$E(N(t)) = 0.$$

The noise $N(t)$ can be represented by

$$N(t) = x(t) \cos \omega_2 t - z(t) \sin \omega_2 t$$

where x and z have zero mean. The random variables $x(t)$ and $z(t)$ are uncorrelated, and

$$R_x(\tau) = R_z(\tau) = \sigma_n{}^2 e^{-\psi|\tau|}.$$

The spectral density of x and z are then

$$S_x(\omega) = S_z(\omega) = \frac{\sigma_n{}^2 \psi}{\omega^2 + \psi^2}.$$

The First Iteration: To obtain the first iteration, we assume that there is no phase jitter present in the VCO (*i.e.*, $\theta_{2_0} = 0$). The output voltage of the VCO (Fig. 5) is then

$$e_{v_0}(t) = \cos \omega_2 t. \tag{13a}$$

The output voltage of the multiplier $e_{m_1}(t)$ is

$$e_{m_1}(t) = G_1 x(t) + G_1(x(t) \cos 2\omega_2 t - z(t) \sin 2\omega_2 t). \tag{13b}$$

The voltage $e_{m_1}(t)$ is then passed through an RC low-pass filter. The bandwidth ω_0 of this filter is much less than the second-harmonic frequency $2\omega_2$. Therefore, in the expression obtained for the difference frequency voltage $e_{d_1}(t)$, the components of $e_{m_1}(t)$ due to the second harmonic $2\omega_2$ can be neglected. (In the experimental model a filter trap is inserted to further reduce the components due to the second harmonic.) Hence, to a first approximation, the spectral density of the difference frequency voltage is

$$S_{e_{d_1}}(\omega) = |H(\omega)|^2 S_{e_{m_1}}(\omega)$$
$$= \frac{1}{1 + \left(\frac{\omega}{\omega_0}\right)^2} \cdot \frac{G_1{}^2 \sigma_n{}^2 \psi}{\omega^2 + \psi^2}$$
$$\simeq \frac{G_1{}^2 \sigma_n{}^2 \omega_0{}^2 / \psi}{\omega^2 + \omega_0{}^2}, \tag{14a}$$

(the bandwidth of the RC filter ω_0 is much less than the bandwidth of the IF filter ψ_0).

The correlation function of e_{d_1} is

$$R_{e_{d_1}}(\tau) = \frac{\omega_0}{\psi} \sigma_n{}^2 G_1{}^2 e^{-\omega_0|\tau|}. \tag{14b}$$

The statistics of $e_{0_1}(t)$ can be found using the relation

$$e_{0_1}(t) = -\zeta e_{d_1}(t) - \omega_a \int_0^t e_{d_1}(\lambda) d\lambda. \tag{15a}$$

The correlation function of e_{0_1} is

$$R_{e_{0_1}}(t, \tau) \simeq \frac{\omega_0 \sigma_n{}^2 G_1{}^2}{\psi} \cdot \left[\zeta^2 e^{-\omega_0|\tau|} + 2\zeta\left(\frac{\omega_a}{\omega_0}\right) + 2\frac{\omega_a{}^2}{\omega_0} t\right] \tag{15b}$$

with

$$t \ll A/\omega_a.$$

Since $e_{d_1}(t)$ is Gaussian, $e_{0_1}(t)$ is also Gaussian with mean zero. The variance of $e_{0_1}(t)$ is, however, an increasing function of time.

The statistics of θ_{2_1} can similarly be obtained using (15a), and the relation

$$\theta_{2_1}(t) = G_2 \int_0^t e_{0_1}(\lambda) d\lambda. \tag{16a}$$

It can be shown that θ_{2_1} is also Gaussian with mean zero, and has a variance, $\sigma_{\theta_{2_1}}$, given by[5]

$$\sigma_{\theta_{2_1}}{}^2(t) = E(\theta_{2_1}{}^2) \simeq \frac{2G_1{}^2 G_2{}^2 \sigma_n{}^2}{4} t \cdot [\zeta^2 + \zeta(\omega_a t) + \tfrac{1}{3}(\omega_a t)^2] \quad (t \ll A/\omega_a). \tag{16b}$$

The Second Iteration: To obtain a second approximation of the output of the multiplier $e_{m_2}(t)$, the input voltage $e_c(t)$ and output voltage $e_{v_1}(t)$ must be multiplied (second harmonic components are again neglected). Then

$$e_{m_2}(t) = G_1 x(t) \cos \theta_{2_1}(t) - G_1 z(t) \sin \theta_{2_1}(t). \tag{17a}$$

If one obtains the statistics of $e_{m_2}(t)$, one sees that for $t \gg 1/\omega_0$, the statistics of e_{m_2} and e_{m_1} are the same.[5] Thus the results obtained in this analysis are valid during the time interval,

$$\frac{1}{\omega_0} \ll t \ll \frac{A}{\omega_a}. \tag{17b}$$

(As a practical matter this means that the results are valid during a time interval of several milliseconds to about 1 hour.)

The results obtained for e_d [(14a) and (14b)] could be obtained if the system operated as an "open-loop" system rather than as a closed-loop system. The open-loop system would be obtained by disconnecting e_0 from the VCO in Fig. 1. The VCO then would transmit a constant frequency signal.

From the preceding analyses it is seen that the APC System analyzed operates as a closed-loop system with respect to an input signal, but as an open-loop system with respect to the noise.

The Response of an APC System to an FM Signal and Narrow-Band Gaussian Noise

In the following discussion, it is assumed that the input voltage,

$$e_c(t) = S \sin \phi_1(t) + N(t).$$

The input voltage $e_c(t)$ was generated by passing the incoming RF signal (and noise) through a narrow-band IF filter. The filter had a center frequency ω_2 rad/sec which is the initial frequency of the VCO, and a bandwidth ψ rad/sec. The input noise is then

$$N(t) = x(t) \cos \omega_2 t - y(t) \sin \omega_2 t$$

where

$$R_x(\tau) = R_z(\tau) = \sigma_n{}^2 e^{-\psi|\tau|}.$$

It is also assumed throughout this discussion that the input signal is unaffected by the IF filter bandwidth.

The Iteration Technique: To obtain a first approximation to the response of the APC System to a signal and noise an iteration technique is utilized. Fig. 6 shows the system representation employed. Initially, the output of the VCO is assumed to be noise free.

Thus

$$e_{v0} = \cos \phi_{20}(t) = \cos (\omega_2 t + \theta_{2_s}(t)) \tag{18a}$$

where $\theta_{2s}(t)$ represents the phase due to the presence of a signal (noise has been neglected in this first approximation). Using this equation a first approximation to the output phase caused by a signal and noise can be obtained; and

$$e_{v1}(t) = \cos (\omega_2 t + \theta_{2s}(t) + \theta_{2n}(t)) \tag{18b}$$

where $\theta_{2_n}(t)$ represents the phase jitter due to the presence of the noise and is the same result which is obtained when only noise is present at the input.

To a first approximation, the system responds "linearly" (since superposition "applies") to the input signal and noise. However, it must be pointed out that the output phase $\theta_{2_s}(t)$ and the phase jitter $\theta_{2_n}(t)$ are not obtained by linearizing the system, but are obtained

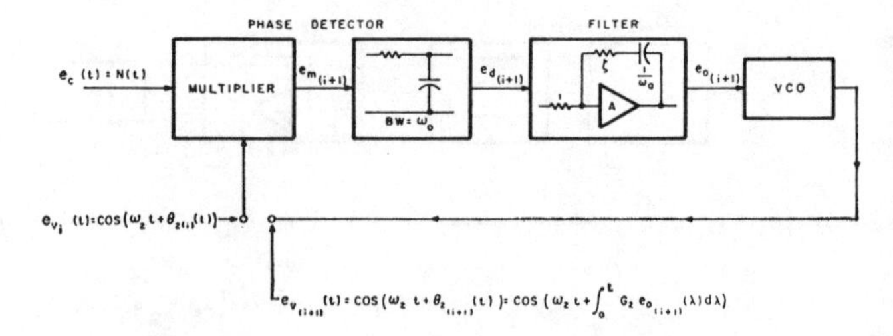

Fig. 6—The iteration technique.

using the nonlinear model. The equation representing $\theta_{2_s}(t)$ and the statistics of $\theta_{2_n}(t)$ are quite different for the linear model than the results obtained here. This pseudo-linearity which occurs is a result of the system responding to the noise as an open-loop system.

This result is quite useful in predicting the response of the APC System to a signal embedded in noise even though second-order corrections to the "iterative response" are not obtained. An output S/N ratio can be formed by comparing the phase jitter caused by the noise to the phase response due to the applied FM signal. Thus,

$$\left(\frac{S}{N}\right)_{\text{output}} = \frac{\theta_{2_s}}{\sigma_{\theta 2_n}} \cdot \tag{19}$$

Although the APC System is responsive to the peak noise and not the standard deviation of the phase jitter, the two are related since the noise is Gaussian. The standard deviation is used throughout this paper to simplify the results.

The standard deviation of the phase jitter is given by (16b), and when normalized using (4), becomes:[2]

$$\sigma_{\theta 2_n}(\tau) \simeq \left[\frac{\sqrt{\omega_a \omega_n}}{\psi}\left(\frac{\sigma_n{}^2}{S^2/2}\right)\tau\left(\epsilon^2 + \epsilon\tau + \frac{\tau^2}{3}\right)\right]^{1/2} \tag{20a}$$

where

$$\tau = \sqrt{\omega_a \omega_n}\, t, \qquad \epsilon = \sqrt{\frac{\omega_n}{\omega_a}}\,\zeta, \qquad G_1 G_2 S = \omega_n \tag{20b}$$

and

$$\frac{1}{\omega_0} \ll t \ll \frac{A}{\omega_a} \cdot \tag{20c}$$

The standard deviation of the phase jitter is proportional to the input N/S ratio, the normalized time τ, the normalizied damping factor ϵ, and the ratio of $\sqrt{\omega_a \omega_n}$ to the IF filter bandwidth (in this sense $\sqrt{\omega_a \omega_n}$ acts as the equivalent bandwidth of the APC system). Several obvious conclusions can immediately be drawn from (20a).

The first is that the greater the input S/N ratio, the smaller the standard deviation of the phase jitter and the higher the probability of synchronization.

The second is that the smaller the value of $\sqrt{\omega_a \omega_n}$, the smaller the standard deviation. However, this term cannot be reduced beyond a lower bound given by

$$\omega_a \omega_n > 2\alpha \tag{21}$$

where α is the slope of the frequency ramp. If $\omega_a \omega_n$ is too close to this lower bound any slight noise perturbation will tend to throw the system out of synchronization or prevent the system from synchronizing.

The standard deviation of the output phase jitter is also proportional to the normalized damping factor ϵ. For critical damping $\epsilon = 2$. During the acquisition period the system should be critically damped for optimum results (see Fig. 11). If the system were underdamped the large overshoots coupled with the phase jitter would tend to decrease the probability of locking. If the system were overdamped, the greater time that is required for acquisition raises the likelihood of noise preventing the system from locking since the phase jitter increases with time. Once the system is locked, the damping factor ϵ should be reduced to reduce $\sigma_{\theta_{2n}}$ and therefore increase the probability of the system remaining locked to the signal.

An Example: The effect of the input S/N ratio and the system parameters on the probability of acquisition can be qualitatively seen using the following example.

Consider a frequency ramp modulated input signal with an amplitude of 10 volts embedded in narrow-band Gaussian noise. The rate of change of the frequency modulation α is 400 rad/sec/sec. The APC loop is closed when the phase error is zero. The frequency error is 22 rad/sec. The system parameters are chosen so that the APC system is critically damped. The IF filter has a bandwidth ψ of 12,000π rad/sec. Then

$$S = 10 \text{ volts}$$

$$\alpha = 400 \text{ rad/sec/sec}$$

and

$$\epsilon = 2.$$

Let

$$\omega_a \omega_n = 425\pi \text{ (rad/sec)}^2,$$

then

$$a = 0.3.$$

A comparison of the output phase $\theta_{2_s}(\tau)$ with the output phase jitter, $\sigma_{\theta_{2n}}(\tau)$, is made in Fig. 7. Using a piecewise linear analysis one finds that the system would synchronize to the incoming signal when $\tau \simeq 0.7$, if no noise were present. Since noise is present one can conclude that the probability of locking is high if the phase θ_{2_s} is greater than θ_{2_n} when $\tau \simeq 0.7$. Thus, for the system parameters chosen, locking is not expected when the input N/S ratio is 20 db. However, locking is quite probable for N/S ratios less than 12 db. These results have been observed experimentally.

In the example chosen the initial phase error is zero. Under actual operating conditions any initial phase

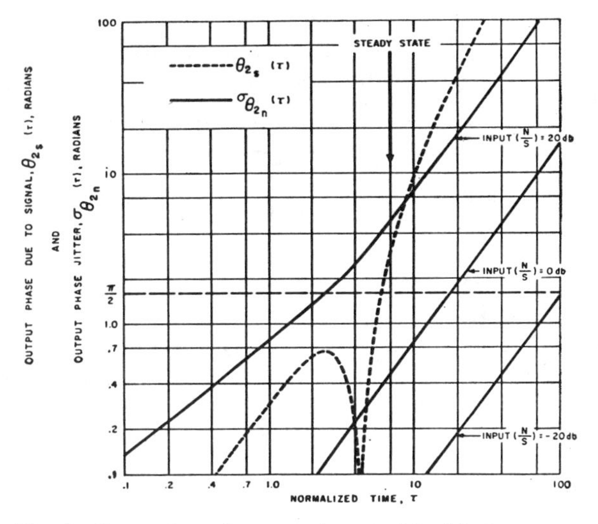

Fig. 7—Comparison of output phase $\theta_{2_s}(\tau)$ with the output phase jitter $\sigma_{\theta_{2_n}}(\tau)$.

error is equiprobable. The time required for locking, of course, depends on the initial phase and frequency error. Thus, the closer the initial conditions compare to the steady-state conditions [in this example: $\phi(0) = -\sin^{-1} 0.3$ (steady state), $\dot{\phi}(0) = 0$ (steady state)] the greater the probability of locking. These results were also verified experimentally.

Experimental Results

The experimental model used to verify and extend the theoretical results is shown in Fig. 8.

The Phase Detector consists of a multiplier, employing collector modulation, and an RC low-pass filter which is used to obtain the difference frequency signal $e_d(t)$. A single-tuned circuit is also used to eliminate the first harmonic distortion. The gain of the Phase Detector was found to be

$$G_1 = 0.025. \tag{22a}$$

The Active Phase Lag filter employed a Chopper stabilized amplifier having an open loop gain,

$$A = -30{,}000. \tag{22b}$$

The Voltage Controlled Oscillator (VCO) used was a simple astable multivibrator having a sensitivity

$$G_2 = 1700\pi \text{ rad/sec/volt}, \tag{22c}$$

at a center frequency of 35 kc. The loop gain (sensitivity) of the system, ω_n is then

$$\omega_n = G_1 G_2 S = 42.5\pi S \text{ rad/sec}.$$

The equation describing the operation of the system (3) then becomes

$$\frac{d^2\phi}{dt^2} + 42.5\pi S \cos\phi \frac{d\phi}{dt} + 42.5\pi S \omega_a \sin\phi = -\alpha \frac{de_m(t)}{dt}. \tag{22d}$$

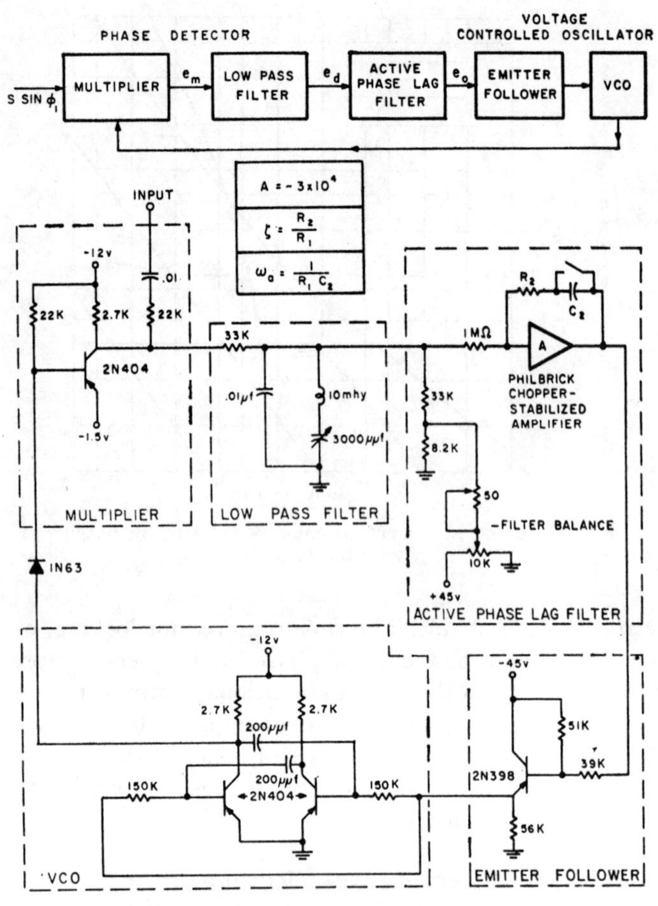

Fig. 8—Experimental model of APC System.

A Comparison of Experimental Results with Theoretical Results when the Input Signal is Modulated by a Frequency Ramp

To determine the transient response of the system to an input signal, the difference frequency voltage $e_d(t)$ was monitored using a recorder. It should be noted from (2a) that $e_d(t)$ is proportional to $\sin\phi$, not to the phase error ϕ.

To compare the experimental results with the theoretical results obtained using the perturbation, piecewise linear, and simple linear techniques shown in Fig. 3, the system coefficients were chosen to be,

$$\epsilon = 0.25$$

and

$$a = 0.157 \text{ radians.} \tag{23a}$$

Choosing a signal strength $S=1$ volt and $\omega_a=1$ rad/sec, α and ζ were calculated using (4).

$$\alpha \simeq 33 \text{ cps/sec}$$

$$\zeta = 6.94 \times 10^{-3}. \tag{23b}$$

The difference frequency signal e_d was recorded for two cases: that of a frequency step ($\alpha=0$), and a frequency ramp with $\alpha=33$ cps/sec. The result when the signal is unmodulated ($\alpha=0$) is shown in Fig. 9. The response shown in the figure is underdamped as expected. When the phase error exceeded $\pi/2$ radians, the time to complete a period of the motion was greater than when the phase error was much less than $\pi/2$ radians. The time per cycle was approximately 0.2 sec when the phase error exceeded $\pi/2$ radians and was approximately 0.11 sec when the phase error was less than $\pi/2$ radians. When the phase error became small, the system seemed to stop locking, as a continual phase jitter was present. This phase jitter was inherent in the input signal source.

When α is nonzero, the variation of the phase error ϕ with time is similar to that shown in Fig. 9, except for the presence of the steady-state phase error

$$\sin\phi_{\text{s.s.}} = -\frac{\alpha}{\omega_a\omega_n}.$$

It is when one observes the locking phenomenon as shown in Fig. 9, that it becomes apparent that the simple linear solution is extremely restrictive. Locking begins when the phase error is much greater than $\pi/2$ radians. This is seen by the dip in the peak of the curve. The dip indicates that $e_d(=\sin\phi)$ becomes less than unity, and therefore the phase error ϕ becomes greater than $\pi/2$. During the time that it takes the maximum phase error to decrease from π radians to 1 radian, the simple linear solution is of little value.

Using the results obtained when the input signal was ramp modulated ($\alpha=33$ cps/sec), a comparison was made between the theoretical and experimental results. The time per cycle and the attenuation (damping) per period for each type of solution is given in Table I. The average time per cycle as shown in Table I is broken into two parts. This was done since the period of the damped oscillations of the system varied as the system locked. This occurred in the experimental and perturbation results, but did not occur in the results obtained using the piecewise linear and the simple linear solution. The "average damping factor" was obtained by considering the phase error ϕ to be approximated by a damped sinusoid of the form

$$e^{-m\tau}\cos\beta\tau. \tag{24}$$

The "average damping factor" m can then be determined.

The results given in Table I clearly indicate that the perturbation technique approximates the experimental results more closely than the piecewise linear or the solution obtained using the simple linear approximation. The experimental and perturbation results would have checked even more precisely if the initial conditions of each would have been closer. The maximum phase error obtained with the perturbation technique was 1.6 radians, while the maximum phase error obtained using

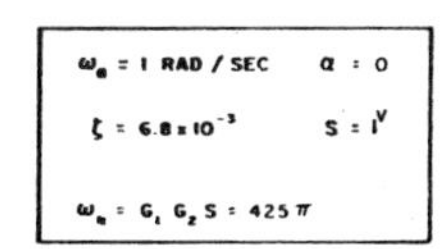

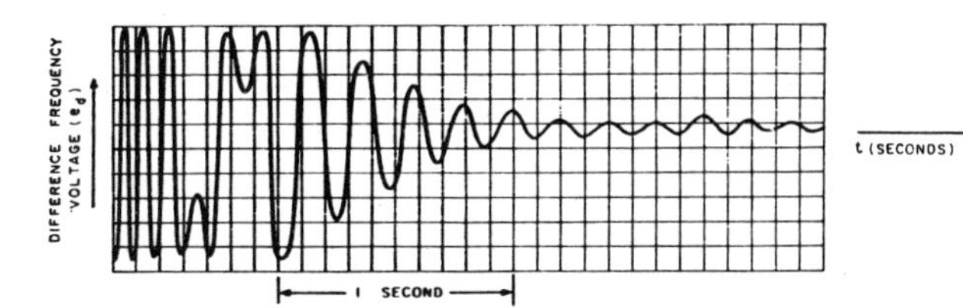

Fig. 9—The response of the APC System to an unmodulated input signal.

TABLE I

	Averaged Normalized Time (τ) per Cycle		m Averaged Attenuation per Normalized Period (Rad)$^{-1}$
	Phase Error <1 Rad	Phase Error <1 Rad	
Experimental	10.4	8	0.0625
Perturbation	7.75	7.4	0.0635
Piecewise Linear	7.85	7.85	0.1
Simple Linear	6.3	6.3	0.125

$\phi(\tau) \sim e^{-m|\tau|} \cos \beta\tau$

the experimental results was 2 radians. This resulted in the discrepancies that occurred in the "time per cycle" readings. The damping per period occurring experimentally is seen to be the same as the damping per period obtained using the perturbation technique. This indicates that while the time per cycle is a function of the maximum phase error, the damping per period is not.

The Experimental Response of an APC System to Colored Gaussian Noise

Before determining the response of the APC System to a signal embedded in noise, the behavior of the system to colored Gaussian noise (when no signal was present) was investigated. The theoretical aspect of this investigation was presented above. That analysis and the analysis of the signal plus noise[6] rely on the fact that the response of the system to noise can be analyzed using the iteration technique. One of the more important results obtained using this technique was that the system acted as an open loop system to the noise. The following discussion describes the experiments performed and shows that the theoretical model is correct.

The colored Gaussian noise used in this experiment had a center frequency of 35 kc and a bandwidth of 6 kc. Using a true rms meter, the relation between the rms input noise $e_{d_{rms}}$, and $e_{0_{rms}}$, were determined using different parameters of the active phase lag filter. The results of this measurement are shown in Table II.

TABLE II

ζ	σ_n (volts)	Measured (volts)		calculated (volts)	
		$e_{d_{rms}}$	$e_{0_{rms}}$	$e_{d_{rms}}$	$e_{0_{rms}}$
6.8×10^{-3}	3.5	0.38	2.8×10^{-3}	0.36	2.4×10^{-3}
47×10^{-3}	3.5	0.38	17.5×10^{-3}	0.36	17.9×10^{-3}
100×10^{-3}	3.5	0.38	40×10^{-3}	0.36	38×10^{-3}
820×10^{-3}	3.5	0.38	330×10^{-3}	0.36	312×10^{-3}

$$e_{d_{rms}}(\text{calc}) = G_1 \sqrt{\frac{1000}{6000}}\,\sigma_n$$

$$e_{0_{rms}}(\text{calc}) = \zeta e_{d_{rms}}$$

$$\frac{1}{11} \leq \omega_a \leq 1$$

It should be observed that the readings are independent of ω_a. The measured results shown in the table are compared to the results calculated assuming the system operates as an open loop system (e_0 disconnected from the VCO). The rms values of e_d and e_0 are therefore given by

$$e_{d_{rms}} = G_1 \sqrt{\frac{\omega_0}{\psi}}\,\sigma_n \tag{25a}$$

and

$$e_{0_{rms}} = G_1\zeta \sqrt{\frac{\omega_0}{\psi}}\,\sigma_n \tag{25b}$$

where $\omega_0 = 2\pi$ (1000)rad/sec, and $\alpha = 2\pi$ (6000)rad/sec. The measured values definitely verify the calculated values.

The measurements were made when the system operated normally in the closed loop position, and again, with the system operating as an open loop system with e_0 disconnected from the VCO. The same values were obtained in each case for $e_{d_{rms}}$ and $e_{0_{rms}}$. This clearly indicates that e_v and e_c (Fig. 5) are uncorrelated, and as a result of this the noise sees an open loop system.

To determine the probability distribution of e_d the voltage was recorded, and 24 amplitude intervals and 1700 samples were taken from the recording. The probability density was determined and plotted on probability paper as shown in Fig. 10. The resulting curve approximates a straight line and therefore approximates the Gaussian distribution found theoretically.

A χ^2 "goodness of fit" test[7] was performed. The results obtained indicate that by chance alone, if the probability distribution is Gaussian, then with a probability of 0.99, the χ^2 test would indicate a worse fit than the one actually observed.

[6] W. Davenport and W. Root, "Random Signals and Noise," McGraw-Hill Publishing Co., Inc., New York, N. Y.; 1958.

[7] H. Cramer, "Mathematical Methods of Statistics," Princeton University Press, Princeton, N. J.; 1946.

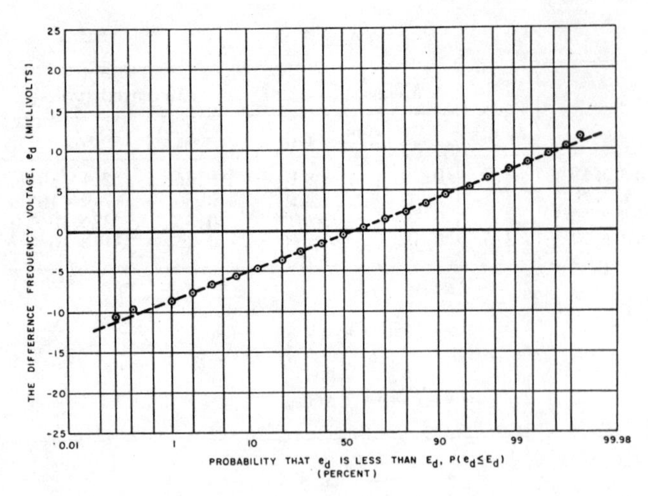

Fig. 10—Experimental verification of the Gaussian response of the difference frequency voltage e_d to Gaussian input noise.

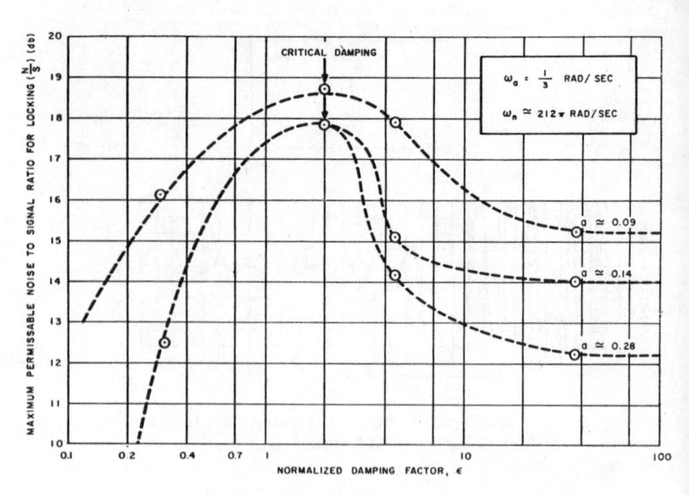

Fig. 11—Experimentally determined maximum permissible N/S ratio for locking to occur.

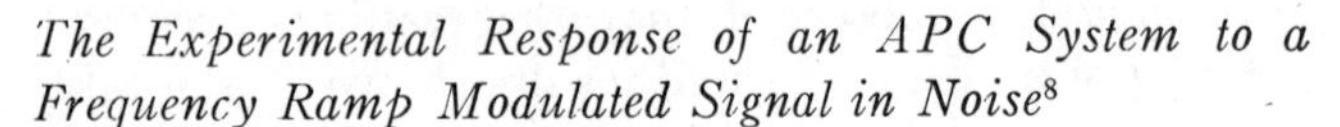

The Experimental Response of an APC System to a Frequency Ramp Modulated Signal in Noise[8]

A frequency ramp modulated signal and narrow-band Gaussian noise were simultaneously applied to the APC System and the effect of varying the normalized damping factor ϵ and the normalized slope of the frequency ramp a on the performance of the APC system was studied. The results are shown in Fig. 11. The maximum N/S ratio was determined by obtaining a specified N/S ratio at the input and noting whether or not the system locked to the signal. The rms value of the input noise was increased in 1-volt steps until the system started to lock only part of the time. The last value of 100 per cent probability of locking yielded the "maximum allowable N/S ratio." Only 2-volts rms above this value and the probability of locking became nil. Fifteen trials were taken at each N/S ratio considered.

It is seen from Fig. 11 that the more rapid the change of frequency (a) the smaller the allowable maximum N/S ratio to insure locking. When a critically damped system was used the variation of input N/S ratio with a was only about 1 db. When the system became severely underdamped ($\epsilon = 0.2$) the variation of input N/S ratio increased to 6 db. When the system was greatly overdamped the maximum value of N/S ratio seemed to approach a constant for a particular a. The underdamped case displayed poorer results since the phase error due to the signal oscillated widely about its steady state value. As this steady state phase error increased (as a increased) a greatly reduced N/S ratio was required, to insure locking. The overdamped system caused the phase jitter to increase since the standard deviation of the noise is a function of ϵ. However, since the system is extremely sluggish it did not follow the noise, and the maximum allowable N/S ratio depended mainly on the steady state phase error ($= -\sin^{-1} a$) rather than on the normalized damping factor.

Fig. 4 clearly shows that if $\omega_a\omega_n$ is held constant, best results occur when the rate of change of frequency α is made as small as possible. However, α is fixed by the conditions of the problem. If $\omega_a\omega_n$ is now increased, the normalized rate of change of frequency a decreases (which tends to improve the performance of the system), but the phase jitter increases (10a). Thus, given a value of α, the value of $\omega_a\omega_n$ used becomes important. A lower bound on this value is

$$\omega_a\omega_n > 2\alpha.$$

Experimentally, values of $\omega_a\omega_n$ between 4α and 9α yield best results.

In conclusion, the best system operation occurs when the system is critically damped ($\epsilon = 2$). Referring to Fig. 11 it is seen that in this mode of operation the maximum N/S ratio permitted, is fairly independent (1 db variation) of a for a wide range of values.

Conclusions

The two analytical techniques presented permit the calculation of the response of an Automatic Phase Control System to FM signals. The comparison made between the theoretical and the experimental results indicates that the perturbation and piecewise linear techniques yield excellent approximations of the second-order nonlinear differential equation studied. The perturbation technique yields extremely good analytical results. However, it can be used only to determine the response of an underdamped system. A piecewise linear solution is easier to obtain than the perturbation solution. However, the linearization of the system results in a loss of accuracy.

The response of the APC System to noise was obtained without "linearizing" the system equation, through use of an iteration technique. It was shown that with an RC low-pass filter and an active phase lag

[8] J. Frazier and J. Page, "Phase lock loop frequency acquisition study," IRE Trans. on Space Electronics and Telemetry, vol. SET-8, p. 210–227; September, 1962.

filter present in the loop, the system responded to noise as an open loop system.

The response of the APC System to an FM signal and noise was obtained using an iteration technique. Employing this technique it was shown that, to a first approximation, the output phase consists of superposition of the output phase obtained when only a signal is present, θ_{2_s}, and the output phase obtained when only noise is present, θ_{2_n}.

The results obtained indicate that the APC System operates best if crtically damped. Although optimum operation occurs during acquisition of the signal if the system is critically damped, the value of damping required should be reduced after acquisition has been accomplished. This is seen from (20a) where reduction of ϵ reduces the output phase jitter.

ACKNOWLEDGMENT

Grateful acknowledgment is made to Prof. M. Schwartz, Polytechnic Institute of Brooklyn, N. Y., for his guidance and criticism of the research and review of this paper.

Appreciation is also expressed to the staff of the Electronics Research Labs., School of Engineering and Applied Science, Columbia University, New York, N. Y., where much of this work was performed. In particular, the author wishes to thank Dr. R. I. Bernstein, Associate Director, for suggesting the topic, and Prof. L. H. O'Neill, Director, G. S. Bodeen, Assistant Director and L. B. Lambert, Laboratory Supervisor, for their encouragement and interest in the work. The author also wishes to thank Dr. L. Abramson, J. Dutt and H. Schachter for their helpful suggestions and discussions.

The Application of Linear Servo Theory to the Design of AGC Loops*

W. K. VICTOR† AND M. H. BROCKMAN†

Summary—**An analytical technique for designing automatic gain control (AGC) circuits is presented. This technique is directly applicable to high-gain high-performance radio receiving equipment. Use of this technique permits the designer to specify the performance of the AGC system completely with respect to step changes in signal level, ramp changes in signal level, frequency response, receiver gain error as a function of receiver noise, etc., before the receiver is constructed and tested. When used in conjunction with the statistical filter theory the technique has been used to synthesize optimal AGC systems when the characteristics of the signal and noise are appropriately defined.**

The mathematical derivation of the closed-loop equations is presented. The resulting expressions are simple and easy to understand by anyone acquainted with linear servo theory. Furthermore, the underlying assumptions used in theory have been tested experimentally, and the close agreement between theory and experiment attests the usefulness of the design technique.

Automatic gain control (AGC) is a closed-loop regulating system which automatically adjusts the gain of a receiver to maintain a constant signal amplitude at the receiver output. The AGC loop is normally capable of operating over a very wide range of signal input levels. When the signal is narrow-band and its amplitude is detected synchronously, the loop is capable of performing efficiently in the presence of wide-band noise. The purpose of this paper is to derive the basic equations of the AGC loop which minimize the mean square error in the estimate of receiver gain when the signal level, noise level, and transient performance are specified.

Fig. 1 is a block diagram showing the principal elements of the AGC loop with the waveform equations at various points in the loop. The desired output of the receiver is unity. The amplitude of the signal $a(t)$ is expressed as a fraction with respect to unity. The gain of the receiver is expressed as (attenuation)$^{-1}$, or $1/a^*(t)$. When $a(t)=1$, $a^*(t)=1$; the gain is unity, and the receiver output is also unity. When $a(t)=0.1$, for example, $a^*(t)=0.1$, the gain is $1/a^*(t)=10$, and the receiver output is unity. The attenuation of the receiver is introduced as a useful concept because it is the attenuation of the receiver that is required to follow the changes in signal level. The variation in attenuation of the receiver is some function of the control voltage b; thus, the receiver may be considered as a voltage-controlled attenuator. This idea is expressed in block diagram form in Fig. 2.

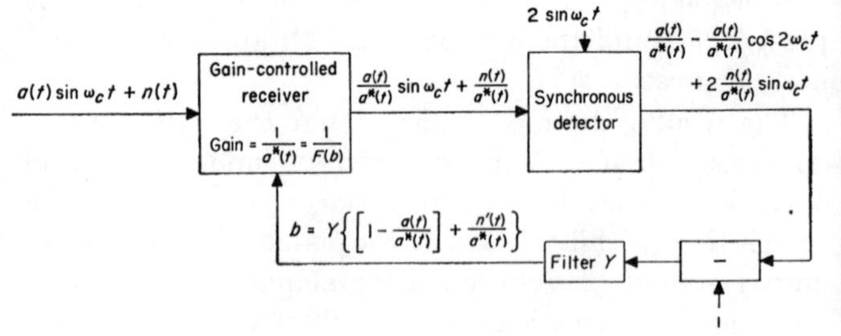

Receiver attenuation $a^*(t) =$

$$F(b) = \text{Function } Y\left\{\left[1-\frac{a(t)}{a^*(t)}\right]+\frac{n'(t)}{a^*(t)}\right\};$$

where

$a(t)$ = amplitude of RF carrier expressed as a fractional part of unity,
ω_c = radian frequency of the carrier,
$n(t)$ = interference of flat spectral density over a range of frequencies about ω_c,
$n'(t) = n(t)[2 \sin \omega_c t]$ = interference of same spectral density as $n(t)$,

$$YX(t) = \int_0^\infty y(\tau)X(t-\tau)d\tau,$$

$$y(\tau) = \text{weighting function of filter} = \frac{1}{2\pi j}\int_{-j\infty}^{+j\infty} Y(s)\epsilon^{s\tau}ds.$$

Fig. 1—Conventional AGC circuit with coherent detection.

Fig. 2 illustrates a recognition of the fact that the output of the AGC loop is the receiver attenuation $a^*(t)$ and that this output signal is required to match the input signal $a(t)$ with a minimum error. The synchronous detector is easily eliminated because it does nothing more than frequency-translate the signal and the noise $n(t)$ from the carrier frequency ω_c to zero frequency, or dc. In proceeding from Fig. 1 to Fig. 2 it should be noted that the two circuits are mathematically equivalent; the solution for the output attenuation $a^*(t)$ is

$$\text{Func } Y\left\{\left[1-\frac{a(t)}{a^*(t)}\right]+\frac{n'(t)}{a^*(t)}\right\}$$

in each case. The diagram is rearranged to provide a better understanding of what actually takes place when the loop is functioning.

The next step in the analysis is to choose a function for the variation of receiver attenuation with control voltage. If b is the control voltage (see Fig. 3), $F(b)$ is chosen to be $10K_A{}^{b/20}$; K_A is a constant associated with the attenuator (or amplifier) and has the dimension db/volt. Although $F(b)$ is highly nonlinear, it should be noted that log $F(b)$ is a linear function.

* Original manuscript received by the IRE, April 6, 1959; revised manuscript received, September 1, 1959. This paper presents the results of one phase of research carried out at the Jet Propulsion Laboratory, California Institute of Technology, under Contract No. DA-04-495-Ord 18, sponsored by the Department of the Army Ordnance Corps.

† Jet Propulsion Lab., California Inst. of Technology, Pasadena, Calif.

Reprinted from *Proc. IRE*, vol. 48, pp. 234–238, Feb. 1960.

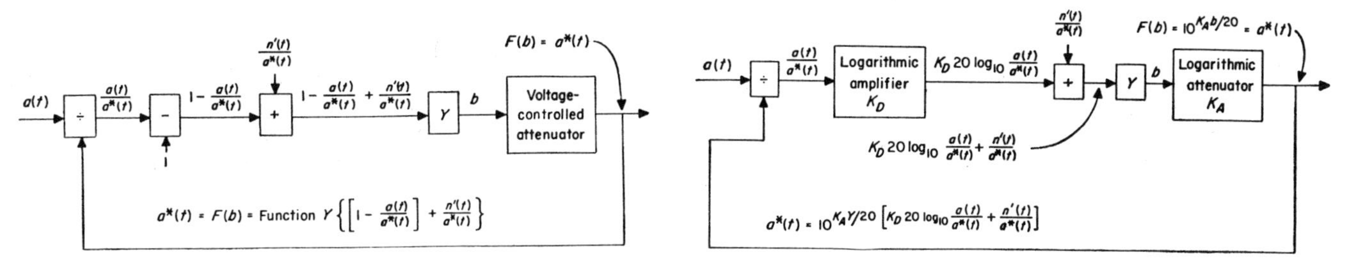

Fig. 2—Modified conventional AGC circuit.

Fig. 3—Nonlinear equivalent AGC circuit.

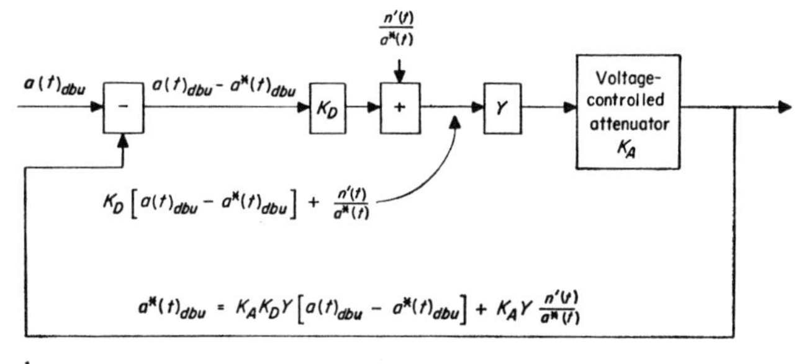

where

$$a(t)_{\text{dbu}} = 20 \log_{10} \frac{a(t)}{1} = \text{amplitude of signal expressed in db with respect to unity}$$

$$a^*(t)_{\text{dbu}} = 20 \log_{10} \frac{a^*(t)}{1} = \text{attenuation of receiver in db with respect to unity}$$

Fig. 4—Linear AGC system.

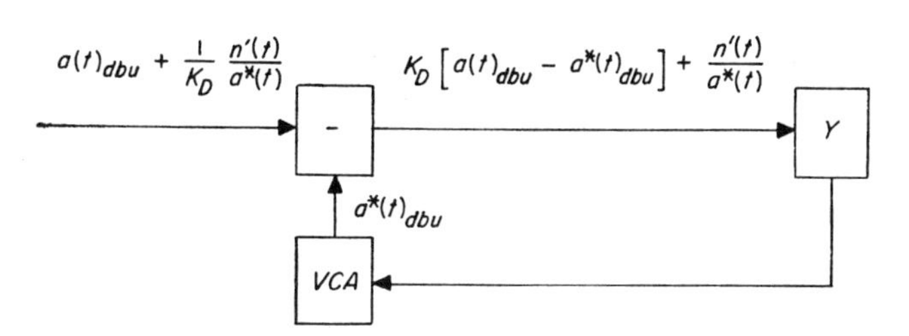

Fig. 5—Simplified linear AGC system.

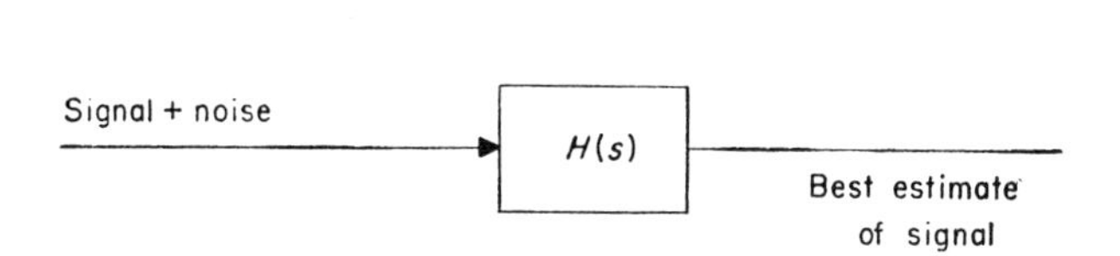

Fig. 6—Standard servo problem.

Having made the decision (see Fig. 3) that the attenuator characteristic should be linear in decibels, the function $1-a(t)/a^*(t)$ is studied and found to be approximately equal to $20 \log_{10} a(t)/a^*(t)$ over the range of 3 db, or 30 per cent variation in $a^*(t)$. Therefore, the differencing function in Fig. 2 can be replaced by the logarithmic amplifier in Fig. 3 without altering the nature of the loop, providing the loop error does not exceed 3 db. (This limitation is similar to the requirement in automatic phase-control systems that the phase error not exceed 30°.) Within this restriction, then, Fig. 3 is a true representation of the AGC loop, and K_D is the constant associated with the logarithmic amplifier in volts/db. The equation of the loop as indicated in Fig. 3 is

$$a^*(t) = 10^{K_A Y/20}\left[K_D 20 \log_{10} \frac{a(t)}{a^*(t)} + \frac{n'(t)}{a^*(t)}\right].$$

This equation can be solved for $a^*(t)$ by taking the logarithm of both sides and, for convenience, expressing the answer in decibels relative to unity. When this is done

$$a^*(t)_{\text{dbu}} = K_D K_A Y[a(t)_{\text{dbu}} - a^*(t)_{\text{dbu}}] + K_A Y \frac{n'(t)}{a^*(t)}\cdot$$

It is now apparent that when the signal level and the receiver attenuation are expressed as a logarithm, the AGC loop becomes a linear system. This system is shown in Fig. 4 and may be simplified still further to the system shown in Fig. 5. The problem has been reduced to the standard servo problem indicated in Fig. 6 and can be solved for the $H(s)$ which gives the minimum rms error in receiver gain.

This problem has been solved using the Weiner methods outlined in a previous paper.[1] The input signal was

[1] E. Rechtin, "The Design of Optimum Linear Systems," Jet Propulsion Lab., California Inst. Tech., Pasadena, Calif., External Publication No. 204; April, 1953.

assumed to be in the form of small step changes in amplitude. The transient error was defined as the infinite time integral of the squared error:

$$\text{transient error} = \int_0^\infty [a(t)_{\text{dbu}} - a^*(t)_{\text{dbu}}]^2 dt$$

and was assumed to be independent of the amplitude noise. The additive noise is assumed to be essentially flat over the spectrum, producing a gain-jitter, σ_N^2, which is

$$\text{gain error due to noise} = \frac{1}{2\pi j}\int_{-j\infty}^{+j\infty} |H(s)|^2 \Phi_N(s) ds.$$

The closed-loop transfer function which minimizes the gain-jitter while holding the transient error to a specified maximum value is of the form

$$H(s) = \frac{1}{1 + \frac{1}{K_B} s}$$

where K_B is a parameter in sec^{-1} which depends upon the amplitude step and the noise spectral density. The solution of the loop equation for filter Y yields $Y(s) = K_B/s$, a pure integrator. If the loop gain is high, $1/s$ may be approximated by $1/(1+\tau s)$, a low-pass filter. Solving for $H(s)$ yields

$$H(s) = \frac{GY(s)}{1 + GY(s)} = \frac{1}{\left(\frac{1}{G} + 1\right) + \frac{\tau}{G} s} \approx \frac{1}{1 + \frac{\tau}{G} s}$$

where G is the dimensionless product of K_D and K_A and is greater than 10.

To demonstrate the usefulness of the theory and the validity of the assumptions made in linearizing the loop, three experiments were performed on the AGC loop of a particular synchronous receiver.

1) The frequency response of the loop was measured using sine-wave variations in the input signal level.
2) The transient response of the loop was measured using exponential changes in the input signal level.
3) The rms error in receiver gain was measured as a function of the input-noise spectral density.

The AGC loop forming a part of this system is similar to the diagrams shown in Figs. 1 through 6. The filter Y is a low-pass filter having essentially a single time constant of 0.4 second. The loop gain has been measured at several different values of input signal level and varied from 66 for a -40-dbm signal level to 38 at -80 dbm. However, over a signal-level range of 3 to 6 db, the gain is essentially constant.

Frequency Response

Using the measured values of gain, the frequency response of the AGC loop was calculated for signal levels of -40, -60, and -80 dbm, and the curves have been plotted in Figs. 7, 8, and 9. The frequency response of the loop was then measured using a sine-wave modulating voltage which attenuated the carrier approximately 2.5 db. The measured points are plotted in Figs. 7, 8, and 9 for comparison with the calculated curves. The experimental data may be observed to agree generally within 1.5 db of the calculated curve.

Transient Response

The transient response of the AGC loop to an exponential change in input signal level of magnitude Δa under the restrictions outlined above can be determined by using transform relation

$$A^*(s) = H(s)A(s)$$

where $A(s)$ = Laplace transform of the input signal and $A^*(s)$ = Laplace transform of the resultant output signal or

$$A^*(s) = \frac{1}{\left(1 + \frac{1}{G}\right) + \frac{\tau}{G} s} \times \frac{\Delta a}{s(1 + \tau_{\text{in}} s)}$$

where τ_{in} = rise time of the input signal.

The solution of this equation expressed as a function of time is

$$a^*(t)_{\text{dbu}} = \frac{\Delta a_{\text{db}}}{1 + \frac{1}{G}} \left[1 - \frac{\frac{1}{\tau_{\text{in}}} e^{-(G/\tau)t} - \frac{G}{\tau} e^{-(t/\tau_{\text{in}})}}{\left(\frac{1}{\tau_{\text{in}}} - \frac{G}{\tau}\right)} \right]$$

where $a^*(t)_{\text{dbu}}$ represents the resultant change in receiver attenuation. The amplitude of the input signal Δa is expressed in decibels. Using the measured values of loop gain, the transient response of the AGC loop was calculated for a 3-db exponential change in signal level at input signal levels of -40 and -80 dbm. The calculated AGC output is plotted in Figs. 10 and 11 as resultant change in receiver attenuation.

The transient response of the AGC loop was then measured by introducing known changes in the input signal level and recording the resultant AGC output (see Figs. 10 and 11). The change in input signal level was accomplished using a current-controlled microwave ferrite attenuator which was varied by a step change in control current. The rise time of the input signal change was 4 to 5 times faster than the rise time of the resultant AGC voltage change. The measured AGC voltage change was expressed as db attenuation change using the measured value of K_A. The experimental results are plotted in Figs. 10 and 11 for comparison with the calculated curves. The experimental data agree with the calculated results to within 0.5 db.

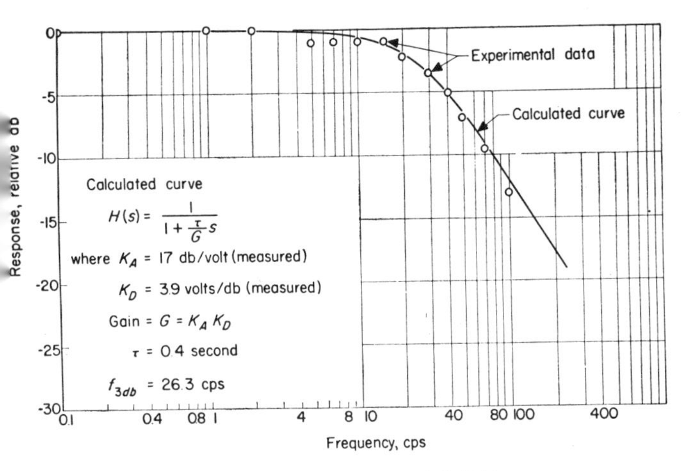

Fig. 7—Frequency response of AGC loop; input signal level = −40 dbm.

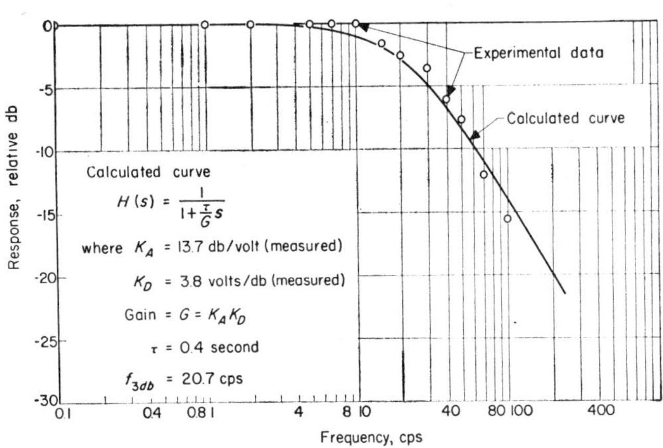

Fig. 8—Frequency response of AGC loop; input signal level = −60 dbm.

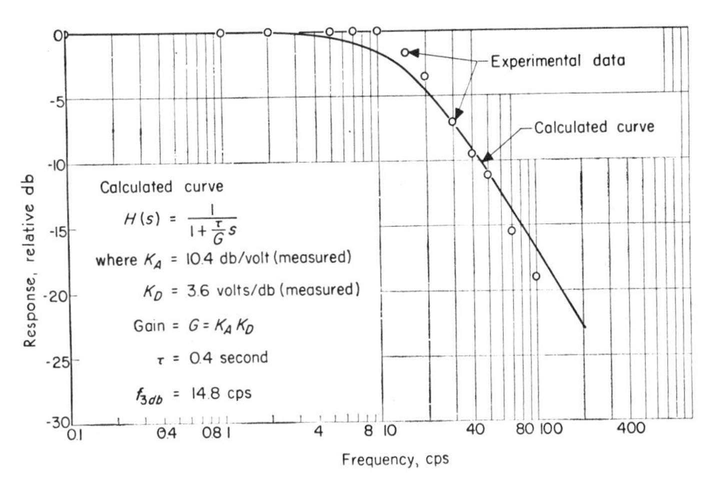

Fig. 9—Frequency response of AGC loop; input signal level = −80 dbm.

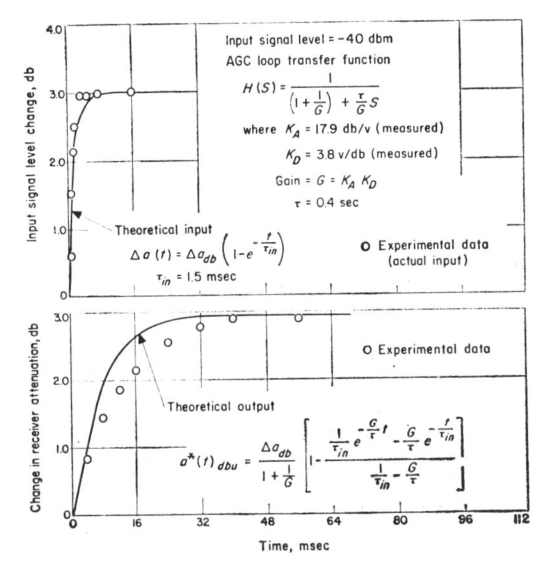

Fig. 10—Transient response of AGC loop.

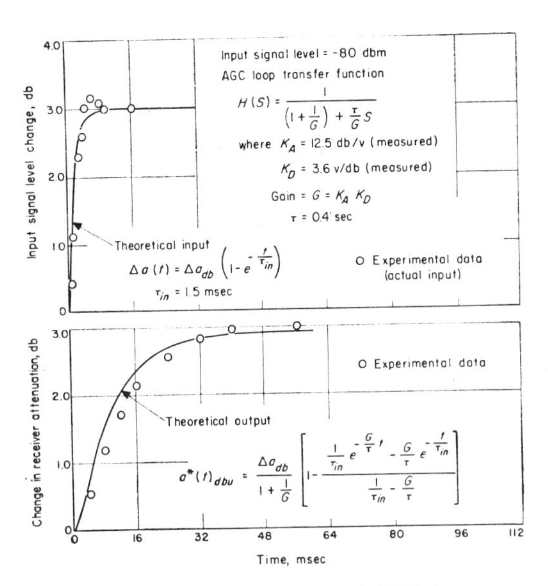

Fig. 11—Transient response of AGC loop.

RMS Error in Receiver Gain

The operation of the AGC loop was analyzed with random noise jamming, and the root-mean-square (rms) error in receiver gain was calculated. An experiment was then performed using a synchronous receiver to determine if the AGC system performed according to the theory. The analytical method is presented first.

The mean square error for the linear system with an error spectral density of $\Phi_\epsilon(\omega)$ is given by

$$\sigma^2 = \frac{1}{2\pi}\int_{-\infty}^{\infty} \Phi_\epsilon(\omega)d\omega. \tag{1}$$

If the system is considered to be distortionless with respect to the signal, the mean square error can be written as

$$\sigma^2 \text{ distortionless} = \frac{1}{2\pi}\int_{-\infty}^{\infty} \Phi_N(\omega)\,|H(j\omega)|^2 d\omega \tag{2}$$

where $\Phi_N(\omega)$ is the noise spectral density at the input in units determined by those of the signal, and $H(j\omega)$ is the system transfer function (dimensionless).

The jamming noise was assumed to have an rms amplitude of N volts and a flat spectral density of

$$\Phi_N(\omega) = \Phi_N(0) = \left[\frac{\Delta a(FS)}{\Delta a'(0)}\right]^2\left(\frac{N}{S}\right)^2 \frac{1}{2B_N}\ \frac{\text{db}^2}{\text{cps}} \tag{3}$$

where

$\Delta a(FS)$ = full-scale value of the gain error curve = 12 volts,

$\Delta a'(0)$ = slope of the error curve in volts/db at zero gain displacement for the signal level under investigation,

S = rms amplitude of the signal in volts, and

$2B_N$ = the effective bandwidth of the input noise.

$$\sigma^2 \text{ distortionless} = \Phi_N(0)\,\frac{1}{2\pi}\int_{-\infty}^{\infty}|H(j\omega)|^2 d\omega \ \text{db}^2$$

$$= \left[\frac{\Delta a(FS)}{\Delta a'(0)}\right]^2\left(\frac{N}{S}\right)^2 \frac{1}{2B_N}\,\frac{1}{2\pi}\int_{-\infty}^{\infty}|H(j\omega)|^2 d\omega \ \text{db}^2 \tag{4}$$

$$= \left[\frac{\Delta a(FS)}{\Delta a'(0)}\right]^2\left(\frac{N}{S}\right)^2 \frac{B_L}{B_N}\ \text{db}^2 \tag{5}$$

where

$$2B_L = \frac{1}{2\pi}\int_{-\infty}^{\infty}|H(i\omega)|^2 d\omega, \tag{6}$$

and the approximate AGC loop transfer function is given by

$$H(j\omega) = \frac{1}{1 + j\omega\dfrac{\tau}{G}} \tag{7}$$

where

τ = 0.4 second,

G = gain of the AGC loop, dimensionless $= K_D K_A$;

where

K_D = AGC detector constant expressed in volts/db, and

K_A = constant associated with the gain of the receiver expressed in db/volt.

The rms error in receiver gain is obtained by taking the square root of (5).

$$\sigma \text{ distortionless} = \frac{\Delta a(FS)}{\Delta a'(0)} \times \frac{N}{S} \times \sqrt{\frac{2B_L}{2B_N}}\ \text{db rms.} \tag{8}$$

Eq. (8) appears in graphical form in Fig. 12 for the receiver under test. Superimposed on the graph are the measured values for comparison purposes. Agreement between measured and calculated values is within 1 db.

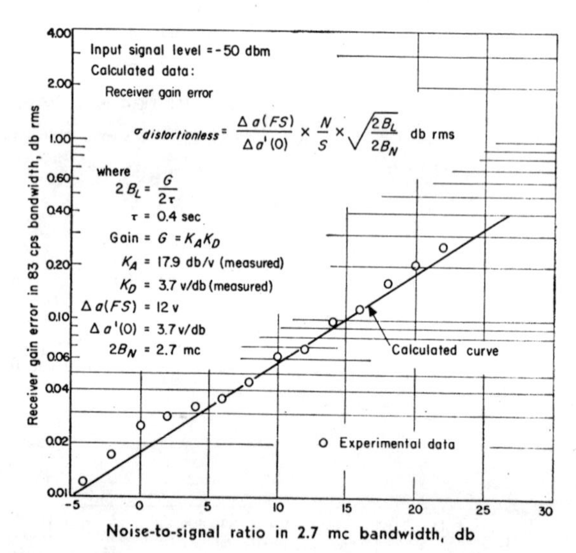

Fig. 12—Receiver gain error vs input noise-to-signal ratio.

Conclusion

AGC systems utilizing synchronous detection may be analyzed with considerable accuracy using the simple theoretical approach outlined here. The assumptions made in linearizing the AGC loop are valid for noise-free and noise-perturbed signals alike, and the analytical technique is a useful design tool.

The ability to achieve this goal is based on the recognition that an almost linear relationship exists between signal level and receiver attenuation when they are both expressed in decibels relative to unity. With the establishment of this fact, more advanced noise theory may be directed toward the synthesis of optimum AGC systems.

Part XI
Digital Phase-Locked Loop

On Optimum Digital Phase-Locked Loops

SOMESHWAR C. GUPTA, SENIOR MEMBER, IEEE

REFERENCE: Gupta, S. C.: ON OPTIMUM DIGITAL PHASE-LOCKED LOOPS, Southern Methodist University, Dallas, Tex. Rec'd 1/6/67; revised 5/15/67 and 10/1/67. Paper 68TP10-COM, approved by the IEEE Communication Theory Committee for publication without oral presentation. IEEE TRANS. ON COMMUNICATION TECHNOLOGY, 16-2, April 1968, pp. 340–344.

ABSTRACT: This paper gives the design procedure of optimum digital filters for analog-digital phase-locked loops. The inputs considered are the step and ramp change in phase. The digital filters can easily be realized on the digital computer or otherwise. Design curves are given to choose proper noise bandwidth, sampling period, and loop parameters. It is shown that the optimum digital-analog phase-locked loop is not the discrete version of the optimum continuous phase-locked loop.

Reprinted from *IEEE Trans. Commun. Technol.*, vol. COM-16, pp. 340–344, Apr. 1968.

Introduction

Phase-locked loop is playing an increasingly important role in modern communication and tracking systems. Ever since the optimum design, based on the Wiener filtering theory, has been presented by Jaffe and Rechtin,[1] there has been a continuous flow of papers on various aspects of the phase-locked loop. No effort will be made here to list such literature as extensive bibliographies exist.

There is now an increasing interest in the digital communications and the use of digital computers within the tracking and communication equipment. This has led to digital-analog and purely digital loops. The digital phase control techniques have been considered by Westlake.[2] Essentially he has taken the standard Rechtin filter and inserted a sampler within the loop. Then he has utilized the z-transform technique to analyze the loop. There is an increasing tendency to discretize the optimum Rechtin filter whenever there is a need to utilize the digital-analog loop.

This paper utilizes the standard optimization of the loop using the z transform and modified z-transform approach.[3]–[8] It is shown that the optimum digital loop is not the discrete version of the analog loop. It should be emphasized that the basic technique presented here should be useful in the design of all types of digital loops.

Statement of Problem

A digital-analog phase-locked loop is shown in Fig. 1.

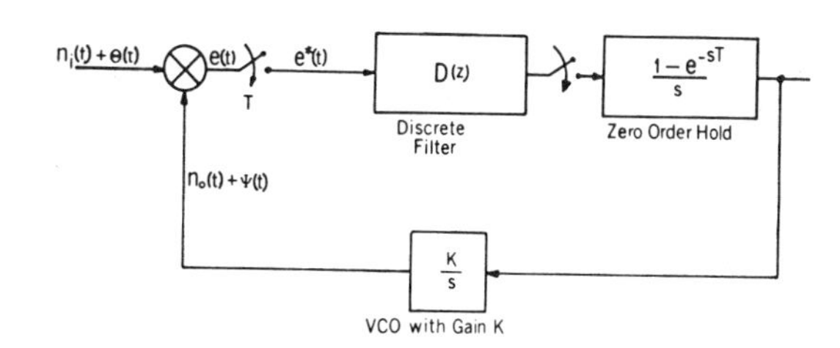

Fig. 1. Digital-analog phase-locked loop.

The phase-locked loop is the same as the continuous case except the optimum filter is replaced by a discrete filter followed by a hold circuit. The discrete filter function can be replaced by a digital computer.

The input to the loop is assumed to be white noise $n_i(t)$ and a phase signal $\theta(t)$. In our study we will have the input phase as a step and a ramp. The noise and the input phase are assumed independent.

The design of the optimum discrete filter is based on the minimization of the function.

$$F = \overline{n_0^2(t)} + \lambda\, \overline{e^2(nT)} \tag{1}$$

where $\overline{n_0^2(t)}$ is the integral squared value of the output noise of the voltage controlled oscillator (VCO) due to the input white noise only. $\overline{e^2(nT)}$ is the sum squared value of the sampled error between the actual phase input and output phase of the VCO in the absence of the noise. Note that it is necessary to consider the error at sampling instants only because that is all the remaining system sees. This is the major difference in the criterion of design between the continuous and the digital loop.

The factor λ in (1) is determined from noise bandwidth considerations as follows. As $D(z)$ is found while minimizing F, the overall transfer function of the loop can be obtained as a function of λ. For any desired noise bandwidth, λ can be found and hence the design of the optimum discrete filter $D(z)$ is complete.

Optimization and Design of the Discrete Filter

In order to obtain F in (1), we evaluate $\overline{n_0^2(t)}$ and $\overline{e^2(nT)}$.

Evaluation of $\overline{n_0^2(t)}$

Considering the input as noise only and letting

$$G(s) = K\frac{1-e^{-sT}}{s^2} \tag{2}$$

we can obtain, using the ordinary and modified z transforms,[3]–[8]

$$\begin{aligned}\Phi_{N_0N_0}(z,m) &= N_0(z,m)N_0(z^{-1},m)\\ &= W(z)W(z^{-1})G(z,m)G(z^{-1},m)\Phi_{N_iN_i}(z)\end{aligned} \tag{3}$$

where

$$W(z) = \frac{D(z)}{1+D(z)G(z)}. \tag{4}$$

Since the input noise is assumed white, we can write

$$\Phi_{N_iN_i}(z) = N(\text{constant}).$$

We can then obtain

$$\begin{aligned}\overline{n_0^2(t)} &= \frac{T}{2\pi j}\int_0^1 dm \int_\Gamma \Phi_{N_0N_0}(z,m)dz\\ &= \frac{1}{2\pi j}\int_\Gamma A(z)W(z)W(z^{-1})\frac{dz}{z}\end{aligned} \tag{5}$$

where

$$A(z) = NT\int_0^1 G(z,m)G(z^{-1},m)dm. \tag{6}$$

Evaluation of $\overline{e^2(nT)}$

In this case we only consider the input phase $\theta(t)$. We can obtain

$$\begin{aligned}\Phi_{EE}(z) &= E(z^{-1})E(z)\\ &= [1-W(z)G(z)][1-W(z^{-1})G(z^{-1})]\Phi_{\theta\theta}(z).\end{aligned} \tag{7}$$

Therefore

$$\overline{e^2(nT)} = \frac{1}{2\pi j}\int_\Gamma [1-W(z)G(z)][1-W(z^{-1})G(z^{-1})]\Phi_{\theta\theta}(z)\frac{dz}{z}. \tag{8}$$

Obtaining the Discrete Filter

Utilizing (5) and (8), we have

$$F = \frac{1}{2\pi j}\int_\Gamma [\lambda + P(z)W(z)W(z^{-1}) - \lambda W(z)G(z)\Phi_{\theta\theta}(z) - \lambda W(z^{-1})G(z^{-1})\Phi_{\theta\theta}(z)]\frac{dz}{z} \tag{9}$$

where

$$P(z) = A(z) + \lambda G(z)G(z^{-1})\Phi_{\theta\theta}(z). \tag{10}$$

Defining

$$P(z) = P^+(z)P^-(z) \tag{11}$$

where the plus part has all poles and zeros inside the unit circle and the minus part has all the poles and zeros outside the unit circle, we can apply the standard[4]–[8] minimization procedure to F to obtain the optimum $W(z)$ as

$$W_0(z) = \frac{z\left[\dfrac{\lambda G(z^{-1})\Phi_{\theta\theta}(z)}{zP^-(z)}\right]_+}{P^+(z)} \tag{12}$$

where $\left[\dfrac{\lambda G(z^{-1})\Phi_{\theta\theta}(z)}{zP^-(z)}\right]_+$ represents that part of the partial fraction expansion of $\dfrac{\lambda G(z^{-1})\Phi_{\theta\theta}(z)}{zP^-(z)}$ whose poles are inside the unit circle.

From (4), we then have the optimum discrete filter $D(z)$ as

$$D(z) = \frac{W_0(z)}{1 - W_0(z)G(z)}. \tag{13}$$

Determination of λ from Noise Bandwidth Considerations

The optimum filter obtained in the foregoing will be a function of λ. We therefore must use one more constraint in order to determine λ. This constraint is traditionally taken in terms of noise bandwidth, which is defined as

$$B_N = \frac{1}{2\pi j}\int_{-j\infty}^{j\infty} H(s)H(-s)ds \quad (\text{Hz}) \tag{14}$$

where $H(s)$ is the transfer function of the system. Since we have a sampler in the system, (14) can easily be written utilizing the modified z transform as[8]

$$B_N = \frac{T}{2\pi j}\int_0^1 dm \int_\Gamma H(z,m)H(z^{-1},m)\frac{dz}{z} \quad (\text{Hz}). \tag{15}$$

In our case

$$H(z,m) = W_0(z)G(z,m). \tag{16}$$

Utilizing the optimum discrete filter, we can obtain λ as a function of B_N and hence, λ can be obtained for any specified noise bandwidth. This completes the design of the discrete filter.

Design of Discrete Filter for Step Change in Phase

Here $\theta(t) = 1(t)$ and $G(s) = K(1 - e^{-sT})/s^2$. Therefore

$$\Phi_{\theta\theta}(z) = \theta(z^{-1})\theta(z) = \frac{z}{z-1}\frac{z^{-1}}{z^{-1}-1} \tag{17}$$

and

$$A(z) = NT\int_0^1 K^2T^2\left[m + \frac{1}{z-1}\right]\left[m - \frac{z}{z-1}\right]dm$$

$$= -\frac{K^2NT^3}{6}\cdot\frac{z^2+4z+1}{(z-1)^2}. \tag{18}$$

We can obtain from (10)

$$P(z) = \frac{-K^2NT^3}{6}\cdot\frac{z^4+2z^3-(6+q)z^2+2z+1}{(z-1)^4}, \text{ where } q = 6\lambda/NT$$

$$= \left\{\frac{-K^2NT^3}{6}\cdot\frac{az^2+bz+c}{(z-1)^2}\right\}\left\{\frac{az^{-2}+bz^{-1}+c}{(z^{-1}-1)^2}\right\} \tag{19}$$

$$= P^+(z)P^-(z)$$

where

$$P^+(z) = \frac{-K^2NT^3}{6}\cdot\frac{az^2+bz+c}{(z-1)^2}$$

$$P^-(z) = \frac{az^{-2}+bz^{-1}+c}{(z^{-1}-1)^2}$$

and

$$a^2+b^2+c^2 = -(6+q)$$

$$ab+bc = 2 \tag{20}$$

$$ca = 1.$$

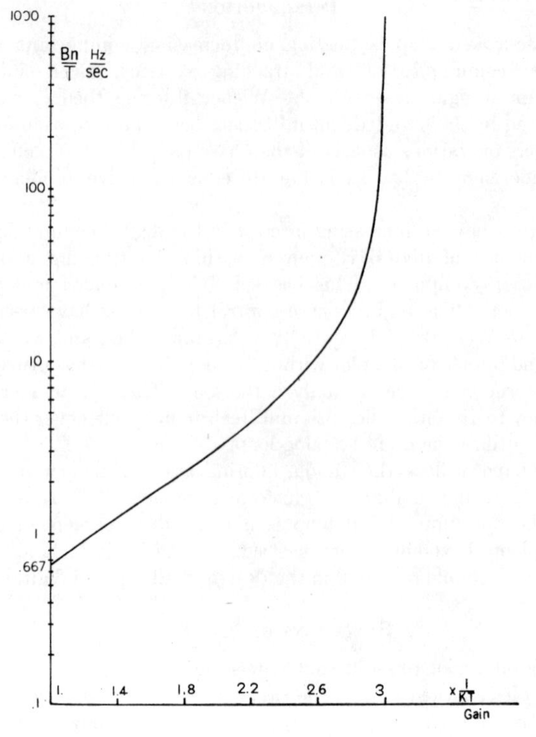

Fig. 2. B_N/T versus gain curve for loop with step change in phase

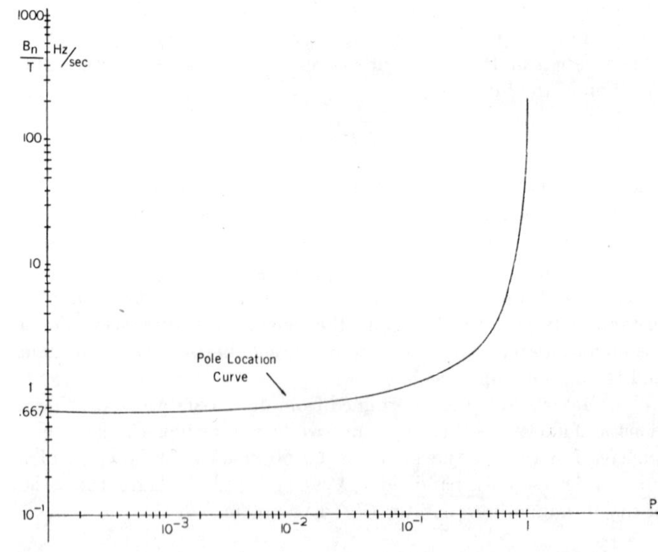

Fig. 3. B_N/T versus location of pole curve for loop with step change in phase.

Utilizing (12), we obtain the optimum $W_0(z)$ as

$$W_0(z) = \frac{z(z-1)}{KT}\cdot\frac{a+b+c}{az^2+bz+c}. \tag{21}$$

We can now obtain the noise bandwidth B_N utilizing (15) and tables of integrals[4],[8] as

$$B_N = \frac{T(2a+b+2c)(a+b+c)}{3(a-c)(a+c-b)} \tag{22}$$

where a, b, and c are previously related by

$$ab+bc = 2, \quad ca = 1. \tag{23}$$

Also

$$D(z) = \frac{W_0(z)}{1 - G(z)W_0(z)} = \frac{\delta z}{z - p} \tag{24}$$

where

$$\delta = \frac{a + b + c}{aKT}, \quad p = \frac{c}{a}.$$

We can now plot B_N/T versus the gain δ as well as the location of the pole p. We must note, however, that the roots of the equation $az^2 + bz + c = 0$ must lie within the unit circle. The conditions herefore are[3]-[5]

$$a + b + c > 0, \quad a - b + c > 0, \quad |c| < |a|.$$

Utilizing (22), (23), and the foregoing constraints of stability, B_N/T is plotted versus gain δ and pole location p in Figs. 2 and 3, respectively, with the aid of digital computer. For any choice of B_N/T which can be chosen on the curve, we can find the gain as a function of VCO gain K and the sampling period T. We can also find the corresponding pole of the compensator. Note that for the optimum loop we cannot have $B_N/T < 0.667$ Hz/s.

In comparison with the continuous loop, we note that the optimum loop is of one degree higher than the continuous loop and the results are to be obtained from the graph.

As an example let two-sided noise bandwidth $B_N = 10^{-2}$ Hz and $T = 10^{-3}$ second. Then $B_N/T = 10$ which gives the gain from Fig. 2 as

$$\delta = \frac{2.56 \times 10^3}{K}.$$

Also $p = 0.75$. Hence the compensator is

$$D(z) = \frac{2.56 \times 10^3 z}{K(z - 0.75)}. \tag{25}$$

This filter is easily synthesized.

Design of Discrete Filter for Ramp Change in Phase

Here we proceed as before except note that $\theta(t) = t \cdot 1(t)$. We obtain

$$P(z) = \frac{-K^2NT^3}{6} \cdot \frac{z^6 - 9z^4 + (16 + r)z^3 - 9z^2 + 1}{(z - 1)^6}, \quad \text{where } r = \frac{6\lambda T}{N}$$

$$= \left\{\frac{K^2NT^3}{6} \cdot \frac{az^3 + bz^2 + cz + d}{(z - 1)^3}\right\}\left\{\frac{az^{-3} + bz^{-2} + cz^{-1} + d}{(z^{-1} - 1)^3}\right\}$$

$$= P^+(z)P^-(z) \tag{26}$$

and

$$W_0(z) = \frac{z - 1}{KT} \cdot \frac{(3a + 2b + c)z + d - b - 2a}{az^3 + bz^2 + cz + d} \tag{27}$$

where

$$\begin{aligned} ab + bc + cd &= -9 \\ ac + bd &= 0 \\ da &= 1. \end{aligned} \tag{28}$$

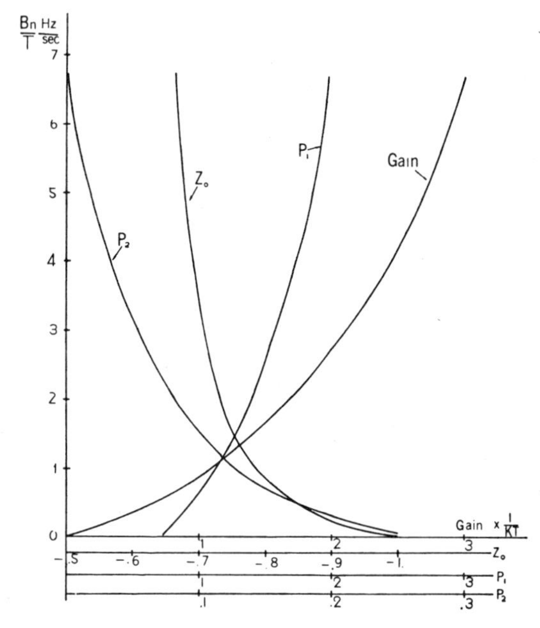

Fig. 4. B_N/T versus gain, pole locations, and zero location curves for loop with ramp change in phase.

The noise bandwidth can similarly be obtained as

$$\frac{B_N}{T} = \frac{\frac{2}{3}(\alpha^2 + \beta^2)Q_0 + \frac{1}{3}\alpha\beta Q_0 - \left\{\frac{1}{3}(\alpha^2 + \beta^2) + \frac{4}{3}\alpha\beta\right\}Q_1 + \frac{1}{3}\alpha\beta Q_2}{(a^2 - d^2)Q_0 - (ab - cd)Q_1 + (ac - bd)Q_2} \tag{29}$$

where

$$\begin{aligned} \alpha &= 3a + 2b + c \\ \beta &= d - b - 2a \\ Q_0 &= a^2 + ac - bd - d^2 \\ Q_1 &= ab - cd \\ Q_2 &= b^2 + bd - ac - c^2. \end{aligned}$$

Also

$$D(z) = \frac{\gamma}{KT} \frac{z(z + z_0)}{z^2 - p_1 z - p_2} \tag{30}$$

where

$$\text{gain} = \frac{\gamma}{KT} = \frac{3a + 2b + c}{aKT} = \frac{\alpha}{aKT}$$

$$\text{zero location } z_0 = \beta/\alpha$$

and pole locations are determined by

$$p_1 = \frac{2a + b + c}{a}, \quad p_2 = \frac{d}{a}.$$

Note that the roots of the equation $az^3 + bz^2 + ca + d = 0$ must be within the unit circle, which implies[3],[4]

$$\begin{aligned} a + b + c + d > 0, \quad a - b + c - d > 0, \\ |d^2 - a^2| > |db - ac|, \quad |d| < |a|. \end{aligned} \tag{31}$$

Fig. 4 shows the plot of B_N/T versus gain, zero location z_0, and p^1 and p_2 while satisfying (28) and (31). The maximum value of

$$\frac{B_N}{T} = 6.67 \quad (\text{Hz/s}).$$

For any choice of B_N/T less than the previous value, all the unknown parameters of the discrete filter are immediately determined.

As an example let two-sided bandwidth $B_N = 10^{-3}$ Hz and $T = 10^{-3}$ second. We have

$$\frac{B_N}{T} = 1 \quad (\text{Hz/s})$$

$$\text{gain } \frac{\gamma}{KT} = \frac{1.1 \times 10^3}{K},$$

where K is the gain of VCO. Also, $z_0 = -0.785$, $p_1 = 1.125$, and $p_2 = 0.1275$. Hence, the discrete filter is

$$D(z) = \frac{1.1 \times 10^3 z(z - 0.785)}{K(z^2 - 1.125z - 0.1275)}. \tag{32}$$

Note that there the loop becomes of the third order compared to the continuous case where it is of second order for a frequency step or ramp change in phase. The previously mentioned discrete filter can easily be synthesized.

Some Comments on General Method of Design for Analog-Digital Phase-Locked Loop

In order to design the overall digital-analog loop as developed in the last few sections, one can proceed as follows.

First an overall desired noise bandwidth is chosen. The choice of T must be made from sampling theorem considerations as well as the restrictions on the ratio B_N/T from Figs. 2, 3, and 4. For a choice satisfying the foregoing, the overall transfer function can be obtained and a damping in the loop can be derived. If the damping is satisfactory, the design is complete. Otherwise a different choice of T is made until all the previous constraints are satisfied. Note that the gain in the loop is inversely proportional to T and therefore consideration to the gain must also be given. Consideration might also have to be given to the synthesis of the discrete compensator.

Conclusion

Optimum discrete compensators have been obtained for a phase-locked loop with input as step and ramp change in phase using the z transform and modified z-transform theory.

The design curves enables one to design the filter for a choice of overall noise bandwidth and sampling frequency. Stability and suitable time response of loops can be maintained. It is found that the optimum digital loops are not the discrete version of continuous phase-lock loops. The order of the loops in the discrete case is one degree higher. It is now possible to use digital computer within the loop while maintaining optimum performance.

The techniques of analysis developed in the paper should have widespread application to the design of optimum digital loops of various kinds, as well as for analyzing loops where there is delay.

References

[1] R. Jaffe and E. Rechtin, "Design and performance of phase-lock circuits capable of near-optimum performance over a wide range of input signal and noise levels," *IRE Trans. Information Theory*, vol. IT-1, pp. 66–76, March 1955.

[2] P. R. Westlake, "Digital phase control techniques," *IRE Trans. Communications Systems*, vol. CS-8, pp. 237–246, December 1960.

[3] S. C. Gupta, *Transform and State Variable Methods in Linear Systems.* New York: Wiley, 1966.

[4] E. I. Jury, *Theory and Application of the z-Transform Method.* New York: Wiley, 1964.

[5] B. Kuo, *Analysis and Synthesis of Sampled Data Control Systems.* Englewood Cliffs, N. J.: Prentice-Hall, 1963.

[6] J. T. Tou, "Statistical design of linear discrete-data control systems via the modified z-transform method," *J. Franklin Inst.*, vol. 271, no. 4, pp. 249–262, 1961.

[7] S. S. L. Chang, *Synthesis of Optimum Control Systems.* New York: McGraw-Hill, 1961.

[8] S. C. Gupta and L. Hasdorff, *Fundamentals of Automatic Control.* New York: Wiley, to be published.

Performance of a First-Order Transition Sampling Digital Phase-Locked Loop Using Random-Walk Models

JACK K. HOLMES, MEMBER, IEEE

Abstract—**A new mechanization of a first-order all digital phase-locked loop (ADPLL) is discussed and analyzed. The purpose of the loop is to provide continuous tracking of the incoming waveform corrupted by the presence of white Gaussian noise (WGN). Based on a random-walk model, solutions are obtained for the steady-state timing-error variance and mean time to slip a cycle. As a result of the mean first-time-to-slip analysis, a difference equation and its solution are obtained that generalize a result of Feller [1]. Using a procedure that appears new, an upper bound on the timing-error bias due to a Doppler shift of the synchronized waveform is also derived. An example, for which the results presented here are applicable, is considered in some detail.**

I. Introduction

ANALOG phase-locked loops (PLL) have long played an important role in modern communication systems. Their theory of operation is well documented in numerous papers throughout the past 15 years. The increased interest in digital communication systems, due primarily to the decreased size and cost and increased reliability, led to the digital phase-locked loop (DPLL). One of the earliest reported DPLL was constructed by adding a sample and hold circuit between the filter and the voltage-controlled oscillator (VCO) [2] in an analog loop. With the advent of integrated circuits, more refined and more nearly all-digital PLL emerged [3], [4]. These loops utilize analog integration for the midphase in-phase (MI) type phase detector, with the remaining functions of the loop being accomplished digitally. A loop reported by Gota [5] was, except for the use of a digital to analog converter, all digital in operation; he did not consider noise in the analysis of his loop.

To date the most common function of the DPLL has been to provide synchronization of a signal. However, this is not the only application, for example, an all-digital phase-locked loop (ADPLL) has been built for FM demodulation [6]. It was designed for potential application in large multiple data set installations, which provide low-speed serial data communications, for time-shared computers. Additional applications of DPLL are summarized in a paper by Gupta [7].

Recently an ADPLL, employing a new simple type of phase detector has been reported [8] for square-wave signal waveforms. The loop was conceived to provide tracking of the subcarrier signal for a command system. Basically, tracking was accomplished by 1) sampling the input waveform at the points in time where the signal transitions or axis crossing occur, 2) accumulating m of these samples, and 3) incrementing the phase of the local reference (clock) in such a direction as to bring the value

Paper approved for publication by the Communication Theory Committee of the IEEE Communications Society after presentation at the 4th Hawaii International Conference on System Science, January 1971. This paper presents the results of one phase of research carried out at the Jet Propulsion Laboratory, California Institute of Technology, under Contract NAS 7-100, sponsored by the National Aeronautics and Space Administration. Manuscript received July 1, 1971.

The author is with the Jet Propulsion Laboratory, Pasadena, Calif. 91103.

Reprinted from *IEEE Trans. Commun.*, vol. COM-20, pp. 119–131, Apr. 1972.

of the accumulation toward zero. In [9] the results are generalized for a broad class of input signals using the same loop.

This paper is concerned with a more comprehensive and complete analysis with additional results not contained in [9].

Solutions for the stationary timing-error variance (mod 2π) and the mean time to the first cycle slip are developed in this paper. As a consequence of the cycle-slip analysis an equation is derived for the mean first slip time that generalizes an equation of Feller [1] to the case when the transition probabilities are state dependent. Using a procedure that appears to be new, an upper bound on the timing error bias (due to a Doppler shift of the waveform to be synchronized to) is derived. An example from which the theory was motivated, a subcarrier synchronization loop for a command system, is then considered in some detail.

This paper is divided into five sections. Section I is the introduction. Section II discusses the operation of the transition tracking loop and introduces the Markov model. In addition, results for the case when the input signal is a square-wave, is developed under the assumption that the presampling filter is sufficiently wide band to neglect waveform distortion. The general case when the transition probabilities are state dependent (i.e., the waveform is not a square wave) is considered in Section III. In Section IV, some results are obtained on Doppler effects, the effect of modulated waveforms (subcarriers) and a multiple sampling procedure. In Section V a specific example of the loops' performance, with modulation, is considered for a command-system subcarrier synchronization loop. In this example the distortion of the presampling filter is considered.

II. Undistorted Signal Case

We now consider the derivation of the results for the steady-state (mod 2π reduced) timing-error variance, as well as the mean first slip time for the first-order digital phase-locked loop. The input waveform is assumed to be a square-wave signal plus white Gaussian noise. The presampling low-pass filter (PSLPF), seen in Fig. 1, is assumed to be sufficiently wide band so that we may neglect the distortion due to its action on the square wave. This wide-band assumption makes the analysis much simpler and in addition allows solutions to be written in a form that allows greater insight into the operation of the loop. The case when the bandwidth is not assumed to be wide is considered in Section III and is the case of greater practical interest.

A. Operation of the ADPLL

The incoming waveform $y(t)$ (Fig. 1) is assumed to be composed of a square wave of amplitude A and period T_s that has passed through a white Gaussian noise channel. The spectral density of the channel noise is denoted by $N_0/2$ (two-sided).

The first element in the loop is modeled as an ideal low-pass presampling filter of W hertz that passes all the essential frequency components of the signal.

The sampler obtains a sufficient number of equally spaced samples per period to adequately represent the signal. From the sampling theorem this requires that the sampling rate be equal to $2W$. Hence there will be $2W/f_s$ samples/period where f_s is the square-wave signal frequency. The samples are then converted by the analog to digital converter to digital format. We assume in what follows that a sufficient number of bits are used for digitizing to introduce negligible error.

The phase detector is formed by the combination of the transition-sample selector and an m sample accumulator. Fig. 2 illustrates a square-wave signal with negligible distortion for which the timing system has an error of τ seconds. The timing error is the time between the negative-going transition and the corresponding sample. We see that the sampled value of the signal of the error shown in the figure with a (+) produces a positive control signal. If the sample was taken after the transition the signal part of the sample would be negative. Hence the transition sampler produces an output, on the average, that has the same sign as the timing error. After the next sample is taken [the one with (−)] the sign of the sample must be changed to provide the correct control signal to the loop. Because noise will cause errors in updating the loop, the performance depends on the sample signal-to-noise ratio. The function of the accumulator (in Fig. 1) is to increase the update signal-to-noise ratio (SNR) by summing m transition samples before updating the timing. This update occurs after M signal periods or $MT_s = T$ seconds. If no transition samples are deleted $m = 2M$, however, with modulation (see Section V) it may be necessary to delete some transition samples due to transition ambiguities so that in general $m \leq 2M$.

The sum of m transition samples are then hard quantized in the sign detector. The output of the sign detector controls the addition or deletion of a clock pulse, which in turn shifts the sampling location by some fixed fraction Δ of one signal period. Since the transition sampler selects the samples the loop assumes are the transition samples, the location of the transition samples are advanced or retarded by Δ according to the sign of the update signal.

The divider following the clock divides the clock rate down to the sample rate, i.e., divides by $R = f_c/2W$ where f_c is the clock frequency and $2W$ is the sampling rate.

If the loop were to utilize many samples centered around the transition region then the phase detector would be a discrete approximation to a matched filter detector. However, for the single sample loop described here the comparison is difficult to make since the amplitude of the signal portion of the single-sample phase detector is independent of timing error, whereas the signal portion of the MI phase detector is proportional to the timing error. Because of this proportionality, in

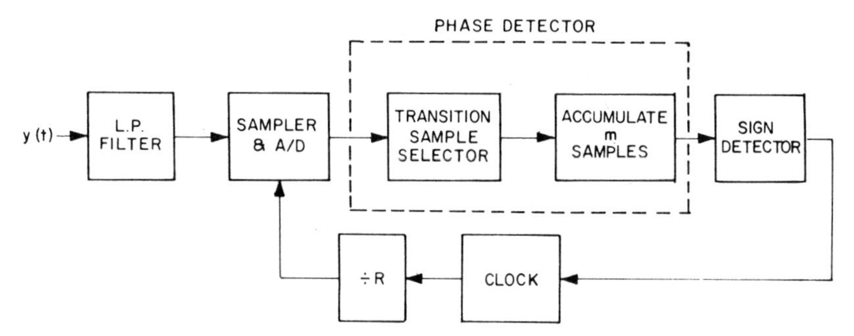

Fig. 1. Block diagram of timing update loop.

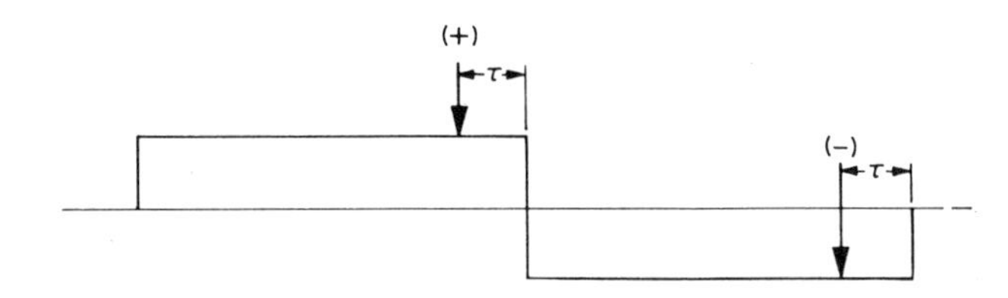

Fig. 2. Square-wave subcarrier with negative timing error shown with two samples/cycle.

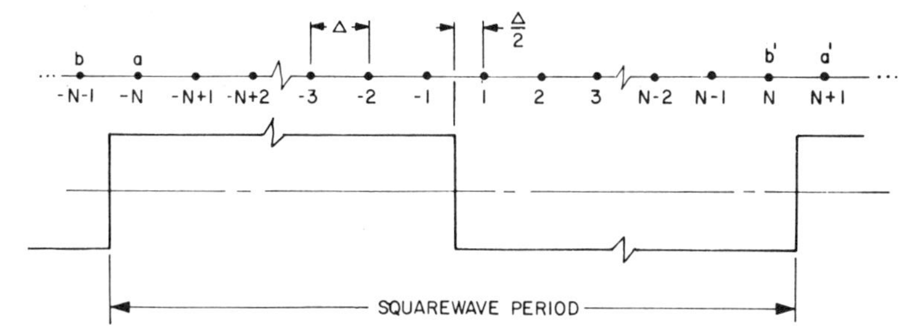

Fig. 3. Error state diagram above and its relationship to the square-wave signal below.

MI phase-detector loops the integration window is decreased to boost timing performance for small timing errors [3]. In the ADPLL discussed here, timing performance is improved by decreasing W to allow some distortion without exceeding the allowable distortion level of the signal waveform.

B. Random-Walk Model

White noise passed through the ideal low-pass filter of bandwidth W and then sampled at the rate of $2W$ samples/second, produces independent samples. Because the updating is determined by the sum of m of these statistically independent samples and since the updating is a fixed fraction of a period Δ, the error state forms a discrete-parameter Markov chain. For any sample taken in the positive region of the square wave the transition probabilities (the probabilities of adding or subtracting Δ from the loop timing) is independent of the sample location. We define the "error state" of the loop to be the state (integer) corresponding to the existing timing error of the loop just before being updated. As a consequence, the Markov chain has a countable number of error states. Fig. 3 illustrates the error-state diagram along with the square-wave signal. The error in timing is measured relative to the negative-going transition of the squarewave. All states are separated by Δ fractions of one period and there are $2N = 1/\Delta$ states in one period. In the case of no noise, since the loop will never make an error, it will eventually oscillate between states -1 and 1 with timing error $\pm\,\Delta/2$.

In actuality the negatively going transition need not be centered between states -1 and 1 but could be closer to either state -1 or state 1. This uncentered effect can be analyzed with a slight modification of the theory but its effect is not significant in the usual case when Δ is small (1/128 of a period) so it will not be pursued further.

It can be shown that since there are a countable number of states, in the limit as the time increases without bound, the long run probabilities all tend to zero and consequently the timing-error variance will be unbounded. To obtain a meaningful result we shall determine the steady-state phase-error distribution modulo-one period just as has been done for the continuous phase-locked loop [10]. We shall refer to this in what follows as mod-2π reduction. It is clear that the modulo-2π-reduced timing-error variance is finite since in this case there are a finite number of states.

C. Steady-State Probabilities and the Timing-Error Variance

We can take into account the fact that errors are reduced modulo 2π by regarding the two outermost error nodes as reflecting states. At a reflecting state N, the probability of going to state $N-1$ is p_N and the probability of remaining at state N is q_N. The reflecting state can be visualized as having a vertical reflecting boundary located halfway between states N and $N+1$. At the nonreflecting states, the transition probabilities p_k and q_k are defined as follows.

p_k Prob(of reducing the error | system is in state k)

q_k Prob(of increasing the error | system is in state k) (1)

and $p_k + q_k = 1$.

Referring to Fig. 3 we see that the transition from state a to state b modulo 2π places the loop error at state b'. However, as far as the timing-error magnitude we can just as well place the loop error at state a since the magnitude of timing error is the same at state b' and state a. Consequently we can model states N and $-N$ as reflecting states.

The state-transition diagram can now be reduced to N total error states from $2N$, by noting that the pairs of states -1 and 1, and -2 and 2, etc., contribute the same squared error. This reduced state transition diagram is illustrated in Fig. 4. In the figure we have used the fact the transition probabilities are state independent (independent of k) since the square-wave amplitude is constant for every state of the figure. Consequently, for example, for states 2 through N the probability of being updated from state k to state $k-1$ is denoted by p.

We have reduced the problem to a state-independent homogeneous finite aperiodic Markov chain. Following convention we shall refer to the chain as a "random walk." Now we shall determine p and q in terms of the signal and noise parameters.

As mentioned previously white noise, passed through the ideal phase-shift low-pass filter (PSLPF) of bandwidth W and then sampled at a rate $2W$ samples/second, produces independent samples. These samples have a variance of N_0W where the spectral density N_0 and the bandwidth W are both one-sided parameters. The square wave is assumed to have amplitude A and to have negligible distortion when passed through the low-pass filter. Hence the sum of m independent samples provides an update statistic SNR of

$$\rho = \mathrm{m}A^2/N_0W. \tag{2}$$

The transition probability p is therefore given by

$$p = \operatorname{erf}[\rho^{1/2}] \tag{3}$$

where

$$\operatorname{erf}(x) = \int_{-\infty}^{x} \frac{1}{\sqrt{2\pi}} \exp(-t^2/2)\, dt.$$

The transition probability q is given by $q = 1 - p$.

Now with the parameters of our random walk defined

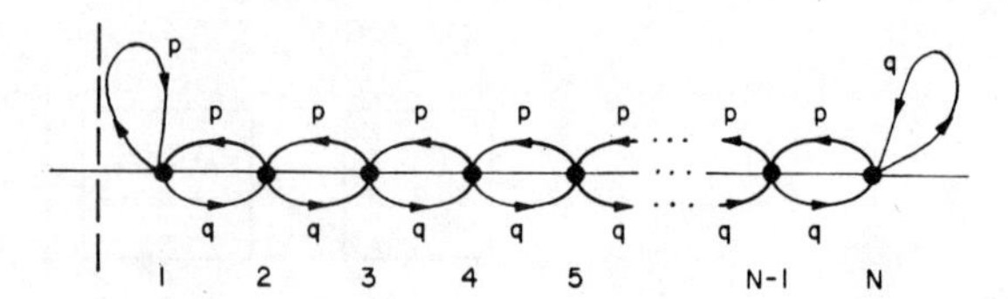

Fig. 4. Reduced state transition diagram showing transition probabilities with reflecting states at each end.

we can now determine the steady-state[1] or the long-run probabilities. Denote by $P_{jk}^{(n)}$ the probability of the event that the loop, starting in state j, is in state k after n updates. Then we define the steady-state probabilities denoted by P_k, by

$$P_k = \lim_{n\to\infty} P_{jk}^{(n)}. \tag{4}$$

It is shown in [11] that these steady state probabilities exist. It is well-known that the steady-state probabilities, for the state transition diagram of Fig. 4, satisfy the following difference equation.

$$P_k = qP_{k-1} + pP_{k+1}, \qquad 2 \le k \le N-1 \tag{5}$$

where p and q are the state-independent transition probabilities. This equation is interpreted as follows. The (steady-state) probability of being in state k is the probability of being in state k given that the system came from state $k-1$, times the probability of being in state $k-1$, plus the probability of being in state k given that the system came from state $k+1$, times the probability of being in state $k+1$. The boundary conditions take into account the reflecting barriers at states 1 and N, i.e.,

$$\begin{aligned} P_1 &= pP_2 + pP_1 \\ P_N &= qP_{N-1} + qP_N. \end{aligned} \tag{6}$$

A general solution to (5) is of the form (see [1])

$$P_k = A + B(q/p)^k \tag{7}$$

Using the boundary conditions and the fact that

$$\sum_1^N P_k = \tfrac{1}{2}$$

yields the steady-state mod-2π-reduced timing-error probabilities

$$P_k = (\tfrac{1}{2}) \frac{1 - q/p}{1 - (q/p)^N} \left(\frac{q}{p}\right)^{|k|-1},$$

$$k = -N, \cdots, -1, 1, \cdots, N \tag{8}$$

where, by symmetry, we have used the fact that $P_k = P_{-k}$. It is shown in Appendix I that this solution is unique. It

[1] They are called the ergodic probabilities by some authors.

is interesting to note that for the case of a zero-amplitude signal, it can be shown from (8) that P_k takes the form

$$P_k = \frac{1}{2N}, \qquad k = -N, \cdots, -1, 1, \cdots, N. \tag{9}$$

So, when the signal is absent, the steady-state mod-2π-reduced timing-error probabilities have a discrete uniform distribution that is the analog of the uniform density obtained in the analog loop under zero-signal conditions.

Before we compute the timing-error variance note that if we modify [12] our loop to allow three possible timing corrections: $+\Delta$, 0, $-\Delta$ where the zero correction occurs with probability r and the $+\Delta$ and the $-\Delta$ corrections are as before ($p + q + r = 1$) then it can be shown that the following difference equation is satisfied

$$P_k = q'P_{k-1} + p'P_{k+1} \tag{10}$$

with the following boundary conditions:

$$P_1 = p'P_2 + p'P_1$$
$$P_N = q'P_{N-1} + q'P_N, \tag{11}$$

where

$$p' = \frac{p}{1-r}, \qquad q' = \frac{q}{1-r}, \qquad p' + q' = 1. \tag{12}$$

Hence we see this problem is formally the same as the one we had just solved. It is only necessary to substitute p' for p and q' for q in (8) to obtain the steady-state timing-error probabilities for the more general case of three possible timing corrections. However, as long as the ratio of p/q (with $r < 1$) is the same the steady-state timing-error probabilities are independent of the value of $r(r < 1)$. For convenience in notation as well as implementation we continue with the original two-level update system.

With the help of (8) it is an easy matter to compute the steady-state timing-error variance. Denoting this variance by σ_{TE}^2 we have

$$\sigma_{TE}^2 = 2 \sum_1^N (k - \tfrac{1}{2})^2 \Delta^2 P_k. \tag{13}$$

Equation (13) can be evaluated in closed form by forming the second derivative of the appropriate geometric series. Letting $\alpha = q/p$ we have

$$\sigma_{TE}^2 = \frac{\Delta^2}{4} + \frac{\Delta^2}{1-\alpha^N}\Bigg\{-N(N+1)\alpha^N + 2 \cdot \left[\frac{\alpha - (N+1)\alpha^{N+1}}{1-\alpha}\right] + 2\left[\frac{\alpha^2 - \alpha^{N+2}}{(1-\alpha)^2}\right]\Bigg\}. \tag{14}$$

Since this equation is rather unwieldy it has been plotted in Fig. 5 as a function of ρ, the update statistic signal-to-noise ratio. The asymptotic values correspond

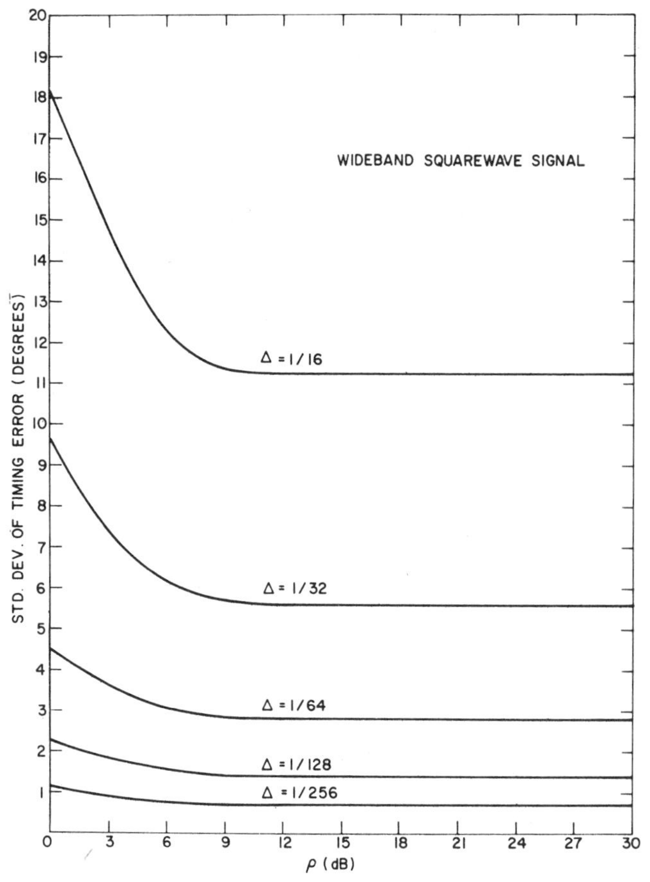

Fig. 5. Timing loop error performance for the wide-band square-wave signal case.

to the limit $\sigma_{TE}^2 = (\Delta/2)^2$ for each value of Δ. It is important to emphasize at this point that allowing W to be wide enough to produce only negligible distortion on the square-wave signal is not an efficient procedure to obtain signal synchronization for a given E/N_0 ratio where $E = A^2T$ and is the energy of the signal. [The ratio E/N_0 and ρ are related by $\rho = (E/N_0)(m/WT)$.] In fact decreasing W to allow distortion improves the performance of the loop. It will be seen in Section III that the timing-error performance is quite acceptable, for reasonable E/N_0 ratios.

D. Mean First Slip Time

Since it is desirable to keep the timing error small at all times, a useful statistic in this regard is the expected time for the timing error to exceed some magnitude (usually π or 2π rad) for the first time. We arbitrarily define a slip to occur when the error first reaches either state $N + 1$ or $-(N + 1)$ given that the loop started at state 1 or -1. (Slips to state $2N + 1$ or state $-2N - 1$ cannot be handled by the state independent model.)

In terms of the random-walk model, we place absorbing barriers at states $N + 1$ and $-(N + 1)$ and start the process at either state 1 or -1. An absorbing state or barrier is characterized by the fact that once the system enters the absorbing state it remains there forever. Because of the symmetry involved and the fact that each step is independent of the previous one we may place a reflecting boundary at the origin as before and reduce the problem to one of $N + 1$ states from one of $2N + 2$ states. Fig. 6 illustrates the state transition diagram used to model the first-passage time problem under consideration. State 1 is reflecting and state $N + 1$ is absorbing.

Rather than derive the probabilities z_n of starting at state 1 and reaching state $N + 1$ in time n and then forming the average

$$T_1^{N+1} = \sum_1^{\infty} n z_n \tag{15}$$

for the mean first slip time we shall use a method indicated in [1].

Let T_k^{N+1} be the mean duration of time it takes to reach state $N + 1$ starting at state k. If the first move is to the right the procedure continues as if the initial position had been $k + 1$. The conditional expectation of the duration assuming the move was to the right is therefore $T_{k+1}^{N+1} + 1$. On the other hand, if the initial move is made to the left the conditional expectation is $T_{k-1}^{N+1} + 1$. Therefore, the mean time T_k^{N+1} must satisfy the following difference equation

$$T_k^{N+1} = qT_{k+1}^{N+1} + pT_{k-1}^{N+1} + 1, \qquad k = 2, \cdots, N \tag{16}$$

with the two associated boundary conditions

$$T_{N+1}^{N+1} = 0$$
$$T_1^{N+1} = qT_2^{N+1} pT_1^{N+1} + 1 \tag{17}$$

The former accounting for the absorbing state at state $N + 1$ and the latter accounting for the reflecting state located at state 1. By direct substitution it can be shown [1] that

$$T_k^{N+1} = \frac{k}{p-q} + A + B\left(\frac{p}{q}\right)^k, \qquad p > q \tag{18}$$

is a solution. Upon evaluating the boundary conditions we obtain

$$T_k^{N+1} = \frac{k - N - 1}{p - q} + \frac{1/q + 1/p - q}{\frac{p}{q} - 1} \cdot \left\{\left(\frac{p}{q}\right)^N - \left(\frac{p}{q}\right)^{k-1}\right\}. \tag{19}$$

It is not hard to show that this is the unique solution (Appendix II). Hence the mean first slip time becomes, upon setting $k = 1$ in (19),

$$T^{N+1} = -\frac{N}{p-q} + \frac{\frac{1}{q} + \frac{1}{p-q}}{\frac{p}{q} - 1}\left[\left(\frac{p}{q}\right)^N - 1\right], \tag{20}$$

where it is assumed $p > q$, since $p = q$ corresponds to no signal. [(20) also holds for $p - q$.] Equation (20) is plotted in Fig. 7 as a function of ρ and the update step size Δ. It is seen that T^{N+1} very rapidly becomes a very large number. It can be shown from (20) that T^{N+1} is asymptotic to exp $[\rho/4\Delta]$, hence the smaller Δ is the larger the mean time to first slip for a given ρ.

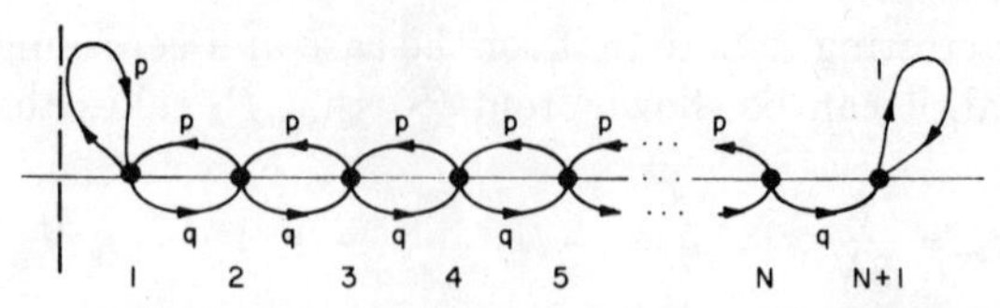

Fig. 6. Reduced state transition diagram for the first-passage time model.

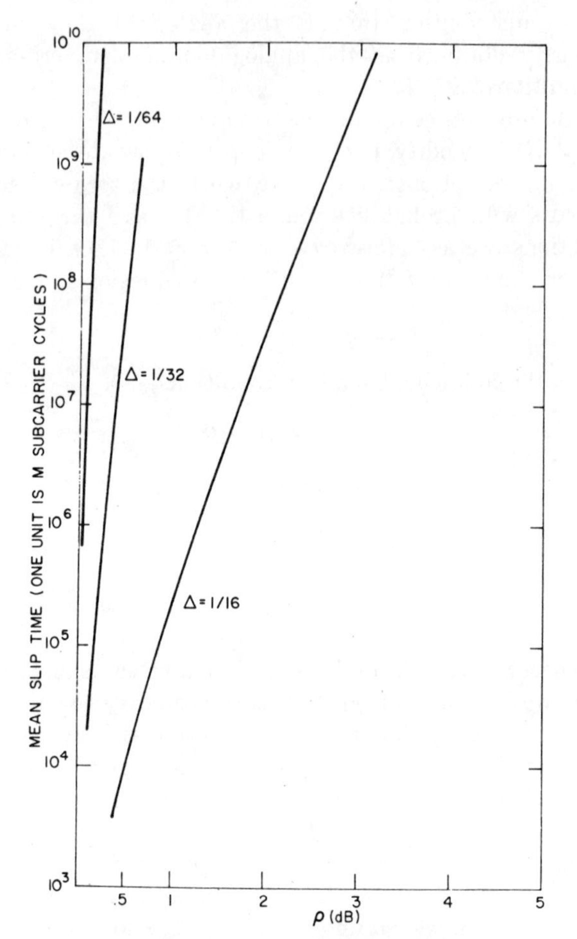

Fig. 7. Mean slip time versus update signal-to-noise ratio for a square-wave subcarrier.

III. General Signal Case

Section II dealt with the wide-band case specialized to a square-wave signal. Under these conditions it was possible to formulate the problem in terms of state-independent transition probabilities and hence a random-walk problem. The performance under wide-band conditions is, of course, not as good as under narrower prefiltering conditions, however the specialization served to illustrate the model and method of analysis unencumbered by the added complexity of the problem in the general case.

The general case of a periodic signal, odd symmetric about the center of every transition, with two zero crossings/period, is now considered. By allowing the low-pass filter bandwidth to be reduced, tracking performance can be significantly improved. In some applications severe distortion, due to a very narrow bandwidth, may not be permissible.

A. General Formulation of the Steady-State Error Probability

In the general case the signal-sample amplitude, at various timing errors, is not independent of the error location as in the square-wave signal case. Hence, the transition probabilities are error state dependent. Again we let f_s denote the signal frequency. The state-transition probabilities for the state-dependent case, in terms of the signal and noise parameters are given by

$$p_k = \text{erf}\,[(mA_k^2/N_0W)^{1/2}], \qquad q_k = 1 - p_k, \tag{21}$$

where A_k is the amplitude of the signal at state k. Now the steady-state probabilities, for the positive states, must satisfy the following second-order difference equation [1]:

$$P_k = q_{k-1}P_{k-1} + p_{k+1}P_{k+1}, \qquad 1 < k < N. \tag{22}$$

The interpretation of (22) is similar to that following (5). The associated boundary conditions for (22) are

$$P_1 = p_1P_1 + p_2P_2$$
$$P_N = q_{N-1}P_{N-1} + q_NP_N. \tag{23}$$

Fig. 8 illustrates the reduced state-transition diagram for the general case.

The solution of (22) along with the boundary conditions can be shown by induction or directly [13] to be

$$P_k = P_1 \prod_{j=1}^{|k|-1} \frac{q_j}{p_{j+1}}, \qquad k = -N, \cdots, -1, 1, \cdots, N \tag{24}$$

with the definition

$$\prod_{j=1}^{0} \frac{q_j}{p_{j+1}} = 1. \tag{25}$$

In (24) P_1 is given by

$$P_1 = \frac{1}{2}\left[1 + \sum_{k=2}^{N} \prod_{1}^{k-1} (q_j/p_{j+1})\right]^{-1}. \tag{26}$$

In Appendix I it is shown that the solution is unique. Once the steady-state probabilities are determined the timing-error variance can be computed.

$$\sigma_{\text{TE}}^2 = 2\Delta^2 \sum_{k=1}^{N} (k - \tfrac{1}{2})^2 P_k. \tag{27}$$

The results for the case of a filtered square-wave signal has been plotted in Fig. 12 and is discussed in Section V.

We again note that if we modify the timing loop to allow three possible timing corrections $+\Delta$, 0, $-\Delta$, where the zero correction occurs with probability r_k in state k,

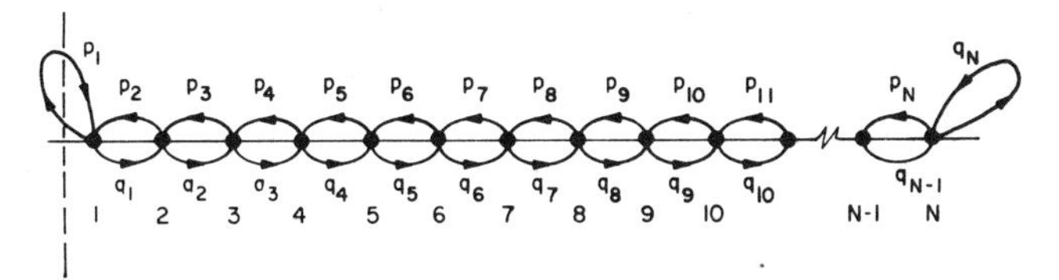

Fig. 8. Reduced state transition diagram in the state-dependent case showing the transition probabilities with reflecting states at each end.

then the following difference equation for p_k is satisfied

$$p_k = q'_{k-1}P_{k-1} + p'_{k+1}P_{k+1} \tag{28}$$

with the following boundary conditions

$$P_1 = p'_1P_1 + p'_2P_2$$
$$p_N = q'_{N-1}P_{N-1} + q'_NP_N \tag{29}$$

where

$$p'_k = \frac{p_k}{1 - r_{k-1}}, \qquad q'_k = \frac{q_k}{1 - r_{k+1}}, \qquad 1 < k < N$$
$$p'_1 = \frac{p_1}{1 - r_1}, \qquad q'_N = \frac{q_N}{1 - r_N}. \tag{30}$$

Therefore this problem is formally the same as the $r_k = 0$ problem solved above except the solution replaces p_k and q_k with p'_k and q'_k as defined in (30). We have

$$P_k = P_1 \prod_{j=1}^{|k|-1} \left(\frac{q'_j}{p'_{j+1}}\right), \qquad k = -N, \cdots, -1, 1, \cdots, N$$
$$P_1 = \frac{1}{2}\left[1 + \sum_{k=2}^{N} \prod_{1}^{k-1} (q'_j/p'_{j+1})\right]^{-1} \tag{31}$$

again the steady-state probabilities only depend on the ratio $q'_j/\,p'_{j+1}$ as mentioned before in the case when the transition probabilities were state independent.

B. Mean First Slip Time for the General Case

In this section a method given in Feller [1] to determine the mean first slip time is modified to allow the transition probabilities to be state dependent. The state transition diagram of Fig. 6 is still applicable except that the transition probabilities are now indexed according to the state they eminate from.

Again let T_k^{N+1} be the mean duration of time it takes to reach state $N + 1$ starting at state k. Further denote the event "incorrect update" as the act of moving Δ units towards state $N + 1$ and "correct update" as the act of moving Δ units towards state 1. Then we note, that starting at state k:

mean duration given an incorrect update

$$= T_{k+1}^{N+1} + 1$$

mean duration given a correct update

$$= T_{k-1}^{N+1} + 1 \tag{32}$$

and

$$P(\text{incorrect update} \mid \text{initial state is } k) = q_k$$
$$P(\text{correct update} \mid \text{initial state is } k) = p_k. \quad (33)$$

Hence combining (32) and (33) we arrive at the difference equation that T_k^{N+1} satisfies:

$$T_k^{N+1} = q_k T_{k+1}^{N+1} + p_k T_{k-1}^{N+1} + 1. \quad (34)$$

With the associated boundary conditions

$$T_1^{N+1} = q_1 T_2^{N+1} + p_1 T_1^{N+1} + 1$$
$$T_{N+1}^{N+1} = 0. \quad (35)$$

The solution is derived in Appendix III for arbitrary k. The mean first slip time, defined in Section II-D is given by

$$T_1^{N+1} = \sum_{l=1}^{N} \left\{ \sum_{j=1}^{l-1} \prod_{i=j+1}^{l} \frac{p_i}{q_i p_i} + \frac{1}{q_l} \right\}, \quad (36)$$

where we define for arbitrary α_j

$$\sum_{j}^{k} \alpha_j = 0, \qquad j > k.$$

In Appendix II it is shown that (36) is the unique solution.

If we had chosen the limits $2N + 1$ and $-(2N + 1)$ in our definition of the mean first slip time then N, in (36), would be changed to $2N$. This allows a slip to be defined to occur when the phase error first leaves the interval $(-\pi, \pi)$, rather than $(-\pi/2, \pi/2)$ as was used in Section II-D.

If we generalize our model to the case of three possible timing corrections at a given state as described previously it can be shown that the mean first slip time is given precisely by (36) except that now for given p_k and q_k, r_k is defined by $r_k = 1 - p_k - q_k$. For example if we let double primes on the transition probabilities denote the two update system and no primes denote the three-level update system then the following holds, as is evident from (36). If

$$p_k > q_k, r_k > 0$$
$$p_k/q_k = p''_k/q''_k$$

then

$$T_k^{N+1} < \hat{T}_k^{N+1} \quad (37)$$

where $\hat{T}_k^{N+1}$ is the mean first time to slip given that $r_k > 0$. In other words, increasing r_k from the value of zero increases the mean first slip time when the ratio of p_k to q_k is kept the same.

IV. Additional Results

In this section we consider a useful generalization of the results in Section III. Up until now we have assumed that no modulation was impressed on the signal to be synchronized. We show that the effect of bit errors is to modify the transition probabilities. Before we consider this generalization we describe a result that yields a bound on the effect of a Doppler shift of the incoming signal relative to the transition sampling rate of the loop.

A. Effect of a Doppler Frequency

If the transition sampling rate, which corresponds to one sample every half-period of the signal, is not equal to exactly twice the frequency of the incoming signal then the Doppler rate is said to be nonzero. In this case the sampling system is not synchronous with the signal transitions. The effect, (assuming the signal is fixed), as viewed on the error-state diagram (See Fig. 3 for example), is to continuously move the error states to the left if the sampling rate is greater than the signal transition rate and to the right if it is less. Consequently stationary probabilities do not exist for this problem since the transition probabilities now depend on time as well as the state. However a bound can be obtained on the bias effect of the Doppler frequency.

In order to obtain our bound we assume, in addition to the transition region being odd symmetric about the negative-going transition, that the signal is shaped such that there exist symmetric timing errors $\pm \tau_0$ (measured from the zero crossing of the transition region) that satisfy the following. The amplitude of the signal is strictly monotonically increasing with timing error up to $\pm \tau_0$ and beyond these points does not increase but may decrease. This is not a severe practical restriction since most physical signals approximate this symmetry condition.

In order to derive the bound we assume that signal transition rate is greater than the sampling rate, in other words, the Doppler is positive. Hence the sampling location will move farther to the right of the zero crossing every sample. Let us assume that, the magnitude of this rate is Δd fractions of a cycle/cycle. As the sample location drifts to the right and the timing error increases, an update signal with higher probability of correct update is generated as long as the error does not exceed the maximum points $\pm\tau_0$. Eventually if Δd is not too large the higher probability of correct updating to the left balances the drift to the right resulting in an equilibrium. The net result being a positive phase-error bias.

In order to obtain an upper bound on the bias, the position of lowest signal amplitude must be selected from all the possible positions that state 1 takes on due to the Doppler induced drift. Hence by inspection, if state 1 is located at the zero crossing the amplitude will be least (zero) for that state. (States -1 and $+1$ must straddle the zero crossing by definition of these states.) Fig. 9 illustrates an example of a model of a filtered square-wave signal that is modeled to have a linear transition region in which state 1 has been located at the zero crossing. This signal will be discussed in greater detail in Section V.

Let $\tilde{p}_k$ denote the probability of moving toward the zero crossing given that the system is in a positive state k in this "shifted" error state configuration. (Here we

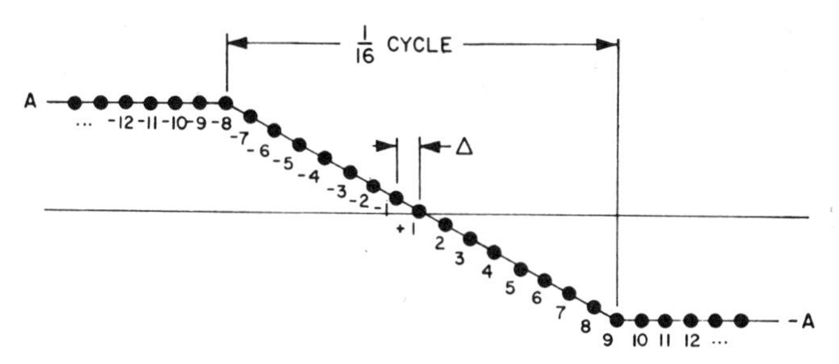

Fig. 9. Worst state configuration of the error-state diagram superimposed on a negative-going transition for the filtered square-wave signal case.

define $\tilde{p}_1$ to be one-half.) Also let $\tilde{q}_k$ be the probability of moving away from the zero crossing given that the system is in a positive state k. If we define

$$i_k = -\Delta \text{ with probability } \tilde{p}_k$$
$$= +\Delta \text{ with probability } \tilde{q}_k \tag{38}$$

and we denote the change of position from state k at time nT to a new position at time $(n+1)T$ by Δd_k, then

$$\Delta d_k = i_k + M\Delta d \tag{39}$$

since the loop updates every M signal periods. Consequently, the mean change at time $(n+1)T$ is

$$\overline{\Delta d_k} = -(\tilde{p}_k - \tilde{q}_k)\Delta + M\Delta d. \tag{40}$$

Since we have picked the error-state configuration with the least "average restoring ability" it follows that if there exists a minimum error "shifted state" k_0 such that

$$(\tilde{p}_{k_0} - \tilde{q}_{k_0}) \geq M\Delta d/\Delta \tag{41}$$

then the upper bound B_d of the magnitude of the Doppler-induced bias is given by (after a little reflection)

$$B_d = k_0\Delta. \tag{42}$$

In terms of the relative Doppler rate $\Delta f_s/f_s$ we have the condition

$$\Delta f/f \leq \frac{(\tilde{p}_{k_0} - \tilde{q}_{k_0})\Delta}{Tf_s}. \tag{43}$$

The sign of the bias is negative (to the left of the transition) if the Doppler decreases the signal frequency and positive if the Doppler increases it. The meaning of the bound is the following. If there exists a k_0 then the average change will not be away from the zero crossing and hence the loop will lock up. If there does not exist a k_0 either the loop will have a large bias or it will not lock up. This bound was applied to a model of a filtered square wave that is described in Section V with $\Delta = 1/256$ and $E_{b/N_0} = 10$ dB and is plotted in Fig. 10 along with some experimental points taken for that system. Equation (43) was used to plot the bound against the relative Doppler shift $\Delta f_s/f_s$.

A reasonable approximation to the bias may be had by plotting $(k_0 - 1)\Delta$ and connecting a line from the origin to the lowest points of the curve $(k_0 - 1)\Delta$ as indicated by the dashed line in Fig. 10.

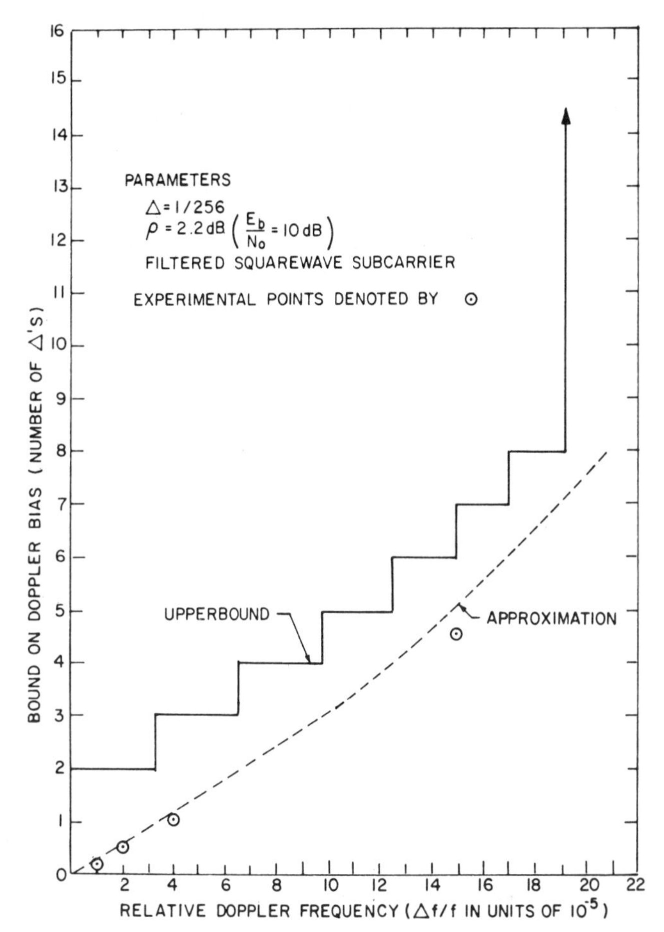

Fig. 10. Upper bound and approximation of Doppler bias versus relative Doppler frequency.

B. Performance With Modulation

Up until now all the results were derived assuming no modulation was present on the signal. With the introduction of modulation it is necessary to modify the loop to employ decision feedback to feedback the sign of the detected bits. The feedback removes the sign of the binary modulation.

The effect of bit errors on the loop is to degrade its timing error performance. To see why the loop is degraded we derive the transition probabilities when modulation is present. Let PE_k denote the bit-error probability, given that the timing error is in a positive state k. Further denote $\hat{p}_k$ as the probability of reducing the timing error taking into account bit errors and given that the error state is k. Let $\hat{q}_k$ denote the probability of increasing the timing error (taking into acount bit errors) given that the error state is k. Then the transition probabilities with modulation are given in terms of the unmodulated ones by

$$\hat{p}_k = (1 - PE_k)p_k + PE_kq_k$$
$$\hat{q}_k = (1 - PE_k)q_k + PE_kp_k. \tag{44}$$

Hence $\hat{p}_k < p_k$ and $q_k > \hat{q}_k$ due to bit errors.

In (44) the correlation between the bit decision statistic and timing update statistic has been neglected. However, usually this will be a negligible effect since the correlation coefficient can be shown to be small if the number of samples/period is large enough, e.g., 16. The correlation coefficient is $\frac{1}{3}$ if 16 samples are used per period. So, for the results derived in Section III to hold with modulation, we must replace p_k and q_k with $\hat{p}_k$ and $\hat{q}_k$. Consider now what happens to PE_k as the system moves away from the transition region, that is, the timing error increases. Since the probability PE_k depends on the autocorrelation function of the modulated waveform, increasing the error decreases the correlation up to a point where it goes negative. The decrease is proportional to the reduction of Eb/N_0. At the point where the correlation goes to zero $PE_k = \frac{1}{2}$ and by (44), $\hat{p}_k = \hat{q}_k = \frac{1}{2}$. Once the modulated waveform correlation function is computed it is not difficult to compute $\hat{p}_k$ and $\hat{q}_k$ and hence the modulated loop performance.

The average bit error rate PE is given by

$$PE = 2 \sum_{k=1}^{N} PE_k P_k. \tag{45}$$

Because of the symmetry involved the negative states were folded over onto the positive ones, which accounts for the 2 in the above equation.

C. *Multiple Transition-Sample Phase Detector*

We now show that the general results derived in Section III for arbitrary signals is applicable to the case when more than one sample is obtained per transition region [14]. Whether the timing error can be reduced by multisamples depends on the ratio A^2/N_0W as well as the signal shape around the transition. We make a reasonable practical assumption that the loop is limited to utilizing the samples obtained by the sampler so that the samples are $1/2W$ seconds apart in time.

In the case of wide-band square waves (undistorted) the multisample scheme, sampling across the transition for a duration of one signal period, is essentially equivalent to the continuous midphase in-phase type of phase detector. That is, they are equivalent in the sense that the output SNR is the same when the timing error is much larger than the time between samples. For small timing errors they are not equivalent as was mentioned in Section II-A.

The effect of the multiple samples is to modify the update signal-to-noise ratio and consequently the transition probabilities. Since the system is digital and there are $2W/f_s$ samples/period available, the most convenient multisampler arrangement utilizes an equal number of samples located symmetrically about the sample the loop thinks is the zero crossing point. This requires $2S + 1$ samples/half-period, where S is an integer so that there are $4S + 2$ samples that may be used per period. These samples may be weighted according to their distance away from the center or zero crossing sample to provide more nearly optimum performance. The basic equations starting at (21) are modified because of the change in update SNR. Let y_j^k, j positive, denote the sample of the signal plus noise taken j samples to the right of the sample taken at state k. When j is negative y_j^k denotes the sample taken j samples to the left (or later) of the one taken at state k. Since the signal part of y_j^k is A_j^k (using the same notation) the transition probability p_k becomes

$$p_k = \operatorname{erf}\left\{\left[\frac{m\left(\sum_{-S}^{S} \gamma_i A_i^k\right)^2}{N_0 W \sum_{-S}^{S} \gamma_i^2}\right]^{1/2}\right\}, \qquad q_k = 1 - p_k \tag{46}$$

where γ_i is the relative weighting of each sample.

This scheme, for $S = 1$, has been investigated for the example considered in Section V with $\gamma_1 = \gamma_{-1}$ and $\gamma_0 = 1$. The value of γ_1 was optimized by the use of a computer to obtain the minimum timing variance (phase error) as a function of the signal-to-noise ratio. The timing error is shown on Fig. 11 for both the single sample/transition and the triple sample/transition. As one might expect, the timing error is reduced with three samples/transition only if the timing error is sufficiently large. For $Eb/N_0 > 1 \gamma_1$ was found to be much less than one in the optimum case, so that the additional processing of adjacent samples was not worthwhile since it reduced the timing variance insignificantly.

V. Low-Data-Rate Command System

The timing system that motivated this paper is now considered (Fig. 1). The input signal is a 60-Hz square-wave subcarrier (SC) biphase modulated by a length-15 pseudonoise (PN) sequence. Each symbol of the PN sequence corresponds to one cycle of the subcarrier. The prefilter has a one-sided bandwidth of 480 Hz. Samples are taken at the Nyquist rate of 960 samples/s and then quantized. This rate provided 16 samples/subcarrier period. Four-bit quantization was used to represent the samples, which introduces negligible degradation of performance. The accumulator adds 22 transition samples/bit. (There are 22 unambiguous transitions in a length-15 PN code.) Positive-going and negative-going transitions are resolved by the PN code generator, which, by means of an internally generated PN code, multiplies the sample by a plus one when the transition is negative going and by a negative one if the transition is positive going. The bit detector determines whether a one or a zero was detected and changes the algebraic sign of the update signal only if a zero was sent.

The filtered PN code modulated subcarrier signal of this system was modeled by a unfiltered square wave with a linear transition region. It was modeled as having a duration of $\frac{1}{16}$ of a subcarrier period. The model agreed quite well with laboratory measurements of the filtered signal.

For modulated signals it is convenient to define the bit-signal energy-to-noise spectral density, assuming a perfectly synchronized digital integrate-and-dump cir-

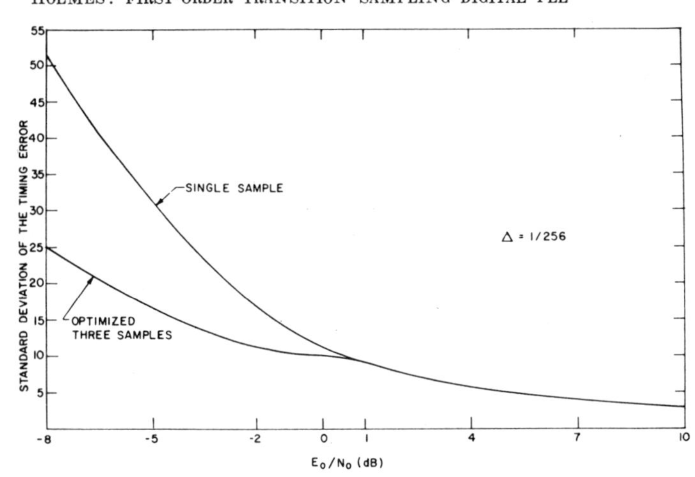

Fig. 11. Timing error for the single sample and the optimized three-sample system for updating.

cuit, by

$$\frac{E_b}{N_0} = \frac{A^2T_b}{N_0}, \tag{47}$$

where T_b is the bit time. Equation (47) neglects the small loss of signal due to the effect of the low-pass filter. Using (24), (26), and (27) of Section III, in conjunction with the modified transition probabilities of (43), the timing-error variance was computed for the case Δ = 1/16, 1/32, 1/64, 1/128, and 1/256 as a function of Eb/N_0. The results are plotted in Fig. 12. Some experimental points obtained from [15] are plotted for $\Delta = 1/256$ and $\Delta = 1/128$.

It is of interest to compare the performance of this loop with that of Cessna [4], which employs a midphase-in-phase type of phase detector. For comparable cases, that is $\Delta = 1/16$, it is seen that for $E_{b/N_0} \geq 10$ dB the phase-error performance is essentially identical. Below 10 dB, Cessna's loop appears to produce a lower phase error. However the phase detector and filter of the loop described here is considerably simpler and is all digital, making it very attractive for uncoded communication systems. In addition the phase error of this loop can always be reduced by reducing Δ.

An important parameter that measures the effectiveness of the timing loop in providing timing is the degradation in the average bit-error probability. Alternately this can be converted to degradation in bit-error SNR. Four values of Δ were considered assuming the Doppler frequency was zero. The values of Δ chosen were 1/16, 1/32, 1/64, and 1/256. The threshold SNR was based on the nominal requirement of a $PE = 10^{-5}$, this led to the value $Eb/N_0 = 10$ dB. The average error probability was computed according to (45) and then converted to the equivalent bit SNR. The difference in decibels between the SNR corresponding to an error rate of 1×10^{-5} and the equivalent SNR determined from (45) was defined to be the degradation.

The results are plotted in Fig. 13. It is seen that an update size of $\Delta = 1/256$ produces a degradation of less

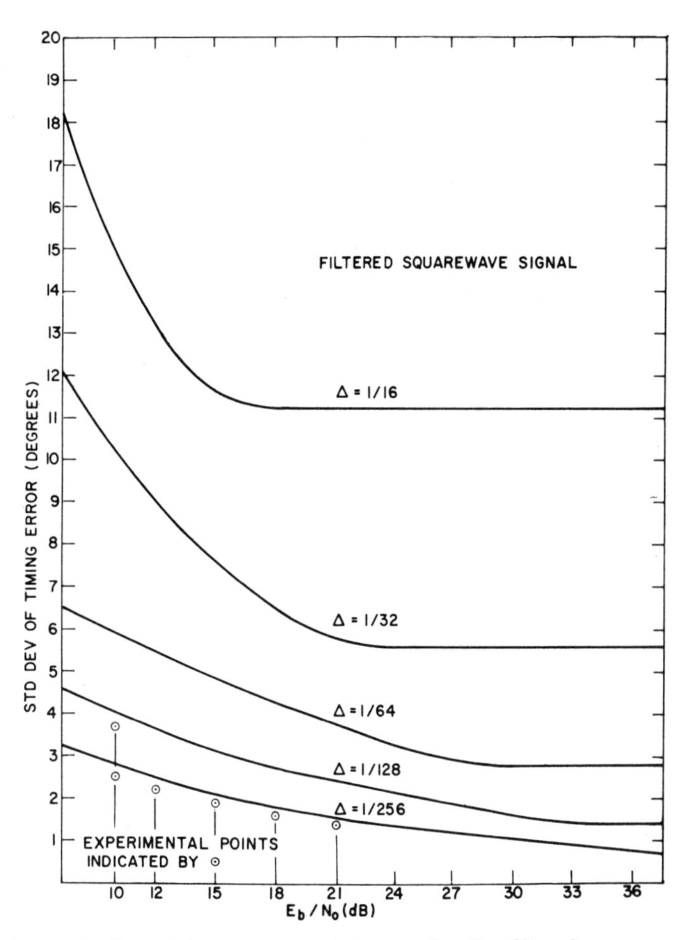

Fig. 12. Timing-loop error performance for the filtered squarewave signal case.

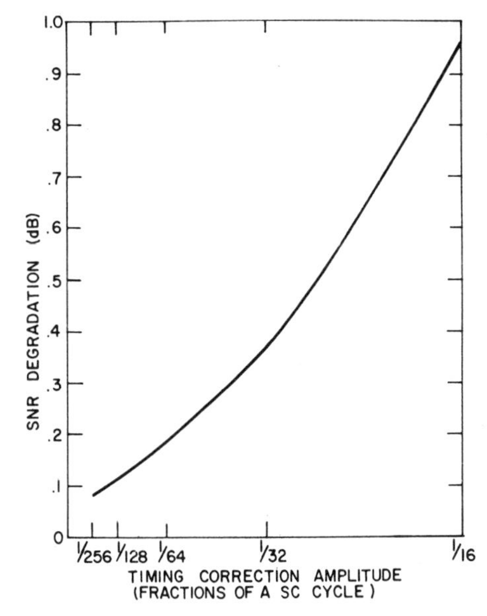

Fig. 13. SNR degradation versus timing update size for PN modulated subcarrier.

than 0.1 dB while the degradation is almost 1 dB with $\Delta = 1/16$.

One parameter that was not derived in the general case but is of interest is the bandwidth of our digital loop. Although the derivation [16] is too long to be included here the result, for one update/bit timing using the filtered square-wave signal model is

$$W_L = \sqrt{2/\pi}\,\frac{\rho^{1/2}\Delta}{\zeta T_b}\ \text{Hz}, \tag{48}$$

where ρ is defined in (2), T_b is the bit time, and ζ is ratio of the transition region duration to the total period of the SC. In Fig. 10, for example, $\zeta = 1/16$ of an SC cycle.

For the system under consideration $T_b = 1/4$ s, $\Delta = 1/256$, $\rho = 1.65$, and $\zeta = 1/16$. Using these values we see that the digital bandwidth, from (48), is 0.256 Hz.

VI. Conclusions

A theory using random-walk techniques has been developed for first-order all-digital phase-locked loops that obtain timing error information by sampling the transition of the signal. Various aspects of performance have been studied such as the steady-state phase-error variance, the mean first time to slip, the bias due to Doppler, and the effect of modulation on synchronization.

The loop lends itself to applications where the advantages of all-digital operation are realized and where simplicity in the construction of the loop is important. These were the guidelines for the development of the loop. There is no claim of optimality for the loop's performance.

The important question of transient analysis has only been partially answered. The mean first slip time has been solved for arbitrary signals but the loop noise bandwidth was given only for the case of a filtered square-wave signal.

The loop was built and its performance was found to agree closely with the theory developed here.

The question arises of whether this approach be applied to the second-order ADPLL. The answer, unfortunately, is that the second-order loop has unbounded memory and therefore is not a simple random walk. However, a vector random walk, which describes the second order loop, can be obtained but it appears to be very difficult to solve.

Appendix I

Uniqueness of the Steady-State Probabilities

We prove that if $q_k > 0$ for at least one k, and

$$P_k = q_{k-1}P_{k-1} + p_{k+1}P_{k+1} \tag{49}$$

with

$$P_1 = p_1P_1 + p_2P_2 \tag{50}$$

$$P_N = q_{N-1}P_{N-1} + q_NP_N \tag{51}$$

$$\sum_1^N P_k = \tfrac{1}{2} \tag{52}$$

then the solution P_k is unique.

Proof: Let two different solutions $\tilde{P}_k$ and $\hat{P}_k$ be solutions of (49) satisfying (50)–(52). Define Δ_k by

$$\Delta_k = \tilde{P}_k - \hat{P}_k \tag{53}$$

then from (50)

$$\Delta_2 = \frac{q_1}{p_2}\Delta_1. \tag{54}$$

By iterating (49)

$$\Delta_k = \prod_{i=2}^{k}\frac{q_{i-1}}{p_i}\Delta_1. \tag{55}$$

Assume $\Delta_1 \neq 0$, then from (55)

$$\sum_{k=1}^{N}\Delta_k = \sum_{k=1}^{N}\prod_{i=2}^{k}\frac{q_{i-1}}{p_i}\Delta_1 \neq 0. \tag{56}$$

But from (52) and (53) we see that we have a contradiction implying $\Delta_1 = 0$, which implies, by (55), that $\Delta_k = 0$ for all k. This implies the solution is unique. If $q_k = 0$ for all k then we also get the unique solution $P_1 = 1/2$, $P_{-1} = 1/2$. Q.E.D.

Appendix II

Uniqueness of the Mean First Slip Time

We prove that if

$$T_{k+1}^{N+1} = q_kT_k^{N+1} + p_kT_{k-1}^{N+1} + 1, \quad 1 < k < N \tag{57}$$

$$T_1^{N+1} = q_1T_2^{N+1} + p_1T_1^{N+1} + 1 \tag{58}$$

$$T_{N+1}^{N+1} = 0 \tag{59}$$

then the solution T_k^{N+1} is unique.

Proof: Let $\tilde{T}_k^{N+1}$ and T_k^{N+1} be different solutions of (57) subject to the boundary conditions in (58) and (59). Let

$$\Delta_k = \hat{T}_k^{N+1} - \tilde{T}_k^{N+1} \tag{60}$$

then from (58)

$$\Delta_1 = \Delta_2 \tag{61}$$

and in general

$$\Delta_k = \Delta_1, \qquad \text{for all } k. \tag{62}$$

From (59) we have

$$\Delta_{N+1} = 0, \tag{63}$$

which implies $\Delta_k = 0$ for all k.

Appendix III

Solution to the Mean First Slip Time

We shall obtain the solution to (57) subject to the boundary conditions of (58) and (59). Starting with (57) and (58) we may iterate to obtain

$$T_{k+1}^{N+1} = T_k^{N+1} - \sum_{j=1}^{k-1}\left(\prod_{i=j+1}^{k}\frac{p_i}{q_iq_j}\right) - \frac{1}{q_k}. \tag{64}$$

Let

$$\alpha_k = \sum_{j=1}^{k-1} \left(\prod_{i=j+1}^{k} \frac{p_i}{q_i q_j} \right) + \frac{1}{q_k} \tag{65}$$

then using (59) it follows that

$$\sum_{k=1}^{N} (T_k^{N+1} - T_{k+1}^{N+1}) = T_1^{N+1} = \sum_{k=1}^{N} \alpha^k. \tag{66}$$

In a similar manner it may be shown that

$$T_k^{N+1} = \sum_{l=k}^{N} \alpha_l.$$

So that from (65) we have

$$T_k^{N+1} = \sum_{l=k}^{N} \left\{ \sum_{j=1}^{l-1} \prod_{i=j+1}^{l} \frac{p_i}{q_i p_j} + \frac{1}{q_l} \right\}. \tag{67}$$

In (67) the following convention was used

$$\sum_{i}^{0} a_i = 0, \qquad \text{any } a_i. \tag{68}$$

It can be shown by some rather tedious algebra that (67) is the solution to (57) and satisfies the boundary conditions of (58) and (59). For example, we now show that the solution satisfies (58). From our solution we can show that

$$T_2^{N+1} - T_1^{N+1} = -\sum_{j=1}^{0} \prod_{i=j+1}^{1} \left(\frac{p_i}{q_i q_j} \right) - \frac{1}{q_1}. \tag{69}$$

Using the conventions of (68) we have

$$T_2^{N+1} = T_1^{N+1} - \frac{1}{q_1}, \tag{70}$$

which is a rearranged form of (58).

References

[1] W. Feller, *An Introduction to Probability Theory and Its Applications*, vol. 1. New York: Wiley, 1957, ch. 14.

[2] P. R. Westlake, "Digital phase control techniques," *IRE Trans. Commun. Syst.*, vol. CS-8, pp. 237–246, Dec. 1960.

[3] W. J. Hurd and T. O. Anderson, "Digital transition tracking symbol synchronizer for LOW SNR coded systems," *IEEE Trans. Commun. Technol.*, vol. COM-18, pp. 141–147, Apr. 1970.

[4] J. R. Cessna, "Steady state and transient analysis of a digital bit synchronization phase-locked loop," in *Proc. Int. Conf. Communications*, vol. 2, pp. 34-15–34-25, June 1970.

[5] H. Gota, "A digital phase-locked loop for synchronizing digital networks," presented at the Int. Conf. Communications, San Francisco, Calif., June 8–10, 1970.

[6] G. Pasternack and R. L. Whalin, "Analysis and synthesis of a digital phase-locked loop for FM demodulation," *Bell Syst. Tech. J.*, vol. 47, pp. 2207–2237, Dec. 1968.

[7] S. C. Gupta, "Status of digital phase locked loops," presented at the Int. Hawaii Conf., Honolulu, Jan. 1970.

[8] J. K. Holmes, "On a solution to a digital first order phase locked loop," in *Proc. UMR-Mervin J. Kelly Communications Conf.*, Rolla, Mo., Oct. 1970.

[9] ——, "Random walk techniques in the solution of first order digital phase locked loops," presented at the Int. Hawaii Conf., Honolulu, Jan. 12–14, 1971.

[10] A. J. Viterbi, *Principles of Coherent Communication*. New York: McGraw-Hill, 1966, ch. 4.

[11] M. Fisz, *Probability Theory and Mathematical Statistics*, 3rd ed. New York: Wiley, 1963, ch. 7.

[12] M. K. Simon, JPL Lab., private communication.

[13] E. Parzen, *Stochastic Processes*. San Francisco: Holden-Day, 1962, pp. 278–281.

[14] J. K. Holmes, "A note on the optimality of the all digital command system timing loop," JPL Lab., Space Program Summary 37-65, vol. 3, Oct. 1970.

[15] A. Couvillion, Jet Propulsion Lab., private communication. second order subcarrier tracking performance," JPL Lab. Rep. TR-1540, Oct. 1971.

Part XII
Applications and Miscellaneous

Analysis of Synchronizing Systems for Dot-Interlaced Color Television*

T. S. GEORGE†, ASSOCIATE, IRE

Summary—**A mathematical analysis is made of two synchronizing systems which might be used in "dot" color television to synchronize the dotting or sampling frequency in the receiver. Synchronizing information is transmitted in bursts of carrier cohered in phase of approximately 3 Mc during line fly-back time. The two systems analyzed are (1) a simple high-Q resonant filter and (2) an oscillator with automatic frequency control (afc).**

In order to maintain sufficient phase accuracy in the sampling frequency, crystal control at the transmitter is necessary. Since the variations of the frequency response of a particular crystal filter with time may be made essentially negligible, the problem resolves itself into calculating the parameters of the synchronizing system to keep below an acceptable value the random phase error due to noise together with the phase error caused by variations in frequency (static phase error) from crystal to crystal. Rather than attempt to minimize the sum of the random and static phase errors, the two have been dealt with separately because of their different character. A fundamental parameter in the calculations is the power carrier-to-noise ratio in the intermediate frequency, the carrier being measured at the sync tips. Calculations are made for values of this ratio of 1, 3, 5, this being considered the critical range.

Results of the calculation show that when the phase error due to noise alone is fixed, the simple resonant filter and the afc with single time constant suffer the same static phase error. If phase errors in the neighborhood of 10° can be tolerated, it appears then that receivers can be designed to operate without manual control of the dotting frequency down to carrier-to-noise ratios of about 2. This should be satisfactory since picture quality at this level is very poor.

IN DESIGNING a "dot interlace" system[1] of color television, synchronizing problems over and above those of ordinary black-and-white television are encountered. A "dot" system consists essentially of a three-channel time division multiplex system with one channel devoted to the transmission of each primary color. This involves sampling each color at the transmitting end of the system at approximately a 3-Mc rate, transmitting information so derived to the receiver, separating this information by resampling at the same 3-Mc rate and applying it to control the three colors displayed on the cathode-ray tube. Thus a sample of each color is obtained on every third pulse. In order to avoid deleterious color distortion, the sampling frequency in the receiver must be closely synchronized in phase with that of the transmitter.

In the receiver we have the problem of first synchronizing the frame, then the line, and finally the dots within each line. The first two of these are common to both color and black-and-white television and since these problems are relatively well understood, less attention will be given to them than to dot synchronizing. There appear to be no smaller tolerances required for frame and line synchronizing in color television than in black-and-white. A continuous-wave 3-Mc wave could be transmitted, picked out by frequency separation, and used for synchronizing purposes. This has obvious disadvantages, however. A better scheme, the one which will be considered here, consists in transmitting a burst of 3-Mc carrier during horizontal sync pulse. This burst together with the sync pulse may be separated from the video in the usual way by clipping. The 3-Mc information may then be separated from the pulse by filtering or by gating. Alternatively, one could first gate out the 3-Mc burst and then filter. The clipped pulse and burst may be applied directly to the conventional pulse afc system without difficulty. However, the pulse must be removed in some way before the burst is applied to a sine-wave afc system.

Assuming that the 3-Mc burst has been in some way separated, it is necessary to consider what mechanism will be used to effect synchronization. Since the tolerances on the stability of the transmitted synchronizing wave will largely determine the mechanism to be used, these will be established. It will be assumed that all synchronizing frequencies in the transmitter are crystal-controlled so that it may be expected that variations in the 3-Mc frequency will not exceed ± 50 cps when switching from station to station. Furthermore, the drift in frequency of a given crystal will be so small as to be negligible. Also, the ratio between the dotting frequency and the line frequency is so chosen that there is no necessity to shift the phase of the dotting frequency at the end of each line or frame. Under these conditions three possible synchronizing devices appear possible: first, a simple high-Q filter (e.g., a crystal filter) which picks out the 3-Mc component, amplifies it, and applies it directly as the sampling wave; second, a locked sine-wave oscillator controlled in frequency by the 3-Mc burst; third, an afc system which controls the frequency of a sine-wave oscillator. Since any sync system must operate satisfactorily under bad fluctuation noise conditions, it must be realized that this requirement will ultimately determine the choice of system.

If, in order to achieve good noise protection, the simple filter device is narrowed down, the point will eventually be reached where the slope of the phase characteristic becomes so large that shifts in frequency such as may be encountered in switching from station to station will cause such an intolerably large static phase error that a manual frequency trimmer must be used. Furthermore, lock-in time on sudden phase transi-

* Decimal classification: R583.13. Original manuscript received by the Institute, May 19, 1950; revised manuscript received, October 9, 1950.

† Philco Corporation, Philadelphia, Pa.

[1] W. P. Boothroyd, "Dot systems of color television," *Electronics*, vol. 22, pp. 88–93; December, 1949.

Reprinted from *Proc. IRE*, vol. 38, pp. 124–131, Feb. 1951.

tions may be too long. This dilemma is common to any sync system in various degrees; that is, if the system is designed to ignore noise fluctuations it tends also to ignore variations from any source.

All three of the possible systems will track variations in phase and frequency with varying degrees of fidelity. The locked oscillator may have some advantage over the simple filter in that its effective bandwidth varies directly as the voltage of the applied synchronizing signal, although its chief merit appears to be in the fact that it is a Q multiplier.[2] Automatic-frequency-control sync would appear to be the most efficient of the three since it separates the functions of phase detection, filtering and amplifying, and generation of the correct synchronizing signal, thus permitting the parameters in each part of the system to be adjusted for optimum performance. The validity of these statements requires some proof and this will now be undertaken. The simple tuned circuit will be compared quantitatively with afc sync, since it is believed that locked oscillator performance lies somewhere between these two.

Since crystal frequency control is assumed used in the transmitter, the problem essentially reduces to the calculation of phase error introduced by random noise, and the static phase error caused by a shift of frequency when station switching. The delay in tracking a phase shift caused by station switching is not believed to be important.

The characteristics of each system which will keep the phase error due to random noise alone below 5° will now be determined, and then the respective static phase errors caused by transients of frequency will be found. This will be done for various signal-to-noise ratios in the intermediate frequency and will furnish a measure of the fidelity of the two systems. The average carrier power-to-noise ratio in the intermediate frequency will be used where the carrier is measured at the sync tips.

If the carrier burst occupies 4 μseconds out of 64 μseconds, when gated, clipped, and filtered, it will yield a sinusoid of amplitude $(C/128)\beta$ where C is the amplitude of the intermediate-frequency carrier during the sync pulse and β is the modulation suppression term derived in Appendix A.

The noise power out of a filter of total bandwidth 2γ will be $2\gamma F_0$ where F_0 is the amplitude of the noise-power spectrum at 3 Mc as derived in Appendix A and 2γ is the equivalent noise bandwidth of the filter. 2γ will turn out to be so small compared with the 4-Mc video bandwidth that the power spectrum may be considered flat across it. When noise is added to a sinusoid, the phase of the sinusoid becomes indeterminate to a degree. It is shown in Appendix C that in such a case, the rms value of the phase error due to noise alone is $\sqrt{\psi_0}/Q$ where ψ_0 is the total noise power added to the sinusoid, and Q is the amplitude of the sinusoid. Thus if it is desired that the rms value of the phase error due to noise alone be less than 5°,

$$\frac{\sqrt{2\gamma F_0}}{Q} < \frac{5}{57.3}.$$

Using the values of F_0 from Appendix A and considering the filter to be a single-tuned circuit it is found that the half bandwidth α must be less than 22, 85, and 135 cps, respectively, for intermediate-frequency signal-to-noise ratios of 1, 3, and 5.

To calculate the static phase error, the filter is again taken to be a single-tuned circuit and the maximum frequency deviation to be ± 50 cps. Then the slope of the phase characteristic at resonance is

$$\left.\frac{d\theta}{d\omega}\right|_{\omega=\omega_0} = \frac{1}{\alpha}$$

where α is the half bandwidth. Thus measuring frequency deviation from resonance, one has $d\theta = d\omega/\alpha$. This yields a phase shift of 2.3, 0.6, and 0.37 radians for intermediate-frequency signal-to-noise ratios of 1, 3, and 5. This amount of phase shift obviously requires an auxiliary manual phase control. If the allowable rms phase error is 10°, then the bandwidths become 88, 340, and 540 cps, respectively, and the static phase shift becomes 33°, 8.4°, and 5.3°. The picture probably becomes unusable for a carrier-to-noise ratio in the intermediate frequency of about 2, so that if these phase errors can be tolerated in the sampling frequency, then the receiver can be operated without auxiliary manual hold control. The transient caused by a sudden step of phase in the input to the filter will have the time constant of the filter. This will occur when switching stations, but it is not considered to be important.

To determine the characteristics of the afc synchronizing system, the configuration of Fig. 1 is assumed.

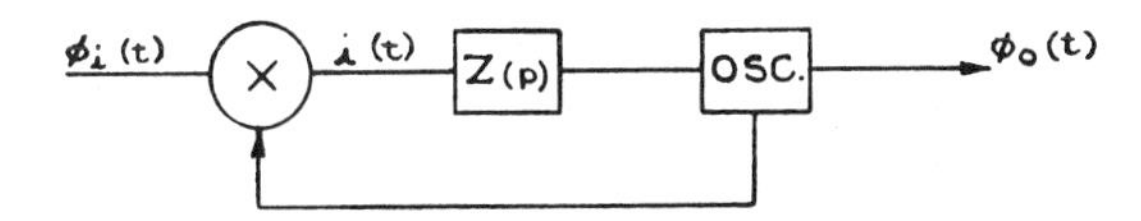

Fig. 1—Automatic-frequency-control servo loop.

The incoming sync information after detection and separation is fed into a phase comparator where its phase is compared with that of the free-running oscillator which determines the frequency of the receiver sync. Thus the incoming phase information is denoted by $\phi_i(t)$. Out of the phase comparator comes an error current $i(t)$ which passes through a filter $Z(p)$ before correcting the oscillator whose output is $\phi_0(t)$. Then out of the phase detector there is obtained

$$i(t) = K_1[\phi_i(t) - \phi_0(t)]$$

where K_1 is the sensitivity parameter of the phase comparator and is taken to be a constant. From the filter $Z(p)$ there is a voltage $e(t)$ which defined by

[2] Kurt Schlesinger, "Locked oscillator for television synchronization," *Electronics*, vol. 22, pp. 112–118; January, 1949.

$$e(p) = Z(p)i(p).$$

It is assumed that the voltage controlling the oscillator is held at a bias which yields the nominal sync frequency so that the voltage applied here results in deviation about that frequency. It is further assumed that the instantaneous deviation of oscillator frequency from the nominal value is directly proportional to $e(t)$.
Then

$$\frac{d\phi_0}{dt} = K_2 e$$

where K_2 is the sensitivity constant of the oscillator.

Since the phase comparator is a nonlinear device, K_1 is not an absolute constant but depends somewhat on the input signal and noise level. However, when measured or otherwise determined for a particular signal and noise, it will not vary much for reasonably small deviations from that value of signal and noise. Presumably, in designing an afc sync system one must first decide the lowest signal-to-noise ratio under which the system is to operate satisfactorily. The design is then optimized for that condition with the expectation that performance will be better when the noise level falls. K_2 will be essentially constant for small values of $e(t)$ for sine-wave oscillators controlled by reactance tubes or for blocking oscillators. Then, solving these equations in the frequency domain, it is seen that

$$\phi_0(p) = \frac{K_1K_2Z(p)}{p + K_1K_2Z(p)}\phi_i(p).$$

If $Z(p)$ is taken as a simple RC low-pass filter,

$$Z(p) = \frac{A_0}{p + \alpha}.$$

Then

$$\phi_0(p) = \frac{A_1}{p^2 + \alpha p + A_1}\phi_i(p)$$

where $A_1 = K_1K_2A_0$. Thus the over-all gain function of the system is given by

$$Z_0(p) = \frac{A_1}{p^2 + \alpha p + A_1}.$$

Now consider the effect on the system of introducing signals such as a step of phase or a step of frequency. These situations will exist when stations are switched. The output phase is given by a Laplace transform, thus

$$\phi_0(t) = \frac{A_1}{2\pi j}\int_{-j\infty}^{j\infty}\frac{g(p)e^{pt}dp}{p^2 + \alpha p + A_1}$$

where $g(p)$ is the spectrum of the input $\phi_i(t)$. In the case of a step of phase of magnitude I_0,

$$\phi_0(t) = \frac{A_1I_0}{2\pi j}\int_{-j\infty}^{j\infty}\frac{e^{pt}dp}{p[p^2 + \alpha p + A_1]},$$

where

$$p^2 + \alpha p + A_1 = (p - p_1)(p - p_2)$$

and

$$p_1 = \frac{-\alpha + \sqrt{\alpha^2 - 4A_1}}{2}, \qquad p_2 = \frac{-\alpha - \sqrt{\alpha^2 - 4A_1}}{2}.$$

These roots are always in the left half plane and may be complex or real depending on $\alpha^2 - 4A_1$ being less than or greater than zero. In the former case, there is no oscillatory approach to the final tracking position while the latter provides a hunting approach. Operation at or near the critically damped position when $4A_1 = \alpha^2$ is generally preferred. In case the system is under-damped, the output is

$$\phi_0(t) = I_0\left[1 - \frac{e^{-(\alpha/2)t}\left(\frac{\beta}{2}\cos\frac{\beta}{2}t + \frac{\alpha}{2}\sin\frac{\beta}{2}t\right)}{\frac{\beta}{2}}\right],$$

where $\beta = \sqrt{4A_1 - \alpha^2}$. This will reduce to the critically damped case when $\beta = 0$. $\phi_0(t)$ can be written as

$$\phi_0(t) = I_0\left[1 - \frac{2\sqrt{A_1}}{\beta}e^{-(\alpha/2)t}\cos\left(\frac{\beta}{2}t - \tan^{-1}\frac{\alpha}{\beta}\right)\right].$$

The output then is as shown in Fig. 2.

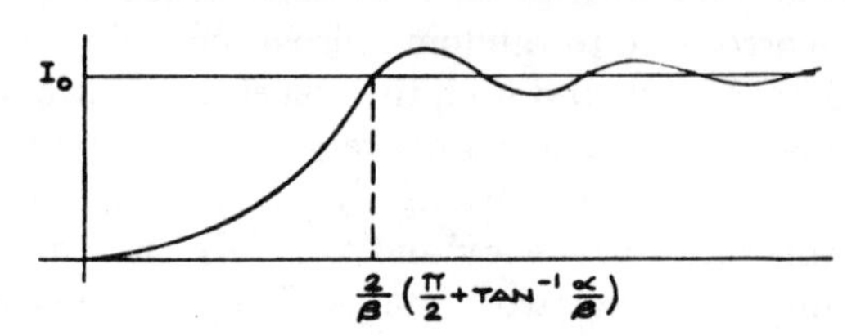

Fig. 2—Output of afc system to a step of phase.

The actual crossing of the line occurs at

$$t = \frac{2}{\beta}\left(\frac{\pi}{2} + \tan^{-1}\frac{\alpha}{\beta}\right)$$

so that if this crossing is taken as a measure of lock-in time, it appears that the lock-in time can be made small by making α large and β large. However, the overshoot becomes larger in such a case, so that a reasonable compromise must be made between fast lock-in time and initial overshoot. By minimizing the error $i(t)/K_1$ in the mean square sense, it can easily be shown to vary as $1/\alpha$ so that in this sense, optimum tracking is obtained by making α as large as possible. All of this is in the absence of noise. It may be further noted that the output actually tracks I_0 regardless of loop gain or filter bandwidth and that tracking time is independent of the magnitude of the phase error. To maintain an oscillatory approach, $4A_1 > \alpha^2$. If for example $4A_1$ is taken to be equal to $2\alpha^2$ and $4\alpha^2$, respectively, the first overshoot is 4 per cent and 16 per cent, which gives an idea of the magnitude of the oscillation. The magnitude of the first overshoot is $I_0(1+e^{-\pi\alpha/\beta})$. These oscillations have been observed experimentally.

Now suppose that the input signal is in correct phase but off in frequency by a fixed amount, due to oscillator drift or station switching. Then beginning at $t=0$, $\phi_i(t)=\omega_0 t$ where ω_0 is the frequency difference. Then $g(p)=\omega_0/p^2$ and

$$\phi_0(t) = \frac{2\omega_0}{\beta} e^{-(\alpha/2)t} \cos\left[\frac{\beta t}{2} - \tan^{-1}\left(\frac{\alpha^2 - 2A_1}{\alpha\beta}\right)\right] + \omega_0 t - \frac{\omega_0\alpha}{A_1}.$$

This last function is asymptotic to $\omega_0 t-(\omega_0\alpha/A_1)$ so that there is a static phase error $\omega_0\alpha/A_1$. This static phase error results in a static shift in the voltage $e(t)$ applied to the oscillator which is just enough to change its frequency by ω_0. One notes that this error is ω_0/RK_1K_2 and is thus independent of the capacity of the filter. The output is as shown in Fig. 3.

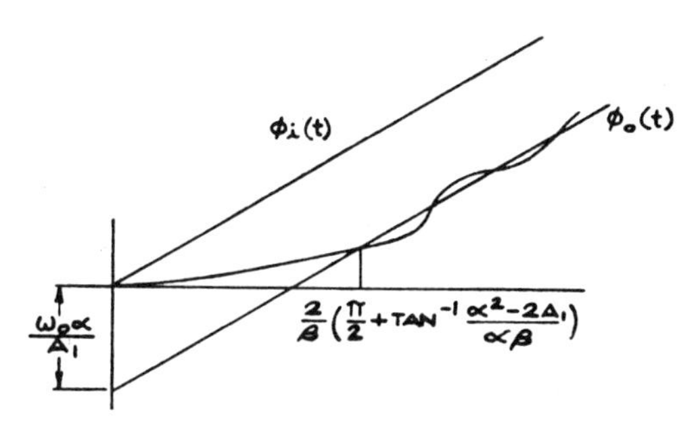

Fig. 3—Output of afc system to a step of frequency.

In general, in designing an afc system, it is necessary to consider the necessity of tracking relatively slow variations in the incoming signal in the presence of noise as well as the more severe transients such as steps of frequency and phase. Having decided upon a typical input signal to be tracked with optimum fidelity in the presence of a given noise level, an attempt would then be made to find the parameters which would minimize the error in some sense. This objective leads naturally to the theory of Wiener[3] in which an attempt is made to find the filter which will minimize the difference between input and output in the mean square sense. In principle, this procedure is straightforward. However, it will generally yield a filter characteristic which is not physically realizable and must therefore be approximated. It is therefore the practice of servomechanists to start with a servo of a given type having adjustable parameters. The mean square error can then be computed directly and the parameters adjusted to minimize it. For our purposes here the simple low-pass filter heretofore assumed is believed adequate. An additional degree of freedom with some improvement in performance can be obtained by starting off with a double time-constant filter. However, this complicates the calculations considerably and will not be gone into here.

The output of the servo as a convolution integral can be given as

$$\phi_0(t) = \int_0^\infty [\phi_i(t-\tau) + \phi_N(t-\tau)]K(\tau)d\tau$$

where ϕ_N is the noise assumed to enter the system with $\phi_i(t)$ and $K(\tau)$ is the characteristic transient of the servo system. The phase error is given by

$$\frac{i(t)}{K_1} = \phi_i(t) - \phi_0(t)$$

$$= \phi_i(t) - \int_0^\infty [\phi_i(t-\tau) + \phi_N(t-\tau)K(\tau)d\tau.$$

Assuming no error in the incoming signal, the phase error becomes

$$\frac{i(t)}{K_1} = -\int_0^\infty \phi_N(t-\tau)K(\tau)d\tau.$$

Thus using Plancherel's theorem, the mean-square phase error due to noise alone is

$$\frac{1}{4\pi}\int_{-\infty}^\infty F_1(\omega)|Z_0(\omega)|^2 d\omega$$

where $F_1(\omega)$ is the phase noise power spectrum and $Z_0(\omega)$ the over-all servo gain function. The above considerations are quite general and apply to any afc system of the type considered here. It may be assumed that the bandwidth of $Z_0(\omega)$ will be so small compared to 4 Mc that $F_1(\omega)$ can be considered flat across it. Therefore, the mean-squared phase error is

$$\frac{1}{4\pi}F_1\int_{-\infty}^\infty |Z_0(\omega)|^2 d\omega = \frac{F_1}{4\pi}\int_{-\infty}^\infty \frac{A_1{}^2\alpha\omega}{(A_1-\omega^2)+\alpha^2\omega^2}$$

$$= \frac{F_1A_1}{4\alpha}.$$

It may be observed here that in trying to minimize this error, the static phase error $\omega_0\alpha/A_1$ is made large so that a compromise must be struck between the two. Since this result will be evaluated for $4A_1=\alpha^2$, little error is caused here by taking the integral to ∞.

Now consider that the afc is applied to the 3-Mc carrier. First the burst is put through a narrow tuned circuit to pick out the 3-Mc sine wave. No phase information is lost by this maneuver. This filter must be wide enough so that an objectionable phase shift does not develop here. Then a sine wave plus noise is put into the phase detector. The power spectrum of the phase noise is, as shown in Appendix A,

$$F_1(\omega) = \frac{2F_0}{Q^2}, \qquad 0 < \omega < \gamma$$

[3] H. M. James, N. B. Nichols, and R. S. Phillips, "Theory of Servomechanisms," Radiation Laboratory Series, vol. 25, McGraw-Hill Book Co., New York, N. Y., chap. 7; 1947.

where Q is the amplitude of the sinusoid, 2γ the bandwidth of the prephase-comparator filter, and F_0 the power spectrum of the video noise at 3 Mc.

Now we wish to compare the simple filter with the afc by constraining them to have the same rms phase noise error and then comparing the ensuing static phase error. If the phase noise errors are equated, it is found that $\alpha_1 = A_1/\alpha_2$ where α_1 is the half bandwidth of the simple filter, and α_2 the half bandwidth of the filter in the afc loop. Then the static phase error for the afc case is $\omega_0\alpha_2/A_1 = \omega_0/\alpha_1$ which is identical with the result for the simple filter, independent of the relationship between A_1 and α_2. If a double time-constant filter is used in the afc loop, it can be shown that for equal static phase errors, the noise error in the double time-constant case can always be made less than that for the single time-constant filter. In such a case, there will then be improvement in performance over the case analyzed here at some reduction in lock-in time.

It will now be shown how the foregoing analysis may be applied to pulsed sync systems, although specific calculations will not be carried through, since such systems are not the primary interest here.

Consider the conventional black-and-white afc pulse sync system. Here the sync information is not provided continuously but intermittently so that there is a little difficulty in defining phase in a sense that may be compared with the continuous case. The position of incoming pulses is determined in a conventional balanced

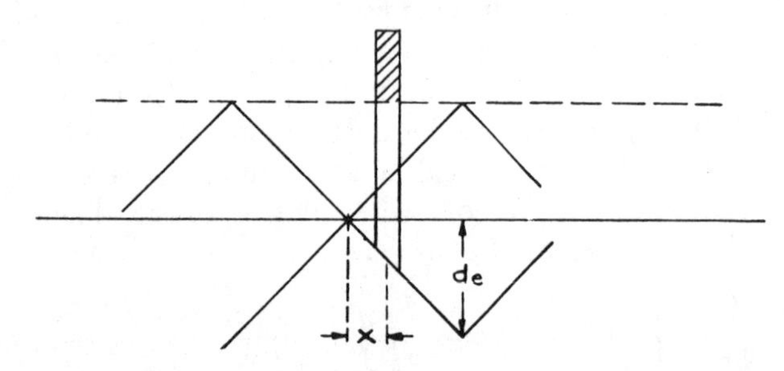

Fig. 4—Balanced pulse phase comparator.

phase detector[4] by feeding back part of the sawtooth output of the oscillator in the fashion shown in Fig. 4. In effect, the pulse is caused to ride on each of two sawteeth which differ only in polarity. In each case the pulse top is then clipped at a fixed (ideally) bias level and the two clipped pulses subtracted. The difference is the output of the phase detector. At the center position there is no output, hence the term balanced-phase detector. If the slope of the sawtooth is d, then the height of the output pulse is $2dx$ where x is the deviation of the pulse from its proper position. If the pulses are to the left of the center point, the polarity of the output (as well as any noise present) is reversed. Thus the output of this phase detector consists of pulses of current of varying amplitude and polarity. If, at time $t=0$, the frequency of the oscillator is correct but there exists a constant phase error, then the input to the servo consists of a sequence of pulses of the same height and polarity.

To make this type of input signal fit in with the previously derived equations describing the servomechanism, it is convenient to observe that the time constant of the over-all servo will be large compared to the time between pulses (of the order of 15 to 1) so that one may without appreciable error spread the area of the input pulse over the entire interval 0 to T. This provides a continuous input to the servo. The area of the error pulse is $2dxW$, where W is its width. The equivalent pulse after spreading then has a height $(2dxw/T)$. x is defined as the phase error. The quantity $2dW/T$ has the dimensions of amperes per unit phase error and is therefore defined as the sensitivity constant K_1 of the phase detector. Under these assumptions all of the preceding results on tracking, lock-in time, and so forth, are valid.

If the sync pulses are not gated, as is generally the case in black-and-white television, there is some difficulty in calculating the noise output of such a phase detector. To design for the worst condition, one would presumably assume the video at black level and then calculate the noise output of the device. This can be done only by numerical methods and will not be gone into here. In practice, the clipping levels of the two diodes in the balanced phase detector are allowed to set themselves. These levels are determined by the pulse signal level, the noise level, the pulse signal position, the diode internal resistance, and the diode plate load resistance. For a given input signal-to-noise ratio and with the input pulse at the neutral position, a solution can be found for the proper plate load resistor to position the clipping level where desired. For small errors in phase then the clipping level will not deviate much from the optimum value. The height of the sync pulse is $C/4$ where $(C/4) > 2de$ in all cases. Ideally, clipping should be done at $(C/4) - de$ since at this level full indication of phase error is obtained while the clipping level is as high as possible to avoid picking up noise. Presumably then, the clipping level is established for the minimum signal at which the device is expected to operate satisfactorily and suffers a design somewhat less than optimum at the higher signal-to-noise ratios. The plate load resistor thus determined affects the value of α in the filter following the phase detector but the capacity there can be varied within limits to adjust α for optimum value in terms of tracking, and so forth.

Appendix A

In the usual sync system, the sync tips are passed into the video and separated there by clipping. The top 25 per cent of the peak carrier may be devoted to hori-

[4] K. R. Wendt and G. L. Fredendall, "Automatic frequency and phase control of synchronization in television receivers," Proc. I.R.E., vol. 31, pp. 7–15; January, 1943.

zontal sync information for about 10 μseconds out of every line of 63.5 μseconds. The usable burst of carrier is probably 4 μseconds long and $C/4$ volts in amplitude where C is the intermediate-frequency carrier amplitude during sync pulse. It is assumed that this burst has been gated so that no noise appears in the sync system except during the burst itself. The noise in the intermediate frequency may be assumed to have a normal first probability density function of the form.

$$W_1(I) = \frac{1}{\sqrt{2\pi\psi_0}} e^{-I^2/2\psi_0}$$

where I is the instantaneous noise current and ψ_0 the average noise power. If the second detector is an envelope tracer, the density function of noise alone in the video is no longer normal but has a Rayleigh distribution

$$W_1(R) = \frac{R}{\psi_0} e^{\frac{-R^2}{2\psi_0}}$$

where R is the voltage of the envelope and ψ_0 the intermediate-frequency noise. If an unmodulated carrier C is added to the intermediate-frequency, the probability density function then becomes

$$W_1(R) = \frac{R}{\psi_0} e^{-(R^2+C^2)/2\psi_0} I_0\left(\frac{RC}{\psi_0}\right) \tag{1}$$

when I_0 is the modified Bessel function. This density function does not depend on the position of the carrier in the intermediate-frequency pass band. To a first approximation, the continuous part of the noise power spectrum $F(f)$ of a sine wave and noise in the video following an envelope is given by[5]

$$F(f) = \pi^2 h_{11}^2 [W(fq - f) + W(fq+f)] + \pi^2 \frac{h_{02}^2}{4} \int_{-\infty}^{\infty} W(x)W(f-x)\alpha x \tag{2}$$

where

$$h_{11} = \frac{1}{2}\left(\frac{y}{\pi}\right)^{1/2} {}_1F_1(1/2; 2; -y)$$

$$h_{02} = (2\pi\psi_0)^{-1/2} {}_1F_1(1/2; 1; -y)$$

$$y = \frac{C^2}{2\psi_0}.$$

${}_1F_1(\alpha; \beta, y)$ is the hypergeometric function. The first part of $F(f)$ is due to the modulation products of signal and noise while the second part is due to the intermodulation of the noise alone. In the case of vestigial side-band transmission, there is only half of the first part of this, namely, $W(fq+f)$ where now f extends over the entire intermediate-frequency and video bandwidth. If $W(f)$ is flat, $= W_0$, the noise power spectrum in the video is as shown in Fig. 5.
Here

$$A = (ch_{02})^2 \left[\frac{BW_0^2}{4C^2}\right] = \frac{W_0}{8} (ch_{02})^2 \frac{1}{\left(\frac{S}{N}\right)_{if}}.$$

The values of $(ch_{02})^2$ and h_{11}^2 are obtained from the curves in the literature.[6] As the signal becomes larger compared with the noise, the contribution due to noise intermodulation drops off until ultimately the noise rides on top of the signal and the spectrum in the video has the same shape as in the intermediate frequency. On the other hand, for zero signal, the spectrum becomes triangular.

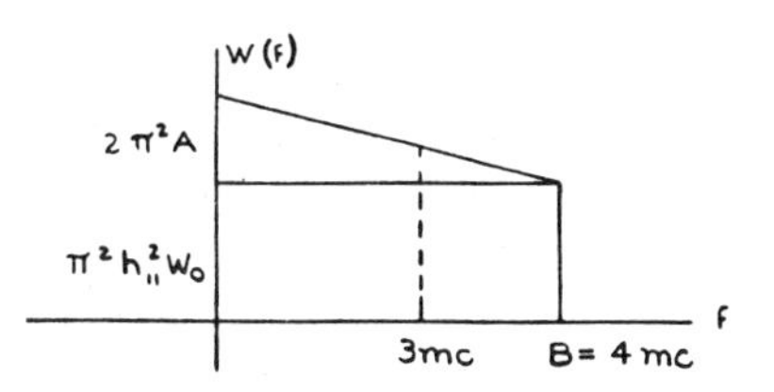

Fig. 5—Power spectrum of sine wave plus noise in the video.

The above shows the power spectrum of the noise alone. As is generally known, when the noise power becomes comparable to that of the signal, the second detector causes a suppression of modulation. Since the modulation index of the carrier burst is small (1,8), the preceding result concerning noise-power spectrum is not altered perceptibly by the addition of the modulation. However, it is necessary to know the extent of modulation suppression and it is convenient, since the modulation index is small, to use the procedure of Ragazzini.[7] The input to the detector can be written

$$e(t) = C \cos \omega_0 t + mC \cos (\omega_0 + \omega_s)t + \sum_0^{m_0} c_n \cos (\omega_n t + \phi_n)$$

where the last term encompasses all noise components with random phase angles ϕ_n. Then the envelope can be written as

$$E(t) \cong [C^2 + m^2C^2 + 2\psi_0 + 2mC^2 \cos \omega_s t]^{1/2}$$

where ψ_0 is the entire intermediate-frequency noise power. This is approximately

$$E(t) = \sqrt{C^2 + 2\psi_0}\left(1 + \frac{mC^2 \cos \omega_s t}{C^2 + 2\psi_0}\right).$$

[5] S. O. Rice, "Mathematical analysis of random noise," *Bell Sys. Tech. Jour.*, vol. 24, pp. 46–157; January, 1945. Rice's constant $1/\pi^2$ has been eliminated here.

[6] See page 148 of footnote reference 5.

[7] J. R. Ragazzini, "The effect of fluctuation voltages on the linear detector," Proc. I.R.E., vol. 30, pp. 277–287; June, 1942.

Thus the detected signal has the amplitude

$$\frac{mC^2}{\sqrt{C^2+2\psi_0}} = \frac{mC}{\sqrt{1+\frac{2\psi_0}{C^2}}}$$

and the modulation suppression is

$$\frac{1}{\sqrt{1+\frac{1}{\left(\frac{S}{N}\right)_{if}}}}.$$

For $S/N_{if}=1$, 3, and 5, the suppression factor becomes 0.71, 0.87, and 0.91.

Now, if in the video, the burst and associated noise is simply gated, it is possible, as is shown in appendix B, to find the signal-to-noise ratio of the 3-Mc sine wave and noise as it leaves the filter. However, some improvement can be made by clipping as well as gating. As shown in Appendix B, the noise spectrum at 3 Mc is, to a first approximation, reduced by the gating factor. The noise may be further reduced by clipping at $3/4C$ since this does not reduce the signal but does reduce the noise. Numerical integration of the second density function (1) shows that the total dc suppressed noise power is reduced from $0.65\psi_0$ to $0.50\psi_0$, $0.83\psi_0$ to $0.52\psi_0$, and $0.93\psi_0$ to $0.61\psi_0$, respectively, for input signal-to-noise ratios of 1, 3, and 5. Thus assuming that the spectral distribution does not change materially in shape, the power spectrum level at 3 Mc can be found.

Utilizing the results of (2), this is to be found to be $0.027W_0$, $0.032W_0$, and $0.037W_0$, respectively. Since ψ_0 the total intermediate-frequency noise $=BW_0$ this may be written

$$\frac{0.027\psi_0}{B}, \quad \frac{0.032\psi_0}{B}, \quad \text{and} \quad \frac{0.037\psi_0}{B}.$$

Appendix B

If we have an ideal gate which opens and closes at intervals T, it may be represented as the time function shown in Fig. 6. To gate noise this time function is

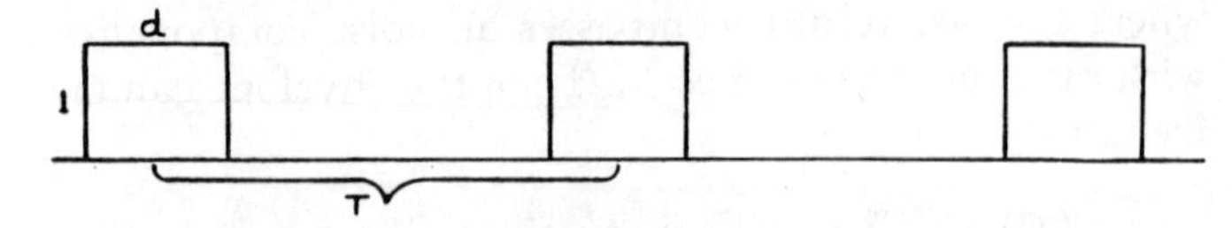

Fig. 6—Gating time function.

multiplied by the noise-voltage time function. It is desired then to find the power spectrum of the product. The autocorrelation of the product is the product of the separate autocorrelations since the two time functions are independent. To each autocorrelation function corresponds, by Fourier transform (Wiener's theorem), the power spectrum of the original time function, Thus,

$$G(\omega) = 4\int_0^\infty R(\tau)\cos\omega\tau d\tau$$

$$R(\tau) = \frac{1}{2\pi}\int_0^\infty G(\omega)\cos\omega\tau dw$$

where $G(\omega)$ is the power spectrum and $R(\tau)$ the unnormalized autocorrelation function. The spectrum of the product of two autocorrelation functions (time functions in the power domain) is given by the complex convolution integral. Thus the power spectrum of the two original time functions is given by

$$G(\omega) = \frac{1}{4\pi}\int_{-\infty}^\infty F_1(\omega-s)F_2(s)ds$$

where $F_1(S)$ is the power spectrum of the noise, say, and $F_2(S)$ is the power spectrum of the gating pulses. For the simple gate assumed, the result is given by

$$F(\omega) = \frac{d^2}{T^2}\sum_{-\infty}^{\infty}\left(\frac{\sin\frac{n\pi d}{T}}{\frac{n\pi d}{T}}\right)^2 G_1\left(\omega-\frac{2\pi n}{T}\right).$$

This sum may be approximated by an integral

$$F(\omega) \cong \frac{d^2}{T^2}\int_{-\infty}^{\infty}\left(\frac{\sin\frac{n\pi d}{T}}{\frac{n\pi d}{T}}\right)^2 G_1\left(\omega-\frac{2\pi n}{T}\right)dn$$

which is valid when the gating frequency is small compared to the bandwidth of the noise, so that successive G's overlap substantially.

If $G_1(\omega)$ is constant and $=G_{11}$

$$F_1(\omega) = \frac{G_{11}d^2}{T^2}\left[1+2\sum_1^\infty\left(\frac{\sin\frac{n\pi d}{T}}{\frac{n\pi d}{T}}\right)^2\right] = \frac{G_{11}d}{T}$$

showing that the noise spectrum and the total noise power are each reduced by the gating factor d/T. By the same argument one may show that for any spectrum $G_1(\omega)$, the total noise power is reduced by d/T.

It may be of some interest to observe here that when gates are used in the sense of samplers of a video signal, the signal-to-noise ratio after gating does not depend on the gating factor d/T in any way, provided that the noise bandwidth is limited to ω_0 prior to sampling where ω_0 is the highest video frequency sampled. This is apparently not generally realized but may be easily shown with the aid of the above formulas.

In the application used here, the gating frequency is small compared with a bandwidth of 4 Mc so that the integral approximation is valid. Under these conditions, it is not hard to show that to a first approximation the spectrum at 3 Mc is also reduced by d/T when the total power is reduced by that amount.

Appendix C

In the case of normally distributed noise in a range of frequencies that is small compared to the center frequency, the noise current may be written after the method of Rice[8] as

$$I_n = \sum_1^N C_n \cos(\omega_n t + \phi_n)$$
$$= \sum_1^N C_n \cos[(\omega_n - q)t + \phi_n + qt]$$
$$= I_c \cos qt - I_s \sin qt$$

where

$$I_c = \sum_1^N C_n \cos[(\omega_n - q)t + \phi_n]$$
$$I_c = \sum_1^N C_n \sin[(\omega_n - q)t + \phi n].$$

Here q is the center frequency of the band,

$$\omega_n = 2\pi f_n, \quad f_n = n\Delta f, \quad C_n{}^2 = 2F(f_n)\Delta f.$$

$F(f)$ is the noise power spectrum. I_c and I_s are normally distributed and each have the rms value $\sqrt{\psi_0}$. If a sinusoid $Q \cos qt$ is added to the noise,

$$I = Q \cos qt + I_n$$
$$= (Q + I_c) \cos at - I_s \sin qt$$
$$= R \cos(qt + \theta)$$

where R is the envelope and θ the phase. Then

$$\theta = \tan - 1 \frac{I_s}{I_c + Q} \cong \frac{I_s}{Q} \quad \text{if} \quad Q \gg \sqrt{\psi_0}$$

and under these conditions the density function for θ may immediately be written as

$$W_1(\theta) = \frac{Q}{\sqrt{2\pi\psi_0}} e^{\frac{-\theta^2 Q^2}{2\psi_0}}$$

Then the rms phase deviation from that of the sinusoid is $\sqrt{\psi_0}/Q$ radians. From the representation of noise in the Fourier series, it is evident that I_s behaves like a noise current whose power spectrum is concentrated in the power part of the power spectrum and is, in fact,

$$F_1(\omega) = F(\omega_q + \omega) + F(w_q - \omega)$$

where $F(\omega)$ is the power spectrum of I_n. Thus if

$$F(\omega) = \begin{cases} F_0 = \dfrac{\psi_0}{B} & \text{for} \quad \omega_q - \dfrac{\beta}{2} < \omega < \omega_q + \dfrac{\beta}{2} \\ 0 & \text{elsewhere,} \end{cases}$$

then

$$F_1(\omega) = \begin{cases} \dfrac{2F_0}{Q^2}, & 0 < \omega < \dfrac{\beta}{2} \\ 0 & \text{elsewhere.} \end{cases}$$

The noise in the video is not normally distributed. However, if a narrow filter at 3 Mc is inserted, randomness is restored and the noise out of the filter can be considered normal, so that the preceding argument can be applied. The filter may be assumed so narrow that the noise spectrum picked out by it will be flat. However, the level of the noise power at 3 Mc must be determined.

Using the results of Appendix B, it can be shown without difficulty that the condition $Q \gg \sqrt{\psi_0}$ is satisfied for the three values of S/N_{if} assumed. A filter 2 kc wide will quite satisfactorily pick out the 3-Mc carrier without the adjacent sidebands which are 15 kc away. This filter width does not give rise to any appreciable static phase error in itself due to changes in frequency and provides a Q suitably larger than $\sqrt{\psi_0}$.

It is of interest to note here that the phase noise-power spectrum has the dimensions of time so that when integrated over ω, it yields phase in radians. The phase correlation may also be easily calculated from the phase power spectrum, being given by Wiener's theorem as

$$R(\tau) = \frac{1}{2\pi} \int_0^\infty F(\omega) \cos \omega\tau dw.$$

In the case of a uniform spectrum of width γ this yields

$$R(\tau) = \frac{F_0\gamma}{4\pi} \frac{\sin \frac{\gamma\tau}{2}}{\frac{\gamma\tau}{2}}$$

showing that phase correlation due to noise does not become small until

$$\tau = \frac{2\pi}{\gamma}.$$

[8] S. O. Rice, "Statistical properties of a sine wave plus random noise," *Bell Sys. Tech. Jour.*, vol. 27, pp. 109–158; January, 1948.

Application of the Phase-Locked Loop to Telemetry as a Discriminator or Tracking Filter*

C. E. GILCHRIEST†

Summary—**This paper studies a system of FM detection suggested by work done in optimum-demodulation theory, and a summary of the problems of ordinary discriminators is made so that possible improvement may be demonstrated readily. This summary points out an important defect of FM discriminators which affects the data accuracy of the telemeter but which is rarely recognized. The system study deals with generalized optimum filtering for transient signals applicable to commutated signals and operators such as discrimination. The solution for particular applications with respect to the phase-locked loop or optimum-demodulation system is offered. With these applications available, noise and transient signals are studied to see how the system fits the practical system. Normalized charts are made which may be applied to any standard telemetering channel that may be designed. A comparison of this system with ordinary discriminators shows a large threshold improvement and a reduction in over-all system error. Graphs of the comparison of these effects are also included. Heuristic suggestions of application are offered for further study.**

I. Introduction

The increasing numbers of complex measurements at improved accuracies required on today's aircraft and guided missiles have forced almost continuous improvements in the state of the telemetering art. Theoretical investigation and evaluation of radio communication techniques by Nichols and Rauch[1] and others stemmed from the necessity of transmitting more information with increased accuracy over narrower communication bandwidths. These investigations have primarily explored and evaluated various forms of pulse and continuous modulation, both frequency and amplitude, of a radio-carrier frequency (or subcarrier). In addition to obtaining meaningful theoretical comparisons between various possible communication systems, these studies have paved the way for a more sophisticated approach to the problem of reducing effective noise power transmitted over a communication channel, and by this reduction improving the maximum data accuracy at reduced transmitted carrier power. This paper attempts to present the theory together with the predicted and experimental performance of such a sophisticated system, as applied to the problem of improving telemetering data, and to compare this new technique to that presently used in telemetering.

II. Optimum Demodulation

The system which is under consideration for system improvement here is the "phase-locked loop." The phase-locked loop (pll) circuit, or automatic phase control (apc), as it is sometimes referred to, is, in effect, one form of correlation detection. Although it is not a new circuit, as pointed out by Lehan,[2] it apparently has not found widely accepted use for the improvements suggested here. Correlation detection is also not new; however, work in this field, reported on by Lehan and Parks,[3] provided the basis for the work being reported here. Their approach was that of "best-fit" for a curve of known character whose errors are random, a problem familiar to data users. The conclusion was that a signal of the nature $e_1(t)$ is best fitted with a signal $e_2(t; a(t))$ under the conditions

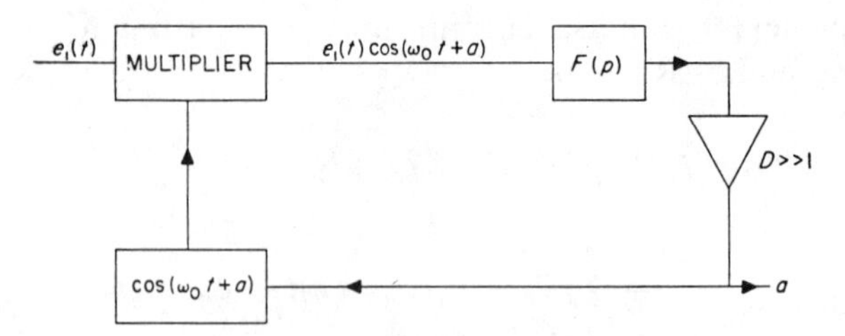

Fig.. 1—Phase-locked loop-implicit solution to (3).

$$F_c(p)\left\{[e_1(t) - e_2(t; a(t))]\frac{\partial e_2(t; a(t))}{\partial a}\right\} = 0. \qquad (1)$$

Rather than burden the reader with any of the details of the solution, he is referred to the original work, while the results for our particular application are simply stated here. For a signal of the FM nature,

$$e_2(t; a(t)) = \sin(\omega_0 t + a), \qquad (2)$$

Lehan and Parks state that

$$F(p)[e_1(t)\cos(\omega_0 t + a)] \cong \frac{a}{D}, \qquad (3)$$

which cannot be solved explicitly for a, but may be solved implicitly by the block diagram of Fig. 1.

Assuming an optimum detection system has been obtained, what then remains to be done? First, it should be shown that a problem exists in the equipment in present use and that this new detection scheme alleviates the situation. Lehan and Parks[3] and Lehan[2] state that threshold improvement is in the offing with the phase-locked loop. Jaffe and Rechtin[4] have further analyzed and

* Manuscript received by the PGTRC, January 3, 1958. This paper presents one phase of research carried out at the Jet Propulsion Lab., California Inst. Tech., under Contract No. DA-04-495-Ord 18, sponsored by the Dept. of the Army, Ordnance Corps.

† Jet Propulsion Lab., California Inst. Tech., Pasadena, Calif.

[1] M. H. Nichols and L. L. Rauch, "Radio Telemetry," John Wiley and Sons, Inc., New York, N. Y., 2nd ed.; 1956.

[2] F. W. Lehan, "Telemetering and information theory," IRE Trans. on Telemetry and Remote Control, vol. TRC-2, pp. 15–19; November, 1954.

[3] F. W. Lehan and R. J. Parks, "Optimum Demodulation," Jet Propulsion Lab., California Inst. Tech., Pasadena, Calif., External Publication 164.

[4] R. Jaffe and E. Rechtin, "Design and performance of phase-locked circuits capable of near optimum performance over a wide range of input signal and noise levels," IRE Trans. on Information Theory, vol. IT-1, pp. 66–76; March, 1955.

Reprinted from *IRE Trans. Telemetry Remote Contr.*, vol. TRC-4, pp. 20–35, June 1958.

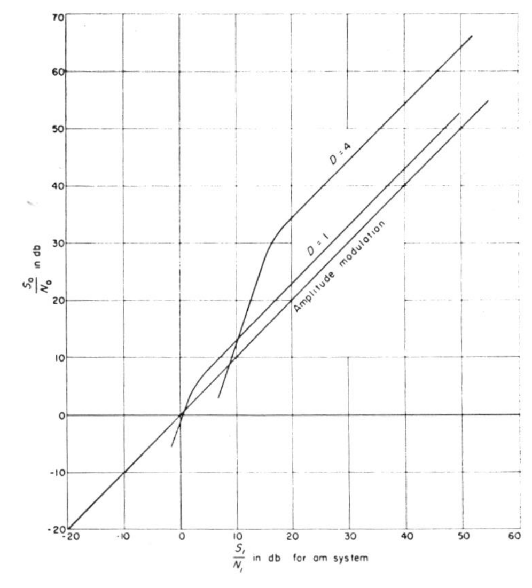

Fig. 2—Signal-to-noise ratio for FM discriminators.

improved the system by optimizing the filtering action of the phase-locked loop. Although greatly advanced, this work was incomplete as applied to some of the required engineering aspects of a telemetry system. Jaffe and Rechtin's work shall be extended and generalized as much as possible, and applied specifically to the telemetry problem.

FM-FM telemetering is perhaps the most widely used of all radio telemetering today, although the present trend is toward a pulse-coded sampled system. There are many modulation schemes, each with virtues unique to themselves, and the trend toward coded modulation is perhaps well-founded. FM-FM systems have suffered under the critical eye of analysis in competition with other schemes.

In discussing some of the limitations of the FM-FM system, it will be shown that one of the reasons for other schemes receiving attention is no longer valid by virtue of the phase-locked loop, and, therefore, conclusions based on these reasons are not valid.

III. Usual Assumed Capabilities of FM

A. Threshold

The threshold of an FM system, relates Goldman,[5] is attributed to the inclusion of the limiter in the system. Its presence seems desirable in order to maintain calibration during fading and to eliminate AM noise and impulse noise.

Commonly used discriminators demonstrate an improvement over the AM system until the signal-to-total-noise ratio reaches a low of about +10 db. This is demonstrated in Fig. 2. This high threshold has caused the FM-FM system to be criticized for use in telemetry because of the large powers required to maintain this signal-to-noise ratio.

[5] S. Goldman, "Certain Aspects of Coherence, Modulation and Selectivity in Information Transmission Systems," AF Cambridge Res. Center, Contract Co. AF 30(635)-2862.

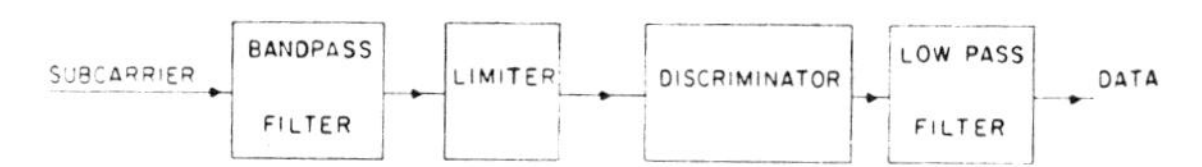

Fig. 3—Block diagram of pulse-counting discriminator.

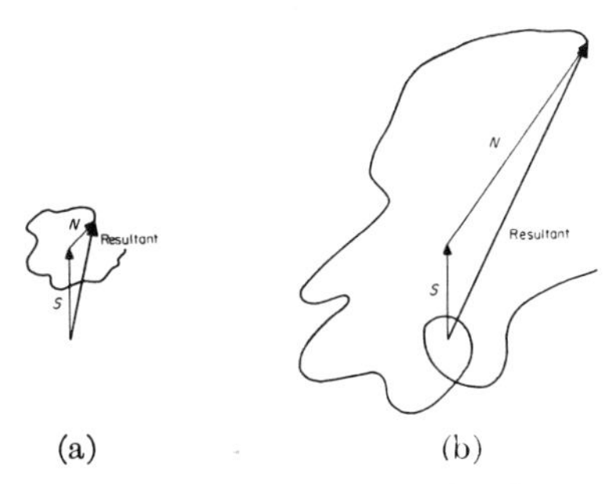

Fig. 4—Signal and noise vector diagram. (a) Low-noise condition. (b) High-noise condition.

B. Signal or Data Suppression

There is an additional degradation in performance of commonly used discriminators which effects data transmission but which is of little concern in ordinary voice communication. This is the suppression of the output signal along with the reduction of the signal-to-noise ratio. This can be explained simply for pulse-counting discriminators, whose block diagram is shown in Fig. 3, by considering the vector diagram of Fig. 4. When the noise level is low, the resultant is essentially that of the signal, and the average zero crossings will be essentially that of the signal. When the noise is high, however, the resultant will be essentially that of the noise, and the average zero crossings will not be influenced greatly by the signal. As anyone with an EPUT meter can verify, a noise-excited band-pass filter will ring about the center-band frequency, and the average count will be this frequency. To cite an example which expresses how this affects telemetry, say that channel 18 is used and data are being sent at the band edge of 75 kc. As noise power is added, the average data-output level will shift more and more toward the center band at 70 kc and is therefore in error. An equivalent statement may be made about a signal of 65 kc. When the data being sent are equivalent to 70 kc, there is no frequency shift effect; an error, which is related to the signal-to-noise ratio and proportional to the actual data, is experienced. This is not particularly noticeable in voice communications but it has a tremendous impact on telemetry instrumentation where accurate data recording is extremely important. This same effect might be explained less aptly for other discriminators by the power constraint of the limiter expressed rather simply by

$$L^2 = s_0^2 + N_0^2 \tag{4}$$

where L is a constant. Thus as noise is increased, the signal level is reduced. This causes a change in the transfer characteristics of the discriminator that follows as shown in Fig. 5, where it is obvious that noise causes a decrease of the signal level. To iterate, this effect is not removed by filtering in the data part of the system. Thus, it is seen that the most important function of the limiter, keeping the calibration constant under all signal levels, can only be approximately realized at high signal-to-noise level conditions.

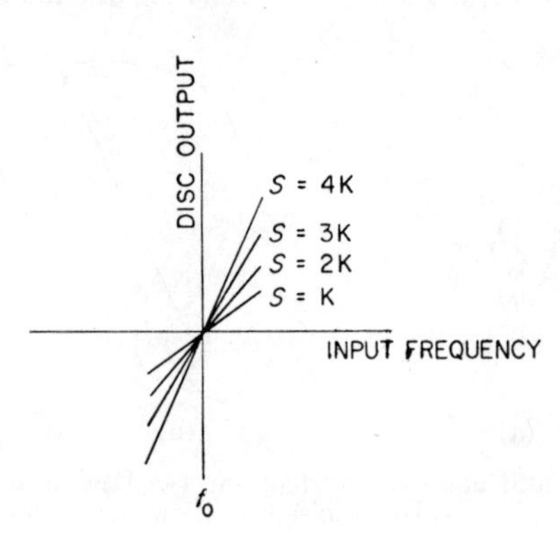

Fig. 5.

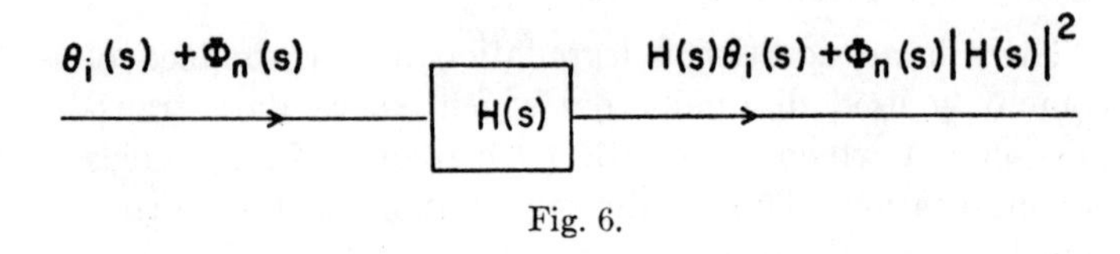

Fig. 6.

IV. Generalized Optimum Filtering of Transient Type Signals and Noise

This section shows the derivation of the optimum filter (see Fig. 6) *for transient signals* and flat noise which will be applicable to most problems in telemetry including ordinary discriminators and phase-locked loops. It extends the work done by Jaffe and Rechtin. Transient signals such as step functions are numerous in telemetry (such as commutated and event data), and it is justifiable to base optimization on them.

The primary design criteria follow closely those of Jaffe and Rechtin. They are: 1) the noise of the data due to noise interference should be minimized, and 2) the transient error between the output and the desired operation on the input for specific inputs should be maintained at a specific amount. Transient error is defined as the infinite time interval of

$$E_T = \int_0^\infty |\text{ input} \times \text{desired operation} - \text{output }|^2\, dt \tag{5}$$

or

$$E_T = \int_0^\infty \left| \int_0^\infty \theta_i(t-\tau)h_1(\tau)\, d\tau - \int_0^\infty \theta_i(t-\tau)h(\tau)\, d\tau \right|^2 dt$$

$$E_T = \int_0^\infty \left| \int_0^\infty \theta_i(t-\tau)[h_1(\tau) - h(\tau)]\, d\tau \right|^2 dt \tag{6}$$

where $h_1(\tau)$ is the weighting function of the desired operation and $h(\tau)$ is the weighting function of the filter or

$$E_T^2 = \frac{1}{2\pi j}\int_{-j\infty}^{j\infty} \left| \theta_i(s)(H_1(s) - H(s)) \right|^2 ds \tag{7}$$

where $H_1(s)$ is the transfer function of the desired operation and $H(s)$ is the transfer function of the filter. Noise output is

$$\sigma_N^2 = \frac{1}{2\pi j}\int_{-j\infty}^{j\infty} | H(s) |^2\, \Phi_N(s)\, ds \tag{8}$$

where $\Phi_N(s)$ is the noise spectral density. The design criteria may be stated symbolically as follows

$$\sigma_N^2 + \lambda^2 E_T^2 = \Sigma^2 = \text{minimum} \tag{9}$$

where λ is a Lagrangian multiplier. Substituting (7) and (8) into (9)

$$\Sigma^2 = \frac{1}{2\pi j}\int_{-j\infty}^{j\infty} [| H(s) |^2\, \Phi_N(s) + \lambda^2 \, | H_1(s) - H(s) |^2 \, | \theta_i(s) |^2]\, ds. \tag{10}$$

Noting that

$$| H(s) |^2 = H(s)H(-s) = H_+H_-$$

and

$$| H_1(s) - H(s) |^2 = (H_1(s) - H(s))(H_1(-s) - H(-s)) = (H_{1+} - H_+)(H_{1-} - H_-),$$

(10) can be written as

$$\Sigma^2 = \frac{1}{2\pi j}\int_{-j\infty}^{j\infty} [H_+H_-\Phi_N(s) + \lambda^2(H_{1+} - H_+)(H_{1-} - H_-) \, | \theta_i(s) |^2]\, ds.$$

Using the standard calculus of variation methods by letting

$$H_+ = H_+ + \epsilon\eta_+ \quad \text{and} \quad H_- = H_- + \epsilon\eta_-$$

$$\Sigma^2 = \frac{1}{2\pi j}\int_{-j\infty}^{j\infty} [(H_+ + \epsilon\eta_+)(H_- + \epsilon\eta_-)\Phi_N(s) + \lambda^2(H_{1+} - H_+ - \epsilon\eta_+) \cdot (H_{1-} - H_- - \epsilon\eta_-) \, | \theta_i(s) |^2]\, ds \tag{11}$$

$H(s)$ now represents the minimizing function. Taking the derivative with respect to the parameter ϵ

$$\frac{d\Sigma^2}{d\epsilon} = \frac{1}{2\pi j}\int_{-j\infty}^{j\infty} \{[(H_+ + \epsilon\eta_+)\eta_- + (H_- + \epsilon\eta_-)\eta_+]\Phi_N(s) + \lambda^2[(H_{1+} - H_+ - \epsilon\eta_+) \cdot(-\eta_-) + (H_{1-} - H_- - \epsilon\eta_-)(-\eta_+)] \, | \theta_i(s) |^2\}\, ds. \tag{12}$$

Evaluating this result at $\epsilon = 0$

$$\left.\frac{d\Sigma^2}{d\epsilon}\right|_{\epsilon=0} = \frac{1}{2\pi j}\int_{-j\infty}^{j\infty} \{[H_+\eta_- + H_-\eta_+]\Phi_N(s) - \lambda^2[(H_{1+} - H_-)\eta_- + (H_{1-} - H_-)\eta_+] \, | \theta_i(s) |^2\}\, ds. \tag{13}$$

Rearranging terms

$$\left.\frac{d\Sigma^2}{d\epsilon}\right|_{\epsilon=0} = \frac{1}{2\pi j}\int_{-j\infty}^{j\infty} \{\eta_+[H_-\Phi_N(s) - \lambda^2(H_{1-} - H_-) \mid \theta_i(s) \mid^2] + \eta_-[H_+\Phi_N(s) - \lambda^2(H_{1+} - H_+) \mid \theta_i(s) \mid^2]\}\, ds$$

or

$$\left.\frac{d\Sigma^2}{d\epsilon}\right|_{\epsilon=0} = \frac{1}{2\pi j}\int_{-j\infty}^{j\infty} \{\eta_+[H_-(\Phi_N(s) + \lambda^2 \mid \theta_i(s) \mid^2) - \lambda^2 H_{1-} \mid \theta_i(s) \mid^2] + \eta_-[H_+(\Phi_N(s) + \lambda^2 \mid \theta_i(s) \mid^2) - \lambda^2 H_{1+} \mid \theta_i(s) \mid^2]\}\, ds.$$

Now let

$$\Phi_N(s) + \lambda^2 \mid \theta_i(s) \mid^2 = \mid \psi(s) \mid^2 = \psi(s)\psi(-s) = \psi_+\psi_- \tag{14}$$

and substituting, the following is obtained

$$\left.\frac{d\Sigma^2}{d\epsilon}\right|_{\epsilon=0} = \frac{1}{2\pi j}\int_{-j\infty}^{j\infty} \{\eta_+[H_-\psi^+\psi^- - \lambda^2 H_{1-} \mid \theta_i(s) \mid^2] + \eta_-[H_+\psi_+\psi_- - \lambda^2 H_{1+} \mid \theta_i(s) \mid^2]\}\, ds. \tag{15}$$

Factoring out appropriate terms

$$\left.\frac{d\Sigma^2}{d\epsilon}\right|_{\epsilon=0} = \frac{1}{2\pi j}\int_{-j\infty}^{j\infty} \left\{\eta_+\psi_+\left[H_-\psi_- - \frac{\lambda^2 H_{1-} \mid \theta_i(s) \mid^2}{\psi_+}\right] + \eta_-\psi_-\left[H_+\psi_+ - \frac{\lambda^2 H_{1+} \mid \theta_i(s) \mid^2}{\psi_-}\right]\right\} ds. \tag{16}$$

For convenience, factor the terms

$$\frac{H_{1-} \mid \theta_i(s) \mid^2}{\psi_+} \quad \text{and} \quad \frac{H_{1+} \mid \theta_i(s) \mid^2}{\psi_-}$$

in the following way

$$\frac{H_{1-} \mid \theta_i(s) \mid^2}{\psi_+} = \left[\frac{H_{1-} \mid \theta_i(s) \mid^2}{\psi_+}\right]_+ + \left[\frac{H_{1-} \mid \theta_i(s) \mid^2}{\psi_+}\right]_- \tag{17}$$

and

$$\frac{H_{1+} \mid \theta_i(s) \mid^2}{\psi_-} = \left[\frac{H_{1+} \mid \theta_i(s) \mid^2}{\psi_-}\right]_+ + \left[\frac{H_{1+} \mid \theta_i(s) \mid^2}{\psi_-}\right]_-. \tag{18}$$

Substituting (17) and (18) into (16) results in

$$\left.\frac{d\Sigma^2}{d\epsilon}\right|_{\epsilon=0} = \frac{1}{2\pi j}\int_{-j\infty}^{j\infty} \left\{\eta_+\psi_+\left[H_-\psi_- - \lambda^2\left[\frac{H_{1-} \mid \theta_i(s) \mid^2}{\psi_+}\right]_+ - \lambda^2\left[\frac{H_{1-} \mid \theta_i(s) \mid^2}{\psi_+}\right]_-\right] + \eta_-\psi_-\left[H_+\psi_+ - \lambda^2\left[\frac{H_{1+} \mid \theta_i(s) \mid^2}{\psi_-}\right]_+ - \lambda^2\left[\frac{H_{1+} \mid \theta_i(s) \mid^2}{\psi_-}\right]_-\right]\right\} ds. \tag{19}$$

If it is noted that

$$\int_{-j\infty}^{j\infty} Z(s)\omega(s)\, ds = 0$$

if $Z(s)$ and $\omega(s)$ are algebraic polynomials having poles only in the same half of the S plane, the above result may be simplified.

$$\left.\frac{d\Sigma^2}{d\epsilon}\right|_{\epsilon=0} = \frac{1}{2\pi j}\int_{-j\infty}^{j\infty} \left\{\eta_+\psi_+\left[H_-\psi_- - \lambda^2\left[\frac{H_{1-} \mid \theta_i(s) \mid^2}{\psi_+}\right]_-\right] + \eta_-\psi_-\left[H_+\psi_+ - \lambda^2\left[\frac{H_{1+} \mid \theta_i(s) \mid^2}{\psi_-}\right]_+\right]\right\} ds \tag{20}$$

$$\left.\frac{d\Sigma^2}{d\epsilon}\right|_{\epsilon=0}$$

will be a minimum if the terms in the brackets reached zero simultaneously or

$$H_-\psi_- - \lambda^2\left[\frac{H_{1-} \mid \theta_i(s) \mid^2}{\psi_+}\right]_- = 0 \tag{21}$$

and

$$H_+\psi_+ - \lambda^2 \left[\frac{H_{1+} \mid \theta_i(s) \mid^2}{\psi_-}\right]_+ = 0 \tag{22}$$

which is true since (21) is the conjugate of (22). The result may be stated simply as

$$H_+ = \frac{\lambda^2}{\psi_+}\left[\frac{H_{1+} \mid \theta_i(s) \mid^2}{\psi_-}\right]_+ \tag{23}$$

or

$$H(s) = \frac{\lambda^2}{\psi_+}\left[\frac{H_1(s) \mid \theta_i(s) \mid^2}{\psi_-}\right]_{\text{Physically Realizable}}. \tag{24}$$

Eq. (24) is the generalized Wiener filter that is necessary for signals such as outlined in the criteria. The next sections show its use in specific examples as applied to FM signals.

V. Solution for the Wiener Optimum Filter for an FM Demodulated Signal

As shown in the previous section, the generalized filter is

$$H(s) = \frac{\lambda^2}{\psi_+}\left[\frac{H_1(s) \mid \theta_i(s) \mid^2}{\psi_-}\right]_{\text{PR}}.$$

In applying this to the filter for the demodulated signal, the new filter is designated as $I(s)$. (See Figs. 7 and 8). The discriminator takes the derivative of the phase input; therefore, the desired operation on the phase is $H_1(s) = S$. The frequency input is $1/S$ while the phase is the integral of this or $1/S^2$. A band-pass filter will be assumed to modify this by $1/(1 + \tau s)$. This is not represented as being the true effect but merely an approximation. If any higher order is assumed, it unnecessarily complicates the solution. It will be shown that an approximation such as this allows improvement which is more realistic.

Therefore

$$I(s) = \frac{\lambda_f^2}{\psi^+}\left[\frac{H_1(s) \mid \theta_i(s) \mid^2}{\psi_-}\right]_{\text{PR}}. \tag{25}$$

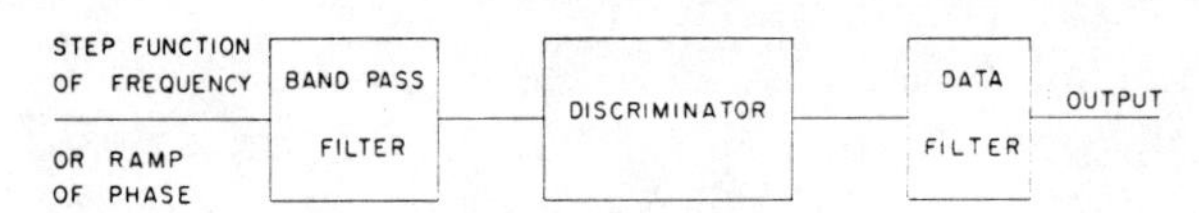

Fig. 7—Assumed system of demodulation.

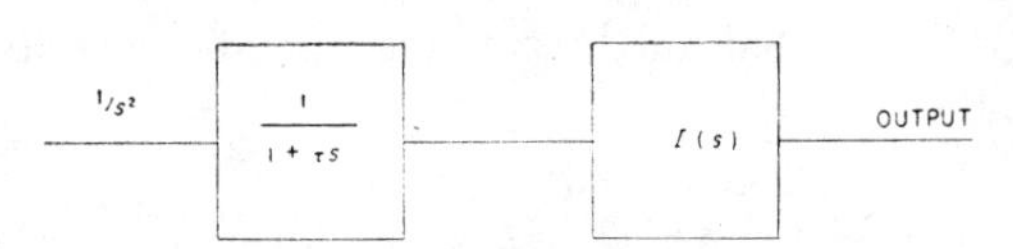

Fig. 8—Transformed equivalent to Fig. 7.

where

$$\theta_i(s) = \frac{\Delta\omega}{s^2}\frac{1}{1+\tau s}$$

or

$$|\theta_i(s)|^2 = \frac{\Delta\omega^2}{s^4}\frac{1}{1-\tau^2 s^2} \tag{26}$$

and

$$H_1(s) = s. \tag{27}$$

Then

$$|\psi(s)|^2 = \psi_+\psi_- = \Phi_N + \lambda_f^2 |\theta_i(s)|^2. \tag{28}$$

Substituting (26),

$$\psi_+\psi_- = \Phi_N + \lambda_f^2 \frac{\Delta\omega^2}{s^4(1-\tau^2 s^2)} \tag{29}$$

$$= \frac{\Phi_N s^4(1-\tau^2 s^2) + \lambda_f^2\Delta\omega^2}{s^4(1-\tau^2 s^2)}$$

$$= \frac{\lambda_f^2\Delta\omega^2 + \Phi_N s^4 - \Phi_N \tau^2 s^6}{s^4(1-\tau^2 s^2)}$$

$$= \frac{\lambda_f^2\Delta\omega^2\left[1 + \frac{\Phi_N}{\lambda_f^2\Delta\omega^2}s^4 - \frac{\Phi_N\tau^2}{\lambda_f^2\Delta\omega^2}s^6\right]}{s^4(1-\tau^2 s^2)}. \tag{30}$$

Let

$$D_0^4 = \frac{\lambda_f^2\Delta\omega^2}{\Phi_N}. \tag{31}$$

Then substituting (31) in (30)

$$\psi_+\psi_- = \lambda_f^2\Delta\omega^2\left[\frac{1 + \frac{s^4}{D_0^4} - \frac{\tau^2}{D_0^4}s^6}{s^4(1+\tau^2 s^2)}\right]. \tag{32}$$

Separating into conjugate parts

$$\psi_+\psi_- = \left[\lambda_f\Delta\omega\frac{(1+\alpha s+\beta s^2+\gamma s^3)}{s^2(1+\tau s)}\right]$$

$$\cdot\left[\lambda_f\Delta\omega\frac{(1-\alpha s+\beta s^2-\gamma s^3)}{s^2(1-\tau s)}\right] \tag{33}$$

where α, β, and γ are the positive real solutions to the simultaneous quadratic equations

$$2\beta - \alpha^2 = 0$$

$$-2\alpha\gamma + \beta^2 = \frac{1}{D_0^4}$$

$$\gamma^2 = \frac{\tau^2}{D_0^4}.$$

Now substituting appropriate parts of (33), (26), and (27) into (29), the following is obtained

$$I(s) = \frac{\lambda_f^2 s^2(1+\tau s)}{\lambda_f\Delta\omega(1+\alpha s+\beta s^2+\gamma s^3)}$$

$$\cdot\left[\frac{s\frac{\Delta\omega^2}{s^4}\frac{1}{1+\tau s}\frac{1}{1-\tau s}}{\frac{\lambda_f\Delta\omega(1-\alpha s+\beta s^2-\gamma s^3)}{s^2(1-\tau s)}}\right]_{PR}. \tag{34}$$

Clearing,

$$I(s) = \frac{s^2(1+\tau s)}{(1+\alpha s+\beta s^2+\gamma s^3)}$$

$$\cdot\left[\frac{1}{s(1+\tau s)(1-\alpha s+\beta s^2-\gamma s^3)}\right]_{PR}. \tag{35}$$

To obtain $[\ ]_{PR}$ use partial fractions or

$$[\ ]_{PR} = \left[\frac{A}{s} + \frac{B}{1+\tau s} + \frac{D}{(\ )} + \frac{E}{(\ )} + \frac{F}{(\ )}\right]_{PR}. \tag{36}$$

Note that the roots corresponding to D, E, and F are in the wrong half plane. Then

$$[\ ]_{PR} = \frac{A}{s} + \frac{B}{1+\tau s} \tag{37}$$

where

$$A = \left.\frac{1}{(1+\tau s)}\frac{1}{(1-\alpha s+\beta s^2-\gamma s^3)}\right|_{s=0} = 1 \tag{38}$$

and

$$B = \left.\frac{1}{s}\frac{1}{(1-\alpha s+\beta s^2-\gamma s^3)}\right|_{s=-(1/\tau)} \tag{39}$$

$$= (-\tau)\frac{1}{\left(1+\frac{\alpha}{\tau}+\frac{\beta}{\tau^2}+\frac{\gamma}{\tau^3}\right)}$$

$$= \frac{-\tau^4}{\tau^3+\tau^2\alpha+\tau\beta+\gamma}. \tag{40}$$

When (37), (39), and (40) are substituted into (35),

$$I(s) = \frac{s^2(1+\tau s)}{(1+\alpha s+\beta s^2+\gamma s^3)}$$

$$\cdot\left[\frac{1}{s} - \frac{\tau^4}{\tau^3+\tau^2\alpha+\tau\beta+\gamma}\frac{1}{1+\tau s}\right]. \tag{41}$$

Clearing,

$$I(s) = \frac{s^2(1 + \tau s)}{(1 + \alpha s + \beta s^2 + \gamma s^3)} \cdot \left[\frac{1 + \tau s - \dfrac{\tau^4}{\tau^3 + \tau^2\alpha + \tau\beta + \gamma} s}{s(1 + \tau s)} \right]$$

$$I(s) = \frac{s\left(1 + \dfrac{\tau(\tau^3 + \tau^2\alpha + \tau\beta + \gamma) - \tau^4}{\tau^3 + \tau^2\alpha + \tau\beta + \gamma} s\right)}{(1 + \alpha s + \beta s^2 + \gamma s^3)}. \tag{42}$$

Eq. (42) may be separated into two parts; the term s appears because of the desired operation $H_1(s)$, which in reality is the discriminator. The remaining filter is the one which must be constructed (see Fig. 9).

It is interesting to note that if $\tau = 0$, the condition that exists if the band-pass filter is disregarded, the filter necessary is of the form of an ordinary galvanometer with a damping of $\sqrt{2}/2$.

$$\overline{I(s)}\bigg|_{\tau=0} = \frac{1}{1 + \dfrac{\sqrt{2}}{D_0} s + \dfrac{s^2}{D_0^2}} \tag{43}$$

where the galvanometer equation using ζ as the damping factor is

$$I(s)_{\text{galvo}} = \frac{1}{1 + \dfrac{2\zeta s}{D_0} + \dfrac{s^2}{D_0^2}}. \tag{44}$$

It is surprising to find that the filters used in most data laboratories are this close to the optimum.

VI. Analysis and Operation of the Phase-Locked Loop

The notation used in this section will be essentially that used by Jaffe and Rechtin[4] with some exceptions so that the results may be compared. A typical system and signals are shown in Fig. 10.

The following definitions are made:

A = the rms input signal level,
C = the rms output signal level of the vco,
$\Phi_N(f_N)$ = the noise spectral density at f_N, the noise frequency.

If the output of the multiplier is inspected closely to see the operation of the loop, the output of the multiplier will be

multiplier output

$$= K_m \sqrt{2}C \cos(2\pi f_0 t + \theta_0)[\sqrt{2}A \sin \cdot (2\pi f_i t + \theta_i) + \sqrt{2}\sqrt{\Phi_N(f_N)} \sin 2\pi f_N t] \tag{45}$$

where K_m is the constant of the multiplier. Note that this relation expresses the peak values.

Further reducing results in

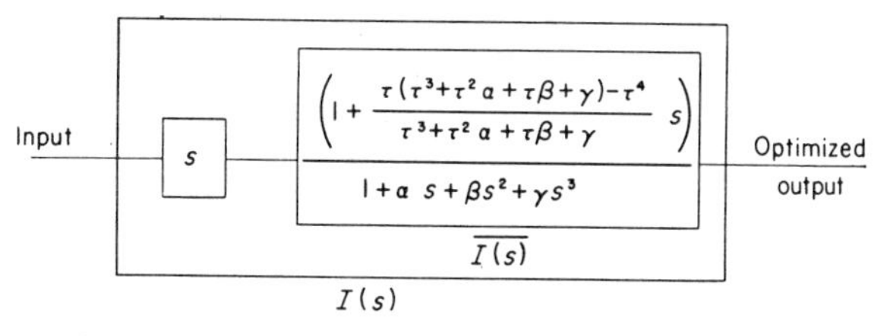

Fig. 9—Optimum filter for FM signals under transient conditions utilizing band-pass filters.

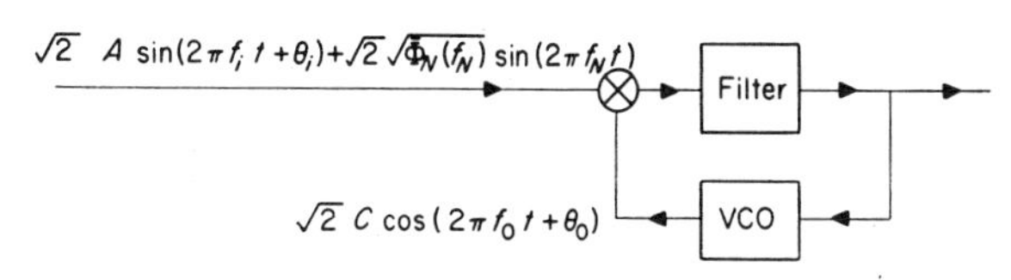

Fig. 10—Typical phase-locked loop with signals.

multiplier output

$$\begin{aligned} &= K_m 2AC1/2[\sin[(2\pi f_0 + 2\pi f_i)t + \theta_0 + \theta_i] \\ &\quad + \sin[(2\pi f_i - 2\pi f_0)t + \theta_i - \theta_0] \\ &\quad + K_m 2C\sqrt{\Phi_N(f_N)}1/2[\sin(2\pi f_0 t + \theta_0 + 2\pi f_N t) \\ &\quad + \sin(2\pi(f_N - f_0)t - \theta_0)]. \end{aligned} \tag{46}$$

The loop response will not be fast enough to follow the higher frequency terms so they may be ignored. Therefore, multiplier output

$$\begin{aligned} &= K_m AC \sin[2\pi(f_i - f_0)t + \theta_i - \theta_0] \\ &\quad + K_m C\sqrt{\Phi(f_N)} \sin[2\pi(f_N - f_0)t - \theta_0]. \end{aligned} \tag{47}$$

Ignore the contribution due to the noise for the present and examine the remaining term.

multiplier output

$$= K_m AC \sin[(2\pi f_i t + \theta_i) - (2\pi f_0 t - \theta_0)]. \tag{48}$$

Let

$$2\pi f_i t + \theta_i = \xi_i$$

and

$$2\pi f_0 t + \theta_0 = \xi_0$$

then

multiplier output

$$= K_m AC \sin(\xi_i - \xi_0). \tag{49}$$

The loop will be unstable unless the sign of the multiplier output is positive. Therefore, the principal value of $(\xi_i - \xi_0)$ must be less than $\pi/2$ which requires that $f_i = f_0$ if t is changing; therefore the loop follows changes of phase (see Fig. 11).

If $\theta_i - \theta_0$ is small,

multiplier output

$$= K_m AC \sin(2\pi(f_i - f_0)t + \theta_i - \theta_0) \cong K_m AC[\theta_i - \theta_0]. \tag{50}$$

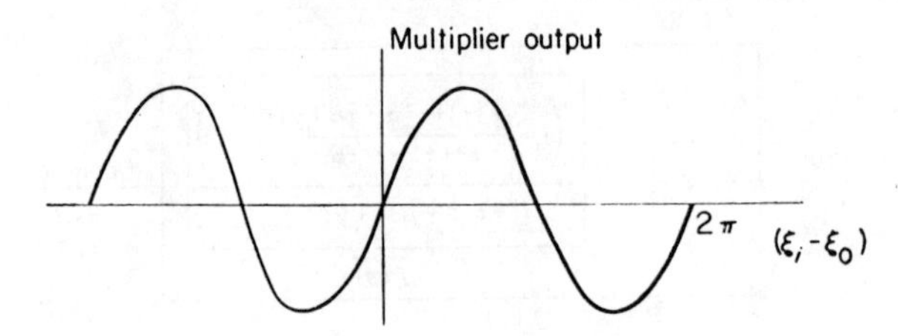

Fig. 11—Multiplier output as function of $(\xi_i - \xi_0)$.

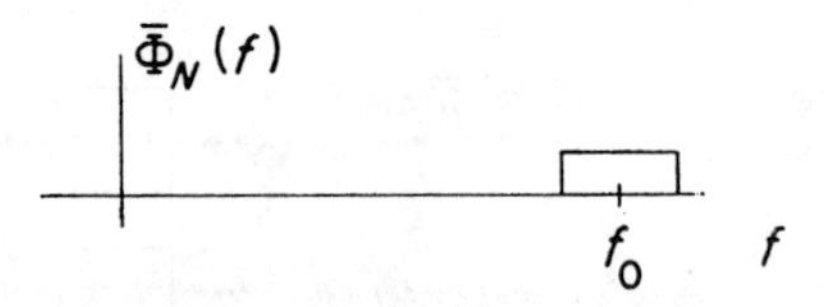

Fig. 12—Typical noise input spectrum.

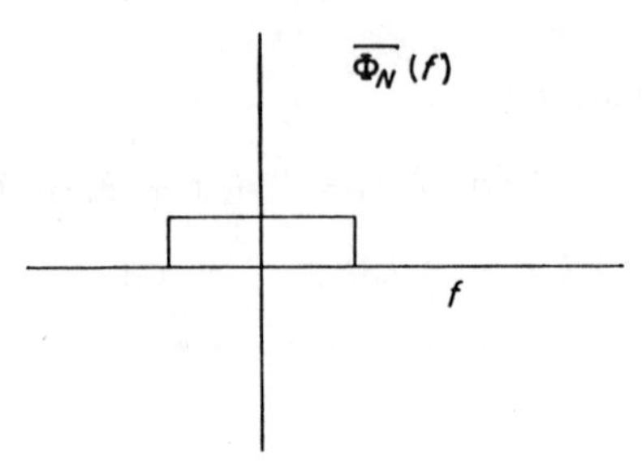

Fig. 13—Multiplier output noise spectrum for typical input spectrum shown in Fig. 12.

The noise output of the multiplier is
multiplier output

$$= K_m C \sqrt{\bar{\Phi}_N(f_N)} \sin [2\pi(f_N - f_0)t - \theta_0)]. \quad (51)$$

Assuming any noise spectrum such as Fig. 12 the vco signal will beat with the "noise" and displace the spectrum along the frequency axis without changing the shape of the spectrum as shown in Fig. 13.

The fact that the spectrum shape and magnitude are not changed is a unique property of the perfect multiplier. A multiplier never allows the noise components to beat upon themselves. If any other type of phase detector were used, such as linear detectors,[6] intermodulation components would be generated which contribute to the overall noise. This is one of the reasons why the multiplier appears in optimum demodulation. The total multiplier output then is
multiplier output

$$= K_m AC \left[\theta_i - \theta_0 + \sqrt{\frac{\bar{\Phi}_N}{A^2}} \right]. \quad (52)$$

To linearize the loop, the input may be replaced with

$$\text{input to pll} = \theta_i(s) + \sqrt{\frac{\bar{\Phi}_N}{A^2}} \quad (53)$$

if an amplifier or $K_m AC$ is inserted within the loop and the multiplier replaced by a different circuit. (Fig. 14).

[6] J. L. Lawson and G. E. Uhlenbeck, "Threshold Signals," M.I.T. Rad. Lab. Ser., McGraw-Hill Book Co., Inc., New York, N. Y., no. 24, pp. 62–63, 159; 1950.

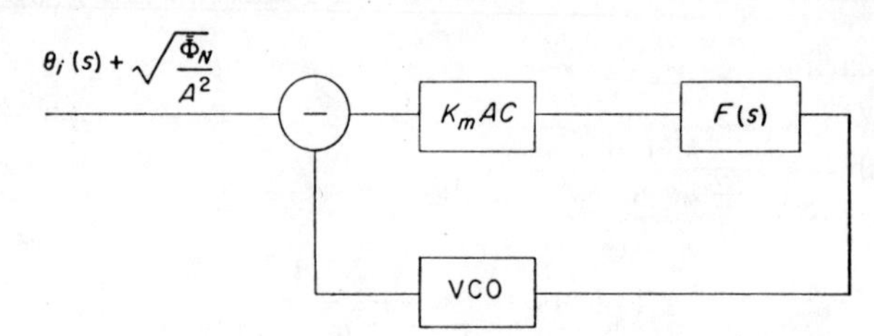

Fig. 14—Linearized phase-locked loop.

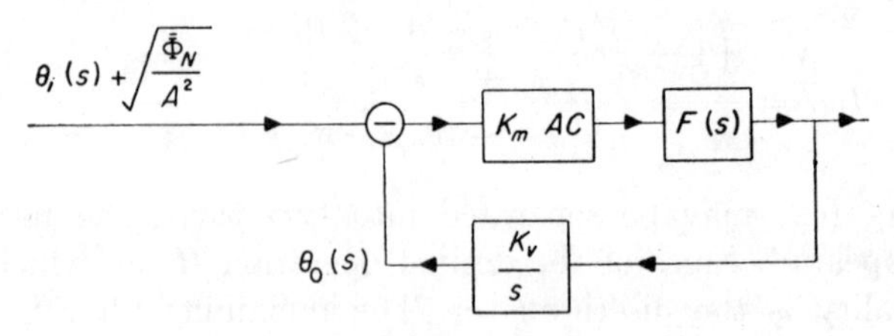

Fig. 15—Equivalent phase-locked loop.

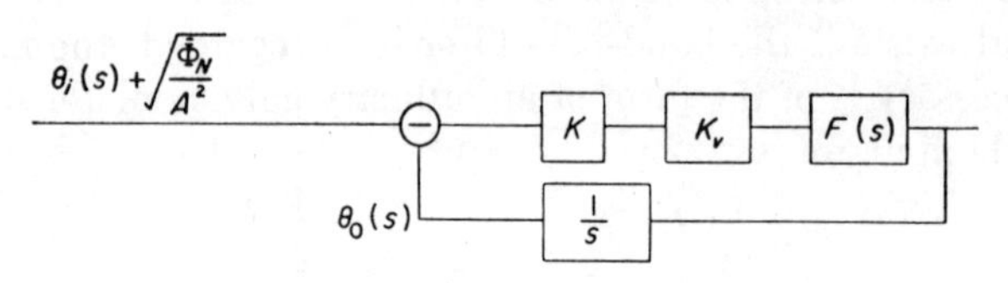

Fig. 16—Phase-locked loop equivalent.

To complete the loop in Laplace notation requires that the vco transfer function be found. A voltage E will cause a frequency change in the output

$$\frac{d\theta_0}{dt} = K_v E(t) \quad (54)$$

where K_v is the sensitivity of the vco

$$\theta_0(t) = \int K_v E(t)\, dt \quad (55)$$

taking the Laplace transform

$$\frac{\theta_0(s)}{E(s)} = \frac{K_v}{s} \quad (56)$$

so that the vco may be replaced by an integrator. The constants K_m, A, and C, may be combined into one constant K so that Fig. 15 can be replaced with Fig. 16.

It is asserted that there are two outputs which can be taken from the phase-locked loop, both of which will be important to us. The loop may take a phase output or a derivative of the phase output. The phase output will be important for the estimation of the threshold, and the derivative of phase will be important as a discriminator output. Since the voltage into the vco is proportional to the frequency, it will be the derivative of the phase or equivalently the demodulated frequency output. This fact makes the phase-locked loop, when used in this manner, equivalent in theoretical signal-to-noise output (if its bandwidth at the derivative output is the same) to any other system under consideration. It is immediately obvious also that the phase output has smaller bandwidth than does the derivative. This can be used to advantage.

Even though a linear system has been derived, the system remains nonlinear, which will limit its usefulness. In the derivation, it was noted that the phase error must remain in the interval

$$-\frac{\pi}{2} < \text{error} < \frac{\pi}{2}$$

otherwise the loop becomes unstable. This maximum error will determine the over-all improvement, if any, over that of ordinary discriminators.

VII. Loop Filter and Operation of the Phase-Locked Loop for Optimum Frequency Filter

This section is included to demonstrate that if the phase-locked loop is to be utilized to the utmost, further optimization must be accomplished. In Section VI it was stated that the error must remain in the interval

$$-\frac{\pi}{2} < \text{error} < \frac{\pi}{2}.$$

Since both the phase-locked loop and ordinary discriminators are in fact discriminators, then both should follow the same theory in the linear regions. It has already been shown that the ordinary discriminators have a threshold due to the wide-band noise at the limiter. A limitation in the phase-locked-loop error is already evident, and nothing has been shown to indicate that the limit will occur at a lower signal-to-noise ratio. If it does not, the optimum demodulation theory has not been of much help. A reduction in bandwidth has been made within the loop; it might be assumed that certainly some improvement might be accomplished as a function of this reduction.

A solution for the optimum filter for the demodulated FM signal has been obtained with which our over-all system must correspond. This was

$$I(s) = \frac{s\left(1 + \dfrac{\tau(\tau^3 + \tau^2\alpha + \tau\beta + \gamma) - \tau^4}{\tau^3 + \tau^2\alpha + \tau\beta + \gamma}s\right)}{(1 + \alpha s + \beta s^2 + \gamma s^3)}. \tag{57}$$

To obtain the system transient phase error the following is evident.

$$\text{error }(s) = \text{input}\left[1 - \frac{1}{s}I(s)\right]$$

$$= \frac{\Delta\omega}{s^2}\frac{1}{1 + \tau s} \tag{58}$$

$$\left[1 - \frac{1}{s}\frac{s\left(1 + \dfrac{\tau(\tau^3 + \tau^2\alpha + \tau\beta + \gamma) - \tau^4}{\tau^3 + \tau^2\alpha + \tau\beta + \gamma}s\right)}{(1 + \alpha s + \beta s^2 + \gamma s^3)}\right]$$

$$= \frac{\Delta\omega}{s^2}\frac{1}{1 + \tau s} \tag{59}$$

$$\left[\frac{1 + \alpha s + \beta s^2 + \gamma s^3 - 1 - \dfrac{\tau(\tau^3 + \tau^2\alpha + \tau\beta + \gamma) - \tau^4}{\tau^3 + \tau^2\alpha + \tau\beta + \gamma}}{(1 + \alpha s + \beta s^2 + \gamma s^3)}\right]$$

$$= \frac{\Delta\omega}{s^2}\frac{1}{1 + \tau s} \tag{60}$$

$$\left[\frac{\left(\alpha - \dfrac{\tau(\tau^3 + \tau^2\alpha + \tau\beta + \gamma) - \tau^4}{\tau^3 + \tau^2\alpha + \tau\beta + \gamma}\right)s + \beta s^2 + \gamma s^3}{(1 + \alpha s + \beta s^2 + \gamma s^3)}\right]. \tag{61}$$

Knowing that the static error is (using the final value theorem)

$$\lim_{s\to 0} s \times \text{error}(s) = \lim_{t\to\infty} e(t) \tag{62}$$

$$\lim_{t-\infty} e(t) = \lim_{s\to 0}\frac{s\Delta\omega}{s(1 + \tau s)}$$

$$\left[\frac{\left(\alpha - \dfrac{\tau(\tau^3 + \tau^2\alpha + \tau\beta + \gamma) - \tau^4}{\tau^3 + \tau^2\alpha + \tau\beta + \gamma}\right) + \beta s + \gamma s^2}{(1 + \alpha s + \beta s^2 + \gamma s^3)}\right], \tag{63}$$

then

$$\lim_{\tau\to\infty} e(t) = \Delta\omega$$

$$\left[\alpha - \frac{\tau(\tau^3 + \tau^2\alpha + \tau\beta + \gamma) - \tau^4}{\tau^3 + \tau^2\alpha + \tau\beta + \gamma}\right]. \tag{64}$$

One important thing to notice from (64) is that

$$\lim_{t\to\infty} e(t)$$

is finite. Another is that the error is reduced by increased bandwidth and that the effect of the band-pass filter makes the error less severe. Without delving too deeply into this, if the error were allowed to be $\pi/2$ when a perfect step function is used ($\tau = 0$), then

$$\lim_{\tau\to\infty} e(t) = \frac{\pi}{2} = \Delta\omega\frac{\sqrt{2}}{D_0} \tag{65}$$

or

$$\frac{\Delta\omega}{D_0} = \frac{\pi}{2\sqrt{2}} = 1.11. \tag{66}$$

The result in (66) seems extreme and certainly shows that there can be no bandwidth reduction to effect any improvement. Usually in a feedback problem the loop gain can be adjusted to make the static error any amount desired, but in this problem the parameters are dictated by (57) which results in (66). If, however, the loop had been designed to minimize the errors in the phase-locked loop instead of the discriminated output, the result would have been quite different. This is the basis for the next section. To elucidate, the loop should be designed to minimize phase error and additional filtering done to accomplish the effect desired for the discriminated output. See Fig. 17, next page.

VIII. Optimum Phase Filter for the Phase-Locked Loop

To solve for the optimum phase filter for the phase-locked loop application, the generalized results of Section IV are used which resulted in the following:

$$H(s) = \frac{\lambda^2}{\psi_+}\left[\frac{H_1(s)\mid\theta_i(s)\mid^2}{\psi^-}\right]_{PR}. \tag{67}$$

Designating the filter for the optimum phase filter as

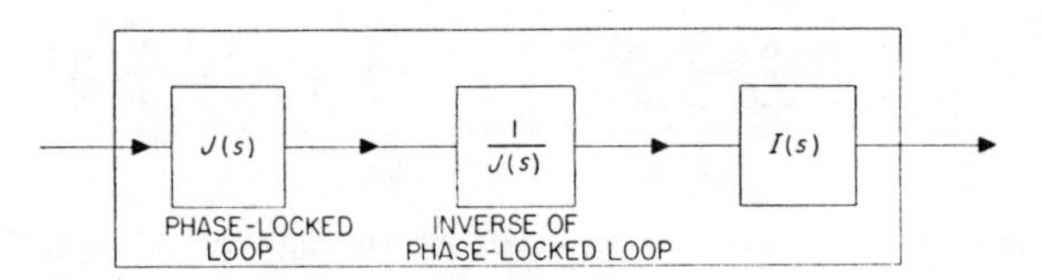

Fig. 17—Optimum system to minimize both phase and frequency errors.

$J(s)$ where

$$\theta_i(s) = \frac{\Delta\omega}{s^2}\frac{1}{1+\tau s} \tag{68}$$

or

$$|\,\theta_i(s)\,|^2 = \frac{\Delta\omega^2}{s^4}\frac{1}{1-\tau^2 s^2} \tag{69}$$

and

$$H_1(s) = 1 \tag{70}$$

which prevail for the conditions of the phase-locked loop. Then

$$J(s) = \frac{\lambda_\theta^2}{\psi_+}\left[\frac{\dfrac{\Delta\omega}{s^4}\dfrac{1}{1-\tau^2s^2}}{\psi^-}\right]_{PR}. \tag{71}$$

$$\psi^+\psi^- = \Phi_N + \lambda_\theta\,|\,\theta_i(s)\,|^2 = \Phi_N + \lambda_\theta\frac{\Delta\omega}{s^4}\frac{1}{1-\tau^2s^2} \tag{72}$$

or from a previous section letting

$$B_0^4 = \frac{\lambda_\theta^2\Delta\omega^2}{\Phi_N} \tag{73}$$

$$\psi^+\psi^- = \left[\lambda_\theta\Delta\omega\frac{(1+\bar\gamma s+\bar\beta s^2+\bar\gamma s^3)}{s^2(1+\tau s)}\right] \cdot\left[\lambda_\theta\Delta\omega\frac{(1-\bar\alpha s+\bar\beta s^2-\bar\gamma s^3)}{s^2(1-\tau s)}\right] \tag{74}$$

where $\bar\alpha$, $\bar\beta$, and $\bar\gamma$ are the positive real solutions to the simulataneous quadratic equations

$$2\bar\beta - \bar\alpha^2 = 0$$

$$-2\bar\alpha\bar\gamma + \bar\beta^2 = \frac{1}{B_0^4}$$

$$\bar\gamma^2 = \frac{\tau^2}{B_0^4}$$

then

$$J(s) = \frac{\lambda_\theta^2 s^2(1+\tau s)}{\lambda_\theta\Delta\omega(1+\bar\alpha s+\bar\beta s^2+\bar\gamma s^3)} \cdot\left[\frac{(1)\dfrac{\Delta\omega}{s^4}\dfrac{1}{1+\tau s}\dfrac{1}{1-\tau s}}{\lambda_\theta\Delta\omega\dfrac{(1-\bar\alpha s+\bar\beta s^2-\bar\gamma s^3)}{s^2(1-\tau s)}}\right]_{PR} \tag{75}$$

$$= \frac{s^2(1+\tau s)}{(1+\bar\alpha s+\bar\beta s^2+\bar\gamma s^3)} \cdot\left[\frac{1}{s^2(1+\tau s)(1-\bar\alpha s+\bar\beta s^2-\bar\gamma s^3)}\right]_{PR}. \tag{76}$$

Now using partial fractions

$$[\ ]_{pr} = \left[\frac{A}{s^2}+\frac{B}{s}+\frac{C}{1+\tau s}+\frac{D}{(\ )}+\frac{E}{(\ )}+\frac{F}{(\ )}\right]_{PR} \tag{77}$$

and noting that the roots corresponding to D, E, and F are in the wrong half plane, then

$$[\ ]_{PR} = \frac{A}{s^2}+\frac{B}{s}+\frac{C}{1+\tau s} \tag{78}$$

where

$$A = \left.\frac{1}{(1+\tau s)(1-\bar\alpha s+\bar\beta s^2-\bar\gamma s^3)}\right|_{s=0} = 1 \tag{79}$$

and

$$C = \left.\frac{1}{s^2(1-\bar\alpha s+\bar\beta s^2-\bar\gamma s^3)}\right|_{s=-(1/\tau)} \tag{80}$$

$$= \frac{\tau^2}{\left(1+\dfrac{\bar\alpha}{\tau}+\dfrac{\bar\beta}{\tau^2}+\dfrac{\bar\gamma}{\tau^3}\right)} \tag{81}$$

$$C = \frac{\tau^5}{\tau^3+\tau^2\bar\alpha+\tau\bar\beta+\bar\gamma} \tag{82}$$

$$B = \frac{d}{ds}\left[\frac{1}{(1+\tau s)(1-\bar\alpha s+\bar\beta s^2-\bar\gamma s^3)}\right]\Bigg|_{s=0} \tag{83}$$

$$= \left.\frac{(\ )(\ )(0)-(1)\dfrac{d}{ds}(1+\tau s)(1-\alpha s+\bar\beta s^2-\bar\gamma s^3)}{(1+\tau s)^2(1-\bar\alpha s+\bar\beta s^2-\bar\gamma s^3)^2}\right|_{s=0} \tag{84}$$

$$=(-1)\left.\frac{\tau(1-\bar\alpha s+\bar\beta s^2-\bar\gamma s^3)+(1+\tau s)(-\bar\alpha+2\bar\beta s-3\bar\gamma s^2)}{(1+\tau s)^2(1-\bar\alpha s+\bar\beta s^2-\bar\gamma s^3)^2}\right|_{s=0} \tag{85}$$

$$= (-1)\frac{\tau-\bar\alpha}{1} = (\bar\alpha-\tau). \tag{86}$$

So that

$$J(s) = \frac{s^2(1+\tau s)}{(1+\bar\alpha s+\bar\beta s^2+\bar\gamma s^3)} \cdot\left[\frac{1}{s^2}+\frac{\bar\alpha-\tau}{s}+\frac{\tau^5}{\tau^3+2\bar\alpha+\tau\bar\beta+\bar\gamma}\frac{1}{(1+\tau s)}\right] \tag{87}$$

$$J(s) = \frac{s^2(1+\tau s)}{(1+\bar\alpha s+\bar\beta s^2+\bar\gamma s^3)} \frac{(1+\tau s)+(\bar\alpha-\tau)s(1+\tau s)+s^2\left(\dfrac{\tau^5}{\tau^3+\tau^3\bar\alpha+\tau\bar\beta+\bar\gamma}\right)}{s^2(1+\tau s)} \tag{88}$$

$$J(s) = \cdot\frac{1+\tau s+(\bar\alpha-\tau)(s)(1+\tau s)+s^2\left(\dfrac{\tau^5}{\tau^3+\tau^2+\tau\bar\beta+\gamma}\right)}{(1+\bar\alpha s+\bar\beta s^2+\bar\gamma s^3)} \tag{89}$$

$$J(s) = \cdot\frac{1+\tau s+(\bar\alpha-\tau)s+\tau(\bar\alpha-\tau)s^2+\left(\dfrac{\tau^5}{\tau^3+\tau^2\bar\alpha+\tau\bar\beta+\bar\gamma}\right)s^2}{(1+\bar\alpha s+\bar\beta s^2+\bar\gamma s^3)} \tag{90}$$

$$J(s) = \cdot \frac{1 + \bar{\alpha}s + \left[\tau(\bar{\alpha} - \tau) + \frac{\tau^5}{\tau^3 + \tau^2\bar{\alpha} + \tau\bar{\beta} + \bar{\gamma}}\right]s^2}{(1 + \bar{\alpha}s + \bar{\beta}s^2 + \bar{\gamma}s^3)} \tag{91}$$

which is the optimum filter that the phase-locked loop must correspond to. When $\tau = 0$, the conditions are the same as assumed by Jaffe and Rechtin and should correspond to $Y(s)$ of their report.

$$J(s) = Y(s) = \frac{1 + \frac{\sqrt{2}}{B_0}s}{1 + \frac{\sqrt{2}}{B_0}s + \frac{s^2}{B_0^2}}.$$

To show that $J(s)$ results in a better system than $I(s)$ alone did, the final value theorem is used on the error of the phase-locked loop constructed to affect $J(s)$. The error in this case is

$$\text{error }(S) = \text{input }[1 - J(S)] \tag{93}$$

$$= \frac{\Delta\omega}{s^2(1 + \tau s)}$$

$$\cdot\left[1 - \frac{1 + \bar{\alpha}s + \left[\tau(\bar{\alpha} - \tau) + \frac{\tau^5}{\tau^3 + \tau^2\bar{\alpha} + \tau\bar{\beta} + \bar{\gamma}}\right]s^2}{(1 + \bar{\alpha}s + \bar{\beta}s^2 + \bar{\gamma}s^3)}\right] \tag{94}$$

$$= \frac{\Delta\omega}{s^2(1 + \tau s)} \times \left[\frac{1 + \bar{\alpha}s + \bar{\beta}s + \bar{\gamma}s^3 - 1}{(1 + \bar{\alpha}s + \bar{\beta}s^2 + \bar{\gamma}s^3)}\right.$$

$$\left. + \frac{\bar{\alpha}s - \left[\tau(\bar{\alpha} - \tau) + \frac{\tau^5}{\tau^3 + \tau^2\bar{\alpha} + \tau\bar{\beta} + \bar{\gamma}}\right]s^2}{(1 + \bar{\alpha}s + \bar{\beta}s^2 + \bar{\gamma}s^3)}\right] \tag{95}$$

$$= \frac{\Delta\omega}{s^2(1 + \tau s)}$$

$$\cdot\left[\frac{\bar{\beta}s^2 + \bar{\gamma}s^3 - \left[\tau(\bar{\alpha} - \tau) + \frac{\tau^5}{\tau^3 + \tau^2 + \tau\bar{\beta} + \bar{\gamma}}\right]s^2}{(1 + \bar{\alpha}s + \bar{\beta}s^2 + \bar{\gamma}s^3)}\right]. \tag{96}$$

Using the final value theorem

$$\lim_{s \to 0} s \times \text{error}(s) = \lim_{t \to \infty} e(t) \tag{97}$$

then

$$\lim_{t \to \infty} e(t) = \frac{s\Delta\omega s^2}{s^2(1 + \tau s)}$$

$$\cdot\left[\frac{\bar{\beta} - \left[\tau(\bar{\alpha} - \tau) + \frac{\tau^5}{\tau^3 + \tau^2\bar{\alpha} + \tau\bar{\beta} + \bar{\gamma}}\right] + \bar{\gamma}s}{(1 + \bar{\alpha}s + \bar{\beta}s^2 + \bar{\gamma}s^3)}\right] \tag{98}$$

$$= 0.$$

Thus the results of this part of the optimization were fruitful.

IX. Transient Errors of the Phase-Locked Loop

In the preceding sections static errors were used to point out the requirements for two different optimizations: the discriminated output and the phase-locked loop. It has already been suggested in Section VII that the optimum system has the inverse of the phase-locked loop, which implies that if the bandwidth of the phase-locked loop is low and if the higher frequencies may be reconstructed, this lower bandwidth in the phase-locked loop would allow operation at much higher signal-to-noise ratios. This is not entirely practical for two reasons. First, the inverse does not always exist in the practical case; second, the phase-locked loop is not entirely linear which requires that the bandwidth be no less than a certain minimum to maintain the phase-locked loop synchronism. The study of this second item is the purpose of this section.

Earlier descriptions of the system accounted for the effect of the band-pass filter. The results of the optimization considering this effect caused the order of the equations to be much higher which complicates the construction of a practical system. Three transient solutions will be shown to see if the more complicated construction problem will yield worthwhile results for the effort exerted. The solutions for α, β, and γ are shown in Table I for systems defined in Table II and were obtained graphically from Figs. 18 and 19. Figs. 20–22 show the time solutions of systems 1, 2, and 3, respectively, with B_0 parametized in τ. (See Table III for value of τ.) The peak values of these curves are important because the loop loses synchronism if they become larger than $\pi/2$. Study of these curves show that the peak value dictates the minimum bandwidth of the loop. Figs. 23 and 24 analyze these peak values in terms of the bandwidth in terms of τ, and the magnitude of the step function for a value of phase error, $\pi/2$. These curves show that an improvement in peak transient values are realized when the bandwidth becomes comparable to τ and larger. Further indicated is the fact that the complication of the system by optimizing accounting for all factors does not do as well as the system designed around the perfect step function used in an actual system which modifies the input by $1/(1 + \tau s)$. This should not be the criterion for discarding one and accepting another since the effect of noise still remains to be studied.

X. Effect of Noise and Bandwidth Upon the Threshold

The last section presented how the peak transient phase error influenced the selection of bandwidth B_0. In the original criterion this was specified by keeping $E_{T\theta}^2$ fixed, but this criterion does not consider the critical operation near $\pi/2$. It was surprising to find that the peak values of the phase error in system 3 were larger than the peak values of system 2. It appears, however, that $E_{T\theta}^2$ of the two systems might be comparable, which would mean that if system 3 were optimum, the noise level would be less and, therefore, some improvement

TABLE I

B_0	$\bar{\alpha}$	$\bar{\beta}$	$\bar{\gamma}$	B_0	$\bar{\alpha}$	$\bar{\beta}$	$\bar{\gamma}$
$\frac{1}{4\tau}$	6.56τ	$21.6\tau^2$	$16.0\tau^3$	$\frac{2}{\tau}$	1.28τ	$0.84\tau^2$	$0.25\tau^3$
$\frac{1}{2\tau}$	3.68τ	$6.74\tau^2$	$4.0\tau^3$	$\frac{4}{\tau}$	0.8τ	$0.32\tau^2$	$0.0625\tau^3$
$\frac{1}{\tau}$	2.12τ	$2.3\tau^2$	$1.0\tau^3$				

TABLE II

System	Input	Filter	Remarks
1	Perfect step function	Designed around perfect step function	More strenuous conditions than exist but system is simpler to build.
2	Step function modified by the effect of the band-pass filter	Designed around perfect step function	Less strenuous conditions and is simpler to build.
3	Step function modified by the effect of the band-pass filter	Designed around step function modified by the effect of the band-pass filter	Optimum system but most difficult to build.

System 1: $$\text{error }(S) = \frac{\Delta\omega}{s^2}\frac{s^2}{B_0^2 + \sqrt{2}B_0 s + s^2}.$$

System 2: $$\text{error }(S) = \frac{\Delta\omega}{s^2}\frac{1}{1+\tau s}\frac{s^2}{B_0^2 + \sqrt{2}B_0 s + s^2}.$$

System 3: $$\text{error }(S) = \frac{\Delta\omega}{s^2}\frac{1}{1+\tau s}\frac{s^2\left[\bar{\beta} - \left[\tau(\bar{\alpha}-\tau) + \frac{\tau^5}{\tau^3 + \tau^2\bar{\alpha} + \tau\bar{\beta} + \bar{\gamma}}\right] + \bar{\gamma}s\right]}{(1 + \bar{\alpha}s + \bar{\beta}s^2 + \bar{\gamma}s^3)}$$

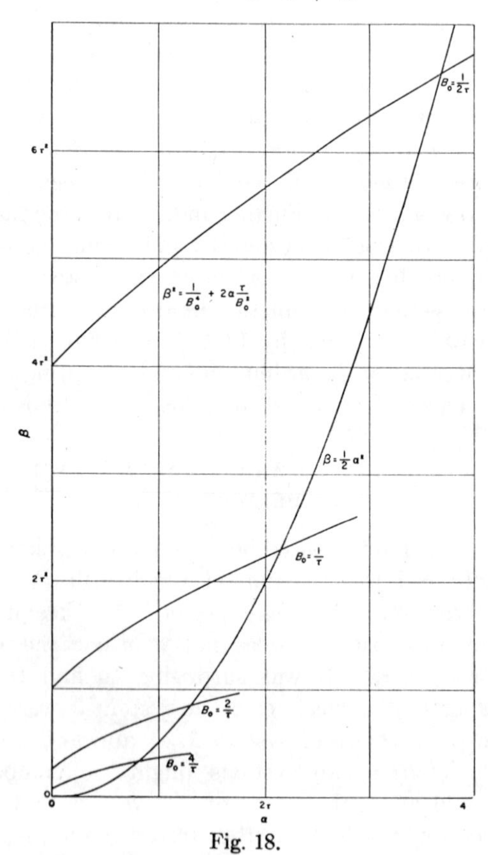

Fig. 18.

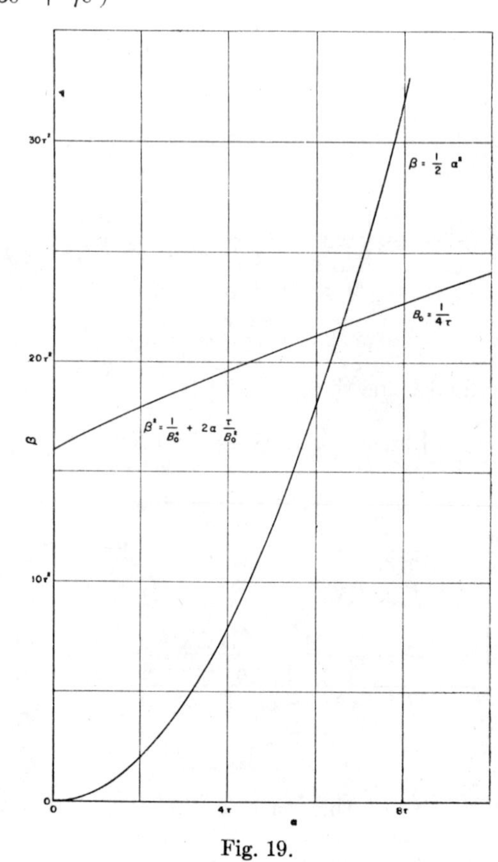

Fig. 19.

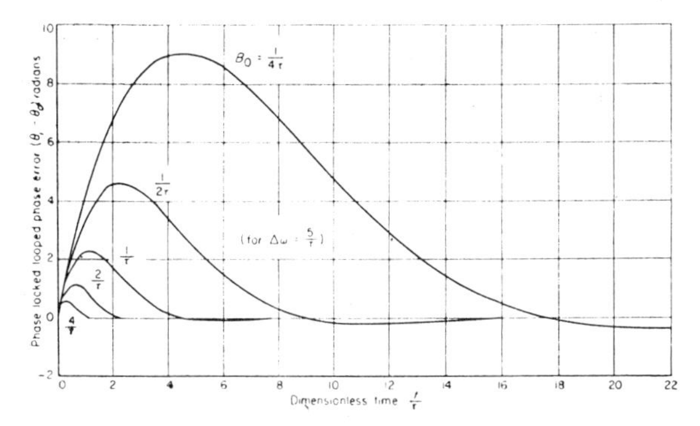

Fig. 20.

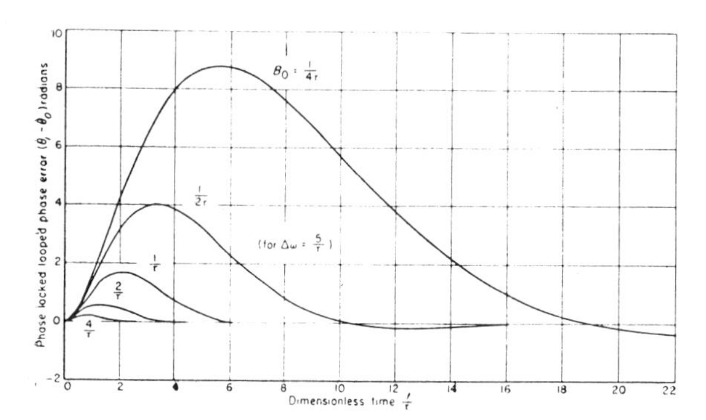

Fig. 21.

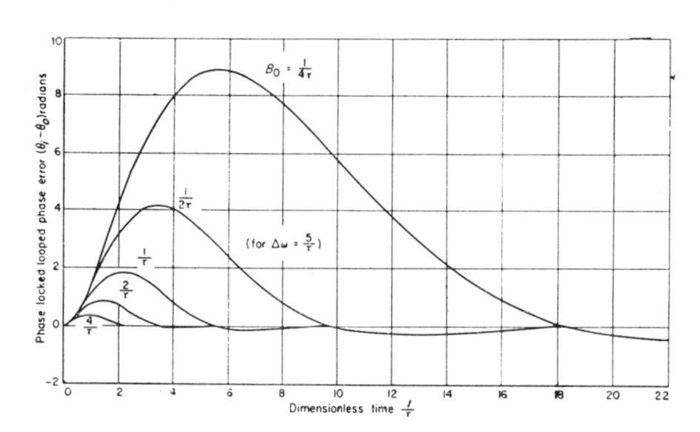

Fig. 22.

would be obtained from the more complicated system.

In considering the noise in each system, it should be explained that $1/(1+\tau s)$ affects only the transient signal and not the noise. This is because some of the sideband power of the signal is stripped by the band-pass filter while the spectrum of the noise is flat within the band-pass and is the same for all systems. The noise filtering of systems 1 and 2 are identical for identical bandwidths but differ from system 3 because it is optimized considering the effect on the transient. The noise is given by

$$\sigma_{N\theta}^2 = \frac{1}{2\pi j}\int_{-j\infty}^{j\infty} |\, J(s)\,|^2 \, \frac{\Phi_n}{A^2}\, ds.$$

For the total noise in each telemetering channel

$$\frac{\Phi_N}{A^2} = \frac{N^2\tau}{A^2 3.3}.$$

For systems 1 and 2

TABLE III

ACCEPTED VALUES OF τ

Channel	τ (seconds)	Channel	τ (seconds)
1	8.8×10^{-3}	10	6.6×10^{-4}
2	6.6×10^{-3}	11	4.8×10^{-4}
3	4.8×10^{-3}	12	3.3×10^{-4}
4	3.8×10^{-3}	13	2.4×10^{-4}
5	2.7×10^{-3}	14	1.6×10^{-4}
6	2.1×10^{-3}	15	1.2×10^{-4}
7	1.5×10^{-3}	16	8.9×10^{-5}
8	1.2×10^{-3}	17	6.8×10^{-5}
9	9.1×10^{-4}	18	5.1×10^{-5}

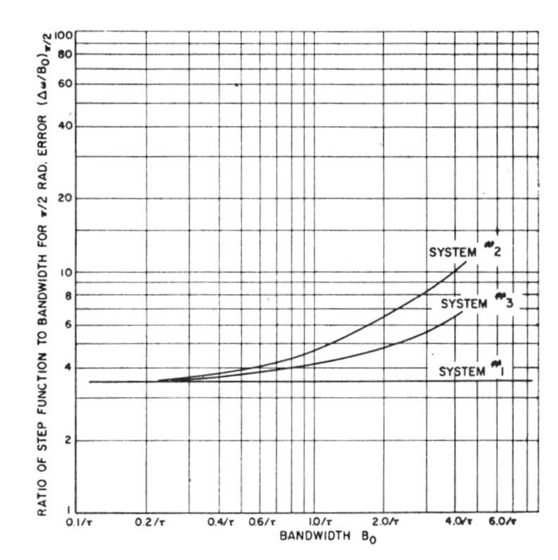

Fig. 23.

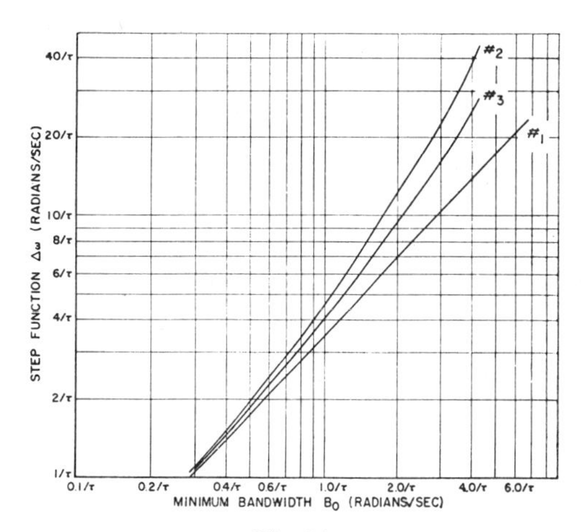

Fig. 24.

$$J(s) = \frac{B_0^2 + \sqrt{2}B_0 s}{B_0^2 + \sqrt{2}B_0 s + s^2}$$

then

$$\sigma_{N\theta}^2 = \frac{1}{2\pi j}\int_{-j\infty}^{j\infty} \left|\frac{B_0^2 + \sqrt{2}B_0 s}{B_0^2 + \sqrt{2}B_0 s + s^2}\right|^2 \frac{N^2\tau}{A^2 3.3}\, ds$$

$$= \frac{1}{2\pi j}\int_{-j\infty}^{j\infty}$$

$$\cdot \frac{B_0^4 - 2B_0^2 s^2}{(B_0^2 + \sqrt{2}B_0 s + s^2)}\,(B_0^2 - \sqrt{2}B_0 s + s^2)\,\frac{N^2\tau}{A^2 3.3}\, ds$$

or[7]

$$\sigma_{N\theta}^2 = \frac{-b_0 + \dfrac{a_0 b_1}{a_2}}{2a_0 a_1} \frac{N^2 \tau}{A^2 3.3}$$

where

$$a_0 = 1 \qquad b_0 = -2B_0^2$$

$$a_1 = \sqrt{2}B_0 \qquad b_1 = B_0^4 .$$

$$a_2 = B_0^2$$

Therefore

$$\sigma_{N\theta}^2 = \frac{N^2 \tau}{A^2 3.3} \frac{3B_0}{2\sqrt{2}}.$$

System 3 noise output is then

$$\sigma_{N\theta}^2 = \frac{1}{2\pi j} \int_{-j\infty}^{j\infty} | J(s) |^2 \frac{N^2 \tau}{A^2 3.3} ds$$

$$\sigma_{N\theta}^2 = \frac{1}{2\pi j} \int_{-j\infty}^{j\infty}$$

$$\cdot \frac{1 + \bar{\alpha}s + \left[\tau(\bar{\alpha} - \tau) + \dfrac{\tau^5}{\tau^3 + \tau^2\bar{\alpha} + \tau\bar{\beta} + \bar{\gamma}}\right]}{1 + \bar{\alpha}s + \bar{\beta}s + \bar{\gamma}s^2}$$

$$\cdot s^2 \Bigg|^2 \frac{N^2 \tau}{A^2 3.3} ds$$

$$= \frac{1}{2\pi j} \int_{-j\infty}^{j\infty}$$

$$\cdot \frac{1 + \left[2\left[\tau(\bar{\alpha} - \tau) + \dfrac{\tau^5}{\tau^3 + \tau^2\bar{\alpha} + \tau\bar{\beta} + \bar{\gamma}}\right] - \bar{\alpha}^2\right]s^2}{(1 + \bar{\alpha}s + \bar{\beta}s^2 + \bar{\gamma}s^3)}$$

$$+ \frac{\left[\tau(\bar{\alpha} - \tau) + \dfrac{\tau^5}{\tau^3 + \tau^2\bar{\alpha} + \tau\bar{\beta} + \bar{\gamma}}\right]^2}{(1 - \bar{\alpha}s + \bar{\beta}s^2 - \bar{\gamma}s^3)} \frac{N^2 \tau}{A^2 3.3} ds.$$

From James, Nichols, and Phillips,[7]

$$\sigma_{N\theta}^2 = \frac{-a_2 b_0 + a_0 b_1 - \dfrac{a_0 a_1 b_2}{a_3}}{2a_0(a_0 a_3 - a_1 a_2)} \frac{N^2 \tau}{A^2 3.3}$$

where

$$b_0 = \left[\tau(\bar{\alpha} - \tau) + \frac{\tau^5}{\tau^3 + \tau^2\bar{\alpha} + \tau\bar{\beta} + \bar{\gamma}}\right]^2$$

$$b_1 = 2\left[\tau(\bar{\alpha} - \tau) + \frac{\tau^5}{\tau^3 + \tau^2\bar{\alpha} + \tau\bar{\beta} + \bar{\gamma}}\right] - \bar{\alpha}^2$$

$$b_2 = 1$$

$$a_0 = \bar{\gamma}$$

$$a_1 = \bar{\beta}$$

$$a_2 = \bar{\alpha}$$

$$a_3 = 1.$$

The variables $\bar{\alpha}$, $\bar{\beta}$, and $\bar{\gamma}$ are shown in the preceding section.

The threshold of the system occurs when the system loses synchronism, or might be defined to mean when synchronism is lost enough of the time to cause a 1 per cent of full bandwidth change in the average. This might be when the rms value ($\sigma_{N\theta}$) of the noise is $\pi/2$ or when three times the rms value of the noise is $\pi/2$. Three $\sigma_{N\theta}$ of a normal distribution assures that less than 1 per cent of the noise peaks exceed $3\sigma_{N\theta}$. The threshold for the three systems are plotted for these two-values as a function of bandwidth in Fig. 25.

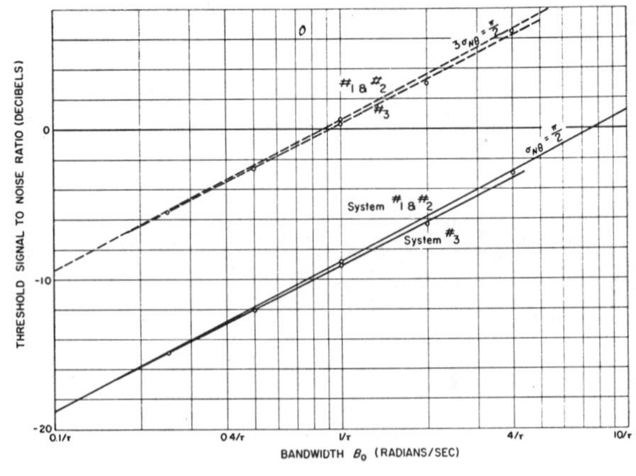

Fig. 25.

From this it can be concluded that there is very little improvement of noise of system 3 over the others and is not worthwhile due to its complication. In fact, system 2 allows a lower bandwidth for a fixed $\Delta\omega$ and will, therefore, have less noise than the corresponding conditions for system 3.

XI. Threshold Comparison of the Phase-Locked Loop and Ordinary Discriminators

It is difficult to make a comparison between the phase-locked loops and ordinary discriminators because the phase-locked loop was optimized against transient signals while ordinary discriminators usually are compared on the basis of sine waves. By plotting the data in the preceding sections in a different manner, it is shown (Fig. 26) that the threshold of the phase-locked loop is 0 db for 3σ and -10 db for σ. This is for the total noise in the channel. The corresponding threshold of ordinary discriminators is fixed by the band-pass filter and is $+10$ db. Therefore

[7] James, Nichols, and Phillips, "Theory of Servomechanisms," M.I.T. Rad. Lab. Ser., McGraw-Hill Book Co., Inc., New York, N. Y., no. 25; 1947.

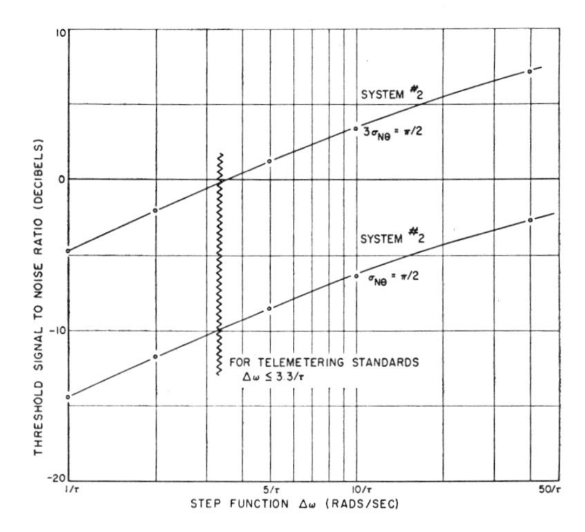

Fig. 26.

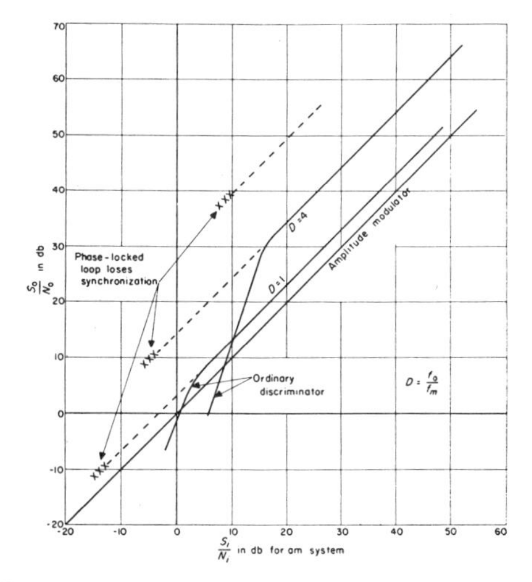

Fig. 27—Threshold level and signal-to-noise ratio comparison of ordinary discriminators and phase-locked loop.

an improvement of somewhere between 10 to 20 db is accomplished.

The exact improvement seems rather nebulous depending upon what is chosen for synch drop-out, σ or 3σ. Fig. 27 shows roughly what this means in terms ordinarily presented. It should be pointed out that B_0 is dictated by $\Delta\omega$ and, therefore, $\Delta\omega$ dictates the improvement, while Do is determined by the channel capacity or $Do \leq 1/\tau$. Figs. 28 and 29, p. 34, show experimental data, comparing the system improvement vs ordinary systems.

XII. General Discussion

From the beginning of this paper, commutated data was implicated. This is one of the most prevalent types of data in use and certainly presents one of the most severe requirements on the operation of the phase-locked loop yet still yields improvement over the ordinary discriminators in use. More threshold improvement is obtained if the magnitude of the step is reduced which allows a reduction in bandwidth but in the long run causes effectively a reduction in deviation ratio and consequently in signal-to-noise ratio. When maximum information bandwidth utilization is not imperative, the bandwidth may be reduced and further improvement in the threshold is obtained. This, however, must be done after the commutator and before the phase-locked loop. Since the data are much lower in bandwidth than the commutation rate, further improvement is obtained by designing the phase-locked loop around the statistical properties of the data if known. Certain techniques of sampled data interpolation allow a reduction in bandwidth of the sampled data. This can be applied to the phase-locked loop so that the phase-locked loop performs this interpolation. This can be accomplished with a decommutation device within the phase-locked loop which switches in a separate filter for each date channel (Fig. 30). The filter performs the function of a memory which returns the vco back to the last known value of the data and, therefore, reduces the magnitude of the individual step functions.

Lock-in time[8,9] is important in phase-locked loop utilization. Fig. 31 shows lock-in time presented in dimensionless units for universal usage with all telemetering channels. For standard use, channel 1 has a lock-in time on the order of 0.1 second while channel 18 is about 1 msec.

Linearity of the phase-locked loop acting as a discriminator is dependent upon the transfer characteristics of the vco within the loop since the vco frequency is identical to the incoming signal. This can be used to advantage since it is the inverse function of what it would be if it were the vco converting the data to the FM signal.

If identical transfer characteristics of the vco at the data source and at the phase-locked loop are used, then the system is linear. This is frequently not the case because of the difficulty in having identical units so a linear discriminator is used. In this case, the phase-locked loop is used only as a filter to obtain best threshold (Fig. 32). This leads to the concept of a tracking filter or a low-pass filter (the phase-locked loop) which follows the signal around yet strips off most of the noise within the band-pass filter.

The phase-locked loop has found other uses in the past and can be applied to others. In particular, it finds use in the carrier of the FM-FM system. To varying degrees of improvement, it may be applied to FM-FM, FM-AM, PCM-AM, PCM-FM, PCM-PM, etc.; discussion of these is best left to a separate paper.

No special problems exist in constructing the system except perhaps for the multiplier. Multipliers were constructed which would operate at frequencies of 400

[8] D. Richmond, "Color-carrier reference phase synchronization accuracy in NTSC color television."

[9] ———. "The dc quadricorrelator: a two-mode synchronization system."

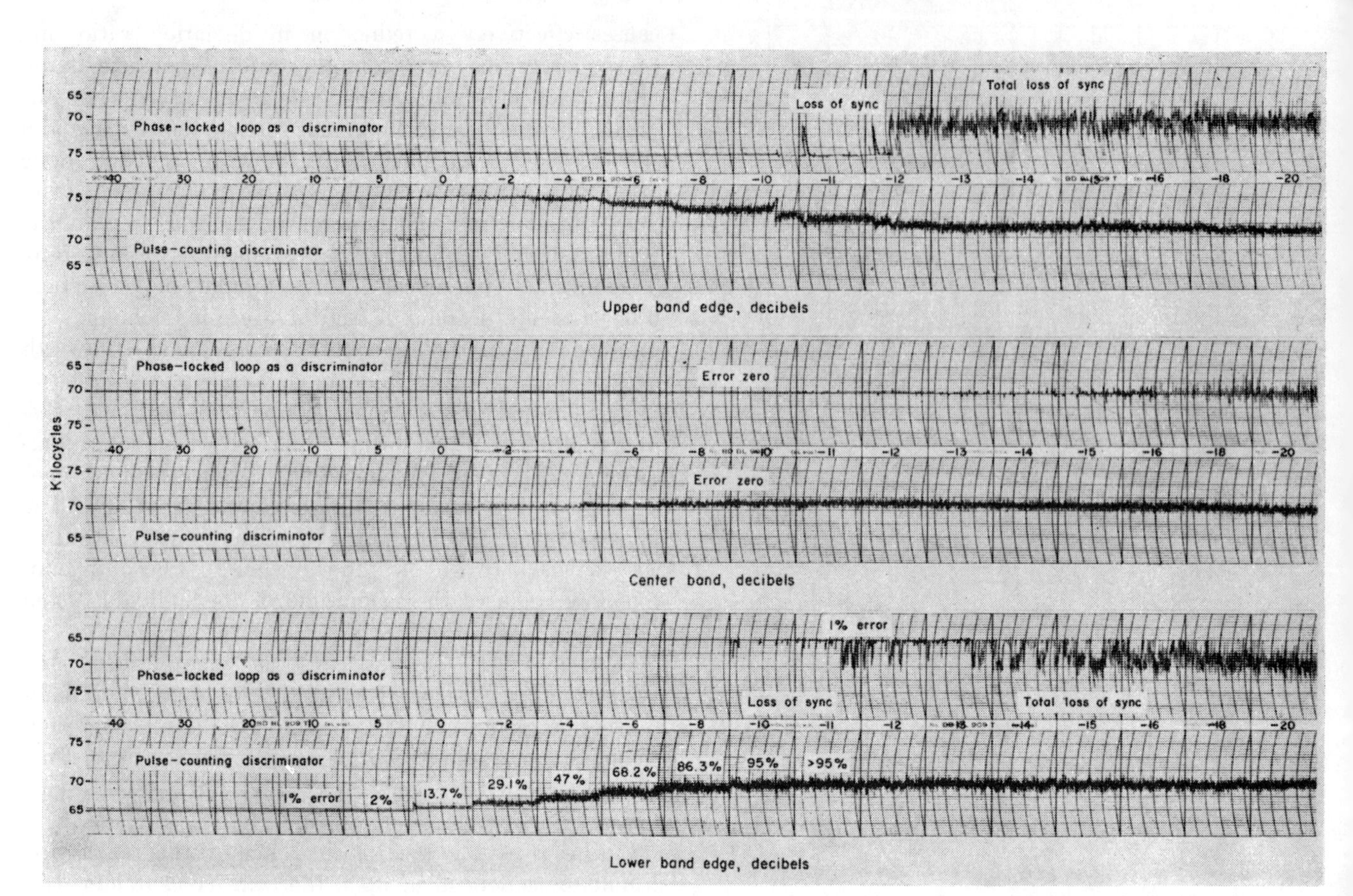

Fig. 28—Comparison of errors and operation of phase-locked loop and pulse-counting discriminator for RDB channel 18 at various signal-to-noise levels.

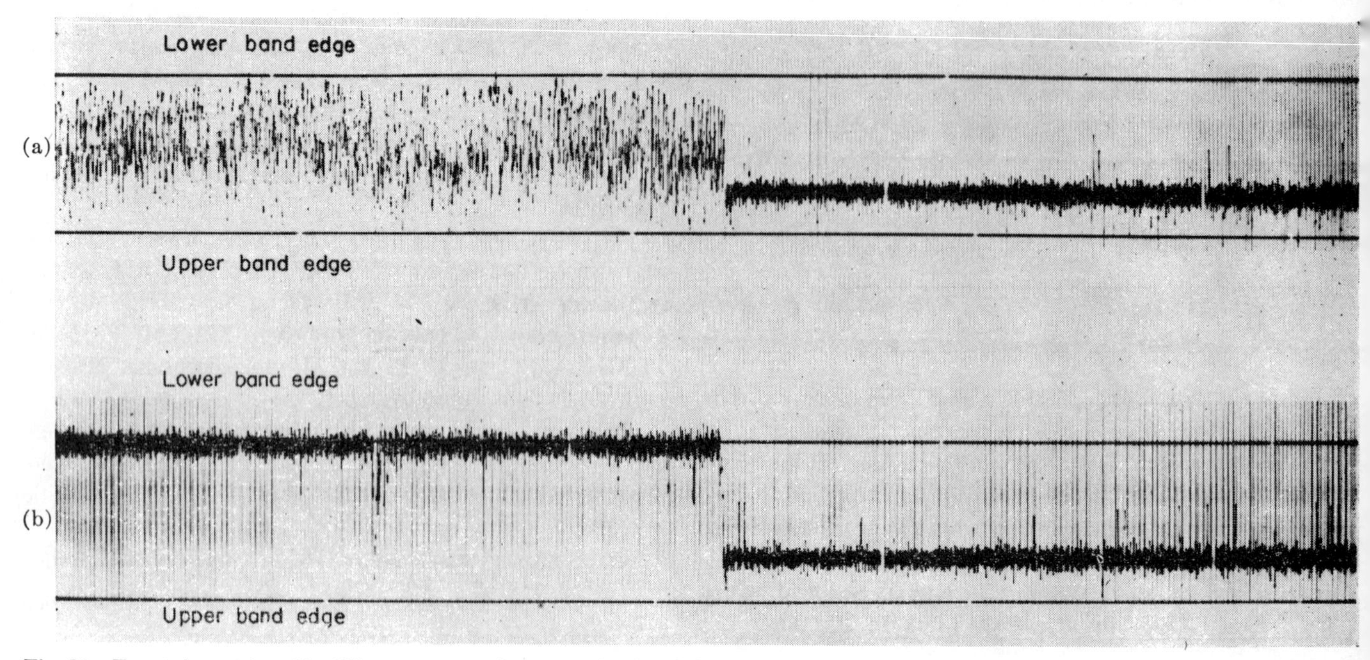

Fig. 29—Event data reduced in different manners from actual noisy flight. (a) Noisy signal through ordinary discriminator. (b) Same signal with phase-locked loop as a tracking filter, preceding ordinary discriminator.

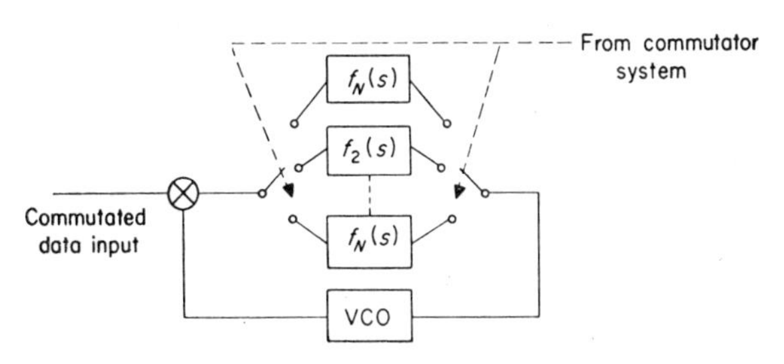

Fig. 30—Device for decommutation, data interpolation, and large threshold improvement.

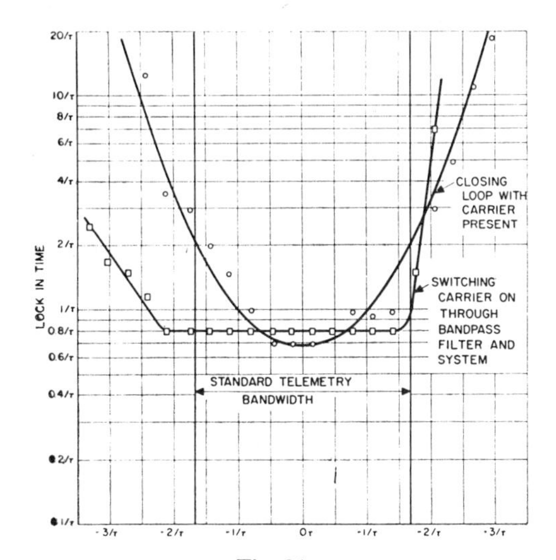

Fig. 31.

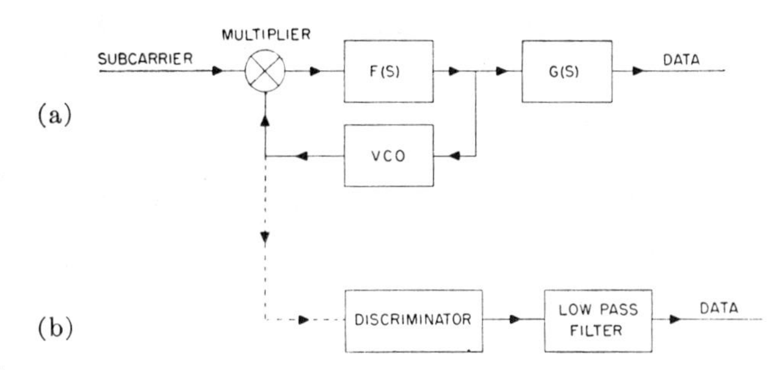

Fig. 32—Block diagram of phase-locked loop (a) as a discriminator or (b) as a tracking filter.

cps to 80 kc which are satisfactory for use in FM-FM subcarrier use. Perfect multipliers are not mandatory but are merely one of the finer points of design. General design criteria work acceptably well with ordinary phase detectors.

Bibliography

[1] Chestnut, H., and Mayer, R. W. *Servomechanisms and Regulating System Design,* New York: John Wiley and Sons, Inc., vol. 1, 1951.

[2] Jensen, G. K., and McGeogh, J. E. "An Active Filter," Radar Techniques Branch Radar Division, Naval Research Laboratory, Washington, D.C., Report No. 4630, November 10, 1955.

[3] Margolis, S. R. "The Response of a Phase-Locked Loop to a Sinusoid-Plus Noise," Jet Propulsion Laboratory, California Institute of Technology, Pasadena, Calif., External Publication No. 348, August 1, 1956.

[4] Rechtin, E. "The Design of Optimum Linear Systems," Jet Propulsion Laboratory, California Institute of Technology, Pasadena, Calif., External Publication No. 204, April, 1953.

Color-Carrier Reference Phase Synchronization Accuracy in NTSC Color Television*

DONALD RICHMAN†, SENIOR MEMBER, IRE

Summary—**The results of an evaluation of the capabilities of the NTSC color-carrier reference signal (the color burst) show this new color television synchronizing signal to be more than adequate; information inherent to the signal permits performance far in excess of that achieved by conventional circuits.**

Phasing information inherent to the burst is considered first with particular regard to measures of accuracy, the required amount of integration, and the extent of the spectral region necessary to translate the burst information.

Properties of elementary passive and active circuits for using the burst in receivers are described along with a determination of the limits of burst synchronization performance for these circuits.

Fundamental considerations in the theory of synchronization show that better performance is obtainable with two-mode systems.

Properties of two-mode systems are considered and lead to an evaluation of the limits of synchronizing performance permitted by the color burst.

The mathematical derivations necessary to support the discussion are presented in the Appendixes.

NTSC COLOR television adds color to a monochrome picture by means of a narrow-band, frequency interleaved carrier color signal which carries one component of the color information in its phase, and another component in its amplitude. It is customary to provide a phase reference in the transmitted signal in order that receivers shall be able to measure the instantaneous phase angle of the carrier color signal so as to reproduce the desired color. This is accomplished by transmitting a short burst of oscillations at color subcarrier frequency during line retrace intervals,[1] at a reference phase which corresponds to the (Y-B) axis.[2]

The color burst carries phasing information. This paper shows how much phasing information is contained in the color burst, and how it may be used.

Analysis of the factors limiting performance shows that, even under extreme conditions of interference and of stabilization requirements, the burst contains adequate information to provide a reliable color-carrier reference signal; in fact, the amount of phasing information in the color signal appears adequate enough so that a customer-operated control relating to color sync should be unnecessary on NTSC color television receivers. Analysis shows that presently used sync instrumentation systems appear capable of meeting but not necessarily exceeding a reasonable measure of the above requirements. However, information existing in the *signal* permits substantially better performance. The real limits of performance and sync systems which more fully utilize the signal information are discussed in this paper. Because of the excess of existing information, a variety of types of circuits can be used.

Several questions may be asked with regard to the amount of phasing information contained in the color burst and its application to provide a reference signal for color demodulation. These are: (a) How closely can the color-carrier reference signal be maintained to the true value, when signals are strong (and hence noise-free) and after transient effects have subsided? (b) How closely can the color-carrier reference signal be maintained in the presence of noise interference? (c) How long will a system or circuit designed to give satisfactory operation on (a) and (b) require to reach a stable mode of operation when stations are switched or a receiver is turned on? (d) How much performance is required in (a), (b), and (c)?

Of these questions, (d) is the most difficult to answer precisely; it depends on many subjective factors and may be obscured by temporary equipment difficulties. In order to provide a standard of comparison for use in this paper, a conservative (pessimistic) estimate has been made, based on past experience.

The answers to the questions are as follows:

(a) With a strong (clean) sync signal, the color-carrier reference signal may be maintained as closely accurate as desired, independent of other factors; in the presence of noise, the *average* phase may be maintained as closely as desired, independent of the required integration and transient characteristics; for example, designs presented later show how the static or average phase of the color-carrier reference signal may be controlled to within five degrees of the true value. Expressed as a time value this is an accuracy of approximately six mμsec. This phase accuracy implies a color fidelity probably substantially better than can be distinguished by the observer.[3]

(b) *The real limitation on performance* is thermal-noise interference, since this type of interference is the most difficult type to reject. It *is* rejected, however, *to any selected measure of reliability* by integration of the synchronization timing information over a suitably long period. Either of two basic types of integrators may be used. These are, one, passive integrators, and two, frequency-and-phase-locked self-oscillating integrators. The analysis presented in this paper shows that, under severe assumptions on the requirements of phase stability and signal-to-noise ratio, the required integration time for passive integrators is of the order of mag-

* Decimal classification: R583. NTSC Technical Monograph No. 7, reprinted by permission of the National Television System Committee from "Color System Analysis," report of NTSC Panel 12.

† Hazeltine Corp., Little Neck, N. Y.

[1] "Recent developments in color synchronization in the RCA color television system," RCA Labs. Report, Princeton, N. J.; Feb., 1950.

[2] Fig. 1 of "Minutes of the Meeting of Panel 14," NTSC; May 20, 1952.

[3] D. L. MacAdam, "Quality of color reproduction," Proc. I.R.E., vol. 39, pp. 468–485; May, 1951.

Reprinted from *Proc. IRE*, vol. 43, pp. 106–133, Jan. 1954.

nitude of 0.005 second, or less than a sixth of a frame period. Locked integrators on the same assumptions require 0.01 second for the integration to take place.

(c) The third requirement, of pull-in or stabilization time, is also limited by the signal-to-noise ratio and the requirement for integration. This may vary considerably with the method of instrumentation, but the limiting or optimum performance with regard to stabilization time is determined by the information carried in the signal; the limit imposed by signal information is found to be (for a reasonable measure of reliability) a few times the integration time discussed above. Later in this report this is shown to be approached under certain conditions by fairly simple passive integrators. It is also shown how locked integrators, characterized by some new forms of automatic frequency- and phase-control loops, may be made to achieve the upper limit of performance. Typical present APC (automatic phase control) circuits fall somewhat short of this limit, but when properly designed can be made to pull in quickly enough so as to appear virtually instantaneous, while permitting most of the burden of frequency stability to be borne by the transmitter.

These facts lead to the conclusion that there is adequate information in the color burst for completely automatic operation, without need for a customer control. The factors leading to this conclusion are presented in the following sequence:

Performance limitations for sync systems which are already synchronized are discussed first, in the section on "Synchronization Accuracy." The reliability of phase difference measurements, and factors relating to the integration time necessary to obtain a specified measure of reliability in the presence of noise are considered.

Then performance limitations of instrumentation systems are discussed with particular regard to the process of synchronization. The basic characteristics of passive and locked integrators are discussed in the section on "Elemental Sync Systems."

Evaluation of ultimate limitations for the signal, and factors leading to new sync systems capable of fully utilizing the signal information are presented in the section on "Theory of Synchronization." Factors of interest are mechanisms of pull-in, the reliability of frequency difference measurements, and the exchange of integration time for a specified measure of reliability in the presence of noise.

Effects of echoes and stability of the gate are briefly discussed.

The conclusions drawn regarding the adequacy of the signal are stated.

Mathematical derivations, which substantiate and illustrate the facts presented in this paper, are presented in several appendexes.

The NTSC Color Synchronizing Signal

Fig. 1 shows the NTSC color synchronizing signal in relation to the video and synchronizing wave form, in the vicinity of one line-retrace interval. It consists of a burst of approximately 9 cycles of sinusoidal wave form at the color-carrier frequency of 3,579,545 ($\pm$.0003%) cps,[4] approximately centered on the portion of the line blanking pulse following each horizontal sync pulse. It is omitted during the nine lines in each field in which the field synchronizing information is transmitted.

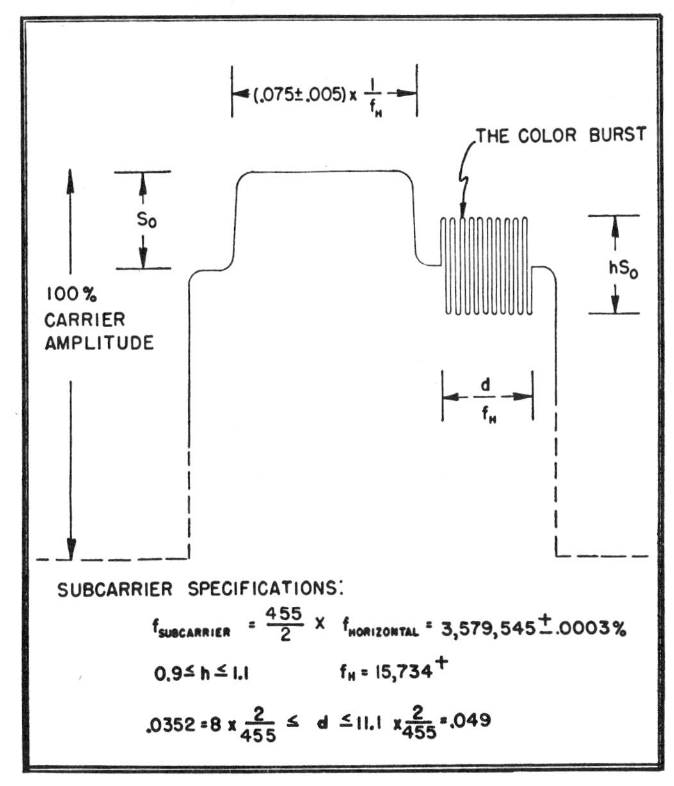

Fig. 1—Wave form during line retrace interval showing horizontal sync pulse and the NTSC burst reference signal.

Parameters of interest which are shown on the figure are:

S_0 = the amplitude of the line and field sync pulses, normally 25% of peak carrier amplitude measured in the video signal.

hS_0 = the peak-to-peak amplitude of the burst, measured in the video signal.

f_H = the line scanning frequency.

d = the duty cycle of the burst.

The color burst is used in the color television receiver to provide a control signal for the generation of a local continuous wave signal at the nominal burst frequency and locked to it in phase.

Synchronization Accuracy

Synchronizing Information

Any time-varying signal can carry timing information, the character of which depends on the distribution of signal energy throughout the frequency spectrum. In

[4] As specified by NTSC in February, 1953. The analysis is not critically dependent on the exact value of the color carrier frequency.

the case of a continuous sine wave, this timing information consists only of phase reference information because it is impossible to identify cycles of the carrier from each other. The same is essentially true of the pulse modulated sine wave which constitutes the burst; envelope information in the burst is not used. It is this phase reference information which is of interest with regard to color-carrier reference phase synchronization.

A signal which passes only through linear noiseless channels may be located in time (or phase) with theoretically unlimited precision. In the presence of noise the data obtained by a time (or *phase*) meter from the signal will fluctuate. This occurs because the timing information which can be extracted from the combination of signal-plus-noise in any specified interval is limited by the signal-to-noise ratio as well as by the statistical characteristics of the signal and noise.

Integration for Signal-to-Noise Ratio Improvement

The fluctuations in the phase data may be smoothed by integration. For example, the instantaneous output of the phase meter may represent the average of all data obtained over some preceding integration period T_M in duration.

Any measuring device which uses any form of integration or memory averages some effective number of independent measurements. One such integrator directly obtains a suitably weighted average (such as the least square error average) of all the data obtained in the preceding period T_M. Such an integrator provides a standard of comparison. Other forms of integrators may then be characterized by their effective integration times, T_M; several practical integrators are described later.

A section of a signal existing in an interval of duration T_M may be expressed as a sum of harmonics of the fundamental frequency $(1/T_M)$; the noise bandwidth associated with each component is equal to the spacing between components, or $1/T_M=f_N$. This means, for example, that if all of the timing information obtained in a period T_M from a signal consisting effectively of a signal sinusoidal component is averaged, that an improvement in reliability is obtained equivalent to that produced by passing the signal through a filter having a noise bandwidth of f_N.

Noise Interference

Noise is specified by its energy content and statistical characteristics. For a flat energy spectrum, taken as an example, impulse noise and white thermal noise represent opposite extremes, since for white thermal noise the relative phases of the several frequency components are completely random and incoherent; for impulse noise the relative phases of all components are related and are not random, although the time of occurrence of any impulse is a random variable.

Noise may be measured in terms of any convenient co-ordinate system into which the signal-plus-noise may be transformed, such as frequency, phase, amplitude, time of arrival, or more complex parameters.

Thermal noise is the most difficult to reject. It may be discriminated against only by averaging; this makes the effective error due to noise vary inversely as the square root of the number of measurements; hence, (for systems with fixed bandwidth) the error varies inversely as the square root of the integration time.

Impulse noise, or noise intermediate between thermal and impulse noise, may be rejected more easily than thermal noise since it represents a signal which can be recognized with a high measure of reliability and removed from the transmission channel.

A synchronizing system is a form of predictor which bases its estimates on past experience. When the input to the system has such a character (such as an improbable amplitude) that it is recognized with high reliability to be a disturbance, it is usually much better to use (at least approximately) the predicted signal as the input to the system for the duration of the disturbance. An equipment system for performing these operations is called an *aperture*. (Aperture systems are now widely used for line and field sync; the same principles are involved in the application to burst sync.)

Since thermal noise represents the most serious (as well as perhaps the most common) limitation to color synchronization performance, it is used in this paper as the measure of interference which must be overcome.

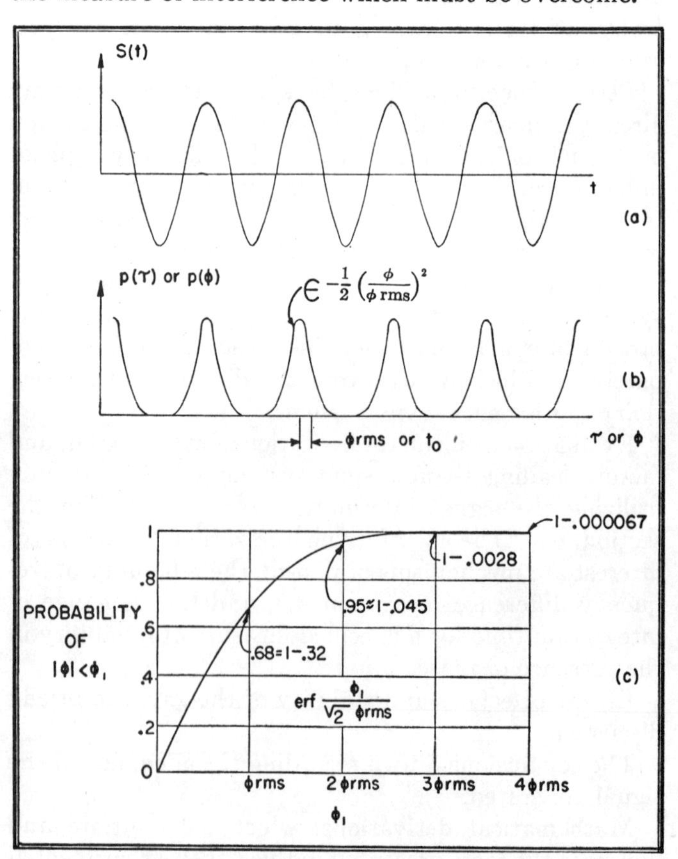

Fig. 2—Timing error distribution.

Measures of Reliability

A section of the burst reference signal is represented as $S(t)$ in Fig. 2(a). The time scale associated with the synchronizing signal may be identified with some representative point in a cycle which is selected as a reference.

The timing accuracy which is obtained for a given signal-to-noise ratio may be expressed in terms of a relative probability density function $p(\tau)$ such as is plotted in the curve of Fig. 2(b). The relative probability density curve permits the determination of the probability that the sync timing answer which results from a single measurement of the sync signal, using all of the information derived from the preceding period T_M, will occur within a specified time or phase interval. This probability is proportional to the area under the curve $p(\tau)$ or $p(\phi)$ within the specified interval. Due to the cyclic nature of the information, the time scale may be replaced by a phase scale. The curve for $p(\phi)$ defines the probability laws for the noise at the output of the synchronizing system. The curve is repetitive at the sync frequency. (The output noise from the sync measuring device has the same basic character from cycle to cycle.) For many signal energy distributions, and particularly for burst synchronization at the levels of output noise which give satisfactory performance, the curve $p(\tau)$ or $p(\phi)$ has very nearly the shape of a normal or Gaussian probability curve represented by the expression

$$\epsilon^{-1/2(t/t_0)^2} \quad \text{or} \quad \epsilon^{-1/2(\phi/\phi_{rms})}$$

in which case the phasing information may be completely described by the rms time error, t_0, or the rms phase error ϕ_{rms}, which may be expected for a specified set of measurement conditions.

For this case of the normal law the absolute probability that any measurement will yield an answer within a specified measure of the true answer may be represented in terms of the rms error. Fig. 2(c), which represents the integral of one lobe of the curve of Fig. 2(b) for the normal law, represents the probability that the magnitude of the phase error at any time is less than some selected phase error ϕ_1. ϕ_1 is measured in multiples of ϕ_{rms}. The curve illustrates that the probability is nearly unity only when ϕ_1 approaches $4\phi_{rms}$, which means that the effective peak value of Gaussian noise is near four times the rms value.[5]

The Sync Accuracy Equation

The parameters which determine the rms time error, t_0 seconds, for burst sync are:

The signal amplitude $\frac{1}{2}hS_0$ volts.
The duty cycle of the gated sine wave d as a fraction.
The rms noise (assumed flat over the band) N_W volts.
The video bandwidth occupied by the signal and noise f_W cycles per second.
The subcarrier frequency f_{SC} cycles per second.
The effective integration time T_M seconds.
The rms phase error ϕ_{rms} in degrees.
Equation (1) relates these parameters

$$\frac{S_0}{N_W} = \frac{1}{\sqrt{df_W T_M}} \frac{1}{t_0 f_{SC}} \frac{1}{\pi h} = \frac{1}{\sqrt{df_W T_M}} \frac{360}{\phi_{rms}} \frac{1}{\pi h} \tag{1}$$

This equation is derived in Appendix A.[6] The physical significance of the several factors in (1) is as follows:

The factor S_0/N_W represents (for example) the smallest ratio of line sync amplitude to rms noise for which $t_0 f_{SC}$ will not exceed a selected arbitrary value. It may be visually estimated if the composite video signal is viewed with a wide band oscilloscope. When $S_0/N_W=1$ the rms noise is equal to sync pulse amplitude. Since S_0 represents 25% carrier amplitude, and since the effective peak value of Gaussian noise is approximately four times the rms value, the condition $S_0/N_W=1$ also corresponds to the "peak" noise being approximately equal to 100% of carrier amplitude.

The factor $t_0 f_{SC}$ represents the fraction of a cycle of phasing error at frequency f_{SC} corresponding to the timing error, t_0. Thus

$$t_0 f_{SC} = \frac{\text{rms phase error in degrees}}{360°} = \frac{\phi_{rms}}{360°}.$$

The factor $df_W T_M$ is the number of effectively independent measurements yielding phase information which may be made in the interval T_M on a signal which is present for only a fraction d of time, and which occupies portions of the bandwidth f_W. The signal is actually present for a period dT_M; the effect of integrating over the period T_M is therefore to reduce the rms error by

$$\sqrt{df_W T_M} = \sqrt{d \frac{f_W}{f_N}},$$

where $f_N = 1/T_M$ is the effective noise bandwidth.

The factor $1/\pi h$ is a constant.

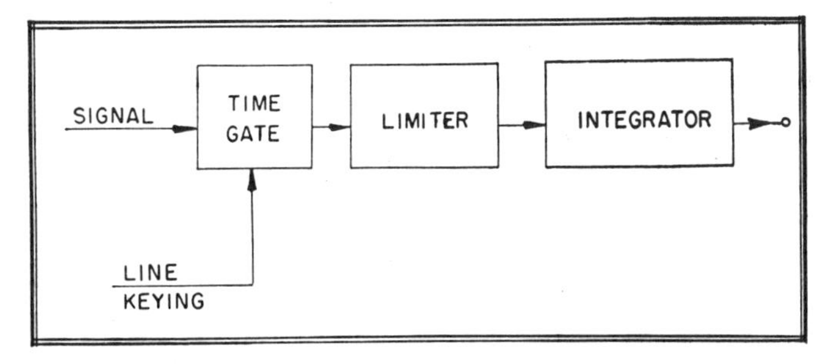

Fig. 3—Typical color-carrier phase reference generation system.

The Required Sync Accuracy

Equation (1) represents the theoretical upper limit of the phasing accuracy which may be derived from the subcarrier burst. A variety of circuits are available which can approach closely to this limit; these circuits are often of the f-rm shown in Fig. 3. The composite

[5] V. D. Landon, "The distribution of amplitude with time in fluctuation noise," Proc. I.R.E., vol. 29, pp. 50–55; Feb., 1941.

[6] D. Richman, "Theoretical limit to time difference measurements," *Proc. NEC*, vol. 5; pp. 203–210; 1949.

video signal is fed to a time-gate which is keyed from line flyback to select the burst, which is then amplitude limited and integrated. Practical integrators are described later.

The sync accuracy equation permits the determination of how much integration is required in order to obtain satisfactory performance under extreme conditions. However, due to the many subjective factors involved it is not possible to specify exactly what is the lowest level of signal-to-noise ratio which will be tolerable from a visual viewpoint;[7] it is equally difficult to specify exactly the largest value of rms phasing error which will not cause visible degradation of the picture. Accordingly, Fig. 4, which is a plot of (1), presents graphically the relations between the relevant factors over a range which probably includes the limiting case of interest. Fig. 4 is based on adverse tolerances presented below.

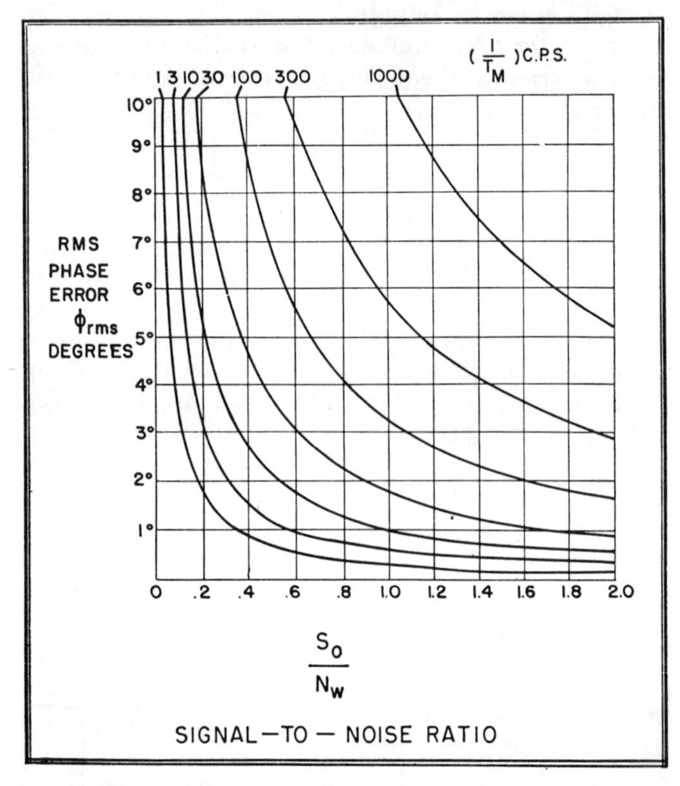

Fig. 4—Phasing accuracy relations for NTSC burst synchronization.

Fig. 4 presents the relation between the rms phase error, $\phi_{\rm rms}$, (in degrees) and the signal-to-noise ratio, S_0/N_W, with the integration time, T_M, (in seconds) as a parameter.

For the case corresponding to the most adverse tolerances, $h=.9$, $d=.0352$, and $f_W=4.3$ mc. Equation (1) then reduces to

$$\phi_{\rm rms}\frac{S_0}{N_W} = .33\sqrt{\frac{1}{T_M}} = \frac{1}{3}\sqrt{\frac{1}{T_M}} \tag{2}$$

which is shown graphically in Fig. 4.

[7] P. Mertz, A. D. Fowler and H. N. Christopher, "Quality rating of television images," PROC. I.R.E., vol. 38, pp. 1269–1283; Nov., 1950.

These curves show that any selected phase accuracy $\phi_{\rm rms}$ can be obtained with decreasing signal-to-noise ratio S_0/N_W if more time T_M is taken for integration of the signal timing information; i.e., if more measurements are integrated in each complete measurement.

(The facts presented later in this paper with regard to the relations between noise integration and other properties of sync systems indicate that the conclusions reached regarding the reliability of the signal are *not* critically dependent upon the assumed values of S_0/N_W and $\phi_{\rm rms}$.)

System Efficiency and the Distribution of Timing Information

The relationships presented above describe the performance of the system when all of the information of the signal is applied usefully. Another parameter which needs to be introduced in order to determine the actual noise bandwidth required is the decoding efficiency, which represents the fraction of the timing information of the signal which is used. Systems with equal noise bandwidths but different decoding efficiencies will give different performance.

In the burst system practical considerations relating to tolerances and to the stability of the gate derived from horizontal sync may result in a gate width r times wider than the narrowest sync burst. Factors relating to this are described later. It results in a requirement of noise bandwidth and integration time such that

$$T_M = \sqrt{r}\, T_{M\,\rm LIMIT}$$

$$f_N = \frac{1}{\sqrt{r}} f_{N\,\rm LIMIT}$$

where

$(1/\sqrt{r})$ is a system efficiency

r = ratio of actual gate width to minimum burst width.

There is another cause of loss of decoding efficiency in sync systems which is of interest. This relates to the relative distribution of timing information in the frequency spectrum. For burst sync systems which are properly designed, effectively all of the information may be used; (common attainment in horizontal sync systems has not been so high).

Fig. 5(a) shows the relative distribution of timing information in the frequency spectrum occupied by the burst. The basis for this curve is discussed in Appendix A. The effective accuracy which can be obtained if only a portion of the information is used may be measured in terms of the ratio of the noise bandwidth required (at any specified signal-to-noise ratio) to the noise bandwidth required if all of the information is used. For example, a problem of interest in receiver design is the relationship between bandwidth in the burst amplification channel and efficiency. If a passband symmetrically tuned about subcarrier frequency is used in this channel then the system efficiency resulting is represented by the curve sketched in Fig. 5(b). The curve depends, of course, on the width of the burst. Even for the narrowest burst a total bandwidth of approximately 600 kc translates nearly all of the timing information.

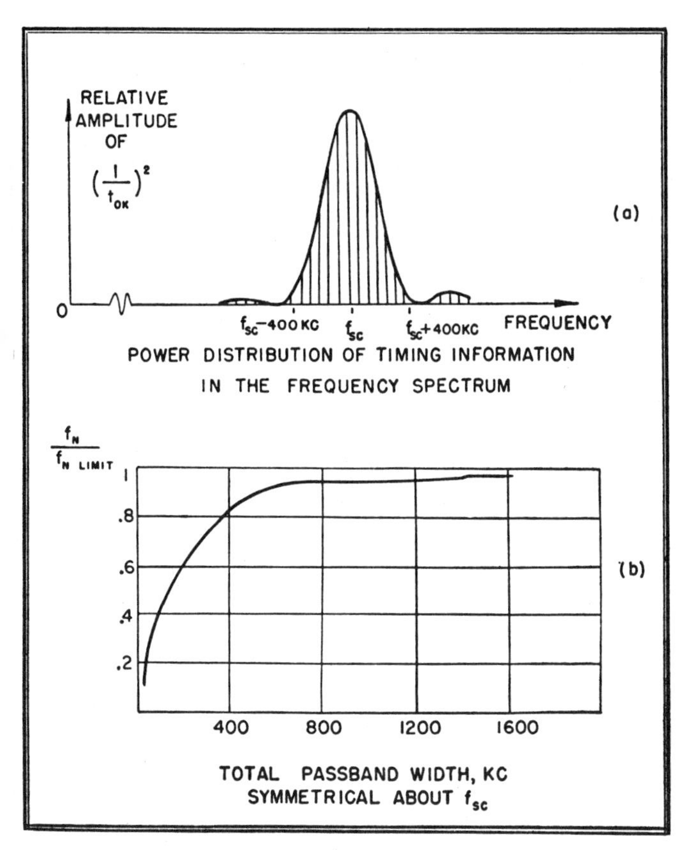

Fig. 5—Frequency distribution and system efficiency of burst sync timing information.

Example: As an illustration suppose the limiting parameter values of interest are approximately $\phi_{rms}=5°$ and $S_0/N_W=1$; these conditions correspond to the point in the center of Fig. 4; then from (2) $T_M \geqq 0.0045$ second. *The required noise bandwidth* for a gate width ratio $r=1.2$ is then approximately $f_N=200$ cycles per second. *This figure is used as a basic design parameter for the practical forms of integrators which will be discussed in this paper.*

Elemental Sync Systems

The function of combining signal information derived over an extended interval of time is accomplished by use of circuits which may broadly be classified as integrators. The performance characteristics of two basic forms of integrators are discussed below. The parameters of interest are:

1. The noise bandwidth and integration time of the system.
2. The static phase accuracy. In general, in systems involving feedback, this varies inversely with a circuit gain parameter and may be made nominally as small as desired.
3. The frequency pull-in range of the system. This is the maximum (single peak) frequency detuning for which the system will automatically achieve the desired final operating condition.
4. The stabilization time T_S; or the time required for all operating characteristics to reach effectively their stabilized conditions. This may consist of one or more definable segments.
5. The phase pull-in time T_ϕ; or the transient time required for the output phase of the system to reach some definable measure of its final conditions.
6. The frequency pull-in time T_F, applicable to systems in which a local signal oscillator must be controlled, or the time necessary for the oscillator frequency to be changed from its initial frequency to some selected reference frequency such as a frequency from which the net differential phase change between sync signal and reference oscillator will not exceed one whole cycle. This overlaps the phase pull-in time T_ϕ.

The first integration system discussed is the Passive Integrator. For this system stabilization consists effectively of a phase transient. The limitations of this system are: practical limitations on how high the circuit Q may be and the possibility of detuning.

These limitations are overcome in the second form of integrator called a Standard APC (Automatic Phase Control) System. In this system the signal is heterodyned against a local carrier at the same frequency permitting the desired filtering to be accomplished by means of a low-pass filter which thus effectively provides unlimited Q. The limitations of this system relate to the difficulties of obtaining synchronization and the long pull-in times which result when narrow noise bandwidths are required.

The real limitations imposed by the signal, and some system fundamentals related to using all of the information in the signal, are presented later.

Passive Integrator

The circuit of Fig. 6(a) shows one form of practical integrator. This is a passive integrator in which the required integration is obtained by use of a high-Q filter. The input signal to the filter consists of time-gated amplitude-limited bursts of sine waves at subcarrier frequency f_{SC}. Because of the gating and limiting, sidebands near f_{SC} (which are separated by integral multiples of f_H) as well as harmonics of f_{SC} which are generated in the preceding limiter, all have effectively the same phase modulation due to noise. The noise bandwidth of the filter needs to be less than or equal to the value of f_N which was computed above. If the filter is approximately equivalent to a single resonant circuit, the noise bandwidth is $f_N=(\pi/2)f_3$ where f_3 is the 3 db bandwidth. The bandwidth f_N is indicated in Fig. 6(b). Thus the filter bandwidth should be approximately $(2/\pi)(200)=127$ cps between 3 db points. The Q desired is $f_{SC}/f_3 \approx 28{,}000$. This requires the use of a crystal filter. Practical crystals in the frequency range of the color subcarrier can achieve the required Q, but up to the present time apparently cannot exceed it by a large factor.[8,9] The sum of transmitter frequency tolerance of ±11 cps and the frequency tolerance of the crystal is comparable with the filter bandwidth. Fig. 6(c) shows how undesirably large static phase shift might result from normal detunings. This is prevented in the system shown in Fig. 6(a) by use of feedback for automatic

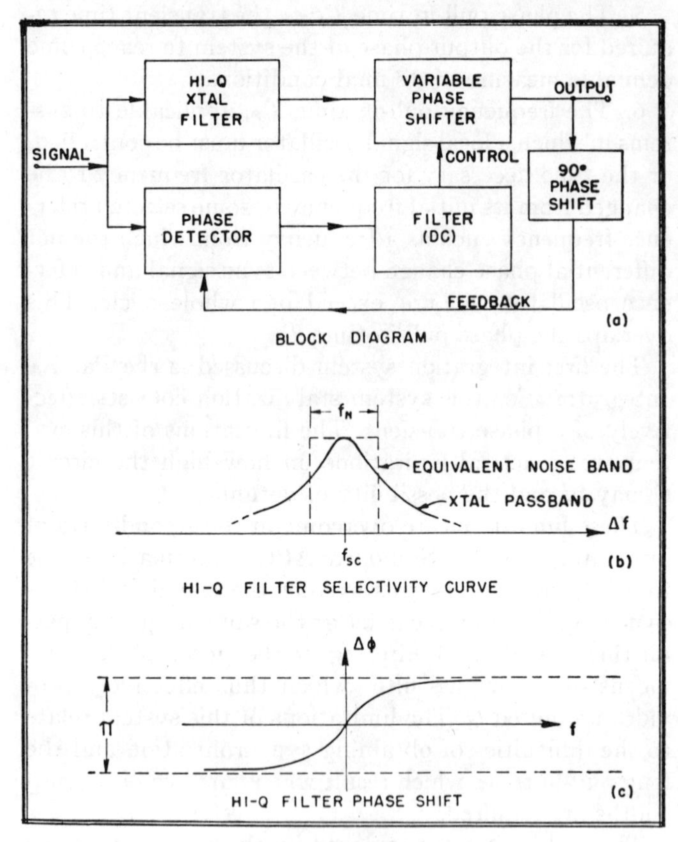

Fig. 6—Passive integrator.

static phase correction. The circuit includes in addition to the high-Q crystal filter a variable phase shifter, a phase detector (which has associated with it a 90° phase shift in one of the signal paths) and a low-pass (dc) filter in the feedback loop for correcting the average phase of the system. Other arrangements are possible; for example, a post-corrector might be used with the feedback signal derived directly from the output of the crystal filter, or a controllable reactance might be coupled to the crystal filter to insure optimum tuning.

The static phase may be maintained as closely accurate as desired by putting a suitably large amount of dc gain in the feedback loop. The signal-to-noise ratio at the output of the system will not be measurably changed if the dc filter is such that the bandwidth of the phase feedback loop is narrower than that of the crystal filter. Design considerations are discussed in Appendix B.

If the crystal stability is comparable to the transmitter frequency stability, the frequency error will be small enough so that rapid phase stabilization will occur when channels are switched. The switching transient is a phase transient and the stabilization time for small detunings will be or the order of a few times the transient time constant of the phase feedback loop. For the crystal bandwidth required, this time is essentially instantaneous. It may be noted however that if appreciable mistuning could occur the gain versus frequency characteristic of the high-Q filter would substantially reduce the amplitude of the correction signal, resulting in considerably increased stabilization time, and effectively reduced loop gain.

Standard Automatic Frequency and Phase Control Locked Integrator

Fig. 7(a) shows the block diagram of a standard automatic frequency and phase control loop. It includes a local reference oscillator, a phase detector which compares the relative phase difference between the sync signal and the oscillator, a filter which partly determines the transfer characteristic of the APC loop as an integrator, and a reactance tube for controlling the oscillator frequency. The loop gain for this system has the dimen-

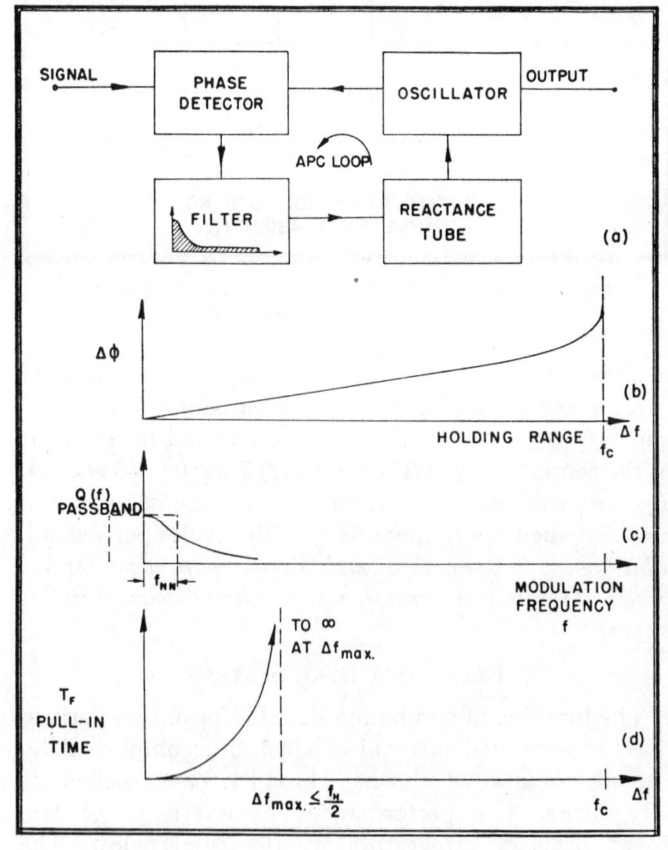

Fig. 7—Standard APC locked integrator.

sions of a frequency, f_C, which is equal to the frequency holding range of the APC system. Included in this characteristic is the dc transmission of the filter. Fig. 7(b) shows the relationship between the static phase error, $\Delta\phi$, and initial oscillator detuning, Δf. By making the holding range much larger than the normal operating range the static phase may be controlled as tightly as desired; here again the price of this control is high loop

[8] W. G. Cady, "Piezoelectricity," McGraw-Hill Book Co., Inc., New York, N. Y.; 1946.

[9] A. W. Warner, "High-frequency crystal units for primary frequency standards," Proc. I.R.E., vol. 40, pp. 1030–1033; Sept., 1952.

gain. Fig. 7(c) shows effective passband characteristic $Q(f)$ of the APC loop as a function of modulation frequency. This is determined largely by the ac transmission of the filter in conjunction with the feedback characteristics of the loop. The noise bandwidth f_{NN} is defined in the normal fashion and indicated on the figure. Since an APC loop phase detector is essentially a synchronous detector and does not distinguish between those noise components which are above or below the local oscillator frequency, then $f_{NN}=f_N/2$, and the effective integration time $T_M=1/2f_{NN}$; the noise bandwidth of the APC loop should not exceed approximately 100 cps for equivalent performance with the high-Q filter.

Fig. 7(d) is a sketch of pull-in time for this loop as a function of Δf. The pull-in range cannot exceed half the gating frequency, i.e. $f_H/2$, and for many designs is substantially smaller. The pull-in mechanism of this loop is not the most efficient one possible. Pull-in times are particularly long near the limit of the pull-in range. The APC loop of Fig. 7(a) is of the same basic type[10] which has achieved essentially universal use in television receivers as an integrator for line frequency synchronizing information. A detailed analysis of the characteristics of this loop is presented in Appendix C and a derivation of the pull-in time relationships is presented in Appendix D.

The pull-in range and time are a function of some design parameters discussed later. It has been found that for optimum design there is a limit to the pull-in performance obtainable with this loop. For these limit designs the following performance is obtained:

(a) The static phase error $\Delta\phi$ may be as small as possible and in fact must be smaller than some specified number in order that pull-in time be minimized.
(b) The pull-in range is equal to $\pm(f_H/2)$.
(c) Except near the limit of pull-in range, the pull-in time and noise bandwidth are very nearly related to the frequency detuning, Δf, by (3)

$$T_F f_{NN} \approx 4\left(\frac{\Delta f}{f_{NN}}\right)^2. \tag{3}$$

This has been used in Fig. 8 to plot the limit of pull-in performance for optimum design standard APC loops. Fig. 8 represents the pull-in time T_F in seconds as a function of the noise bandwidth f_{NN} in cycles per second. The range of f_{NN} in this log-log plot is from 10 to 1,000 cps with the approximate normal required bandwidth of 100 cps in the center of the graph. Pull-in times ranging from less than one-tenth to approximately one second appear instantaneous and may be characterized as "good." Pull-in times between 1 and 10 seconds are acceptable but probably close to the limit of adequate performance and have been designated "fair." Pull-in times in excess of 10 seconds are definitely "poor."

The relationship between f_{NN} and T_F is shown for several values of Δf. For example an optimum design unit having a noise bandwidth of 100 cycles will require 4 seconds to pull in from 1,000 cycles detuning. This indicates that such a sync system should be adequate for completely automatic phase control but that it apparently does not have an excess of available performance; for example, if the noise bandwidth needed to be reduced to 50 cycles, then 32 seconds would be required to pull in 1 kc.

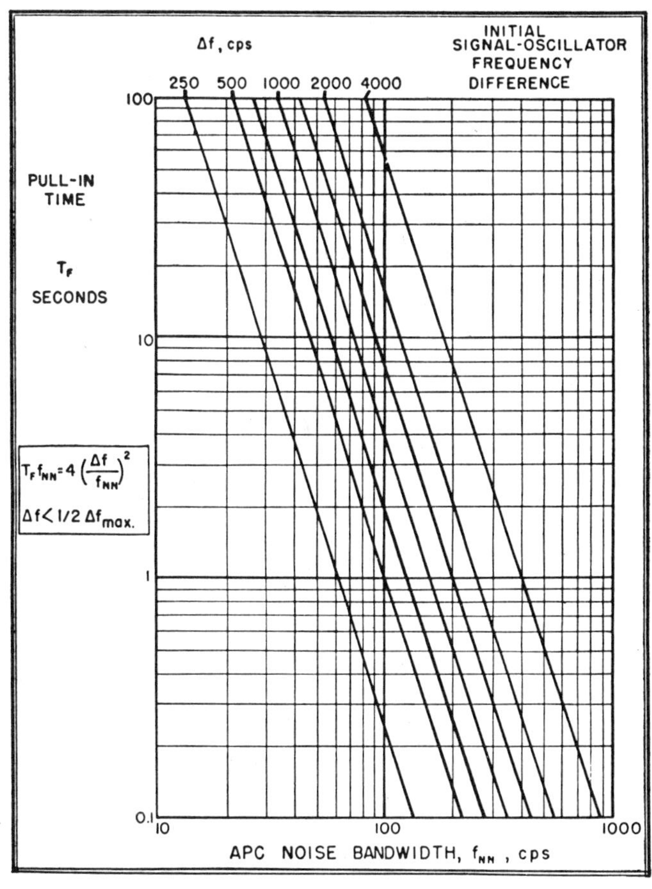

Fig. 8—Standard APC optimum pull-in performance.

Pull-In Performance Attainable with a Standard APC System

Not all designs of APC circuits will achieve the limits of performance discussed with respect to Fig. 8. In fact, partly due to economic limitations, the majority of past designs have fallen short of the limit. Accordingly, Figs. 9 and 10 are presented as a basis for demonstrating the pull-in limitations of the Standard APC System. The curves are expressed in terms of what are believed to be the parameters of interest to the user, specifically the noise bandwidth f_{NN}, the initial frequency difference Δf, and the frequency stabilization time T_F. The dimensionless parameters $T_F f_{NN}$, and $\Delta f/f_{NN}$, are used as ordinate

[10] K. R. Wendt and G. L. Fredendall, "Automatic frequency and phase control of synchronization in television receivers," Proc. I.R.E., vol. 31, pp. 7–15; Jan., 1943.

and abscissa. Two different parameters, designated m and K, which are discussed in Appendix C, appear. The parameter m varies inversely as the dc loop gain for fixed noise bandwidth. The figure shows that increased dc loop gain (smaller m) and hence tighter static phase control permit wider pull-in range and a closer approximation to the minimum pull-in time curve. The parameter K which is a damping coefficient (discussed in

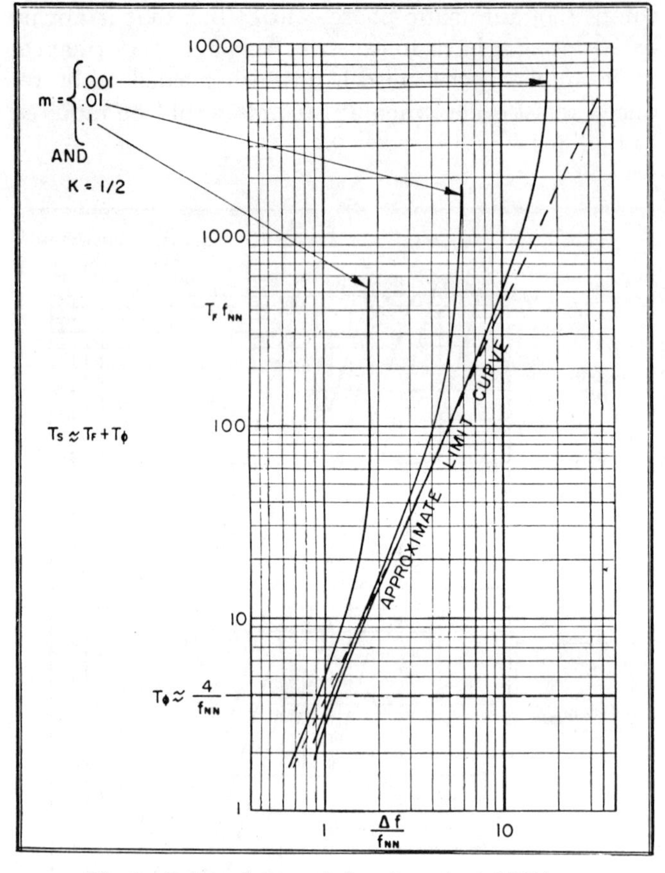

Fig. 9—Pull-in characteristics of standard APC loop.

Appendix C) determines the level of the limit curve as indicated in Fig. 10. Over part of its range of variation the parameter K permits an exchange of minimum pull-in time for pull-in range. The maximum increase, however, is limited to a 50% increase in frequency pull-in range, over designs which approach the optimum pull-in time limit curve.

The mathematics upon which these curves are based is presented in Appendexes C and D. Appendix C introduces and presents the relevant relations between the parameters of the Standard APC System. Derivation of the pull-in time equation and discussion of the pull-in phenomenon is presented in Appendix D.

Theory of Synchronization

Improved Sync Systems

The systems described thus far permit a level of performance which appears to satisfactorily meet the requirements for burst synchronization but do not appear to have a large excess of performance. The signal itself permits substantially better performance.[11] This will be shown below by considering the limitations of the systems presented thus far and introducing the factors which lead to full utilization of the signal information. This leads to a sync system which appears capable of efficiently using all of the timing and synchronizing information in the signal. Then an implementation of this system is described which appears applicable to NTSC color television receivers to produce what may be ideal performance at no substantial cost increase.

Finally, the approximate upper limit of performance capability for the signal is evaluated numerically. The limitations on the previous system relate to the severe restrictions interrelating noise bandwidth and pull-in time. There appear to be a variety of new sync systems which can overcome this limitation. Several varieties have been instrumented and found practical. However, the potentialities of the NTSC burst sync system are perhaps most clearly demonstrated by examining what may be the upper limit of performance.

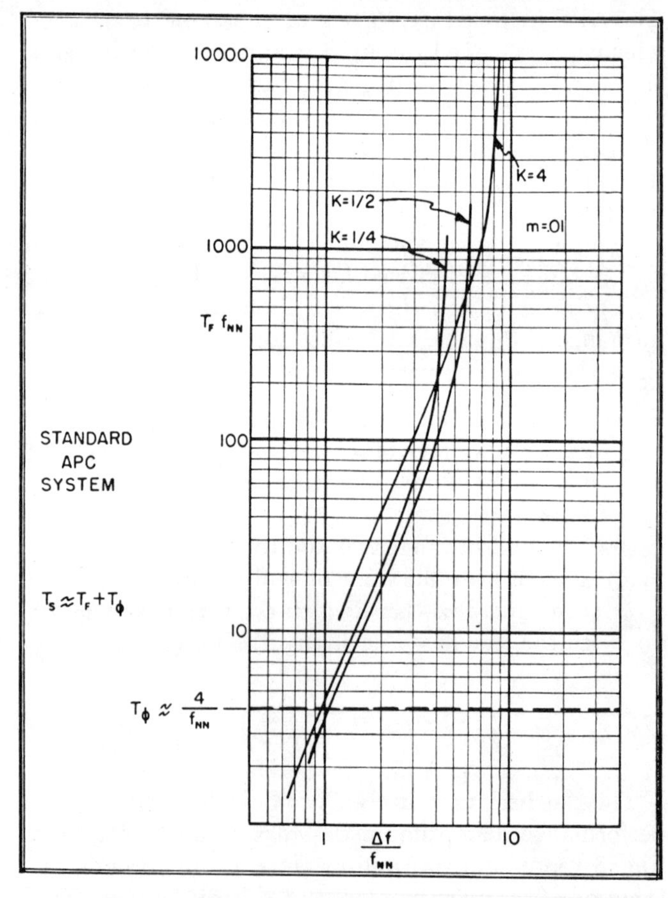

Fig. 10—Effect of variations in the parameter K.

Two Mode Systems

There are two separate and distinct modes of performance of sync systems. These relate to (a) the phase stability attainable after the system has achieved a stable synchronized operating condition, which has been

[11] D. Richman, "Theory of synchronization applied to NTSC color television," IRE Convention Record, Part 4; 1953.

discussed in some detail earlier in this paper, and (b) the performance associated with the system achieving that final state. Each of these modes has fundamental physical restrictions and characteristics associated with it. The full measure of performance permitted by the signal can be achieved by a system which makes these two modes of operation as independent as possible of each other and of each other's limitations.

Some systems use the same mechanism for hold-in and pull-in. The Standard APC System falls into this category. It is inefficient in its use of signal information. Other types of systems use a multiplicity of mechanisms, usually two.[12] One mechanism is designed for stable performance after synchronization, the second mechanism is designed to produce synchronization. *Such a device must have within it the inherent ability to extract from the signal the necessary information with regard to the mode of performance which is required.* For example, it should not confuse noise which may be present when the system is synchronized with a beatnote indicative of a lack of synchronism.

Factors Relating to Frequency Pull-In

There are two basic factors which relate to frequency pull-in. The first problem is concerned with the mechanism whereby a frequency difference is recognized in the presence of strong signals and a control voltage generated which can be utilized for pull-in. The second problem relates to the ability of the mechanism associated with pull-in to discriminate against noise interference.

Frequency Recognition

This separation of the requirements of the system leads to the following principle. *The real limitation of a synchronization system with respect to frequency pull-in is the ability of the system when out of sync to recognize a frequency difference and distinguish it from noise.*

This sets the *real upper limit of performance.* If the frequency determination is effectively linear, then after a time delay which permits the frequency difference to be measured to within a suitable measure of reliability, the reference oscillator may be switched instantaneously by the proper amount to insure synchronization. A system for accomplishing this may be called an *ideal sync system.* Just as with phase measurements this reliability is obtained by integrating the frequency difference information for an adequately long period of time. The shortest stabilization time consistent with reliable performance is therefore determined by the integration time necessary to measure a frequency difference with a suitable measure of reliability.

The Pull-In Control Effect

Fig. 11 represents the generated control effect for pull-in for two important synchronization systems. Fig. 11(a) relates to the frequency pull-in characteristic of a standard APC loop. The generated control voltage for pull-in is shown as a function of instantaneous applied frequency difference Δf. If the frequency is within a range roughly two-thirds that of the noise bandwidth, pull-in (as explained in Appendix D) is effectively instantaneous. The system never slips a cycle; a dc voltage for frequency control is generated which is proportional to the frequency difference. For larger values of Δf the system slips cycles but by virtue of the feedback in the APC loop generates a dc component of control voltage which varies in the inverse fashion with frequency difference indicated in Fig. 11(a). This inefficient control effect may be compensated for in this system by very high ratios of dc to ac loop gain ($1 \gg m$) but at the expense of the long pull-in times indicated by Fig. 9 and 10. An automatic frequency control system[13] containing a linear frequency difference detector[14] which generates a control voltage proportional to the frequency difference for all frequency differences of interest as indicated in Fig. 11(a) provides a more efficient indication of large frequency differences.

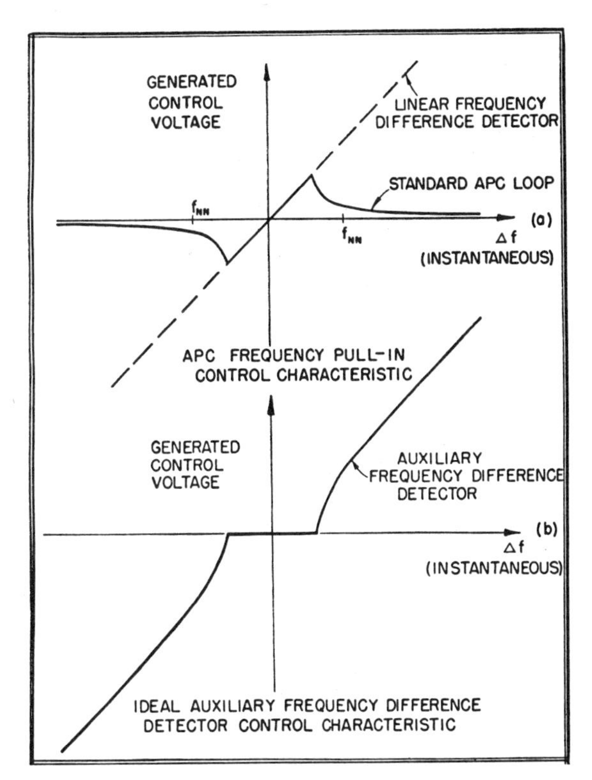

Fig. 11—Synchronization control characteristics.

Improved performance may be achieved by supplementing the APC system with an "Ideal Auxiliary Frequency Difference Detector," the control characteristic of which is shown in Fig. 11(b). Such an auxiliary detector can provide a suitable control effect for nearly optimum pull-in performance and as indicated by the flat

[12] Fundamentals relating to systems analyzed here have been applied to automatic gain control circuits as well as to sync systems.

[13] C. Travis, "Automatic frequency control," Proc. I.R.E , vol. 23, pp. 1125–1141; Oct., 1935.

[14] C. F. Shaeffer, "The zero-beat method of frequency discrimination," Proc. I.R.E., vol. 30, pp. 365–367; Aug., 1942.

portion of the curve will *automatically turn itself off when synchronization has been achieved;* this occurs when the frequency difference is reduced to within the linear sloping portion of the curve of Fig. 11(a), within which range the standard APC loop can produce effectively instantaneous pull-in.

A Sync System which Efficiently Uses the Signal Information

Fig. 12 represents the block diagram for a sync system having the auxiliary frequency detection control characteristic described with regard to Fig. 11(b). It includes a Standard APC System such as was shown earlier in Fig. 7 and in addition an auxiliary frequency difference detector which supplements the pull-in performance of the APC system.

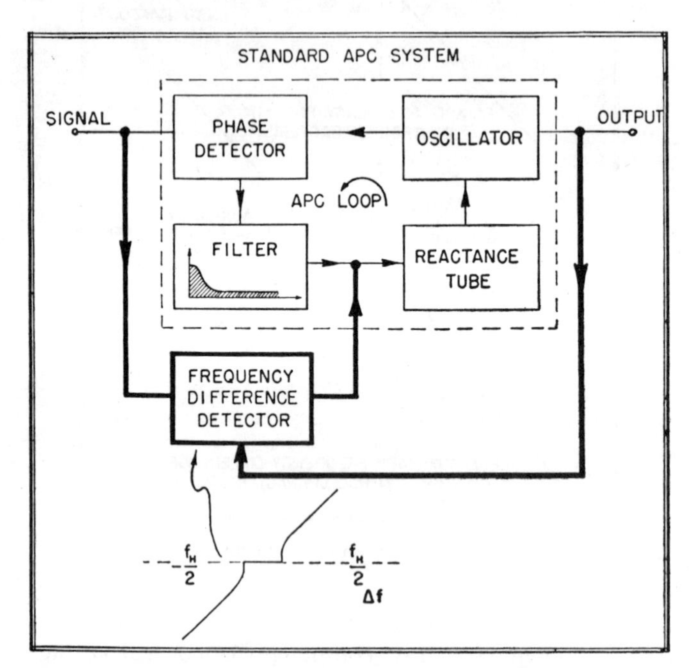

Fig. 12—A synchronization system capable of using total signal information at maximum efficiency.

The idealized upper limit performance described earlier under "Frequency Recognition" may be achieved by means of a suitable interconnection circuit. However, with the stepped characteristic of Fig. 11(b) an essentially direct connection is feasible. The composite system functions as a form of automatic frequency control system when out of sync and as an automatic phase control system when in sync; the auxiliary frequency difference detector turns itself off automatically by virtue of the shape of its control characteristic.

The ideal switched system has pull-in time equal for all frequency differences.

AFC systems normally require high loop gain and are characterized by a pull-in time constant. In some instrumentations of the system of Fig. 12 a loop gain of approximately unity (or a little more for tolerance purposes) may be adequate if the frequency difference detector includes a small amount of delay in its output. As soon as the oscillator is brought near the frequency of the sync signal, the high-gain APC system becomes operative, and the frequency difference detector is automatically inactivated.

The Quadricorrelator: A Frequency Difference Detector

In order to illustrate in more detail the problems and characteristics associated with the achievement of effectively upper limit performance, a form of circuit arrangement is introduced here which appears capable of using elements already present in color television receivers operable on NTSC standards to achieve the ideal frequency difference detection described above. This form of circuit will be called a quadricorrelator in this paper. Analysis of the performance characteristics of the quadricorrelator presented in Appendix E shows that when preceded by a limiter it comes within a few db in signal-to-noise ratio of using all of the signal information for signal-to-noise ratios of interest here. When the limiter is omitted from the system, the quadricorrelator is an efficient frequency detector; the extra noise due to amplitude modulation disappears after pull-in.[15] It is a true frequency difference detector since it is not subject to tuning errors. The excess of available over required noise discrimination suggests that the limiter can be omitted.

There is no real purpose to accomplishing pull-in much more rapidly than perhaps a few tenths of a second. The simple quadricorrelator instrumentations appear (on this basis) to give effectively optimum performance.

A block diagram of a basic form of a quadricorrelator is shown in Fig. 13. Its elements are a pair of synchronous detectors which are fed with reference signals in quadrature with each other so that the phase detector outputs represent "in phase" and "quadrature" components of the applied sync signal. These output signals are then limited in maximum frequency to (for example) $f_H/2$ by filters as indicated in Fig. 13(a). The output of one of the synchronous detectors goes through a differentiating circuit which provides a 90° phase shift through the passband. The two signals are then heterodyned in another synchronous detector, and the output is integrated in a narrow band filter; a low-pass filter is shown. This filter exchanges brevity of integration time for reliability of frequency measurement. The resulting output signal is proportional to the frequency difference (as explained below) and is applied through an interconnection circuit to the controlled oscillator of the APC system of the receiver.

The mechanism by which the frequency difference is determined may be explained as follows: Assume that a frequency difference Δf exists between the sync signal and the local reference oscillator. The input noise may

[15] J. G. Chaffee, "The application of negative feedback to frequency modulation systems," PROC. I.R.E., vol. 27, pp. 317–331; May, 1939.

be considered for simplicity as the sum of two noise-modulated signals in quadrature with each other at the *oscillator* frequency. The output from one synchronous detector will contain a sine beatnote (see Fig. 13(b)) and the noise along one reference axis. The output of the other synchronous detector will contain a cosine beatnote (see Fig. 13(c)) and the noise along an axis in quadrature with the first reference axis. These two *noise* voltages are completely independent of each other.

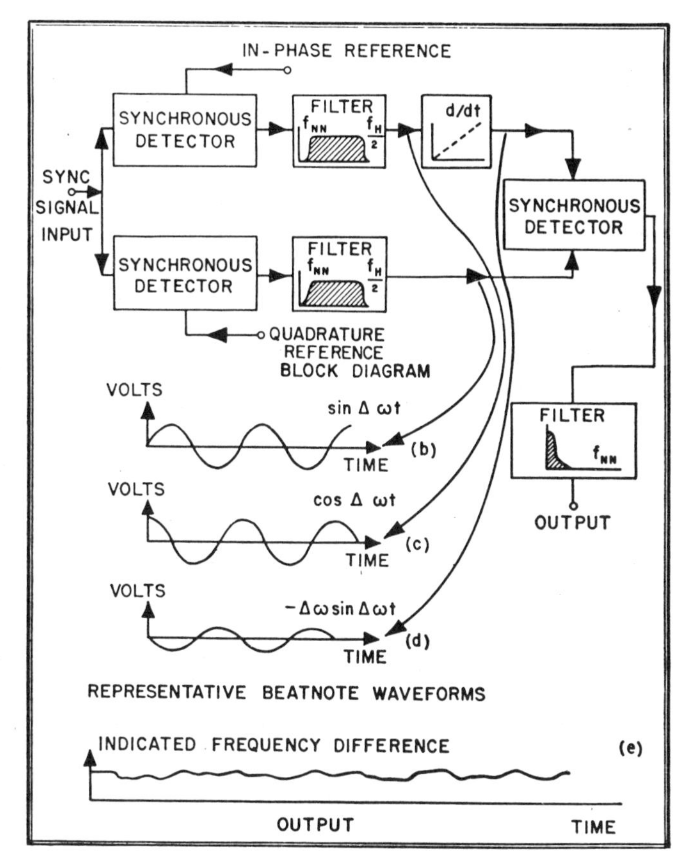

Fig. 13—Basic quadricorrelator.

The cosine beatnote is converted by differentiation to a sine beatnote having an amplitude which is proportional to its frequency, as indicated in Fig. 13(d); its associated noise is differentiated but the two noise voltages are still independent of each other. The output of the cross-multiplying synchronous detector will contain *a dc term proportional to and polarized according to the frequency difference.* In addition the output contains random noise; this noise output is discussed in Appendix E.

The quadricorrelator provides a convenient means for measuring a frequency difference with any selected measure of reliability in the presence of noise by integration of the frequency difference information.

An Illustrative Receiver Operable on NTSC Standards

Fig. 14 shows a partial block diagram of an NTSC color television receiver which includes color difference synchronous detectors, a Standard APC System, and a quadricorrelator for frequency difference detection. The composite video signal is fed to a pair of synchronous detectors for deriving the color difference video signal. The $R-Y$ synchronous detector output may be fed through an amplifier gated during line retrace to a filter, a reactance tube, and an oscillator, the output of which is fed back in the normal fashion to both synchronous detectors. These elements comprise a Standard APC System as described earlier. The gated outputs of both synchronous detectors are fed to a pair of filters as indicated. These may be bandpass filters having low-frequency cutoffs near the noise bandwidth of the APC system and having high frequency cutoffs not higher than half line frequency, as shown in Fig. 14. The differentiating circuit may be included in either beatnote translation path. The third synchronous detector and filter as indicated complete the elements of the quadricorrelator, the output of which is fed to the reactance tube. *The low-frequency attenuation characteristics of the filters in the two channels make the quadricorrelator have an essentially zero transmission characteristic for small beatnote frequency differences.*

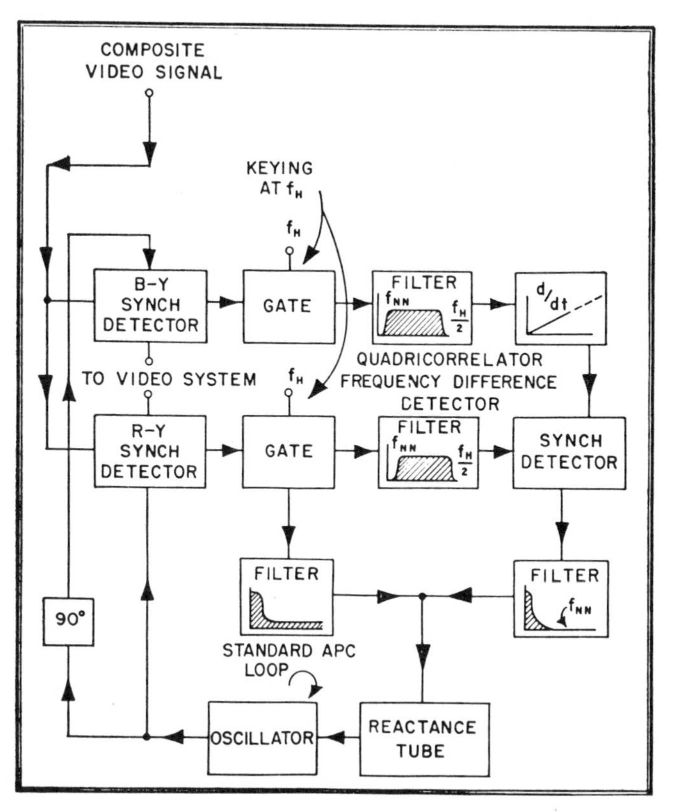

Fig. 14—Application of Fig. 12 sync system to NTSC color television receiver.

The circuit arrangement presented in block form in Fig. 14 provides one means for realizing the sync system of Fig. 12. Since this composite frequency- and phase-control system is essentially free of the previous limitations between noise bandwidth and pull-in time, the over-all system can be more readily designed for desired performance.

There appears to be a variety of frequency difference detector circuits and of linear and nonlinear interconnection arrangements which can be used to approach upper limit performance in the burst system.

The excess of performance inherent to these arrangements appears exchangeable for receiver economy and long term reliability.

The Approximate Limit of Performance Permitted by Signal Information

There are three requirements on the sync system.

(1) The static phase error shall not exceed some selected value, say 5°. It is shown in Appendexes B and C that for both passive and locked integrators this may be accomplished by use of adequately high loop gain.

(2) The rms phase error shall not exceed some selected value, say 5°, for signal-to-noise ratios at least as high as the approximate lowest level for which monochrome video picture information is acceptable; this is approximately $S_0/N_W=1$.

The required noise bandwidth for the APC system is

$$f_{NN} = \frac{1}{2T_M} \approx \frac{1}{2}\left(3\phi_{\text{rms}}\frac{S_0}{N_W}\right)^2$$

from (2). (The effect of excess gate width is small and is neglected here for simplicity.)

(3) The stabilization time shall not be annoyingly long. For example, pull-in times shorter than 1 second are acceptable.

The minimum integration time required for frequency difference detection yielding an rms frequency error f_{rms} is shown in Appendix E to be

$$T_I = \sqrt{\frac{2}{d}}\,\frac{1}{\pi h}\,\frac{N_W}{S_0}\sqrt{\frac{f_H}{f_W}}\,\frac{1}{f_{\text{rms}}} \tag{4}$$

for the signal. This is based on a pull-in range of $\pm(f_H/2)$.

It is shown in Appendexes C and D that the linear portion of the curve of Fig. 11(a) extends to a value of Δf approaching $2f_{NN}/\pi$; the control effect is strong to near f_{NN}. Then, if for frequency differences between approximately $(2/\pi)f_{NN}$ and $f_H/2$, the error in frequency difference measurement is less than $(2/\pi)f_{NN}$, pull-in will occur in time T_I. The more severe of the following two requirements then determines the frequency pull-in time T_F.

$$T_I \geqq \frac{\pi}{2f_{NN}} \quad \text{Approximately} \tag{5}$$

$$f_{\text{rms}} \leqq \frac{1}{4}\left(\frac{2}{\pi}f_{NN}\right) = \frac{f_{NN}}{2\pi}. \tag{6}$$

Combining (4) and (6),

$$T_I \geqq 2\sqrt{\frac{2}{d}}\,\frac{N_W}{S_0}\sqrt{\frac{f_H}{f_W}}\,\frac{1}{f_{NN}}\,\frac{1}{h}. \tag{7}$$

The same adverse tolerances used in obtaining (2) may be used here. If $d=.0352$, $h=.9$, $S_0/N_W=1$, $f_W=4.3$ mc, and $f_H=15734+$cps, then (7) becomes

$$T_I \geqq \frac{1}{f_{NN}}.$$

Thus, the required frequency pull-in time is of the order of magnitude of $1/f_{NN}$ or $(\pi/2)(1/f_{NN})$. After frequency pull-in, phase pull-in occurs. (Both occur effectively simultaneously in the continuous feedback system.) The time for phase pull-in is normally less than

$$T_\phi \approx \frac{4}{f_{NN}}. \tag{8}$$

The constant in (8) depends on the shape of the passband determining f_{NN}.

Then, the stabilization time, T_S is given by

$$T_S \approx T_F + T_\phi. \tag{9}$$

Since the required value of f_{NN} was found earlier to be 100 cps, pull-in times of the order of .05 second are possible. This is considerably shorter than is required, indicating that the information inherently contained in the signal is substantially in excess of what is required.

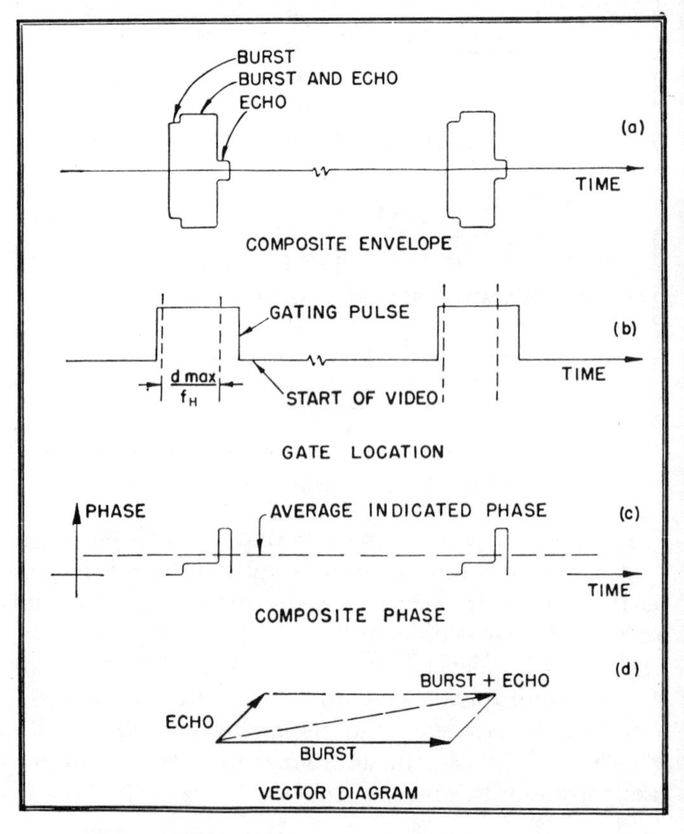

Fig. 15—Effect of echoes on the NTSC color burst.

Other Topics

Effects of Echoes

Some sketches relative to a discussion of the effect of echoes on burst sync are presented in Fig. 15. Fig. 15(a) shows one possible representation of a burst to which an echo has been added. Parameters of interest are the relative delay, the relative amplitude, and the relative phase. If the time-gate exceeds the burst width on the lagging end as indicated in Fig. 15(b) combined signals may be used to operate the burst sync system. In this case the indicated phase as a function of time is as shown in Fig. 15(c) while Fig. 15(d) is a vector diagram representing the signals of interest. Phase angles of interest are indicated for the burst phase, for the phase of the sum of the burst and echo, and for the phase of the echo. The average phase is not necessarily equal to any one of these but may often be near the phase of burst plus echo. The phase of burst plus echo is the correct reference phase for low detail large area colors. For this reason it appears possible that some extra gate width as indicated in Fig. 15(b) may give a useful and efficient exchange of noise immunity for performance in the presence of echoes. However, the existence of high order correlation between widely separated picture elements[16] may be uncommon enough to make this effect relatively unimportant.

A complete discussion of the effect of echoes in the NTSC system is beyond the scope of the present paper.

Effect of Stability of the Gate

The gate is conveniently obtained from horizontal flyback. The effect of gate stability depends on two factors: the stability of the horizontal sync system which produces the gate; and the relative widths of the gating pulse and the burst, which determines the extent to which noise jitter of the gate can be cross-modulated into the burst channel.

The fundamental physical considerations which have been presented and discussed above with regard to burst synchronization are also true of horizontal synchronization although the shape of the spectral distribution for horizontal sync introduces some additional complications. The static phase may be controlled as closely as desired, limited ultimately by transmission tolerances. The stability may be held to any desired level still permitting effectively instantaneous pull-in.

The effect of cross-modulation when it occurs is to increase the noise power for those low-frequency components to which the horizontal sync system is responsive. The horizontal sync system appears to contain more information than it needs. Stability of the gate is a design consideration but it is not a real limitation of the burst sync system.

Conclusions

The discussion above has shown that standard sync systems appear capable of completely automatic synchronization for NTSC burst sync (although without a large *excess* of performance). In the presence of strong signals the burst sync system is capable of yielding a color-carrier reference having a reliability completely determined by the gain in the receiver sync system, while noise is rejected by integrating the timing information for a suitably long period. An *effective* integration time of the order of 1/200th of a second appears appropriate. Passive integrators using controlled crystal filters, appear capable of meeting the requirements on Q, frequency stability, and rapidity of stabilization. The Standard APC System, when designed for near limit performance, appears capable of providing adequate and usable performance. This means that for reasonable operating tolerances, synchronization will always occur, and with adequate synchronization accuracy.

Improved sync systems which overcome the ultimate limitations of the standard APC sync system have been presented along with a discussion of factors leading to improvement and of the upper limit of performance permitted by the signal. These indicate that the requirement of a high order of noise immunity does not limit synchronization performance in the manner and to the degree experience with previous circuits had indicated. A large excess of attainable as compared to apparently necessary performance appears to exist.

The NTSC color-carrier reference phase synchronization signal contains adequate information for reliable performance down to levels of signal-to-noise ratio where the signals are no longer usable in picture content. A variety of circuits can provide satisfactory performance.

Appendix A

Phase of a Sine Wave Plus Random Noise

Derivation of the Equation

The analysis of the theoretical limits to phasing accuracy may be based on the properties of a signal composed of a sine wave plus random noise.[17] The information of each frequency component may be determined separately and then all of the information may be combined.

The problem is solved here first for a continuous (ungated) sine wave.

The probability density distribution of amplitude coefficients for a sinusoidal signal plus two-dimensional Gaussian noise is shown in Fig. 16. The signal is

$$S(t) = S \cos \omega_{SC}t. \tag{A-1}$$

The noise may be written as

$$N(t) = a(t) \cos \omega_{SC}t + b(t) \sin \omega_{SC}t \tag{A-2}$$

where $a(t)$ and $b(t)$ are time-varying parameters, each having a Gaussian distribution, and defined by the mean square values shown below.

$$\overline{a^2} = \overline{b^2} = \overline{N^2}. \tag{A-3}$$

This equality results from the fact that by symmetry, $\overline{a^2} = \overline{b^2}$ while the total noise power

$$\overline{N^2} = \overline{(a \cos \omega_{SC}t + b \sin \omega_{SC}t)^2}$$

[16] E. R. Kretzmer, "Statistics of television signals," *Bell Sys. Tech. Jour.*, vol. 30, pp. 751–767; July, 1952.

[17] S. O. Rice, "Mathematical analysis of random noise," *Bell Sys. Tech. Jour.*, vol. 23, p. 282–332, July, 1944; vol. 24, pp. 46–156, Jan., 1945.

$$= \tfrac{1}{2}\overline{a^2} + \tfrac{1}{2}\overline{b^2}. \qquad \text{(A-4)}$$

For the above case it is possible to express the probability distribution of phase angles for the combination of signal and noise, relative to the phase of the signal. This, however, leads to a cumbersome expression.[18]

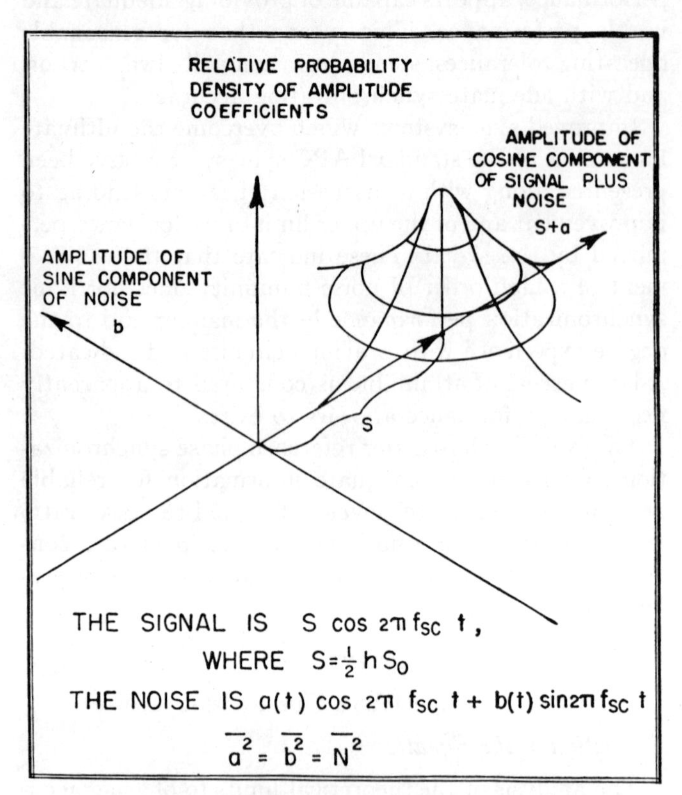

Fig. 16—Probability density distribution of a sine wave and random noise.

It is more convenient to use the simplified vector diagram shown in Fig. 17.[6] Here S represents the signal, and a and b represent the cosine and sine (in-phase and quadrature) components of noise.

Then if ϕ is the phase error,

$$\phi \approx \tan\phi = \frac{b}{S+a} \approx \frac{b}{S} \qquad \text{(A-5)}$$

or, very nearly, since $b_{\text{rms}} = N$,

$$\phi_{\text{rms}} = \frac{N}{S}. \qquad \text{(A-6)}$$

This equation is a good approximation if N/S is not large; in the case where the sync measuring system is primarily responsive to the noise in quadrature with a reference signal controlled by a long time constant of integration, it is accurate enough.

Then, since

$$N = \frac{N_W}{\sqrt{f_W T_M}} = \text{noise in the noise bandwidth} \qquad \text{(A-7)}$$

$f_N = 1/T_M$, and since $S = \tfrac{1}{2}hS_0$, we obtain

$$\phi_{\text{rms}} = 2\pi f_{SC} t_0' = \frac{\dfrac{N_W}{\sqrt{f_W T_M}}}{\tfrac{1}{2}hS_0}. \qquad \text{(A-8)}$$

The above equation applies for a continuous sine wave which is not gated. However, because the signal is present only a fraction d of the time, the integration is only $\sqrt{d}$ times as effective, and hence $t_0 = t_0'/\sqrt{d}$. Therefore, by substitution, the following upper limit relationship is obtained.

$$\frac{S_0}{N_W} = \frac{1}{\sqrt{d f_W T_M}} \cdot \frac{1}{f_{SC} t_0} \cdot \frac{1}{\pi h}. \qquad \text{(A-9)}$$

This is (1), presented earlier.

If the signal plus noise is passed through a limiter, the output of the limiter is approximately

$$S \cos \omega_{SC} t + b(t) \sin \omega_{SC} t$$

for signal-to-noise ratios of interest. Thus, the limiter aids in achieving the upper limit, without improving it.

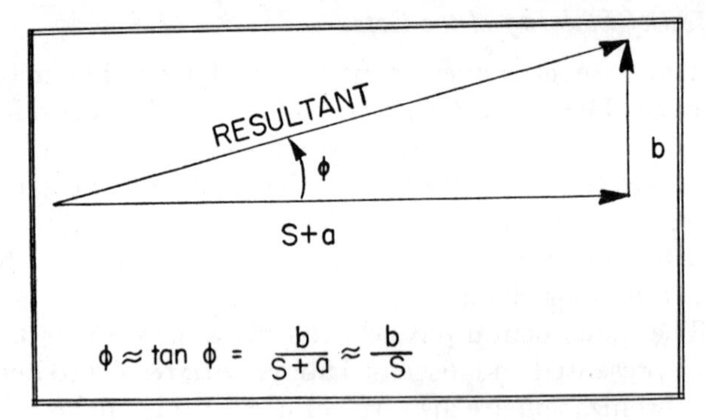

Fig. 17—Simplified vector diagram.

When not all of the signal spectrum is used, the rms error will exceed the limiting value of t_0 computed above.

The burst may be represented by the following Fourier series

$$S(t) = \sum_{k=-k_1}^{k_2} S_k \cos \omega_k t \qquad \text{(A-10)}$$

where

$$S_k = \left(\frac{1}{2} hS_0\right) \cdot d \cdot \left(\frac{\sin dk\pi}{dk\pi}\right) = \frac{hS_0 \sin dk\pi}{2k\pi} \qquad \text{(A-11)}$$

and

$$\omega_k = \omega_{SC} + k\omega_H. \qquad \text{(A-12)}$$

Each of these carries timing information; the error associated with the measurement of any component is very nearly Gaussian. For such a case, the Principle of Least Squares[19] may be applied. Then[20]

$$\frac{1}{t_0^2} = \sum_{k=-k_1}^{k_2} \left(\frac{1}{t_{0k}^2}\right) = \sum_{k=-k_1}^{k_2} \left(\frac{S_k}{N_W}\right)^2 (f_W T_M)(2\pi f_k)^2$$

[18] D. Middleton, "Some general results in the theory of noise through non-linear devices," *Quart. Appl. Math.*, vol. V, p. 471; Jan., 1948.

[19] R. B. Lindsay and H. Margenau, "Foundations of Physics," John Wiley & Sons, Inc., New York, N. Y., chap. IV, pp. 159–187; 1936.

[20] D. Richman, "Frame synchronization for color television," *Electronics*, vol. 25, pp. 146–152; Oct., 1952.

$$= \sum_{k=-k_1}^{k_2} \left(\frac{S_0}{N_W}\right)^2 (f_W T_M)(f_{SC} + kf_H)^2 \left(\frac{h \sin dk\pi}{k}\right)^2. \quad \text{(A-13)}$$

The factor $(1/t_{0_k}^2)$ has been plotted in Fig. 5(a) as the information per component. The effective accuracy, $1/t_0$ varies as the square root of the area under the curve, for any bandwidth. Although there is an optimum weighting, the weighting is not critical in the vicinity of the correct weighting. This is a general characteristic of integration systems.

Appendix B

Passive Integrators

This appendix presents some equations relevant to the performance of the phase stabilized integrating filter shown in Fig. 6(a).

The basic loop parameters are as follows:

(1) The transfer characteristic of the high Q filter is $F(f)$

$$F(f) \approx \frac{1}{1 + j2\dfrac{f - f_{SC}}{f_{SC}} Q} = \frac{1}{1 + j\dfrac{2\Delta f}{f_3}}$$
$$= \frac{1}{1 + j\pi \dfrac{\Delta f}{f_N}} = F(\Delta f) \quad \text{(B-1)}$$

(2) The phase detector sensitivity, for nominal full amplitude input is $\partial E/\partial\phi$

(3) The passband characteristic of the low pass (dc) filter is $Y(f)$, where $Y(0) = 1$. Let

$$Y(f) = \frac{1}{1 + j2\pi fT}. \quad \text{(B-2)}$$

(4) The sensitivity of the phase shifter (assumed broad band) may be represented as

$$\frac{\partial\phi}{\partial E}.$$

(5) The loop gain is G

$$G = \frac{\partial\phi}{\partial E}\cdot\frac{\partial E}{\partial\phi}.$$

(6) The static phase error which would result if there were no feedback is $\Delta\phi_0$

$$\Delta\phi_0 = \arctan\left(-\frac{\pi\Delta f}{f_N}\right) \approx -\frac{\pi\Delta f}{f_N}. \quad \text{(B-3)}$$

The Static Phase Error with Feedback

The static phase error with feedback is $\Delta\phi$

$$|F(\Delta f)|\,\Delta\phi\cdot G\cdot Y_0 = \Delta\phi_0 - \Delta\phi = \Delta\phi_{corr}$$
$$\frac{\Delta\phi}{\Delta\phi_0} = \frac{1}{1 + |F(\Delta f)|\,GY_0} \quad \text{(B-4)}$$

or, since for normal operation $F(\Delta f) \approx 1$ (very nearly) and $Y_0 \equiv Y(0) = 1$

$$\frac{\Delta\phi}{\Delta\phi_0} = \frac{1}{1 + G} \quad \text{(B-5)}$$

Since $\Delta\phi_0 < 90°$, a loop gain of $G > 17$ makes $\Delta\phi < 5°$ always.

The Effect of the Feedback Loop Upon Noise Performance

When noise is present, the phase detector output produces a noise output, which, after filtering by $Y(f)$ produces extra phase modulation noise.

The equation written earlier can be rewritten in terms of the ***phase correction***, $\Delta\phi_{corr}$, since

$$\Delta\phi_{0\ effective} = \Delta\phi_{corr} + \Delta\phi. \quad \text{(B-6)}$$

$\Delta\phi_{0\,effective}$ is the equivalent phase modulation to produce the actual phase detector noise output.

Then

$$[\Delta\phi_0(p) - \Delta\phi_{corr}(p)]\cdot G\cdot Y(p) = \Delta\phi_{corr}(p)$$

or

$$\frac{\Delta\phi_{corr}(p)}{\Delta\phi_0(p)} = \frac{GY(p)}{1 + GY(p)} = \frac{G}{1+G}\left[\frac{1}{1 + p\dfrac{T}{1+G}}\right]. \quad \text{(B-7)}$$

The signals to the phase detector are

(1) The original composite signal+noise, unfiltered.

(2) The filtered signal, with a narrow band of noise having a very small rms value.

Cross beats of signal upon noise produce considerably larger output than the beatnote between noise components, which are therefore negligible.

The output noise may be expressed as a phase:

$$\frac{b(t)}{S} \approx \Delta\phi_{01}(t) \quad \text{or} \quad \frac{b(p)}{S} \approx \Delta\phi_{01}(p)$$
$$\frac{b(p)\cdot F(\Delta p)}{S} \approx \Delta\phi_{02}(p)\,. \quad \text{(B-8)}$$

The total phase noise is $\Delta\phi_{0\,effective} = \Delta\phi^0{}_1 - \Delta\phi_{02}$ since, if the filter $F(f)$ were removed, the phase detector output would be identically zero.

$$\Delta\phi_0(p) = \frac{b(p)}{S}[1 - F(\Delta p)]. \quad \text{(B-9)}$$

There is little noise energy below approximately $f_N/2$ appearing at the phase detector output.

Since the transfer characteristic for this noise is

$$\frac{\Delta\phi_{corr}}{\Delta\phi_0} = \frac{G}{1+G}\left[\frac{1}{1 + p\dfrac{T}{1+G}}\right] \quad \text{(B-10)}$$

(which corresponds to a low pass filter), the following design condition may be employed to insure that the effective Q of the crystal filter will not be degraded by the feedback

$$\frac{1+G}{2\pi T} < \frac{f_N}{2}$$

or
$$T > \frac{1+G}{\pi f_N}. \quad \text{(B-11)}$$

Transient Analysis

The response to a step in differential phase $\Delta\phi_0$, is $\Delta\phi(p)$ or $\Delta\phi(t)$

$$\frac{\Delta\phi(p)}{\Delta\phi_0} = \frac{1}{p}\left[\frac{1+pT}{1+FG+pT}\right]$$
$$= \frac{1}{pT}\left[\frac{1}{\frac{1+FG}{T}+p}\right] + \frac{1}{\frac{1+FG}{T}+p} \quad \text{(B-12)}$$

$$\frac{\Delta\phi(t)}{\Delta\phi_0} = \int_0^t \left[\frac{1}{T}\epsilon^{-[1+FG/T]t}\right] dt + \epsilon^{-[1+FG/T]t}$$
$$= \frac{1}{1+FG}\left[1-\epsilon^{-[1+FG/T)]t}\right] + \epsilon^{-[1+FG/T]t}$$
$$= \frac{1}{1+FG} + \frac{FG}{1+FG}\epsilon^{-[1+FG/T]t} \quad \text{(B-13)}$$
= steady state + transient response.

For $FG \gg 1$, the transient term is negligible for $t > T$.

The time for the phase error to settle down to twice its final value may be computed, as a measure of stabilization time.

The total transient time consists effectively of an amplitude and phase transient of the high Q filter plus the transient time of the feedback loop. The transient is effectively completed in three times the time constant of the filter. Since the noise bandwidth is f_N, the time is

$$T_A \approx 3\left(\frac{1}{4f_N}\right). \quad \text{(B-14)}$$

This overlaps with the phase loop transient time, which, neglecting amplitude effects, would be

$$T_\phi = \frac{1}{\pi f_N} lnG \approx \frac{1}{\pi f_N} ln\left[\frac{\pi\Delta f_{\max}}{f_n\Delta\phi_{\max}} - 1\right] \quad \text{(B-15)}$$

which is based on

$$\frac{\Delta\phi}{\Delta\phi_0} = \frac{1}{1+G}\left[1 + G\epsilon^{-(1+G)(t/T)}\right]$$
$$G = \frac{\Delta\phi_0}{\Delta\phi_\infty} - 1 \approx \frac{\pi\Delta f_{\max}}{f_N\Delta\phi_{\max}} - 1$$
$$T = \frac{1+G}{\pi f_N}.$$

These two pull-in times overlap.

Appendix C

Performance Characteristics of the Standard APC Loop

This appendix presents a description of the operating characteristics of a standard APC system. The basic parameters of the APC loop are defined. The independence of the primary parameters $\Delta\phi$ (the static phase error) and f_{NN} (the APC loop noise bandwidth) is shown; these parameters characterize the performance after the system has stabilized. The limitations of pull-in are discussed and some formulas which are derived later in Appendix D are introduced. The simple relation presented earlier for pull-in time is then obtained.

The formulas derived may be applied for designs based on any convenient set of assumed criteria.

The Basic APC Loop Parameters

(1) The output voltage ΔE of the phase detector, and the phase difference $\Delta\phi$ between the reference oscillation and the signal are related by the control characteristics. When both signals are sinusoidal,

$$\Delta E = \mu \sin \Delta\phi \quad \text{(C-1)}$$

where ΔE is a voltage developed at the phase detector output in response to a phase difference $\Delta\phi$ between signal and reference oscillation. For operation at or very near balance,

$$\frac{\partial E}{\partial\phi} = \mu \quad \text{volts per radian.}$$

(2) The transfer characteristic of the feedback loop filter is denoted by

$$N(\omega) = \frac{\text{output voltage}}{\text{input voltage}}$$

(3) The sensitivity of the reactance tube is denoted by

$$\beta = \frac{\partial f}{\partial E} \quad \text{cycles per second per volt.}$$

(4) The factor $|\mu\beta| \equiv f_c$ is a characteristic parameter of the loop; the time constant $t_c \equiv 1/2\pi f_c$ is the transient time constant of the loop when $N(\omega) \equiv 1$. (This may be verified from (C-3) for $Q(\omega)$ presented later.)

(5) The static phase error, $\Delta\phi$, which results from a "free-running" frequency difference, Δf, between signal and local oscillator may be found from the preceding relations:

$$-\sin\Delta\phi = 2\pi\cdot\Delta f\cdot t_c = \frac{\Delta f}{f_c}. \quad \text{(C-2)}$$

Although (C-2) contains the appropriate signs, it is the magnitudes of the above quantities which are of interest in design work.

(6) The phase following ratio for an APC loop is

$$\frac{\text{phase variation of output phase}}{\text{phase variation of input phase}} = Q(\omega) = \frac{N(\omega)}{N(\omega)+pt_c}. \quad \text{(C-3)}$$

This is the small signal form of the differential equation which characterizes the APC loop. It is used to determine the response of the APC system to noise, after the system is synchronized.

(7) The noise bandwidth of the APC system is f_{NN}. Consistent with the usual practice, this is defined as

$$f_{NN} = \int_0^\infty |Q(\omega)|^2 df = \int_0^\infty Q(\omega)Q(-\omega)df. \quad \text{(C-4)}$$

Representative network configurations for $N(\omega)$ are shown in Fig. 18(a). For each of these networks

$$N(\omega) = \frac{1 + xpT}{1 + (1 + x)pT} \quad \text{(C-5)}$$

where $T = RC$ and $p = j2\pi f = j\omega$. Then

$$Q(\omega) = \frac{1 + xpT}{1 + p(t_c + xT) + p^2(1 + x)t_cT} \cdot \quad \text{(C-6)}$$

This equation suggests one manner in which the meaning of the phase transfer ratio and noise bandwidth of an APC loop may be readily visualized. Fig. 18(b) represents a network having a voltage transfer characteristic which is identical with $Q(\omega)$ given above. If a voltage

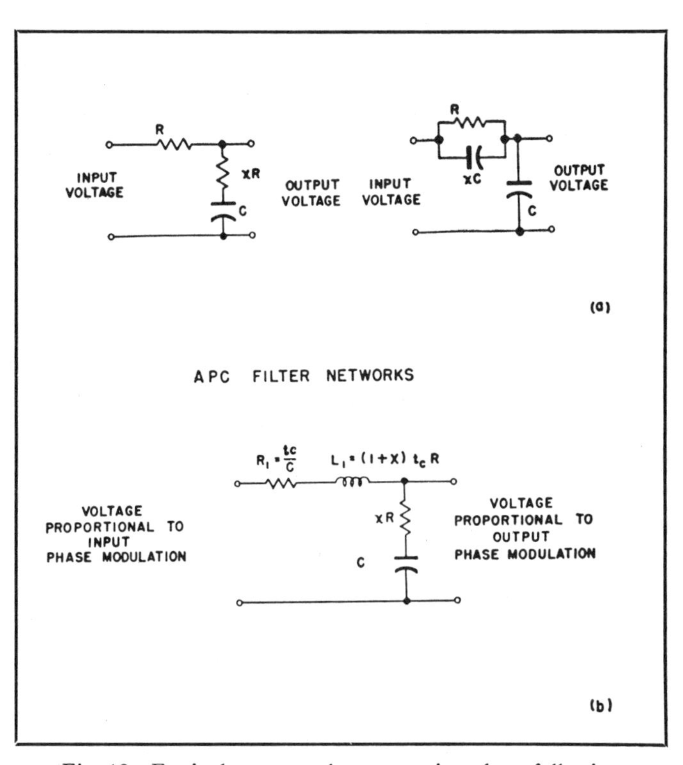

Fig. 18—Equivalent network representing phase following ratio of an APC loop.

proportional to the phase modulation of the synchronizing signal (by noise or any other disturbance) is applied to the input of the network of Fig. 18(b), the output voltage is proportional to the phase modulation of the reference oscillator of the APC loop. The shape of the (low frequency) passband described by $Q(\omega)$ defines the small signal transient response of the loop as well as the noise bandwidth.

(8) The ratio of ac gain/dc gain through the network $N(\omega)$ is

$$m \equiv \frac{x}{1 + x} \quad \text{(C-7)}$$

(from [C-5] when $pT \gg 1$). The parameter m determines the pull-in range of the APC system, when certain other parameters are specified. It is convenient therefore to express the synchronous performance in terms of m.

Also, the term xT/t_c appears often. This is written as

$$y \equiv \frac{xT}{t_c} \cdot \quad \text{(C-8)}$$

Then, rewriting the earlier expressions in terms of these parameters,

$$N(\omega) = \frac{1 + pyt_c}{1 + p\frac{y}{m}t_c} \quad \text{(C-9)}$$

$$Q(\omega) = \frac{1 + pyt_c}{1 + pt_c(1 + y) + p^2\frac{y}{m}t_c^2} \cdot \quad \text{(C-10)}$$

The noise bandwidth is found by integration (at the end of this Appendix C), using the definition presented earlier, to be

$$f_{NN} = \frac{1}{4t_c}\cdot\frac{1 + my}{1 + y} \cdot \quad \text{(C-11)}$$

(9) In order to prevent resonant ringing on noise impulses, $Q(\omega)$ should have a moderately flat graph. Since the denominator of $Q(\omega)$ contains a quadratic expression, it is convenient to define a damping coefficient, K, which is defined by the following equation:

$$(1 + y)^2 = K\cdot\frac{4y}{m} \cdot \quad \text{(C-12)}$$

Then $K = 1$ corresponds to equal roots or critical damping, $K > 1$ corresponds to overdamping and makes $Q(\omega)$ approach the shape of the single (RC) low pass filter, and $K < 1$ tends to give $Q(\omega)$ a high resonant rise.

Fig. 19 shows the shapes of $|Q(\omega)|$ and $|Q(\omega)|^2$ for several values of K, and subject to the simplifications $y \gg 1$, and $my \approx 4K$, derived below. A value of K close to 1 gives best performance.

The Synchronous Performance of the APC System

The basic equations relating to the synchronous performance of the APC system have been presented above. These are

$$-\sin\Delta\phi = 2\pi\cdot\Delta f\cdot t_c \quad \text{(C-2)}$$

$$f_{NN} = \frac{1}{4t_c}\cdot\frac{1 + my}{1 + y} \quad \text{(C-11)}$$

$$m = \frac{4Ky}{(1 + y)^2} \cdot \quad \text{(C-12)}$$

Since both tight static phase and narrow noise bandwidth are desired, it is possible to define a figure of merit for the system as $|(\sin\Delta\phi)/(\Delta f)|\cdot f_{NN}$; the smaller this product is, the better the over-all performance. However, relations above show that any arbitrarily selected figure of merit may be obtained by proper design, since, combining the above relations,

$$\left|\frac{\sin\Delta\phi}{\Delta f}\right|\cdot f_{NN}=\frac{\pi}{2}\left[\frac{1+\dfrac{4Ky^2}{(1+y)^2}}{1+y}\right]. \quad \text{(C-13)}$$

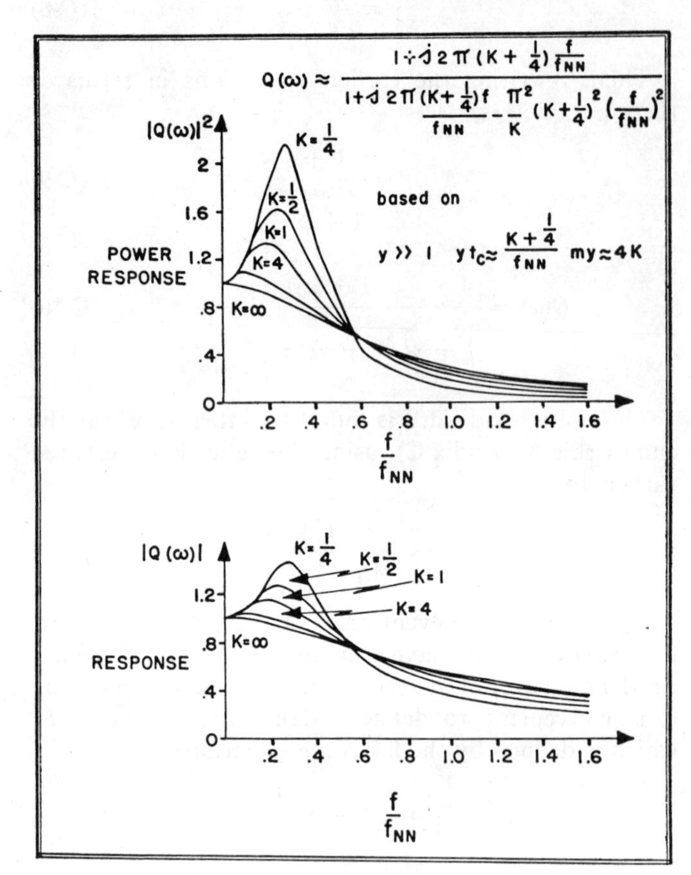

Fig. 19—APC loop small signal modulation response.

For the limiting case of a single time constant filter, $y=0$, and then

$$\left[\left|\frac{\sin\Delta\phi}{\Delta f}\right|\cdot f_{NN}\right]_{y=0}=\frac{\pi}{2}. \quad \text{(C-14)}$$

Thus for the simplified filter the static phase shift and noise bandwidth are interdependent.[21,22] However, for the filters of Fig. 18a, the parameters can be designed for whatever figure of merit is required for synchronous operation.

The above relations may usefully be written in simpler form, since, for the design ranges of interest, $m\ll 1$ and $y\gg 1$; then, very nearly $4K=my$ and hence

$$4f_{NN}t_c=m\frac{1+my}{m+my}\approx m\left(\frac{1+4K}{4K}\right). \quad \text{(C-15)}$$

This equation will be used in expressing the pull-in performance of the system conveniently.

[21] T. S. George, "Analysis of synchronizing systems for dot-interlaced color television," Proc. I.R.E., vol. 39, pp. 124–131; Feb., 1951.

[22] K. Schlesinger, "Locked oscillator for television synchronization," *Electronics*, vol. 22, pp. 112–118; Jan., 1949.

The figure of merit may be written as

$$\left|\frac{\sin\Delta\phi}{\Delta f}\right| f_{NN}=\frac{f_{NN}}{f_c}\approx\frac{\pi}{2}\cdot m\left(\frac{1+4K}{4K}\right). \quad \text{(C-16)}$$

The Transient (Pull-In) Performance

The pull-in behavior of the APC system is investigated in detail in Appendix D. The significant conclusions are as follows: The pull-in performance is expressible in terms of the relations between the parameters

$$\left(\frac{T_F}{xT}\right)\equiv\left(\frac{T_F}{yt_c}\right)$$

and

$$\left|\frac{\Delta f}{mf_c}\right|.$$

Fig. 20 shows the relation between these parameters.

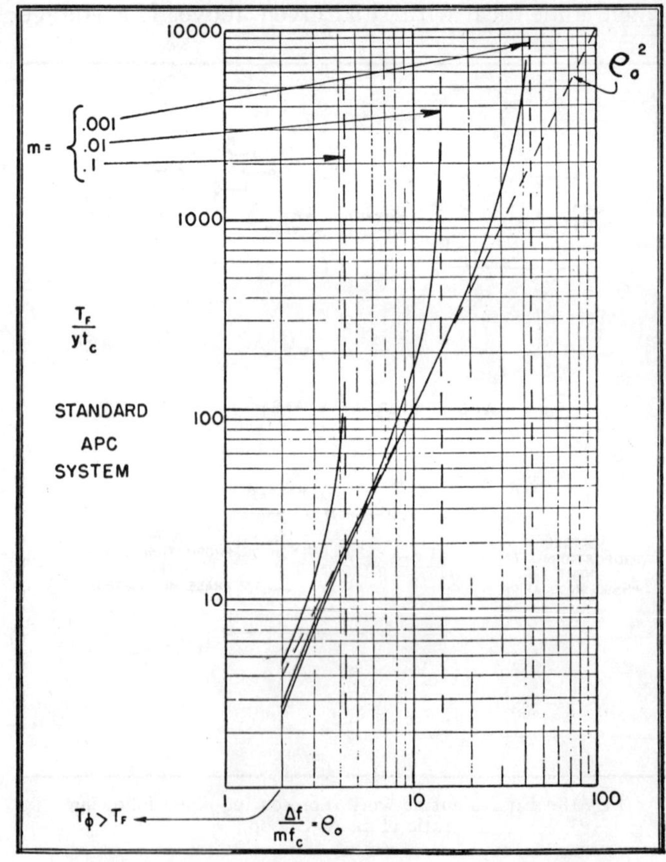

Fig. 20—Universal frequency pull-in characteristics.

The following approximation to the data represented by Fig. 20, based on (D-29), has been found useful in design work, it can also be solved for Δf:

$$T_F\approx xT\frac{\left(\dfrac{\Delta f}{mf_c}\right)^2}{1-\dfrac{\Delta f^2}{2f_c\cdot mf_c}}.$$

If $|\Delta f/mf_c| \leq 1$ the frequency pull-in is effectively instantaneous ($T_F=0$) but a short period is required for the phase to approach closely its stable value. If $|\Delta f/mf_c| > 1$, the system can slip cycles; often the slip is a great many cycles as this pull-in mechanism is fairly inefficient. The pull-in range is limited to the region

$$\left|\frac{f}{mf_c}\right| < \sqrt{\frac{2}{m} - 1} \qquad \text{(C-17)} \quad \text{(D-25)}$$

or

$$\Delta f_{\max} = f_c\sqrt{2m - m^2} \approx \sqrt{2f_c \cdot mf_c}$$

Then,

$$|\sin \Delta\phi| \approx |\Delta\phi| \leq \left|\frac{\Delta f_{\max}}{f_c}\right| = m\sqrt{\frac{2}{m} - 1} \approx \sqrt{2m}. \qquad \text{(C-18)}$$

If $m < (1/250)$ the phase angle after pull-in will always be less than 5°. However, not all of the pull-in range is normally used. If

$$|\Delta\phi| < \frac{1}{2}\Delta f_{\max},\ m < \frac{1}{62} \text{ makes } |\Delta\phi| < 5°.$$

When operation is *well within* the pull-in range the *frequency pull-in* time, T_F, which is defined as the time for the oscillator to be pulled from $|\Delta f|$ to within mf_c of the frequency of the color burst, approaches very nearly the relation

$$\frac{T_F}{yt_c} = \left(\frac{\Delta f}{mf_c}\right)^2. \qquad \text{(C-19)} \quad \text{(D-28)}$$

By making m smaller and f_c larger it is possible to extend the pull-in range far enough so that the gated nature of the signal provides the only real limitation on pull-in; the range is $|\Delta f| < (f_H/2)$. The pull-in time is then expressed by the square law relation above, except near the limit of the pull-in range. Furthermore, making m smaller improves the synchronous figure of merit.

The pull-in relations may be expressed in terms of f_{NN}, since

$$yt_c = my\frac{t_c}{m} \approx 4K \cdot \frac{1}{4f_{NN}}\left(\frac{1 + 4K}{4K}\right) = \frac{K + 1/4}{f_{NN}} \qquad \text{(C-20)}$$

and

$$mf_c = \frac{m}{2\pi tc} \approx \frac{1}{2\pi} \cdot 4f_{NN}\frac{4K}{1 + 4K} = \frac{2}{\pi}\left(\frac{K}{K + 1/4}\right) \cdot f_{NN} \qquad \text{(C-21)}$$

the following equation results

$$T_F f_{NN} = \lambda^2\left(\frac{\Delta f}{f_{NN}}\right)^2 \qquad \text{(C-22)}$$

where, when f_c is large enough so that $\Delta f_{\max} \gg \Delta f$,

$$\lambda^2 \equiv \left(\frac{\pi}{2}\right)^2 \frac{(K + 1/4)^3}{K^2} \geq 4.2. \qquad \text{(C-23)}$$

The approximate value 4 has been used in Figs. 8 and 9. Fig. 21 shows graphically the relation between K and λ^2. The curve has a minimum at $K = 1/2$.

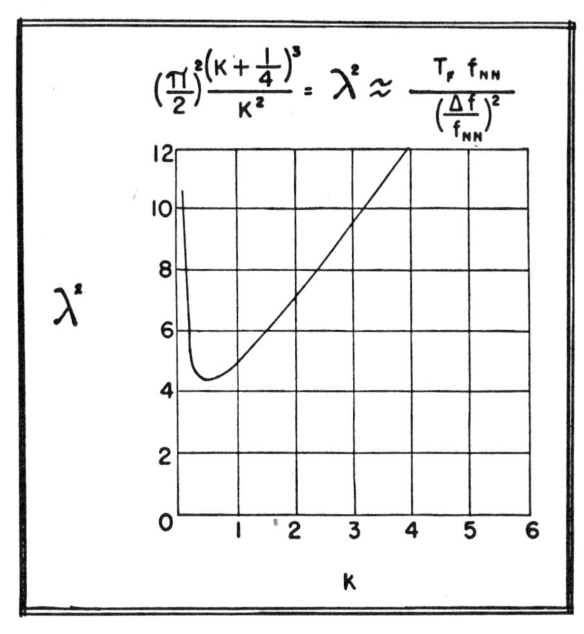

Fig. 21—Graph showing the relation between the damping coefficient, K, and the constant in the APC limit curve equation.

In view of the shape of the curve, and the normal tolerance variations of practical circuits, a value of K near 1 seems desirable. This gives good small signal transient response also. The problem of optimum design is discussed in more detail in a reference.[23]

Derivation of the Noise Bandwidth

The integration is performed as follows. Since

$$Q(p) = \left[\frac{1 + pyt_c}{\frac{m}{y} + pt_c(1 + y)\frac{m}{y} + p^2t_c^2}\right]\left(\frac{m}{y}\right), \qquad \text{(C-10)}$$

then

$$|Q^2| = Q \cdot Q^* = \left(\frac{m}{y}\right)^2 \frac{1 + y^2\theta^2}{(\theta^2 + \theta_\alpha^2)(\theta^2 + \theta_\beta^2)} \qquad \text{(C-24)}$$

where

$$\theta = \omega t_c \qquad \text{(C-25)}$$

and

$$\theta_{\alpha,\beta} = \frac{1}{2}\left\{(1+y)\frac{m}{y} \pm \sqrt{\left[(1+y)\frac{m}{y}\right]^2 - 4\frac{m}{y}}\right\} \qquad \text{(C-26)}$$

but

[23] D. Richman, "APC color sync for NTSC color television," IRE Convention Record, part 4; presented March 23, 1953.

$$\frac{1+y^2\theta^2}{(\theta^2+\theta_\alpha^2)(\theta^2+\theta_\beta^2)}=\left[\frac{1-y^2\theta_\alpha^2}{\theta^2+\theta_\alpha^2}-\frac{1-y^2\theta_\beta^2}{\theta^2+\theta_\beta^2}\right]\cdot\frac{1}{\theta_\beta^2-\theta_\alpha^2}. \quad \text{(C-27)}$$

The above is substituted in (C-4) to give

$$\int_0^\infty Q^2(ft_c)d(ft_c) = t_cf_{NN} \quad \text{(C-28)}$$

$$=\frac{\left(\frac{m}{y}\right)^2}{\theta_\beta^2-\theta_\alpha^2}\int_0^\infty\left[\frac{1-y^2\theta_\alpha^2}{\theta_\alpha^2+(2\pi ft_c)^2}-\frac{1-y^2\theta_\beta^2}{\theta_\beta^2+(2\pi ft_c)^2}\right]d(ft_c).$$

Then, since d/dx arctan $x=1/(1+x^2)$ and arctan $0=0$, and arctan $\infty=\pi/2$.

$$t_cf_{NN}=\left(\frac{m}{y}\right)^2\left(\frac{\dfrac{1-y^2\theta_\alpha^2}{4\theta_\alpha}-\dfrac{1-y^2\theta_\beta^2}{4\theta_\beta}}{\theta_\beta^2-\theta_\alpha^2}\right). \quad \text{(C-29)}$$

This is simplified as follows. Since

$$\theta_\alpha+\theta_\beta=\frac{m}{y}(1+y) \quad \text{(C-30)}$$

and

$$\theta_\alpha\cdot\theta_\beta=\frac{m}{y},$$

then

$$t_cf_{NN}=\left(\frac{m}{y}\right)^2\cdot\frac{1}{4}\left[\frac{\dfrac{\theta_\beta-\theta_\alpha}{\theta_\beta\theta_\alpha}+y^2(\theta_\beta-\theta_\alpha)}{(\theta_\beta-\theta_\alpha)(\theta_\beta+\theta_\alpha)}\right] \quad \text{(C-11)}$$

$$=\frac{1}{4}\frac{m^2}{y^2}\left[\frac{\dfrac{y}{m}+y^2}{\dfrac{m}{y}(1+y)}\right]=\frac{1}{4}\left(\frac{1+my}{1+y}\right).$$

This is the desired result.

Appendix D

Transient Performance of the APC Loop

This appendix provides a description and derivation of formulas relating to pull-in characteristics and pull-in time of APC loops. Exact analysis of a simplified APC loop provides useful formulas and a basis for understanding some of the phenomena relating to pull-in. This then suggests a simple approximate method for reducing the differential equation of the loop to a form which is readily solvable for the pull-in time. The results are plotted and discussed.

The Simplified Loop

The simplest form of APC network is the one for which $N(\omega)=$ a constant. See Fig. 22(a).

The basic equations are:

$$N(\omega)=m$$

$$Q(\omega)=\frac{m}{m+pt_c}$$

$$f_{NN}=\frac{\pi}{2}mf_c \quad \text{(D-1)}$$

$$|\sin\Delta\phi|=\left|\frac{\Delta f}{mf_c}\right|\leq 1.$$

The differential equation of the loop is

$$m\cdot\omega_c\sin\phi=\frac{d\phi}{dt}-\Delta\omega. \quad \text{(D-2)}$$

The same equation has been shown applicable for directly synchronized oscillators.[24]

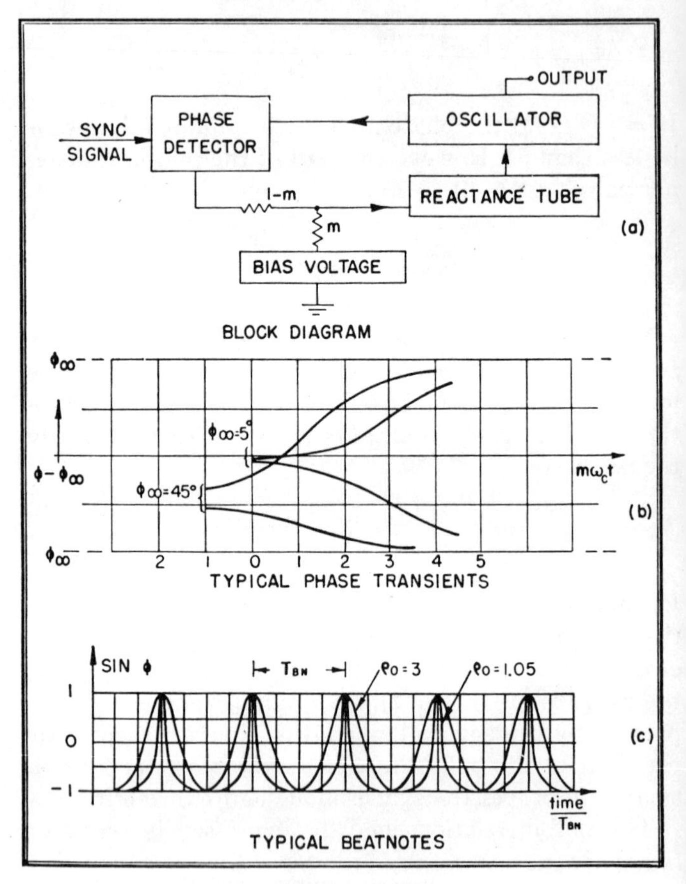

Fig. 22—Basic APC system.

This equation is equivalent to

(Filter transfer characteristic) · (Phase detector output)
= (Rate of change of phase difference)
− (Initial angular frequency difference).

The equation may be rewritten as

$$dt=\frac{d\phi}{\Delta\omega+m\omega_c\sin\phi}. \quad \text{(D-3)}$$

It has two solutions, depending on whether $\Delta\omega/m\omega_c$ is greater than or less than 1. Boundary conditions are

[24] R. Adler, "A study of locking phenomena in oscillators," Proc. I.R.E., vol. 34, pp. 351–357; June, 1946.

$$t = 0 \qquad \frac{d\phi}{dt} = \Delta\omega \qquad \phi = \phi_0$$

$$t = \infty \qquad \frac{d\phi}{dt} = 0 \qquad \phi = \phi_\infty = \arcsin\left(\frac{-\Delta f}{mf_c}\right). \tag{D-4}$$

Equation (D-3) is directly integrable.[25]

The pull-in range is $\Delta f \leq mf_c$. Within the pull-in range the phase stabilizes according to the following equation, which is the integral of (D-3) under this condition:

$$m\omega_c t \cos\phi_\infty = \ln \left| \frac{\tan\frac{\phi}{2} - \cot\frac{\phi_\infty}{2}}{\tan\frac{\phi}{2} - \tan\frac{\phi_\infty}{2}} \right| \cdot \left| \frac{\tan\frac{\phi_0}{2} - \tan\frac{\phi_\infty}{2}}{\tan\frac{\phi_0}{2} - \cot\frac{\phi_\infty}{2}} \right| \tag{D-5}$$

where

$$\frac{\Delta\omega}{m\omega_c} = \rho = -\sin\phi_\infty \qquad (|\rho| < 1) \tag{D-6}$$

and

$$-\sqrt{1-\rho^2} = \cos\phi_\infty. \tag{D-7}$$

Typical phase transients are shown in Fig. 22(b). Phase is plotted relative to ϕ_∞ with a scale calibrated in units of $m\omega_c t$. The starting point on any curve is determined by $\phi_0 - \phi_\infty$.

An approximate time constant of stabilization is

$$\frac{-1}{m\omega_c \cos\phi_\infty} = \frac{1}{\sqrt{(m\omega_c)^2 - (\Delta\omega)^2}},$$

however, *the actual stabilization time is a function of the initial phase.*

Outside the pull-in range $\rho > 1$, and the phase as a function of time is defined by the following equation, which is the integral of (D-3) for this condition:

$$\frac{m\omega_c t\sqrt{\rho^2 - 1}}{2} = \arctan \left\{ \frac{\rho \tan\frac{\phi}{2} + 1}{\sqrt{\rho^2 - 1}} \right\} \Bigg|_{\phi_0}^{\phi}. \tag{D-8}$$

This represents a cyclic variation characterized by its wave form and its fundamental frequency, f_{BN}.

Fig. 22(c) shows examples of the cyclic relationship between $\sin\phi$ and t, $\rho_0 = \Delta f / mf_c$ being specified as 1.05 or 3. The time scale is normalized to the beatnote period $T_{BN} = 1/f_{BN}$. The period T_{BN} is such that t increases by T_{BN} when ϕ increases by 2π, and is found from the following relation:

$$\frac{m\omega_c T_{BN}\sqrt{\rho^2 - 1}}{2} = \pi \quad \text{when} \quad \Delta\phi = 2\pi. \tag{D-9}$$

Then

$$T_{BN} = \frac{2\pi}{m\omega_c\sqrt{\rho^2 - 1}} = \frac{1}{\sqrt{(\Delta f)^2 - (mf_c)^2}}. \tag{D-10}$$

This is an important relationship. It states for example, that if in the APC loop block diagram presented above the bias is adjusted so that the effective open loop frequency difference is $\Delta f (> mf_c)$, the operating beatnote frequency difference is $\sqrt{(\Delta f)^2 - (mf_c)^2}$. If the bias is a slowly varying function of time (as compared to f_{BN}), the above relationship accurately describes the variation of f_{BN} with time.

The dc bias or average dc potential developed at the reactance tube input may be determined from the above relationships. It may be expressed in terms of its effect on frequency.

Integrating the differential equation over a cycle, and dividing by the period

$$\frac{1}{T_{BN}} \oint m\omega_c \sin\phi dt = \frac{1}{T_{BN}} \oint \frac{d\phi}{dt} dt - \frac{1}{T_{BN}} \oint \Delta\omega dt \tag{D-11}$$

or

$$m\omega_c \overline{\sin\phi} = \frac{2\pi}{T_{BN}} - \Delta\omega \tag{D-12}$$

and therefore, dividing by 2π,

$$mf_c \overline{\sin\phi} = \sqrt{(\Delta f)^2 - (mf_c)^2} - \Delta f. \tag{D-13}$$

This is plotted in Figs. 11(a) and 23(a) which represents magnitude of the developed bias as a function of Δf.[26] In the standard loop shown later in which the bias battery is replaced by a capacitor it is proportional to the control effect which causes pull-in.

Fig. 23(a) shows that $m\omega_c \overline{\sin\phi}$, the average angular frequency shift, is a maximum when $\Delta\omega/m\omega_c = 1$ and decreases beyond that point, approaching zero asymptotically. When $(\Delta\omega/m\omega_c) < 1$, the phase does not shift 2π radians in a finite time. Enough bias is produced however, to shift the angular frequency by $\Delta\omega$. This bias is represented by the straight line portion, as discussed with regard to Fig. 11(a).

The Standard APC Loop

The standard APC loop is shown in Fig. 23(b). For the network shown,

$$N(p) = \frac{1 + pyt_c}{1 + p\frac{y}{m}t_c} = m\frac{1 + pyt_c}{m + pyt_c} = m + \frac{1 - m}{1 + p\frac{y}{m}t_c}$$

[25] H. B. Dwight, "Tables of Integrals and Other Mathematical Data," The Macmillan Co., New York, N. Y., Integral 436.00; 1947.

[26] In experimental work this characteristic may be measured in terms of f_{BN}. From (D-10), above, $f_{BN}^2 + (mf_c)^2 = (\Delta f)^2$.

= wideband direct transfer component
+ long time-constant integration component
= resistive component
+ capacitive component. (D-14)

The differential equation in operational form is

$$N(p)\omega_c \sin\phi = p\phi - \Delta\omega \quad \text{(D-15)}$$

which may be written as

$$m\omega_c \sin\phi = p\phi - \Delta\omega - \frac{1-m}{1+p\frac{y}{m}t_c}\omega_c \sin\phi. \quad \text{(D-16)}$$

The term

$$\Delta\omega + \frac{1-m}{1+p\frac{y}{m}t_c}\omega_c \sin\phi \equiv \omega_I \quad \text{(D-17)}$$

is the Fourier transform of a time function representing effective instantaneous impressed frequency difference.

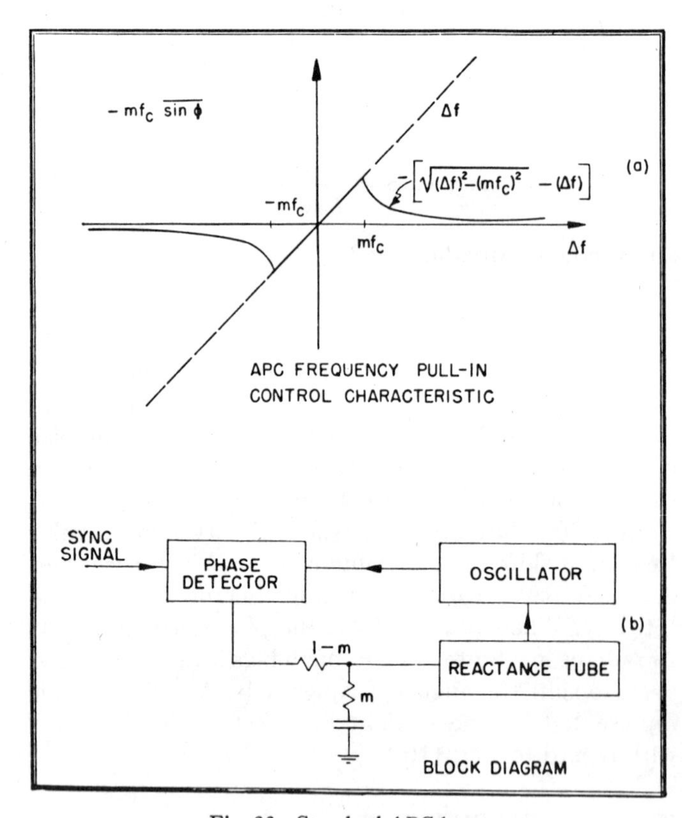

Fig. 23—Standard APC loop.

When this loop is turned on, or has a signal applied to it, the transient of stabilization lasts for a period of time which depends on both the initial phase and the frequency difference. However, the initial phase has only a small effect on the pull-in time and may be neglected for simplicity; the phase transient time, T_ϕ is rapid compared to the frequency pull-in time, T_F. Fig. 22(b) substantiates that for high dc loop gain ($\phi_\infty \ll 90°$) normally $m\omega_c T_\phi < 10$. Then, using (C-15),

$$10 > m\omega_c T_\phi = 4f_{NN}T_\phi \frac{K}{K+\frac{1}{4}}\cdot \quad \text{(D-18)}$$

If $K=\frac{1}{2}$, $T_\phi < (15/4f_{NN})$, while if $K=4$, $T_\phi < (2.7/f_{NN})$.

If the frequency difference is such that $\Delta f < mf_c$ the resistive component of loop feedback is adequate to ensure pull-in. The analysis presented above for simplified loop shows the system never slips a complete cycle.

A definition of frequency pull-in time T_F, and phase pull-in time T_ϕ is desirable; the following are useful.

If the system never slips a cycle, then the transient is defined as phase pull-in and measured in terms of the phase pull-in time, T_ϕ.

If the system slips cycles, then the period of time from the instant of switching or excitation until a definable point is reached from which the phase slip does not exceed a cycle is T_F, the frequency pull-in time.

When the initial frequency difference is such that $\Delta f > mf_c$, the long time integration component of feedback must be relied upon for pull-in.

The time constant $(y/m)t_c = y/(2\pi m f_c)$ is long compared to the loop time constant, t_c/m, since $y \gg 1$. Because of this long time constant, the average bias across the capacitor which may result from an unsymmetrical beatnote wave form from the phase detector will not change rapidly with time. It is not unreasonable therefore to integrate the differential equation for this APC loop over a cycle of beatnote.

Then

$$\overline{m\omega_c \sin\phi} = \frac{2\pi}{T_{BN}} - \overline{\omega_I} = \sqrt{(\overline{\omega_I})^2 - (m\omega_c)^2} - \overline{\omega_I} \quad \text{(D-19)}$$

$$\overline{\omega_I} = \Delta\omega + \overline{\frac{1-m}{1+p\frac{y}{m}t_c}\omega_c \sin\phi}. \quad \text{(D-20)}$$

At this point it is necessary to recognize clearly the nature of the signal circulating in the APC loop. There are two components; there is a *cyclic* component produced as a result of the average frequency difference, and having a harmonic composition which is a function of the frequency difference and hence of time during pull-in; there is a low frequency *drift* component which represents the slow change in frequency difference which constitutes pull-in. It has been shown earlier that the generated frequency shift, $\omega_c \overline{\sin\phi}$, varies in an inverse manner with $\Delta\omega$ or $\overline{\omega_I}$; thus, frequency changes slowly except when $\overline{\omega_I}$ is very near $m\omega_c$; stated another way, almost all of the pull-in time is accrued under the condition that the rate of change of the beatnote frequency is not comparable to the beatnote frequency. Therefore, very nearly

$$\oint \frac{1-m}{1+p\frac{y}{m}t_c}\omega_c \sin\phi\, dt \approx \frac{1-m}{1+p\frac{y}{m}t_c}\oint \omega_c \sin\phi\, dt \quad \text{(D-21)}$$

and

$$\overline{\frac{1-m}{1+p\frac{y}{m}t_c}\omega_c\sin\phi} \approx \frac{1-m}{1+p\frac{y}{m}t_c}\omega_c\overline{\sin\phi}. \quad \text{(D-22)}$$

The term $\omega_c\overline{\sin\phi}$ may be eliminated from the above equations, giving a first order differential equation in $\overline{\omega_I}$, the average angular frequency difference.

$$\overline{\omega_I} - \Delta\omega = \frac{1-m}{1+p\left(\frac{y}{m}t_c\right)} \cdot \frac{1}{m}\left[\sqrt{(\overline{\omega_I})^2-(m\omega_c)^2}-\overline{\omega_I}\right]. \quad \text{(D-23)}$$

This may be written more conveniently, dividing through by $m\omega_c$, writing $\rho=\overline{\omega_I}/m\omega_c$ and $\rho_0=\Delta\omega/m\omega_c$ and by operating on both sides with the differential operator, $1+p(y/m)t_c$.

Then

$$\left(1+p\frac{y}{m}t_c\right)(\rho-\rho_0) = \frac{1-m}{m}\left(\sqrt{\rho^2-1}-\rho\right)$$

or

$$\rho-\rho_0+\frac{y}{m}t_c\frac{d\rho}{dt} = \frac{1-m}{m}\left(\sqrt{\rho^2-1}-\rho\right).$$

Transposing $\rho-\rho_0$ and separating the variables

$$\frac{dt}{\frac{y}{m}t_c} = \frac{d\rho}{\rho_0-\rho+\frac{1-m}{m}\left(\sqrt{\rho^2-1}-\rho\right)}. \quad \text{(D-24)}$$

This equation may be directly integrated (between the limits $\rho=\rho_0$ and $\rho=1$) to yield T_F.

The integration is accomplished with the aid of a change of variable which permits the application of some tabulated integrals. The equations obtained are cumbersome; they are presented at the end of this appendix; they were used for the computations on which the several graphs presented are based. Fig. 20 presents the universal pull-in curves for the standard APC system. The following simplified analysis obtains the significant conclusions, in simpler form.

The limiting pull-in range may be determined as the condition which makes the required pull-in time become infinite. This occurs when the denominator of the above integrand has a real root. It will only occur when

$$\rho_0\left(\equiv\frac{\Delta f}{mf_c}\right) \geqq \sqrt{\frac{2}{m}-1}. \quad \text{(D-25) (C-17)}$$

A simple approximate solution for the "limit-curve" may be obtained by eliminating from the equation the factor which produces the above limitation. (Specifically, the small term $(\rho_0-\rho)$ in the denominator is omitted.)

Then, if $m\ll 1$, approximately

$$\frac{dt}{yt_c} \approx \frac{d\rho}{\sqrt{\rho^2-1}-\rho} = -\left(\sqrt{\rho^2-1}+\rho\right)d\rho \quad \text{(D-26)}$$

and hence, integrating from $\rho=\rho_0$ to $\rho=1$,

$$\frac{T_F}{yt_c} = \frac{\rho_0^2-1}{2}+\frac{\rho_0\sqrt{\rho_0^2-1}}{2} - \frac{1}{2}\ln\left|\frac{\rho_0+\sqrt{\rho_0^2-1}}{1}\right|. \quad \text{(D-27)}$$

Except for ρ_0 near 1, this is closely equal to

$$\frac{T_F}{yt_c} \approx \rho_0^2 = \left(\frac{\Delta f}{mf_c}\right)^2 \quad \text{(D-28) (C-19)}$$

which is the equation presented earlier.

The pole at $\rho_0^2=(2/m)-1$ can be included, writing the simplified equation as

$$\frac{T_F}{yt_c} = \frac{\rho_0^2}{1-\frac{m}{2-m}\rho_0^2} \quad 1<\rho_0^2<\left(\frac{2}{m}-1\right). \quad \text{(D-29)}$$

The exact integration of (D-24) is accomplished with the aid of the following substitution:

$$z=\sqrt{\rho^2-1}-\rho$$

whence,

$$-\rho=\frac{1+z^2}{2z} \quad \text{and} \quad \frac{d\rho}{dz}=-\frac{1}{2}\left(\frac{z^2-1}{z^2}\right).$$

Then

$$\frac{T_F}{yt_c} = \int_{z_0}^{z_1}\frac{-\left(z-\frac{1}{z}\right)dz}{(2-m)z^2+2m\rho_0 z+m}. \quad \text{(D-30)}$$

The limits are

$$\left|\begin{array}{l}\rho=1\\ \rho=\rho_0\end{array}\right. \quad \text{and} \quad \left|\begin{array}{l}z_1=-1\\ z_0=\sqrt{\rho_0^2-1}-\rho_0.\end{array}\right.$$

Referring to H. B. Dwight, "Tables of Integrals and Other Mathematical Data,"[25] Integrals #160.01, #160.11 and #161.11 are used.

Then

$$\frac{T_F}{yt_c} = -\left\{\frac{1}{2(2-m)}\ln\left|(2-m)z^2+2m\rho_0 z+m\right| - \frac{1}{2m}\ln\frac{z^2}{(2-m)z^2+2m\rho_0 z+m} + \left[\frac{2m\rho_0}{2}\left(\frac{1}{m}-\frac{1}{2-m}\right)\frac{2}{\sqrt{4(2-m)m-(2m\rho_0)^2}} \cdot\arctan\frac{2(2-m)z+2m\rho_0}{\sqrt{4(2-m)m-(2m\rho_0)^2}}\right]\right\}\Bigg|_{z_0=\sqrt{\rho_0^2-1}-\rho_0}^{z_1=-1}$$

$$= \frac{-1}{2(2-m)} \ln \left| \frac{2(1-m\rho_0)}{(2-m)z_0^2 + 2m\rho_0 z_0 + m} \right|$$

$$+ \frac{1}{2m} \ln \left(\frac{1}{z_0^2}\right) \left[\frac{(2-m)z_0^2 + 2m\rho_0 z_0 + m}{2(1-m\rho_0)} \right]$$

$$- \left\{ 2\rho_0 \left(\frac{1-m}{2-m}\right) \frac{1}{\sqrt{m(2-m) - (m\rho_0)^2}} \right. \quad \text{(D-31)}$$

$$\cdot \left[\arctan \frac{m\rho_0 - 2 + m}{\sqrt{m(2-m) - (m\rho_0)^2}} \right.$$

$$\left. \left. - \arctan \frac{m\rho_0 + (2-m)z_0}{\sqrt{m(2-m) - (m\rho_0)^2}} \right] \right\}.$$

Appendix E

Reliability of Frequency Difference Detection

This Appendix presents some mathematical derivations relating to the reliability of frequency difference detection.

The relations between rms frequency error and integration time are derived for

(a) the signal

(b) quadricorrelator frequency difference detector preceded by limiter

(c) quadricorrelator frequency difference detector alone.

Basic Signal Characteristics

The combination of signal and noise may be expressed in the following alternate forms (omitting for the moment the time gate factor)

$$S \cos \omega_{SC} + a(t) \cos \omega_{SC}t + b(t) \sin \omega_{SC}t \quad \text{(E-1)}$$

in which the noise is related to the color subcarrier frequency, or

$$S \cos \omega_{SC}t + a_0(t) \cos \omega_0 t + b_0(t) \sin \omega_0 t \quad \text{(E-2)}$$

in which the noise is expressed relative to the local oscillator frequency.

After limiting, the signal can be expressed as

$$S \cos (\omega_{SC}t + \phi(t)). \quad \text{(E-3)}$$

The phase modulation due to noise is $\phi(t)$.

$$\phi(t) = \arctan \frac{b(t)}{S + a(t)} \approx \frac{b(t)}{S} \quad \text{(E-4)}$$

as a first order approximation.

Then

$$\frac{d\phi}{dt} = \frac{S\dfrac{db}{dt} + \dfrac{d}{dt}(ab)}{(S+a)^2 + b^2}. \quad \text{(E-5)}$$

As a second order approximation, the relationship

$$\phi(t) = \int \frac{d\phi}{dt} dt \approx \frac{b(t)}{S} + \frac{a(t)b(t)}{S^2} \quad \text{(E-6)}$$

can be used, since $(S+a)^2 + b^2 \approx S^2$ for signals of interest.

The instantaneous frequency of the amplitude limited signal is

$$f(t) = \frac{1}{2\pi}\left[\omega_{SC} + \frac{d\phi(t)}{dt}\right]. \quad \text{(E-7)}$$

The rms frequency error due to noise is f_{rms}.

$$f_{\text{rms}} = \frac{1}{2\pi}\left[\frac{d\phi(t)}{dt}\right]_{\text{rms}}. \quad \text{(E-8)}$$

The signal amplitudes will also be useful in this analysis.

Then

$S = \frac{1}{2}hS_0$ = amplitude of a burst

$Sd = \frac{1}{2}hS_0 d$ = average amplitude of the component at the burst frequency with gate duty cycle d.

The rms value of $b(t)$ is the square root of the noise power. If *effectively* passed through a filter of bandwidth f_H, and gated with a duty cycle d, the noise power per unit time is $d(N_W^2/f_W)f_H$ and hence, the first order approximation for ϕ_{rms} is

$$\phi_{\text{rms}} \approx \frac{b_{\text{rms}}}{S} = \frac{N_W}{\frac{1}{2}hS_0}\sqrt{\frac{f_H}{df_W}}. \quad \text{(E-9)}$$

These relations are useful in evaluating the relation between integration time and reliability of the best possible frequency difference detector which might be used for the signal.

To relate reliability to time, the signal information may be averaged over a period T_I, and the rms value of the average then has improved reliability by virtue of integration. As in the case of phase information, it is convenient to use a rectangular time aperture for a standard of comparison for integrators.

Then

$$f_{\text{rms}} = \frac{1}{T_I}\left[\int_0^{T_I} [f(t) - f_{SC}]dt\right]_{\text{rms}} \quad \text{(E-10)}$$

$$= \frac{1}{T_I}\left[\int_0^{T_I} \frac{1}{2\pi}\frac{d\phi}{dt}dt\right]_{\text{rms}}$$

$$= \frac{1}{2\pi T_I}[\phi(T_I) - \phi(0)]_{\text{rms}}$$

$$= \frac{\sqrt{2}}{2\pi T_I}\phi_{\text{rms}}$$

and therefore, using the first order approximation above,

$$f_{\text{rms}} \approx \sqrt{\frac{2}{d}} \cdot \frac{1}{h\pi T_I} \cdot \frac{N_W}{S_0} \cdot \sqrt{\frac{f_H}{f_W}}. \quad \text{(E-11)}$$

The term $(S_0/N_W)\sqrt{f_W}$ is the signal-to-*noise-density* ratio.

The factor $(1/T_I)\sqrt{f_H}$ has the dimensions of $(\text{frequency})^{3/2}$; such terms normally result in frequency modulation noise analysis due to the triangular spectrum of the noise.[27]

The second order approximation is

$$f_{\text{rms}} \approx \frac{1}{\sqrt{2}\,\pi T_I}\left[\left(\frac{b_{\text{rms}}}{S}\right)^2 + \left(\frac{(ab)_{\text{rms}}}{S^2}\right)^2\right]^{1/2} \quad \text{(E-12)}$$

Note here that rms values add in quadrature.

The second term varies as

$$\left(\frac{N_W}{S_0}\sqrt{\frac{f_H}{f_W}}\right)^2$$

and for signal-to-noise ratios which at present give satisfactory monochrome video signals is small compared to the first term.

Quadricorrelator with Limiter

The quadricorrelator is shown in Figs. 13 and 14. The quadrature reference is R_Q.

$$R_Q = \sin \omega_0 t. \quad \text{(E-13)}$$

The in-phase reference is R_I.

$$R_I = \cos \omega_0 t. \quad \text{(E-14)}$$

The cosine beatnote is the beatnote between the input signal and R_I. This is conveniently expressed as

$$\frac{2}{S}\overline{[S\cos(\omega_{SC}t + \phi)]\cos\omega_0 t} = \cos\left[[\omega_0 - \omega_{SC}]t - \phi(t)\right]. \quad \text{(E-15)}$$

The sine beatnote is then

$$\frac{2}{S}\overline{[S\cos(\omega_{SC}t + \phi)]\sin\omega_0 t} = \sin\left[[\omega_0 - \omega_{SC}]t - \phi(t)\right]. \quad \text{(E-16)}$$

The derivative of the cosine beatnote is

$$-\left[\omega_0 - \omega_{SC} - \frac{d\phi}{dt}\right]\sin\left[[\omega_0 - \omega_{SC}]t - \phi(t)\right]. \quad \text{(E-17)}$$

Let

$$\omega_0 - \omega_{SC} \equiv \Delta\omega \equiv 2\pi\Delta f. \quad \text{(E-18)}$$

Then, the indicated frequency, which is the integrated output from the product of the signals expressed in (E-16) and (E-17), as multiplied in the output synchronous detector of the quadricorrelator, is, with due regard to signs,

$$f(t) = \frac{1}{\pi T_I}\int_0^{T_I}\left[\Delta\omega - \frac{d\phi}{dt}\right]\sin^2\left[\Delta\omega t - \phi(t)\right]dt. \quad \text{(E-19)}$$

The polarity of the indicated frequency may be reversed (when so required) by transferring the differentiating circuit from the cosine channel to the sine channel.

Then, since $\sin^2 x = \frac{1}{2} - \frac{1}{2}\cos 2x$, and since

$$\left[\Delta\omega - \frac{d\phi}{dt}\right]\cos 2[\Delta\omega t - \phi]dt = \cos 2[\Delta\omega t - \phi]\cdot d[\Delta\omega t - \phi]$$

we obtain

$$f(t) = \Delta f + \frac{1}{2\pi T_I}\int_{t=0}^{t=T_I} d\phi - \frac{1}{2\pi T_I}\int_{t=0}^{t=T_I}\cos 2\eta\, d\eta \quad \text{(E-20)}$$

where

$$\eta \equiv \Delta\omega t - \phi(t).$$

Thus the output noise consists of two components: the first represents the frequency noise of the signal; it could be measured as output noise if ω_{SC} were ***known***. The second represents extra noise introduced by the measurement of an unknown frequency in this circuit. Then

$$f_{\text{rms}} = \frac{1}{\sqrt{2}\,\pi T_I}\left[\phi_{\text{rms}}^2 + \text{extra noise}^2\right]^{1/2}. \quad \text{(E-21)}$$

The extra noise is evaluated as follows:

$$\frac{1}{2\pi T_I}\int_{t=0}^{t=T_I}\cos 2\eta\, d\eta = \frac{1}{2\pi T_I}\left[\tfrac{1}{2}\sin 2\eta(T_I) - \tfrac{1}{2}\sin 2\eta(0)\right]. \quad \text{(E-22)}$$

Two effects are indicated:

(a) Due to the use of a rectangular time aperture, an extraneous "sampling distortion" term appears unless $\Delta\omega T_I$ = a multiple of 2π, which is therefore assumed for simplicity.

(b) The output noise has the character of random noise which is passed through a nonlinear amplifier having a gain proportional to the sine of the input. This crushes the noise peaks and reduces the rms value.

Then, since $\sin^2 x < x^2$

$$\left[\frac{1}{2\pi T_I}\int_{t=0}^{t=T_I}\cos 2\eta\, d\eta\right]_{\text{rms}} < \frac{\sqrt{2}}{2\pi T_I}\phi_{\text{rms}}. \quad \text{(E-23)}$$

If there is substantial integration ($f_H T_I \gg 1$), the two noise components approach complete independence and add in quadrature, hence at worst,

$$f_{\text{rms}} \approx \frac{1}{\pi T_I}\phi_{\text{rms}}. \quad \text{(E-24)}$$

Thus, the quadricorrelator, with a limiter, measures a frequency difference to within a few db of the ultimate reliability permitted by signal information. It has no "detuning" error. The stepped characteristic may be introduced to give

[27] M. G. Crosby, "Frequency modulation noise characteristics," Proc. I.R.E., vol. 25, pp. 472–514; April, 1937.

$$f_{\text{rms}} \approx \frac{1}{\pi T_I} \phi_{\text{rms}} \sqrt{\frac{f_H - 2f_{NN}}{f_H}} \cdot \quad \text{(E-25)}$$

Since $f_H \gg 2f_{NN}$, the simpler equations above are adequate.

Quadricorrelator without Limiter

The input signal is

$$S \cos \omega_{SC} t + a_0(t) \cos \omega_0 t + b_0(t) \sin \omega_0 t. \quad \text{(E-2)}$$

The cosine beatnote is proportional to

$$\cos \Delta\omega t + \frac{a_0(t)}{S} \cdot \quad \text{(E-26)}$$

The sine beatnote is

$$\sin \Delta\omega t + \frac{b_0(t)}{S} \cdot \quad \text{(E-27)}$$

The derivative of the cosine beatnote is

$$-\Delta\,\omega \sin \Delta\omega t + \frac{1}{S}\frac{da_0(t)}{dt} \cdot \quad \text{(E-28)}$$

The quadricorrelator output is

$$f(t) = \frac{1}{\pi T_I}\int_0^{T_I}\left[\Delta\omega \sin \Delta\omega t - \frac{1}{S}\frac{da_0}{dt}\right] \cdot \left[\sin \Delta\omega t + \frac{b_0}{S}\right] dt$$

$$= \Delta f - \frac{1}{\pi T_I}\int_0^{T_I}\frac{1}{S}\frac{da_0}{dt}\sin \Delta\omega t dt + \frac{1}{\pi T_I}\int_0^{T_I}\frac{\Delta\omega}{S} b_0 \sin \Delta\omega t dt - \frac{1}{\pi T_I}\int_0^{T_I}\frac{1}{S^2} b_0 \frac{da_0}{dt} dt. \quad \text{(E-29)}$$

The evaluation of the several terms is aided by integration by parts:

$$-\frac{1}{\pi T_I}\int_0^{T_I}\frac{1}{S}\frac{da_0}{dt}\sin \Delta\omega t dt = -\frac{1}{\pi T_I}\sin \Delta\omega t \frac{a_0}{S}\Big|_0^{T_I} + \frac{1}{\pi T_I}\int_0^{T_I}\frac{\Delta\omega}{S} a_0 \cos \Delta\omega t dt. \quad \text{(E-30)}$$

Then

$$\left[-\frac{1}{\pi T_I}\sin \Delta\omega t \cdot \frac{a_0}{S}\Big|_0^{T_I}\right]_{\text{rms}} = \frac{1}{\pi T_I}\cdot\frac{a_{0\text{rms}}}{S} \cdot \quad \text{(E-31)}$$

The term

$$\frac{1}{\pi T_I}\int_0^{T_I}\frac{\Delta\omega}{S}\left[b_0(t)\sin \Delta\omega t + a_0(t)\cos \Delta\omega t\right] dt \quad \text{(E-32)}$$

now appears. This is a two-dimensionally noise modulated sine wave, of the type shown in Fig. 16. The bandwidth of the noise, however, is such that it heterodynes with the carrier to produce a dc component. Then, the integral of this term has the rms value

$$\frac{\Delta\omega}{\pi S}\frac{N_{\text{rms}}}{\sqrt{f_H T_I}} = \frac{\Delta\omega}{\pi S}\frac{a_{0\text{rms}}}{\sqrt{f_H T_I}} \quad \text{(E-33)}$$

since there are $f_H T_I$ effective harmonic components. The remaining term is evaluated as follows, integrating by parts:

$$\left[\frac{1}{\pi T_I}\int_0^{T_I}\frac{a_0}{S^2}\frac{db_0}{dt}dt\right]_{\text{rms}} = \left[\frac{1}{\pi T_I}\int_0^{T_I}\frac{b_0}{S^2}\frac{da_0}{dt}dt\right]_{\text{rms}} = \frac{1}{\sqrt{2}}\left[\frac{1}{\pi T_I}\int_0^{T_I}\frac{1}{S^2}\frac{d}{dt}(a_0 b_0) dt\right]_{\text{rms}} = \frac{1}{\pi T_I}\frac{(a_0 b_0)_{\text{rms}}}{S^2} \cdot \quad \text{(E-34)}$$

Then

$$[f(t) - \Delta f]_{\text{rms}} = f_{\text{rms}} = \frac{1}{\pi T_I}\left[\left[\left(\frac{a_{0\text{rms}}}{S}\right)^2 + \left(\frac{(a_0 b_0)_{\text{rms}}}{S^2}\right)^2\right] + \left[\frac{2\Delta f}{\sqrt{f_H T_I}}\cdot\frac{a_{0\text{rms}}}{S}\right]^2\right]^{1/2} \cdot \quad \text{(E-35)}$$

These terms add in quadrature as they represent independent random variables. The first bracketed term is of similar form as, but 3 db larger than, the second order signal approximation presented in (E-12), and is nearly equal to

$$\frac{1}{\pi T_I}\frac{a_{0\text{rms}}}{S} \cdot$$

The extra noise due to amplitude modulation appears in the last term of (E-35). The ratio of the AM component of noise to the FM component of noise is near

$$\frac{2\pi \Delta f T_I}{\sqrt{f_H T_I}} \cdot \quad \text{(E-36)}$$

When Δf is small, the quadricorrelator without a limiter approaches the limit of performance permitted by the signal. When Δf approaches $\frac{1}{2} f_H$, a poorer signal-to-noise ratio is obtained. The time T_I must be selected so that f_{rms} does not exceed some selected value, when Δf is the nominally maximum design value for pull-in range.

Equation (E-35) shows that the operation of pulling in results in a large reduction of output noise from the quadricorrelator.

TABLE OF SYMBOLS

Symbol	*Description*
a, $a(t)$	cosine component of noise at the frequency of the sync signal
a_0, $a_0(t)$	same at frequency of oscillator
b, $b(t)$	sine component of noise at the frequency of the sync signal
b_0, $b_0(t)$	same at frequency of oscillator
C	capacitor, Fig. 18
d	the duty cycle of the burst
f_c	the dc loop gain of an APC system, equal to the peak frequency holding range
f_{BN}	beatnote frequency, appearing in Appendix D
f_H, $f_{\text{HORIZONTAL}}$	the line-scanning frequency, 15,750 cps
f_N	effective noise bandwidth of a phase-detection system
$f_{N_{\text{LIMIT}}}$	the value of f_N if all of the phase information is used
f_0	frequency of the local reference oscillator
f_{NN}	the noise bandwidth of an APC loop
f_{rms}	root-mean-square frequency error of a frequency-difference detector
f_{SC}, $f_{\text{SUBCARRIER}}$	the subcarrier (color carrier) frequency
$f(t)$	indicated frequency difference
f_W	video bandwidth occupied by signal and noise
f_3	the 3 db pass-band width of a high Q filter
$F(f)$	transfer characteristic of high Q filter
$F(\Delta f)$	transfer characteristic of high Q filter measured in terms of frequency difference
G	loop gain of phase-control system of Appendix B
h	the ratio of the peak-to-peak amplitude of the burst and the line and field sync pulses
k	index number used in Appendix A
K	a damping coefficient relating to the pass band of an APC loop
m	the resistive divider ratio of a standard APC filter, the ratio of ac gain over dc gain through the network $N(\omega)$
$N(t)$, N	noise signal as a function of time
N_W	the root-mean-square noise in the entire video pass band, assumed flat over the band
$N(\omega)$, $N(p)$	transfer characteristic of the filter of an APC loop
p	$j\omega$, $j2\pi f$, or d/dt as appropriate
$p(\tau)$	the relative probability density distribution function for timing data
$p(\phi)$	the relative probability density distribution function for phasing data
$Q(f)$, $Q(\omega)$	the effective modulation pass band transfer characteristic of an APC system after synchronization
r	ratio of actual gate width to minimum burst width
R	resistor, Fig. 18
R_I	in-phase reference signal
R_Q	quadrature reference signal
S	amplitude of color burst
S_k	amplitude of kth frequency component
S_0	the amplitude of the line and field sync pulses
$S(t)$	the synchronizing signal as a function of time
T	time constant of low-pass filter $Y(f)$
T	time constant RC of Fig. 18
T_A	transient response time of high Q filter in Appendix B
T_{BN}	beatnote period
t_c	characteristic time constant of an APC loop
T_F	frequency pull-in time
T_I	integration time of a frequency-difference detector
T_M	effective integration time
$T_{M_{\text{LIMIT}}}$	the required value of T_M if all of the phase information is used
t_0	root-mean-square time error
t_0'	root-mean-square error if the color synchronizing signal were present all of the time
t_{0_k}	rms timing error of the k'th frequency component
T_S	stabilization time of a synchronizing system
T_ϕ	phase pull-in time
x	constant relating to Fig. 18(a)
y	parameter defined by (C-8)
$Y(f)$	pass-band characteristic of the low-pass filter in the phase-control system of Fig. 6, Appendix B
z	variable of integration used in Appendix D
β	sensitivity of the reactance tube of an APC loop
ΔE	output voltage of the phase detector of an APC loop
Δf	frequency difference between oscillator and sync signal
$\Delta f_{\max}$	maximum frequency difference from which pull-in will occur
$\Delta\phi$	the static phase error of an APC loop
$\Delta\phi$	as used in Appendix B—phase error of phase feed-back system
$\Delta\phi_{\text{corr}}$	the phase correction produced by the phase feed-back system of Appendix B
$\Delta\phi_{\text{corr}}(p)$	frequency spectrum of $\Delta\phi_{\text{corr}}$
$\Delta\phi_0$	static phase error of high Q filter
$\Delta\phi_{0_{\text{effective}}}$	equivalent phase modulation (re. B-6)
$\Delta\phi_0(p)$	frequency spectrum of $\Delta\phi_{0_{\text{effective}}}$
$\Delta\phi_{0I}(p)$	frequency spectrum of beat between filtered signal and direct noise at the phase detector of Fig. 6(a) (re. B-8)
$\Delta\phi_{02}(p)$	frequency spectrum of beat between the direct signal and filtered noise at the phase detector of Fig. 6(a) (re. B-8)
$\Delta\omega$	angular frequency difference
η	parameter used in Appendix E
θ, θ_α, θ_β	parameters defined and used in Appendix C
λ^2	see (C-23) and Fig. 21
μ	transfer gain of phase detector of an APC loop
$\rho=\dfrac{\Delta f}{mf_c}$	normalized frequency difference defined in Appendix C
ρ_0	initial value of ρ
τ	time scale for a probability density
ϕ	phase angle
ϕ_1	a phase variable relating to Fig. 2(c)
ϕ_{rms}	root-mean-square phase error
ϕ_0	initial phase
ϕ_∞	static phase difference due to frequency detuning
ω_c	$2\pi f_c$
ω_H	$2\pi f_H$
ω_I	instantaneous angular frequency difference (re. D-17)
ω_k	kth angular frequency
ω_0	$2\pi f_0$
ω_{SC}	$2\pi f_{SC}$

Phase-Locked Loops with Signal Injection for Increased Pull-In Range and Reduced Output Phase Jitter

PETER K. RUNGE

Abstract—Second-order phase-locked loops with signal injection (PLLI's) have orders of magnitude larger pull-in ranges than conventional second-order PLL's with identical natural frequencies and damping factors. The PLLI utilizes a little-known characteristic of the Van der Pol oscillator, and for practically no additional hardware the pull-in range is increased sufficiently to eliminate in many cases the need for crystal control of the voltage-controlled oscillator (VCO). In nonlinear systems (e.g., a retiming chain of a digital repeater) where the in-phase and in-quadrature noise components are correlated, the PLLI in addition features a significant reduction in output phase jitter. A 44.6-MHz PLLI, designed for timing recovery in a fiber optic repeater, has 8.4-kHz conventional minimum jitter bandwidth. To reduce time jitter accumulation in a chain of repeaters, the damping factor was increased to 4 resulting in a Q of 1300. The pull-in range of this loop is 9.6 MHz compared to 18 kHz for the same loop without signal injection. The static phase error is ±0.6° for ±0.3 MHz oscillator drift. By injecting with 18° phase lag, the PLLI can take advantage of the correlation between amplitude and phase noise components and the total output phase jitter is reduced by 10.5 dB due to destructive interference.

Paper approved by the Editor for Wire Communication of the IEEE Communications Society for publication after presentation in part at the National Telecommunications Conference, San Diego, CA, December 2-4, 1974. Manuscript received December 15, 1975.

The author is with Bell Laboratories, Holmdel, NJ 07733.

INTRODUCTION

ANALOG phase-locked loops (PLL's) have been an important building block with many applications in communication and other systems for many years. For a review paper on this subject see Gupta [1]. High-Q PLL's, however, have always suffered from their rather limited pull-in range. The PLL by itself is a poor frequency acquisition device [2] and a number of schemes have been devised to improve the pull-in range at the expense of additional hardware.[1] A very popular solution to the acquisition problem in phase-controlled oscillators is to impose bounds on the drift range of the voltage-controlled oscillator (VCO) by crystal control. The required VCXO's are then even more expensive.

[1] Numerous schemes have been suggested to increase the pull-in range of a conventional PLL. All require a considerable amount of additional hardware. See, e.g., [3], [4].

Reprinted from *IEEE Trans. Commun.*, vol. COM-24, pp. 636–644, June 1976.

Lahti and Beling [5] first realized the benefit of a synchronizing signal on the acquistion performance of an FM detector. Others followed suit.[2] For some unknown reason, however, the principle of signal injection was never fully utilized to solve one of the PLL's basic shortcomings.

We will demonstrate that signal injection into the VCO of a PLL can increase the pull-in range by more than two orders of magnitude. This is a consequence of a little known characteristic of the Van der Pol oscillator. The unsynchronized, but driven, oscillator produces a discrete Fourier component at the injecting frequency. The phase of this component is $\pm\pi/2$ out of phase with the injected current and changes sign with the frequency difference. In conjunction with the phase detector (PD) of the PLL this mechanism effectively serves as a frequency detector and increases the pull-in range of the complete phase-locked loop with injection (PLLI). The PLLI requires a multiplier-type PD (e.g., a diode quad) to detect the discrete frequency component. The additional path for the signal injection, however, consists of only a few passive components. Signal injection, therefore, is a very economic solution to a long-standing problem of the acquisition performance of PLL's.

Previous analyses of the PLL acquisition concerned themselves with the relative pull-in range in terms of the lock range [9], [10]. However, the question still remains of how large the lock range can be designed in practice to insure proper PLL performance.

We have analyzed the pull-in performance of PLLI and offer an unambiguous expression for its pull-in range in terms of component limitations.

The phase angle of the injection becomes an important parameter of the PLLI. It controls mainly the conversion of input amplitude noise to output phase noise, which vanishes for in-phase injection. However, by injecting with a slight phase lag, the PLLI can utilize existing correlation between input amplitude and phase noise components to achieve a considerable reduction in total output phase noise without penalty in transient response time. Conventional PLL's are insensitive to cross correlation in the input noise components and therefore cannot take advantage of it. This type of correlation was found to exist in the nonlinear retiming chain of a digital repeater and the PLLI offers a genuine reduction in total output phase noise by at least 10.5 dB.

PLL'S WITH SIGNAL INJECTION

A. Fundamental Equations

Fig. 1 shows a block diagram of the PLLI. The following complex input voltage is assumed:

$$\bar{V}_{in} = (V_1 + v_c(t)) \exp\{j(\omega_i t + \theta_{in}(t))\} + jv_s(t) \exp\{j(\omega_i t + \theta_{in}(t))\} \quad (1)$$

where V_1 is the amplitude of the RF carrier at ω_i and $\theta_{in}(t)$ its phase, which is allowed to be a slowly varying function of time. $v_c(t)$ and $v_s(t)$ are the in-phase and in-quadrature noise components, respectively, and are both assumed to be random variables of wide-sense stationary processes. A current proportional to the input voltage is injected directly into the tank circuit of the VCO with the complex transconductance

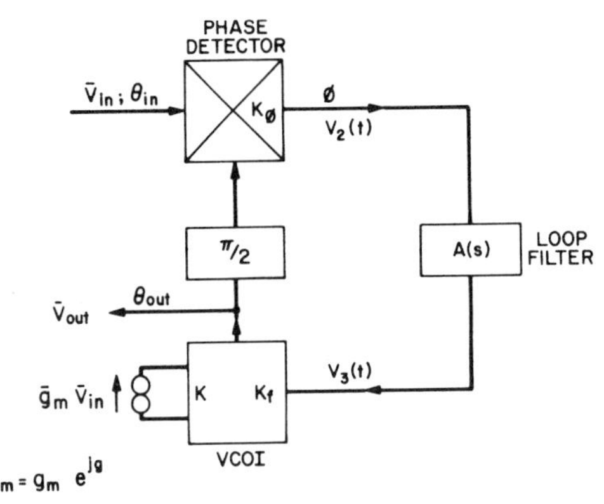

Fig. 1. Block diagram of a PLL with current injection into the oscillator.

$$\bar{g}_m = g_m \exp\{jg\}. \quad (2)$$

The phase angle g of the injection will become an important parameter. Adler [11] has derived the differential equation of the output phase $\theta_{out}(t)$ of a Van der Pol oscillator which was driven by an injected voltage. Adler's analysis applied to the VCO with current injection yields

$$\frac{d\theta_{out}(t)}{dt} = K\left[\left(1 + \frac{v_c(t)}{V_1}\right) \sin(\phi(t) + g) + \frac{v_s(t)}{V_1} \cos(\phi(t) + g)\right] - \omega_i + \omega_r + K_f V_3(t) \quad (3)$$

with the injection constant K

$$K = \frac{g_m V_1}{V_0 2C} \quad (4)$$

where V_0 is the free running voltage amplitude and C is the capacitance of the oscillator. ω_r is the rest frequency of the VCO and K_f its frequency constant in rad/V.

The phase error of the PLLI is

$$\phi(t) = \theta_{in}(t) - \theta_{out}(t). \quad (5)$$

With a $\pi/2$ phase shift between VCO output and the sinusoidal PD, the output voltage of the PD is

$$V_2(t) = K_\phi \left(1 + \frac{v_c(t)}{V_1}\right) \sin\phi + K_\phi \frac{v_s(t)}{V_1} \cos\phi. \quad (6)$$

The input and output voltages of the loop filter are related by

$$V_3(t) = \int_0^t a(t - t_0) V_2(t_0)\, dt_0 \quad (7)$$

[2] References [6]-[8] also suggest a synchronized loop in conjunction with a phase-locked oscillator. None, however, achieves a comparable increase in pull-in range.

with $a(t)$ the inverse Laplace transform of the filter transfer function

$$A(s) = A_0 \frac{1 + s\tau_2}{1 + sA_0\tau_1}. \quad (8)$$

Equation (8) describes an active loop filter with dc gain A_0 and time constants τ_1 and τ_2 of the feedback network. Equations (3), (5), (6), and (7) completely describe the PLLI. To solve this set of nonlinear integro-differential equations, two common approximations are made. To study the loop dynamics, a large phase error $\phi(t)$ is assumed and all noise components are neglected. To study the noise performance, on the other hand, the loop is assumed to be at lock, and the phase error therefore is a small quantity allowing appropriate linearization of the equations.

B. Loop Dynamics

$v_c(t)$ and $v_s(t)$ are assumed identically zero. Furthermore, assuming that the input phase is constant, $\theta_{in} \neq f(t)$, the above set of equations is solved for the phase error

$$\frac{d^2\phi}{dt^2} + \frac{d\phi}{dt}\left[2\omega_{n0}\xi \cos\phi + \frac{1}{A_0\tau_1}\right] + \omega_{n0}^2 \sin\phi - \frac{\omega_i - \omega_r}{A_0\tau_1} = 0 \quad (9)$$

with the natural frequency ω_{n0} of the PLLI

$$\omega_{n0} = \sqrt{\frac{K_f K_\phi}{\tau_1}} \quad (10)$$

Note that ω_{n0} is independent of the dc gain A_0 of the active loop filter with transfer function (8). The damping factor

$$\xi = \xi_0 + \xi_i \cos g \quad (11)$$

consists of two terms:

$$\xi_0 = \frac{\omega_{n0}}{2} \tau_2 \quad (12)$$

is the damping factor of a conventional PLL;

$$\xi_i = \frac{K}{2\omega_{n0}} \quad (13)$$

is an additive term proportional to the magnitude of the current injection, and it is the only contribution in (9) due to the injection. ($\xi_i = 0$ and $\xi_0 \neq 0$, therefore, describe a conventional PLL.) Two approximations have been made in deriving (9),

$$K \ll A_0 K_f K_\phi, \quad (14)$$

which is always justified, since the dc loop gain A_0 can be made arbitrarily large in a PLLI as shall be explained later. Secondly, the injection angle g is sufficiently small so that

$$\cos(\phi + g) \approx \cos\phi \cos g. \quad (15)$$

We now proceed to discuss the pull-in range of the PLLI. Equation (9) is identical to the differential equation for a second-order PLL with imperfect integrator that has the following pull-in range [2]:

$$\omega_i - \omega_r < 2\omega_{n0}\sqrt{\xi\omega_{n0}\tau_1 A_0}. \quad (16)$$

Clearly, the dc gain A_0 should be made as large as possible to achieve a large pull-in range. In conventional PLL's ($K = 0$) the gain is limited by the combined offset voltages V_{offset} of the phase detector and the active loop filter to avoid saturation of the active filter.

In addition, a large pull-in range requires a wide-band operational amplifier and high quality components in the feedback network to maintain the characteristic of (8) over a wide band. In a practical high-Q PLL, these two effects tend to limit the pull-in range to approximately 1 to 10 times the loop natural frequency. Some other means, usually crystal control, are then applied to keep the free-running frequency of the VCO within the narrow pull-in range.

The PLLI, however, does not depend on the loop filter for its pull-in. A free-running, but driven, Van der Pol-type oscillator emits a discrete frequency component $V_{\omega i}$ at the injecting frequency that is always $\pm\pi/2$ out of phase with respect to the injected current, which we will proceed to prove.

It is assumed that the active loop filter is saturated such that $V_3(t) \neq f(t)$ in (3). The oscillator rest frequency ω_r, therefore, is changed to $\omega_0 = \omega_r + K_f V_3$. Equations (3) and (5) can then be reduced to Adler's equation

$$\frac{d(\phi + g)}{dt} = -K \sin(\phi + g) + \omega_i - \omega_0. \quad (17)$$

Adler gave the complete solutions for the above differential equation [11]. The solution for the unlocked but driven state is

$$\tan\frac{\phi(t) + g}{2} = k + \sqrt{1 - k^2}\tan\left[\frac{\Delta\omega(t - t_0)}{2}\sqrt{1 - k^2}\right], \quad \text{for } |k| < 1 \quad (18)$$

where

$$k = \frac{K}{\omega_i - \omega_0}. \quad (19)$$

The output phase of the VCOI according to (18) is a periodic function of time. The output voltage, therefore, has a discrete Fourier spectrum and the spectral component at the injecting frequency can be found following the analysis of Armand [12]

$$\bar{V}_{\omega_i} = -j\frac{k}{1 + \sqrt{1 - k^2}} V_0 \exp\{j(\omega_i t + \theta_{in} + g)\}. \quad (20)$$

Fig. 2 shows amplitude and phase of this Fourier component

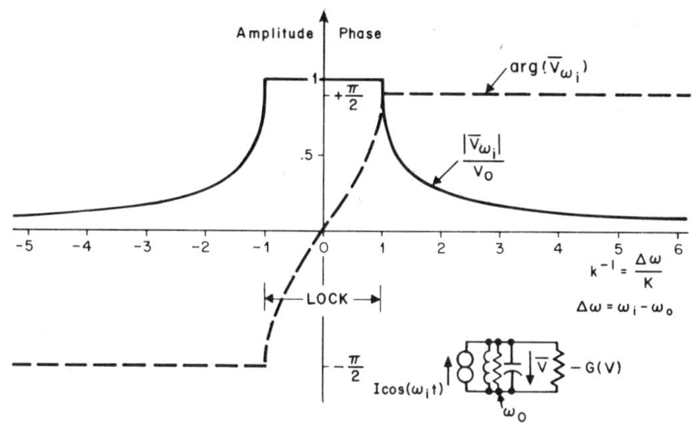

Fig. 2. Amplitude and phase of the Fourier component at ω_i as a function of $\Delta\omega/K$ for a Van der Pol oscillator with injected signal at ω_i.

as a function of k^{-1} for in-phase injection ($g = 0$). For $|k| < 1$, the phase of this Fourier component is offset by $\pm\pi/2$ from the input phase and the phase offset changes sign with the frequency difference $\Delta\omega$. The driven but unlocked VCOI exhibits frequency discriminator characteristics. This effect will increase the pull-in range of the complete PLLI significantly. The amplitude and phase behavior as indicated in Fig. 2 was studied experimentally by investigating a VCO with a network analyzer. The phase behaved exactly as shown in Fig. 2. The measured amplitude characteristic is shown in Fig. 3 in good agreement with (20).

To initiate acquisition, the PD output voltage has to overcome the offset voltage V_{offset} that initially caused the assumed saturation of the active loop filter. The new pull-in frequency is then given by

$$|\omega_1 - \omega_0| \leqslant \omega_{n0}\xi_i M \cos g \tag{21}$$

with ξ_i from (13) and the figure of merit of the PD

$$M = \left| \frac{K_\phi}{V_{\text{offset}}} \right| \tag{22}$$

which is a quantity in the order of 500.

The pull-in range is now determined by the lowest value of (16) or (21). Since saturation of the active loop filter is no longer limiting the pull-in, A_0 in (16) can be made arbitrarily large, and the pull-in is given by (21). The pull-in range then is a function of the figure of merit of the phase detector (22) and the magnitude and angle of the injection. To maximize the pull-in range, ξ_0 from (12) should be made zero (the loop filter then has pure low-pass characteristic) and the pull-in frequency becomes

$$|\omega_1 - \omega_0| \leqslant \omega_{n0}\xi M, \qquad \xi_0 = 0. \tag{23}$$

The following experiment was performed to demonstrate the large increase in pull-in range. An experimental PLLI was built for a fiber optic $T3$ rate repeater (44.6 Mbits/s) with

$$\omega_{n0} = 16.7k \text{ rad/s}; \qquad M = 450; \qquad \xi = 4$$

$$Q = 1300; \qquad A_0 = 10^5.$$

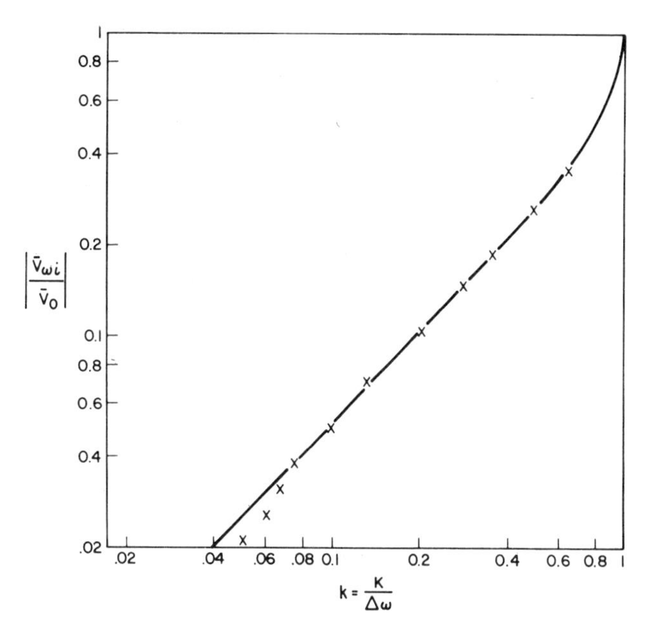

Fig. 3. Measured magnitude of Fourier component at ω_i as a function of normalized frequency difference.

The signal path for the injection consisted of a simple potentiometer plus an R-C phase shifter between PLLI input and VCOI. The PLLI could be changed into a PLL and vice versa by simply varying the magnitude of the injection and τ_2 in the loop filter (12). A direct comparison between the two versions of phase-locked loops was possible because no other component was altered. The PLLI had a pull-in range of 9.6 MHz. The PLL, however, required a reduction in dc gain and then exhibited a maximum pull-in range of only 18 kHz.

The temperature drift of the VCO was 500 kHz over the range $-40°$C to $80°$C. Therefore, the loop without injection would require crystal control, whereas the loop with injection can accommodate the temperature drift.

An additional benefit of larger dc gain A_0 is a smaller static phase error ϕ_0 for a frequency offset $\Delta\omega$ between VCO rest frequency and input frequency, from (9)

$$\sin\phi_0 = \frac{\Delta\omega}{K_\phi K_f A_0}. \tag{24}$$

The pull-in time is obtained from (9) in an analysis similar to that in [2]

$$T_{\text{pull-in}} \cong \frac{\Delta\omega^2}{4\xi\omega_{n0}^3}.$$

C. Linear Noise Analysis

The PLLI is assumed to be a lock and the phase error ϕ small to allow a first-order Taylor approximation of the trigometric functions in (3) and (6). Neglecting terms of second order, taking the Laplace transform, and combining equations (3), (5), (6), and (7) we obtain

$$\theta_{\text{out}}(s) = H(s)\left[\theta_{\text{in}}(s) + \frac{v_s(s)}{V_1}\right] + G(s)\frac{v_c(s)}{V_1} \tag{25}$$

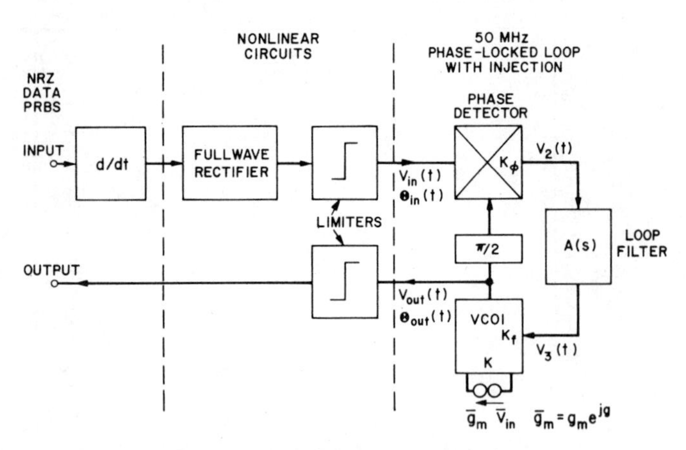

Fig. 4. Block diagram of a PLLI and preceding circuits in a retiming chain of a digital repeater.

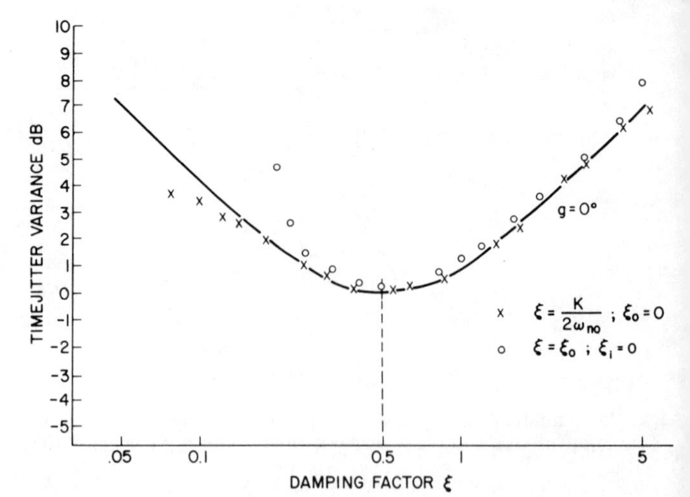

Fig. 5. Measured variance of output time jitter as a function of the damping factor ξ. ×: $\xi = (K/2\omega_{n0})$ (with signal injection, $g = 0°$); ○: $\xi = \xi_0$ (no injection).

with the closed loop transfer functions

$$H(jx) = \frac{1 + j2\xi x}{1 + j2\xi x - x^2}; \qquad G(jx) = \frac{2x\xi_i \sin g}{1 + j2\xi x - x^2} \tag{26}$$

where $x = \omega/\omega_{n0}$ and ξ and ξ_i from (11) and (13), respectively. For in-phase injection ($g = 0$), $G(jx)$ becomes zero, and the PLLI behaves like a conventional second-order PLL with natural frequency ω_{n0} and damping factor ξ from (11). Its integrated noise bandwidth is

$$B_L = \int_{-\infty}^{\infty} |H(jx)|^2 \frac{\omega_{n0}}{2\pi}\, dx = \frac{\omega_{n0}}{2}\left[\xi + \frac{1}{4\xi}\right] \text{(Hz)}. \tag{27}$$

To evaluate the noise performance of a PLLI, the output time jitter was measured for an input signal typical for a timing chain in digital repeaters. The received data have a raised cosine nonreturn-to-zero (NRZ) pulse shape and its spectrum therefore does not contain a discrete frequency for the PLL to lock on to. The signal is differentiated and rectified (Fig. 4) to regain a discrete frequency component at the baud rate and limited to assure stable loop parameters. The limiter at the output of the PLLI suppresses amplitude fluctuations.

The input signal is a band-limited pseudorandom bit stream (PRBS) with a $2^{20} - 1$ word length. Fig. 5 shows the measured variance of the output time jitter (normalized by $\langle|\Delta t|^2\rangle = 1.6\ 10^{-21} s^2$) versus the damping factor ξ from (11) for $g = 0$. The PLLI ($\xi_0 = 0$) is again compared with the PLL ($\xi_i = 0$) and with the theoretical noise bandwidth (solid line) of (27). For $\xi > 0.3$ both loops have indeed the same noise bandwidth. The PLLI version, of course, features orders of magnitude larger pull-in range.

The output phase noise (25) becomes a function of the input amplitude noise in addition to the input phase noise for out of phase injection ($g \neq 0$). Possible cross correlation between the two noise components then, of course, has to be taken into account in (25). This type of correlation is very likely to exist but difficult to analyze for a signal that was processed in nonlinear circuits (Fig. 4). A nonlinear element (e.g., a diode), whose reactance is amplitude dependent, will introduce cross correlation between amplitude and phase fluctuations. We do not attempt to predict the statistical properties of the noise, but rather make the most general assumptions about the spectral densities associated with the individual noise components. As we shall see later, the PLLI will then allow us to very easily and accurately measure the actual noise properties including the cross correlation terms.

We assume the following spectral densities to be associated with the noise components: amplitude noise density S_{cc}; phase noise density S_{ss}; cross-correlation density S_{sc} and S_{cs} with the necessary condition $S_{sc} = S_{cs}{}^*$; input phase jitter density ψ.

Integrating (25) with $\xi_0 = 0$ yields

$$\gamma = \frac{\langle|\theta|^2\rangle(\xi, g)}{\langle|\theta|^2\rangle(0.5, 0)} = \frac{\xi}{\cos g}\left[1 + \frac{1}{4\xi^2} + (A-1)\sin^2 g + B \sin 2g\right] + C \tan g. \tag{28}$$

Equation (28) is the total output phase jitter variance normalized to the jitter variance at the conventional minimum ($\xi = 0.5, g = 0$), with the relative noise spectral densities A, B, C,

$$A = \frac{S_{cc}}{\psi + S_{ss}}; \qquad B = \frac{\text{Re}\,(S_{cs})}{\psi + S_{ss}};$$

$$C = \frac{\omega_{n0}}{\psi + S_{ss}} \frac{d}{d\omega} \text{Im}\,(S_{sc}). \tag{29}$$

Equation (28) possesses an absolute minimum at $\xi_{\min}$ and $g_{\min}$. From

$$\frac{\partial\gamma}{\partial\xi} = 0 : \xi_{\min}^2 = \frac{1}{4[1 + (A-1)\sin^2 g_{\min} + B \sin 2g_{\min}]} \tag{30}$$

and from

$$\frac{\partial \gamma}{\partial g} = 0 : \tan g_{\min} = 2B + \frac{1}{\xi_{\min} \cos^2 g_{\min}} \cdot \left[C + \frac{\sin g_{\min}}{4\xi_{\min}} + A\xi_{\min} \sin g_{\min} [1 + \cos^2 g_{\min}] \right]. \tag{31}$$

The functional value at the minimum is

$$\gamma_{\min} = \frac{1}{2\xi_{\min} \cos g_{\min}} + C \tan g_{\min}. \tag{32}$$

Explicit solutions of $\xi_{\min}$ and $g_{\min}$ as a function of the coefficients A, B, and C could not be found. In most cases, however, A, B, and C are not known and it is desirable to calculate them from measured parameters at the absolute minimum.

Given $\gamma_{\min}$, $\xi_{\min}$, and $g_{\min}$ ($\xi_0 = 0$), we find

$$A = \frac{1 + \dfrac{1}{4\xi_{\min}{}^2 \cos^2 g_{\min}} - \dfrac{\gamma_{\min}}{\xi_{\min} \cos g_{\min}}}{\sin^2 g_{\min} [1 + \tan^2 g_{\min}]} \tag{33}$$

$$B = \frac{\dfrac{1}{4\xi_{\min}{}^2} - 1 - (A - 1) \sin^2 g_{\min}}{\sin 2 g_{\min}} \tag{34}$$

$$C = \frac{\gamma_{\min} \cos g_{\min} - \dfrac{1}{2\xi_{\min}}}{\sin g_{\min}}. \tag{35}$$

Thus, by two simple measurements of the variance of the output phase jitter performed at the conventional minimum and at the absolute minimum, we can determine $\gamma_{\min}$, $\xi_{\min}$, and $g_{\min}$. From (33)-(35) we can then calculate the relative spectral densities and with (29) and (27) the absolute spectral densities of amplitude and phase noise including the densities of their cross correlation. This is a very interesting feature that makes the PLLI useful as a measuring device.

Experimentally, an absolute minimum was found at $\xi_{\min} = 4$ and $g_{\min} = -18°$ (Fig. 6) of magnitude $\gamma_{\min} = -4.5$ dB. From (33)-(35) we can now calculate the relative spectral power densities of the input noise:

$$A = 8.75; \qquad B = 2.93; \qquad C = -0.69. \tag{36}$$

The measured phase jitter variance at the conventional minimum is $\langle |\theta|^2 \rangle (0.5, 0) = 1.26 \; 10^{-4}$ rad^2 and with $B_{L\min} = 8.35$ kHz the relative power densities of the input noise (36) may be converted to absolute spectral densities. It should be pointed out that C (which is the only frequency dependent noise power density) is only a correction factor of small-magnitude compared to A and B, which are frequency independent.

In Fig. 6 we have plotted the normalized time jitter variance as a function of the damping factor with the injection angle $g = 0$ and $-18°$. The solid lines represent (28) with the above constants. $g = 0$ represents the PLLI with a characteristic identical to that of a conventional second-order PLL. The minimum at $\xi = 0.5$ is what we have called the conventional minimum. The curve for $g = -18°$ passes through the absolute minimum where theory and experiment were matched. The agreement outside this minimum, however, is also very good.

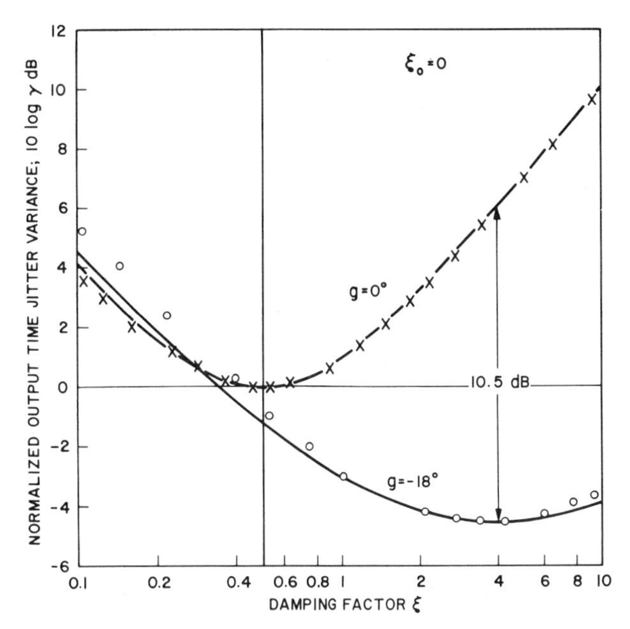

Fig. 6. Measured variance of output time jitter as a function of the damping factor for injection angles $g = 0°$ and $-18°$.

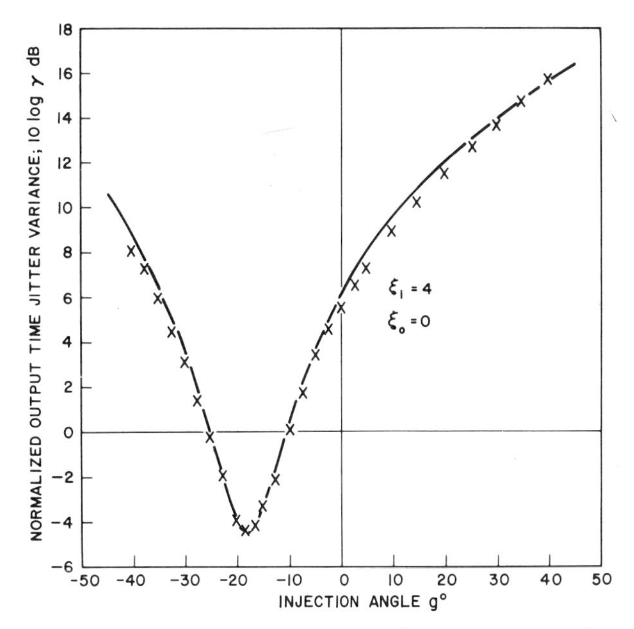

Fig. 7. Measured variance of output time jitter as a function of the injection angle for $\xi_i = 4$.

In timing recovery, PLL's should be overcritically damped to minimize jitter buildup in long repeater chains [13]. Damping factors in the range $\xi \cong 4$ to 10 are therefore desirable. Inspecting Fig. 6, we find that injection at $g = -18°$ offers a genuine reduction in output phase jitter variance of 10.5 to 13.5 dB.

In Fig. 7 the normalized time jitter variance is plotted versus the injection angle g, leaving $\xi_i = 4$ constant. Theory and measurement again agree very well. To demonstrate the variation of the output phase jitter over a larger range of the parameters ξ and g, we have included the computer printouts in Figs. 8-11 using the previously determined coefficients A, B, and C.

A plot of the spectral density of the output phase jitter as a function of frequency illustrates the effect of destructive interference. From (25) we obtain

$$\frac{|\theta_{out}(x)|^2}{\psi(0)+S_{ss}(0)} = \frac{1+4x^2(\xi^2+A\xi_i^2\sin^2 g+2B\xi_i\xi\sin g+C\xi_i\sin g)}{1+2x^2(2\xi^2-1)+x^4} \tag{37}$$

with $\xi=\xi_0+\xi_i\cos g$ from (11).

Equation (37) is plotted in Fig. 12 for $\xi=4$ and 10 ($\xi_0=0$) for the equivalent conventional PLL ($g=0$) and for the PLLI ($g=-18°$). The effect of destructive interference on the output spectral density is very pronounced. Spectral components around the natural frequency are suppressed in excess of 10 dB. It should be emphasized that the plots for $g=-18°$ do not represent transfer functions of the PLLI but output spectral densities. The transfer function $H(s)$, which determines the loop's dynamic behavior as pull-in time and phase transient response time, is altered only insignificantly by injecting at $-18°$ compared to 0° since it depends on the term $\cos g$ only (26). The output jitter, however, is reduced significantly by utilizing the cross correlation in the noise components, which usually remains unnoticed in the case of conventional PLL's.

D. Unwanted Injection in a Conventional PLL

Experimenting with the PLLI at 50 MHz, we noticed that signal injection into the VCO is very often obtained unintentionally by coupling or by reverse passage through interconnecting networks. Since the $\pi/2$ phase shifter between VCO and PD is missing in a conventional PLL, the angle of this unwanted injection is about $g\cong\pm\pi/2$, because a natural phase shift of $\pm\pi/2$ exists at the PD. The interference between amplitude and phase noise, that was utilized in the PLLI leading to destructive cancellation of certain noise components, may, of course, also lead to constructive buildup. This noise buildup will be a maximum at $g=+\pi/2$, entirely within the possibility of the above unwanted injection. To demonstrate the effect, we have calculated the spectral densities of the output jitter according to (37) for the following case.

$\xi_0=4$ (overcritically damped conventional second-order PLL)

$g=+\frac{\pi}{2}$ (worst case)

$\xi_i=0;0.2;0.4;0.6;0.8;1;1.2$

and spectral densities A, B, and C as before. Fig. 13 shows the

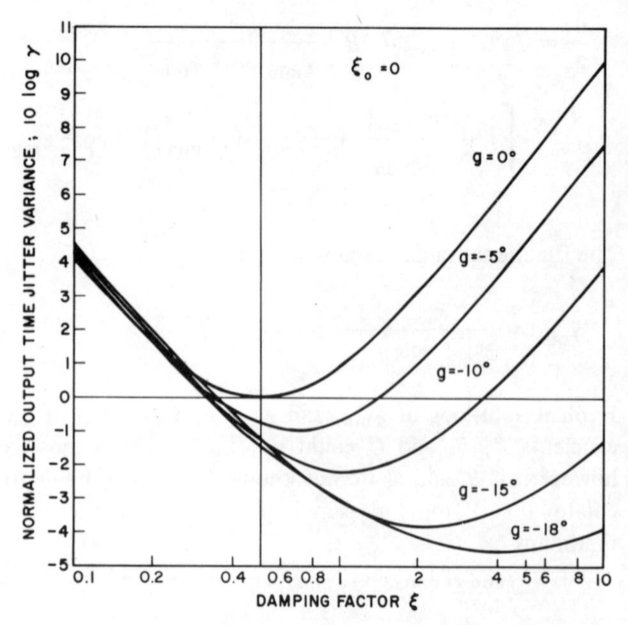

Fig. 8. Variance of output time jitter as a function of the damping factor ξ with the injection angle g as a parameter. Spectral densities as before.

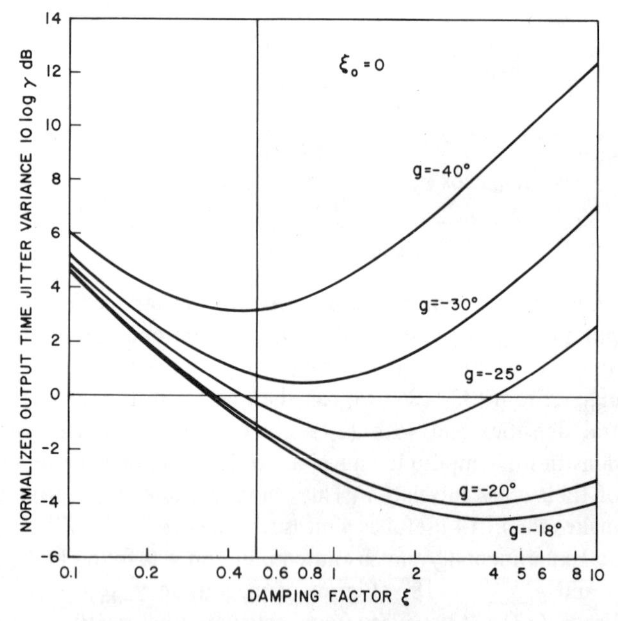

Fig. 9. Variance of output time jitter as a function of the damping factor ξ with the injection angle g as a parameter. Spectral densities same as before.

computer plot. The conventional PLL ($\xi_i=0$) has a jitter "peak" of only 0.2 dB. 5 percent additional unwanted injection already results in a jitter peak of about 1.3 dB, which in many cases cannot be tolerated.

CONCLUSION

We have shown theoretically and proven experimentally that signal injection into the VCO can increase the pull-in range of a PLL by several orders of magnitude without sacrifice in output jitter performance. The PLL with injection

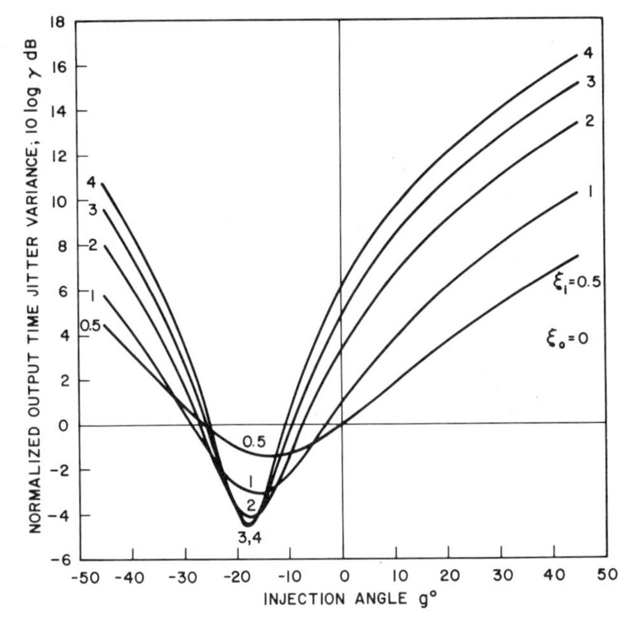

Fig. 10. Variance of output time jitter as a function of the injection angle g with the damping factor ξ as a parameter. Spectral densities same as before.

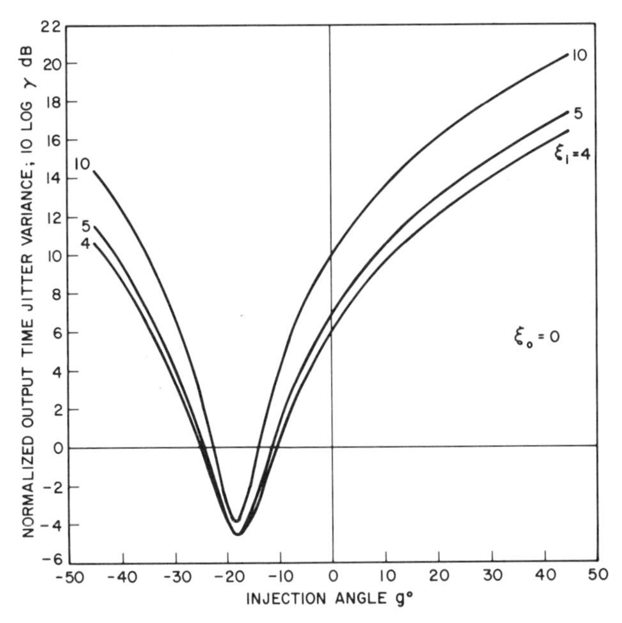

Fig. 11. Variance of output time jitter as a function of the injection angle g with the damping factor ξ as a parameter. Spectral densities same as before.

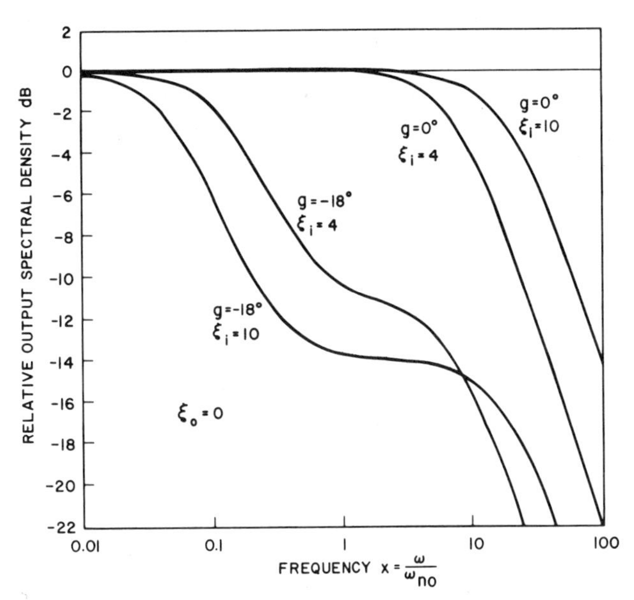

Fig. 12. Relative spectral density of the output phase jitter as a function of the normalized frequency.

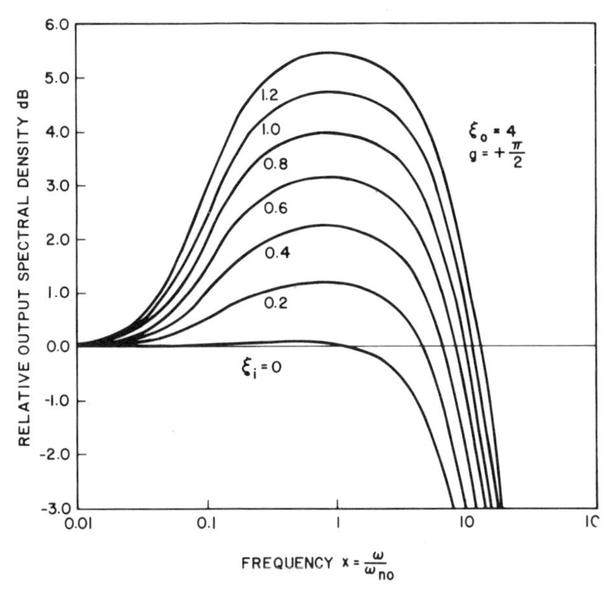

Fig. 13. Relative spectral density of output phase jitter as a function of normalized frequency for a conventional PLL with damping factor $\xi_0 = 4$ and unwanted signal injection at $g = +\pi/2$ and ξ_i from 0 to 1.2 in steps of 0.2 (5 percent per step).

(PLLI) tolerates a much higher dc loop gain since out-of-lock saturation no longer limits the pull-in range. As a side benefit, the static phase error is reduced significantly. Signal injection requires practically no additional hardware and eliminates the need for crystal control of the VCO.

We have presented a detailed noise analysis for a PLLI and have proven that a PLLI can have a significant reduction in output jitter compared to a conventional PLL with the same loop parameters, provided the input "amplitude" and "phase" noise components are correlated. The experiments confirm that this type of correlation is indeed present in the digital input signal to the PLL when the common nonlinear operation to regain a tone is performed on the input data stream. We were able to measure the spectral densities including the cross correlation density of the input noise components and have achieved a 10.5-dB reduction in total output jitter variance compared to a conventional PLL.

Furthermore, we have shown that this type of correlation in the noise component can severly degrade the performance of a conventional PLL in the retiming chains of a digital repeater if the PLL has some unwanted injection. In the worst case, an increase of the damping factor due to unwanted injection by only 5 percent results in a jitter peak of about 1.3 dB. This cannot be tolerated in long repeater chains.

ACKNOWLEDGMENT

The author would like to thank Mrs. D. Vitello for the computer graphs and Mr. A. R. McCormick for his assistance with the experiments.

REFERENCES

[1] S. C. Gupta, "Phase-locked loops," *Proc. IEEE,* vol. 63, pp. 291–306, Feb. 1975.
[2] A. Y. Viterbi, *Principles of Coherent Communication.* New York: McGraw-Hill.
[3] D. Richman, "Color-carrier reference phase synchronization accuracy in NTSC color television," *Proc. IRE,* vol. 42, pp. 106–133, Jan. 1954.
[4] A. Acampora and A. Newton, "Use of phase subtraction to extend the range of a phased locked demodulator," *RCA Rev.,* p. 577, Dec. 1966.
[5] A. W. Lahti and T. E. Beling, "Phase-locked-loop coherent FM detector with synchronized reference oscillator," U.S. Patent 3 189 825, filed Mar. 29, 1962.
[6] K. Murakami, "A new phase lock demodulator with injection locking," *Electron. Commun.,* vol. 52, p. 119, Feb. 1969.
[7] B. N. Biswas, "Combination injection locking with indirect synchronization technique," *IEEE Trans. Commun. Technol.,* vol. COM-19, pp. 574–576, Aug. 1971.
[8] B. N. Biswas and P. Banerjee, "Range extension of a phase-locked loop," *IEEE Trans. Commun. Technol.,* vol. COM-21, pp. 293–296, Apr. 1973.
[9] J. F. Oberst, "Generalized phase comparators for improved phase-locked loop acquisition," *IEEE Trans. Commun. Technol.,* vol. COM-19, pp. 1142–1148, Dec. 1971.
[10] L. J. Greenstein, "Phase-locked loop pull-in frequency," *IEEE Trans. Commun.,* vol. COM-22, pp. 1005–1013, Aug. 1974.
[11] R. Adler, "A study of locking phenomena in oscillators," *Proc. IRE,* vol. 34, pp. 351–357, June 1946; also, *Proc. IEEE,* vol. 61, pp. 1380–1385, Oct. 1973.
[12] M. Armand, "On the output spectra of unlocked driven oscillators," *Proc. IEEE,* vol. 57, pp. 798–799, May 1969.
[13] E. Roza, "Analysis of phase-locked timing extraction circuits for pulse code transmission," *IEEE Trans. Commun.,* vol. COM-22, pp. 1236–1249, Sept. 1974.

On the Selection of Signals for Phase-Locked Loops

JACK J. STIFFLER, MEMBER, IEEE

REFERENCE: Stiffler, J. J.: ON THE SELECTION OF SIGNALS FOR PHASE-LOCKED LOOPS, Raytheon Company, Sudbury, Mass. Formerly with Jet Propulsion Laboratory, Pasadena, Calif. Rec'd 12/1/67. Paper 67TP1212-COM, approved by the IEEE Communication Theory Committee for publication after sponsored presentation at the 1967 IEEE International Conference on Communications, Minneapolis, Minn. IEEE TRANS. ON COMMUNICATION TECHNOLOGY, 16-2, April 1968, pp. 239–244.

ABSTRACT: The performance of phase-locked loops and related signal tracking devices is dependent upon the cross-correlation function between the input signal and a locally generated signal. This paper is concerned with the problem of selecting both the input and local signals so as to minimize the tracking error due to additive noise.

Introduction

PHASE-LOCKED loops have been analyzed at great length and from many points of view. Almost invariably, however, the analysis begins by prescribing the waveform (usually a sinusoid) to be tracked by the loop. Rarely is the waveform itself treated as a system variable capable of being altered to improve the loop performance. This paper considers some of the aspects of this problem of signal selection for phase-locked loops.

A block diagram of a phase-locked loop is shown in Fig. 1. The received signal plus noise $Ax(t + \tau) + n(t)$ is multiplied by the locally generated signal $y(t)$. If the loop bandwidth is narrow relative to the reciprocal of the periods of the signals $x(t)$ and $y(t)$, the error signal for a given τ is determined, in the absence of noise, entirely by the average of the product $Ax(t + \tau)y(t)$; i.e., by the quantity

$$\lim_{T\to\infty} \frac{1}{T} \int_{-T/2}^{T/2} Ax(t + \tau)y(t)dt = A\rho_{xy}(\tau) \qquad (1)$$

where $\rho_{xy}(\tau)$ is the cross correlation of the signals $x(t + \tau)$ and $y(t)$. The noise term $n'(t) = n(t)y(t)$ is easily shown to be approximately white and Gaussian, regardless of the signal $y(t)$, if $n(t)$ is white and Gaussian (as will be assumed here). The noise contribution to the error signal is therefore independent of $y(t)$ so long as $y(t)$ is constrained to have a fixed power. For convenience we let both $x(t)$ and $y(t)$ represent unity power

$$\lim_{T\to\infty} \frac{1}{T} \int_{-T/2}^{T/2} x^2(t)dt = \lim_{T\to\infty} \frac{1}{T} \int_{-T/2}^{T/2} y^2(t)dt = 1. \qquad (2)$$

The signal design problem for phase-locked loops then reduces to the problem of specifying the optimum cross-correlation function $\rho_{xy}(\tau)$. In the next section we demonstrate that, at least for the first-order loop, the optimum function $\rho_{xy}(\tau)$ is a square wave.

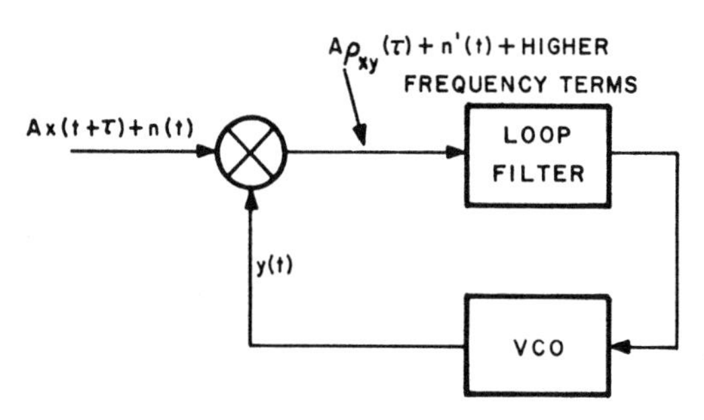

Fig. 1. The phase-locked loop.

The Optimum Function $\rho_{xy}(\tau)$

The phase-locked loop attempts to track a null in the cross correlation between the received and locally generated signals. At least heuristically a rather strong case can be made for a square wave as the optimum cross-correlation function for such a device. If the locally generated signal "leads" the received signal, the maximum possible positive error signal would be available to inform the device of this fact. Similarly, if the local signal "lags" the received signal, the error signal would be maximally negative. This proposition, however, seems to be difficult to prove for a general phase-locked loop. The analysis of the performance of a phase-locked loop operating on the basis of a square-wave correlation function clearly precludes the use of the linear model of the loop. Yet a nonlinear analysis of the phase-locked loop has been accomplished only in certain special cases.[1]–[3]

In this section, we use the results of the analysis of the exact model of the first-order loop to demonstrate the optimality of the square-wave autocorrelation function for this particular loop.

The density function of the phase error ϕ due to additive white Gaussian noise in a first-order loop has been shown to be[1]

$$f(\phi) = \frac{e^{\alpha g(\phi)}}{\int_{-\pi}^{\pi} e^{\alpha g(\phi)}\, d\phi}, \quad -\pi \le \phi \le \pi \qquad (3)$$

where

$$g(\phi) = \int^{\phi} \rho_{xy}\left(\eta \frac{T}{2\pi}\right) d\eta$$

and where $\alpha = A^2/N_0B_L$ is the loop signal-to-noise ratio and $\rho_{xy}(\tau)$ is the correlation function between the received and locally generated signals. The optimality of a square-wave cross-correlation function will be demonstrated in the following sense: If $P_0(|\phi| > \phi_0)$ is the probability of ϕ exceeding ϕ_0 in absolute amplitude when $\rho_{xy}(\eta T/2\pi)$ is a square wave and $P_1(|\phi| > \phi_0)$ is the probability of the

Reprinted from *IEEE Trans. Commun. Technol.*, vol. COM-16, pp. 239–244, Apr. 1968.

same event when $\rho_{xy}(\eta T/2\pi)$ is some other function of η, then

$$P_0(|\phi| > \phi_0) \leq P_1(|\phi| > \phi_0), \quad -\pi \leq \phi \leq \pi \tag{4}$$

for all ϕ_0. Moreover, it will be shown that (4) will always hold with strict inequality for some values of ϕ_0.

The proof begins with the observation that $|\rho_{xy}(\eta T/2\pi)| \leq 1$ and that the optimum cross-correlation function will be an odd function of η (i.e., if a particular correlation for positive $\eta = \eta_0$ results in an optimum performance, precisely the negative of this correlation would also result in an optimum performance when $\eta = -\eta_0$). As a consequence, when $-\pi$ is chosen as the lower limit of integration for the integral defining $g(\phi)$, all such functions $g(\phi)$ must exhibit the following properties:

$$g(-\pi) = g(\pi) = 0$$

$$g(\phi) = g(-\phi) \tag{5}$$

$$0 \leq g(\phi) \leq \pi - |\phi| \equiv g_0(\phi), \quad |\phi| \leq \pi$$

where $g_0(\phi)$ is the function $g(\phi)$ when the correlation function is a square wave. Moreover, because $|\rho_{xy}(\eta T/2\pi)| \leq 1$, the function $\epsilon(\phi) = g_0(\phi) - g(\phi)$ cannot decrease as $|\phi|$ decreases. Thus, $\epsilon(\phi)$ is a non-negative nonincreasing function of $|\phi|$. But, for arbitrary $g(\phi)$

$$f(\phi) = \frac{e^{\alpha g(\phi)}}{\int_{-\pi}^{\pi} e^{\alpha g(\phi)}\, d\phi} = \frac{e^{\alpha g_0(\phi)}\, w(\phi)}{\int_{-\pi}^{\pi} e^{\alpha g_0(\phi)}\, w(\phi) d\phi} \tag{6}$$

where $w(\phi) = e^{-\alpha\epsilon(\phi)}$ is a non-negative nondecreasing function of $|\phi|$ with

$$0 < w(\phi) \leq 1$$

and

$$w(\pi) = 1.$$

Therefore

$$P_1(|\phi| < \phi_0) = \frac{\int_{-\phi_0}^{\phi_0} e^{\alpha g_0(\phi)}\, w(\phi) d\phi}{\int_{-\pi}^{\pi} e^{\alpha g_0(\phi)}\, w(\phi) d\phi}$$

$$= \left[1 + \frac{\int_{|\phi|>\phi_0} e^{\alpha g_0(\phi)}\, w(\phi) d\phi}{\int_{|\phi|<\phi_0} e^{\alpha g_0(\phi)}\, w(\phi) d\phi}\right]^{-1}$$

$$\leq \left[1 + \frac{w(\phi_0)\int_{|\phi|>\phi_0} e^{\alpha g_0(\phi)} d\phi}{w(\phi_0)\int_{|\phi|<\phi_0} e^{\alpha g_0(\phi)} d\phi}\right]^{-1}$$

$$= P_0(|\phi| < \phi_0) \tag{7}$$

which was to be proved. Clearly, unless $w(\phi)$ is a constant, the inequality in (7) will be a strict inequality. But if $w(\phi)$ is a constant, $w(\phi) = 1$, $g(\phi) = g_0(\phi)$, and the correlation function is a square wave.

The Optimum Performance Attainable With a First-Order Loop

When $g(\phi) = g_0(\phi)$ as defined in (5), the density function of the phase error is

$$f(\phi) = \frac{\alpha e^{\alpha(\pi - |\phi|)}}{2(e^{\alpha\pi} - 1)}. \tag{8}$$

Thus the mean-squared phase error is

$$\sigma^2 = \int_{-\pi}^{\pi} \phi^2 f(\phi) d\phi = \frac{1}{e^{\alpha\pi} - 1}\left\{\frac{2e^{\alpha\pi}}{\alpha^2} - \left(\pi^2 + \frac{2\pi}{\alpha} + \frac{2}{\alpha^2}\right)\right\} \tag{9}$$

which becomes, when α is large,

$$\frac{2}{\alpha^2} = 2\left(\frac{N_0 B_L}{A^2}\right)^2. \tag{10}$$

The mean-squared error using a sinusoid, in contrast, is approximately $1/\alpha$ for large α. Accordingly, when the mean-squared error is small, the optimum loop error is proportional to the square of that of the conventional sinusoidal loop.

Another interesting parameter is in the frequency of cycle slipping f_s. This quantity is given by the formula[1]

$$f_s = \frac{2}{T(\pi)} \tag{11}$$

where

$$T(\pi) = \frac{\alpha}{4B_L}\int_0^{\pi}\int_{\phi}^{\pi} \exp[\alpha(g(\phi) - g(x))]dx d\phi$$

with $g(\phi)$ as previously defined.

Using the $g(\phi) = g_0(\phi)$ of (5) we find that

$$f_s = \frac{8B_L}{\pi}\left[\frac{e^{\alpha\pi} - 1}{\alpha\pi} - 1\right]^{-1} \tag{12}$$

which becomes, for large α,

$$8B_L\, \alpha e^{-\pi\alpha}.$$

For comparison, the same loop using sinusoidal signals yields

$$f_s = \frac{2B_L}{\pi^2}\frac{1}{\alpha I_0^2(\alpha)} \sim \frac{4B_L}{\pi} e^{-2\alpha} \tag{13}$$

the approximation valid for large signal-to-noise ratios α.

Direct Approximation of the Square-Wave Correlation Function

The preceding section demonstrated some significant advantages in choosing two signals $x(t)$ and $y(t)$ having a square-wave correlation function for use in conjunction with a first-order loop. Unfortunately, a square-wave correlation function is well known to be mathematically impossible. In this section, we attempt to specify two signals $x(t)$ and $y(t)$, having a cross-correlation function which approximates as closely as possible a square wave.

To accomplish this, of course, we must use some mathematically tractable measure of "closeness." For this reason we choose the mean-square measure of distance, i.e., if $\rho_0(\tau)$ denotes a square wave of period T and $\rho(\tau)$ an arbitrary correlation function, then the distance separating them is

$$\sigma_e^2 = \frac{1}{T}\int_0^T (\rho(\tau) - \rho_0(\tau))^2 d\tau. \tag{14}$$

It is this distance we wish to minimize.

Without loss of generality, let both the received signal $x(t)$ and the locally generated signal $y(t)$ have equal periods T. Then

$$x(t) = \sum_n a_n e^{j\omega_n t}, \quad \omega_n = \frac{2\pi n}{T} \tag{15}$$
$$y(t) = \sum_n b_n e^{j\omega_n t}$$

and the cross-correlation function $\rho_{xy}(\tau) = \rho(\tau)$ can be written

$$\begin{aligned}\rho(\tau) &= \frac{1}{T}\int_0^T x(t)y(t+\tau)\,dt \\ &= \sum a_n b_n^* \, e^{j\omega_n \tau}\end{aligned} \tag{16}$$

where b_n^* is the complex conjugate of b_n. The ideal cross-correlation function $(\rho_0(\tau))$ can be expressed in the form

$$\rho_0(\tau) = \sum c_n e^{j\omega_n t} = \begin{cases} 1, & \frac{-T}{4} < \tau < \frac{T}{4} \\ -1, & \frac{T}{4} < \tau < \frac{3T}{4} \end{cases} \tag{17}$$

where

$$c_n = \begin{cases} (-1)^{(|n|-1)/2}\dfrac{2}{\pi|n|}, & |n| \text{ odd} \\ 0, & |n| \text{ even} \end{cases}$$

Thus, we wish to choose a_n and b_n so as to minimize the mean-squared difference σ_e^2 between $\rho(\tau)$ and $\rho_0(\tau)$:

$$\begin{aligned}\sigma_e^2 &= \frac{1}{T}\int_0^T \left(\sum_n a_n b_n^* \, e^{j\omega_n \tau} - \sum_n c_n e^{j\omega_n \tau}\right)^2 d\tau \\ &= \sum_m \sum_n \frac{1}{T}\int_0^T d_m d_n \, e^{j(\omega_n \tau + \omega_m \tau)} d\tau = \sum_n |d_n|^2\end{aligned} \tag{18}$$

where

$$d_n = a_n b_n^* - c_n.$$

Accordingly

$$\sigma_e^2 = \sum (|a_n|^2 \, |b_n|^2 - a_n b_n^* c_n - a_n^* b_n c_n^* + |c_n|^2) \tag{19}$$

and letting $a_n = \alpha_n e^{j\theta_n}$ and $b_n = \beta_n e^{j\phi_n}$ with α_n and β_n both real and positive, we have

$$\sigma_e^2 = \sum (\alpha_n^2 \beta_n^2 - 2\,\alpha_n \beta_n c_n \cos(\theta_n - \phi_n) + c_n^2). \tag{20}$$

Clearly, regardless of the values of α_n and β_n, the optimum choice of the phase angles θ_n and ϕ_n satisfies the relationship

$$\cos(\theta_n - \phi_n) = \begin{cases} 1, & c_n > 0 \\ -1, & c_n < 0 \end{cases}. \tag{21}$$

Under this condition

$$\sigma_e^2 = \sum (\alpha_n \beta_n - |c_n|)^2. \tag{22}$$

This expression is obviously minimized by equating the product $\alpha_n \beta_n$ to $|c_n|$. But to do so necessarily violates the condition that $x(t)$ and $y(t)$ both represent unit power, i.e., that

$$\sum_n \alpha_n^2 = \sum_n \beta_n^2 = 1. \tag{23}$$

In fact, if $\alpha_n \beta_n = |c_n|$ either $\sum_n \alpha_n^2 = \infty$ or $\sum_n \beta_n^2 = \infty$ or both. Hence, we must explicitly impose this additional constraint on α_n and β_n. In particular, we must minimize the quantity

$$\sum (\alpha_n \beta_n - |c_n|)^2 + \lambda_1 \sum \alpha_n^2 + \lambda_2 \sum \beta_n^2 \tag{24}$$

where λ_1 and λ_2 are defined to satisfy the conditions (23).

This minimization procedure can be simplified by the following observation: First of all, the product $\alpha_n \beta_n$ will never exceed $|c_n|$ for any n when α_n and β_n are chosen optimally, since were this the case, the mean-squared error could always be decreased by decreasing that particular product and increasing $\alpha_{n'} \beta_{n'}$ for some n' for which $\alpha_{n'} \beta_{n'} < |c_{n'}|$. (At least some of the products $\alpha_n \beta_n$ must be less than $|c_n|$ because of conditions (23).) Now letting $\alpha_m^2 + \beta_m^2 = K_m$ and differentiating (22) with respect to α_m we obtain

$$2(\alpha_m \beta_m - |c_m|)\,\beta_m\,[1 - \alpha_m^2/\beta_m^2] \tag{25}$$

a positive quantity when $\alpha_m > \beta_m$ and a negative quantity otherwise. Thus for any K_m, the mean-squared error is decreased by decreasing α_m when it exceeds β_m and increasing it when it is smaller than β_m. Evidently, the optimum situation obtains when $\alpha_m = \beta_m$ for all m. Thus we wish to select the set of coefficients $\{\alpha_m\}$ minimizing the expression

$$E^2 = \sum (\alpha_n^2 - |c_n|)^2 + \lambda \sum \alpha_n^2. \tag{26}$$

Differentiating, we find

$$\frac{\partial E^2}{\partial \alpha_m} = 4\,\alpha_m \left[\alpha_m^2 - |c_m| + \frac{\lambda}{2}\right] \tag{27}$$

and E^2 is an increasing function of α_m for $\alpha_m^2 > |c_m| - \lambda/2$ and a decreasing function of α_m otherwise. We conclude, therefore, that since α_m is real

$$\alpha_m = \begin{cases} \left(|c_m| - \dfrac{\lambda}{2}\right)^{1/2}, & |c_m| \geq \dfrac{\lambda}{2} \\ 0, & |c_m| < \dfrac{\lambda}{2} \end{cases} \tag{28}$$

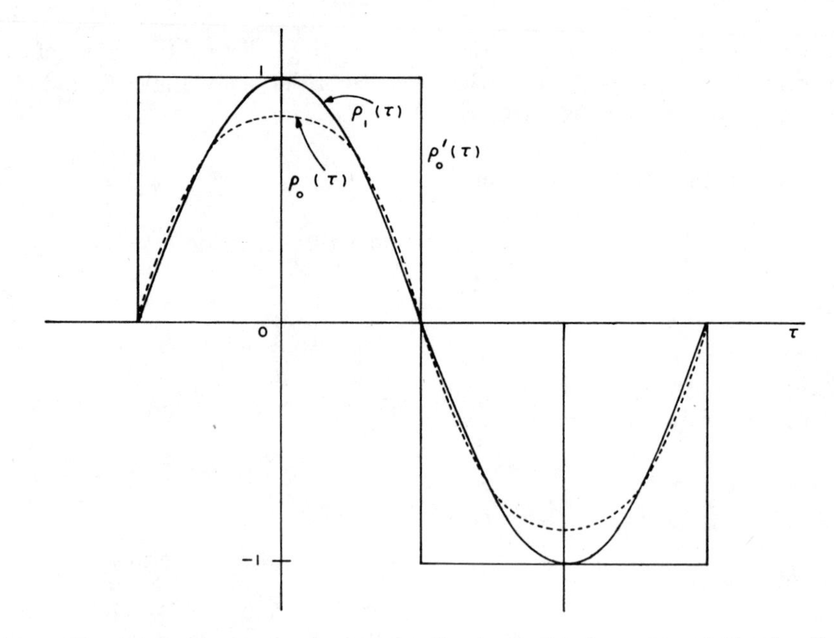

Fig. 2. The optimum mean-square approximation to the ideal cross-correlation function.

And since

$$\sum_m \alpha_m^2 = 1$$

$$2 \sum_{\substack{m=1 \\ m \text{ odd}}}^{N} \left(|c_m| - \frac{\lambda}{2}\right) = 1 \tag{29}$$

where N is the largest integer for which

$$|c_N| > \frac{\lambda}{2}. \tag{30}$$

(Note the $|c_m|$ terms are monotonically decreasing functions of $|m|$.) Thus

$$\frac{\lambda}{2} = \frac{2}{N+1} \sum_{\substack{m=1 \\ m \text{ odd}}}^{N} |c_m| - \frac{1}{N+1}. \tag{31}$$

Substituting (28) into the inequality (27), we find that $N = 3$ and that

$$\alpha_{-1} = \alpha_1 = \left(\frac{1}{4} + \frac{2}{3\pi}\right)^{1/2} \cong 0.68$$
$$\alpha_{-3} = \alpha_3 = \left(\frac{1}{4} - \frac{2}{3\pi}\right)^{1/2} \cong 0.2 \tag{32}$$

and

$$\rho(\tau) = 0.924 \cos \frac{2\pi\tau}{T} - 0.076 \cos \frac{6\pi\tau}{T}.$$

This cross-correlation function is plotted in Fig. 2 along with the ideal function $\rho_0(\tau)$ and the function

$$\rho_1(\tau) = \cos \frac{2\pi\tau}{T}$$

which results when all of the available power is placed in the fundamental. It is evident from (32) and from Fig. 2 that little is to be gained, under the criterion used in this section, in using other than a single sinusoid as a tracking signal. The mean-square criterion, of course, is quite arbitrary and different conclusions might result from different criteria.

Indirect Approximation of the Square-Wave Autocorrelation Function

A second approach fortunately yielding more fruitful results is to try to approximate the *performance* attainable using an ideal square-wave cross-correlation function rather than attempting to approximate the correlation function itself. One way of doing this is suggested by the following observation: Suppose a phase-locked loop is designed to track a sinusoid with some quite small mean-squared tracking error σ_1^2 rad². Then the signal-to-noise ratio α must be large and, to a good approximation,

$$\sigma_1^2 = \frac{1}{\alpha} = \frac{N_0 B_L}{A^2}. \tag{33}$$

In the preceding derivations, the loop gain was held fixed regardless of the actual signals used. (This can be justified, incidentally, in a number of ways. Heuristically, the loop gain is chosen as a compromise between the desire to limit the effect of the additive noise and the desire to emphasize any variations in the signal phase. Since neither the noise nor the signal phase statistics are altered by changing the signal waveform, the gain should remain constant, independent of the signals actually used. This conclusion is justified from another point of view shortly.)

Now suppose, instead, the frequency of the sinusoid which the loop is to track is increased by a factor of n. If the loop gain is kept fixed, the loop bandwidth will increase in direct proportion to the signal frequency. This follows because the VCO output *phase* $\theta_1(t)$ is still given, in terms of its input voltage $e(t)$, by

$$\theta_1(t) = K \int e(t)dt \text{ radians.}$$

But, since nothing has changed except the signal frequency, this is still a measure of phase in terms of the original period of T_0 seconds, i.e., 2π radians correspond to T_0 seconds. The effective loop gain, referred to the new frequency is therefore $K' = nK$ radians per volt second and the bandwidth is $B_L' = nB_L$ (with $B_L = AK/4$ the bandwidth of the original loop.) The variance $(\sigma_2')^2$ of the phase error of this loop is therefore approximately

$$(\sigma_2')^2 = \frac{N_0 B_L'}{A^2} = \frac{n}{\alpha}. \tag{34}$$

This is the phase-error variance referred to the new signal having a period only $(1/n)$th that of the original sinusoid. Since it is presumably the magnitude of the tracking error which is of concern and not the magnitude relative to the period of the particular signal being tracked, the two variances can be compared only if they are referred to the same period. In this case, the variance of the higher frequency loop can be written

$$\sigma_2^2 = \frac{1}{n^2}(\sigma_2')^2 = \frac{1}{n\alpha} \tag{35}$$

where 2π radians correspond to one period of the original sinusoid.

The phase-error variance, therefore, is inversely proportional to the frequency of the signal being tracked. This suggests the possibility of reducing the tracking error to an arbitrarily small value simply by increasing the signal frequency. This, of course, cannot be done. The variance (34) does increase with n, and while, as we have just argued, this variance is not the parameter of fundamental concern, it is a measure of the ability of the loop to remain in lock. When this quantity becomes too large, the loop begins to slip cycles and (35) is no longer meaningful. There must therefore be some maximum acceptable value for n.

To determine this maximum allowable n, we refer back to (13) for the frequency of cycle slipping in a phase-locked loop tracking a sinusoid. For the loop tracking the higher frequency sinusoid, we have

$$f_s \approx \frac{4nB_L}{\pi} e^{-2\alpha/n}. \tag{36}$$

Now, if f_s is to be small ($f_s \ll 1$) we find

$$\frac{4nB_L}{\pi} \ll e^{2\alpha/n}$$

and

$$\log_e(1/f_s) \approx \frac{2\alpha}{n}.$$

Accordingly,

$$\max n \approx \frac{2\alpha}{\log_e(1/f_s)} \tag{37}$$

where f_s is the maximum acceptable cycle-slipping frequency.

Using this value of n in (35) we obtain

$$\sigma_2^2 \approx \frac{\log_e(1/f_s)}{2\alpha^2} = \frac{\log_e(1/f_s)}{4}\sigma_0^2 \tag{38}$$

where σ_0^2 is the variance of the ideal loop (10). The phase-error variance, then, can be made to exhibit the same dependence on the signal-to-noise ratio as that of the ideal loop simply by increasing the frequency of the signal being tracked. Note that the constant $\log_e(1/f_s)/4$ is only logarithmically dependent upon f_s and hence will generally be relatively small. If $f_s = 10^{-4}$ Hz, for example, (an average of about one cycle slipped every 3 hours)

$$\sigma_2^2 \approx 2.3\ \sigma_0^2.$$

As a further justification for keeping the loop gain constant, we mention that when the loop is operating in the linear region, as it will be under the assumptions made here, the linear model can be used to also determine the effect of jitter in the received signal phase. In so doing, we find the mean-squared phase error due to input phase fluctuations to be also inversely proportional to n. Increasing the frequency reduces the magnitudes of the tracking error due to the input phase jitter and that due to the additive noise by equal factors, a statement which would not be true were the loop gain allowed to vary with the signal frequency.

This same approach, incidentally, can be taken with higher order loops when the linear model is used. The maximum acceptable value of n can be determined, for example, by a condition on the phase-error variance as referred to the period of the signal being tracked. In contrast with that of first-order loops, the bandwidth of higher order loops does not increase in direct proportion to the frequency when the loop gain is held constant. In particular, it is generally found that B_L is proportional to n^r where, as before, n is the factor by which the frequency is increased and r is a constant in the range $0 < r < 1$. Essentially the same argument as used previously for the first-order loop then leads to a phase-error variance which is inversely proportional to $\alpha^{2/r}$. The second-order loop, for example, designed to track a frequency step transient, yields a value of $r = 1/2$; the phase-error variance in the loop tracking the optimized frequency in this case is inversely proportioned to α^4.

Conclusion

It is shown that, at least for the first-order loop, the optimum cross-correlation function between the received signal and the locally generated signal is a square wave. While it seems difficult to approximate satisfactorily this correlation function with physical signals, the performance attainable in this ideal situation can be quite closely approximated by using sinusoids and properly selecting their frequency.

It should be emphasized that the frequency chosen in this manner is a function of the loop signal-to-noise ratio.

As this signal-to-noise ratio changes, the loop performance will deviate from that attainable with an ideal loop.

A second disadvantage of the approach outlined here is that, as a consequence of increasing the signal frequency, the number of lock-in points in a given interval also increases. Specifically, if the frequency is increased by a factor of n, the number of lock-in points in a given interval increases by this same factor. If the recovered signal is to be used to generate a local reference for the lower frequency signal, this n-fold ambiguity must somehow be resolved. Numerous techniques are available for accomplishing this, however, and the amount of additional power required to do so is generally quite small. In some cases, as for example when delay-locked loops are used,[4] the effective signal period can be reduced by a factor of n without introducing any additional ambiguities. The same conclusion as to the advantage of decreasing the (effective) signal period and as to the amount by which this can be done with impunity apply to delay-locked loops as well as ordinary phase-locked loops.[5] In this case, the cost of reducing the effective signal period is in the increased amount of time required to bring the loop into lock.

References

[1] A. J. Viterbi, "Phase-locked loop dynamics in the presence of noise by Fokker-Planck techniques," *Proc. IEEE*, vol. 51, pp. 1737–1753, December 1963.

[2] V. I. Tikhonov, "The effects of noise on phase-lock oscillator operation," *Avtomat. i Telemekh.*, vol. 22, no. 9, 1959.

[3] V. I. Tikhonov, "Phase-lock automatic frequency control application in the presence of noise," *Avtomat. i Telemekh.*, vol. 23, no. 3, 1960.

[4] J. J. Spilker, Jr., and D. T. Magill, "The delay-lock discriminator—an optimum tracking device," *Proc. IRE*, vol. 49, pp. 1403–1416, September 1961.

[5] J. J. Stiffler, "Phase-locked loop synchronization with non-sinusoidal signals," Jet Propulsion Lab., Pasadena, Calif., Space Programs Summary 37–38, vol. IV, pp. 227–232, August 1964.

Author Index

Subject Index

R

S

T

U

V

W

Editors' Biographies

William C. Lindsey (S'61–M'63–SM'73–F'74) is Professor of Electrical Engineering at the University of Southern California, Los Angeles. He has been a frequent consultant to government and industry and currently provides his services through LinCom Corporation. He has published numerous papers on varied topics in communication theory and holds several patents. He has written two books: *Synchronization Systems in Communication and Control* (Englewood Cliffs, NJ: Prentice-Hall, 1972) and *Telecommunication Systems Engineering*, coauthored with M. K. Simon (Englewood Cliffs, NJ: Prentice-Hall, 1973).

Dr. Lindsey serves on Commission C, Signals and Systems of the International Scientific Radio Union (URSI).

Marvin K. Simon (S'60–M'66–SM'75–F'78) was born in New York, NY, on September 10, 1939. He received the B.S. degree in electrical engineering from the City College of New York, New York, NY, in 1960, the M.S. degree in electrical engineering from Princeton University, Princeton, NJ, in 1961, and the Ph.D. degree from New York University, New York, NY, in 1966.

During the years 1961 to 1963 and again from 1966 to 1968 he was employed at Bell Laboratories, Holmdel, NJ, where he was involved in theoretical studies of digital communication systems. Coherent with his doctoral studies between 1965 and 1966 he was employed as a full-time Instructor of Electrical Engineering at New York University. Since 1968 he has been with the Applied Communications Research Group of the Jet Propulsion Laboratory, Pasadena, CA, where his research has explored all aspects of the synchronization problem associated with digital communication systems, in particular, the application of phase-locked loops. More recently, his interests have been somewhat diverted toward the study of modulation techniques for efficient spectrum utilization. In addition to his technical contributions in other areas, Dr. Simon has published over 30 journal papers on the subject of synchronization of digital communication systems and is coauthor with W. C. Lindsey of the textbook *Telecommunication Systems Engineering* (Englewood Cliffs, NJ: Prentice-Hall, 1973). He also serves as a consultant to private industry and the government in the above-mentioned areas of specialization, and has participated in short courses offered by these same installations and leading universities throughout the country.

Dr. Simon is a member of Eta Kappa Nu, Tau Beta Pi, and Sigma Xi. In the past he has served as Editor of Communication Theory for the IEEE TRANSACTIONS ON COMMUNICATIONS and is currently Chairman of the Communication Theory Committee of the IEEE Communications Society. His most recent responsibility is the Technical Program Chairmanship of the 1977 National Telecommunications Conference.